International Congress on Energy 2016 (ICE)

AA003042

Topical Conference at the 2016 AIChE Annual Meeting

San Francisco, California, USA
13 - 18 November 2016

Volume 1 of 2

ISBN: 978-1-5108-3420-0

Printed from e-media with permission by:

Curran Associates, Inc.
57 Morehouse Lane
Red Hook, NY 12571

Some format issues inherent in the e-media version may also appear in this print version.

Printed by Curran Associates, Inc. (2017)

AIChE
120 Wall Street, FL 23
New York, NY 10005-4020

Phone: (800) 242-4363
Fax: (203) 775-5177

www.aiche.org

Additional copies of this publication are available from:

Curran Associates, Inc.
57 Morehouse Lane
Red Hook, NY 12571 USA
Phone: 845-758-0400
Fax: 845-758-2633
Email: curran@proceedings.com
Web: www.proceedings.com

TABLE OF CONTENTS

VOLUME 1

VOLUME 2

468388 Control of Fe-Ni Nanoparticle Activity for the Oxygen Evolution Reaction (OER) and Methanol Electrooxidation

Abstract:

The electrolysis of water for hydrogen generation and the electrochemical oxidation of alternative fuels such as methanol for alkaline fuel cells currently relies on the use of precious metal catalysts. However, the scarcity and high cost of these precious metals lead to the search for earth-abundant alternatives with similar electrocatalytic properties. Key challenges in the development of catalysts for the oxygen evolution half reaction (OER) of water electrolysis remain; these challenges include reducing the overpotential of the reaction and developing non-precious metal catalysts that are stable and active with minimal mass transport limitations. For direct electrooxidation of alternative fuels such as methanol, non-precious metal alternatives to precious metals such as platinum and palladium are often much less active. Among the non-precious metals, multimetallic nickel-based catalysts show particular promise for their high stability and high activity for the oxygen evolution reaction (OER) half-reaction of water electrolysis. In particular, it has recently been demonstrated by Trotochaud, Burke, Boettcher, and co-authors that iron incorporation into nickel hydroxide and cobalt hydroxide catalysts causes a significant increase in the OER activity of the catalyst [1-3]. It is now understood that the iron is in fact the active site of such bimetallic catalysts [2]. Additionally, nickel has exhibited high activity for methanol oxidation, particularly when in the form of a bimetallic catalyst such as NiPd [4]. However, promising active catalysts that are comprised entirely of non-precious metals such as nickel and iron have not been often reported for alternative fuel electrooxidation. Furthermore, there is a need for the development of nanostructured catalysts that can address mass transport limitations inherent in bulk or film-based catalyst materials [2]. As such, there continues to be an opportunity in alkaline electrocatalyst development to discover and understand nanostructured catalyst materials for both OER and alternative fuel electrooxidation.

In this presentation, our results on the development and optimization of an FeNi hydroxide core-shell nanoparticle catalyst will be discussed for both OER and methanol electrooxidation. Through control of critical synthesis parameters, such as the addition rate of reducing agent, ratio of reducing agent to iron, and reaction time for nanoparticle synthesis, we have demonstrated the ability to tune nanoparticle catalyst activity for either OER or methanol electrooxidation. Results for maximum OER current as a function of the rate of addition of reducing agent sodium borohydride illustrate the sensitivity of catalyst performance to changes in synthesis parameters. Characterization of nanoparticle structure and phase will be discussed and correlated to electrochemical performance. Results suggest that tuning of the crystalline structure and order are critical to electrocatalyst performance optimization, as well as the presence of oxide and hydroxide.

References

[1] L. Trotochaud, S.L. Young, J.K. Ranney, S.W. Boettcher, Nickel-iron oxyhydroxide oxygen-evolution electrocatalysts: The role of intentional and incidental iron incorporation, J. Am. Chem. Soc., 136 (2014) 6744-6753.

[2] M.S. Burke, L.J. Enman, A.S. Batchellor, S.H. Zou, S.W. Boettcher, Oxygen evolution reaction electrocatalysis on transition metal oxides and (oxy)hydroxides: Activity trends and design principles, Chemistry of Materials, 27 (2015) 7549-7558.

[3] M.S. Burke, M.G. Kast, L. Trotochaud, A.M. Smith, S.W. Boettcher, Cobalt-iron (oxy)hydroxide oxygen evolution electrocatalysts: The role of structure and composition on activity, stability, and mechanism, J. Am. Chem. Soc., 137 (2015) 3638-3648.

[4] A. Dutta, J. Datta, Energy efficient role of Ni/NiO in PdNi nano catalyst used in alkaline DEFC, J. Mater. Chem. A, 2 (2014) 3237-3250.

475202 Transition Metal-Nitrogen-Carbon Electrocatalysts for Oxygen Reduction Reaction

Abstract:

UNM has developed the Sacrificial Support Method (SSM) as a main approach for templated synthesis of hierarchically structured electrocatalysts materials. In this method the catalysts precursors are being absorbed on, impregnated within or mechanically mixed with the support (usually mono-dispersed or meso-structured structured silica), thermally processed (pyrolyzed) and then the silica support is removed by etching (in KOH or HF) to live the open frame structure of a "self-supported" material that consists of the catalysts only. Such hierarchical structures are advantageous in enhancement of the fuel cell performance since they correspond to the different levels of transport in the corrugated electrode matrixes. A wide variety of materials can be made by these methods in which not only the composition but also the microstructure can be varied. It is the combination of these attributes - control over microstructure at a number of different length scales and composition, simultaneously - that is extremely important to the performance of the electrocatalyst materials in a fuel cells.

The makeup and structure of the active site/sites of the Platinum Group Metal-free (PGM-free) ORR electro-catalysts, including geometry (coordination) and chemistry (composition and oxidation state) remain contentious even after 50 years of research. There is an emerging agreement that iron and nitrogen functionalities, displayed on the surface if the carbonaceous substrate/support, govern ORR activity. This is often combined with a broadly accepted hypothesis that micro-porous surface area plays a critical role forming the active site. Candidate structures participating in ORR include multitudes of nitrogen defect motifs in the carbon matrix of different degrees of graphitization, with metal incorporated as metal nano-particles, corresponding (native) oxides and/or as atomically dispersed, oxidized metal species, linked (coordinated) to nitrogen defects in carbonaceous matrix in a variety of possible configurations. These consist of in-plane defects such as graphitic nitrogen and Fe-coordinated to three or four nitrogen atoms and a plurality of possible edge sites such as pyridinic, pyrrolic, quaternary and $Fe-N_2$/$Fe-N_x$ sites. In this study, we have analyzed more than 45 distinct M-N-C electrocatalysts synthesized from three different families of precursors, such as (i) N-containing polymer, (ii) N-chelating macrocycles and (iii) N-containing small organic molecules (amines). Catalysts were evaluated and analyzed structurally using exactly the same protocol for deriving structure-to-property relationships. We have identified possible active sites participating in different ORR pathways: (1) metal-free electrocatalysts support partial reduction of O_2 to H_2O_2; (2) pyrrolic nitrogen acts as a site for partial O_2 reduction to H_2O_2; (3) pyridinic nitrogen displays catalytic activity in reducing H_2O_2 to H_2O; (4) Fe coordinated to N ($Fe-N_x$) serves as an active site for $4e^-$ direct reduction of O_2 to H_2O. The ratio of the amount of pyridinic and $Fe-N_x$ to the amount of pyrrolic nitrogen serves as a rational design metrics of M-N-C electrocatalytic activity in oxygen reduction reaction occurring through the preferred $4e^-$ reduction to H_2O.

There are two main hypotheses in the literature: one claiming that nitrogen functionalities on/in carbon-based support are directly responsible for their ORR activity and the second suggesting that nitrogen groups serve as the coordinating environment for metal ions, serving as reactive centers for the ORR. Theoretical calculations have confirmed that incorporation of nitrogen atoms in the carbon matrix enhances its electronic properties, but nitrogen atoms are not catalytically active towards a direct reduction of oxygen. Our mechanistic and structural studies have shown that multiple (yet similar) chemical functionalities most likely co-exists in practical M-N-C catalysts, displaying engagement in various steps of the ORR and thus obscuring the "simple model" of the "most active "***active site***".

471580 Electrocatalytic Generation of H_2O_2: Carbon Based Material Synthesis and Device Design for Portable Low Cost Water Purification

Abstract:

Hydrogen peroxide, produced at a rate of 3 million tons/year, is one of the most important industrial chemicals widely used as a bleaching and oxidation agent.[1] Compared to chlorine which is the most commonly used water purification agent at the moment, hydrogen peroxide does not generate harmful byproducts after decomposition, and the higher oxidation potential makes it a more favourable reagent for treatment of water pollutants.[2] Hydrogen peroxide is mainly produced in large-scale facilities using a batch anthraquinone oxidation process, facing the challenge of high transportation and capital cost as well as explosive hazard during storage. For further applications in developing world a selective electrochemical catalyst towards H_2O_2 production is important and a prototype fuel cell/electrolyzer-solar cell device needs to be designed to enable continuous, small-scale and decentralized production of H_2O_2.

Through DFT calculation, we have predicted that different types of nitrogen-doped (N-doped) carbon or carbon with defects are able to selectively catalyze a 2-electron transfer oxygen reduction. We synthesized different types of carbon based materials and tested them along with some commercially available mesoporous carbons. Our results suggest that carbon defect sites without heteroatom doping could serve as the active center. The best catalyst by far is able to catalyse the reaction with no over-potential under alkaline condition with over 90% selectivity.

In this study, prototype devices, both fuel-cell and electrolyzer setups have also been designed and tested. Early results in a fuel-cell setup shows that H_2O_2 has higher affinity towards the aqueous phase as compared to the gas phase. Also H_2O_2 can be easily decomposed in contact with transition metals cations and reductive organic compounds. Considering the fact that electrocatalytic H_2O_2production will have the largest impact in the developing world, itâ€™s

desirable to have a simple material manufacturing process and low material cost. All these considerations lead to the current electrolyzer design using plastic with no Nafion membrane and precious metal catalysts involved.

In our electrolyzer design, the produced H2O2 is up-concentrated in the electrolyte. At the current stage, the electrolyzer cell is able to operate with 50 mA overall current at pH 13 with an applied cell potential 1.7 V and faradaic efficiency close to 100%. The device can also be connected to a PV system as the energy supply. A ten-minute operation will able to accumulate 10ml solution with the concentration around 500mg/L.

1. Campos-Martin, J. M.; Blanco-Brieva, G.; Fierro, J. L. G., Hydrogen peroxide synthesis: An outlook beyond the anthraquinone process. *Angew Chem Int Edit* **2006,** *45*(42), 6962-6984.

2. Chu, S.; Majumdar, A., Opportunities and challenges for a sustainable energy future. *Nature* **2012,** *488* (7411), 294-303.

473513 Niobium Doped Molybdenum Disulfide Catalyst for CO_2 Reduction Reaction

Abstract:

The world population consumes an average of 15 trillion watts of power, 85% of which comes from burning fossil resources including petroleum, natural gas, coal, and wood.[1] Despite their low cost and high quality, the use of fossil fuels is unavoidably coupled to the release of many compounds such as carbon dioxide (CO_2) that may adversely affect the environment such as acidifying oceans and climate change.[2,3] Thus, controlling CO_2 emission is essential using either renewable energy sources or remediation techniques. Among various remediation techniques, electrochemical reduction of CO_2 is known as the best approach since it not only reduces the amount of existing CO_2 in the atmosphere but also recycle the inert CO_2 molecule into energy-rich products (e.g., syngas). Layered transition metal dichalcogenides (TMDCs) have recently shown a significant potential to use as the catalyst for a wide range of electrochemical energy conversion and storage systems such as hydrogen evolution and CO_2 reduction[4,5]. However, it has been always a challenge to tailor the electronic structure of TMDCs to achieve highly promoted catalysts.

In this report, we have discovered a superior catalytic performance of Niobium (Nb) doped vertically aligned MoS_2 (VA-MoS_2) as an energy efficient and highly active catalyst for syngas production. Our results indicate that with only 5% Nb concentration at the edge atom structures, MoS_2 exhibits more than two orders of magnitude higher activity than Ag nanoparticles and one order of magnitude higher than pristine MoS_2 at low overpotentials (-0.2 V vs RHE).[6] Additionally, the product characterization results reveal that this promoted catalyst is selective for syngas (CO and H_2) formation at the potential range of -0.164 V to -0.764 V vs RHE. The effect of doping on the structure of pristine MoS_2 was studied by synthesis and testing

different concentrations of dopant for the CO_2 reduction reaction. Different characterization methods such as scanning transmission electron microscopy (STEM), X-ray photoelectron spectroscopy (XPS), electron energy loss spectra (EELS), energy-dispersive X-ray spectroscopy (EDX) and in-situ differential electrochemical mass spectrometry (DEMS) were applied to fully elucidate the effect of dopant on the electronic structure of MoS_2 and its catalytic activity and selectivity.

References:

(1) Davis, S. J.; Caldeira, K.; Matthews, H. D. Future CO2 Emissions and Climate Change from Existing Energy Infrastructure. *Science.* **2010**, *329*, 1330–1333.

(2) Haszeldine, R. S. Carbon Capture and Storage: How Green Can Black Be? *Science.* **2009**, *325*, 1647–1652.

(3) Lastoskie, C. Caging Carbon Dioxide. *Science.* **2010**, *330*, 595–596.

(4) Behranginia, A.; Asadi, M.; Liu, C.; Yasaei, P.; Kumar, B.; Phillips, P.; Foroozan, T.; Waranius, J. C.; Kim, K.; Abiade, J.; *et al.* Highly Efficient Hydrogen Evolution Reaction Using Crystalline Layered Three-Dimensional Molybdenum Disulfides Grown on Graphene Film. *Chem. Mater.* **2016**, *28*, 549–555.

(5) Asadi, M.; Kumar, B.; Liu, C.; Phillips, P.; Yasaei, P.; Behranginia, A.; Zapol, P.; Klie, R. F.; Curtiss, L. A.; Salehi-Khojin, A. Cathode Based on Molybdenum Disulfide Nanoflakes for Lithium–Oxygen Batteries. *ACS Nano,* **2016**, 10 (2), pp 2167–2175.

(6) Asadi, M.; Kumar, B.; Behranginia, A.; Rosen, B. a; Baskin, A.; Repnin, N.; Pisasale, D.; Phillips, P.; Zhu, W.; Haasch, R.; *et al.* Robust Carbon Dioxide Reduction on Molybdenum Disulphide Edges. *Nat. Com.* **2014**, *5*, 4470.

460933 Reutilization of Coal Fly Ash for the Production of Highly Beneficial Products

Abstract:

Coal fly ash (CFA), generated during the combustion of coal in power generation plant, is an industrial waste that continues to increase as a result of the growth in power demand. This large amount of industrial waste can cause environmental and human health problems if it is not treated properly. Research on the potential reuse of CFA has therefore been undertaken intensively worldwide. From those perspectives, this work aims to reduce the problems related to the disposal of CFA by converting them into beneficial products, including zeolite, geopolymer and fertilizer.

Zeolites: Hydrothermal alkaline reaction was employed for converting CFA to zeolites. Two different types of alkali (i.e. NaOH and KOH) were used as activating agents. Different Si/Al ratio, time and temperature used for the hydrothermal reaction were investigated for their effects on the type and amount of zeolites synthesised. It was found that, by using NaOH as activating agent, a mixture of Na-A and hydroxy-sodalite was mostly formed. On the other hand, a different mixture of merlinoite and perlialite was formed when KOH was used as the activating agent. The obtained zeolites were then employed for dye removal from wastewater and it was found that zeolite obtained from NaOH has the highest adsorption capability of 16.26 mg/g.

Geopolymer: In order to produce CFA-based geopolymer, CFA was firstly treated with the aqueous solution of Na_2SiO_3 and NaOH. Concentration of Na_2SiO_3, SiO_2/Na_2O mole ratio, ash/alkaline solution mass ratio, temperature and curing time were varied to study their effects on heat resistance and compressive strength of geopolymers. It was found that the compressive strength of geopolymer increases with Na_2SiO_3 concentration, temperature and curing time, whereas heat resistance of geopolymer is insignificantly affected by varying those parameters (dry weight loss between 2.5 and 4.9%). Moreover, it was found that geopolymers produced have higher compressive strength (1115 MPa) and better heat resistance as compared to the hardened Ordinary Portland Cement (OPC) (400 MPa).

Fertilizer: CFA consists of all the elements present in soil except organic carbon and nitrogen, making it becomes a valuable raw material for producing fertilizer. In order to produce fertilizer, CFA was mixed with another two components: (i) wood compost, which contains significant amounts of basic nutrients for the plants i.e. carbon, hydrogen oxygen nitrogen and sulfur; and (ii) biochar, which can provide a physical domain for bacteria and fungi, including beneficial mycorrhizal species and also enhance the water-holding capacity of the soil. From the ICP analysis, the fertilizer produced contains significant amounts of useful nutrients for plants, including phosphorus and potassium as primary macronutrients; calcium and magnesium as secondary macronutrients; cobalt, copper, iron, manganese, sodium, and zinc as micronutrients. Moreover, the final product also contains a significant amount of silica (~18.9 wt. %). Last but not least, it was found that the fertilizer obtained can significantly enhance the quality of soil.

468869 Characterizing Catalytic Waste Gasification - Effects of Substrate and Catalyst

Abstract:

This research is primarily aimed to extend proof-of-concept gasification experiments as a suitable alternative for waste management and sustainable technology. While initially intended to make more efficient use of resources in space exploration missions by producing high value reaction products from normally unused or wasted materials, the same principles can be applied to municipal waste as well.

The experiments performed examine gasification dynamics for different substrates promoted by different catalysts. It utilizes a high pressure batch scale reactor which allows examining controlled multi-phase catalytic isothermal oxidation of solid and melting substrates in the presence high purity air. The reaction time was varied between 2 and 24 hours and a preliminary overall kinetic model was formulated based on conversion, selectivity, and yield determinations. The studied substrates, Polyethylene and Cellulose, are both long chain organic polymers, and make up a substantial portion of both space and municipal waste composition. Although similar in nature, these substrates exhibit marked differences as it pertains to gasification, and were therefore selected as model substrates. Namely, cellulose contains oxygen as part of its molecular structure, while polyethylene does not. Since the technology investigated is aimed at working to relatively low temperatures (below 350 ^{o}C), cellulose remains as a solid particulate at the investigated reaction temperature while polyethylene melts.

Following the reaction phase, both the liquid and solid reaction residues, along with the gaseous products, are collected for characterization and analysis. Gas quantification is performed using a Gas Chromatograph. While thermal characterization (DSC) and Scanning Electron Microscopy (SEM) are used to characterize solid and liquid residuals.

Preliminary results suggest that catalytic gasification provides potential control for the liquid-phase oxidations. Under appropriate conditions, oxidation was found to follow the incomplete oxidation pathway which resulted in significant yields of hydrogen and methane, which are potentially high value products as they can be used for energy recovery.

Comparisons between the gasification rates for the two substrates are drawn. Similarly, comparative studies of different catalysts are also presented. The results are used in conjunction with a kinetic model for the gas-phase to formulate a methodological model for the gasification of solid waste. The results will be presented along with a preliminary extension for a continuous gasification process.

459628 Experimental and Numerical Study on Combining Anaerobic Digestion and Gasification for Potential Energy Recovery from Biomass and Sewage Sludge

Abstract:

The purpose of this study is to improve energy recovery and realize full utilization of municipal solid wastes (MSWs). This paper proposes a novel concept of a two-stage MSWs disposal process that combines the anaerobic digestion and biomass gasification. Both experimental and numerical studies have been conducted to investigate the performance of an integrated system for the energy recovery from MSWs based on anaerobic digestion and gasification. The experiments are performed at two stages. At the first stage, anaerobic digestion of MSWs, such as grass, leaves and animal manure, is conducted by mixing with anaerobic sludge. At the second stage, the digested solid residue is co-gasified with woody biomass for syngas production. The effect of inoculum to substrate ratio on anaerobic digestion and wood chips to digested solid residue ratio on gasification performance are investigated respectively. A combined kinetic model is then built for prediction and optimization of biogas and syngas yields in this two-stage process. The deviation between the numerical and experimental results obtained was lower than 10%. Both experimental and numerical results show that full utilization of MSWs can be realized to produce energy through anaerobic digestion followed by gasification process.

461746 Hydrothermal Carbonization of Organic Fraction of Municipal Solid Waste and Corresponding Digestate Via Anaerobically Digestion

Abstract:

Steam autoclaving is a technology that has been around since the 80â€™s and can efficiently divide the biogenic and non-biogenic fractions of Municipal Solid Waste (MSW) for best re-use and upgrading. Autoclaving eliminates hazardous hand sorting of raw garbage and in the process of separating the pulp fibers, sterilizes the output and allows for sequestration of pollutants that could otherwise leach into the groundwater or the atmosphere over time. The organic fraction is recovered by trommel screen (accepts <3/8â€) and is comprised of a moist pulp matrix embedded with ash and debris. The food waste from MSW is solubilized and absorbed onto the pulp matrix, creating a unique feedstock for further thermochemical processing (e.g., hydrothermal carbonization, HTC) biological processing (e.g., anaerobic digestion, AD), or a combination of both.

Hydrothermal carbonization (also called thermal hydrolysis, or wet torrefaction) is a treatment process which converts moist feedstocks into homogenized, carbon rich, and energy dense solid fuel, called hydrochar. One of the main advantages of HTC compared to other thermochemical treatment processes is the use of moisture as reaction medium and reactant, thereby precluding the need for drying prior to treatment. Thermodynamic properties of water change greatly in the subcritical region from 180-280 °C, and as a result, subcritical water behaves as a non-polar solvent and mild acid and base catalyst simultaneously. When subjected to HTC, biomass undergoes rapid hydrolysis and other reactions, releasing oxygen-containing volatiles and producing a highly hydrophobic hydrochar. Aqueous products typically include organic acids, sugars, and furans.

In this study, the autoclaved organic fraction of municipal solid waste pulp (OFMSW) and the digestate from OFMSW pulp after AD were processed by HTC at 200, 250, and 300 °C for 30 min and 2 h. The focus of this work was to evaluate the potential for producing an energy-dense, stable, solid hydrochar by HTC from either OFMSW or its digestate. The fate of inorganics, especially the heavy metals in hydrochar is discussed. In addition, the aqueous products were analyzed to identify the chemicals produced and the chemical changes during HTC. A carbon balance was performed from the solid and liquid phase analyses. Moreover, two conceptual MSW treatments consisting of autoclaving with or without AD, but including HTC were proposed and mass and energy balances were performed. The proximate, ultimate, FTIR, and fiber analyses show that the fuel value of the hydrochar increases with increasing HTC temperature. Inorganics have a strong propensity for the hydrochar matrix, and are found only in small amounts in the aqueous phase. The fuel content is as high as 44.7 MJ kg^{-1} (dry ash-free basis) when HTC digestate was treated at 300 °C for 2h. A cascaded design including AD and HTC yields a higher net energy production than HTC or AD alone.

461951 Minimizing Municipal Solid Waste Accumulation through the Application of Appropriate Technology

Abstract:

Along with population and urbanization in developing countries, the amount of municipal solid waste generated is also increasing. Although programs and initiatives to recycle and manage waste can often be found in the major population centers, these technologies are slow to spread to or are not yet present in the rural areas. Heavily populated urban slums are also lacking in the infrastructure needed to collect and manage trash, particularly plastic packaging. Solving the problems caused by this accumulation of municipal solid waste (MSW) derived plastic is a global challenge. Plastic waste accumulation in landfills and oceans is detrimental to ecosystems as well as public health. In developing regions particularly, where MSW handling infrastructure is limited or unavailable, plastic is often simply left on the ground where it eventually makes it way to the ocean. Because these sources are widespread, the solutions to dealing with this plastic must also be widely dispersed.

Fortunately, many plastics can easily be converted into a fuel oil that can be directly used as a substitute for petroleum derived kerosene or diesel. Plastics commonly found in MSW such as High Density Polyethylene (HDPE), Low Density Polyethylene (LDPE), Polypropylene (PP), and Polystyrene (PS) are all viable raw materials for this process. Plastics such as polyethylene terephthalate (PET) and poly vinyl chloride (PVC) are unfortunately not suitable for this process, due to the presence of oxygen and chlorine respectively in the polymers. The chemistry of converting plastic waste to a hydrocarbon fuel oil is simple and well established. Since, the molecules of these plastics are only made up of carbon-hydrogen chains, when thermally heated to temperatures of approximately 430°C (Wong, et al., 2015), the hydrocarbon chains break, decomposing the polymer, and yielding a hydrocarbon gas, which is then condensed to obtain the fuel oil product. However, industrial processes for carrying out this chemistry are typically not available in the developing regions that are the source of much of the plastic that ends up in the ecosystem. To address this problem, solutions based on combining the principles of appropriate technology with green process engineering are needed.

Appropriate technology (AT) is a design philosophy first described in 1973 by E.F. Schumacher in his book *Small is Beautiful*. According to Hazeltine and Bull in their 1999 book "*Appropriate Technology: Tools, Choices, and Implications*", it refers to "technological choices and applications that are small scale, decentralized, labor-intensive, energy efficient, environmentally sound, and locally controlled" (Hazeltine, 1999). AT is simply technology suitable for a specific region, designed to meet specific needs of certain individuals or communities. It differs from industrial scale processes in the sense that AT is specifically intended to be used in rural or underdeveloped regions. AT is low-cost, small-scale, easy to construct, and can be operated by individuals with limited formal technical education. The purpose of AT is not to reproduce industrial technology on a small scale; instead, it is to design specific solutions based upon the resources available in a given region or for a given community.

It is our assertion that appropriate technology is the most viable option for solving the global problem of waste plastic accumulation in developing regions. Therefore, we will make the case that through the utilization of a low-cost easy to use processor for converting waste plastic to fuel oil, a sustainable solution for reducing the volume of waste plastic entering the ecosystem can be achieved. We will address the challenges for implementing such a strategy and discuss how these challenges can be met.

460695 Struvite Precipitation for Nitrogen Removal in Horse Manure for Gasification: Experimental Study and Economic Analysis

Abstract:

Animal manure is one of the most common kinds of waste in farming areas. Instead of direct disposing or burning, gasification is a more promising and cleaner way to harvest the energy from waste and reduce pollution at the same time. From previous study, addition of horse manure and chicken manure to woodchips could improve syngas production up to 40%. However, the nitrogen content inside the manure, taking up around 2-3 wt% in the sample, may course environmental or explosive concern for gasification. On the other hand, retrieving nitrogen from animal manure can be a significant source as fertilizers. From FT-IR tests, nitrogen was found to be in the form of ammonia mainly in the manure samples. Thus, struvite precipitation is considered as an excellent approach to remove ammonia and produce struvite at the same time. In this work, horse manure from Singapore Turf Club was studied. Elemental analysis, FT-IR and TGA tests for the manure samples were conducted as feedstock characterization. Firstly, ammonia was leached out from manure samples using water at different ratios. After leaching, ammonia concentration could reach around 100 mg/L. For precipitation process, optimal experimental conditions, such as molar ratios of Mg^{2+} and PO_4^{3-} sources, mixing time and pH, were studied. Performance of different kinds of Mg^{2+} and PO_4^{3-} sources were compared. In the end, economic analysis of struvite precipitation process was also conducted to find the optimal solution for the whole process. The objective of this work is to find out the optimal experimental conditions for fully utilization of manure waste in a most cost-effective and high-yield way.

454626 Highly Selective Conversion of Biogenic Waste Materials to "Biogenic Formic Acid" As a Green Platform Chemical

Abstract:

The selective oxidation of complex, water-insoluble and wet waste biomasses from second and third generation to formic acid including effective catalyst recycling is reported. Additionally, the relevance and limits of potential contaminants are illustrated by different experimental approaches. By using a very robust homogeneous polyoxometalate catalyst in aqueous solution, molecular oxygen as oxidant and an acid as solubilizer, it is possible to convert different lignocellulosic and algae feedstock into formic acid and pure carbon dioxide. The applied green oxidation system benefits from its low reaction temperature (below 100 °C) and its very selective nature. The upcoming requirements for sustainable production of chemicals will lead to an increasing substitution of fossil fuels by biorenewable feedstock. This is mainly due to the fact that almost all use of hydrocarbon fossil fuels in production processes is directly or indirectly linked to carbon dioxide (CO_2) emissions that contribute to global warming. However, biomass conversion is usually characterized by complex feedstock compositions leading to sophisticated processes with many parallel and consecutive reactions involved. This often results in low yields of the designated products. Consequently, only few processes for directly converting biomass into value-added chemicals have been established in academic research. The high oxygen content of biomass requires typically a number of reductive conversion steps in biomass valorization sequences leading almost unavoidably to drawbacks like formation of tar or gluey byproducts. Moreover, the ultimate goal of producing valuable chemicals in high yields by direct conversion of complex biomasses, such as e.g. lignocellulosic substrates or algae is still far ahead. Major difficulties arise from the fact that such substrates are not well soluble in conventional solvents and are very resistant to chemical and biological transformations.

Other important considerations for the chemical use of biogenic feedstocks arise from conflicts with the use of biomass as food and animal nutrition. To characterize the potential for such conflicts, the broad range of biogenic feedstock conversion processes has been classified into three generations: The first generation of biomass conversion processes deals with feedstock that is in direct conflict with the supply of food. Examples are sugars, starch and vegetable oils. Conversion processes that utilize lignocellulosic biomass like wood or herbage do not use feedstock that is in direct conflict with food supply. However, these biomasses grow on farmland and are therefore in indirect conflict with food supply. The use of microorganisms, such as algae, to produce fuels or chemicals is denoted as third generation biomass conversion processes. These processes are preferred as future biomass conversion processes as no direct or indirect conflict with food and nutrition applications arises from a large scale, industrial realization.

Formic acid (FA) is a basic chemical that is widely used in chemical, textile, leather, pharmaceutical, rubber and other industries. Recently, FA is discussed in addition as fuel for direct FA fuel cells, and as hydrogen storage material. FA can be easily and selectively decomposed to hydrogen and CO_2through metal catalysed processes under very mild conditions. In addition, thermal decomposition of FA (above 373 K) forms CO and water. Thus, FA can be regarded as a liquid syngas equivalent and conversion of biomass to FA can be regarded as up-grading an abundant natural resources into a green energy and synthesis equivalent.

Oxidative conversion of wet biomass to formic acid using homogeneous POM catalysts has been developed by our group in the recent five years (OxFA-process). The concept essentially overcomes all major problems of classical biomass gasification or reforming processes. The interesting difference of the OxFA route compared to reductive biomass valorization approaches is that all thermally induced formation of solid or gluey by-products is completely avoided. Moreover, all heteroatoms in the biomass are converted into their highest oxidized, water-soluble forms (e.g. sulfates, nitrates) in contrast to the formation of H_2S and ammonia encountered under reductive conditions.

The OxFA-process is mildly exothermic and operates under mild temperature conditions of typically below 373 K. As water is applied as solvent, biomass of different origin, composition and humidity can be applied without drying. Formic acid is produced as the only product in the liquid phase avoiding tedious liquid phase separation processes. The only side-product is pure carbon dioxide that forms in the gas phase and can be re-used for the production of new biomass substrate (e.g. for the production of algae in an algae farm). As the OxFA process converts a very broad range of biogenic raw materials into only two products that separate nicely into gas and liquid phase its simplicity and robustness are clear advantages compared to other biomass valorization technologies. The OxFA process applies homogeneous polyoxometalate (POM) complexes as redox catalysts. POMs are characterized by strong Brønsted acidity, high proton mobility, fast multi-electron transfer, high solubility in various solvents and excellent resistance against hydrolytic or oxidative degradations in solution. The named properties make this class of catalysts very attractive for the oxidative conversion of biomass in aqueous media. In particular, their tunable redox potential and acidity offers enormous potential for their future use in oxidation chemistry. To increase the efficiency of the OxFA process for the conversion of water-insoluble, complex biomass, we recently introduced the use of sulfonic acids as additives. The latter have been found to act as efficient depolymerization catalysts, solubilizers and promotors. These findings have paved the way for the oxidative processing of e.g. lignocellulosic biomass, a class of substrates that is notoriously difficult to convert in aqueous phase reforming.

Oxidation experiments for expanding the substrate scope

In this contribution, we significantly expand on the scope of biogenic substrates that can be processed with the OxFA-process. Therefore, we first studied a range of second generation feedstocks like cane trash, pomace, fruit pulp, bark from hard and softwood, leafs, straw, grass clippings and aerated material. As representatives for third generation biomasses, we tested green algae like *chlorella* and *ulva lactuca*, brown algae like *ascophyllum nodosom*, blue-green algae like *spirulina* or *cyanobacter* and *chondrus crispus*as red algae. To enlarge the list of feedstocks even to the treatment of problematic and contaminated biological waste materials, we also applied effluent sludge and deinking sludge from a paper manufacturing plant under typical OxFA process conditions. Additionally, we used railway sleepers and beech condensates as substrates for the process. The effluent sludge contained about 80 wt% organic materials with a high nitrogen content of about 10 wt%. This substrate also contained large amounts (ca. 10 wt% in total) of inorganic materials and salts like calcium oxide and different phosphates. The applied deinking sludge was of even more inhomogeneous composition. Besides 35 wt% of organic material, it contained many different heavy metal salts like copper, lead, cadmium and nickel compounds. Furthermore, this substrate contained highly volatile halogenated hydrocarbons, like chloroethane and vinyl chloride as well as polycyclic and polychloric hydrocarbons. The applied railway sleepers were characterized by a high content of aromatic compounds. The applied beech condensate was also rich in aromatic compounds.

Based on our previous findings, we applied for the screening experiments of the named new substrates molecular oxygen as the oxidant, the polyoxometalate $H_5[PV_2Mo_{10}O_{40}]$ (HPA-2) in aqueous solution as a catalyst and para-toluenesulfonic acid (TSA) as an additive and solubilizer. The conversion of all substrates was performed in a 600 mL Hastelloy autoclave with a gas entrainment impeller and additional cooling to deal with the exothermicity of the reaction. Effects of different liquid volumes in the reactor (100 mL of reaction mixture in 600 mL of reactor volume) can be excluded as the reaction takes place only in the liquid phase where the oxygen is entrained by the impeller. In particular, there is no further reaction in the gas phase. All biogenic feedstocks were processed for 24 hours at 90 °C under 30 bar oxygen pressure. The initial composition of the reaction mixture in each oxidation experiment was 100 g of water as a solvent, 1.74 g (0.9 mmol) of HPA-2 polyoxometalate as a catalyst, 1.90 g (11 mmol) of TSA as additive/solubilizer and 3.30 g of the substrate under investigation. To enable the calculation of yields based on the comparable basis of carbon atoms available in the substrate, we performed an elemental analysis of each substrate prior to our catalytic experiments (results see Table 1).

The results in Table 1 clearly show that the applied homogeneous oxidation system is able to convert all applied 2nd and 3rd generation biomasses to predominantly formic acid and carbon dioxide. Under the applied conditions no CO formation from FA decomposition and no catalytic FA decomposition to $CO_2$and hydrogen were detected (see Table 1, entry 19). The yield in FA was found to depend strongly on the structure of the substrate and the number of oxygen-functionalities in the carbon-framework. Among the tested new substrates, the best raw material for producing formic acid was found to be pomace with 55 % FA-yield (Entry 1) followed by cane trash with 49 % FA-yield (Entry 2) and fruit pulp with 45 % FA-yield (Entry 3). Note that all these latter three origin from commercial and industrial biomass conversion processes making the here presented technology an interesting add-on to widely practiced processes of the food industry.

Table 1: Oxidative conversion of complex, water-insoluble biomass of 2nd and 3rd generation in presence of the additive TSA using a HPA-2 POM-catalyst.

Entry	Substrate	Molecular composition[a]	Combined Yield[b] $FA+CO_2$ [%]	FA-Yield[b] [%]	$FA:CO_2$-selectivity[b] [%]
1	Pomace	$C_{0.24}H_{1.59}O_1$	97.3	54.6	

56:44

2

Cane trash

$C_{0.92}H_{2.04}O_1$

97.5

48.9

50:50

3

Fruit pulp

$C_{1.00}H_{1.92}O_1$

99.0

45.3

46:54

4

Spruce chips

$C_{1.30}H_{1.98}O_1$

89.5

34.6

39:61

5

Poplar splint

$C_{1.32}H_{1.98}O_1$

66.0

30.8

47:53

6

Grass clippings

$C_{1.17}H_{2.05}O_1$

74.6

29.9

40:60

7

Chondrus crispus

$C_{0.71}H_{1.30}O_1$

48.3

21.7

45:55

8

Chlorella

$C_{1.69}H_{2.92}O_1$

53.7

21.6

40:60

9

Oak bark

$C_{1.21}H_{1.87}O_1$

56.0

21.1

38:62

10

Willow bark

$C_{1.34}H_{2.15}O_1$

59.5

16.1

27:73

11

Cyanobacter

$C_{0.72}H_{1.32}O_1$

27.4

15.8

58:42

12

Spirulina

$C_{1.70}H_{2.99}O_1$

40.6

15.7

39:61

13

Birch bark

$C_{1.39}H_{2.29}O_1$

41.0

15.5

38:62

14

Ascophyllum nodosum

$C_{0.82}H_{1.42}O_1$

29.2

14.8

51:49

15

Straw

$C_{1.04}H_{1.62}O_1$

53.1

11.9

22:78

16

Nettle leafs

$C_{1.10}H_{1.98}O_1$

54.2

11.0

20:80

17

Ulva lactuca

$C_{0.75}H_{1.53}O_1$

16.4

7.6

46:54

18

TSA

$C_7H_8O_3S$

-

-

-

19

FA

CH_2O_2

-

-

-

20

Neat pomace[c]

$C_{0.24}H_{1.59}O_1$

52.0

12.5

24:76

21

Neat cyanobacter[c]

$C_{0.72}H_{1.32}O_1$

14.8

4.3

29:71

Reaction conditions: 3.3 g substrate, 1.9 g (11 mmol) additive and 1.74 g (0.9 mmol) catalyst dissolved in 100.0 mL H_2O, 30 bar O_2, 90 °C, 24 h, 1000 rpm; [a] Determined via C, H, N, S elemental analysis; [b] Yield and selectivity determined by means of ^{1}H-NMR using benzene as an external standard according to n (FA)/n(C-atoms feedstock); [c] Reference experiments without TSA additive.

Interestingly, also substrates from the third generation like red algae (Chondrus crispus) or green algae (Chlorella) gave surprisingly high FA-yields. Note, that for these substrates the selected reaction time of 24 hours was definitely too short, so not all the substrate was converted to FA and CO_2. From our experience, it can be expected that prolonged reaction times would lead to a final FA yield for these substances close to the FA selectivity obtained after 24 h reaction time. These experiments could also confirm the very positive effect of the toluenesulfonic acid (TSA) additive and solubilizer on the POM-catalyzed biomass oxidation. Experimental evidence is provided by the reactions with and without the TSA additive (compare Table 1, entries 1 and 11 against entries 20 and 21). For pomace, the experiment without TSA (Entry 20) showed only very little FA formation (Y (FA) =12.5 %) compared to a FA-yield of 54.6 % (entry 1) with TSA for otherwise identical conditions. For cyanobacter, the experiment without TSA (entry 21) yielded only 4.3 % FA after 24 h reaction time, in contrast to the experiment with the TSA additive in entry 11 (15.8 % FA-yield). Based on these results, it is obvious that the TSA additive does not only act as a solubilizer but also as reaction accelerator and selectivity enhancer. The latter can be explained by the solubilizing effect of the TSA additive giving the homogeneous POM catalyst more access to the functional groups within the carbon-framework. As these groups are normally carbonyl- or hydroxy-groups this leads to an increase of the FA-yield based on our previous findings with different funtionalized C_2-model compounds. These results are in good agreement with previous findings concerning the suitability of the TSA additive for the transformation of weekly water soluble biomass substrates using homogeneous POM catalysts in water. A baseline experiment using pure TSA as the substrate showed again no oxidation neither to FA nor to CO_2(see Table 1, entry 18).

In an additional set of experiments, we tested conversion of chemically contaminated substrates in the POM catalyzed oxidation. For this purpose, we processed effluent sludge, railway sleeper, beech condensate and deinking sludge under typical OxFA conditions. In these experiments, the initial composition of the reaction mixtures contained 100 g of water as a solvent, 1.74 g (0.9 mmol) of HPA-2 as a catalyst and between 1.85 g and 10 g (100 mmol) of the substrate dependent on its water content. The additive TSA was not used in these experiments as the unknown composition of the applied â€œreal lifeâ€ substrates was expected to open additional but unknown reaction pathways for TSA (e.g. salt formation with metal ions). This would lead to consecutive reactions and uncontrollable loss of the additive which would have made interpretation of the obtained results rather difficult. For all experiments of this series the reaction temperature was kept at 90 °C in order to prevent thermal FA decomposition. The oxidation reactions were carried out at 30 bar oxygen atmosphere with stirring of 1000 rpm for 24 h reaction time. The obtained results are summarized in Table 2.

Table 2: Oxidative conversion of chemically contaminated biomass substrates with the HPA-2 catalyst.

Entry	Substrate	Molecular composition[a]	Combined Yield[b] FA+CO_2 [%]	FA-Yield[b] [%]	FA:CO_2-Selectivity[b] [%]
1	Effluent sludge	$C_{6.31}H_{11.35}O_{6.63}N_{1}$	46.1	15.7	34:66
2	Railway sleeper	$C_{1.25}H_{1.90}O_{1}$	39.5	11.2	29:71
3	Beech condensate	$C_{1.21}H_{1.86}O_{1}$	2.5	1.0	41:59
4	Deinking sludge				

$C_{0.38}H_{0.47}O_1$

-

-

-

Reaction conditions: 100 mmol substrate (1.85 g for entry 1; 5 g for entry 2; 10 g for entry 3 and 5.35 g for entry 4) depending on molecular weight; 1.74 g (0.9 mmol) catalyst dissolved in 100.0 mL H_2O; 30 bar O_2, 90 °C, 24 h, 1000 rpm. [a] Determined via C, H, N, S elemental analysis; [b] Yield and selectivity determined by means of ^{1}H-NMR using benzene as an external standard according to n(FA)/n(C-atoms feedstock).

It is obvious that the HPA-2 catalyst also works with contaminated feedstock and is highly robust against many impurities. Thus effluent sludge with the highest amount of organic and oxidisable matter yields nearly 16 % FA at a combined FA + CO_2 yield of 46 % after 24 h reaction time. The high nitrogen content within the sludge from a wastewater treatment plant has a negative effect on the conversion rate but does not prevent the oxidation reaction. Also for railway sleepers containing high amounts of aromatic compounds, the oxidation to FA and CO_2 is observed. A FA + CO_2combined yield of 40 % is observed after 24 h resulting with FA selectivity of 29 %. In contrast to the latter two substrates, highly diluted beech condensate and deinking sludge show almost no oxidative conversion after 24 h reaction time.

Defined poisoning experiments for the POM catalyst

In order to elucidate the complete lack of reactivity with the deinking sludge substrate we performed a couple of poisoning experiments to learn more about the sensitivities of the applied POM oxidation system. Our first hypothesis was that the heavy metal content in the applied deinking sludge (Palm GmbH & Co. KG paper manufacturing company) could be responsible for the observed catalyst deactivation. Such heavy-metal cations are expected to form water-insoluble and inactive complexes with the anion of the polyoxometalate catalyst. To confirm this hypothesis, we added on-purpose water-soluble salts of higher transition metals like copper, lead, cadmium and nickel as contaminants to a glucose oxidation experiment. Moreover, similar contamination experiments were performed with other typical components of deinking sludge, such as e.g. gypsum, kaoline, TiO_2 or ZrO_2. For these experiments, 5 g (25 mmol) of glucose as a water-soluble substrate and 1.74 g (0.9 mmol) of the catalyst dissolved in 100 mL water, together with 8 mmol (0.63-2.18 g) of the impurity were processed under 90 °C, 30 bar oxygen with 1000 rpm for 6.5 h in a 600 mL high-pressure autoclave. All results are displayed in Table 3; the baseline result for a non-contaminated glucose oxidation experiment is shown as entry 9.

Table 3: HPA-2 catalysed glucose oxidation - defined poisoning experiments with typical impurities from paper manufacturing to elucidate the lack of catalytic activity for deinking sludge as substrate.

No.	Impurity	Potential source of origin	Combined Yield[a] FA+CO_2 [%]	FA-Yield[a] [%]	FA:CO_2-Selectivity[a] [%]

1

$CuSO_4$

printerâ€™s ink

39.8

-

0:100

2

$CdCl_2$

printerâ€™s ink

87.6

44.4

51:49

3

$NiCl_2$

printerâ€™s ink

n.d.

n.d.

n.d.

4

$PbCl_2$

printerâ€™s ink

6.2

4.4

71:29

5

Gypsum

filler

89.9

				47.5	53:47
6	Kaoline	filler	93.7	49.2	53:47
7	TiO_2	filler	72.3	44.4	61:39
8	ZrO_2	paper varnish	83.0	38.8	47:53
9	-	-	100	46.2	46:54

Reaction conditions: 5 g (25 mmol) glucose as a substrate; 1.74 g (0.9 mmol) HPA-2 as a catalyst dissolved in 100.0 mL H_2O; 8 mmol (0.63-2.18 g) of the impurity depending on molecular weight; 30 bar O_2, 90 °C, 6.5 h, 1000 rpm. [a] Yield and selectivity determined by means of ^{1}H-NMR using benzene as an external standard according to n(FA)/n(C-atoms feedstock).

The results in Table 3 clearly indicate a strong influence of Pb- and Cu-salts on the catalytic activity for glucose oxidation. Other heavy metal ions, like Cd^{2+}, appear to be of little effect. Interestingly, the presence of copper ions does completely inhibit the formation of FA from glucose oxidation, only CO_2 is detected as a product. This goes along with the observation that copper ions form water-insoluble complexes with the catalyst heteropolyanion in form of small blue particles at the bottom of the reaction vessel. Unfortunately, the oxidation result with the added $NiCl_2$ could not be quantified by the usual NMR-method due to the strong deformation of the spectra due to the paramagnetic nature of Ni (II) in the sample. Nevertheless, also for this case, the formation of FA could be qualitatively confirmed by the help of ^{13}C-NMR. In contrast to the experiments with Cu, no formation of water-insoluble crystals was observed in the case of $NiCl_2$ addition. Finally, the addition of Pb (II) species to the aqueous glucose solution resulted in an immediate color change from orange to a blue-green solution. The ^{51}V-NMR spectra of the obtained solution strongly indicate complexation of the heteropolyanion with the Pb^{2+} cations. In addition, the formation of yellow crystals at the bottom of the reaction vessel was observed thus removing the active catalyst from solution. In general, ^{51}V-NMR spectra of the reaction solutions containing late transition-metal cations show significant changes after addition of Cu-, Ni- and Pb-salts. In contrast, the ^{51}V-NMR spectra with added fillers do not show significant changes of the polyoxometalate catalyst. Addition of these water-insoluble fillers, like gypsum, kaoline, titanium dioxide and zirconia, resulted in an increase of the viscosity of the reaction mixture; however, this had very little effect on the observed glucose oxidation reaction.

Our contribution gives strong support to the scientific and economic attractiveness of homogeneously catalyzed biomass oxidation to formic acid by polyoxometalate catalysts. We could expand the substrate scope for the reaction considerably by converting a very wide range of substrates including many complex water-insoluble biogenic feeds and also contaminated materials. While waste from the food industry, effluent sludge and even railway sleeper was found to be effectively converted under typical OxFA reaction conditions (90 °C, 30 bar O_2, 24 h reaction time), a clear and so far undetected limitation has been found for metal ion containing feedstocks. Whenever contaminated biological feed with metal ions forming stable water-insoluble complexes with the catalytically active polyoxometalate ion were present in the applied feedstock, the reactivity of the oxidation system was strongly reduced or almost completely eliminated. The favorable role of the TSA additive for supporting depolymerization and solubilisation of 2nd and 3rd generation was impressively underlined by comparative oxidation experiments with pomace and cyanobacter. From all results of this study we anticipate a high potential of POM-catalyzed biomass oxidation for future decentralized biomass conversion processes of wet and contaminated substrates. As formic acid is a chemical product used in high amounts by farmers (e.g. in the preparation of silage) and waste streams from farming or food manufacturing represent very attractive feedstocks, the technology appears particularly attractive if linked to agricultural activities or small to medium sized food industries.

456032 Systematic Framework for Carbon Dioxide Capture and Utilization Processes to Reduce the Global Carbon Dioxide Emissions

Abstract:

In the year 2013, 9.5 billion metric tons of carbon dioxide gas was emitted into the air, and each year this amount is increasing [1]. Carbon dioxide emissions are of particular concern as they represent 80% of greenhouse gas emissions and therefore are a large contributor to global warming. Among the two approaches that are currently being investigated, carbon capture and storage (CCS) and carbon capture and utilization (CCU) [1] to address this issue, the later approach is more promising as it reuses captured carbon dioxide, as a fuel, reactant, solvent, and others, to produce valuable products. There is not only a need for technologies for capture and utilization, via conversion, but also there are numerous questions that need to be resolved. For example, which higher value chemicals can be produced, what are their current demands and costs of production, and, how much of the captured carbon dioxide would be utilized? Also, how much carbon dioxide would be indirectly generated due to the capture and utilization efforts? Can the regulated carbon dioxide emission reduction targets be achieved only through the CCU and/or CCS efforts? Therefore, there is a need for a systematic computer-aided framework through which the issue of global carbon dioxide emissions can be investigated in terms of different available capture-utilization technologies, solution methods, and benefit scenarios, with the objective to determine more sustainable solutions within an appropriate application boundary. The framework would need to provide, amongst other options: useful data from in-house databases on carbon dioxide emission sources; mathematical models from a library of process-property models; numerical solvers from library of implemented solvers; and, work-flows and data-flows for different benefit scenarios to be investigated.

It is useful to start by developing a prototype framework and then augmenting its application range by increasing the contents of its databases, libraries and work-flows and data-flows. The objective is to present such a prototype framework with its implemented database containing collected information-data on various carbon dioxide emission sources and available capture-utilization technologies; the model and solution libraries [2]; and the generic 3-stage approach for determining more sustainable solutions [3] through superstructure (processing networks) based optimization – adopted for global carbon dioxide emission problems. More specifically, the prototype framework and the three-stage approach adopted for systematic and sustainable design of carbon capture and utilization processes incorporates (i) process synthesis stage, the determination of a processing path from a network of alternatives; (ii) process design, the design and analysis of a process or generated processing path in terms of process "hot-spots" or deficiencies to set design targets for improvement; and (iii) innovative and more sustainable design, determination of solutions matching the design targets. In this way, the starting point is an analysis of the current carbon dioxide emission status and the end point is an analysis of the more sustainable solutions in terms of one or more carbon dioxide benefit scenarios.

The developed framework contains a database constructed based on a developed knowledge representation structure, which has collected data on carbon dioxide emission sources, capture technologies, carbon dioxide-based reaction paths, and known utilization technologies leading to various products. The database currently has 5 carbon dioxide emission sources, 3 capture technologies, 10 conversion routes leading to 8 products, and over 50 reaction paths for many more products. The framework also integrates a software tool (the Super-O interface [2]) containing a library of process models and links to numerical solvers that enables the generation of processing networks (superstructure) and the determination of the optimal processing route. Finally, the framework has access to, a collection of tools for analysis, such as economic analysis, sustainability and life cycle assessment, and links to detailed process simulation (process simulators).

Following the work-flow and data-flow implemented in the framework, data for the carbon capture and utilization alternatives are retrieved from the database and linked to form a network using the interface tool, Super-O. For each alternative, models for economic and environmental impacts are available; the promising utilization products considered are methanol, dimethyl carbonate, and succinic acid. The optimal utilization process (or processes) is determined via a superstructure-based method (using solvers in GAMS). Then, the carbon dioxide emission values for the utilization routes are compared with the current industrial production routes for the listed products in order to identify those that have zero or negative emission values (objective function). Subsequently, the identified utilization processes are designed rigorously and further improved with unique technology and integration options. Using this approach, the utilization processes for methanol, dimethyl carbonate and succinic acid give the following interesting result: by converting at least 70% (methanol), 40 % (dimethyl carbonate) and 20% (succinic acid) of the current industrial processes to the corresponding utilization processes, the emissions from these processes can be neutralized, representing over 100 million metric tons of emissions that are neutralized. Note that the issue of product demand increase has not been considered. Also, on a global scale, this reduction of carbon dioxide emissions represents a very small but very valuable reduction. A more complete sustainability analysis is needed to enlarge the boundary of the analysis; for example, the boundary can change in terms of reduction of non-renewable resources and the use of renewable energy sources, as well as more efficient and sustainable designs of current production routes. While a combination of methods is needed to globally address carbon dioxide emissions and other aspects of sustainability, this systematic approach to designing carbon capture and utilization processes shows the capacity for such processes to reduce emissions and improve sustainability while producing valuable products. The current prototype framework with its implemented methods and tools is a small but important step. Collaboration and integration of data, methods and tools is necessary to provide a more sustainable solution to the global carbon dioxide emission problem.

References

[1] Yuan, Z., Eden, M.R., Gani, R., 2016. Toward the Development and Deployment of Large-Scale Carbon Dioxide Capture and Conversion Processes. Ind. Eng. Chem. Res. 55, 3383–3419. doi:10.1021/acs.iecr.5b03277

[2] Bertran, M.-O., Frauzem, R., Zhang, L., Gani, R., 2016. A generic methodology for superstructure optimization of different processing networks, in: Kravanja, Z. (Ed.), Proceedings of the 26th European Symposium on Computer Aided Process Engineering – ESCAPE 26. Elsevier B.V., Portorož, Slovenia.

[3] Babi, D.K., Holtbruegge, J., Lutze, P., Gorak, A., Woodley, J.M., Gani, R., 2015. Sustainable process synthesis–intensification. Comput. Chem. Eng. 81, 218–244. doi:10.1016/j.compchemeng.2015.04.030

462685 Design of Energy Saving CO2 Separation Process Using Circulating Fluidized Bed

Abstract:

By fossil fuel combustion, large amounts of greenhouse gases (GHGs) are generated and emitted into the atmosphere. The reduction of carbon dioxide (CO2) emission has become a major target in efforts to suppress global warming. Carbon capture and storage (CCS) has attracted attention in the past decades to reduce greenhouse gas (GHG) emissions. CO2 separation from gas mixtures requires huge amounts of energy. In fact, CO2 separation from gas mixtures is one of the most energy intensive process in CCS. Therefore it is necessary to reduce the energy consumption of CO2 separation for propagation of CCS.

The most widely used method among CO2 separation is a chemical absorption using amine based solution such as monoethanolamine (MEA). In the conventional CO2 separation by absorption using amine based solution, the thermal energy is supplied from the bottom of regenerator which utilizes reaction heat for regeneration of the solutions and heat of vaporization for stripping. To reduce the supply of the thermal energy, authors have developed CO2 gas separation processes based on self-heat recuperation. Applying the self-heat recuperation to thermal processes, not only the latent heat but also the sensible heat of the process stream can be circulated into the processes without any heat addition, leading to minimizing exergy destruction of heat transfer and reduction of the process energy consumption. However, the absorption using solution physically and chemically requires additional energy for separation of liquid/gas, vaporization for stripping.

In this paper, an innovative energy saving CO2 separation process using chemical absorption was proposed for energy saving point of view. In this process, metal oxide particles were selected as absorbents not for requiring separation energy of liquid/gas. These particles were fluidized by feed gas mixture for increasing segregation effect in the bed and heat transfer rate from chemical reaction to a heat pump system. Heat of absorption was recuperated and supplied to the heat of thermal decomposition through the heat pump by following exergy recuperation. In addition, some of the heat was circulating with absorbents. The energy saving performance of the proposed process was evaluated by simulations.

464620 Systems Design and Economic Analysis of Direct Air Capture of CO2 through Temperature Vacuum Swing Adsorption on Metal Organic Frameworks

Abstract:

The rising levels of CO_2 concentration in the atmosphere poses a serious threat to global climate stability. To mitigate this threat, efforts needs to be made for reduction of CO_2 levels. One of the strategy is to capture CO_2 from stationary, point, sources. These point sources accounts for approximately only one-third of the global emissions. Another strategy is to capture CO_2 directly from ambient air which, if successfully implemented, could result in capture of CO_2 from disperse emission sources. In contrast to CO_2 capture from stationary sources, Direct Air Capture (DAC) plants are location independent and can be scaled more easily. Thus, the DAC plant can be set up near a sequestration or utilization site, eliminating the need for CO_2transportation infrastructure.

Our work proposes comparison of two metal organic frameworks, namely, MIL-101(Cr)-PEI and mmen-Mg_2(dobpdc) [1] for DAC through temperature vacuum swing adsorption (TVSA). Isotherm parameter estimation for CO_2 adsorption was carried out for both the adsorbents using published and original data and a kinetic study was performed through empirical correlations [2] to estimate the mass transfer rates for the process . The adsorbents are grown as films inside 900 cpsi (cells per square inch) monolith. To achieve a temperature swing, steam is used as a stripping agent during the desorption step. In this thermal desorption operation, it is believed that the amines groups present in the metal organic frameworks are prone to oxidative degeneration at higher temperatures[3]. To overcome this challenge, we propose a five step process: adsorption, evacuation, pressurization, desorption and cooling. In the proposed process, evacuation and cooling steps are implemented through vacuum swing to remove the oxygen in the channels when the temperature is high. In order to model this system, partial differential algebraic equations have been implemented in gPROMS, which is a commercial dynamic process modeling and optimization software. The linear driving force (LDF) model is used to approximate the rate of CO_2adsorption [2].

This talk will discuss the energy requirements and overall economics of this DAC process. Different components which contribute to consumption of energy are identified and the TVSA model is simulated to estimate the net energy requirement for MIL-101-PEI-50 and mmen-Mg_2(dobpdc). Operating costs during the five steps (adsorption, evacuation, pressurization, desorption and cooling) and capital costs for metal organic frameworks and monoliths have been estimated for both the adsorbents. It has been shown that lifetime of the adsorbent is critical in estimating the net cost and a sensitivity analysis of the adsorbentâ€™s lifetime has been performed to analyze the DAC economics for both the adsorbents.Â We have identified sensitive parameters such as gas flow rate, cycle time, film thickness etc. which effects the overall energy requirements and net economics of the DAC process.

1. McDonald, T.M., et al., *Capture of carbon dioxide from air and flue gas in the alkylamine-appended metal-organic framework mmen-Mg2(dobpdc).* J Am Chem Soc, 2012. **134**(16): p. 7056-65.

2. Y. Y. Li, S.P.P., B. D. Crittenden, *Zeolite monoliths for air separation Part 2: Oxygen Enrichment, Pressure Drop and Pressurization.* Trans IChemE, 1998. **76**: p. 931-941.

3. Heydari-Gorji, A. and A. Sayari, *Thermal, Oxidative, and CO2-Induced Degradation of Supported Polyethylenimine Adsorbents.* Industrial & Engineering Chemistry Research, 2012. **51**(19): p. 6887-6894.

466081 New Performance Indicators for Adsorbents Used in CO_2 Capture Swing Adsorption Processes

Abstract:

As adsorption process has increasingly received attention in CO_2 capture area, a number of studies have suggested a variety of adsorbents such a metal-organic frameworks (MOFs), zeolitic imidazolate frameworks (ZIFs) and many modified versions of conventional adsorbents. However, because of their limited scope of the studies, they have evaluated their adsorbents by using conventional performance indicators: working capacity and selectivity. The performance indicators rarely consider the situation of the adsorbents' practical application in the adsorption process. In order to evaluate the adsorbents with the consideration of their eventual application, e.g., pressure swing adsorption (PSA) process, dynamic simulations with a rigorous model of the adsorption process are needed, but the simulations cannot be carried out conveniently by the developers of new adsorbents. Furthermore, the high computation cost caused by the nonlinearity and cyclic operation of the PSA system hinders the evaluation of adsorbentsâ€™ suitability for their process-level application. For those reasons, this work derives simple formulas for the process-level performance of an idealized PSA system. The efficiency and the CO_2 purity of produced gas in the idealized PSA system are defined as performance indicators for the evaluation of a target adsorbent. For further simplicity, the performance indicators are derived in the explicit form of an analytic solution. The derived explicit equations are suggested as formulas of the new performance indicators for the evaluation. With the formulas, the performance of a target adsorbent in the PSA process can be obtained with simple calculation only from lab-scale measurement data such as adsorption isotherm data. Because the suggested formulas are based on the idealized PSA process where the theoretically maximum performance of each adsorbent is achieved, the suggested way is fairer than comparing the simulations of different adsorbents with some prior design and operating conditions which may be preferred by particular adsorbents. In order to show the usefulness of the formulas, a case study involving several real adsorbents is presented. The results from the formulas about the new indicators are then compared both to the simulation results of a rigorous dynamic model and to the conventional performance indicators.

Acknowledgement: The authors would like to acknowledge the financial support from the grant (G01160051) funded by Saudi Aramco-KAIST CO2 management center.

466627 Optimal Retrofit of a Post Combustion CO_2 Capture Process Using Reduced Superstructure and Rate-Based Models

Abstract:

We propose a superstructure-based methodology for optimal retrofit of a CO_2 capture pilot plant using rigorous rate-based model for the reactive distillation. The optimal retrofit for existing plants is particular challenging because rigorous process models used for describing the currently operating process often result very complex optimization problems.[1] Herein, we preselect process alternatives implemented in the superstructure based on thermodynamic analysis. Thermodynamic analysis identifies unit operations that have high energy saving potential and process alternatives improving these unit processes are considered in the superstructure. Consequently, the problem size of the superstructure can be substantially reduced and optimization can be performed using commercially available process simulators and optimization solvers. The optimum process configuration and operating conditions are obtained stochastically using a genetic algorithm. Near global optimum solution is obtained by solving the problem multiple times with different crossover fraction. Results indicate that the optimum retrofit process includes three stages solvent recirculation, lean vapor recompression, and mechanical vapor recompression. Thermal energy and total energy consumption in the optimal retrofit process are reduced about 59% and 27%, respectively.

471576 Dynamic Modeling with Uncertainty Quantification of Solid Sorbent Based CO_2 Capture Processes

Abstract:

Post-combustion CO_2 capture is a potential approach to mitigate buildup of CO_2 concentration in the environment. Significant research is currently being carried out all over the world to reduce the penalty for CO_2 capture. Due to the large amount of CO_2 that needs to be captured, significant scale-up is needed, starting from the lab scale investigation of a potential capture technology to its commercial implementation for large-scale power plants. Since scale-up of novel chemical technologies generally extends over 20-30 years, it is imperative that the time required be significantly reduced for the accelerated commercialization of the potential technologies. Rigorous process models of these novel capture processes can be very useful for this purpose. However, process models suffer from uncertainty. The parameters for submodels are typically calibrated deterministically for each submodel. Further, the models are imperfect and the experimental data are not only imperfect, but are often inadequate. The resulting parametric and model-form uncertainties can lead to high discrepancy in the model outputs. Therefore, process models should be developed with quantified uncertainty, which can also be instrumental in identifying where additional experimental data should be collected and/or where different or modified model form should be considered to reduce the discrepancy.

With these motivations, a 1-D dynamic, non-isothermal and non-adiabatic fixed bed adsorption reactor model is developed for novel solid sorbent-based CO_2 capture processes. Rigorous models for mass transfer, heat transfer and chemical kinetics, if applicable, of both physisorption- as well as chemisorption-dominated processes are developed. A Bayesian calibration approach is used for quantifying parametric uncertainty. Model-form discrepancy is described by a set of orthogonal functions. Posterior distribution of the parameters as well as the model-form discrepancies are propagated through the process model to obtain the bounded confidence on the modelâ€™s predictions of variables of interest such as bed pressure drop, breakthrough curve, and bed temperature profile. Our study shows that the output discrepancy strongly depends on the operating conditions and helps to identify approaches that can be used to reduce the scale-up uncertainty of the novel CO_2 capture processes.

468716 CO_2 Capture and Conversion to Chemicals Via Syngas: Rigorous Modeling, Intensification, and Superstructure-Based Process Synthesis

Abstract:

Recent global warming and climate change are attributed to increased CO_2 emission from fossil fuels. Significant efforts have been made in the past to develop CO_2 capture and sequestration (CCS) technologies. However, large-scale CCS is still not deployed in reality for reasons including high cost, technological barriers and uncertainty in geological storage. A promising alternative to CCS is to convert CO_2 to value-added chemicals via syngas (CO and H_2), which is an intermediate for many hydrocarbon-based fuels and chemicals. As much as 2 Gt/yr of fuels and 200 Mt/yr of chemicals can be produced from CO_2 utilization [1].

In this work, we explore alternative technologies and routes for the thermochemical conversion of CO_2 to syngas and then syngas to various oxygenates. We postulate a flowsheet superstructure which includes layers with various alternatives for separation, conversion and upgrading. We allow both pure and dilute raw materials, such as CO_2 from flue gas, methane from stranded sources, oxygen from air and water, and hydrogen from renewables and other sources. A novel feature of our superstructure is that it also includes various intensified alternatives. One such example is the inclusion of a tri-reforming membrane reactor which, instead of using pure CO_2, uses the mixed flue gas as feed, produces syngas, and separates nitrogen from the product at the same time. Therefore, it does not need a separate CO_2 capture section at the upstream. For each alternative, we develop a rigorous model detailing the transport and kinetics to accurately predict the process performance. We use a pseudo-homogenous model for plug-flow reactors (PFR) [2] to describe various alternatives for CO_2 reforming (e.g., steam methane reforming, dry-reforming, tri-reforming, partial oxidation, and variants of combined reforming). Using the rigorous simulation platform, we also develop efficient surrogate models [3] for conversion as it changes with reactor type, design and operating conditions. The replacement of the ODE models with their algebraic surrogates allows us to pose the overall synthesis problem as a mixed-integer nonlinear optimization (MINLP) problem, which we solve to optimality using ANTIGONE [4]. We consider different objectives such as maximizing the overall utilization of CO_2, minimizing the total annualized cost, and maximizing profit. We also take into account the auxiliary carbon dioxide emissions associated with various processing tasks in the network. In this presentation, we will discuss our overall synthesis framework and the optimization results.

References:

[1] Quadrelli, E. A.; Centi, G.; Duplan, J.-L.; Perathoner, S. Carbon Dioxide Recycling: Emerging Large-Scale Technologies with Industrial Potential. *ChemSusChem.* 2011, 4:1194â€“1215.

[2] Aboosadi, Z. A.; Jahanmiri, A. H.; Rahimpour, M. R. Optimization of tri-reformer reactor to produce synthesis gas for methanol production using differential evolution (DE) method. *Applied Energy*, 88(8):2691â€“2701, 2011.

[3] Bajaj, I.; Hasan, M. M. F. Effective Sampling, Modeling and Optimization of Constrained Black-Box Problems. Accepted for publication in proceedings of ESCAPE26, Slovenia, 2016.

[4] Misener, R.; Floudas, C. A. ANTIGONE: algorithms for continuous/integer global optimization of nonlinear equations. *Journal of Global Optimization.* 59 (2-3), 503-526, 2014.

463231 Techno-Economic and Life Cycle Analysis of Chemical Production from Biomass Feedstocks

Abstract:

Techno-economic and Life Cycle Analysis of Chemical Production from Biomass Feedstocks

Abhay Athaley[1], Praneeth Annam[1], Basudeb Saha[2], Marianthi Ierapetritou[1]

[1] Department of Chemical and Biochemical Engineering, Rutgers - The State University of New Jersey

[2.]Catalysis Center for Energy Innovation & Department of Chemical & Biomolecular Engineering, University of Delaware

This work focuses on the techno-economic analysis and life cycle analysis on the production of chemicals from lignocellulose feedstocks including three types of hydrolysis of lignocellulose to produce sugar mixtures to convert to furan derivatives and the production of p-Xylene. The development of efficient and economically sustainable routes for biobased chemical production has raised a lot of interest especially for the platform chemicals such as hydromethylfurfural (HMF), furfural, levulinic acid, etc.[1,2] The aforementioned chemicals can be produced from glucose or xylose solutions. Different type of hydrolysis is used to produce sugar mixtures such as enzymatic hydrolysis, thermo-chemical hydrolysis etc.[3,4,5] It is found that it is more economically and environmentally attractive to manufacture HMF and furfural from C5 and C6 mixtures that are produced from hydrolysis of 2nd generation (lignocellulosic) biomass.[6,7]

The acceptance of bio-products in the market place depends on their competitiveness compared with petroleum-based products not only in terms of economics but also in terms of process sustainability[8]. In our previous work we evaluated the economics and environmental impacts of bio-products such as p-Xylene[9-11]; however this study compares different hydrolysis processes and the integrated flowsheet to produce p-Xylene. This is significant since different hydrolysis processes produce different mixtures of C5 and C6. Present day, water and energy systems are interconnected with each other- water-carbon-energy nexus. Hence life cycle analysis is also carried out in this study to check the competitiveness of the above process in terms of sustainability.

This work uses techno-economic analysis and life cycle assessment to design and evaluate the production alternatives. The detailed process flowsheet is developed and simulated using Aspen Plus.[12]Three different hydrolysis steps are used to produce sugar mixtures: 1) Dilute Acid Hydrolysis 2) Concentrated Acid Hydrolysis and 3) Hydrolysis of Biomass using Molten Salt Hydrates (add references to each one). At the next stage heat integration is carried out and economic analysis is done. Also sensitivity analysis is carried out to compare the cost of p-Xylene with fluctuating cost of raw materials and to compare the cost of p-Xylene when HMF is considered as by-product to produce different chemicals. It is seen that utility cost is directly related to the solvent requirement and thus related to price of p-Xylene and also the cost of p-Xylene increases when the amount of HMF considered as by-product increases. Finally, life cycle analysis is carried out to check for process sustainability with respect to carbon production and water consumption. High requirement of water is needed for the process and thus sustainability related to water consumption is also presented.

References:

1. Bozell JJ, Petersen GR. Technology development for the production of biobased products from biorefinery carbohydrates-the US Department of Energy's "Top 10" revisited. *Green Chemistry.* 2010;12(4):539-554.

2. Dutta S, De S, Saha B, Alam MI. Advances in conversion of hemicellulosic biomass to furfural and upgrading to biofuels. *Catalysis Science & Technology.* 2012;2(10):2025-2036.

3. Harris JF, Laboratory FP. *Two-stage, dilute sulfuric acid hydrolysis of wood: an investigation of fundamentals*: U.S. Dept. of Agriculture, Forest Service, Forest Products Laboratory; 1985.

4.Weydahl K.R., Process for the production of Alcohol, 2012, Patent application no. 13/318302.

5. Deng W., Kennedy J.R., Tsilomelekis G., Zheng W., Nikolakis V., Cellulose Hydrolysis in Acidified LiBr Molten Salt Hydrate Media, *Industrial & Engineering Chemistry Research*;2015;54:5226-5236.

6. Chheda JN, Roman-Leshkov Y, Dumesic JA. Production of 5-hydroxymethylfurfural and furfural by dehydration of biomass-derived mono- and poly-saccharides. *Green Chemistry.* 2007;9(4):342-350.

7. Zhang J, Lin L, Liu S. Efficient Production of Furan Derivatives from a Sugar Mixture by Catalytic Process. *Energy & Fuels.* 2012/07/19 2012;26(7):4560-4567.

8.Hawkins ACG, David A.; Buelter, Thomas; Wade, James; Meinhold, Peter; Peters, Matthew W.; Gruber, Patrick R.; Evanko, William A.; Aristidou, Aristos A., Assignee, Inventors; Gevo Inc. USA, assignee. METHODS FOR THE ECONOMICAL PRODUCTION OF BIOFUEL PRECURSOR THAT IS ALSO A BIOFUEL FROM BIOMASS 2009.

9. Lin Z, Ierapetritou M, Nikolakis V. Aromatics from Lignocellulosic Biomass: Economic Analysis of the Production of p-Xylene from 5-Hydroxymethylfurfural. *AIChE Journal.* 2013;59(6):2079-2087.

10. Lin Z, Nikolakis V, Ierapetritou M. Alternative Approaches for p-Xylene Production from Starch: Techno-Economic Analysis. *Industrial & Engineering Chemistry Research.* 2014/07/02 2014;53(26):10688-10699.

11. Lin Z, Nikolakis V, Ierapetritou M. Life Cycle Assessment of Biobased p-Xylene Production. *Industrial & Engineering Chemistry Research.* 2015/03/04 2015;54(8):2366-2378.

12. Torres AI, Daoutidis P, Tsapatsis M., Continuous Production of 5- hydroxymethylfurfural from fructose: a design case study. Energy & Environmental Science, 2010;3(10):1560-1572.

467521 Carbon Footprint Analysis of Biomethane from Anaerobic Digestion of a Food Waste / Manure Mixture in Colorado

Abstract:

This study determined the life cycle greenhouse gas (GHG) emissions of bio-methane produced from anaerobic digestion of a mixture of food waste and cow manure in Colorado. Included in the system boundary for this analysis are long distance semi-truck transport of collected food waste from the Denver metropolitan area, avoid landfilling (and composting in a separate scenario) emissions for the food waste, short haul transport of manure waste from local dairy farms, anaerobic digestion process and biogas purification, distribution of the bio-methane into the natural gas pipeline grid, and combustion of bio-methane. The carbon accounting included both fossil and biogenic carbon. The effects of co-products solid compost sold to retail gardening market and nutrient rich liquid digestate distributed to local farm fields were accounted for in the analysis.

After including emission credits for avoiding landfill emissions of CO_2 and CH_4, bio-methane emits over its life cycle -3.5 kg CO_2 equivalents / kg bio-methane, thus reducing global warming effect with each unit of production of bio-methane. This favorable emission intensity is compared to that of fossil methane of between 3.3 to 5.1 kg CO_2 equivalents / kg fossil methane. Effects of life cycle system parameters in the model on GHG results are investigated in scenario analyses and also in time dependent analysis of landfill emissions.

469717 Multistage Torrefaction of Biomass and in Situ Catalytic Upgrading to Hydrocarbon Biofuels and Specialty Biochemicals

Abstract:

Thermochemical conversion of lignocellulosic biomass via pyrolysis and catalytic upgrading of bio-oil is touted as an environmental panacea, capable of producing fuel, energy, and value-added bio-based chemicals with an improved environmental profile compared to conventional fossil routes. However, bio-oil contains a host of compounds with myriad chemical functionalities, making selective catalytic upgrading to fuel and/or chemicals challenging and a major technical bottleneck. This work explores an innovative design strategy, in which the primary constituents of lignocellulosic biomass (i.e. hemicellulose, cellulose, lignin) are selectively decomposed via a series of torrefaction and pyrolysis reactors. Staged thermal decomposition produces several fractionated bio-oil streams comprised primarily of light oxygenates, furan derivatives, and phenolic species respectively. The purposed multistage design permits tailored catalytic upgrading of the fractionated bio-oil streams to hydrocarbon fuels and/or specialty bio-based chemicals, and thus holds promise for improved environmental and economic performance over the current state of the art (i.e. fast pyrolysis and hydroprocessing fuel platforms).

This work performs a prospective life cycle assessment to determine the environmental profile of concurrent hydrocarbon biofuel and specialty biochemical production via a multistage torrefaction and *in situ* catalytic upgrading platform. Several multistage design strategies consisting of different catalytic strategies and targeted platform chemicals (i.e. 2-methylfuran, tetrahydrofuran, and cyclopentanone) are considered. Detailed material and energy balances for multistage systems are developed using a combination of experimental data and ASPEN process simulations. Several critical sustainability metrics including life cycle greenhouse gas (GHG) emissions and Energy Return on Investment (EROI) are used to compare the environmental performance of multistage systems, and benchmark against a base-case fast pyrolysis and hydroprocessing platform. Preliminary results reveal that multistage systems have the capacity of achieving over 75% GHG reductions relative to petroleum diesel, and produce high quality hydrocarbon transportation fuels and specialty bio-chemicals. The influence of different LCA allocation schemes and coproduct scenarios for the produced biochar on the environmental profile of hydrocarbon biofuels and biochemicals will be discussed.

469925 Optimizing Land Use Change and Life Cycle Greenhouse Gas Emissions of Biofuels

Abstract:

Biofuels are a potential source of renewable liquid fuel, in contrast to the nonrenewable fossil fuels we rely on today. Another driver for the interest in biofuels is the possibility that life cycle greenhouse gas (GHG) emissions for biofuels might be smaller than fossil fuels, providing an environmental benefit in addition to the benefits of a secure and renewable fuel supply. However, there still exist GHG emissions sources throughout the biofuel life cycle, and many of them do not exist at a reasonable level for fossil fuels. For example, there are cultivation GHG emissions associated with cultivating the biomass feedstocks, land use change (LUC) GHG emissions incurred by transforming land into cropland to grow the biomass feedstocks, etc. GHG emissions from LUC due to biofuels are perhaps the most difficult to characterize and quantify; they are not released from a smokestack or a tailpipe. A host of researchers propose different characterization and quantification methods to calculate GHG emissions from LUC due to biofuels, but there is still significant disagreement on just how many GHG emissions from LUC biofuels are responsible for.[1,2] In other words, no one has been able to show which mix of biofuel production pathways and biomass sourcing strategies are best to minimize life cycle GHG emissions including LUC emissions. It is imperative to reduce uncertainty on this issue as current biofuels legislation and policies are based on these estimates. Furthermore, previous studies explore fixed biofuel production scenarios, e.g. 14 Bgal/year of bioethanol from corn, and are not able to identify better pathways with diminished GHG emissions from LUC. Thus, not only are current policies and biofuel production and biomass sourcing strategies based on uncertain LUC GHG emissions estimates, but they are also not founded on optimized stratagems.

In this work, we develop a first-ever LUC life cycle optimization (LCO) model that incorporates cutting-edge characterization and calculation methods for bioethanol LUC GHG emissions within an optimization framework. This modeling framework identifies an optimal bioethanol production and biomass sourcing strategy that minimizes LUC GHG emissions by selecting any combination and size of 14 bioethanol production pathways and 5 biomass feedstocks. We choose bioethanol production pathways that are either already commercialized or are commercialization-ready, using data from technoeconomic analyses/reports from the U.S. DOE or the peer-reviewed scientific community.[3-6] LUC and direct processing GHG emissions are considered as well as excess electricity production credits from processing. We calculate LUC GHG emissions from exhaustive data bases for soil organic carbon (SOC), carbonaceous biomass cover, foregone CO_2 sequestration by forested land, and the net primary productivity (NPP) of each land type in each region around the world.[7-10] The CGE is integrated into the LCO modeling framework with the land carbon data to determine LUC GHG emissions around the world due to economic forces arising from increased bioethanol production in the US. Thus, for the first time, cutting-edge LUC characterization and calculation methods are integrated into an LCO modeling framework. Furthermore, this study presents the first time an integrated CGE-LCO model has been developed and utilized.

We consider a case study of US bioethanol production levels ranging from 10 – 100 Bgal/year. We find that minimized LUC GHG emissions from our results are lower than previous estimates, indicating that these previous estimates are at least possible. Corn stover is always preferred, as LUC due to its utilization is negligible. Switchgrass is then selected when the sustainably harvestable corn stover resource base is exhausted. Two-stage dilute acid pretreatment followed by distillation or dilute acid pretreatment followed by pervaporation are preferred technologies for corn stover conversion. Dilute acid pretreatment followed by distillation is always the selected bioethanol production pathway for switchgrass. Forested land is largely preserved around the world whenever possible, and pastureland is by far preferred for conversion to cropland. However, some of the recent LUC GHG emissions estimates for biofuels are close to our minimized value, indicating that some might be optimistic. Nonetheless, if the optimal bioethanol production and biomass sourcing strategies produced by our work are followed, then life cycle GHG emissions including LUC GHG emissions of bioethanol can be managed.

References

1. Qin Z, Dunn JB, Kwon H, Mueller S, Wander MM. Influence of spatially dependent, modeled soil carbon emission factors on life cycle greenhouse gas emissions of corn and cellulosic ethanol. *GCB Bioenergy.* 2016.

2. Plevin RJ, Beckman J, Golub AA, Witcover J, O'Hare M. Carbon accounting and economic model uncertainty of emissions from biofuels-induced land use change. *Environmental science & technology.* 2015;49(5):2656-2664.

3. Pfromm PH, Amanor-Boadu V, Nelson R, Vadlani P, Madl R. Bio-butanol vs. bio-ethanol: a technical and economic assessment for corn and switchgrass fermented by yeast or Clostridium acetobutylicum. *Biomass and Bioenergy.* 2010;34(4):515-524.

4. Zhu Y, Gerber MA, Jones SB, Stevens DJ. *Analysis of the Effects of Compositional and Configurational Assumptions on Product Costs for the Thermochemical Conversion of Lignocellulosic Biomass to Mixed Alcohols – FY 2007 Progress Report,*. Richland, Washington: PNNL;2009.

5. Zhu Y, Jones SB. *Techno-economic Analysis for the Thermochemical Conversion of Lignocellulosic Biomass to Ethanol via Acetic Acid Synthesis.* Richland, Washington: PNNL;2009.

6. Dal-Mas M, Giarola S, Zamboni A, Bezzo F. Strategic design and investment capacity planning of the ethanol supply chain under price uncertainty. *Biomass and Bioenergy.* 2011;35(5):2059-2071.

7. Change IPoC. *2006 IPCC guidelines for national greenhouse gas inventories*: Intergovernmental Panel on Climate Change; 2006.

8. Monfreda C, Ramankutty N, Hertel TW. Global agricultural land use data for climate change analysis. *Economic analysis of land use in global climate change policy.* 2009;14:33.

9. West TO, Brandt CC, Baskaran LM, et al. Cropland carbon fluxes in the United States: increasing geospatial resolution of inventory-based carbon accounting. *Ecological Applications.* 2010;20(4):1074-1086.

10. Gibbs H, Yui S, Plevin R. *New estimates of soil and biomass carbon stocks for global economic models*: Center for Global Trade Analysis, Department of Agricultural Economics, Purdue University;2014.

470623 Developing a Systemic Methodological Framework for the Sustainability Assessment of Biobased Fuels and Chemicals

Abstract:

In the last two decades, there has been some effort in academia and industrial practice to provide tools that can deal with the challenges of designing sustainable chemical processes. Taking into account the diversity of the sustainability objectives, data availability and, therefore, modelling detail, it is inevitable that the structure of the developed tools varies significantly, ranging from simple indicators to holistic frameworks for an overall process assessment. These sustainability frameworks have also been applied in the field of biorefineries, in particular for biofuels, bioethanol, methanol and syngas production, incorporating process integration and life cycle assessment (LCA), exergy and inherent safety analysis, and life cycle energy and environmental analysis. However, perhaps the most striking difference between biobased and conventional chemical production in terms of sustainability assessment is the availability of process data, even for single biobased production lines, not to mention for integrated biorefinery concepts.

From this perspective, the aim of this study is (i) to present the methodological steps for the development of a model-based database of LCA related sustainability impacts in biomass based production, (ii) to demonstrate the use of this database in single- and integrated multi-production lines and (iii) to provide a first generation of surrogate models that correlate available input to desirable output process parameters and assessment metrics with computationally inexpensive algebraic forms.

In particular, this study presents detailed cradle-to-gate inventory analysis in biomass based production using protocols for process flowsheeting and filling of data gaps (e.g., for cradle-to-gate production steps and waste treatment of process effluents), considers various environmental impact allocation methods for the resulting joint and co-production biorefinery systems, applies a wide list of sustainability impact assessment metrics and tests multiple surrogate modelling approaches (i.e., from simple linear regression to partial least squares, Kriging and neural network approaches) on the basis of model generated data and sampling plans.

The overall methodological framework comprises an accessible, modular and expandable database of biobased production lines, including mass and energy balances, other process related information and various allocation approaches for the calculation of life cycle inventories in single- and multi-production systems. The resulting data structures are integrated with a reporting mechanism which offers information about the environmental impacts expressed by relevant indicators proposed by life cycle impact assessment methods such as the RECIPE and Cumulative Energy Demand methods and provides an environment for benchmarking alternative choices.

Molreover, metrics of "green chemistry" assessing the efficiency of processes (yield, mass intensity, E-factor etc.) as well as other process-related variables are extracted from the data structures aiming at the development of a first generation of surrogate models to provide guidance for a fast screening of innovative biobased processes and products.

The data structures, assessment results and surrogate modelling performance is demonstrated for a series of platform chemicals (e.g., syngas, sugars and lignin) and biofuels (e.g., biodiesel, biogas, and alcohols), starting from diverse biomass sources (e.g., wood chips, wheat straw, vegetable oil) and biomass availability scenarios, in single and integrated multi-production lines.

471048 Assessing Sustainability By Life Cycle Assessment Versus Techno-Ecological Synergy: A Case Study for Biofuel Production

Abstract:

Limitations on resource availability and the increasing consumption patterns have necessitated the quest for sustainable development. Many methods and tools have been developed to assess and design for sustainability. Attempts have been made to improve process efficiency to reduce resource consumption and waste generation. Paradoxically, even though the environmental impacts associated with per unit product have been reduced significantly, our ecosystem has been degraded further. This is mainly due to the following two reasons, 1) problems may shift outside the selected boundary, and 2) the role of ecosystem goods and services for supporting human activities is largely ignored despite their essential role in sustaining human activities. Therefore, for a methodology to claim its contribution to guiding sustainable decision making, both of these aspects need to be accounted for.

The most widely used sustainability assessment method is Life Cycle Assessment (LCA). It is used to estimate and assess the environmental impacts of resources used and emissions created during the main stages of a product's life cycle. Even though this method can potentially expand the system boundary to include the upstream and downstream effects, it largely ignores the role of ecological goods and services. LCA encourages decisions that do â€œless badâ€. Thus, the results from LCA are comparative in nature. The alternative that has the lowest environmental impacts is considered sustainable. This can potentially lead to misleading results and perverse decisions [1].

Techno-Ecological Synergy (TES) framework aims to assess and encourage synergies between systems in technological and ecological spheres [1]. The methodology considers systems at multiple spatial scales. Technological systems, ranging from individual processes to supply chains and life cycles can be evaluated along with their relevant supporting ecological systems. The key concept for TES is that resource uses and emissions are permitted but must stay in line with the capacity of relevant ecosystems. The necessary but not sufficient condition to be satisfied for reaching the goal of sustainability is to maintain the demand for ecosystem services within the supply at the largest ecological scale. By directly considering the role of ecosystem services, different insights and novel ecological solutions can be obtained.

The potential benefits achievable with TES methodology are illustrated by an application to a biofuel life cycle, including the production of selected direct inputs into farming, the farming phase and the ethanol manufacturing. Three ecosystem services have been considered to imply the general applicability of TES, namely carbon sequestration, air regulation and water provisioning. From this multi-scale and multi-objective analysis, we can identify processes that demand more from nature than can be supplied. This insight can be used to understand trade-offs and synergies between various ecosystem services, and identify opportunities for technological improvements and ecological restoration. Thus, by using TES approach, we not only have the option to reduce resource inputs and emissions as with the LCA approach, but also the option of increasing the supply of ecosystem services that can be achieved by proper land use. This solution might be potentially more cost-effective and environmentally friendlier than solutions developed by traditional methods such as LCA.

References:

[1] Bakshi, Bhavik, Guy Ziv, and Michael Lepech. "Techno-ecological synergy: A framework for sustainable engineering." Environmental Science & Technology 49.3 (2015): 1752-1760.

460852 Engineering Macroscale Thermoelectric Transport By Chemical Modulation of Nanoscale Interfaces

Abstract:

Optimizing the performance of thermoelectric materials has historically proven difficult as their essential transport parameters are cross-coupled in bulk transport equations. However, new classes of hybrid inorganic/organic materials are emerging which present exciting opportunities as their transport behavior can strongly deviate from classical models. In this talk, I will present recent achievements in demonstrating new, interface-driven mechanisms for enhancing and modulating thermoelectric transport in nanoscale hybrid systems. Highlighted research will include p- to n-type conversion of transport type in Tellurium nanowires, subphase alloy scattering to boost power factor in Tellurium nanowires, phase transition anomalies in Cu_2Se nanocrystal arrays, and unexpectedly low thermal conductivity in hybrid perovskite nanowires.

472306 Doped Semiconducting Polymers As Solution-Processable Thermoelectric Materials

Abstract:

Due to inherently low thermal conductivity and tunable electronic properties, semiconducting polymers are very promising solution-processable-materials for thermoelectric applications. Polymeric semiconductors are insulators in its pristine state, but are rendered conductive through chemical doping. Here, we studied the thermoelectric properties of poly(2,5-bis(3-tetradecylthiophen-2-yl)thieno[3,2-*b*]thiophene (PBTTT-C_{14}) doped either from solution or from the vapor phase. Overall, vapor doped films yield large Seebeck coefficients at high electrical conductivity (Power Factors ~ 100 Î¼Wm^{-1}K^{-2}). It important to note that the introduction of the chemical dopants can perturb the molecular packing and the morphology relative to the neat film. By leveraging various synchrotron X-ray scattering techniques that probe both the local and long range order, we discovered key structural features that can be correlated to the enhancement in the Seebeck coefficient and electrical conductivity for vapor doped films. Specifically, the introduction of the chemical dopant must limit the structural perturbation at small length scales while maintaining long-range correlation of polymer chains essential to efficient macroscopic charge transport. Lastly, to further elucidate the charge transport mechanism driving the thermoelectric properties, we report on the temperature-dependent measurements of both the Seebeck coefficient and electrical conductivity.

468413 Thermal and Thermoelectric Transport Coefficients in Graphene

Abstract:

Thermal and thermoelectric transport coefficients

in graphene

Enrique Muñoz, Physics Institute, Pontificia Universidad Católica de Chile, Vicuña Mackenna 4860, Santiago, Chile.

Abstract

Thermal, electronic and thermoelectric transport coefficients in graphene were studied. The effect of electron-phonon scattering processes, as well as the presence of C^{13} isotopes over the thermoelectric properties of extrinsic graphene were considered. Electron-phonon interaction is described within the deformation potential approximation. Electrical and thermal resistivity, as well as the thermopower, were calculated within the Bloch theory approximations. Analytical expressions for the different transport coefficients were obtained from a variational solution of the Boltzmann transport equation. The phonon-limited electrical resistivity $\rho e{-}ph$ shows a linear in temperature dependence at high temperatures, and follows a $\rho e{-}ph \quad T^4$ at low temperatures, in agreement with experiments. The phonon-limited thermal resistivity at low temperatures exhibits a T dependence and achieves a nearly constant value at high temperatures. The predicted Seebeck coefficient at very low temperature is $Q(T) \quad \pi^2 kB\, T/(3eEF)$, which shows a $n^{-1/2}$ dependence with the carrier density n, in agreement with experiments. On the other hand, the presence of a random distribution of C^{13} isotopes at fixed concentrations was studied as an strategy to reduce the lattice thermal conductivity, with the corresponding enhancement of the figure of merit for

thermoelectric applications.[3]

[1] E. Muñoz, Journal of Physics: Condensed Matter 24 (2012), 195302.

[2] E. Muñoz, J. Lu and B. I. Yakobson, Nano Letters 10 (2010), 1652.

[3] E. Muñoz, submitted (2016).

463547 Carrier Scattering at Alloy Nanointerfaces Enhances Power Factor in PEDOT:PSS Hybrid Thermoelectrics

Abstract:

This work demonstrates the first method for controlled growth of heterostructures within hybrid organic/inorganic nanocomposite thermoelectrics. Using a facile, aqueous technique, semimetal-alloy nanointerfaces are patterned within a hybrid thermoelectric system consisting of tellurium (Te) nanowires and the conducting polymer poly(3,4-ethylenedioxythiophene): poly(styrene-sulfonate) (PEDOT:PSS). Specifically, this method is used to grow nanoscale islands of $Cu_{1.75}Te$ alloy subphases within hybrid PEDOT:PSS-Te nanowires. This technique is shown to provide tunability of thermoelectric and electronic properties, providing up to 22% enhancement of the systemâ€™s power factor in the low-doping regime, consistent with preferential scattering of low energy carriers. This work provides an exciting platform for rational design of multiphase nanocomposites and highlights the potential for engineering of carrier filtering within hybrid thermoelectrics via introduction of interfaces with controlled structural and energetic properties.

465380 Combining Density Functional Theory Calculations, Supercomputing, and Data-Driven Methods to Design New Thermoelectric Materials

Abstract:

Density functional theory (DFT) simulations solve for the electronic structure of materials starting from the Schrödinger equation. Many case studies have now demonstrated that researchers can often use DFT to design new compounds in the computer (e.g., for batteries, catalysts, and hydrogen storage) before synthesis and characterization in the lab. In this talk, I will focus on how DFT calculations can be executed on large supercomputing resources in order to generate very large data sets on new materials for functional applications. First, I will briefly describe the Materials Project, an effort at LBNL that has virtually characterized over 60,000 materials using DFT and has shared the results with over 17,000 registered users. Next, I will talk about how such data can help discover new materials, describing how preliminary computational screening led to the identification and confirmation of a new family of bulk AMX2 thermoelectric compounds with measured zT reaching 0.8. I will outline future plans for how such data-driven methods can be used to better understand the factors that control thermoelectric behavior, e.g., for the rational design of electronic band structures, in ways that are different from conventional approaches.

450621 Phase Diagram, Microstructure and Thermoelectric Properties

Abstract:

Thermoelectric materials and devices have attracted intensive research and development interests due to their ability to convert waste heat directly into electricity. Energy usage efficiency could then be improved which is a very critical concern nowadays. Good thermoelectric properties require low thermal conductance but good electrical conductance. Phonon scattering caused by nanoprecipitates has been reported to be an efficient mechanism to enhance the thermal resistance but causing no significant hindrance on electrical conductance. Microstructures of thermoelectric materials are essential for their thermoelectric properties. Phase diagrams provide information of phase equilibria, and are fundamentally important for processing routes selection and understanding of materials microstructures.

Ternary thermoelectric materials systems, Bi-Te-In, Ag-Pb-Te and Co-Sb-Ga, are used as examples to illustrate the relationships between phase diagrams, microstructures and thermoelectric properties. Two kinds of ternary alloys were prepared. Ternary alloys of one kind were equilibrated at pre-determined temperatures, and the equilibrium phases were determined. Alloys of the other kind were quenched, and their primary solidification alloys were determined. Isothermal sections of ternary phase equilibria were proposed based on the results of determined ternary equilibrium phases together with the information obtained from their constituent binary phase diagrams, while the liquidus projections were determined based on results of the primary solidification phases.

According to the determined liquidus projections, ternary alloys with compositions at the ternary eutectics in the Bi-Te-In and Ag-Sb-Te systems, L (Bi- 13.0at%In-77.0at%Te)=Bi_2Te_3+Te+In_2Te_5, and L (Ag-40.at%Sb-36.at%Te)=$AgSbTe_2$ +Ag_2Te+δ-Sb_2Te, were prepared. Their microstructures were determined, and their thermoelectric properties were measured. Nanoprecipitates of Ag_2Te phase were observed. This ternary eutectic Ag-Pb-Te alloy has very low thermal conductivity (0.3w/mK at 415K), presumably due to the nano-sized lamellar structure and the nanoprecipitates enhancing phonon scattering. Very fine microstructures with nano- whisker-type

precipitates were observed in the as-solidified ternary Bi-Te-In eutectic. These nanoprecipitates contribute to the alloys' thermoelectric properties.

Ternary alloys in the Bi-Te-In and Co-Sb-Ga systems with compositions close to the ternary solubility limits of Bi_2Te_3 and $CoSb_3$ phases were also prepared. The thermoelectric properties of the Ga-doped $CoSb_3$ alloys were improved with the concentration of Ga doping and the figure of merit increased to 0.7 with a 0.1at% Ga doping. Similar results were observed in the In-doped Bi_2Te_3 alloys. A small amount alloying would enhance the phonon scattering and reduce lattice thermal conductivity. However, secondary phases would form if the doping concentrations are higher than the solubility limits. Since the secondary phase might not have good thermoelectric properties, its existence would deteriorate the thermoelectric properties of the alloys. The phase diagrams are thus fundamentally important.

459105 Shaping the Spectrum of Thermal Radiation Using Nanostructured Materials for Efficient Thermophotovoltaic Power Generation

Abstract:

By manipulating the structure of materials at the nanoscale, we can tune the spectrum of thermal radiation. In doing so, we can selectively enhance or suppress certain modes linked to either desirable or undesirable properties. Such spectrally-selective transport holds promise for more efficient solar and thermal power generation.

Here, we enhance spectrally-selective radiative transport using nanostructured materials to improve the efficiency of thermophotovoltaics (TPVs). TPVs are solid-state heat engines that convert energy from a high temperature radiant source—the emitter—into excited charge carriers in a semiconductor diode—the cell. TPVs are a promising technology for applications in concentrated solar power and distributed combined heat and power because they are solid-state, scalable and compatible with a variety of high-temperature heat sources. To achieve high efficiencies, it is well known that spectral control over thermal radiation is critical. Nevertheless, several questions remain: 1) *What spectral features are most important in TPVs when accounting for parasitic losses and non-idealities?, and 2) How can nanostructured materials be designed to meet the stringent spectral and thermal requirements in TPVs?*

To address these questions, we developed radiative transport and energy conversion models to identify the most important spectral features in realistic TPVs. Based on these models, we show that suppressing radiative exchange between the emitter and the cell at energies below the bandgap of the cell (i.e., at long wavelengths) is critical for high efficiency. This is explained by the fact that, at the optimal operating temperature, the emissive power below bandgap is much higher than above the bandgap. Operation at higher temperatures is undesirable because of the increased high-temperature radiative losses and excited charge carrier thermalization in the cell.

Based on these insights, we developed thin-film and nanostructured materials for TPV applications where spectrally-selective transport is based on the interaction of thermal radiation with wavelength-scale structures. The structures are chosen to minimize radiative transport at energies below the bandgap and then optimized to achieve high efficiencies in the presence of realistic losses. By suppressing radiative exchange at low energies, we demonstrate a solar TPV device that is an order of magnitude more efficient than previous solar TPVs and has the potential to exceed the limit of single-junction solar cells. Further, we show how a scalable and thin-film cell design promotes spectral selectivity using a back-side surface reflector to return sub-bandgap radiation back to the hot emitter. Spectroscopy measurements confirm that the reflectance below the bandgap for these thin-film cells is significantly higher than for cells on bulk substrates. Based on these spectral properties we show that the heat engine efficiency can be significantly higher than the state-of-the-art, exceeding 30%.

This work provides a detailed understanding of the relative importance of spectral features in TPVs and offers strategies for how to best engineer thin-film and nanostructured surfaces to shape the spectrum of thermal radiation for TPVs. With integration of thermal storage, TPVs can become an important part of distributed energy systems and may lead to increased adoption of intermittent renewables.

466335 Enzymatic Production of Fructo-Oligosaccharides from Inexpensive and Abundant Substrates Using a Membrane Reactor System

Abstract:

The concept of zero waste in the sugar industry has encouraged the development of technologies that promote utilization of byproducts, including the enzymatic conversion of waste feed stocks into nutraceutical products. Here we investigated a membrane reactor for the production of fructo-oligosaccharides (FOS) with immobilized enzyme (Pectinex Ultra SP-L) system using pretreated molasses as substrate with cleaning efficiency and less fouling as performance tools. The resulting spectra showed a high FOS yield of 63% with 89% conversion from available sucrose. So the EMR system with innovative continuous separation of enzyme was proved to fit for the production of FOS.

Reference: Amad Ur Rehman A, et al, Separation Science and Technology, DOI: 10.1080/01496395.2016.11667740

461271 Consolidated Process for Biobutanol Production from Cellulose By Two Step Fermentation of Different Clostridia Strains

Abstract:

n-Butanol is an important industrial chemical and also has been proposed as an alternative biofuel superior to ethanol. Biological production of *n*-butanol is promising because it can use renewable biomass as feedstock and is environmentally friendly. Conventional biorefinery of lignocellulosic biomass needs separate processes, including feedstock pretreatment, cellulase/hemicellulase production, biomass enzymatic hydrolysis, sugar fermentation, and product recovery, which is not cost-effective due to the complex process and subsequently high equipment capital requirement. Consolidated process only use one tank for enzyme production, biomass hydrolysis, and sugar fermentation, which will greatly simplify the process and reduce the equipment investment. In our research, *Clostridium cellulovorans*, a natural cellulose/hemicelluloses utilizing, but non-butanol producing bacterium, was metabolically engineered for *n*-butanol production from cellulose by overexpressing enzymes in the butanol-producing pathway. The obtained mutants could produce above 2 g/L *n*-butanol from cellulose. However, the *n*-butanol productivity was very low and *n*-butanol titer was hard to be further improved, which could be attributed to the poor specific cellulose hydrolysis activity of the enzyme system from *Clostridium cellulovorans*, as well as limited amount of the secreted enzymes under the circumstances of metabolic burden and butanol stress. To unblock this bottleneck, a novel consolidated process was developed by applying another clostridia strain, *Clostridium thermocellum*, with naturally high cellulose hydrolyzing activity. There two clostridia strains have similar nutrient requirement and could grow in the same cellulose medium. Due to the difference in optimal growth temperature (55°C vs. 37°C), a two step fermentation was necessary. The first step of fermentation under 55°C is a dedicated enzyme production phase by *Clostridium thermocellum*, in which maximal cellulose hydrolyzing activity of the whole culture system is the target of process optimization. The second step under 37°C could be believed as a simultaneous enzymatic hydrolysis and *n*-butanol production phase by *Clostridium cellulovorans* mutant strain. By combining the advantages of these two clostridia strains, this novel consolidated process showed promising potential for cost-effective biobutanol production from lignocellulosic biomass.

Bioprocess Development for Pleuran production by *Pleurotus ostreatus* using submerged cultivation system in Semi-Industrial scale.

El Enshasy H.A[1,2*], *Johari Masri, M.H*[3], *Malek R.A*[1], *Othman, N.Z.*[1], *Leng, O.M*[4], *Aziz, R*[1],

1. Institute of Bioproduct Development, Universiti Teknologi Malaysia, Universiti Teknologi Malaysia, Johor Bahru, Malaysia.
2. City of Scientific Research and Technology Applications, New Burg Al Arab, Alexandria, Egypt.
3. RD Department, All Cosmos Industries Sdn. Bhd., Johor Bahru, Malaysia, Institute of Bioproduct Development, Universiti Teknologi Malaysia, Skudai, Malaysia,
4. Harita Go Green, Johor Bahru, Malaysia and Institute Bioproduct Development, Universiti Teknologi Malaysia, Skudai, Malaysia

(*, corresponding author: Email: henshasy@ibd.utm.my, Phone: +601117739884)

Abstract

Nowadays, the demand of new natural and safe pharmaceutical production of immunomodulatory and anticancer properties is very high. For many years it have been reported that *Pleurotus ostreatus* is characterized by pleuran production, a β-1(1,3/1,6)-D-glucan biopolymer. This polysaccharide has binding affinity to carbohydrate receptors on immune cells and thus have good immunomodulatory activity. When cells cultivated in submerged culture system, they are able to production the pleuran as exopolysaccharide (EPS) in the fermentation medium. In this work, a complete industrial platform was developed to increase the pleuran production using submerged cultivation system in semi-industrial scale. At first, different media formulations were tested for polysaccharide production. The medium of highest EPS production was optimized using statistical approach in shake flask level. After this step, cultivations were conducted in 16-L stirred tank bioreactor for bioprocess industrialization and optimization of different bioprocessing engineering parameters to develop complete industrial platform for this polysaccharide in Semi-industrial scale.

Introduction

Mushrooms have been used since centuries as healthy food and medicine in different cultures. This based on its high nutritional value as rich source of proteins, carbohydrates, lipids, minerals, and vitamins (1-3). Recent research highlighted the medicinal values of macrofungi were highlighted in different research as antibiotic, antioxidant, antitumore/anticancer, immunomodulator, and antiatherosclerosis agents (4). Of different types of medicinal mushrooms known for their high value metabolites, *Pleurotus ostreatus* (oyster mushroom) is world known mushrooms based on its high nutritional value compounds make it one of the mostly used mushroom in food market (5). In addition, this mushroom is characterized by the ability to produce polysaccharides of high medicinal value with many bioactive properties. Pleuran (CAS # 159940-37-1) is one of most important polysaccharides which belong to β-1(1,3/1,6)-D-glucan biopolymer. Like other hexose polymers, it has the general chemical formula of $(C_6H_{10}O_5)_n$ (6). This polysaccharide is characterized by its high immunomodulating and anticancer properties (7-9). Pleuran is traditionally produced by mushroom cultivated in green house under controlled temperature and humidity. However, this method has many disadvantages such as long time cultivation (more than 3 months), low yield of bioactive compound and the ease of contamination by microorganism in this open cultivation system. Therefore, alternative method using submerged cultivation is applied to decrease the production time and increase productivity. In addition, extraction of pleuran from mushrooms in green house require many downstream processes, using a lot of solvent, and many steps of purification. In submerged culture system, the required polysaccharides can be obtained directly through simple purification of the exopolysaccharides from the fermentation broth by direct precipitation followed by drying or ultrafiltration to isolate the required high molecular weight polysaccharides followed by freeze drying. However, the different methods used for pleuran production is presented in figure 1. This work is focused on the development of new cultivation medium which support high yield of pleuran production in both shake flask and in bioreactor culture. Two approaches have been applied for medium optimization: One Factor At Time (OFAT) method and statistical optimization. Furthermore, cultivations were conducted in pilot plat 16-L stirred tank bioreactor to improve the production of pleuran production in semi-industrial scale.

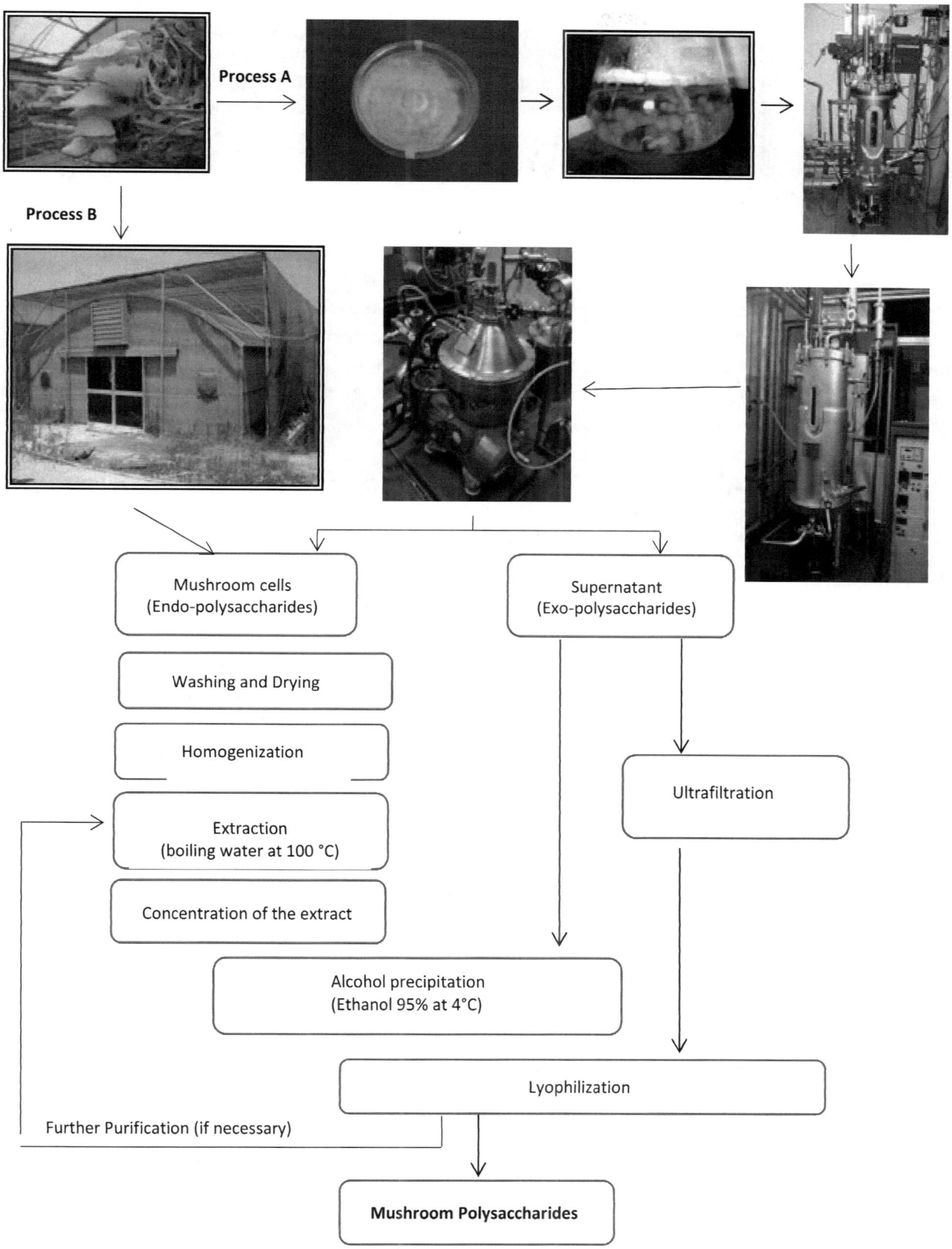

Figure 1. Production process for Pleuran production by *P. ostreatus* using submerged mycelium fermentation (Process A) or solid state fermentation (Process B). Process A started by mushroom

cell propagation on agar plate followed by submerged cultivation of cells in shake flask as intermediate step for inoculum preparation for bioreactor culture. Cultivation in bioreactor carried out by scaling up the process from small to large bioreactor followed by centrifugation to separation cells from the culture broth. If polysaccharides produced extracellular, extraction is usually carried out either through direct precipitation using cold ethanol followed by lyophilization to obtain the crude extract. Other new solvent free method was also developed to separate high molecular weight compounds directly by ultrafiltration followed by direct lyophilization of the polysaccharides (10). For endo-polysaccharides, they will be extracted using the conventional scheme of polysaccharides extraction from mushroom fruiting bodies (Process B). In case of Process using fruiting bodies as starting materials (Process B), cells were first washed and dryed followed by homogenization before many hours extraction using boiling water. The water extract of crude polysaccharide will be concentrated using vacuum extraction method followed by alcohol precipitation using cold ethanol. The precipitate is then lyophilized. In this process, further purification could be also done by repeating the steps between water extraction step and lyophilization. (11).

Materials and Methods

Microorganism

Pleurotus ostreatus NRRL 366 was used in this study. This mushroom was initially obtained from ARS culture collection, Agriculture Research Service, Peoria, USA. Once received from the microbial cell bank, it was subcultured on PDA agar medium (PDA, Oxoid, UK) After inoculation, the inoculated agar plates were incubated at 26 °C for 12 days and stored thereafter in a refrigerator at 5 °C for further use. Cells were activated every 2 months to ensure the cell viability during storage.

Vegetative growth medium and inoculum preparation

Cells were first activated by incubation on PDA agar plate for 12 days, harvested in sterile saline solution, and used to inoculate YM medium composed of (g/L): glucose, 10.0, yeast extract, 3.0, malt extract, 3.0; peptone, 5.0 (pH 5.5). After 7 days cultivation, the grown cells were harvested, homogenized using Omni homogenizer (Omni International, USA) for better inoculum homogenization and prevent the cell aggregation to large pellet.

Production medium

The initial medium used for EPS production before optimization was composed of (g/L): glucose, 60.0; peptone, 2.0; KH_2PO_4, 0.46; K_2HPO_4, 1.0; $MgSO_4.7H_2O$, 0.5 (pH 5.5),. This medium was modified from El Enshasy et al. (11). Glucose was sterilized separately and added to the fermentation medium before inoculation. This medium was further optimized in this study using RSM method in this study.

Shake flask and bioreactor cultivation
In shake flask experiment, 250 mL Erlenmeyer flasks of 50 mL working volume were used, incubated for 14 days at 26 °C on rotary shaker incubator at 200 rpm. Samples in form of two flasks were taken at different time intervals every day for analysis. For bioreactor cultivation, 16-L stirred tank bioreactor (BioEngineering, Wald, Switzerland) with working volume of 8L was used in this study. The stirrer speed with adjusted to 200 RPM and aeration rate of 1 v/v/min.

Determination of glucose and EPS
Glucose concentration in culture broth was determined using glucose lactate analyzer (YSI 2700), which based on enzymatic method. The centrifuged sample was mixed first with cold ethanol in ration (1:3, v/v), kept for 20 hours at 4 °C for polysaccharides precipitation. The precipitated EPS was then centrifuged at 10,000g for 20 min for full separation. The precipitated polysaccharides was then washed consequently with three solvent system (ethanol, acetone, and ethyl ether). The obtained white precipitate was then lyophilized for 3 days, and weighted.

Results and Discussion

Medium optimization for high EPS production by *P. ostreatus*
Set of experiments were conducted to improve the medium components to increase EPS production. Two approaches were applied to achieve this target using one factor at time method (OFAT) and statistical design method using response surface methodology (RSM). The results of medium optimization is summarized in table 1.

Table 1. Medium composition obtained after from different optimization methods to increase EPS production by *P. ostreatus*

Components	**Un-optimized**	**One factor at time (OFAT)**	**Statistical Optimization (RSM)**
Glucose (g/L)	60.0	100.0	105.0
Yeast extract (g/L)	2.0	3.0	5.2
$(NH_4)_2SO_4$ (g/L)	-	2.5	2.6
K_2HPO_4 (g/L)	1.0	1.0	2.2
$MgSO_4.7H_2O$ (g/L)	0.2	0.2	0.2
Peptone (g/L)	1	-	-
EPS (g/L)	**2.13**	**3.33**	**3.65**

As shown optimization of culture medium using statistical optimization increased the volumetric production of pleuran up to 3.65 g/L (almost 71% higher than those value obtained in unoptimized medium. However, the difference between the polysaccharides production in medium optimized using OFAT and RSM was only 9.6% higher in case of RSM culture.

Batch cultivation of *P. ostreatus* in 16-L stirred tank bioreactor

After medium optimization in shake flask level, cultivations were conducted in 16-L stirred tank bioreactor for bioprocess industrialization. Figure 2 shows the growth profile of mushroom cells when cultivated in bioreactor level. As shown, cells grew with low growth rate in the first 48 hours followed by higher growth rate until 240 hours, reaching maximal biomass of 15.22 g L^{-1}. During that time, glucose consumed in culture with rate of about 0.37 g L^{-1} h^{-1}. On the other hand, the maximal volumetric and specific EPS production was 3.75 g L^{-1} and 0.24 g g^{-1}, respectively.

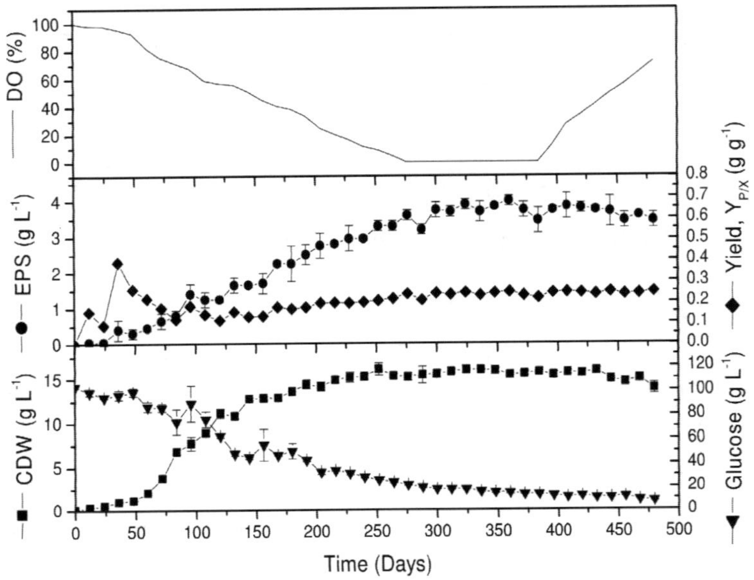

Figure 2. Cell growth, glucose consumption, DO, and EPS production profile during cultivation of *P. ostreatus* in 16-L stirred tank bioreactor.

Fed-batch cultivation of *P. ostreatus* for high EPS production in 16-L stirred tank bioreactor

Based on the results of batch cultivation, fed-batch cultivation was designed using constant feeding approach with either glucose alone or full medium feeding. Figure 3 shows the growth and EPS production profile in cultures fed with two different feeding strategies. Feeding started after 240 hours to prolong cell growth and EPS production up to 384 hours. As shown, the biomass in full medium fed culture reached maximal value of about 23.81 g L^{-1} which was about 16% higher than the value in glucose fed culture. However, the maximal EPS production was also obtained in full medium fed culture and reached 4.5 g L^{-1}, which is about 20% compared to glucose fed culture (which gave the same volumetric production as in batch culture). It is also worthy to note that the specific EPS production in both culture was almost the same. This indicates that the increase of EPS in complete medium fed-culture was due to the increase in biomass not due to higher cell productivity.

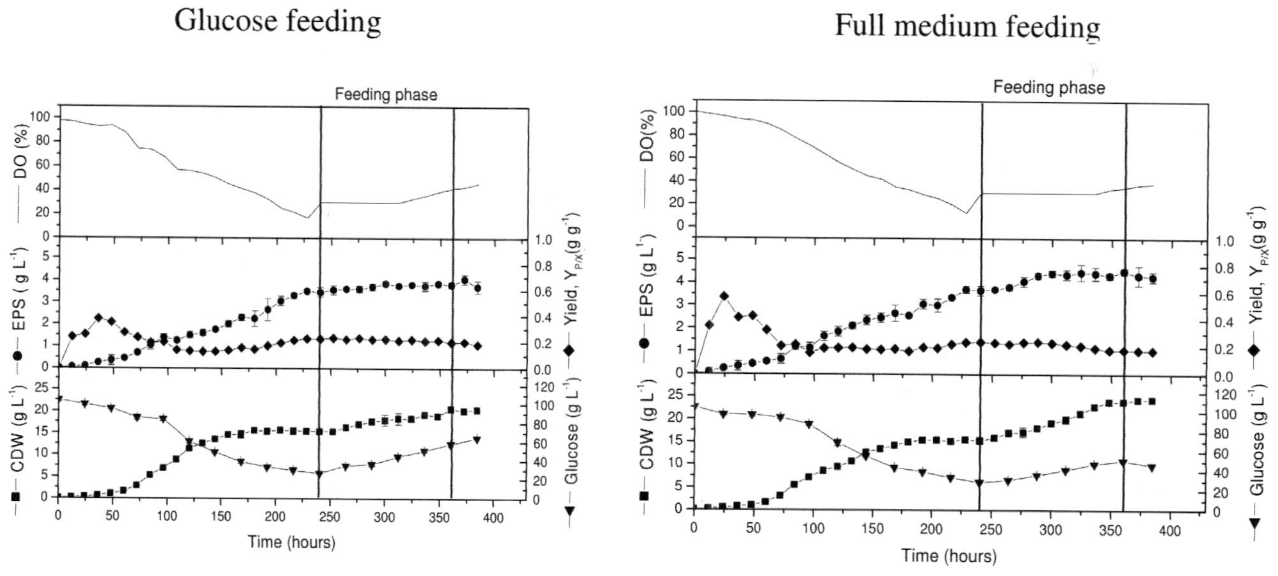

Figure 3. Chang in cell growth, glucose concentration, DO, and EPS production profiles during cultivation of *P. ostreatus* in 16-L stirred tank bioreactor using constant fed-batch cultivation strategy. In this cultivation, two different feeding solutions were applied, either glucose feeding only or complete medium feeding.

Table 2. Comparison between cell growth and EPS production when cultivating *P. ostreatus* in fed batch culture using different feeding solutions.

Parameters for Fed-Batch Cultivation	Exponential time in feeding period (240 hour to 360 hour)	
	Glucose feeding	Full medium feeding
Biomass, X (g L^{-1})	20.491	23.81
Specific growth rate, μ (h^{-1})	0.00243	0.0037
EPS production (g L^{-1})	3.8	4.5
Yield coefficient product over cell mass $Y_{p/x}$ (g g^{-1})	0.185	0.189

Conclusion

This study demonstrates a complete industrial platform for pleuran production in semi-industrial scale. Medium optimization was carried out using two approaches (OFAT) and (RSM), which lead to the development of high volumetric EPS production. The process was scaled up in 16-L stirred tank bioreactor and further optimized by applying fed-batch cultivation strategy. The results clearly demonstrate that fed batch cultivation using complete medium is more suitable than feeding with glucose only. However, the increase of EPS production was due to the higher biomass production and not to the increase in cell productivity.

Acknowledgement
This research was financially supported by Harita Go Green Sdn. Bhd., authors would like to thank also Research Management Centre (RMC), Universiti Technology Malaysia for their support.

References

1. Smith J. E, Rowan, N.J., Sullivan R. (2002). Medicinal mushrooms: Their therapeutic properties and current medical usage with special emphasis on cancer treatment. USA. University of Strathclyde.
2. Pushpa, H.; Purushothama, K.B. Nutritional analysis of wild and cultivated edible medicinal mushrooms. *World Journal of Dairy and Food Science.* 5, 140-144.
3. Chang, S.T., Buswell, J.A. Mushroom nutraceuticals. *World Journal of Microbiology and Biotechnology.* 12, 473-476.
4. Chen, J. and Seviour, R. (2007). Medicinal importance of fungal β-(1→3), (1→6) glucan. *Mycological Research.* 111, 635-652.
5. Maftoun, P., Johari, H., Soltani, M., Malik, R., Othman, N.Z., El Enshasy, H.A. (2015). The edible mushroom *Pleutotus* spp.: I. Biodiversity and nutritional values. *International Journal of Biotechnology for Wellness Industries.* 4, 67-83.
6. El Enshasy, H.A. Maftoun, P., Abd Malek, R. (2012). Pleuran: Immunomodulotor polysaccharide from *Pleurotus ostreatus*, structure, production and application *In.* Mushrooms: Types, Properties and Nutrition (Andres, S. and Baumann, N. Ed.). Nova publisher, NY, USA). pp 153-172.
7. Sun, Y., Liu, J. (2009). Purification, structure and immunological activity of a water soluble polysaccharide from the fruiting body of *Pleurotus ostreatus. Bioresource Technology.* 100, 983-986.
8. Levi, I., Friesem, D., Geresh, S., Hadar, Y., Schwarz, B. (2006). An aqueous polysaccharide extract from the edible mushroom *Pleurotus ostreatus* induces anti-proliferative and pro-apoptotic effect on HT-29 colon cancer cells. *Cancer Letters.* 244, 61-70.
9. El Enshasy, H. (2010). Immunomodulators. In *The Mycota: Industrial Applications,* Esser K., Bennett J.W., Eds.; Springer Verlag: Berlin, Heidelberg, Vol. 10, pp. 165-194.
10. El Enshasy, H. (2006) New method for mushroom cell cultivation and production of polysaccharides/ peptidoglycan in bioreactors. Egyptian Patent No. 24637. (Dated 7/6/2006).
11. El Enshasy, H.A. and Hatti-Kaul, R. (2013). Mushroom immunomodulators: unique molecules with unlimited applications. *Trends in Biotechnology.* 31, 668-677.
12. El Enshasy, H., Daba, A.; El-Demellawy, M.; Ibrahim, A.; Elsayed, S., El Badry, I. (2010). Bioprocess development for large scale production of anti-cancer exopolysaccharide by *Pleurotus ostreatus* in submerged culture. *Journal of Applied Science.* 10(21): 2523-2529.

464571 Process Engineering of Clostridium Cellulovorans for Butanol Production from Biomass

Abstract:

Clostridium cellulovorans can utilize a wild range of low-value lignocellulosic biomass as substrates because of its efficient cellulosome complex. The objective of this study was to achieve a robust lignocellulosic butanol production using metabolically engineered *C. cellulovorans* mutant by the process engineering of consolidated bioprocessing (CBP). The metabolomics was used to create an intracellular metabolite profiling of *C. cellulovorans* and analyze the interaction between cellular metabolism and fermentation process parameters. The bioreactor operation parameters and fed-batch process parameters were specifically analyzed. The changes in core metabolites responsible for carbon, energy and redox balance at different fermentation conditions were identified. The CBP was engineered by optimizing the seeding density, pH, cellulose feeding and trace element. The butanol production was improved to titer of >3 g/L and selectivity of > 0.4 g/g from cellulose. The direct fermentation of corn cob generated butanol titer of > 1.5 g/L.

466640 Ethanol Production on Date Syrup Using Yeast: Process Optimization Study for Pilot Scale Production

Abstract:

Yeast is one of a most promising species for ethanol production using various substrates. In the present study bioethanol production from commercially available date syrup was evaluated for yield and process optimization parameters using Central Composite Design (CCD). Date syrup as substrate is quite economic in this part of the world. The ethanol production is specifically chosen to support the industry during falling crude prices globally on one hand, and if successfully produced & economically optimized for large scale production over date syrup; the ethanol production can become a boon to the otherwise struggling date palm agro-economic sector in region; on the other hand. Fermentation temperature, concentration of date syrup sugar, nitrogen source, C/N ratio, feed flow rate, pH, was optimized through CCD in a pilot scale fermentor. The data was compared with production over standard media. Using date syrup, growth kinetics and yields were found to be near double to that of the standard media. With a controlled flow rate; above 97% sugar conversion was achieved. Urea was used as supplemental Nitrogen source for pilot level production and the efficiency of the conversion process was recorded.

Bioprocess Strategies to Optimize Thermostable Phytase Production By *Escherichia coli* B121 (DE3) When Induced with Lactose

Nor Zalina Othman[1], Solleh Ramli[1],Roslinda Abd Malek[1], Mohamad Roji Sarmidi[1], Rajni Hatti-Kaul[3], Hesham A. El Enshasy[1,2]

[1]*Institute of Bioproduct Development, Universiti Teknologi Malaysia, 81310 Skudai Johor, Malaysia;* [2]*Bioprocess Development Department, Mubarak City for Scientific Research and Technology Applications (MuCSAT), New Burg Al Arab, Alexandria, Egypt;* [3]*Department of Biotechnology, Center for Chemistry and Chemical Engineering, Lund University, Box 124, S-221 00 Lund, Sweden.*

Introduction

IPTGis used as a molecular mimic of allolactose, a lactose metabolite that triggers transcription of the *lac* operon. Unlike allolactose, the sulphur atom create a chemical bond which is non-hydrolyzed by the cell, preventing the cell from degradation by the host cells that makes the IPTG concentration remains constant (Fernández-Castané *et al.*, 2012). IPTG is however toxic and comparably expensive (Donovan *et al.,* 1996). Nowadays, lactose has been considered as an alternative inducer in fermentation of recombinant *E.coli*. Since the product yields is compatible with IPTG as inducer however it depends on the strategy of induction to increase heterologus protein expression (Weng *et al.*, 2006; Abdel-Hamid *et al.*, 2014). Induction of promoter gene by lactose during foreign protein expression has a limiting step. It is clearly observed that IPTG can bind directly to the *lac*-repressor molecules and express the phytase protein after one hour of induction. However, the expression of phytase was delayed almost an hour compared induction with IPTG and the amount of lactose needed as inducer should be further optimized for cultivation in different scale (Tran *et al.*, 2010). In case of lactose as inducer, a few steps are needed before induction process. This is due to the time needed by lactose to be transported into the cell by the present of lactose permease and processed to allolactose by β-galactosidase before binding to the repressor (Becker *et al.*, 2012; Santillán *et al.*, 2007). Furthermore, lactose can be metabolized as energy sources. Further catabolic activity of allolactose may produce glucose and galactose that can contribute to increase the metabolic overflow due to excess carbon sources during induction period (Santillán and Mackey, 2008). As a result of metabolite burden, the high concentrations of inducer that are often used in an effort to fully induce the *lac* promoter do not necessarily lead to maximal expression of a target protein. This is due to high residual lactose concentration in the broth led to accumulation of glucose by hydrolysis of excess lactose during post-induction phase (Lim *et al.*, 2004). Therefore the objective of this study to optimize the induction strategy whenlactose as inducer with addition of additive for *E. coli* BL21 (DE3) which growth in glycerol minimal medium to increase productivity of phytase expression and leakage of expression protein from periplasmic space to outside of cell membrane in semi-scale bioreactor.

Materials and Methods

Microorganism

E.coli BL21 (DE3) harboring pE10C2 with the leader peptide that carries thermophile phytase gene from *Bacillus* sp. MD2 under the control of *lac*-O promoter to express thermostable phytase enzyme was used for this research. The strain was originally from of Biotechnology, Center for Chemistry and Chemical Engineering, Lund University and was delivered and stored in Wellness Industries Culture Collection (WICC), Institute of Bioproduct Development, Universiti Teknologi Malaysia. Cells were first propagated on LB agar plate containing 100 $\mu g mL^{-1}$ amphicillin and were incubated at 30 °C for 24 hours. The arisen colonies were harvested by glycerol solution (50 %) and put in series of 2 mL cryogen vials. The vials were frozen at -20 °C for 24 hours followed by further storage as working cell bank at -80 °C for further use.

Inoculum preparation and cultivations conditions in shake flask

E. coli BL21 (DE3) cells were grown on a LB-agar plate containing 100 µg mL^{-1} of ampicillin at 30 °C for 24 h. A single colony was picked from the plate and cultured for 5h in 50 ml LB broth containing 100 µg mL^{-1} of ampicillin at 30°C. The shake flasks cultivations were carried out in 250 ml Erlenmyer flasks containing 50 ml broth and incubated in a rotary shaker (4230 Innova, New Brunswick, NJ, USA) at 200 rpm. In all shake flask experiments, the media was supplemented with 100 µg mL^{-1} of amphicillin. Inoculation was performed by transferring 5 mL of seed culture to 50 mL medium in a 250 mL Erlenmeyer flask.

Production medium in shake flask cultivation and bioreactor

Medium that used in the shake flask cultivation and the batch culture for bioreactor 16-L was glycerol minimal medium modified from studied of Spexard *et al.*, (2010). For batch cultivation of *E. coli* BL21 (DE3), carbon source was prepared and autoclaved separately in 1000 ml and added aseptically to the bioreactor before inoculation.For the fed-batch pre-induction growth phase, the feeding solution was started using DO-stat feed of a solution containing:yeast extract, 10 g L^{-1}; glycerol, 500 g L^{-1}, trace elements, 10 mL L^{-1}, 1 M $MgSO_4$, $7H_2O$, 50 ml L^{-1}. During the post-induction phase, feeding of glycerol was constantly fed at 15 g L^{-1} h^{-1}. For the induction phase, the induction solution for one-pulse induction strategy and the feeding solution were prepared as batch cultivation. Induction was aseptically carried by using 30 mM of lactose with 10 mM $CaCl_2$ and 0.65 % (w v^{-1}) of glycine which were sterilizing separately, mixed together and added for batch cultivation when the cell mass approximately 6-8 g L^{-1}.

Optimization of induction strategies using RSM

The shake flasks cultivations were carried out in 250 ml Erlenmyer flasks containing 50 ml broth and incubated in a rotary shaker (4230 Innova, New Brunswick, NJ, USA). Induction was aseptically carried when OD_{600nm} of each flask approximately 0.6-0.8. The sample analysis was done after 6 hours of post-induction phase.A Box-Behnken design with a total of 30 experiments was run to further optimize the significant variables. Box-Behnken design was used for optimisation strategy for enhancing leakage of phytase from periplasmic space to the extracellular medium. Concurrently, the correlation of lactose, $CaCl_2$ and glycine concentration to the periplasmic and extracellular phytase activity was studied. Each factor in the design was studied at three different levels that was low, intermediate and high value. Each variable are denoted by the numbers (-1), (0) and (+1), respectively as shown in Table 1; Lactose (**A**), $CaCl_2$ (**B**) and glycine(**C**).

Table 1: Experimental range and levels of factors influencing induction strategy of *E.coli* BL21(DE3) for production of phytase in a Box-Behnken design.

Factors	Low(-1)	Centre (0)	High (+)
Lactose (mM)	15	25	35
$CaCl_2$ (mM)	5	10	15
Glycerine (% w/v)	0.25	0.65	1

Bioreactor operation

Cultivations with growing cells were carried out in a 16-L stirred-tank bioreactor (BioEngineering, Wald, Switzerland). The bioreactor is equipped with pH probe, oxygen probe, foam sensor, and stirrer of two–four bladed Rushton turbines. During the experiments, controlled the pH, temperature, aeration rate and the agitation speed were controlled at 7, 30 °C, 2 v v^{-1} min^{-1} and 400 rpm, respectively. The dissolved oxygen was adjusted to 100% at the beginning of cultivation and remains controlled at 30 % along the cultivation process. Antifoam reagent (Silicone antifoam, Sigma, USA) was added to suppress the foaming when necessary.

Analysis

Phytase activity assay: Determination of phytase activity

The extracellular phytase activity was determined in the cell-free supernatant obtained after centrifugation of the sample at 8,000 g for 10 min at 4 °C. The cell pellet was washed with physiological saline and the remaining cell pellet was re-suspended in 0.1 M Tris–HCl buffer, pH 7, containing 5 mM $CaCl_2$. The periplasmic content was extracted by using the osmotic-shock method. All the assays were performed in triplicates. The phytase activity was determined from the soluble protein sample using supernatant obtained from extracellular and periplasmic fraction. Base on the method studied by Tran *et al.*,(2010), a mixture of appropriate diluted enzyme sample and sodium phytate (1.5 mM) was dissolved in 100 mM Tris-HCl buffer, pH 7 containing 5 mM $CaCl_2$. Then, the solution was incubated at 70 °C for 20 min. The reaction is stopped by adding an equal volume of 15% trichloroacetic acid (TCA) solution. The precipitate obtained is removed by centrifugation, and 500 µl of the clear supernatant transferred to an Eppendorf tube followed by addition of 500 µl ammonium molybdate reagents (containing 1:4 mixtures of 2.7% $FeSO_4$ and 1.5% ammonium molybdate in 5.5 % H_2SO_4). After 5 min incubation at room temperature, absorbance at 700 nm was read and used to calculate the concentration of P_i from a standard curve made using sodium dihydrogen phosphate solution in the concentration range of 0-600 µmol/l P_i. All the samples were assayed in triplicates. One unit of phytase activity was defined as the amount of enzyme that releases 1 µmole of P_i per minute under the assay conditions.

Acetic acid, glycerol and lactose determination

For analyzing acetic acid, glycerol and lactose in culture supernatant by HPLC. HPLC separation was performed with Waters liquid chromatography (Milford, MA, USA) equipped with Waters 2690 Separation Module and Waters 996 Photodiode Array Detector (DAD). The data was stored and processed by HPLC Empower software. The data was stored and processed by HPLC Empower software. Chromatographic separations were performed on an ROA-Organic Acid H+ (Phenomenex, USA) column (300 x 7.80 mm, 8 micron). As the mobile phase 0.005N H2SO4 was used at a flow rate of 0.6ml/min. The column was kept in column heater at 40 °C. For detection, an optimum wavelength was selected for the acetic acid is 210 nm.

Result and discussion

Optimization of induction strategyfor extracellular and periplasmic phytase activity in the shake flask cultivation.

The aims of this section to optimize excretion of phytase outside the cell membrane towards maximal total phytase activity. The ANOVA for the extracellular and periplasmic phytase activity are shown in Table 2. The calculated value of F were compared to the critical value of $F_{(p-1,\ n-p,\alpha)}$. The ANOVA showed that the calculated value of F for the extracellular and periplasmic phytase activity that was 3.09 and 6.63, respectively were greater than the critical values for the F distribution table of $F_{(3,26,0.05)}$=2.934. If the calculated value of F exceeds the $F_{(3,26,0.05)}$, the null hypothesis is rejected at the level of significance, and it is inferred that the coefficient estimates are not all zero and the variation verified that the model is significantly greater than the unexplained variation (Wonacott and Wonnacott, 1990). Therefore the p-value for all of the response is below 0.05 which means that at the 95 % confidence level of the models were significant.The determination coefficient of R^2 which implies that the sample variation of above 95 % for both extracellular and periplasmic activity was attributed to the concentration of lactose, $CaCl_2$ and glycine. This study signifies a good correlation between the experimental data and the predicted values which overall below than 5% was not explained by the model (Varnalis *et al.*, 2004).

Table2: Analysis of variance (ANOVA) of the response variables

Response	Source	DF *[1]	SS*[2]	MS*[3]	F value, F_{cal}	Prob(p)>F	r^2(%)	F (0.05)
Extracellular phytase activity (Y_1)	Regression Residual Total	3 26 29	27.623 77.544 105.167	9.208 2.982	3.09	0.045	97.15	2.934
Periplasmic phytase activity (Y_2)	Regression Residual Total	3 26 29	1.069 0.826 1.895	0.356 0.032	6.63	0.002	98.80	2.934

[1]-Degree of Freedom,[2]- Sum of Square, [3]-Mean Square

Confirmation test of optimization study between non-optimized and optimized induction strategy

Through Box-Behnken design of optimization plot for induction strategy consists of lactose, $CaCl_2$ and glycine of 27.68 mM, 11.32 mM and 0.65 % [w v^{-1}], respectively. The extracellular phytase activity (7.86 U mL^{-1}) and total phytase activity (8.70 U mL^{-1}) obtained in fermentation used the optimal induction strategy that increased after 6 hour induced with lactose by 36.45 % and 29.65 %, respectively. The increment of extracellular phytase activity that excreted outside the cell membrane after statistically optimized induction strategy supplemented with glycine increased up to 90.34 % within 6 hours of the post-induction phase when compared to un-optimization induction strategy only 85.84 % (Table 3).The lost of cell membrane permeability by supplemented with glycine increased the uptake of lactose which reduced the time required for transport of lactose into the cell (Shuman and Silhavy, 2003). In this study, the consumption rate of lactose is not significant but the enzyme production rate was more significant based on glycine alteration on the membrane permeability (Puertas *et al.*, 2010; Ding *et al.*, 2010).This in consequently causes the accumulation of phytase in the cytoplasm and the periplasmic space and consequently secreted outside the cell membrane.

Table 3: The comparison study of the optimized induction strategy with and without addition of ampicillin in the shake flask cultivation.

Parameters	Addition of ampicillin (100 µg mL^{-1})		Without-ampicillin (100 µg mL^{-1})
Induction strategy variables			
Lactose [mM]	30	25.68	25.68
$CaCl_2$ [mM]	10	11.32	11.32
Glycine % [wv^{-1}]	-	0.65	0.65
Growth kinetics and phytase production			
X_{max} [g l^{-1}]	1.79	1.59	1.86
µ [h^{-1}]	0.11	0.12	0.16
$P_{max\ (Exo)}$ [U mL^{-1}]	5.76(12hr)	7.86 (6 hr)	6.50(6hr)
$P_{max\ (Total)}$ [U mL^{-1}]	6.71(12hr)	8.70 (6 hr)	7.05(6hr)
$Y_{Totalphytase/cell\ mass}$ [U g^{-1}]	6,414.91	10,875.00	5875.00
Q_{Exo} [U L^{-1} h^{-1}]	578.00	1,310.00	1,083.33
Q_{Total} [U L^{-1} h^{-1}]	671.00	1,450.00	1,175.00
% excretion outside the cell membrane	83.50	90.34	92.19

X_{max}: maximal cell dry weight, $P_{exo\ or\ total}$: maximal production of extracellular,P_{total}: maximal production of phytase (as sum of periplasmic, cytoplasmic and extracellular activity), $Y_{p/x}$: unit of enzyme produced per unit cell mass (at the time of maximal total phytase activity), Q_p [U L^{-1} h^{-1}]: volumetric phytase production rate

Hence, to study effects of ampicillin for plasmid stability cultivation without and addition of ampicillin were carried out. The cultured grew faster when the cultivation broth not supplemented with ampicillin with the specific growth rate of 0.16 h^{-1}.The maximal cell mass production was increased about 16.98 % (1.86 g L^{-1}) when compared cultured with ampicillin. The growth rate of 0.16 h^{-1} gave high acetic acid production (max=1.55 g L^{-1}). This contributed to rapid declined of plasmid lost after 2 hours of induced with lactose. Finally contributed to the decrease of total phytase activity to 18.96 % with 92.19 % was excreted outside the cell membrane.

Fed-Batch Cultivation to Increase Production of Thermostable Phytase in 16-L Stirred Tank Bioreactor

From the batch cultivation of *E. coli* BL21 (DE3) in the 16-L bioreactor, the addition of lactose as the only carbon source in the fermentation broth in the post-induction phase was not only for cell growth but also for heterologous protein expression by the host cells as also reported by Li *et al.* (2012). Using lactose as the only carbon source in this phase was not efficient for cell growth and protein expression (Lim *et al.*, 2004). Meanwhile, when induced with lactose only, 4.78 U mL^{-1} of total phytase activity was reached after 6 hours of the post-induction phase. Therefore, due to limitations of the carbon source and dissolved oxygen, there was a significant accumulation of acetic acid in the fermentation broth of up to 1.51 g L^{-1}. Therefore, the fed-batch cultivation was carried out by glycerol feeding at the rate of 15 g L^{-1} h^{-1}. From Figure 1, it is shown that this mode of glycerol feeding can maintain the cell growth with the specific growth rate of 0.08 h^{-1}. After 5 hours lactose induction, both of lactose and glycerol was actively consumed in this phase.

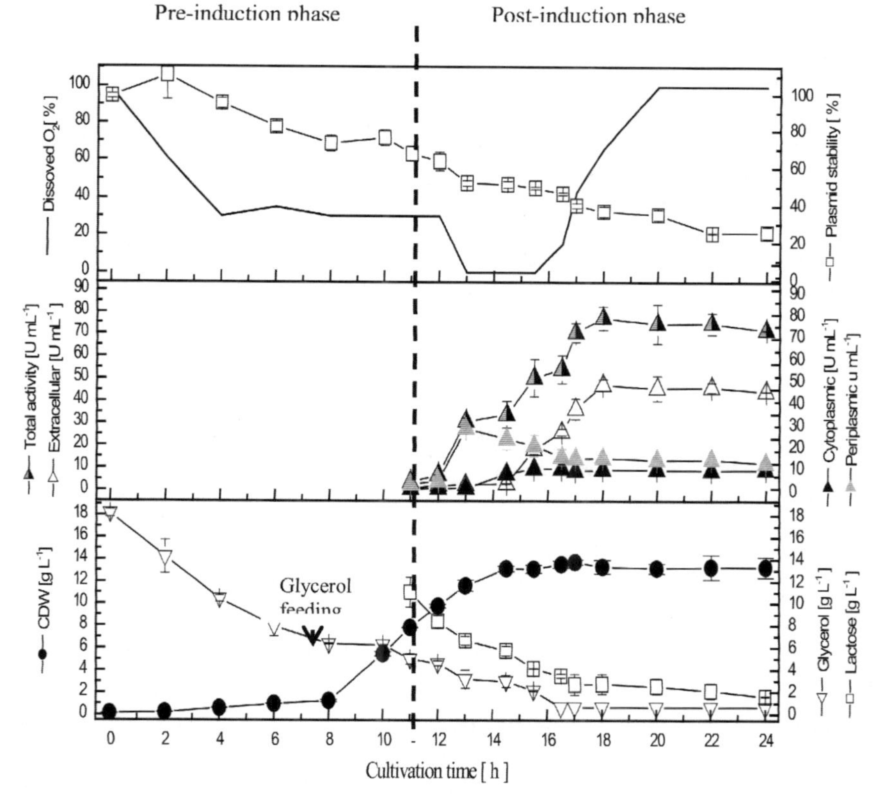

Figure 1: The cell mass production, phytase expression, distribution of phytase activity, residual glycerol/lactose, acetic acid production and plasmid stability in fed-batch cultivation of *E. coli* BL21 (DE3) induced with one-pulse lactose with 15 g L^{-1} hr^{-1} of glycerol feeding during the post-induction phase (----start of induction).

The residual of glycerol was almost 0.1 g L^{-1} and lactose was 3.46 g L^{-1} after 5 hours induced (Figure 1). The presence of glycerol in the post-induction phase contributes, to the carbon stress response of *E. coli* during cultivation (Zhang *et al.,* 2009; Martinez-Gomez *et al.*, 2012).

Conclusion

In this study revealed factor that significantly influenced cell mass, extracellular phytase activity and accumulation of acetic acid by *E. coli* BL21 (DE3) in the optimization of induction strategies was lactose. In this study, more than 90 % of the total phytse activity was excreted outside the cell membrane by addition of glycine also contributed to the highest volumetric and productivity induction with lactose. This work has demonstrated the use of the second order polynomial equation to determine conditions leading to the maximum extracellular phytase activity and minimize acetic acid production. The new developed of medium composition and induction strategy can be used for the study in 16 L bioreactor without the addition of ampicillin.

Acknowledgement

The authors would like to thanks the MOHE and Universiti Teknologi Malaysia (UTM) for Contract research grant no R.J130000.7344.andR.J130000.7646.4C101 .

References:

1. Becker,A.N., Peters, P.J., James Maher III, L. (2012) Mechanism of promoter repression by Lac repressor-DNA loops. Nucleic Acids Research. 1-11.
2. Ding, R., Zhaofeng, L. , Chen, S., Wu, D. and Wu, J., Chen, J. (2010).Enhanced Secretion of Recombinant α-Cyclodextringlucosyl Transferase from *E. coli* by Medium Additives. *Process Biochemistry,* 45: 880–886.
3. Donovan, R.S., Robinson, C.W. and Glick, B.R. (1996). Review: Optimizing Inducer and Culture Conditions for Expression of Foreign Proteins under the Control of the *lac* promoter.*Journal Industrial Microbiology*, 16:145-154.
4. Fernandez-Castane, A., Vine, C.E., Caminal, G.and Lopez-Santin, J. (2012). Evidencing the Role of Lactose Permease in IPTG uptake by *Escherichia coli* in Fed-Batch High Cell Density Cultures. *Journal Biotechnology*, 157:391–398.
5. Abdel-Hamid, A. M., Attwood, M.M. and Guest, J. R. (2001). Pyruvate oxidase Contributes to the Aerobic Growth Efficiency of *Escherichia coli.Microbiology,* 147:1483–1498.
6. Li, M., Zhang, X., Agrawal, A. and San, K. Y. (2012). Effect of Acetate Formation Pathway and Long Chain Fatty Acid Coa-Ligase on the Free Fatty Acid Production in E. coli Expressing Acy-ACP Thioesterase From *Ricinus communis. Metabolic Engineering,* 14(4): 380-387.
7. Lim, H.K., Kim, S.G., Jung, K.H. and Seo, J.H. (2004). Production of the Kringle Fragments of Human Apolipoprotein(α) by Continuous Lactose Induction Strategy. *Journal of Biotechnology,*108(3):271-278.
8. Martinez-Gomez, K, Flores, N., Castañeda, H. M, Martínez-Batallar G, Hernandez-Chavez, G., Ramirez, O. T., Gosset, G., Encarnación, S. and Bolivar, F. (2012). New insights into *Escherichia coli* Metabolism: Carbon Scavenging, Acetate Metabolism and Carbon Recycling Responses during Growth on Glycerol. *Microbial Cell Factories,* 11:46.
9. Pei, X.L., Wang, Q.Y., Li, C.L., Qiu, X.F., Xie, K.L., Huang, L.F., Wang, A.M., Zeng, Z.W. andXie, T. (2011). Efficient Production of a Thermophilic 2-Deoxyribose-5-Phosphate Aldolase in Glucose-limited Fed-batch Cultivations of *Escherichia coli* by Continuous Lactose Induction Strategy. *Applied Biochemistry Biotechnology*, 165:416–425.
10. Puertas, J. M., Ruiz, J., de La Vega, M. R., Lorenzo, J., Caminal, G. and Gonzalez, G. (2010). Influence of Specific Growth Rate Over the Secretory Expression of Recombinant Potato Carboxypeptidase Inhibitor in Fed-batch Cultures of *Escherichia coli. Process Biochemistry,* 45(8):1334–1341.

11. Santillán, M., and Mackey, M.C.(2008). Quantitative Approaches to the Study of Bistability in the lac operon of *Escherichia coli. Journal Royal Society Interface,* 5:S29–S39.
12. Shuman, H. A. and Silhavy, T. J. (2003). The Art and Design of Genetic Screens: *Escherichia coli. Nature Reviews Genetics,* 4: 419-431.
13. Santillán, M., Mackey, M. C. and Zeron, E.S. (2007). Origin of Bistability in the *lac* operon. *Biophysiology Journal*, 92:3830–3842.
14. Spexard, M., Beshay, U., Risse, J.M., Miksch, G., Flasche, E. (2010). Screening for conditions of enhanced production of a recombinant *b*-glucanase secreted into the medium by *Escherichia coli.* Biotechnol. Lett .32:243–248.
15. Tran, T.T., Mamo, G., Mattiasson, B. and Hatti-Kaul, R. (2010). A Thermostable Phytase From *Bacillus sp.* MD2: Cloning, Expression and High-Level Production in *Escherichia coli. Journal Industrial Microbiology Biotechnology,*37(3):279-87.
16. Weng, Y.P., Hsu, F.C., Yang, W. S. and Chen, H.P. (2006). Optimization of the Over expression of Glutamate mutase S Component under the Control of T7 System by using Lactose and IPTG as the Inducers. *Enzyme and Microbial Technology,* 38: 465–469.
17. Wonnacott, T.H., and Wonacott, R.J. (1990). Introductory statistics. New York: John Willey and Sons.
18. Varnalis, A., Brennan, J., MacDougall, D., Gilmour, S. (2004). Optimization of high temperature puffing of potato cubes using response surface methodology. Journal Food Engineering, 61: 153-163.
19. Zhang, B. H., Mao, X. O., Wang, Y. J. and Hu, X. Q. (2009). Optimization of Culture Conditions for High-level Expression of Dextransucrase in *Escherichia coli. Journal of Food, Agriculture and Environment,* 7(3&4):75 - 78 .

474193 Laccase: Fermentation Process Intensification and Efficient Purification By Magnetic Separation

Abstract:

A promising farnesol addition strategy was developed to enhance the laccase production from *Trametes versicolor*, and the laccase fermentation was successfully scaled up to a 2 m^3 stirred-tank reactor. The addition of farnesol resulted in an increased expression of the laccase (*lac*) gene and the *RhoA* gene. The *RhoA* gene correlated with hyperbranched mycelia, which facilitated the secretion of the intracellular laccase. An emphasis is highlighted on direct capture of laccase from *T. versicolor* fermentation broth using magnetic mesoporous silica nanoparticles with metal affinity ligands. The Cu^{2+}-chelated magnetic mesoporous silica nanoparticles with an average pore size of 14.5 nm provide 60.6-fold purification of laccase and 114.6% recovery yield of enzyme activity. The simple and efficient bioseparation method has great potential to be used for the robust and inexpensive preparation of purified laccase at the industrial scale.

471297 A Biosensor for Human Milk Oligosaccharide Detection

Abstract:

Human milk oligosaccharides (HMOs) are prebiotic compounds found in breast milk that favor the growth of beneficial bacteria in the infant gut. Often these compounds are derivatives of lactose or n-acetyllactosamine that contain either a fucose or sialic acid moiety. While there are many efforts underway to synthesize these beneficial complex carbohydrates, the current standard for their detection is low-throughput HPLC-MS. We present a high-throughput, fluorescence-based detection system for fucosylated and sialylated human milk oligosaccharides in vivo in *E. coli.* Such high-throughput systems facilitate directed evolution for increased in vivo HMO production. Additionally, the reporter enables the use of HMOs as orthogonal inducers of cellular circuits in synthetic biological applications.

466525 Compositional Effect of Varying Feedstocks on Heavy Oil Upgrading By Supercritical Water

Abstract:

Crude oil prices have shown big fluctuations around 150 and 35 $/barrel during 2008-2016. Such a volatile and unpredictable oil price tendency, which is being drastically affected by political issues, forces energy policy makers to focus on alternative energy conversion technologies for energy security. On the other hand, one third of Worldâ€™s recoverable oil reserves are in a form of heavy oil. Therefore developing alternative oil upgrading technologies with improved conversion and tunable selectivity providing low coke and improved light liquid yield are of desire. However, the variations in feedstock chemistry due to the source of the heavy oil affect economics of investments to a great extent. Supercritical fluids can be used for the development of alternative and innovative heavy oil upgrading processes and supercritical water (SCH_2O) having decreased dielectric constant and increased ionic product values is a potential candidate for the reactive extraction and thereby upgrading of heavy oils in an intensified process scheme. These altered physicochemical properties of water enable it to mimic hydrocarbons for reactive extraction purposes and also to increase [H_3O^+] concentrations for triggering acid-catalyzed cracking reactions. Domestic tar sand samples together with feedstocks of state-of-the art heavy oil upgrading units (Delayed Cokers - DCU), namely 50/70 bitumen from 2 different refineries in Turkey, 70/100 pen bitumen, and by product of Base Oil Unit, PDA bitumen, were used for the experiments. The reactions were carried out at 440 °C and 30MPa for 1 hour with water/oil ratio of 2 in a home-made 10 cm^3, SS316, bomb reactor. Characterization was done by GPC, H-NMR, DSC, TGA, FTIR, XRD, GC-Simdist, API, SEM images and SARA analysis. The coke-liquid-gas yield, coke type and boiling point distribution of liquids were shown to be influenced from the feedstock chemistry. Varying conversions with scattered selectivity values have been obtained. Modeling studies will be carried out to derive a tool for the prediction of conversion and selectivity of the reactions for a given feed under certain conditions as a part of a complementary study.

467087 *Product Formation and Kinetics of the Non-Isothermal Hydrothermal Liquefaction of Soy Protein Isolate*

Abstract:

James Sheehan - Graduate Student

Dr. Phil Savage – Adviser

The Pennsylvania State University, Department Chemical Engineering

Product Formation and Kinetics of the Non-isothermal Hydrothermal Liquefaction of Soy Protein Isolate

Hydrothermal liquefaction (HTL) is a thermochemical process capable of producing biocrude oil from biomass. HTL of various biomass feedstocks including microalgae, bacteria, lignocellulose and model compounds has been studied extensively under subcritical temperatures and long holding times (>60 min).[1–7] More recently, HTL of microalgae under rapid heating conditions and short reaction times (<1 min) has achieved higher biocrude yields achieved than HTL of microalgae under subcritical conditions and longer holding times.[8,9] Protein is a major biochemical constituent of microalgae and we performed HTL of soy protein isolate to understand better the incorporation of protein derived products into biocrude.

We conducted non-isothermal, fast HTL of soy protein isolate (SPI) at temperatures ranging 140 °C to 500 °C and residence times of 10 s to 300 s. SPI solids rapidly transform into aqueous phase products and biocrude. The highest biocrude yields (38-40wt%) were obtained within 45 s to 120 s and at temperatures ranging from 375 °C to 435 °C. The largest recovery of atomic N in the aqueous phase products was observed prior to the formation of substantial yields of biocrude. Ammonia formation was significant when the hydrothermal reaction medium reached supercritical conditions. Over 50% of the atomic N was observed as ammonia under such conditions. We developed a kinetic model and determined Arrhenius parameters that describe the fast HTL of SPI. The results of the modeling are shown in Figure 1. As can be seen, the proposed reaction network and optimized Arrhenius parameters can accurately model the major product yields from fast HTL of SPI.

Figure 1: Experimental (discrete points) and model predicted yields (continuous curves) for fast HTL of SPI at set point temperatures 400°C.

References

(1) Valdez, P. J.; Nelson, M. C.; Wang, H. Y.; Lin, X. N.; Savage, P. E. *Biomass and Bioenergy* **2012**, *46*, 317–331.

(2) Valdez, P. J.; Nelson, M. C.; Faeth, J. L.; Wang, H. Y.; Lin, X. N.; Savage, P. E. *Energy & Fuels* **2014**, *28* (2013), 67–75.

(3) Biller, P.; Ross, A. B. *Bioresour. Technol.* **2011**, *102*, 215–225.

(4) Teri, G.; Luo, L.; Savage, P. E. *Energy & Fuels* **2014**, *28*, 7501–7509.

(5) Changi, S.; Zhu, M.; Savage, P. E. *ChemSusChem* **2012**, *5*, 1743–1757.

(6) Luo, L.; Sheehan, J. D.; Dai, L.; Savage, P. E. *ACS Sustain. Chem. Eng.* **2016**.

(7) Toor, S. S.; Rosendahl, L.; Rudolf, A. *Energy* **2011**, *36* (5), 2328–2342.

(8) Faeth, J. L.; Valdez, P. J.; Savage, P. E. *Energy and Fuels* **2013**, *27*, 1391–1398.

(9) Hietala, D. C.; Faeth, J. L.; Savage, P. E. *Bioresour. Technol.* **2016**, *214*, 102–111.

James Sheehan - Graduate Student

Dr. Phil Savage – Adviser The Pennsylvania State University, Department Chemical Engineering

Product Formation and Kinetics of the Non-isothermal Hydrothermal Liquefaction of Soy Protein Isolate

-

Hydrothermal liquefaction (HTL) is a thermochemical process capable of producing biocrude oil from biomass. HTL of various biomass feedstocks including microalgae, bacteria, lignocellulose and model compounds has been studied extensively under subcritical temperatures and long holding times (>60 min).[1–7] More recently, HTL of microalgae under rapid heating conditions and short reaction times (<1 min) has achieved higher biocrude yields achieved than HTL of microalgae under subcritical conditions and longer holding times.[8,9] Protein is a major biochemical constituent of microalgae and we performed HTL of soy protein isolate to understand better the incorporation of protein derived products into biocrude.

We conducted non-isothermal, fast HTL of soy protein isolate (SPI) at temperatures ranging 140 °C to 500 °C and residence times of 10 s to 300 s. SPI solids rapidly transform into aqueous phase products and biocrude. The highest biocrude yields (38-40wt%) were obtained within 45 s to 120 s and at temperatures ranging from 375 °C to 435 °C. The largest recovery of atomic N in the aqueous phase products was observed prior to the formation of substantial yields of biocrude. Ammonia formation was significant when the hydrothermal reaction medium reached supercritical conditions. Over 50% of the atomic N was observed as ammonia under such conditions. We developed a kinetic model and determined Arrhenius parameters that describe the fast HTL of SPI. The results of the modeling are shown in Figure 1. As can be seen, the proposed reaction network and optimized Arrhenius parameters can accurately model the major product yields from fast HTL of SPI.

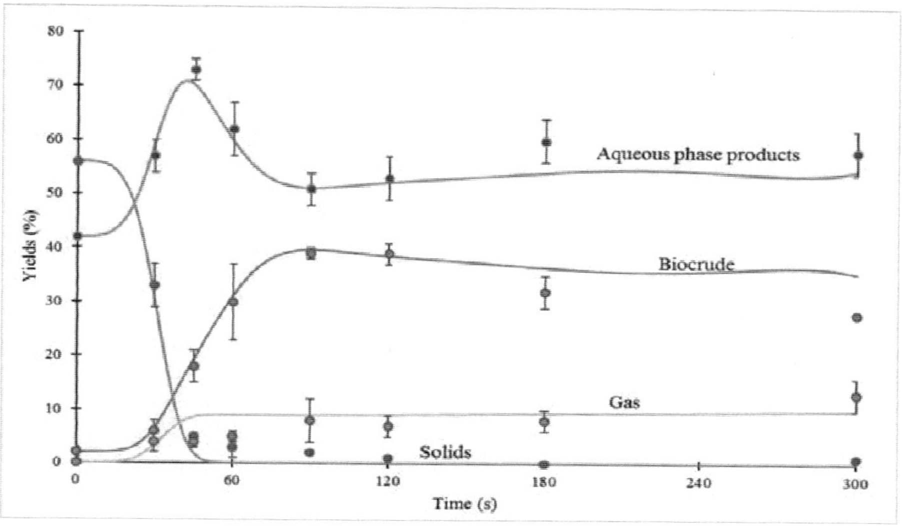

Figure 1: Experimental (discrete points) and model predicted yields (continuous curves) for fast HTL of SPI at set point temperatures 400°C.

References

(1) Valdez, P. J.; Nelson, M. C.; Wang, H. Y.; Lin, X. N.; Savage, P. E. *Biomass and Bioenergy* **2012**, *46*, 317–331.

(2) Valdez, P. J.; Nelson, M. C.; Faeth, J. L.; Wang, H. Y.; Lin, X. N.; Savage, P. E. *Energy & Fuels* **2014**, *28* (2013), 67–75.

(3) Biller, P.; Ross, A. B. *Bioresour. Technol.* **2011**, *102*, 215–225.

(4) Teri, G.; Luo, L.; Savage, P. E. *Energy & Fuels* **2014**, *28*, 7501–7509.

(5) Changi, S.; Zhu, M.; Savage, P. E. *ChemSusChem* **2012**, *5*, 1743–1757.

(6) Luo, L.; Sheehan, J. D.; Dai, L.; Savage, P. E. *ACS Sustain. Chem. Eng.* **2016**.

(7) Toor, S. S.; Rosendahl, L.; Rudolf, A. *Energy* **2011**, *36* (5), 2328–2342.

(8) Faeth, J. L.; Valdez, P. J.; Savage, P. E. *Energy and Fuels* **2013**, *27*, 1391–1398.

(9) Hietala, D. C.; Faeth, J. L.; Savage, P. E. *Bioresour. Technol.* **2016**, *214*, 102–111.

467990 An Investigation on Short Residence Time Macro Algae Hydro Thermal Liquefaction in a Batch Reactor

Abstract:

AIChE annual meeting, November 2016 San Francisco

An investigation on short residence time macro algae hydro thermal liquefaction in a batch reactor

Arash Izadpanah [a], Niccolˇ Le Brun [a], Klaus Hellgardt[a]

[a] Department of Chemical Engineering, Imperial College London, South Kensington Campus, London, SW7 2AZ, United Kingdom

Bio crude oil production from macro algae may become a key technology to tackle tomorrowÕs renewable fuels challenge. In order to seed this technology it is important to identify and evaluate the right process pathways to convert macro algae to biofuels. In conventional algal biofuels processing, drying of the algae is a costly procedure, which could be bypassed by using Hydrothermal Liquefaction (HTL) technology with substantial advantages. For an efficient implementation of a continuous HTL process, the effect of process variables on the conversion and selectivity towards liquid biofuels needs to be understood. This is often a challenge because the timescale of the HTL process is of the order of minutes and it is difficult to be captured properly by ordinary methods. In this study we employed a molten salt bath to facilitate rapid heat transfer to the batch reactor and thus study HTL for precisely controlled residence times between 1 and 25 min. Under experimental conditions the conversion of macro algae to bio crude oil was found to be complete within 15 min reaching a maximum and then decreasing slightly (23% maximum yield after 11 min at 390 ¡C). Using different analytical techniques including CHNS-O, GC-MS, Simulated Distillation and Size Exclusion chromatography the bio crude from macro algae is characterized and compared with North Sea crude oil.

Keywords: Macro algae, Hydrothermal liquefaction, Conversion rate, Filtration, Centrifugation

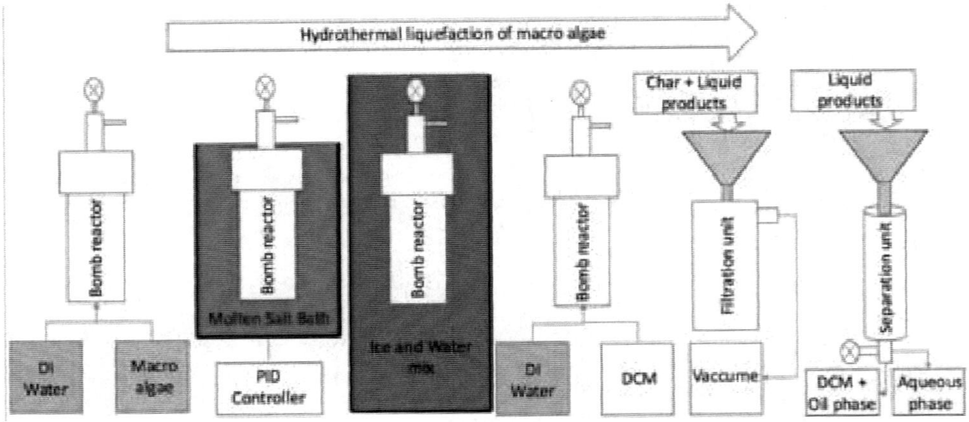

Figure 1:Graphical summery of the experiment

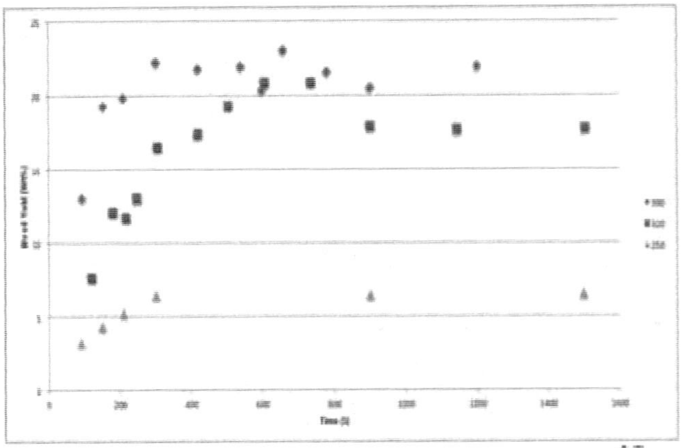

Figure 2: Bio crude oil yield as a function of residence time at 250 ℃, 320 ℃, 390 ℃.

459274 Hydrothermal Liquefaction (HTL) of High-Ash Algal Biomass: The Effect of Ash Contents in HTL Reactions

Abstract:

Previous studies demonstrate that the high ash contents appeared to inhibit the formation of biocrude oil in the hydrothermal liquefaction (HTL) processes. Mixed-culture algal biomass from wastewater systems (AW) were screened into three sizes (<106 μm, 106-180 μm and >180 μm) and their ash contents were measured. AW biomass with different ash contents was converted into biocrude oil via HTL at 300 °C for a 60 minute reaction time, which is the previously determined optimum condition for producing biocrude oil from AW. Compared to AW biomass before screen pretreatments, the ash contents of AW of 106-180 μm and <106 μm were respectively decreased from 53.3 % to 38.5% and 41.8 %, while the biocrude oil yields remained almost constant. When the ash contents were reduced from 53.3 % to 38.5 % in the AW biomass, the HHV of biocrude oil was increased from 27.5 MJ/kg to 32.3 MJ/kg and the fraction of light oil (boiling point of 100-300 °C) was increased from 31% to 49%. On the other hand, GC-MS analyses of algal biocrude oil and aqueous products show that the ash contents could promote denitrogenation, catalyze the formation of hydrocarbons, as well as mitigate the recalcitrant compounds in aqueous products under the HTL processes. The above results indicate that algal biomass with certain amounts of ash contents can be converted into biocrude oil with reasonable quality and quantity. In order to quantify the limit of the ash contents under the HTL processes, further HTL conversions with model algal feedstock and representative ash contents were conducted in this study. Elemental and thermogravimetric analyses of the resulting biocrude oil both demonstrate that when the ash content in the algal feedstock was below 40%, the HHV and boiling point distribution of the algal biocrude oil could be hardly affected. This fact substantiates the feasibility of using high-ash algal biomass as HTL feedstock and diminishes the necessity of multi-step pretreatments and modifications of high-ash algal biomass for biofuel application.

456792 Simultaneous Solid, Phosphorus Recovery and Biocrude Production through Hydrothermal Liquefaction of Algae

Abstract:

Subcritical water has become a highly regarded method for converting biomass, primarily algae, to an energy dense biocrude product. This conversion technique, more commonly referred to as hydrothermal liquefaction (HTL), still faces challenges of nutrient recovery, producing a practical biocrude substitute, and generating value-added coproducts before becoming a sustainable and commercially viable method. Recovering the phosphorus (P) in the solid biochar could be beneficial for secondary applications.

Previous work has shown that the inorganic content of the growth media, in particular Ca, P, and Si, and pH during harvesting dictate the fate of P in the HTL products. This paper expands on the idea of solid P-recovery via *in situ* synthesis of calcium phosphates present in the biochar, as seen previously in mixed culture, wastewater-cultivated algae. HTL was performed on diatoms, a unique group of algae that harness Si naturally within their cell walls, and mixed cultures of diatoms and *Chlorella kessleri*. Powder x-ray diffraction (XRD) characterized both the starting algal solids and resulting HTL biochar while acid digestions and inductively coupled plasma optical emission spectrometry (ICP) closed the P balance of HTL reaction.

In addition, it is hypothesized the algal carbohydrates, proteins, and lipids also play a role in the morphology and structure of the biochar. Model compounds were individually reacted via HTL, in the presence and absence of calcium phosphates, and calcium and phosphate ions. The resulting biochar, aqueous coproduct, and biocrude were characterized. Gas chromatography mass spectrometry, CHN/O elemental analysis, and simulated distillation analysis of the biocrude product were analyzed to identify potential catalytic upgrading properties of the calcium phosphate biochar.

453154 Supercritical Water Treatment of Fractions of Crude Oil: Quantification of the Products

Abstract:

The Arabian Heavy (AH) crude oil was fractionated into distillate ($\leq C_{20}$) fraction and bottoms fraction. The distillate and the bottoms fraction of crude oil were treated with supercritical water (SCW) in a batch reactor setup. The gaseous and liquid phase products were analyzed using gas chromatography (GC-MS, GCxGC, FID, SCD). The change in structural composition of maltenes and asphaltenes in the bottoms fraction before and after supercritical water treatment was investigated using proton NMR, FTIR and Raman spectroscopy.

The distillate fraction primarily consisted of n-alkanes ($\leq C_{20}$) plus a minor fraction of alkyl aromatics and sulfur compounds. SCW treatment of the distillate fraction resulted in cracking of higher carbon number alkanes to lower carbon number alkanes, and production of gaseous species such as methane, ethane and propane.

SCW treatment of the bottoms fraction caused much more dramatic changes. Much of the heavy components in the crude oil cracked to form compounds in the gasoline and diesel range which are observed by GCxGC. The higher carbon number aliphatics ($\geq C_{20}$) and long chain ($\geq C_9$) alkyl aromatic compounds in the bottoms fractions underwent cracking and produced C_6-C_{18} alkanes and C_3-C_8 alkylbenzenes. Supercritical water treatment of the bottoms fraction cracked the heavy sulfur compounds, and resulted in increased recovery of lighter sulfur compounds such as alkylated benzothiophene and dibenzothiophenes. The SCW-treated bottoms also showed increased recovery of polycyclic aromatic hydrocarbons such as naphthalene. Despite being carried out at temperatures similar to those used in conventional thermal coking processes, the SCW process did not form significant amounts of coke.

The maltenes and asphaltenes were separated from the bottoms fraction and the SCW-treated bottoms fraction. The proton NMR of the SCW-treated maltenes fraction showed a decrease in the protons corresponding to aliphatic CH_2 and CH_3 groups. There was also an increase in protons in single ring and polycyclic aromatic hydrocarbons. The proton NMR results support the observations made by the GCxGC measurements.

The proton NMR, FTIR and Raman spectroscopy on untreated and SCW-treated asphaltenes showed changes in the structural composition of asphaltenes post-SCW treatment. The proton NMR experiments showed a decrease in the protons which correspond to aliphatic CH_2 and CH_3 groups. The decrease in aliphatic CH_2 and CH_3 functional groups could be due to scission of the long alkyl chains attached to the aromatic core structure of the asphaltenes. The proton NMR results also showed 49 % increase in the mono-aromatic protons and 1 % decrease in di-aromatic and poly-aromatic protons, which suggests formation of new aromatic rings and/or perhaps conversion of maltenes containing single aromatic rings into asphaltenes (e.g. by the cleavage of alkyl tails, affecting the solubility).

FT-IR spectroscopy showed similar results to proton NMR, wherein an increase in aromatic and heteroatom content (O-H/N-H) was observed in the SCW-treated asphaltenes. Raman spectroscopy also showed evidence of chemical rearrangement and higher aromatic content in the SCW-treated asphaltenes when compared to asphaltenes from untreated bottoms fraction.

Overall, the SCW treatment of the crude oil fractions resulted in cracking of heavier hydrocarbon components into more valuable lighter molecules, which could be used in transportation fuel or as chemicals.

Acknowledgement: This research was supported through a Research Agreement with Saudi Aramco, a Founding Member of the MIT Energy Initiative.

458338 Supercritical Water Process - Finding a Way to Accomodate into Petroleum Industry

Abstract:

Supercritical fluid is not new to petroleum industry. Solvent extraction process utilizes supercritical fluid for enhancing selectivity and yield. However, supercritical water has not been applied in industrial scale due to its harsh condition. With progress of material and process technologies, supercritical water has been studied and evaluated in various areas to resolve technical challenges of conventional technologies. Recent research activities on heavy oil upgrading by supercritical water resulted in serious consideration of accommodating such a nonconventional way in petroleum industry. This presentation will introduce potential application areas of supercritical water technology as well as its implication in petroleum industry.

463912 Effect of Date Pit Based Additive on the Thermal Stability of Water-Based Drilling Fluid: Experimental Studies and Mathematical Modeling

Abstract:

Successful completion of oil and gas wells depends to a large extent on the properties of the drilling fluid used during the operation. The ability of drilling fluids to perform their fundamental functions is measured by certain performance properties including fluid loss prevention, stability under various temperature and pressure operating conditions, and stability against corrosion. Varieties of additives (such as polymers) are usually added to drilling fluid formulation in order to fulfill these property requirements. However, the poor thermal stability of such additives makes it difficult for them to be used in high temperature and high pressure drilling operations. Naturally occurring composites (such as date pit) are a new set of materials that are currently generating interest in varieties of applications. Date pit is a green, readily available and cheap material with superior thermal stability as compared to the polymers. Therefore, the objective of this work is to investigate the effect of date pit particles on the thermal stability of water â€" based drilling fluid formulation.

Infusion method was used to prepare date pit particles/poly (2-acrylamido-2-methyl-1-propanesulfonic acid) (PAMS) composite. Commercial Bentonite (Alumino Silicate Hydrate Montmorillonite) was used as the base material for preparing the water based drilling formulation. The formulation was prepared by adding 22.5g of Bentonite and the composite additive in 400ml of pure water. The percentage weight of the composite additive that was used ranges from 0.5 to 10 wt percent. Thermal stability of the formulations with and without date pit particles was evaluated using thermogravimetric analysis within a temperature range of 30 and 600°C under N_2 environment. Particle size analysis was done using laser light diffraction. FTIR study was also performed to evaluate the interaction between the additives and the bentonite.

In our previous studies, it was observed that the addition of 10wt% composite increased the rheological properties (PV, YP and YVR) by 80, 33 and 35% respectively, decreased the filter cake thickness by 35% and reduced the amount of fluid loss by 44%. The results from the present study revealed that the thermal degradation temperature of pure bentonite, date pit and PAMPS are 455, 250 and 160°C, respectively. Formulations with 10wt% date pit additive can be used up to a temperature of 300°C with only a 8% drop in the thermal stability while the thermal stability of the formulation with only PAMPS (without date pit) additive decreased by 53% at the same temperature as compared to the formulation without additive. Therefore, the prepared composite demonstrated some potential for use as a drilling fluid additive in high temperature environment. This observation is an eye opener into the suitability of date pit based drilling fluid formulation for high temperature and high pressure drilling fluid operations. The research outcome is expected to provide useful information regarding the use of readily available cheap waste materials for oil and gas exploration.

475008 Biocarbon Applications- Current Status and Future Directions

Abstract:

Biobased composites for lighter auto-part applications are the wave of the transportation industries. The corporate average fuel economy (CAFE) standard requires 54.5 mpg fleet average by the year 2025. Carbonaneous and lighter biocomposites are being researched and the resulting new innovations show their potential applications in injection moulded parts for car uses. Around 20 wt.% weight saving is achieved for such newly engineered biocomposites as contrast to talc filled counterparts. Carbon black, a petro-based colorant and filler is being used predominantly in plastics and rubber-based industries that accounts to around $ 20 billion market value globally. A more eco-friendly alternative, named as bio-carbon (from pyrolyzed biomass) is attracting recent attention as a filler/reinforcement in automotive and other plastic/rubber-based industries. Although a low content of ~2-4% of carbon black is used as colorant in many plastic-based products, bio-carbon with appropriate modification can be incorporated comparatively at a considerable higher percent (even 30 to 40 wt. %) in specific plastic based composite materials. Such biobased composites show promising potential as possible substitute for talc/short glass fibre reinforced polymer composites.

Acknowledgements: This research is financially supported by (1) the Ontario Ministry of Agriculture, Food, and Rural Affairs (OMAFRA)-University of Guelph Research Program(Project # 030055, 200399); (2) the Natural Sciences and Engineering Research Council (NSERC), Canada, Discovery Grants (Project # 401111); (3) Agriculture and Agri-Food Canada and Competitive Green Technologies (Project #051910, 052881) and (4) the Ontario Research Fund, Research Excellence Program; Round-7 (ORF-RE07) from the Ontario Ministry of Research and Innovation (MRI), currently known as the Ontario Ministry of Research, Innovation and Science (MRIS) (Project # 052644 and # 052665).

467117 Development of a Pilot-Scale Phytate Extraction System from Ethanol Coproducts

Abstract:

Introduction

Commercial production of fuel ethanol involves breaking down the starch present in corn to simple sugars, followed by the bioconversion of sugar molecules to ethanol catalyzed by yeast. The downstream processing involves the recovery of ethanol by distillation and the residual coproducts are further processed to serve as animal feed. In the dry-grind process, which accounts for over 70% of the total plants in the United States, the residual components are separated after the fermentation and distillation.

When coproducts are fed as an energy source to animals (above 15 to 20% of diet), protein and phosphorus are overfed. The environmental impact of feeding protein and phosphorus-rich materials to animals is related to one of the most pervasive forms of pollution from modern agriculture, as animal manure is applied to the soil. Phytate is the primary storage form of phosphorus and inositol in plants. The bioavailability of phosphorus bound to phytate is low for non-ruminants, and thus the phytate-derived phosphate ingested by these species (e.g. poultry and swine) is largely excreted, resulting in both nutritional deficiencies and environmental pollution. On the other hand, phytate is a widely applied valuable chemical as human nutrition, pharmaceutics, cosmetics, chelating chemical, and the raw material to produce inositol (Vitamin B8). It is primarily produced from rice bran in Asia and imported to the US

This creates a great opportunity for the US ethanol industry because extracting phytate as a new product from corn ethanol coproducts, knowing that corn germ has up to 5% phytate in its composition, can create additional revenue while increasing the feeding value of coproducts and decrease the phosphorus excretion in the animal manure.

Methods

Samples from four companies in the Midwest (Iowa, Illinois, and Minnesota) were obtained. Samples are stored at -20 °C prior to analysis and use.

Total-P and Phosphate-P are determined using P-ash method, in which digested samples are analyzed with a colorimetric assay of P in the digests. Samples are first dried at 105 °C until constant dry weight, and subsequently, they will be ashed at 550 °C. Concentrated HNO_3 (7 mol L^{-1}) will be used to dissolve the inorganic ash residue. Spectrophotometric assays are then performed with the molybdenum blue method. An enzymatic method derived from De Boland, Garner, & O'Dell (1975) is used for determination of phytate phosphorus.

Five macroporous resins (IRA series) and the fine resin (AG-1 series) were evaluated for the phytate adsorption capacity, high selectivity, high purity after desorption and low maintenance requirements. All of the resins selected for this study are alkali-type and have strong phytate adsorption capacity. The selection of the resin is one of the most important factors to be considered in the extraction process because all of the extraction process designs are based upon the resin selection.

The adsorption isotherms predict the maximum adsorption capacity of each resin. They are expressed in amount of substrate adsorbed in equilibrium onto the resin (qe) vs. the residual substrate in solution in equilibrium (Ce). Freundlich, Langmuir and Dubinin-Radushkevich adsorption models were used to estimate isotherm parameters.

Results and discussion

WS, containing undissolved corn fragments and residual yeast cells, present a moisture content of 87.94%. The TS, after the solid-liquid separation from WS, has the moisture content of 95.41%. The CDS, concentrated from TS through series of evaporators in order to reduce its moisture content, achieves the moisture content of 70.48%. On regards to the liquid fraction, mainly due to the increase in suspended solids, viscosity and osmotic pressure, the remaining water in CDS was difficult to remove by evaporation (Belyea et al., 2006), achieving moisture content levels close to 70%. The CDS is then mixed with the wet distillers grains to produce WDGS (M.C. 49.75%), which is then further dried to produce DDGS, at the moisture content of 14.92%. Usually, limited by cost in transportation for marketing, DDGS has a low moisture content ranging from 10% to 15%, which also prevents it to become moldy and unusable. Evaporation and drying are one of the most energy intensive steps in the corn ethanol processing.

WS has a total phosphorus concentration of 12.02 mg/g, of which 40% is found as phytate-phosphorus, at a concentration of 4.81 mg/g. After the first solid-liquid separation, the concentrations of total and phytate phosphorus (23.57 mg/g and 10.11 mg/g, respectively) in the TS are over doubled. The water removal step on TS to produce CDS maintains similar levels for total phosphorus (23.77 mg/g) and a decrease on phytate phosphorus (9.41 mg/g). WDGS and DDGS showed similar concentrations of total phosphorus (10.11 mg/g and 10.59 mg/g, respectively), whereas DDGS had higher levels of phytate than WDGS (4.50 mg/g and 3.27 mg/g, respectively). Comparing the results reported at literature (Alkan-Ozkaynak et al., 2010), similar values were found, especially for fractions of TS (19.4 mg/g), CDS (18.8 mg/g), and WS (11.0 mg/g) in means of total phosphorus.

Phytate is found most concentrated in the small-size range of particles in TS, indicating that it is possible to solubilize or aggregate these small particles. Selecting the appropriate size range of particles is crucial in order to process only the most concentrated fraction, which will allow the most efficient adsorption operation. The fractions of 0.10 mm and below account for about 90% of the total phytate found in TS. Screening experiments were conducted to evaluate the pressure drop required to process TS and 0.10 mm-filtered TS. Our results showed that having a filtration step prior to the adsorption process may be energetically feasible, since this step decreases the required pressured drop of the adsorption process by 55%.

Thin stillage is a phytate-rich ethanol co-product and is the liquid fraction that is being evaluated as the raw material in the phytate extraction. It also has significant presence of phosphate and other types of phosphorus-based groups. Reactive-phosphorus (i.e. free orthophosphate) was also adsorbed onto the resin. The characteristics of the most effective resin are those with the capability to adsorb most of the phytate and the least amount of phosphate, when determining resin selectivity. IRA-400 (0.128 mg phytate/mg resin) and IRA-900 (0.126 mg phytate/mg resin) resins adsorbed the most phytate per mass of resin, but IRA-900 (0.733 phytate phosphorus / total phosphorus) showed more selectivity compared to the IRA-400 (0.661 phytate phosphorus / total phosphorus) resin from this screening experiment.

In order to reduce the operating cost of the adsorption process and to achieve technical feasibility, the optimal resin should have an extended lifespan. A study was conducted to evaluate the regeneration capacity of the 5 macroporous (IRA series) resins using NaCl (2 M) as desorbent and NaOH (150 g/L at a loading rate of 4 %) as a regenerating agent for 4 cycles of operation. The equilibrium concentrations (q_e) of phytate-P (mg/g resin) in all the four cycles for all the resins tested, were within 2.5 standard deviations from the mean, which is an acceptable deviation for this study. The adsorption- desorption capacity and selectivity were similar for all 4 cycles.

References

Alkan-Ozkaynak, A., Karthikeyan, K., & Roa-Espinosa, A. (2010). Reducing phosphorus concentration in animal feed coproducts from the corn distilling industry. *Transactions of the ASABE, 53*(4), 1287-1294.

Belyea, R., Rausch, K., & Tumbleson, M. (2004). Composition of corn and distillers dried grains with solubles from dry grind ethanol processing. *Bioresource Technology, 94*(3), 293-298.

De Boland, A. R., Garner, G. B., & O'Dell, B. L. (1975). Identification and properties of phytate in cereal grains and oilseed products. *Journal of Agricultural and Food Chemistry, 23*(6), 1186-1189.

474771 New Biocomposite Materials from Poly(lactic acid) Blended with Poly(ethylene glycol) and Biobased Carbon

Abstract:

Biobased and biodegradable polymeric materials continue to be a fundamental step towards achieving an environmentally-friendly society. Therefore, poly(lactic acid) (PLA) being the most common biopolymer in the marketplace today is useful as the base component in the formation of future compostable consumer products. The current downsides of PLA being its toughness and low thermal stability are the key reason for its limited applications and a major focus for research and optimization.

In this study, poly(lactic acid) based composites prepared through melt blending with poly(ethylene glycol) (PEG) and biobased carbon were tested for their thermal and mechanical properties. The addition of poly(ethylene glycol) acted as a plasticizer, via greater chain mobility and as a crystallization promoter, by the reduction in energy necessary for chain folding during spherulite formation. The biobased carbon was beneficial in establishing nucleation sites for the poly(lactic acid) upon cooling, increasing the crystallization relative to the neat material. The poly(D-lactic acid) (PDLA) was essential in the development of the PLA stereocomplex between the L- and D- enantiomers. The crystallinity of the polymeric composites was elevated when PEG was included in the system. The heat deflection temperature of the poly(lactic acid) with PEG and PDLA was hindered by the low glass transition temperature of the samples. The mechanical properties of the composites samples are still under characterization and will be incorporated into the results prior to event. It is expected that the tensile strength and Youngâ€™s modulus will be lowered with the inclusion of poly(ethylene glycol), while the elongation at break will be increased. Further investigation of the nothched Izod impact strength will be conducted and optical microscopy for visual representations of the crystallization of the composites. Also, scanning electron microscopy (SEM) will be used to determine the morphology and dispersion of the components throughout the composite in order to correlate all findings.

Acknowledgment: *This research is financially supported by (1) the Ontario Ministry of Agriculture, Food, and Rural Affairs (OMAFRA)-New Directions Research Program (Project # 049528) and (2) the Ontario Ministry of Economic Development and Innovation (MEDI), Canada, Ontario Research Fund, Research Excellence Round 4 program (ORF-RE04) (Project # 050231 and 050289).*

474990 Effect of Butyl Glycidyl Ether Model Co-Monomer on Poly (glycerol succinate) Network and Dynamics for the Design of Multifunctional Hyperbranched Polymers

Abstract:

An original approach has been developed to easily design functional materials from poly (glycerol succinate) (PGS). This strategy is related to the introduction of an epoxidized functional agent during the polyesterification between the glycerol and succinic acid. The impact of this epoxide group on the polymerization chemistry and its hyperbranched network has been modeled by using the butyl glycidyl ether (BGE) as co-monomer. The potential reactions have been confronted with the units topology revealed by 2D NMR correlations. The regioselectivity and the stoichiometry of the system were modified *in situ* by the kinetic control of parallel reactions. As a result, we observed a delay of the gelation and an increase of the polyesterification conversion. The resulting hyperbranched polymers (HBPs) obtained just after gelation present a temperature of glass transition (T_g) of -3.9 °C for PGS and -16.1 °C for Poly(glycerol-co-butyl glycidyl ether) succinate (PGS-co-BGE). This difference was related to the dynamic motion of the BGE butyl tails which play the role of spacer between the polymer chains during the relaxation process. The dependency of the effective activation energy ($E_{î±}$) through the T_g was calculated by using the advanced isoconversional method. The variation of $E_{î±}$ was associated to a competition between the cooperative chains motions and the sequestration effect of the hydrogen bonds. This behavior can be potentially generalizable to the plastic glass containing a critical amount of secondary interactions.

Acknowledgment: *This research is financially supportedÂ by (1) the Ontario Ministry of Agriculture, Food, and Rural Affairs (OMAFRA)-New Directions Research Program (Project # 050155); and (2) OMAFRA-University of Guelph Bioeconomy-Industrial Uses Theme (Project # 200283).*

475007 Value Added Biodegradable Materials from Industrial Wastes

Abstract:

Biodegradable plastics and composites can be a solution to the plastic waste pollution issue. With regards to practical biodegradability, the term compostability is a standardized way to identify plastics and composites that undergo biodegradation more than 90 % of their weight in a specific time frame; six month. At the Bioproducts Discovery and Development Centre (BDDC), University of Guelph, research is going on to develop compostable composites formulations from biodegradable plastics such as PLA, PBS, etc., and biomass such as agricultural residues, perennial grasses, and industrial co-products/by-products. This approach has two advantages. First, it can create new outlets for the under-valued biomass and bring new revenue to the farmer. Second, it can result in faster biodegradation of the composite in comparison with the neat plastic, as previous studies at the BDDC confirm that. With the compostability in mind, the mechanical and rheological performance of the composite material is an important factor to consider. Green surface treatments of the biomass and the use of processing aids and compatibilizers are effective ways to enhance or keep balanced the properties of the composite material. To date, there is no commercial compatibilizer based on biodegradable plastics. At the BDDC, maleated bioplastic compatibilizers were produced customized towards the bioplastic matrix and successfully utilized for improving the mechanical properties of the compostable composite materials. On the other hand, hybridization of the fillers of different types of biomass was another approach pursued in this study to benefit from different aspects of different fillers at the same time. The amount of compatibilizer and different fillers was optimized to achieve a balance in mechanical and physical properties. The produced compostable composites were successfully certified as 100 % compostable and used in food packaging applications.

Acknowledgements: *This research is financially supported by (1) the Ontario Ministry of Agriculture, Food, and Rural Affairs (OMAFRA)- University of Guelph Bioeconomy Industrial uses Research Program (Project # 200245, 200245); (2) Agriculture and Agri-Food Canada (AAFC) and Competitive Green Technologies (CGTech) through AgriInnovation (Program Project # 051909, 051910, 052880); and (3) and the Ministry of Economic Development and*

Innovation (MEDI), Ontario Research Fund - Research Excellence Round - 4 (Program Project # 050231 and 050289).

467847 Indirect Liquefaction of Biomass to Transportation Fuels Via Mixed Oxygenated Intermediates

Abstract:

This paper presents a comparative techno-economic analysis of five conversion pathways from biomass to gasoline-, jet-, and diesel-range hydrocarbons via indirect liquefaction with specific focus on pathways utilizing oxygenated intermediates (derived either via thermochemical or biochemical conversion steps). The four emerging pathways of interest are compared with one conventional pathway (Fischer-Tropsch) for the production of the hydrocarbon blendstocks. The processing steps of the four emerging pathways include: biomass-to-syngas via indirect gasification, gas cleanup, conversion of syngas to alcohols/oxygenates, followed by conversion of alcohols/oxygenates to hydrocarbon blendstocks via dehydration, oligomerization, and hydrogenation. We show that the emerging pathways via oxygenated intermediates have the potential to be cost competitive with the conventional Fischer-Tropsch process. The evaluated pathways and the benchmark process generally exhibit similar fuel yields and carbon conversion efficiencies. The resulting minimum fuel selling prices are comparable to the benchmark at approximately \$3.60 per gallon-gasoline equivalent, with potential for two new pathways to be more economically competitive. Additionally, the coproduct values can play an important role in the economics of the processes with oxygenated intermediates derived via syngas fermentation. Major cost drivers for the integrated processes are tied to achievable fuel yields and conversion efficiency of the intermediate steps, i.e., the production of oxygenates/alcohols from syngas and the conversion of oxygenates/alcohols to hydrocarbon fuels.

465137 Incorporating Unsteady-State Fermentation Models in Aspen Plus: Fed-Batch and *in Situ* Gas Stripping Simulation and Analyses

Abstract:

Process simulation can predict the most informative experiments to conduct for process design, control and optimization. Traditional refinery processes have been extensively simulated and analyzed using the thermodynamic models and separation unit operations in process simulation tools, such as Aspen Plus. However, these essential tools for optimizing refinery processes have not been fully extended to biorefineries as a result of the inherent unsteady-state nature of fermentation processes. One of the approaches used to reduce substrate and product inhibition to fermentation microorganisms and increase fermentation productivity is fed-batch fermentation and *in situ* product recovery techniques, such as gas stripping. In the fed-batch fermentation, a bioreactor is charged with a relatively low concentration of substrate and as the substrate is consumed, fresh substrate is supplied at a rate that keeps the substrate concentration in the reactor below the inhibitory levels to the microorganisms. Gas stripping as a chemical separation method allows for the selective removal of the volatile components from the aqueous fermentation solution using a gas bubbled through the fermentation broth.

In our previous study, the batch acetone-butanol-ethanol (ABE) fermentation was used as a model system to simulate an unsteady-state batch ABE fermentation using a Fortran user kinetics subroutine linked to the batch reactor (RBatch) in Aspen Plus as a starting point for a biorefinery. This study extends the Aspen Plus batch model to fed-batch fermentation and *in situ* gas stripping with the aim of comparing the trends in the simulation to their corresponding laboratory fed-batch and *in situ* gas stripping experiments. An existing ABE fermentation model was reformulated to describe cell growth and inhibition using Monod kinetics, making it sensitive to changes in the fermentation environment as a result of fresh substrate addition (fed-batch) and product accumulation. This study provides a platform that can be used to guide integrated fed-batch fermentation and *in situ* gas stripping experiments and assess the accuracy of literature data. Additionally, a universally accepted process simulator, Aspen Plus, can be used in the simulation of fermentation processes linked with separation unit operations to provide strategies for the design and optimization of biorefineries.

463187 Techno-Economic Analysis of Direct Coal-Biomass to Liquids (CBTL) Plants with Shale Gas Utilization and CO_2 Capture and Storage (CCS)

Abstract:

Direct coal liquefaction (DCL) technology is a promising technology for producing transportation fuels from non-petroleum sources. However, a large amount of H_2 is required in the DCL system because of the low H/C ratio in coal. A recent study shows that about 0.48 tonne of CO_2 is released per barrel of transportation fuels produced in the Shenhua DCL plant, where about 80% of CO_2 is produced in the gasification-based H_2 production unit. Addition of biomass and applying carbon capture and storage (CCS) technologies are two possible solutions to reduce the carbon footprint, but would lead to higher capital and operating costs. Furthermore, H_2 can also been produced from shale gas instead of coal with less capital investment and CO_2 emission, as shale gas is currently abundant in most of the continental US. Even though co-processing of coal and biomass and inclusion of CCS have been widely studied for indirect liquefaction and power generation system, limited studies have been conducted for direct liquefaction processes. With this motivation, we will present a techno-economic study of direct coal-biomass to liquid (CBTL) plants with shale gas utilization and CCS.

In this study, the direct CBTL plants is designed based on the catalytic two-stage liquefaction process investigated by HTI, where inline hydrotreating is considered.[1] Four different plant configurations, named SMR_CCS, SMR_VT, CG_CCS, CG_VT, are considered in this work based on different extent of CCS and H_2 sources. In the SMR_CCS and SMR_VT processes, hydrogen is produced from shale gas steam methane reforming (SMR) and partial oxidation of liquefaction residue. In the CG_CCS and CG_VT processes, hydrogen is produced from co-gasification (CG) of

coal/biomass/residue. High extent of CCS is considered in the SMR_CCS and CG_CCS processes, while all CO_2 is vented (VT) in the SMR_VT and CG_VT processes. Both chemical and physical absorption processes are considered for carbon capture depending on the sources and partial pressure of CO_2 in the CO_2-containing stream. The process model and economic model are developed mainly in Aspen Plus® and Aspen Process Economic Analyzer® (APEA®), respectively, and are validated by comparing with the data obtained from the open literature.

In this presentation, we will focus on the following aspects: (1) development of plant-wide process model and rigorous economic model in Aspen Plus®, Aspen Exchanger Design and Rating®, APEA® and Excel, (2) sensitivity studies to analyze the impact of key design parameters (i.e. biomass/coal ratio, H_2 sources, extent of CCS) and investment parameters (i.e. price of coal, biomass, shale gas, project life, plant contingency and plant capacity) on key economic measures, such as net present value (NPV), internal rate of return (IRR), break-even oil price (BEOP) and equivalent oil price (EOP), (3) analysis of the potential environmental credits due to use of biomass and CCS, (4) comparison between indirect and direct CBTL plants with or without CCS.

Reference

(1) Comolli, A.G., Lee, L.K., Pradhan, V.R., Stalzer, T.H., Harris, E.C., Mountainland, D.M., Karolkiewicz, W.F. and Pablacio, R.M., Direct Liquefaction Proof-of-Concept Facility, Technical Progress Report POC Run 01, Contract No. AC22-92PC92148, Hydrocarbon Research Inc., 1995.

469616 Techno-Economic (TEA) and Life Cycle Analysis (LCA) of the Pyrolysis-Bioenergy-Biochar Pathway for Carbon-Negative Energy

Abstract:

Techno-economic (TEA) and Life Cycle Analysis (LCA) of the Pyrolysis-Bioenergy-Biochar Pathway for Carbon-Negative Energy

Wenqin Li.[b], Qi Dang [ab], Wright M. M.[ab], Brown, R. C.[ab], Laird, D.

[a]Bioeconomy Institute

[b]Department of Mechanical Engineering

[c]Department of Agronomy

Iowa State University

The objective of this project is to evaluate the economic and environment impacts of the Pyrolysis-Bioenergy-Biochar Pathway for carbon-negative energy. Biochar, a carbon rich coproduct of pyrolysis, has the potential to improve soil quality by sequestering carbon thus helping to achieve a carbon negative energy goal. In addition to biochar, fast pyrolysis of biomass can also produce biofuels, bio-power and chemicals. Techno-economic Analysis (TEA) and Life Cycle Analysis (LCA) results will vary with the selection of pyrolysis products, feedstock materials, and facility locations.

Economic analysis was conducted on two pyrolysis product utilization scenarios: biochar & biofuel and biochar & bio-power. In the biochar & biofuel scenario, we estimate that the minimum fuel-selling price (MFSP) varies between \$2.91 and \$3.03 per gallon if biochar can be sold for \$60 to \$20 per tonne. On the other hand, if the biofuel can be sold at market prices of \$2.00 to \$5.00 per gallon, the biochar can be sold for \$364 to -\$637 per metric tonne. The biochar & bio-power scenario yields minimum electricity-selling prices of 9.85 to 10.07 ¢/kWh for biochar prices of \$60 to \$20 per tonne. Biochar can be sold for -\$336 to \$400 per tonne at electricity prices of 12 to 8 ¢/kWh. These prices do not include environmental credits or incentives. More pyrolysis product combinations scenarios will be assessed and compared.

Biochar properties and yields vary from different types of biomass. Regression models of relationship between biochar & biofuel yield and biomass carbon, ash & oxygen contents have been built based on series of fast pyrolysis experimental yields data. The regression model was validated for different types of feedstock and used to predict the yield for various feedstocks. Process models for fast pyrolysis of various feedstocks will be built in Aspen Plus™, with the products prediction data. Mass and energy balance and subsequent techno-economics analysis will be assessed and compared to elect the best economic performance feedstock.

The pyrolysis facility locations might affect various factors, such as feedstock cost, feedstock type, capital cost, and operating cost. Biochar type and the demand for biochar will also vary with the plant locations. Three specific land areas of US are investigated depending on the selection of pyrolysis facility locations: Upper Mississippi River Basin (UMRB), California, and U.S. Southeast. The capital cost, operating cost and minimum fuel selling price was estimated to indicate the economic benefits of employing integrated Pyrolysis-Bioenergy-Biochar pathway in various regions. Life cycle analysis will also be conducted using GREET to evaluate the GHG emissions of the integrated Pyrolysis-Bioenergy-Biochar production facility in different regions.

467412 Techno-Economic and Environmental Life Cycle Assessments of Hydrocarbon Biofuel from Loblolly Pine

Abstract:

Thermochemical conversion of wood to biofuel via fast pyrolysis is regarded as a promising conversion pathway for producing biofuels. The process involves rapid thermal degradation of wood in the absence of air at a temperature of approximately 500-550°C with a short residence time of less than 2 seconds. Hydrocarbon biofuel is then produced through catalytic upgrading of the pyrolysis bio-oil product. Drying and size reduction of wood are major contributors to energy consumption of this process. Torrefaction, often referred to as mild pyrolysis, tends to enhance bio-oil properties by reducing water content, minimizing acidity, and reducing size reduction energy. The development of a two-stage fast pyrolysis process that involves a torrefaction pretreatment step prior to pyrolysis was investigated as an approach to minimize the energy consumption associated with size reduction. The impacts of different torrefaction temperatures have on the cost of production of hydrocarbon biofuel as well as the environmental impacts was investigated by carrying out techno-economic analysis (TEA) and life cycle assessment (LCA), respectively, with two-stage processing compared to the one-stage pyrolysis-based processing and to fossil fuels. Effects of the use of renewable co-products such as char and non-condensable gases from pyrolysis to replace fossil energy inputs (natural gas for process heat) to the process are explored in scenario analyses.

Aspen Plus process simulation package was used to model the two stage torrefaction-fast pyrolysis process with catalytic upgrading of pyrolysis bio-oil to hydrocarbon fuel blends. The effect of torrefaction severity on product composition, yield of pyrolysis bio-oil and subsequently yield of hydrocarbon fuel were obtained from the literature.[1-6] The effect of torrefaction on the energy requirement for size reduction was also included in our model based on literature data.[7] Using these data sources, mass and energy balances were obtained, and used in sizing the equipment, with equipment prices estimated from a number of sources such as the Aspen Economic Process Analyzer, previous

works and equipment vendors. A Discounted Cash Flow Rate of Return spreadsheet was used to obtain the gate cost of production, and data obtained from the simulation also served as inputs for the LCA carried out using the LCA software SimaPro 8.0 with greenhouse gas emissions (GHG) as the impact category.

From our model simulations, having a torrefaction step as a pretreatment step prior to fast pyrolysis reduces the cost of bio-oil production from loblolly pine, by reducing energy cost associated with size reduction. Results from the model also show a reduction in GHG emissions associated with the bio-oil production and catalytic upgrading as a result of the torrefaction pretreatment step. The effect of substituting renewable energy (char) for process heat demand rather than natural gas yielded a cost of production decrease from $4.26/gallon hydrocarbon fuel to $4.15/gallon for the one-step pyrolysis-based process. Full results on the TEA and LCA for all scenarios will be presented.

References

1. Park, J., Meng, J., Lim, K.H., Rojas, O.J. & Park, S. Transformation of lignocellulosic biomass during torrefaction. *Journal of Analytical and Applied Pyrolysis* **100**, 199-206 (2013).

2. Westerhof, R.J. et al. Stepwise fast pyrolysis of pine wood. *Energy & fuels* **26**, 7263-7273 (2012).

3. Jones, S. et al. (National Renewable Energy Laboratory (NREL), Golden, CO., 2013).

4. Vispute, T. Pyrolysis oils: characterization, stability analysis, and catalytic upgrading to fuels and chemicals. (2011).

5. Huber, G.W., Chheda, J.N., Barrett, C.J. & Dumesic, J.A. Production of liquid alkanes by aqueous-phase processing of biomass-derived carbohydrates. *Science* **308**, 1446-1450 (2005).

6. Furimsky, E. Catalytic hydrodeoxygenation. *Applied Catalysis A: General* **199**, 147-190 (2000).

7. Phanphanich, M. & Mani, S. Impact of torrefaction on the grindability and fuel characteristics of forest biomass. *Bioresource technology* **102**, 1246-1253 (2011).

458183 Novel Process of Biofuel Production from Acid Oil and Hydrous Bioethanol Using Ion-Exchange Resin; I. Experimental Optimization of Operating Conditions

Abstract:

Introduction

Liquid biofuels offer attractive advantages over other forms of biomass energy resources in terms of high energy density and ease of storage and transport. Bioethanol is useful as a gasoline additive (up to 10%) and biodiesel, or fatty acid methyl ester (FAME), is also used as a fuel alternative (100%) to petroleum diesel. However, there has been no significant increase in the production of these fuels in the past few years. The reason is that there is no profitability for producers owing to high production cost and no benefits for consumers because of unstable product quality and lower fuel economy and power. Moreover, emergence of next-generation automobiles will lower the demand for gasoline. The water content of ethanol used as the gasoline additive should be less than 0.5 wt% and the dewatering process requires a large energy input. Thus, the necessity of anhydrous ethanol as gasoline additive fades and new technology which can use hydrous ethanol is desired. On the other hand, the diesel fuel continues to be used for heavy duty on-road commercial transportation, non-road construction and various types of ocean vessels. If new technology to economically produce high quality biodiesel can be developed, the necessity of biodiesel utilization will increase.

In this research (Part I), biodiesel is produced in the form of fatty acid ethyl ester (FAEE) using hydrous ethanol by esterification of free fatty acid (FFA). The FAEE has higher cetane number and fuel economy and power than the traditional fatty acid methyl ester (FAME) and FAEEâ€™s low temperature fluidity is also improved[1]. However, the conventional technologies for esterification of FFA never use hydrous ethanol because the water in the reaction solution must be completely removed to obtain a high conversion. We have reported to produce FAME at more than 95% conversion by esterification of acid oil using a cation-exchange resin catalyst without dewatering and adding excess alcohol[2]. This means that the catalytic activity of the resin is not inhibited by water. Thus, using the cation-exchange resin catalyst the batch and continuous esterification experiments with hydrous ethanol are performed. The maximum water content in hydrous ethanol to obtain high conversion is clarified.

Materials and methods

The cation-exchange resin, Diaion PK208LH (Mitsubishi Chemical Co., Ltd., Tokyo, Japan) with a high catalytic activity is used[2]. The resin was supplied in an activated H-form swelled with water. For pretreatment, the resin was packed with the column and ethanol is supplied to remove water in the resin before the esterification. In the batchwise esterification, the preheated mixed-solution at the molar ratio of oleic acid and ethanol, 1:3, and 33 wt% of the resin were added to the glass bottle. The bottle was well shaken in a thermal bath at 50 °C. At specific time intervals, samples of the solution were collected and the concentrations of the reactants and products were analyzed. The water content in ethanol is set at 0, 5, and 10 wt%.

In the continuous esterification, the column packed with the resin was kept constant at 50 °C. For the pretreatment, ethanol with a certain water content was supplied similarly to the batchwise system. Then, the mixed solution of oleic acid and the ethanol at the same molar ratio as that in the batchwise system was supplied to the bottom of the column by a pump. Under each water content in ethanol, the residence time through the resin bed was regulated by changing the feed flow rate. The effluent solutions from the top of the column were collected and the concentrations of the reactants and the products in the solution were also determined.

Results and Discussion

In the batchwise system, using anhydrous ethanol for both pretreatment and esterification, the complete conversion was reached after 50 h. Using ethanol with 5wt% water content, the esterification rate became slower and more than 90% conversion was obtained at 120 h. After the pretreatment using ethanol with 5wt% water content, the water content in the resin was higher than that in the resin pretreated with anhydrous ethanol, so that the lipid soluble reactant, FFA, was hard to transfer into the resin to lower the esterification rate.

In the continuous system, the conversion was determined after steady state was reached under each condition. Using anhydrous ethanol, the complete conversion was obtained at the residence time through the bed of 10 h. Using ethanol with 5wt% water content, more than 95% conversion was attained at the residence time of 35 h and the complete conversion was also obtained at 60 h. Using ethanol with 10wt% water content, more than 90% conversion was obtained at 60 h. In the continuous system, the esterification rate became slower with increasing water content in ethanol, but its effect was smaller than that in the batchwise system. This was because the water formed by esterification was accumulated in the batch system, whereas the water was removed by continuous flow in the continuous system. Therefore, the continuous esterification using the column reactor packed with the resin is very effective for utilization the hydrous ethanol.

Lower esterification rate leads to an increase in energy consumption, thus contributes negatively from a life cycle perspective by inducing more environmental impacts associated with energy. On the other hand, increased water content means lower duty for distillation of ethanol. In another part of this research (Part II), to allow a system-wide optimization, modeling of the system with a scope that contains all the relevant product life cycles is performed. Using the data acquired from the experiments with continuous configuration, the abovementioned trade-off is highlighted using a case study with a Japanese example.

Reference

1)N.Shibasaki-Kitakawa et al., Kagaku Kogaku Ronbunshu, 42 (2016) 30-36

2)N.Shibasaki-Kitakawa et al., Fuel, 139 (2015), 11-17.

<u>**Economic and Environmental Assessment of Methane to Ethylene Via Oxidative Coupling**</u>

Abstract ID# 469784

Kenneth O'Neill and Benjamin J. Davis, The Cooper Union (New York, NY)

One possible separation train for the purification of ethylene from reactor effluent is proposed, with interest in meeting economic and environmental (via LCA) performance targets. The system consists of two flash tanks and four distillation columns which serve to purify polymer-grade ethylene and reactor recycle. The reaction for producing ethylene is the oxidative coupling of methane on nanowire catalysts, and low conversions mean a large percentage of the methane enters the separation train and must be separated. The four columns following the two initial flash tanks are a carbon monoxide removal column for the separation of carbon monoxide from methane for reactor recycle, a demethanizer column for separation of the remaining methane from the heavier hydrocarbons, a C_2-splitter for the separation of ethylene from ethane, and a carbon dioxide removal tower which operates at vacuum to break the azeotrope which forms between ethylene and carbon dioxide to obtain polymer grade ethylene. The purified ethylene distillate from the carbon dioxide removal tower is 99.481 % pure and the separation system has purchase and bare module costs of \$776,220 and \$3,300,000, respectively (based on 2015 CEPCI values).

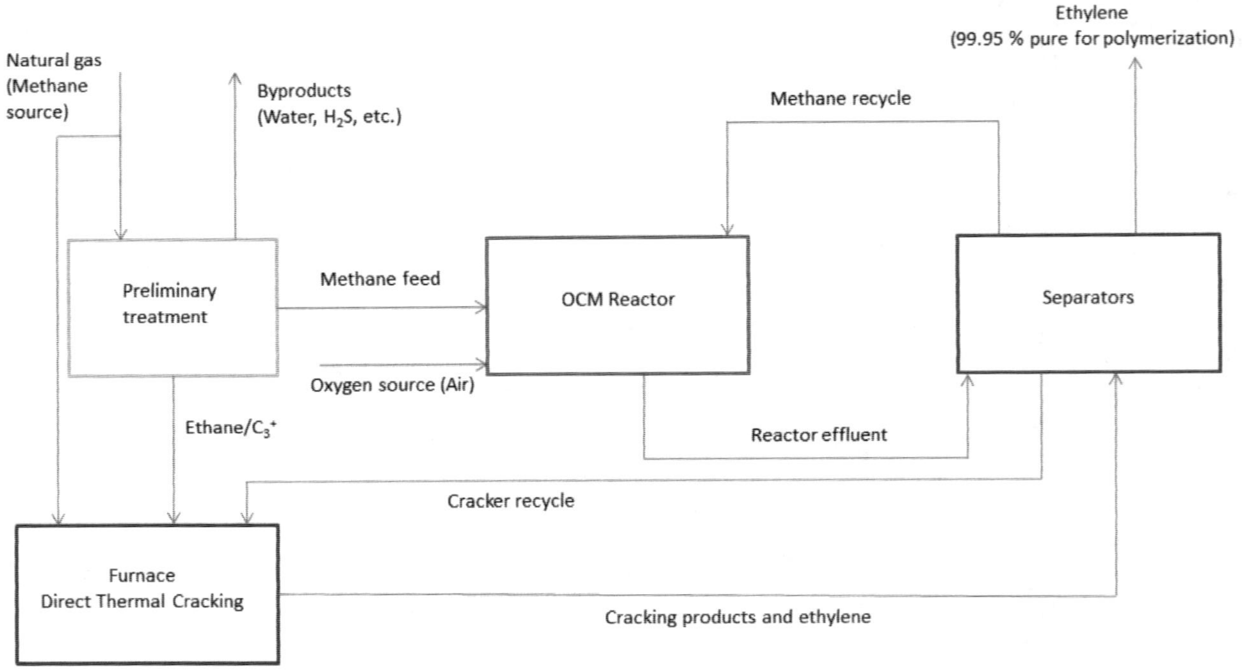

Figure 1: BFD for Siluria's OCM process integrated into an existing ethane cracker

Approximately 90,000 tons of ethylene were produced in United States (US) gas plants in 2014[i], making it one of the world's largest volume commodity chemicals. Ethylene is valuable as a precursor to a variety of products, including low and high density polyethylene, vinyl chloride monomer (VCM), ethylene glycol, and styrene; more than 30 million tons of these commodities were sold in 2012[ii]. Because of its high volume, finding cheaper feedstocks is one of the few ways to have an impact on the ethylene production market. Traditionally ethylene is made from cracking ethane or naptha, which are less abundant species in natural gas and crude oil, respectively. As of February 16, 2016, the sales price of natural gas was \$1.95 per MMBTU,[iii] while the total sales of ethylene based polymers decreased 4 % to \$69.4 billion in 2014[iv].

Due to the importance of ethylene in the chemical industry, means of revolutionizing and optimizing its production are constantly sought after. Siluria Technologies has created and patented nanowire catalysts[v] and processes[vi] related to the oxidative coupling of methane (OCM) to make ethylene; this is a more advantageous route because methane is cheaper and more abundant than naptha and ethane. A simplified flowsheet from their website is shown below as Figure 2.

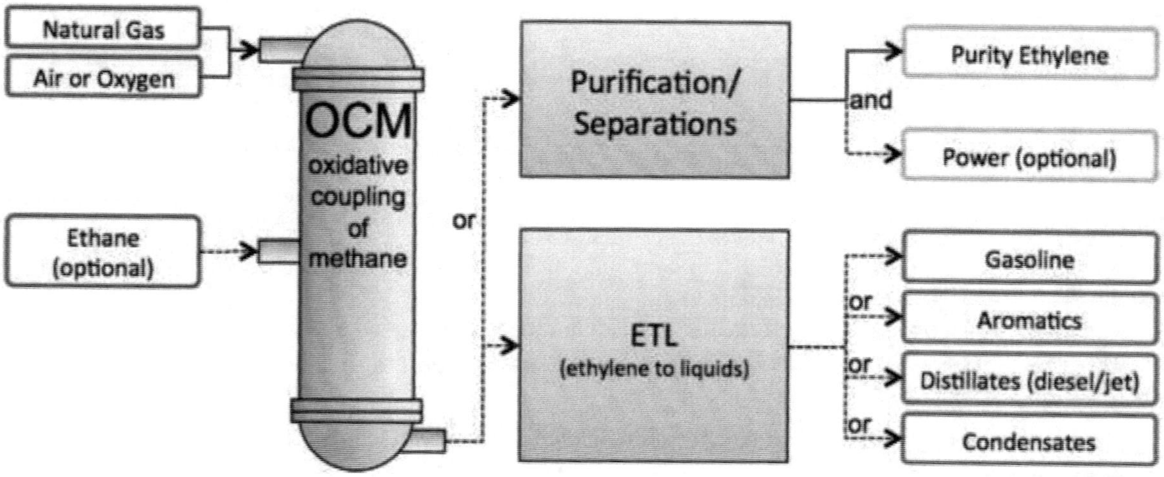

Figure 2: Simplified flowsheet of Siluria's OCM process[vii]

In the Siluria process, the feed to the OCM reactor consists of a methane source (natural gas) and an oxygen source (air or O_2 in the Siluria process). The reactor effluent can then be routed to one of two steps: a separation train for the recovery of high-purity ethylene, or to an ethylene to liquids process to produce gasoline/aromatics/distillates/condensates. The present investigation attempts to propose a more efficient means of separating the reactor effluent for the acquisition of high-purity ethylene.

Traditionally, methane conversions in OCM reactors with ordinary catalysts have been found to vary between 1 % and a theoretical upper bound around 40 % according to Bhasin *et al.* and Lunsford *et al.*, with the remaining 60 % of methane either unreacted, burned, or coked. Selectivities for ethane, ethylene, and acetylene (C_2) tend to inversely with methane conversion, from highs of 90 % to some reports of as low as 1 % have been reported[viii,ix]. At 799 °C, one catalyst patented by Siluria exhibits methane conversión around 21 % with a 70 % selectivity for C_2; on a similar catalyst over an SiO_2 support, the conversion is about 28 % with a 60 % selectivity for at 800 °C. In this study the properties of the former catalyst were used.

With the goal of having an industrially viable OCM system, Siluria has one of two options: construct a stand-alone facility focused on OCM-based ethylene production or integrate the process into a preexisting ethylene cracking facility. With integration into an existing ethylene cracker in mind, the preliminary design investigated herein was one in which the OCM reactor produces about 40 metric tons/day of product (40,000 kg/day).This corresponds to overall plant production of 740 metric tons/day, with 5.4 % of the cracking facility's ethylene from OCM and the remainder from the cracker.
(700 metric tons/day from the cracker).

The primary goal of the research thus far has been to separate the OCM reactor effluent. The difficulty in the separation of high-purity ethylene comes from the complicated OCM reactor effluent mixture. This mixture was assumed to consist of the following species:

Species and Boiling Points	
Component	**Normal Boiling Point (K)**
N_2	77.35
CO	81.15
O_2	90.15
CH_4	111.5
C_2H_4	169.15
C_2H_6	184.15
C_2H_2	189.15
CO_2	194.15
C_3H_6	225.15

Table 1: Components in OCM reactor effluent in order of increasing boiling points

The following assumptions were used in simplifiying the number of components in the reactor effluent:

- Air served as the oxygen source
- Hydrogen was considered light enough to be separated relatively easily
- Water's distinct chemical differences from the hydrocarbon mixture allowed for the assumption that it too could be easily removed
- Propylene was used as a proxy for hydrocarbons of chain length 3 or more, e.g. propanes, butanes, etc., due to those species' relatively higher boiling points allowing for relatively easy separations

In addition to the similarities in properties between the hydrocarbon components, these compounds are also soluble in one another, making separations via distillation more difficult than a simple boiling point difference, making determining appropriate operating conditions for the required separation units difficult. These four simplifying assumptions, along with the results of a material balance around the reactor given below, served as the basis for the simulation.

Component	Mass Flow Rate (MT/day)	Mass %	Molar Flow Rate (kmol/day)	Mole % (OCM Effluent)
CH_4	205.4	42	12800	57
O_2	60.1	0.12	18.8	.083
N_2	226	46	8070	36
CO	3.97	8.2E-01	142	0.63
CO_2	12.3	2.5	279	1.2
C_2H_6	28.1	5.8	938	4.2
C_2H_4	10.079	2.1	360	1.6
C_2H_2	0.041	8.4E-03	1.58	0.0070
C_3H_6	0.082	1.7E-02	1.95	0.0086
Total	487	100	22600	100

Table 2: Mass and molar flow rates and percents for the feed to the separation process simulated via ASPEN Plus V8.0

Figure 3 shows the current separation process as simulated in ASPEN Plus V8.0. The Peng-Robinson equation (P-R EOS) of state was used for the simulation because it works well for light gases[x].

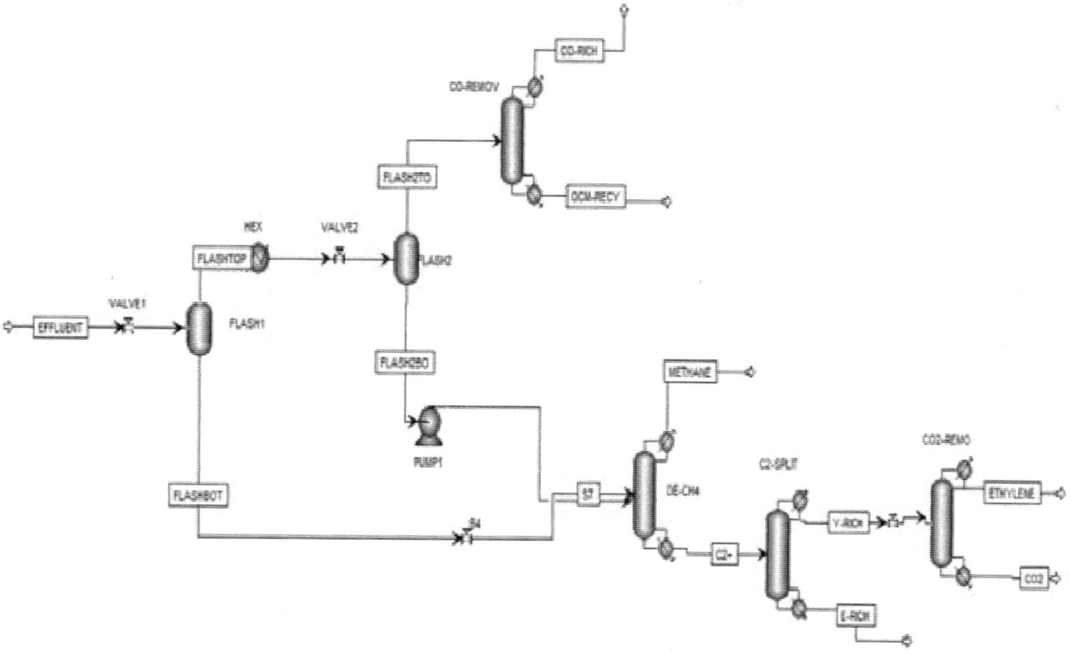

Figure 3: Separation process flowsheet from ASPEN Plus V8.0

The rationale for the flowsheet above was to aim for a good initial separation of light and heavy components, where the ideal case has all of the methane and carbon monoxide leaving as vapor with oxygen and nitrogen, and the remaining components leaving as liquid bottoms from flasg 1. The vapor overhead from the initial flash is cooled, throttled, and flashed again at atmospheric pressure to condense the remaining ethylene; the vapor product is sent to a cryogenic distillation column for the separation of carbon monoxide from the methane and oxygen to be sent as reactor recycle (above the higher flammability limit of methane in air).

The specifications on the individual units are as follows in Table 3 and Table 4.

	Carbon Monoxide Removal	Demethanizer	C2-Splitter	Carbon Dioxide Removal
# of Theoretical Stages	30	10	50	30
Stage Efficiency	0.79	0.77	0.79	0.9
Actual Stages	38	13	63	32
Feed Stage (from top)	17	8, 8	15	24
Reflux Ratio	9	3	3.5	11
Distillate:Feed Ratio	0.14	0.505	0.4	0.544
Condenser P (atm)	2	9.5	9.5	0.08
Condenser T (K)	84	148	220	135
Reboiler P (atm)	2.18	9.555	9.8	0.11
Reboiler T (K)	113	227	241	143

Table 3: Distillation column specifications

Unit	Flash 1	Flash 2	Valve 1	Valve 2	Pump
P or ΔP (atm)	10	2.1	10	1	9.75
T (K)	165	120	-	-	-

Table 4: Flash tank, valve, and pump specifications

From the heavy components (20 % of the feed), CO_2 would preferably be removed as early as possible due to its abundance in the mixture, but due to its high solubility in hydrocarbons and its azeotrope with ethylene, vacuum distillation was necessary to separate it from the ethylene product. The bottoms of the two flash tanks are fed into the demethanizer column to separate the remaining methane (the bottoms of the second flash tank would be pumped back up to 9.75 atm). The distillate from the demethanizer is 99 % pure methane, with the remainder oxygen and nitrogen (above the higher flammability limit) and can be recycled back to the reactor. The bottoms of the tower containing all C_2^+ species would then be sent to a C_2-tower for the separation of ethylene from ethane. Finally, the ethylene and carbon dioxide rich distillate from the C_2-tower would be sent to a vacuum distillation column to break the azeotrope and get a 99.95 % pure ethylene product to be used for polymerization to various grades of polyethylene.

The ethylene separated in the CO_2-removal tower was 99.481 % pure, close to the required purity limit, and would need to be sent through an adsorber prior to polymerization. The CO-Removal tower is processing 90 % of the process flow at cryogenic temperatures and will most likely be the most expensive unit in the system with the consideration of utilties. In addition, the vacuum distillation column for breaking the azeotrope between ethylene and CO_2 will require expensive auxiliary equipment to generate vacuum in the column. The CO-free methane recovered from the bottoms of the CO-Removal tower and the remaining methane recovered as overhead from the demethanizer can be recycled to the OCM Reactor to decrease feed costs. Ethane, acetylene, and other heavier products can be used as feedstock for a traditional cracking furnace to further increase ethylene yields. The current focus of the study is to optimize the designs of the distillation columns in terms of utilities required for the heat exchangers and the power requirements for the vacuum pump.

The columns were sized with 14 inches between trays, 3 feet at the top of the column for vapor disengagement, and 10 feet at the bottom for liquid holdup and structural support. The reflux drums were sized in the same way as the flash tanks previously and the pumps were priced based on the operating pressures and necessary powers.

Distillation Columns				
Specification	CO-Removal	Demethanizer	C2-Splitter	CO2-Removal
Theoretical Trays	30	10	50	25
Tray Efficiency	0.8	0.8	0.8	0.8
Actual Trays	38	13	63	32
R (mole ratio)	9	3.5	3.5	11
Diameter (ft)	6	2.5	3	4.3
Height (ft)	70	32.5	107.5	61
Condenser Duty (MMBTU/h)	-14.3	-1.7	-1.4	-1.5
Reboiler Duty (MMBTU/h)	10.2	2.6	1.4	1.5
Cond. Utility (K)	83.7	147.9	220.4	135.1
Reboil. Utility (K)	112.7	226.9	240.9	142.6
P condenser (atm)	2	9.5	9.5	0.08
P reboiler (atm)	2.18	9.55	9.8	0.11
Purchase Cost	$154,000	$24,100	$243,000	$83,000
Bare Module Cost	$334,000	$58,800	$701,000	$177,000
Reflux Drums				
L/D	5	5	5	5
Volume (ft^3)	40.33	13.87	9.64	4.52
D (ft)	5.71	1.96	1.36	0.64
L (ft)	28.53	9.81	6.82	3.20
Thickness (in)	0.07	0.08	0.05	0.00
Purchase Cost	$23,100	$4,910	$3,390	$2,650
Bare Module Cost	69500	15300	10200	7960
Pumps				
	Demethanizer Pump		Vacuum Pump	
Purchase Cost	$6,970		$6,900	
Bare Module Cost	$27,800		$20,700	

Table 5: Sizing and costing of the distillation columns, their corresponding reflux drums, and the two pumps required in the flowsheet

Finally, after a possible optimal design is made, a pinch analysis for heat integration will be performed to determine the number heat exchangers and amount of utilities required, a full cost-analysis, and LCA will be done. When this is complete, a comparison as a stand-alone plant versus a supplementary ethylene production facility to a traditional ethane cracking plant. Alternative means of separating the CO_2—including membrane separation and the deposition of solid CO_2—will be investigated.

References

[i] U.S. Energy Information Administration. "U.S. Gas Plant Production of Ethane-Ethylene." https://www.eia.gov/dnav/pet/hist/LeafHandler.ashx?n=pet&s=metfpus1&f=a. Feb. 2, 2016. Web.

[ii] Esposito, Frank. "American Chemistry Council reports 2012 resin statistics." http://www.plasticsnews.com/article/20130220/NEWS/130229994/american-chemistry-council-reports-2012-resin-statistics. Feb. 2, 2016. Web.

[iii] U.S. Energy Information Administration. "Natural Gas." http://www.eia.gov/naturalgas/. Feb. 14, 2016. Web.

[iv] UN Comtrade Database. "Polymers of ethylene, in primary forms." http://comtrade.un.org/.Feb. 8, 2016. Web.

[v] United States Patent Application Publication, Scheret *al*. Publication No.: US 2012/0041246 A1, Feb. 16, 2012.

[vi] United States Patent Application Publication, Iyeret *al*. Publication No.: US 2014/0012053 A1, Jan. 9, 2014.

[vii]http://siluria.com/Technology/Overview

[viii]Bhasin, M. M., and Keller, G. E. "Synthesis of Ethylene via Oxidative Coupling of Methane I. Determination of Active Catalysts."*Journal of Catalysis* 73.1 (1982): 9-19. Print.

[ix] Clearfield, Abraham, Hinson, Paul G., and Lunsford, Jack H. "The Oxidative Coupling of Methane on Chlorinated Lithium-doped Magnesium Oxide."*Journal of the Chemical Society*. 20.1 (1991): 1430-1432. Print.

[x] Henley, Ernest J., Junior D. Seader, and D. Keith Roper. *Separation Process Principles*. 3rd ed. Hoboken, NJ: Wiley, 2011. Print.

472330 Highly Efficient Formate Production By Hydrogenation of Captured CO_2

Abstract:

Climate change resulting from the emission of CO2 has become a widespread concern in the recent years. Carbon Capture and Storage (CCS) is a promising technology that has the potential to address the 40% of emission emanating from large-point sources such as power plants. The utilization of CO2 as a raw material in the synthesis of chemicals and liquid energy carriers offers a direct way to mitigate the increasing CO2 build-up but catalytic reduction of CO2 to fuels or value added chemicals is still one of the greatest challenges. Recently, increasing efforts have been devoted to the transformation of CO2 into value added substances such as formic acid or formates, which are highly attractive as commodity chemicals used as preservative and antibacterial agents, as well as clean sources for H2 generation. Herein, an effective low-temperature process of hydrogenating liquid ammonia or amine captured CO2 over supported palladium nano-catalysts, which can leverage the existing industrial infrastructure to produce value-added formate by utilizing the CO2 derivatives in the commercial CCS process, has been developed. For the first time, we found that the yield of formates is highly dependent on the ethyl carbonate ion intermediates in a solution of ammonium carbonate or amine-CO2 adduct. We further designed the rechargeable â€œhydrogen batteryâ€ based on the redox equilibrium between liquid ammonia captured CO2 and the corresponding ammonium formates, where hydrogen was stored in chemical bonds. The stored hydrogen can be catalytically liberated under mild conditions as the feed to fuel cells for regenerating electricity.

Allothermal gasification of lignite and biomass with integrated syngas cleaning and CO_2 removal

Peter Treiber, Michael Neubert, Jonas Leimert, Juergen Karl
Friedrich-Alexander-University, Erlangen-Nuremberg, Germany

Introduction

For the production of synthetic natural gas (SNG) from lignite or biomass in general the following process steps are required [1].

- Gasification of the solid feedstock (allothermal or autothermal)
- Syngas cleaning from particles, tars, and off-gases
- Methanation
- Optional product gas conditioning for grid injection

For small- and mid-scale, decentralized plants the main challenge is to keep the plant complexity to a minimum while still reaching the requirements for gas grid injection. The Heatpipe Reformer technology is a suitable way for decentralized SNG production [2], [3]. A hydrogen rich syngas with a nearly ideal composition for methanation can be produced in the Heatpipe Reformer using allothermal steam gasification. A syngas cleaning unit with integrated removal of CO_2 would reduce the plant complexity and provide syngas for methanation without elaborate gas cleaning steps.

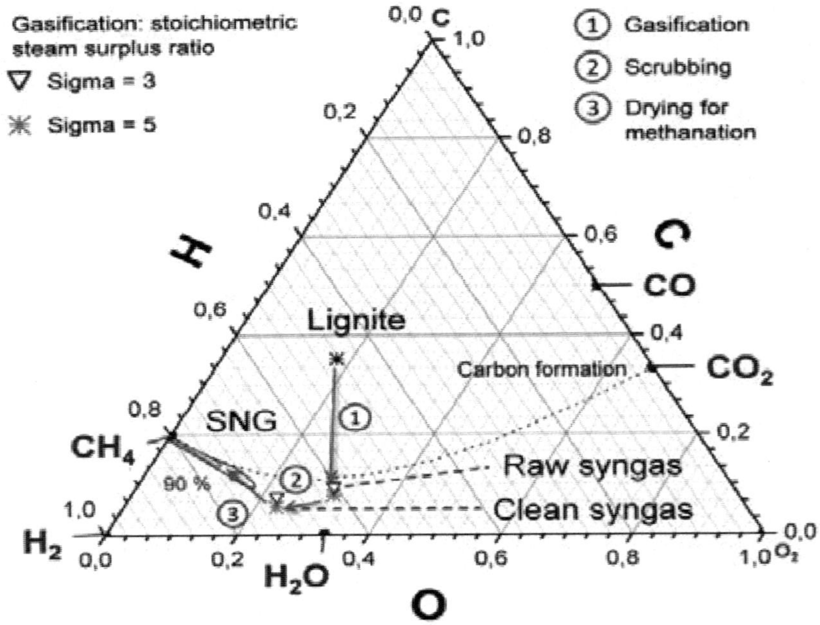

Figure 1: Coal-to-SNG process chain in the ternary diagram

With the EU research project CO2freeSNG2.0 (RFCR-CT-2013-00008), funded by Research Fund for Coal and Steel (RFCS), the research focus of the Chair of Energy Process Engineering at the Friedrich-Alexander-University Erlangen-Nürnberg for the subjects of gasification, syngas cleaning, and methanation is intensified [4]. In addition to the formerly proposed hot gas cleaning [5], [6] a hot carbonate scrubbing unit based on the Benfield process [7], [8] is integrated in the process chain for the combined bulk removal of CO_2, sulfur, and tar species. The proposed process chain and its structure in the ternary diagram are depicted in Figure 1. The tests stands and the results of the parameter variation tests are presented in the following chapters.

5 kW Bench scale process chain

For more flexibility and a reduced test time frame, a process chain for the production of SNG was designed and manufactured in the bench scale for 5 kW fuel input. The schematic setup of the single test rigs and the possible combinations are shown in Figure 2.

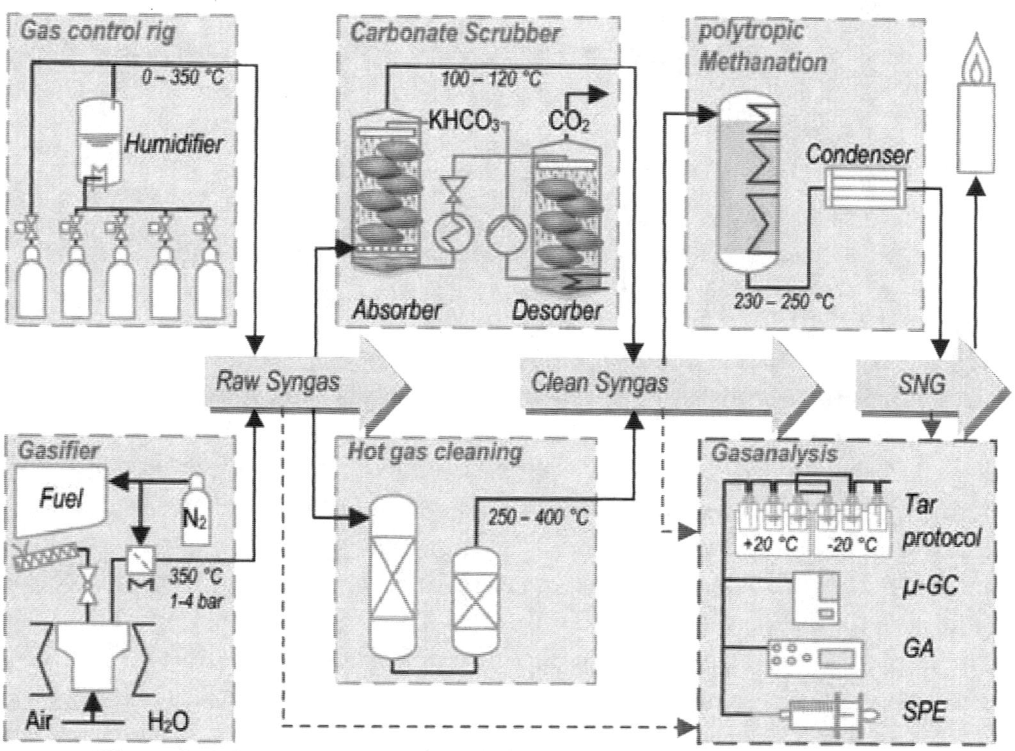

Figure 2: Schematic layout of the EVT 5 kW Coal-to-SNG process chain

Besides real syngas from coal or biomass also bottled gas mixtures can be used in the test rigs. The whole process chain is designed for an operating pressure of 5.2 bar_a. Furthermore, gas sampling ports have been integrated between each of the process steps to analyze gas composition, tar loading and sulfur species in the gas streams.

The basis for the bench scale SNG production is the 5 kW bench scale gasifier. Lignite or biomass can be gasified in a pressurized, steam blown fluidized bed. The regulation of the fluidization behavior as well as operating pressure is integrated in the fully automated test rig which allows for long term tests of over 100 h. The raw syngas from the gasifier passes a hot gas filter where particles and alkali are removed from the gas stream. Subsequently the syngas is passed to the next test rig (scrubber or methanation unit with hot gas cleaning) reducing the pressure between each test stand by 200 mbar to enable an independent pressure regulation for each unit. At the end, the syngas or product gas is disposed over a lab scale flare. The setup and function schematics of the bench scale gasifier are shown in Figure 3.

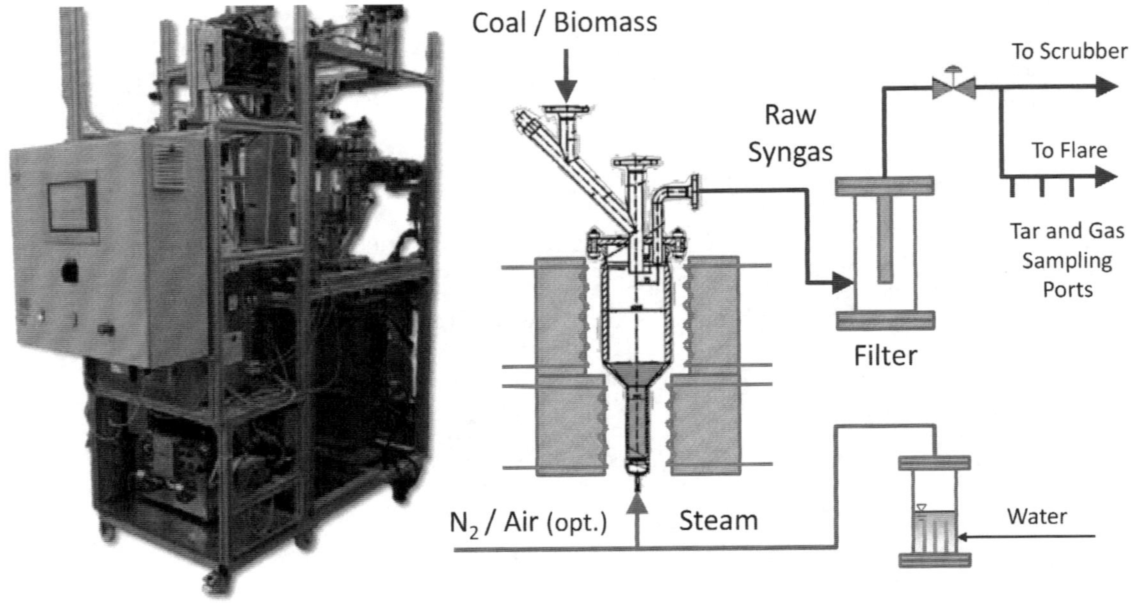

Figure 3: 5 kW bench scale gasifier: setup (left) and function schematics (right)

After the gasifier, the raw syngas is passed on to the bench scale carbonate scrubber (Figure 4) at a pressure of 5.2 bar_a. In the absorber, the raw syngas passes in counter current flow to the solvent over a random packing of 6 mm Raschig rings and exits the column at the top. In contrast to other absorption processes, the solvent is not cooled before entering the absorber (as in amine scrubbing) of the hot potassium carbonate process but it is directly drawn from the reboiler and recycled into the absorber at temperatures only slightly below boiling point (105°C). Therefore, the energy-intensive cooling of the solvent is not needed. Furthermore the regeneration is supported by a partial evaporation of water in the solvent, lowering the partial pressure of CO_2 already at the top inlet of the desorber column. In addition to the absorption of CO_2, also hydrogen sulfide is absorbed and heavier tar components are captured in the solvent flow due to condensation. Therefore, the carbonate scrubbing unit can be considered as bulk removal option for CO_2, H_2S, and heavier tar components.

Table 1: Carbonate scrubbing reactions

$$K_2CO_3 + CO_2 + H_2O \rightleftharpoons 2\,KHCO_3 \qquad (1)$$

$$K_2CO_3 + H_2S \rightleftharpoons KHS + KHCO_3 \qquad (2)$$

The capture of tar components is mainly based on condensation while the sour gas removal is based on chemisorption with potassium carbonate following the chemical reactions listed in Table 1. The reduced reboiler duty and the omitted solvent cooling are the main advantages of the carbonate scrubbing process compared to other scrubbing processes.

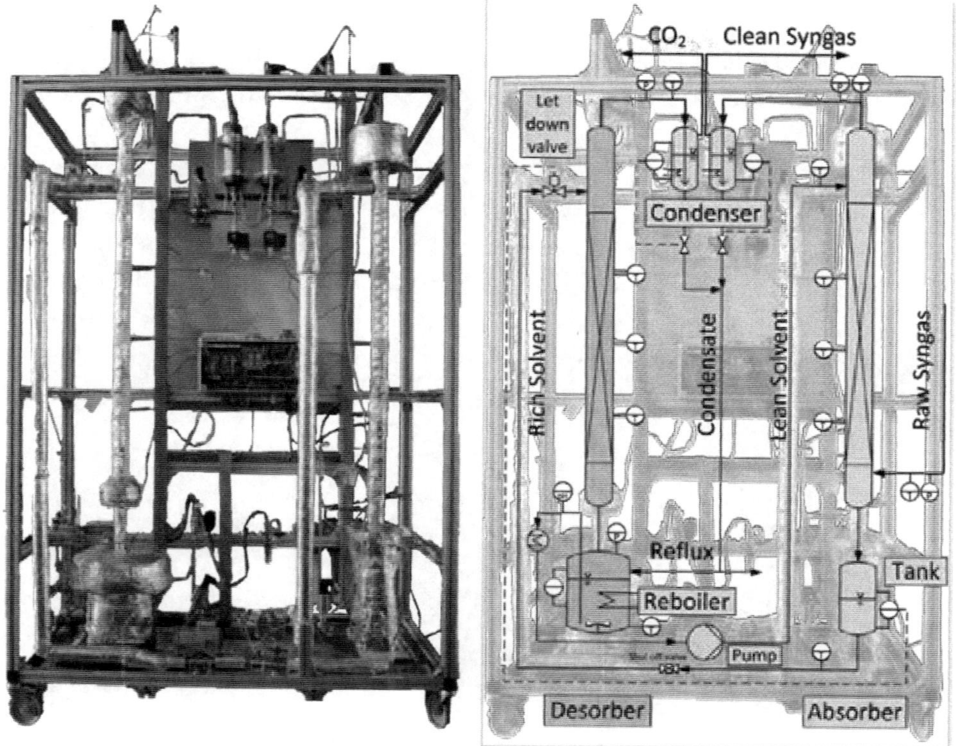

Figure 4: 5 kW bench scale carbonate scrubber: setup (left) and function schematics (right)

100 kW Pre-pilot process chain

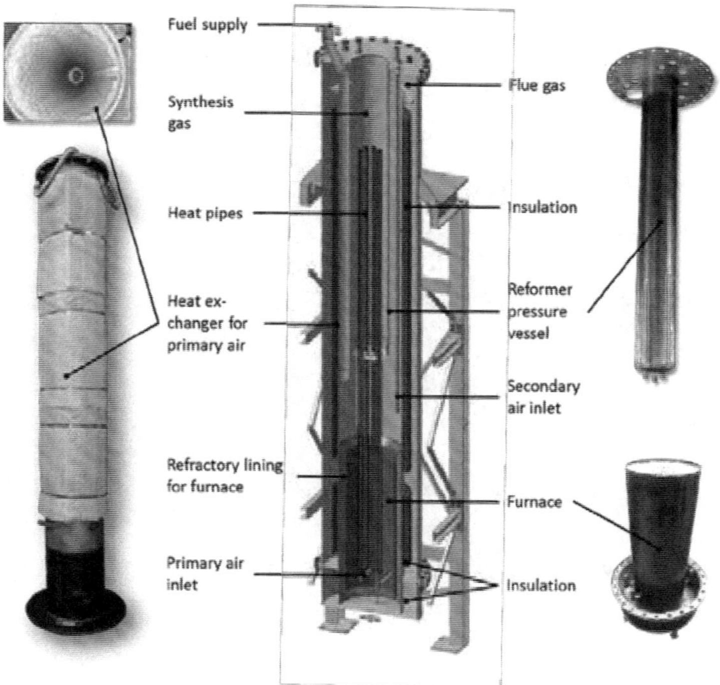

Figure 5: Setup of the 100 kW Heatpipe Reformer (HPR) [3]

Since the 5 kW bench scale test stands are rather prone to heat losses and edge effects (carbonate scrubber columns) the process chain at a 100 kW pre-pilot scale is also investigated. Similar to the bench scale process chain a pressurized, fluidized bed steam gasifier provides the raw syngas. Figure 5 shows the setup of the Heatpipe Reformer. The atmospheric combustion chamber generates the necessary heat which is transported via liquid metal heatpipes to the pressurized gasification reactor at approx. 850°C.

Like in the bench scale process chain, the raw syngas from the Heatpipe Reformer passes over a hot gas filter for the removal of fine coke as well as ash particles. After the filter the syngas enters the absorber column at the bottom inlet. The pre-pilot scale scrubbing unit is shown in Figure 6. The columns (packing height: absorber 3 m and desorber 4 m, diameter: both 100 mm) are packed with 10 mm Raschig rings and have been dimensioned by prior rate based simulations with ASPENplus^{Tm}. The solvent is let down by a throttle valve into a flash drum at the top of the desorber.

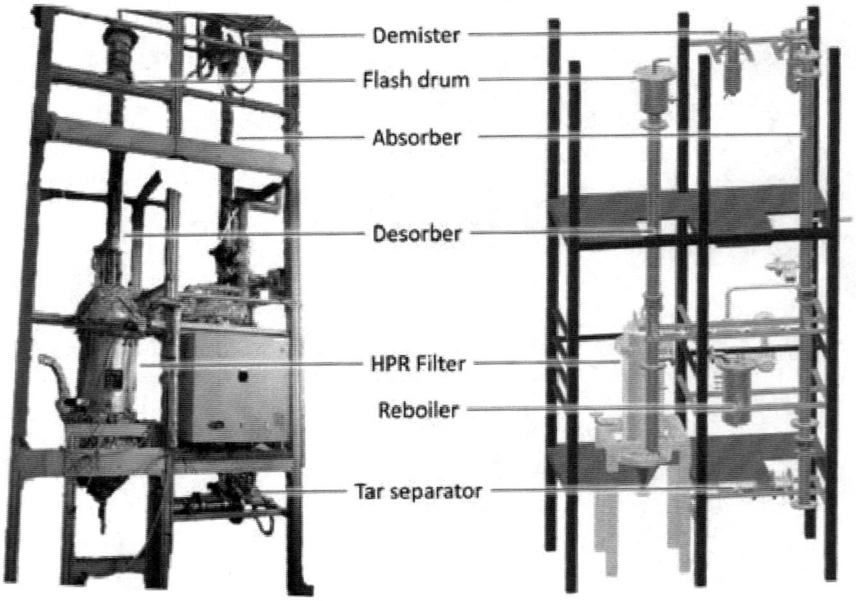

Figure 6: Setup of the 100 kW Benfield scrubber

In contrast to the bench scale scrubber, the reboiler unit of the 100 kW scrubber is not located at the column bottom but is a separate unit which is fed by an additional pump to prevent the enrichment of potassium carbonate in the reboiler. Additionally, a tar separator by means of a coalescence-mesh was integrated in the solvent flow before the main solvent pump. Therefore, a lighter and a heavier tar fraction (compared to the solvent) shall be removed from the solvent flow during regular operation. The total removal of light hydrocarbons is not intended since prior tests have shown that they can be converted in the methanation step [9], [10].

At the top of the absorber column the clean syngas passes a demister and can be either provided for a methanation reactor or disposed in a thermal afterburner. As proposed by Benson and Field [8], [11], [12] a 30 wght.-% potassium carbonate solution shall be used as solvent. Due to reduced solvent volume, the variation of solvent concentration and possible additives is predominantly investigated at the bench scale test rig.

Results

Allothermal steam gasification

In Figure 7 the syngas composition after the gasifier at a stoichiometric steam surplus of $\sigma = 5$ is shown. The relatively high steam surplus causes a syngas which is rich in hydrogen and carbon dioxide.

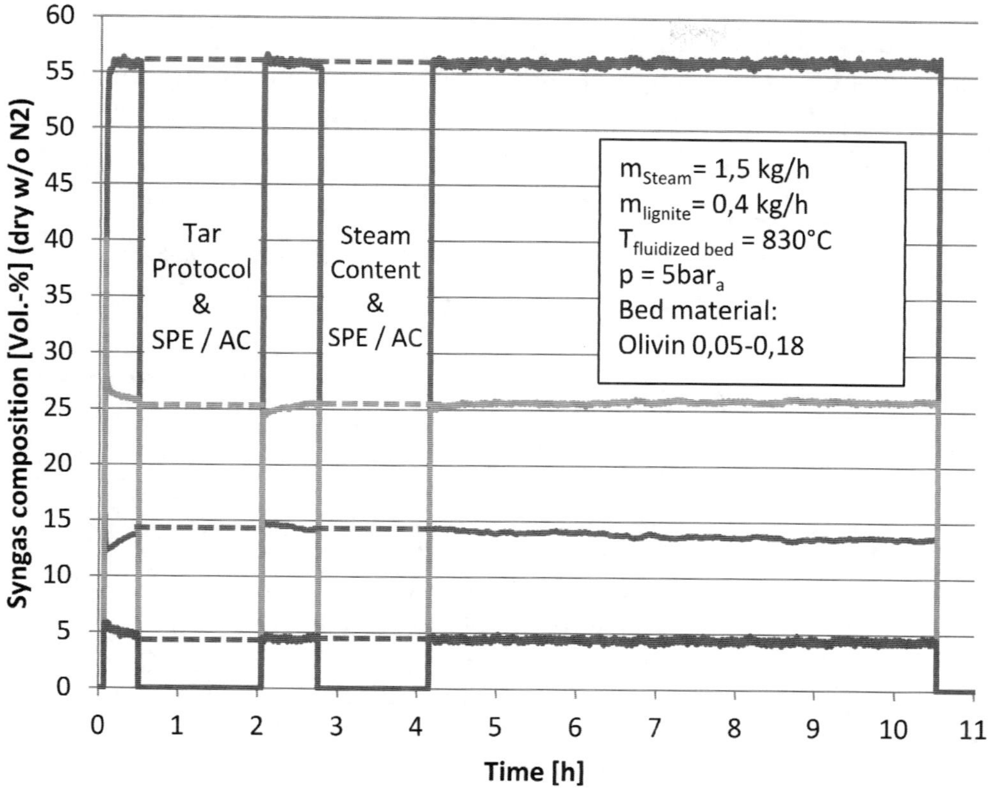

Figure 7: 10-h gasification test at the bench scale gasifier; syngas composition

During the test shown in Figure 7 a steam content measurement and a tar content analysis by means of the tar protocol [13] were performed. The results of the tar protocol measurement are depicted in Figure 8.

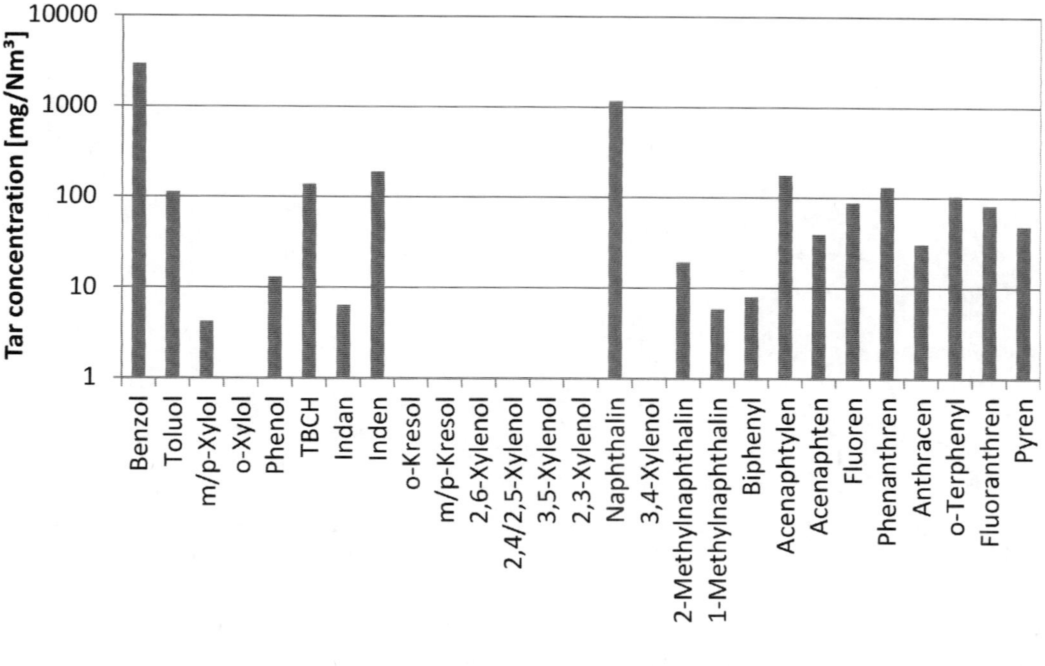

Figure 8: 10 h gasification test, tar concentration in the raw syngas by means of the tar protocol [13]

Integrated syngas cleaning

The influence of the solvent temperature at the absorber inlet as well as the ratio between the solvent and syngas mass flow (liquid-to-gas ratio – L/G) are depicted in Figure 9.

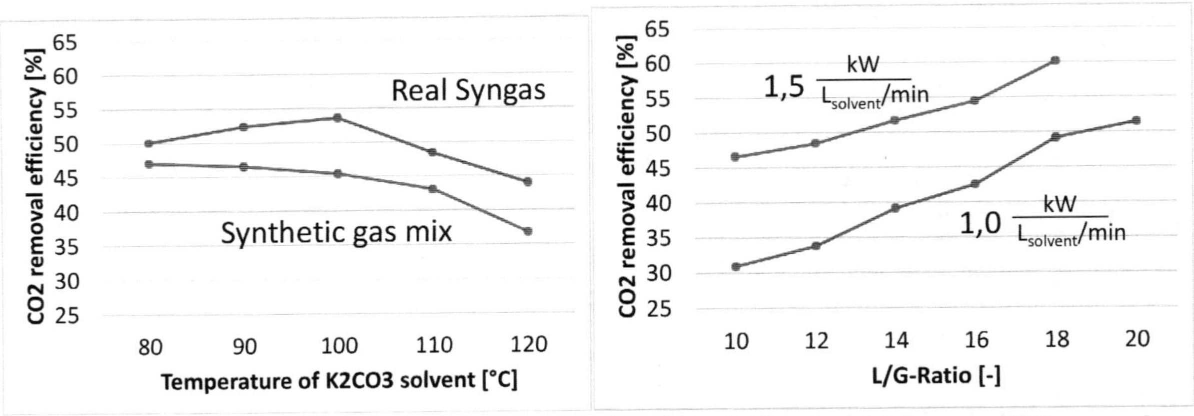

Figure 9: Influence of the solvent temperature (left) and the L/G ratio (right) on the CO2 removal efficiency at the 5 kW bench scale scrubber

At temperatures below 100 °C the CO_2 removal efficiency decreases due to the hindered mass transfer of CO_2 to the solvent and the carbonate reaction rate. At higher temperatures, the solvent loading capacity decreases due to the limited solubility of CO_2.

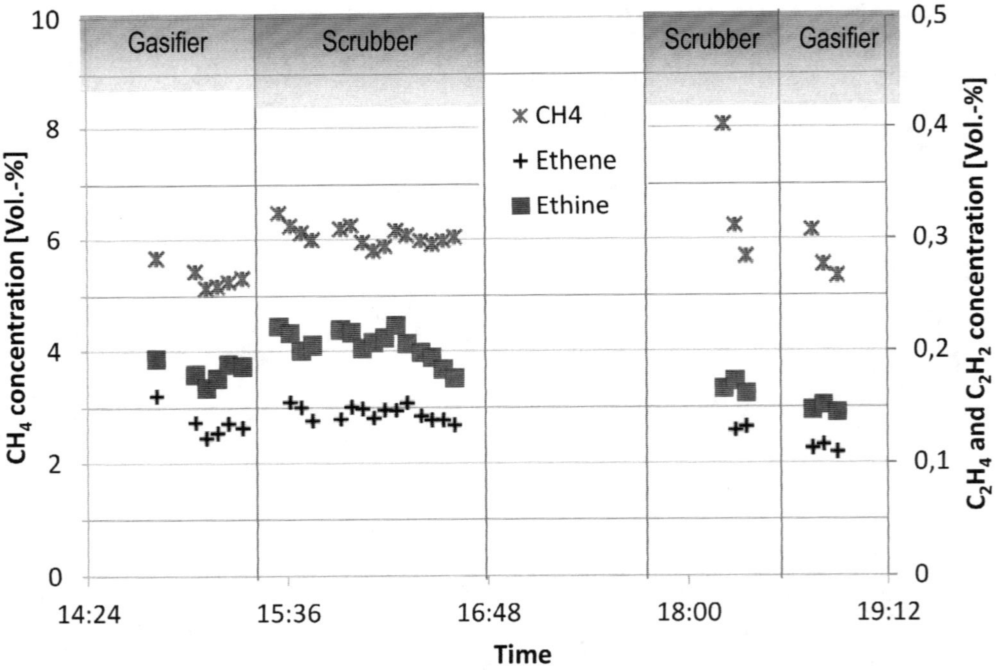

Figure 10: µ-GC measurements after the bench scale gasifier($\sigma = 5$, $T_{gasifier} = 830°C$, $p = 5.2$ bar$_a$) and scrubber ($L/G = 17$, $T_{Solvent} = 100°C$, $p = 5.0$ bar$_a$)

Furthermore, a higher removal efficiency of the real syngas could be demonstrated since a slip stream of the raw syngas was used for gas analysis compared to the synthetic tests with bottled gases reducing the syngas volume entering the absorber column. The variation of the solvent flow at constant gas flow proofed an increase in removal efficiency at higher L/G-ratios. An additional increase in the specific reboiler heat duty showed a positive influence on the solvent regeneration and thus an increasing CO_2 removal efficiency.

Compared to the prior simulations, the expected removal efficiency for CO_2 of 90% could not be reached in the bench scale tests. Since the test rig has a relatively small column diameter and the Raschig ring diameter was therefore rather large the scrubber showed dominant edge effects prevented the necessary mass transport in the absorber column. Final tests with the 100 kW pre-pilot test stand are expected to reach the prior simulated removal efficiency.

Besides the direct measurements for the CO2 removal efficiency, also the removal efficiency for other gas components was investigated with a µ-GC. The results shown in Figure 10 demonstrate that light hydrocarbons like ethene and ethine are not influenced by the carbonate scrubbing process and can be measured in constant concentrations in the clean syngas.

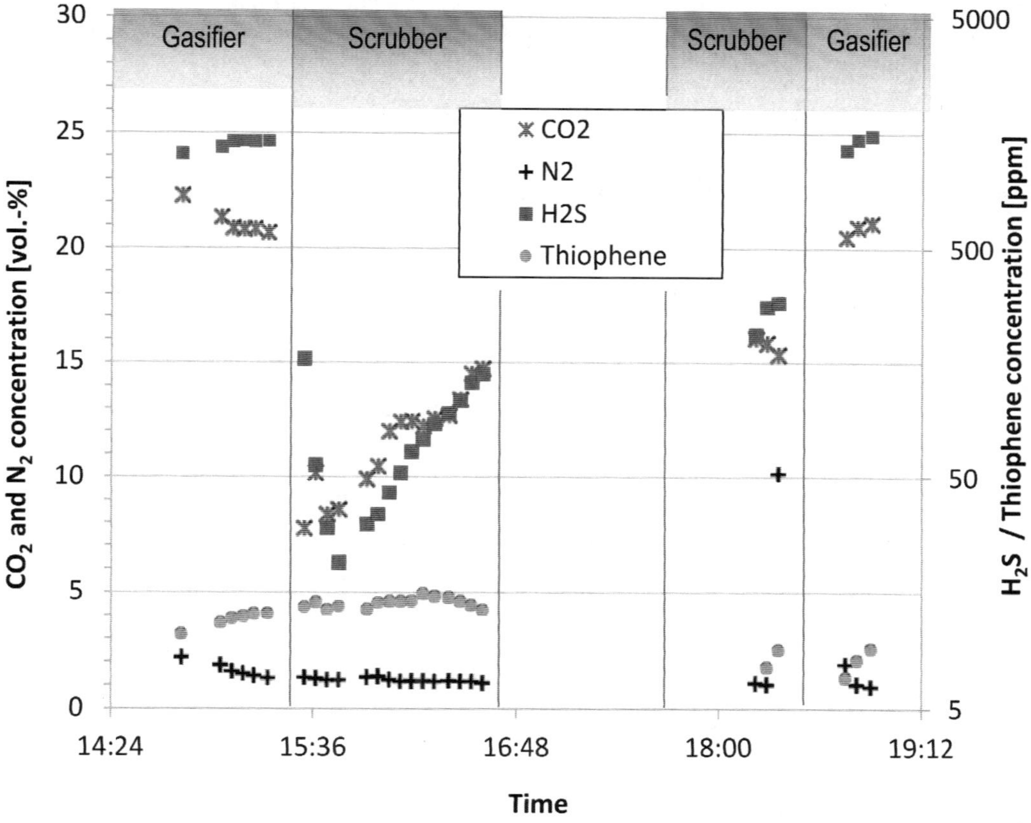

Figure 11: µ-GC measurements after bench scale gasifier and scrubber

The results of the µ-GC measurements in the first half of Figure 11 show the solvent loading process with increasing CO_2 and H_2S content. Since the bench scale scrubber is pressurized and preheated using nitrogen as inert gas, the solvent is stripped to a minimum content of CO_2 and H_2S. In the middle section the product gas after the methanation unit was analyzed showing no content of sulfur species. The final tests demonstrate higher removal efficiencies for hydrogen sulfide than for CO_2 in stationary operation. However the concentration of thiophene was hardly influenced by the carbonate scrubbing unit.

Conclusion

Within the EU research project CO2freeSNG2.0 (RFCR-CT-2013-00008) the process chain for the production of substitute natural gas (SNG) from lignite or biomass for 5 kW und 100 kW fuel input have been constructed. The syngas produced by allothermal steam gasification is treated by means of a hot carbonate scrubbing process and with the same process the bulk share of CO_2 is removed from the syngas stream. First tests could already proof the feasibility of the proposed gas cleaning system which is based on the Benfield process. Besides CO_2 also the bulk share of H_2S was removed from the raw syngas stream while lighter hydrocarbons remained in the gas stream due to higher scrubber temperature. Thus, they are available for conversion in the subsequent methanation step. Organic sulfur components such as thiophene, however, were hardly removed by the carbonate scrubber. Therefore, the proposed process chain demonstrates a bulk syngas cleaning technology for smaller decentralized SNG plants, where conventional syngas cleaning systems are not be suitable due to the high investment costs. In future tests, increased removal efficiency for CO_2 in the 100 kW process chain shall be targeted and the influence of the scrubbing process on tar and sulfur components in the syngas stream will be investigated more thoroughly.

Acknowledgements

The presented test stands and results have been funded by the European Commission and the Research Fund for Coal and Steel (RFCS) as part of the research project CO2freeSNG2.0 (RFCR-CT-2016-00008).

References

[1] J. Karl, *Dezentrale Energiesysteme: Neue Technologien im liberalisierten Energiemarkt*. 2012.

[2] J. Karl, “Biomass heat pipe reformer—design and performance of an indirectly heated steam gasifier,” *Biomass Convers. Biorefinery*, vol. 4, no. 1, pp. 1–14, Oct. 2013.

[3] J. M. Leimert, P. Treiber, and J. Karl, “The Heatpipe Reformer with optimized combustor design for enhanced cold gas efficiency,” *Fuel Process. Technol.*, vol. 141, pp. 68–73, Jan. 2016.

[4] P. Treiber and J. Karl, “CO2freeSNG2.0,” 2013. [Online]. Available: http://www.co2freesng20.eu/. [Accessed: 24-Mar-2014].

[5] C. Baumhakl, J. Karl, T. Kienberger, A. Schweiger, D. Buchholz, S. Bajohr, E. I. Koytsoumpa, and S. Karellas, *Substitute natural gas from coal with internal sequestration of CO 2 (CO2freeSNG) - Final Report*. European Union, 2013.

[6] C. Baumhakl, “Substitute Natural Gas Production with direct Conversion of Higher Hydrocarbons,” phdthesis, 2014.

[7] H. E. Benson, J. H. Field, and R. M. Jimeson, “CO2 absorption: Employing hot potassium carbonate solutions,” *Chem. Eng. Prog.; (United States)*, vol. 50:7, Jul. 1954.

[8] H. E. Benson and J. H. Field, “Method for separating CO2 and H2S from gas mixtures,” US Patent 2,886,405, 1956.

[9] T. Kienberger, C. Zuber, K. Novosel, C. Baumhakl, and J. Karl, “Desulfurization and in situ tar reduction within catalytic methanation of biogenous synthesis gas,” *Fuel*, vol. 107, pp. 102–112, May 2013.

[10] C. Baumhakl and J. Karl, “Efficient Bio-SNG Production from Thermal Gasification with Hot-gas-cleaning Technologies,” in *21st European Biomass Conference and Exhibition*, 2013.

[11] H. E. Benson and D. H. McCrea, “Removal of acid gases from hot gas mixtures.” Google Patents, 10-Jul-1979.

[12] H. E. Benson and J. H. Field, “New data for hot carbonate process,” *Pet. Refin.*, vol. 39, no. 4, 1960.

[13] “DIN CEN/TS 15439:2006-08,” 2006. [Online]. Available: http://www.din.de/de/mitwirken/normenausschuesse/nmp/din-spec/wdc-beuth:din21:84563381. [Accessed: 01-Apr-2016].

471466 Biocatalytic Polymer Material for Partial Oxidation of Methane to Methanol

Abstract:

An industrial process for the selective partial oxidation of methane under mild conditions would be highly valuable for controlling emissions to the environment and for utilizing vast new sources of natural gas. Methane monooxygenase enzymes perform the difficult activation of methane under nearly ambient conditions, suggesting a biochemical process for methane conversion could be more efficient, less capital intensive, and more easily downscaled compared to current industrial practice. However, the industrial use of enzymes for large-scale applications, especially those involving gas-to-liquid reactions, has been limited by the low enzyme loading and poor mass transfer characteristics of stirred-tank reactors. We have successfully embedded the enzyme particulate methane monooxygenase (pMMO) within a 3D printed hydrogel to create a mechanically robust and gas permeable material that overcomes mass transfer limitations to convert methane to methanol. We studied the kinetics, reactant dependence, and product inhibition of the partial oxidation of methane by pMMO, as these parameters are not well understood and are critical to optimization of our biocatalytic hydrogel material design.

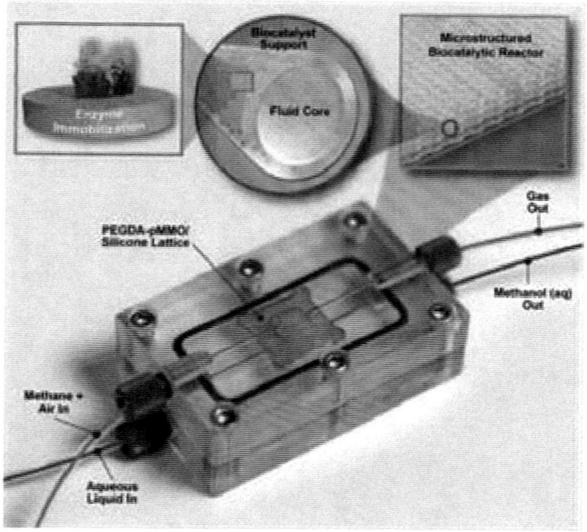

Figure 1: Schematic of enzyme immobilization within hydrogel in a structured polymer scaffold to create a gas-permeable biocatalytic material in a flow-through bioreactor.

Methods: pMMO was immobilized in poly(ethylene glycol) (PEG) hydrogel during crosslinking by UV-initiated photopolymerization. The activity of pMMO was measured by a methanol assay at 45°C using NADH as the reducing agent; methane was injected into the reaction headspace and the methanol produced was quantified by GC-MS. Kinetic studies were conducted by stopped flow analysis with varying pMMO, NADH, dissolved oxygen, and dissolved methane concentrations.

Results and Discussion: Immobilization of pMMO within PEG resulted in a biocatalytic material with nearly 100% physiological enzyme activity. By increasing the surface area to volume ratio of the material using 3D printing, we were able to achieve high volumetric methanol productivity of 2 g/L/hour. Additionally, the hydrogel material can be cured within a poly(dimethyl siloxane) scaffold as part of a flow-through reactor, shown in Figure 1, which is capable of continuous methanol production. Finally, we have determined the dependence of methanol production rate on gas phase reagents, which will facilitate optimization of reactor design parameters such as enzyme loading, membrane thickness, gas-to-liquid volume fractions, surface area to volume ratio, and flow rates.

This work was performed under the auspices of the U.S. Department of Energy by Lawrence Livermore National Laboratory under Contract DE-AC52-07NA27344.

473137 XTL Process with Low Overall GHG Emissions

Abstract:

A patent-pending process that can be used for the production of liquid fuels from biomass/coal based syngas (H_2:CO â ‰¤1) over iron-based catalysts while completely suppressing the formation of CO_2 in the FTS process will be presented. Based on the most optimistic data available from Sasol, the fraction of carbon in the original coal that results in the formation of liquid hydrocarbons is only 16.9 % using iron-based catalysts. The data from research employing this new technology have shown that this fraction can be improved to 51.1 %. This is achieved by co-feeding CO_2 (produced in the upstream processes) such that the CO_2:syngas ratio in the reactor is above 3.33:1 and conducting the FTS at conditions above the critical pressures and temperatures of CO_2. The liquids selectivity moved towards higher hydrocarbons (C_{10}- $C_{22}$with peak content in the range of C_{13}-C_{15}). The ScCO_2 â€“ FTS process also takes advantage of the strong solvent yet tunable power of ScCO_2 and allows for pressure tunable fractionation of the products. A final but important addition to the ScCO_2 â€“FTs process is the addition of product gas recycle capabilities that renders the following two unique advantages. (A) recycle helps in improving the throughput (reduced reactor size). (B)Advantages of ScCO_2FTS are further enhanced by the recycle of HC gases (along with the unreacted feed) since HCs like hexane and pentane become supercritical. The results from the various stages of development of this technology will be discussed.

474331 Simulation of Hydrogen and Acetylene Production By Methane Pyrolysis in Thermal Plasma

Abstract:

With the recent shale gas disposal and increase of biogas production, acetylene production from methane could play an important role in near future. Indeed, gas pyrolysis by thermal plasma can produce competitive hydrogen and acetylene for PVC production. Besides, acetylene could be converted to ethylene and ethylene to synfuel which can be an attractive gas to liquid solution to produce transportable fuel [1]. A closer process has been already massively set-up and used in Germany before and after Second World War to produce synthetic rubber and hydrogen for coal hydrogenation process [2]. Recently, we have done many experiments converting coal or liquid hydrocarbons to acetylene by thermal plasma with satisfied performances [3]. However, acetylene is unstable at medium temperature and can convert to tar and oil in about 2 ms and then, the pyrolysis products should be cooled rapidly. Cold gas/liquid hydrocarbons can be used as quenching media to increase the yield of ethylene, i.e. a desired by-product. In order to improve the process efficiency, a deep study has been carried out based on experimental and numerical simulation with methane used as feedstock.

The experimental results show that the methane conversion through both plasma devices is higher than 90% and leads mainly to acetylene and hydrogen with high yields. Also, few-layers graphene have been obtained in most of the cases. This investigation suggests that such a simple technique is easy to operate and suitable for mass production of acetylene and few-layer GNs in a scalable process. To improve and have a better understanding of our lab experimental set-up, a numerical model has been established. The plasma is considered using a MHD model. The reaction kinetics and carbon nanoparticle formation have been considered in a sophisticated model. A parametric study has been made on plasma power and hydrogen/methane ratio. The plasma parameters and the hydrogen ratio could be adjusted to improve the acetylene formation and the graphite nanosheets formation

474574 Enabling C_1 Chemistry through Atmospheric Pressure Corona Discharge (non-thermal) Plasmas

Abstract:

Efficient implementation of C_1 chemistry has been a major goal in industrial chemistry for over a century – currently, the major methane to liquids processes involves reforming to syngas followed by catalytic CO hydrogenation. Our group has developed novel technology using atmospheric pressure corona discharge (non-thermal) plasmas implemented in microreactors. In this work we have demonstrated C_1 coupling to C_2 mixtures (ethylene and/or ethane and excess hydrogen, depending on the gas stream composition) with energetic efficiencies in the order of 70%, as well as to longer hydrocarbons. This presentation will summarize recent developments from our lab in this area.

469126 Physical Solvent Mixtures for CO_2 Absorption at Simulated Pre-Combustion Conditions

Abstract:

We have developed physical solvent mixtures for CO_2 absorption at simulated pre-combustion conditions. In this study, the gas solubility and mass transfer coefficient of CO_2, H_2, CH_4 in a physical solvent, an ionic liquid, and other mixtures were measured in a Continuous Stirred Tank Reactor (CSTR) by the step-pressure dynamic method. The solvent performance and their mass-transfer properties in a gas-liquid system are important for engineering design and unit operation. The absorption of CO_2, and H_2 in physical solvents and other mixture were investigated. Effects of pressure on the volumetric coefficient K_La were studied with various gas liquid systems in the pressure range 0-30 atm. Experiment data of the mass transfer coefficient of CO_2 in physical solvents and mixtures matched predictions. Based on the experiment observation, physical properties and the kinetics of mixed solvents followed ideal solution mixing rules. Results of this study provided a base for further investigation on solvent mixture for gas absorption.

460582 High Performance, Bio-Derived Solvents for Pre-Combustion Capture

Abstract:

While Selexol/DMPEG is a well-established process/solvent for CO2 removal in applications such as natural gas processing and syngas purification, and is expected to be a leading candidate for pre-combustion CO2 capture applications, the viscosity of DMPEG is several times that of water. As such, DMPEG may face challenges when used within hollow fiber membrane contactors (HFMCs) and other advanced devices designed to promote high mass transfer rates. Recently, we have developed new types of physical solvents from bio-derived sources that exhibit much lower viscosities than DMPEG, low volatility and have a greater absorption capacity for CO2 at given temperature and pressure conditions. As such, this class of solvents appears highly promising for further scale up and demonstration projects. This presentation will discuss solvent production, physical property characterizations, molecular simulation and process modeling.

471696 New Water-Lean Amine-Based Solvents for Carbon Dioxide Capture with Low Regeneration Energy

Abstract:

Carbon dioxide capture and storage (CCS) of anthropogenic CO_2 released from large point sources such as coal fired power plants is required in order to mitigate adverse environmental effects like climate change. Toward this goal numerous CO_2 capture processes have been developed, the most advanced technology utilizes aqueous solutions of primary and secondary alkanolamines such as MEA or diethanolamine (DEA) for chemical absorption of CO_2. The regeneration energy of these aqueous amine solvents is very high due to the high heat capacity of water. To address this problem water-lean transformational solvent systems have been developed. As part of an ongoing program studying multiple solvent chemistries and combining the key findings from different solvent classes, we have developed a new class of water-lean non-aqueous amine-based solvents with potential to be energy efficient CO_2 capture materials. These amines bind CO_2 at standard temperature and pressure (STP) to form liquid carbamate salts with high gravimetric capacity ~ 20 wt%.Â These solvents have low thermal regeneration temperature (70-100 Â°C) and low CO_2â€“rich solvent viscosity compared to other water-lean materials. The regeneration temperature can be lowered to 60 Â°C by applying polarity swing assisted regeneration (PSAR)Â making it feasible to utilize lower grade heat from the power plant for CO_2 stripping resulting in energy savings.

461574 Molecular Design Strategies to Reduce the Viscosity of Non-Aqueous Carbon Capture Solvents

Abstract:

In power generation, high viscosities encumber the use of post-combustion non-aqueous CO_2 capture solvents. Using single molecule CO_2 binding organic liquids (CO_2BOLs) as example, the key molecular features that determine viscosity and CO_2 uptake kinetics are identified. The close proximity of the alcohol and amine sites involved in CO_2 binding result in the concerted formation of a Zwitterion containing both an alkyl-carbonate and a protonated amine. Capture is controlled by viscosity since the small binding energy barrier suggests that diffusion is rate limiting. The viscosity of CO_2BOLs increases exponentially as a function of CO_2 loading. Inter-molecular hydrogen bonding between positively and negatively charged functional groups ultimately determines viscosity. The hydrogen-bonding network can be controlled by chemically tuning the solvents to favor an internal hydrogen bonding structure. Additionally, a molecular design strategy to significantly reduce viscosity by shifting the proton transfer equilibrium toward a non-charged acid/amine species, as opposed to the ubiquitously accepted zwitterionic state, is presented. Based on knowledge obtained from molecular simulation, a reduced model was developed that predicts CO_2BOL viscosity given a few structural parameters and CO_2 loading. In collaboration with experiment, candidate compounds were synthesized to measure their viscosities and verify the model.

467008 Molecular Simulation of CO2 Absorption in the Ionic Liquid [P2228][2-Cnpyr] Using Reaction Ensemble Monte Carlo

Abstract:

CO_2 capture using monoethanolamine (MEA) suffers from amine degradation, solvent emissions, and large heat requirements to regenerate the solvent. Ionic liquids (ILs) are an attractive alternative to MEA due to their thermal stability, minimal vapor pressure, and tunable properties. Gurkan et al. [1] designed a pyrrolide-based anion that binds CO_2 more weakly than MEA, but still absorbs CO_2 in near equimolar amounts at standard conditions. Further IL optimization is complicated by the exceptionally large number of possible cation-anion pairs. Molecular simulation could be a valuable tool to screen proposed ILs for CO_2 absorption, if it were possible to simulate both the condensed phase IL environment and the C-N bond formation as CO_2 binds to the pyrrolide anion. Quantum mechanical methods can accurately model bond formation and breaking, but are limited to small system sizes representative of the gas phase environment. Classical simulations, on the other hand, can be used to model the bulk liquid at a molecular level, but cannot describe covalent bond formation.Â Weâ€™ve developed a methodology for predicting CO_2 absorption in ILs using reaction ensemble Monte Carlo (RxMC) that utilizes both quantum and classical simulations. Liquid phase configurations are sampled using traditional translation, rotation and regrowth Monte Carlo moves. The equilibrium composition is sampled using reaction moves that delete reactant molecules and insert product molecules. The acceptance probability of reaction moves depends on the liquid-phase change in potential energy due to molecular insertions and deletions and the gas-phase change in free energy dueÂ to the reaction. The latter is determined by computing the potential of mean force using ab initio molecular dynamics. We investigate the absorption of CO_2 in the ionic liquid triethyloctylphosphonium 2-cyanopyrrolide, [P2228][2-CNpyr], and present CO_2 absorption isotherms at a range of temperatures.

[1] Gurkan, B., Goodrich, B.F., Mindrup, E.M., Ficke, L.E., Massel, M., Seo, S., Senftle, T.P., Wu, H., Glaser, M.F., Shah, J.K., Maginn, E.J., Brennecke J.F. and Schneider W.F., "Molecular design of high capacity, low viscosity, chemically tunable ionic liquids for CO2 capture." The Journal of Physical Chemistry Letters 1.24 (2010): 3494-3499.

472611 A Novel CO_2 Absorption Process Enabled By Dedicated Biphasic Solvents: Solvent Screening and Process Development

Abstract:

Biphasic solvent-based absorption processes have recently been of growing research interest as an alternative technology to the conventional absorption processes for post-combustion CO_2 capture. Such processes employ biphasic solvents that can form dual liquid phases during CO_2 absorption, with the absorbed CO_2 concentrated in one phase and lean in the other. The CO_2-concentrated liquid phase is the only part of solvent to be regenerated, resulting in a reduced mass for CO_2 stripping, an elevated stripping pressure associated with high CO_2 loading, and thus reduced energy consumption.

We are presently developing a novel absorption process enabled by a dedicated new class of biphasic solvents to further improve the performance of biphasic absorption processes by maximizing absorption kinetics and minimizing the concern of high solvent viscosity associated with high CO_2 loading. Totally 34 solvent blends have been formulated and experimentally screened with respect to phase transition behavior, absorption kinetics, and vapor-liquid equilibrium performance. Preliminary results revealed that the phase transition, in terms of its dependence on CO_2 loading and volumetric distribution between two phases, could be tuned with selection of solvent components or composition. The solvents could be formulated to have faster absorption rates than that of the benchmark 5M monoethanolamine (MEA) aqueous solution. The loading capacities of the solvents for CO_2 stripping were 40 to 250% higher than 5M MEA and the stripping pressures generally ranged between 5 and 10 bar under the regenerated lean solution conditions at 120 °C.

In this presentation, an overview of the novel biphasic absorption process will be provided, the preliminary results of

the biphasic solvent screening study will be updated, and their indications to the process techno-economic performance will be discussed.

471201 Deterministic and Stochastic Mass Transfer Models for CO_2 Capture Processes

Abstract:

For accelerating deployment of novel CO_2 capture technologies, rigorous process models can be very valuable to reduce the risk and uncertainty in scaling up the process by skipping intermediate scales. For such process models, it is critical that the uncertainties in the key submodels be quantified. For chemical solvent-based post-combustion CO_2 capture technologies, mass transfer models play an important role especially in estimating the performance of the absorber. With this motivation, comprehensive mass transfer models of a chemical solvent based CO_2 capture process with quantified uncertainties is developed in this work.

The typical approach in the literature to development of mass transfer models is sequential. In chemical solvent based CO_2 capture systems such as for the MEA-H_2O-CO_2 system, very fast reaction kinetics make it difficult to obtain measure of the mass transfer flux and identify the associated parameters with the mass transfer models. Assuming the diffusivity model to be satisfactory, mass transfer coefficients are typically calculated from the experimental data from the wetted wall column (WWC) experiments. The experimental data from packed columns are the used to obtain the interfacial area model assuming the mass transfer coefficient models are still valid. As the WWC and packed column operate in different flow and operating regimes, it is not necessary that the mass transfer coefficient models and their parameters are still optimal for the packed column. In addition, assumption of accuracy of the diffusivity model is also questionable. Overall, a suboptimal mass transfer model is obtained. To circumvent these issues, simultaneous regression of parameters for the mass transfer coefficient model, diffusivity model, and interfacial area model is carried out in this work for the MEA-H_2O-CO_2 system by simultaneously using the experimental data from the packed columns and wetted wall columns. Since such simultaneous regression is not feasible in existing state-of-the-art simulation software, a computation tool named Framework for Optimization, Quantification of Uncertainty and Surrogates (FOQUS) developed as part of U.S. DOE's Carbon Capture Simulation Initiative is leveraged to accomplish this objective.

Following the deterministic parameter regression, rigorous uncertainty quantification is performed for the mass transfer models. The best guess for parametric uncertainty (priors) is then cast in the framework of a Bayesian inference methodology to obtain more informed posterior parametric uncertainty. The computational expense for using the Bayesian approach to these models can be prohibitive. Therefore, response surface models are generated as computationally inexpensive surrogate models with sufficient accuracy by using the input-output data. Finally, the posterior distributions in all parameters are propagated through the process model to obtain the uncertainty bounds in key operating variables. Our results show how the uncertainty in process models vary depending on the operating conditions. Our study also indicates possible approaches to reducing uncertainty in mass transfer models.

456618 Mechanism of Furfural Hydrodeoxygenation on Metal/Metal Oxide Catalysts

Abstract:

Selective C-O bond activation at low temperatures is essential for oxygen removal to produce second generation biofuels and chemicals from biomass. Constraints, imposed by energy scaling relationships, make monofunctional catalysts (metals, oxides, and carbides) inefficient. At the same time, recently discovered multifunctional metal/metal oxide catalysts (e.g., Rh/MoO_x, Ir/VO_x) exhibit unusual activity and selectivity for C-O bond hydrogenolysis at temperatures < 200°C. Herein, we utilize density functional theory calculations, first principles microkinetic modeling, and electronic structure analysis to shed light on the site cooperativity in the Ru/RuO_2 catalyst, which enables up to the 76% yield of 2-methyl furan in catalytic transfer hydrogenolysis of furfural. We find the facile C-O scission to be associated with a formation of a conjugation-stabilized furfuryl radical intermediate on RuO_2 oxygen vacancies. Three types of catalytic sites required for catalytic transfer hydrogenolysis, are identified. Furthermore, we elucidate electronic factors responsible for the high catalyst activity, and establish correlations between binding energy descriptors and the catalyst activity for a variety of reducible oxides. The mechanistic picture is fully consistent with isotopic labeling experiments and kinetic studies, and the insights gained might prove useful in catalyst selection and reaction pathway design for hydrodeoxygenation of biomass.

471366 Kinetic Analysis of Hydrogenation and Hydrogenolysis of Complex Furans over Supported Noble Metals

Abstract:

Upgrading furan-based compounds through mild hydrotreatment is an important step to produce biomass-derived commodity chemicals and transportation fuels. Biomass derived functionalized furans, such as furfural or 5-hydroxyfurfural, are used in a variety of carbon â€" carbon bond coupling chemistries to produce products in the jet, diesel and lubricant range has also been established [2-4]. These products typically retain the furan functional group, which require further hydrotreatment in order to produce hydrocarbons or valued-added oxygenates.

The kinetics and mechanism of hydrogenating the furanyl moiety over carbon-supported noble metals were previously studied using 2,5-dimethylfuran (DMF) as a model substrate, in liquid phase at conditions removed of both internal and external mass transfer effects (333K - 363K, p_{H2} = 0.4-0.7MPa). Of the catalysts studied, Pt/C had the greatest activity for C-O hydrogenolysis, producing 2-hexanone as the major product, whereas Pd/C exclusively saturated the aromatic ring into its tetrahydrofuran analog. We proposed a reaction network, supported by a Langmuir Hinshelwood kinetic model, in which not only are ring opening and ring saturation of DMF competitive pathways, but the three main products observed: 2,5-dimethyltetrahydrofuran, 2-hexanone, and 2-hexanol result directly from hydrogenation of DMF over Pt/C and not from hydrogenation in series.

We apply the findings made from our model system to understand the hydrogenation behavior of more complex, substituted aromatic furans with carbon numbers in the diesel range, such as bismethylfuranylmethane (BMFM), a C_{11} compound containing two adjacent furan moieties. Similar to DMF, ring opening and ring saturation of the aromatic ring of BMFM occurred in parallel, with the former pathway preferred at elevated temperatures and low p_{H2} conditions. This behavior was determined to be due to the high activation energy and a low rate order dependence on

p_{H2} associated with the ring-opening step of DMF. Contrary to DMF, BMFM hydrogenation activity over Pt/C was sluggish when compared to activity over Pd/C, in which rate of conversion occurred 8-fold faster to produce not only ring saturated cyclic ethers, but also significant quantities of ring-opened alcohol products (~30% yield). C-O bond hydrogenolysis over Pd/C occurred selectively on the interior, aromatic C-O bond as opposed to the more accessible exterior C-O bond, accounting for 96% of the ring opened products. These differences in hydrogenation activity of BMFM suggest that geometry of the furanic substrate plays a dominant role in the reaction network.

References

[1] Y. Román-Leshkov, J.N. Chheda, J.A. Dumesic, Phase Modifiers Promote Efficient Production of Hydroxymethylfurfural from Fructose, Science, 312 (2006) 1933-1937.

[2] A. Corma, O. de la Torre, M. Renz, Production of high quality diesel from cellulose and hemicellulose by the Sylvan process: catalysts and process variables, Energy Environ. Sci., 5 (2012) 6328.

[3] G. Li, N. Li, Z. Wang, C. Li, A. Wang, X. Wang, Y. Cong, T. Zhang, Synthesis of High-Quality Diesel with Furfural and 2-Methylfuran from Hemicellulose, ChemSusChem, 5 (2012).

[4] M. Balakrishnan, E.R. Sacia, A.T. Bell, Syntheses of Biodiesel Precursors: Sulfonic Acid Catalysts for Condensation of Biomass-Derived Platform Molecules, ChemSusChem, 7 (2014) 1078-1085.

466351 ZrO_2 Is Preferred over TiO_2 As Support for the Ru-Catalyzed Hydrogenation of Levulinic Acid to Î³-Valerolactone

Abstract:

In the last few years, Î³-valerolactone (GVL), readily obtained by the hydrogenation of levulinic acid (LA), has attracted the attention of both academic researchers and industries because of its possible use in different applications related to the chemical industry (e.g., as fuel additives, food ingredients, nylon intermediate or as a solvent). Numerous approaches have been examined for LA hydrogenation, with the main focus in most cases on obtaining high GVL yields by variation of either the active phase or the catalyst support. Studies focusing on catalyst stability are much more limited in number. Indeed, while high activity/selectivity for GVL can relatively easily be achieved with most of the reported catalysts, as well as with commercial ones (e.g., Ru/C), stable performances especially under more industrially-relevant conditions is much harder to achieve. Long-term exposure to relatively harsh, polar conditions, as well as the use of non-pure, real LA feeds containing impurities from previous process steps indeed pose significant challenges to catalyst stability.

Herein, we present a study of catalyst stability and performances for the hydrogenation of LA to GVL of a Ru/ZrO_2 catalyst [1-2]. Catalyst performance has been compared with commercial Ru/C and a self-synthesized Ru/TiO_2 catalyst, which can generally considered as benchmark catalysts for this process. All catalysts show high GVL yields under standard conditions. Large differences were seen in catalyst stability; though, Ru/ZrO_2 maintanied its stability and high GVL yields were obtained even after five recycling tests. Remarkably, in the fresh Ru/ZrO_2 catalyst, Ru was found to be fully atomically dispersed on the fresh catalyst even at 1 wt.% Ru loading, with some genesis of Ru nanoparticles being observed upon recycling. Notably and in strong contrast to the benchmark catalysts (Ru/C and Ru/TiO_2), the Ru/ZrO_2 catalyst can efficiently perform the hydrogenation of LA even in the presence of impurities in

the LA feed, e.g., H_2SO_4, salts (Na_2SO_4, NaCl, and H_3PO_4), and organic compounds/contaminants derived from fractions of the biomass, which have been reported as impurities existing in real streams. To better understand these remarkable results with regards to catalytic performance, an extensive catalyst characterizatin study was performed to provide more insights into the relation between structure, performance, and long-term stability of the catalysts.

[1] J. Ftouni, A. MuÃ±oz-Murillo, A. Goryachev, J. P. Hofmann, E. J.M. Hensen, L. Lu, C. J. Kiely, P. C.A. Bruijnincx, B. M. Weckhuysen. ZrO_2 is preferred over TiO_2 as support for the Ru-catalyzed hydrogenation of levulinic acid to Î³-valerolactone,Â *submitted*.

[2] J. Ftouni, P. C.A. Bruijnincx, B. M. Weckhuysen. Method for preparing a chemical compound using a ruthenium metal catalyst on a zirconium oxide support in the presence of a contaminant, *PCT/EP2016/054031*.

465502 C–O Bond Hydrogenolysis of Biomass–Derived Cyclic Ethers over Ni–Based Catalysts

Abstract:

Catalytic deoxygenation reactions play a crucial role for upgrading of biomass–derived oxygenates to high value chemicals. Among several deoxygenation processes, C–O hydrogenolysis has emerged as a viable technology to convert oxygenates such as sugar polyols and furanic derivates into target molecules. The key element to selectively cleave the C–O bond is the development of an efficient catalytic system, while minimizing the side reactions and carbon loss. Recent experimental studies suggested that rhodium and iridium catalysts modified by some oxophilic metals such as Re, Mo, W and V oxides display high activity and selectivity in the hydrogenolysis of glycerol[1], tetrahydrofurfuryl alcohol (THFA)[2] and tetrahydropyran–2–methanol (2–THPM)[2-3] to their corresponding α,ω–diols. However, due to the high cost and depleting resources of noble metals, it is necessary to develop catalytic systems based on non–noble metals. Among non–noble metals, nickel–based catalysts are widely used in hydrogenation and hydrogenolysis reactions because they exhibit good catalytic performance and are inexpensive. It was reported that C–O bond cleavage can be occurred over Ni catalysts at temperatures above 473K for both furan and tetrahydrofuran rings[4-5] . Therefore, in this contribution we focus on synthesis of Ni–based catalysts, suitable for ring opening of cyclic compounds derived from furanic conversions, particularly THFA. The effect of Ni and different support properties on their performance as THFA hydrogenolysis catalysts was examined under liquid phase reaction. The synergistic roles of the metal and acid sites of Ni/HZSM–5 and the reaction mechanism of THFA hydrogenolysis as well as their respective reaction intermediates and products are described and discussed. Ring opening of THFA into 1,5–PDO takes place over both nickel metallic sites and HZSM–5 acidic sites, followed by ring closure into tetrahydropyran (THP) product over acidic sites up to 70% yield.

[1] Y. Nakagawa, M. Tamura, K. Tomishige, *J. Mater. Chem. A.* **2014**, *2*, 6688-6702.

[2] Y. Nakagawa, M. Tamura, K. Tomishige, *ACS Catal.* **2013**, *3*, 2655-2668.

[3] T. Buntara, S. Noel, P. H. Phua, I. Melián Cabrera, J. G. de Vries, H. J. Heeres, *Angew. Chem. Int. Ed.* **2011**, *50*, 7083-7087.

[4] S. Sitthisa, D. E. Resasco, *Catal. Lett.* **2011**, *141*, 784-791.

[5] J. Lee, Y. Xu, G. W. Huber, *Appl. Catal., B.* **2013**, *140*, 98-107.

469628 A Novel Pathway to 1,5-Pentanediol without the Use of Noble Metals

Abstract:

Recent focus in biomass conversion has shifted from the production of low value fuels to high value chemicals.Â Î±,Ï ‰-diols are important chemicals for their uses as polyurethane precursors, resins, and plasticizers. The petroleum industry produces 138 and 2,500 kilotons/year of 1,6-hexanediol and 1,4-butanediol, respectively. This market size and a projected market growth rate of 8% per year allow for additional production from biorenewable sources without market saturation [1,2].

We have developed a new route to synthesize 1,5-pentanediol in high yields from the hemicellulose fraction of lignocellulosic biomass. Previous work focused on direct hydrogenolysis of the fully hydrogenated form of furfural, tetrahydrofurfuryl alcohol, to 1,5-pentanediol with RhRe catalysts [3,4]. Our new multi-step process uses a combination of heterogeneous acids and supported base metal catalysts.Â The ability to use base metal catalysts stems from the improved reactivity of a key reaction intermediate.

In this presentation, we will focus on mechanistic studies used to determine why the new reaction intermediate is more active than tetrahydrofurfuryl alcohol and why base metals can be utilized for this reaction system. Experiments were performed with nuclear magnetic resonance (NMR) spectroscopy and *in situ* attenuated total reflectance Fourier transform infrared spectroscopy (*ATR-FTIR*) to help identify reaction intermediates.

[1] Lux Research, Bio-based Materials and Chemicals Intelligence Service

[2] MarketandMarkets Report on 1,6-Hexanediol

[3] Chia, M. et al. *Journal of the American Chemical Society.* **2011**, *32*, 12675-89.

[4] Koso, S. et al. Chemical Communications. **2009**, 15, 2035-2037.

465036 Evaluation and Characterization of Carbon-Supported Nobel Metals for the Hydrodeoxygenation (HDO) of Acetic Acid

Abstract:

Hydrodeoxygenation (HDO) of organic acids (i.e. acetic acid, propanoic acid, fatty acids) can yield desirable products such as hydrocarbon and alcohols [1–4]. The HDO reaction pathways consist of hydrogenation, decarbonylation and decarboxylation [5]. This study has explored the catalytic chemistry of HDO of acetic acid over noble metals (Pt, Pd, Ru and Rh) on activated carbon. The reactions are evaluated in a continuous plug-flow reactor operated between 200-400ºC under atmospheric pressure with concentrations of 1% acetic acid/20%H_2/balance He. The gaseous products were analyzed by an on-line gas chromatograph. Transition metals supported on activated carbon yield primarily alkane products in the HDO of acetic acid, making them promising candidates for production of liquid fuels from organic acid feeds. The activation energies ranging from 19.7-20.3 kcal/mole and kinetics involving the reaction rate orders with respect to acetic acid (α=0 .07-0.41) and H_2 (β=0.15-.40) have been examined. The intrinsic activity based on the turn over frequency (TOF) has been evaluated for each catalyst, suggesting that Ru and Pt are the most active metals. The activity has been observed to decrease with time on stream, and post-reaction analysis by x-ray diffraction (XRD), scanning transmission electron microscopy (STEM) and temperature programmed oxidation (TPO) suggests that a combination of sintering and carbon deposition during reaction is responsible. The probable mechanism for HDO of acetic acid over these catalysts is discussed in light of these findings and those in the literature.

[1] Z. Wang, G. Li, X. Liu, Y. Huang, A. Wang, W. Chu, X. Wang, N. Li, Aqueous phase hydrogenation of acetic acid to ethanol over Ir-MoOx/SiO2 catalyst, Catal. Commun. 43 (2014) 38–41. doi:10.1016/j.catcom.2013.09.007.

[2] N. Joshi, A. Lawal, Hydrodeoxygenation of acetic acid in a microreactor, Chem. Eng. Sci. 84 (2012) 761–771. doi:10.1016/j.ces.2012.09.018.

[3] H. Olcay, L. Xu, Y. Xu, G.W. Huber, Aqueous-Phase Hydrogenation of Acetic Acid over Transition Metal Catalysts, ChemCatChem. 2 (2010) 1420–1424. doi:10.1002/cctc.201000134.

[4] Z. He, X. Wang, Required catalytic properties for alkane production from carboxylic acids: Hydrodeoxygenation of acetic acid, J. Energy Chem. 22 (2013) 883–894.

[5] Y.K. Lugo-José, J.R. Monnier, C.T. Williams, Gas-phase, catalytic hydrodeoxygenation of propanoic acid, over supported group VIII noble metals: Metal and support effects, Appl. Catal. Gen. 469 (2014) 410–418. doi:10.1016/j.apcata.2013.10.025.

457028 Direct Conversion of Methane to Methanol and Ethanol

Abstract:

The enormous scale of methane reserves has motivated significant research activities focused on its conversion to fuels and chemicals. Since large amounts of natural gas are located in remote areas and transporting gases in pipelines is difficult, processes for converting methane into denser products are highly desirable. To address this challenge, we developed a bi-functional catalyst that is capable of activating the C-H bond in methane to form surface methyl groups. Some of these methyl groups undergo surface reaction that lead to the formation of higher alkyl chains. In the presence of steam, methanol and ethanol are formed by hydrolysis of the surface alkyl groups. Oxygen has to be co-fed to provide a thermodynamic driving force for this reaction. In addition to alcohols, carbon dioxide and hydrogen are formed as by-products. Importantly, formation of alcohols occurs at 450 °C under steady state conditions with a turnover frequency of at least 50 h^{-1}. This is a significant improvement from previous studies, in which a high-temperature calcination step was required for every turnover. The identification of promising catalysts and suitable reaction conditions will be discussed. Spectroscopic studies are performed to identify the active sites and the synergistic way in which they enable the reaction.

470752 Aqueous-Phase Phenol Hydrogenation over Platinum and Rhodium: Electrocatalysis Versus Thermal Catalysis

Abstract:

A major future need is providing renewable energy without emitting greenhouse gases. Catalysis can play an integral role in the chemical processes required to produce transportation fuels and store energy. One method to make renewable fuels is to convert biomass to hydrocarbon fuels. Bio-oil (modeled here using phenol) is an especially promising feedstock, which can be hydrogenated and then deoxygenated to yield fuel grade hydrocarbons.[1] Although hydrogen generally comes from steam methane reforming, which emits carbon dioxide, renewable electricity can be used to drive carbon-neutral (i) electrochemical hydrogen production which is then used for thermal hydrogenation (TH), or (ii) direct electrochemical hydrogenation (ECH) of the organic substrate. Both of these routes require high rates for industrial application, and have the advantage that greenhouse gases are not emitted, and renewable energy can be stored as transportation fuels. ECH has the further advantage that there is no need to store hydrogen as a gas (other than as a side-product of ECH). TH and ECH of phenol on rhodium and platinum have similar activation barriers due to a shared rate determining step of adsorbed H adatoms reacting with adsorbed phenol.[2] The conditions (hydrogen pressure or applied potential) influence the reaction rate, but how they affect the rate and how the conditions of TH and ECH compare is not well-understood. Additionally, the high rates needed for industrial application are not currently achievable due to catalyst deactivation at high temperatures.[2] Similar behavior for gas-phase benzene TH is due to the poisoning of surface sites by dehydrogenated benzene.[3] We show here that a similar effect exists for TH of phenol, and that ECH and TH rates are controlled by the surface coverages of the reactants, which are in turn controlled by either applied potential or hydrogen pressure. We also discuss differences in ECH and TH and propose solutions to improve reaction rates.

On Pt, TH activity decreases from 60 to 80 °C, when the hydrogen pressure (P_{H2}) is below 1 bar. At 10 bar, the rate of reaction is greatly increased and the reaction takes on Arrhenius behavior from 60 to 80 °C. Based on the third-order in hydrogen, the decreased rate from 60 to 80 °C must be from poisoning species derived from dehydrogenated phenol, which are removed when the hydrogen pressure is sufficiently high (between 1.5 to 6 bar in this work) to rehydrogenate them. At pressures sufficiently high to remove the dehydrogenated phenol, the order in hydrogen is first-order, indicating a dependence on the hydrogen coverage. At even greater high pressures the hydrogen order decreases to zero, indicating further changes in the hydrogen coverage do not greatly influence the rate. The applied potential has a similar effect on the reaction rate for ECH, where initial increases in potential (from 0.05 to -0.05 V vs. RHE) greatly increase the rate of reaction, but from -0.05 to -0.45 V vs. RHE the rate increases slowly, after which it plateaus. The similarities in the maximum rate achieved for TH and ECH at a given temperature indicate the rate is controlled by the coverage and surface reaction of adsorbed phenol and an adsorbed hydrogen adatom. By comparing the rates of ECH and TH we are able to see conditions where the coverages must be similar, and thus can empirically relate applied potentials to pressure. For example, under this study it appears that 6 bar P_{H2} results in similar coverages as -0.45 V vs. RHE, and 1 bar P_{H2} is similar to -0.3 V vs. RHE.

We also present *in situ* characterization using X-Ray Absorption Spectroscopy (XAFS) on Pt nanoparticles in the presence and absence of phenol under applied potential. Through analysis of the near-edge (XANES) portion of XAFS we detect H-adsorption on Pt nanoparticles at potentials known to fully saturate the Pt surface with hydrogen. We see that the addition of phenol to the electrolyte solution lowers the H-adsorption signal, indicating phenol displaces hydrogen, lowering the hydrogen coverage. As higher applied potentials are applied, in the presence of phenol, the surface coverage of hydrogen increases. This supports our hypothesis from the reaction data that we see increased ECH rates at higher applied potentials because of an increase in hydrogen coverage, which reacts with adsorbed phenol. From the extended X-ray absorption fine structure (EXAFS) we see that the introduction of phenol also results in interaction between Pt and carbon atoms, which we believe to be adsorbed phenol. The Pt-C distance matches calculations for phenol adsorption on platinum,[4] and cannot be attributed to oxygen due to the absence of surface oxide from our XANES results. The strength of the Pt-C interaction is inversely related to the H-adsorption signal from XANES, supporting our conclusion that phenol displaces hydrogen on the surface, and that the applied potential can control the H-adsorption, and indirectly control the phenol coverage. Because the H-adsorption appears to affect phenol adsorption, we believe that phenol and hydrogen compete for sites on platinum, which is supported by electrochemical results that show phenol inhibits the H adsorption from protons at low applied potentials.[2]

References

(1) Zhao, C.; He, J.; Lemonidou, A. A.; Li, X.; Lercher, J. A. *J. Catal.* **2011**, *280* (1), 8.

(2) Song, Y.; Gutiérrez, O. Y.; Herranz, J.; Lercher, J. A. *Appl. Catal. B Environ.* **2016**, *182*, 236.

(3) Chou, P.; Vannice, M. A. *J. Catal.* **1987**, *107*, 140.

(4) Yoon, Y.; Rousseau, R.; Weber, R. S.; Mei, D.; Lercher, J. A. *J. Am. Chem. Soc.* **2014**, *136* (29), 10287.

472437 Highly Efficient Conversion of Terpenoids to High Density Bio- Jet Fuels

Abstract:

The market demand for renewable aviation fuels is consistently increasing due to the growing concerns about global climate change. Bio- jet fuels produced from hydro-treating vegetable oils or Fischerâ€“Tropsch synthesis of bio-syngas are mainly straight chain alkanes with low octane numbers and volumetric densities and cannot meet the specification of petroleum based jet fuels. Herein, we develop the novel process to upgrade biomass derived terpenoids, with various cyclic hydrocarbon structures, e.g., 1,8-cineole, to produce cycloalkanes with high thermal stabilities, energy densities, and octane numbers, which can be blended to produce high-quality bio- jet fuels. We found that the effective deoxygenation of 1,8-cineole underwent three main steps: C-O ring opening, dehydration, and hydrogenation. We have designed multi-functional catalytic systems for the conversion of terpenoids to cyclic alkanes in a â€œone-potâ€ process. The conversion of monoterpenoid reached ~100% with ~100% selectivity to cycloalkanes.

456461 Organosolv and Kraft Lignin: Fractionation and Conversion to Melt Spun and Electrospun Carbon Fibers

Abstract:

The global market for carbon fiber is growing rapidly as potential uses in energy and energy savings currently are expanding. The low cost, potential abundance, and highly aromatic carbon content of lignin, compared to polyacrylonitrile (PAN) which is the main precursor for making carbon fiber, make it an attractive precursor for making low-cost carbon fiber.

Melt spinning and electrospinning are two common methods for making fibers with different characteristic and applications. While melt spun fibers can be used in structural and composite applications, electrospun fibers have potential applications in energy storage and filtration. However, these spinning methods required lignins with different characteristics. For example, electrospinning requires high molecular weight and soluble lignin, while melt spinning requires low molecular weight and fusible lignin. Technical lignins, such as kraft and organosolv, are heterogeneous and contains different molecular weight fractions and functionality. Ultrafiltration and sequential solvent extraction are methods which have been employed for separation of different fractions from lignin.

We performed a sequential solvent extraction, using methanol and methanol/methylene chloride mixture to fractionate softwood kraft lignin and organosolv switchgrass lignin. The solvent extraction resulted in obtaining three lignin fractions, according to their solubility. Lignin samples were evaluated by thermal analysis (DSC and TGA), different spectroscopy method (FTIR and NMR), elemental and compositional analysis. Finally, the lignin fractions were used for making melt spun and electrospun carbon fibers. Softwood kraft lignin had the highest ash content and was not fusible. Although switchgrass lignin was fusible, continuous spinning was not possible. Both switchgrass and softwood kraft lignins were not completely soluble in electrospinning solvent (DMF/methanol). After solvent extraction, the methanol soluble fraction had low T_g and thermal decomposition temperature, while the methanol/methylene chloride soluble fraction had high T_g and thermal decomposition temperature (117 vs 181°C for softwodd kraft lignin and 107 vs 161°C for switchgrass lignin) which could indicate low and high molecular weight of these fractions. Methanol soluble fraction of both lignins was used for melt spinning. The methanol soluble lignin fraction was able to continuously spin fibers free of defects, which is a significant improvement over the raw lignin. The methanol/methylene chloride fractions from both lignin demonstrate good solubility in electrospinning solution and performed very well in electrospinning process. Continus meltspinning for about 40 min was achived with both lignins and meltspinnning was performed for about 8 h. Carbon fibers were evaluated by SEM and tensile properties of fibers were measured. Diameter of melt spun carbon fibers were in the range of 15-30 Î¼m; while average diameter of elctrospun carbon fibers were in the range of 350-700 Î¼m, depend of concentration of electrospinning solution. Based on tensile properties data tensile strnght of meltspun switchgrass carbon fibers were 378 MPa and it was 441 MPa for meltspun softwood kraft carbon fibers. Results indicate lignin fractionation can provide different fractions with different characteristics and applications and can play a role like refinery for oil industry.

465028 Chemically Modified Lignin Surfactant: Synthesis, Characterization and Their O/W Interfacial Properties

Abstract:

Lignin-based surfactant was prepared through chemical modification. This bio-based surfactant demonstrated to be a good emulsifier for W/O emulsion. Herein, grafting of butyric anhydride (BA), 3-(trimethoxysilyl)propyl methacrylate (MPS) and 2-bromoisobutyryl bromide (BIBB) onto kraft lignin to change its amphiphilicity was conducted. The chemical reactions were confirmed by FTIR, ^{1}H NMR, GPC, TGA analysis, and ^{31}P NMR. Lignin modified with MPS and BIBB could only dissolve partially in organic monomer due to the low grafting ratio. In contrast, over 95% of the hydroxyl group conversion was achieved when butyric anhydride was used. The butyric anhydride modified lignin resultant (lignin-B) was found to be completely soluble in oil monomers. According to the interfacial tension measurement, lignin-B was able to decrease the interface energy between water and styrene from 35 $J{\cdot}m^{-2}$ to 15 $J{\cdot}m^{-2}$. As expected, stable water-in-oil emulsion was prepared with lignin-B as emulsifier. The average diameters of emulsion droplets were 499.4 nm and 363.9 nm with a lignin-B content of 0.5 wt % and 5 wt % respectively. No separated aqueous phase was observed for 30 days. These results suggest lignin-B has promising potential to be used as a bio-based surfactant or emulsifier.

466557 Reactive Diluents Prepared from Lignin Model Compounds:the Effect of Structure and Impurites on the Properties of Vinyl Ester Resins

Abstract:

Vinyl ester (VE) resins are utilized to produce polymer matrix composites for use in a wide range of applications due to possessing relatively high moduli, strengths, and glass transition temperatures (T_gs) while maintaining low weight and cost. VE resins often contain high concentrations of a petroleum-based reactive diluent (RD), such as styrene. In addition to lowering the resin viscosity, styrene is able to improve the overall polymer performance by increasing monomeric reactivity and extent of cure; however, styrene has been designated as a HAP, potential carcinogen, and a VOC. Renewable RDs with similar performance to petroleum-based RDs are desired to mitigate the aforementioned hazards as well as to facilitate the transition from petrochemical feedstocks to bio-based feedstocks.

Lignin is an abundant renewable resource with a high aromatic content that has the potential for the development of VE resins, including RDs. RDs that can be derived from lignin, including phenyl methacrylate (PM), 2-methoxyphenyl methacrylate (MG), and 4-propyl-2-methoxyphenyl methacrylate (M4PG), were synthesized and investigated as potential styrene replacements.Â A commercial VER was blended with each lignin-derived RD, and styrene, in 7:3 and 1:1 weight ratios and cured to â‰¥ 94 %, based on near-IR spectral analysis.Â The substituents on the aromatic ring of the lignin-derived RD are shown to have small effects on viscosity, thermogravimetric, and thermomechanical properties of the cured resins, where, in general, the smaller the RD, the higher the T_g and the lower the viscosity.Â Increasing purity of the RDs through chromatographic separations substantially increases the T_g and degradation temperatures of the resins. The thermogravimetric and thermomechanical properties of all lignin-derived RDs are directly comparable to that of styrene.Â These results suggest that PM, MG, and M4PG are effective bio-based RD replacements for styrene.

472760 Fine Fractionation of Lignin By Molecular Weight Using Supercritical Fluids

Abstract:

As lignin is the only major constituent of biomass with aromaticity, it has been proposed as a renewable biopolymer replacement for petroleum-derived applications such as the polyols in polyurethane foams, the phenols in phenol–formaldehyde resins, and the polyacrylonitrile in carbon fibers. However, increasing evidence indicates that the molecular weight of the lignin should best be "tuned" for a given product application. In this study, concentrated acetic acid–water mixtures were used to dissolve most of a low-ash (1-2%) lignin recovered from a kraft black liquor via the SLRP process. Supercritical carbon dioxide (CO_2) was then added as an anti-solvent to the acetic acid–water–lignin mixture in increments of increasing pressure. Both the molecular weight and metals impurities of the lignin fractions that precipitated from solution were found to be a strong function of the pressure of the supercritical CO_2 being added to the acetic acid–water mixtures. The precipitated lignin fractions were characterized by GPC, ^{1}H, ^{13}C, and HSQC-NMR, and unexpected relationships between molecular weight and functional-group composition were obtained. Based on these limited results, we conclude that supercritical CO_2 can be an energy-efficient, alternative method for fractionating lignin in terms of molecular weight, while holding other molecular properties constant.

469139 Application of Metamaterials and Rectenna for Capture of Blackbody Radiation

Abstract:

Metamaterials, a new synthetic set of materials with properties that do not exist on their own in nature, typically consist of various composite materials such as metals and dielectrics. Their repeating geometric structure and material set provide the metamaterial with a negative index of refraction. The negative index of refraction induces a primary function of metamaterials: their ability to manipulate electromagnetic waves. In this study, metamaterials are designed and fabricated to manipulate blackbody radiation into a well-organized, functional source of electromagnetic waves that are rectified to a DC voltage.

The metamaterial-rectenna unit is designed to convert heat into electricity using the optical properties of the blackbody radiation emitted from heat source. The blackbody radiation is optically manipulated by the metamaterial. In general, as the temperature of the object increases, the wavelength and intensity of the blackbody radiation increase. The metametarial-rectenna units fabricated in this study are designed to capture electromagnetic radiation with a frequency of 1 THz.

The constructed device uses a rectenna as the rectifying device to convert the alternating current (AC) induced from the electromagnetic waves emitted as blackbody radiation to a direct current (DC). The rectenna, containing a metal insulator metal (MIM) tunneling diode, utilizes the ordered electromagnetic radiation from the metamaterial. The metamaterial consists of copper and a photosensitive polymer in an alternating pattern that focuses the electromagnetic radiation over onto the rectenna. The metal of choice, copper, is selected due to its high thermal and electrical conductivity. SU-8 is selected as the photosensitive polymer due to its well-characterized optical properties as well as use in photolithography. The metamaterial is fabricated using various techniques of photolithography, including wet and dry etching of metals and oxides, thin-layer deposition of metals using evaporation, and electrodeposition.

In order to test the functionality of the metamaterial, a heat source applies heat at a set temperature to the metamaterial. Using test pads that have been built into the structure and connected to the rectenna, the diodes are characterized using four-point probe IV measurements. The thermal measurements utilize two of these probes to detect any voltage that is generated by the rectenna-metamaterial device when exposed to a controlled heat source. In this study, a proof of concept design of the metamaterial-rectenna units is achieved and a plan for future development is suggested.

469647 Testing Materials and Devices for Electromagnetic Energy Capture and Conversion to Electricity

Abstract:

The subject of this paper is terahertz (THz) recteenna. Rectenna are devices based on two fundamental components. There is an antenna that captures the photonic energy and a rectifier that transforms alternating current (AC) into a directional current (DC) for use in power systems. Research on our team is presently focused on developing efficient rectenna devices capable of capturing electromagnetic (EM) radiation is the form of infrared (IR) emitted waste heat. Materials-based solutions used in this regard include tailoring antenna geometry and composition and coupling metamaterials into the receiver. The antenna are coupled through a feedpoint to a metal-insulator-metal (MIM) type diode. MIM diodes possess the characteristic of being well-suited for high-speed aplications.

This bulk of this paper will center on testing strategies which are not trivial in the 30 - 300 THz region. Both free-space testing and thermal testing results will be presented. In order to truly quantify overall efficiency numbers, our reserach team uses computational finite element modeling (FEM). Thie paper will present results from FEM thermal transport models, optoelectronic models, and nanofabrication models. In summary, results will be provided about rectenna device performance at 1, 5, and 30 THz.

459954 Incorporation of Photo-Responsive Membrane Protein Species into Nanostructured Silica for Light-Driven Ion Transport

Abstract:

Membrane proteins are versatile biomolecules with diverse functionalities that impart sensing, signaling, transport, or catalytic properties that support the viabilities of biological cells. Such functionalities are highly selective to particular ions or molecules and often occur at high rates, which would be attractive for technological applications, such as chemical or biological sensing, separations, bioanalytics, and energy conversion. To effectively exploit membrane proteins for technological purposes often requires their incorporation into synthetic host materials that enable the proteins to function stably and be integrated into macroscopic devices. One interesting example is the membrane protein proteorhodopsin, which functions as a light-driven H^+-ion-pump that might be harnessed for photochemical energy conversion. Synthetic host membranes that contain macroscopically aligned proteorhodopsin species are expected to generate bulk ion gradients across host materials under illumination. Here, we present a solution-based synthetic protocol that allows high concentrations (up to 15 wt%) of active orientationally ordered proteorhodopsin species to be incorporated within nanostructured silica membrane hosts. Synthesis conditions and compositions were selected to stabilize proteorhodopsin molecules in the presence of the structure-directing surfactant and soluble network-forming silica species that co-assemble to form nanostructured silica host matrices, as established by small-angle X-ray diffraction analyses. Multidimensional solid-state NMR spectra show that proteorhodopsin molecules incorporated within nanostructured silica hosts retain native-like structures, though with some interesting differences. The optical absorbance behaviors of proteorhodopsin within the synthetic hosts are analogous to the photochemical reaction cycle of proteorhodopsin in native-like environments associated with the active transport of H^+-ions. The synthesis protocol is expected to be general and has been adapted to incorporate other functionally active membrane proteins within nanostructured silica host membranes. The versatile nanostructured silica-surfactant host materials open opportunities to integrate membrane proteins and their diverse functionalities into synthetic semi-permeable membranes.

461940 Suppression of Infrared Absorption in Nanostructured Metals By Controlling Faraday Inductance and Electron Path Length

Abstract:

While metals are widely used for electrodes in many optoelectronic devices, the device efficiency is limited by optical losses due to metals. As thinner devices are desired for reduced cost without sacrificing performance, the metal electrodes can be nanostructured to concentrate light in the photoactive material. A popular approach for the light concentration is to excite surface plasmon polaritons (SPPs). However, in such SPP-based devices, strong metal absorption is a major source of losses. The problem of this parasitic absorption becomes critical when a high device efficiency is desired. Moreover, if the device operates over a broad band as in solar cells, metal absorption needs to be controlled over the whole spectrum of interest and this poses a significant scientific and engineering challenge. Here, we investigate the loss suppression in metal nanocoil structures in the infrared range. Surprisingly, we find that the nanostructured metals can behave almost like a vacuum with a negligible optical loss in the infrared region. The loss is reduced in comparison to flat films by more than an order of magnitude over most of the very broad spectrum between short and long wavelength infrared. Our detailed analysis shows that this suppression of optical losses is due to both a large Faraday inductance and a long electron path. These two factors render the conduction electron in the metal effectively heavier than a hydrogen molecule. To investigate the usefulness of this effect in infrared detectors, we include a photoactive material in the coiled metal structures. In this case, the fraction of absorption in the active material increases by two orders of magnitude compared to non-coiled structures. Moreover, we demonstrate that the coiling of the infrared detector nanostrips increases the photoresponsivity by 15 times without additional electrical losses. These findings could benefit many metal-based applications that require low loss such as photovoltaics, photoconductive detectors, solar selective surfaces, infrared-transparent defrosting windows, and other metamaterials.

454155 Light Harvesting in Dye Sensitized Solar Cell Based on Consensitizer in Core-Shell Nanofiber Configuration Reducing Charge Recombination

Abstract:

Dye sensitized solar cell offers the potential of being a low-cost, high-efficiency photovoltaic device. However, the power conversion efficiency is limited as it cannot utilize all photons of the visible solar spectrum. A novel design of a core-shell photoanode is presented herein where a thin-shell of infrared dye is deposited on the core of sensitized TiO_2 nanofiber. Specifically, the ruthenium based dye (N719) sensitized TiO_2 nanofiber is wrapped by a thin-shell of copper phthalocyanine (CuPc). In addition to broadening the absorption spectrum, this core-shell configuration further suppresses the electron-hole recombination process. Instead of adopting the typical Förster resonance energy transfer, upon photons being absorbed by the infrared dye, electrons are transferred efficiently through a cascade process from the CuPc to the N719 dyes, the conduction band of TiO_2, the FTO electrode and finally the external circuit. This results in a high power conversion efficiency of 9.48% for the device. The proposed strategy provides an alternative method for enhancing the performance of DSSC for low-cost renewable energy in the future.

469311 Nanocrystal Doping Stabilizes the Perovskite Phase of Cesium Lead Iodide

Abstract:

All-inorganic, perovskite-phase cesium lead iodide is a promising material for photovoltaics. However, under ambient conditions it rapidly transforms to a non-functional, so-called yellow phase. Here we demonstrate more than two orders of magnitude improvement in the stability of nanostructured perovskite-phase cesium lead iodide thin films through a nanocrystal doping approach. The doped, functional perovskite phase is synthesized by co-assembling cesium lead iodide nanocrystals with cesium lead chloride dopant nanocrystals. The resulting doped nanocrystal solid is subsequently fused into a polycrystalline thin film by chemically-induced, room-temperature sintering. This process ensures nanometer-scale mixing even at compositions that exceed the bulk miscibility of the two constituents. Spectroscopy and X-ray diffraction indicate that some of this chloride is further dispersed during sintering and a polycrystalline mixed phase is formed. By comparison to DFT-calculated values, the relative change in band gap and the lattice contraction are shown to be consistent with a Cl:I ratio of a few percent in the mixed phase.

460501 Metal Oxide Electron-Selective Layers for Inverted Perovskite Solar Cells By Atomic Layer Deposition

Abstract:

Perovskites are promising candidates for low cost, high efficiency light absorbers with a large, tunable band gap. Recent advances in photostability and thermal stability from the introduction of mixed-cation lead mixed-halide perovskites in the form of cesium formamidinium (CsFA) perovskites has lead to significant interest in the development of perovskite-silicon tandem solar cells. Improvements in the deposition of transparent contacts, stability of organics within the device structure and blocking of moisture to the perovskite absorber must be made to realize highly efficient and stable perovskite-silicon devices. The improved thermal stability of the CsFA perovskite has opened up new processing options through atomic layer deposition (ALD) for the conformal coating of metal oxide materials on top of perovskite absorbers. In this work, we explore TiO_x and SnO_x ALD thin films deposited on top of CsFA perovskite devices in an inverted structure as dual-purpose layers for electron-selectivity and degradation protection. We will discuss the effects of ALD processing conditions on TiO_x and SnO_x electronic properties as well as on CsFA perovskite stability. We achieve stable >15% power conversion efficiency with 1.6 eV perovskite devices through maximum power point tracking and demonstrate the effectiveness of ALD metal oxide layers as buffer layers for the sputtering of indium tin oxide contacts for semi-transparent devices with >14% efficiency.

467509 Transparent Conductive Oxide Nanocrystals Coated with Insulators By Atomic Layer Deposition

Abstract:

Thin films comprised of transparent conducting oxide (TCO) nanocrystals are attractive for a number of optoelectronic applications. However, it is often observed that the conductivity of such films is very low when in contact with air. It has recently been demonstrated, somewhat surprisingly, that filling in films comprised of initially insulating TCO nanocrystals with another insulator by atomic layer deposition (ALD) dramatically increases the conductivity by many orders of magnitude. This work aims to elucidate the mechanism by which the ALD coating increases the conductivity. We used an *in situ* electrical probe to examine the effect on sheet resistance of removing two adsorbed oxygen species (physisorbed molecular water and chemisorbed hydroxide) and compared this result to thin films comprised of ZnO nanocrystals coated with Al_2O_3 and also HfO_2 by ALD. Although both insulating infills decrease the sheet resistance and increase the stability of the films, there is a stark discrepancy between the two. From the *in situ* measurements, it was found that coating with Al_2O_3 removes both physisorbed water and chemisorbed hydroxide resulting in a net reduction of the ZnO nanocrystals. Coating with HfO_2 only removes physisorbed water, which was confirmed by Fourier transform infrared spectroscopy. A similar phenomenon was observed when coating thin films comprised of Sn-doped In_2O_3 nanocrystals with Al_2O_3 by ALD; suggesting Al_2O_3 can be used to reduce and stabilize metal-oxide nanocrystals in general.

464634 Microstructure Development in $Cu_2znsn(S_xSe_{1-x})_4$ Thin Films during Annealing of Colloidal Nanocrystal Coatings

Abstract:

Copper zinc tin sulfoselenide is a promising material for thin film solar cells because it comprises inexpensive and abundant elements. Thin polycrystalline films of $Cu_2ZnSn(S_xSe_{1-x})_4$, or CZTSSe, may be formed by annealing nanocrystal coatings cast from colloidal dispersions in a selenium and/or sulfur atmosphere. This is a promising low-cost and high-throughput route for fabrication of CZTSSe-based thin film solar cells. During annealing, the nanocrystal films can transform into polycrystalline thin films with micrometer size grains. While benchmark efficiencies of CZTSSe solar cells have risen sharply in the last two decades, continued improvements in the solar cell performance and cost reduction require a fundamental understanding of the factors that influence the film microstructure. We have been investigating the microstructure evolution of polycrystalline CZTSSe thin films through systematic control of parameters such as chalcogen (sulfur or selenium) vapor pressure, annealing temperature, and heating rate in a closed system. In this talk, we will contrast microstructure development in CZTSSe films formed by annealing CZTS nanocrystal coatings in selenium vapor with microstructure development in CZTS films formed by annealing CZTS nanocrystal coatings in sulfur vapor. There are significant differences in the mechanisms of microstructure development despite the similar approaches. These differences are due to differences in thermodynamic phase equilibria of sulfur and selenium as well as due to differences in chemical reactions of sulfur and selenium with the carbon containing ligands that surround the nancrystals: these ligands are used to stabilize the nanocrystal dispersions and often end up in the coatings and films even after annealing. For example, annealing CZTS nanocrystal coatings at high selenium pressure (*e.g.*, 450 Torr) leads to the formation of a bilayer microstructure containing a microcrystalline CZTSSe layer on top of a nanocrystalline layer that is rich in carbon. In contrast, films annealed with sulfur do not show a distinct carbon-rich layer and most of the carbon volatilizes from the film during annealing. We suggest that carbon remains in the film due to the formation of a non-volatile carbon selenide polymer, the presence of which inhibits nanocrystal sintering and leads to the formation of a large-crystal top layer. The microstructure of films annealed in the presence of selenium liquid is influenced by liquid phase sintering. These films exhibit monomodal coarsening and densification into a compact film without the formation of a bilayer structure.

Multiscale study of the Self-Organized Vertical Concentration Profile of PEDOT:PSS for Work Function Optimization

Min Huang*
School of Chemical Science and Engineering, Tongji University
1239 Siping Road, Shanghai, China 200092

ABSTRACT

Conventional conducting polymers, poly(3,4-ethylenedioxythiophene):poly(styrenesulfonate) (PEDOT:PSS) has been widely used as the hole injection layer (HIL). It would be preferable to design a gradient profile of work function near the interface in vertical for individual applications. In the literature, tetrafluoroethylene-perfluoro-3,6-dioxa-4-methyl-7-octene-sulfonic acid copolyme (PFI) and nano Nickel Oxide were experimentally added into the PEDOT:PSS layer to create a self-organized vertical concentration profile. In this contribution, the mechanisms of the self-organized gradient effect were studied using multiscale analysis and dissipative particle dynamics (DPD) simulation. The Flory-Huggins parameters and intramolecular bonding parameters were obtained through reverse mapping of molecular dynamics simulations. DPD simulations with different additive molecular structure and concentration were performed, their influences on the self-organized concentration profile were discussed.

Keywords: multiscale, dissipative particle dynamics, PEDOT:PSS : PFI, NiOx nano particle, work function

1. INTRODUCTION

Transparent conductive polymer pss:pedot is very widely used in many applications such as organic solar cells[1,2], OLED[3], and recently perovskite solar cells[4]. it is often inserted in between indium tin oxide (ITO) and the active layer to serve as the hole injection layer (HIL). The work function of ITO is usually at about 4.8eV. Work function of the active layers according to their applications, maybe 5.8 eV. The work function of PSS:PEDOT can vary strongly according to its composition, different PSS:PEDOT ratio[5]. Therefore, in practice, optimized PSS:PEDOT ratio may be chosen to form a cascaded step-changed profile in work function. However, in some applications, such as OLED, this commonly used PEDOT:PSS has been found to have serious problems due to its inefficient hole-injection, inefficient electron-blocking, and substantial quenching of excitons close to the PEDOT:PSS. It was proposed that a smooth gradient profile can solve or at least lessen these problems. Many experimental studies aiming towards this goal, solvent surface treatment[6], NiOx nano particles[7], PFI diblock copolymer[8]

Modeling of the interface and additives interact with PEDOT:PSS is very important for mechanisms understanding, future device architecture designing and structure optimization. Classical molecular dynamics is a very powerful tool for it can provide atomistic detail of the interface, but it also has an equally great disadvantage, it can only apply to very small size dimension and short time duration, far from the length and time scales in real sense. This disadvantage maybe overcome through recently matured multiscale simulations. In a multiscale simulation mesoscale simulations were

* minontongji@tongji.edu.cn; phone 86 021 65982232; fax 86 021 65982287

performed to provide a bridge to relate microscopic view from molecular dynamics to the dimension and time in real sense. There are two popular mesoscale simulation tools, Coarse grain dynamics which uses parameters from direct mapping of the molecular dynamics, therefore, retained most features of molecular dynamics. The other method is the so called Dissipative particle dynamics (DPD). DPD was developed in the early 90's[9] and reformulated by Espanol[10], which combines coarse graining with a branch of thermodynamics, the Flory-Huggins theory for polymer[11,12,13]. Flory-Huggins theory for polymer mix states that the free energy of a polymer mix including three terms:

$$\left(\frac{\Delta G_{mix}}{RT}\right)\left(\frac{V_r}{V}\right) = \frac{\phi_A}{N_A}\ln\phi_A + \frac{\phi_B}{N_B}\ln\phi_B + \chi_{AB}\phi_A\phi_B \quad (1)$$

the first two terms is entropy of mix from ideal solution, the third term led by the interaction parameter χ_{AB} represents enthalpic interaction between monomer segments and/or solvents which can be calculated form cohesive energy density. The non-ideality of entropy of mix can be incorporated through chain flexibility using intramolecular bonding parameters. According to second law of thermodynamics, the energy will be at its minimum when the system reaches its equilibrium. Therefore, in this way, the DPD simulation has a thermodynamic constrain built in, the simulation process will go towards nothing but thermodynamic final state.

2. MODELS AND METHODS

Dissipative particle dynamics[14] (DPD) is a mesoscale dynamic simulation tool, therefore, atomistic details are not considered. This allowing it to give access to longer time and length scales than classical molecular dynamics (MD) simulations. In DPD simulation different beads with similar volume size and mass must be used to represent polymer monomers in a wormlike structure.

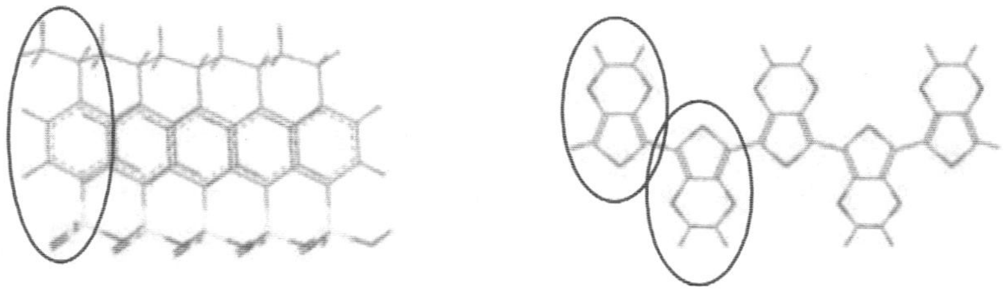

PSS PEDOT

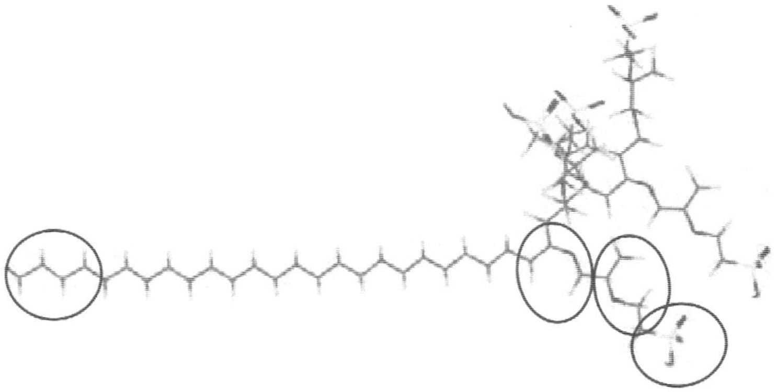

PFI

Figure 1. DPD beads representation of PEDOT, PSS and PFI

All monomers in this study were optimized using DFT, B3LYP[15,16,17] functional and 6-311G++2d,2p basis set[18]. The definition of PSS:PEDOT and PFI beads are shown in Figure 1. The polymer chain length is also very important parameter. In a condensed phase, a polymer chain may fold along its chain length, therefore forming many contact points with self and/or with other chains. As a result, much short chain model is needed. Previous studies, through solubility simulations, found that 15 or more monomers will be suffice. In this study, 25 were used. When block copolymer was involved, the molecular configuration also needs to be considered—there may be differences in degree of polymerization in each branch or alternating structure may exist.

2.1 Flory-Huggins Repulsion parameter

Enthalpic contribution χ_{ij} is obtained through a series of single and bi-component simulation boxes assembled according to bulk average concentration. Molecular Dynamics simulations were then performed. CFF force field was used for PSS:PEDOT:PFI system while UFF force field was used in PSS:PEDOT:NiOx system. Average cohesive energy density were extracted according to equation. Flory-Huggins interaction parameters χ_{ij} and DPD repulsion parameters were calculated accordingly.

$$\chi_{ij} = \left(\frac{\Delta E_{ij}}{RT}\right) V_{brad} = \left(\frac{\Delta E_{ij}}{8.314 \times 298} 6.022 \times 10^{23} \times 10^{-30}\right) V_{brad} = \left(2.43 \times 10^{-10}\right) \Delta E_{ij} V_{brad} \qquad (2)$$

where V_{bead} is the volume of a bead.

$$\Delta E_{ij} = \phi_i \left(E_{coh}\right)_i + \phi_j \left(E_{coh}\right)_j - \left(E_{coh}\right)_{ij} \qquad (3)$$

where E_{coh} represents cohesive energy.

$$\chi_{ij} = \left(\frac{\Delta E_{ij}}{RT}\right) V_{bead} \qquad (4)$$

Once the Flory-Huggins parameters were decided, the DPD repulsion parameters of different beads can then be obtained according to Groot[19]

$$\alpha_{ij} = \frac{\chi_{ij}}{0.306} + \alpha_{ii} \qquad (5)$$

2.2 Intramolecular bonding parameter

Simulation box contains all related components was assembled according to again the bulk average concentration. The generated trajectory was converted to obtain intra-molecular distances and angles distributions, the first peak was used to extract bond stretching and bending distribution function for each component, then convert to spring constants.

2.3 DPD simulation

The DPD simulation box were 10x10x20 DPD units for PEDOT:PSS:NiOx and 20x20x30 DPD units for PEDOT:PSS:PFI. Chain length of 25 repeated monomer beads were used, of which PEDOT(monomer):PSS equals to 1:2.5 weight ratio according to the literature[20]. Different PFI concentration were used in experiments in the same literature, low PFI concentrations have been published elsewhere[21], in this contribution, concentration of weight ratio from 5 and 8.5 of PFI will be presented. The concentration of NiOx nano particle is not as clearly defined in the literature, therefore, 0.1 and less were used.

In the experiment, PEDOT:PSS and PFI were in appearance, well mixed in the solution according to the standard device fabrication procedures. However, monomers confined to polymer chains do not have the same freedom as small molecules, total random mix condition at microscopic level is seldom met. The mixture will always be in some level of semi-ordered form. Therefore, DPD simulation boxes with stacking and random PEDOT:PSS and PFI initial conditions were generated side by side to mimic the processes change from ordered to semi-ordered and form random to semi-ordered form; and verify the results for each other. For PEDOT:PSS:NiOx, only random initial condition was used.

3. RESULTS AND DISCUSSES

3.1 DPD parameters

DPD parameters of PSS:PEDOT:PFI system were published in somewhere else[12]. Obtained PSS:PEDOT:NiOx DPD repulsion parameters are listed in Table 1, PEDOT and NiOx nano particle have strong attraction, while PSS and NiOx nano particles repulse mildly. Water molecule attracts to PEDOT and repulse to PSS mildly. Water attracts to NiOx nano particle very strongly. Their intramolecular bond stretching and bending results are listed in Table 2.

Table 1. PSS:PEDOT:NiOx DPD repulsion parameters α_{ij}

ij	EDOT	SS	NiOx	H_2O
EDOT	25			
SS	24.12	25		
NiOx	17.0	28.24	25	
H_2O	23.3	27.8	13.58	25

From Table 2, adding water to the PSS:PEDOT chain flexibility of both polymer have only little change. When NiOx is added, PEDOT chain become twice as rigid. This rigidness will contribute negatively to mixing and tend to form phase segregation.

Table 2. PSS:PEDOT Intramolecular bonding parameters

System	Interactions	AAMD Simu	
	Bond length	ks (kcal/mol/A^2)	r(A)
PEDOT:PSS	EDOT-EDOT	0.8158	4.4201
	SS-SS	0.0436	5.8071
PEDOT:PSS:H2O	EDOT-EDOT	0.6085	4. 3002
	SS-SS	0.0566	6.8432
PEDOT:PSS:NiOx	EDOT-EDOT	1. 6998	4.5438
	SS-SS	0.0179	9.1388
	Bend angle	k_θ (kcal/mol/rad^2)	θ_0 (degree)
PSS:PEDOT	EDOT-EDOT-EDOT	4.0538	128.7380
	SS-SS-SS	0.3159	31.1367
PSS:PEDOT:H2O	EDOT-EDOT-EDOT	4.4385	140.1322
	SS-SS-SS	0.1548	27.7342
PSS:PEDOT:NiOx	EDOT-EDOT-EDOT	2.1044	125.7708
	SS-SS-SS	0.3983	45.8840

3.2 DPD simulation results

DPD simulation from both initial conditions reached similar final state in all the PFI concentrations in study. The effect of concentration of different kinds on the vertical profiles from random initial condition are shown in Figure 2. PSS:PEDOT:PFI forms phase separation with PFI concentrated on the top surface, leading to self-organized gradient profiles for PSS and PEDOT.

PEDOT:PSS:PFI 1:5:5 1:2.5:5 1:2.5:8.5

Figure 2. DPD simulation results of the self-organized gradient effect of ITO-PEDOT:PSS:PFI, random initio condition, where grey, blue, cyan and fulvous beads represent ITO, PEDOT, PSS and PFI respectively.

The same mechanism causing PEDOT to have steeper profile than PSS towards the surface, therefore, large eV value can be expected on PFI rich surface.

The effect of NiOx concentration on the vertical profiles are shown in Figure 3. From Table1 and 2, NiOx nano particles cling to PEDOT polymer chains, causing PEDOT chain to be more stretched and rigid, thus easier to form phase separation. At the same time, the strong attraction between NiOx nano particles and PEDOT pulls PEDOT together to form more condensed PEDOT phase, thus more condensed PSS phase on the opposite direction which lead to the self-organized PEDOT and PSS vertical concentration profile, which, through optimization, can be used to achieve gradient work function in vertical direction.

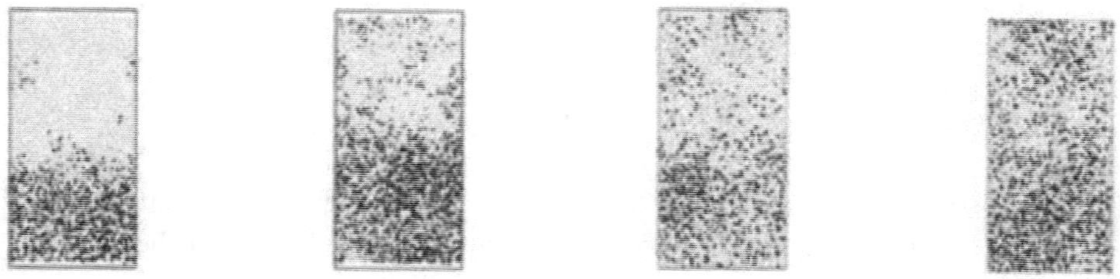

PEDOT:PSS:NiOx 1:5:0.1 1:2.5:0.05 1:5:0.04 1:2.5:0.025

Figure 3. DPD simulation result of the self-organized gradient effect of PEDOT:PSS:NiOx. Where blue, cyan and yellow beads represent PEDOT, PSS and NiOx respectively.

4. CONCLUSIONS

PSS:PEDOT:PFI forms phase separation in a wide PFI concentration region, with PFI concentrated on the top surface, leading to self-organized gradient profiles for PSS and PEDOT, The same mechanism causing PEDOT to have steeper profile than PSS towards the surface, therefore, large eV value can be expected on PFI rich surface. PSS:PEDOT:NiOx takes a different mechanisms. NiOx nano particles cling to PEDOT polymer chains, causing PEDOT chain to be more stretched and rigid, thus easier to form phase separation. At the same time, the strong attraction between NiOx nano particles and PEDOT pulls PEDOT together to form more condensed PEDOT phase, thus more condensed PSS phase on the opposite direction.

The model can be used for future numerical experimenting on effects of composition, structure configuration and additives for potentially tailored design and structure optimization.

REFERENCES

[1] Sun, K., Zhang, S., Li, P., Xia, Y., Zhang, X., Du, D., Isikgor, F. H. and Ouyang, J., "Review on application of PEDOTs and PEDOT:PSS in energy conversion and storage devices," J Mater. Sci: Mater Electron. 26, 4438–4462 (2015)

[2] Nie W., Tsai, H., Asadpour, R., Blancon, J.-C., Neukirch, A.J., Gupta, G., Crochet, J.J., Chhowalla, M., Tretiak, S., Alam, M.A., Wang, H.-L. and Mohite, A.D., "High-efficiency solution-processed perovskite solar cells with millimeter-scale grains," Sci. 367, 522-525 (2015)

[3] Wang, G.-F., Tao, X.-M. and Wang, R.-X., "Fabrication and characterization of OLEDs using PEDOT:PSS and

MWCNT nanocomposites," Compos. Sci. Technol. 68, 2837–2841 (2008)
[4] You J, Hong Z, Yang YM, Chen Q, Cai M, Song TB, Chen CC, Lu S, Liu Y, Zhou H, Yang Y, "Low-temperature solution-processed perovskite solar cells with high efficiency and flexibility", ACS Nano, 8(2):1674-80 (2014)
[5] Petrosino, M., Vacca, P., Miscioscia, R., Nenna, G., Minarini, C. and Rubino, A., "Effect of PEDOT:PSS ratio on the electrical and optical properties of OLEDs," Proc. SPIE 6593, 659310 (2007)
[6] I. Cruz-Cruza, M. Reyes-Reyesa, M.A. Aguilar-Frutisb, A.G. Rodrigueza, c, R. López-Sandoval, "Study of the effect of DMSO concentration on the thickness of the PSS insulating barrier in PEDOT:PSS thin films", Synthetic Metals, 160:1501–1506 (2010)
[7] Anil Kanwat, William Milne, Jin Jang, "Vertical phase separation of PSS in organic photovoltaics with a nickel oxide doped PEDOT: PSS interlayer", Solar Energy Materials and Solar Cells, 132:623–631 (2015)
[8] Chin, B.D., Choi, Y., Baek, H.-I. and Lee, C., "High efficiency phosphorescent OLEDs based on the heterostructured light emission and charge injection layers," Proc. SPIE 7415, 741510 (2009)
[9] P. J. Hoogerbrugge and J. M. V. A. Koelman. Simulating microscopic hydrodynamic phenomena with dissipative particle dynamics. Europhysics Letters, 19(3):155–160, JUN 1 1992
[10] P. Español and P. B. Warren. Statistical-mechanics of dissipative particle dynamics. Europhysics Letters, 30(4):191–196, MAY 1 1995
[11] Flory, P.J., "Thermodynamics of High Polymer Solutions," J. Chem. Phys. 10, 51-61 (1942)
[12] Huggins, M.L., "Some Properties of Solutions of Long-chain Compounds," J. Phys. Chem. 46, 151-158 (1942)
[13] Huggins, M.L., "Theory of Solutions of High Polymers," J. Am. Chem. Soc. 64, 1712-1719 (1942)
[14] Groot, R.D., and Warren, P.B., "dissipative particle dynamics: bridging the gap between atomistic and mesoscopic simulation," J. Chem Phys. 107, 4423-4435, (1997)
[15] Becke, A.D., "Density-functional exchange-energy approximation with correct asymptotic behavior," Phys. Rev. A 38, 3098-3100 (1988)
[16] Chengteh, L., Weitao, Y. and Robert, G.P., "Development of the Colle-Salvetti correlation-energy formula into a functional of the electron density," Phys. Rev. B 37 (2), 785-789 (1988)
[17] Vosko, S. H., Wilk, L. and Nusair, M., "Accurate spin-dependent electron liquid correlation energies for local spin density calculations: a critical analysis," Can. J. Phys. 58, 1200-1211 (1980)
[18] Ditchfield, R., Hehre, W.J., Pople, J.A., "Self-consistent molecular-orbital methods. IX. an Extended Gaussian-type basis for molecular-orbital studies of organic molecules," J. Chem Phys. 54, 724-728 (1971)
[19] Groot, R.D. and Madden, T.J., "Dynamic Simulation of Diblock Copolymer Microphase Separation," J. Chem. Phys. 108(20), 8713-8724, (1998)
[20] Lee, T.-W., Chung, Y., Kwon, O., and Park, J.-J., "Self-Organized Gradient Hole Injection to Improve the Performance of Polymer Electroluminescent Devices," Adv. Funct. Mater. 17, 390–396 (2007)
[21] Min Huang, "Multiscale study of the self-organized gradient effect of novel hole injection material PEDOT:PSS:PFI", Proc. of SPIE Vol.9941:994116, Organic Light Emitting Materials and Devices XX, 2016 SPIE Optics & Photonics, San Diego, California, USA, Aug. 29, 2016

472603 Study of Charge Transfer Dynamics in Spray Deposited Cu2ZnSnS4 (CZTS) Photoelectrodes for Performance Improvement

Abstract:

Thin film solar cell based on Cu_2ZnSnS_4 (CZTS) has the potential for large scale electricity production using solar radiation while all of the elements are earth abundant, inexpensive and environment friendly. The state of art CZTS solar cell has reached 12.6% efficiency while the commercial production will require at least 15% for economic feasibility. Here, we approach to understand the charge transfer dynamics in CZTS using ultrafast laser spectroscopy and focus on improving the efficiency of CZTS photovoltaics through the incorporation of a mesoscopic hole acceptor, nickel oxide. CZTS photoelectrode is prepared using spray deposition of precursor solution, and characterized using SEM, TEM, XRD, XPS and Raman Spectroscopy. The performance of the material is tested in three electrode photo electrochemical cell.

463080 Highly Efficient Solar Cells Made with $Cu_{1-X}K_XInSe_2$ Alloys: A Foundation for Engineering K in $Cu(In,Ga)Se_2$

Abstract:

For the first time, Cu-KF-In-Se co-deposition was used to form $Cu_{1-x}K_xInSe_2$ (CKIS) films over the full composition range (0 < x < 1). Compared to K-free $CuInSe_2$, the CKIS films exhibited increased majority carrier concentrations and wider band gaps, while 0 < x < 0.22 compositions showed increased minority carrier lifetimes. Thin film solar cells were fabricated using soda-lime glass/Mo/CKIS/CdS/intrinsic-ZnO/Al:ZnO/Ni/Al stacks, where a Ni/Al finger grid with partial shading was used for the top contact. Photovoltaic (PV) performance was optimal at x ~ 0.07, which resulted in an officially-measured 15.0%-efficient deviceâ€"equal to the world record $CuInSe_2$ efficiency (grown at 575 Â°C), despite being grown at 500 Â°C, without the three stage process, and without device stack optimization. The devices were characterized with quantum efficiency, temperature- and light-intensity-dependent current-voltage, capacitance-voltage, and admittance spectroscopy. These data were then used to calculate charge carrier recombination rates in the bulk absorber, the depletion region, and at the interface. Bulk recombination was reduced for x > 0. Interface recombination was reduced at x ~ 0.07, while x â‰¥ 0.22 compositions had poorer performance due to interface recombination, relative to x ~ 0. These trends are consistent with both surface and bulk recombination affecting apparent lifetimes. The results showed that large K compositions deteriorated PV performance in CKIS alloys, while nevertheless, $Cu(In,Ga)Se_2$ solar cells with record efficiencies have exhibited large K compositions at the absorber/buffer interface. Processing techniques were employed to address this apparent discrepancy in material quality at high K content. Increasing the substrate's Na composition was found to increase the extent of CKIS alloy formation, relative to $CuInSe_2$ + $KInSe_2$ formation. Similarly, decreased substrate temperature was found to favor CKIS formation, relative to $CuInSe_2$ + $KInSe_2$. Substrate Na and temperature can therefore be used to engineer K bonding in $Cu(In,Ga)Se_2$ absorbers to enhance both initial and long-term PV power generation. Material growth

methods such as these may also be used to investigate the differences between CKIS alloys and Cu-K-In-Se films with very low $KInSe_2$ content.

468083 Study of Electron Transport Mechanism in Dye-Sensitized Solar Cell with the Effect of Morphology, Crystalline Structure and Electron Mobility

Abstract:

Dye-sensitized solar cells (DSSCs) are one of the most promising developments in photovoltaic devices. To obtain high efficiency DSSCs, extensive research has been carried out to consider the effect of dyes, electrolytes, and alternatives to titania.

The performance of photovoltaic devices could be improved by using nano-composites which can provide high electron mobility and efficiently collect photo-generated electrons through a better understanding of their fundamentals. One-dimensional nanomaterials including Single-walled carbon nanotubes (CNT) and metals, such as gold and silver, exhibit very high electron mobility.

However, the incorporation of such materials into DSSCs for efficient photovoltaic performance is still challenging. Also the fundamental of the above system is not well understood.

This study focuses on the synthesis of core-shell nano-composites made of TiO_2, various metals, and CNTs. Several fabrication techniques including chemical vapor deposition, electrochemical deposition, spin-coating, and electro spinning were applied to create various nanostructured composite materials such as one dimensional nanorods, nanowires, and nanotubes. Designing core-shell nanostructures of TiO_2in DSSC help us investigate and compare electron transfer properties of the different core materials.

The Mechanistic study of the DSSCs were carried out by using the following techniques: UV-vis scoptrocopy, time-resolved photoluminescence spectroscopy, impedance analyzer in combination. In addition, characterization methods such as TEM, SEM, XRD, and Raman spectroscopy were used. These studies will help to better understand the principle of DSSCs and better design efficient DSSC devices.

473712 Solventless Synthesis of Zeolitic-Imidazole Framework ZIF-8 Membranes Via Crystal-Specific Sintering Phenomenon

Abstract:

ZIFs, constructed by interconnecting divalent metal nodes (e.g., Zn, Co, Cd or their combinations) and imidazole linkers in periodic manners, have been one of the most intensively studied metal-organic frameworks (MOFs) primarily due to their exceptional thermal/chemical stabilities compared to other MOFs.[1] In combination with the ZIFsâ€™ stabilities, their pore apertures with molecular level dimension have led them to be considered as promising material candidates for energy-efficient membrane-based separations.[2]

When prepared, ZIF membranes have polycrystalline nature like zeolite membranes. Therefore their performances, considering the materials are properly chosen, depend on membrane microstructures, most importantly grain boundary structure. Since the membrane microstructures are generally governed by membrane processing technique, current research focus on ZIF membranes has been mainly about developing new processing techniques.[3] As results, newly designed methods, which are broadly classified into in-situ and secondary growth, enabled significant improvements in membrane performance, synthesis simplification, and potential scalability.[2]In spite of all these advancements, however, since the majority of the current techniques are still solution-based syntheses, there are still issues inevitably encountered such as producing wastes (e.g., environmentally harmful solvent, excessively-used expensive ligand, metal precursor) and putting a limitation on reducing membrane thickness which is very important for large quantity gas processing. Therefore, there are still needs of developing synthesis protocols which are sustainable and allow membranes with thin thickness.

Here, we report a new synthesis strategy which allows the formation of continuous ultrathin ZIF-8 films / membranes (~300 nm) in a solventless condition. This new approach is on the basis of our new discovery that intentionally-designed defective ZIF-8 crystals display sintering phenomenon without crystal degradation under specific vapor environment, leading to crystal merging and subsequently grain growth.

[1] K. S. Park, Z. Ni, A. P. CÃ´tÃ©, J. Y. Choi, R. Huang, F. J. Uribe-Romo, H. K. Chae, M. Oâ€™Keeffe, O. M. Yaghi, *Proceedings of the National Academy of Sciences* **2006**, *103*, 10186-10191.

[2] aJ. Yao, H. Wang, *Chemical Society Reviews* **2014**, *43*, 4470-4493; bB. Chen, Z. Yang, Y. Zhu, Y. Xia, *Journal of Materials Chemistry A* **2014**, *2*, 16811-16831.

[3] H. T. Kwon, H.-K. Jeong, A. S. Lee, H. S. An, J. S. Lee, *Journal of the American Chemical Society* **2015**, *137*, 12304-12311.

472934 Microplasmas for Substrate-Independent Deposition of Nanostructured Metals and Oxides

Abstract:

A general, substrate-independent method for plasma deposition of nanostructured, crystalline metals and oxides will be presented. The technique uses a flow-through, micro-hollow cathode plasma discharge (supersonic microplasma jet) with remote ring anode to deliver a highly-directed flux of growth species to the substrate. A diverse range of nanostructured materials, e.g., metals (Cu, Ni, Pd), oxides (CuO, SnO_2, NiO, Fe_2O_3, Co_xO_y), doped oxides ($Fe_xNi_{1-x}O$), and spinels ($NiFe_2O_4$) with different morphologies (nanoparticles, dense columnar films, or hierarchical nanostructures) can be deposited at room temperature on virtually any substrate (conductors, insulators, plastics, fibers, and patterns) in a conformal fashion. The effects of deposition conditions, substrate type, and patterning on film morphology, nanostructure and surface coverage will be highlighted. The microplasma synthesis approach presented herein provides a general and tunable method to deposit a variety of functional and hierarchical metal/oxide materials on many different surfaces. High surface area, conversion-type CuO electrodes for Li-ion batteries are demonstrated as a proof-of-concept example.

472002 Hierarchically-Structured Porous Carbon Films By Multiscale Templating and Interfacial Engineering

Abstract:

The promise of inorganic membranes such as carbon molecular sieves (CMS) to revolutionize continuous high-selectivity gas and liquid separations has been quelled by persistent struggles to break through commercial performance barriers associated with the well-known trade-off between permeability and selectivity. Respective synthetic challenges common to CMS and other inorganic membrane materials include the need to minimize defect-free membrane thickness while simultaneously tailoring membrane texture and function. In addition, whereas strategies exist for scaling polymeric membranes (e.g., hollow fiber spinning), the lack of corresponding strategies for facile areal scaling of inorganic membranes, commonly prepared as two-dimensional sheets, challenges their ability to meet stringent demands for industrial separations productivity. In this talk, we will describe our synthetic efforts to realize a class of ultra-thin, three-dimensionally corrugated, self-supporting carbon thin films with tunable meso- and micro-structure toward novel CMS membranes. We will describe a synergistic strategy that exploits nanoparticle (silica)-mediated continuous porous thin film formation and concomitant interstitial and interfacial confinement of polymerization and pyrolysis as a new strategy for simultaneously tuning thickness, geometry, and microstructure of selective carbon features.[1] Specifically, convective co-assembly of sacrificial silica nanoparticle templates with molecular carbon precursors results in tunable, 3D-interdigitated mesopore topologies with extensive specific surface areas.[2] Upon template sacrifice, the resulting films offer two types of independent but interdigitated pore structures contacted across ultra-thin (i.e., nanometer-scale) three-dimensionally distributed CMS pore walls. This 3D interdigitated pore structure represents a significant area-to-volume enhancement that holds promise as a platform for molecular separations. Beyond mesoscale structuring, we also find that template-mediated interfacial phenomena provide a new handle for simultaneously tuning molecular-scale structural characteristics (e.g., sp^2 hybridized carbon content) that ultimately impact film texture. This work aims to demonstrate how sacrificial, template-mediated hierarchical structuring may advance the viability of CMS membranes, and, more generally, may serve as a new framework for structuring other inorganic membrane materials.

References

[1] Z. Tian, M. A. Snyder*, "Hard-templating of symmetric and asymmetric carbon thin films with three-dimensionally ordered mesoporosity," *Langmuir,* 30 (2014) 9828-9837.

[2] Z. Tian, M.A. Snyder*, "Nanocasting of carbon films with interdigitated bimodal three-dimensionally ordered mesopores by template-replica coassembly," *Langmuir*, 30 (2014) 12411-1242

454878 Effect of the Chain Length of Alkylamine on Film Formation from Alcohol-Soluble Copper Complex Ink

Abstract:

Metal-organic decomposition (MOD) ink has a promising application for printed electronics technology. We synthesized self-reducible conductive ink with copper formate (Cuf) and primary alkylamines as ligand to form alcohol-soluble complex. Highly conductive copper films could be obtained on both rigid and flexible substrates at low sintering temperature under nitrogen atmosphere. The decomposition mechanism was also studied. It was found that amines could not only activate the dissociating of carboxyl from Cuf and thus decrease the decomposition temperature, but also retard the subsequent growth of copper nuclei. These two effects were dependent on the alkyl chain length of amines. With the increase of the alkyl chain length, the decomposition rate of copper-amine complex was slowed down, while the capping effect was enhanced resulting in narrow copper particle size distribution and less void space. The film from Cuf-Octylamine (Cuf-Octyl)showed the lowest resistivity, which was 10.12μΩ·cm sintered at 180 . Using blended amines with long-chain and short-chain could balance the dual effects, contributing to the formation of dense packed copper films with less organic residues and lower resistivity. The resistivity of copper film from Cuf -Octylamine-Butylamine (Cuf-Octyl-Butyl) ink was 5.25μΩ·cm sintered at 180 . Moreover, the solubility in alcohols and stability of Cuf-Octyl-Butyl were both improved compared to Cuf-Octyl. By adjusting the viscosity and surface tension, the prepared ink could meet the requirement of inkjet printing.

471516 Understanding of Diffusion Pathway of Cyclohexane through Nanoscale MFI Zeolite

Abstract:

Understanding of diffusion pathway of cyclohexane through nanoscale MFI zeolite

Xiaoduo Qi, Vivek Vattipalli, Wei Fan

Department of Chemical Engineering, University of Massachusetts Amherst, 686 N Pleasant Street, Amherst, 01002

Zeolites have been widely used in the fields of shape selective catalysis and separation due to their unique molecular sieving property. During catalytic reactions, reactants diffuse into pore network, react at active site, and diffuse out of pore network. To minimize the effects of internal mass transfer limitation on the overall reaction rate, nanoscale zeolite crystals have been designed and synthesized.[1] However, within small zeolite crystals, diffusion study has shown that the apparent diffusivity appears to be orders of magnitude smaller than that for large zeolite crystals[2]. This discrepancy of apparent diffusivity has led to the phenomena called "Surface Barrier", which is a general term to describe a mass transfer resistance near the surface of zeolites. For large zeolite particles, since the internal microspore diffusion is dominant, such a surface resistance is often negligible. While for small zeolite particles, the surface resistance can contribute significantly to the overall mass transfer rate.

To explain the origin of the surface barrier, pore blockage and pore narrowing mechanisms have been proposed[2]. However, with such hypotheses, Kinetic Monte Carlo simulation suggested 99.9% of the micropores on external surface have to be blocked[3]. In this presentation, we propose a new "Pore Re-entering" mechanism. According to this mechanism, molecules undergo a surface diffusion on the external surface during the desorption step. The surface diffusion is much slower than the Knudsen diffusion. Diffusion on the external surface might lead to re-entering of micropore openings, which significantly increase apparent diffusion length for the molecules. In order to demonstrate the existence of such effects, silcalite-1/silica nanoparticle mixtures were developed where silicalite-1 particles were isolated from each other by 35 nm solid silica nanoparticles. Mass transport study using cyclohexane suggested that the diffusion of cyclohexane is still controlled by the micropore diffusion with the same activation energy, but much faster in the mixtures, providing a strong evidence for the existence of the Pore Re-entering effect. The observation provides critical information for the rational synthesis of hierarchical porous materials for the applications in separation and catalysis. [1] Chang, C.-C.; Teixeira, A. R.; Li, C.; Dauenhauer, P. J.; Fan, W., Enhanced Molecular Transport in Hierarchical Silicalite-1. Langmuir 2013, 29, (45), 13943-13950.

[2] Teixeira, A. R.; Chang, C.-C.; Coogan, T.; Kendall, R.; Fan, W.; Dauenhauer, P. J., Dominance of Surface Barriers in Molecular Transport through Silicalite-1. The Journal of Physical Chemistry C 2013, 117, (48), 25545-25555.

[3] Teixeira, A. R.; Qi, X. D.; Conner, C. W.; Mountziaris, T. J.; Fan, W.; Dauenhauer, P. J., 2D Surface Structures in Small Zeolite MFI Crystals. Chemistry of Materials 2015 27 (13), 4650-4660

469390 Neutron Reflectometry Investigation of Hydrogen in Plasma Treated Hydrogen Doped Nanoporous TiO_2 Thin Films for Water Splitting Photocatalysis

Abstract:

Although hydrogen has long been accepted as the species responsible for photocatalytic activity of hydrogenated titania (also referred to as â€œblack titaniaâ€ because of its strong light absorbing properties), direct observation of this species has been elusive. In this work, we use neutron reflectometry (NR) to detect the hydrogen in plasma treated hydrogen doped nanoporous titania thin films. The cubic ordered nanoporous TiO_2 thin films were prepared by a surfactant-templated sol-gel method and were treated with hydrogen plasma, an approach hypothesized to capitalize on the high degree of disorder in the material and the high energy of the plasma species to achieve efficient hydrogen doping. UV-vis absorbance spectra indicated that the incorporated hydrogen atoms made TiO_2 films black with the enhancement of visible light absorption, and XPS analysis showed peak for Ti^{3+} state. The presence of hydrogen in hydrogen doped nanoporous titania (H-TiO_2) films is suggested by the scattering length density (SLD) profiles obtained from NR. The SLD of H-TiO_2 is much lower than that of TiO_2 film due to the negative SLD of hydrogen. Thus, for the first time, NR showed that the dramatic change in visible light absorbance of H-treated black TiO_2 is accompanied by significant hydrogen uptake and not just Ti^{3+} generation. The photocatalytic activity of hydrogen doped TiO_2 (H-TiO_2) films was evaluated by chronoamperometry. The effect of light sources on photoelectrochemical performance was explored using UV (365 nm) and blue (455 nm) LEDs. In photoelectrochemical water splitting, the H-TiO_2 showed about 27 times and 8 times higher photocurrent under UV and blue LEDs, respectively, compared to undoped TiO_2 films. Electrochemical impedance spectroscopy showed that better charge transfer is obtained in plasma treated hydrogen doped titania compared to undoped titania films. This study suggests that the dramatic change in visible light absorbance of hydrogenated black TiO_2 is due to significant hydrogen uptake, and plasma-induced doping is a promising strategy to enable the efficient incorporation of heteroatoms into surfactant-templated TiO_2 films, thereby leading to significant enhancement in photoelectrochemical hydrogen production.

462156 Synthesis of Tin(II) Monosulfide Nanoplates: A Potential 2D Material

Abstract:

Two dimensional (2D) layered materials such as graphene, metal dichalcogenides, and black phosphorus are of increasing interest because of their unique electronic properties. Tin(II) monosulfide (SnS) has the same crystal structure as black phosphorus and is a semiconductor with a 1.3 eV direct band gap, a 1.1 eV indirect band gap, and a high absorption coefficient (>10^4 cm^{-1}) in the visible range of the electromagnetic spectrum. SnS has potential applications in photovoltaics, photocatalysis, thermoelectrics, and lithium ion batteries. We present a facile chemical synthesis of SnS nanoplates via thermal decomposition of a single precursor in a coordinating solvent. Specifically, tin(IV) diethyldithiocarbamate ($Sn(dedtc)_4$) dissolved in oleic acid is injected into hot (300-340 °C) oleylamine to initiate rapid nucleation and subsequent growth of SnS after decomposition. Oleylamine complexes with $Sn(dedtc)_4$ and induces its decomposition to form SnS. Examination of the reaction product after 60 minutes at 340 °C with scanning electron microscopy shows thin plates that are 1 to 10 µm wide. Atomic force microscopy of these plates dispersed onto silicon show that the plates are 3-60 nm thick. Energy dispersive X-ray spectroscopy (EDS) indicates that the Sn-to-S ratio is one within the accuracy of EDS quantification. The crystal structure of the plates is confirmed to be orthorhombic (a=0.433 nm, b=1.119 nm, c=0.398 nm) Herzenbergite α-SnS phase with X-ray diffraction. When the plates are >1 µm wide, X-ray diffraction (XRD) from SnS plates cast onto substrates shows only the (040) diffraction as the plates settle on the substrate with their [010] axis aligned with the substrate normal. This textured XRD from plates cast on flat substrates and transmission electron microscopy indicate that SnS plates grow preferentially in the [100] and [001] directions with the [010] axis direction. Raman spectroscopy and confocal imaging show Raman scattering peaks at 96, 161, 186 and 217 cm^{-1}, all consistent with SnS. At low temperatures (300 °C) and short synthesis times (3 minutes) we also observe the presence of SnS_2 with both XRD and Raman spectroscopy. Raman spectroscopy shows a strong peak at 312 cm^{-1}. SnS_2 disappears at later times. This suggests that decomposition of $Sn(dedtc)_4$ first produces SnS_2 which is subsequently reduced, likely by oleylamine, to SnS. The SnS plate sizes depend on the oleylamine concentration, time, and temperature. Plate sizes increase with decreasing oleylamine concentration and increasing temperature. Plates also become larger and thicker with time. SnS nanoplate dispersions in toluene exhibit an optical absorption feature in the visible range of the electromagnetic spectrum, which we surmise to be of plasmonic origin. The nanoplate dispersions in toluene respond to electric fields.

461321 MoS_2-Passivated Bilayer Phosphorene Phototransistors

Abstract:

Despite its unique and promising properties such as high carrier mobility and conspicuous in-plane anisotropy that generated a tremendous amount of excitement for the past few years, BP has been suffering from its one significant drawback that may dim its potential as an active material in the next-generation nanoelectronics: severe crystal deterioration upon exposure to oxygen and water. Field-effect transistors (FETs) based on a MoS_2-passivated black phosphorus (BP) van der Waals (vdW) heterostructure could provide an effective solution to tackling this important challenge, while enhancing photoresponse of the device and retaining the conductivity of a BP channel almost intact. In this work, we fabricate 4 MoS_2-passivated BP FETs with varying BP thicknesses, 1.5 (bilayer), 5, 13, and 20 nm, to explore the effect of the MoS_2 passivation layer on the device stability and electrical characteristics both in the dark and under illumination (Î» = 600 nm). When in contact with a trilayer MoS_2, photoluminescence intensity of the bilayer BP crystal decreases by 25%, suggesting that a built-in electric field forms at the BP-MoS_2 p-n interface that could help dissociate photo-generated electron-hole pairs, thereby reducing the probability of the recombination events. The effectiveness of a few-layer MoS_2 as a vdW protection layer is tested by exposing BP-MoS_2 vdW vertical heterostructures to the air for up to 3 weeks as well as by annealing at a high temperature (350°C) in an inert argon environment. Then, the heterostructure FETs are fabricated so that direct comparisons between exposed and MoS_2-passivated BP regions on the same flake can be made. In general, the MoS_2 passivation layer gives rise to only an insignificant decrease in the transport characteristics in the dark, with the trend where its impact decreases as the thickness of the underlying BP crystal increases. Upon illumination, the MoS_2-passivated regions of the bilayer BP device shows enhanced photoresponse compared with the exposed regions by 75 % with no applied gate voltage. Also, a possible photogating effect is observed in the bilayer BP device while photoconductivity dominates the photoresponse process in the thicker devices. Thus, 2D MoS_2 2D thin films may aid in the realization of the future

transparent, flexible BP electronic and optoelectronic devices by acting not only as an atomically thin passivation layer but as a photoresponse enhancer.

464739 Process Simulation and Economic Analysis of Producing Liquid Transportation Fuels from Biomass

Abstract:

Although the global energy sector is still driven by fossil fuels, renewable and more environmental friendly energy sources need to be explored due to the insecurity of crude oil supply and the environmental problems associated with fossil fuel consumption. The conversion of biomass to liquid transportation fuels is a promising alternative. A process is developed in ASPEN Plus to simulate the conversion of biomass to produce gasoline, diesel, and kerosene at varying design/operation conditions. Biomass is dried and gasified to generate synthesis gas, which is converted to a mixture of hydrocarbons via Fischer-Tropsch synthesis (FTS). Given the same biomass feedstock, three FTS technologies including conventional FTS, once-through FTS, and supercritical FTS are comparatively studied. Hydrocarbons from FTS are upgraded to liquid transportation fuels which meet all necessary physical property standards through several upgrading technologies (hydrocracking, hydrotreating, isomerization, catalytic reforming, and alkylation). In conventional FTS, the produced $C_1/C_2/C_3$ light gas in the hydrocarbon upgrading process is reformed to produce additional syngas and then recycled back to the FTS reactor along with the fresh syngas to improve the overall carbon conversion rate. For once-through FTS, the $C_1/C_2/C_3$ light gas from hydrocarbon upgrading process and the unreacted syngas from the FTS reactor are directed to a gas turbine for co-production of electrical power. For supercritical FTS, a supercritical fluid is used as the reaction medium to improve/tailor the performance of the FTS reactions.

This study first investigates the product distribution and CO_2 emission of the biomass conversion processes associated with the three FTS technologies. Then heat and power integration is performed for these three biomass conversion process topologies and a given set of operating conditions. Finally, a detailed economic analysis is performed using Aspen Process Economic Analyzer and unit cost functions obtained from literature.

466069 Modeling Plasma Gasification of Biomass with Thermodynamic and Kinetic Approach in Series

Abstract:

Modeling Plasma gasification of biomass with thermodynamic and kinetic approach in series

Babita K. Verma, Rajesh Elangovan, T. Renganathan, S. Pushpavanam*

Department of Chemical Engineering, Indian Institute of Technology Madras,

Chennai 600036, India

*Corresponding author: spush@iitm.ac.in, Tel: +91-44-22574161

Plasma gasification is a promising green technology in which carbon based feed-stocks are converted to synthesis gas. In this process, a plasma arc is used to supply heat to the reactor to sustain the endothermic gasification reactions. Plasma gasification operates at temperatures above 2500 K. Here the reaction rates are very high and this leads to lower residence time as compared to conventional gasification [1]. This drastically reduces the size of gasifiers operating at atmospheric pressure. These advantages motivate the development of accurate mathematical models for plasma gasifiers which can help to analyze the effect of various parameters on the performance.

There are two ways of modeling: Equilibrium and kinetic approach. The purely equilibrium approach does not need information on all reactions which occur at the high temperatures in the reactor here when using the non-stoichiometric approach. The purely kinetic approach requires the consideration of heterogeneous reactions, resistances associated with gas-solid interaction and the formation of radicals. In the present work, the gasifier is modeled as consisting of a high temperature zone followed by a quenching zone where the reactor contents lose heat to the ambient. In the former the reactions are very fast and hence it can be assumed to be at thermodynamic equilibrium. In the quenching zone the reactant mixture cools down and the composition of the gases vary along the reactor length at a finite rate determined by the reaction kinetics. This motivates the reactor to be modeled as being made of two regions namely a plasma gasifier zone (PGZ) which is simulated using a thermodynamic non-stoichiometric approach and a quenching zone (QZ) which is simulated based on a kinetic approach. The schematic representation of the plasma gasifier model [2] is shown in Figure 1

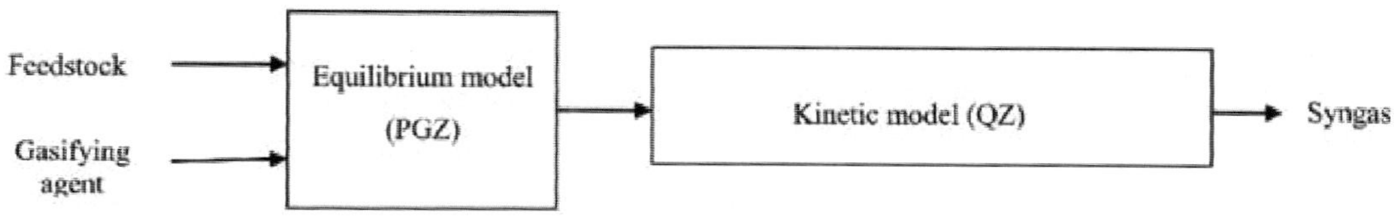

Figure 1. Schematic representation of the plasma gasifier model

The effect of various parameters – gasifier temperature, steam flow rate, air flow rate and feed moisture content – on the performance of the gasifier is studied. Figure 2 shows a typical result i.e. the effect of the gasifier zone temperature on the heat duty and the cold gas efficiency existing in PGZ and QZ region.

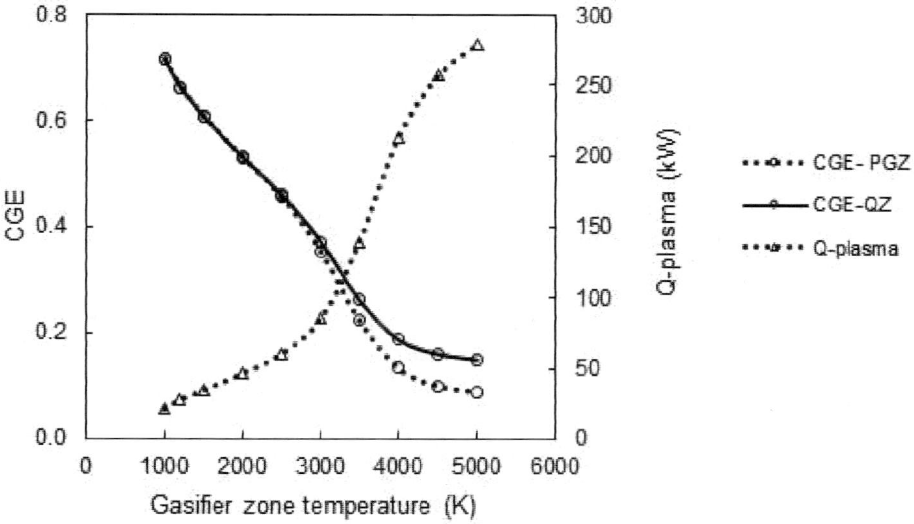

Figure 2: Effect of temperature on heat duty and CGE exiting PGZ and QZ region (Air=13.25kg/hr, Steam=3.90 kg/hr, Moisture=10%)

References:

[1] S. J. Yoon and J. G. Lee, "Syngas Production from Coal through Microwave Plasma Gasification : Influence of Oxygen , Steam , and Coal Particle Size," *Energy & Fuels*, vol. 26, pp. 524–529, 2012.

[2] M. Gorokhovski, E. I. Karpenko, F. C. Lockwood, V. E. Messerle, B. G. Trusov, and a. B. Ustimenko, "Plasma technologies for solid fuels: experiment and theory," *J. Energy Inst.*, vol. 78, no. 4, pp. 157–171, Dec. 2005.

16AIChE
Annual Meeting, San Francisco, CA

Sustainable Solar Fuel Production by a Mixed Metal Oxide Based Thermochemical H_2O/CO_2 Splitting Cycle

(Principal Author)
Rahul Bhosale
Assistant Professor
Department of Chemical Engineering
College of Engineering, Qatar University, Doha, Qatar
rahul.bhosale@qu.edu.qa

(Other Authors)
Parag N. Sutar
Postdoctoral Fellow
Department of Chemical Engineering
College of Engineering, Qatar University, Doha, Qatar
parag.sutar@qu.edu.qa

Fares A. AlMomani
Associate Professor
Department of Chemical Engineering
College of Engineering, Qatar University, Doha, Qatar
falmomani@qu.edu.qa

Ivo Alxneit
Senior Scientist
Solar Technology Laboratory
Paul Scherrer Institute, Villigen – PSI, Switzerland
ivo.alxneit@psi.ch

Prepared for Presentation at
American Institute of Chemical Engineers
2016 Annual Meeting
San Francisco, California
November 13-18, 2016

Sustainable Solar Fuel Production by a Mixed Metal Oxide Based Thermochemical H_2O/CO_2 Splitting Cycle

Rahul Bhosale
Department of Chemical Engineering
College of Engineering
Qatar University, Doha, Qatar
rahul.bhosale@qu.edu.qa

Parag Sutar[1], Fares AlMomani[1], Ivo Alxneit[2]
[1]Department of Chemical Engineering
College of Engineering
Qatar University, Doha, Qatar

[2]Solar Technology Laboratory
Paul Scherrer Institute, Villigen – PSI, Switzerland

Keywords: Solar Fuel, Mixed Metal Oxide, Thermochemical Cycles, H_2O/CO_2 Splitting, Barium Ferrite

Abstract

This paper reports the synthesis of Ba-ferrite via sol-gel method for the production of solar fuel such as H_2 or syngas via thermochemical H_2O and/or CO_2 splitting cycle. Sol-gel synthesis of Ba-ferrite nanoparticles includes dissolution of metal precursors in ethanol, addition of propylene oxide as a proton scavenger to achieve gel formation, aging of the gel, gel drying, and subsequent calcination in presence of air. The effect of addition of propylene oxide (PO) as a proton scavenger on physico-chemical properties of sol-gel derived Ba-ferrite was investigated in detail. Various characterization techniques such as powder X-ray diffraction (XRD), BET surface area analyzer (BET), and scanning electron microscope (SEM) were utilized in this study.

1. Introduction

Metal oxide based thermochemical production of solar fuels such as solar H_2 or solar syngas (a mixture of H_2 and CO) via a two-step H_2O and CO_2 splitting cycle is considered as one of the most promising new technologies for the storage of solar energy in terms of chemical fuels [1-9]. In past, various metal oxide based thermochemical cycles have been investigated towards H_2O and CO_2 splitting reactions. In recent years, this research has been focused towards mixed ferrite materials.

Different ferrite materials have been investigated which includes: Ni-ferrite, Zn-ferrite, Co-ferrite, Mn-ferrite, Sn-ferrite, Ni-Zn-ferrite, Ni-Mn-ferrite, Ni-Sn-ferrite, and others [10-29]. Although these materials showed promising results in terms of solar fuel production, material stability (grain growth due to sintering) due to thermal cycling is an issue which poses limitation for the ferrite based thermochemical cycles to be a commercial process.

Therefore, in this paper we attempted to dope Ba in the crystal structure of iron oxide to synthesize Ba-ferrite. As an initial step only 10% doping was attempted. To achieve this sol-gel method was employed. Effect of addition of propylene oxide as proton scavenger and aging time on physico-chemical properties of Ba-ferrite and iron oxide mixture was studied. Various analytic techniques such as powder X-ray diffraction (XRD), BET surface area analyzer (BET), and scanning electron microscope (SEM) were utilized in this study.

2. Materials and Methods

Materials: The precursors and chemicals required for the synthesis of Ba-doped ferrite nanoparticles i.e. barium nitrate and iron nitrate, ethanol, and propylene oxide (gelation agent) were purchased from Alfa Aesar and Sigma Aldrich and used without any pre-treatment.

Synthesis of Ba-doped ferrite nanoparticles: The Ba-doped ferrite nanoparticles were derived via sol-gel method. At first, nitrate precursors of barium and iron were dissolved in ethanol (in respective weight ratios) by using a sonic bath. After the dissolution, different amounts of propylene oxide (as a proton scavenger) were added drop-wise and the gel formation was achieved. The Ba-doped ferrite gel was aged for multiple hours at atmospheric conditions. After aging, the gel was dried at 100°C for 2 h using a temperature–controlled hot plate. The dried powder obtained was further calcined at 600°C. The obtained calcined powder was stored in a dry atmosphere and then further characterized to determine the phase composition, crystallite size (CS) and specific surface area (SSA). The process flow diagram for the sol-gel synthesis of Ba-ferrite is shown in Fig. 1.

Characterization of sol-gel derived Ba-doped ferrite nanoparticles: The physico-chemical characterization of the sol-gel derived Ba-doped ferrite was carried out by using different characterization techniques. The phase composition and crystallite size (CS) was determined by using powder X-ray diffractometer. Crystallite size was calculated by using the Scherrer's equation. The specific surface area (SSA) of the sol-gel derived Ba-doped ferrite_was estimated using the BET surface area analyzer via the adsorption/desorption isotherms obtained after degassing the calcined powder at 200°C for 12 h.

3. Results and Discussion

The effect of addition of propylene oxide as a proton scavenger on the gel time, phase composition, CS, SSA and PV of the sol-gel derived Ba-ferrite was studied in detail. At first, the variation in the gel time with respect to the amounts of propylene oxide added was examined and the obtained results are presented in Fig. 1 (10 g of metal precursors + 20 ml of ethanol). As shown in Fig. 2, as the amounts of propylene oxide added increases the time required for gel formation decreases.

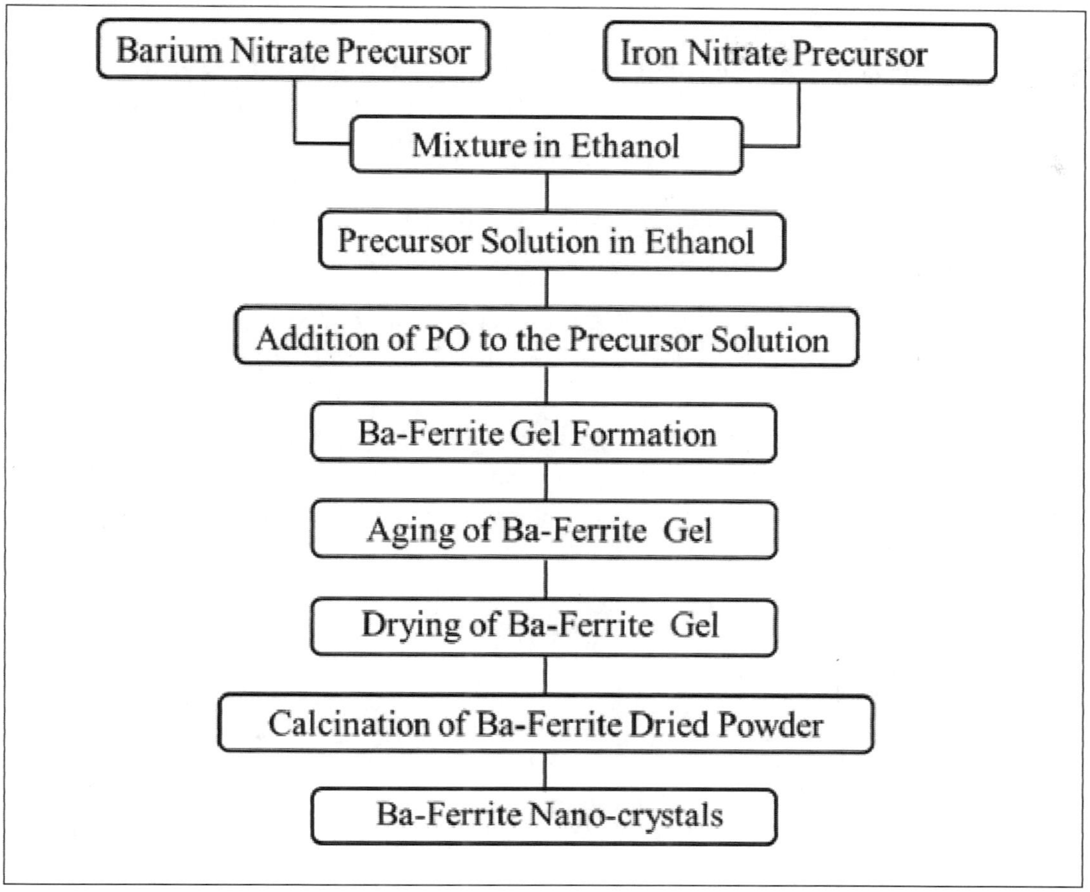

Fig. 1. Process flow diagram of sol-gel synthesis of Ba-ferrite.

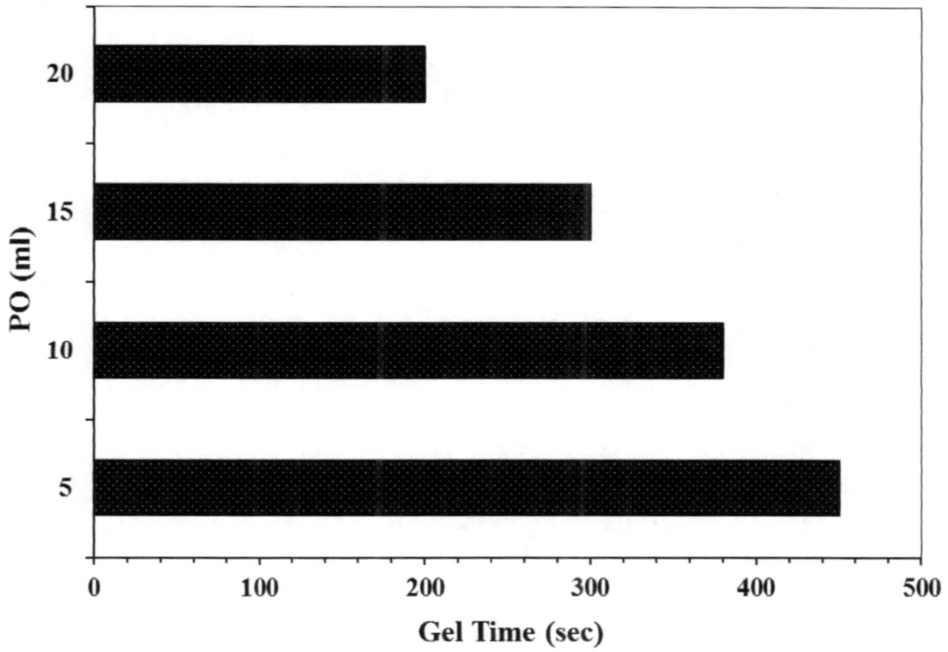

Fig. 2. Effect of PO volume used on gel time of Ba-ferrite.

By adding 5 ml of PO, the gel formation can be achieved in 630 sec. As the amount increase upto 10 ml, the gel time decreases to 565 sec. Furthermore increase in the amount of propylene oxide added induces a quick gel formation. For instance, when 15 and 20 ml of propylene oxide was added the gel was formed in 490 and 400 sec, respectively. Addition of propylene oxide as a proton scavenger helps in freezing the molecular chains and reducing the time required for gel formation.

In addition to the gel time, the phase composition and the CS of the Ba-ferrite gel prepared by adding different amount of propylene oxide and aged for 24 h was also determined. The aged gel was dried at 100°C and further calcined at 600°C in presence of air. The calcined powder was further characterized using a powder X-ray diffractometer. These XRD patterns indicate that due to the addition of different amounts of proton scavenger the phase composition of the Ba-ferrite does not altered. For instance, the XRD of the calcined Ba-ferrite powder is reported in Fig. 3.

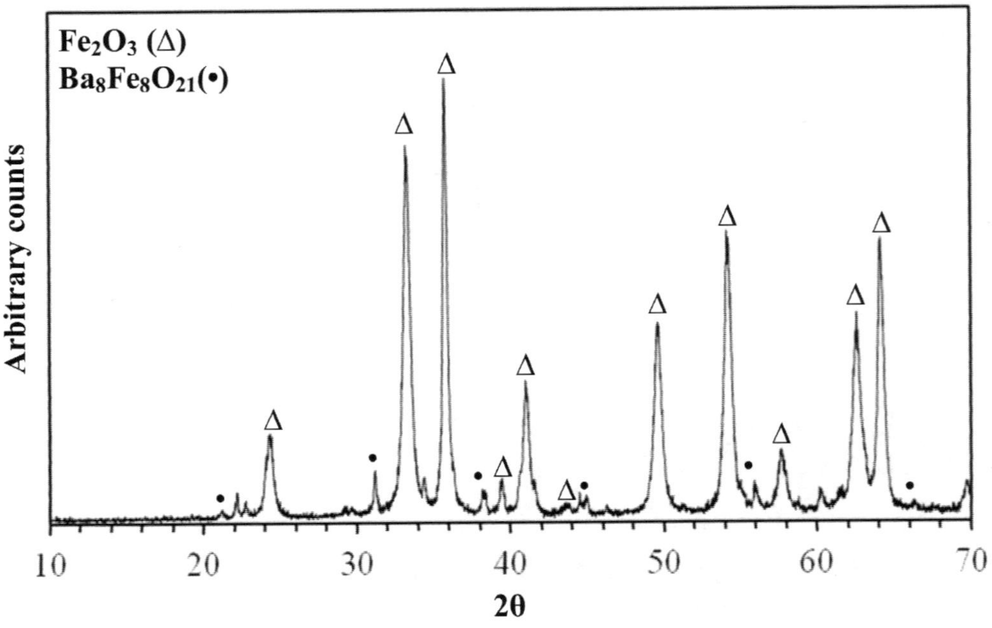

Fig. 3. Powder XRD pattern of Ba-ferrite (10% doping).

Furthermore, by utilizing results obtained via X-ray diffraction, the CS of the Ba-ferrite was calculated by using Scherrer's equation. Negligible variation in the CS of the Ba-ferrite as a function of propylene oxide added as a proton scavenger was observed. For instance, at PO = 5 ml the CS = 22 nm and at PO = 20 ml the CS = 21.45 nm. These results further indicate that the Ba-ferrite nanoparticles are formed.

After analysing the effect of amount of propylene oxide added as a proton scavenger on gel time, phase composition and C_S, the variation in the SSA and P_V was also explored. As shown in Fig. 4, the SSA of the sol-gel derived Ba-ferrite was observed to be increased with the increase in the amount of propylene oxide used. Initially, when 5 ml of propylene oxide was added, the

SSA of the derived Ba-ferrite was observed to be equal to 10 m^2/g. As the amount of propylene oxide added increases to 10, 15, and 20 ml, the SSA was also increased by a factor of 19, 27, and 35 m^2/g, respectively.

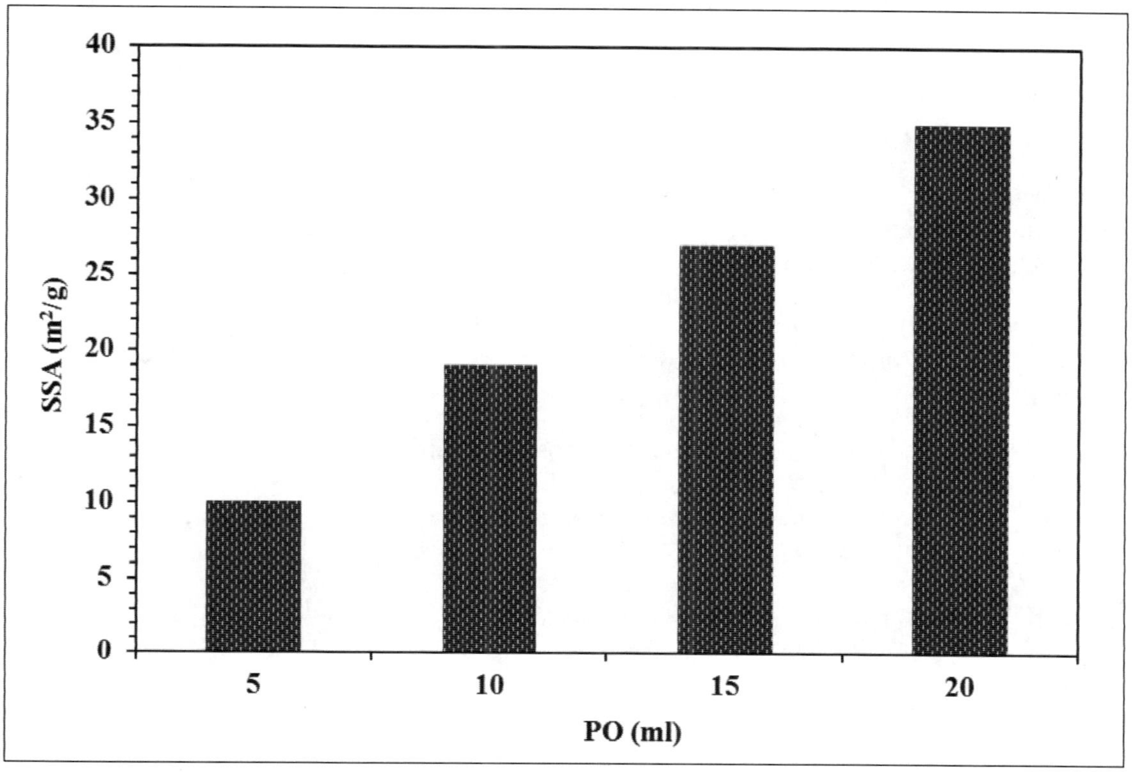

Fig. 4. Effect of PO volume used on SSA of Ba-ferrite.

4. Conclusion

The synthesis of Ba-ferrite (10% doping) was successfully carried out using sol-gel method. The effect of PO addition on physico-chemical properties of Ba-ferrite such as phase composition, gel time, CS, and SSA was studied. It was observed that with the increase in the PO concentration, the gel time decreases and SSA increases. In contrast to this, the phase composition and CS remained unchanged.

5. Acknowledgments

This publication was made possible by the NPRP grant (NPRP8-370-2-154) from the Qatar National Research Fund (a member of Qatar Foundation). The statements made herein are solely the responsibility of author(s).

References

[1] R. R. Bhosale. A. Kumar, F. AlMomani, I. Alxneit, [Sol-gel derived CeO_2–Fe_2O_3 nanoparticles: Synthesis, characterization and solar thermochemical application], *Ceram. Int*, **2016**, in press. doi:10.1016/j.ceramint.2016.01.042.

[2] J. Scheffe, A. Steinfeld, [Oxygen exchange materials for solar thermochemical splitting of H_2O and CO_2 : a review], *Mater. Today*, 2014, 17, 341 – 348.

[3] D. Dardor, R. R. Bhosale, S. Gharbia, A. AlNouss, A. Kumar, F. AlMomani, [Solar carbon production via thermochemical ZnO/Zn carbon dioxide splitting cycle], *J. Emerg. Trends Eng. Appl. Sci*, **2015**, 6, 129 – 135.

[4] D. Dardor, R. R. Bhosale, S. Gharbia, A. Kumar, F. AlMomani, [Solar thermochemical conversion of CO_2 into C via SnO_2/SnO redox cycle: Thermodynamic study], *Int. J. Eng. Res. Appl*, **2015**, 5, 134 – 140.

[5] R. R. Bhosale. A. Kumar, F. AlMomani, U. Ghosh, D. Dardor, Z. Bouabidi, M. Ali, S. Yousefi, A. AlNouss, M. S. Anis, M. H. Usmani, M. H. Ali, R. S. Azzam, A. Banu, [Solar co-production of samarium and syngas via methanothermal reduction of samarium sesquioxide], *Energy Convers. Manage*, **2016**, 112, 413 – 422.

[6] R. R. Bhosale. A. Kumar, F. AlMomani, S. Al-Muhtaseb, R. Gupta, I. Alxneit, [Assessment of $Ce_xZr_yHf_zO_2$ based oxides as potential solar thermochemical CO_2 splitting materials], *Ceram. Int*, **2016**, in press. doi:10.1016/j.ceramint.2016.02.100

[7] A. M. Turkey, [Catalytic oxidation of CO over pure and doped Cu-Ni-O System], *J.Applicable.Chem*, 2016, 5 (1), 219 – 230.

[8] V. K. Mahajan, G. H. Sonawane, [Improved photocatalytic activity of CeO_2 coupling ultrasound for Eosin-Y degradation], *J.Applicable.Chem*, 2015, 4 (5), 1500 – 1506.

[9] B. Suryanarayana, V. Raghavendra, K. ChandraMouli, [Structural, electrical and magnetic properties of $Ni_{0.5}Zn_{0.5}Fe_2O_4$ nanoscale particles by co-precipitation method], *J.Applicable.Chem*, 2015, 4 (4), 1237 – 1242.

[10] C. S. Thatte, M. V. Rathnam, M. S. Kumar, [Synthesis, characterization and application of Chitosan based Schiff base-transition metal complexes (Mn,Cu,Co,Ni)], *J.Applicable.Chem*, 2013, 2 (5), 1192 – 1206.

[11] B. R. Gaikwad, P. P. Khirade, D. V. Kurmude, A. B. Shinde, A. A. Pandit, K. M. Jadhav, [Comparative study of the structural and magnetic properties of magnesium ferrite prepared by ceramic and sol-gel auto combustion technique], *J.Applicable.Chem*, 2013, 2 (5), 1275 – 1280.

[12] M. R. Devi, C. Kannan, [Synthesis of unusual large pore Mg substituted alumino phosphate ($MgAlPO_4$) mesoporous molecular sieve and its catalytic activity], *J.Applicable.Chem*, 2013, 2 (3), 398 – 404.

[13] C. Agrafiotis, C. Pagkoura, A. Zygogianni, G. Karagiannakis, M. Kostoglou, A. G. Konstandopoulos, [Hydrogen production via solar-aided water splitting thermochemical cycles:

Combustion synthesis and preliminary evaluation of spinel redox-pair materials], *Int. J. Hydrogen Energy*, **2012**, 37, 8964 – 8980.

[14] N. Gokon, T. Kodama, N. Imaizumi, J, Umeda, T. Seo, [Ferrite/zirconia-coated foam device prepared by spin coating for solar demonstration of thermochemical water-splitting], *Int. J. Hydrogen Energy*, **2011**, 36, 2014 –2028.

[15] R. R. Bhosale, R. V. Shende, J. A. Puszynski, [H_2 generation from thermochemical water splitting using sol-gel derived Ni-ferrite], *J. Energy Power Eng,* **2010**, 4, 27 – 38.

[16] R. R. Bhosale, R. V. Shende, J. A. Puszynski, [H_2 generation from thermochemical water-splitting using sol-gel synthesized Zn/Sn/Mn-doped Ni-ferrite], *Int. Rev. Chem. Eng*, **2010**, 2, 852 – 862.

[17] F. Fresno, T. Yoshida, N. Gokon, R. Fernandez-Saavedra, T. Kodama, [Comparative study of the activity of nickel ferrites for solar hydrogen production by two-step thermochemical cycles], *Int. J. Hydrogen Energy*, **2010**, 35, 8503 – 8510.

[18] M. Roeb, N. Gathmann, M. Neises, C. Sattler, R. Pitz-Paal, [Thermodynamic analysis of two-step solar water splitting with mixed iron oxides], *Int. J. Energy Res*, **2009**, 33, 893 – 902.

[19] R. R. Bhosale, R. P. Khadka, J. A. Puszynski, R. V. Shende, [H_2 generation from two-step thermochemical water-splitting reaction using sol-gel derived $Sn_xFe_yO_z$], *J. Renewable Sustainable Energy*, 2011, 3, 063104-1 – 063104-12.

[20] R. R. Bhosale, R. V. Shende, J. A. Puszynski, [Thermochemical water-splitting for H_2 generation using sol-gel derived Mn-ferrite in a packed bed reactor], *Int. J. Hydrogen Energy*, **2012**, 37, 2924 – 2934.

[21] F. Fresno, R. Fernandez-Saavedra, M. Gomez-Mancebo, A. Vidal, M. Sanchez, M. Rucandio, A. Quejido, M. Romero, [Solar hydrogen production by two-step thermochemical cycles: Evaluation of the activity of commercial ferrites], *Int. J. Hydrogen Energy*, **2009**, 34, 2918 – 2924.

[22] M. Allendorf, R. Diver, N. Siegel, J. Miller, [Two-step water splitting using mixed-metal ferrites: thermodynamic analysis and characterization of synthesized materials], Energy Fuels, **2008**, 22, 4115 – 4124.

[23] R. R. Bhosale. A. Kumar, L.J.P. van den Broeke, S. Gharbia, D. Dardor, M. Jilani, J. Folady, M. Al-Fakih, M. Tarsad, [Solar hydrogen production via thermochemical iron oxide–iron sulfate water splitting cycle], *Int. J. Hydrogen Energy*, **2015**, 40, 1639 – 1650.

[24] T. Kodama, N. Gokon, R. Yamamoto, [Thermochemical two-step water splitting by ZrO_2-supported $Ni_xFe_{3-x}O_4$ for solar hydrogen production], *Sol. Energy*, **2008**, 82, 73 – 79.

[25] J. R. Scheffe, J. Li, A. W. Weimer, [A spinel ferrite/hercynite water-splitting redox cycle], *Int. J. Hydrogen Energy*, **2010**, 35, 3333 – 3340.

[26] J. E. Miller, M. Allendorf, R. Diver, L. Evans, N. Siegel, J. Stuecker, [Metal oxide composites and structures for ultra-high temperature solar thermochemical cycles], *J. Mater. Sci*, **2008**, 43, 4714 – 4728.

[27] T. Kodama, Y. Kondoh, R. Yamamoto, H. Andou, N. Satou, [Thermochemical two-step water splitting by ZrO_2-supported $Ni_xFe_{3-x}O_4$ for solar hydrogen production], *Sol. Energy*, **2008**, 78, 623 – 631.

[28] R. R. Bhosale. A. Kumar, F. AlMomani, I. Alxneit, [Propylene oxide assisted sol–gel synthesis of zinc ferrite nanoparticles for solar fuel production], *Ceram. Int*, **2016**, 42, 2431 – 2438.

[29] M. Neises, M. Roeb, M. Schmucker, C. Sattler, R. Pitz-Paal, [Kinetic investigations of the hydrogen production step of a thermochemical cycle using mixed iron oxides coated on ceramic substrates], *Int. J. Energy Res*, **2010**, 34, 651 – 661.

[30] S. Lorentzou, A. Zygogianni, K. Tousimi, C. Agrafiotis, A. G. Konstandopoulos, [Advanced synthesis of nanostructured materials for environmental applications], *J. Alloys Compd*, **2009**, 483, 302 – 305.

[31] R. R. Bhosale. A. Kumar, F. AlMomani, [Solar thermochemical hydrogen production via terbium oxide based redox reactions], *Int. J. Photoenergy*, **2016**, vol. 2016, Article ID 9727895, 9 pages, 2016. doi:10.1155/2016/9727895.

[32] S. Lorentzou, A. C. Agrafiotis, A. G. Konstandopoulos, [Aerosol spray pyrolysis synthesis of water-splitting ferrites for solar hydrogen production], *Granular Matter*, **2008**, 10, 113 – 122.

[33] H. Kaneko, T. Miura, A. Fuse, H. Ishihara, S. Taku, H. Fukuzumi, Y. Naganuma, Y. Tamaura, [Rotary-type solar reactor for solar hydrogen production with two-step water splitting process], *Energy Fuels*, **2007**, 21, 2287 – 2293.

Ti-base alloy coking behavior during steam cracking of ethane

Stamatis A. Sarris, Marie-Françoise Reyniers, Kevin M. Van Geem, Guy B. Marin
Ghent University, Laboratory for Chemical Technology, Technologiepark 914, 9052 Gent, Belgium.

Keywords: Ti-base alloy, carburization, steam cracking, pretreatment, cracking, thermal cracking, oxidation, ethane, coke formation, aging, jet stirred reactor, cyclic aging

* Corresponding author: Technologiepark 914, 9052 Gent, Belgium;
Kevin.VanGeem@Ugent.be

Introduction

One of the dominant processes to manufacture light olefins is steam cracking of hydrocarbons. The major problem occurring during this process is the coke deposition on the inner wall of the tubular crackers. The formed coke layer reduces the available reactant cross-sectional area causing elevated pressure drop along the reactor. These promote bimolecular over monomolecular reactions, decreasing the olefin selectivity. As a result, more energy is required to maintain the cracking severity and the desired selectivity[1, 2]. At a certain point, either when the metallurgic limits are reached or the pressure drop over the reactor is excessive, the operation is ceased and decoking of the reactor is necessary; typically for 48 h. The latter affects negatively the process finance[3]. In an effort to mitigate coke formation several anti-coking technologies have been evolved, namely 3-D reactor technologies[4-6], application of additives[7-10] and use of advanced surface technologies[11-17].

The present work, testing the cracking behavior of a Ti-base alloy, belongs clearly in the last technologies category. Titanium is used most of the times to enrich the composition of a high temperature alloy in concentrations of 0 to 10 wt %[18] aiming at an increased rigidity. However, to the best of the knowledge of the authors, it has not been used as base material for crackers, therefore no quantitative coking behavior observations are available. In order to evaluate the coking performance of this alloy under ethane steam cracking conditions, an experimental study was performed in a jet-stirred reactor thermogravimetric set-up. As a first step, the optimum pretreatment of Fe-Ni-Cr alloys was applied and compared with a pretreatment increased by 100 K temperature. This, based on the available literature[19-22], aimed at a better oxidation of the surface, thus at an improved coking behavior of the material (see Figure 1). According to that a factor 8 increase in the absolute weight of oxides is expected by increasing the temperature from 1173 K to 1273 K.

The target was partially achieved, the results proved decreased coking rates of the Ti-base alloy after the new treatment in comparison with the standard treatment used for Fe-Ni-Cr alloys, at the expense of the pronounced formation of Carbon oxides. However, the coking performance was still not better than the industrially representative alloy, while the tested Ti-base alloy coupons were partially or fully broken in pieces or slices after application of cyclic aging and cooling down back to ambient conditions. Therefore, the tested Ti-base alloy was not characterized as industrially interesting.

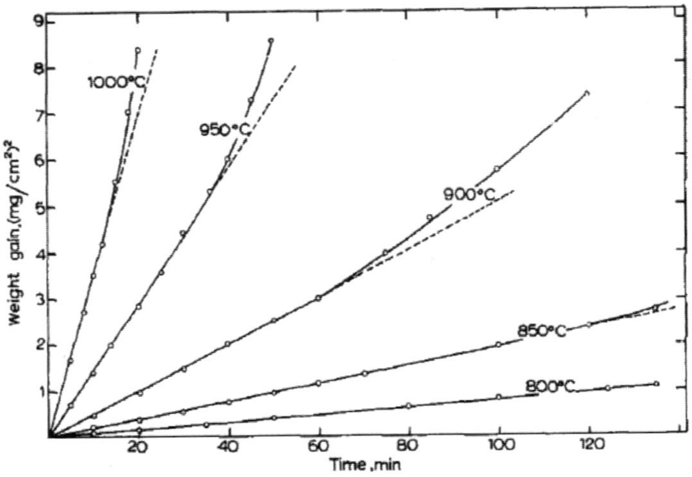

Figure 1. Effect of temperature on the oxides gain over time

Experimental Set-up

A picture of the jet-stirred reactor (JSR) is illustrated in Figure 2. The reactor is entirely made of quartz, including the nozzles and walls, minimizing coke formation on the inner surface. A small flat coupon is placed in the JSR, suspended from the arm of an electro-balance. Typically, the samples are cut by electro-erosion from the internal surface of industrial tubes to dimensions 10 mm x 8 mm x 1 mm. The coupons are polished before assembly to a Ra roughness of 2.6 μm.

The electro-balance can measure the weight difference of the coupons on-line with an accuracy of 1 μg s^{-1}, giving the advantage of actual coking rate values over time. The reactor effluent is quenched to prevent further cracking and coking and its composition is measured online with the aid of two gas chromatographs (GC). Nitrogen is used as internal standard, while the detection limit ranges from methane to naphthalene, satisfactory for Ethane cracking[23].

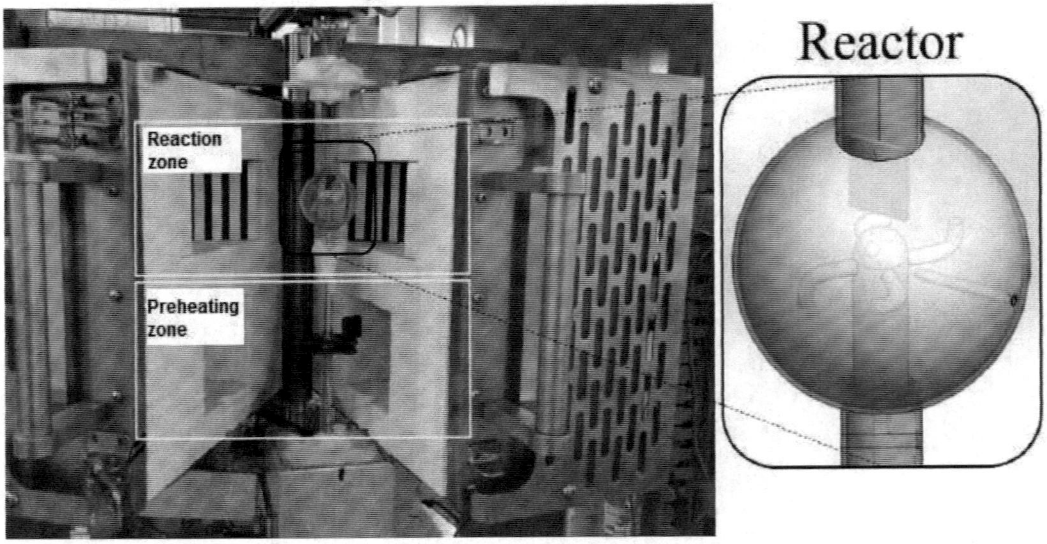

Figure 2. Jet stirred reactor used for coking behavior evaluation during steam cracking

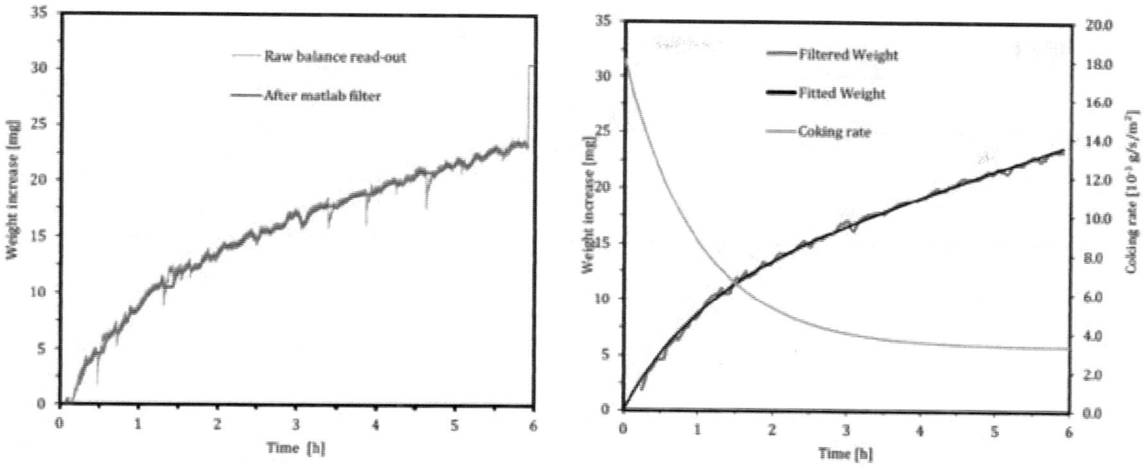

Figure 3. JSR data processing.

A graphical visualization of the data processing as executed by the MATLAB program is shown in Figure 3. In this work, the mean measured value between the initial 15 and 60 minutes is determined as initial coking rate characterizing the catalytic coking mechanism. The asymptotic or pyrolytic coking rate is related to the radical coking mechanism and is reported as the mean measured coking rate between the 5th and 6th hour of cracking.

The samples extracted after the experiment, are ex-situ analyzed by means of SEM and EDX. Similarly to Figure 4, surface and cross sectional analyses are performed aiming at a better understanding of the phenomena occurring during and before cracking.

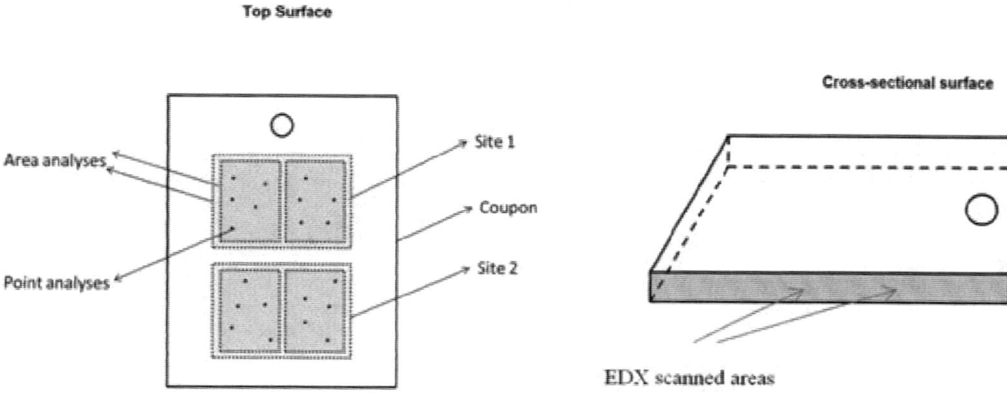

Figure 4. SEM and EDX analysis performed for the JSR samples

Experimental Results

The coking performance of the new Ti-base alloy under ethane steam cracking conditions was evaluated in the JSR set-up of LCT. Initially, the effect of the in-situ pretreatment was evaluated under blank ethane steam cracking conditions.

As mentioned above, a higher temperature pre-oxidation - 1273 K instead of 1173 K - than the typically applied for Fe-Ni-Cr alloys can increase significantly the absolute weight of oxides formed on Titanium[20, 21]. Based on previous work the formation of a uniform and thick oxide layer can improve the anti-coking behavior of a material[11, 12]. However, to the best of the knowledge of the authors, no coking rates results are available for Ti-base alloys.

Table 1 and

Table 2 summarize the experimental conditions and sequences under the two different pretreatments.

Table 1. Conditions of the two different applied pretreatments

Pretreatment	Id	Steps 1	Steps 2	Steps 3
Standard pretreatment	*A*	$6.7 \cdot 10^{-3}$ Nl s^{-1} of Air; 14h; T=1023 K	1:1 $8.15 \cdot 10^{-3}$ Nl s^{-1} of Air:N_2; 30 min; T=from 1023 K to 1173 K	$8.15 \cdot 10^{-3}$ Nl s-1 of Air and $6.67 \cdot 10^{-6}$ kg s^{-1} Steam; 15min ; T=1173 K
High Temperature Preoxidation + Steam Treatment at higher temperature	*B*	$6.7 \cdot 10^{-3}$ Nl s^{-1} of Air; 14h; T=1273 K	1:1 $8.15 \cdot 10^{-3}$ Nl s^{-1} of Air:N_2; 30 min; T=from 1023 K to 1273 K	$8.15 \cdot 10^{-3}$ Nl s^{-1} of Air and $6.67 \cdot 10^{-6}$ kg s^{-1} Steam; 15min ; T=1273 K

Table 2. Experimental sequence during the two performed pretreatments.

	1023 K Air	1023->1173 K N_2+Air	1173 K Air+H_2O	1173 K C_2H_6 +H_2O	1023->1173 K N_2+Air	1173 K Air +H_2O	1173 K C_2H_6 +H_2O	1023->1173 K N_2+Air	1173 K Air +H_2O	1173 -> 273 K He	
A	In situ Pre-oxidatio n 14 h	mild Pre-oxidation 30 min	Steam Treatme nt 15 min	1st coking cycle of 6 h	Decoking 30min	Steam Treatme nt 15 min	2nd-8th coking cycle of [2nd, 3rd and 8th of 6 h, 4-7th of 1 h]	Decoking 30 min	Steam Treatme nt 15 min	cooling down	time
	1273 K Air	1023->1273 K N_2+Air	1273 K Air+H_2O	1173 °C C_2H_6 +H_2O	1023->1273 K N_2+Air	1273 K Air +H_2O	1173 K C_2H_6 +H_2O	1023->1273 K N_2+Air	1273 K Air +H_2O	1273 -> 273 K He	
B	In situ Pre-oxidatio n 14 h	High T Pre-oxidation 30 min	Steam Treatme nt 15 min	1st coking cycle of 6 h	Decoking at higher T 30 min	Steam Treatme nt 15 min	2nd-8th coking cycle of [2nd, 3rd and 8th of 6 h, 4-7th of 1 h]	Decoking at higher T 30 min	Steam Treatme nt 15 min	cooling down	time

Table 3 illustrates the most important mass yields during ethane cracking. For a better comparison the effluent results of a typically used material are also included in the presented yields. A factor 10 and 300 increased values of CO, for the typical and higher temperature pretreatment respectively, were measured over the 8 cracking cycles in comparison with the Fe-Ni-Cr alloy. For CO_2, the effect was less pronounced; only after application of the high temperature pre-oxidation (b) an increase of a factor 3 was observed. In addition, the increase of Carbon oxides is followed by an increase of H_2. A minor decrease of CH_4 is also noticed, however it is still within the error margin of measured values. The observations support the idea that Ti and/or Ti-oxides present on the surface have a catalytic effect towards coke gasification. These results already prove that the Ti-base alloy is significantly worse than the reference material.

Table 3. Averaged (over the 8 cracking cycles) mass product yields. Ethane steam cracking: F_{HC}= 29.18 x 10^{-6} kg s^{-1}, δ= 0.33 kg_{H2O} $kg^{-1}{}_{HC}$, $T_{reactor}$= 1173.15 K, P= 101.35 kPa, F_{H2O}=9.72 x 10^{-6} kg s^{-1}.

Components [wt%]	**Fe-Ni-Cr alloy A**	**Ti-base alloy A**	**Ti-base alloy B**
H_2 [±0.07]	4.24	4.39	4.45
CO_2 [±0.003]	0.006	0.005	0.019
CO	0.05 [1]	0.56 [2]	1.87 [3]
CH_4 [±0.21]	6.99	6.97	6.91
C_2H_6 [±0.52]	30.13	30.19	30.06
C_2H_4 [±0.25]	49.86	49.61	49.57
C_3H_8 [±0.03]	0.11	0.11	0.11
C_3H_6 [±0.02]	0.74	0.70	0.69
C_2H_2 [±0.05]	1.38	1.46	1.41
1,3-C_4H_6 [±0.07]	2.00	1.93	1.89
Benzene [±0.13]	2.46	2.53	2.53

[1] [±0.008], [2] [±0.07], [3] [±0.35]

Nevertheless, coking rates results are also processed and summarized in Table 4. For a qualitative comparison the results of reference material during continuous addition of DMDS are presented as well. Depending on the cracking cycle the Ti-base alloy performs significantly worse than the reference material. The increased temperature pretreatment improves the coking rates of the Ti-base alloy. However, this improved coking behavior still keeps the Ti-base alloy out of competition against the blank reference coking behavior. In terms of anti-coking performance, the Ti-base alloy is placed between the reference during steam cracking with continuous addition of DMDS and the reference with no DMDS.

Additionally, the reference material maintains a rather stable coking behavior over cyclic aging, while the sample of the Ti-base alloy shows increased coking after several cycles.

Table 4. Calculated initial (average 15min-60min) and asymptotic (average 5h-6h) coking rates. Ethane steam cracking: F_{HC}= 29.18 x 10^{-6} kg s^{-1}, δ= 0.33 kg_{H2O} $kg^{-1}{}_{HC}$, $T_{reactor}$= 1173.15 K, P= 101.35 kPa, F_{H2O}=9.72 x 10^{-6} kg s^{-1}.

Alloy Conditions	Fe-Ni-Cr alloy Blank	Ti-base alloy Blank	Ti-base alloy Blank	Fe-Ni-Cr alloy CA
Pretreatment	A	A	B	A
Cracking Temperature (K)	1173	1173	1173	1173
dilution	0.33	0.33	0.33	0.33
CA DMDS (ppmw S per HC)	0	0	0	41
Cc	Coke formed (mg)			
1	1.29	5.48	3.66	7.55
2	1.26	9.01	4.99	8.38
3	1.41	7.43	4.82	8.66
4	0.60	3.57	2.93	4.12
5	0.67	3.03	2.17	4.55
6	0.69	2.97	2.13	4.53
7	0.47	2.95	2.32	4.51
8	1.19	7.69	5.35	12.46
Cc	Initial coking rate [10^{-6} kg/s/m^2]			
1	0.71	1.73	1.23	3.03
2	0.67	3.26	1.81	4.74
3	0.86	3.65	2.59	4.23
4	0.85	3.80	3.11	5.85
5	0.96	3.91	2.76	6.45
6	0.97	3.80	2.70	6.42
7	0.67	3.79	2.79	6.39
8	0.63	3.27	2.46	5.29
Cc	Asymptotic coking rate [10^{-6} kg/s/m^2]			
1	0.22	1.06	0.61	1.53
2	0.22	1.58	0.83	1.43
3	0.23	0.93	0.53	1.61
8	0.21	1.25	0.74	2.47

The most discouraging result obtained by the Ti-base Alloy coupons was their behavior after cooling down to ambient conditions. Most specifically, the coupons broke either in slices or in smaller pieces. In Figure 5, an indication of the thickness decrease after cracking and of the thickness of the slices is given. Clearly, thick oxides are formed during oxidation of Ti, making the coupon more brittle as material towards mechanical stresses. The latter leads to the formation of cracks due to the mechanical stresses evolved during cooling down. An additional example of a cracked coupon is given in Figure 6. For a better understanding of this phenomenon SEM and EDX analysis are performed.

The cross sectional analyses of the two samples treated at the two different temperatures reveal part of the phenomenon occurring in this study (Figure 7). In the standard temperature, 1173 K, 0.1 to 0.2 μm thick oxides are formed and spalling off the coupon from both sides leaving no oxide covering the surface. At 1273 K, thick oxides are formed leading to oxidation of almost the whole coupon volume. Clearly the oxides that are formed are quite brittle. Thus, they are separating from the core of the coupon. In the case of the increased temperature oxidation, the core is too thin, therefore the crack is propagating causing breakage of the coupon in smaller pieces.

Combining the above experimental results with the formation of cracks on the samples' surface can give a fruitful insight on the phenomenon. Based on the increased values of carbon oxides and H_2 differences of the yields, an increased catalytic surface of the Ti-base alloy samples can be assumed. That most probably derived from the presence of the surface cracks. As a result both coke formation and coke gasification were catalyzed.

Figure 5. Thickness values of the Ti-base alloy before cracking (left), after cracking (middle) and of a broken slice (right), pretreated at 1173 K.

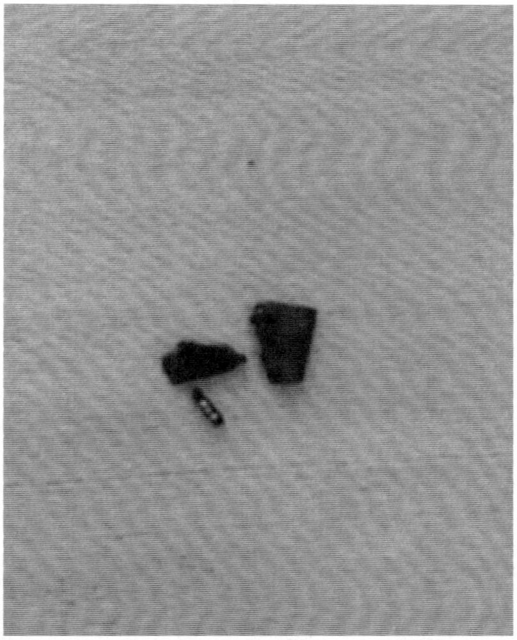

Figure 6. Broken Ti-base alloy coupon after cracking pretreated at 1273 K.

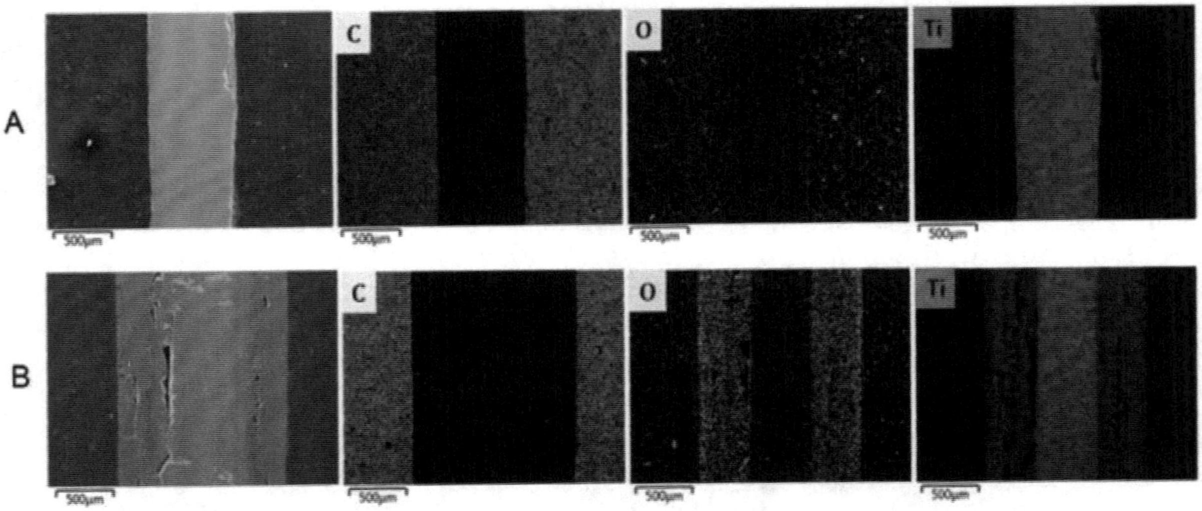

Figure 7. SEM and EDX cross sectional analysis of the two pretreatments. A is referring to temperature of 1173 K and B to 1273 K.

Conclusions

A newly tested Ti-base alloy was evaluated under industrially relevant ethane steam cracking conditions. Outrageously increased values of CO, both for the standard and higher temperature pretreatment, were measured over the 8 cracking cycles in comparison with a typical Fe-Ni-Cr alloy under the same experimental conditions. Notable increase of the CO_2 was observed between the higher temperature pre-oxidation (b) and the typical pre-oxidation used for Fe-Ni-Cr (a) when applied to the Ti-base alloy. No significant difference was observed for the CO_2 values of Ti-base alloy in comparison with the reference material in the standard treatment.

The increase of Carbon oxides is followed by a slight increase of H_2, supporting the idea that Ti and Ti-oxides are catalytic towards coke gasification. The significant differences in terms of gas phase composition can be explained by the increased available surface interacting with the gas phase. Clearly at the points of crack formation coke formation and gasification was catalyzed, affecting the gas phase product distribution.

Additional surface analysis will lead to a better understanding of the crack formation phenomenon observed in these alloys. Judging by these experimental results, the use of the Ti-base alloy in an industrial cracker is considered highly risky and certainly not beneficial in terms of coking or carbon oxides formation.

References

1. Rao, M. R.; Plehiers, P. M.; Froment, G. F., The coupled simulation of heat transfer and reaction in a pyrolysis furnace. *Chemical engineering science* **1988,** 43, (6), 1223-1229.
2. Stefanidis, G.; Merci, B.; Heynderickx, G. J.; Marin, G. B., CFD simulations of steam cracking furnaces using detailed combustion mechanisms. *Computers & chemical engineering* **2006,** 30, (4), 635-649.
3. Heynderickx, G.; Schools, E.; Marin, G., Optimization of the decoking procedure of an ethane cracker with a steam/air mixture. *Industrial & engineering chemistry research* **2006,** 45, (22), 7520-7529.
4. Brown, D. J., Internally finned radiant coils: A valuable tool for improving ethylene plant economics. In *6th EMEA Petrochemicals Technology Conference*, London, UK, 2004.
5. Albano, J.; Sundaram, K.; Maddock, M. *Application of extended surfaces in pyrolysis coils*; New York, NY; American Institute of Chemical Engineers: 1988.

6. Schietekat, C. M.; van Goethem, M. W. M.; Van Geem, K. M.; Marin, G. B., Swirl flow tube reactor technology: An experimental and computational fluid dynamics study. *Chemical Engineering Journal* **2014,** 238, (0), 56-65.
7. Reyniers, M.-F. S. G.; Froment, G. F., Influence of Metal Surface and Sulfur Addition on Coke Deposition in the Thermal Cracking of Hydrocarbons. *Industrial & Engineering Chemistry Research* **1995,** 34, (3), 773-785.
8. Wang, J.; Reyniers, M.-F.; Marin, G. B., Influence of Dimethyl Disulfide on Coke Formation during Steam Cracking of Hydrocarbons. *Industrial & Engineering Chemistry Research* **2007,** 46, (12), 4134-4148.
9. Wang, J.; Reyniers, M.-F.; Marin, G. B., The influence of phosphorus containing compounds on steam cracking of n-hexane. *Journal of Analytical and Applied Pyrolysis* **2006,** 77, (2), 133-148.
10. Wang, J.; Reyniers, M.-F.; Van Geem, K. M.; Marin, G. B., Influence of Silicon and Silicon/Sulfur-Containing Additives on Coke Formation during Steam Cracking of Hydrocarbons. *Industrial & Engineering Chemistry Research* **2008,** 47, (5), 1468-1482.
11. Muñoz Gandarillas, A. E.; Van Geem, K. M.; Reyniers, M.-F.; Marin, G. B., Influence of the Reactor Material Composition on Coke Formation during Ethane Steam Cracking. *Industrial & Engineering Chemistry Research* **2014,** 53, (15), 6358-6371.
12. Muñoz Gandarillas, A. E.; Van Geem, K. M.; Reyniers, M.-F.; Marin, G. B., Coking Resistance of Specialized Coil Materials during Steam Cracking of Sulfur-Free Naphtha. *Industrial & Engineering Chemistry Research* **2014,** 53, (35), 13644-13655.
13. Verdier, G.; Carpentier, F., Consider new materials for ethylene furnace applications: An innovative metallurgy solves maintenance issues. *Hydrocarbon processing* **2011,** 90, (5), 61-62.
14. Wolpert, P.; Ganser, B.; Jakobi, D.; Kirchheiner, R., Using steam; tube with uniform cross-section; pyrolysis; antideposit agents; interior fins; swirling gas flow. In Google Patents: 2005.
15. Jakobi, D.; van Moesdijk, C.; Karduck, P.; von Richthofen, A., Tailor-made materials for high temperature applications: New strategies for radiant coil material development. *CORROSION 2009* **2009**.
16. Jakobi, D.; Karduck, P.; von Richthofen, A. F., The High-Temperature Corrosion Resistance of Spun-Cast Materials for Steam-Cracker Furnaces-A Comparative Study of Alumina-and Chromia-Forming Alloys. *CORROSION 2013* **2013**.
17. Asteman, H.; Hartnagel, W.; Jakobi, D., The Influence of Al Content on the High Temperature Oxidation Properties of State-of-the-Art Cast Ni-base Alloys. *Oxidation of Metals* **2013,** 80, (1-2), 3-12.
18. Petrone, S. S. A.; Mandyam, R. C.; Wysiekierski, A. G., Surface alloyed high temperature alloys. In Google Patents: 2000.
19. Kumar, S.; Sankara Narayanan, T. S. N.; Ganesh Sundara Raman, S.; Seshadri, S. K., Thermal oxidation of Ti6Al4V alloy: Microstructural and electrochemical characterization. *Materials Chemistry and Physics* **2010,** 119, (1–2), 337-346.
20. Kofstad, P.; Anderson, P. B.; Krudtaa, O. J., Oxidation of titanium in the temperature range 800–1200°C. *Journal of the Less Common Metals* **1961,** 3, (2), 89-97.
21. Kofstad, P., High-temperature oxidation of titanium. *Journal of the Less Common Metals* **1967,** 12, (6), 449-464.
22. Kumar, S.; Narayanan, T. S. N. S.; Raman, S. G. S.; Seshadri, S. K., Thermal oxidation of CP-Ti: Evaluation of characteristics and corrosion resistance as a function of treatment time. *Materials Science and Engineering: C* **2009,** 29, (6), 1942-1949.
23. van Goethem, M. W. M.; Barendregt, S.; Grievink, J.; Verheijen, P. J. T.; Dente, M.; Ranzi, E., A kinetic modelling study of ethane cracking for optimal ethylene yield. *Chemical Engineering Research and Design* **2013,** 91, (6), 1106-1110.

469612 Fermentation Design and Gas Transfer Considerations for Biochemical Methane Conversion

Abstract:

Fermentation Design and Gas Transfer Considerations for Biochemical Methane Conversion

Kyle Stone, Matthew Hilliard, Q. Peter He, and Jin Wang

Department of Chemical Engineering, Auburn University, Auburn, AL, 36849, USA

Methane is an essential component of the global carbon cycle and one of the most powerful greenhouse gases (GHGs) with a global warming potential over 20 times that of CO_2. Yet, methane is a rich source of carbon and energy that is utilized for heat, fuel synthesis, and valued chemical production. In addition, there is a rapid expansion of global methane supply predominantly because of increased production of natural gas [1] and biogas [2]. Due to concerns about climate change, abundance of supply, and favorable economic conditions, there has been growing interest in biological processes as a sustainable and efficient way of converting methane to valuable products while reducing GHG emissions.

These biological processes utilize the largely aerobic bacteria called methanotrophs that consume methane as their sole source for carbon and energy. Due to their unique metabolism, methanotrophs can convert methane at ambient temperatures and pressures to many different products. Specifically, methanotrophs have the natural ability to produce methanol, valued proteins, biopolymers, lipids (for biodiesel), organic acids, sucrose, and ectoine [3,4]. With genetic modification, methanotrophs can also produce non-native products such as carotenoids, isoprene, 1,4 butanediol, and farnesene [3,4]. Several of these products can potentially be simultaneously generated with specifically designed processes [4].

However, all processes with methanotrophs face several challenges, such as clean sources of methane, biocatalyst modification, and process design. For process design, the most influential factor is efficient gas-liquid transfer (i.e., methane and/or oxygen gas transfer to the liquid medium where cells reside). The gas-liquid transfer of the poorly soluble methane and oxygen can be strongly affected by reactor design and mass transfer enhancing agents in the liquid medium.

In this work we review recent progress about effective gas transfer for bioprocesses with methanotrophs. One focus will be on the quantification of methane transfer rate and cellular uptake rate. We will also provide a detailed review and discussion of the bioreactor configurations and the influence of these systems on gas-liquid transfer. Finally, we will discuss gas-liquid transfer enhancement with promoting agents. Recommendations will be provided and potential research areas for methane conversion through methanotrophs will be discussed.

[1] International Energy Statistics: Natural gas proved reserves, (2015). http://www.eia.gov/cfapps/ipdbproject/IEDIndex3.cfm?tid=3&pid=3&aid=6.

[2] Biogas Opportunities Roadmap, U.S. Department of Agriculture, U.S. Environmental Protection Agency, U.S. Department of Energy, 2014.

[3] P.J. Strong, S. Xie, W.P. Clarke, Methane as a resource: can the methanotrophs add value?, Environ. Sci. & Technol. (2015).

[4] P.J. Strong, M. Kalyuzhnaya, J. Silverman, W.P. Clarke, A methanotroph-based biorefinery: potential scenarios for generating multiple products from a single fermentation, Bioresour. Technol. (2016).

460680 How Do Electrolyte Cations Affect Activity and Selectivity of the Electrochemical Reduction of CO_2 over Ag and Cu?

Abstract:

The electrochemical conversion of water and CO_2 to renewable fuels offers a sustainable route to support future energy demands and store intermittent solar energy. The efficiency of such conversion over metals is strongly affected by electrolyte cations. Previous attempts to explain this phenomenon have not been successful. We show here that cation size affects the activity and selectivity of Ag and Cu for CO_2 reduction, and we propose that the observed effects can be attributed to hydrolysis of hydrated cations in the immediate vicinity of the cathode. With increasing cation size, the pKa for hydrolysis decreases and is sufficiently low for K^+ Rb^+, and Cs^+ for these hydrated cations to serve as buffering agents. Buffering lowers the pH near the cathode leading to an increase in the local concentration of dissolved CO_2. The consequences of these changes are an increase in cathode activity, a decrease in faradaic efficiencies for H_2 and CH_4 and an increase in faradaic efficiencies for CO, C_2H_4, and C_2H_5OH in full agreement with experimental observations for CO_2 reduction over Ag and Cu. The present study further reveals that hydrolysis of hydrated cations can be used to increase the activity and selectivity of any proton-transfer reaction.

472181 Understanding the Influence of [EMIM] Cl on the Suppression of the Hydrogen Evolution Reaction on Transition Metal Electrodes

Abstract:

Ionic liquids have recently been suggested to suppress the hydrogen evolution reaction (HER), as well as increase selectivity and activity for CO_2 reduction to high value chemicals and fuels. To learn more about the origin of this effect, we have studied the influence of the ionic liquid 1-ethyl-3-methylimidazolium chloride ([EMIM] Cl) on HER on Ag, Cu and Fe electrodes in acidic and basic media. In the absence of CO_2, HER is suppressed on all three electrodes in acidic media; no such suppression is observed in basic media. Furthermore, we used ^1H NMR spectroscopy to identify compounds that are formed as the [EMIM] Cl breaks down at both the working and counter electrodes under electrochemical reaction conditions.

463736 Effects of Temperature and Gas-Liquid Mass Transfer on the Operation of Small Electrochemical Cells for the Quantitative Evaluation of CO_2 Reduction Electrocatalysts

Abstract:

In the last few years, there has been increased interest in electrochemical CO_2 reduction (CO2R). Many experimental studies employ a membrane separated, electrochemical cell with a mini H-cell geometry to characterize CO2R catalysts in aqueous solution. Gas products are measured periodically by sampling the effluent CO_2, which is continuously sparging the cell, while liquid products are measured by batch sampling of the electrolyte. The sensitivity of this approach for the detection of liquid products can be increased by increasing the catalyst surface area to electrolyte volume ratio (*S/V*) of the cell. However, we show that operating cells with high *S/V* at high current densities can have subtle consequences, especially for the cell temperature and electrolyte CO_2 concentration.

Both effects were evaluated quantitatively in high *S/V* cells using Cu electrodes and a bicarbonate buffer electrolyte. Electrolyte temperature is a function of the current and total voltage passed through the cell as well as the cell geometry. Even at very high current density, 20 mA cm^{-2}, the measured temperature rise was less than 4 °C and a decrease of <10% in the dissolved CO_2 concentration is predicted. In contrast, limits on the CO_2 gas-liquid mass transfer into the cells produce much larger effects. By using the pH in the cell to measure the dissolved CO_2 concentration, significant undersaturation of CO_2 is observed, even at more modest current densities of 10 mA cm^{-2}.

Importantly, undersaturation of CO_2 produces large changes in the faradaic efficiency product distribution observed on Cu electrodes, with H_2 production becoming increasingly favored as the CO_2 undersaturation worsens. We show that the size of the CO_2 bubbles being introduced into the cell is critical for maintaining the equilibrium CO_2 concentration in the electrolyte, and we have designed a high S/V cell that is able to maintain the equilibrium CO_2 concentration at current densities up to 15 mA cm^{-2}.

464666 Surface Structure Engineering of Cu Thin Films for Electrochemical CO_2 Reduction

Abstract:

Global dependence on fossil fuels as energy sources and the alarming increase of greenhouse gas emissions has necessitated the development of carbon-free and carbon-neutral renewable energy sources for the future. The sequestration of CO_2 emissions along with subsequent electrochemical reduction into fuel products, forms a carbon-neutral synthetic fuel cycle which could potentially be streamlined into existing fuel infrastructures. To date, only Cu has displayed any propensity as a catalyst to electrochemically reduce CO_2 into longer chain hydrocarbons. Previous studies on Cu singles crystals have shown large facet sensitivities for electrochemical CO_2 reduction product selectivity. The aim of our research is to utilize these surface structure motifs to design active and selective catalysts that are amenable to device integration.

To this end, we use physical vapor deposition to synthesize Cu thin films in both low and high Miller index orientations for electrochemical CO_2 reduction. X-ray pole figures demonstrate that Cu thin films can be epitaxially grown in the <100>, <111>, and <751> orientations by utilizing the orientation of the single crystal substrate. The different Cu thin film orientations were subsequently tested for their CO_2 reduction activity and selectivity using electrochemical measurements in tandem with gas phase (gas chromatography) and liquid phase (nuclear magnetic resonance) product detection methods. Here, we will demonstrate how surface structure engineering can guide selectivity for C-C coupling and oxygenate formation during electrochemical CO_2 reduction.

458210 Electrochemical Reduction of CO_2 over Phase-Segregated Cuag Bimetallic Electrodes with Enhanced Oxygenate Selectivity By CO Spillover

Abstract:

The use of solar energy to produce fuels and commodity chemicals from water and carbon dioxide (CO_2), present either in the atmosphere or dissolved in seawater, represents an opportunity to obtain these products in a sustainable manner. One approach to this goal would be to convert solar energy into electrical energy using a photovoltaic device and to then utilize the generated electricity to drive an electrochemical cell for CO_2 reduction (CO_2R).[1–4] Within such an electrochemical cell water is oxidized at the anode and CO_2 is reduced at the cathode. The electrocatalyst used as the cathode will largely define the overall efficiency of the process as well as the composition of the generated products. To date, extensive research has shown that the only metal capable of producing hydrocarbons and oxygenated products from CO_2 with Faradaic efficiencies (FEs) in excess of 1% is Cu.[5] The principal products formed over Cu with high FEs (> 10%) are H_2, $HCOO^-$, CO, CH_4, C_2H_4, and C_2H_5OH.[6,7] The distribution of these products is sensitive to many variables, including the surface morphology of the Cu cathode, the electrolyte composition, and the applied potential. While alloying Cu with other transition metals has been identified as a promising method of tuning the product distribution obtained during CO_2R,[8] no studies in the current literature have observed a positive effect of doing so on the FEs of potential fuels.

Herein we report our investigations of electrochemical CO_2 reduction over phase-segregated CuAg bimetallic electrodes, which we have observed to be the only metallic electrocatalysts yet discovered that are more selective for the formation of multi-carbon oxygenates than hydrocarbons. The selectivity of oxygenates relative to hydrocarbons is found to scale with the relative distribution of Cu and Ag facet terminations

present at the electrode surface at steady state. The unique selectivity observed over these bimetallic electrodes is explained in terms of CO spillover from the Ag phase to the Cu phase, which results in an enhanced steady state coverage of CO adsorbed to the Cu facet terminations present at the electrode surface. The enhanced coverage of this key reaction intermediate on Cu increases the rate of C-C coupling at the expense of C-H bond formation, resulting in an improved selectivity to multi-carbon oxygenated products at the expense of hydrogen and hydrocarbons.

References

1. Jitaru, M., Lowy, D. A., Toma, M., Toma, B. C. & Oniciu, L. Electrochemical Reduction of Carbon Dioxide on Flat Metallic Cathodes. *J. Appl. Electrochem.* **27,** 875–889 (1997).

2. Gattrell, M., Gupta, N. & Co, A. A Review of the Aqueous Electrochemical Reduction of CO2 to Hydrocarbons at Copper. *J. Electroanal. Chem.* **594,** 1–19 (2006).

3. Hori, Y. in *Mod. Asp. Electrochem.* (Vayenas, C. G., White, R. E. & Gamboa-Aldeco, M. E.) 89–189 (Springer, 2008). doi:10.1007/978-0-387-49489-0_3

4. Whipple, D. T. & Kenis, P. J. A. Prospects of CO2 Utilization via Direct Heterogeneous Electrochemical Reduction. *J. Phys. Chem. Lett.* **1,** 3451–3458 (2010).

5. Hori, Y., Kikuchi, K. & Suzuki, S. Production of CO and CH4 in Electrochemical Reduction of CO2 at Metal Electrodes in Aqueous Hydrogencarbonate Solution. *Chem. Lett.* 1695–1698 (1985).

6. Hori, Y., Murata, A. & Takahashi, R. Formation of Hydrocarbons in the Electrochemical Reduction of Carbon Dioxide at a Copper Electrode in Aqueous Solution. *J. Chem. Soc. Faraday Trans. 1* **85,** 2309–2326 (1989).

7. Kuhl, K. P., Cave, E. R., Abram, D. N. & Jaramillo, T. F. New Insights into the Electrochemical Reduction of Carbon Dioxide on Metallic Copper Surfaces. *Energy Environ. Sci.* **5,** 7050–7059 (2012).

8. Hansen, H. A., Shi, C., Lausche, A., Peterson, A. & Nørskov, J. K. Bifunctional alloys for the electroreduction of CO_2 and CO. *Phys. Chem. Chem. Phys.* 9194–9201 (2016). doi:10.1039/C5CP07717F

464322 Interaction of Anion-Exchange Ionomers with Carbon Dioxide Reduction Electrocatalysts

Abstract:

Electrocatalytic reduction of carbon dioxide may provide a path to generating carbon-neutral liquid fuels and feedstocks that at present are derived from petroleum. Practical devices for carbon dioxide reduction will require membrane-electrode assemblies, similar to those used in many commercial hydrogen fuel cells, to maximize energy efficiency. Membrane-electrode assemblies consist of gas diffusion layers and catalyst layers on either side of a membrane. The catalyst layer must facilitate transport of electrons, gases such as carbon dioxide and products, and ionic species such as hydroxide to and from the surface of nanoparticle catalysts. Ionic transport is achieved using ionomers which must be in direct contact with the catalyst surface; despite this, the influence of the ionomer on the underlying catalystâ€™s activity and selectivity for electrochemical carbon dioxide reduction is poorly understood. In this presentation, we will discuss how anion-exchange ionomers influence transition metal electrocatalysts for carbon dioxide reduction. Using electrochemical and spectroscopic studies, we are able to elucidate the molecular interaction between the ionomer and the catalyst surface. Our results provide insights into the design of new anion-exchange ionomers for electrochemical carbon dioxide reduction.

470521 Electrochemical Reduction of Carbon Dioxide Using Solid Oxide Electrolysis Cells

Abstract:

Extensive use of fossil fuels and consequential high levels of carbon dioxide (CO_2) emissions are major contemporary challenges.[1] The most direct way to mitigate this process is to activate the reverse chemical pathways in which CO_2 is reduced into high-energy molecules.[2] The electrochemical reduction of CO_2 to CO has attracted increasing attention, since CO represents a valuable intermediate to the production of synthetic fuels using established processes, such as Fischer-Tropsch.[3-6] One approach to electrochemically reduce CO_2 to CO is using solid oxide electrolysis cells (SOECs). SOECs are solid-state electrochemical systems that, in principle, can facilitate the electrochemical reduction of CO_2 to CO with potentially very high rates due to the favorable kinetics at high operating temperatures.[7] While electrochemical reduction of CO_2 using SOECs offers a great deal of promise, these systems are limited by the inability to operate near the thermodynamic reversible potentials due to the activation overpotential losses.[1] In the present work, we combine experimental and theoretical techniques to determine the factors that lower the overpotential losses in SOECs during electrolysis of CO_2. We find that the conventional Ni-yttria stabilized zirconia (Ni-YSZ) composite cathodes exhibit significant overpotential losses toward electrochemical reduction of CO_2. Incorporation of gadolinium-doped ceria (GDC) improves the SOEC performance. We have conducted controlled electrochemical studies to investigate the influence of gadolinium-doped ceria in the electrochemical reduction of CO_2.

REFERENCES

1. S.D. Ebbesen, S.H. Jensen, A. Hauch, M.B. Mogensen, Chem. Rev. 114 (2014) 10697-10734.

2. B.P. Sullivan, K. Krist, H. Guard, Elsevier, 2012.

3. C. Sanchez-Sanchez, V. Montiel, D. Tryk, A. Aldaz, A. Fujishima, Pure Appl. Chem. 73 (2001) 1917-1927.

4. K. Xie, Y. Zhang, G. Meng, J.T. Irvine, Energy Environ. Sci. 4 (2011) 2218-2222.

5. E. Iglesia, Appl. Catal., A. 161 (1997) 59-78.

6. D. Wilhelm, D. Simbeck, A. Karp, R. Dickenson, Fuel Process. Technol. 71 (2001) 139-148.

7. C. Graves, S.D. Ebbesen, M. Mogensen, Solid State Ionics, 192 (2011) 398-403.

476892 Overview of U.S. DOE Environmental and Waste Cleanup Programs

Abstract:

The Department of Energy has a liability for the safe cleanup of the environmental legacy brought about from decades of nuclear weapons development and government-sponsored nuclear energy research. The Departmentâ€™s Office of Environmental Management (EM) is responsible for this cleanup function and over the 25-plus years of its existence have safely cleaned up 91 sites, leaving 16 sites, with these remaining 16 sites posing the greatest technical and programmatic challenges. Four of the remaining 16 sites are tank waste sites where wastes resulting from reprocessing of spent fuel are stored in over 200 large underground tanks, containing approximately 100 gallons of highly radioactive and chemically toxic wastes. EM has a nominal $6 Billion annual budget with approximately 35% of the budget targeted to safe management, treatment and disposition of tank waste as well as closure of the tanks

The four sites include the Savannah River Site, Hanford, the Idaho National Laboratory (INL) and the West Valley Demonstration Project (WVDP). Progress in waste retrieval, treatment and disposition vary among the sites, with the treatment mission essentially complete at WVDP and not yet substantially started at INL and Hanford. To accomplish this tank waste cleanup mission, EM and its field sites utilize approximately 5,000 contractors employed by various corporations, usually in the form of limited liability partnerships. In addition, EM and its contractors rely on specific technical expertise from national laboratories, outside technical experts, and universities to provide assistance. This is particularly true in the technology development area, in which partnerships with other agencies, e.g., NASA for robotics, and university partnerships are being utilized. This talk will address these aspects that are enablers to achieving the EM tank waste mission.

469614 Adapting Power Generation Lessons Learned to Industrial CO_2 Capture

Abstract:

The Southern States Energy Board (SSEB) is adapting lessons learned in CO_2 capture and storage from power generation to promote commercialization and dissemination of clean energy technologies within the refining and petrochemical industries. The U.S. Department of Energy has partnered with seven Regional Carbon Sequestration Partnerships (RCSP) to advance innovation and research in carbon capture and sequestration. SSEB manages the Southeast Regional Carbon Sequestration Partnership (SECARB), including its Phase III Anthropogenic Test. The Anthropogenic Test is a fully integrated CO_2 capture, transportation, injection, storage and monitoring project that began injecting in August 2012 and concluded injection in September 2014. The project currently is in a 3-year post injection monitoring phase. The post-combustion CO_2 capture facility is a 25 megawatt equivalent advanced amine system, located at Alabama Power Companyâ€™s Plant Barry fossil based power generation facility.

470738 Highly Efficient Warm Gas Carbon Capture System for IGCC Power Plants

Abstract:

TDA is developing a novel sorbent that removes CO_2 via physical adsorption from synthesis gas. The relatively strong affinity of the sorbent to CO_2 enables effective operation at temperatures up to 300°C (well above the dew point of synthesis gas stream generated by commercial gasifiers). However, because the sorbent and the CO_2 do not form a true covalent bond, the energy needed to regenerate TDAâ€™s sorbent (5.4 kcal per mol of CO_2) is much lower than that observed for either chemical absorbents (e.g., 29.9 kcal/mol CO_2 for sodium carbonate) or amine-based solvents (e.g., 14.2 kcal/mol CO_2 for monoethanolamine). TDAâ€™s sorbent can be regenerated isothermally and CO_2 can be recovered at pressure (as high as 150 psia). Thus, the energy needed to regenerate the sorbent and compress the CO_2 for sequestration is significantly lower than that for any other technology reported to date. The high surface area and favorable porosity of the sorbent also provides a unique platform to introduce additional functionality, such as active groups to catalyze the water-gas-shift (WGS) reaction or to remove trace metals (e.g., Hg, As). Hence, TDAâ€™s CO_2 sorbent could be used to do a warm gas capture of CO_2 above the dew point of the synthesis gas in an Integrated Gasification Combined Cycle (IGCC) power plant. TDA's pre-combustion CO_2 capture system achieves a net plant efficiency of 34% on a coal high heating value basis when operated on bituminous coal, which is significantly higher than that can be achieved with the Selexol^{TM} scrubber at 31.6% (state-of-the-art pre-combustion carbon capture technology). The levelized cost of electricity including the transport, storage and monitoring costs for CO_2 is calculated as \$92.9/MWh for TDA's system, in comparison to \$105.2/MWh for Selexol^{TM}. In a current DOE sponsored project, TDA is preparing to carryout field tests with the warm gas CO_2 capture unit at 0.1 MWe scale. In this paper we present the results from the system analysis, bench-scale evaluations, and the fabrication and testing of the 0.1 MWe field test unit.

467400 Development and Optimization of Pressure/Temperature Swing Adsorption for Post-Combustion Carbon Capture

Abstract:

Development and Optimization of Pressure/Temperature Swing Adsorption for Post-Combustion Carbon Capture

K.T. Leperi,[1] F. You,[2] R.Q. Snurr[1]

[1]Northwestern University, Evanston, IL, USA

[2] Cornell University, Ithaca, NY, 14850

With fossil fuels expected to be a significant portion of the world's energy mix for the near future, it is important to minimize CO_2 emissions from current power plants through carbon capture and sequestration (CCS). In post-combustion CCS, the CO_2 is separated from the power plant flue gas emissions, which contain mainly N_2 and CO_2. Among the more promising technologies for accomplishing this separation with minimal energy are adsorption based processes. Only a few studies have focused on temperature swing adsorption (TSA) compared with pressure/vacuum swing adsorption (P/VSA) for separation of CO_2 from N_2. This has been primarily due to long cycle times associated with the heating and cooling of the bed for TSA. However, potential benefits of TSA over PSA are larger adsorbent working capacities for temperature differences compared to pressure differences and the lower cost of using thermal energy to regenerate the bed compared to using electricity for compressors and vacuum pumps. Therefore, it is important to determine not only the optimal regeneration method for CO_2 capture, but also the optimal cycle utilizing these regeneration methods.

In this work, various PSA, TSA and Pressure/Temperature Swing Adsorption (P/TSA) processes are simulated utilizing our previously developed model.[2] This model consists of a system of partial differential algebraic equations incorporating mass and energy balances, pressure drop across the column, competitive multi-site Langmuir isotherms and the linear driving force model. This system of PDAEs is solved by discretizing the spatial dimension utilizing the finite volume method coupled with the weighted essential non-oscillatory (WENO) scheme to generate a system of ODEs, which is solved in MATLAB. The non-dominated sorting genetic algorithm (NSGA-II) is then used to determine the optimal operating parameters for capturing the CO_2 from the flue gas at the lowest cost while maintaining the desired CO_2 purity and recovery. The objective of this work is to compare the different methods for regenerating the adsorbent and identify the ideal cycle for minimizing the cost of capturing CO_2 from flue gas. Each cycle is investigated with zeolites 13X and 5A along with Ni-MOF-74 in order to determine potential influences the adsorbent has on the determination of the ideal cycle and regeneration method.

1. MIT. *The Future of Coal*; 2007. http://web.mit.edu/coal/.

2. Leperi KT, Snurr RQ, You F. Optimization of Two-Stage Pressure/Vacuum Swing Adsorption with Variable Dehydration Level for Postcombustion Carbon Capture. *Ind. Eng. Chem. Res.* 2016;55:3338-3350.

455673 An Investigation of SO_2 Interference on Separation Performance of Amine-Containing Facilitated Transport Membrane for CO_2 Capture from Flue Gas

Abstract:

Post-combustion carbon capture is a key approach to control CO_2 emissions from power plants using fossil fuels. Treating real flue gas may be problematic as it contains not simply just CO_2 and N_2. The minor contaminants, such as SO_2 and O_2, may be the major factors that can limit the long-term stability of a membrane. The typical flue gas after flue gas desulfurization (FGD) treatment is at its discharge temperature of 57°C with 40 – 50 ppm SO_2. The SO_2 level can be further reduced to even 1 ppm by a caustic scrubber, i.e., 20 wt.% NaOH. With the presence of SO_2 in the feed gas, SO_2 could preferably react with amine carriers and hinder the permeation of CO_2 in amine-containing facilitated transport membranes. Possible amine degradation and competitive reactions raise the concerns on the membranes for use in real flue gas. Amine-containing facilitated transport membranes were tested at 102 and 57°C, respectively. SO_2 concentrations of 0.7 – 5 ppm were employed in our transport study. Unstable membrane performance was observed for the membrane tested at 102°C in the presence of SO_2. It was found that amines reacted with SO_2 to form bisulfite and sulfite products irreversibly at 102°C. On the contrary, the membrane showed a stable separation performance in the presence of 1 – 3 ppm SO_2 at 57°C. Moreover, an O_2 concentration of 3% was also added in the feed gas to simulate the real flue gas conditions. The membrane also showed a stable separation performance upon exposure to 3 ppm SO_2 and 3% O_2 for > 100 hours. In addition, infrared spectra were collected with respect to time by using an in-situ FTIR, and the growth of sulfite characteristic peaks was identified. The spectra results were consistent with the observed membrane separation performance, which is an affirmation of the SO_2 interference of the amines in the membrane.

The stable membrane performance at 57°C with SO_2 at ppm levels and O_2 at 3% indicates the applicability of the developed amine-containing membrane for use in CO_2 capture from flue gas.

460857 Engineering Enhanced Permeability and Selectivity in Hybrid CO_2 Capture Membranes

Abstract:

Despite ongoing advances in renewable energy, it remains true that ~90% of global power generation is still derived from carbon-based sources. It is therefore imperative to develop energy-efficient approaches to sequestering CO_2, facilitating decarbonization of energy production. CO_2 capture (pre- or post-combustion) using separations membranes represents an attractive alternative to amine looping, as membranes do not require temperature- or pressure-swings to achieve their performance. Historically, membranes for CO_2 sequestration have been dominated by pure polymers, due to ease of processability and scaling. However, within the solution-diffusion model for pure polymers, it has been challenging to increase CO_2 permeability as molecular diffusion and solubility are typically inversely related. In this talk, I present recent research into developing new molecular transport mechanisms in hybrid organic/inorganic membranes by introduction of chemically modulated inorganic nanocrystals and metal-organic frameworks (MOFs) into classical polymers forming altogether new classes of membranes that offer enhanced solubility, diffusion, and mechanical robustness. I will also briefly discuss strategies for H_2 separations as well.

467335 Two Dimensional Nanosheets-Embedded Thin Film Composite Membrane for CO_2 Cature Applications

Abstract:

The development of cost-effective CO_2 capture technologies for coal power plants will reduce the adverse impact of CO_2 emissions on global climate. Amine absorption technologies have been extensively studied and are regarded viable for post-combustion CO_2 capture. However, the energy comsuption for absorption technologies is still much higher than minimum requirement for large scale post-combustion flue gas. It seems clear that the membrane-based gas separations have presented a suitable alternative to the traditional separation processes for CO_2 capture, particularly for the treatment of flue gas streams including more than 15% CO_2. Membrane systems offer relevant gains in terms of flexibility, reliability, modularity and that they need energy intensity generally lower, or, at least, comparable with that of absorption technology. For competition with existing technologies, most of all, membrane materials with high permeability and high selectivity should be further developed with feasible processibility for real membrane platforms such as hollow fibers or thin-film composite membranes. According to recent report, the membrane systems can compete with existing technologies when membrane shows high CO_2 permeance (~4000 GPU, 1 GPU = 1×10^{-10} cm^3(STP)/cm^2.sec.cmHg) and high CO_2/N_2 selectivity (~50). However, to achieve this goal, many factors should be considered: selective layer thickness, gutter layer, and the porosity of support membrane in the form of thin-film composite membrane. In this study, we have developed high permeable, selective thin-film composite membrane for post-combustion CO_2 capture. Selective layers consist of high CO_2 permeable polymer and two-dimensional nanosheets. The nanosheets as dispersed phase in continuous polymer matrix help enhance CO_2/N_2 selectivity (~60) without any loss of CO_2 permeability. Also, such nanosheets at low content help prevent CO_2-induced plasticization, often leading to dramatic decrease in gas mixture. To develop high permeable, selective thin-film composite membranes, we have analyzed the effect of selective layer thickness, types of gutter layer (or gutter layer thickness), and support membrane properties, and these results were used for preparing optimized thin-film composite membranes. These thin-film composite membranes exhibited high CO_2 permeance and high CO_2/N_2 selectivity, even in the presence of water, which is close to the membrane criteria suitable for membrane-based CO_2 capture system. We have also prepared plate-to-frame modules, based on these membranes, showing high CO_2 flux of 5 m_3/h, for pilot plant demonstration.

462467 CO_2 Capture from Power Plant Flue Gas By Polaris^{TM} Membranes: Update on Field Demonstration Tests

Abstract:

To mitigate the harmful effects of global climate change, CO_2 in power plant flue gas must be captured and either sequestered or utilized in processes such as enhanced oil recovery. Membrane technology is an attractive CO_2 capture option because of advantages such as energy-efficient passive operation, tolerance to acid gases and oxygen, no hazardous chemical emissions or handling issues, minimal water requirements, and no steam use requiring modifications to the existing boiler.

Working with the US Department of Energy, MTR has developed new membranes and process designs to recover CO_2 from power plant flue gas. MTR Polaris membranes have CO_2 permeances ten times higher than standard commercial membranes, which greatly reduces the cost of a membrane capture system. These membranes are combined with a novel process design that uses incoming combustion air to sweep membranes and recycle CO_2 to the boiler. Design calculations estimate that this membrane process can capture 90% of the CO_2 in flue gas as a supercritical fluid using approximately 25% of the plant power, at a cost of \$40-\$50/ton of CO_2 captured. This translates to an increase in the levelized cost of electricity of about 50%.

In 2010, MTR completed a three-month field demonstration of the membrane CO_2 capture process with Arizona Public Service (APS) at their Cholla coal-fired power plant. The pilot system used commercial-scale membrane modules, and removed 90% CO_2 from a post-FGD flue gas slipstream containing 1 ton CO_2/day. In late 2011, the system was moved to the National Carbon Capture Center (NCCC) operated by Southern Company in Wilsonville,

AL, for further demonstration. Overall, the system has an accumulated run time of over 10,000 hours, and the membrane modules have exhibited stable performance. With experience learned from these tests, MTR has designed a scale-up small pilot unit to capture 20 ton CO_2/day from a power plant slipstream (equivalent to 1 MW_e-scale coal-fired power generation). The system was initially commissioned in August 2014 and testing concluded in June 2015. During testing, advanced modules demonstrating higher packing density, lower pressure drop, and cost savings were validated. In fall 2015, the 20 ton CO_2/day small pilot unit was moved to Babcock & Wilcoxâ€™s (B&W) research facilities in Barberton, OH for a fully integrated recycle test of the small pilot unit with a 1.8 MW_{th} coal-fired research boiler. The field demonstration to be conducted at B&W during the summer of 2016 will be the first time membrane operation with real flue gas and boiler operation with CO_2-laden air have been conducted in a single test. Technical results to date for the field tests, details of an upcoming hybrid membrane-absorption field test at the SRP pilot plant located on the Pickle Research Campus at UT-Austin, and future plans on technical development will be discussed in this presentation.

472304 Selective Hydrogenation of Bio-Oil Model Compounds over Molybdenum Carbide Supported Catalysts

Abstract:

The increasing complexity and cost of drilling for oil, the need to decrease global net CO2 emissions, and the desire to continue using liquid transportation fuels have motivated research to produce liquid fuels from biomass. Fast pyrolysis is a thermochemical conversion process in which biomass is reacted at high temperatures (near 500°C) at atmospheric pressure for very short residence times (on the order of seconds) in the absence of added molecular oxygen. Bio-oil from conventional fast pyrolysis is corrosive, has poor heating value, is thermally unstable, cannot be blended with conventional fuels without the addition of surfactant, and can become very viscous with aging[i]. Selective hydrogenation can improve the properties of bio-oil by increasing its thermostability and energy density, and potentially reducing acidity and therefore improving the ability to blend it with conventional fuels.

^{1}H and ^{13}C NMR analyses of pyrolysis bio-oils have shown that typical functional groups include aldehydes, ketones, hetero-aromatics, aliphatics, methoxys, alcohols, alkenes, dibenzenes, esters, benzyls, esters, and carboxylic acids[ii]. Considering the molecular character of pyrolysis bio-oil, crotonaldehyde was used as the model compound for work described in this paper as it is a simple aldehyde and can be used to probe a variety of different reaction pathways. The catalysts are based on Mo carbide (Mo_2C). Molybdenum carbide is active for a variety of hydrogenation reactions including the hydrogenation of CO to hydrocarbons (Fischer-Tropsch reaction)[iii] and naphthalene to tetralin[iv]. In addition, this material can possess, depending on preparation and treatment conditions, multiple types of surface sites (e.g. acidic, basic and metallic sites)[v],[vi],[vii] and is stable under conditions of fast pyrolysis[5]. Transition metal and alkali promotion of Mo_2C has been shown to cause shifts in the selectivity for a number of reactions, including shifting the selectivity for CO hydrogenation (Fischer-Tropsch reaction) to alcohols and heavier hydrocarbons, depending on the promoter[viii],[ix]. The goal of this work was to understand how metal promotion of Mo2C-based catalysts could be used to shift selectivity of crotonaldehyde hydrogenation.

The Mo_2C catalysts were synthesized by a temperature-programmed reaction method as described previously[x]. Metal salt was added to the native (unpassivated) Mo_2C catalyst by incipient wetness in an oxygen-free inert atmosphere. Then, the material was dried at 110°C for 2 hrs in H_2, reduced in H_2 at 450°C for 4 hrs, quenched to room temperature, and passivated in 1% O_2/He for at least 6 hrs. All of the catalysts were characterized by x-ray powder diffraction, N_2 physisorption, and temperature programmed desorption of NH_3 and CO_2. The crotonaldehyde hydrogenation experiments were performed in a flow reactor system equipped with a bubbler to introduce vapors of crotonaldehyde, and the reactor effluent was analyzed with a gas chromatograph equipped with flame ionization and thermal conductivity detectors. The catalyst was loaded into a quartz u-tube reactor and supported on quartz wool. Prior to the rate and selectivity measurements, the catalysts were reduced in 15% CH_4/H_2 for 4 h at 590 °C. All reactions were performed between 275-350°C in atmospheric pressure. The reaction mixture consisted of 0.7 mol% crotonaldehyde, 4.2 mol% H_2, and 95.1% N_2. The H_2/acetic acid molar ratio (6) was chosen to achieve approximately 2 times the stoichiometric amount of H_2required for complete saturation and deoxygenation of crotonaldehyde to butane.

The unpromoted Mo_2C favored formation of butyraldehyde and butene (ca. 40% each). Since butene is believed to be a secondary hydrogenation-dehydration product of crotyl alcohol[xi], there is not a clear preference for a specific pathway on the unpromoted catalyst. Selectivities to butyraldehyde, butene, and butadiene decreased with increasing potassium coverage, while selectivity to furan increased with potassium coverage. At coverages equivalent to 50% of a monolayer (ML), the furan selectivity was ca. 80%, while the selectivities to the other three products were each less than 10%. There was no significant further increase in the selectivity upon increasing the potassium loading, suggesting that the effect of potassium promotion was maximized at ~50% ML.

Alkali promotion is known to affect the acid-base properties of a catalyst, and acid-base properties are also known to affect selectivity[6,xii]. Acid and base site densities of the catalyst surfaces were quantified via temperature-programmed desorption of NH_3 and CO_2, respectively. While acid site densities decreased with increasing potassium loading, base site density increased with potassium promotion. The highest base site density (~$4.3x10^{14}$ molecules/cm^2) was for the K/Mo_2C catalyst with 1.1ML-equivalent coverage of potassium; the highest acid site density was on the bulk Mo_2C and was $1.5x10^{14}$ molecules/cm^2. There was a strong correlation between the furan production rate and the base site densities, implicating base sites the location of the rate-determining step. Furan appears to be produced from crotyl alcohol intermediate, indicating that potassium promotion enhanced the hydrogenation of crotonaldehyde to crotyl alcohol.

The model compound results presented in this paper demonstrate the ability to tune the product selectivity of molybdenum carbide catalysts by adding potassium. Tuning product selectivity may prove invaluable in upgrading and stabilizing bio-oils. Performing simultaneous carbon-carbon coupling and hydrogenation-dehydration reactions could ultimately lead to an upgraded bio-oil with lesser oxygen content, greater hydrogen content and a higher average carbon number.

---

[i] (a) Czernik, S.; Bridgwater, A.V.; *Energy & Fuels*, **2004**, *18*, 590-598; (b) Zhang, Q.; Chang, J.; Tiejun, W.; Yung, X. *Energy Conversion and Management*, **2007**, *48*, 87-92.

[ii] Mullen, C.A.; Strahan, G.D.; Boateng, A.A. *Energy & Fuels*, **2009**, *23*, 2707-2718.

[iii] Schaidle, J.A.; Thompson, L.T. *Journal of Catalysis*, **2015**, *329*, 325-334.

[iv] Pang, M.; Liu, C.; Xia, W.; Muhler, M.; Liang, C. *Green Chemistry*, **2012**, 14, 1272-1276.

[v] Bej, S. K.; Bennett, C. A.; Thompson, L. T., *Applied Catalysis A: General,* **2003**, *250*, 197-208.

[vi] Bej, S. K.; Thompson, L. T., *Applied Catalysis A: General,* **2004**, *264*, 141-150.

[vii] Sullivan, M. M.; Held, J. T.; Bhan, A., *Journal of Catalysis,* **2015**, *326*, 82-91.

[viii] Park, K.Y.; Seo, W.K.; Lee, J.S. *Catalysis Letters*, **1991**, *11*, 349-356.

[ix] Griboval-Constant, A.; Giraudon, J.-M.; Leclerq, G.; Leclercq, L. *Applied Catalysis A: General*, **2004**, *260*, 34-45.

[x] J.A. Schaidle, N.M. Schweitzer, O.T. Ajenifujah, L.T. Thompson, J. Catalysis 289 (2012) 210-217.

[xi] Zanella, R.; Louis, C.; Giorgio, S.; Touroude, R. *Journal of Catalysis*, **2004**, *223*, 328-339.

[xii] Kotarba, A.; Adamski, G.; Piskorz, W.; Sojka, Z.; Sayag, C.; Djega-Mariadassou, G. D. *Journal of Physical Chemistry B*, **2004**, *108*, 2885-2892.

471968 Guaiacol Deoxygenation Using Methane over Pt-Bi Catalysts: Reaction Pathways and Kinetics

Abstract:

The deoxygenation of guaiacol, a typical molecule from the pyrolysis of ligno-cellulosic biomass, was investigated using methane as the reductant to overcome the high cost issue of H_2 as the reductant for the hydrodeoxygenation (HDO) process. In our prior work, a Pt-Bi/C catalyst was used to deoxygenate guaiacol and methane was found to exhibit as good deoxygenation performance as H_2. In the present work, by varying the space velocity, kinetic data was acquired. Following similar reaction pathways as in the case using H_2, rate constants and activation energies for each sub-reaction in the network were obtained. Owing to more difficult activation of C-H bond in CH_4, as compared to H-H bond in H_2, the kinetic results show higher reaction barriers when methane is used as the reductant. This work provides a new approach for bio-oil upgrading using methane as the reductant instead of hydrogen.

481344 Fluidized Bed Catalytic Glycerol Hydrodeoxygenation to 1,3-Propanediol

Abstract:

1,3 propandiol (1,3-PDO) is a monomer for commercial polyesters like Corterra® (Shell) and Sorona® (DuPont). Glycerol is an attractive feedstock for this monomer because of its low cost. Here we propose a gas-phase hydrodeoxygenation process in which we atomize a dilute solution of glycerol to a fluidized bed of of Pt/WO3/Al2O3 operating above 240 °C and atmospheric pressure. Gas-solids fluidized beds have excellent heat transfer properties that vaporize the droplets within milliseconds thereby minimizing thermal degradation of the glycerol. After 2 h at 260 °C, 1,3-PDO yield reached 14%. The other products included 1,2-PDO (18%), 1-propanol (28%), and 2-propanol (15%). We postulate that glycerol first dehydrates to acrolein, then rehydrates to 3-hydroxypropanal followed by the hydrogenation to 1,3-PDO. The yield of other compounds (acrolein, ethylene glycol, propane, and acetone) increased at higher operating temperatures.

454718 Reaction Pathways for the Hydrodeoxygenation of Anisole and Guaiacol over Zn-Pt Bimetallic Catalysts

Abstract:

There is an increasing interest in the use of lignin as a renewable feedstock for the production of high-value aromatic compounds; however, due to its high oxygen content, selective hydrodeoxygenation (HDO) of the aromatic oxygenates produced from lignin depolymerization is required. Metal alloys containing a group 10 metal and a more oxyphillic metal, such as Zn, have been shown to be highly selective for this class of reactions, at least under some conditions. Although, the mechanism by which alloying enhances selectivity is still poorly understood. To provide mechanistic insight, in this study we have investigated the reaction of typical lignin-derived oxygenates, anisole and guaiacol, on single-crystal Pt and Zn-Pt model catalysts by temperature programmed desorption (TPD) and high resolution electron energy loss spectroscopy (HREELS). For Zn-free Pt it was found that anisole and guaiacol interact strongly with the surface via the aromatic ring, which promotes undesired ring hydrogenation. Adding Zn to the Pt surface was found to inhibit bonding via the aromatic ring and promoted a stronger interaction with the oxygens. This in turn facilitated selective C-O bond scission without hydrogenating the aromatic ring. These results suggest that Zn-Pt alloys may be an effective catalyst for HDO of lignin-derived aromatic oxygenates with low activity for ring hydrogenation. This hypothesis was then tested and verified by investigating the reaction of anisole or guaiacol with H_2 over high surface area carbon-supported Pt and Zn-Pt catalysts.

459948 Novel Bio-Oil Hydrodeoxygenation Catalysts Based on Strong Electrostatic Adsorption

Abstract:

Fast pyrolysis of biomass produces an intermediate crude oil (â€˜bio-oilâ€™) with potential to become fungible transportation fuel. Catalytic hydrodeoxygenation (HDO) has currently gained attention as a technique for bio-oil upgrading, due to its ability to convert highly oxygenated and reactive species to hydrocarbons. Most studies of HDO catalysts are process and not catalyst oriented; noble metal catalysts are typically used in conjunction with common supports, and important variables which affect catalyst performance are often not decoupled during preparation, including catalyst particle size, hydrothermal stability, and precision of bimetallic formations.

We hypothesize that a precisely synthesized combination of bimetallic base metal catalysts can possess requisite activity and selectivity for pyrolysis oil HDO. We studied the hydrotreatment of bio-oil from switchgrass (regular and catalytic fast pyrolysis) in a batch reactor, using a series of active and hydrothermally stable HDO catalysts. Strong electrostatic adsorption was used to prepare single and bimetallic combinations (Pt, Ru, PtRu, Ni, Cu, NiCu) on carbon, mesoporous alumina, and metal-doped SBA-15. Particle size was determined by XRD to fall within 3 â€“ 5 nm with narrow distribution, and in many cases, as small as 1.0 nm. Catalyst stability was determined by examining the support crystal structure before and after reaction. The selectivities of each catalyst toward deoxygenation, hydrogenation, and coke inhibition were compared, and this information was used to identify optimum HDO catalysts.

460969 Biomass Pyrolysis Interfaced with Honeycomb Structured Upgrading Reactor for the Production of High Quality Biofuels

Abstract:

A reactor interfaced a pyrolysis zone with a doubly stacked honeycomb catalytic upgrading zone was designed and evaluated for biomass conversion to liquid fuels. The pyrolysis zone was operated through a bubbling bed with fine sand particles as the heat carrier. Two zeolite-coated cordierite honeycomb (ZCCH) catalyst matrices separately prepared using ZSM-5 of different Si/Al ratios were stacked in the upgrading zone. The results showed that ZCCH catalytic upgrading following the steam pyrolysis of corncob effectively reduced the oxygen content and viscosity of bio-oil. The yield, HHV and stability of the bio-oil were increased by the presence of a moderate amount of ethanol. The HHV of bio-oil was found to linearly correlate with the oxygen concentration in the oil. Models of bio-oil energy yield from corncob based on unit energy yield and yield in ethanol equivalent (YEE, in L ethanol) were developed according to the oxygen content in the produced bio-oils.

471424 Upgrading of Biomass Pryolysis Vapors of Modified ZSM-5: Effects of Metal(s) Loading and Pretreatment

Abstract:

Ga/ZSM-5 and Ni/ZSM-5catalysts with varied nickel loadings were evaluated for their ability to produce aromatic hydrocarbons by upgrading of pine pyrolysis vapors. The effect of catalyst pretreatment by hydrogen reduction was also investigated. Results indicate that the addition of gallium or nickel increases the yield of aromatic hydrocarbons while simultaneously increasing the conversion of oxygenates, relative to ZSM-5, and these effects are more pronounced with increasing nickel loading. Additionally, while initial activity differences were observed between the oxidized and reduced forms of nickel on ZSM-5 (i.e., NiO/ZSM-5 vs. Ni/ZSM-5), the activity of both catalysts converge with increasing time on stream. These reaction results, coupled with characterization of pristine and spent catalysts, suggest that the catalysts reach similar active states during catalytic pyrolysis, regardless of pretreatment, as NiO under goes *in situ* reduction to Ni by biomass pyrolysis vapors. This reduction of NiO to Ni was confirmed by reaction results and characterization by NH_3 TPD, TPR, and XRD. This finding is significant in that the ability to reduce or eliminate the need for a pre-reaction H_2 reduction of Ni-modified zeolite catalysts could reduce process complexity and operating costs in a biorefinery-based vapor phase upgrading process to produce biomass-derived fuels and chemicals. The ability to monitor catalyst activity in real-time with a molecular beam mass spectrometer used to measure uncondensed, hot pyrolysis vapors, allows for improved understanding of the mechanism for improved activity with Ni-addition to ZSM-5, which is attributed to the ability to prevent deactivation by deposition of coke and capping of zeolite micropores.

467798 Cross Metathesis of Unsaturated Carboxylic Acids to Bioterephthalic Acid Intermediates

Abstract:

Terephthalic acid is essential for the preparation of a variety of polyester fibers and is largely produced *via* the catalytic oxidation of petroleum-derived *p*-xylene [1]. The ongoing exchange of naphtha crackers with shale gas crackers is causing a shortage in some aromatics species, such as *p*-xylene [2]. As a consequence, alternative and sustainable routes to produce terephthalic acid from biomass derivatives, e.g. bio-sourced sorbates [3], are highly desirable. In the last decade, Ru-catalyzed olefin metathesis, and in particularly cross-metathesis, has become an important tool for the formation of carbon–carbon bonds and the formation of valuable organic molecules [4].

We have investigated the ruthenium-catalyzed cross metathesis of sorbates with acrylates to form di-ester muconates. In a later step, the reaction between alkyl muconates and ethylene affords the aromatization to dialkyl terephthalate. The effect of alkyl substituents and solvents was investigated over various organometallic ruthenium-based catalysts to increase conversion and selectivity towards the alkyl muconates. Conversions of up to 20% were achieved within the first hour of reaction with very low amounts (0.1 mol.%) of ruthenium-based catalyst containing a *N*-heterocyclic carbene (NHC) ligand. If a solvent was used, the cross metathesis products were formed in higher yields with the use of more polar solvents, *i.e.* CH_2Cl_2. With the aim to develop a greener process, the acrylate reactant was used in excess with no other solvent added. In this case the highest selectivity to the desired product was observed (~ 40%). By using different alkyl substitutents the steric effect of the substitutents was investigated. Using reactants (sorbates and acrylates) with smaller alkyl substitutents the yield of cross metathesis products was favored, possibly due to easier accessibility of the reactants to the catalyst metal site. Upon further development, this route for obtaining di-ester muconates, may provide a sustainable alternative for producing biomass derived precursors to polyethylene terephthalate (PET).

References

[2] C. Luke Williams, Chun-Chih Chang, Phuong Do, Nima Nikbin, Stavros Caratzoulas, Dionisios G. Vlachos, Raul F. Lobo, Wei Fan, and Paul J. Dauenhauer. ACS Catal. 2 (2012) 935

[4] Georgios C. Vougioukalakis, and Robert H. Grubbs. Chem. Rev. 110 (2010) 1746

460301 Ultralight, Reusable Cellulose Diacetate Aerogels for Selective Fluid Sorption

Abstract:

Uncontrolled oil or chemical spillage into fresh water can trigger a cascade of events including toxic loading and contamination of food resources, all of which are heavily detrimental to the ecosystem[1,2]. Widespread technologies proposed to deal with oil spills include oil skimming, *in-situ* burning, mechanical containment and utilization of dispersants, solidifiers and degrading microorganisms[3-8]. However, these methods are either inefficient or environmentally unfriendly. The need for efficient oil spill cleaning technologies was imminent in 2010 deep water horizon spill, one of the largest man-made disasters in the recent history, where oil gushed out in the Gulf of Mexico for straight 87 days, pouring out over 200 million gallons of crude oil[9]. In this type of scenario, the use of sorbents is attractive as they are easy to deploy and do not generate byproducts. There are three categories of sorbent materials: organic (milkweed, wood chips, rice straw, among others), inorganic (organo-clay, perlite and sand, zeolites, etc.) and synthetic (non-woven polypropylene mats, for example).[10-12] However, they (1) suffer from low sorption capability (typically less than 10 g per gram of solid support); (2) are difficult to recover (they are not buoyant), and (3) are not biodegradable.

In recent years, aerogels have gained attention as sorbent materials because of their light weight and high pore volume. In this study, highly porous (99.7% air volume) and ultra-light (4.3 mg/ml density) cellulose ester aerogels were synthesized for unprecedented water uptake (45-90 g/g) while affording wet strength and mechanical robustness (maximum compressive stress and strain of 350 kPa and 92%, respectively). The high compression strains are generally achieved with carbon nanofiber aerogels but the 92% strain is higher than those of nanocellulose aerogel reported in literature. The compressive stress of 350 kPa is 100 times higher than the reported cellulosic aerogels (see Figure 1). The aerogels were further adjusted towards hydrophobicity and oleophilicity via chemical vapor deposition with an organo-silane species. The as-modified aerogel exhibited high oil retention (20-30 g/g aerogel) while maintaining mechanical integrity for fast oil cleanup from aqueous media under marine conditions.

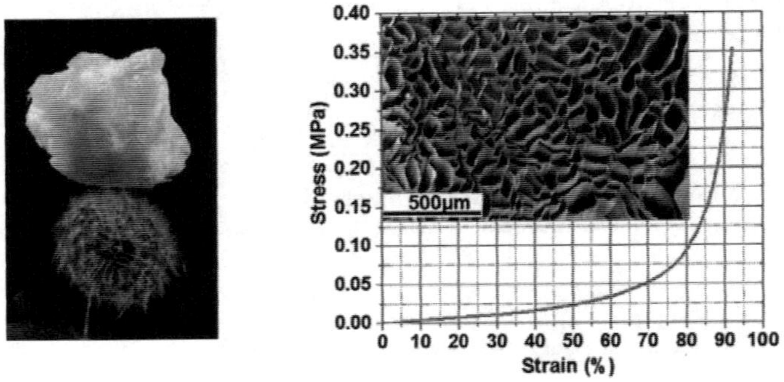

a) b)

Figure 1: a) Image of a 2 w/v% aerogel (density 4.3 mg/ml) on a dandelion leaf; b) Compressive stress-strain profile for 4 w/v% aerogels with the maximum compressive stress of 350 kPa and maximum strain of 92%. Inset: SEM image showing the radial cross-section of 4 w/v% CDA aerogel with a "honeycomb" like structure.

References:

1. Aguilera F, Méndez J, Pásaroa E, Laffona B. Review on the effects of exposure to spilled oils on human health. *J Appl Toxicol*. 2010;30(4):291-301. doi:10.1002/jat.1521.

2. Burton NHK, Musgrove a. J, Rehfisch MM, Clark N a. Birds of the Severn Estuary and Bristol Channel: Their current status and key environmental issues. *Mar Pollut Bull*. 2010;61(1-3):115-123. doi:10.1016/j.marpolbul.2009.12.018.

3. Fingas M. Oil Spill Science and Technology. *Elsevier, Amsterdam*. 2011;a.

4. Fingas M. "In-situ burning" in Oil Spill Science and Technology. In: *Elsevier*. Gulf publishing company, New York NY; 2011:737-903.

5. Michel J, Adams EE, Addassi Y, et al. *Oil Spill Dispersants. Efficacy and Effects*. Washington, DC; 2005.

6. Rosales PI, Suidan MT, Venosa AD. A laboratory screening study on the use of solidifiers as a response tool to remove crude oil slicks on seawater. *Chemosphere*. 2010;80(4):389-395. doi:10.1016/j.chemosphere.2010.04.036.

7. Correa M, Padron E, Petkoff I. The San Rafael De Laya oil spill : A Case of cleanup and remediation in Venezuela. In: *International Oil Spill Conference*.; 1997:932-935.

8. Aldhous P, Hecht J. Beware long-term damage when cleaning up oil spills. *New Sci*. 2010;10(206):2765. doi:10.1016/S0262-4079(10)61464-9.

9. Graham B, Reilly WK, Beinecke F, Boesch DF, Garcia TD, Murray C a. *Deep Water: The Gulf Oil Disaster and the Future of Offshore Drilling*.; 2011.

10. Teas C, Kalligeros S, Zanikos F, Stournas S, Lois E, Anastopoulos G. Investigation of the effectiveness of absorbent materials in oil spills clean up. *Desalination*. 2001;140(3):259-264. doi:10.1016/S0011-9164(01)00375-7.

11. Adebajo MO, Frost RL, Kloprogge JT, Carmody O, Kokot S. Porous Materials for Oil Spill Cleanup : A Review of Synthesis. *J Porous Mater*. 2003;(10):159-170. doi:10.1023/A:1027484117065.

12. Carmody O, Frost R, Xi Y, Kokot S. Adsorption of hydrocarbons on organo-clays-Implications for oil spill remediation. *J Colloid Interface Sci*. 2007;305(1):17-24. doi:10.1016/j.jcis.2006.09.032.

464405 Exploring Cellulose Nanofibrils As Renewable Resource for Metal-Free Heteroatoms-Doped Carbon Electrocatalyst

Abstract:

With increasing development toward sustainable energy technologies for the future, a cost-effective electrochemical reaction has become a critical aspect for its practical applications. Replacing traditional metal-based electrocatalysts with carbon-based electrocatalyst offers both economical and environmental benefits. Cellulose nanofibrils, as biomass resources, have drawn a great interest as promising raw materials for preparing carbonaceous materials. In addition, enhanced electrochemical properties of carbon materials could be achieved by incorporating heteroatoms dopants. In this work, metal-free heteroatoms-doped carbons derived from carbon nanofibrils are explored as electrocatalyst for oxygen reduction reaction. The characteristics of doped carbons were investigated and their electrocatalytic activities were demonstrated.

459777 Cellulose Nanofibers from Recycled Pulp: Production, Characterization and Application to Reinforce Recycled Paper

Abstract:

Papermaking is a sustainable sector where paper recycling has been recognized of a great importance [1, 2]. In Europe, 54% of the paper industryâ€™s raw material comes from recovered paper which corresponds to a paper recycling rate of 72% [3]. However, one of the main problems in the utilization of secondary fibers is to maintain the quality levels of the products that is continuously increasing following customer demands. On the other hand, paper demand has decreased due to both economic recession and replacement of paper by other information supports, causing cost pressures in the paper industry. Although natural and synthetic strength additives are commonly used in recycled paper, poor tensile strength is still the main source of customer complaints to paper manufacturers. In this context, new strategies to improve interfiber bonding must be explored and the use of cellulose nanofibers (CNF) are alternative additives to increase mechanical properties of paper with some advantages, such as their renewable nature, biodegradability and potential high availability.

CNF have gained increasing attention due to their high strength and stiffness combined with their low weight. These properties make CNF promising for applications in fields such as papermaking, composites, packaging, electronic devices, coatings, biomedicine and automotive. Among all of the applications, their use in the papermaking industry can improve paper quality and many studies have proved that the addition of CNF in mass to the pulp suspension increases the mechanical properties of paper [4]. CNF are currently produced from wood, mainly from eucalyptus one, but the reality is that the CNF market is still very limited [5]. The main drawbacks to use CNF in large volume industrial sectors, such us papermaking, is the cost of the product due to the high amount of energy consumption during CNF processing, the difficulties in producing uniform nanocellulosic particles, the dewatering difficulties and the problems to pump at high concentration [6]. A potential solution is to produce the CNF on-site from the mill pulp to avoid drying and transport costs, as well as dilution and dispersion problems.

Therefore this paper focuses on the on-site production of CNF from a cellulose source available at the recycled paper mill and its direct application to reinforce the recycled paper.

The production of CNF was optimized by studying different TEMPO-mediated oxidation levels before the homogenization mechanical process. The obtained CNF were characterized and several doses were added to the recycled pulp to evaluate their effect in terms of paper strength enhancement.

CNF were obtained from recycled newsprint pulp by TEMPO-mediated oxidation using 2.5, 5, 10 and 15 mmol of NaClO per each gram of pulp [7]. Once the pulp was oxidized, a cleaning process was performed through filtration steps using distillate water to reach a pH value around 7. Finally, homogenization at 600 bar was carried out in a PANDA PLUS 2000 laboratory homogenizer manufactured by GEA Niro Soavi (Italy).

Characterization of the oxidized cellulose pulp was performed by measuring the amount of carboxilate groups, as an indicator of the oxidation degree achieved after TEMPO oxidation, by conductimetric titration. CNFs were characterized by determination of the nanofibrillation yield, the cationic demand, the transmittance and the polymerization degree. Nanofibrillation yield was measured by centrifugation of a diluted CNF suspension (0.1% wt) at 4500 g for 30 min in order to isolate the nanofibrillated fraction in the supernatant from the non-fibrillated fraction deposited in the sediment. Cationic demand was measured by colloidal titration of the diluted suspension at 0.05%, with 0.001 N polyDADMAC, using a particle charge detector, Mütek PCD04, manufactured by BTG Instruments GmbH (Herrsching, Germany). Transmittance of the CNF suspensions diluted at 0.1% were measured between 400 and 800 nm of wavelength using a Cary 50Conc UV-Visible spectrophotometer manufactured by Varian Australia PTI LTD. Polymerization degree was calculated from the limiting viscosity number (intrinsic viscosity) of CNF suspensions, which was determined following the international standard ISO5351/1 with cupriethylendiamine as solvent [8].

Pulps were prepared through disintegration of 20 g of dry recovered paper in 2000 mL of water, by using a Messmer pulp disintegrator (Mavis Engineering Ltd, London). The recovered paper with the correspondent amount of CNF was left to soak at least 24 h before disintegration to favor swelling. A three-component retention system was added to the pulp (1.25 mg/g of coagulant, 0.75 mg/g cationic polyacrylamide as flocculant and 1.7 mg/g hydrated bentonite clay based on industrial recommendations). The pulp was used to prepare handsheets with basis weight of 60 g/m^2 in a normalized handsheet former Rapid-Köthen (ISO 5269/2, DIN 54 358). These handsheets were physically and mechanically characterized by using an AUTOLINE 300 from Lorentzen & Wettre (Stockholm, Sweden).

Results showed that cheaper CNF can be prepared on-site from recycled paper through chemical pretreatment followed by homogeneization treatment. The amount of carboxilic groups of the oxidized pulp increased linearly with the TEMPO-mediated oxidation level from 0.09 to 1.04 mmoles/ g o.d. pulp, for non-oxidized and 15 mmoles NaClO/g o.d. TEMPO-mediated oxidized pulp, respectively. Maximum CNF yield was achieved with 15 mmoles NaClO/g o.d. up to 77%. Cationic demand and transmittance also increased with TEMPO-oxidation level. The highest oxidized CNF achieved a 29% increase of tensile strength index by adding 3 wt% of CNF to the recycled pulp.

ACKNOWLEDGEMENTS

The authors wish to thank the Economy and Competitiveness Ministry of Spain for the support of the project NANOSOLPAPELREC with reference CTQ2013-48090-C2-1-R.

REferences

1. Miranda, R., Bobu, E., Grossmann, H., Stawicki, B. Blanco, A., 2010. "Factors influencing a higher use of recovered paper in the european paper industry". Cellulose Chemistry and Technology. 44(10), 419-430.

2. Blanco, A., Miranda, R. Monte, M.C., 2013. "Extending the limits of paper recycling: improvements along the paper value chain". Forest Systems. 22(3), 471-483.

3. CEPI, 2014. "Key Statistics. European Pulp and Paper Industry. http://www.cepi.org/topics/recycling".

4. Taipale, T., Osterberg, M., Nykanen, A., Ruokolainen, J. Laine, J., 2010. "Effect of microfibrillated cellulose and fines on the drainage of kraft pulp suspension and paper strength". Cellulose. 17(5), 1005-1020.

5. Chirayil, C.J., Mathew, L. Thomas, S., 2014. "Review of recent research in nano cellulose preparation from different lignocellulosic fibers". Reviews on Advanced Materials Science. 37(1-2), 20-28.

6. Osong, S.H., Norgren, S. Engstrand, P., 2016. "Processing of wood-based microfibrillated cellulose, and applications relating to papermaking: a review". Cellulose. 23(1), 93-123.

7. Saito, T., Kimura, S., Nishiyama, Y. Isogai, A., 2007. "Cellulose nanofibers prepared by TEMPO-mediated oxidation of native cellulose". Biomacromolecules. 8(8), 2485-2491.

8. Balea, A., Merayo, N., Fuente, E., Delgado-Aguilar, M., Mutje, P., Blanco, A. Negro, C., 2016. "Valorization of corn stalk by the production of cellulose nanofibers to improve recycled paper properties". Bioresources. 11(2), 3416-3431.

472899 Phenolic Catalyzed Hydroxyl Radical Oxidation for Cellulose Nanofiber Production

Abstract:

Hydroxyl radicals (•OH) can change the surface chemical properties of cellulose to produce charged carboxylates in production of cellulose nanofibers (CNF). While ozone gas is an efficient oxidizing agent, ozone alone is relatively ineffective in oxidizing cellulose surfaces. We show that the addition of phenols from lignin and its residual byproducts are effective hydroxyl radical promoters in aqueous solution. Our preliminary results indicate that lignin-derived phenolics are effective catalysts to activate ozone for oxidation of the cellulose surface thereby improving cellulose dispersion in liquid media. Changes in the zeta-potential of the •OH radical modified CNF suggested a colloidal stability after the chemical treatment. The initial 7% yield of CNF from the starting never dried pulp feedstock increased after removing the surface modified CNF by centrifugation/filteration and re-applying the chemical treatment to the remaining cellulose. The •OH radical-cellulose reaction needs to be carefully controlled to inhibit cellulose structural degradation which affects CNF characteristics. We propose a model that predicts surface characteristics as a function of phenolic catalyst loading, ozone consumption, and mechanical treatment.

473470 Performance of Hematite Oxygen Carrier for Hydrocarbon Conversion

Abstract:

Hematite has been studied extensively as an oxygen carrier for chemical looping combustion. This study investigates a Canadian hematite under reduction and re-oxidation multi-cycle experiments at a variety of conditions including temperature, space velocity, cycle length, and number of cycles. Post-run characterization is presented to identify which property changes may be most strongly linked to change in performance.

467048 Characterization of Ilmenite As Oxygen Carrier during Chemical Looping Combustion and Reforming

Abstract:

<span" roman="Roman"" new="New"> Fossil fuel usage is considered a major source of CO_2 emissions, the major contributor to greenhouse effect. Carbon capture and storage (CCS) is considered a promising approach to mitigate CO_2 emissions from such industrial facilities. CCS involves separation of carbon dioxide from the flue gas, transportation to a storage location and subsequent long-term storage. Among these steps, CO_2 capture is the major challenge due to issues relating to the technical feasibility and cost. One of these technologies is chemical looping process (CLP). In the CLP process, an oxygen carrier (a metal oxide) is circulated between two fluidized-bed reactors; a fuel reactor where the oxygen carrier is reduced according to equation (1) and (2), and an air reactor where the reduced oxygen carrier is oxidized according to equation (3). The main advantage with CLP, compared to other conversional conversion processes, is that CO_2 is inherently separated from the other flue gas components, and therefore reducing the economic cost of CO_2separation.

<span" roman="Roman"" new="New">$CH_4 + 4Me_xO_y \rightarrow CO_2 + 2H_2O + 4Me_xO_{y-1}$(1)

<span" roman="Roman"" new="New">$CH_4 + Me_xO_y \rightarrow CO + 2H_2 + Me_xO_{y-1}$(2)

<span" roman="Roman"" new="New">$Me_xO_{y-1} + 1/2O_2 \rightarrow Me_xO_y$(3)

<span" roman="Roman"" new="New"> CanmetENERGY has two on-going research programs on CLP, one is pressurized chemical looping combustion (PCLC) for the production of power and steam, the other one is pressurized chemical looping reforming (PCLR) for hydrogen production. High reactivity and proper flow behaviour of the oxygen carrier are crucial in developing both PCLC and PCLR technologies. Among all the low-cost materials including ores and industrial residues, ilmenite, a nature-occurring mineral material, has been paid much more attentions for their potential in CLP due to its desired thermodynamical properties resulting in 100 % theoretical combustion efficiency. Crushed ilmenite ore, henceforth referred to as ilmenite, is considered a leading candidate for oxygen carrying in CanmetENERGY's pressurized chemical looping (PCL) processes since it is relatively inexpensive and possesses good physical attributes. However, as an iron-based oxygen carrier, the reactivity of ilmenite is lower than Mn-, Co-, Ni- and Cu-based ones. Lower reactivity will result in large bed inventory (8-10 times that required for the synthetic Fe-based oxygen carriers) and high pressure drop when used in CFB, and unconverted gas fuels in the outflow of fuel reactor. In addition, there is evidence of attrition, sintering and agglomeration of ilmenite in certain CLP conditions, which negatively impact its reactivity and operating behaviour. This study focuses on introducing nonporous (or mesoporous) dopants of CeO_2, ZrO_2, NiO or Al_2O_3into the ilmenite (mainly immobilized on their surface) to improve the reactivity of the ilmenite-based OCs towards methane combustion and reforming. On the other hand, when mesoporous dopants are employed, the agglomeration of ilmenite particles could be eased.

<span" roman="Roman"" new="New"> A pressurized thermogravimetric analyzer (PTGA) and a pressurized fixed-bed reactor (PFBR) are used for testing the doped ilmenite particles at various temperatures (850-1050°C), gas compositions, pressures, and cycle stages. These particles are then analyzed by X-ray fluorescence (XRF), X-ray diffraction (XRD), and scanning electron microscopy with energy dispersive X-ray spectroscopy (SEM-EDX). The results from the measurements and discussion on the OC performance are outlined in this paper.

469032 The Role of Support in Increasing Agglomeration Resistance of Cu-Based Oxygen Carriers

Abstract:

Chemical looping combustion (CLC) is a promising alternative to the conventional, *i.e.* combustion based, fossil fuel conversion processes. In CLC, a solid oxygen carrier is used to transfer oxygen from air to a carbonaceous fuel. This indirect combustion route allows for effective CO_2 capture since a sequestrable stream of CO_2 is inherently produced without any need of energy intensive CO_2 separation step. From a thermodynamic point of view, CuO is arguably the most promising oxygen carrier candidate for CLC. However, the main challenge associated with the use of CuO for CLC is its structural instability as the Tammann temperatures of Cu (405 ºC), Cu_2O (481 ºC) and CuO (526 ºC) are lower than the typical operating temperatures of the CLC process. Thus, irreversible microstructural changes are expected for CuO-based oxygen carriers during CLC operation. To avoid the thermal sintering and agglomeration of CuO, it is commonly stabilized by a high Tammann temperature ceramic, *e.g.* Al_2O_3, $MgAl_2O_4$, MgO or ZrO_2. However, the role of support in suppressing agglomeration remains unclear to date because a high Tamman temperature of the support does not always ensure a high resistance to agglomeration. For example, the Tammann temperatures of ZrO_2 and $MgAl_2O_4$ are 1214 ºC and 929 ºC, respectively. Gayán *et al.*[1] identified that compared to $MgAl_2O_4$ (40 wt. %) a higher weight fraction of ZrO_2 (60 wt. %) was required to manufacture agglomeration-resistant oxygen carriers. In this work, we investigate the mechanism through which a support decreases the agglomeration tendency of Cu/CuO. For this, well-defined surfaces comprising thin films of Cu/CuO and a support (Al_2O_3, Y_2O_3 ZrO_2, MgO, SiO_2) were prepared *via* magnetron sputtering. The cyclic redox stability of the synthesized materials was evaluated in a thermo-gravimetric analyser (TGA) at 400 °C or 900 °C using 10 vol. % H_2 in N_2as the fuel and air for re-oxidation. Energy dispersive X-ray (EDX) spectroscopy on the cross-sections of the thin films revealed that Cu

atoms have the tendency to diffuse outward through the films of the support material during the redox reactions. The support that inhibits the outward movement of Cu/CuO is postulated to provide the highest agglomeration resistance.

1. Gayán P, Adánez-Rubio I, Abad A, de Diego LF, García-Labiano F, Adánez J. Fuel. 2012;96:226-38.

462983 Effects of Structural and Surface Promoters on Manganese-Containing Oxides in Cyclic Redox Reactions

Abstract:

Mn-containing oxides are frequently investigated as low cost and environmentally benign oxygen carriers for chemical looping combustion. The current study investigates Mn-based mixed oxides for two distinct redox applications, i.e. chemical looping with oxygen uncoupling (CLOU) for coal combustion and Chemical Looping based Oxidative Dehydrogenation (CL-ODH) for ethylene production from ethane.

In terms of CLOU, perovskite-structured oxygen carriers with a general formula of $A_xAâ€™_{1-x}Mn_yB_{1-y}O_3$ (A/Aâ€™ = Sr, Ca, Ba; B = Fe, Ni) are investigated. Although $CaMnO_3$ has the tendency to lose its oxygen carrying capacity through irreversible transition to spinel and Ruddlesden-Popper phases, small amount of Sr or Fe dopants enhances the redox stability of $CaMnO_3$. Moreover, the initial oxygen release temperature is significantly lowered, enabling facile oxygen donation and more effective coal combustion. DFT calculations were performed to interpret the dopant effect. It was determined that the aforementioned dopants changes the spatial locations, charge and vacancy formation energy for oxygen anions. Satisfactory correlations among vacancy formation energy, energy barrier for vacancy migration, free energy for phase transition, and redox activity and stability of Mn-containing oxygen carriers are observed.

With respect to CL-ODH, a number of Mn-containing model oxides are investigated. In this cyclic redox scheme, a Mn-based redox catalyst is used to for ethane ODH. Since air is used to regenerate the redox catalyst, CL-ODH avoids the air separation plant. Mg_6MnO_8 based model redox catalysts are investigated in the current study. The effect of various promoters are determined along with the corresponding reaction pathway and underlying mechanisms. It was determined that unpromoted Mg_6MnO_8 redox catalysts are highly selective for CO_2 formation, making it a potentially suitable oxygen carrier for chemical looping combustion. In comparison, promoters such as Na and W significantly improve the selectivity and ethylene yield for the ODH reaction. The overall ODH reaction proceeds through parallel gas phase cracking reaction and selective H_2 combustion on the surface of the promoted redox catalysts.

470135 Morphological Evolution of Transition Metal Nanostructures with Oxygen Dissociation and Migration during Chemical Looping Processes

Abstract:

Transition metal are heavily used in chemical looping technologies because of their high oxygen capacity and high thermal reactivity. These oxygen activities result in the oxide formation and oxygen vacancy formation that affect the nanoscale crystal phases and morphologies within these materials and their subsequent bulk chemical behaviour. In this study, two selected earlier transition metals manganese and cobalt, as well as two selected later transition metals copper and nickel, are investigated with their undergoing cyclic redox reactions. We found Co microparticles exhibited increased CoO impurity when oxidized to Co_3O_4 upon cyclic oxidation; CuO redox cycles prefer to be limited to a reduced form of Cu_2O and an oxidized form of CuO; Mn microparticles were oxidized to a mixed phases of MnO and Mn_3O_4, which causes delamination during oxidation. For Ni microparticles, a dense surface were observed during the redox reaction. The atomistic thermodynamics methods and density functional theory (DFT) calculations are carried out to investigate the effect of oxygen dissociation and migration on the morphological evolution of nanostructures during the redox processes. Our results indicate that the earlier transition metals (Mn and Co) tend to have stronger interaction with O_2 than the later transition metals (Ni and Cu). Also, our modified BrÃ¸nstedâ"Evansâ"Polanyi (BEP) relationship for reaction energies and total reaction barriers reveals that reactions of earlier transition metals are more exergonic and have lower oxygen dissociation barriers than those of later transition metals. In addition, it was found that for these transition metal oxides the oxygen vacancy formation energies increase with the depth. The oxide in the higher oxidation state of transition metal has lower vacancy formation energy, which can facilitate to form defective nanostructures. The fundamental understanding of these metal oxide reactions is essential in directing the development of metal oxide-based oxygen carriers for chemical looping applications.

468911 Pt Modified Fe_2O_3@CeO_2 Core/Shell Structures for Effective Thermochemical Water Splitting

Abstract:

Pt modified Fe_2O_3@CeO_2 core/shell structures for effective thermochemical water splitting

Davood Hosseini, Paula Macarena Abdala, Sung Min Kim and Christoph R. Müller

Laboratory of Energy Science and Engineering
Department of Mechanical and Process Engineering, ETH Zurich, Switzerland

Currently H_2 is produced on the large scale largely via the steam reforming of natural gas or the gasification of coal. To yield high-purity H_2, energy intensive purification and separation steps are required. Using chemical looping based approaches, these energy intensive separation steps could be avoided potentially [1]. Chemical looping based H_2 production is a two-step process. In the first step a synthesis gas (*i.e.* a mixture of H_2 and CO) reduces an oxygen carrier, typically a transition metal oxide. In the second step, the reduced oxygen carrier is re-oxidized with steam producing, after the condensation of un-reacted steam, a pure stream of H_2.

In this context, Fe_2O_3 is one of the most promising oxygen carriers owing to its favorable thermodynamics, wide availability and for being environmentally benign. However, at high temperatures (~ 700 – 1000°C) pure, *i.e.* unsupported, Fe_2O_3 deactivates after only a few cycles of reduction and oxidation if reduced fully to metallic Fe [2,3]. One approach to improve the redox stability of Fe_2O_3 is to support it on a high Tammann temperature material, *e.g.* Al_2O_3, ZrO_2, $MgAl_2O_4$ or CeO_2 [1]. The synthesis gas used in the reduction process can be produced *via* the dry reforming of methane (DRM) using an appropriate catalyst, *e.g.* Ni or Pt.

In this work, we report the development of a bi-functional material that allows the integration of the dry reforming of methane into the thermochemical water splitting cycle. The material is composed of a Fe_2O_3@CeO_2 core/shell structure, decorated with Pt nanoparticles. In this work the redox characteristics of the newly developed oxygen carriers were determined at 800 °C using both a thermo-gravimetric analyzer and a fixed bed reactor. To elucidate the effect of the addition of Pt nanoparticles on the reduction of iron oxide, in situ XRD studies were performed.

References

[1] Adánez J., Abad A., García-Labiano F., Gayán P., de Diego L.F. (2012), *Progress in Energy and Combustion Science* **38**, 215.
[2] Bohn C.D., Cleeton J.P., Müller C.R., Chuang S.Y., Scott S.A., Dennis J.S. (2010), *Energy Fuels* **24**, 4025.
[3] Bohn C.D., Müller C.R., Cleeton J.P.E., Hayhurst A.N., Davidson J.F., Scott S.A., Dennis J.S. (2008), *Ind. Eng. Chem. Res.* **47(20)**, 7623.

- Abstract Text

457210 Study on the Performance ofÂ Fe_2O_3/MgO/Al_2O_3 in Chemical-Looping Hydrogen Generation

Abstract:

Fe_2O_3/Al_2O_3, prepared by the impregnation method, was used as the oxygen carrier in Chemical-Looping Hydrogen Generation. The evaluation performance indicated that the oxygen carrier was active that methane was totally oxidized, and the hydrogen yield was notably high at first. However, carbon deposition was high when reacted with methane and it affected the purity of H_2. The addition of MgO to Fe_2O_3/ Al_2O_3 as a promoter was tested and characterized by BET, SEM, TPR, XRD and XPS methods. Fe_2O_3- MgO/ $LaNiO_3$ was evaluated under the same evaluation conditions as Fe_2O_3/Al_2O_3. The evaluation performance indicated that Fe_2O_3- MgO/ $LaNiO_3$ remained low carbon deposition even after 100 cycles. TPR and XPS characterization illustrated that MgO enhanced the dispersity of Fe_2O_3.

470919 Investigating Synergetic Effects Between Deposited PGM Nanoparticles and Transition Metal Carbide Nanotube Support through Surface Characterization and Device Performance

Abstract:

Platinum group metals (PGM) supported on transition metal carbide (TMC) surfaces (PGM/TMC) have been the focus of significant research interest, due to the similar electronic and geometric structures between TMC and Pt.[1] This presentation focuses on probing a new class of electrocatalyst fabrication using high purity Î²-molybdenum carbide nanotube (Mo_2C) as a support and atomic layer deposition (ALD) as the PGM deposition technique. Lattice spacing analysis and Pt binding energy shift in X-ray photoelectron spectroscopy (XPS) results showed strong interaction between deposited Pt particles and the TMC support. This interaction was correlated with enhanced hydrogen evolution (HER) and hydrogen oxidation reaction (HOR) activity. In addition, significantly increased durability was observed in constant potential electrolysis (CPE) and potential cycling. Finally, the Pt/Mo_2C catalyst performance was systematically evaluated in a device setting for different Pt loadings by applying it in anode and cathode of a proton exchange membrane fuel cell (PEMFC), with accelerated degradation test (ADT) being carried out between 0.6V to 1 V for 30,000 cycles at a scan rate of 500 mV s^{-1}.[2] Significantly higher power density was observed in MEAs with Pt/Mo_2C catalysts than 20% Pt/C was observed after the vigorous ADT.

References

1. R. B. Levy and M. Boudart, *Science*, 1973, 181, 547â€“549.
2. B. Han, C. E. Carlton, A. Kongkanand, R. S. Kukreja, B. R. Theobald, L. Gan, R. O'Malley, P. Strasser, F. T. Wagner and Y. Shao-Horn, *Energy Environ. Sci.*, 2014, **8**, 258â€“266.

470659 Bimetallic Ru-Pt and Pt-Co Fuel Cell Catalysts Prepared By Strong Electrostatic Adsorption and Electroless Deposition

Abstract:

The methods of Strong Electrostatic Adsorption (SEA), and its incipient wetness analog, Charge Enhanced Dry Impregnation (CEDI), have been shown to yield supported metal nanoparticle catalysts with high dispersion and narrow size distribution. Catalysts prepared by SEA and CEDI therefore are desirable as seeds for addition of secondary metal using Electroless Deposition (ED), as the prepared bimetallic catalysts should be of similar dispersion as the base catalyst. This study demonstrated the use of these methods to prepare two series of carbon supported bimetallic catalysts containing Pt which were then characterized and evaluated with cyclic voltammetry. The first system used Pt as the base metal, prepared by SEA on XC72R carbon, with Ru as the secondary metal added by ED (Pt@Ru/C). The second system used Co as base metal, prepared by a modified CEDI method on Carbon Black, with Pt added by ED (Pt-Co/C)

XRD, HAADF-STEM, and XEDS characterization showed that the Pt@Ru/C catalysts did not have alloying of the component metals and that the catalysts have bimetallic surface composition. The resulting Pt-Co/C catalysts had particles with irregular morphology that were larger than the seed Co particles. These were determined to have thin alloyed Pt-Co phases and domains of pure Pt. Evaluation of these catalysts showed higher mass activities, with respect to platinum content, compared to commercially available monometallic and bimetallic catalysts. This enhancement in performance is associated with the electronic interaction between Pt and Ru on the catalyst surface and the lattice contraction for Pt-Co alloys.

465251 Understanding the Effects of pH and Alkali Cations on H/OH Adsorption and Hydrogen Oxidation on Transition Metal Surfaces

Abstract:

Understanding how the complex interface between a catalytic electrode and electrolyte affects the rate and mechanism of electrocatalytic reactions is crucial to designing more active and selective electrocatalysts. As an example of this effect, it is now well known that the rate of the hydrogen oxidation reaction on a platinum electrode is orders of magnitude faster in acid than in base [1, 2] and that the rate varies continuously as a function of pH [3]. Sheng et al. have shown that the dependence of the rate on pH is correlated with what is believed to be a shift in the binding energy of hydrogen with pH, as measured by cyclic voltammetry [3]. We have recently shown using density functional theory (DFT) simulations that the low potential peaks measured by cyclic voltammetry on Pt(100) and Pt(110) are actually the co-adsorption or competitive adsorption of hydrogen and hydroxide, and that the shift of this peak with pH may not be solely related to the binding energy of hydrogen. We further show, in conjunction with our prior work on alkali cation specific adsorption [4], that potassium can specifically adsorb to Pt(111), Pt(100), and Pt(110) electrode surfaces at high pH and weaken the adsorption of hydroxide [5]. We see similar results on stepped surfaces of platinum, in particular Pt(553) and Pt(533), where both the binding strength of hydrogen on the step in the gas phase and in an electrochemical environment match only when competitive adsorption of hydroxide is considered. We extended this analysis of hydrogen and hydroxide adsorption across transition metal surfaces and find a similar effect of adsorbed alkali cation on hydroxide adsorption. Hydrogen and hydroxide competitive adsorption occurs on most surfaces which bind both species more strongly than Pt(111). Recent work by Zheng et al. has shown that the pH dependence of the hydrogen oxidation reaction and of the adsorption peaks measured by cyclic voltammetry are not unique to platinum, and occur across transition metals [6]. This has significant implications in the understanding of the hydrogen oxidation mechanism, where it has been proposed that adsorbed hydroxide may play a role in dictating the rate [7]. We will discuss our work using DFT to model hydrogen oxidation reaction mechanisms as well as other important conclusions we can draw (such as more accurately calculating electrode surface area or the surface restructuring which occurs with

platinum electrodes under electrochemical â€œconditioning/cleaningâ€) with our improved understanding of hydrogen and hydroxide adsorption on transition metal surfaces.

[1] Sheng, W.; Gasteiger, H.A.; Shao-Horn, Y. Hydrogen oxidation and evolution reaction kinetics on platinum: Acid vs Alkaline Electrolytes. *Journal of the Electrochemical Society*. **2010**, 157, B1529-B1536.

[2] Durst, J.; Siebel, A.; Simon, C., HaschÃ©, F.; Herranz, J.; Gasteiger, H.A. New insights into the electrochemical hydrogen oxidation and evolution reaction mechanism. *Energy Environ. Sci.* **2014**, 7, 2255-2260.

[3] Sheng, W.; Zhuang, Z.; Gao, M.; Zheng, J.; Chen, J. G.; Yan, Y. Correlating Hydrogen Oxidation and Evolution Activity on Platinum at Different pH with Measured Hydrogen Binding Energy. *Nat. Commun.* **2015,** *6:5848.*

[4] Mills, J.N.; McCrum, I.T.; Janik, M.J.; Alkali cation specific adsorption onto fcc(111) transition metal electrodes. *Phys. Chem. Chem. Phys.* **2014**, 16, 13699-13707.

[5] McCrum, I.T.; Janik, M. J. pH and alkali cation effects on the Pt cyclic voltammogram explained using density funtional theory. *J. Phys. Chem. C.* **2016**, 120, 457-471.

[6] Zheng, J.; Sheng, W.; Zhuang, Z.; Xu, B.; Yan, Y. Universal dependence of hydrogen oxidation and evolution reaction activity of platinum group metals on pH and hydrogen binding energy. *Sci. Adv.* **2016**, 2, e1501602.

[7] Strmcnik, D.; Uchimura, M.; Wang, C.; Subbaraman, R.; Danilovic, N.; van der, V.; Paulikas, A. P.; Stamenkovic, V. R.; Markovic, N. M. Improving the Hydrogen Oxidation Reaction Rate by Promotion of Hydroxyl Adsorption. *Nat. Chem.* **2013,** *5*, 300-306.

466793 Nanoparticle Catalysts Supported on Substitutionally Doped Graphene: Effects on Activity and Stability for Hydrogen Oxidation

Abstract:

Fuel cells represent an important part of a renewable future hydrogen-based energy grid. In particular, hydroxide exchange membrane fuel cells (HEMFCs) offer the advantage of non-noble metals being more stable against dissolution than in the acidic environment. However, the rate of the hydrogen oxidation reaction (HOR) decreases by approximately two orders-of-magnitude when switching from acid to base. As a result, developing more efficient and cheaper HOR catalysts is imperative to advancing the current state-of-the-art fuel cells. In a recent publication, we have demonstrated that Ni nanoparticles supported on nitrogen-doped carbon nanotubes (N-CNT) exhibit a threefold increase in the exchange current density relative to a pristine CNT support, and a factor of 30 increase relative to an amorphous carbon support (Zhuang, Z., *et al.*, *Nature Communications* **2016**, *7*, 10141). However, a fundamental study on the synergistic interaction between the supported nanoparticle and doped CNT has yet to be done.

Herein, using graphene as a two-dimensional analog for the CNT support used experimentally, we study from first-principles the impact of graphene and doped graphene supports on the hydrogen oxidation reaction occurring on nanometer-sized catalysts. We consider nickel, copper, and silver nanoparticle compositions, and nitrogen-doped, boron-doped, and phosphorous-doped graphene supports. To understand quantitatively the effect of substitutional doping on the nanoparticle/support interaction, we study: 1) the charge transfer between the nanoparticle and support as a function of dopant and dopant location, 2) the modulation of the nanoparticle's d band center due to the presence of a dopant, 3) the interaction of metal adatoms with graphene and doped-graphene, 4) the hydrogen oxidation activity of the nanoparticle/support systems, and 5) the oxidative stability of the supported copper and nickel nanoparticles.

467551 Impact of Combining Metal Nanoparticle Catalysts and Semiconductor Photo-Electrodes on Photo-Catalytic Performance

Abstract:

Semiconductors have found significant utility in the field of photo-electrochemistry as excellent optical absorbers. Metal nanoparticles are typically loaded onto these surfaces to add catalytic function to the system. The addition of metal nanoparticles also changes the optical properties of these systems since metal nanoparticles strongly interact with light through interband transitions and plasmonic phenomena. For a semiconductor substrate that is a good absorber and has long charge carrier diffusion lengths, the interaction of light with metallic nanoparticles will result in a loss of optical efficiency.

In this contribution, we will discuss our work on developing strategies for combining semiconductors and metal nanoparticle catalysts to simultaneously improve the catalytic and optical efficiencies of a photo-electrochemical device. We will present the results of our experimental and theoretical efforts based on a model system consisting of planar silicon electrodes and platinum nanoparticles as composite photo-electrodes for the hydrogen evolution reaction.

462154 Trends in Hydrogen Evolution Reaction Activity Among Metal Modified Carbide Thin Films and Powders

Abstract:

Water electrolysis, coupled with renewable energy sources, is an important technology for producing CO_2-free hydrogen gas that can be used for energy storage or CO_2 conversion. One of the major barriers to widespread electrolyzer use is the high cost of platinum-group metals required to catalyze the hydrogen evolution reaction (HER) and the oxygen evolution reaction (OER) at the electrolyzer cathode and anode, respectively. In the current study, metal-modified transition metal carbides were investigated as inexpensive replacements for bulk Pt for the HER. Transition metal carbides were synthesized as thin film electrocatalysts and modified with monolayer coverage of precious and non-precious metals. The HER activities of these catalysts were measured electrochemically in both acidic and basic electrolyte, and activity trends were established based on density functional theory (DFT) calculated hydrogen binding energy (HBE). Pt-modified niobium carbide showed promise as a low-cost alternative to bulk Pt for the HER, and it also showed impurity tolerance in the presence of common tap water ions. Thin film results were then extended to industrially relevant high surface area molybdenum carbides (Mo_2C) with dispersed metal nanoparticles. Pt-Mo_2

techniques for well-defined high surface area catalysts in order to facilitate the use of DFT to help screen for active HER catalysts.

461163 Experimental Determination of Water Orientation Effects on Kinetic Barriers to Alkaline Hydrogen Oxidation and Evolution

Abstract:

It has long been recognized that the reaction rates of the hydrogen oxidation and hydrogen evolution reactions (HOR and HER) are slower in basic than acidic electrolytes, even though the surface intermediate of adsorbed hydrogen is independent of solution pH. Understanding the root of this observation is critical to designing catalysts for a multitude of electrochemical reactions with relevance to energy conversion and storage. In this work, we undertake a fundamental investigation relating interfacial water structure to the reaction barrier for hydrogen underpotential adsorption and to the HER/HOR kinetics.

The pH-dependence of HOR/HER kinetics has been attributed to a variety of reasons. The research groups of Gasteiger and Yan have both suggested that in base, hydroxide ions stabilize the Pt-H bond for stronger binding and slower catalysis [1,2]. Central to this hypothesis is an experimentally measured shift with pH in the peak potential for hydrogen underpotential deposition on the (100) and (110) steps on the Pt surface. Although it is unclear why pH would stabilize adsorbed hydrogen, stronger binding would push platinum further to the right of the volcano peak and decrease the overall activity. Markovic et al proposed instead that in base, the water recombination/dissociation step of HOR/HER requires specific adsorption of hydroxide ions and therefore optimal binding to two adsorbates [3]. One observation that has been less discussed than either hydrogen or hydroxide adsorption is the effect of pH not only on the apparent hydrogen adsorption energy, but the kinetic barrier to adsorption. Koper et al attributed this observation to a pH-dependent water orientation at the surface, and hypothesized that "H-down" in acid vs "H-up" in base resulted in slower transfer of hydrogen to and from the surface [4]. The same pH-dependence of water orientation was recently supported computationally by Rossmeisl's group[5].

In this work, we examine specifically the hypothesis that water orientation governs the rate of hydrogen adsorption and thus the overall HER/HOR kinetics. The native oxide formed on chromium metal is known from colloidal literature to have a potential of zero charge (Epzc) of 1.22V vs. RHE[6]. We hypothesize that if water orientation truly governs the alkaline HOR/HER rate, the negative oxide surface charge on chromium oxide will induce an H-down water orientation adjacent to platinum nanoparticles supported on the surface. This will result in faster hydrogen adsorption kinetics and thus, higher activity for alkaline HOR/HER than extended platinum surfaces.

To test this hypothesis, platinum nanoparticles were synthesized via pulsed electrodeposition on chromium oxide. The nanoparticle loading was controlled by the duration of deposition pulse, with longer times corresponding to higher coverage as well as larger particles. The hydrogen adsorption kinetics were measured by cyclic voltammetry in the hydrogen underpotential deposition region. Varying the sweep rate results in greater peak separation that can be used to extract rate constants for the adsorption reaction via traditional electroanalysis[7].

A typical voltammogram of Pt/CrOx in 0.1M KOH is shown in Fig. 1a. As the scan rate increases, kinetic barriers cause greater peak separation (dEp) for hydrogen adsorption on the 100 and 110 facets of platinum. For lower loadings with shorter pulse periods, both smaller currents and smaller peak separations are observed due to the lower Pt surface area. The effects of surface area can be accounted for by normalizing current to Pt surface area as measured by Hupd coulometry. The resulting surface-adsorption Tafel plot is shown in Fig 1b. The peak potential separation dEp is much greater for the larger nanoparticles with higher loading. This difference suggests that smaller particles have qualitatively more rapid adsorption kinetics due to the advantageous water orientation induced by negatively charged CrOx.

The results of this study contribute to resolving a long-standing paradox in electrocatalysis and surface science by highlighting the importance of kinetic barriers, as well as adsorption energies, and suggest the ability to modulate alkaline hydrogen activity through metal-support interactions. Future work will discuss the relation of hydrogen underpotential deposition kinetics to HER/HOR activity and the effects of different oxide supports.

(1) Durst et al. Energy Environ. Sci. 2014, 2255.

(2) Sheng et al. Nat. Commun. 2015, 5848.

(3) Strmcnik et al, Nat. Chem. 2013, 1.

(4) Van Der Niet et al. Catal. Today 2013, 105.

(5) Rossmeis et al. Phys. Chem. Chem. Phys. 2013, 10321.

(6) McCafferty, E. Electrochim. Acta 2010, 1630.

(7) Angerstein-Kozlowska et al. J. Electroanal. Chem. Interfacial Electrochem. 1979, 1.

459138 Co-Oxidation of CO and Hydrocarbons on Pd/Ceria-Zirconia/Al2O3 Three-Way Catalysts: Experiments and Modeling

Abstract:

Co-oxidation of CO and Hydrocarbons on Pd/Ceria-Zirconia/Al2O3

Three-way Catalysts: Experiments and Modeling

Wendy Lang[1], Michael P. Harold[1*], Yisun Cheng[2], Carolyn Hubbard[2], Manish Sharma[2], and Paul Laing[2]

[1]*Department of Chemical and Biomolecular Engineering, University of Houston, Houston, TX*

[2]*Research and Innovation Center, Ford Motor Company, Dearborn, MI*

*mharold@uh.edu

Three-way catalyst (TWC) light-off performance is critical in cost effectively meeting future emission standards. Improvements in catalyst light-off (LO) should include catalyst modifications that minimize inhibition effects of exhaust species such as CO and hydrocarbons (HCs), lowering precious group metal (PGM) loading, or using a lower cost PGM such as Pd. Alumina (Al_2O_3) and zirconia (ZrO_2) supports provide high surface area and stability. Furthermore, ceria (CeO_2) supports for PGMs provide the key benefit of oxygen storage, compensating for deviations from stoichiometric TWC operating conditions by supplying oxygen for CO and HC oxidation under rich transient conditions. Studies have demonstrated that inclusion of ceria in PGM catalysts enhances CO and HC conversion and minimizes self-inhibition; this improved activity is mainly attributed to ceria containing more efficient oxygen activation sites at the PGM-support interface relative to other support materials such as Al_2O_3 [1-4].

Monolith catalysts having washcoats with 1 wt% Pd, 3 g/in^3 loading, 400 cpsi and formulations Pd/Al_2O_3, Pd/CZO, and Pd/CZO/Al_2O_3 (17 wt% CZO, balance Al_2O_3) were synthesized. Catalyst performance studies were conducted to elucidate the role of Pd and support (CZO, Al_2O_3) compositions and loadings on inhibition during CO and C_3H_6 LO under near-stoichiometric conditions. Transient and steady-state bench scale reactor studies with simulated exhaust gas mixtures were conducted to extract oxidation kinetics, compare LO behavior, and develop a predictive model for understanding and optimizing catalyst performance.

Catalyst activities were compared via temperature-programmed oxidation experiments. Figure 1 compares CO LO curves using feed concentrations of 0.5% CO and 1% CO ($\lambda = 1.01$ in both cases). Using all three catalysts, LO temperatures increase with feed CO concentration, showing that CO oxidation is self-inhibiting. With increasing CZO content, LO temperatures decrease, demonstrating the promotional effects of using ceria.

Figure 2 shows C_3H_6 LO using feed concentrations of 250 ppm C_3H_6 and 500 ppm C_3H_6 ($\lambda = 1.01$). Like CO, C_3H_6 is self-inhibiting. The shoulder feature in the LO curves using Pd/Al_2O_3 and Pd/CZO/Al_2O_3 is speculated to be due to a change in the oxidation state of Pd and/or the accumulation of inhibiting carbonaceous surface species. LO begins at lower temperatures and the shoulder becomes less prominent with increasing CZO content and is absent with Pd/CZO.

Comparing the individual oxidations of 1% CO and 500 ppm C_3H_6 in Figures 1 and 2 to LO behavior during their co-oxidation (1% CO + 500 ppm C_3H_6) in Figure 3, LO temperatures for both species are higher in the mixture, i.e. CO and C_3H_6 are mutually-inhibiting. During co-oxidation, CO lights off before C_3H_6 and the C_3H_6 shoulder does not appear, suggesting that CO inhibits C_3H_6 adsorption and the accumulation of carbonaceous species on the catalyst. Once CO lights off and temperatures increase, C_3H_6 oxidation proceeds. Significantly lower LO temperatures are achieved using Pd/CZO, and higher conversions of CO and C_3H_6 are reached at lower temperatures using catalysts containing CZO.

Differential kinetics studies were conducted to quantify reaction orders and activation energies. The reaction order with respect to CO was found to be approximately −1 using Pd/Al_2O_3 and $Pd/CZO/Al_2O_3$ and 0 to slightly negative using Pd/CZO, consistent with literature values [5-7]. The activation energies for CO oxidation was found to be 125-149 kJ/mol using Pd/Al_2O_3 and $Pd/CZO/Al_2O_3$, and 29-44 kJ/mol using Pd/CZO. Using the Pd/Al_2O_3 catalyst, the reaction order with respect to C_3H_6 was found to be approximately −1 and activation energy 77 kJ/mol.

Ongoing work involves the use of a mechanistic-based kinetic model incorporated into a low-dimensional monolith model [8-9]. The global reactor models are used with CO and C_3H_6 oxidation on Pd with steps involving ceria. The 1+1 dimensional model formulation comprises coupled species balances with transverse-average equations having external and internal mass transfer coefficients. As shown in Figure 4, the model verifies the experimental LO of CO on Pd/Al_2O_3. For example, it validates additional experimentally-observed oxidation behavior such as the switch in order of ignition between individual and mixture oxidations on Pd/Al_2O_3 and $Pd/CZO/Al_2O_3$ as well as the promotional effect of ceria.

Steady-state and transient CO and C_3H_6 oxidation experiments using fresh Pd/Al_2O_3, Pd/CZO, and $Pd/CZO/Al_2O_3$ monolith catalysts were conducted to compare LO behavior and kinetic parameters (i.e. reaction orders and activation energies). CO and C_3H_6 individual and mixture LO behavior was improved with increasing catalyst ceria content. A reactor model confirms experimentally-observed oxidation behavior and predicts catalyst performance. Finally, experiments involving other hydrocarbons including acetylene and toluene are being conducted to investigate inhibition during LO, extract kinetic parameters for oxidation reactions, and provide a basis for validating modeling simulations.

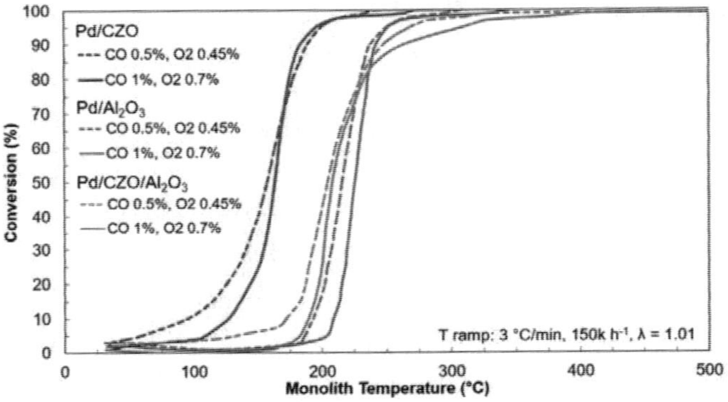

Figure 1: CO conversion (%) versus monolith temperature using fresh Pd/CZO, Pd/Al_2O_3, and $Pd/CZO/Al_2O_3$ catalysts for feed concentrations of 0.5% CO (λ = 1.01) and 1% CO (λ = 1.01).

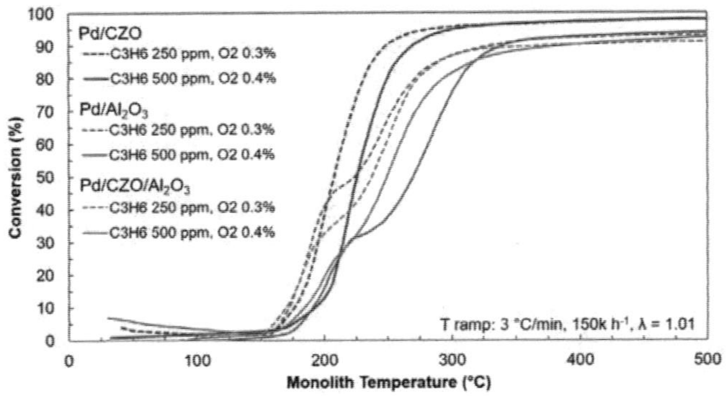

Figure 2: C_3H_6 conversion (%) versus monolith temperature using fresh Pd/CZO, Pd/Al_2O_3, and $Pd/CZO/Al_2O_3$ catalysts for feed concentrations of 250 ppm C_3H_6 (λ = 1.01) and 500 ppm C_3H_6 (λ = 1.01).

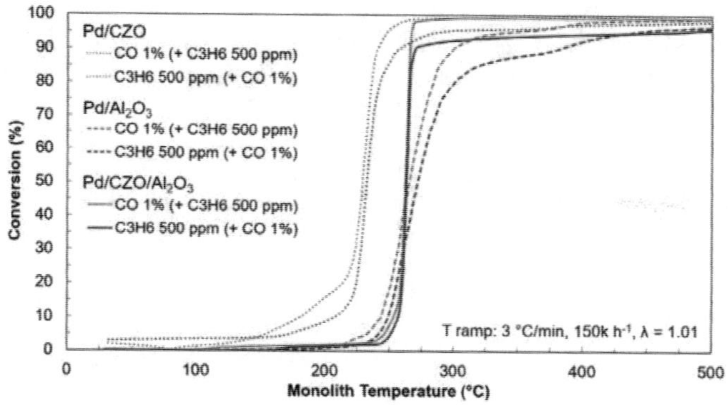

Figure 3: CO, C_3H_6 conversion (%) versus monolith temperature using fresh Pd/CZO, Pd/Al_2O_3, and Pd/CZO/Al_2O_3 catalysts for feed concentration of 500 ppm C_3H_6, 1% CO ($\lambda = 1.01$).

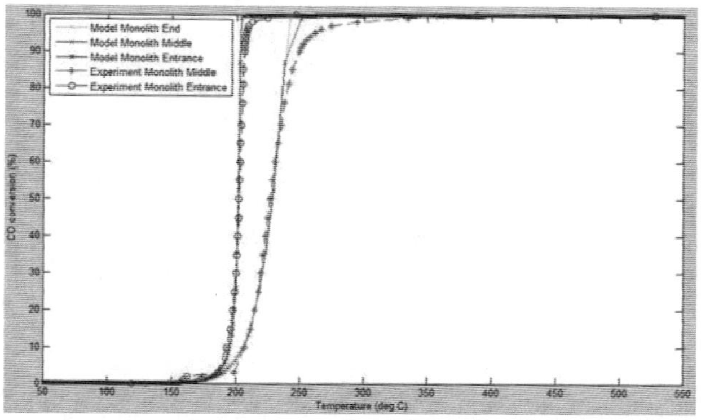

Figure 4: CO conversion (%) versus temperature using fresh Pd/Al_2O_3 catalysts for experimental and simulated feed concentration of 1% CO ($\lambda = 1.01$).

References

[1] Fernandez-Garcia, M.; Martinez-Arias, A.; Iglesias-Juez, A.; Hungria, A.B.; Anderson, J.A.; Conesa, J.C.; and Soria, J. *App. Cat. B* 2001, 3**1**, 39-50.

[2] Martinez-Arias, A., Fernandez-Garcia, M.; Iglesias-Juez, A.; Hungria, A.B.; Anderson, J.A.; Conesa, J.C.; and Soria, J.. *App. Cat. B* 2001, **31**, 51-60.

[3] Sharma, S.; Hegde, M.S.; Das, R.N.; and Pandey, M. *App. Cat. A* 2008, **337**, 130-137.

[4] Harmsen, J.M.A.; Hoebink, H.B.J.; and Shouten, J.C. *Chem. Eng. Sci.* 2001, **56**, 2019-2035.

[5] Yao, Y-.F. Y. *J. Catal.*, 1984, **87**, 152-162.

[6] Cant, N.W.; Hicks, P.C.; and Lennon, B.S. *J. Catal.* 1978, **54**, 372-383.

[7] Rainer, D.R.; Koranne, M.; Vesecky, M.; and Goodman, D.W. *J. Phys. Chem. B* 1997, **101**, 10769-10774.

[8] Joshi, S.; Harold, M.P.; Balakotaiah, V. *AIChE J.,* 2009, **55,** 1771-1783.

[9] Raj, R.; Harold, M.P.; Balakotaiah, V. *Chem. Eng. J.*, 2015, **281**, 322-333.

456585 Interaction Between Highly Dispersed Ionic Pd2+ and Mn-Ce Solid Solution Support for Low Temperature CO Oxidation

Abstract:

Interaction between highly dispersed ionic Pd^{2+} and Mn-Ce solid solution support for low temperature CO oxidation

Chao Wang, Cun Wen, Erdem Sasmaz, Jochen Lauterbach*

Department of Chemical Engineering, University of South Carolina, Columbia, SC, 29208, U.S.A.

*lauteraj@cec.sc.edu

Supported Pd catalysts are widely used in automobile three-way catalysts (TWC) and diesel oxidation catalysts. While Pd catalysts perform well for CO and hydrocarbon emission control during continuous engine operation, 80â€“90% of CO and hydrocarbons are emitted into the atmosphere during the cold start phase of an automobile when the catalyst temperature is low [1]. Hence, promotion of catalytic activity at low temperatures is desired to achieve better removal of pollutants. Pd-CeO_2 catalyst is a widely studied category of supported Pd catalysts for low temperature CO oxidation. By doping Mn or both Mn and Sn into the CeO_2 lattice and forming a solid solution, we found that Pd-Ce based catalysts show excellent CO oxidation activity at even ambient temperature [2]. Metal-support interaction has been proposed as one of the most important factors that determine the Pd-CeO_2 catalyst activity at low temperature, while doping a secondary metal into the CeO_2 support may improve and change such interaction between Pd and Ce based supports. So far, the dynamic Pd structure change and interaction with the Ce based solid solution supports before and during reaction are not yet fully understood.

In this study, Pd supported on Mn doped CeO_2 solid solution (Pd-MC) and pure CeO_2 are synthesized and compared to evaluate metal-support interaction. CO oxidation tests show that Pd-MC achieves almost 100% CO conversion at 50°C, which is much higher than Pd-CeO_2 (40%). The high activity of Pd-MC can be attributed to the formation of more highly dispersed ionic Pd^{2+} on MC and strong oxygen transfer interaction between Pd species and MC support. Detailed dynamic changes of Pd sites on MC and CeO_2 before and during reaction were characterized and compared by a series of ex-situ and in-situ spectroscopic experiments, including X-ray photoelectron spectroscopy (XPS), X-ray absorption spectroscopy (XAS), and diffuse reflectance infrared Fourier transform spectroscopy (DRIFTs). H_2 temperature-programmed reduction (H_2-TPR) and CO-O_2 pulse were used to investigate the effect of metal-support interaction on the redox properties of the catalysts. In addition, for practical implementation of low temperature CO oxidation catalysts, effects of hydrothermal treatment, steam, hydrocarbon, and space velocity on the catalytic activity were evaluated on Pd-MC.

References

1. M. Shelef, R. W. McCabe, Catal. Today 62, (2000) 35-50.

2. C. Wang, E. Sasmaz, C.Wen, J. Lauterbach, Catal. Today 258 (2015) 481â€“486.

472730 Effect of Rare-Earth Metal Dopants on Ceria Catalysts for CO Oxidation

Abstract:

Ceria has been intensively studied in a field of catalysis with a variety of applications, such as automotive emission control, catalytic combustion, and hydrocarbon reforming, and electrocatalytic reactions. Especially, ceria has been used as an active component in three-way catalysis (TWC) where pollutant emission from internal combustion engines is effectively reduced, due to the nature of outstanding redox behavior being amenable for adsorbing and giving off the oxygen from the atmosphere. Although most researchers have investigated transition metals-doped ceria, rare-earth metal-doped ceria nanocatalysts for CO and soot oxidation may be better support materials.

In this study, we examine CeO_2 (ceria) doped by various rare-earth metals to search for the suitable CO oxidation activity. We synthesized rare-earth metal (RE=Pr, Sm, Nd, and La) doped CeO_2 nanoparticles with two different RE concentrations, $Ce_{0.9}RE_{0.1}O_{2-\delta}$ (RE=10%) and $Ce_{0.8}RE_{0.2}O_{2-\delta}$ (RE=20%), by EDTA-citrate complexing method which is capable of producing highly pure and homogeneous nano-powders. Then, the nanoparticles were characterized by X-ray diffraction (XRD), Scanning Electron Microscopy (SEM), and Transmission Electron Microscopy (TEM). The surface area was also evaluated by Brunauer-Emmett-Teller (BET) measurement. The catalytic efficiency of RE-doped CeO_2 for CO oxidation was measured by gas chromatography (GC). Our results will be helpful for effectively exploring the RE dopant candidates of ceria and the optimal concentration for the enhanced catalytic activity.

456809 Carbon Monoxide Oxidation and Preferential Oxidation on Pt Alloy Nanoparticle Catalyst with Engineered Surface

Abstract:

Carbon monoxide oxidation and preferential oxidation (PROX) are two important reactions for environmental control and for applications requiring the use of clean hydrogen, for instance hydrogen fuel cell technology and ammonia synthesis. It is because that the typical hydrogen production processes by reforming hydrocarbons and diesel fuels generate CO as a byproduct, which is notorious poisoning species and readily deactivating many catalysts. There is thus a need for selective removal of CO before hydrogen can be used for these applications. We report the study of Pt alloy nanoparticles with well-engineered surfaces as low temperature CO oxidation and PROX catalyst. In particular, the researched octahedral Pt3Ni alloy nanoparticle catalysts, which are exclusively enclosed by the (111) surface, exhibit both high PROX activity and 100% selectivity from RT all the way till around 100 °C. The much promoted CO oxidation kinetics on octahedral Pt3Ni was attributed to significantly decreased Ea comparing to spherical Pt3Ni and Pt (~40.0 kJ/mol vs. 56.9 and 66.3 kJ/mol) and the presence of both Pt and Ni active sites on the engineered (111) surface, which can work synergistically to activate and react CO and O2. The 100% PROX selectivity was associated with the unique (111) surface geometry of the octahedral Pt3Ni and the big difference in the adsorption energy of CO and H2 on the surface, which help with suppressing H2 from adsorption and oxidation.

481309 Sensitivity of XANES and XES to the local environment: Multiple adsorption of H_2O and NH_3 on Cu-SSZ-13

Abstract:

Sensitivity of XANES and XES to the local environment: Multiple adsorption of H_2O and NH_3 on Cu-SSZ-13

Renqin Zhang[1], Hui Li[1], Christopher Paolucci[2], Atish A. Parekh[3], Janos Szanyi[4], Feng Gao[4], Trunojoyo Anggara[2], Hui Li[2], William F. Schneider[2], Fabio Ribeiro[3] and Jean-Sabin McEwen[1*]

1. Washington State University, Pullman, WA; 2. University of Notre Dame, Notre Dame, IN; 3. Purdue University, West Lafayette, IN; 4. PNNL, Richland, WA

*js.mcewen@wsu.edu

Nitrogen oxides (NO_x) are one of the main air pollutants present in the exhaust from diesel engines, which are popular for vehicle transportation due to their efficiency and durability [1]. However, NO_x emission control is a challenge in these "lean-burn" engines [2]. As is well known, the selective catalytic reduction (SCR) of NO_x with NH_3 is a reaction between the NO, NO_2, and O_2 oxidants and the NH_3 reductant to form N_2 and H_2O. Copper-exchanged small-pore micro-crystalline materials with the chabazite structure (Cu/CHA), such as Cu-SSZ-13, display excellent catalytic activity and hydrothermal stability in SCR of NO_x, as has been shown in a number of recent studies [3, 4]. Although Cu-SSZ-13 have been commercialized as diesel after-treatment catalysts, the fundamental chemical and physical properties need to be characterized in order to aid in the design of new and better catalysts [5].

X-ray absorption spectroscopy (XAS) is a versatile tool to determine the oxidation state and the local structure of Cu in Cu/CHA. Cu K-edge X-ray absorption near edge spectra (XANES) has been widely used to study the properties of Cu-SSZ-13 because of its high activity and selectivity in the NH_3 SCR of NO_x. It is demonstrated that XANES is extremely sensitive to the local structure of Cu ions [6, 7]. However, the exact relationship between XANES features and local structure is still unclear. In this work, we examine how the geometry of a Cu^+ ion with a linear configuration and Cu^{2+} ion with a square planar configuration in Cu-SSZ-13 correlate with the XANES features. When water and ammonia binds to a Cu^+ ion with a linear configuration, a strong intensity peak around 8983 eV in Cu K-edge XANES appears. When two ammonia molecules bind at a Cu^+ ion, it is also found that the intensity of the peak around 8983 eV decreases as one decreases the N-Cu-N bond angle from 180 to 100 degree (see Figure 1a). By analyzing the corresponding PDOS, it is found that the splitting of 4p state correlates with the N-Cu-N bond angle and the corresponding decrease in the K-edge intensity. Concerning the case of Cu^{2+} ions with a square planar configuration bounded to four ammonia molecules, it is found that there are two feature peaks around 8986 and 8993 eV in their Cu K-edge XANES. By studying this simple model, we find that changing the N-Cu-N angle between one of the bonded NH_3 and the Cu-centered ion from 180 to 100 degree results in a decrease in the intensity of both peaks around 8986 and 8993 eV. It is also found that the broken of square planar configuration results in the delocalization of 4p state.

An important limitation of XANES is that it is difficult to distinguish the nitrogen and oxygen atoms in the proximity of the metal center. X-ray emission spectroscopy (XES) could overcome this limitation. It is reported that the kbeta″ peak in XES of a Cu-SSZ-13 sample presents a blue shift with the formation of Cu-N bond [8], as shown in Figure 1b. This can be understood by the fact that the kbeta″ peak originates from the ligand 2s to a metal 1s crossover transition and the nitrogen 2s level lies higher in energy than the one for oxygen. By comparing the XES results between H_2O and NH_3 adsorbed on Cu-SSZ-13, it is found that the kbeta″ peak of Cu-O bond appears at lower energy than that of Cu-N bond. In addition, the coordination number of Cu ions significantly affects the energy difference between the position of the kbeta″ peak. By performing a density of states (DOS) analysis, these results can be understood at the electronic level. Overall, our careful elucidation of the spectroscopic properties of Cu ions in Cu-SSZ-13 via theoretical techniques can assist us in interpreting the XANES and XES experimental results so as to gain a deeper understanding of the SCR of NO_x with NH_3 throughout its catalytic cycle.

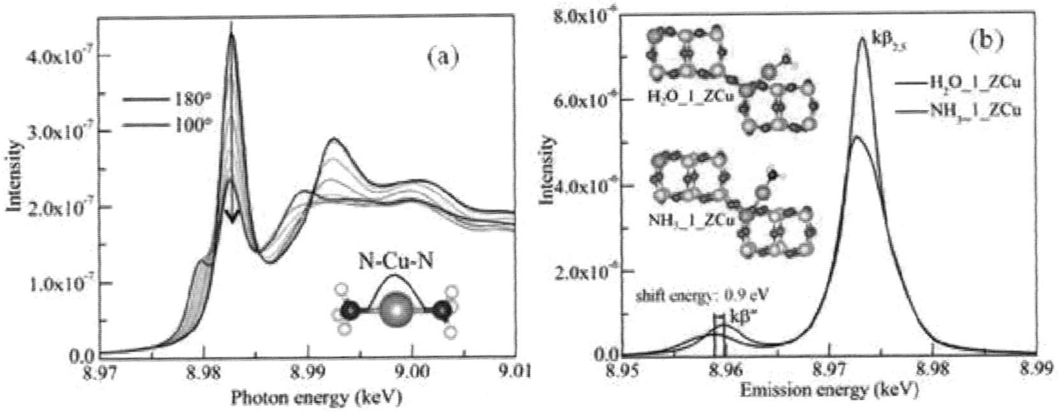

Figure 1. (a) Variation of Cu K-edge XANES for 2 NH_3 adsorbed on Cu^+-SSZ-13 while changing the N-Cu-N angle (as shown insert in the right bottom) from 180 degrees (black line) to 100 degrees (blue line). (b) Computational XES of Cu ions in H_2O and NH_3 adsorbed on Cu^+-SSZ-13. Two emission lines, kbeta" and $kbeta_{2,5}$, are presented.

References

[1] U. Deka, I. Lezcano-Gonzalez, B.M. Weckhuysen, A.M. Beale, ACS Catal., 3 (2013) 413

[2] S.i. Matsumoto, CATTECH, 4 (2000) 102

[3] J.H. Kwak, R.G. Tonkyn, D.H. Kim, J. Szanyi, C.H. Peden, J. Catal., 275 (2010) 187

[4] L. Ma, Y. Cheng, G. Cavataio, R.W. McCabe, L. Fu, J. Li, Chem. Engin. J., 225 (2013) 323

[5] F. Gao, J. Kwak, J. Szanyi, C.F. Peden, Top. Catal., 56 (2013) 1441

[6] R. Zhang, J. Szanyi, F. Gao, J.-S. McEwen, Catal. Sci. Technol., 6 (2016) 5812

[7] R. Zhang, K. Helling and J.-S. McEwen, Catal. Today, 267 (2016) 28

[8] F. Giordanino, E. Borfecchia, K.A. Lomachenko, A. Lazzarini, G. Agostini, E. Gallo, A.V. Soldatov, P. Beato, S. Bordiga and C. Lamberti, J. Phys. Chem. Lett., 5 (2014) 1552

469446 Mercury Oxidation over Cu-SSZ-13 Catalysts Under SCR Conditions for Power Plant Applications

Abstract:

Many combustion-based systems, such as power plants and automobiles, utilize selective catalytic reduction (SCR) as a pollution control technology to reduce the emissions of nitrogen oxide (NO_x) species. In power plants, the reduction of the NO_x occurs via the injection of NH_3 over a V-(Mo or W)/Ti oxide catalyst. Recently, the promulgation of the Mercury and Air Toxic Standards regulates mercury, requiring a 90% reduction of Hg emissions. While activated carbon injection is one solution to limit Hg pollution, it can be costly. An alternative is the catalytic oxidation of the elemental Hg to Hg^{+2} via HCl, where the oxidized pollutant is readily water soluble and easily captured downstream in the flue gas desulfurization unit. In this way, the traditional SCR material acts as a bifunctional catalyst limiting both NO_x and Hg emissions. While previous studies have examined the oxidation activity of the commercial SCR catalyst, they often display limited activity, falling short of the required emissions limits. To address this, multiple new materials have been proposed for Hg oxidation. In this study, small pore zeolite Cu-SSZ-13 was investigated for its possible application in coal-fired power plants as a new bifunctional SCR and Hg oxidation catalyst.

In this study, SSZ-13 and Cu-exchanged SSZ-13 are compared to a commercial SCR catalyst with respect to their Hg oxidation activity under a variety of flue gas compositions and temperatures. A simulated flue gas environment is created through the combustion of methane in air in a high-temperature furnace. Other pollutants common to coal-combustion flue gas, such as HCl, SO_2, NO, and NH_3 are injected downstream of the furnace but upstream of the packed-bed reactor housing the catalyst. Concentrations of elemental and oxidized mercury were measured using a PS Analytical mercury analyzer to determine the amount of mercury oxidation. Initially, the flue gas composition was varied to examine the effect of individual flue gas components on Hg oxidation with the catalysts kept at 375°C, an intermediate SCR temperature. The compositions investigated include flue gas only (CO_2+H_2O+N_2+O_2), flue gas with HCl, with HCl+SO_2 and with HCl+NO+NH_3. Finally, the temperature effect was studied by varying the temperature between 300-450°C, representative of standard operating temperatures in SCR units for coal-fired power plants.

459898 Thiophenic Species Removal over a New Zn-Offretite Zeolite Via Selective Adsorption for FCC Processes

Abstract:

This study considers the sulfur selective adsorption of a new Zn-Offretite zeolite based additive under FCC conditions. This Zn-Offretite (Zn-OFF) zeolite was prepared via direct inclusion of zinc in the OFF synthesis. The goal is to obtain a sorbent suitable for thiophenic chemical species removal via selective adsorption under FCC conditions. Physicochemical characterizations combined with reactivity studies were developed to understand the adsorption process. It was proven that the introduction of Zn in the OFF zeolite considerably increases acidity and Lewis/Brönsted acid site ratio. These two properties promote the thiophenic selective adsorption in the OFF structure. The valuable performance of the Zn-OFF additive was demonstrated in a CREC Riser Simulator unit under typical FCC conditions. Reactivity studies showed that the Zn-OFF additive is a suitable sorbent for thiophenic species adsorption with an optimum operation at 5s and 530°C. It was found that the Zn-OFF in combination with a FCC commercial catalyst can reach 51 wt% of 2-methyl-thiophene reduction with only 2wt% of sulfur left as coke. In addition, and when the FCC catalyst was blended with 5-15wt% of 3.5Zn-OFF, it yielded a 210% coke reduction. On the basis of the above, it is demonstrated that the Zn-OFF additive can contribute to both selective adsorption of sulfur containing species and lower coke. This reduced coke may become a significant Zn-OFF additive feature for mitigating SOx emissions in the FCC regenerator.

468495 Wash-Coat Development for Lean-Burn Engine-Exhaust Aftertreatment: A Novel Way of Incorporation of Binder to Pd Supported on Sulfated Zirconia

Abstract:

Wash-coat development for Lean-burn Engine-exhaust Aftertreatment: A Novel Way of Incorporation of Binder to Pd Supported on Sulfated Zirconia

Sreshtha S. Majumdar, Gokhan Celik, Anne-Marie Alexander, Umit S. Ozkan*The Ohio State University, Columbus, Ohio 43210, U.S.A.

*ozkan.1@osu.edu

As the environmental regulations are getting more stringent, it is of great importance to develop a catalytic aftertreatment system for natural gas-fired lean-burn engines to reduce the emissions of nitrogen oxides (NO_x), unburned hydrocarbons (CH_4, C_2H_6, and C_3H_8) and carbon monoxide efficiently. We have developed a dual-catalyst aftertreatment system which utilizes the hydrocarbons present in the exhaust streams of lean-burn engines for reducing NO_xemissions. The dual-catalyst aftertreatment system is a physical mixture composed of a reduction catalyst, palladium supported on sulfated zirconia (Pd/SZ), and an oxidation catalyst, cobalt supported on ceria. Such a catalytic system which utilizes methane in the exhaust stream offers several advantages considering that the emissions of air-pollutant greenhouse gases are controlled in a single unit without a need of injecting and handling an external reducing agent such as ammonia [1-3]. Development of a catalytically active washcoat for monolith cores is essential for the practical use of this dual-catalyst system. However, if the washcoat adheres poorly to the monolith core, the aftertreatment unit will suffer from the irreversible loss of the catalytically active phase. In order to improve the adhesive properties, conventionally, binders are used in a catalyst slurry.

In this study, we have aimed to improve the adhesive properties at the molecular level. For this purpose, alumina, as a binder, was incorporated in situ to the sol-gel medium of Pd/SZ prior to the gelation during synthesis. Samples prepared by this novel approach have shown superior differences in terms of catalytic performance than the samples prepared by the conventional method. Addition of binder before gelation to the sol-gel medium also resulted in changes of textural and structural properties of binder-free samples, and adhesive properties of the washcoat as shown by N_2 physisorption, in-situ X-ray diffraction (XRD) calcination, ^{27}Al-NMR, laser RAMAN and infrared spectroscopy, scanning electron microscopy (SEM) and ultrasonication, Water and sulfur tolerance as well as hydrothermal stability of the binder-incorporated catalyst will be presented as well.

[1] B. Mirkelamoglu, M. Liu, U.S. Ozkan, Catalysis Today, 151 (2010) 386-394.

[2] P. Gawade, A.-M.C. Alexander, R. Silver, U.S. Ozkan, Energy and Fuels, (2012).

[3] P. Gawade, B. Bayram, A.-M.C. Alexander, U.S. Ozkan, Applied Catalysis B: Enviromental (2012).

469481 Carboxylation of Propylene Oxide to Propylene Carbonate in Slurry and Trickle Bed Reactors

Abstract:

Carboxylation of Propylene oxide to Propylene carbonate in slurry and trickle bed reactorS

Pallavi Bobba, Xin Jin, Bala Subramaniam and Raghunath V. Chaudhari

Center for Environmentally Beneficial Catalysis, Department of Chemical & Petroleum Engineering, University of Kansas,
1501 Wakarusa Dr., Lawrence, KS 66047

CO_2 is naturally available carbon source and emitted from industrial processes, automobiles and petrochemical refineries. Catalytic conversion of CO_2 to value added chemicals provides an alternative green, cheap and sustainable synthesis route for industrial chemicals, otherwise produced using toxic reagents such as phosgene. An important example is the carboxylation of propylene oxide (PO) to propylene carbonate (PC), a key intermediate for dimethyl carbonate and polycarbonates. These cyclic carbonates are widely used as aprotic solvents, antifoaming agents, antifreeze, plasticizers and monomer for various commodity polymers. Several studies are known on synthesis of cyclic carbonates by carboxylation of epoxides employing homogeneous and heterogeneous catalysts. Heterogeneous catalysts consisting of metal oxides, zeolites, polymer supported quaternary onium salts, ion exchange resins, and polymer supported ionic liquids have been studied with the goal of improving catalytic activity and selectivity. However, no efforts have been made to understand the intrinsic kinetics of carboxylation and the related reaction engineering studies. Here, we report an experimental study on (a) intrinsic kinetics of carboxylation of PO to PC using ion exchange resin catalyst in a batch slurry reactor, and (b) modeling of a trickle bed reactor with experimental validation.

The experiments for kinetic studies were carried out in a stirred pressure reactor with 100 cm^3 capacity with provisions for control of agitation speed, temperature and sampling of liquids. In these experiments, the CO_2 pressure in the reactor was kept constant by continuous supply through a CO_2 reservoir using a constant pressure regulator such that the temporal reaction progress was followed from the pressure decrease in the reservoir. At the end of each experiment, liquid products were analyzed for PO and PC to assess the material balance. The effects of catalyst loading, PO concentration, pressure and temperature were studied as shown in Figures 1 and 2. Kinetic analysis of these data will be presented along with discrimination of rate models and estimation of rate parameters.

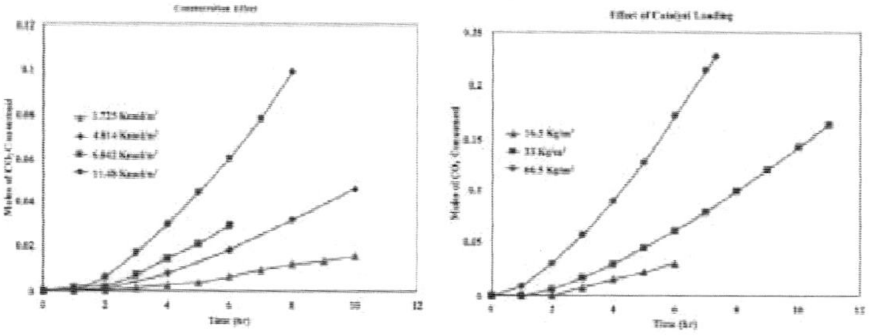

Figure 1. Effect of Concentration and Catalyst loading on CO_2 consumption as a function of time at 368.15K and 1.4 MPa

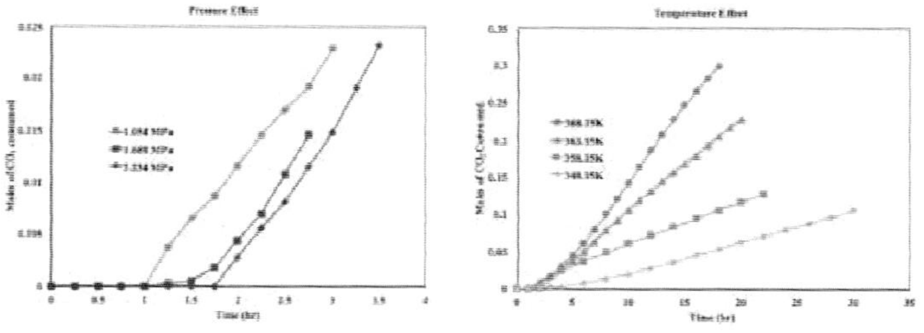

Figure 2. Effect of Pressure at 368.15K and Temperature at .14 MPa on CO_2 consumption.

For trickle bed experiments, a high-pressure trickle bed reactor of 2.5 cm diameter and 10 cm length was used with down flow of gas (pure CO_2) and liquid (PO in PC) phase. The effect of gas and liquid velocity, inlet PO concentration and temperature was studied. Steady state conversion and selectivity were observed for different reaction conditions. The analysis of reactant/products was carried out using GC. Typical results showing the effect of liquid velocity on PO conversion are shown in Figure 3. Detailed mathematical model for the trickle bed reactor based on the kinetics determined in the slurry reactor will be presented incorporating the effect of mass transfer and the wetting of catalyst particles.

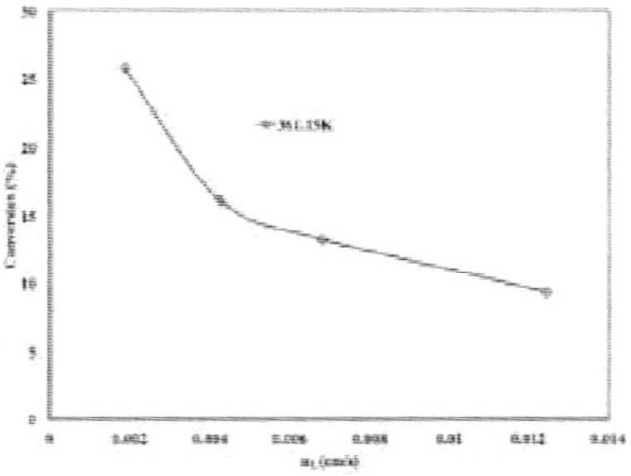

Figure 3. Conversion as function of liquid velocity at 361.15K at CO_2 velocity of 1.867 cm/s and pressure of 0.4 MPa

474206 Nanorod-like CH3NH3PbI3 for Planar Heterojunction Perovskite Solar Cell with Improved Performance

Abstract:

Hybrid organic-inorganic perovskite solar cell has recently attracted significant attention owing to the rapid improvement of the power conversion efficiency (PCE) as well as low cost manufacturing process. Great efforts involving high-quality perovskite film formation methodologies, compositional engineering of perovskite materials, and device architecture together with interface engineering have been made over the last six years, leading to a PCE leap from 3.8% in 2009 to the current world record of 22.1% in 2016 (NREL chart, http://www.nrel.gov/ncpv/images/efficiency_chart.jpg, Accessed 13.03.2016, 2016). The photo-to-electron conversion in perovskite solar cell involves several key steps, including light absorption by perovskite layer, charge transport in perovskite capping layer, and charge separation at n-type oxide/perovskite and perovskite/HTL interfaces. To achieve high photovoltaic performance, the morphology of perovskite is critically vital particularly for the plannar structured PSC, where the crystalline perovskite layer has to be self-standing without the support of a mesoporous scaffold.

This work reports a facile one-step HMT-assisted solution deposition method for reproducible fabrication of a high-quality nanorod-like $CH_3NH_3PbI_3$ thin films via controling over the $CH_3NH_3PbI_3$ crystallization in the presence of ammonia agent. SEM analysis review that a dense and grain-boundary-free perovskite capping layer can be prepared with orienting alignment of perovskite grains that fully cover the ammonia-modified ZnO substrate. XRD results demonstrate a high purity of $CH_3NH_3PbI_3$ thin film without no PbI_2 impurity. The perovskite film produces an appreciable enhancement in both photovoltaic performance and reproducibility as comparison with pristine perovskite, leading to an average PCE of 14.87±0.78%. The introduction of ammonia agent as nucleation-controlling and crosslinking agents and interface modifier provides a facile and effective approach for processing high purity and grain-boundary-free metal-halide perovskite thin film and tune the oxide/perovskite interface. The simplicity, ammonia

agent-assisted and low temperature solution method can extend to construction of other perovskite-based hybrid optoelectronic devices.

473102 Lead-Free, Hybrid, Organic-Inorganic Halide for Light Harvesting

Abstract:

We will present the synthesis and optoelectronic properties of a lead-free, organic inorganic halide structures and show their integration as light harvesters in solar cells. Organic inorganic lead halide perovskite based solar cells have gained attention as a promising contender with nearly 20 % power conversion efficiencies. However, the environmental liabilities with the use and disposal of lead has led to the development of several non-lead based hybrid halides. The natural choice of many researchers has been tin (Group 14 , period 5) and to some extent bismuth (group 15, period 6). Although relatively new, the instability of tinâ€™s 2+ oxidation state has been a concern in spite of good semiconducting behavior. Here we will report the successful synthesis of a Ti based semiconductor with tunable bandgap based on the choice of the organic moiety. We will present and discuss the optical characteristics (ranging from 1.5 to 2.06) as well as the I-V characteristics of the of the various structures. While some of the synthesized products were clearly perovskites (based on XRD data), there are couple of structures whose crystalline structures have not yet been identified. The authors believe the XRD patterns to be a result of layered structures and expect to have it deciphered to present. The response of solar cell containing the hybrid halides coated onto a TiO_2 or ZnO will be presented.

463784 Improving Electron Transport in Nanostructured TiO_2 Electrode

Abstract:

Titanium dioxide (TiO_2) is one of the most widely used semiconductors in photovoltaics and photocatalysis because it is nontoxic, abundant, stable and photoactive. However, the wide bandgap, low electron mobility and short minority carrier diffusion length of TiO_2 limit its quantum efficiency in these applications. In this work, we present a solution chemical approach for making TiO_2 nanostructures for improving the electron transport in nanostructured TiO_2 electrodes.

461988 Symmetry-Breaking in Light-Trapping Nanostructures on Silicon for Solar Photovoltaics

Abstract:

While various materials have been investigated for photovoltaics, solar cells based on crystalline silicon (c-Si) dominate the current photovoltaics market. To reduce the cost of c-Si cells, wafer manufacturing companies have produced competitively priced thin c-Si films, ranging from a few microns to tens of microns, using a kerfless process. In such thin-film c-Si cells, light absorption becomes poorer than in thick films and light trapping is crucial to increase the photovoltaic efficiency. Han et al. have demonstrated that, among various light-trapping schemes, symmetry breaking in photonic nanostructures can approach the Lambertian light-trapping limit very closely. However, fabricating symmetry-breaking nanostructures in a scalable, cost-effective, manufacturable manner remains elusive. Here, we introduce a new approach to systematically break the symmetry in photonic nanostructures on c-Si surface. Using our approach, we fabricate low-symmetry inverted nanopyramid structures. Our method makes use of low-cost, manufacturable wet etching steps on c-Si(100) wafers without relying on expensive off-cut wafers. Our experiment and computational modeling demonstrate that the symmetry breaking can increase the Shockley-Queisser efficiency from 27.0 to 27.9% for a 10-micron-thick c-Si film. Further, our computation reveals that this improvement would increase from 28.1 to 30.0% with over-etching for a 20-micron-thick c-Si film.

464829 Computational and Kinetic Considerations for Morphology Prediction of Donor-Acceptor Oligomers for Organic Photovoltaics

Abstract:

Organic photovoltaic (OPV) solar cells are inexpensive, flexible, lightweight devices for sustainably generating electrical power. The efficiency and stability of OPV devices depends on their morphology, which we currently have poor synthetic control over. In this work, we perform molecular dynamics simulations of neat benzodithiophene-theinopyrrolodione (BDT-TPD) oligomers used in OPVs to evaluate the degree to which current simulation techniques are capable of predicting morphology in BDT-TPD films. The morphologies of simulated BDT-TPD volumes are characterized with simulated grazing-incidence X-ray scattering (GIXS) experiments and compared against experimental GIXS measurements.Â We find that the morphologies predicted from instantaneously quenched simulations better match experiments than structures that are allowed to equilibrate. This result highlights the competition between thermodynamics and kinetics in the structural evolution of polymer thin films.Â Further, we find that approximating the aromatic rings in BDT-TPD with rigid bodies rather than combinations of bond, angle, and dihedral constraints, results in better computational efficiency but poorer agreement with experimental structure. We determine the primary structural features in films of BDT-TPD are the stacking of oligomer backbones and the self-organization of backbone stacks into â€œribbonsâ€' with long-range spatial correlations. We conclude with a discussion of the shortcomings of current modeling tools and potential directions for improvement.

469339 Enhancing Dye Sensitized Solar Cell J-V Behavior By Integrating Nanoscale Polymer Films

Abstract:

Dye sensitized solar cells (DSSCs) are inexpensive, are easy to fabricate, have tunable optical properties such as color and transparency, and can be integrated into buildings and apparel easily. A DSSC is composed of three main components: (1) a photosensitive dye which converts sunlight into electrons, (2) a mesoporous TiO_2 photoanode which provides a large surface area for dye adsorption and a pathway for electron transport, and (3) a liquid electrolyte comprised of an iodide-triiodide redox couple which is used to regenerate the dye and acts as a hole transport medium. A major limitation of the current DSSC design is the liquid electrolyte, which is prone to leakage and evaporation along with being corrosive to metal contacts such as silver. Furthermore, electron recombination loss is significant at the TiO_2-electrolyte interface. Polymer electrolyte DSSCs can overcome these disadvantages and can potentially enhance the cell's photocurrent-voltage behavior [1]. However, due to the nanostructured, mesoporous nature of the TiO_2 photoanode and the effects of viscosity on fluid transport in such a structure, conventional polymer deposition techniques such as spin casting and dip coating lead to ineffective pore penetration. Often only the top 2 µm out of a >10 µm photoanode layer is in contact with the polymer with the remaining photoanode below completely void of polymer. This severely reduces device performance because a large portion of the adsorbed dye is not accessible and cannot be regenerated. To ensure optimal performing DSSCs, there needs to be intimate contact between the polymer electrolyte and the TiO_2 throughout the entire photoanode layer to ensure the redox couple can reach the adsorbed dye.

To overcome the limitations of liquid methods for incorporating polymer electrolytes into the mesoporous TiO_2, we employ a novel polymerization technique called initiated chemical vapor deposition (iCVD). iCVD is a solvent-free polymerization method that relies on gas-to-surface reactions at low pressures in which the reagents for polymerization, the monomer and initiator, are heated to vapors that can easily penetrate into the mesoscale voids of

the TiO

occurs conformally inside the pore surfaces. This makes it especially useful for cases that require penetration of sub-micron and nanometer-sized pores as issues of wettability and surface tension with liquid processing are absent. In addition, the lack of any solvent medium removes the possibility of liquids getting trapped in the photoanode, which may degrade performance, and are difficult to remove [2].

In this work, iCVD is used for the synthesis and integration of polymer electrolyte thin films within the mesoporous TiO_2 photoanode of DSSCs. Fourier transform infrared spectroscopy (FTIR), gel permeation chromatography (GPC), and X-ray photoelectron spectroscopy (XPS) analyses confirm that the polymers formed are stoichiometric in composition expected of linear homopolymers and chemically identical to ones formed with liquid solution methods. The surface reaction kinetics and mass transport have been found to depend on the fractional surface saturation of monomer, , which is an adsorption parameter that provides a measure of the surface availability of the monomer. By carefully controlling , conformal coatings throughout the TiO_2 photoanode have been achieved. To gain a better understanding on the effect of the polymer electrolyte on DSSC behavior, current-voltage, UV-VIS absorption, incident photon-to-current efficiency (IPCE) and electrochemical impedance spectroscopy (EIS) measurements have been employed. Experimental results indicate that DSSCs with ultrathin polymer electrolyte thin films integrated on the photoanode pore surface can enhance cell performance. For example, by applying poly(1-vinylimidazole) (PVIZ), power conversion efficiencies are found to be 27% higher than liquid-electrolyte DSSCs due to a simultaneous increase in short circuit current J_{sc} and open circuit voltage V_{oc} [3]. To our knowledge, this is the first reported demonstration of PVIZ as a polymer electrolyte in DSSCs.

1. Smolin, Y. Y.; Nejati, S.; Bavarian, M.; Lee, D.; Lau, K. K. S.; Soroush, M. *Journal of Power Sources* **2015,** 274, 156-164.

2. Nejati, S.; Lau, K. K. *Nano letters* **2010,** 11, (2), 419-423.

3. Kuba*, A. G.; Smolin*, Y. Y.; Soroush, M.; Lau, K. K. S. *Chemical Engineering Science. Special Issue: Energy Conversion and Storage* **2016,** (*in press*.) http://dx.doi.org/10.1016/j.ces.2016.05.007 (*equal contribution)

468024 Photocatalytic Photosystem I/PEDOT Composite Films Prepared By Vapor Phase Polymerization

Abstract:

Photosystem I (PSI) is an integral membrane protein complex contained within the chloroplast granum of plants and cyanobacteria. The product of over a billion years of systematic evolution, PSI is adapted to the earth's climate and sun's radiant energy profile. PSI transduces blue-violet and red light, sustaining 1.1eV of stored energy in <100ps by way of a cascade of irreversible internal reactions. Internal quantum efficiency for this process is nearly 100%, making PSI a model photodiode. As such, PSI has garnered interest as a macromolecule for use in photovoltaic and photocatalytic systems in a variety of roles. Recently, there have been efforts to include PSI within composite films, whereby the high potential and photocatalytic efficiency of PSI is coupled with an electrically conductive matrix.

We describe a versatile technique to fabricate photocatalytic composite films containing PSI and a chosen intrinsically conductive polymer (ICP). Iron (III)—a friedel crafts catalyst—is added to protein solutions by way of water soluble $FeCl_3$. Drop casting of these solutions on a substrate of choice is followed by contact with low temperature (40°C) monomer vapor of a well-studied ICP, poly(3,4-etylenedioxythiophene) PEDOT. In minutes, a ~200 nm thick polymer/protein composite is grown. Composite films retain the high conductivity of the used ICP and the characteristic absorbance and photocatalytic capabilities of PSI multilayers. Vapor phase processing allows versatility in choice of underlying substrate.

455709 Energy Innovation - Challenges to Commercialization of New Energy Technology

Abstract:

Meeting growing global demand for energy in the 21st century in an environmentally responsible manner will require a significant escalation in innovation and the development of new, clean energy sources. Innovation is defined as invention plus commercialization, so that the innovation process isn't complete until the invention has been turned into a useful product that provides real world benefits. Innovation in energy is complex, has a high rate of failure, takes a long time. Commercializing new energy technology requires numerous technical and business disciplines, partnerships with large energy entities, support from service organizations and most importantly, significant investment.

This presentation will explore the energy commercialization process and how a technology accelerator, such as the Houston Technology Center, can reduce the risk of failure and speed up the innovation process by helping energy entrepreneurs navigate the commercialization process, develop support networks, acquire skills and resources, develop strategic relationships and attract funding.

463641 Energy and Nanomaterials: Interfacial Intersection

Abstract:

With energy at the forefront of the national economy and security energy-directed applications are of particular interest. Not surprisingly, energy and materials are intimately related. Many forms of energy utilization, conversion and storage and generation are dominated by interfacial chemistry. Therein nanomaterials as an interfacial modifier can play a critical role in these processes.

Specific energy applications we have explored include the following:

1) Storage: Increased energy density in Li ion batteries and supercapacitors using carbon nanotubes

2) Efficiency: Reduced friction using nanolubricants between moving parts

3) Transfer: Improved thermal management using nanofluids in heat transfer applications

4) Conservation: Lightweight polymeric composites incorporating nanotubes, nanoclays and graphene oxide for vehicle composites

5) Control: Gas sensors based on nanoscale metal oxide semi-conductors for process control and monitoring

6) Generation: Catalysts and photocatalysts using nanostructured oxides for accelerated charge transfer and minimal recombination losses

Highlights in each application will be presented.

Though synthesis of a host of carbon and metal oxide nanomaterials has been demonstrated, their integration into practical applications remains highly challenging. This talk will provide an overview of their synthesis, integration and value in energy storage, conservation, transfer, efficiency, control and generation. Alternative, scalable synthesis approaches such as flame synthesis and associated laser-based optical diagnostics will be touched upon.

470168 Carbon Nanotubes in Real World Applications – A Perspective on Translational Challenges and Industry Progress

Abstract:

It's been 25 years since the discovery of the carbon nanotube (CNT). CNTs were lauded as a breakthrough materials, with potential to revolutionize virtually every industry from energy to health. Early research indicated that many of the challenges of working with these materials were related to scaling of high purity, high perfection materials and directing their organization into controlled morphologies. Over the past decade, there has been a decline in carbon nanotube suppliers and decrease in press releases, accompanied by a shift in attention to newer materials such as graphene and other two-dimensional materials. Yet, many companies have been quietly incorporating CNTs into applications. This talk will focus on the state of the art for CNTs in applications, including their use in functional coatings, energy storage, and light-weighting. Our group's efforts at Battelle are in translational science – using new science discoveries to solve a broad range of industrial problems. The dominant theme in this type of research is how to control structures at interfaces to enable functionality, durability, and stability, while minding aspects of scalability and manufacturability. The strategic fit of a new material into an existing device is often a major hurdle to be managed. A case study from our group will be used to highlight translation of a CNT coating into a vehicle application -- a low-power, lightweight ice protection system for unmanned aerial vehicles (UAVs). Icing is a serious problem of UAVs, due to their flight characteristics, limited power budget, and restrictions on added weight and size. Our approach is to integrate an electrically-powered heater coating between the standard coating system and selectively power to keep flight surfaces clean. Due to their high conductivity and ability to be spray-applied, CNTs are an excellent candidate for creating heaters that can be retrofitted onto existing aircraft. Successful deployment of these systems required balancing several other requirements, primarily related to power, weight, and lifetime. Through fundamental study of conduction mechanisms in CNT networks, their environmental response, and fatigue behavior, approaches for creating stable, functional systems were identified. We will discuss our path to managing power budget, stability, and durability to create a system that has demonstrated successful performance in a simulated field environment. The key lessons learned from this case study will be summarized and generalized for others going through the product development process. We expect to see growth in the number of applications of CNTs, as the utility of CNTs are demonstrated in real-world environments.

473378 Solution Processed Inorganic Solar Cells

Abstract:

SOLUTION PROCESSED INORGANIC SOLAR CELLS

Caleb Miskin and Rakesh Agrawal
School of Chemical Engineering
Purdue University, West Lafayette, Indiana, USA

With the abundant availability of renewable energy from the sun, one can easily obtain all the power needed for society to continue its growth. What is holding the progress from moving forward is not the capability, but rather the economics. Thin-film solar cells create one step towards the reduction of costs by utilizing less material, but they have difficulty in scalability. To assist in scalability and further reduce costs, we have turned to the development of solution processed solar cells.

In our solution-based approach, we have taken a two prong approach. In the first approach, we synthesize nanoparticles of candidate compound semiconductors and disperse them in solvents to formulate inks. In the second approach, we are developing a new chemistry based on amine-thiol mixtures to dissolve precursor compounds that could not be dissolved earlier in common solvents to prepare molecular precursor inks. These inks are then used to print micron thick films on suitable substrates and heat treated to grow large grains suitable for good optoelectronic properties. Our group is studying the dissolution mechanism for molecular precursors, mechanism for grain growth in nanoparticle films during annealing under different environments and is characterizing final devices for their optoelectronic performance. We are also using the solution chemistry to synthesize nanoparticles of some new and promising semiconducting materials and their thin films. Some examples of the semiconducting materials being explored in our lab are: Cu(In,Ga)(S,Se)2, Cu2Zn(Sn,Ge)(S,Se)2, Cu3P(S,Se)4, Cu3As(S,Se)4, ZnSnP2 , SnS etc.

In this talk, we will highlight some of the important findings of our groupâ€™s research including fabrication of

Cu(In,Ga)(S,Se)2 and Cu2Zn(Sn,Ge)(S,Se)2 solar cells with power conversion efficiencies of 15% and 9.4% respectively. The excitement of this routeâ€™s science and technology along with its challenges will be presented.

471833 Processing as Viable Strategy for Forming High Performance Lithium Ion Battery Electrodes

Abstract:

Future applications of energy storage devices such as lithium ion batteries will require substantial improvement in electrochemical performance. Key metrics will be higher specific capacity leading to high energy and power densities, high Coulombic efficiency, and low capacity fade over several hundred to several thousand charge-discharge cycles.

Developments in this area can be divided into two distinct parts (i) efforts to discover new materials, or (ii) devise novel strategies for placing already known materials into optimal configurations. The latter strategy will be reviewed in this presentation. Vapor phase and colloidal processing techniques to create nanostructured anodes will be described, and the performance of half cells created by these methods will be analyzed. Several challenges still remain, and these will be discussed.

461202 Nanostructured Block Copolymers for Lithium Batteries and Biofuels Purification

Abstract:

Nanostructured block copolymers enable the design of membranes with optimized transporting channels for a variety of applications. This presentation addresses the use of these membranes to transport lithium ions in polymeric lithium battery electrolytes, and to separate biofuels such as ethanol and butanol from dilute fermentation broths where they are produced. Previous studies of electrolytes showed an inverse correlation between ionic conductivity and shear modulus: increasing conductivity invariably resulted in a decrease in the shear modulus. Similarly, previous studies of alcohol-selective membranes showed an inverse correlation between alcohol separation factor (ratio of alcohol concentration in the exiting stream to that of the incoming stream) and total flux: increasing the separation factor invariably resulted in a decrease in total flux. Block copolymer membranes do not suffer from these inverse correlations. Practical implications of our discoveries for enabling safe rechargeable batteries and second generation biofuels will be discussed. A large fraction of energy and materials that we use today are derived from industries that consume non-renewable resources. The two projects described here may enable a gradual transition toward to a more sustainable approach.

474724 High Performance LT-SOFC Based on a Strontium Iron Cobalt Molybdenum Oxide Based Ceramic Anode Support

Abstract:

High Performance LT-SOFC Based on a Strontium Iron Cobalt Molybdenum Oxide Based Ceramic Anode Support

Ke-Ji Pan, Colin Gore, Lei Wang, Luis Correa, Thomas Langdo, and Bryan Blackburn

Redox Power Systems, LLC

College Park, MD 20742

Abstract

Solid Oxide Fuel Cells (SOFCs) consisting of a ceramic anode have several advantages when compared to a traditional Ni-cermet based anode. The main advantages include better red-ox cycling stability and chemical stability with the use of hydrocarbon fuels. However, ceramic materials, such as Nb or La doped $SrTiO_3$, are much less conductive than Ni. Furthermore, such materials usually display appreciable conductivity only after a high temperature reduction step, thus practically limiting their use to higher operating temperatures. Unfortunately, the catalytic activity of ceramic anodes is not typically as good as Ni. At low operating temperatures fuel oxidation kinetics are hindered and must be improved through the infiltration of catalysts. In addition, ceramic anode-supported SOFCs have more fabrication difficulties than Ni-cermet cells, such as the potential for chemical reactions between cell components, thermal mismatch, and shrinkage mismatch. Therefore, until now there have been very few reports of ceramic anode-supported LT-SOFCs. In this work, we have successfully synthesized new strontium, iron, cobalt, molybdenum oxide ceramic anode materials with different compositions (SFCM#1-#5) and fabricated anode-supported SOFCs using GDC as the electrolyte. SFCM displays a conductivity greater than 30 $S*cm^{-1}$ between 450 ¡ãC to 650 ¡ãC without the need for a high temperature reduction step, as shown in Figure 1. Button cells have been fabricated and demonstrated a power density greater than 0.4 W/cm^2 at 500 ¡ãC and have shown good stability in H_2/CH_4 mixtures for several hundred hours. We have scaled up the size of the cells to greater than 5 cm by 5 cm (Figure 2) and will discuss additional performance results and future challenges as we work towards integrating the ceramic anode cells into an LT-SOFC stack.

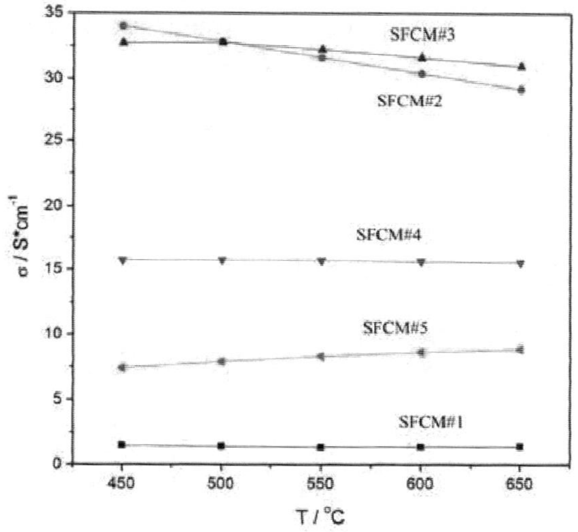

Figure 1. Conductivities of various SFCM compositions in 10% H_2 / 90% N_2

Figure 2. Scaled-up SFCM ceramic anode SOFC.

Key Words: LT-SOFC, ceramic anode, high performance, long term stability

459447 Improved Activity and Stability of Ni-Based Anode for Direct Methanolâ€“Fueled Solid Oxide Fuel Cells

Abstract:

Mo is utilized to modify the Ni- $Ce_{0.8}Sm_{0.2}O_{1.9}$ (SDC) anode. The NiO-MoO_3-SDC anode materials are prepared by impregnation technique, which are reduced to SDC and Ni-Mo alloy phases in H_2 atmosphere. Solid oxide fuel cells (SOFCs) with Ni-Mo-SDC composite anodes are fabricated and tested with methanol as fuel. The addition of Mo improves the catalytic activity for methanol pyrolysis and the resistance to carbon deposition of Ni-SDC anode. The electrochemical performance test demonstrates that the addition of Mo into Niâ€“SDC anode effectively improves performance and stability of the single cell with methanol as fuel at 700 °C. The anode with a mole ratio of Mo to Ni of 0.03:1 (Ni-3Mo-SDC) exhibits the lowest polarization resistance. The cell with that anode and SDC-carbonate composite electrolyte exhibits a maximum power density of 680 mW cm^{-2} at 700 °C, as shown in Fig. 1. The stability of the cell is enhanced with the increase of the content of Mo in the anode.

In order to further improve the cell performance and stability, rare earth elements (La, Pr and Sm) are introduced into Ni-3Mo-SDC. The effects of La, Pr and Sm addition on the catalytic activity for methanol pyrolysis and coking resistance of the Ni-3Mo-SDC anode are examined. The electrochemical performance of the SOFC is enhanced with introducing La and Sm into the anode. The cell with La addition in the anode shows a maximum power density of 765 mW cm^{-2} at 700 °C with methanol as fuel. The addition of La, Pr and Sm into Ni-3Mo-SDC anode improves the stability of the cell, which is mainly attributed to the decreased amount of carbon deposition with a high graphitization degree.

460851 Low Temperature Sintering of Cathode-Supported Micro-Tubular Solid Oxide Fuel Cells Using Sintering Aids

Abstract:

Micro-tubular solid oxide fuel cells (SOFCs) with typical diameters ranging from sub-millimeter to a few millimeters offer several advantages such as facile sealing, high volumetric power density, and better endurance against thermal cycling and thermal shock. Cathode-supported micro-tubular SOFCs, in which a thick cathode layer serves as the structural support, allow for flexible cell-to-cell connections using inexpensive transition metals and also provide better stability under redox conditions. Additionally, cathode-supported cells with a one-end-closed configuration make it possible to use pressurized oxygen as the oxidant, which leads to better power generation performance. However, the fabrication of cathode-supported micro-tubular SOFCs has been found to be challenging because of the high co-sintering temperatures (â‰¥1400 Â°C) required to obtain a dense electrolyte layer on top of the cathode support. At such high co-sintering temperatures, electrically insulating secondary phases are formed due to chemical reactions between cathode and electrolyte materials. Moreover, the high co-sintering temperature gives rise to coarsening of cathode particles and a very low porosity of the cathode support. Thus, it is desirable to reduce the sintering temperature of the electrolyte layer to achieve high-performance cathode-supported SOFCs. In the present work, we fabricated micro-tubular SOFCs with strontium-doped lanthanum manganite (LSM) cathode supports by multi-step dip coating and low temperature co-sintering techniques. NiO and Fe_2O_3 were used as sintering aids to obtain a dense yttria-stabilized zirconia (YSZ) electrolyte at lower co-sintering temperatures. The role of the sintering aids on densification behavior of the electrolyte and the properties of the fabricated micro-tubular SOFCs will be discussed in detail.

458928 Impact of Molecular- and Mesoscales on Macroscopic Thermodynamic and Transport Properties in Perfluorosulfonic-Acid Membrane Using Multiscale Modeling

Abstract:

Performance of polymer-electrolyte-fuel-cells (PEFCs) links strongly to thermodynamic and mass-transport properties of the polymer membrane. The quintessential PEFC-membrane material is perfluorosulfonic-acid (PFSA) polymer, which phase separates into nanoscale hydrophilic water-filled domains through which ions and water transport and into hydrophobic polymer-matrix domains that provide structural integrity and durability. Although considerable progress has been made to reduce reactant mass-transport limitations in PFSA membranes, mass-transport is still a critical area of concern. Aqueous electrolyte transport across a phase-separated membrane is inherently a multiscale problem with aqueous-dissolved species moving through nanoscale domains that are connected to form a transport network at the intermediate mesoscale. To guide the optimization of PEFC material design and operation, we decouple the impact of micro and macro length-scale on mass transport in PEFC membranes.

Specifically, molecular-scale interactions between aqueous-electrolyte and polymer in hydrophilic domains of PFSA membranes are modeled with mean-field local-density theory aligned with an experimentally consistent 3D domain geometry. Our molecular-scale description of membrane conductivity accounts for solvation, electrostatics, solvent dielectric saturation, finite size, and confinement effects. The microscale framework is validated against atomistic simulations and, subsequently, up-scaled to predict macroscale conductivity and solvent absorption by accounting for the interconnectedness of the hydrophilic domains. Excellent agreement is found with experimental data. Our proposed multiscale model for membrane conductivity and solvent sorption provides a new tool to explore potential avenues for improving PFSA membrane performance.

474510 Analysis and Multi-Objective Optimization of Solid Oxide Fuel Cellâ€“Gas Turbine Hybrid Cycle

Abstract:

Abstract:

Chemical process optimization problems often have multiple and conflicting objectives, such as profit, capital cost, operating cost, production cost, energy consumptions, environmental impacts and safety. In such cases, Multi-Objective Optimization (MOO) is useful in finding many optimal solutions, to understand the quantitative trade-offs among the objectives, and also to obtain the optimal values of decision variables. Natural gas, syngas or biogas can be converted into power, heat and electricity, using combustion engine, gas turbine or Solid Oxide Fuel Cell (SOFC). SOFC with Gas Turbine (GT) has shown higher thermodynamic performance, due to better heat integration. Hence, this hybrid conversion system leads to better utilization of natural resource, reduced environmental impacts, and more profit.

SOFC-GT system is divided into five sub-systems: (1) fuel processing using steam reformer, (2) SOFC, (3) anodic GT, (4) cathodic GT, and (5) CO_2 compression. The unconverted fuel from the anodic side of SOFC is combusted in a burner, and then it is used in anodic GT to produce electricity. The SOFC-GT has been simulated in BELSIM-Vali (version 4.7.0.3) flowsheeting software. This study optimizes performance of SOFC-GT system for minimization of both levelized electricity cost and annualized capital cost per kWh, simultaneously. The MOO problem has 12 decision variables, from all five sub-systems of SOFC-GT. The MOO of SOFC-GT system is performed using OSMOSE, which has four integrated parts: (1) stochastic MOO program which provides the values of decision variables, (2) to pass values of decision variables to BELSIM Vali for simulating SOFC-GT system, (3) to obtain temperatures and flow rates of important streams from BELSIM Vali and perform heat integration, and (4) performance evaluation or calculations of objective functions for SOFC-GT system. For optimal SOFC-GT designs, the composite curves for maximum amount of possible heat recovery indicate good performance of the system, and the optimal designs do not require any external hot utility. The SOFC-GT system separate the CO_2, which can be stored and used in other applications. Finally, first law energy and exergy efficiencies of optimal SOFC-GT system are significantly better compared to other conversion systems. The results of this study will be presented at 2016 AIChE Annual Meeting.

Keywords:

Multi-Objective Optimization, Solid Oxide Fuel Cell, Gas Turbine, Heat Integration.

References:

1. Caliandro P., Tock L., Ensinas A.V, Maréchal F., Thermo-economic optimization of a solid oxide fuel cell - gas turbine system fuelled with gasified lignocellulosic biomass. Energy Conversion and Management 2014; 85: 764-773.
2. Facchinetti E., Favrat D., Maréchal F., Design and optimization of an innovative solid oxide fuel cell - gas turbine hybrid cycle for small scale distributed generation. Fuel Cells 2014; 14(4): 595-606.
3. Facchinetti E., Favrat D., Maréchal F., Hybrid cycle SOFC-inverted gas turbine with CO_2 separation. Patent Number - US9228494 B2, 2016.
4. Maréchal F., Palazzi F., Godat J. and Favrat D., Thermo-economic modelling and optimization of fuel cell systems, Fuel Cells 2004; 5(1): 5-24.
5. Palazzi F., Autissier N., Maréchal F., Favrat D., A methodology for thermo-economic modeling and optimization of solid oxide fuel cell systems. Applied Thermal Engineering 2007; 27(16): 2703-2712.

ONSITE

Operation of a Novel SOFC-Battery Integrated Hybrid Generator for Telecommunication Energy Systems

erich erdle

2016 AIChE Annual Meeting
San Francisco - November 13-18, 2016

1

contents

- ❑ general project information - project data and consortium
- ❑ the energy situation within the telecommunication industry
- ❑ a glimpse on trends in the telecommunication industry
- ❑ needs of the telecommunication industry
- ❑ ONSITE's mission
- ❑ ONSITE's achievements so far
  - system design
  - key power sources sofc and NaNiCl-battery
  - first proof of concept test
  - commissioning of sofc generator
- ❑ concluding remarks

2

general project information

- overall objective:
  construction and operation of a containerized a hybrid sofc/NaNiCl-battery (SNC) system that generates more than 10 $kW_{electric}$ at high efficiency (>40% electrical, >85% total) and economically competitive costs to prove the concept
- budget: 5.525.540 Euro (3.012.0338 Euro funding from FCH JU)
- duration: 3 years and 9 months
- partners:

Consiglio Nazionale delle Ricerche · ATA · Bonfiglioli Vectron · efceco · ERICSSON · FIAMM Energy Storage Solutions · Hes·so · HTceramix · JEN

3

the situation

- telecommunication industry consumes up to 1% of the total electricity of a country
- electricity cost is a dominant cost factor, reduction via more efficient power generation is important
- the users of telecommunication services request 100% availability
- the electric grid is not 100% reliable in many countries
- reliability is key, not only for the customers but also for the operators, blackouts cause big financial losses

Country	Network	Energy Consumption	% of Country Total Energy Consumption
USA	Verizon 2006	8.9 TWh	0.24%
Japan	NTT 2001	6.6 TWh	0.7%
Italy	Telecom Italia 2005	2 TWh	1%
France	France Telecom-Orange 2006	2 TWh	0.4%
Spain	Telefonica 2006	1.42 TWh	0.6%

source: Steve Roy: Energy Logic for Telecommunications; Emerson

AVERAGE COST FOR ONE HOUR OF POWER INTERRUPTION.

Industry	Amount (U.S.)
Cellular communications	$41,000
Telephone ticket sales	72,000
Airline reservation system	90,000
Semiconductor manufacturer	2,000,000
Credit card operation	2,580,000
Brokerage operation	6,480,000

Source: U.S. Department of Energy

4

ONSITE FCH efceco

the illustration of the challenge (1)

- additionally telecommunication companies
  - will have to serve a steadily increasing number of customers in the future
  - and they have to offer many new energy-intense services (e.g. high speed and high volume data communication)
- in 2010 there were 5.3 billion mobile subscribers (that's 77 % of the world population)
- this number has been growing daily
- at the same time, geographically, more areas have to be covered, especially areas where no or only unstable electric grid is available
- thus, there is need for more power and for reliable, grid-independent power

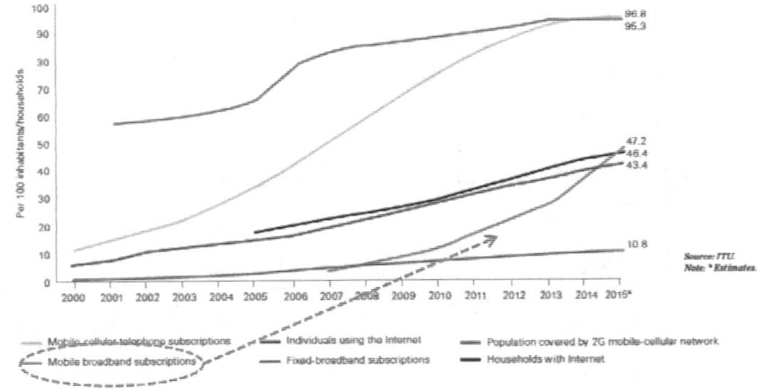

5

ONSITE FCH efceco

the illustration of the challenge (2)

3G mobile-broadband coverage is extending rapidly and into the rural areas

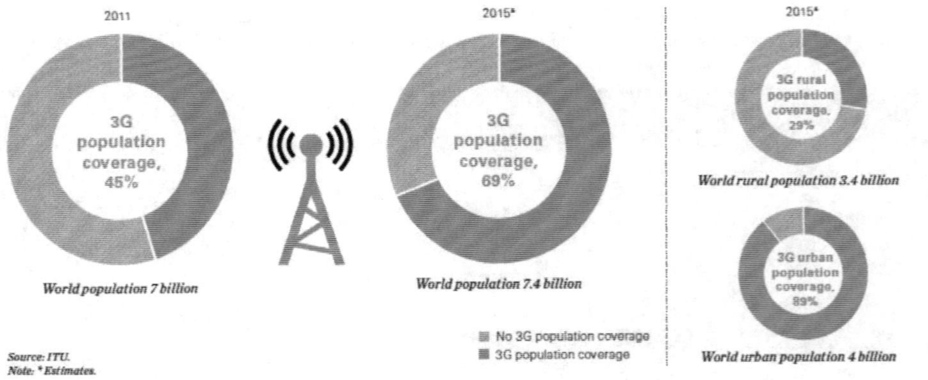

6

needs of telecommunication industry

situation (source: a white paper from EMERSON):

- the cost of and the demand for energy are on the rise, but stable utility power is not always available, especially in remote areas and underdeveloped parts of the world
- despite these challenges, demand for wireless solutions around the world is increasing, prompting service providers to seek and wisely use multiple energy sources
- hybrid energy technologies are emerging as an answer
- there is no "one-size-fits-all" solution. Differences in climate, social environments, site objectives and usage expectations often differ from one site / region to the next, and thus, what works well in one application may be inadequate for another.

needs:

- reduction of environmental impact and CO_2 footprint
- consequent reduction of energy consumption (and OPEX in total), i.e. more efficient systems

the "ONSITE" technical concept aims at

- supply for both on-grid and off-grid Information Technology (IT) plants
- supply for both small Radio Base Stations (RBS) (1-10 kW) and medium sized Data Centers (50 kW +)
- integration with Renewable Energy Systems (RES, e.g. solar PV, wind) minimizing their unpredictability
- integration of the IT plants into the future smart grids/smart cities

7

ONSITE's mission

the "ONSITE" project

- shall accomplish a proof-of-concept of an intelligent supply system (> 10 $kW_{electric}$) for telecommunication equipment and infrastructures (e.g. radio base stations, data centers)
- features a novel hybrid concept that integrates a SOFC (Solid Oxide Fuel Cell) and high temperature Sodium-Nickel-Chloride (SNC) batteries
- aims at
  - better utilization of primary energy
  - higher power generation efficiency
  - use of the waste heat from the fuel cell for heating or cooling purposes
  - significant reduction of the impact on the environment and the climate
- thanks to its multi fueling capability SOFC can provide off-grid power for telecommunication services in remote areas with no or limited electric and/or gas grid, e.g. by operating on LPG

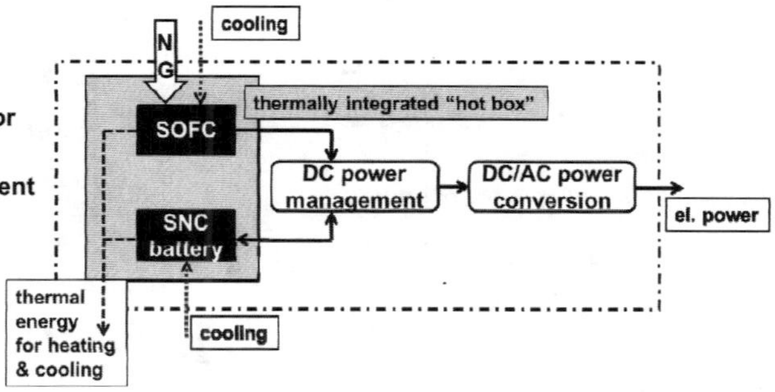

8

ONSITE FCH efceco

system design - electrical scheme

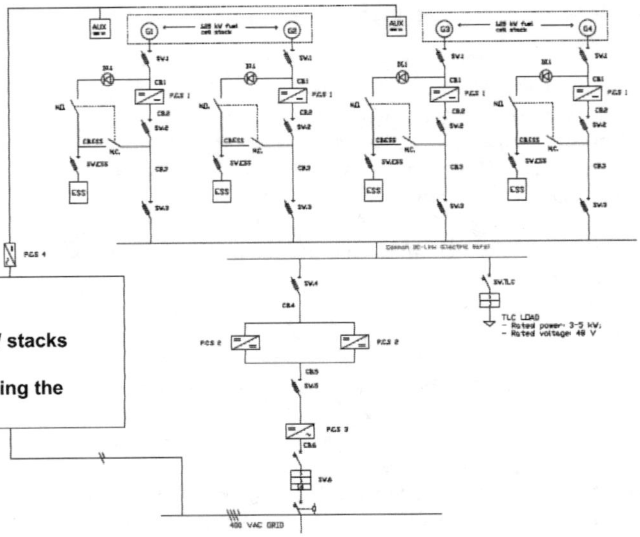

- the system will consist of two subsystems
- each subsystem comprises
  - a 2,5 kW SOFC Hotbox containing 2 1,25 kW stacks and 2 batteries with 3 kW power each
  - each stack has its own DC/DC converter taking the stack voltage to battery level

9

ONSITE FCH efceco

SOFC generator features

feature	values
SOFC stacks operating voltage range	25 to 44 Vdc
fuel	natural gas
nominal power	2.5 kW
SOFC generator operating voltage (regulated)	400 Vdc
operating ambient temperature	0 to +60 °C
SOFC stack operating temperature	750 ÷780 °C
exhaust temperature	< 80°C
temperature of recovered heat	45-65°C

10

SNC battery characteristics

characteristics	values
nominal voltage	48 Vdc
operating voltage range	42 to 56 Vdc
nominal capacity	200 Ah
maximum discharge current	50 A
operating temperature	270 °C
operating ambient temperature	- 20 to +60 °C

achievements: first proof of concept test (1)

first successful test of SOFC/SNC hybrid generator

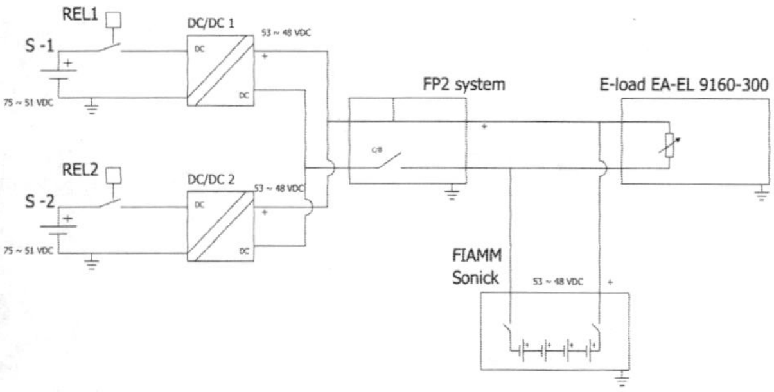

12

achievements: first proof of concept test (2)

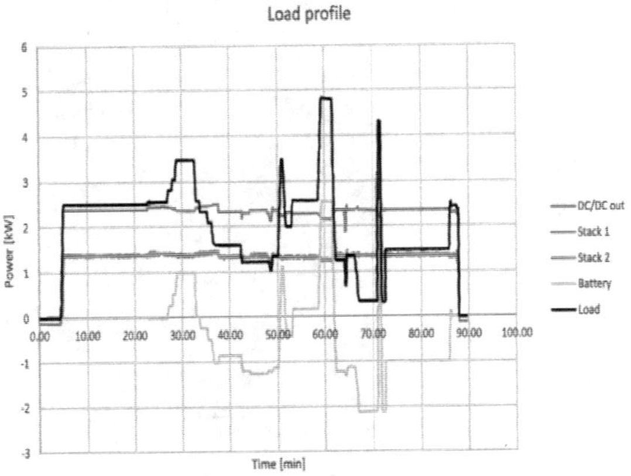

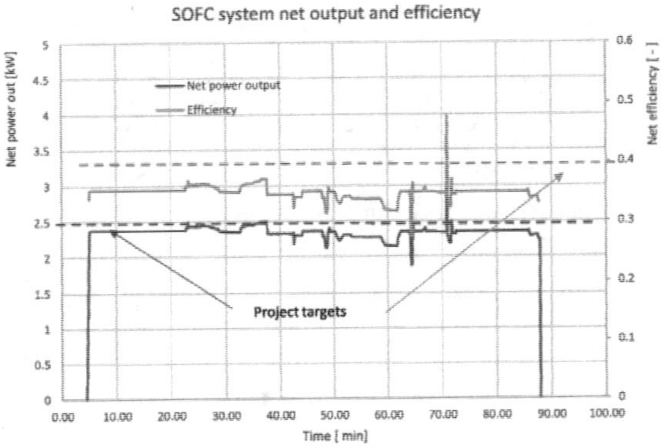

13

commissioning test of sofc system (1)

sofc system running on natural gas

upper compartment: sofc stacks

lower compartment: bop

14

ONSITE FCH efceco

commissioning test of sofc system (2)

480 hours endurance test of sofc system, electrical efficiency approaching 40%, total efficiency above 85%

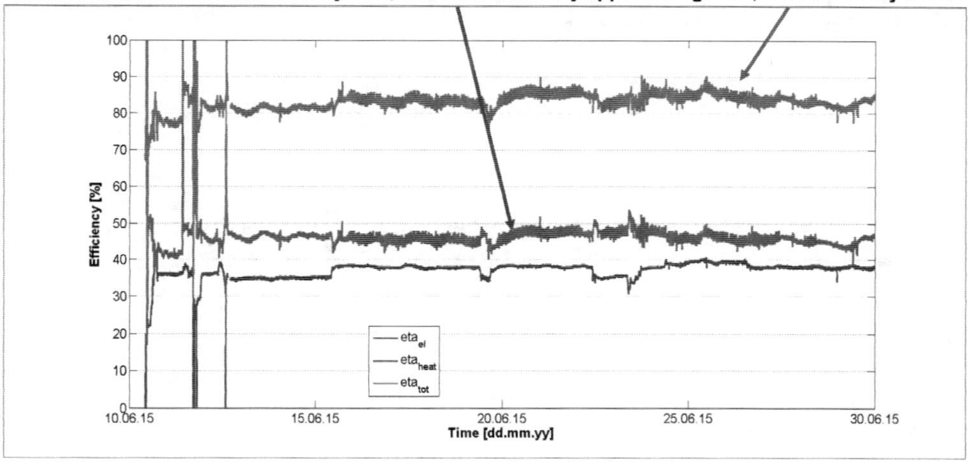

15

ONSITE FCH efceco

commissioning test of sofc system (3)

load variation test of sofc system

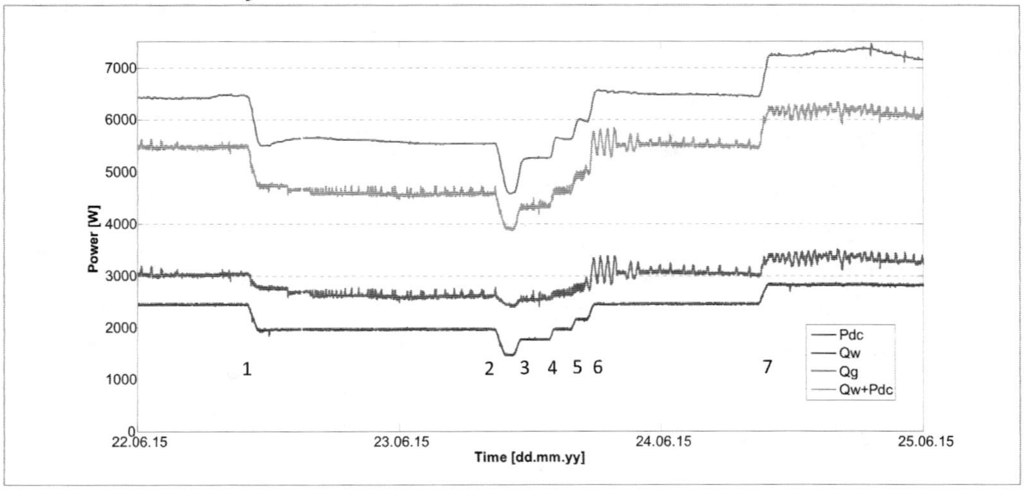

16

commissioning test of sofc system (4)

- electrical efficiency of the stacks and of the system
- parasitic losses and efficiency of DC/DC converter have to be and will be improved

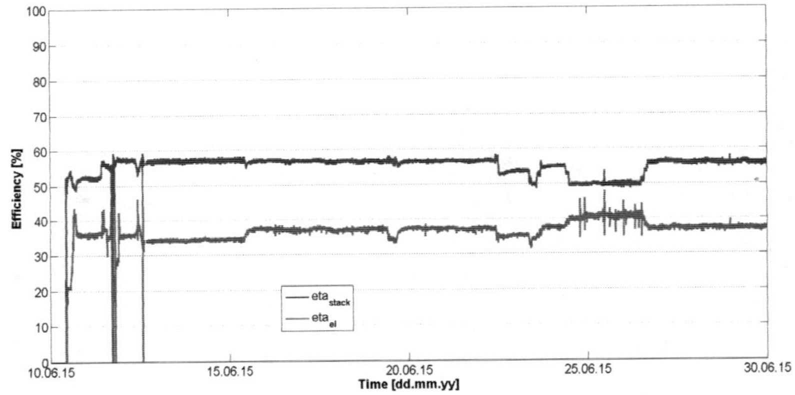

17

commissioning test of sofc system (5)

electric, thermal and total efficiency and electrical and thermal power at different loads

P_{el} [W]	P_{th} [W]	eff_{el}%	eff_{th}%	eff_{tot}%
2000	2700	38%	45%	83%
2500	3000	38%	45%	83%
2800	3400	40%	48%	88%

project goals approached and met at high load, respectively

18

concluding remarks

- power demand in the telecommunication industry is globally growing rapidly
- reliability of power supply has to be improved
- decentralized power generation systems can contribute to solve these issues
- ONSITE is pursuing a new approach by
  - continuous operation instead of backup power concept
  - replacing the battery packs for backup in off-grid locations by a hybrid generator
- due to that and by using a sofc the system can also provide heat and be embedded in smart grids

- Acknowledgements:
  - Funding from the European Union's Seventh Framework Programme for research, technological development and demonstration via its Fuel Cells and Hydrogen Joint Undertaking under grant agreement no. 325325 is gratefully acknowledged
  - I want to thank
    - the entire consortium of ONSITE for the excellent collaboration
    - in particular my co-authors V. Antonucci, G. Brunaccini, M. Ferraro and F. Sergi from CNR ITAE as well as S. F. Au from HTceramix

19

Thanks for listening!

Questions?

www.onsite-project.eu

www.efceco.com

20

472984 Catalytic Bitumen Partial Upgrading Under Methane Environment

Abstract:

Heavy oil and bitumen extracted from oil sands such as the ones abundant in Canada cannot be directly sent to existing refineries mainly due to its high viscosity. Traditionally, hydrotreating is widely employed to break down the long carbon chain and complicated polycyclic molecular structure of heavy crude to lower the viscosity, and remove the impurities such as sulfur, nitrogen and metals simultaneously. However, such treatment consumes expensive hydrogen and requires extremely high operating pressure. Replacing hydrogen with methane, the principal component of natural gas, and lowering the operating pressure would significantly decrease the cost of oil refining. The activation and conversion of methane has been a great challenge due to its symmetric molecular structure and high C-H bond energy. The presence of higher hydrocarbons, which is rich during the upgrading process of heavy oil, would significantly enhance the conversion of methane. In the present work, methane is activated with ZSM-5 based catalysts loaded with active metals and upgrade the heavy oil at low pressure. The charged catalyst witnesses significant viscosity reduction, excellent product stability and compatibility, decreased total acid number, averaged molecular weight, asphaltene content, and increased H/C atomic ratio of the product. The reaction between the heavy oil molecules and methane is studied by choosing several representative model compounds including n-butylbenzene to analog the feedstock components in heavy oil. The study of the reaction is executed from multiple aspects with several analytic instruments, including the GC-MS analysis of products, which provides the composition information, and Diffuse Reflectance Infrared Fourier Transform (DRIFT) Spectroscopy, which detects the formation of organic species on the surface of the catalyst during the upgrading process. ^{13}C and ^{2}D enriched methane are also employed to track the catalytic pathways. Several key products such as ethylbenzene, propylbenzene, isopropylbenzene and butylbenzene, which are the saturation products of styrene, are separated and purified. Their ^{1}H NMR, ^{13}C NMR, FTIR and GC-MS spectra reveal how the CH_x and H are added into unsaturated bonds. This work clearly demonstrates the feasibility of upgrading heavy oil by directly using cheap natural gas on zeolite-supported catalyst at moderate pressure. This

mechanistic study would also benefit the rational design of catalyst for direct upgrading of heavy crude oil using natural gas in the future.

472197 Molecular Characterziation and Modeling for Feedstocks Including Heavy Oil and Reactors

Abstract:

Simulation of refinery-wide processes helps the petroleum industry to design, optimize and operate the refineries profitably and safely, while meeting all necessary regulatory requirements. The traditional pseudo-component approach developed over 30 years ago is still widely used today to represent the feedstocks and products in these process models. This approach uses correlations and heuristics to provide thermodynamic and other properties required in the simulation. For refinery reactor models, kinetic lumps are used to represent molecular classes that undergo reactions to produce the desired products. These kinetic lumps are determined from either finger-print data of the feed or correlations that are based on other properties (such as PNA or sulfur contents) of the feed. Both the pseudo-component and kinetic lump representations do not contain molecular level information. Due to the increasing demands of environment regulations, changing feedstocks, higher margins, etc., the petroleum industry is very interested in improving the accuracy and understanding of the refinery process models and has been exploring the use molecular information for quite some time. In this presentation, we will describe the molecular approach recently developed to leverage molecular knowledge to increase the accuracy of process simulation. The molecular method was applied to characterize crude oils using traditional laboratory assay data (quantitatively match) as well as analytical data such as Fourier Transform Ion Cyclotron Resonance Mass Spectrometry (FT-ICR-MS) (qualitatively match). The resulting molecular compounds, compositions and associated properties are then used to calculate the properties of the process streams. The molecular information are also used directly to represent kinetic lumps composition used in refinery reactor models.

471382 An Innovative Technique for Oxy-Cracking of Quinolin-65 As a Model Heavy Hydrocarbon Compound

Abstract:

Oil refining processes typically involves a coking process that results in a carbon rich solid waste hydrocarbon byproduct called petroleum coke (petcoke). The production rate of the petcoke is over 20 million tonnes per day. Recently, petcoke from Canadian oil sands in U.S. refineries has led to the appearance of large stockpiles adjacent to urban communities in Detroit and Chicago, because limited markets exist for this commodity. The petcoke has strong effects on human and animal health. The main threat to urban populations in the vicinity of petcoke piles is most likely fugitive dust emissions in the form of fine particulate matter. Current industrial processes for treating these kinds of such complicated waste hydrocarbons like direct gasification and combustion have environmental drawbacks centralized mainly in the high production of gaseous CO_2.

In this study, we propose a nano-catalyzed oxy-cracking process that works at mild operating temperature and pressure for high solubilisation of petcoke in relatively low proportions of water. This would open the possibilities for an efficient disposition of petcoke which could be converted at low processing costs into a transportable stream or, through their selective adsorption, be anchored on active catalysts for converting them into lighter products, or alternatively be used efficiently as hydrogen generating source via catalytic steam gasification at much lower temperature than the conventional gasification process.

To understand the mechanism and the kinetics of such innovative technology, Quinolin-65 (Q-65) was selected as a model-oxy-cracking compound mimicking the solid waste hydrocarbons in aqueous-based solutions. In the first phase of this study, the reaction took place in a non-catalyzed batch process operated at optimized oxygen partial pressure and temperature. The preliminary results enabled us to propose a reaction pathway based on the radical reaction mechanism to describe the kinetic behavior of the oxy-cracked Q-65. Hence, kinetic experimental results indicated that the Q-65 oxidation undergoes a parallel consecutive reaction in which an oxidative decomposition took place in the first step producing different aromatic intermediates as confirmed by the NMR analysis. Further, the intermediates were oxy-cracked consecutively into different families of organic acids and small amount of solubilized CO_2 as confirmed by the inorganic carbons analysis. To support our experimental findings, a theoretical computational modeling is conducted. Therefore, the complete radical reaction mechanism will be discussed and elaborated.

465284 Upgrading of Petroleum Vacuum Residue Using Supercritical Hydrocarbon Solvents in Batch and Continuous Systems

Abstract:

As demands for transport fuels and lighter oils have increased, petroleum industries have been forced to refine low-quality extra heavy oils. Among the heavy feedstocks, petroleum vacuum residue (VR), which is the heaviest fraction of crude oil, has high boiling points over 520 °C and contains about 5 wt. % of sulfur, 0.5 wt. % of nitrogen, and organometallic compounds such as nickel and vanadium. Generally, heavy feedstocks can be converted into light products by either carbon rejection process or hydro-conversion process. However, it is difficult for extra heavy oils to be converted into liquid fuels by the conventional upgrading processes due to the high concentration of impurities and asphaltenes.

In the study, supercritical upgrading process (SUP) was applied to upgrade vacuum residue (VR) to light products with small amount of coke formation in batch and continuous systems. Since the physical and transport properties at near-critical region of fluid are altered sensitively, the parametric study for the SUP was required to elucidate optimum operating conditions, depending on various hydrocarbon solvents. The upgrading results were evaluated on residue conversion, coke formation, and product distribution (naphtha, middle distillate, vacuum gas oil, and residue) at various operating conditions; temperature (350-450 °C), hydrogen partial pressure (0-3.45 MPa), reaction time (5-30 min). In addition, further study on a continuous system was conducted on the basis of the results from the batch system. In the continuous system, residence time worked as one of key operating parameters to suppress the coke formation.

470536 The Binary Pyrolysis of Sulfur Compounds with Alkyl Aromatics and n-Alkanes

Abstract:

Petroleum refiners need reliable kinetic models for the optimization of oil residue conversion processes. Modeling of chemical reactions involving heavy ends of petroleum is complicated. Structures and mixtures are very complex, and dissimilar depending on the origin of crudes.

Countless species differ in reactivity should be accounted for in these models, which in turn needs explicit molecular detail of the feed. During conversion many structural changes take place, high boiling compounds broken down into smaller fragments; all sulfur as well as metals such as vanadium and nickel are removed and the most of the aromatics turn into naphthenes which further turn into paraffins and iso-paraffins.

Pure hydrocarbon model compounds pyrolysis gives important information about the reaction kinetics. The purpose of model compounds pyrolysis and its components is to reveal the actual chemical kinetics of the molecules that play a fundamental role during the conversion process. Binary pyrolysis of pure hydrocarbons obtained from the reaction of sulfur compounds reactions will provide important information about heavy oil conversion reactions.

In this study, the binary pyrolysis of decilmercaptane, thiophene n-tetradecane and butyl benzene was investigated at reaction temperatures of of 400-500 °C temperature for reaction time ranging from 30 min to 180 min. In the second phase with the same reaction conditions the binary compounds pyrolysis was carried out. Reaction rate constants and arrhenius parameters for the pyrolysis of binary mixtures obtained from the reaction model were compared with pure compounds.

468016 Use of Unconventional Resources in State of the Art Oil Refining Processes

Abstract:

Conventional oils which are readily accessible and homogenous compared to the unconventional oils are running out. This scarcity draws growing attention to unconventional oils in order to meet the energy requirement of the world. On the other hand, unconventional oils are heavy and complex structures which are tightly trapped between or bound to sand or rock; therefore, new challenges emerge to convert these oils into more valuable products.

In this study, thermal upgrading of oil sand and heavy bitumen derived from domestic reserves is studied to reveal the processibility of these feedstocks in delayed coker units. In addition to that, these feedstocks are blended with highly aromatic refinery by-products to evaluate the effect of aromaticity on coke and liquid yield as well as coke morphology. Feedstock of the delayed coker unit from a refinery is used as a reference for both upgrading and characterization parts.

Oil sands are subjected to solvent extraction method with toluene to separate sand and hydrocarbon samples. After that, these unconventional feedstocks are evaluated for their elemental composition, SARA content, number and weight average molecular weight, specific gravity, dynamic viscosity and S-value (an indicator for stability of asphaltene molecules and coke morphology). Thermal upgrading reactions are conducted at 500 °C for 90 minutes in a tubing bomb reactor having a 10 mL internal volume. Both neat feedstocks and the ones blended with refinery by-products are used in reactions. Coke and liquid products are characterized by means of scanning electron microscopy (SEM) and gas chromatography (GC-SIMDIS), respectively. Moreover, liquid products are further analyzed for their elemental composition, SARA content, molecular weight, specific gravity and dynamic viscosity.

The characterization results indicated that oil sands and heavy bitumen can be used in delayed coker reactor systems but they should be blended with conventional oils to meet design requirements of the units in terms of sulfur and asphaltene content or specific gravity. Furthermore, thermal upgrading reactions demonstrated that unconventional and reference feedstock give nearly same coke yield (~20% wt.) but their liquid product distributions are quite different from each other. Addition of refinery by-products provides around 10% decrease in coke yield which can be attributed to the prevention of phase separation during thermal upgrading reaction by aromatic compounds which can dissolve maltene fractions and asphaltene precursors. Thereby, polycondensation reactions are retarded and coke yield is decreased. Moreover, the S-value results and microstructures from SEM analysis indicated that addition of highly aromatic refinery by-products affects the coke morphology, which specifies the value of coke and occurrence of hot-spot problems in delayed coker units.

Knowing the fact that liquid products are more valuable as compared to solid coke samples whose annual production capacities may reach up to million tons/year for a single refinery, decreasing coke yield and changing coke morphology with easy-to-implement solutions is a powerful tool to increase the profit of oil refineries in a competitive global market.

449940 Developing a Dead Heavy Oil Viscosity Correlation for Lower Fars, Kuwait

Abstract:

Abstract

Viscosity of the produced fluid is an important property for oil and gas industry. It is used in calculations of fluid flow in reservoirs, designing production tubing and flow lines. Accurate estimation of heavy oil viscosity is the most important, as it is one of the key parameters in evaluating a given prospect for the possibility of cold oil production and planning the type of thermal Enhanced Oil Recovery (EOR) methods such as hot water, steam cycling, steam flooding, steam assisted gravity drainage or in-situ combustion. Viscosity of crude oil is usually measured at reservoir temperature at various pressures using bottom-hole sample or surface recombined sample.

Occasionally, the sample is difficult to obtain or becomes unreliable. In this case calculation methods are used to estimate the crude oil viscosity as a function of easily measured production information such as API gravity or detailed compositional data. Many correlations have been presented to estimate crude oil viscosity for crude oils of a certain field or region. Application of these correlations to other parts or regions results in large error in estimated viscosity because each crude oil has its own physico-chemical characteristics that makes it different from other crudes.

In this study, the accuracy of the well-known correlations are tested against heavy oil viscosity data collected from Lower Fars formation in Kuwait. Later, a new correlation is presented to estimate the dead oil viscosity of Lower Fars. Accuracy and reliability of the proposed model is compared to published correlations such as Beal (1946), GlasØ (1980), Kartoatmodjo and Schmidt (1994), and Al-Omair et al. (2014).The results showed that the new correlation is able to predict the dead oil viscosity with mean absolute relative error of 0.16, correlation coefficient equal of 96%, and standard deviation of 0.17.

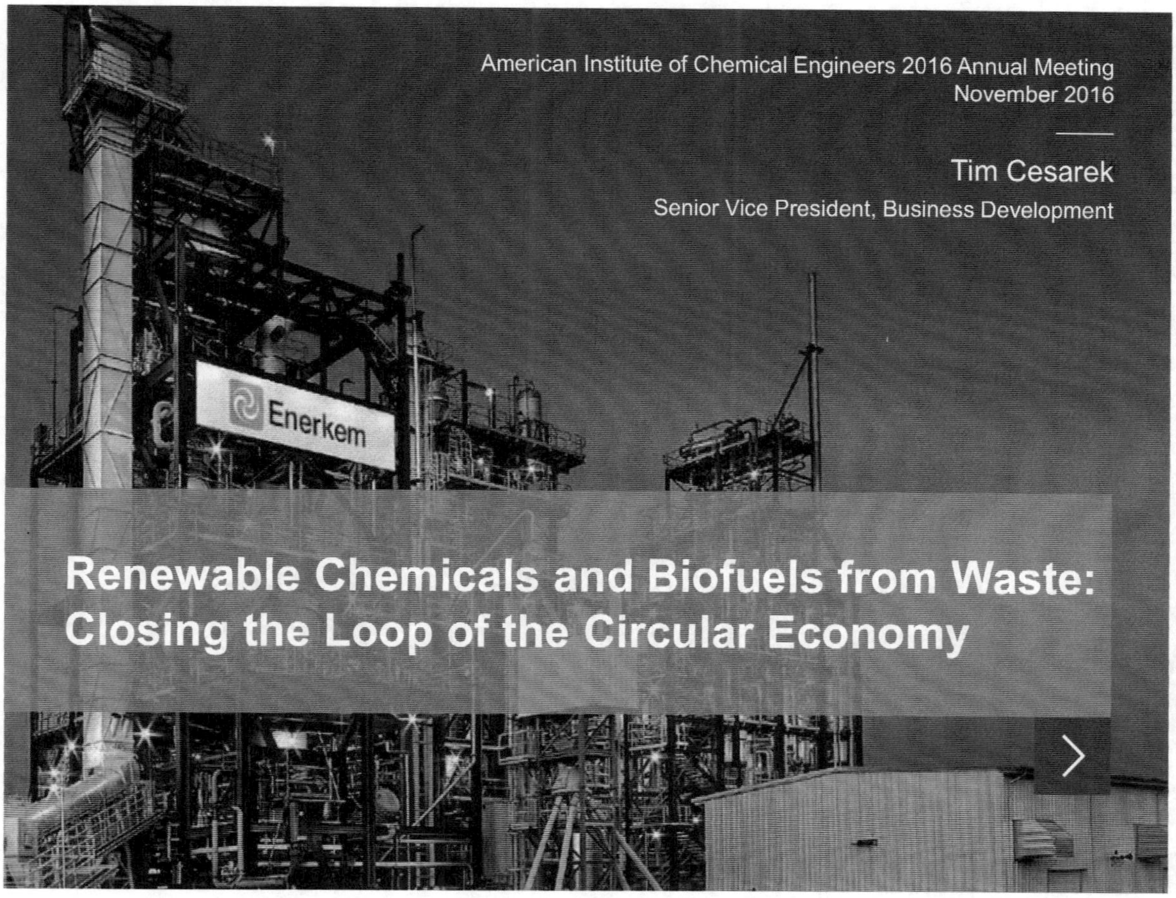

American Institute of Chemical Engineers 2016 Annual Meeting
November 2016

Tim Cesarek
Senior Vice President, Business Development

Renewable Chemicals and Biofuels from Waste: Closing the Loop of the Circular Economy

Enerkem at a glance

- MSW-based biofuels and renewable chemicals producer
- World's first full-scale commercial MSW biorefinery beginning operations in Edmonton, Alberta
- $400M invested to date to move from R&D to commercial stage
- 200 employees
- New alternative to landfilling and incineration

We're building the bioeconomy. |

Enerkem

World's first commercial MSW-to-biofuels and chemicals facility

ENERKEM ALBERTA BIOFUELS

Capacity:	38 million litres per year (i.e. 1 X standard Enerkem system)
Feedstock:	25-year agreement with City of Edmonton for 100,000 dry tonnes of MSW per year
Products:	Biomethanol, cellulosic ethanol

Methanol sale and delivery

Proprietary thermochemical technology and process

Feedstock preparation	Gasification	Cleaning and conditioning process	Catalytic synthesis and product purification
Sorting, shredding, drying (if required) and feeding	Conversion of carbon-rich residues into synthetic gas	Primary syngas purification	Conversion of chemical-grade syngas into final renewable products

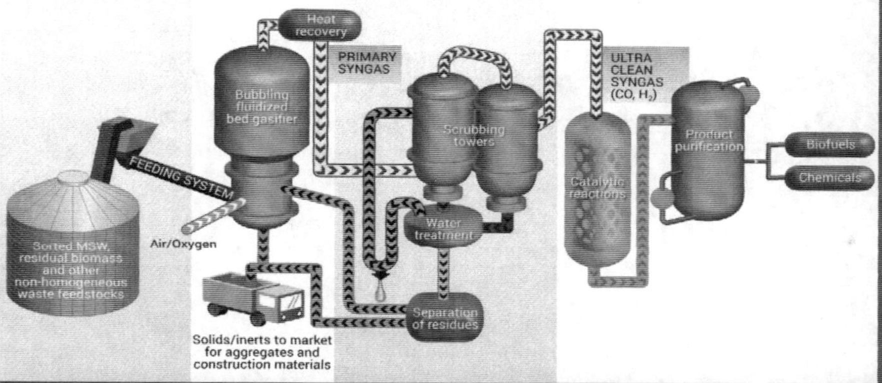

Heterogenous feedstock	Key IP Low severities → Low OPEX → Low CAPEX	Key IP Gas purification & separation (chemical-grade CO, H_2 & CO_2 streams)	CO, H_2 & CO_2 reacted with catalyst to produce fuels & chemicals	Product optionality

Low-severity technology capable of converting heterogeneous feedstock to clean syngas, the fundamental building block for all Enerkem's products

We're building the bioeconomy. | © Enerkem, 2016

Enerkem

Key market drivers for waste as feedstock

- Increased scarcity of urban landfill airspace and societal desire for waste diversion
- Circular economy or "cradle-to-cradle" approach
- Low cost unconventional feedstocks
- Renewable fuels mandates around the world
- Consumer pull for renewable and biobased products
- Focus on carbon footprint and GHG emissions reduction

We're building the bioeconomy. | © Enerkem, 2016

Enerkem

Large market potential

MSW IN THE WORLD

RECYCLED / COMPOSTED

85% TO LANDFILL

1.3 BILLION
METRIC TONS OF MSW
GENERATED PER YEAR

548 MILLION
METRIC TONS OF MSW
SUITABLE FOR ENERKEM'S TECHNOLOGY PLATFORM

THE POTENTIAL:
261 BILLION
METRIC TONS OF METHANOL
USING ENERKEM

Source: World Bank, 2012

We're building the bioeconomy. | Enerkem

VANERCO
Enerkem's second MSW to biofuel facility – Montreal Canada
Product: Biomethanol with ethanol expansion
Capacity: 45,000 MT; 69 million litres per year (18 million gallons) of biomethanol (1 train Enerkem plant)
Feedstock: 115,000 dry tons of RDF per year – C$20/MT RDF tipping fee
Location: Varennes industrial park (South Shore Montreal)
Engineering Start: Underway
Procurement Start: Q1 2017
Construction Start: Q3 2017
Operations Start: Q3 2018
Enerkem

Biofuels from waste increase energy diversity and reduce GHG emissions

Trends towards Low Carbon Transportation Fuels

Gasoline, Gasoline Blendstock or Replacement	Definition	Carbon Intensity (including land use effect)
CARBOB	CARBOB	99.78
Corn ethanol*	Derived from corn	75.97
Cellulosic Ethanol*	Derived from corn stover	41.05
Sugarcane Ethanol*	Derived from Brazilian sugarcane	56.66
Electricity	CA grid electricity	105.16
Hydrogen	Compressed H2 from central reforming of NG	151.01

** California Air Resources Board temporary CIs for fuels with indeterminate CIs*

Source: California ARB, July 31, 2015 (Proposed re-adoption of the LCFS: Third 15-Day Modified Regulation Order, Table 6. Tier 2 Lookup Table for Gasoline and Diesel and Fuels that Substitute for Gasoline and Diesel and Table 7. Temporary FPCs for Fuels with Indeterminate CIs)

Using waste as a feedstock for the chemical industry

Waste-to-Chemicals public-private partnership in Europe

AkzoNobel

Enerkem

AIR LIQUIDE

van Gansewinkel

AVR.

Clean Tech Delta

Port of Rotterdam

Innovation Quarter

Gemeente Rotterdam

provincie Zuid Holland

rotterdam partners

We're building the bioeconomy. | © Enerkem, 2016 Enerkem

Location of Project in Port of Rotterdam

We're building the bioeconomy. | © Enerkem, 2016 Enerkem

Product pathways of syngas and methanol

CO + H2 (syngas) and methanol are key building-block molecules of chemical industry

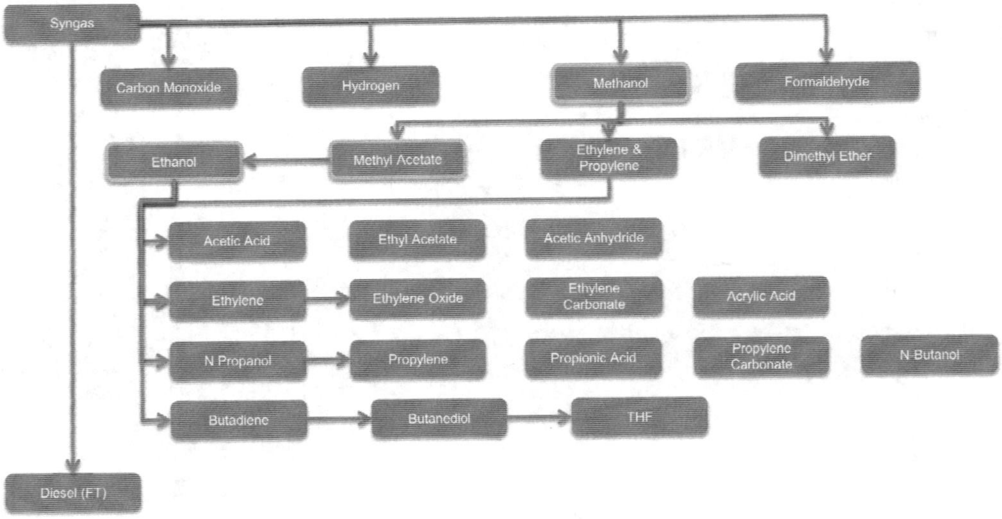

Multi-product platform being developed

Enerkem's platform is being expanded to new products. Acrylic Acid and Acetic Acid are the next steps:

Chemical building blocks in Enerkem's synthesis gas

CO

H_2

CO_2

Product	Global Market Size	Applications
Acrylic Acid	>$14 billion[1]	Industrial Coatings, Paints, Fiber
Acetic Acid	>$4.3 billion[2]	Solvents, Pesticides, and Coatings, Chemical Intermediate
Ethyl Acetate	~$1.5 billion[2]	Coatings, Solvents, Ink
Butanol	>$2.4 billion [3]	Solvent, Chemical Intermediate, Fuel
Propanol	>$3.5 billion [3]	Fuel Additive, Plastic Intermediate, Solvents

Sources:
(1) ICIS, in USD
(2) ICIS; Company estimates
(3) Argus DeWitt

Target growth areas for global partnerships

- Strategic partnerships with leading industrial groups
- Selection based on market attractiveness:
  - public policies
  - tipping fees
  - proximity to petrochemical infrastructure
  - population

474725 Compact Steam Reformers for Power and Hydrogen Generation

Abstract:

Precision Combustion, Inc. (PCI) is developing high heat flux, compact, thermally-integrated steam reformer technology. The innovation consists of intimately coupled catalytic endothermic and catalytic exothermic sections implemented via novel metal mesh (MicrolithÂ®) supported catalyst. This results in high heat flux for improved product selectivity due to uniform temperature distribution, and flameless operation results in lower peak operating temperatures for increased material safety. These compact reformers are being implemented for fuel cell power generation and H_2 generation. Compared to conventional designs, PCIâ€™s approach offers improved performance, durability and power density. This is facilitated by the use of high surface area, low pressure drop MicrolithÂ® substrates coated with proprietary catalyst formulations. Additionally, implementation of flameless catalytic oxidation avoids high flame temperatures, increases thermal uniformity, distribution, durability and control. The burner component can utilize system waste heat, anode exhaust or supplemental fuel (as needed) to drive the endothermic steam reformer for efficiency advantages. Performance of the steam reformer has been rigorously evaluated for hundreds of hours with conventional as well as renewable fuels, including at high pressures for enabling hydrogen separation and purification. Data from these tests will be presented and discussed, demonstrating stable, equilibrium limited performance without coking or catalyst degradation. The sulfur tolerance studies of the burner and the steam reforming catalyst will also be presented.

PCI has extensive experience with balance of plant components and controls development. The BOP components e.g. fuel, air and water pumps and controllers are separately packaged, resulting in a compact system. The technology is scalable. The reformer was examined to generate SOFC-quality reformate and was coupled with a water-gas-shift reactor to generate HT-PEM, or PEM-quality (with additional CO cleanup) reformate. Additionally, integration with a hydrogen membrane separator resulted in production of high purity hydrogen. Alternatively, it could also be coupled with a pressure swing adsorber. Data from the reformer as well as from integrated testing with H_2 separator operating on HRJ and S8 will be presented, demonstrating the potential for a compact H_2 generation system. These reformers offer a solution towards compact, fuel flexible power generation systems for various military and industrial applications, as well as for on-site H_2 generation systems.

458513 Development of Smart Ni Based Nano-Oxyhydrides for Hydrogen Production from Bioethanol

Abstract:

Development of smart Ni based nano-oxyhydrides for hydrogen production from bioethanol

L. Jalowiecki-Duhamel[1,*], Y. Romani,[1] W. Fang[1,2], C. Pirez[1], S. Paul[1], H. Jobic[1,3], F. Dumeignil[1]

[1]*Univ. Lille, CNRS, Centrale Lille, ENSCL, Univ. Artois, UMR 8181 - UCCS - Unité de Catalyse et Chimie du Solide, Lille, 59000, France*
E-mail: louise.duhamel@univ-lille1.fr
[2]*Key Laboratory of Medicinal Chemistry for Natural Resource - Ministry of Education, School of Chemical Science and Technology, Yunnan University, Kunming, 650091, China*
[3]*IRCELyon Institut de Recherches sur la Catalyse et l'Environnement de Lyon, 69626 Villeurbanne Cedex, France.*

At the world level, people are concerned by their health and even life degraded by environmental pollution. Therefore, the potential benefits of a hydrogen economy coming from renewable resources are creating a large consensus and increased attention is focused on hydrogen production technologies. So far, hydrogen has been produced from ethanol by steam reforming, partial oxidation, oxidative steam reforming, and autothermal reforming reactions. The endothermic steam reforming reaction extracts more hydrogen atoms from ethanol and water, however, additional energy supply is needed. One alternative way of supplying heat is to add oxygen or air to the feedstock and simultaneously to burn a portion of ethanol. [1-3] Moreover, in complement, some chemical energy can be also brought by a "smart" catalyst, such as an oxyhydride. [4-7]

Well tuning the preparation, formulation and conditions applied, different series of nickel based nano-oxyhydride (Ni-Ce-(Al)-O and Ni-Mg-Al-O) catalysts were developed for the highly efficient and sustainable H_2 formation from ethanol in the presence of oxygen and water with low energy input. The chemical energy released from the reaction between hydride species stored in the nano-materials and O_2, is used to obtain continuous complete conversion of ethanol with simultaneously production of H_2 with an oven temperature at only 50°C (Figs 1 and 2). The influence of different parameters on the activity and selectivity was analyzed, such as the reaction temperature, water-oxygen-ethanol feed compositions, as well as Ni content and in-situ pre-treatment in H_2 of the catalysts. Moreover, different physico-chemical characterizations were performed allowing a discussion on active site and mechanism.

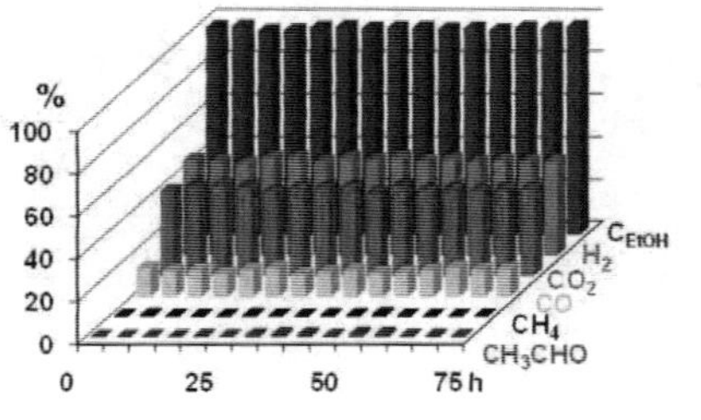

Fig. 1: Ethanol conversion and products distribution on Ni-Ce-O based nano-oxyhydride.

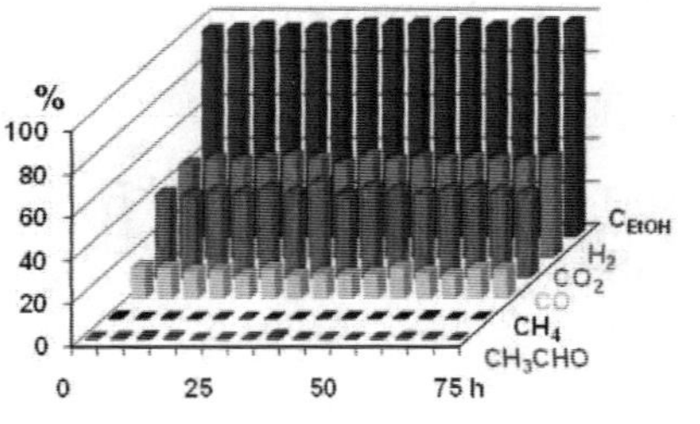

Fig. 2: Ethanol conversion and products distribution on Ni-Mg-Al-O based nano-oxyhydride.

1. G. A. Deluga, J. R. Salge, L. D. Schmidt, X. E. Verykios, *Science* **30**, 5660, 993 (2004).
2. V. Subramani, C. Song, *Catalysis* **20**, 65 (2007).
3. L. V. Mattos, G. Jacobs, B. H. Davis, F. B. Noronha, *Chem. Rev.* **112**, 4094 (2012).
4. W. Fang, C. Pirez, S. Paul, M. Capron, H. Jobic, F. Dumeignil, L. Jalowiecki-Duhamel, *ChemCatChem* **5**, 2207 (2013).
5. C. Pirez, M. Capron, H. Jobic, F. Dumeignil, L. Jalowiecki-Duhamel, *Angew. Chem. Int. Ed* **50**, 10193 (2011).
6. C. Pirez, W. Fang, M. Capron, S. Paul, H. Jobic, F. Dumeignil, L. Jalowiecki-Duhamel, *Appl. Catal. A* **518**, 78 (2016).
7. W. Fang, C. Pirez, S. Paul, M. Jiménez-Ruiz, H. Jobic, F. Dumeignil, L. Jalowiecki-Duhamel, *Int. J. Hydrogen Energy* (2016) DOI: 10.1016/j.ijhydene.2016.04.019.

- Abstract Text

466889 Exploring Transition Metal Carbides and Phosphides for Ex-Situ Catalytic Fast Pyrolysis

Abstract:

Hexagonal Fe_2N type molybdenum carbide (Î²-Mo_2C) has been explored as biomass upgrading catalyst in recent decades for model compounds via hydrodesulfurization (HDS), hydrodenitrogenation (HDN), and hydrodeoxygenation (HDO). Molybdenum carbide has also been shown to be a superior catalyst support, for example, for platinum in comparison to carbon in electrocatalytic water splitting. However, there have not been any studies to investigate the effects of supporting biomass upgrading catalysts such as nickel phosphide (Ni_2P) on molybdenum carbide. Additional carbide-phosphide interfacial sites thus generated have the potential to be of tremendous importance in biomass conversion catalysis. One of the main stumbling blocks for biomass derived pyrolysis oils finding wider use in energy and chemical feedstock generation is the presence of significant amount of oxygen, unlike in fossil reserves. While model compound studies using phosphide and carbides have shown encouraging HDO activities, only a limited number of studies have explored the HDO catalysis on abundant biopolymers such as lignin and cellulose. This presentation will be based on the ex situ catalytic fast pyrolysis of lignin and cellulose in the presence of molybdenum carbide supported nickel phosphide. Particular emphasis will be devoted to the characterization of the interfacial sites and the effect of different Ni_2P loadings on catalysis. Stability of the carbide support during the synthesis of the composite material will also be discussed. In addition, a detailed analysis of product fragments that originate due to the presence of interfacial catalyst sites will be presented via comparison with catalysis results from Mo_2C and Ni_2P individually.

469915 Microwave-Assisted Fast Pyrolysis (MW-FP) of Several Lignocellulosic Feedstocks

Abstract:

Using fast pyrolysis processes, renewable biomass resources can be converted to bio-oils that retain more than 60% of the original energy content, and these oils can then be upgraded to motor fuels or other commodity chemicals. In order to attain economic feasibility, feedstocks for commercial-scale fast pyrolysis must be obtained from a diverse set of low-cost resources for which the conversion performance is well-understood. This presentation summarizes the development of a microwave-assisted fast pyrolysis (MW-FP) reactor that employs a 3 kW microwave beam to heat samples of 1 to 2 grams at rates as fast as 200 °C/s. Liquid, solid and gas yields from as many as 8 samples per day can be measured and analyzed to determine product qualities. Because the individual experiments can be conducted in minutes rather than hours and only small quantities of material are required, MW-FP is advantageous for parametric studies exploring the effects of feedstock material, conversion temperature, and heating rate on fast pyrolysis product yields. Direct measures of product yields show that oil yield increases with volatile matter and lignin contents and decreases with potassium and sodium contents. Contents of total ash, calcium and magnesium did not appear to measurably affect oil yield on a dry, ash-free material basis. Results from the MW-FP reactor have been calibrated to floor-scale fluidized bed fast pyrolysis tests to ensure reliability of the conclusions.

Biomass to Energy – A novel process

By Rakesh Gupta
Lead Process Engineer, Lockheed Martin Energy
Lockheed Martin Corporation
Moorestown, NJ

Background

Biomass is defined as organic material derived from living organisms especially plants. It is an abundant source of renewable energy and chemicals. It is widely available in the United States and worldwide. Its use is being encouraged worldwide as a source of energy, especially in the area of power production. Several government programs have initiated subsidy programs to help develop the power generation from renewable sources. Since biomass is a renewable source, it can be a significant source of fuel for power generation.

Wood is the largest source of biomass. It can come from crop residues, forest thinning, milling residues, agricultural process residues and even municipal solid waste (MSW).

Biomass constituents include organics such as cellulose and lignin. They also contain water and other components such as chlorine, nitrogen, phosphorus, sulfur, silicon and trace metals.

The most important characteristic of a biomass is its heating value. This value is reduced by the presence of water and ash. Although, water presence in the biomass can also be beneficial by assisting in the steam reformation process discussed later.

Lockheed Martin is committed to developing renewable energy sources and has committed significant resources towards its development. Lockheed Martin has partnered with Concord Blue, licensor of a novel advanced gasification technology to convert biomass to energy. The purpose of this partnership is to develop a scalable and fully integrated solution. The Concord Blue process utilizes a patented method of "staged" reforming to convert organic material into a hydrogen rich syngas. This syngas is cleaned of contaminants and it can be either separated into valuable chemicals or fed to a combustion engine or turbine for power generation.

These power plants can range from 250 kW up to 20 MW in size.

Technology

There are many ways to use biomass as a source of energy for power generation. The most common method is to use it as a co-fuel with other sources such as coal, natural gas, etc. However, biomass can also be gasified using air or oxygen or pyrolyzed using a heat source. Some systems have a combination of pyrolysis and combustion/oxidation processes.

Syngas, short for synthetic gas, is the generation of gas from an organic fuel such as MSW, or woodchips by pyrolysis. In this process, the organic waste is gasified by addition of heat. The gases produced have a higher hydrogen to carbon ratio than the source material. The gases also contain carbon monoxide, hydrogen, methane, tars and other contaminants.

Pyrolysis involves the decomposition of organic compounds into gases and solids. This is done in the absence of oxygen. When the biomass is initially fed to the pyrolyzer, the volatile components are released almost immediately. The remaining char is converted to H_2 and CO in slower reactions.

Steam reformation process involves converting the volatile components such as methane and other organics into hydrogen with endothermic and reversible reactions. This process requires addition of steam and energy. These reactions can be repeated for higher hydrocarbon chains also, although each of those reactions have their own energy requirements.

In the Concord Blue staged steam reformation process, syngas is generated with components such as hydrogen, methane, carbon monoxide and carbon dioxide along with other tars and contaminants.

The raw syngas is cooled and cleaned in a scrubber system and then fed to a combustion engine to generate power, but ultimately the syngas could have other uses as well. The components of the syngas could be separated into valuable chemicals or converted into biofuels.

The process generates many valuable byproducts such as char that may be used for agricultural applications, steam for plant use and heat that could be used for applications such as hot water for district heating. Biochar produced from the pyrolysis process can also be used as a source of fuel for the pyrolysis process. It can also be used as activated carbon feedstock.

Syngas generated from the conversion process can also be converted into synthetic crude oil which can be further refined into clean synthetic fuels and chemicals such as kerosene, diesel, jet fuel, lubricants, etc. It can also be blended into crude oil and processed in a refinery. The practice involves using the Fischer-Tropsch process.

Suppliers

Gasifiers are provided by many suppliers such as GE, Conoco, Shell, Siemens, MHI and KBR. Shell and Siemens systems are oxygen blown using slagging type burner. MHI and KBR technologies use air or oxygen, do not use a burner and are non-slagging type.

The technology licensed by Lockheed Martin from Concord Blue utilizes staged reforming as a method of generating synthetic gas. The organic material is initially pyrolyzed and then reformed using steam and heat.

Pilot Plant

In order to demonstrate this technology, Concord Blue's patented method of "staged reforming", Lockheed Martin has built a pilot scale project at its Owego, NY facility. Lockheed Martin selected wood chips as the initial fuel source, since they are easily obtained in consistent size, they are available in abundant supply and they provide a consistent source of organics. Literature search showed wood chip samples from New York state have heating values of 5,600 Btu/lb. to 7,800 Btu/lb., moisture content of 28 – 42 % and ash content from 0.4 to 1%.

The size of this of this project is 250 kWe. It has the capability of producing renewable / sustainable power using various solid wastes, organic wastes, non-organic residues and woody biomass feedstocks such as clean wood chips and refuse derived fuel (RDF), while showcasing the technology. The feedstock will initially be woody biomass with an average moisture content of 37%.

The following is a simple sketch of the Concord Blue technology.

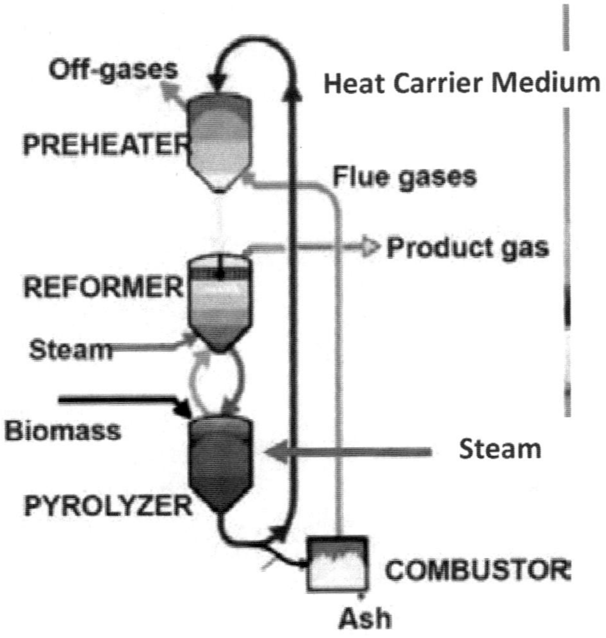

Figure 1 – Concord Blue Staged Reforming Technology
(Institute of Wood Technology and Wood Biology (HTB), Hamburg, Germany)

The biomass is fed to the pyrolyzer at temperatures above 900 °F. The gases generated are reformed in a steam reformer to increase the hydrogen concentration in the syngas. Heat is supplied to the reformer and the pyrolyzer by a heat carrier medium. The heat carrier medium is heated in the preheater as shown above in Figure 1.

The 250 kW Technical Assessment and Demonstration Facility's subsystems include:

- Feedstock Storage and Handling - The biomass is received in specially equipped trucks. It is transferred to the pyrolyzer using a number of conveyors. All feedstock materials are processed off-site by a feedstock provider and transported to the plant. Feedstock processing includes sorting, shredding to size, and removing all ferrous and non-ferrous metals as well as all inert material such as rocks, stones, ceramics, etc.

- Staged Reforming Equipment - The primary function of the Staged Reforming System is to gasify the incoming feedstock to generate syngas for downstream combustion in the Internal Combustion Engine. Biomass is fed to the Pyrolyzer where it gasified in a slow and controlled manner. Gas from the Pyrolyzer is fed to the Reformer where higher temperature reactions take place in the absence of oxygen. It is in this high temperature reformer where the water gas shift reaction occurs. It consumes water and energy to increase the heating value of the syngas by generating hydrogen. This is also where tars are cracked with water to produce hydrogen and carbon monoxide. This step improves the quality of the syngas by reducing the amount of tars that end up downstream into the syngas cleanup equipment potentially fouling piping and the gas cleaning system, thereby reducing gas flow to the generator

- Burner - The Burner is a vendor supplied package, that is designed to combust natural gas, syngas and other fuel sources, such as spent biodiesel, in order to generate hot flue gas to supply heat to Staged Reforming System's Pre-Heater. Biodiesel is atomized into the burner using compressed air. Fresh outside air, using combustion air blower provides the oxygen source for combustion within the burner. During start-up, natural gas is used to fuel the burner and bring the flue gas up to operating temperature. Note that in other installations, the char generated by the process will be combusted to provide the heat for the Staged Reforming System.

- Syngas Cleaning - Raw syngas from the Staged Reforming System Reformer contains a number of contaminants not suitable for running an IC engine, such as acid gases (i.e. H2S, HCl, HF, COS), ammonia, particulate matter, and tars. (The presence and concentration of contaminants is affected by the type of fuel used.) The function of Syngas Processing and Conditioning system is to clean and condition the raw syngas from the Staged Reforming Process so that it conforms to the syngas quality requirements of the internal combustion (IC) engine. The first step in the syngas scrubbing process is the quenching of the hot gases. The gases then pass through a venturi scrubber to remove particulates. The system includes a scrubber to remove ammonia and another scrubber to remove H2S. The syngas pressure is boosted to meet the requirements of the engine. Also, the gases are reheated to reduce the humidity of the gases.

- Flue Gas Cleaning will be added in the future depending upon initial results of environmental stack testing.

- Syngas Storage / IC Engine-Generator - The function of the Power Generation and Gas Storage system is to convert the cleaned syngas into electrical energy through the IC Engine. The gas storage system provides a mechanism to even out the composition of the syngas and provides a few minutes of storage to allow for graceful shutdown of the engine, when required. A single 250 kWe internal combustion engine vendor package is used to convert the cleaned syngas into electrical energy.

- Flare – The flare is used to burn off any off-specification syngas or cleaned syngas in case the engine is not operating. The flare provides safe combustion of the syngas to prevent harmful toxins and combustible materials from emitting to the atmosphere. The flare is designed to maintain a constant flame at the flame tip. Nitrogen is used as the purge gas for the flare velocity seal.

- Wastewater Treatment – The wastewater treatment system collects the water discharge from the scrubbing system. Currently, it is shipped offsite for treatment. In other applications, it would be treated on-site and returned to the plant for process water uses.

- Cooling Water System – The quenched gases in the syngas scrubbing system require cooling for the engine to combust them. They are cooled using a closed circuit cooling tower.

- Compressed and Instrument Air System – The compressed air system provide the air required for the operation of the instruments, valves and the atomizing air for the burner.

- Natural Gas – Natural gas is currently used as the fuel source for heating the heat carrier medium.

- Nitrogen – Nitrogen is used in the process to purge out the system before generating syngas. It is also used in a number of components to prevent mixing of syngas with air.

- Biodiesel – Biodiesel is used as the absorbent for the tars generated in the process. The spent biodiesel is used as a fuel in the burner.

A simplified flow diagram of the system is as follows:

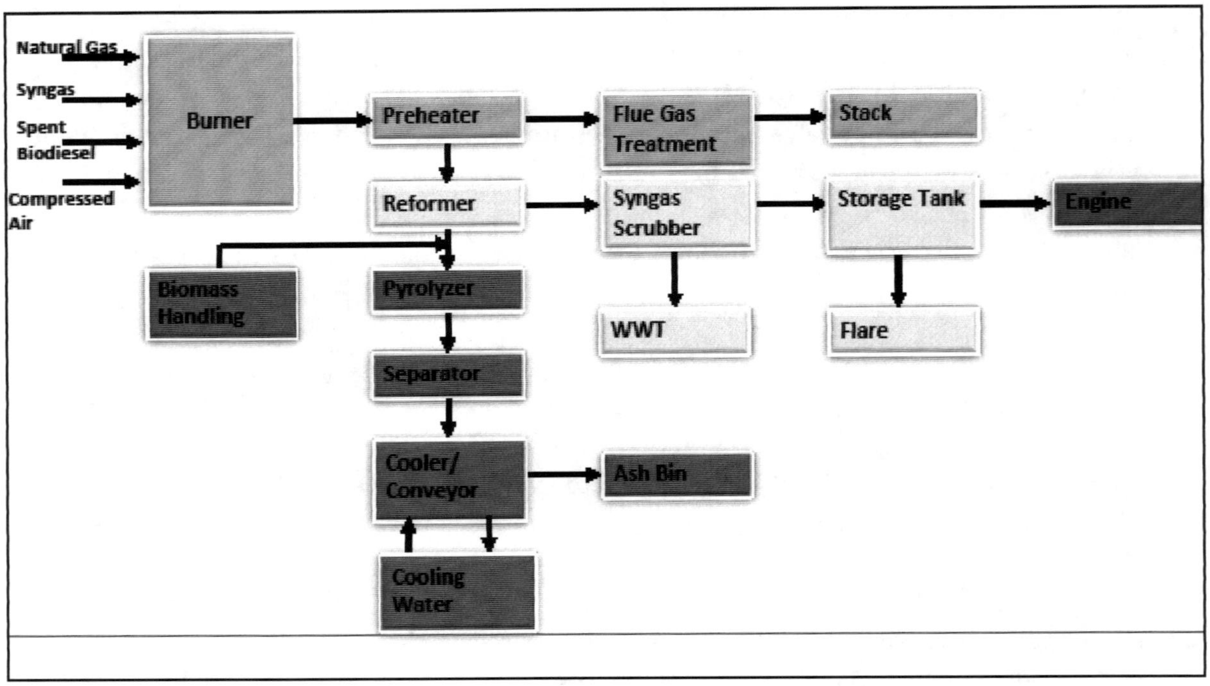

Figure 2 - Block Flow Diagram of the Owego pilot plant

The following is conceptual design of the pilot plant:

Figure 3 – Initial concept of the pilot plant

The following Figure 4 is a picture of the installed system.

Figure 4 – Owego pilot plant

Commercial Scale Plant –Herten, Germany

This is a full scale plant that will be built to generate more than 5 MW of power using wood chips as a fuel. It is similar to the pilot plant described above, except fuel drying has been added to the process.

Other energy-saving strategies are being implemented in this design, including use of biochar as a heat source for the Staged Reforming System, and use of heat from engine exhaust as a heat source for the biomass dryer.

The pictures below Figures 5 & 6 show a conceptual view of the original site plan.

Figure 5 – Herten Biomass Plant Conceptual View

Figure 6 – Herten original site plan

Results

The Owego pilot plant has completed the subsystems commissioning. System level commissioning underway. In the last month, we have made syngas that meets the quality requirements of the engine and have run the engine briefly. Plans are under way to complete the tuning and optimization of the process. Complete results of the tests will be available later this year.

The Herten, Germany project is still under design and will begin construction later this fall.

References

1. "Elemental Analysis of Wood Fuels", Final Report, June 2013, NYSERDA Report 13-13

GENERATION OF HYDROGEN FROM SOLID FEEDSTOCK USING THE HEATPIPE REFORMER TECHNOLOGY WITH IN-SITU HYDROGEN SEPARATION BY NICKEL MEMBRANES

Jonas M. Leimert, Jürgen Karl
Institute of Energy Process Engineering, Friedrich, Alexander-University Erlangen-Nuremberg, Germany

Introduction

The Heatpipe Reformer was developed as an allothermal gasification system at TU Munich during the EU project ``BioHPR" (Ref. ENK5-CT-2000-00311). The concept was commercialized by Agnion Inc., which built a demonstration plant in Pfaffenhofen, Germany in the scale of 500 kW and two commercial plants in the 1 MW scale in Grassau, Germany and Auer, Italy. Results from these plants are given by Gallmetzer in [1].

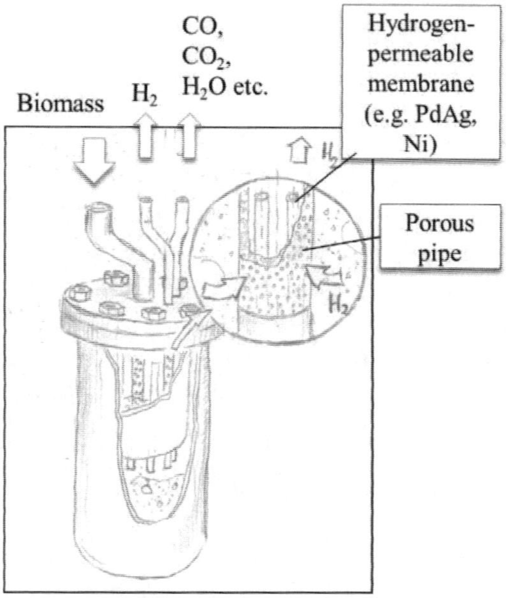

Figure 1: Sketch of Heatpipe Reformer with integrated hydrogen separation by membranes

Figure 1 shows a process scheme of the Heatpipe Reformer: In contrast to other dual fluidized bed gasification processes, the gasification takes place in a pressurized reformer chamber at 2-10 bar and 800°C. The gasification system can process both biomass and coal as feedstock. The pressurized synthesis gas allows usage for SNG synthesis, combustion in engines or gas turbines. Alternatively, the hydrogen in the synthesis gas can be separated for coupled heat and hydrogen generation. This could be achieved by a conventional process chain using a downstream CO shift reactor and hydrogen purification systems. As proposed by Karellas et al. and Schäfer the hydrogen could also be separated directly in the reformer chamber [2], [3]. This system enhances hydrogen generation due to the shift of the water-gas reaction to the product side caused by hydrogen removal:

$$CO + H_2O \rightleftharpoons CO_2 + H_2$$

For this application dense hydrogen permeable membranes with a solution-diffusion mechanism can be used, providing hydrogen with highest purity. Such dense hydrogen permeable membranes have been covered extensively over the years. In the case of palladium membranes the temperature regime of these investigations was in the range of 200-600°C restricted by the kinetics of hydrogen permeation at lower temperatures and the poor durability of Pd/Ag composite membranes at higher temperatures. Also membrane systems with other materials like Niobium, Tantalum, Titanium, Vanadium or Nickel have been proposed with varying limitations regarding temperature, pressure and atmospheres. For the Heatpipe Reformer at FAU-EVT the membrane system has to withstand temperatures of 750-900°C and pressures of up to 5 bar.

State-of-the-art

The approach at the Institute of Energy Process Engineering is the usage of dense nickel capillaries for hydrogen separation due to their low cost and high durability at the desired process temperature. Nickel membranes were already used in the Heatpipe Reformer system at FAU-EVT for hydrogen degassing of liquid metal heat pipes as described in [4]. In this application the nickel capillaries were operated for 2500 h at 800 – 870°C in sodium atmosphere without failure.

This section introduces the mathematical background of hydrogen permeation and the analysis of turnover rates and hydrogen yields. It also shows the state-of-the-art of hydrogen permeable membranes for high-temperature applications.

Hydrogen permeation and process analysis

The permeation of diatomic gases like hydrogen through a dense body is divided into several steps: The adsorption of the hydrogen on the surface, the dissociation into protons and dissolution into the metal, which is sometimes referred to as solution of the hydrogen. The hydrogen diffuses through the metal as protons, which is in most cases the rate-controlling step of the whole process, finally recombines and desorbs from the solid [3], [5], [6].

The following equation describes the hydrogen molar flow through a dense membrane n_{H2}, it depends on the permeability P of the membrane material, membrane area A, thickness Δx and the square root of the pressure difference between retentate and permeate p_1, p_2 respectively.

$$\dot{n}_{H_2} = -P \cdot \frac{A}{\Delta x} \cdot (p_1{}^{0.5} - p_2{}^{0.5}) \quad (1)$$

The permeability itself is a temperature activated material property and can be calculated with the Arrhenius equation using the permeability constant P_0 permeability activation energy E_A and gas constant R at the respective temperature T:

$$P = P_0 \cdot e^{-\frac{E_A}{R \cdot T}} \quad (2)$$

To describe the hydrogen separation in relation to the amount of treated gas, the permeate flux is divided by the maximum amount of hydrogen, which is possible to generate from the feed. The result is the dimensionless hydrogen separation rate Y_M.

$$Y_M = \frac{\dot{n}_M}{4\,\dot{n}_{CH4} + \dot{n}_{CO} + \dot{n}_{H2}} \quad (3)$$

The feedstock in gas phase reaction consists mainly of CO and CH_4. To describe the course of the reaction, the CO and CH_4 turnover X_{CH4} / X_{CO} is introduced:

$$X_i = \frac{\dot{n}_{i,out}}{\dot{n}_{i,feed}} \quad (4)$$

State-of-the-art of high temperature hydrogen permeable membranes

Figure 2 shows a summary of the literature study of high temperature membrane systems. Additionally to the permeability values of Pd based membranes mentioned above, also values from Akamatsu et al. [7], Hatlevik et al. [8], Yan et al. [9], Kusakabe et al. [10], Fernandez et al. [11], Guo et al. [12], Mardilovich [13] and Xomeritakis and Lin [14] are included. Palladium composite membranes show high hydrogen fluxes nearly reaching the physical boundary, which is shown by the hydrogen flux calculated from pure palladium. However, many membranes show fluxes well below this ideal value covering up to three orders of magnitude. The temperature limit for these membranes lies around 650°C making an application in gasification processes very challenging. Further problems of Pd-composite membranes in synthesis gas applications are coke formation, which can lead to membrane failure as carbon atoms diffuse into the Pd lattice leading to cracking [15]. They also show a degradation in hydrogen flow when exposed to synthesis gas components like CO, H_2O or sulfur components like H_2S [15]–[18].

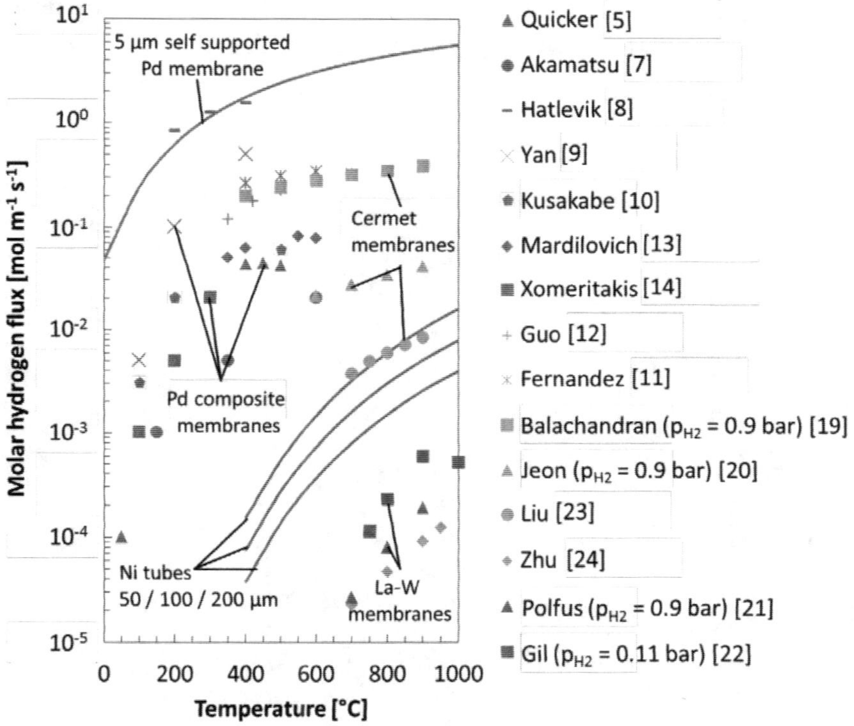

Figure 2: Literature values of membranes for high temperature hydrogen separation calculated from published permeability values at P_{H2} = 1 bar unless otherwise stated

Future membrane solutions could come from cermet or La-W membranes as described in [19]–[24], which combine high molar fluxes with temperature stability and tolerance for many gas components. However, the transfer from laboratory to an industrial application could again pose problems. Recent literature describes only small, disk shaped membranes.

Nickel membranes show in the temperature regime of gasification hydrogen fluxes in the same order of magnitude as some of the Pd and cermet membranes while showing high stability. Considering low nickel prices, this material could be easily used in process engineering without an elaborate membrane fabrication process. When small tube diameters are used, high membrane areas can be accomplished easily by constructing membrane bundles.

Gasification experiments

Fig. 3 shows a CAD sketch of the 100 kW Heatpipe Reformer at EVT-FAU. The reformer pressure vessel is integrated into the combustion chamber, described in a previous publication [25]. Eight heat pipes, filled with sodium, transfer the heat, which is needed for gasification, from the combustor into the reformer. The fluidized bed furnace is made of refractory concrete to withstand the temperatures and abrasive conditions of the fluidized bed. The primary air inlet is located in the lower part of the furnace. Before entering the combustion chamber the primary air is preheated. Therefore, the air is lead through an integrated double walled heat exchanger with a pillow-like structure located at the external wall of the freeboard. The secondary air is provided by four steel pipes with a nominal diameter of DN25 and is blown into the freeboard radially. The flue gas is led out of the combustor at the top with a temperature of about 200-400°C after preheating primary and secondary air.

The reformer is filled with olivine with a particle size of 0.063 - 0.180 mm from Sibelco Europe with a bed height of 1.1 m. Pressurized steam and lignite is fed into the gasifier through a stand pipe. This ensures a thorough mixing of steam, lignite and olivine. The synthesis gas leaves the gasifier with a temperature of about 500-700°C through an upper outlet.

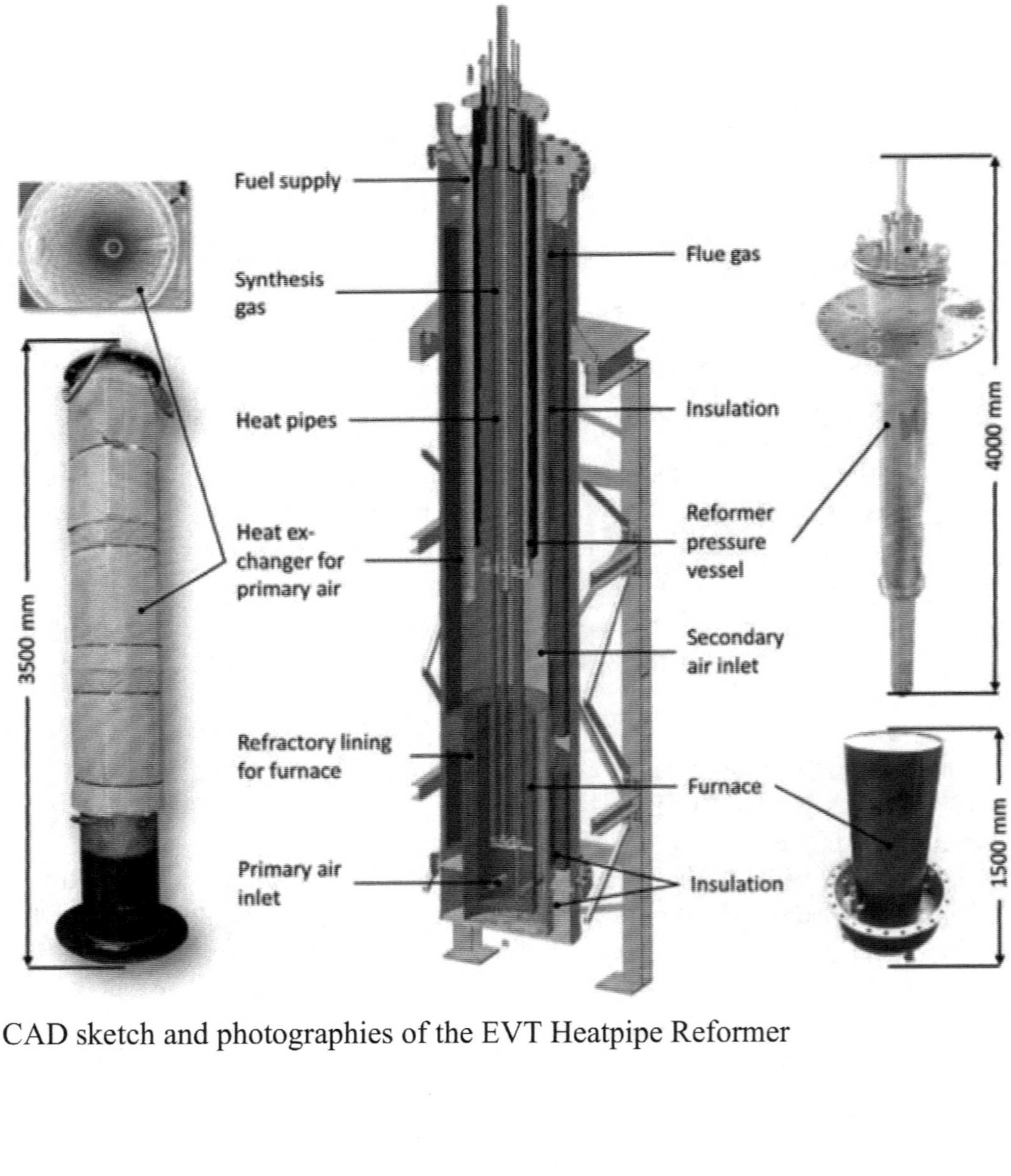

Figure 3: CAD sketch and photographies of the EVT Heatpipe Reformer

Experimental setup

The Heatpipe Reformer described above is integrated in a process chain which guarantees the fuel, air and steam supply as well as the conditioning and analysis of the synthesis gas. The piping and instrumentation diagram of this process chain is shown in Figure 4.

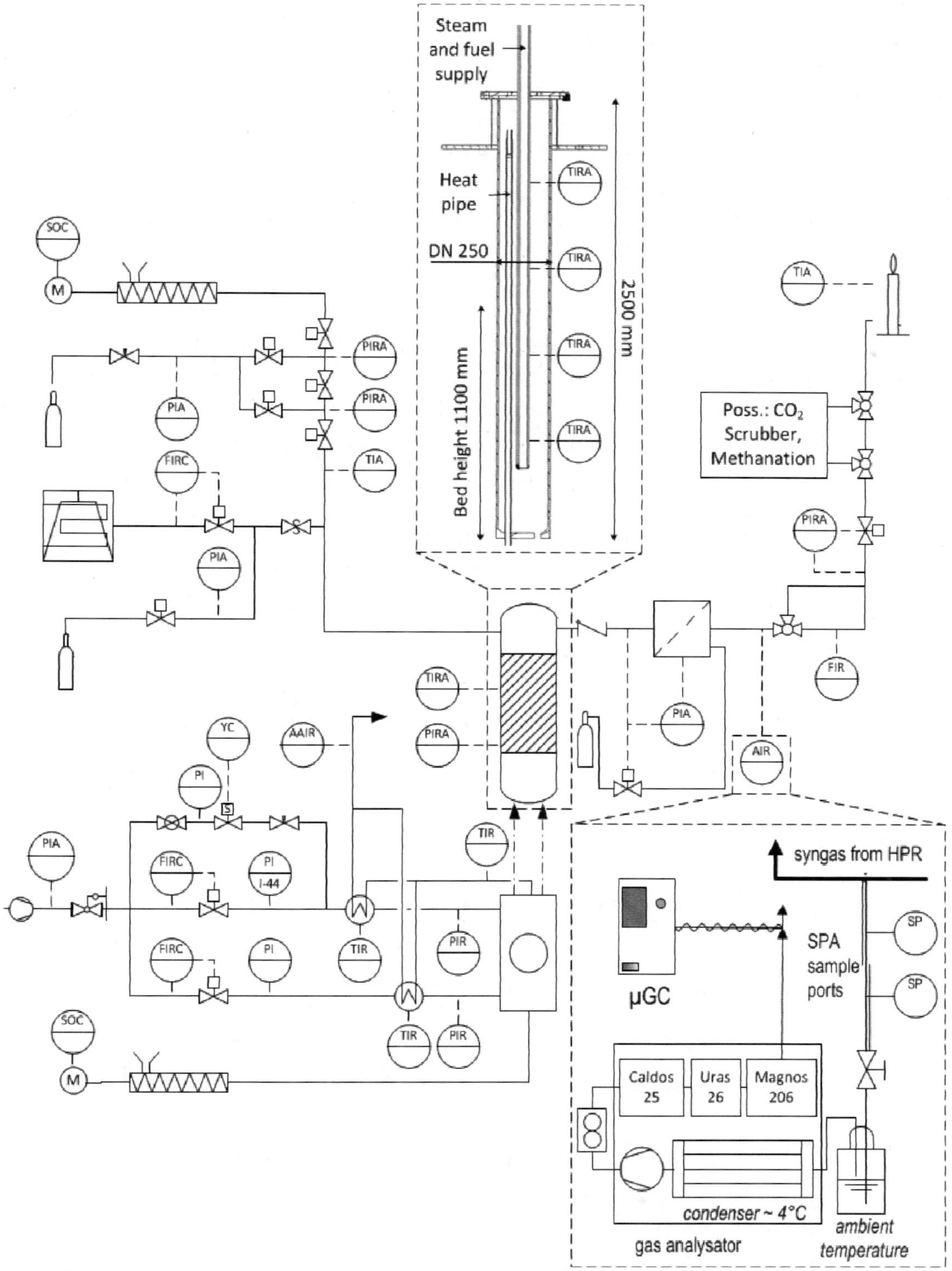

Figure 4: Piping and instrumentation diagram of the Heatpipe Reformer process chain and the gas and tar analysis system

For the combustion, the fuel input is ensured by a screw conveyor. The inlet of primary and secondary air into the combustion chamber is regulated with mass flow controllers fed by compressed air. The lignite supply to the reformer also takes place by a screw conveyor. To pressurize the lignite a lock hopper system with three ball valves is installed above the reformer. The lignite is pressurized by using nitrogen, which also acts as a tracer component for subsequent measurement of the synthesis gas flow. Following the lock hopper system the steam inlet is mounted. The steam supply is guaranteed by a steam generator, it is controlled with a Coriolis flow meter.

After leaving the gasifier the synthesis gas is directed into a candle filter, which removes olivine, ash and char particles. Downstream the candle filter the gas and tar analysis of the synthesis gas takes place, which is described in the following section. Afterwards a pressure-sustaining valve is mounted, which adjusts the reformer pressure. At the end of the process chain the flammable components of the synthesis gas are disposed of in the thermal afterburning.

Results

The measurement campaign in the presented work had a total duration of 72 h with 24 h of gasification operation. After reaching a safe reactor temperature above 800°C the gasification was started. During the measurement campaign five operation points were tested. Table 1 shows the main parameters of the operating points. During operation the fuel input into the gasifier as well as the steam mass flow was kept constant at each operation point while the heat duty of the combustor was adjusted to maintain steady state operation at gasification temperatures between 800 and 840°C.

Table 1: Operation points of the lignite measurement campaign

Name	Pressure [bar]	Heat duty Reformer [kW]	Heat duty furnace [kW]	Steam mass flow [kg s^{-1}]
OP1	1.3	10.8	26.2	4.0
OP2	1.3	15.1	28.9	5.7
OP3	1.5	19.4	34.0	7.3
OP4	4.0	10.8	26.8	4.0
OP4	4.0	15.1	33.2	5.7

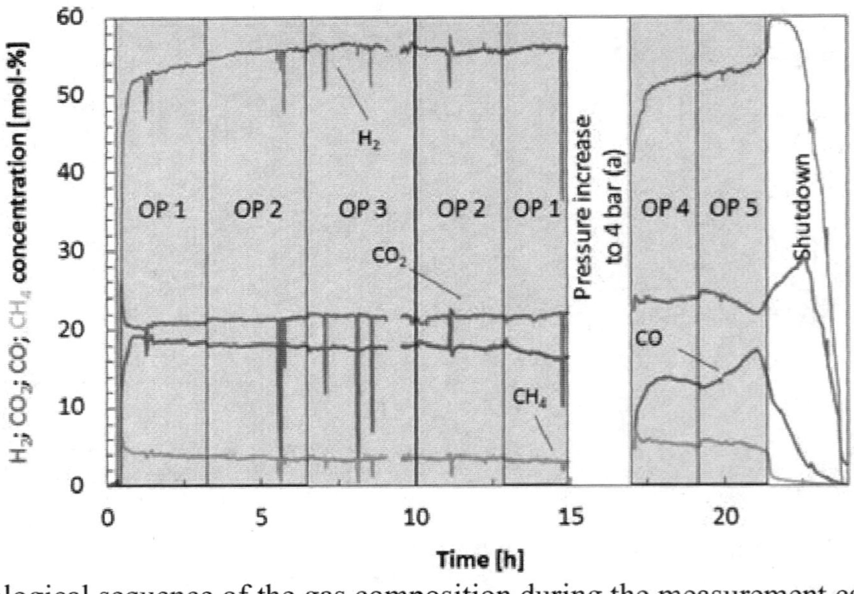

Figure 5: Chronological sequence of the gas composition during the measurement campaign

Figure 5 shows the volumetric concentration of the main gasification products, namely H_2, CO_2, CO and CH_4, as well as chronological sequence of the different operation points. The concentrations remain stable over the measurement campaign after reaching steady state after approx. 5-10 h, which can can be monitored best with the H_2 concentration. Tar content of the synthesis gas was declining with operation temperate of the gasifier and was in the range of 3-5 g/Nm³.

The gasifier showed constant hydrogen contents of 55 % in the synthesis and could safely be operated at 4 bar (a), allowing high hydrogen partial pressures for the membrane separation with nickel membranes. The medium gas composition at OP4 and OP5 was used as a bottle-mixed feed for the following membranes experiments

Membrane experiments

After single membrane experiments for preliminary tests, published in [26] and [27], the gathered information were used for a layout of a membrane bundle for the separation of approx. 1 l/min pure hydrogen. For this purpose 50 Ni 201 tubes with diameter 2.0x0.1 mm from Shaanxi Corp. were used. The following sections cover the measurement techniques for the membrane bundle experiments as well as the analysis of the influence of hydrogen separation on reaction equilibria.

Experimental setup

Figure 6 shows the membrane test rig for the membrane bundle experiments. A mass flow controller was used for the measurement of the hydrogen flow through the membrane bundle. Nitrogen acted as a tracer component to measure the dry volumetric flow rate and as a gas barrier between membrane sealing and measurement cell to prevent condensation at the temperature-sensitive sealing.

To determine the influence of the hydrogen separation on gas composition, the hydrogen permeate flow was fed back to the retentate after the measurement cell and analyzed using a µ-GC "Agilent 490" for the components CO, CO_2, N_2 and CH_4. Although measurement of H_2 is not possible as helium is used as carrier gas in the µ-GC system, it can be calculated as the difference to 100 %.

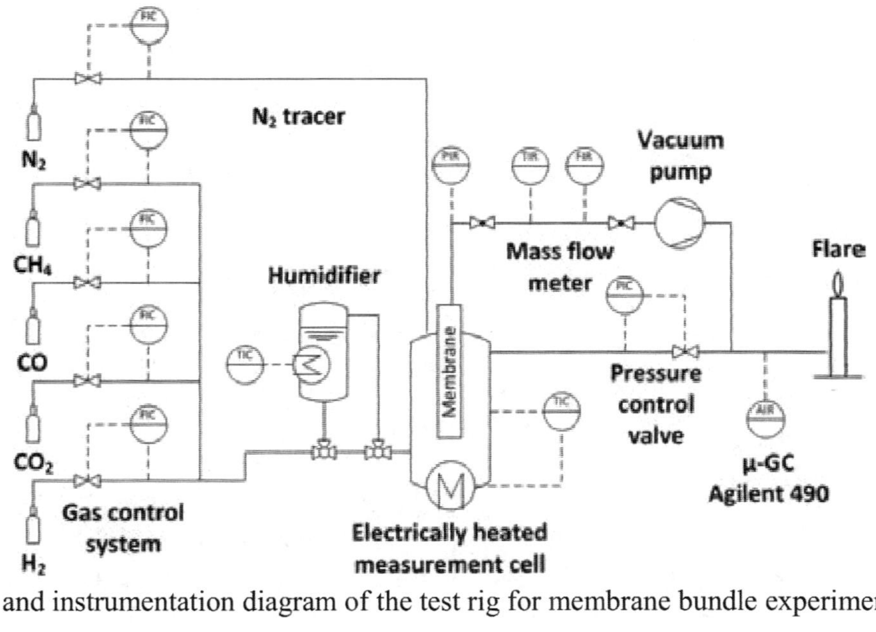

Figure 6: Piping and instrumentation diagram of the test rig for membrane bundle experiments

Results

The gas composition determined in the gasification experiments was used as feed for the measurement cell in the membrane test rig. The composition was 55 % H_2, 24 % CO_2, 16 % CO, 5 % CH_4 with a steam / ratio of 2, which resembles a steam content of 32 %. Figure 7 shows the results from these measurements. The gas composition at a hydrogen separation rate of zero does not resemble the feed composition as the gas mixture reacts due to the high temperatures and the presence of the catalytic nickel surface. Hydrogen separation has a significant influence on the turnover of CH_4 and CO. The CO turnover is negative at low hydrogen separation rates as CH_4 is converted to CO and H_2. As a result, the H_2 and CO_2 content rises due to the ongoing conversion of the feedstock. As proposed in the introduction, it is possible to influence the conversion of the synthesis gas to hydrogen by using nickel membranes.

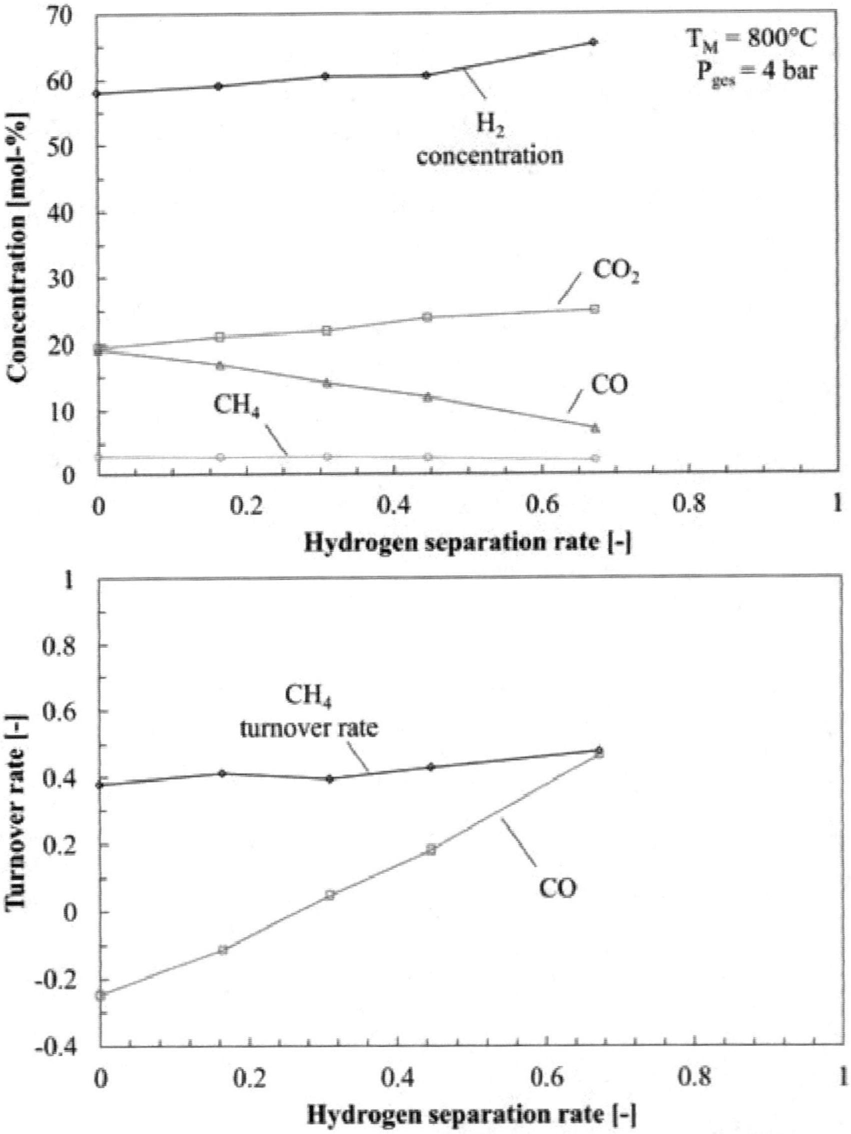

Figure 7: Resulting gas composition and turnover rates from hydrogen separation from bottle-mixed synthesis gas, Membrane temperature: 800°C, pressure 4 bar, gas composition (dry base): 55 % H_2, 24 % CO_2, 16 % CO, 5 % CH_4; steam content 32 %

Conclusion

The paper shows results from the gasification tests with the Heatpipe Reformer at FAU-EVT as well as the results from membrane experiments with a nickel membrane bundle using synthesis gas. As outlined in the state-of-the-art section, the temperature regime of gasification is very challenging for existing membrane systems such as Palladium composite membranes. The usage of nickel membranes was demonstrated and evaluated as an alternative to these systems.

Our experimental results show the good performance of the Heatpipe Reformer system and the suitability of the produced synthesis gas for hydrogen generation. The membrane results show, that the hydrogen yield and the CH_4 and CO turnover can be raised significantly by using nickel membranes.

Ongoing development at FAU-EVT concentrates on the completion and demonstration of a 100 kW SNG process chain with long term experiments. Scientific development is focused on advanced gas cleaning systems, e.g. membranes or scrubbing processes, for a sustainable usage of solid feedstock in energy process engineering

Bibliography

[1] G. Gallmetzer, P. Ackermann, A. Schweiger, T. Kienberger, T. Gröbl, H. Walter, M. Zankl, and M. Kröner, "The agnion Heatpipe-Reformer: operating experiences and evaluation of fuel conversion and syngas composition," *Biomass Convers. Biorefinery*, vol. 2, no. 3, pp. 207–215, May 2012.

[2] S. Karellas, E. Kakaras, T. Papadopoulos, C. Schäfer, and J. Karl, "Hydrogen production from allothermal biomass gasification by means of palladium membranes," *Fuel Process. Technol.*, vol. 89, no. 6, pp. 582–588, Jun. 2008.

[3] C. Schäfer, "Diffusionseigenschaften bestimmter Metalle bei der Hochtemperatur-Wasserstoffabtrennung," Technische Universität München, 2010.

[4] J. M. Leimert, M. Dillig, and J. Karl, "Hydrogen inactivation of liquid metal heat pipes," *Int. J. Heat Mass Transf.*, vol. 92, pp. 920–928, 2016.

[5] P. Quicker, "Dehydrierung von Ethylbenzol zu Styrol," FAU Erlangen-Nuremberg, 2000.

[6] H. Möllenhoff, H. J. Schröter, and H. Jüntgen, "Hydrogen Permeation through Metallic Heat Exchanger Materials during Steam Gasification of Coal," *Zeitschrift für Physikalische Chemie*, vol. 147, no. 1_2. pp. 273–284, 1986.

[7] K. Akamatsu, M. Nakane, T. Sugawara, T. Hattori, and S. Nakao, "Development of a membrane reactor for decomposing hydrogen sulfide into hydrogen using a high-performance amorphous silica membrane," *J. Memb. Sci.*, vol. 325, no. 1, pp. 16–19, Nov. 2008.

[8] Ø. Hatlevik, S. K. Gade, M. K. Keeling, P. M. Thoen, A. P. Davidson, and J. D. Way, "Palladium and palladium alloy membranes for hydrogen separation and production: History, fabrication strategies, and current performance," *Sep. Purif. Technol.*, vol. 73, no. 1, pp. 59–64, May 2010.

[9] S. Yan, H. Maeda, K. Kusakabe, and S. Morooka, "Thin Palladium Membrane Formed in Support Pores by Metal-Organic Chemical Vapor Deposition Method and Application to Hydrogen Separation," *Ind. Eng. Chem. Res.*, vol. 33, pp. 616–622, 1994.

[10] K. Kusakabe, S. Yokoyama, S. Morooka, J.-I. Hayashi, and H. Nagata, "Development of supported thin palladium membrane and application to enhancement of propane aromatixation

on Ga-silicate catalyst," *Chem. Eng. Sci.*, vol. 51, no. 11, pp. 3027–3032, 1996.
[11] E. Fernandez, J. A. Medrano, J. Melendez, M. Parco, J. L. Viviente, M. van Sint Annaland, F. Gallucci, and D. A. Pacheco Tanaka, "Preparation and characterization of metallic supported thin Pd-Ag membranes for hydrogen separation," *Chem. Eng. J.*, 2015.
[12] Y. Guo, H. Zou, H. Wu, L. Zhou, H. Liu, and X. Zhang, "Preparation of palladium membrane by bio-membrane assisted electroless plating for hydrogen separation," *Int. J. Hydrogen Energy*, vol. 39, no. 13, pp. 7069–7076, 2014.
[13] P. P. Mardilovich, Y. She, Y. H. Ma, and M.-H. Rei, "Defect-free palladium membranes on porous stainless-steel support," *AIChE J.*, vol. 44, no. 2, pp. 310–322, Feb. 1998.
[14] G. Xomeritakis and Y.-S. Lin, "CVD synthesis and gas permeation properties of thin palladium/alumina membranes," *AIChE J.*, vol. 44, no. 1, pp. 174–183, 1998.
[15] A. Basile, A. Julianelli, T. Longo, S. Liguori, and M. De Falco, "Pd-based Selective Membrane State-of-the-Art," in *Membrane reactors for hydrogen production processes*, M. De Falco, L. Marrelli, and G. Iaquaniello, Eds. Springer London, 2011, pp. 20–56.
[16] A. Kulprathipanja, G. O. Alptekin, J. L. Falconer, and J. D. Way, "Pd and Pd-Cu membranes: Inhibition of H2 permeation by H 2S," *J. Memb. Sci.*, vol. 254, pp. 49–62, 2005.
[17] H. Amandusson, L. Ekedahl, and H. Dannetun, "The effect of CO and O 2 on hydrogen permeation through a palladium membrane," *Appl. Surf. Sci.*, pp. 259–267, 2000.
[18] C. Chen, "Sulfur tolerance of Pd/Au alloy membranes for hydrogen separation from coal gas," Worcester Polytechnic Institute, 2012.
[19] U. Balachandran, T. H. Lee, C. Y. Park, J. E. Emerson, J. J. Picciolo, and S. E. Dorris, "Dense cermet membranes for hydrogen separation," *Sep. Purif. Technol.*, vol. 121, pp. 54–59, Jan. 2014.
[20] S. Y. Jeon, M. B. Choi, C. N. Park, E. D. Wachsman, and S. J. Song, "High sulfur tolerance dual-functional cermet hydrogen separation membranes," *J. Memb. Sci.*, vol. 382, no. 1–2, pp. 323–327, 2011.
[21] J. M. Polfus, W. Xing, M.-L. Fontaine, C. Denonville, P. P. Henriksen, and R. Bredesen, "Hydrogen separation Membranes based on dense ceramic composites in the La27W5O55.5–LaCrO3 system," *J. Memb. Sci.*, vol. 479, pp. 39–45, Jan. 2015.
[22] V. Gil, J. Gurauskis, C. Kjølseth, K. Wiik, and M.-A. Einarsrud, "Hydrogen permeation in asymmetric La28 − xW4 + xO54 + 3x/2 membranes," *Int. J. Hydrogen Energy*, vol. 38, no. 7, pp. 3087–3091, Mar. 2013.
[23] M. Liu, W. Sun, X. Li, S. Feng, D. Ding, D. Chen, M. Liu, and H. C. Park, "High-performance Ni–BaZr0.1Ce0.7Y0.1Yb0.1O3−δ (BZCYYb) membranes for hydrogen separation," *Int. J. Hydrogen Energy*, vol. 38, pp. 14743–14749, 2013.
[24] Z. Zhu, W. Sun, Y. Dong, Z. Wang, Z. Shi, Q. Zhang, and W. Liu, "Evaluation of hydrogen permeation properties of Ni-Ba(Zr0.7Pr0.1Y0.2)O3−δ cermet membranes," *Int. J. Hydrogen Energy*, vol. 39, no. 22, pp. 11683–11689, Jul. 2014.
[25] J. M. Leimert, P. Treiber, and J. Karl, "The Heatpipe Reformer with optimized combustor design for enhanced cold gas efficiency," *Fuel Process. Technol.*, vol. 141, pp. 68–73, 2016.
[26] J. M. Leimert and J. Karl, "Nickel membranes for in-situ hydrogen separation in high-temperature fluidized bed gasification processes," *Int. J. Hydrogen Energy*, vol. 41, no. 22, pp. 9355–9366, 2016.
[27] J. M. Leimert, A. Bartolf, and J. Karl, "Hydrogen concentration measurement in high temperature applications using hydrogen permeation," *Int. J. Hydrogen Energy*, vol. 41, no. 32, pp. 14360–14367, 2016.

469525 Performance of Transition Metal Catalysts Based on Initial Predictor for Fischer Tropsch Reaction

Abstract:

Performance of transition metal catalysts based on initial predictor for Fischer Tropsch reaction

Authors: Sumegha Godara[1], Suraj Gyawali[1] and Dr. Daniela S. Mainardi[2]

[1]Institute for Micro-manufacturing, Louisiana Tech University

[2]Institute for Micro-manufacturing, Chemical Engineering, Louisiana Tech University

All Fischer-Tropsch mechanisms known to date begin with the adsorption of carbon monoxide followed by its dissociation on a given catalysts surface. Cluster models of Co, Fe, Ni, Pd, Pt, and Ru have been explored using Density Functional Theory. Carbon monoxide adsorption energies were calculated for adsorption on all possible catalytic sites, and the most preferred CO adsorption sites were found in each case. An initial predictor that can be used to anticipate potentially effective catalysts was identified as a percentage difference, using CO adsorption energy results in combination with CO dissociation energies from the previously found sites. A greater catalysts performance is expected when that percentage difference is maximized. On pure clusters, the predictor indicates that ruthenium is expected to be the best catalyst followed by cobalt, in very good agreement with the current knowledge in this field. Hydrogen assisted CO adsorption and formation of hydrocarbons is under study.

Thus, this work presents a mechanism to quickly explore the natural potential of a catalyst material to break the CO bond. The predictor presented here can guide the synthesis of new catalysts, involving modifications of known and

currently used catalysts for the Fischer-Tropsch process, and investigations on catalysis in general.

474745 Session Keynote - Co-Extrusion: Advanced Manufacturing for Energy Devices

Abstract:

The development of mass markets for large-format batteries, including electric vehicles (EVs) and grid support, depends on both cost reductions and performance enhancements to improve their economic viability. Palo Alto Research Center (PARC) has developed a multi-material, advanced manufacturing process called co-extrusion (CoEx) to remove multiple steps in a conventional battery coating process with the potential to simultaneously increase battery energy and power density. CoEx can revolutionize battery manufacturing across most chemistries, significantly lowering end-product cost and shifting the underlying economics to make EVs and other battery applications a reality.

PARCâ€™s scale-up of CoEx for electric vehicle (EV) batteries builds on a solid base of experience in applying CoEx to solar cell manufacturing, deposition of viscous ceramic pastes, and Li-ion battery chemistries. In the solar application, CoEx has been deployed commercially at production scale where multi-channel CoEx printheads are used to print viscous silver gridline pastes at full production speeds (>40 ft/min). This operational scale-up provided invaluable experience with the nuances of speed, yield, and maintenance inherent in taking a new technology to the factory floor. PARC has leveraged this experience, adapting the CoEx process for Lithium-ion (Li-ion) battery manufacturing. To date, PARC has worked with Li-ion battery materials and structured cathodes with high-density Li-ion regions and low-density conduction regions, documenting both energy and power performance. Modeling results for a CoEx cathode show a path towards a 10-20% improvement in capacity for an EV pouch cell. Experimentally, we have realized a co-extruded battery structure with a Lithium Nickel Manganese Cobalt (NMC) cathode at print speeds equivalent to conventional roll coating processes. The heterogeneous CoEx cathode enables improved capacity in thick electrodes at higher C-rates. The proof-of-principle coin cells demonstrate the feasibility of the CoEx technology and a path towards higher energy and higher power EV pouch cells.

463732 Instability of Reversible Pump Turbines and Its Physical Origin

Abstract:

Currently, reversible pump turbines are being developed worldwide in the pumped hydro energy storage power plants. One of the paramount technical problems of the reversible pump turbines is the instabilities during the off-design working conditions. For example, in generating modes, significant pressure fluctuation has been observed in the reversible pump turbines with prominent systematic vibrations. In the present paper, several examples of the measured instabilities of reversible pump turbines are given for both the prototype and model type turbines. Based on those experimental data, the basic features of the instabilities are revealed together with the paramount influencing parameters (e.g. guide vane opening). Numerical simulations has also been performed to further investigate the physical origins of the instability. Finally, some recent progress are also briefly discussed for relieving the instability.

473631 Simplified Model of a Redox Flow Battery for Deriving Design and Dispatch Strategies in a Resource Planning Tool

Abstract:

Energy resource planning tools (ERPTs) that integrate an energy storage technology with renewable energy sources require representative models for the energy storage resources. ERPTs may involve use of optimization framework to select the optimal mix of resources and dispatch strategies. Such optimization problems can involve a yearlong time frame and several energy resources. Hence use of linear models is desirable for faster convergence of the optimization problem.

Flow battery technology has emerged as one of the potential storage candidates which can be used for stationary energy storage systems. The flexibility of scaling energy and power independently may give these systems an advantage over other energy storage technologies for applications where the energy is high and power demand is low. Flow battery systems exhibit nonlinear behavior during charging and discharging which is typical of electrochemical energy storage systems. Hence there is a need to develop a simplified (preferably linear) model without undue compromise of the underlying physics of the batteries.

Rather than modeling a flow battery as an ideal, finite source of energy with constant energy efficiency, a simple model incorporating the electrochemical behavior of a redox flow battery was developed. This model enables the optimizer to (i) choose energy and power independently (design decision) and (ii) access the performance of the chosen battery size (dispatch decision) in order to derive the optimal energy mix. The talk will also focus on quantification of errors associated with the simple model compared with the nonlinear representation.

Acknowledgements

The work presented herein was funded by the Office of Naval Research.

474162 Selection of Ion-Exchange Membrane for Vanadium Redox Flow Battery

Abstract:

The increasing share of renewable energy sources in the total energy production emphasizes the need for a reliable and cheap stationary energy storage in order to compensate for hardly predictable non-stabilities in output power of photovoltaics or wind turbines on the power transmission level. Vanadium redox flow battery (VRFB) appears as a promising solution for the stationary energy storage as it offers high efficiency (80% DC-DC), versatile arrangement of decoupled power (kW) to capacity (kWh), extended durability and fast demand response. However, the broader commercialization of the technology is still obstructed by relatively high investment costs.

Ion-exchange membrane is a key component of VRFB cell as it directly influences the power and the efficiency of the stack - the resistance of the typically used cation-exchange membrane counts for up to 50 % of the total internal resistance of the battery stack. In the same time, price of the membrane may exceed 25 % of the overall costs. Thus, the identification of highly conductive, low-cost membrane with minimal permeability for vanadium ions and sufficient durability in acidic electrolytes is vitally needed for VRFB developers.

In our contribution we present the results of a broad systematic study focused on the effect of membrane properties (charge, ion-exchange capacity, thickness, method of preparation and preconditioning) on VRFB operation. The various types of commercially available ion-exchange membranes were characterized with the respect to the properties relevant for VRFB operation such as: i) through-plane ionic conductivity in the environment of VRFB cell, ii) permeability for vanadium ions of different oxidation states, iii) performance in VRFB single-cell and iv) durability in the presence of oxidizing V^{V}electrolyte.

From the results obtained from permeation measurement it is evident that the effective diffusion coefficient of V^{3+} ions is significantly lower than of VO^{2+}ions for all the studied membranes, which corresponds with the differences in the solvation numbers of the ions. Generally higher conductivities were found for cation-exchange membranes when compared to anion-exchange ones, which is caused by the different mobility of the transported ions (hydronium vs. sulfate ions). According to our assumptions, the conductivity and selectivity of membrane significantly depend on the concentration of ion-exchange groups in the membrane. However, these properties are also influenced by the specific inner arrangement of hydrophobic and hydrophilic domains in the material. The observed trends were also confirmed by the characterization of membranes in VRFB single-cell. Based on these results we identified optimum membranes for VRFB with the respect to the energy efficiency of charging discharging cycle and maximal power output. Perfluorinated cation-exchange membranes are the solution when high power and long durability is required. Anion-exchange membranes show lower vanadium crossover and thus appears more suitable for the application when low maintenance is required.

463885 Energy Storage for the Enhancement of Penetration Level of Wind Energy: A Case Study

Abstract:

In the past five years, wind energy increases significantly in China. In terms of the equipped capacity of wind turbine, China serves as the top country in the world. However, serious problems occur due to the connection of the wind energy with the electrical grid. For example, the fluctuation nature of the wind energy causes significant instabilities in the grid, leading to the rejection of the wind energy. In some local area of China, the rejection rate of the wind energy could reach 40% or even more. To solve this dilemma, it is quite necessary to develop the energy storage facilities (in terms of both large-scale and small-scale capacities). In the present paper, effects of the energy storage on the penetration level of wind energy are investigated based on the wind farms in the western parts of the Inner Mongolia province of China. The detailed analysis are given based on the quantitative data obtained from the government and industries. The importance of the energy storage are categorized into several groups and several emerging trends are briefly discussed.

474663 An Investigation of a Polysulfide Polyiodide Aqueous Redox Flow Battery

Abstract:

Redox flow batteries (RFBs) have attracted considerable academic and industrial interest based on their favorable combination of performance and safety in the widespread integration of renewable, non-dispatchable energy sources and in the improvement of energy efficiency of the electric grid. However, at present, RFBs are broadly considered too expensive for widespread deployment. A cost-effective aqueous RFB based on alkaline metal polysulfide (PS) and polyiodide (PI) is shown here for the application grid-scale energy storage. The proposed RFB applies earth-abundant elements, which shows great potential to meet the stringent cost target for RFBs ($100/kWh). The OCV of the proposed RFB is ~ 0.88 V with a theoretical energy density as high as ~ 46 Wh/L that is ~84% higher than that of all vanadium redox flow batteries. Galvanostatic and potentiostatic studies were carried out on both the negative and positive sides, which sheds light on the electrochemistry of sulfide and iodide species, respectively. The cell performance of the proposed PS-PI chemistry was further studied using a flow battery architecture operated at 5 mL/min and 20 degree Celsius. The resultant RFB displayed stable performance at 15 mA/cm^2 for over 150 hours (25 cycles) with relatively small capacity decay. A system diagnosis after the cycling experiment revealed that membrane (sodiated Nafion membrane) degradation is the primary reason for the significantly decreased energy efficiency of the cell after 25 cycles.

459286 Comparative Techno-Economic Analysis of Bio-Hydrocarbon Production from Pyrolysis-Derived Bio-Oil Via Hydroprocessing and Zeolite Cracking

Abstract:

This work examines the comparative techno-economic analysis of two pathways for the upgrading of fast pyrolysis-derived bio-oil into bio-hydrocarbons, *viz.* hydroprocessing and zeolite cracking. The hydroprocessing pathway entails a two-stage hydrodeoxygenation process, followed by a hydrocracking process, which is coupled with a steam reforming process of the aqueous phase of the bio-oil to supplement hydrogen requirements for hydroprocessing. The zeolite upgrading pathway features a single stage regenerator, assumed to be in a full combustion mode and fitted with a catalyst cooler to tackle excess heat due to the accompanying excess coke formation typically associated with this route. The two production pathways were modelled in Aspen Plus® process simulator based on a hypothetical 72 t/day fast pyrolysis plant. Aspen Process Economic Analyser® was used for equipment sizing and cost estimation. The respective minimum fuel selling prices (MFSPs) were estimated by the discounted cash flow method, assuming the plant operates for 20 years at a 10% annual discount rate. The models of the two upgrading pathways were compared in terms of energy efficiency and profitability with regards to their relative MFSPs. Furthermore, sensitivity analysis and uncertainty analysis via Monte-Carlo simulations were conducted for the two upgrading pathways to examine the effect of variations in economic parameters on profitability in a deterministic and probabilistic manner, respectively. The energy efficiencies of the hydroprocessing and the zeolite cracking upgrading routes were 68% and 52%, respectively with corresponding MFSPs of £6.25/GGE and £7.20/GGE. Sensitivity analysis revealed that the MFSPs of both upgrading pathways are mainly sensitive to variations in fuel yield, operating cost, and income tax. Furthermore, uncertainty analysis *via* Monte-Carlo simulations resulted in probability distributions for hydroprocessing and zeolite cracking with expected MFSPs of £6.42/GGE and £8.30/GGE and standard deviations of £1.05 and £1.39, respectively. The results provide evidence to support past research regarding the economic viability of the production of biofuels from hydroprocessing and zeolite cracking. The results obtained from this work suggest that the upgrading of pyrolysis derived bio-oil *via* hydroprocessing is more techno-economically viable than the zeolite upgrading route.

464269 Economics of Biofuels and Bioproducts from an Integrated Pyrolysis Biorefinery

Abstract:

This presentation describes a novel integrated biorefinery based on fast pyrolysis and subsequent bio-oil upgrading to yield diverse renewable product portfolios. Three portfolios are considered: hydrocarbon biofuels (gasoline and diesel), biobased products (bioasphalt, calcium acetate, and dextrose), and hydrocarbon chemicals (aromatics, olefins, and fuel oil). Based on the economic assumptions and historical market prices of the products, internal rate of returns (IRRs) of the biorefinery (2000 dry metric tons per day) over 30 years are calculated to evaluate the performance of the proposed biorefinery. Given the historical low, average, and high prices of the products, a total of twenty-seven scenarios are explored and the maximum IRR for each scenario is determined. Results indicate that upgrading bio-oil to biobased products generates the maximum IRR of 67.9% if the biobased products can be marketed at a historically high price level (between \$257 and \$1491 per metric ton) and an IRR of 42.5% at an average price level (between \$240 and \$950 per metric ton). The capital and operating costs of producing biobased products are \$166.5 and \$80.6 million. Under some market conditions, the biorefinery should produce hydrocarbon chemicals or biofuels to increase the IRR. IRRs of less than 10% are found under market prices of \$820 to \$909 per metric ton for biofuels, \$222 to \$416 per metric ton for biobased products, and \$172 to \$1356 per metric ton for hydrocarbon chemicals indicating potential financial risks under low market product prices.

470751 A Kinetic Model for Updraft Biomass Gasification By Aspen Plus

Abstract:

Biomass gasification is now widely accepted as a potential and environment-friendly energy resource. In this work, a kinetic model for updraft biomass gasifiers is presented. The model is developed in Aspen Plus V8.6 by solving the detailed gasification kinetics and it approximates the composition of syngas by two steps. The biomass pyrolysis step is accomplished by stoichiometric reactions and the biomass combustion/gasification step is simulated by plugging in heterogeneous and homogeneous reaction kinetic information. Two lab-scale updraft gasifiers experiments available in literature are used to test the kinetic model (K-model) and a widely used minimizing Gibbs Energy model (MGE-model). The syngas composition predicted by K-model agrees well with experimental measurements. K-model is also more accurate than MGE-model especially for the prediction of tar. Moreover, the gasification of new biomass samples at different equivalence ratio (ER), gasification temperature and air/steam ratio are also investigated. Simulation results show that gasification process can be improved if extra heat is available to the gasifier or when it is operated under the desired ER range.

463896 A Multiscale, Multiphysics Modelling Framework for the Processes Involved in Consolidated Bioprocessing

Abstract:

Current biofuel production in the world is primarily from starch or sugar sources, such as corn and sugar cane. This has led to concerns about competition affecting prices as demand for both fuel and food increases. In order to overcome this, focus has turned to lignocellulose resources. The production of biofuels is currently carried out by processes such as separate hydrolysis and fermentation that convert biomass over a series of stages. These have proven to be inefficient and unattractive in terms of cost effectiveness. In order to make biofuels economically viable and competitive with oil-based equivalents, it is necessary to improve the efficiency and cost effectiveness of these processes. One of the ways to achieve this is by consolidated bioprocessing (CBP). CBP involves utilizing microorganisms to produce the cellulases needed to digest the biomass. whilst simultaneously metabolising the sugars produced into desirable products, such as ethanol, all in one reactor. This removes the need for extra reactors, heaters, pumps and other auxiliary equipment. The metabolism of the sugars as they are produced has the potential to reduce product inhibition of the enzymatic hydrolysis step. However, there is not a single organism that is able to both produce the enzymes to hydrolyse the biomass and ferment the sugars to products at a desirable rate and yield. Therefore, research has focused on engineering desirable traits into organisms to create an ideal CBP microorganism, such as modifying *Saccharomyces Cerevisiae* to produce hydrolytic enzymes, or adding fermentative pathways to naturally cellulolytic organisms.

We have developed a multiscale, multiphysics model for the production of valuable fuels and chemicals by consolidated bioprocessing. The thermophile, *Geobacillus Thermoglucosidasius*, has been used as the microorganism for the work. A metabolic reconstruction of the key pathways in a genetically modified strain was carried out. This was then combined with kinetic expressions of intracellular reactions to create a dynamic metabolic flux model that allowed us to visualise fluxes through the organism to products, cell growth and enzyme production. A hydrolysis

model describing the breakdown of cellulose into cellobiose using Langmuir adsorption and competitive product inhibition was developed. This, combined with the metabolism model, has allowed us to follow carbon from the biomass through to the end product. A global sensitivity analysis was carried out on model parameters to identify key factors, as well as to locate possible sources of uncertainty in the model. By identifying the most important rate limiting steps we can focus future work on the areas that will have the greatest impact on performance.

463346 Pd Particle Stabilization on Carbon Supports Under Hydrothermal Conditions

Abstract:

Carbon, as a catalyst support, has been shown to provide improved hydrothermal stability relative to common metal oxide supports such as silica and alumina, making it suitable for aqueous phase biomass conversion to chemicals and fuels. However, for supported metal catalysts, not only the stability of the carbon support, but also the stability of the supported metals needs to be considered. The stability is likely dependent on the carbon surface chemistry. The current work systematically examined this premise. Three carbon coated silica supports were synthesized at different temperatures to provide a range of oxygen content and functionality followed by the deposition of Pd. Through detailed and quantitative XPS and ^{13}C NMR characterization, lower temperature synthesized carbons gave a range of oxygen functional groups, while higher temperature carbons had more aromatic groups. The hydrothermal stability was tested using treatment in water at 170°C for 24 hours. Additionally, aqueous phase hydrogenation of furfural was performed in both batch and flow reactor systems to further characterize the relative stabilities of the catalysts. The low temperature carbon with more oxygen functional groups led to better Pd dispersion and improved hydrothermal stability.

478951 Study the Effect of Metal Salts on Glucose Isomerization by Means of Raman and Reaction Induced Difference Infrared Spectroscopy

Abstract:

Study the effect of metal salts on glucose isomerization by means of Raman and Reaction Induced Difference Infrared Spectroscopy

469719 Vacancy-Mediated Hydrodeoxygenation of Furfuryl Alcohol over Metal Oxide

Abstract:

Hydrodeoxygenation (HDO) of biomass-derived molecules is a necessary step in their utilization as fuels or chemical intermediates. Processes in which high-pressure molecular hydrogen is used suffer from the disadvantages of cost and safety. Ru/RuOx/C materials can successfully catalyze the HDO of furfural [1] and HMF [2] in the presence of isopropanol, without the need for external hydrogen, as the alcohol acts as a hydrogen donor and solvent.

Theoretical (DFT and MKM) and experimental studies of this system show that the HDO of furfural to methylfuran (MF) proceeds via a Meerwein-Ponndorf-Verley transfer hydrogen transfer from isopropanol to furfural [3]. The furfuryl alcohol produced in this step is subsequently deoxygenated over an oxygen vacancy on the RuO_2 surface [4]. Vacancies on the RuO_2 surface form by the reaction of H_2 produced from the dehydrogenation of isopropanol on the Ru surface with surface O atoms of RuO_2. However, nucleation of the vacancies causes the complete reduction of RuO2, resulting in the suppression of the HDO reaction and poor recyclability of the catalyst [2]. As such, the discovery of a stable catalyst is of great importance for the viability of this process.

In this work, we generalize and extend our conclusions to different oxides and metals that have been reported in the literature as active for HDO reactions, such as MoO3 [5] and Pt [6]. We combine experimental measurements of the HDO rate over a range of oxides in a batch reactor with a theoretical investigation of oxygenated species on these materials. These investigations yield an excellent correlation of the rate of HDO with intrinsic properties of the materials. Based on these investigations, we are able to propose and successfully employ a rationally-designed active and stable catalyst system for the HDO of furfural to MF.

References:

1. Panagiotopoulou, Vlachos. Appl. Catal A: General, 480 (2014) 17-24

2. Jae, et al. ChemCatChem, 6 (2014) 848-856

3. Gilkey, et al. ACS Catal. 5 (2015) 3988-3994

4. Mironenko and Vlachos. J. Am. Chem. Soc. (2016) Accepted

5. Prasomsri, et al. *Energy Environ. Sci.*, 6 (2013) 1732-1738

6. Luo, et al. Catal. Lett. 146 (2016) 711-717

470395 Influence of Water and Solvents on the Production of Levoglucosenone and 5-(Hydroxymethyl)Furfural from Cellulose

Abstract:

Cellulose is the most abundant renewable source of carbon on the earth. 5-hydroxymethylfurfural (HMF) as a major decomposition product from cellulose hydrolysis is one of the most promising platform chemical. Anhydro-sugars could also be obtained from cellulose depolymerization via pyrolysis or dehydration under anhydrous conditions. Levoglucosenone (LGO) is an anhydro-sugars that has potential as a chiral building block, which was used for the synthesis of a wide variety of high value pharmaceuticals. Both HMF and LGO could be converted to 1,6-hexanediol, which could serve as a precusor in polymer industry. Water, which is ubiquitus in biomass feedstock and solvents both play very important role on the product selectivites during the formation of HMF and LGO from cellulose.

In the absence of water, cellulose is first dehydrated to form anhydro-sugar (Levoglucosan), which could be further dehydrated to LGO. The final product distribution could be controlled by the addition of water to the reaction system. Cellulose will be hydrolyzed to glucose, which could be further dehydrated to HMF. The selectivity to HMF will gradually increase with increasing water content in the solvent. A series of solvents have been screened for the production of HMF and LGO from cellulose, it has been shown that the use of polar, aprotic solvents is very promising.

473767 Transformation of Chitin into Value-Added Chemicals

Abstract:

Chitin is the world second most abundant biopolymer widely existed in the exoskeleton of shellfish and insects containing 7 wt% of biologically fixed nitrogen. This sustainable biomass has an underestimated but remarkable potential for the production of industrially important chemicals especially nitrogen containing compounds. To date, the direct conversion of chitin into high value chemicals with high yield has yet to be demonstrated. [1] In this presentation, we will show case some of our efforts in the catalytic transformation of chitin into *N*-containing chemicals (Scheme 1). Several strategies, including dehydration, liquefaction, hydrogenolysis and hydrothermal treatment, will be described in detail. The direct, one-pot conversion of chitin into a N-containing furan derivative with 7.5 % yield was achieved in both organic and ionic liquid solvents, by using boric acid and metal chlorides/HCl additives.[2, 3] The boron-sugar complex intermediate formed during the reaction was probed by NMR analysis, which is the key intermediate in the reaction. Chitin liquefaction, which transforms solid biomass into oils, was demonstrated in ethylene glycol under the catalysis of sulphuric acid.[4] After liquefaction, the crystalline structure of chitin was efficiently damaged. The total yield of liquefied monomers was around 30 %. Moreover, chitin monomer has been converted to its corresponding amide/amino substituted sugar alcohols, over noble metal catalysts in the presence of hydrogen in water.[5] By using Ru/C catalyst, 71.9% C6 polyols (N-containing) 8.7% N-acetylmonoethanolamine (NMEA) and 6.1% C4 polyols were obtained. [6] Kinetic studies revealed four major parallel reaction pathways that lead to these various products. In an effort to enhance the product yields, a comparative, systematic investigation of chitin treatment by various methods was studied and the structure-reactivity correlation has been identified. The yield of dehydration product was increased to 4-fold as previous by ball mill treatment. This series of work demonstrated that chitin is indeed an attractive starting materials for the manufacturing of certain value added chemicals.

REFERENCES

[1] N. Yan and X. Chen, Nature, 524 (2015) 155.

[2] X. Chen, S.L. Chew, F.M. Kerton and N. Yan, Green Chem., 16 (2014) 2204.

[3] X. Chen, Y. Liu, F.M. Kerton and N. Yan, RSC Adv., 5 (2015) 20073.

[4] Y. Pierson, X. Chen, F.D. Bobbink, J. Zhang and N. Yan, ACS Sustainable Chem. Eng., 2 (2014) 2081.

[5] F.D. Bobbink, J. Zhang, Y. Pierson, X. Chen and N. Yan, Green Chem., 17 (2015) 1024.

[6] X. Chen, Y. Gao, L. Wang, H. Chen and N. Yan, ChemPlusChem, 80 (2015) 1565.

472530 Direct Carbon-Carbon Coupling of Furanics with Acetic Acid over BroNsted Zeolites

Abstract:

Torrefaction of biomass at mild thermal conditions can selectively decompose hemicellulose, which yields a blend of furanic and carboxylic acid species that are difficult to separate. The acidity introduced by acetic acid facilitates the polymerization of furanic species at room temperature, creating obvious storage and transportation challenges[1]. One promising route to eliminate these acids was by converting them to ketones via decarboxylative ketonization over HZSM-5[2], but this results in the loss of carbon in the form of is CO_2. Here we report direct acylation of methylfuran with acetic acid as acylating agent over HZSM-5 zeolite to produce acetyl methyl furan with very high selectivity. A reaction mechanism is proposed involving the dehydration of acetic acid as rate-determining step based on a combination of temperature programmed desorption (TPD) studies, reaction kinetics, and Denisty Functional Theory (DFT) calculations. TPD experiments to generate surface acyl groups followed by pulses of methylfuran demonstrate the direct C-C coupling of the two species to form acetyl methyl furan. The apparent activation energy for acylation was estimated to be 24.5 kJ/mole over the temperature range of 220-250 °C at under steady state flow conditions while the reaction behaves as first order with respect to acetic acid. DFT calculations reveal the generation of acyl species requires a true activation energy barrier of 104 kJ/mole, with a corresponding first order an apparent barrier of 16 kJ/mole. The barrier for the formation of acetyl methyl furan is low due to its ability to stabilize the charge of the acyl group in the transition state, and thus the overall barrier is determined by the rate of acyl formation. The presence of water is inevitable in all the streams produced from biomass, and the introduction of water in the feed reavealed an inhibition on acylation reaction rate by an order of -0.7 with a positive effect on catalyst stability. This study illustrates a new path of C-C coupling to produce higher value chemicals from biomass without sacrificing carbon as CO_2in the process.

Reference

1. Stocker, M. Biofuels and Biomass-To-Liquid Fuels in the Biorefinery: Catalytic Conversion of Lignocellulosic Biomass using Porous Materials. *Angewandte Chemie-International Edition* **47**, 9200-9211, (2008).
2. Gumidyala, A., Sooknoi, T. & Crossley, S. P. Title: Selective ketonization of acetic acid over HZSM-5: The importanceof acyl species and the influence of water. *Journal of Catalysis*, (Accepted).

467604 Supported Cu Electrocatalysts for Electrocatalytic Hydrogenation and Hydrogenolysis of Furfural

Abstract:

Furfural (FF) is a C_5 biomass-derived platform chemical that can be obtained by hydrolysis and dehydration of lignocellulose. By chemical hydrogenation and hydrogenolysis (CH) of FF, furfural alcohol (FA), an industrial adhesive intermediate, and 2-methyl furan (MF), a promising biofuel, can be produced[1, 2]. Traditionally, CH of FF provides significant selectivity towards to FA and MF on Cu-based catalysts, but it requires externally supplied hydrogen gas and high temperature or/and high pressure[3].

In contrast to CH, electrochemical hydrogenation and hydrogenolysis (ECH) allows for the conversion of FF to FA and MF without external hydrogen gas, high temperature or high pressure. This process becomes more viable when excess electricity from renewable sources is used for ECH of FF.

Results of previous studies showed that strongly acidic electrolytes produced both MF and FA [4], while higher pHs mainly produced FA [5] on bulk metallic electrodes. In ECH of FF, Cu foil electrodes also showed remarkable selectivity (c.a. 80%) to MF in 0.5 M H_2SO_4 [4]. Also, substantial selectivities of FA with carbonate pH 10 buffer (80%) on a Cu foil [6] and with 0.2 M NH_4Cl (79%) on a Ni foil [5] have been demonstrated.

We have experimentally shown further relationships between products and reaction conditions in acidic electrolytes on Cu foil electrodes where product yield significantly depended on the current density when equivalent charge was transferred to the system. However, product selectivity depended on pH, not anions of electrolytes.

Here, we developed high surface area supported Cu electrocatalysts to investigate the impact of morphology on selectivity and activity in comparison with Cu foils in 0.2 M NH_4Cl and 0.5 M H_2SO_4 electrolytes. Electrocatalysts were characterized utilizing XRD, SEM, TEM and XPS to discern the differences in catalytic performance. Product analysis was performed with gas chromatography mass spectrometry (GC-MS).

References

[1] J.N. Chheda, G.W. Huber, J.A. Dumesic, Liquid-phase catalytic processing of biomass-derived oxygenated hydrocarbons to fuels and chemicals, Angew. Chem.-Int. Edit. 46 (2007) 7164-83.

[2] J.-P. Lange, E. van der Heide, J. van Buijtenen, R. Price, Furfural—A promising platform for lignocellulosic biofuels, ChemSusChem 5 (2012) 150-66.

[3] Y. Nakagawa, M. Tamura, K. Tomishige, Catalytic reduction of biomass-derived furanic compounds with hydrogen, ACS Catalysis 3 (2013) 2655-68.

[4] P. Nilges, U. Schröder, Electrochemistry for biofuel generation: production of furans by electrocatalytic hydrogenation of furfural, Energy & Environmental Science 6 (2013) 2925-31.

[5] Z. Li, S. Kelkar, C.H. Lam, K. Luczek, J.E. Jackson, D.J. Miller, C.M. Saffron, Aqueous electrocatalytic hydrogenation of furfural using a sacrificial anode, Electrochimica Acta 64 (2012) 87-93.

[6] P. Parpot, A.P. Bettencourt, G. Chamoulaud, K.B. Kokoh, E.M. Belgsir, Electrochemical investigations of the oxidation-reduction of furfural in aqueous medium: Application to electrosynthesis, Electrochimica Acta 49 (2004) 397-403.

469287 Rational Synthesis and Evaluation of Pd Bimetallic Catalysts for Furfural Rearrangement to Cyclopentanone in Aqueous Phase

Abstract:

Furfural, a product of acid-catalysed hydrolysis of hemicellulose in lignocellulosic biomass, is one of most promising biochemical platforms. With multiple active functional groups, furfural (FFA) can be catalytically transformed into a series of chemicals. Cyclopentanone (CPO) is one of downstream products from furfural hydrogenation which is widely used as solvents and fine chemical starting materials for fungicides, fragrances and pharmaceuticals [1].Â Production of CPO is a good way to develop further utilization of biomass-based furfural.

Hydrogenation of furfural has been intensively reported in literature, however the rational method to prepare bimetallic catalysts for this reaction hasnâ€™t been covered much. In this work, the aqueous phase hydrogenation of FFA to CPO was investigated on supported bimetallic catalysts (Pd-Ni, Pd-Cu and Pd-Co) prepared by simultaneous Strong Electrostatic Adsorption (SEA) and incipient wetness (IW). In SEA method, hydroxyl groups populate on the surface of support are charged by adjusting pH value in the solution away from its neutral point which is termed as Point of Zero Charge (PZC), metal precursors with opposite charges will be adsorbed. Unlike incipient wetness, the strong electrostatic interaction between metal complexes and supports renders highly dispersed metal nanoparticles. X-ray diffraction, elemental mapping and H_2 chemisorption were performed to confirm the high dispersion (<1.5 nm) and intimate contact between Pd and second metal.

Furfural hydrogenation reaction was carried out in 57 mL deionized water with 3 g of FFA at 150 °C, 500 PSIG (P_{H2}= 430 PSIG) for 3 hours, products were sampled at an interval of 20 min and analyzed by GC. Reaction results showed enhanced catalytic performance with Pd bimetallics prepared by SEA compared to their IW counterparts: the yield of CPO increased by more than 5 times; nano-alloy survived and slightly sintered (~2.5 nm) in SEA catalysts while Ni and Co might have leached out from IW ones after reaction.

References

[1] M. Hronec, K. Fulajtarove, I. Vavra, T. Sotak *et al.*, App. Cataly. B : Environmental, 181 (2016) 210-219.

472910 Polymeric Solid Acid Catalysts for Efficient Hydromethyfurfural and Levulinic Acid Production from Biomass Feedstock

Abstract:

Polymeric solid acid catalysts were synthesized and optimized for the production of reducing sugars, HMF and levulinic acid from biomass feed stock. The catalysts consist of a poly(sulfonic) acid chain and a poly(ionic liquid) chain grafted on a solid substrate. Crystalline cellulose as well as dilute acid, base and steam pretreated biomass samples from National Renewable Energy Laboratory were investigated with our novel tunable catalysts in ionic liquid solvent and ionic liquid/water, ionic liquid/GVL solvent mixtures. Over 95% of total reducing sugar (TRS), over 70% HMF and over 50% levulinic acid yields were obtained from pure cellulose samples at 105°C, 135°C and 155°C respectively. For the pretreated biomass samples, different yields were obtainded from three different pretreatment methods. Yields of over 80% TRS, over 60% HMF/Furfural, and over 40% levulinic acid were achieved. GVL cosolvent was found to facilitate the conversion process whereas the opposite is true for water. Our results demonstrate the high catalytic activity and versatility of these unique polymeric acid catalysts.

473494 Performance of a Chemical Looping Combustion with Oxygen Uncoupling Process Development Unit

Abstract:

Chemical looping combustion (CLC) is a technology to combust solid fuels while intrinsically separating the greenhouse gas CO_2. This eliminates the energy cost associated with gas separation, making CLC an attractive technology for CO_2 separation. CLC involves circulating a metal-based â€œoxygen carrierâ€ between two reactors in which it is oxidized by air and then reduced by a fuel, in essence combusting the fuel. In the variant of CLC known as chemical looping with oxygen uncoupling (CLOU), the oxygen carrier releases gas-phase oxygen as O_2once in the fuel reactors reducing environment, allowing for a gas-solid reaction to occur. Particles are then circulated to the air reactor and reoxidized. The natural mineral ilmenite presents a cost-effective oxygen carrier material with demonstrated strength and cyclability, while copper allows for the desired oxygen uncoupling behavior of the metal oxide during CLOU operation.

The University of Utah has developed a 10 kW_{th} bench scale dual fluidized bed chemical looping combustion system. The goal of the study presented here was to assess system performance and compare conventional chemical looping combustion, known as indirect gasification-CLC or iG-CLC, with CLOU chemical looping. In this study two different fuels, coal and coal char, were processed and the degree of fuel conversion, CO_2 capture efficiency and CO_2 purity were evaluated. Performance of the unit itself is also evaluated by changing variables such as temperature, flow rate, bed inventory, and circulation rate to determine optimal performance.

469774 Recent Operating Experiences from Netl's 50kW$_{th}$ Circulating Chemical Looping Combustion Test Facility

Abstract:

In order to balance future energy needs with the goal of reducing greenhouse gas emissions, sustainable technology options for fossil fuel power generation are needed. Chemical Looping Combustion (CLC) is widely recognized as a potential option to reduce greenhouse gas emissions at a cost that is significantly lower than a conventional power plant with a conventional CO_2 absorption system. Although the general concept has a long history, CLC has numerous technical issues that must be addressed in order to prepare this technology for a commercial debut. Researchers at the Department of Energy's National Energy Technology Laboratory are using a small 50 kW$_{th}$ circulating test facility to address some of these issues. This paper will describe the test facility and recent operating experience using iron-based oxygen carrier materials. Operating experience has indicated very good gas phase mass balances, as well as, some operational issues associated with solids handling and control. This paper will describe some of the advantages and disadvantages associated with operations using various "commodity" oxygen carrier

472261 The Syngas Chemical Looping Processes for Hydrogen and Electricity Co-Generation with Carbon Capture: Pilot Plant Development and Operation

Abstract:

Chemical looping technologies have evolved into a promising alternative for the efficient conversion of carbonaceous fuels to electricity and/or high value chemicals with minimal carbon emissions. With the exponential growth of research and publications in this field, chemical looping has expanded to encompass power and chemical production with in-situ gas separation. The Ohio State University (OSU) has successfully developed the syngas chemical looping (SCL) and coal direct chemical looping (CDCL) processes from laboratory studies to two fully integrated pilot scale demonstration facilities for the conversion of gaseous and solid fuels, respectively, to high purity H_2 and heat with nearly 100% CO_2 capture. The unique counter-current moving bed reactor design allows the OSU SCL process to achieve full fuel conversion to CO_2/H_2O while minimizing the solid circulation rate and has the capability of producing high purity H_2 for further chemical production.

The present paper summarizes key developments of the 250 kW_{th} â€" 3MW_{th} SCL pilot plant constructed at the National Carbon Capture Center (NCCC). An overview of the process design, cold flow model studies, and construction will be provided. Operational challenges overcome in the SCL pilot plant included the development of auxiliary equipment such as a pilot-scale raw gas burner and baghouse. An advanced high temperature, high pressure solid flow measurement device, electrical capacitance volume tomography (ECVT), was also tested for instantaneous solid flow control in this chemical looping system. The SCL pilot plant has been successfully demonstrated to achieve nearly full syngas conversion. Start-up and operational results including solid circulation studies, process heat up, syngas conversion, and solids analysis will be presented and discussed.

462528 CFD Simulations of a Chemical Looping with Oxygen Uncoupling System: Aiding in Reactor Design

Abstract:

Chemical looping with oxygen uncoupling (CLOU) is a carbon-capture technology that utilizes a metal oxide as an oxygen carrier to separate oxygen from air. The metal oxide oxidizes in an air reactor by the air used to fluidize the reactor and is transferred to another reactor where releases gaseous O_2 into a reactor where fuel, such as coal, is fed. The reduced oxides are then transferred back to the air reactor, completing the loop. Previous experimental and simulated work has been performed on systems less than 1 MW. CLOU kinetics have been introduced into a CLOU dual reactor configuration. These simulations have been used to explore the effects of fuel properties, process properties, and combustion kinetics.

From these simulations design changes to the reactor have been explored including decreasing fluidization volumetric flow rate, altering fluidizing gases, inclusion and design of a carbon stripper. The fluidization volumetric flow rate needed to be decreased proportional to the volatile content of the fuel, this is to maintain proper fluidization regime. The fluidizing gases in the air to fuel reactor loopseal needed to maintain high enough oxygen content as to not introduce oxygen uncoupling in the loopseal resulting in decrease in oxygen transferred to the fuel reactor. From simulations a simple carbon stripper was designed to be used in the system.

473130 Mediated Oxycombustion with Integrated Uncoupled Oxygen Supply (MOBIUS)

Abstract:

In order to sustain the global growth experienced currently, the appetite for energy is going to continually increase. To meet this demand, it is projected that combustion of carbonaceous fuels driven electricity production will continue to play a major role. Coupled with this demand for energy is an increase in the public awareness regarding the need to maintain clean water and air. Therefore, to improve the sustainability of such processes, it is necessary to enhance the efficiency of these power plants and to find reliable environmental compliant solutions. Two such solutions are the Oxy-combustion process that uses oxygen diluted with CO_2 rather than air and the Chemical Looping Combustion (CLC) that employs metal oxide/metal redox couples. There are, however, bottlenecks for both processes. In the former, separate oxygen separation units operated by cryogenic separation or membrane separation becomes a necessity while the latter incurs the technical difficulty of hot solids materials handling as well as performance issues such as reduced conversion rates when dealing with solid fuels. In recent years, there has been a push towards direct carbon fuel cells (DCFC) where the electricity is generated directly in a fuel cell by the direct use of carbonaceous fuels. The primary technological hurdles of DCFC lies in the a) identification of a suitable metal/metal oxide anode system to allow the facile transport of the oxygen from the SOFC electrolyte to the carbon and have a sufficiently high open circuit voltage and b) a compatible electrolyte for the chosen material such the rates of oxidation are high without causing deterioration in the electrolyte by processes such as corrosion. We present a combustion system â€“ Mediated Oxycombustion with Integrated Uncoupled Oxygen Supply (MOBIUS) that synergistically combines the effective aspects of the three above mentioned technologies along with innovations to overcome the hurdles faced by these technologies. We will discuss the development of the combustion system components. The key component of this design is a trifunctional membrane reactor based on suitably doped ceria that facilitates the oxygen transport. We will present the impact of doping on the oxygen uptake on the O rich side, its transport through the membrane and

uncoupling at the O lean side.

467303 Soot Nucleation and Consumption in Oxy-Coal Systems

Abstract:

With the rising concern of carbon dioxideâ€™s effects on climate change, research involving oxy-fuel combustion has increased significantly. The environment of an oxy-fuel system can vary greatly from that of conventional air-fired systems; as a result, models to describe system physics must be adapted to account for these differences. This work focuses on some of the differences found in the oxy-coal environment and the effects this has on soot nucleation and consumption.Â The transformation of tar into soot particles along with the growth of those particles are explored and models are presented. Oxidation rates are explored and models optimized to experimental data published in the literature. Model parametersâ€™ uncertainty are also computed and presented. Gasification rates are considered and modeled in a similar way with their respective uncertainties. Effects of gasification, which is typically neglected in conventional systems, are also explored in simulation of an oxy-fuel combustor at the University of Utah; results are presented.

469046 Investigation of NO_x Chemistry in Oxy-Combustion Flue Gas

Abstract:

Oxy-coal combustion technology can be a promising solution to meet the challenge of reducing carbon emissions from power plants while continuing with the fossil fuel usage. Implementation of such technology requires the removal of nitrogen (N_2) from the oxidizer and introduction of a pure oxygen (O_2) stream and a recycled flue gas stream into the boiler. In absence of N_2, the flue gas stream generated from such system will be low in volume and highly concentrated in carbon dioxide (CO_2) which in turn will facilitate the capture and sequestration process. Moreover, oxy-coal combustion process has shown promise in reducing nitric oxide (NO) emissions. Lower NO emissions are expected in the oxy-combustion process due to the absence of N_2 in the oxidizer and reduction of recycled NO in the flame zone through the reburn mechanism. In the reburn process, recycled NO interacts with fuel fragments forming different intermediates. Depending on the combustion conditions, these intermediates can form $N_2$and thus reduce the NO emission. To gain a better understanding of this reduction process under oxy-combustion conditions and to evaluate the influences of different combustion parameters, a comprehensive study focused on nitrogen chemistry in oxy-combustion is crucial.

The present study aims to provide speciation data of nitrogen oxides (NO_X) and to reveal the reaction pathways of recycled NO destruction under oxy-combustion conditions. To do so, the oxy-coal combustion environment has been simulated in a lab-scale reactor with a time-temperature history representing a plant boiler. Flue gas samples collected from different temperature points of the reactor have been analyzed by applying Fourier Transform Infrared (FTIR) spectroscopy for NO_X speciation. The influence of different combustion parameters, i.e., equivalence ratio (0.8-0.98), O_2 percentage (28-34%) and recycled NO concentration (800-2000 ppmv) has also been investigated in the course of the study. Moreover, the combustion experiments have been simulated by conducting kinetic modeling with Chemkin to evaluate the performance of reaction mechanisms reported in the literature. Sensitivity analysis has also been performed to identify the destruction routes of recycled NO. The data reported in the present study can aid in predicting NO_x emissions from power plants operating under oxy mode and to evaluate the performance of combustion mechanisms in the literature.

459266 On the Role of Sodium Content of Fine Ash Particles in Determining Deposition Rates Under Air- and Oxy-Coal Combustion Conditions

Abstract:

During the combustion of coal, the sodium content in the ash is commonly thought to be the "bad actor" in determining ash deposition rates on boiler heat transfer surfaces. We have conducted 16 tests in which deposition rates of three coals, burned under air fired and oxy-fired combustion conditions, were measured. A 100kW down-fired laboratory combustor, coupled with a specially designed deposition probe was employed. Emphasis was on deposition rates of the tightly bonded "inside" deposits, rather than on the loosely bound "outside" deposits, which could not be weighed precisely. Surprisingly, the measured deposit rates correlate *neither* with the sodium content of the coal, *nor* with the sodium content of the fine ash aerosol (PM1 or PM0.6). Instead they correlate well with the *overall* flue gas concentration of PM1, the alkali composition of which varied significantly over the 16 tests. This suggests a mechanism of deposit adhesion that depends simply on the presence of submicron particles of *any* composition. Additional research is needed to confirm this, because this conclusion disagrees with nearly all existing models for ash deposition and has implications for simulations predicting boiler performance for coal or biomass fired fuels.

467596 NO_x Formation in Syngas/Air Combustion at Elevated Pressure

Abstract:

Combustion of high hydrogen content (HHC) fuels is central to the implementation of integrated gasification combined cycle (IGCC) systems or to other configurations that would optimize power production while minimizing emissions. Gas turbines utilizing HHC fuels for power generation applications will be required to meet stringent pollutant emission standards, particularly with respect to nitrogen oxides (NO_x). One of the effective NO_x reduction technologies for gas turbines is lean premixed (LPM) combustion method. Accurate numerical prediction of NO_x from such combustors using multi-dimensional models has only been of limited success. This is to some extent due to the complexity of NO_x formation chemistry in LPM combustion. Experimental data is necessary for generating accurate computational models of NO_x production in combustion. Improvement of the model predictions could be accomplished by validation against data from well-defined experiments to design a very low emission process with optimum efficiency. For gas turbine conditions, reliable high-pressure data is needed to both validate and improve reaction mechanisms to simulate NO_x chemistry.

In this study, NO_x formation in post-flame gases of syngas combustion was studied at elevated pressures. A high-pressure burner facility well suited to the study of reaction mechanisms was designed and fabricated. A series of experiments were performed to investigate the NO_x formation from lean premixed syngas/air combustion with various stoichiometries (equivalence ratio between 0.5-1.0 and H_2/CO ratio between 0.25-1.0). Acquiring an accurate temperature profile throughout the system is critical for making accurate predictions via kinetic simulations. Flame temperature and two-dimensional temperature distribution were measured under various conditions. Detailed NO_x speciation measurements in the post-combustion zone were conducted by Fourier transform infrared spectrometer (FTIR). Additionally, computational fluid dynamics (CFD) simulations with detailed chemical kinetics were employed

to predict the temperature and species profiles.

462880 Formation and Decomposition Kinetics of HONO and HNO2

Abstract:

The HNO_2 system â€“ cis-HONO, trans-HONO, and HNO_2 â€“ are important intermediates in the combustion of nitrogen. These isomers are typically formed either from H-abstraction or disproportionation. To date, most experimental and theoretical work on the HNO2 potential energy surface has focused on the chemically activated reaction H + NO_2â†' OH + NO. The isomerization and decomposition kinetics of the HNO2 isomers has received considerably less attention.

This talk will present recent theoretical work on the isomerization and thermal decomposition of cis-HONO, trans-HONO, and HNO_2. For the stationary points, geometry optimization and normal mode analysis was done using CCSD(T)/cc-pVQZ. Five post-coupled cluster methods were used to further improve the electronic energy. The tight transition state for isomerization between trans-HONO and HNO2 was computed using conventional transition state theory with rigid-rotor harmonic oscillator approximations. For the three H + NO_2pathways and the two OH + NO pathways, variable reaction coordinate transition state theory was used. The active space was 14 electrons in 11 orbitals.

The final part of the talk will focus on the relative importance of including all three isomers in kinetic models versus the simplicity of treating all three as a single, lumped species.

447496 Experimental Study on NO Reduction By Iron with Methane in Wet Simualted Flue Gas

Abstract:

Nitrogen oxide (NO) is a serious pollutant emission from coal-fired boilers and other industrial plants. In the current SCR and SNCR technologies, the reducing agent is ammonia, which is expensive and requires special handling and storage as well as a sophisticated metering system to avoid NH_3 slip. There is a great incentive to use natural gas as the reductant in stationary SCR/SNCR units rather than NH_3. A novel and effective method to reduce NO by methane over iron and/or iron oxides was recently proposed by the authors and experimental results in dry simulated flue gas proved this method is very promising.

However, there is water vapor in the flue gas in real conditions. In the SCR-NH_3 technology, many of the catalysts will lost their activity when there is water vapor. The resistance of this method to vapor need to be demonstrated.

In this paper, the effect of water vapor on NO reduction by methane over iron was experimentally investigated at 300~1100°C in a ceramic tubular flow reactor. The iron samples after reaction were characterized by XRD, SEM and XPS. Results demonstrated that water vapor had a small effect on NO reduction by methane over iron. In N_2 atmosphere, water vapor involved in the oxidation of iron and caused decreased NO reduction efficiencies when 2.5~7% water vapor was added into the reactor, as compared to that when vapor was not added. When the water vapor increased from 2.5% to 7%, however, NO reduction efficiency increased with the increase of water vapor due to the porous surface of the iron sample caused by the oxidizing effect of water vapor on iron. Methane was involved in the reduction of the iron oxides leading to the formation of a dense layer of Fe_3O_4 and FeO, which inhabited the interaction of NO and metallic iron and caused the decreased NO reduction as compared to that without methane. In simulated flue gas atmosphere, water vapor increased the NO reduction by methane over iron. When the excess air ratios in reaction zone and burnout zone were SR_1=0.7 and SR_2=1.2 respectively, the NO reduction efficiencies were found to be 96.7% and 90.6% respectively at 1050 °C when there were 7% and 0% water vapor. SO_2 caused a slight decrease of NO reduction. Experimental data showed that more than 90% of NO was reduced by 1.14% methane over iron at 1050°C in a durable test over 50 hours in simulated flue gas atmosphere containing 7% H_2O and 0.02% SO_2.

The results proved that the present method for NO reduction by iron with methane is very effective and is not influenced by vapor.

464644 Inferential–MODEL Based FLARE SET POINT Determination

Abstract:

Recent EPA rule (40 CFR parts 60 and 63) finalizes single minimum combustion zone net heating value (NHVcz) operating limit as 270 BTU/scf during any 15–minute average period for flares subject to Petroleum Refinery Maximum Achievable Control Technology standards to ensure 98% destruction efficiency (DE) or 96.5% combustion efficiency (CE) or higher at all times. In addition, refinery flares meet the flare tip velocity requirements (max. 400 ft/s), visible emission requirements (20% opacity limits) and net heating value dilution parameter (NHV_{dil}) minimum operating limit of 22 BTU/ft^2 and corrected vent gas heat content of 1212 BTU/scf for hydrogen to demonstrate the compliance. This study aims at developing robust inferential flare models to express DE, CE and opacity as a function of operating variables and to determine set point for steam and make up fuel at the incipient smoke point operation for NHVcz ≥ 270 BTU/scf and $NHV_{dil} \geq 22$ BTU/ft^2. Multivariate statistical distribution and response surface models were developed for steam and air assisted flares burning propylene, propane, natural gas, methane, ethylene and typical refinery vent gas mixtures. Curve fitting tool box in MATLAB and Minitab were used in modeling. In this study, soot emissions, visible emission scale, DE, and CE test data from 1983 to 2014 (including 1983/1984/1985 EPA, 2010 TCEQ, 2009/2010 Marathon Detroit/Texas City, and 2014 Carleton University) were analyzed. The input variables include measurable variables like exit velocity, lower heating value, flare tip diameter, vent gas species (double bond, carbon to hydrogen molar ratio and molecular weight), and operating variables (NHVcz, NHV_{dil}, steam/air assists, and fuel). Response surface models developed for CE and opacity showed high coefficient of determination (R^2) values 0.91 and 0.95 respectively for steam assisted flare data.

470281 Multiscale Optimization and Intensification of Natural Gas Separation and Storage

Abstract:

Natural gas is currently the fastest growing energy commodity and the cleanest fossil fuel with abundant reserves. Methane is the predominant component of natural gas. Significant efforts have been undertaken in the past for the separation, storage and transportation of methane. However, the following challenges remain: (i) materials with high selectivity for methane separation from natural gas often do not have high-deliverable capacity for storage, and (ii) materials with high-deliverable capacity for methane storage often do not have high selectivity for separation. In this work, we develop a combined separation and storage (CSS) technology to separate and selectively store methane at the same pressure in the same vessel, thereby essentially eliminating the need for a separate purification process. The proposed method is based on microporous materials (e.g., zeolites). While the state-of-the-art research considers materials for methane separation or storage separately [1], we propose an intensified technology enabled by advanced materials which simultaneously separate and store methane from natural gas.

Using a multi-scale systems engineering approach, we have designed a CSS process that takes a mixture of $CH_4/CO_2/N_2$ as feed, and stores CH_4 inside an adsorbent-packed column while venting N_2 from the system. A non-linear algebraic partial differential equation (NAPDE) model describing the conservation equations and other equations pertaining to adsorption isotherm, mass transfer driving force and pressure drop across the bed is used to study the process. In order to obtain optimal operating conditions, the detailed process model is completely discretized in both space and time into a system of nonlinear equations, whose size increases with the level of discretization. The discrete decisions related to stepwise cyclic operation together with the nonlinear equations result in a large-scale mixed integer nonlinear programming (MINLP) problem. To aid the convergence of the model, simulation-based warm starts are provided to the optimization framework. Optimization is performed to select microporous adsorbents from the database of existing materials [1] and a rank ordered list of candidate materials for different process conditions is generated. Preliminary studies reveal that for achieving simultaneous separation and storage, both purity and storage capacity of CH_4 have to be balanced by identifying the best candidate material and the corresponding operating conditions.

Our multiscale framework has the capability to simultaneously optimize the material and the process. Additionally, we will demonstrate that the framework can be used to obtain optimal properties of hypothetical materials, which can serve as a guide for future experimental efforts to design novel materials for combined gas separation and storage.

References:

[1] First, E. L.; Hasan, M. M. F.; Floudas, C. A. Discovery of Novel Zeolites for Natural Gas Purification through Combined Material Selection and Process Optimization Approach. *AIChE Journal*, 60(5), 1767â€"1785, 2014.

[2] Siderius, D.W., Shen, V.K., Johnson III, R.D. and van Zee, R.D., Eds. â€œNIST/ARPA-E Database of Novel and Emerging Adsorbent Materialsâ€, *NIST Standard Reference Database Number 205*, National Institute of Standards and Technology, Gaithersburg MD, 20899, http://adsorbents.nist.gov.

472821 Ideas Based Synthesis of Fossil Fueled Power Generation/Hydrogen Production Flowsheets with No Carbon Dioxide Emissions

Abstract:

In this work, the Infinite-DimEnsionAl State-space (IDEAS) conceptual framework is employed as a process synthesis activity in the formulation of a flowsheet that utilizes natural gas, carbon, and coal combinations to co-produce hydrogen and formic acid, without any carbon dioxide emissions. IDEAS formulations result in a linear program that allows for a broad search of the state-space in order to identify intensified flowsheets. Employing combustion, steam methane reforming (SMR), pressure swing adsorption, and formic acid production from CO2 and H2, a carbon dioxide emission free flowsheet can be synthesized. Thermodynamic analysis, heat and power integration studies, and operating costs/revenue studies are all presented to determine the designâ€™s operating revenue to cost ratio.

467387 A Heuristic-Based Model for Gasification-Based Plants with Flexible Feedstock/Product in Carbon Constrained Scenarios

Abstract:

Gasification is one of the most versatile thermochemical conversion technologies for carbonaceous feedstocks. The inherent feedstock flexibility of gasification technology as well as the broad range of end use products makes it an attractive option in coal and biomass conversion. The commercially available technologies for pre-combustion CO_2 capture from syngas are extremely capital intensive. The economic impact of a carbon tax on a gasification based plant significantly depends on the quality of feed, technology, and application of interest.

In this study, the impact of the sulfur, ash, moisture content, and heating value are considered for a wide range of coals and biomass. Heuristics, continuous and discrete design variables and cost functions are used to build various modules for major process units including gasification island, acid gas removal, water gas shift reactor, and application specific processes. A heuristic-based algorithm is used to identify the near-optimum modules (nodes) of gasifier and downstream subsystems to determine the process route with the maximum net present value under various carbon tax schemes for a given set of inputs and constraints.

This model examines a myriad of alternative pathways to determine the most economical match of carbonaceous feedstock composition, gasifier choice, and syngas processing subsystems to a process application of interest including electricity, hydrogen, methanol and Fischer-Tropsch hydrocarbon production.

470303 Thermodynamic Modeling and Uncertainty Quantification of CO_2-Loaded Aqueous MEA Solutions

Abstract:

The US DOE's Carbon Capture Simulation Initiative (CCSI) has a strong focus on the development of state of the art process models to accelerate the development and commercialization of post-combustion carbon capture system technologies. One of CCSI's goals is the development of a "gold standard" process model that will serve as a definitive reference for benchmarking the performance of MEA-based CO_2 capture systems. The focus of this specific work is the development of the thermodynamic framework, an essential element of the physical properties package for a reactive separation system, for the gold standard model. A stochastic model of the thermodynamic framework is also included in this work, in which a Bayesian inference methodology is used to quantify the parametric uncertainty.

The thermodynamic framework is developed in Aspen Plus®, and the complete e-NRTL activity coefficient model is used as a starting point. The Aspen Plus® physical properties regression system is used to calibrate the parameter values, including data for VLE of binary MEA-H_2O and ternary MEA-H_2O-CO_2 systems as well as heat of absorption and heat capacity of the ternary system. Due to the large number of parameters in the full e-NRTL model, a parameter screening methodology based on the Akaike Information Criterion (AIC) is developed. The use of a reduced model is not only useful in avoiding over parameterization in the deterministic model, but also improves the feasibility of the uncertainty quantification (UQ) procedure. A Bayesian inference methodology is used to develop the stochastic model, in which prior distributions estimated from the deterministic regression output and the experimental data are used to derive informative posterior distributions. To reduce the computational expense of the Bayesian approach, response surface models are generated as computationally inexpensive surrogate models with sufficient accuracy by using the input-output data. The Markov Chain Monte Carlo (MCMC) algorithm with Gibbs Sampling is used in the Bayesian inference procedure.

The thermodynamic framework is implemented into an Aspen Plus® model of the CO_2 absorption process, consisting of both absorber and regenerator columns, which are modeled as rate-based columns with kinetic reactions. The reaction kinetics used in this work are written in terms of the activity-based reaction equilibrium constants in order to ensure consistency with the thermodynamic model. Case studies are performed, separately for the absorber and regenerator columns, in which prior and posterior parameter distributions are propagated through the process models. This results in estimates of the uncertainty in output variables, specifically CO_2 capture efficiency in the absorber and outlet solvent CO_2 loading from the regenerator, due to the thermodynamic model. The study shows that the uncertainty due to the VLE model can vary widely depending on the operating conditions, but does reduce considerably for the posterior distributions in comparison to the priors for all the cases studied. The study also helps to identify how the uncertainty due to the thermodynamic model can be reduced further.

467832 Design and Advanced Optimization of a Natural Gas Combined Cycle Power Plant with CO_2 Capture

Abstract:

As the penetration of intermittent renewable energy sources into the electrical grid keeps increasing, traditional fossil-fired power plants would be expected to cycle their load much more frequently than in the past. However, current fossil-fueled generation plants are not optimally designed for cycling. With this motivation, a rigorous model of a natural gas combined cycle (NGCC) plant with CO_2 capture is developed. The plant is then designed for optimal load-following characteristics.

First, a steady-state model of the NGCC plant with a triple-pressure reheat cycle steam generator is developed in Aspen Plus. A stage-by-stage model of the steam turbine is developed in Fortran and implemented as a user model in Aspen Plus for estimating the performance of the triple-pressure steam turbine with multiple steam addition and extraction points including the large LP steam extraction for the CO_2 capture process. A mixture of piperazine-promoted methyldiethanolamine is used as the solvent for CO_2 capture. A rate-based model for this solvent system is developed and validated using available experimental data. While traditional optimization approaches in this area have looked mainly into maximizing efficiency, this work is focused on optimizing a cost function that considers both the operating costs as well as the capital costs required for improved load-following characteristics. As an example, even though increased steam temperature and pressure increases the overall efficiency, it increases steam-side oxidation of the boiler tubes especially the high pressure superheater tubes resulting in lower thermal conductivity, which leads to higher skin temperatures. This effect not only leads to higher corrosion for the tubes but also results in lower strength of the tube materials thereby necessitating a design with greater tube thickness. The oxide scale also has a different thermal expansion coefficient compared to the base material leading to stress buildup that can result in spalling especially during startup and shutdown of the plant. Furthermore, creep and fatigue characteristics of the plant under design conditions, during operations under elevated conditions, and under thermal cycling should be optimally designed for load-following. An evolutionary optimization technique is setup by bridging optimization software with Aspen Plus to maximize the net present value. Pareto curves in the presence and absence of the CO_2 capture process are generated showing the trade-offs between the operating and capital costs.

462335 Natural Gas to Liquid Transportation Fuels Utilizing Chemical Looping Technologies for Syngas Generation: Process Synthesis and Global Optimization

Abstract:

Even with advances in electric vehicles and improved vehicle fuel economy, the demand for liquid fuels in the transportation sector is projected to remain steady through 2040 [1]. Liquid fuels produced from domestic feedstocks would partially offset this demand and could ultimately reduce dependence on foreign crude imports, enhancing national energy independence. A potential feedstock is natural gas due to its high hydrogen to carbon ratio and methane-rich composition, increasing the overall carbon conversion to liquid fuels instead of CO_2. Natural gas to liquid fuels (GTL) processes may also be economically promising due to an abundance of inexpensive natural gas in the United States (2015 spot price: $2.62/MMBtu [2]).

Previous multi-scale engineering [3] work by Baliban et al. [4] demonstrated that optimal GTL processes can be economically competitive with petroleum refineries. The primary natural gas conversion technologies investigated were syngas generation via autothermal reforming or stream reforming, partial oxidation to methanol, and direct oxidative coupling to olefins, before further upgrading to liquid fuels via methanol synthesis or Fischer-Tropsch processes. Natural gas conversion is low through partial oxidation and oxidative coupling routes, and while it is greater through traditional reforming, high pressure operation limits the overall conversion.

However, recent developments in chemical looping [5, 6] as an alternative for syngas generation from natural gas have significantly expanded the potential GTL capabilities. Chemical looping can be operated at low pressure and offers close to complete conversion of natural gas in a single pass. A high concentration of syngas can be produced without using pure oxygen, eliminating the need for additional air separation and syngas conditioning units. This could greatly reduce the capital and operational costs associated with generating syngas in GTL processes and improve the overall process efficiency.

In this work, two chemical looping technologies are incorporated into a GTL process superstructure [4, 7, 8, 9] as alternatives to reforming for syngas generation. All process technologies are rigorously modeled and together with other important process components, such as hydrogen and oxygen production, wastewater treatment, and light gas handling, form a large-scale nonconvex mixed-integer nonlinear model (MINLP). A deterministic global optimization branch-and-bound algorithm is used to solve the MINLP [10]. Simultaneous heat and power integration is also performed to minimize utility cost [11]. This process synthesis framework provides an illuminating means to systematically compare chemical looping technologies against other competing technologies in a GTL process.

Several case studies are examined to highlight the prospective benefits of chemical looping technologies over other conversion routes. The effect of plant capacity and production ratios on the overall profit of the GTL process is analyzed. Major topological decisions on process technologies will be discussed. Economic and environmental trade-offs will also be presented.

[1] U.S. Energy Information Administration. Annual Energy Outlook 2015 with projections to 2040. Available at: https://www.eia.gov/forecasts/aeo/pdf/0383%282015%29.pdf. Accessed May 2016.

[2] U.S. Energy Information Administration. "Henry Hub Natural Gas Spot Price". Available at: http://www.eia.gov/dnav/ng/hist/rngwhhdm.htm. Accessed May 2016.

[3] Floudas, C. A.; Niziolek, A. M.; Onel, O.; Matthews, L. R. Multi-scale systems engineering for energy and the environment: Challenges and opportunities. AIChE Journal 2016, 62 (3), 602-623.

[4] Baliban, R. C.; Elia, J. A.; Floudas, C. A. Novel Natural Gas to Liquids Processes: Process Synthesis and Global Optimization Strategies. AIChE Journal 2013, 59 (2), 505-531.

[5] Luo, S.; Zeng, L.; Xu, D.; Kathe, M.; Chung, E.; Deshpande, N.; Qin, L.; Majumder, A.; Hsieh, T.-L.; Tong, A.; Sun, Z.; Fan, L.-S. Shale gas-to-syngas chemical looping process for stable shale gas conversion to high purity syngas with a H2:CO ratio of 2:1. Energy & Environmental Science 2014, 7 (12), 4104–4117.

[6] de Diego, L. F.; Ortiz, M.; García-Labiano, F.; Adánez, J.; Abad, A.; Gayán, P. Hydrogen production by chemical-looping reforming in a circulating fluidized bed reactor using Ni-based oxygen carriers. Journal of Power Sources 2009, 192 (1), 27–34.

[7] Baliban, R. C.; Elia, J. A.; Floudas, C. A. Biomass and Natural Gas to Liquid Transportation Fuels: Process Synthesis, Global Optimization, and Topology Analysis. Industrial & Engineering Chemistry Research 2013, 52 (9), 3381-3406.

[8] Onel, O.; Niziolek, A. M.; Elia, J. A.; Baliban, R. C.; Floudas, C. A. Biomass and Natural Gas to Liquid Transportation Fuels and Olefins (BGTL+C2_C4): Process Synthesis and Global Optimization. Industrial & Engineering Chemistry Research 2015, 54 (1), 359-385.

[9] Niziolek, A. M.; Onel, O.; Elia, J. A.; Baliban, R. C.; Floudas, C. A. Coproduction of liquid transportation fuels and C6_C8 aromatics from biomass and natural gas. AIChE Journal 2015, 61 (3), 831-856.

[10] Baliban, R. C; Elia, J. A; Misener, R.; Floudas, C. A Global optimization of a MINLP process synthesis model for thermochemical based conversion of hybrid coal, biomass, and natural gas to liquid fuels. Computers & Chemical Engineering 2012, 42, 64-86.

[11] Baliban, R. C.; Elia, J. A.; Floudas, C. A. Simultaneous process synthesis, heat, power, and water integration of thermochemical hybrid biomass, coal, and natural gas facilities. Computers & Chemical Engineering 2012, 37, 297-327.

470988 Optimal Design and Operation of Hybrid CO2 Capture Systems

Abstract:

This work presents a mixed integer nonlinear programming model (MINLP) for the optimal design and operation of carbon capture plants. A mix of first principle and surrogate-based models have been developed to characterize the process design and operation.
Coal-based power generation is one of the most important sources of energy in the US. As such, the importance of cost effective carbon capture and storage (CCS) technologies to mitigate the environmental issues of power generation plants is obvious. The most significant technologies for carbon capture include solvents, membrane and solid sorbent systems. This work analyzes and compares the solid adsorption and membrane systems on the basis of minimizing the cost of electricity.
In solid sorbent capture systems the process is mainly driven by the interaction of the solid sorbent particles with the flue gas (mainly consisting of N_2, CO_2, and H_2O). The main process units of the solid sorbent systems are multiphase reactors, such as moving bed and fluidized beds.
In gas permeation separation systems, the CO_2 is separated by high CO_2 selective membranes. The main driving force of the process is the partial pressure between the feed and the permeate side. A typical membrane separation plant consists of multi-stage membrane configurations and compressors.
Given the modeling complexity and the difficulty of formulating algebraic models for such processes, previous studies are generally limited to either process-simulation[1] based frameworks or simpler models for mathematical optimization. This work proposes a mathematical optimization framework in which both first principles-based models and surrogate models[2] have been developed to obtain the optimal plant design and operation to capture at least 90% of the CO_2. The use of commercial process simulators to develop detailed, carefully tuned models have been exploited in the preparation and characterization of surrogate models (ALAMO[2]). Model validation and sample generation play a very important role in this mathematical framework.

The modeling framework considers a set of adsorbers (with two possible operating technologies), regenerators (with two possible operating technologies), membrane stages, mixers, splitters, compressors, pumps and heat exchanger units which can be installed or not. Design decisions determine the plant layout, by selecting the optimal number of parallel trains, number of adsorbers and regenerators, number of membranes, equipment design (diameter, length, etc.), heat exchangers, mixers, splitters and configuration of the system. At the operating level, the decisions involve the optimal flows, temperatures, concentrations of gas and solids. For the adsorbers and regenerators, the solids bed length, flux velocity and other parameters are also considered. While, the membrane characteristics are to be determined by the permeability, permeate fraction, pressure ratio, among others.
[1] Eslick, J. C., Tong C., Lee A., Dowling A. W., Mebane D. S. (2015). Optimization under uncertainty with rigorous process models. Proceedings AIChE Annual Meeting 2015.
[2] Cozad, A., N. V. Sahinidis and D. C. Miller, Learning surrogate models for simulation-based optimization, AIChE Journal, 2211-2227, 2014.

464491 An Investigation into the Effects of Impurities in Captured CO_2 stream on Safety and Environmental Indicators within Transportation Pipeline and Downstream Storage

Abstract:

Abstract

The realms of carbon dioxide pipeline transportation and its subsequent storage have long been stirring many incentives for multidisciplinary research across the world. However, there are several aspects of these processes yet to be explored in detail. Amongst these, the coexistence of certain impurities in transportation pipeline and/or storage facilities along with carbon dioxide is of utmost significance.

In our previous works we defined the indices corresponding to the safe operation of transport and storage processes with regard to the concentration of specific impurities (factors). Along with these impurities, three safety indices (responses) were considered i.e. pipeline corrosion, acute and chronic toxicities. However, these studies were focused on the effects of individual impurities alone and lacked any actual combined effects beyond additive ones. Therefore, little has been established on how these so-called impurities would affect these indices when present simultaneously within such processes.

In this study, the interaction effects of several impurities including water, oxygen, NO_2, SO_2, Cl and several heavy metals (i.e. Pb, Hg, Cd and As) on the above safety indices have been investigated in order to define the resulting safe operational envelopes. The interactions were revealed by initially developing a full factorial DoE framework and then implementing an additive approach to represent the combined indices in each run. These interaction were next amalgamated in order to produce the corresponding 3D and contour plots which highlight the boundaries between the safe and unsafe operating conditions.

This study sheds further light on how the combined effects of various impurities present in the captured CO_2 stream would affect the transportation as well as the downstream infrastructures. This is expected to potentially bring about complementary considerations. A further study to bridge between the identified safe operating conditions and the capture cost is now under investigation. This would aim to successfully provide the decision makers and engineers with a better understanding of the associated costs linked to the required safety considerations as well as the underlying trade-offs.

457329 Keynote - Modeling Electrochemical Interfaces: An Approach from First Principles

Abstract:

Electrified interfaces are central for electrocatalysis, batteries and molecular electronics. Experimental characterization of these complex interfaces with atomic resolution is highly challenging. First principles modelling could provide a link between the measurable quantities and an atomic scale understanding. However, such simulations are far from straightforward. Although approaches that include the effect of the potential and the electrolyte have been proposed, detailed validation has been scarce and "indirect" since atomically resolved experimental studies of systems that can be convincingly simulated are scarce.

We introduce in this lecture the adsorption of pyridine on Au(111) as a convenient and relevant model: the adsorption mode of pyridine switches as a function of the electrochemical potential. We demonstrate that the primitive surface-charging model gives qualitatively correct results at a low cost. For quantitative agreement, however, the model needs to include a more realistic description of the electrical double layer. Approximating the latter through the linearized Poisson-Boltzmann equation leads to a quantitative improvement, lowering the error in the transition potential from 1 V to acceptable 0.3 V. Hence, we demonstrate the qualitative usefulness of the surface charging method and the excellent agreement that can be obtained by slightly more sophisticated electrolyte models.

References

1) Stephan N. Steinmann and Philippe Sautet "Assessing a First Principles Model of an Electrochemical Interface by Comparison with Experiment" J. Phys. Chem. C 120, 5619–5623 (2016)

2) Stephan N. Steinmann, Carine Michel, Renate Schwiedernoch, Jean-Sebastien Filhol, Philippe Sautet* Modelling the HCOOH/CO2 Electrocatalytic Reaction: When Details Are Key, ChemPhysChem 16, 2307 –

2311 (2015)

467825 Feature Engineering of Machine-Learning Chemisorption Models for Bifunctional Electrocatalyst Design

Abstract:

Transition metal alloys have shown great potential for catalyzing many chemical and electrochemical reactions. Attributed to the versatility of transition metals on the periodic table, the catalyst surface reactivity can be manipulated by varying its geometric characteristics, metal ligands, and extrinsic factors in the vicinity of active sites [1]. However, it is very time-consuming and costly to search for highly optimized alloy composition and geometry by high-throughput experiments and/or quantum-chemical calculations.

In this talk, we will introduce a new computational catalyst design framework, which integrates machine-learning algorithms with the descriptor-based design approach for rapid screening of transition-metal catalysts [2-3]. By engineering numerical representation of surface metal atoms using easily accessible features such as the local electronegativity and the effective coordination number that are dependent on the surroundings of an adsorption site, together with the intrinsic properties of active metal atoms including the electronegativity, ionic potential, and electron affinity, the machine-learning model optimized with ~500 *ab initio* adsorption energies on bimetallic alloys can capture complex, non-linear adsorbate/substrate interactions with the root mean squared errors (RMSE) ~0.12 eV. To validate the model, we applied the model to search for high-performance {111}-terminated bifunctional catalysts for electrocatalytic oxidation of methanol. The *CO and *OH adsorption energies represent important efficiency metrics to describe the electrocatalytic reactivity. We show that our machine-learning-augmented model exhibits outstanding prediction power in screening for the bifunctional, multimetallic surfaces with desired adsorption properties for both *CO and *OH on different adsorption sites. Compared with the traditional high-throughput computational and experimental trial-and-error approach, the machine-learning chemisorption models have great potential in accelerating the discovery of interesting catalytic materials.

[1] A. Vojvodic and J. K. Nørskov, "New design paradigm for heterogeneous catalysts," *Natl. Sci. Rev.*, p.

nwv023, Apr. 2015.

[2] X. Ma, Z. Li, L. E. K. Achenie, and H. Xin, "Machine-Learning-Augmented Chemisorption Model for CO2 Electroreduction Catalyst Screening," *J. Phys. Chem. Lett.*, vol. 6, no. 18, pp. 3528–3533, Sep. 2015.

[3] Z. Li, X. Ma, and H. Xin, "Feature Engineering of Machine-Learning Chemisorption Models for Catalyst Design," Catalysis Today, 2015 (Accepted).

469263 Effect of Lanthanum Doping and Chlorine Addition on Strontium Titanate Perovskites for Electrocatalytically-Assisted Oxidative Dehydrogenation of Ethane

Abstract:

Oxidative dehydrogenation (ODH) of alkanes is an important catalytic reaction that is widely studied. It can be used to convert less valuable alkanes, such as ethane, to more valuable olefins, such as ethylene. However, the further oxidation of olefins to carbon dioxide and carbon monoxide limits the selectivity of the process [1]. Oxide ion-conducting electrolyte reactors allow for controlling oxygen availability to the reaction and thus, improve the selectivity for olefin formation [2, 3]. The reactor setup is similar to solid oxide fuel cells with an important difference: instead of generating power as in solid oxide fuel cells, an external current is applied to control the oxygen flux through the electrolyte. When the current is applied, the oxygen molecules at the cathode side are ionized to O^{-2} ions. Oxygen ions conduct through the oxygen ion-conducting electrolyte to the anode, where they react with ethane to form ethylene. Since the amount of oxygen conducted through the electrolyte can be controlled by the current applied, this can help regulate the alkane conversion and selectivity.

In this study, strontium titanate type of perovskites were tested as anode catalysts for electro ODH reaction. La-doped strontium titanates were synthesized via modified Pechini route. The effect of lanthanum doping and chlorine incorporation on these catalysts was investigated. The characteristics of the catalysts were investigated using N_2 physisorption, ambient and *in-situ* X-ray diffraction (XRD), X-ray photoelectron spectroscopy (XPS), diffuse reflectance infrared spectroscopy (DRIFTS), and electronic conductivity measurements. Besides Electro-ODH experiments to test the activities, impedance measurements were also taken using to better understand the cell characteristics and resistances. Oxygen flux measurements were also performed.

Electro-ODH tests on the cell with $La_{0.2}Sr_{0.8}TiO_{3Â±Î´}Cl_{Ïƒ}$ anode showed that ethane conversion and ethylene production rates increase with increasing current. Hydrogen production rates were also observed to be very low, indicating oxidative dehydrogenation is taking place. It was also observed that Cl addition increases the ethylene selectivity and La doping increases the conductivity of strontium titanate.

1. Kung, H.H., and Kung, M.C., Applied Catalysis A: General 157, 105 (1997).

2. Coronas, J., Menendez, M., and Santamaria, J. Industrial & Engineering Chemistry Research 34, 4229 (1995).

3. Sundmacher K., Rihko-Struckmann, L.K., Galvita, V., Catalysis Today 104, 185 (2005).

461301 Mechanistic and Trend Analyses of Selective Furfural Electroreduction on Transition Metals from First-Principles Methods

Abstract:

Furfural is an important biomass-derived platform compound. Electrocatalytic hydrogenation and deep reduction, via the C-O bond cleavage, of furfural on transition metals can be developed into a sustainable strategy to obtain valuable chemicals and fuels. Further, the complex chemical nature of furfural presents challenges to understand the electroreduction pathways and the product selectivity, which are still being debated.

Density functional theory (DFT) calculations were performed on the close-packed (111) facets of Cu, Ag, Pb and Ni to investigate furfuryl alcohol (FA) and 2-methylfuran (MF) formations in acidic conditions. Two FA formation pathways via 2-methoxyfuran (mh6) and 2-hydroxymethylenefuran (mh7), and three MF formation pathways were considered in this study. The computational hydrogen electrode (CHE) model was applied to determine free energies of the reactions under an external applied potential. The energy barriers of elementary reduction steps were estimated using the BrÃ¸nstedâ€"Evansâ€"Polanyi (BEP) relationship. On Cu, Ag, Pb and Ni, it has been found that the pathway via mh6 is competitive for FA formation. On Ni, the pathway via mh7 is also competitive. For MF formation, pathways involving mh6 are more competitive on all considered metals. In order to gain insights into the product selectivity, micro-kinetic modeling was performed based on the mechanism derived from DFT calculations. The modeling confirms that FA and MF are formed via parallel pathways. The turnover frequencies of FA and MF formations were also obtained, and Cu is found to be the most active for MF production, which is supported by electrochemical measurements.

464406 NO Electrochemical Reduction on Pt(100) from First Principles

Abstract:

Electrochemical denitrification is a promising technology for the removal of toxic nitrate and nitrite species from groundwater due to this processâ€™s environmental compatibility, energy efficiency, safety, and product selectivity.[1] The adsorbed NO is generally considered to be a selectivity-determining species for the reaction in acid electrolytes.[2]However, molecular-level insights into the mechanism are challenging to obtain from experiments alone, and atomic-scale studies using density functional theory (DFT) can thus provide important information about general reaction scheme of NO electrocatalysis. These basic building blocks, determined on the most common electrocatalyst for this reaction, platinum, may, in turn, serve as a starting point for future trends-based analysis on different transition metal surfaces.

While Pt(111) terraces are the lowest in energy of any single crystal Pt surface, terraces of Pt(100) have shown unusual activity and selectivity patterns for the electroreduction of NO.[3] Therefore, in this work, we use periodic, self-consistent DFT calculations to clarify the adsorption structures and energetics of NO and its reaction intermediate species on the Pt(100) surface. Solvation energies and protonation kinetics are calculated using water-incorporated systems in the DFT analysis.[4]In addition, we implement a rigorous kinetic Monte Carlo (kMC) approach that takes into account coverage-dependent interactions and activation barriers and that as that produces potential-dependent values of current from multiple reaction pathways as a function of voltage.

In our simulation, the most favorable pathway is the ammonia formation. The mechanism consists of two consecutive proton-electron transfer steps after the adsorption of NO (denoted the EE mechanism), with favorable formation of NOH* as a first intermediate. The Tafel slope, which was obtained by plotting peak potentials at different scan rates in kMC simulations, supports the EE mechanism rather than the chemical dissociation steps. Also, due to the fast

protonation, the probabilities of forming N_2 or N_2O are negligible. Finally, implications of these results for both NO and nitrate electroreduction on other transition metal surfaces will be discussed.

[1] M. Duca and M. T. M. Koper, Energ. Environ. Sci. 5 (2012) 9726.
[2] A. Cuesta, M. Escudero, Phys. Chem. Chem. Phys. 10 (2008) 3628.
[3] V. Rosca, M. T. M. Koper, J. Phys. Chem. B 109 (2005) 16750.
[4] A. Clayborne, H. J. Chun, R. B. Rankin, J. Greeley, Angew. Chem. Int. Ed. 54 (2015) 8255.

472512 Flame Synthesis and High Pressure High Temperature Annealing of Anatase TiO_2 for Increased Photocatalytic Activity

Abstract:

Titanium dioxide (TiO_2) is of particular interest in photocatalysis due to its photoactivity under ultraviolet illumination and its stability under catalytic reaction conditions. As most of incident solar radiation is visible light, many researchers have sought a visible-light active photocatalyst by modifying TiO_2 in order to reduce the band gap energy (BGE). It is known that the lattice structure, particle size, dopant concentration, and supported metal nanoparticles affect the BGE of TiO_2. Through low pressure flame spray pyrolysis, carbon doping, high pressure high temperature annealing, and supporting of metals we have successfully decreased the BGE and increased the photocatalytic activity of anatase phase TiO_2.

High pressure high temperature annealing (HPHT) was conducted at 200 °C in a Parr reactor under 250 psig of hydrogen or nitrogen. We have observed that, compared to the parent sample, HPHT increases particle size, decreases the BGE, causes a slight darkening of color, and increases the absorbance throughout the UV and visible spectrum. The increase in UV and visible light absorbance indicates that HPHT samples will have a greater photocatalytic activity than un-modified anatase. This was seen via an increase in hydrogen production during methane steam reforming (MSR). Hydrogen yield increased from a negligible amount over anatase to 1.05 % and 1.12 % under UV and visible light irradiation MSR over HA-Anatase at 700 °C.

It was found that supporting copper on anatase, 1%Cu(O)/Anatase, decreased the band gap from 3.22 eV to 3.13 eV. By combining the two modification methods, HPHT and metal supporting, we aim to decrease the BGE further and therefore increase the photocatalytic activity under UV and visible light irradiation. An order of operations variation was performed by annealing in hydrogen either before or after metals were supported and the sample calcined. All

variations of HPHT and copper supporting yielded BGE lower than unmodified anatase, with HPHT hydrogen annealed copper supported on anatase, HA-1%Cu(O)/Anatase, having the smallest BGE of 3.07 eV. EPR analysis indicates that HPHT annealing under hydrogen or nitrogen increases Ti^{3+} centers in the samples. Since it has been reported that oxygen vacancies play an important role in absorption and active sites, we hypothesize that the HPHT samples have a higher photocatalytic activity than the unmodified parent sample partially due to the effect of the Ti^{3+} centers.

A third area of interest has been the low pressure flame synthesis of carbon doped TiO_2. Various compositions and flow rates of the premixed gases are introduced to the low pressure synthesis chamber and ignited to produce a flat flame. The low pressure flame spray pyrolysis method produces various polymorphs of TiO_2 which show promise in photocatalytic conversion of methanol at moderately low temperatures. Flame temperature profiles and carbon loading are predicted via computational analysis using a SPIN code to ensure that the flame of the temperature is comparable across samples made from different pre mixed gases. Photocatalytic activity of the carbon doped samples is determined via methanol oxidation reactions under UV illumination.

467273 Photoelectrochemical Oxidation of Biorenewable Alcohols By Nitroxyl Radical Catalysts

Abstract:

Transforming primary alcohols to acids is a fundamental synthetic process of interest in both the academic and industrial worlds, but it typically requires stoichiometric oxidants which are costly and generate potentially toxic waste. On the other hand, conversion of biomass-derived chemicals in a *photoelectrochemical* (PEC) cell is one of the most environmentally-friendly chemical production routes available, by taking advantage of renewable carbon sources from the earth, renewable photons from the sun, and renewable electrons from wind, geothermal, or photovoltaic sources.

Stable nitroxyl radicals such as TEMPO are a well-known class of homogeneous oxidation catalysts with widespread use in industrial and laboratory applications.[1] Furthermore, TEMPO is easily oxidized to its active form, an oxoammonium cation, at moderate electrode potentials and is very selective for C OH or C=O oxidations. It was recently reported that the overpotential required to drive TEMPO oxidation can be significantly reduced in a PEC reactor, where energy is harvested from visible light with an earth-abundant bismuth vanadate ($BiVO_4$) semiconductor electrode.[2] However, external electrical energy was still required to enhance electron-hole separation and achieve reasonable photocurrents. Additionally, the electron efficiency for TEMPO oxidation was <100% owing to competition with water oxidation. Extensive work has been done to improve the performance of $BiVO_4$ photoanodes for solar water splitting including nanostructuring, doping, and the addition of an electrocatalyst layer, but optimization to promote TEMPO oxidation has not been explored.

Previously, we have studied selective electrocatalytic oxidation of alcohols and 5-hydroxymethylfurfural (HMF) on supported noble metal catalysts.[3-4] Now, we have investigated electrochemical and PEC systems for efficient TEMPO-mediated oxidation of key biomass-derived chemicals including HMF, glycerol, and

ethylene glycol. $BiVO_4$ modified with electrocatalyst layers facilitated TEMPO oxidation at overpotentials >400 mV lower than unmodified $BiVO_4$, while significantly suppressing water oxidation. The novel photoelectrocatalytic system achieved nearly 100% efficiency to TEMPO-mediated oxidations and bulk electrolysis gave high yields of organic acid products. In this talk, factors that tune the competition between TEMPO and water oxidation, including photoanode structure, applied potential, and other reaction conditions will be discussed.

References:

(1) Rafiee, M.; Miles, K. C.; Stahl, S. S., JACS (2015), 137 14751−14757.

(2) Cha, H. G.; Choi, K. S., Nature Chem. (2015), 7 328–333.

(3) Chadderdon, D. J. et al., Green Chem. (2014), 16 3778–3786.

(4) Chadderdon, D. J. et al., ACS Catal. (2015), 5 6926−6936.

457859 Effect of Membrane on the Performance of a Hydrogen-Vanadium Regenerative Fuel Cell

Abstract:

Hydrogen-vanadium regenerative fuel cells offer a feasible solution for storing electrical energy from the grid or directly from renewable energy sources such as wind and solar. Due to the relatively high cost of vanadium ($ per kg), the hydrogen-vanadium flow battery is attractive over the all-vanadium system due to the 50% reduction in the amount of vanadium solution required. Furthermore, by replacing the negative vanadium electrode with a hydrogen electrode the effect of self-discharge by the crossover of unwanted ions, mostly the smaller vanadium II and III ions, is expected to be greatly reduced. Also, by having a two-phase system, gaseous negative electrode and aqueous positive electrode, crossover species can be easily separated and recovered. In this presentation, the performance of electrospun perfluorosulfonic acid (PFSA)/polyphenylsulfone and PFSA/polyvinylidene fluoride (PVDF) composite membranes, relative to PFSA membranes such as Nafion, with regards to the self-discharge rate and electrical performance of a hydrogen-vanadium regenerative fuel cell will be discussed.

464994 Dynamics of Anhydrous Proton Transport on the Surface of Functionalized Graphene

Abstract:

Proton transport phenomena has been of tremendous importance for the application of proton transport membrane fuel cells for their efficiency and clean emissions. We are specifically focused with the applications of proton transport at interfaces for a better design of new materials for proton exchange membrane (PEM) fuel cells in anhydrous conditions. Study of transport of protons over the surface of hydroxylated graphene in anhydrous conditions is performed using density functional theory. The hypothesis is that a facile anhydrous proton transport will occur on a surface that has a continuous network of hydrogen bonded hydroxyl groups showing "grotthuss-like" mechanism. Simple configurations of 1-D single file and a 2-D hydrogen bonded network of hydroxyl groups on graphane membrane will sufficiently facilitate the transport of protons. Dynamical properties and energetics related to the transport of the excess proton is studied. In this work, facile transport of excess proton over the surface of functionalized graphene has been confirmed with computed energy barriers using nudged elastic band technique and their respective mechanism pathways will be discussed. Ab-initio molecular dynamics (AIMD) simulations are used to determine the dynamical properties of 1-D single file and 2-D network of fully hydroxylated graphene surface with one excess proton concentration. Correlations influencing the dynamical properties for the above patterns of hydroxylation are investigated. Stability of the functionalized graphene material for varying degree of hydroxylation, morphological defects and also excess proton concentrations will be reported. A monte-carlo based lattice model of 1-D hydrogen bonded network system mimicking the proton hopping mechanism observed from the AIMD simulations is developed. This is to support and verify the behavior of normal diffusion of excess proton transport for larger system size and longer time domain.

472540 Indirect and Direct Observation of Ionomer Colloidal Systems with Applications to Fuel-Cell Catalyst Layers

Abstract:

Manufacturing methods for fuel cell catalyst layers are critically important for achieving high material utilization and performance. Typically, electrode systems are cast from an 'ink' or 'colloidal' state, where the solid material is suspended in a solution containing a solvent and a polymer. For polymer-electrolyte fuel cells, a perfluorinated sulfonic-acid (PFSA) ionomer, typically Nafion, is used in the ink and catalyst layer, where its roles are to act as a binding agent and conduit for proton conduction to the reaction site.[1] Recently, it has been suggested that the ionomer film that surrounds the platinum reaction site may contribute to increased resistances that limit performance at low Pt loadings.[1,2] Thus, it is of great interest to understand the factors controlling the film formation in catalyst layers to optimize material proeprties and cell performance. The nature of the dispersion and casting process could have significant influence on the morphology and properties of the dispersion-cast membranes.[3-5] To isolate these effects, the morphology of PFSA dispersed in various solvents (glycerol, water/2-propanol, and NMP) was reported previously using small angle neutron scattering (SANS), where catalyst layers from the different solvents exhibit different performance.[6] The understanding of *local* 'ink' structures as a function of active material (polymer and solid) volume fraction and across solvent space and length scales can aid in understanding bulk properties and the impact of ink formulation in cast systems. In this talk, a direct observation of the colloidal suspension of the ionomer in different solvents will be described using confocal laser scanning microscopy (CLSM). CLSM has long been utilized as a tool for characterizing the structure of biphasic colloidal gels. We combine dynamic light scattering, ultra-small angle x-ray scattering, in-situ saxs (during printing) and confocal microscopy to understand PFSA size and structure in dilute (1,3, and 5 wt%) ionomer solutions. Moreover, a novel we explore how the structure evolves during printing processes using synchrotron based small angle x-ray scattering methods. Finally, we contrive a model 'ink' composed of attractive and repulsive silica spheres and PFSA suspended in different solvents (water, NMP, and isopropyl-alcohol) and use CLSM to

gain a greater understanding about solid|PFSA interactions in model inks.

1. A. Z. Weber and A. Kusoglu, *Journal of Materials Chemistry A*, **2**, 17207 (2014).

2. A. Kongkanand and M. F. Mathias, *The Journal of Physical Chemistry Letters*, 1127 (2016).

3. T. T. Ngo, T. L. Yu and H.-L. Lin, *Journal of Power Sources*, **225**, 293 (2013).

4. C.-H. Ma, T. L. Yu, H.-L. Lin, Y.-T. Huang, Y.-L. Chen, U. S. Jeng, Y.-H. Lai and Y.-S. Sun, *Polymer*, **50**, 1764 (2009).

5. Y. S. Kim, C. F. Welch, R. P. Hjelm, N. H. Mack, A. Labouriau and E. B. Orler, *Macromolecules*, **48**, 2161 (2015).

6. C. Welch, A. Labouriau, R. Hjelm, B. Orler, C. Johnston and Y. S. Kim, *Acs Macro Letters*, **1**, 1403 (2012).

466036 Multiscale Modeling of Polymer Electrolyte Membrane for Fuel Cell

Abstract:

In this study, we present a multiscale modeling of nanophase-segregation of the nanophase-segregated structures and transport properties of polymer electrolyte membranes for fuel cell. Molecular interactions between entities in full-atomistic molecular dynamics (MD) simulations are described by force field that is developed using quantum mechanical density functional theory (DFT).

The equilibrium structures are obtained from MD simulations, providing detailed information of local structures through pair correlation functions. The extents of nanophase-segregation in this full-atomistic structures are evaluated from the structure factor analysis. To scale up the simulations from full-atomistic modeling to meso-scale coarse-grained modeling, we employ Chi parameter as a coarse-grained energy parameter, which is used for coarse-grained simulations such as MesoDyn and dissipative particle dynamics (DPD) simulation. Using such multiscale modeling approach, we investigate various polymeric materials including Nafion, dendrion, PEEK, and so on.

453357 High Anion Conduction in Partially Fluorinated Multiblock Copolymers

Abstract:

High Anion Conduction in Partially Fluorinated Multiblock Copolymers

Lisha Liu[*†], John M. Ahlfield[‡], and Paul A. Kohl[‡]

[†]School of Materials Science and Engineering,

[‡]School of Chemical and Biomolecular Engineering

Georgia Institute of Technology

Atlanta, Georgia 30332-0100

Anion Exchange Membranes (AEM) are used in fuel cells, electrolyzers and electrodialysis devices. Alkaline conditions provide a route to overcoming fundamental issues with acid-based devices. The fuel cell and electrolyzer issues include the high cost of platinum catalysts, complex water management, and sluggish electrochemical reactions. However, the performance of AEM fuel cells is not as good as that of Proton Exchange Membrane (PEM) fuel cells partially because of the limitations of current anion exchange membranes, such as low ionic conductivity, poor stability at high pH, and high water uptake.

A series of partially fluorinated multiblock copoly(arylene ether)s with long head-group tether were synthesized for use in AEM fuel cells and electrolyzers. The multiblock copolymers were synthesized via polycondensation of hydroxyl-terminated oligomers and fluoro-terminated oligomers. The resulting multiblock structure has one hydrophilic block and one hydrophobic block. It was designed so that nanophase-separation occurs and efficient conductive channels are formed with low water uptake. Multiblock copolymers with different block lengths and ion exchange capacity (IEC) were synthesized to maximize ion conductivity and explore the relationship between chemical structures and membrane properties. Table 1 shows the membrane properties. A non-linear relationship was found between the number of head-groups on a hydrophilic block and the conductivity. Doubling the number of head-groups more than doubled the hydroxide conductivity. Hydroxide conductivity as high as 119 mS/cm at 80°C have been observed with a specific size block size: $X_{5.4}Y_7$, where X is the hydrophobic block and Y is the hydrophilic block. The hydrophobicity of the backbone has allowed synthesis of polymers with minimal water uptake. The ratio of conductivity-to-water uptake shows that less water is absorbed compared to conventional materials. The number of waters per ion within the polymer is as low as 4. This is reflected in the measurement of the amount of free-water and bound-water. No conductivity loss was observed after soaking the membrane in 1M NaOH solution at 60°C for over 600 hours.

Financial support from US Office of the Deputy Assistant Secretary of the Army for Defense Exports and Cooperation (DASA-DE&C) is gratefully acknowledged.

Table 1. Summary of properties of membranes

Block copolymer	Molecular Weight (GPC)	IEC (Ion Exchange Capacity) (meq/g)	OH^- Conductivity (mS/cm)				Water Uptake
			R.T.	40°C	60°C	80°C	(%)
Y_8-1	18K	1.18	13.12	17.91	24.34	36.05	35.85
$X_{3.1}Y_{3.6}$-1	88.6K	0.66	16.41	26.37	35.55	51.50	5.56
$X_{5.4}Y_7$-1	68.2K	0.73	14.17	21.08	27.94	34.72	8.00
$X_{5.4}Y_7$-2	68.2K	1.30	38.21	58.29	96.07	119.70	50.77
$X_{3.1}Y_8$-2	55.9K	1.56	23.13	45.41	66.41	94.03	26.67
$X_{3.1}Y_{3.6}$-2	66.0K	1.19	25.79	41.75	59.03	85.00	25.00
$X_{5.9}Y_5$-2	59.0K	1.10	22.14	33.56	50.29	66.70	19.57

1. X-hydrophobic block; Y-hydrophilic block; subscripted number-block lengths; 1 or 2-tether amount)
2. IEC (Ion Exchange Capacity) was calculated via ^1H NMR results.
3. OH^- conductivity was measured by four-probe conductivity cell.
4. Water uptake was measured at room temperature.

465994 Anion Exchange Membranes for Fuel Cell Applications Based on Advanced Cations

Abstract:

The potential of anion exchange membrane (AEM) fuel cells to provide inexpensive compact power from a wider variety of fuels than is possible with a proton exchange membrane (PEM) fuel cell, has continued to drive the research interest in this area. Alkaline catalysis in fuel cells has been demonstrated with non-precious metal catalysts, and with a variety of fuels beyond H_2 and methanol. Alkaline fuel cells (AFCs), based on aqueous solutions of KOH, have serious drawbacks associated with system complexity and carbonate formation. Anion exchange membrane (AEMs) fuel cells have a number of advantages over both PEM fuel cells and traditional AFCs; however, although anionic conductivity in AEMs can be comparable to PEMs the chemical stability of membrane attached cations in hydroxide is still not always sufficient for practical applications. Recently, it has been recognized that a number of advanced cations, may give AEMs the needed chemical stability. In some circumstances simple trimethyl benzyl ammonium cations are stable up to 60°C allowing us to being to study hydroxide and water transport in these systems. We use in-plane conductivity and multi-nuclear PFGSE to measure self-diffusion coefficients of the water and where possible, the ion, i.e. F^-, carbonate and bicarbonate. Together with temperature and RH dependent SAXS we couple this information with the morphological changes in the materails.

By their nature these organic cations form dipoles, which have a tendency to interact, see Figure below for a trimethylbenzyl cation with bromide for illustrative purposes. For maximum ion transport the cation should be distributed along the polymer chain, however, cation clustering is a common phenomena. It occurs in phase separated diblock polymers when an attempt is made to raise the IEC, no net increase is observed in ionic conductivity above a certain IEC. In many other less structured polymers it occurs in the fully humidified state when the distributed disordered cations order in agglomerates at a disorder/order T_d. This T_d is often observed in the operating temperature

range expected for a device, in DMA, DSC, and broadband electric spectroscopy. Interestingly, even though the SAXS clearly shows an agglomeration on the nano-scale, little difference is seen in the bulk conductivity with temperature. However, steric hindernce effects are expected to discourage aggregation, and may allow the cation to tune all properties of the AEM. As these phenomena are so prevalent in AEM materials the implications and science behind them are being investigated by us and will be discussed in this presentation. The implications of more complex cations proposed for stability in AEMs will be described.

462223 Sustanion™ Membranes for Electrolyzers and Fuel Cells

Abstract:

This paper will discuss the properties of Sustanion™ membranes for electrolyzers and fuel cells. The membranes are based on a styrene backbone that has been functionalized with Imidazolium groups and other additives. Different formulations show conductivities between 30 and 70 mS/cm in KOH and are unexpectedly stable. A CO_2 electrolyzer has been run for 6 months at room temperature with no significant loss in performance. We observe CO_2 reduction currents up to 600 mA/cm^2 at 3.25 V. The closest competitor needs to go to 6.5 V to obtain similar currents. An alkaline water electrolyzer has been run for 200 hrs at 60 C in 1 M KOH. We observe currents of 1 A/cm^2 at voltages less than 2V. Again, 3 time higher than the closest competitor. We are expecting to offer these membranes for beta-testing by various academic groups. Please let us know if you are interested.

460806 Design and Development of Electrolyte Membranes for Solid Alkaline Fuel Cells

Abstract:

Solid alkaline fuel cells (SAFCs) are another system that holds the potential to achieve high-energy conversion efficiency without Pt catalysts. Although most of metal catalysts can be used under alkaline environment, development of durable electrolyte membranes in alkaline media is the key for this technology. We are systematically designing and developing new materials 1,2) from the molecular level to the device level based on organic chemistry3) and quantum chemical simulation4). We have designed a new anionic polyelectrolyte membrane consisting of highly cross-linked and packed ion-exchange sites with aromatic backbone. Each aromatic monomer unit has three ion-exchange sites and the sites can cross-linked with other monomer units. The monomer was polymerized in pores of a porous membrane substrate and the membrane showed high ionic conductivity, exceptionally low fuel cross-over and good chemical stability. Those functionalities can satisfies the anticipated requirements for a viable solid state alkaline fuel cells.

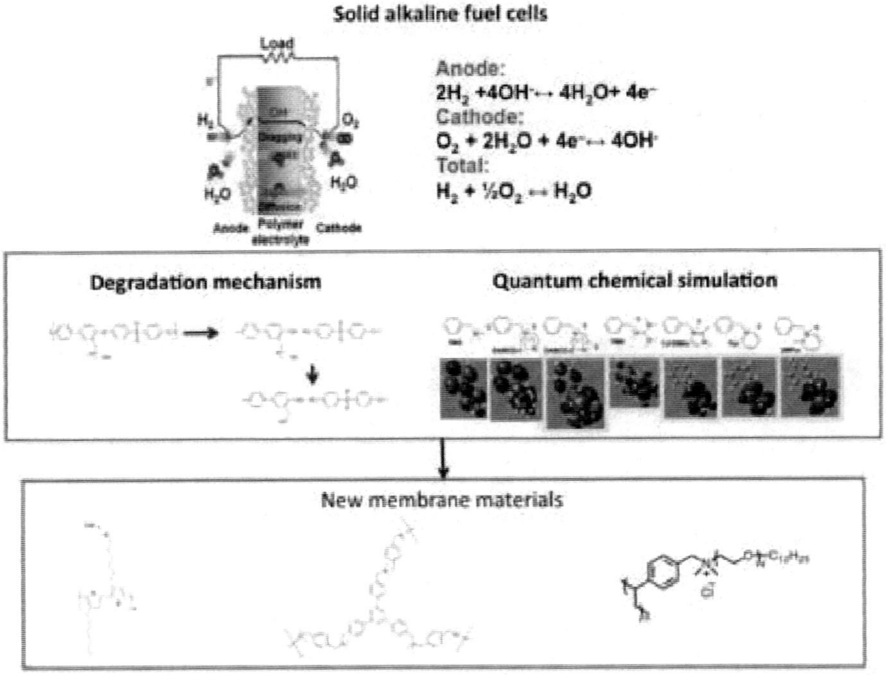

References

1.) S. Miyanishi, T. Fukushima, T. Yamaguchi, ***Macromolecules***, 48(8), 2576-2584 (2015)

2.) G. S. Sailaja, S. Miyanishi, T. Yamaguchi, ***Polym. Chem.***, 6, 7964-7973 (2015)

3.) S. Miyanishi, T. Yamaguchi, ***Phys. Chem. Chem. Phys.***, in press (2016)

4.) K. Matsuyama; H. Ohashi, S. Miyanishi, H. Ushiyama, T. Yamaguchi, ***RSC Advances***, 6, 36269-

36272 (2016)

466840 Membrane Electrolytes with Magnetic Field Aligned Graphene Oxide Nanosheets for Alkaline Fuel Cells

Abstract:

This study aims to elucidate the magnetic graphene oxide (GO) nanofiller alignment in polymer matrix on the resulting filmsâ€™ ionic conduction and methanol barrier properties. Magnetic iron oxide (Fe_3O_4) nanoparticles were fabricated and decorated onto the GO nanosheets. The magnetic GO-Fe_3O_4 nanofillers were mixed with quaternized polyvinyl alcohol (QPVA) to form polymer composite. By applying magnetic field on parallel sides along in-plane direction during drying and film formation step, the GO-Fe_3O_4 nanosheets were aligned in an orientation parallel to the direction of magnetic field and connected each other into elongated domains. The QPVA composite membranes were doped with alkali to become hydroxide-conducting electrolytes for direct methanol fuel cell application. The magnetic field-aligned QPVA/GO-Fe_3O_4 membrane showed higher ionic conductivity and lower methanol permeability compared to nonaligned QPVA/GO-Fe_3O_4 membrane. The aligned QPVA/GO-Fe_3O_4 composite exhibited a peak power density of 172.4 mW cm^{-2} at 60°C, 35.6% higher than that without magnetic field treatment.

463585 Highly Thermal-Stable and Functional Cellulose Nanocrystals and Nanofibrils Produced Using Fully Recyclable Organic Acids

Abstract:

Here we report the production of highly thermal stable and functional cellulose nanocrystals (CNC) and nanofibrils (CNF) by hydrolysis using concentrated organic acids. Due to their low water solubility, these solid organic acids can be easily recovered after hydrolysis reactions through crystallization at a lower or ambient temperature. When dicarboxylic acids were used, the resultant CNC surface contained carboxylic acid groups which facilitate functionalization and dispersion in aqueous processing. Carboxylic acid group content was 0.1 – 0.4 mmol/g for CNC produced from a bleached eucalyptus kraft pulp (BEP) using oxalic acid at concentrations of 50 – 70 wt%. Onset thermal degradation temperature for the CNC was increased to 322°C from 274°C for the feed BEP fibers, compared with 218°C for CNC produced from the same feed fibers using conventional concentrated sulfuric acid hydrolysis. The low strength (high pKa) of organic acids also resulted in CNC with longer lengths of 200 – 400 nm and higher crystallinity than those produced using mineral acids. Fibrous cellulosic solid residue (FCSR) collected from acid hydrolysis was an excellent feedstock for producing CNF through subsequent mechanical fibrillation with low energy input. The ability to recover the organic acids using a conventional and commercially proven crystallization method makes these organic acids uniquely suited for sustainable and green production of cellulose nanomaterials. The resultant CNC and CNF with high thermal stability and large aspect ratio are excellent for bio-composite applications.

467780 Two-Step Process to Create "Roll-Off" Superamphiphobic Paper Surfaces

Abstract:

The fabrication of anti-wetting surfaces has gained increased attention during recent years; of particular interest is the achievement of superamphiphobic (superhydrophobic and superoleophobic) materials. For anti-wetting applications such as disposable non-stain clothes, low cost medical testing strips or packages for liquids and food, paper has many merits over polymeric or glass containers such as breathability, low cost, recyclability and texture. In this study, the fabrication of paper which repels organic liquids with a surface tension as low as 27.5 mN/m (n-hexadecane) is reported. Paper was exposed to a 30 minute oxygen plasma followed by a 15 minute vapor-phase silanization with fluoroalkyl trichlorosilane. The resulting surfaces cause all test liquids (water, diiodomethane, motor oil and hexadecane) to "bead up" and readily roll off the treated surface. Unlike the widely used approach that relies on the deposition of nanoparticles to achieve superamphiphobic surfaces, this method results in good adhesion between coating and the paper substrate. Anti-wetting behavior and surface stability were determined by contact angle measurements and surface XPS analysis after dynamic impact tests. The surface modification process is simple and does not produce any waste liquid. Furthermore, the superamphiphobic paper has high gas permeability due to pore volume enhancement by plasma etching, but maintains the mechanical flexibility and strength of untreated paper.

467826 3D Printed Nanocellulosic Materials and Their Composite

Abstract:

Cellulose nanocrystals and fibers, which can be extracted from the cell wall of trees, offer a sustainable and greener alternative for petroleum products. Studies have shown nanocellulosic polymer composite materials can exhibit light weight and high barrier properties. These properties have been exploited to form barriers films, reinforcing films, platforms for printable electronics, and many other cellulosic products. Nevertheless, it is still a challenge to find compatible nanocellulose-curable resins composite, and design a facile and scalable approach for fabricating composite materials into functional structures. In this work, we introduced cellulose nanocrystals into an acrylate based photo-curable resin, then projection stereolithography was utilized to 3D print nanocellulose composite structures. A custom built direct ink write 3D printer was also utilized to 3D print cellulose nanocrystals and nanofiber, which resulted in pure cellulosic 3D aerogels after freeze dry processing. The mechanical properties of the resultant 3D printed structures were characterized. This work established the printing platform for generating nanocellulosic 3D structure and aerogel, which enabled future possibility for incorporating targeted functionalities into these structures.

466469 Modifying the Surface Properties of Wood Using ATRP Grafting Polymerization

Abstract:

Wood is a well-known bio-based material which can be used for a wide range of applications, with a major focus on its utilization as a building material. In order to profit from its natural properties and to develop smart and innovative materials, a variety of chemical modifications can be used. Cabane et al. have shown that wood can be modified by in-situ grafting of synthetic polymers enhancing the natural properties of wood and making it competitive with other available materials [1]. We recently extended the modification of wood through the grafting of fluorinated acrylates by surface-initiated ATRP (Fig 1A). ATRP is a living polymerization technique allowing for a better control over grafted polymer chain length. The polymer chains are grafted from the wood biopolymers through a two-step reaction. In the first step the ATRP initiator is grafted onto the wood surface (via esterification), and it is used in the second step to grow the polymer chains.

Several techniques were used to characterize the wood properties after the polymerization. Amongst others, the most outstanding results obtained were in regard to the wettability. Through the functionalization of wood with a fluorinated polymer, and following the examples of superhydrophobic surfaces found in nature, we have acquired a renewable material that is both superhydrophobic and oleophilic. The contact angles of water on the wood surface measured before and after modification increase drastically with values raising from 73° to 180 ° (Fig 1B). This increase is both due to the chemical modification and the surface roughness acquired by the combination of the both the natural roughness of wood and the grafted polymer chains.

Moreover, quaternary amine acrylates were grafted onto the wood in order to further study the wettability given by the combination of the natural structure of wood and the chemical and physical properties of the polymer introduced. Both quaternary amines and fluorinated acrylates have been separately and successfully grafted into wood (Fig. 1C) offering different wettabilities (Fig 1D). In further work we would like to extend the methods used to copolymerize fluorinated and quaternary amine acrylates, combining the properties of both polymers. By this copolymerization it would be possible to obtain a material that is both hydrophilic and oleophobic [2]. This new material could be used to extract the oil present in contaminated water resources.

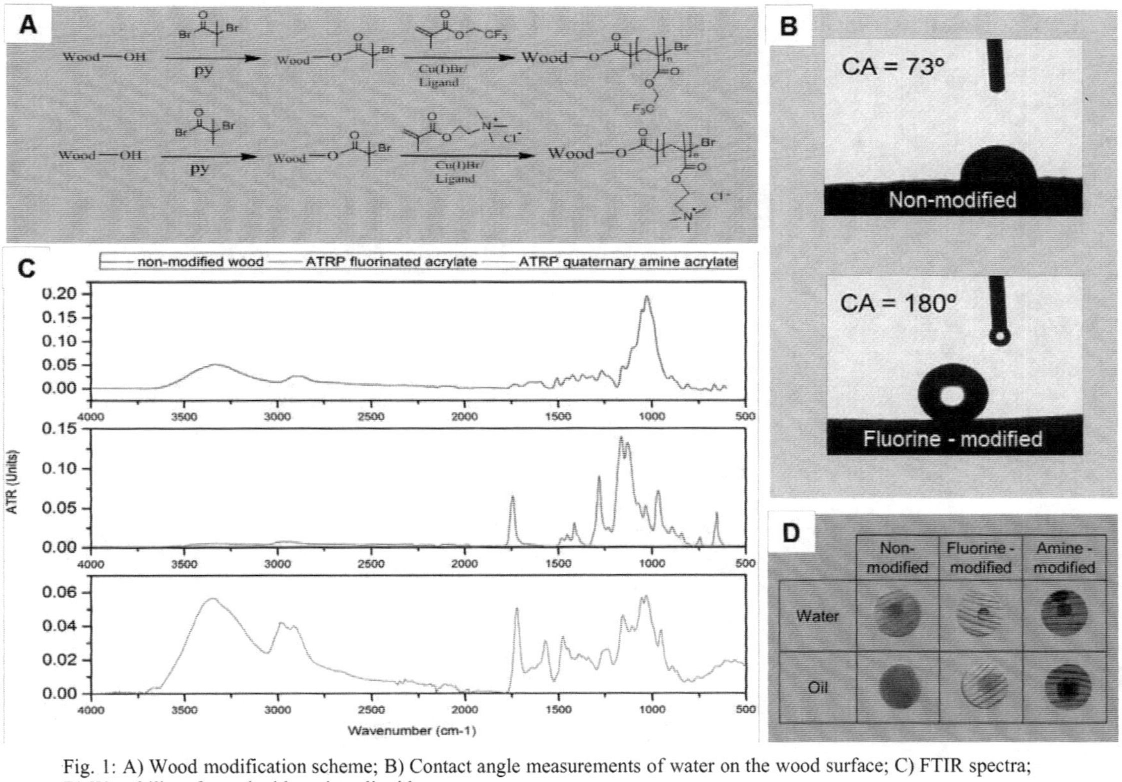

Fig. 1: A) Wood modification scheme; B) Contact angle measurements of water on the wood surface; C) FTIR spectra; D) Wettability of wood with various liquids.

[1]. E. Cabane, T. Keplinger, V. Merk, P. Hass, I. Burgert, *ChemSusChem* **2014**, 7, 1020-1025.

[2]. P. Brown, O. Atkinson, J. Badyal, *ACS Applied Materials & Interfaces* **2014**, 6, 7504-7511.

458027 Synthesis of Cellulose-Based Injectable Hydrogel Composite for pH-Responsive Drug Delivery

Abstract:

Abstract: Localized drug delivery, and prolonged, stimulus-driven release are approaches to maximize the function of drugs and minimize their side effects. This study developed a novel cellulose-based injectable hydrogel composite system containing pH responsive poly (ethylene oxide)-block-poly (2-(diisopropylamino) ethyl methacrylate) (PEO-b-PDPA) copolymer micelles for localized delivery and pH-triggered release. First, covalently in situ generated hydrogels were constructed by mixing hydrazide-modified carboxymethyl cellulose (CMC-NH_2) with oxidized carboxymethyl cellulose (CMC-CHO). The resultant hydrogels could be well tuned in the aspects of gelation time, degradation profile, and mechanical properties. Then, the pH-responsible copolymer PEO-b-PDPA was introduced to enhance the loading of hydrophobic substances in hydrogels as well as a pH-triggered release, which was demonstrated by fluorescent spectroscopy using Nile Red as a model drug. In addition, the results demonstrated that the hydrogel itself could act as a second diffusion barrier to reduce the drug release rate. The cell viability tests showed that both CMC-NH2 and hydrogels possessed excellent biocompatibility, while CMC-CHO was cytotoxic at a relative high concentration.

464660 Preparation of Whole Biomass Aerogels from Douglas Fir Using Molten Salt Hydrate As Solvent

Abstract:

Aerogel is a material with porous network structure, low density, high specific surface area, and low thermal conductivity. Aerogel has been used for thermal insulator, supercapacitor, storing medium, filter material for extremely fine particles, drug delivery systems, and drug coating. In the present study, whole biomass aerogels with high porosity, large surface area and low-density open pore 3D structure were prepared from Douglas fir without hemicellulose removal and delignification process. Lithium bromide molten hydrate was used to dissolve â€“ regenerate the cellulose or biomass to form cellulosic network structure. The results indicated that the density of the aerogels prepared from Douglas fir whole biomass ranged from 4.6 to 29 mg/cm^3 with porosity from 99.7% to 98.1%. The aerogel properties changes with biomass pretreatment, different reaction conditions, and biomass loading were investigated.

473457 so$_2$ Interactions with Pt, Pd, and Pd-Pt/Al_2O_3 Catalysts

Abstract:

Typical natural gas engine exhaust will contain uncombusted CH_4 that will need to be oxidized. Due to their high catalytic activity for the complete oxidation of methane, Pd-based catalysts are the typical choice for such oxidation systems. These catalysts can be exposed to trace amounts of sulfur compounds present in the engine exhaust. This sulfur exposure can lead to degradation in activity. The focus of this study was how sulfur interacts with Pd-based oxidation catalysts, as well as bimetallic Pt/Pd systems since Pt has been shown to have some resistance to sulfur poisoning. Temperature programmed desorption and oxidation were used to probe these interactions, as was DRIFTS to characterize surface species. The results show that the temperature, at which sulfur species are desorbed from sulfur-treated, mono and bimetallic Pd and Pt/Al_2O_3 catalysts, is dependent on the Pd-to-Pt mole ratio of the catalyst as well as the precious metal-crystallite particle size.

463008 Insights into the Deactivation of Cu/SAPO-34 By Low- and High-Temperature Hydrothermal Treatment

Abstract:

Cu/Chabazite (CHA) catalysts, Cu/SSZ-13 and Cu/SAPO-34, have been successfully commercialized for selective catalytic reduction (SCR) of NOx in lean-burn engine exhausts due to their superior activity, N_2 selectivity and durability. High-temperature (> 600 °C) hydrothermal aging treatments typically cause performance decrease for these catalysts, rationalized from partial structure degradation and loss of Cu-sites in active forms. However, recent studies have shown that the exposure of Cu/SAPO-34 to water vapor at *low temperatures* (< 100 °C) also result in an irreversible decline in NH_3-SCR activity [1, 2]. This may be attributed to structure degradation since SAPO-34 is known to hydrolyze in the presence of water vapor at such low temperatures. However, transformation of active Cu-sites to less active forms cannot be ruled out.

To gain insights into the similarity and difference between low- and high-temperature deactivation mechanisms, model Cu/SAPO-34 catalysts with varying Cu loadings were prepared using a one-pot method. These catalysts were further hydrothermally treated at either 70 or 800 °C. Effects of hydrothermal treatments on catalytic performance were examined using standard SCR and non-selective NH_3 oxidation kinetics. The zeolite framework integrity (crystallinity, porosity, coordination environments, Si-O-Al bonds, acidity) were characterized by XRD, N_2 physisorption, solid state NMR, DRIFTS and NH_3-TPD. A particular focus was given to changes in the local environments of Cu species during hydrothermal aging, examined using two-dimensional (2D) pulsed EPR. Finally, an attempt is made to elaborate the deactivation mechanisms that correlate the variation of NH_3 storage sites (acid sites) and redox sites (Cu^{2+} sites) caused by low- and high-temperature hydrothermal treatments.

[1] Wang, J., Fan, D., Tie, Y., et al., *J. Catal. 322*:84-90, 2015

[2] Leistner, K., Olsson, L., *Appl. Catal. B: Environ.* 165: 192-199, 2015

464314 Lean-Rich Switching over a Modified Three-Way Catalyst:Â Experiments and Modeling

Abstract:

Lean-Rich Switching Over a Modified Three-Way Catalyst: Experiments and Modeling

Mengmeng Li, Sam Malamis, Michael P. Harold*, William Epling**

Department of Chemical and Biomolecular Engineering, University of Houston, Houston, TX 77204

** mharold@uh.edu, ** wsepling@central.uh.edu*

The most recent CAFE standard for automobile requires an average fuel economy of 54.5 miles per gallon by 2025 [1]. At the same time, more stringent emission rules enacted by the EPA for 2017-18 will require an 80% reduction in non-methane organic gases (NMOG) plus NOx from current Tier 2 Bin 5 levels [2]. Gasoline vehicles that operate under both lean-burn and stoichiometric combustion hold promise since lean-burn combustion is more fuel efficient than conventional stoichiometric combustion. However, elimination of NOx ($NO+NO_2$) by the traditional three-way catalyst (TWC) is not possible under lean conditions. Advanced NOx reduction strategies must be applied to exploit the more efficient lean-burn engine combustion such as the three-way catalyst with a NOx storage function (TWNSC). The TWNSC stores NO_x storage during lean operation and reduces the trapped NOx during stoichiometric operation. To date, there have been only a few studies that report TWNSC. In this study a predictive TWNSC model is developed to predict TWNSC performance and in so doing provide operational insight and optimization.

A model of a catalytic monolith channel was developed that incorporates a TWNSC kinetic model into a low-dimensional model formulation [3–5]. The convection-diffusion-reaction equations are averaged in the transverse direction which replaces the transverse gradient term with an algebraic term that contains an internal mass transfer coefficient. Overall heat and mass transfer coefficients are used to represent the heat and mass transfer in the transverse direction between the bulk of fluid and solid phase. The TWNSC kinetics consists of a combination of elementary steps and global reactions. Following our previous experimental study of NOx reduction by propylene over a multifunctional NOx trap catalyst [6], the hybrid model includes the kinetics of the oxidation of CO, H_2, NO and hydrocarbons, of reduction of NO/NO_2 by CO, H_2 and representative hydrocarbons, and of storage of NOx and O_2. Fig. 1 shows the reaction network including 10 main paths that described the chemistry in qualitative terms.

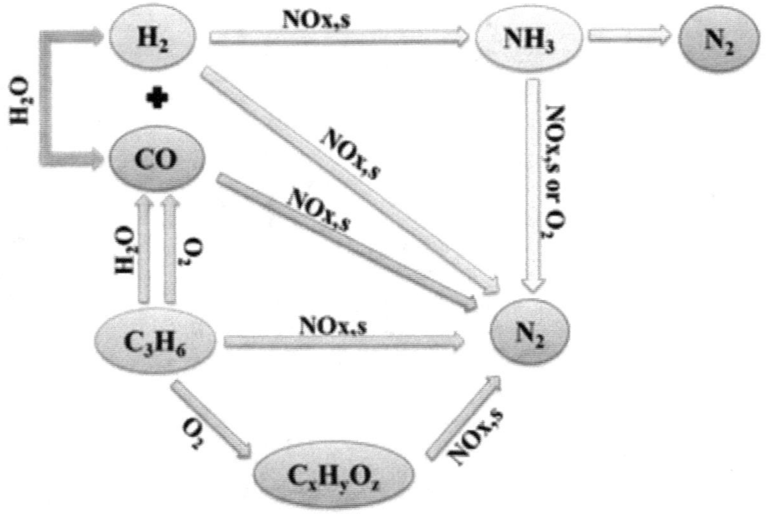

Figure 1. Simplified reaction network of NOx reduction using propylene as reductant[6]

A step-wise approach was used to develop kinetics for the oxidations of CO, H_2 and propylene. First, global kinetics are fitted tuned to transient light-off experiments following Raj et al. [5]. For example, the Langmuir-Hinshelwood expression is given by

$$R_{co} = \frac{k_1 X_{CO} X_{O_2}}{(1 + K_{CO} X_{CO})^2} \quad (1)$$

Fig. 2 shows the fitted light-off curve for CO oxidation and the comparison between experimental and model data. The model shows a reasonable fit and captures the dynamical feature of CO TPO (temperature programmed oxidation) experiment.

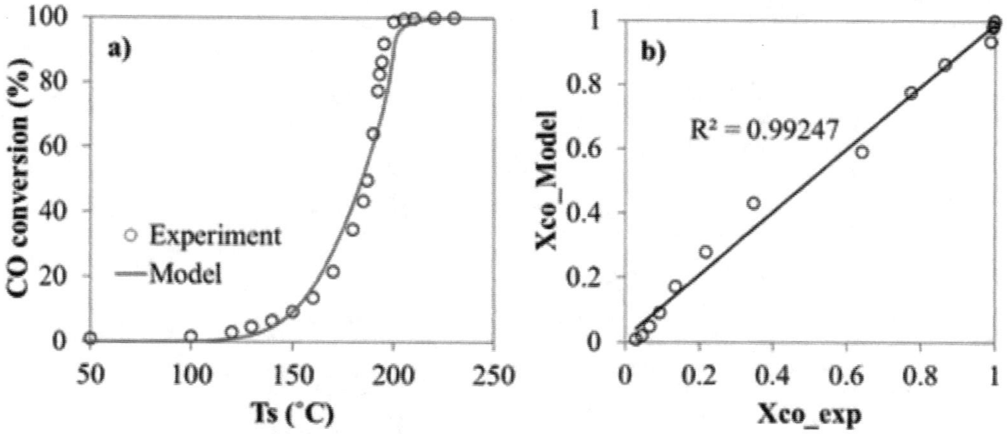

Figure 2. (a) Fitted light-off curve of CO oxidation (b) comparison between experimental and model-predicted conversions. [Conditions: 0.5% CO, 0.5% O; temperature ramp: 3°C/min]

The rate expressions for the co-oxidation of CO, H_2, and C_3H_6 include coupling effects. These are combined with a dual-site NOx storage model and incorporated into species balances.

We will show how the model is effective in capturing the main trends in the lean-rich switching data. This study advances the understanding of TWNSC catalyst and provides guidance for optimizing catalyst formulation and operation strategies.

Reference

[1] http://www.nhtsa.gov/About+NHTSA/Press+Releases/2012/Obama+Administra tion+Finalizes+Historic+54.5+mpg+Fuel+Efficiency+Standards.

[2] T. Johnson, SAE Int. J. Engines. 6 (2014) 699–715.

[3] S.Y. Joshi, M.P. Harold, V. Balakotaiah, AIChE J. 55 (2009) 1771–1783.

[4] D. Bhatia, R.D. Clayton, M.P. Harold, V. Balakotaiah, Catal. Today. 147 (2009) 250–256.

[5] R. Raj, M.P. Harold, V. Balakotaiah, Chem. Eng. J. 281 (2015) 322–333.

[6] M. Li, V.G. Easterling, M.P. Harold, Catal. Today. 267 (2016) 177–191.

453155 Importance of Exhaust Hydrocarbon Speciation in the Studies of Low-Temperature Emission Control Technologies for Next-Generation Fuel-Efficient Vehicles

Abstract:

As the fuel economy regulations being mandated all over the world require development of engines with improved fuel efficiency, the energy content in the exhaust will generally decrease and the resulting lower exhaust temperatures present significant challenges in catalytic converter performance, especially during cold-start periods. This, coupled with the more stringent future emission standards (e.g., LEV-III/Tier 3 regulations in U.S.) and recent adoption of green-house gas regulations including CH_4, makes catalytic oxidation of exhaust hydrocarbons (HCs) at low temperatures critically important.

Hydrocarbons in engine exhaust generally consist of unburned fuel components and partial combustion products. The latter includes light (C_1-C_3) saturated and unsaturated HCs such as methane, ethane, ethene and propene. Unburned fuel components typically consist of higher molecular-weight (>C_4) alkanes and aromatics. For gasoline engines, it has been commonly observed that C_5 and C_8 alkanes as well as C_7 and C_8 aromatics (toluene and xylene) are among the most abundant unburned fuel components in the exhaust. Regarding the nature of C_5 and C_8 alkanes, detailed GC-based speciation studies of gasoline exhaust HCs show that alkane isomers, such as iso-pentane and iso-octane, are the predominant species, with the concentrations of the corresponding n-alkanes (i.e., n-pentane and n-octane) being much lower.

The direct consequence of lower exhaust gas temperatures is that some of the HC species in engine exhaust are likely to remain unconverted even after the converter is fully warmed-up, thus presenting considerable challenges in HC emission control for future fuel-efficient engines. Thus, it is of practical importance to identify the types of HC species that are likely to contribute significantly to tailpipe HC emissions in cooler exhaust from next-generation gasoline engines and understand how the conversions of those HCs are affected by the type & loading of noble metals (Pd vs Rh), O_2 concentration level and catalyst aging. To that end, we conducted laboratory reactor experiments aimed at investigating the catalytic oxidation of various HCs in gasoline engine exhaust (alcohol, n-alkanes, alkane isomers, alkene and aromatics) over modern commercial three-way catalysts with variable noble metal loadings before and after aging. Another approach being considered for improved cold-start HC emission control is the passive HC adsorber technology, which is designed to store HCs during engine cold-start and oxidize (convert) the stored HCs as they are released at higher temperatures. As such, the effectiveness of this technology depends on the interplay between the adsorption, desorption and oxidation characteristics of specific HC species of interest. With this background in mind, we evaluated commercial HC adsorber catalysts for their cold-start HC emission reduction potential in a feedstream containing paraffins, olefins and aromatics.

This presentation will first discuss proper choice of "representative" HCs in gasoline engine exhaust based on HC speciation data and then examine their oxidation activities over degreened and aged commercial three-way catalysts as a function of noble metal loading. The oxidation activities were measured using a feedstream which contains not only the HC of specific interest and O_2 but the other usual exhaust constituents, such as CO, H_2, NO, CO_2 and H_2O as well. This would allow us to characterize the reactivity of each HC species under realistic conditions, namely, in the presence of the promotion effects of $H_2$and the inhibition effects of CO and NO. Also discussed in this talk are the performance characteristics of HC adsorbers in a feed containing various types of HCs including olefins, aromatics and paraffins. It was found that these types of HCs can be controlled to very different extents, due to variations in their adsorption, desorption and light-off (oxidation) properties. We also investigated the effects of hydrothermal aging as well as noble metal loading, oxygen concentration and adsorber volume on the HC conversion performance of the HC adsorber catalysts. These results will be discussed to illustrate the capabilities and limitations of the current HC adsorber technology, and to point ways toward improvement in the effectiveness of adsorber systems.

471850 Rapidly Pulsed Reductants in Diesel NOx Reduction with Lean NOx Traps: Spatial Distribution of Species

Abstract:

Lean NOx Trap (LNT) is a deNOx method using a NOx storage and reduction (NSR) catalyst that can be a desirable choice for NOx reduction on light duty diesel vehicles with smaller engines (< 2.0 L). A LNT used in a conventional way with long lean (60 s) – short rich (5 s) cycles obtains poor NOx conversions at high temperatures (T>400°C) and space velocities. Rapidly Pulsed Reductants (RPR) is a method for injecting hydrocarbons in rapid pulses ahead of a LNT which has shown the potential to achieve greater than 90% NOx conversion at temperatures as high as 600°C. This approach has recently been discussed by Toyota and named Di-Air [1]; it appears to be useful for smaller diesel vehicles that do not easily accommodate SCR technologies.

Four sets of parameters have been found important in RPR operation: 1- flow field and reductant mixing uniformity [2], 2- pulsing parameters, 3- the reductant type [2], and 4- catalyst composition, PGM type/loading, NOx storage material, and oxygen storage capacity (OSC). This parameter space has been explored using a synthetic bench reactor equipped with a novel high frequency injection and mixing system capable of pulsing reductants over the catalyst at up to 100 Hz.

The main findings from our previous studies on the effect of aforementioned parameters are; (a) The NOx conversion is maximized using nearly perfect plug flow pulses with minimal axial mixing [2], (b) The pulses of injected hydrocarbon need to be large enough in magnitude to generate a rich mixture during the pulse in order to achieve high NOx conversion. However, the cycle-averaged overall A/F ratio was always net lean. (c) The NOx conversion is dependent on the type of reductant at different operating temperatures [2]. (d) There exists an optimal operating condition, in terms of pulse amplitude and frequency that maximizes NOx conversion and minimize the fuel penalty associated the pulsed reductant.

In the current study, the spatial distribution of species along the catalyst have been explored using spatially resolved capillary inlet mass spectrometry (Spaci-MS) [3]. The experiments were conducted at near optimal operating conditions found from previous studies, with maximal NOx conversion and minimal byproduct (CO, N_2O, ...) formation. Ethylene was used as the main reductant on a typical Pt/Rh/BaO/Al_2O_3 LNT sample (no Ce). Experiments were conducted under reductant pulsing conditions to find the spatial distribution of the catalyst bed temperature (T/C) and the concentration of some of the important reactants and products, such as, C_2H_6, O_2, CO, CO_2, H_2O, H_2, NO, and NH_3 in the temperature range of 150°C to 600°C. Due to the limited response time of the instruments and high frequency pulsing approach, transient measurements were not possible of products and cycle-averaged measurements were performed. As an example, Fig.1 shows the cycle-averaged NO concentration vs temperature and axial location.

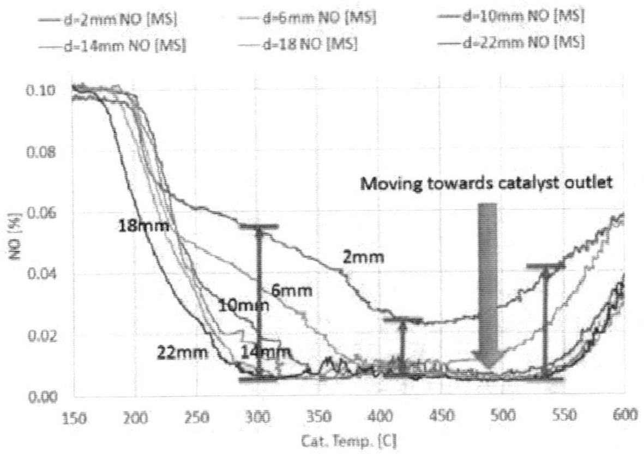

Figure 1 NO concentration vs temperature, measured at different axial locations of a LNT monolith with 25mm of length. Experimental conditions are: 30,000h^{-1} space velocity, P = 2s, 15% pulse duty cycle, and $\lambda_{pulse} = 0.73$.

As can be seen, the length of catalyst necessary to achieve high or complete NOx conversion is dependent on the operating temperature regime. At 300°C significant NOx conversion is seen 2 mm into the catalyst and complete conversion occurs 18 mm into the catalyst. At 420°C NOx conversion is high only 2 mm into the catalyst and complete at 6 mm into the catalyst, about one fourth of its length. At 540°C we find a new regime, where NOx storage capacity becomes the limiting factor over much of the length of this catalyst.

It should be mentioned that, the Spaci measurements are prone to variation due to the restriction effects of the capillary inside the channel. We have observed considerable deviation (10-20%) of these measurements from the bulk concentration at the outlet of catalyst. Thus, while these data are not perfectly quantitative, they are very useful in determining where in the catalyst product distributions change.

As described in the caption of Fig. 1, the spatial distribution provides insight into the limiting factors and different operating regimes. In addition, it enables us to locate the formation and consumption of important reaction byproducts, such as NH_3, CO and H_2. For instance, it was observed that NH_3 was formed within the first few millimeters of the catalyst, and then consumed moving downstream. Furthermore, formation of CO and H_2 was observed, around 350-400°C as an indication of steam reforming, which occurred within the first 2-6mm of the catalyst. A simple balance showed that the generated CO was then consumed moving downstream, mainly by the water gas shift reaction, forming H_2 which was observed at the catalyst outlet.

In summary, Spaci-MS is a useful technique which provides spatially resolved concentration and temperature distributions to help better understand the overall reactivity data. In the context of the RPR process, we have employed this information for determining the governing factors at different operating regimes, as well as important reaction pathways.

References

[1]Bisaiji, Yuki, et al. "Development of Di-Air-A New Diesel deNOx System by Adsorbed Intermediate Reductants."

[2] Reihani, A; Corson, B; Hoard, JW; Fisher, GB; Smirnov, E; Roemer, D; Theis, J; Lambert, C. “Rapidly Pulsed Reductants in Diesel NOx Reduction by Lean NOx Traps: Effects of Mixing Uniformity and Reductant Type.”

[3] Choi, Jae-Soon, William P. Partridge, and C. Stuart Daw. "Spatially resolved in situ measurements of transient species breakthrough during cyclic, low-temperature regeneration of a monolithic Pt/K/Al 2 O 3 NO x storage-reduction catalyst."

A Rational Pathway for Generic Three-way Catalyst Model Development in Stoichiometric Natural Gas Engines

Jian Gong, Di Wang, Junhui Li, Neal Currier and Aleksey Yezerets

Corporate Research and Technology, Cummins Inc., 1900 McKinley Avenue, Columbus, IN 47201, United States

Abstract

Three-way Catalysts have been widely used on stoichiometric natural gas engines for CO, hydrocarbon and NOx emission control [1]. A generic TWC model is highly desirable for design of aftertreatment systems, development of advanced control strategies and on-board diagnostics (OBD), and TWC architecture selection. However, there are many challenges in developing a generic TWC model. First, TWCs typically operate under dynamic conditions (e.g. fast lean/rich dithering). Second, TWC chemistry consists of a complex reaction network. In this work, a rational pathway in Figure 1 has been established to develop a generic TWC model that is capable of capturing the complicated chemistry as well as transient behaviors. This TWC model includes an OSC sub-model [2] and PGM related kinetic sub-model which is sensitive to catalyst state. The former is calibrated based on dynamic oxygen storage capacity (OSC) characterization performed on a bench reactor [3]. The latter is calibrated base on a complex steady state light-off performance protocol developed by the authors. The full model with calibrated kinetics is finally validated by dithering tests performed on a transient reactor and engine test data. This approach is found to be very capable and effective for generic TWC model development.

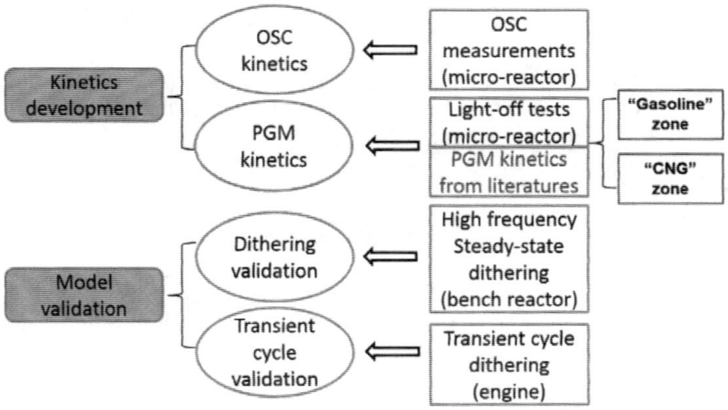

Figure 1: A schematic of the rational pathway for TWC model development

References

1. Gandhi, H.S., Graham, G.W., and McCabe, R.W., "Automotive exhaust catalysis," J. Catal. 216(1-2):433–442, 2003, doi:10.1016/S0021-9517(02) 00067-2.
2. Gong, J., Wang, D., A., Li, J., Currier, N., Yezerets, A., "Dynamic Oxygen Storage Modeling in a Three Way Catalyst for Natural Gas Engine: A Dual-site and Shrinking-core Diffusion Approach", Applied Catalysis B: Environmental, 2016 (in press), DOI: 10.1016/j.apcatb.2016.11.005
3. Wang, D., Gong, J., Junhui, L., Kamasamudram, K., Currier, N., Yezerets, A., "Controlling factors of surface and subsurface Oxygen Storage Capacity (OSC) of Three-Way Catalysts (TWCs) used in natural gas aftertreatment systems," 2015 Annual AIChE Meeting, Salt Lake City, UT, November 10, 2015.

460174 Effects of Cu Density in Cu-SSZ-13 Zeolites on Low-Temperature NOx Selective Catalytic Reduction with NH_3

Abstract:

Effects of Cu density in Cu-SSZ-13 Zeolites on Low-Temperature NOx Selective Catalytic Reduction with NH_3

Ishant Khurana[1], Atish A. Parekh[1], Jonatan D. Albarracin-Caballero[1], Christopher Paolucci[2], Arthur Shih[1], John R. Di Iorio[1], W. Nicholas Delgass[1], Jeffrey T. Miller[1], William F. Schneider[2], Rajamani Gounder[1], Fabio H. Ribeiro[1]

[1]School of Chemical Engineering, Purdue University, 480 Stadium Mall Drive, West Lafayette, IN 47907, USA

[2]Department of Chemical and Biomolecular Engineering University of Notre Dame

182 Fitzpatrick Hall, Notre Dame, IN 46556, USA

NOx selective catalytic reduction (SCR) using NH_3 as a reductant over Cu-SSZ-13 zeolites is a commercial technology used to meet emissions targets in lean-burn and diesel exhaust. Low temperature (473 K) NOx SCR on Cu-SSZ-13 occurs via a Cu(II)/Cu(I) redox cycle [1] on two types of isolated Cu sites, both divalent Cu^{2+} cations compensating paired framework Al atoms and monovalent $[CuOH]^+$ complexes at isolated framework Al atoms, which are solvated by NH_3 during reaction [2]. Both types of cationic Cu sites were identified by infrared, UV-Visible and *operando* X-ray absorption (XAS) spectroscopies and also from their exchange stoichiometry with surface protons, the residual number of which was quantified by NH_3 titration.

Under standard SCR conditions, in which O_2 is used as a four-electron oxidant in the Cu(I)/Cu(II) redox cycle, we have identified two different kinetic regimes that are characterized by different apparent orders, activation energies, and steady-state Cu(I)/Cu(II) distributions. The prevalence of standard SCR kinetic regimes depends strongly on the Cu density (determined by the Si/Al and Cu/Al) of the Cu-SSZ-13 catalysts. One regime corresponds to SCR cycles limited by Cu(I)->Cu(II) oxidation with O_2, which we probe in *operando* by varying reaction parameters (temperatures, gas pressures) while collecting steady-state kinetic data and XAS spectra, and in transient oxidation experiments of reduced Cu(I)-SSZ-13 catalysts using either O_2 and NO_2. These experiments reveal differences in the kinetics and Cu site requirements between O_2 and NO_2-catalyzed Cu(I)->Cu(II) oxidation, which respectively are four-electron and one-electron oxidants that propagate SCR cycles under "standard" and "fast" conditions. We provide evidence that O_2-oxidation requires two Cu sites and two NO molecules to be oxidized, while NO_2-oxidation occurs at single, isolated Cu sites. As a result, standard SCR rates depend on Cu cation proximity and distribution in Cu-SSZ-13 zeolites when Cu(I) oxidation steps are kinetically-relevant, but are independent of Cu density otherwise.

[1] C. Paolucci, A.A. Verma, S.A. Bates, V.F. Kispersky, J.T. Miller, R. Gounder, et al., Isolation of the Copper Redox Steps in the Standard Selective Catalytic Reduction on Cu-SSZ-13, (2014) 12022–12027.

[2] C. Paolucci, A.A. Parekh, I. Khurana, J.R. Di Iorio, J. Albarracin, A. Shih, et al., Catalysis in a Cage : Condition-Dependent Speciation and Dynamics of Exchanged Cu Cations In SSZ-13 Zeolites, J. Am. Chem. Soc. (2016) 1–63.

473866 The Effects of Catalyst Aging on Sulfur Degradation over Fully Formulated Lean NO_x Trap Catalysts

Abstract:

Lean NOx trap catalysts are subjected to both hydrothermal aging and sulfur deactivation in real world environments. Studies have shown that hydrothermal aging can increase noble metal particle size and subsequently reduce the ability of the catalyst to oxidize NO. This results in reduced NOx storage, and ultimately reduced NO_x conversion. Furthermore, literature has observed that the degree of sulfur deactivation varies with changes in catalysts parameters such as barium loading[1], this is an indicator that changing catalyst morphology via another mechanism such as hydrothermal aging could also change the degree to which the catalyst is sulfated. Because, both catalyst aging and deactivation due to sulfur occur in real world environments it is important to understand the effects of these two mechanism in conjunction.

In the following study a model Pt/Ba/CeZr/Al_2O_3 catalyst was subjected to hydrothermal aging. The hydrothermally aged catalysts were subjected to 1 hour sulfation in a lean gas environment. It was found that while hydrothermal aging initially resulted in variations in, NOx uptake, total NOx conversion, as well as ammonia and N_2O production, after sulfation, the cycling trends changed, most dramatically during the rich phase. Figure 1 shows that prior to sulfation ammonia production upon rich gas introduction is greatest for the catalyst aged at 700°C and the least for the catalyst aged at 600°C. After sulfation, ammonia production for the 600°C and 700°C aged catalysts are increased substantially, while the 800°C catalyst shows only a small increase in ammonia production and as a result has the lowest total ammonia production. The effects of hydrothermal aging and sulfation have also been characterized by BET, CO-Chemisorption, XRD and TEM and used to identify the dominant factors that have resulted in the observed deviations upon lean/rich cycling.

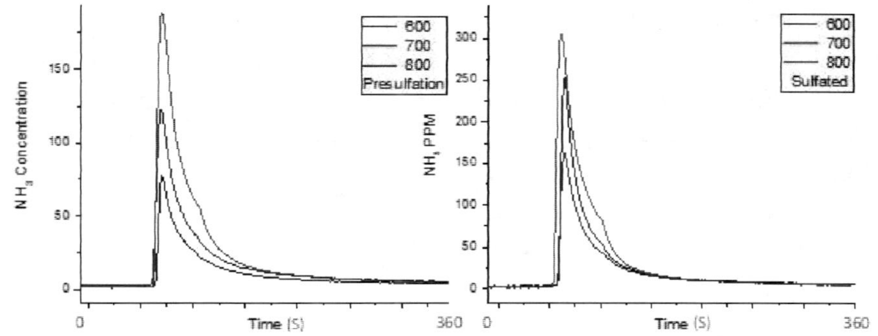

Figure 1 shows the variation in ammonia production after Lean/rich cycling at 300°C after presulfation (left) and post sulfation (right)

Acknowledgements

This project has received funding from the *European Union's Horizon 2020 research and*

innovation programme under grant agreement No 636380.

1. Kim, D. H.; Szanyi, J.; Kwak, J. H.; Szailer, T.; Hanson, J.; Wang, C. M.; Peden, C. H. F., Effect of Barium Loading on the Desulfation of Pt-BaO/Al2O3 Studied by H2 TPRX, TEM, Sulfur K-edge XANES, and in Situ TR-XRD. *The Journal of Physical Chemistry B* **2006,** *110* (21), 10441-10448.

459082 A Comparative Study of ZSM-5 and BEA-Zeolites for Hydrocarbon Trap Applications Under "Cold-Start" Conditions

Abstract:

Regulations on vehicle emissions are becoming more stringent worldwide due to increasing concerns of the impact of air pollution on environment and public health. Significant attention has been paid to develop methods to treat automobile exhausts for the regulated emissions such as hydrocarbons especially during engine cold starts. An effective solution is to employ suitable porous materials which can trap and retain hydrocarbons (HCs) temporarily until emission control catalysts are lit off (i.e., exhaust gas temperature reaches high enough for catalysts to become active) [1,2]. For this purpose, zeolites have been found to be a preferred type of materials due to their affinity to HCs and relative thermal stability [3]. The goal of this work was to understand key factors controlling hydrocarbon trapping and release performance of zeolites. Gained insights can, in turn, help to design HC traps with enhanced performance suitable for addressing future emission control challenges related to low temperature exhausts expected from more efficient engines. Zeolites were evaluated using propylene (C_3H_6) as a model HC compound in simulated exhaust conditions relevant to vehicle cold starts. The specific objective was to investigate in detail trends observed between C_3H_6 trapping-release performance, zeolite framework structure (ZSM-5 vs. BEA), acidity (high vs low), exchanged cation (e.g. H^+ vs. Ag^+ vs Pd^+), synthetic procedure (incipient wetness vs. ion exchange) and gas composition. Our results show that the amount of C_3H_6 adsorbed on H-zeolites significantly decreased in the presence of water, likely due to competitive adsorption and inhibition of adsorption sites. The results demonstrated that ion-exchanged metal is necessary to achieve increased storage and release of C_3H_6 in the presence of H_2O. Increased metal loading led to increased C_3H_6 adsorption ability and decreased desorption temperature. Most of stored C_3H_6 was not released as C_3H_6, but as various products of oxidation and cracking catalyzed by zeolites. We will discuss potential strategies to design enhanced HC traps based on the gained insights.

Acknowledgements. Research sponsored by the U.S. Department of Energy, Vehicle Technologies Office.

References

[1] A. Iliyas, M.H. Zahedi-Niaki, M. Eić, S. Kaliaguine, Micropor. Mesopor. Mat. 103 (2007) 171-177.

[2] J.-H. Park, S.J. Park, I.-S. Nam, G.K. Yeo, J.K. Kil, Y.K. Youn, Micropor. Mesopor. Mat. 101 (2007) 264-270.

[3] T. Kanazawa, Catal. Today 96 (2004) 171-177.

467005 Combined Experimental and Modeling Approach to Triboelectric Charging of Polyethylene Powders

Abstract:

The charging of objects caused by their mutual friction is called triboelectric charging or triboelectrification. Triboelectrification accompanies not only our daily activities like brushing hair or walking in shoes, but also several industrial processes. And because the charging is a surface-related phenomenon, it is most apparent in systems with high specific surface like powders (polymers, pigments, flour, drugs). For example, in fluidized-bed polymerization reactor, the polymer particles collide with other particles and with the device wall, thereby generating the excess electrostatic charge on their surfaces. Charged particles, in general, are attracted to the metal walls of devices due to electrostatic induction. Moreover, oppositely charged particles attract each other. Such behavior contributes to the unwanted particle agglomeration and fouling, which can impair the production and manufacturing of polymers. Although triboelectric charging has been documented for centuries, its mechanism hasn't been fully explained yet. Thus we combined the cascade method apparatus (that is a slide followed by the Faraday's pale) with Discrete Element Method (DEM) based model of colliding particles, in order to study the charging of polyethylene (PE) powder particles. Using the apparatus, we measured not only the saturation charge of PE particles, but also the charging dynamics as a function of industrially relevant parameters (temperature, air humidity, material of the slide, presence of sorbent). The charging was less pronounced at low humidities and low temperatures. Moreover, we observed significant charging even for the contact of 'identical' materials (PE particle with PE slide). Our DEM model of particle-wall contact charging was complemented by the capacitor model of charging. Another model was used for the prediction of particle-particle contact charging, in which Surface States theory was implemented. The model predicted fast charging for systems of a wide particle size distribution and/or spatial segregation of 'small' and 'big' particles. Such a result correlates well with the observations of charging in natural and industrial granular systems. Moreover, the model predicted more pronounced charging at high temperatures, which correlates well with our measurements. Combining particle charging and elasticity measurements with modeling, we further address the question of whether

the charging depends only on the contact area of colliding objects (which correlates with the spontaneous release of electrons during collision) or also on the amount of dissipated mechanical energy during the contact (which correlates with the excitation energy necessary for the electron transfer). Our study thus contributes to a better understanding of triboelectric charging fundaments and, at the same time, may provide valuable information for polyolefin industry

461607 The Influence of Surface Chemistry on the Electrostatic Properties of Particles

Abstract:

Background

Electrostatic phenomena are frequently observed whenever powder comes in contact with different surfaces, particularly when the motion of particles is involved. The resulting charge accumulation and its potential prolonged retention become a serious concern when insulating materials, such as pharmaceutical powders, are involved in the process. Due to a large number of factors influencing the process of charge generation, quantification and prediction of electrostatic properties is still one of the greatest challenges in this field.

Aims

The aim of this work is to assess the influence of surface chemistry on the ability of a material to acquire and subsequently dissipate electrostatic charge. The main hypothesis is that surface functional groups can strongly dictate the overall charging behaviour of the material and therefore could provide some indication of materials' electrostatic properties.

Methods

The surface of glass beads, a model material in this study, was modified using a silanisation reaction to produce chemically functionalised samples. Contact angle measurements were used to confirm the extent of surface modification of the samples. Electrostatic properties were studied using a purpose build instrument, which incorporates a capacitive probe[1] within a Dynamic Vapour Sorption instrument. The modified glass beads were tribocharged by contact with a stainless steel surface and their maximum potential as well as decay curves were measured at 25 °C at three different humidity levels, i.e. 20% RH, 50% RH and 75% RH to evaluate the impact of environmental conditions on charge dissipation.

The influence of surface hydrophobicity on the electrostatic behaviour was further investigated for mannitol crystals. Mannitol was recrystallized from water at four supersaturation levels (σ = 3%, 13%, 23% and 33%) to produce needle-shaped crystals with varied aspect ratio and hence different overall hydrophobic character. Strong correlation between crystal shape and crystal surface chemistry for mannitol has been determined previously.[2] The electrostatic properties of the crystals were determined using the capacitive probe method and compared with industrially available instrument called Charge Decay Time Analyser.

Results

Successful surface modification of glass beads produced four samples with distinct surface chemistries, which exhibited differing degrees of surface hydrophobicity. Electrostatic measurements showed that the magnitude of charging followed the trend in hydrophobicity with the most hydrophobic particles acquiring the largest charge. Up to six times larger charge was accumulated by fluoro functionalised glass beads, acquiring charge of -11.2 ± 2.8 V/mg, compared to hydroxyl functionalised sample with overall charge of -1.9 ± 0.3 V/mg. In addition, the unique properties of heteroatoms, such as electronegativity, seem to have further impact on the overall charge adopted as observed by significant charging of the fluoro functionalised sample and charge inversion in case of amino functionalised surface. For all surfaces, increasing the external humidity resulted in faster charge dissipation; however, the actual effect of humidity on charge decay rate seems to be surface dependent. Under insulated charge conditions, increase in humidity from 20% RH to 75% RH resulted in lowering charge half-life time from 21.4 min to 4.9 min for hydrophilic sample and from 45.1 min to 13.3 min for amino functionalised surface.

Crystallisation of mannitol from solutions under varied conditions produced a range of crystals – from thin, long needle crystals at low supersaturation to short wide plate-like crystals at high supersaturation. These showed to have different surface energy properties and hence different electrostatic characteristics.

The data generated was repeatable and reproducible indicating that by controlling particle size, shape and chemistry the variation in electrostatic charge can be limited, but also shows that the capacitive probe setup used for electrostatic measurements is reliable and capable of detecting electrostatic charge of powders.

Conclusion

The results obtained show that surface chemistry and hence surface hydrophobicity play an important role in powder electrostatics. It was shown that materials having different chemical groups on the surface can exhibit different electrostatic behaviour. This includes not only the overall charge acquired under the same conditions, but also charge polarity and charge decay characteristics. Such observations could be beneficial in predicting a material's propensity towards charge accumulation during processing based on its chemical structure and molecular arrangement within a crystal lattice.

Acknowledgments

The authors acknowledge Dr Daryl Williams and Dr Jin Wang Kwek for their comments and useful discussions.

References

(1) Kwek, J. W.; Jeyabalasingam, M.; Ng, W. K.; Heng, J. Y. Y.; Tan, R. B. H., Comparative Study of the Triboelectric Charging Behavior of Powders Using a Nonintrusive Approach. *Ind. Eng. Chem. Res.* **2012,** *51* (50), 16488-16494.

(2) Ho, R.; Wilson, D. A.; Heng, J. Y. Y., Crystal Habits and the Variation in Surface Energy Heterogeneity. *Crystal Growth & Design* **2009,** *9* (11), 4907-4911.

470762 Catastrophic Failure of Agitated Filter-Driers Due to Triboelectric Charging of Certain APIs

Abstract:

Agitated filter driers, typically consisting of a dished vessel with a perforated plate, and jacketed walls, bottom dish, and impeller, are commonly used in the process of solid API isolation. This work will present the investigation of the physical processes underlying a catastrophic failure during the drying of a certain API at the pilot plant scale. Prior to the failure, the powder resistance to flow would increase dynamically during the course of the agitation resulting in an increase in the agitator torque (as measured by the hydraulic pressure required to maintain the rotational speed set point). Eventually, this increase in flow resistance would lead to a solid body-like rotation of the powder bed, translating the agitator's mechanical energy, through the non-slip contact at the dryer bottom, into shear on the filter cloth sufficient to tear it out of its retaining o-ring.

Additional experimentation at scale showed that this dynamic increase in flow resistance was repeatable across multiple batches, and was also evident in other APIs and intermediates. Further investigation identified tribo-electric charging as the underlying cause for this behavior. Characterization of the dependence of the phenomenon on the process parameters and cake properties allowed for the development of a procedure for the early detection of potential "problem" compounds at the lab-scale.

COMPARISON OF MAGNETICALLY ASSISTED IMPACTION COATING (MAIC) WITH TRADITIONAL MIXING TECHNIQUES FOR THE ADDITION OF A SILICA FLOW AID

Tim Freeman[1], William A. Hendrickson[2], Charles R. Bowman[2], Christopher J. Rueb[2], Robert G. Bowman[2], Katrina Brockbank[1], Jamie Clayton[1]

[1]Freeman Technology, Tewkesbury, Gloucestershire, UK
[2]AVEKA Inc., Woodbury, Minnesota, USA

Keywords*: Solids Flow, Handling, and Processing, Mixing, Process Design & Development

Introduction

Many powders have poor basic flow properties, they block in hoppers and die feed frames, exhibit inconsistent or pulsatile discharge rates, adhere to equipment surfaces or do not mix readily with other materials. Additionally, many powders tend to cake on storage or during transport due to changes in environmental conditions and/or consolidation. As such, flow aids are frequently added to a blend to enhance flow properties and/or inhibit caking.

In order to optimise the influence of a flow aid, it must be dispersed in a fine, uniform layer on the surface of the substrate particles. However, most flow aids, for example fumed silica or magnesium stearate, consist of very fine particles that have a tendency to form agglomerates. The mixing/coating process employed can therefore have a major influence on how efficiently the flow aid is distributed. The process must be intense enough to disperse the agglomerated flow aid without damaging the particles of the substrate or the flow aid itself. This is particularly true for many food and pharmaceutical ingredients which are often relatively friable and easily deformed by extreme mechanical forces [1].

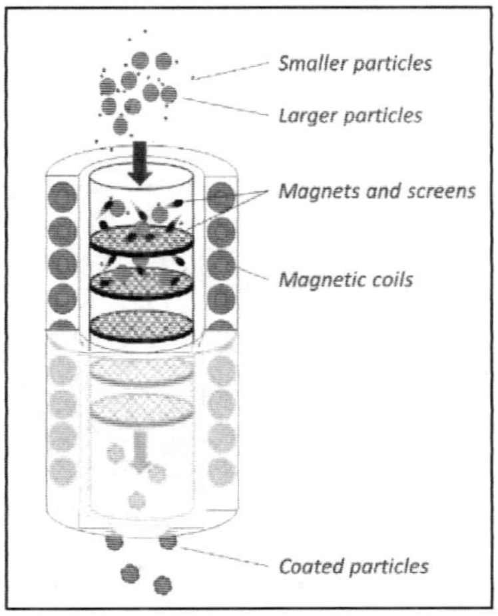

Figure 1 - Magnetically Assisted Impaction Coating (MAIC) process

Magnetically Assisted Impaction Coating (MAIC) has been demonstrated as a technique for coating host particles without causing major changes to particle shape and size [1,2]. This system relies on a pounding process in which an oscillating magnetic field is used to accelerate and spin large magnetic particles (600-800 µm, though size may vary with application), causing the magnetic particles to fluidise, through which the host and guest particles (flow aid) are passed in a continuous fashion (fig 1). The agitated magnetic particles impart energy to the host and guest particles, and coating is then achieved by means of impaction or peening of the guest particles onto the host particles as collisions occur between particles and between particles and the vessel wall [1].

Materials and Methods

Jet milled citric acid (Hawkins Chemicals, jet milled by *AVEKA*) was pre-blended with 0.05%, 0.5%, and 2.0% (w/w) fumed silica (Aerosil 200 Pharma, Evonik) using a V-blender with intensifier bar for 5 minutes. The blends were then fed through the MAIC process (US Patent 5962082 [2]). The samples were run through the MAIC (*AVEKA*, USA) in increasing mass percent of silica, so as to minimize the impact of potential contamination of subsequent samples. Prior to this, citric acid was passed through the MAIC without any treatment with fumed silica, and without being placed in the V-blender. In addition to the MAIC-treated samples, two v-blended samples not subjected to MAIC were created, with 0.5% and 2.0% (w/w) fumed silica. The untreated citric acid made for the seventh and last of the evaluated samples.

After being randomized, the samples were labeled without any indication of processing method (V-blend only or MAIC-treated) or silica loading, which facilitated a blind study of the flow characteristics. All packaged samples were then shipped from Minnesota, USA by air to Tewkesbury, UK. Each sample was then evaluated for flow on an FT4 Powder Rheometer (Freeman Technology, UK) before the identity of each sample was revealed and matched to the relevant data.

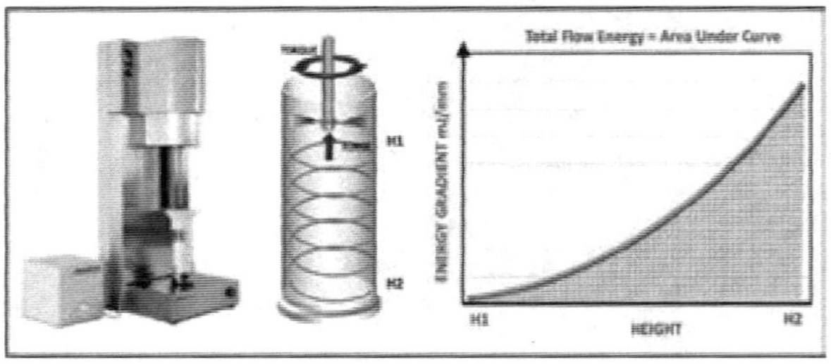

Figure 2 – Measurement of the flow energy using the FT4 Powder Rheometer

The FT4 measurements included dynamic flow, bulk and shear properties. Dynamic properties, including Basic Flowability Energy (BFE) (fig. 2), were obtained by means of a patented measurement principle that evaluates the resistance to the motion of a specially shaped twisted blade passing through a precise volume of the sample along a prescribed path. The required torque and force are recorded and converted into a flow energy [3]. The repeatability of all measurements was enhanced by the use of a conditioning cycle which removes packing history and operator-induced variability.

Results and Discussion

This study compared MAIC with V-blending through comparison of the flow properties of the substrate after the addition of silica. In addition, this study also investigated the effect of silica concentration on samples blended via MAIC.

MAIC vs. V-Blending

The MAIC blends at both silica concentrations (0.5% and 2.0% (w/w)) presented lower BFE, Aerated Energy (AE), Cohesion, Consolidation Index (CI) and Pressure Drop values indicating improved flow properties compared with the V-blended samples (fig. 3).

BFE measures the resistance to flow in a dynamic, constrained forced flow environment, for example feeding via a screw conveyer. Whilst all the silica blends exhibited improved flow compared to the raw material, the MAIC blends exhibited the largest decrease in flow energy, suggesting the greatest improvement in flow. This is due, in part, to the fact that MAIC creates a more uniform coating with the flow aid than can be achieved by V-blending alone [4].

The silica blends also exhibited a decrease in Specific Energy (SE) compared to the raw material suggesting reduced mechanical interlocking and friction. However, there were limited differences between the MAIC and V-blended samples at 0.50% (w/w) silica; although at the higher load of 2.0% (w/w) silica, the MAIC blend presented a greater improvement in flow compared with the V-blended sample.

The lower AE values presented by the MAIC samples not only indicate that they were more responsive to the introduction of air, but also demonstrated a clear reduction in the absolute cohesion compared with the V-blended samples. This is also indicated by the lower cohesion values obtained during the shear tests suggesting that the MAIC samples were also less resistant to flow at higher (applied) loads. This behaviour has been previously observed on a batch MAIC process, and used to improve the ability of a powder to fluidize [5].

Powders can also undergo compaction when exposed to external vibrational forces, for example during transport or due to proximity to (other) process equipment, resulting in reduced flowability. The V-blended samples both exhibited much larger increases in flow energy with notably higher Consolidation Index (CI) values following tapping, demonstrating that these samples were much more sensitive to vibrational forces and indicating reduced flow when exposed to external forces. Overall these results suggest that the MAIC blends exhibited improved flow properties at low, moderate and high stress regimes, and may be more stable during storage.

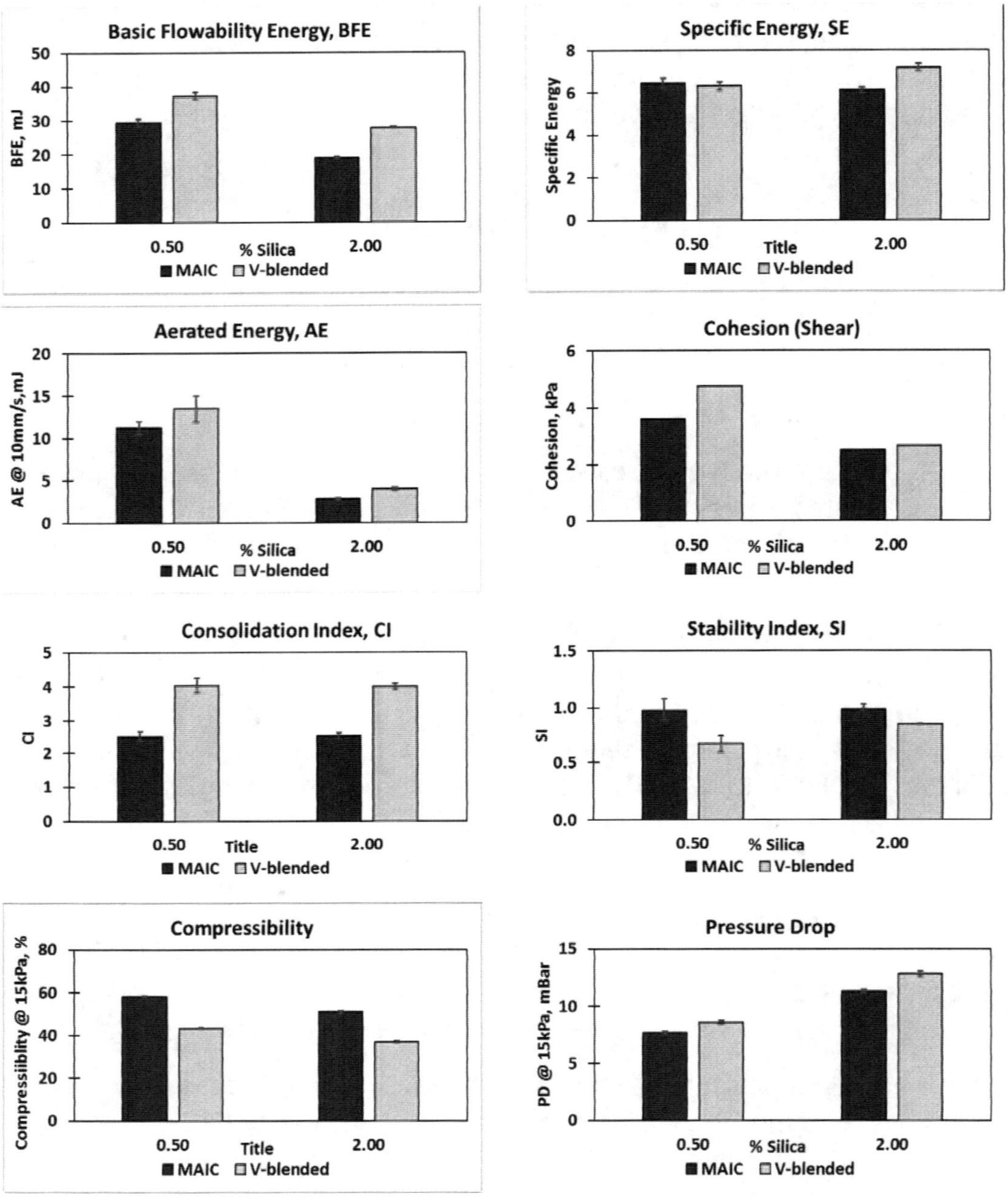

Figure 3 – FT4 Powder Rheometer Results for MAIC and V-Blended samples with 0.50% and 2.0% (w/w) fumed silica

The FT4 dynamic flow tests also provide an indication of the physical stability of the powder by measuring the change in flow energy over repeat test cycles. Whilst the MAIC blends were demonstrated to be physically stable over 7 test cycles, the V-blended samples presented a marked decrease in flow energy with Stability Index (SI) values less than 1, indicating that resistance to flow decreased as the sample was tested. This suggests that the fumed silica was not fully dispersed at the start of the test, with the mixing action of the blade as it traversed the powder bed resulting in a more uniform and wider dispersion of the fumed silica over the duration of the test. This in turn resulted in the decrease in the measured flow energy. It should be noted that these measurements were made after the samples were shipped via air from the USA to the UK. This may have generated stress conditions in the powder samples prior to testing which the MAIC process appears less affected by.

While the MAIC samples clearly showed reduced cohesion during flow, they were also demonstrated to contain more entrained air compared to the V-blended samples with lower Bulk Density (in both the Dynamic Flow and Compressibility tests) and Compressibility values. Flow aids improve particle flow behaviour by acting as a surface lubricant, reducing mechanical interlocking and friction, and/or increasing interparticular spacing, thereby removing agglomerates. Increased particle spacing may partially account for the increases in entrained air, however, the bulk of the entrained air is more likely to be due to improved separation of both the host and guest particles. Whilst this will allow for a more uniform distribution of the fumed silica on the surface of the citric acid particles as more of the host surface will be exposed, it will also cause the powder to mix with air which will subsequently be trapped in the powder bed.

% Loading

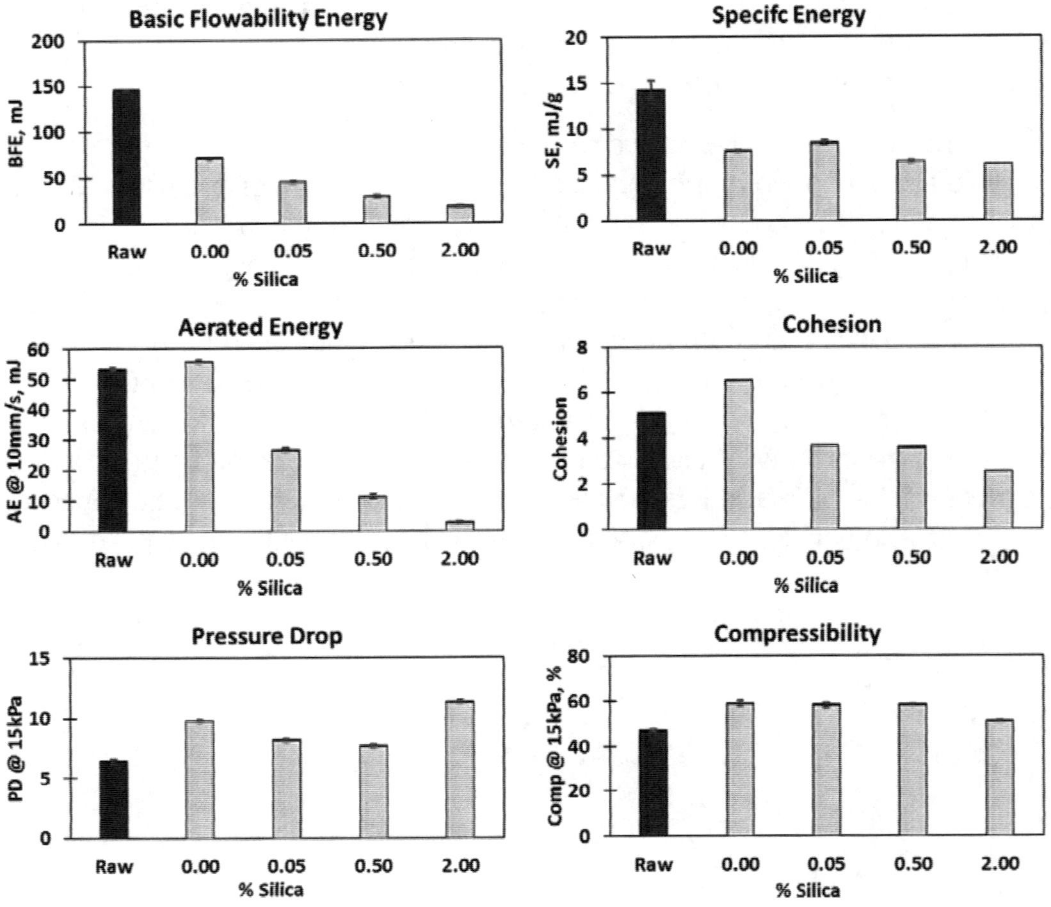

Figure 4 – FT4 Powder Rheometer Results for MAIC blended sample with increasing % silica (w/w)

As demonstrated by the results in Fig. 4, the MAIC blends exhibited a clear decrease in the BFE with increasing silica loading, with the 2.0% (w/w) presenting the lowest BFE. However, a marked decrease in flow energy was also demonstrated between the raw sample and the 0% (w/w) blend with a similar decrease demonstrated by the SE. This indicates that the MAIC process can improve reduce the resistance to flow in a low-moderate dynamic regime without the addition of silica. Indeed, there were only minor reductions in SE with addition of silica at higher loads (at 0.05% (w/w) the SE was actually slightly higher), suggesting the mechanical interlocking and friction had minimal influence following MAIC.

Improvements in the flow properties of the raw material following MAIC may be due to a number of factors. As previously mentioned, both the Bulk Density and Compressibility values indicate that the treated samples contain more entrained air, suggesting improved separation of the particles and wider interstitial spacing. The higher Pressure Drop values also suggest less efficient particle packing. Improved separation may result in the breakage / reduction of liquid and solid bridging, whilst wider interstitial spacing will decrease short range forces, such as van der Waals forces (effective range of approximately 100 nm) [6], and mechanical interlocking and friction. Collisions between the agitated magnetic particles and substrate may also result in the loss of surface asperities, smoothing the particle surfaces and again reducing mechanical interlocking and frictional effects, although SEM has shown limited changes in the particle physical properties. It may also be that in the absence of flow aid, the MAIC process attaches the fine particle fraction (of the citric acid) to the larger citric acid particles in the bulk, not only removing the smaller and therefore typically more cohesive particles from the bulk but also increasing interparticular spacing between the larger particles.

With respect to the effect of silica loading on the particle packing properties (as demonstrated by Compressibility and Permeability), the silica concentrations appear to have minimal effect below 2.0% (w/w) silica and even at 2.0% (w/w) there is only a slight improvement. This may be due to the reduced cohesion allowing for improved particle packing, or excess silica filling interstitial spaces. Either way, the number/size of air pockets and channels which allow for the passage of air would be reduced, explaining why the Pressure Drop increases at 2.0%(w/w) silica, whereas it presented minor decreases at lower concentrations.

Interestingly, when the effects of particle packing are negated, either by fully aerating the powder bed or through compaction (shear), the 0% (w/w) blend is marginally more cohesive compared to the raw sample with a slightly higher Aerated Energy (AE) and Cohesion values. With the addition of silica both tests present a clear improvement in flow properties with the lowest AE / Cohesion values presented at 2.0% (w/w) silica. The Aeration test presents a similar trend to the BFE with a clear reduction in the AE with increasing silica loading whereas Cohesion presents limited variation between the 0.05% (w/w) and 0.5% (w/w) blends despite a 10-fold increase in silica concentration (this is also shown by other shear parameters including Angle of Internal Friction and Flow Function). Indeed, a 10-fold increase in silica did not result in similar magnitude of change in flow properties for any of the measured parameters, as such while higher concentrations may exhibit the largest improvements, as long as the powder is fit for purpose at lower silica concentrations, further benefits may not outweigh the added costs and complexities of increasing the silica concentration. Furthermore, previous studies have shown that continuing to increase the flow aid concentration can be detrimental to flow as once the point of surface saturation is reached, the flow aid remains loose in the powder bulk adding to the total cohesion [7].

Conclusion

Although both the V-blended and MAIC samples exhibited a clear improvement in flow properties compared to the raw materials, the results were more pronounced for the MAIC samples, which were shown to be less cohesive across a range of stress regimes and modes of flow. The MAIC also resulted in a more conditioned uniform blend that is more resistant to changes in its flow properties, even after extensive shipping and handling. In terms of sample loading, whilst the 2% (w/w) silica blend was shown to be the least cohesive in a number of tests, the marked increases in silica concentrations (from 0.05% (w/w) to 2.0% (w/w)) did not yield similar magnitudes of improvement in the flow properties. As such, once the added costs and complexities of substantially increasing the silica concentration are considered, a lower silica loading may be more beneficial as these still return a marked improvement in flow behaviour. Interestingly, even without the addition of silica, the MAIC process appear to have some positive effects on the flow properties, however the reason for this requires further investigation. Overall, it can be concluded that MAIC shows an advantage over V-blending when adding flow aid.

[1] M. Ramlakhan, C.Y. Wu, S. Watano, R.N. Dave, R. Pfeffer, Dry particle coating using magnetically assisted impaction coating: modification of surface properties and optimization of system and operating parameters, Powder Technology, 112 (2000) 137-148.

[2] W.A. Hendrickson, J. Abbott, Process for applying liquid coatings to solid particulate substrates, US Patent 5962082 A (1997)

[3] R. Freeman, Measuring the flow properties of consolidated, conditioned and aerated powders — A comparative study using a powder rheometer and a rotational shear cell, Powder Technology, 174 (2007) 25-33.

[4] J. Yang, A. Silva, A. Banerjee, R.N. Dave, Dry particle coating for improving the flowability of cohesive powder, Powder Technology, 158 (2005) 21-33

[5] R. Pfeffer, C.H. Nam, R.N. Dave, G. Liu, J.A. Quevedo, Q. Yu, C. Zhu, System and method for nanoparticle and nanoagglomerate fluidisation, US Patent 7658340 B2, (2010)

[6] J.Q. Feng, D.A. Hays, Relative importance of electrostatic forces on powder particles, Powder Technology, 135–136 (2003) 65-75.

[7] T. Freeman, J. Yin, M. Delancy, J. Clayton, Characterising and optimising the use of flow additives in powders and powder formulations, AIChE Annual Meeting (2014)

471620 Surface Chemical Modification and Its Effect on Dynamic Flow Properties

Abstract:

Surface Chemical Modification and its Effect on Dynamic Flow Properties

C.G. Jange; R. P. K. Ambrose

Department of Agricultural and Biological Engineering, Purdue University, West Lafayette, IN

Powder flow properties are crucial parameters in handling and processing operations, including discharge from hoppers and silos, transportation, mixing, compression and packaging. Particle size, shape, density, and particle size distribution are some of the factors that influence the flow properties. The interparticle forces are highly dependent on surface properties of particles and may impact flow behaviour. van der Waals, electrostatic and capillary forces, and mechanical interlocking are the possible interparticle interaction that could exist between particles. This research quantified the effect of surface chemical composition on the powder flowability. Qualitative and quantitative understanding on the effect of surface chemical modifications on interparticle interactions and flow behavior remain poor in published literature. To study the general effect of surface composition on bulk flow properties of powders, samples of uncoated and coated glass beads with size range of 125 µm and 180 µm were used. Flow property measurements were performed using the FT4 Powder Rheometer with pre-consolidation stress from 1kPa to 9kPa. The samples were coated with gluten, starch, and fat before flow property analysis. The average values of unconfined yield strength (UYS), major consolidation stress (, cohesion index (C), flow function coefficient (ff_C), and angle of internal friction, for 125 µm size uncoated glass beads were 0.25kPa ± 0.03, 2.50kPa ± 0.31, 0.09 ± 0.01, 20.27 ± 3.96, 18.66 ± 0.32, respectively. For 180 µm samples, the values were 0.23kPa ± 0.06, 2.35kPa ± 0.02, 0.08 ± 0.02, 30. 46 ± 10.58, 19.66 ± 0.52, respectively. Results indicated 'easy flowing' characteristics for both size ranges followed by a slight increase in ff_C for particles in the range of 180 µm. Particle size of coated and uncoated glass beads were measured using G3-ID-Raman Spectrometer (Malvern Instruments, UK). In completion of this study, we expect to quantify the influence of surface chemical composition on the dynamic flow properties of particulate material.

Key words: Surface Composition; Flowability; Interparticle Forces

472544 Effect of Surface Roughness and Particle Size on Wall Friction

Abstract:

Friction between particulate materials and the process, storage and transfer equipment wall surface is of critical importance for design and operation of solids processing systems. A typical example would be hoppers. The wall friction angle between the particulate material and the hopperâ€™s material of construction is measured and used along with the particle-particle friction to calculate the required outlet diameter of a hopper for unobstructed flow (i.e. without forming solid bridge). The type of the material, the surface finish (i.e. the degree of roughness), and the coating used inside the equipment may vary greatly and affect the wall friction. Despite the vast amount of experimental data collected for various surfaces and particulate materials over the last couple of decades, the effect of surface roughness on wall friction is unpredictable and depends on the particulate material. In this study, we investigate the effect of wall roughness systematically for model materials such as glass spheres and polymer particles. The effect of particle size will be also presented.

465034 Extracting Kinetic Parameters from TGA Datasets - Ensuring Optimal Outcomes for Cellulose Pyrolysis

Abstract:

Thermal gravimetric analysis (TGA) has been used for decades to study the thermal decomposition of whole biomass and its separated component phases, cellulose, hemicellulose and lignin. Extraction of kinetic parameters from TGA datasets, however, has resulted in a wide range of activation energies and associated pre-exponential factors. A proposed global optimization strategy that incorporates a Nelder-Mead simplex algorithm was used to find *optimized* kinetic parameters for the thermal decomposition of crystalline cellulose. Visualization of the objective function topology illustrates the difficulty in locating the global optima. A number of plotting strategies are also illustrated as useful tools for the collective comparison of fit outcomes. A systematic study of the effects of search strategy, objective function choice and sample temperature uncertainty was shown to unify the historical findings.

466394 The Influence of Potassium Content and Heating Rate on Biomass Pyrolysis

Abstract:

A non-isothermal one-dimensional model has been developed to describe biomass pyrolysis at fast heating rate (600-10^4 K s^{-1}), high temperatures (up to 1500°C) and is valid for different biomass particle sizes (< 10mm). The model was developed to estimate the yields of volatile gas and char. The model relies on the concept applied in fast pyrolysis of cellulose through the formation of an intermediate liquid (so called metaplast) which reacts further to char and gas. The kinetics of the fast pyrolysis was described by the Broido-Shafizadeh scheme.

The influence of particle size and shape was included in the model. Cylindrical representation of a biomass particle shape was chosen to be the most suitable in the pyrolysis model. The evolution of devolatilization time required for the complete pyrolysis showed that the particles with a mean diameter < 0.45mm may be considered as thermally thin at high heating rates. The predicted results by one-dimensional model are in agreement with the experimental work, and emphasize a key role of intra-particle heat conduction in biomass particles > 0.45mm.

The potassium influence on the char yield was implemented in the model based on the experimental results in the wire mesh and drop tube reactors with respect to the stronger catalytic effect of potassium on the char yield at low and intermediate heating rates compared to pyrolysis at high heating rates. The heating rate and potassium content affected significantly the char yield as evidenced from the experimental data obtained in the wire mesh and drop tube reactors. Thus, the model including these two parameters provides an acceptable fit of char yield to the experimental data. The present results showed that the proposed kinetic model for the fast biomass pyrolysis is relatively simple and predicts reasonably accurately the char yield of woody and herbaceous biomass particles < 10mm using one fixed set of kinetic parameters valid for woody and herbaceous biomass.

471557 Synthesis and Characterization of Maleic Acid â€“ Aluminum Catalyst for Glucose Conversion

Abstract:

Maleic acid (MA) and $AlCl_3$ self-assemble into catalytic complexes (Al-$(MA)_2$-$(OH)_2$(aq)) with improved selectivity for converting glucose to value-added chemicals in water. Use of maleic acid and $AlCl_3$ compared to HCl and $AlCl_3$ shows an increase in selectivity (2.6 to 11) of fructose conversion to hydroxymethylfurfural at 180°C. We report the production of Al-$(MA)_2$-$(OH)_2$(aq) formed by reaction of maleic acid and $AlCl_3$ and confirmed by high pressure liquid chromatography (HPLC) and mass spectrometry. Reaction of 100 mM maleic acid and 100 mM $AlCl_3$ at 100°C for 24 hours shows a modest 25.9% production of Al-$(MA)_2$-$(OH)_2$(aq) with some degradation of maleic acid to malic acid. Increasing beginning concentrations to 500 mM maleic acid and 500 mM $AlCl_3$ shows an increase in selectivity (0.34 to 0.47) for Al-$(MA)_2$-$(OH)_2$(aq) over malic acid production. Storage of the refluxed liquid at 4°C yields formation of precipitate crystals containing 92.6% catalyst complex. Alteration of initial reflux concentrations to 500 mM maleic acid and 250 mM $AlCl_3$ shows an increase in Al-$(MA)_2$-$(OH)_2$(aq) production and purity (18.2% to 20.2%) in Al-$(MA)_2$-$(OH)_2$(aq) present in solution. Forming crystals by precipitation shows a change in morphology of formed crystals and an increase in purity to 94.6% compared to the equal molar reflux liquid. Rinsing crystals with deionized water removes residual maleic acid and malic acid yielding crystals with 99.2% purity catalyst complex. Future work will include a study on x-ray crystallography and a study on complex catalytic conversion of glucose including kinetics and conversion yield.

472913 Catalytic Hydropyrolysis of Milled Lignins Derived from Hardwood, Softwood and Herbaceous Biomass Using Bifunctional Catalysts

Abstract:

Catalytic hydropyrolysis of milled lignins derived from hardwood, softwood and herbaceous biomass using bifunctional catalysts

Yuan Xue[a], Ashokkumar Sharma[b], Xianglan Bai[a], Feng Cheng[c], Catherine E. Brewer[c]

[a] Department of Mechanical Engineering, Iowa State University

[b] Bioeconomy Institute, Iowa State University

[c] Department of Chemical & Material Engineering, New Mexico State University

Lignin is the second most abundant natural polymer after cellulose and accounts for 15-30% of biomass. Lignin is also abundantly produced by paper, pulping industries and biorefineries as a low value by-product. In fact, the phenyl-propane-based lignin polymer can be a potential source of aromatic hydrocarbons. Nevertheless, lignin is also known for its recalcitrant for thermal decomposition and catalytic deoxygenation due to its structural complexity and high char and coke yields. In this work, catalytic hydropyrolysis of lignin and its model compounds was studied using several different types of bifunctional zeolite catalysts at atmospheric pressure. Three milled lignins were derived from red oak (hardwood), loblolly pine (softwood) and corn stover (herbaceous) to represent the natural lignins in different biomass species. Phenol and guaiacol were used to represent lignin-derived monomers. Catalysts tested are HZSM-5, Ni/ZSM-5, MoO_3/ZSM-5, Fe-HBeta, Ru/Fe-HBeta and Pd/Fe-HBeta, respectively. It was found that the hydrogen environment effectively increased the hydrocarbon yields from the lignins compared to the helium environment for all the catalysts. The catalytic coke with dramatically reduced yields and the aromatic hydrocarbons with an increased selectivity for benzene, toluene and xylene, were observed with the modified HZSM-5 catalysts. The model compound study shows that MoO_3/ZSM-5 and Pd/Fe-HBeta effectively hydrodeoxygenate phenolic monomers, whereas Ru/Fe-HBeta and Ni/ZSM-5 exhibit hydrocracking ability.

460000 Inhibition Effect during Catalyzed Biomass Char Gasification in Steam and Carbon Dioxide and Its Reversal

Abstract:

Gasification is a chemical process in which a gasifying agent reacts with carbonaceous materials to produce gases including high value CO and H_2. Gasification of renewable resources is a promising and environmentally friendly solution to the increasing global demand for energy and chemicals. The process takes place at high temperatures where biomass undergoes pyrolysis to rapidly form the biomass char. The char continuously reacts with the gasifying agent as the rate limiting step in the process. The gas-phase composition influences the gasification rate and is one of many variables that can be used for process improvement. Observations of gasification with low cost gases, steam (H_2O) and carbon dioxide (CO_2), have been reported extensively. The gasification reactivity behavior appears to vary among reports showing either the presence or lack of a synergistic effect for mixtures of these gases. The difficulty of predicting or explaining the optimum gasification rate stems from a lack of understanding the fundamental gasification mechanism in mixtures of steam and CO_2.

This study focused on understanding the mechanism in the evolution of gasification reactivity during conversion of char with CO_2, steam, and their mixtures. Bagasse char as starting material was prepared by pyrolysis of potassium-rich Brazilian bagasse in an inert gas pressurized entrained flow reactor. The char was then exposed to varying gasifying environments in a thermogravimetric analyzer to undergo full conversion. The initial gasification reactivity was higher in steam than in CO_2. However, overall and final reactivities were higher in CO_2 than in steam. Gasification inhibition was observed as a continuously decreasing reactivity with conversion with steam that was not observed with CO_2. On the other hand, CO_2 gasification rate showed a continuous increase with char conversion. The underlying mechanism was explored further by studying the transient behavior under exposure to step changes in

model gas compositions of CO_2, steam, H_2, and O_2. It is found that the active sites for both CO_2 and steam gasification are initially the same. However, as the char gasification progresses in steam, the active sites are likely to be blocked by in-situ hydrogen product formation (â€œinhibitionâ€), which does not happen during CO_2 gasification. On the contrary, concentration of active sites is increased as the char gasification progresses in CO_2 due to the increasing K/C ratio. In this study, several methods were demonstrated for the reversal of the inhibition effect. A generalized mechanism, consistent with experimental observations, is proposed. The findings provide a key insight into the source of the elusive synergistic effect during gasification with steam and CO_2.

453986 Acetone Condensation over Ion-Exchanged Hydroxyapatite Catalysts

Abstract:

Biomass can be used to produce platform molecules such as ethanol, acetone, and hydroxymethylfurfural, but further processing is needed to convert renewable feedstocks into useful products. One promising solution is bimolecular coupling of alcohols or aldehydes to form higher-chained commodity chemicals and transportation fuels. Recent reports in literature have shown that hydroxyapatite (HAP; $Ca_5(PO_4)_3OH$) is efficient at catalyzing a variety of coupling reactions, such as aldol condensation, Knoevenagel condensation, and Guerbet coupling; however, the mechanism and site requirements for these reactions remain a point of debate. [1,2]

In this work, we investigate the gas-phase condensation of acetone to mesityl oxide over HAP as a model reaction. Using a post-synthesis ion-exchange method, a series of divalent cation-substituted HAP catalysts were prepared to understand how different elements influence reaction rates. Initial characterization by XRD and BET suggest that the surfaces of exchanged-HAP catalysts are enriched with the introduced dopant while retaining an apatitic structure. To overcome equilibrium limiations, catalysts were tested in a fixed bed reactor using a physical mixture of HAP and Pd/SiO_2 to hydrogenate unsaturated products with H_2. Our results show that the nature of the cation in HAP plays a key role in governing catalytic activity. Isotopic labeling and KIE experiments reveal that enolate formation and C-C coupling are fast and reversible under the range of reaction conditions tested. The rate-limiting step occurs after the formation of diacetone alcohol, which is an intermediate in the reaction. A relationship between key descriptors of HAP structure and microkinetic parameters of acetone condensation will be discussed.

[1] Hanspal, S.; Young, Z. D.; Shou, H.; Davis, R. J. *ACS Catal.* **2015**, 5, 1737â˘'1746.

[2] Ho, C. R.; Shylesh, S.; Bell, A. T. *ACS Catal.* **2016**, 6, 939â˘'948.

461524 U.S. Power System Outlook: A Range of Perspectives from a Suite of Standard Scenarios

Abstract:

U.S. Power System Outlook: A Range of Perspectives from a Suite of Standard Scenarios

Wesley J. Cole and James Richards

Strategic Energy Analysis Center, National Renewable Energy Laboratory, Golden, CO, USA

The U.S. electricity system has undergone significant change over the last decade with declines in coal powered generation and increases in natural gas and renewable energy generation [1]. Many of these changes were largely unexpected. For example, before the shale gas revolution, natural gas prices were expected to be much higher than they are today [2], and solar photovoltaic capital costs have declined much more rapidly that most anticipated (see e.g., [3] vs. [4]). While continued regulation, such as the Clean Power Plan, and expected cost declines in renewable energy costs, may support a continuation of these trends, there is still tremendous uncertainty in terms of what a future power sector will look like. This work aims to identify a range of possible future electric system generation mixes via scenario analyses of some key sensitivities in the power sector. We present scenario results from the National Renewable Energy Laboratory's suite of Standard Scenarios, coupled with future technology cost analysis, to better understand how the power sector might evolve under various futures. Specifically, we consider electricity sector evolution with various natural gas price trajectories, technology costs, and policy drivers. We show how the capacity mix and resulting CO_2 emissions change as different assumptions are applied. The range of outcomes of these possible futures can provide insight into which drivers can have substantial impact on how the system develops.

References

[1] J. Logan, K. B. Medlock III, and W. C. Boyd, "A Review of Sector and Regional Trends in US Electricity Markets: Focus on Natural Gas," National Renewable Energy Laboratory, Golden, CO, NREL/TP-6A50-64652, 2015.

[2] EIA, "Annual Energy Outlook 2010," U.S. DOE Energy Information Administration, Washington, DC, DOE/EIA-0383(2010), 2010.

[3] Black & Veatch, "Cost and Performance Data for Power Generation Technologies," Black & Veatch Corporation, Overland Park, KS, 2012.

[4] NREL, "NREL Annual Technology Baseline (ATB)," 2015. [Online]. Available: http://www.nrel.gov/analysis/data_tech_baseline.html. [Accessed: 21-Apr-2015].

457798 Economics and Dynamic Flexibility of Concentrated Solar Power Technologies

Abstract:

Most techno-economic studies of concentrated solar power (CSP) systems only consider levelized cost of electricity (LCOE) to quantitatively compare economics and performance. This is problematic as LCOE neglects the time varying value of electricity and additional revenue opportunities from providing ancillary services. As consequence, LCOE-centric analyses undervalue the flexibility provided by thermal energy storage and misrepresent the true economics of owning and operating CSP systems [1, 2]. For example, a CSP plant providing 10 MW of regulation capacity (an ancillary service) for all hours of 2015 in the California energy market would have received $500,000 in capacity payments alone. Similarly, shifting 10 MW_e of generation from the average price (30 $/MWh) to the 1% most extreme prices (97 to 1,621 $/MWh) yields additional revenues of $400,000/yr. Thus selection of the appropriate economic and performance metrics is paramount when evaluating CSP systems and comparing against other energy technologies.

Beyond economics, cooling system water usage is increasing in importance. CSP systems, in particular, operate in water-sparse areas and thus water conservation is priority if the technology is to be competitive. Cooling systems influence overall system efficiency both due to primary thermal (Rankine cycle) efficiency impacts and parasitic loading associated with cooling system operation, which in turn impact system economics. Most CSP optimization literature, however, focuses only on economic metrics and does not consider water usage.

Compressive analysis of market-based revenue potential and water usage for CSP systems is especially complex because it depends on the selected operating policy. In several studies, revenue estimation is formulated as an optimization problem: manipulate the operational policy (e.g., mass and energy flows) to maximize market revenue subject to physical constraints [3 - 5]. Broadly speaking, CSP operational methods may be classified into two categories: in revenue-focused studies (e.g., [3 - 5]), emphasis is placed on capturing market rules and start-up and shutdown physical restrictions (e.g., minimum up/down times). Simplified correlations (e.g., piecewise linear approximations) are commonly used to model the solar collector and Rankine cycle, and 1 hour time discetizations are considered. In contrast, control focused studies (e.g., [6 - 8]) use much finer time discretizations (e.g., minutes) and consider nonlinear (partial) differential algebraic equation models to capture CSP system physics. These studies neglect market rules and start-up/shutdown limitations, and typically seek to maximize solar energy production or similar metrics. As such, neither of these approaches are capable of analyzing revenue opportunities available at multiple timescales from day-ahead (hour frequency) and real-time markets (minutes frequency).

We present a unified framework that considers market rules, start-up/shutdown restrictions and detailed system physics to analyze multiple CSP performance metrics, especially profits and water usage. The electricity market mathematical model is sufficiently detailed to evaluate revenues from selling/providing both energy and ancillary services, including regulation and (non)-spinning reserve capacity, in the California day-ahead and real-time markets. The CSP physics models include mass and energy balances and nonlinear equipment performance correlations. They are sufficiently complex to resolve temperature profiles in the thermal storage tanks and study the impact of storage dynamics on system performance. Revenue estimation and operational policy determination are formulated as mixed integer nonlinear programs (optimization problems). We present a decomposition strategy that uses physics inspired relaxations to efficiently find approximate solutions that are feasible for all of constraints and locally optimal with respect to a fixed start-up/shut-down schedule. The framework is implemented in Julia/JuMP (an open-source programming language/modeling environment) and leverages large-scale and efficient numerical optimization solvers such as IPOPT and Gurobi.

References:

[1] P. L. Joskow, Comparing the cost of intermittent and dispatchable generation technologies, American Economic Review: Papers & Proceedings 100 (3) (2011) 238-241.

[2] E. Lizarraga-Garcia, A. Ghobeity, M. Totten, A. Mitsos, Optimal operation of a solar-thermal power plant with energy storage and electricity buy-back from grid, Energy 51 (2013) 61-70.

[3] S. H. Madaeni, R. Sioshansi, P. Denholm, How thermal energy storage enhances the economic viability of concentrating solar power, Proceedings of the IEEE 100 (2) (2012) 335-347.

[4] J. L. Gonzalez, I. Dimoulkas, M. Amelin, Operation Planning of a CSP Plant in the Spanish Day-ahead Electricity Market, in: 11th International Conference on the European Energy Market (EEM), 2014, pp. 1-5.

[5] J. Usaola, Operation of concentrating solar power plants with storage in spot electricity markets, IET Renewable Power Generation 6 (1) (2012) 59 - 66.

[6] K. M. Powell, T. F. Edgar, Modeling and control of a solar thermal power plant with thermal energy storage, Chemical Engineering Science 71 (2012) 138 - 145.

[7] K. M. Powell, J. D. Hedengren, T. F. Edgar, Dynamic optimization of a hybrid solar thermal and fossil fuel system, Solar Energy 108 (2014) 210-218.

[8] E. F. Camacho, A. J. Gallego, Optimal operation in solar trough plants: A case study, Solar Energy 95 (2013) 106-117.

460426 Leveraging Storage and Hybridization to Maximize Renewable Utilization

Abstract:

In the pursuit for renewable power generation, a sentiment of competition between renewables and fossil fuels has emerged among much of the population. Renewables are carbon free and sustainable, but lack the reliability of fossil fuels, which can be dispatched on demand. Hybridization is often overlooked as at least an interim solution. By hybridizing a system, the benefits of each type of technology is realized, with the emissions benefits of renewable generation and the reliability of a fossil fuel system. In this work, a hybrid solar thermal and natural gas concept is explored. A central receiver solar concentrating system is used to pre-heat compressed air before it enters a conventional natural gas combined cycle power generation system. High temperature packed-bed thermal energy storage (pre-gas combustion) is used to increase the solar share of the plant. With the plant operating in hybrid mode, the flexibility of the storage and the supplemental natural gas are used in an optimization-based control scheme to maximize the amount of solar energy collected. This configuration demonstrates that by using a fossil fuel with a variable source of renewable energy, a certain synergy is achieved, with the system flexibility being leveraged to maximize the amount of solar energy harvested. The study presented is based on experimental and simulation results and highlights many of the feasibility challenges associated with achieving such a concept at grid scale.

463938 State of Health Estimation Method Design for Energy Storage System of Lithium Ion Battery and Comparative Study : From Cell to Demonstration Sites

Abstract:

State of health Estimation method design for Energy Storage System of Lithium ion battery and comparative study : from cell to demonstration sites

Keonhee Park, Ph.D. candidate
Seoul National University

Recently, Increasing demands of renewable energy and necessity of stable electricity supply are accelerating ESS(Energy storage system) market. ESS, which can stores the power and supplies electricity when it is needed, is possible to increase efficiency of electricity-usage and still has converted from production-consumption paradigm of electricity system to production-storage-consumption paradigm of that so far. ESS is comprehensive system including EMS(Energy Management system), BMS(Battery management system), PCS(Power conversion system). BMS has important role for sustainable ESS operation. Main state variables of BMS are SOC(State of charge) and SOH(State of health). SOC means current residual capacity based on present nominal capacity. And SOH is a prominent indicator to determine degradation of battery. Performance of repeated cycled battery decreased and it should be replaced at an appropriate time for usability. Imprecise SOH can affected SOC accuracy which is important variable for on-line monitoring. However, SOH accurate estimation is challenging due to few obtainable output variables for the battery system and the lack of understanding for electrochemical phenomena in cell. On-line SOH estimation method is needed since a lot of ESS application sites are isolated. This study proposes the modified online SOH estimation method, and compares estimation results of related SOH estimation methods. The results are validated with

experimental cell data and demonstration operation data. This method uses the equivalent circuit model parameters and recursive least square method. The cell experimental data is given by Korea testing certification(KTC) and demonstration data is obtained by 2 sites, which are interconnected with PV power system. The one is installed in micro-grid system in the island and the other is integrated with plant site. From the test results show maximum 2% accuracy in cell data and 5% accuracy in demonstration data. The accuracy of proposed method is dependent on input data quality highly. Also we can estimate the life time of battery system and implement easily without complex change of BMS algorithm.

464641 The Effect of Centralized Energy Storage on an Integrated Distributed Photovoltaics/CHP System for District Power, Heating and Cooling

Abstract:

The rise in availability of Distributed Energy Resources (DERs) has spurred utility companies and grid operators to search for methods to incorporate these resources (i.e., photovoltaics and energy storage) in the energy infrastructure in a manner that keeps the customers connected to the grid without jeopardizing grid stability and reliability. In the past decade, the explosive spread of photovoltaic generation (both rooftop and centralized) has lowered energy demand from conventional generation methods (i.e., power plants) during the afternoon. The result is the famous "duck curve" for conventional electricity demand, with medium demand in the early morning, low demand in the afternoon (when the sun is shining), and a sharp increase to high demand in the evening when the sun sets [1]. While the growth trend for solar photovoltaic generation has important environmental benefits, it can –if not properly managed – lead to an over-generation situation. In this case, demand during the day can be satisfied largely using solar power; however, the increased evening demand, corroborated with a natural drop in solar photovoltaic generation, may require very high power plant ramp rates. Most current power plants are unlikely to be able to withstand such rapid changes in generation rate. Motivated by this, significant research efforts have been expended towards defining strategies for controlling photovoltaic generation and grid-level storage systems. However, there are very few published works exploring the influence of photovoltaics and energy storage on the *design* of future power generation plants.

In our previous work [2], we provided a framework for simultaneous optimization of design and operation strategy for a CHP plant, providing the ability to select equipment sizes as well as plant operation modes to meet the projected electricity, heating, and cooling demand of a residential neighborhood. Based on realistic data collected from Pecan Street Research Inc., our results revealed that in the seasons with low electricity and cooling demand (winter and spring, considering a predominantly cooling climate), rooftop photovoltaic generation may exceed the neighborhood demand, while in the warmer seasons, the cooling capacity can often exceed the system's needs.

Motivated by the above, in this paper, we study the same system operating in an islanded (i.e., grid-disconnected) mode, and consider the impact of a centralized energy storage facility on system design and operations. We discuss an extension of our formulation for simultaneously optimizing the design and operation of the CHP plant and the centralized energy storage to minimize the capital and marginal costs of the CHP plant including transition penalties accrued by the equipment turning on and off. Using the aforementioned data collected from a neighborhood located in Austin, TX, we investigate the benefits of energy storage relative to the level of photovoltaic integration from the neighborhood, as well as identify economic benefits from photovoltaic integration and energy storage in terms of plant savings.

References:

[1] CAISO (2016). *What the duck curve tells us about managing a green grid.* https://www.caiso.com/Documents/FlexibleResourcesHelpRenewables_FastFacts.pdf

[2] Ondeck, A.D., Baldea, M., Edgar, T.F. *Simultaneous Optimization of Design and Operation Strategies for CHP Systems.* Computers & Chemical Engineering. Manuscript in preparation.

468128 Reducing the Cost of Operational Water in Military Water Systems: A Data-Driven Modeling and Optimization Study

Abstract:

Reducing the Cost of Operational Water in Military Water Systems: A Data-Driven Modeling and Optimization Study

Corey James, Michael Webber and Thomas Edgar

Department of Chemical Engineering

The University of Texas at Austin, 1 University Station C0400, Austin, TX 78712

email: cmjames@utexas.edu

Military water systems are tasked to provide safe and clean water to support soldiers, their families and mission related activities. Due to the historic low price and relative abundance of water, there has been little motivation to ensure that water systems minimize water waste after treatment. Currently, water is treated for disinfection as needed to meet demand. The concentrations of common disinfecting agents (chloramines or chlorine) begin to exponentially decay the moment they are placed in the water [1]. A system that treats water without rigorously considering unique demand data and the temperature dependent nature of chloramine decay will inevitably waste water. Waste is created when excess water is treated and placed into the system that exceeds demand. The volume that exceeds demand resides in the system long enough to run out of "lifetime", where the concentration of chlorine in the treated water decays below the legal safe threshold. This water must be flushed, or wasted, in order to ensure a safe water source. Embedded in the water waste is the energy and cost required to pump and disinfect that volume of water. A system that is not optimized to account for chlorine decay could cause thousands of gallons of water wasted on a monthly basis, not to mention the embedded energy and cost.

Approximately 300 million gallons of potable water was wasted in 2013 due to routine maintenance in Texas' 10 largest cities [2]. Water wasted due to decayed chlorine levels is unknown, but likely similar. Lack of optimization of an appropriate objective function in water systems is the main cause of wasted water and energy related to lifetime. Military installations have unique water requirements with regard to emergencies, base resiliency and fires. Volume and pumping constraints within the model are vital to reducing the water and power consumed. Awareness of water shortages, installation of water conserving fixtures/appliances and an increase in stewardship over the past decade created an overall reduction (~13%) in per capita and overall potable water consumption [3]. Unfortunately, the reduction in consumption creates a new unintended consequence for the water system. Water systems are not optimized for the new reduced flow and further waste is created when excess treated water that reaches its disinfectant lifetime is purged. A substantial amount of studies have examined optimizing both residential and commercial water consumption networks, but rigorous studies optimizing water use and related power consumption on military installations are not available .

Results on the optimal production of treated water to minimize waste and energy consumption (pumping costs) while meeting demand and quality constraints utilize data from Fort Carson, Colorado and The Pecan Street Project. The unique demand data and constraints of a typical military system are used to create a time-resolved water profile for an installation, which is then used to examine the fully burdened cost, energy consumption and waste of less than optimal water systems through rigorous multi-tiered optimization and control. Financial, social, economic and environmental impacts are analyzed as part of the fully burdened cost.

References

[1] Peter J. Vikesland, Kenan Ozekin and Richard L. Valentine (2001). "Monochloramine Decay in Model and Distribution System Waters". *Water Resources* 35(7): 1766-1776. <>[2] Amarillo Globe-News, "Cities wasted nearly 300M gallons of water." 4-July-2014.

[3] M.A. Maupin, J.F. Kenny, S.S. Hutson, J.K. Lovelace, N.L. Barber and K.S. Linsey (2014). "Estimated use of water in the United States in 2010". *U.S. Geological Survey Circular* 1405(56).

469581 Energy Demand Management in Process Systems Subject to Time-Varying Electricity Prices: A Decomposition-Based Approach

Abstract:

As the availability and pricing of electricity vary during the day due to intermittency in renewable resources or real-time electricity tariff policies, the chemical industry, especially those with power-intensive processes, needs to rethink operating strategies to maximize economic viability. The implementation of demand response (DR) strategies, which focuses on demand side activities and has been in the center of all pragmatic energy policy decisions, is the key to minimizing production costs and maximizing profits. The economic potential of DR for industrial processes has been acknowledged in numerous studies (see, for example, [1]-[3]).

In developing DR strategies, the target lies in scheduling the production rates, production times and inventory levels at each time step to achieve specific objectives. Therefore, process control-related activities are closely linked to the responsive demand augmentations given that control actions involve real-time manipulation of selected process variables to hold key product quality and production rates near their desired target values. Moreover, as the plant operating schedule varies, process control facilitates the dynamic transition periods between different operating modes. Therefore, it becomes imperative to develop design and operating methodologies that can effectively integrate production scheduling and control functions (e.g., see [4]-[5] for some recent comprehensive reviews of the literature on the integration of scheduling and control.

In a prior work [8], we posed the DR problem through a mixed-integer nonlinear programming (MINLP) formulation where optimal operating policies (modes) were found in response to varying electricity prices and intermittent on-site renewable generation. A key assumption in that work was that the control policies were predetermined and no attempt was made to optimize the transition periods. By relaxing that assumption, the goal of the present work is to offer a new DR formulation for the integration of scheduling and control subject to time-varying electricity prices. The formulation simultaneously takes into account the production scheduling with respect to minimizing production cost, the optimization of controller parameters to realize the objective of reducing the time and lost product during transitions, and the integration of scheduling and control to achieve a more economical operating policy. Instead of formulating one complex and unsolvable MINLP problem, we propose a novel decomposition scheme that generates smaller sub-problems that can often be solved more easily. The proposed strategy is implemented within a receding horizon optimization framework, in which only the first transition is carried out with optimally tuned controller parameters and a subsequent scheduling problem is then reformulated with the new updated storage level corresponding to the optimized transition time. A case study involving a continuous chemical reactor is used to demonstrate the contributions and efficacy of the developed approach.

References:

[1] M. Paulus and F. Borggrefe, "The potential of demand-side management in energy-intensive industries for electricity markets in Germany," *Applied Energy*, **47**, 1492-1503, 2011.

[2] S. Mitra, I. E. Grossmann, J. M. Pinto and N. Arora, "Optimal production planning under time-sensitive electricity prices for continuous power-intensive processes," *Comput. Chem. Eng.*, **38**, 171-184, 2012.

[3] Y. Wang and L. Li, "Time-of-use based electricity demand response for sustainable manufacturing systems," *Energy*, **63**, 233-244, 2013.

[4] S. Engell S and I. Harjunkoski, "Optimal operation: Scheduling, advanced control and their integration," *Comput. Chem. Eng.*, **47**, 121-133, 2012.

[5] M. Baldea and I. Harjunkoski, "Integrated production scheduling and process control: A systematic review," *Comput. Chem. Eng.*, **71**, 377-390, 2014.

[6] C. Tong, A. Palazoglu, N. H. El-Farra and X. Yan, "Energy demand management for process systems through production scheduling and control," *AIChE Journal*, **61**, 3756-3769, 2015.

473688 New World of Opportunity Crudes

Abstract:

The refining industry has changed noticeably over the past several years with an ample supply of opportunity crude oils available, which include light tight oils and heavy sour crudes, and sometimes high TAN crudes. Opportunity crudes have been around for years, but only until recently has the abundance of these discounted crudes shifted the behavior of refiners.

With the use of opportunity crude oils comes new processing challenges; Automation technology is advancing to meet these challenges to ensure refiners are successfully alleviating these new issues. Advances in real time crude analysis and crude unit monitoring and control allow refiners to maximize opportunity crude processing while mitigating issues like accelerated fouling and corrosion.

Refiners on the coast will typically have access to more opportunity crudes than those landlocked. Because crude oils can be very different in properties, refiners try to match the required crude oil composition for their refinery configuration; usually by blending two or more crude oils. To make matters more challenging, sometimes crude oils purchased to process are crudes that the refiner has no experience processing.

Crude blending operation has become more important with the ability to establish consistent crude feed properties. Steady operation enables better optimization of the crude unit by pushing against the appropriate constraints which often includes maximizing the quantity of discounted opportunity crude oils blended.

This presentation will highlight on new operating challenges, and show a couple examples where refiners have made modifications in both technology and behavior to adapt to the new world of opportunity crudes.

463572 Hydrocarbon Mixture Bubble Points in Nanoporous Silica Monoliths

Abstract:

Shale reservoirs are now becoming the focus of new oil and gas exploration and development. These shale reservoirs usually have lower permeability and porosity with a pore size distribution ranging between 1-20 nm. It is hypothesized that fluids in the confined systems (shale reservoirs) have different thermo-dynamic properties than conventional reservoirs. However, because of the difficulty of the realization of the pore structure of shales, there have been few experimental studies. In my research, two different nanoporous silica based monoliths were synthesized with similar pore size ranges as the shale reservoirs, represented in the porous system of shale reservoirs. Monoliths can be synthesized into a desired shape by putting them in a customized container. Because of this advantage, the synthesized monoliths have only nano-sized pores, not macropores. Two nanoporous monoliths were synthesized by modifying the synthesizing procedures that could be used to synthesize hexagonal and cubic mesoporous silica powders, which have different morphologies. From comparing the saturation pressure with and without nanoporous media, we can confirm the hypothesis that nano-sized pores have an effect on saturation pressure. Also, we can see the different morphologies of the two monoliths have an effect on saturation pressure. These synthesized monoliths have been characterized by X-ray diffraction (XRD), nitrogen adsorption/desorption isotherm (BET), and transmission electron microscopy (TEM), which show their physicochemical properties. The phase diagram of methane and decane mixture has been modeled with the Winprop module from the CMG reservoir simulator. The experimentally measured saturation pressure matches with the simulated result. However, the measured saturation pressure changes with nanoporous media.

455711 Forecasting Gas-Oil Ratios and Solution Gas Production from Liquid Rich Shale Reservoirs

Abstract:

The basic issue: To enhance our knowledge about the production mechanisms of shale oil reservoirs and better exploit their economic potential, it is important to be able to forecast production of the secondary phase, namely gas.

Why it is important: Stakeholders often focus on oil forecasts when it comes to shale oil reservoirs, ignoring the equally important solution gas produced from these plays. Lack of production data, complicated flow mechanisms of liquid rich shale reservoirs, production pattern of producing gas-oil ratios among other factors, make the task of forecasting solution gas production difficult.

What we are proposing: The objective of this paper is to present a new method for estimating producing gas-oil ratios (GOR) and solution gas production from shale volatile oil reservoirs. The method is data driven while observing the fundamental equations that govern hydrocarbon production over time. It entails the following steps:

Step 1 – Generate representative collection of data: To develop the functional form that describes the evolution of GOR over time for a producing well, run a large number of representative simulations over a fairly long production time (e.g. 30 years) so that a variety of wells in various reservoirs under diverse producing conditions can be well represented in the sample.

Step 2 – Identify natural GOR shapes through PCA: Stack GOR profiles on top of each other, to construct a matrix of dimension perform principal component analysis (PCA) on , and then approximate as , where . The vectors can be thought of as natural shapes over time, whose linear combination can reproduce – with reasonable approximation – every conceivable GOR profile over time.

Note:Steps 1 and 2 will have to be completed perhaps only once or at worst a limited number of times, since the natural GOR shapes they produce will be universal.

Step 3 – Fit a linear combination of natural GOR shapes to data over limited time: Given GOR data for a limited production history over time , identify estimates by finding .

Step 4 – Use the model of Step 3 for future predictions: The value of GOR at a future time along with a corresponding confidence interval can be estimated using standard formulas as where is the information matrix.

Case study: A commercial compositional simulator was used to simulate 30 years of production from a shale volatile oil reservoir with 4 different reservoir fluids. The reservoir permeability and porosity were 0.001 md and 6% respectively. Several values for the following parameters were considered in a total of the simulations:

1. 1. Critical gas saturations between 5% and 20%;
2. 2. Constant bottomhole pressures (BHP) between 500 and 1500 psi; and
3. 3. Initial reservoir pressures between 3000 and 5000 psi.

PCA on the GOR data produced by these simulations revealed (through cross-validation) that a small number of principal components adequately capture the GOR profiles observed over the 30-year period. To test how well the resulting model can forecast future production from production data over limited time, we considered a horizontal well of length 5000 ft, with 20 hydraulic fractures of 150 ft half-length, spaced 250 ft apart. We simulated producing GOR over 30 years, fit the model to the corresponding simulated data over periods from 6 months to 3 years, and then compared GOR forecasts by the model to GOR values produced by the simulator into the future. The future profiles produced by the PCA model matched those produced by the simulator with reasonable level of accuracy. The PCA model was also found to match the simulator on solution gas production, calculated by integrating producing GOR over cumulative oil production.

Conclusions: This work presents a novel approach of forecasting production from liquid rich shale plays, based on a well established statistical approach. It bypasses the need to invent an ad hoc functional form for GOR over time that would be both simple for computation and realistic enough to capture real data. The proposed approach can help improve reserves estimation, reservoir management, and project economics.

467020 Solvent Extraction for Bitumen Recovery from Oil Sands Ores

Abstract:

Unconventional oils, including oil/tar sands, shale oil, heavy oil, etc., have been considered as an alternative of traditional crude oils due to their huge reserves (about two thirds of the total oils) buried around the world. It is estimated that there are about 1700 trillion barrels of bitumen contained in the oil sands ores in Alberta, Canada(Chew, 2014; Gray et al., 2009). To release this kind of petroleum, hot water-based extraction (HWBE) has been proposed and applied in industry which creates large volumes of wet tailings with fine solids. To avoid this environment pollution, the solvent extraction has been attracting increasing attention from industry due to its high extraction efficiency, less water and energy consumption, and high compatibility for different kinds of ores.

In the present study, solvent extraction of oil sands has been investigated using pure organic solvents, including toluene, tetrahydrofuran (THF), cyclohexane, n-heptane, n-hexane, ethyl acetate, and etc. Results of bottle tests showed that the bitumen recovery increased with the increase of the number of carbon and the cyclic hydrocarbons (toluene, THF and cyclohexane) performed better than alkane hydrocarbons (n-heptane, n-hexane, ethyl acetate and acetone). To analysis the bitumen in the extracted phase more detailedly, it has been divided into four fractions, named as saturates, aromatics, resins, and asphaltenes (SARA). Results show that increasing the aromaticity of the solvent is beneficial for the recovery of heavy fractions (e.g., resins and asphaltenes) of bitumen from the oil sands.

To further explore the solvent extraction process, solubility simulation in COSMOtherm software was applied and four typical molecules were selected as the representatives of SARA. A similar trend of bitumen recovery with different solvents was obtained as an order of: THF â‰¥ toluene > cyclohexane > n-heptane > n-hexane > ethyl acetate. Firstly, the efficiency of solvent extraction is found to be highly dependent on the polarity of the solvent. The solvents with too low (i.e. n-hexane and n-heptane) or too high (i.e. acetone and methanol) polarity had a poor solubility to bitumen. The polarity of the solvent will cause certain effect to the bitumen recovery because some special polar groups (i.e. groups containing oxygen, nitrogen, sulfur, and etc.) affected the properties of bitumen components. Secondly, the enhancement of solvent aromaticity makes the recovery of aromatics, resins and asphaltenes increase. The reason for these may be that the heavy bituminous constituents contain aromatic or cyclic structures. From the solubility point of view, the more similarity of microstructure between bitumen components and solvent, the more bitumen was recovered. Finally, after careful analysis, it is found that the dispersion force of the solvent plays a dominate role (weight > 55%) in dissolving bitumen fractions, followed by hydrogen-bonding force (weight at 0~26%) and polar force (weight at 0~18%). This finding allows us to consider more detailed influence factors when selecting the potential solvents for bitumen recovery process and solubility simulation may be a relatively efficient and accurate previous prediction of solvents screening.

Accordingly, the solvent extraction process would be a promising method for the recovery of unconventional oil from the ores. This study would improve the knowledge base for bitumen recovery mechanism and serve for future work on the solvent extraction technology applications.

References:

Chew, K.J., 2014. The future of oil: unconventional fossil fuels. Philos Trans A Math Phys Eng Sci 372, 20120324.

Gray, M., Xu, Z., Masliyah, J., 2009. Physics in the oil sands of Alberta. Phys. Today 62, 31-35.

464457 Short Residue to Useful Products: Different Pathways

Abstract:

With the depletion of the oil reserves, the oil is getting heavy day by day. The utilization of the heavy oil in the refinery process leads to the generation of high amount of short residue from the vacuum distillation units. The present study is about finding the different pathways for the utilization of the short residue to the formation of useful products. The conversion of short residue to useful product also helps in increasing the economic viability of the refinery processes. This research work initially studies about the quality of the various crudes available in the Indian petroleum industry. The different processes involved for the upgradation are categorized as Carbon Rejection Process and Hydrogen Addition Process. The present work further discussed on the selection of the specific process depending on the quality of the short residue left in the refinery.

452776 Production of Light Olefins from Protein-Rich Microalgae By Hydrothermal Liquefaction and Sequential Catalytic Cracking

Abstract:

In recent years, biomass resources have gained increasing attention as sustainable feedstock for production of energy fuels and chemicals [1]. Algae is a very promising source as the third-generation biomass, following after the first-generation biomass (edible crops, sugars, and starches) and the second-generation biomass (inedible lignocellulose). Algal biomass can be converted into biofuels such as biogas, bioethanol, biodiesel, and bio-oils by anaerobic digestion, fermentation, gasification, liquefaction, pyrolysis, and so on [2]. Although hydrothermal liquefaction (HTL) is considered the most promising method for conversion of wet microalgae to often called â€˜bio-oilâ€™, industrial viability of microalgae bio-oil production still has a lot of challenges to overcome [3]. In this study, we propose a unique method to produce light olefins such as ethylene, propylene, and butene from protein-rich microalgae.

Dry microalgae cells of Spirulina sp. was purchased from DIC LIFETEC Co., Ltd. (Tokyo, Japan). Spirulina sp. consisted of 4.7 wt% moisture, 6.1 wt% ash, 12.6 wt% carbohydrates, 7.8 wt% lipids, and 68.8 wt% protein. It contained 53.0 wt% C, 8.0 wt% H, 11.5 wt% N, 0.9 wt% S, and 26.6 wt% O (dry ash free). All other materials were purchased from commercial suppliers and used as received without any pretreatment or purification. Hydrothermal reaction experiments were carried out in a 100-mL hastelloy autoclave to obtain both water soluble and insoluble fractions. A fixed bed reaction system was used for catalytic cracking of water soluble fraction. The products were analysed using a gas chromatograph, high-performance liquid chromatographs, a total organic carbon analyser, a total nitrogen analyser, an elemental analyser, and etc. The yields of the products (mol%/C) were calculated based on the carbon content.

We demonstrated the hydrothermal liquefaction of *Spirulina sp.* at from 250 °C to 300 °C and sequential catalytic cracking using a HZSM-5 zeolite catalyst at 600 °C. We show that we obtained light olefins as major products along with aromatics and light alkanes as minor products in the presentation. The possible conversion pathways of carbohydrates, lipid, and protein to bio-oils in our system are also proposed. This research provides a useful production method of light olefins from protein-rich microalgae.

References

[1] a) J. N. Chheda, G. W. Huber and J. A. Dumesic, Angewandte Chemie-International Edition 2007, 46, 7164-7183; b) A. J. Ragauskas, C. K. Williams, B. H. Davison, G. Britovsek, J. Cairney, C. A. Eckert, W. J. Frederick, J. P. Hallett, D. J. Leak, C. L. Liotta, J. R. Mielenz, R. Murphy, R. Templer and T. Tschaplinski, Science 2006, 311, 484-489.

[2] L. Brennan and P. Owende, Renewable & Sustainable Energy Reviews 2010, 14, 557-577.

[3] a) D. López Barreiro, W. Prins, F. Ronsse and W. Brilman, Biomass and Bioenergy 2013, 53, 113-127; b) Y. Guo, T. Yeh, W. Song, D. Xu and S. Wang, Renewable and Sustainable Energy Reviews 2015, 48, 776-790.

454715 Selective Adsorption of Au (III) By Epichlorohydrin/Thiourea Modified Porous Alginate Beads

Abstract:

Selective adsorption of Au (III) by epichlorohydrin/thiourea modified porous alginate beads

Xiangpeng Gao,[a] Yan Zhang,[a*] Yuming Zhao[b]

[a] Department of Process Engineering, Memorial University of Newfoundland

[b] Department of Chemistry, Memorial University of Newfoundland

Abstract

Sodium alginate is an anionic polysaccharide distributed widely in the cell walls of brown algae, which can be extracted by treating algae with sodium carbonate solution and recovered by precipitation. Sodium alginate and its derivatives can be used as biosorbents to retain toxic metal ions, or recover precious metal ions from industrial wastewater. However, the dissolution of sodium alginate in aqueous solution has limited its application as a sorbent. Moreover, powder form of sodium alginate is not suitable to be packed in a column for applications at preparative scale because they may cause clogging problems, a huge pressure drop or serious hydrodynamic limitations. Therefore beads are regarded as the best form of biopolymer based sorbents for metal uptake from aqueous solutions. So far, most of the studies in biosorption area focused on the sorption of monometallic system. A major complication in the application of biosorption to industrial conditions is the presence of other metal ions in the wastewater to be treated. The effect of the co-existing metal ions on the selective adsorption of the meal of interest may be different under different scenarios. In this study, porous alginate beads were prepared by modifying epichlorohydrin and thiourea with sodium alginate, and then dropped into calcium chloride solution to form beads. The epichlorohydrin/thiourea modified alginate beads were soaked in tetraethyl orthosilicate (TEOS) sol and finally freeze dried to generate pores. Selective adsorption of Au (III) from mixed metal solutions under different acidic conditions was studied. Mathematical models were used to describe the adsorption isotherm and kinetics of Au(III). FT-IR, SEM-EDS, and XPS analyses were used to characterize the sorbent and identify the functional groups and potential biosorption mechanisms that contributed to the selective adsorption of Au (III).

Keywords: Selective adsorption; Alginate; Porous biosrobent

461820 Hydrothermal Carbonization of FOOD Wastes from LOCAL Eatery

Abstract:

Since 2014, it has been estimated that the world has produced nearly 2 billion tons of garbage also known as municipal solid waste (MSW) and it is further projected that this generated waste will be increased by nearly five times by 2050 due to large economic and urban growth of developing countries. In the U.S, a typical garbage disposal fee of $20-50 per month, varies from county-to-county, applies to every household. Garbage contains mainly food waste and the composition can add up to 70 %. Except from some cases, where food waste garbage has been composted or used as animal feed, it ends up in a landfill, where it contributes to environmental pollution via greenhouse gases (mainly methane and carbon dioxide). Large metropolitan cities like Los Angeles, New York etc. are facing serious waste disposal challenges with the lack of adequate lands. Although garbage has huge potential for energy production, converting garbage to energy in both efficiently and economically is yet to be discovered.

Hydrothermal carbonization (HTC) is an exceptional treatment process for biofuel production, as residual water is the reaction media for HTC. Therefore, the wet biomass need not be dried prior to the HTC. Subcritical water at around 200-260 °C has the highest ionic constant, which means it is quite reactive. As a result, within 5 minutes of reaction with biomass, solid-hydrophobic biofuel is formed along with liquid and gas byproducts. The solid product, also called hydrochar, is quite hydrophobic, friable, and more importantly energy dense; an increase in energy density means that less mass is required to produce the same amount of energy. Pelletization of the hydrochar allows for more energy densification as well as better storage of the fuel. Moreover, the exothermic HTC reaction ensures the minimum energy input for biofuel production. Researchers around the world have been developing HTC over the last decade for woody biomass and agricultural residues.

The main goal of this study is to apply HTC to local eatery wastes for producing clean and sustainable energy. Waste disposal issues will be automatically resolved with this proposed process. In this study, a 2L batch Parr reactor was used for HTC for recycle, compostable, and trash bins wastes at 200, 230, and 260 °C at their water saturation pressure. Solid hydrochar, HTC process liquid, and gaseous products were analyzed for their physicochemical properties. Based on batch HTC results, a simple techno-economic analysis was performed using ChemCAD 6.0 simulation. The simulation results indicate that higher HTC temperature is favorable for solid fuel production, however, such high-pressure and high-temperature system requires high capital investment.

Production of Jet Fuel and Butadiene from Lignocellulose and Lignocellulosic Platform Chemicals

Cui Xingkai, Zhao Xuebing*, Liu Dehua

Institute of Applied Chemistry, Department of Chemical engineering, Tsinghua University

Correspondence: zhaoxb@mail.tsinghua.edu.cn

Introduction

With the reduction of fossil energy reserves, increasing concerns of environmental pollutions and greenhouse gases emissions, more and more attentions have been payed to developing green and sustainable chemicals and fuels. Lignocellulose, which is generated numerously every year in nature, is a feasible alternative to fossil resources. Three strategies are usually applied to produce fuels from lignocellulose: (i) Pyrolysis to produce bio-oil and fuels are produced by bio-oil hydrogenation processing (ii) Gasification to produce syngas and then F-T Synthesis to get fuels; (iii) Pretreatment, hydrolysis and dehydration to generate platform chemicals, which can be converted to liquid fuels by bio-processing or chemical processing [1].

A process employing solid acid catalyst preparation, dehydration, C-C coupling techniques and hydrodeoxygenation (HDO) is being developed to convert lignocellulosic platform chemicals (including furfural, hydroxymethylfurfural (HMF) and 2,3-butanediol) to methyl ethyl ketone (MEK), butadiene and jet fuel. First, lignocellulosic materials, such as bagasse and straw, are converted to solid acid by Incomplete carbonization and H_2SO_4 sulfonation subsequently [2]. Next, the prepared lignocellulose based catalyst is used to catalyze 2,3-butanediol dehydration reaction that can produce methyl ethyl ketone (MEK) at low temperatures and butadiene at high temperatures. Further, by undergoing aldol condensation[3] and hydrodeoxygenation (HDO) subsequently, MEK and furfural (or HMF) can be converted to C9-C14 (or C10-C16) linear and branched hydrocarbons that meet the requirements of jet fuels. The aldol condensation kinetics are also investigated to make molecular size distributions in the final hydrocarbons controllable.

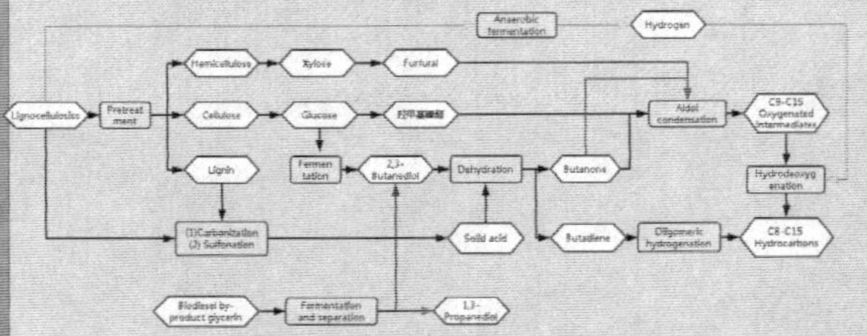

Figure 1. Research framework

Results and Discussions

Preparation of lignocellulosic-based solid acid

Sugarcane bagasse and lignocellulosic residues after pretreatment, simultaneous saccharification and fermentationis (SSF) are used as feedstocks for solid acid preparation. Firstly, the feedstocks are incompletely carbonized with protection of nitrogen. Secondly, materials from the first step are sulfonated by concentrated sulfuric acid.

Table 1. Changes in the composition at different carbonization temperatures

(Carbonization time: 0.5h, nitrogen flow rate: 200mL / min)

Temperature(℃)	285	315	345	375	405	435	465
Mass before carbonization(g)	10.72	12.54	11.28	10.23	11.46	11.77	12.12
Mass after carbonization(g)	7.91	7.14	4.45	3.19	3.21	3.1	2.9
The percentage reduction in mass(%)	26.21	43.06	60.55	68.82	71.99	73.66	76.07
Cellulose content (w/w%)	38.8	47	36.5	1.38	1.29	1.09	1.23
Hemicellulose content(w/w%)	27.5	10.8	3.14	0	0	0	0

- At lower temperatures, hemicellulose removal is dominant
- When the temperature is higher than 345 ℃, almost all the cellulose and hemicellulose is removed.

Dehydration of 2,3-butanediol to produce methyl ethyl ketone

Dehydration reaction is carried out in a round-bottomed flask with magnetic stirring, methyl ethyl ketone is separated online with a condenser. Bagasse-based solid acid is used as catalyst. Results are displayed in table 2.

In order to to increase yield of methyl ethyl ketone, lower carbonization temperature, longer carbonization time and shorter sulfonation time should be adopted. For bagasse raw material, optimal conditions for preparing the solid acid is: carbonization temperature 345 ℃, carbonization time 1.5h and Sulfonation time 20min. In addition, the catalyst exhibit good reuse stability. 82% activity remains after 4 reuses.

Table 2. Different conditions for preparing solid acid and comparison of their catalytic performance

(Reaction conditions: 2.5g catalyst, 50g 2,3-butanediol, reaction temperature is 168 ℃, reaction time 6h, stirring speed 300rpm)

Carbonization temperature(℃)	Carbonization time(h)	Sulfonation time(h)	Conversion of 2,3-butanediol(%)	Selectivity of MEK(%)	Yield of MEK (%)
315	0.5	8	61.97	50.89	31.53
345	0.5	8	63.82	73.37	46.83
375	0.5	8	67.81	60.17	40.80
405	0.5	8	69.33	54.00	37.44
435	0.5	8	66.29	53.76	33.98
345	0.17	8	56.60	77.61	43.94
345	1	8	63.06	73.42	46.30
345	1.5	8	61.78	81.82	50.56
345	1.5	0.33	68.55	76.47	52.42
345	1.5	0.5	60.44	81.83	49.46
345	1.5	1	55.24	86.12	47.57
345	1.5	3	56.34	84.10	47.38

Aldol condensation of methyl ethyl ketone and furfural

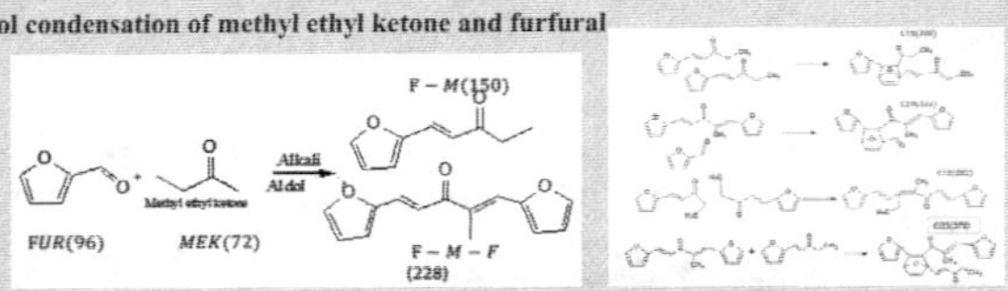

Figure 2. Mechanism of aldol condensation and side reactions

Aldol condensation is carried as two-phase reaction, the organic phase take tetrahydrofuran as solvent, furfural and methyl ethyl ketone are dissolved; the aqueous phase is aqueous sodium hydroxide solution that serves as catalyst; reaction is carried out in a shaker at 300 rpm, the reaction time was 24h, samples taken are analyzed by gas chromatography.

Table 3. Aldol condensation results of furfural and methyl ethyl ketone

Mass ratio of organic to aqueous phase	Furfural in organic phase (w%)	Molar ratio of furfural to MEK	Molar ratio of NaOH to furfural	Reaction temperature(℃)	8h				24h			
					Conversion of furfural (%)	Selectivity of F-M(%)	Selectivity of F-M-F(%)	Selectivity of heavy condensates (%)	Conversion of furfural (%)	Selectivity of F-M(%)	Selectivity of F-M-F(%)	Selectivity of heavy condensates (%)
1:1	13	2:1	2.3	25	65.64	75.36	21.21	3.43	79.00	47.93	40.45	11.61
1:1	13	1:1	2.3	25	89.02	86.91	10.07	3.03	96.09	78.80	13.52	7.68
1:1	13	2:5	2.3	25	94.16	98.01	1.99	0	99.43	92.0	3.60	4.41
1:1	13	1:1	1	25	96.48	85.06	12.48	2.46	99.64	81.81	14.12	4.06
1:1	20	1:1	2.3	25	89.31	80.90	9.82	9.28	96.44	66.01	12.67	21.32
3:1	13	1:1	1	25	84.31	88.49	7.03	4.47	88.97	78.75	9.22	12.03
1:1	13	1:1	1	40	99.27	84.91	12.37	2.71	100	82.28	12.04	5.68
1:1	13	1:1	1	55	100	79.75	15.35	4.90	100	75.47	14.47	10.06

To improve the conversion of furfural and selectivity of F-M and F-M-F, and control selectivity of heavy condensates, following strategies need to be adopted: lower mass ratio of organic to aqueous phase, lower furfural fraction in organic phase, lower molar ratio of furfural to MEK and lower molar ratio of NaOH to furfural.

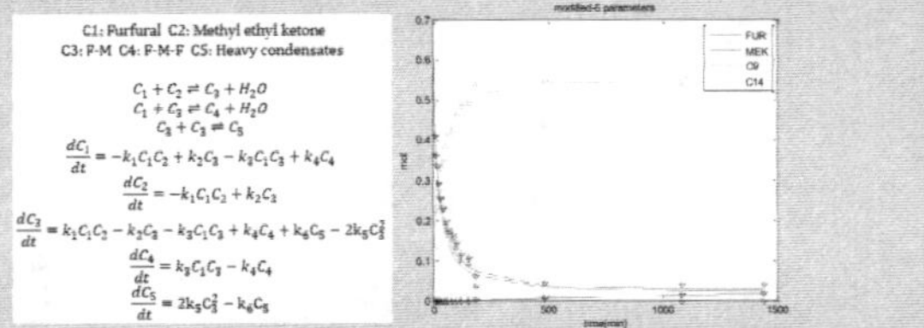

Figure 3. Kinetic modelling of aldol condensation

(simulation work is based on reaction conditions in the 5th row in table 3)

Conclusions

- Bagasse-based solid acid exhibit good catalytic activity and reuse stability in converting 2,3-butanediol to methyl ethyl ketone
- With catalysis of NaOH solution, aldol condensation of furfural and methyl ethyl ketone gives high furfural conversion, and distribution carbon chain length in condensates can be adjusted by reaction conditions.

References

1. John R. Regalbuto. Cellulosic Biofuels--Got Gasoline? *Science,* 2009, 325, 822.
2. Wenyong Lou et al. A highly active bagasse-derived solid acid catalyst with properties suitable for production of biodiesel. ChemSusChem 2012, 5, 1533-1541.
3. Rong Xing et al. Production of jet and diesel fuel range alkanes from waste hemicellulose-derived aqueous solutions. *Green Chem.*, 2010, 12, 1933–1946.

465838 Macroalgae Pretreatment Using Peg-Linked Dicationic Acidic Ionic Liquids

Abstract:

Macroalgae has become a sustainable source of carbohydrates for the production of bio-fuel and bio-based chemicals. Compared with lignocellulosic biomass, macroalgae has higher photosynthetic efficiency, higher yields, lower lignin content and requires a simpler environment for its cultivation. Saccharification of macroalgae is commonly performed in a pressurized system in the presence of an acid catalyst. But ionic liquids (ILs) have been gaining popularity as effective and environmentally benign solvents for biomass as it enables hydrolysis to be performed at atmospheric conditions.

Task specific ILs are currently being explored in biomass processing owing to their versatility and multi-functionality (Olivier-Bourbigou et al., 2010). Particularly, Bronsted acidic ILs could replace traditional mineral acids such as sulfuric and hydrochloric acids. Typically, mono-cationic ILs are used to initially dissolve the biomass forming a homogeneous solution which is subsequently hydrolyzed using either enzymes or an acid catalyst. Meanwhile, dicationic ILs have various advantages over mono-cationic ILs such as higher thermal stability, decreased toxicity and multi-functionality. Particularly, ether-functionalized ILs tend to exhibit lower viscosities than their aliphatic counterparts. Moreover, the oxygen atoms embedded in the ether chains could act as hydrogen bond acceptors promoting biomass dissolution (Zhao et al., 2008). Dicationic ILs with acidic counter-anions are scarcely reported in biomass processing.

In this work, dicationic acidic ILs (DAILs) with varying poly-ethylene glycol (PEG) linker chains were synthesized, characterized and utilized for the conversion of *Gelidium amansii*, a red macroalgae into its corresponding monomeric sugars. The structures of PEG-DAILs having an acidic counter-anion (i.e. $[2HSO_4]^-$) were confirmed using

spectroscopic techniques. Results showed the liquid form of PEG-DAILs, which are thermally stable up to $T_{peak} <$ 350 ºC. PEG-DAILs with short ether chains (e.g. Di-, Tri- and Tetra-ethylene glycol) exhibited faster reaction rates than those with long ether chains (e.g. PEG400 and PEG600). Parameters such as particle size, biomass loading, reaction temperature and time were varied to achieve optimal sugar yields. Results showed $[Tri\text{-}EG\text{-}(MIm)_2]\ 2HSO_4$ as the best PEG-DAIL for maximum sugar release from *G. amansii*. Overall findings show the potential of PEG-DAILs for one-pot dissolution and hydrolysis of macroalgae to sugars.This work was supported by the National Research Foundation of Korea (NRF) grant funded by the Ministry of Education (No. 2009-0093816) and by the Ministry of Science, ICT & Future Planning (No. 2015R1C1A2A01054605).

References:

Olivier-Bourbigou, H., Magna, L., Morvan, D., *Appl. Catal. A-Gen.*, 373 (2010) 1.

Zhao, H., Baker, G. A., Song, Z., Olubajo, O., Crittle, T., Peters, D., *Green Chem.*, 10 (2008) 696.

467504 Fast Pyrolysis Bio-Oil As Precursor of Thermosetting Resins

Abstract:

The use of biomass-based resources is currently a hot topic for both academic and industrial research as an alternative to mitigate the strong dependence on fossil carbon in the chemical and petrochemical industries. The U.S. Department of Energy (DOE) and the U.S. Department of Agriculture (USDA) have prioritized the development of bioenergy and bio-products; and they have the goal to produce 18% of the current U.S. chemical commodities from biomass by 2020, and 25% by 2030. The macromolecular chemistry based on lingocellulosic feedstock represents one of the answers to the quest for polymeric materials capable of replacing their fossil-based counterparts. In particular, fast pyrolysis bio oil is an excellent candidate for the bio-sourcing of polymeric resin. Their competitive cost, worldwide availability and built-in functionality have catapulted it use as a source of macromonomers for polymer applications. The development of this area has been carried out mainly through the reaction and functionalization of hydroxyl groups in the bio oil to produce thermosetting resins such as: epoxy, phenolic, polyurethanes, etc. Thus, the main purpose of this project is to cover the major aspects related to the chemical synthesis, physical-chemical characterization and study of thermo-mechanical properties of bio-based resin, where highly functionalized components are synthetized by chemical modification. Results showed that this new bio-based polymeric systems display interesting properties that are close to their commercial counterparts.

463504 Micro/Nano Lignocellulosic Fibrils (MNLCF) Aerogels from Coconut and Oil Palm Tree Residuals and Application for Environmental Remediation

Abstract:

The extensive utilization of oil palm and coir plantations worldwide has led to a substantial generation of residual biomass[1]. Therefore, an integrated utilization of related biomass streams is a necessity that, in fact, can become a source of valuable natural fibers. In addition, replanting operations and environmental issues can be avoided this way. The lignocellulose sources used in this study, include coir fibers and oil palm empty fruit bunches (EFB), both of which have high microfibril angle (~45^0) that imparts high breaking strain and low elastic modulus, making them useful for producing ropes, mats and fishing nets[2,3]. Coir and EFB fibers are cheap and a renewable resource but a vast supply is wasted due to a lack of better utilization. These fibers have high lignin content, approximately 43% and 36% dry weight in coir and EFB, respectively. Lignin is an undesirable component in paper manufacturing as it causes yellowing of the paper and hence, coir and EFB fibers do not find their use in high-volume papermaking. In fact, the current research on nanocellulose also focuses on producing lignin-free cellulose nanofibers which leaves a high carbon footprint on the environment[4,5]. Currently, lignin is utilized for their high calorific value which is a result of a cross-linked phenolic structure[6]. Recently, lignin has been shown to act as cementing material between cellulose nanofibers[7]. In addition, it has also been demonstrated that the porosity, hydrophilicity and the barrier properties of nanopapers or films of nanocellulose can be tuned via lignin content.

A high microfibril angle and a high lignin content, limits the applications of coir and EFB fibers. In the present study, we approach this problem by isolating micro/nano fibrils from the source fiber, thus foregoing effects related to the microfibril angle. We also target retaining the native lignin of coir and EFB by subjecting them to mild pulping conditions via hydrothermal treatment (see Table 1). A recently reported micro-emulsion method is used to facilitate defibrillation of the fibers[8]. The micro-emulsions swell and delaminate different structures of the fibers by disrupting intra-molecular hydrogen bonding. The micro/nano lignocellulosic fibrils (MNLCF) are obtained through microfluidization. A detailed characterization of the fiber defibrillation process is carried out via composition analysis, scanning electron microscopy (SEM), atomic force microscopy (AFM) and X-ray diffraction (XRD). The migration of lignin from cell wall to the defibrillated fibers is studied via confocal microscopy. The thermal properties of MNLCF is studied via thermogravimetric analyses (TGA) and differential scanning calorimetry (DSC). Our detailed analyses provide useful information to identify potential applications for the produced MNLCF. As a proof of concept, water-resistant and durable aerogels are synthesized from MNLCF via freeze drying (see Figure 1) and tested for adsorption of Zn(II), Pb(II) and Cd(II) salts.

Table 1: Chemical composition of fiber precursor before and after the treatment. All values are in %

Sample	Extractives	Lignin	Cellulose	Hemicellulose
Coir (raw material)	0.5	43	30	23
Coir (Hydrothermal treatment)	<0.1	34	56	7
Coir (microemulsion treatment)	<0.1	31	56	5
EFB (raw material)	0.8	36	34	20
EFB (Hydrothermal treatment)	<0.5	29	58	4
EFB (microemulsion treatment)	<0.5	24	67	3

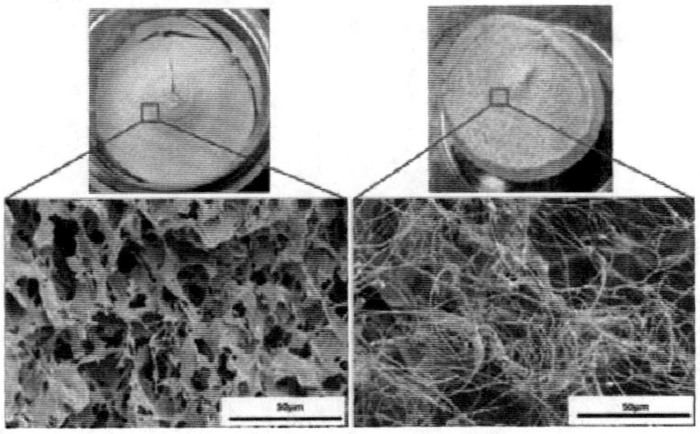

Figure 1: As-prepared aerogels from EFB (left) and coir (right) micro/nano lignocellulosic fibrils (MNLCF) along with their SEM images. The scale bar corresponds to 50μm

References:

1. Ferrer A, Vega A, Rodríguez A, Jiménez L. Acetosolv pulping for the fractionation of empty fruit bunches from palm oil industry. *Bioresour Technol*. 2013;132(November 2015):115-120. doi:10.1016/j.biortech.2012.12.189.

2. Khalil HPSA, Alwani MS, Omar AKM. Chemical composition, anatomy, lignin distribution, and cell wall structure of Malaysian plant waste fibers. *BioResources*. 2006;1(2):220-232. doi:10.15376/biores.1.2.220-232.

3. Satyanarayana KG, Kulkarni AG, Rohatgi PK. Structure and Properties of Coir Fibres. *Proc Indian Acad Sci Ser Eng Sci*. 1981;V 4 PT4(February):419-436. doi:10.1007/BF02896344.

4. Abraham E, Deepa B, Pothan L a., et al. Extraction of nanocellulose fibrils from lignocellulosic fibres: A novel approach. *Carbohydr Polym*. 2011;86(4):1468-1475. doi:10.1016/j.carbpol.2011.06.034.

5. Deepa B, Abraham E, Cordeiro N, et al. Utilization of various lignocellulosic biomass for the production of nanocellulose: a comparative study. *Cellulose*. 2015;22(2):1075-1090. doi:10.1007/s10570-015-0554-x.

6. Vanholme R, Demedts B, Morreel K, Ralph J, Boerjan W. Lignin Biosynthesis and Structure. *Plant Physiol*. 2010;153(3):895-905. doi:10.1104/pp.110.155119.

7. Rojo E, Peresin MS, Sampson WW, et al. Comprehensive elucidation of the effect of residual lignin on the physical, barrier, mechanical and surface properties of nanocellulose film. *Green*. 2015:1853-1866. doi:10.1039/c4gc02398f.

8. Carrillo CA, Laine J, Rojas OJ. Microemulsion Systems for Fiber Deconstruction into Cellulose Nano fi brils. *Appl Mater Interfaces*. 2014;6:22622-22627. doi:10.1021/am5067332 |.

469073 High Concentration Glucose Production from CELF Pretreated Corn Stover Using Wild Type Clostridium Thermocellum Supplemented with β-Glucosidase

Abstract:

Glucose production from cellulosic biomass can provide a powerful platform on which to build a sustainable fuels industry. Today expensive cellulases are needed to further breakdown pretreated biomass into sugars with high yields. In response, researchers have turn towards biological saccharification of glucan using *C. thermocellum*, a highly cellulolytic anaerobe possessing cellulosomes capable of glucan saccharification at high rates and yields. However, *C.thermocellum* does not produce much β-glucosidase, a necessary enzyme for cellobiose hydrolysis into glucose. Additionally, Co-solvent Enhanced Lignocellulosic Fractionation (CELF) has been shown to be very effective at producing glucan solids from corn stover that are highly susceptible to enzymatic hydrolysis. Thus, the objective of this research is to develop a process by which *C. thermocellum* supplemented with β-glucosidase can achieve high yields of glucose when fed CELF pretreated corn stover solids at high concentrations (>100g/L). Advantages of the process include: 1) reduction of expensive fungal cellulases, 2) no need for genetic engineering, and 3) high glucose concentrations. CELF pretreatment prepared corn stover for biological conversion, and the resulting washed solids were loaded into 50 mL bottles at a 50-100 g/L glucan loading for incubation with *C.thermocellum* supplemented with β-glucosidase at 60°C with a shaking speed at 180rpm. Some samples were also supplemented with sodium azide to suppress cell metabolism and enhance glucose yields. Results will be compared for active and inactive *C.thermocellum* cultures and to results from conventional fungal hydrolysis at high and low loadings at 50°C, pH 5, and 150 rpm.

472224 Switchgrass Solubilization By Mixed Methanogenic Enrichments with Comparison to Pure Cultures of *Clostridium Thermocellum*

Abstract:

To meet anticipated global demand for low carbon liquid transportation fuels, second generation biofuels will likely be produced from lignocellulose, a low cost and abundant plant biomass resource. However, the lignocellulose in plant cell walls is difficult to convert. Increasing lignocellulose conversion will significantly enhance biofuel production and lower capital & operational cost. Microbially-mediated fermentation of lignocellulose is a promising approach. Diverse, mixed microbial enrichment cultures provide an indication of the achievable extent of biomass solubilization. Yet there have been few fundamental studies of grass fermentation under the conditions anticipated for a high-rate, low cost industrial process, including very high solids concentration, short residence times and high temperatures. Motivated by this perspective, we have initiated study of thermophilic switchgrass fermentation using mixed culture anaerobic enrichments at decreasing residence times & increasing grass concentrations.

Inoculum was collected from a two-stage anaerobic digester (Vermont Technical College) fed with manure and food waste. Triplicate semi-continuous, anaerobic digesters were operated at 55°C and 30g/L mid-season switchgrass concentration for 214 days, during which two of them were allowed to come to steady-state sequentially at residence times (RT) of 20 days, 10 days, 5 days and 3.3 days, with the other one running as control at RT = 20 days. This feed rate was maintained by regular replacement of 1/10 of reactor contents to achieve the nominal residence times. Whereas we expected a shift from methanogenesis to acidogenesis as RT decreased to 5 days, surprisingly, even at RT=3.3 days stable methane production was still seen and acids inhibition was not observed, which indicates that the digesters were still healthy and balanced. Total carbohydrate solubilization at RT = 20, 10, 5 and 3.3 days was 69.1%, 62.8%, 54.9% and 50.0% respectively. In addition to these continuous trials, batch studies compared the performance of mixed enrichment cultures to pure cultures of *Clostridium thermocellum*, where solubilization was similar despite

the more limited enzyme diversity of the pure culture species.

16S rRNA analysis was used to characterize microbial populations dynamics. It was found that Firmicutes dominated each of the reactors, representing between 54-96% of the sequence reads, and these populations increased with reduced residence time. In addition, Synergistetes, Euryarchaeota etc. taxa were also well represented and relatively stable. By using bioinformatics tools, species present in the consortia were also matched to closely related cultured species including *Clostridium clariflavum* that have cellulolytic activities. Plate culture isolation further revealed that *Clostridium clariflavumÂ* was one of the major cellulolytic species existing in the mixed culture consortia. Data at feed concentrationsÂ â‰¥Â 75 g/L, approaching industrial conditions, will also be presented.

453000 Substituent Effect of Phenolic Aldehydes Inhibition on Alcoholic Fermentation By *Saccharomyces Cerevisiae*

Abstract:

Phenolic compounds significantly inhibit microbial fermentation of biomass hydrolysates. To understand the quantitative structure-inhibition relationship of phenolic aldehydes on alcoholic fermentation, the effect of eleven different substituted benzaldehydes on the final ethanol yield was examined. The results showed that the degree of phenolic benzaldehydes inhibition was strongly associated to the position of phenolic hydroxyl groups, but not the number of phenolic hydroxyl groups. It was observed that ortho-substituted 2-hydroxybenzaldehyde resulted in 15-20 fold higher inhibition than the meta- or para-substituted analogues of 3- and 4-hydroxybenzaldehydes. By correlating the molecular descriptors to inhibition potency in yeast fermentation, we found a strong relationship between log *P* (octanol/water partition coefficient) of aldehydes and EC_{50} value. The most inhibitory 2-hydroxybenzaldehyde has a highest log *P* and possesses an ortho –OH group capable of forming an intramolecular hydrogen bond, which can potentially increase the cell membrane permeability and their toxicity. The results also indicated that the calculated free energy change between phenolic aldehydes and amino acids can be used to predict their structure-inhibitory activity relationship.

467613 Enzymatic Process for Making Plasticizers from High Oleic Soybean Oil

Abstract:

Epoxidized soybean oil (ESBO) is a chemically modified soybean oil derivative that is used to make plastics pliable as well as for making bio-based polyurethanes for coatings, epoxide-based resins, and high temperature lubricants. While traditional conversion methods use peroxyacetic acid to catalyze the reaction, lipases can also catalyze epoxidation reactions. In this paper we show conversion of soybean oil and high-oleic content soybean oil using immobilized lipase derived from *Candida antarctica* without the need for additional solvents, such as toluene, or added free fatty acids. At 35°C, soybean oil can be epoxidized with >90% yield within 15 hours, while using peroxyacetic acid only obtains 70% yield at 50°C after 10 hours of reaction time as side reactions reduced the selectivity of the epoxidation. Reaction kinetics for individual unsaturated fatty acids will be presented, as well as characterization of the enzymatic ESBO compared to the peroxyacetic acid conversion with both soybean oil and high-oleic content soybean oil. Recyclability of the immobilized enzyme is discussed in the context of potential commercial applications as a continuous process.

473034 Biomass Blending & Densification: Impacts on Biochemical Conversion Performance

Abstract:

A blended feedstock strategy has been proposed to address the high cost of feedstock production and reduce risk to the biomass supply chain. The use of blended and densified feedstocks in biochemical conversion pathways instead of conventionally-ground biomass from a single source addresses several challenges in the current biomass supply chain, including availability, transportation, storage, cost, quality, and supply variability. Blending provides a mechanism to reduce risk to the supply chain and reduce feedstock cost by coupling location-specific resource utilization with biofuel production. Preliminary results suggest blending and densification provide a viable path forward to a low-cost, consistent biomass supply. Recent work has investigated the impacts of blended and densified feedstocks on sugar yields in biochemical conversion pathways. Recent studies have demonstrated that densification does not negatively impact biomass recalcitrance; reactivity was equivalent for ground and pelleted corn stover. Recent studies provide a glimpse into the potential for preprocessing options, like blending and densification, to provide benefits to not only logistics, but downstream conversion. However, the impact blending and preprocessing may have on variability and conversion performance has not been fully vetted. The cumulative impact of densification and blending on biochemical conversion processes will be discussed.

466246 *Tuning Ionic Liquids for Low Cost: Applications in Lignocellulose Deconstruction*

Abstract:

Ionic liquids (ILs) have proven to be highly tunable 'designer solvents' capable of a wide range of exciting chemistries. However, industrial application at large scale is hampered by high solvent cost. This cost is, however, a tunable feature of the solvent itself – provided the ion selection is handled with a careful eye aimed at limiting synthetic complexity. Lowering the solvent cost will increase the attractive opportunities of ILs for bulk processing of lower cost end products – including such applications as biofuels. One of the key challenges in biorefining is the initial separation or deconstruction of lignocellulosic feedstock into separate components. ILs offer unique advantages in this area, due to their unusual thermochemical properties. However, there are serious concerns about the economic viability of their use due to the very high cost of most ionic liquids (> $50/kg). We have overcome this by redesigning the IL based deconstruction process to use low-cost, acidic ILs for lignin dissolution rather than cellulose dissolution, yielding filterable cellulose and a dissolved lignin for precipitation or conversion to high-value chemicals. We have found that processability of the cellulose is high and lignin recoveries near quantitative.

We use a range of 'protic' ILs, the family typically used in IL industrial processes, because their simple acid-base chemistry results in a simple and cheap synthesis, with a cost (< $1/kg) similar to common organic solvents such as acetone or toluene. This presentation will discuss how ionic liquids can be 'tuned' to control cost structure of the final solvent, and what implications this will have for the chemical processes involved. The impact of the solvent on large-scale applications, such as biomass pretreatment, will be discussed, with a focus on performance and process considerations such as how the ILs maintain solvent stability under long-term processing conditions, that they can be recovered and continue to exhibit very good performance after multiple reuses. These properties highlight that the ILs have the flexibility to be useful for a variety of downstream chemical processing techniques, and for use in other applications as well.

471708 Investigation of Lignin Streams Generated during Ionic Liquid Pretreatment of Lignocellulosic Biomass

Abstract:

Lignin is a complex three-dimensional amorphous polymer derived from paracoumaryl, coniferyl, and synapyl alcohols, cross-linked together by different a number of linkages. In plant cell walls, lignin fills the spaces between cellulose and hemicellulose, and it acts like a 'glue' that holds the lignocellulose matrix together. Despite years of research, the lignin structure is still not well understood as a function of feedstock and process. The ability of certain ionic liquids (ILs) to dissolve cellulose and/or lignin enables the possibility of commercial lignin upgrading and valorization to improve the overall biorefinery economics. However, to date depolymerization mechanisms of lignin in ILs is not well understood. In this study, the changes in the chemical composition of lignin after pretreatment with IL were investigated. Three different type of biomass, namely switch grass, eucalyptus and pine were utilized in this study. The relative changes in the common β-O-4, β-β and β-5 inter-unit linkages of lignin during pretreatment process were investigated using solution state two dimensional (2D) nuclear magnetic resonance (NMR). The changes in the size and molecular weight distribution of lignin were analyzed using size exclusion chromatography (SEC) and mass spectrometry.

Swelling and Dissolution of Cellulosic Fibers

Mohammad Ghasemi, Marina Tsianou, Paschalis Alexandridis*

Department of Chemical and Biological Engineering
University at Buffalo, The State University of New York (SUNY)
Buffalo, New York 14260-4200, USA

* E-mail: palexand@buffalo.edu

Abstract

The dissolution of cellulose is a critical step for the efficient utilization of this renewable resource as a starting material for the synthesis of high value-added functional polymers. Important aspects of cellulose dissolution such as solvent-induced decrystallization and chain disentanglement are not well-understood. We have developed a phenomenological model that captures the phenomena governing the dissolution of semicrystalline polymers. This model allows the quantification of two important features, i.e., solvent effectiveness in cellulose (i) decrystallization and (ii) chain disentanglement, the balance of which controls the mechanism and kinetics of cellulose dissolution. The insights obtained from this study improve the understanding of cellulose-solvent interactions underlying decrystallization and disentanglement and their contributions in controlling the kinetics of cellulose swelling and dissolution.

Introduction

The processing of cellulose into desired products typically involves its swelling and/or dissolution.[1] A major barrier to cellulose solubility is the recalcitrance of cellulose, which originates from a partial crystalline structure and extended noncovalent interactions.[2] Only a few solvent systems are effective for direct dissolution of cellulose, and they operate under rather strict conditions of composition and temperature.[3] It is generally thought that the capacity of the solvent to break the intra- and inter-molecular hydrogen bond network of cellulose is a key factor in its dissolution.[4] However, it was recently suggested that, since cellulose has amphiphilic properties, hydrophobic interactions might play a significant role in cellulose solubility.[5]

The homogenous swelling and dissolution of cellulose fibers is shown schematically in Figure 1. When solvent diffuses into the cellulose crystalline domains they are gradually disrupted become amorphous, and the fiber swells.[6] After a certain induction time, cellulose chains gradually become disentangled and move out of their network into the bulk solution.[7] Therefore, the dissolution of cellulose involves two inter-related phenomena: decrystallization and chain disentanglement.[7] If a solvent can destroy the crystalline network of cellulose but is not capable to disentangle the chains, then the crystallinity of cellulose is reduced, however, no dissolution happens, as in the case of 1-N-butyl-3-methylimidazolium chloride ([bmim]Cl) at 70 °C.[8] On the other hand, if a solvent can only disentangle the chains, then only the amorphous regions of cellulose can dissolve.

In order to gain insights into the swelling and dissolution of cellulose, a microscopic description of macroscopic experimental information has been provided by using a newly developed phenomenological model [7] that captures the phenomena governing the dissolution of semicrystalline cellulose. Our model fits well experimental data [8] for swelling and dissolution of cotton fibers in the ionic liquid [bmim]Cl.

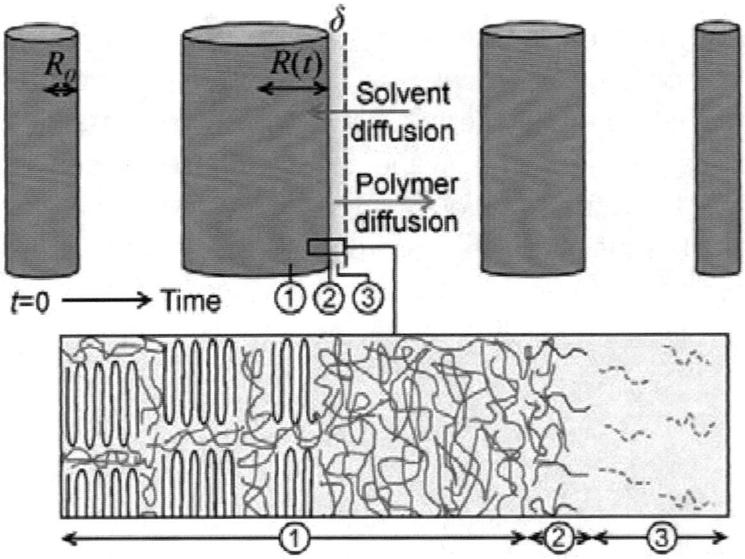

Fig. 1: *Schematic representation of swelling and dissolution of cellulose fiber: (1) fiber interior: crystalline domains (black) and amorphous domains (blue, light shade); (2) fiber-solvent interface at which disentanglement occurs; (3) external diffusion boundary layer at which disentangled cellulose chains (red and dashed) diffuse toward the bulk solution.*[7]

Methods

In the model the cellulose fiber is considered as a long semicrystalline polymer cylinder (consisting of two components; crystalline and amorphous) with a uniform cross section area and with properties that may vary along the cross section.[7] The model describes different steps involved in cellulose dissolution, e.g., solvent diffusion, transformation from crystalline to amorphous domains, specimen swelling, cellulose chain disentanglement, and diffusion of disentangled chains toward the bulk solution, using appropriate transport phenomena and kinetics equations.[7] The ability of solvent molecules to transform the crystal lattice into amorphous and to disentangle cellulose chains is captured by the decrystallization rate constant (k_{dec}) and the disentanglement rate (r_{dis}), respectively. These two are the only parameters fitted to experimental data; all other model parameters are independently obtained.

Results and Discussion

The impact of cellulose decrystallization in the dissolution process is not the same at all decrystallization constant values: at some conditions decrystallization has a strong effect, while at other conditions its effect is negligible (Figure 2a). For example, decreasing k_{dec} from 2.5×10^{-1} to 2.5×10^{-3} s^{-1} has little impact on the total dissolution time (obtained when fraction of dissolved fiber = 1), however, decreasing k_{dec} below 2.5×10^{-3} s^{-1} has a significant impact on the time evolution of fiber dissolution. For $k_{dec} \leq 2.5\times10^{-4}$ s^{-1}, the contribution of solvent molecules to the decrystallization of cellulose is less than their contribution in dissociating the chains, and so the dissolution process is mainly controlled by decrystallization. Because of this, a small change in the k_{dec} value results in a considerable variation in the time evolution of the fiber radius, crystallinity, and solubility, whereas for $k_{dec} \geq 2.5\times10^{-3}$ s^{-1} these fiber characteristics are not affected much by k_{dec}.

When the cellulose-solvent interactions that affect chain disentanglement are stronger than those that influence cellulose decrystallization, the fiber remains crystalline up to almost the end of the dissolution process. This condition happens for $r_{dis} \geq 1\times10^{-9}$ m/s (Figure 2b). Therefore, for $r_{dis} \geq 1\times10^{-9}$ m/s, solvent molecules can disentangle the amorphous domains well, while the decrystallization rate is not high enough to transform the crystalline domains at the fiber surface to amorphous (decrystallization control). However, when $r_{dis} \leq 5\times10^{-10}$ m/s the decrystallization rate of cellulose is relatively greater than the rate of disentanglement, and so the overall kinetics of fiber dissolution are mainly controlled by the chain disentanglement.

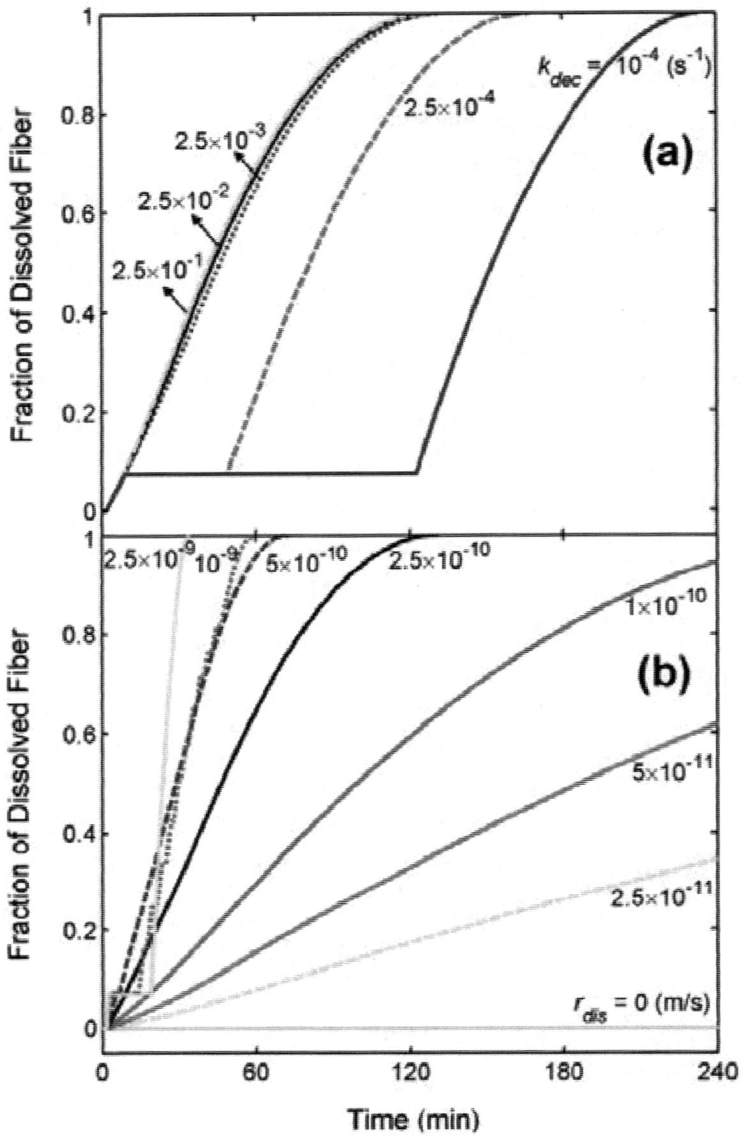

Fig. 2: *Effect of (a) k_{dec} and (b) r_{dis} on the time evolution of the fraction of the fiber that has dissolved.*[7]

The above approach allows us to evaluate the conditions of cellulose dissolution in a variety of systems in the context of the relative contributions of decrystallization and disentanglement. Accordingly, the dissolution conditions can be organized into four regimes: (1) $k_{dec} \leq 2.5\times10^{-4}$ 1/s and $r_{dis} \leq 5\times10^{-10}$ m/s. At these conditions the ability of solvent in both cellulose decrystallization and chain disentanglement is not good, the solubility is limited and the process can be controlled by either decrystallization or disentanglement, depending on the k_{dec} and r_{dis} values. (2) $k_{dec} \leq 2.5\times10^{-4}$ 1/s and $r_{dis} \geq 1\times10^{-9}$ m/s. Here the cellulose-solvent interactions which affect disentanglement are much stronger than interactions that impact decrystallization (decrystallization control).

(3) $k_{dec} \geq 2.5\times10^{-3}$ 1/s and $r_{dis} \leq 5\times10^{-10}$ m/s. At this condition the solvent is more effective in disturbing cellulose intermolecular interactions that result in decrystallization than the interactions that disentangle the chains (disentanglement control). (4) $k_{dec} \geq 2.5\times10^{-3}$ 1/s and $r_{dis} \geq 1\times10^{-9}$ m/s. Here the rates of both decrystallization and disentanglement are very high, the solvent is very aggressive toward cellulose dissolution, and the mechanism of the dissolution process can be controlled by either decrystallization or disentanglement, depending on the k_{dec} and r_{dis} values.

Conclusions

When the solvent effectiveness in the disruption of cellulose crystals is relatively lower than its ability to disentangle the chains (e.g., for LiCl/DMAc[9]), the kinetics of dissolution are controlled by decrystallization. Furthermore, conditions that facilitate cellulose chain disentanglement, in addition to increasing the rate of dissolution, can result in faster decrystallization. The solvent effectiveness in chain disentanglement is the only factor that determines the decrease of the cellulose fiber radius. In cases where the fiber dissolution rate is higher than the decrystallization rate (e.g., in [bmim]Cl at 80 °C[8]), the dissolution of cellulose is mostly controlled by the solvent ability to disentangle the chains. Such knowledge allows the construction of operating condition diagrams for the design of dissolution processes in the biorefinery.

Acknowledgements

We thank the U.S. National Science Foundation (Grant CBET-1159981) for supporting research on cellulose dissolution in our laboratory.

References

(1) Wang, H.; Gurau, G.; Rogers, R. D., Ionic liquid processing of cellulose. *Chem. Soc. Rev.* **2012,** *41* (4), 1519-1537.

(2) Himmel, M. E.; Ding, S.-Y.; Johnson, D. K.; Adney, W. S.; Nimlos, M. R.; Brady, J. W.; Foust, T. D., Biomass recalcitrance: Engineering plants and enzymes for biofuels production. *Science* **2007,** *315* (5813), 804-807.

(3) Liebert, T., Cellulose solvents - remarkable history, bright future. *ACS Symp. Ser.* **2010,** *1033*, 3-54.

(4) Keshk, S. M. S., Effect of different alkaline solutions on crystalline structure of cellulose at different temperatures. *Carbohydr. Polym.* **2015,** *115*, 658-662.

(5) Medronho, B.; Romano, A.; Miguel, M. G.; Stigsson, L.; Lindman, B., Rationalizing cellulose (in)solubility: Reviewing basic physicochemical aspects and role of hydrophobic interactions. *Cellulose* **2012,** *19* (3), 581-587.
(6) El Seoud, O. A., Understanding solvation. *Pure Appl. Chem.* **2009,** *81* (4), 697-707.
(7) Ghasemi, M.; Alexandridis, P.; Tsianou, M., Cellulose dissolution: Insights on the contributions of solvent-induced decrystallization and chain disentanglement. *Cellulose* **2016**.
(8) Jiang, G.; Huang, W.; Wang, B.; Zhang, Y.; Wang, H., The changes of crystalline structure of cellulose during dissolution in 1-butyl-3-methylimidazolium chloride. *Cellulose* **2012,** *19* (3), 679-685.
(9) Ramos, L. A.; Morgado, D. L.; Gessner, F.; Frollini, E.; El Seoud, O. A., A physical organic chemistry approach to dissolution of cellulose: Effects of cellulose mercerization on its properties and on the kinetics of its decrystallization. *Arkivoc* **2011,** *7*, 416-425.

471942 Towards a New Paradigm in the Ionic Liquid Pretreatment for the Production of Lignocellulosic Biofuels: Technoeconomic Insights

Abstract:

Ionic liquids (ILs) have been receiving increasing attention in biomass pretreatment applications (Konda et al., 2014 and 2015) as some ionic liquids are shown to be very efficient in biomass pretreatment. Among several benefits of ionic liquids in general, they are often regarded as *green solvents* due to their negligible volatility. More importantly, given the availability of numerous anions and cations, it is often possible to *tune* their properties to suit specific application(s) by appropriately selecting the combination of anion and cation. Specific to biomass pretreatment, some ionic liquids are shown to be feedstock agnostic (Li et al., 2013) which is a very desirable characteristic especially given the uncertainty with feedstock supply and quality (e.g., seasonal variations). Such a technology can be potentially disruptive and can play a critical role in enabling large-scale cellulosic biorefineries.

To tap into the full potential of ionic liquids, however, it is important that the processes are designed such that they are economically viable as well as environmentally sustainable. These criteria demand for minimal/reduced use of any other chemicals (e.g., co-solvents, water) and energy footprint in the entire process. Furthermore, given the highly integrated nature of these cellulosic biorefineries, these pretreatment technologies need to be developed in such a way that they can be â€˜*plugged-in*â€™ while minimizing any negative impact on the other sections of the process (including downstream). Realizing such a process is often challenging as there exists significant trade-offs amongst these criteria. This emphasizes the need for greater insights into the mass and energy balances of such complex processes. To this end, with the help of virtual biorefinery models, technoeconomic analysis (TEA) offers invaluable guidance to nurture technological advances. Equipped with technoeconomic insights, scientists at Joint BioEnergy Institute (JBEI) are harnessing the tunable nature of ILs, together with the novel process design and operational concepts, to develop advanced pretreatment technologies. Our previous TEA work (Konda at al., 2014) emphasized the importance of high biomass loading, high IL recovery, low IL and water usage to ensure the prospective cellulosic

biorefineries can indeed be economically sustainable. A recent advancement in this direction includes the development of a novel integrated process that is capable of high yields under high-gravity (HG) conditions, while minimizing the use of water and IL. TEA revealed that the continued developments with the proposed integrated process could potentially reduce minimum fuel selling price (MFSP) to below $3/gal.

References:

- Konda et al. (2014). Understanding cost drivers and economic potential of two variants of ionic liquid pretreatment for cellulosic biofuel production. Biotechnology for Biofuels 7: 86.
- Konda et al. (2015). Towards a cost-competitive ionic liquid based pretreatment: Insights from technoeconomic analysis. AIChE Annual Meeting, November 2015, Salt Lake City, UT.
- Li et al. (2013). Comparing the recalcitrance of eucalyptus, pine, and switchgrass using ionic liquid and dilute acid pretreatments." BioEnergy Research 6, no. 1 (2013): 14-23.

458154 Effect of Alcohol Structure on the Kinetics and Mechanism of Acid Catalyzed Etherification and Unimolecular Dehydration over Tungstated Zirconia

Abstract:

Branched and linear ether molecules have attracted recent interest as diesel additives and lubricants that can be built from various structures of biomass-derived alcohols. In this study, a comprehensive investigation of the effect of alcohol structure on the mechanism and kinetics of liquid phase etherification and unimolecular dehydration reactions was performed. In this investigation, 12 wt% WO_x/ZrO_2 was identified as an effective and green solid acid catalyst for selective etherification of primary alcohols in the liquid phase, achieving ether selectivities of >96% for C_6-C_{12} linear alcohols at 120 °C, even though at this temperature unimolecular dehydration to alkenes is thermodynamically preferred. The proposed mechanisms for Brønsted acid-catalyzed unimolecular dehydration and etherification were supported by isotopic labeling and by measuring kinetic isotope effects. The catalyst was characterized with XRD, BET, and DRIFTS FTIR with pyridine adsorption. It is proposed that, while the desired etherification reaction proceeds over a small number of strong Brønsted acid sites, the surrounding Lewis acid sites help bring reactant molecules to the surface to facilitate the bimolecular reaction. The effects of alcohol structure including linear alcohol length, carbon chain branching positions, and carbon chain branching size, were probed by measuring the apparent kinetics for dehydration reactions over tungstated zirconia. Trends in apparent activation energies and selectivities were consistent with the proposed mechanism, and can elucidate feasible pathways to produce valuable ether products from biomass-derived alcohols.

470138 Pathways to Selectively Form Individual Aromatic Products from Ethanol

Abstract:

Ethanol derived from the fermentation of biomass can be catalytically converted into higher molecular weight hydrocarbons and oxygenates to produce advanced biofuels and lubricants. The Guerbet reaction, which involves dehydrogenation, aldol condensation, and hydrogenation, is one pathway for this process. Cascades of subsequent self- and cross-condensation reactions between ethanol and $C_{\geq 4}$ alcohol derived intermediates provide a wide-range of possible higher carbon number (C_n) products. Successful production of these products requires high selectivities for dehydrogenation and condensation reactions to avoid the loss of carbon by forming undesirable products (e.g., C_1-, C_2-
form specific molecules which can serve as fuels and platform chemicals.

Here, we show how bifunctional catalysts, comprised of amphoteric oxides (TiO_2, hydroxyapatites, or hydrotalcites) and Cu clusters promote these networks of reactions by measuring formation rates of more than 30 distinct products as functions of reactant concentrations, temperature (523 - 623 K), and the ratios of the numbers active sites for hydrogen transfer and those for condensation reactions. The selectivities with which the C_2-units of ethanol are incorporated into growing chains is limited by the selectivity of both catalytic functions, while the isomeric structure of the products is determined by the concentrations of the unsaturated oxygenates present on the oxide surfaces and differences between the deprotonation energies of C-atoms α to the carbonyl groups of these species. Reactions among product mixtures that predominantly contain alcohol and aldehyde reactants form broad distributions of fatty alcohols whose product selectivities follow predictions from step-growth polymerization models. These diverse mixtures, however, are only suitable as fuels.

Aldol condensations directly form unsaturated aldehydes, which exist at much higher concentrations than

their saturated analogs in the absence of effective hydrogen transfer catalysts. Large differences in the deprotonation energies between α C-atoms for aldehydes and unsaturated aldehydes allow these product mixtures to be exploited to form a very small set of C_8 aromatic species (ortho- and para-tolualdehydes). The structure of the resulting products effectively halts further condensation reactions and provides a unique opportunity to make a desirable platform chemical within this complex network of reactions. The isomeric structure of the C_8 aromatics are sensitive to the presence or absence of steric hindrance about the acid-base site pairs that catalyze the reaction. Coupling hydroxyapatite, which lacks microporosity produces predominantly ortho-C_8 products with minor (< 10%) amounts of the para-isomers. These selectivities can be inverted, however, when the reactant stream include ethylene (derived from in situ dehydration of ethanol or added intentionally). The competing pathways to ortho- and para-tolualdehydes reflect the relative rates of aldol and Diels Alder additions between hexadienal and either acetaldehyde or ethylene co-reactants.

These findings provide evidence that the Guerbet reaction proceeds by dehydrogenation and aldol condensation steps and elucidates and the complex sets of surface reactions that dictate the formation rates of certain categories of products. This knowledge will help guide further research into reaction schemes to convert alcohols into larger oxygenates and hydrocarbons.

452415 Combined Experimental and Theoretical Study of Ethanol Catalytic Conversion to 1,3-Butadiene on MgO

Abstract:

Biomass can serve as a sustainable and renewable carbon source. Ethanol is relatively inexpensive and widely available and can its catalytic upgrading has been proposed as a viable route of indirect biomass valorization. 1,3-butadiene, the most important monomer for synthetic rubber, has been produced via catalytic processing of ethanol during World War II by USSR and USA, using Lebedev and Ostromislensky processes, respectively. The former utilized catalytic conversion of ethanol to 1,3-butadiene in one-step over MgO/SiO_2 catalystswhile the latter utilized a two-step process with the first step of ethanol dehydrogenation to acetaldehyde over Cu/SiO_2 catalysts and transforming acetaldehyde into 1,3-butadiene over a tantalum-based catalyst during the second step. Recent abundance of shale gas resulted in a different catalytic cracker product distribution dominated by ethylene. This caused a worldwide shortage of C4 hydrocarbons such as 1,3-butadiene. Since ethanol can be produced using variety of biomass processing routes, including fermentation and gasification, it recently reemerged as the green route to catalytically form 1,3-butadiene but improvements in yields are necessary.

In this study, we identified the reaction intermediates and mechanisms of ethanol on MgO surface by combining both theoretical and experimental study by means of DFT, *in-situ* FTIR and TPSR. A complex mechanism involving acetaldehyde, crotyl species as well as unexpected pathway via ethylene are discussed.

459233 Aldol Condensation of Acetaldehyde over Titania, Hydroxyapatite and Magnesia

Abstract:

Aldol condensation is a well-known reaction that results in the formation of a C-C bond and is used in the production of important chemicals such as 2-ethylhexanal, isophorone, and crotonaldehyde among others. This work aims to elucidate the mechanism of aldol condensation of acetaldehyde over three materials: anatase titania (TiO_2), hydroxyapatite (HAP), and magnesia (MgO). Reactions were carried out in a fixed bed reactor with a total system pressure of 220 kPa at temperatures between 533 and 633 K and acetaldehyde partial pressures between 0.05 and 50 kPa. Crotonaldehyde was the only product observed over all three catalysts and severe catalyst deactivation occurred at acetaldehyde partial pressures of 5 kPa or greater. The aldol condensation reaction over all three catalysts was first order at low acetaldehyde partial pressure and approached zero order at high acetaldehyde partial pressure. No kinetic isotope effect (KIE) was observed with fully deuterated acetaldehyde reacting over TiO_2 or HAP implying C-H bond activation is not kinetically relevant. These measurements are consistent with a mechanism in which adsorption and desorption steps are kinetically significant during the reaction. Characterization of the catalysts by adsorption microcalorimetry of acetaldehyde and ethanol and diffuse reflectance Fourier transformed infrared spectroscopy of adsorbed acetaldehyde, crotonaldehyde and acetic acid revealed a very high reactivity of these catalysts, even at low temperatures.

467972 Single Step Conversion of Short Carboxylic Acid Mixtures Obtained from Fermentation to Aromatics and α-Olefins over a Cu/ZrO2 Catalyst

Abstract:

Biomass represents an attractive feedstock for producing petrochemical substitutes. However, the high variety of functional groups found in biomass often reduces selectivity and yields during biomass conversion reactions. To overcome this problem, biological funneling can be used to drastically reduce the amount of functional groups. Using bioprocessing, biomass polysaccharides, which contain many functional groups can be converted to carboxylic acid mixtures that contain a single functional carboxylic acid group.

In this work, short carboxylic acids obtainable from a fermentation process were converted into aromatic compounds and/or alpha-olefins in a single step catalytic process with a near theoretical carbon yield. The experiments were carried out from 623 K to 673 K under hydrogen atmosphere using Cu/ZrO_2 catalyst. The combination of copper and zirconia oxide allowed full conversion of carboxylic acids via cascade C-C coupling reactions resulting in higher molecular weight (C8-C15) products, which form a water immiscible organic phase, and light alpha-olefins in the gas phase. Remaining oxygen functionalities are selectively removed by hydrogenation-dehydration reactions over copper while avoiding the hydrogenation of unsaturated C-C bonds reducing hydrogen consumption. Moreover, by tuning reaction conditions, selectivity can be controlled to favor the increased production of short alpha-olefins or heavier products.

454772 Mechanistic Insights on the Formation of Maleic Anhydride from the Oxidation of Levulinic Acid over Supported Vanadates

Abstract:

With the realization that, even though biomass derived transportation fuels present a great solution to the diminishing oil reserves, they cannot compete with their petrochemical equivalents, we turn our attention to bio-based commodity chemicals. Those have the potential to compete with the existing market and present opportunities for the utilization of lignocellulosic biomass. Along those lines, we present an alternative catalytic pathway for the production of maleic anhydride (MA) through the gas phase oxidative cleavage of the methyl carbon in levulinic acid (LA). We previously showed that over a vanadia phase supported on silica, high yields of MA (71%) can be achieved. We were also intrigued to observe that the methyl carbon cleavage does not occur on monofunctional methyl ketones. Here, we try to probe the nature of this unique chemistry, by employing a number of different ketones and keto-carboxylic acids in order to decide if this unique selectivity is due to steric effects or the dual functionality of LA.

458085 P-Toluic Acid from Biomass-Derived Coumalic Acid: A Reaction Network Analysis

Abstract:

The Diels-Alder/decarboxylation/dehydrogenation sequential reaction of biomass-derived coumalic acid (CMA) with propylene provides a renewable one-pot reaction pathway to *p*- toluic acid (pTA), an intermediate that can be used to produce terephthalic acid. A detailed reaction network of the formation of toluic acid (TA) was established by varying the key parameters including temperature, solvent, and reaction time, showing that high yields of TA (85 %) are possible at complete conversion of CMA and that the *para* to *meta* selectivity of TA can be controlled by adjusting reaction temperature and time. Changing the reaction media from non-polar (toluene) to polar aprotic (Î^{3}-valerolactone) revealed a strong solvent effect with an increase in the formation of TA from 58 % to > 85 %. Performing the reaction of CMA and propylene with and without the presence of catalyst gave additional information about byproduct formation demonstrating the importance of the catalyst presence to quickly dehydrogenate TA intermediates and to prevent unwanted byproduct formation. Product yield and selectivity can be further optimized when water free reaction conditions are applied as accelerated decomposition rates of CMA were observed by adding small amounts of water (5 vol-%) to the solvent.

464791 Rational Design of Nanoflower Shaped Bimetallic Ptmn Nanocatalysts for Tartronic Acid Synthesis from Biomass

Abstract:

Tartronic acid is considered as one of the important industrial building blocks for a variety of everyday chemicals such as detergents and plastics, as well as valuable pharmaceutical products. Conventionally, tartronic acid is synthesized from methanol and CO *via* multiple energy intensive and toxic steps involving cyanide agents. Significant formation of waste by-products and poor yield are major issues plaguing the traditional route. Catalytic oxidation of renewable biomass feedstocks would undoubtedly provide alternative approaches for tartronic acid production.

Several mono and bimetallic nanocatalysts have been proposed for catalytic glycerol oxidation to tartronic acid, which are often obtained from seed or polymer template-based methods and exhibit ordered core-shell and alloy structures. Most of existing nanocatalysts however show limited performances for oxidation activity and tartronic acid selectivity. In this presentation, we describe a new completely disordered morphology of PtMn nanocatalysts. Unique bud- and cauliflower-shaped bimetallic PtMn nanoclusters were synthesized using lattice mismatch induced self-assembly technique without adding polymer stabilizers. The influence of Pt/Mn ratios on the morphologies of bimetallic nanoclusters was studied systematically under various synthetic conditions. Due to the large lattice mismatch between Pt and Mn metals, anisotropic growth bimetallic crystals was obtained with high surface index numbers. The novel bimetallic nanocatalysts display synergistic an eight-fold enhancement in catalytic activity for aqueous phase oxidation of glycerol to tartronic acid compared with monometallic Pt catalysts.

474591 Upgrading Corn Stover Pretreatment Wastes into Biodiesel Via Oleaginous *Rhodococci*

Abstract:

The derivation of advance biofuels and specialty oils from agricultural waste streams via oleaginous bacteria is a rapidly developing biotechnology primed for both bioenergy and biochemical sectors. However, finding viable effluents and promising microbes that produce an array of tangible oils for subsequent targeting is a current challenge. Here, we examine and compare alkaline and alkaline-peroxide pretreatments of corn stover as a feedstock for three strains of wild-type *Rhodococci*. In comparison, *R. opacus* DSM 1069 and PD 630 exhibited aggressive degradation of low molecular weight aromatics and glucose monomers compared to *R. jostii*, resulting in heightened viability and improved yields. Interestingly, all three strains yielded oils that could be implemented as a fungible diesel post-transesterification or utilized as a specialty oil for novel biochemical/bio-pharmaceutical applications. Overall, this work presents a novel method for waste stream optimization for bio-refineries.

474502 Biological Design of Lignin Conversion

Abstract:

The utilization of lignin for fungible fuels and chemicals remains one of the most imminent challenges in modern biorefinery. The fundamental challenge in lignin degradation lies in the highly recalcitrant structure that requests redox (oxidation-reduction) reactions to break down the chemical linkages. First, we addressed the challenges by revealing the chemical and biological mechanisms for synergistic lignin degradation via a bacterial and enzymatic system, which significantly improved lignin consumption, cell growth and lipid yield. Chemical analysis suggested that laccase, *R. opacus* cell and Fenton reaction reagents promoted the degradation of different functional groups, elucidating the chemical basis for the synergistic effects. Second, we further improved lignin depolymerization by establishing a new enzyme-mediator system to enhance the electron transfer in enzymatic lignin depolymerization to achieve more than 35% solubilization of kraft lignin without extreme temperature, pH, and pressure. The study has manifested that an efficient enzyme-mediator system can be exploited to establish a bioprocess to solubilize lignin, cleave lignin linkages, modify the structure, and produce substrates amenable to bioconversion, all of which are vital for improving current processing and conversion methods for lignin based biofuels and biomaterials. Third, based on the aforementioned studies, consolidated lignin conversion was engineered in both *Rhodococcus opacus* and *Pseudomonas putida* using systems biology-guided biodesign. Efficient ligninase secretion system has been established through proteomics-guided optimization of promoter/rbs, secretion peptide, and others. The engineered strains have led to significantly increased cell growth and bioproduct yield on lignin substrates. Overall, the biological and chemical design of lignin conversion has opened new avenues for lignin conversion toward various bioproducts.

459510 Selective Conversion of Guaiacol to Substituted Alkylphenols in Supercritical Ethanol over MoO_3

Abstract:

Recently, transition metal phosphides, nitrides, carbides and oxides have been intensively employed in biomass conversion{Ma, 2015 #41}.[1-4] The earth-abundant MoO_3 is regarded as a representative of promising alternative catalysts.[2] However, the bottleneck for the development of HDO technique over MoO_3 lies in the consumption of hydrogen and high temperature. In this work, the conversion of guaiacol to various substituted alkylphenols is investigated using MoO_3 in supercritical ethanol at a low temperature of 280 °C in inert nitrogen. Furthermore, the preliminary reaction pathways are proposed and the molybdenum active state is explored.

The conversion of guaiacol was carried out in a batch reactor at a temperature, in 260-300 °C, for a desired reaction time, 1-5 h over a MoO_3 catalyst. The product sample was injected neat into a GC-MS and a GC-FID instrument for analysis. Several possible reaction intermediates (phenol, catechol, anisole, 4-ethylphenol and 2-isopropylphenol) were employed as reactants to understand the transformation path of guaiacol. XRD, XPS and Raman spectroscopy were used to characterize the catalyst samples.

Fig. 1 exhibits that substituted phenols such as ethylphenols, iso-propylphenols, butylphenols (tert-, iso-) and tert-amylphenol were obtained as the dominating products with a high combined selectivity of 86% and at a conversion of 99% over MoO_3 under 280 °C for 4 h in supercritical ethanol.

To explore the reaction pathways of guaiacol conversion, control experiments with phenol, anisole and catechol as the starting materials under the same reaction condition were carried out. Catechol gave a complete conversion while

phenol and anisole did not. Besides, a similar product distribution compared to guaiacol as the reactant is achieved with catechol. Considering phenol was never found in the guaiacol conversion process, we propose that catechol is formed as the intermediate via demethylation of guaiacol and followed by direct conversion to alkylphenols without the formation of phenol. Subsequently, we employed 4-ethylphenol and 2-isopropylphenol as the reactants and obtained higher alkylphenols as the products. There were no isopropyl, tertiary (iso) butyl or tertiary pentyl derived aliphatic compounds or n-propyl, n-butyl, n-pentyl substituted phenols detected in the liquid and gaseous products of guaiacol conversion. Thus, we speculate that the higher alkylphenols are from consecutive substitution of lower alkylphenols with methyl and ethyl groups produced by ethanol medium, instead of direct substitution with corresponding substituent groups or isomerization with interrelated alkylphenols (Fig. 2).

The Mo (3d) energy region of used MoO_3 is shown in Fig. 3. As can be seen, it contains large amounts of Mo^{5+} (56.4%) with the $3d_{5/2}$ and $3d_{3/2}$ bands located at 230.9 eV and 234.1 eV, respectively. Moreover, molybdenum trioxide (Mo^{6+} state) and molybdenum dioxide (Mo^{4+} state) have been proved to be inactive in control experiments. Therefore, Mo^{5+} was considered to be responsible for the excellent performance of guaiacol conversion.

468094 Enabling Microbial Utilization of Thermally Depolyermized Lignin Monomers

Abstract:

The lignin component of biomass is an attractive source of carbon and energy for the microbial production of fuels and chemicals. This adding of value (*valorization*) to lignin by biological conversion can increase the overall economic viability of biorenewable fuel and chemical production. Thermal depolymerization of biomass through fast pyrolysis shows promise in terms of its ability to recover aromatic monomers from the lignin component of the biomass. Specialized microbes, such as *Pseudomonas putida* KT2440, can utilize many of these aromatic monomers when provided in pure culture [1,2]. However, microbes can be negatively affected by toxic impurities in biomass derived streams, and lignin rich pyrolysis fractions are insoluble in water and therefore inaccessible to microbes in liquid culture medium. Here we addressed these toxicity and insolubility issues by developing and testing an emulsion formula which granted *P. putida* KT2440 access to the phenolic monomers. The optimized emulsion formula contained 18.2 MΩ water, a 70:30 ratio of Tween20® to Span80® surfactant, and the lignin rich pyrolysis fraction. *P. putida* KT2440 was able to use this emulsion formulation as sole carbon source. Additionally, studies with a pure aromatic monomer, p-coumarate, indicated that the emulsion procedure at least partially protects *P. putida* KT2440 from the substrate toxicity of this monomer. To minimize surfactant cost, the surfactant amount was minimized to 7 wt% of the total lignin rich pyrolysis fraction weight. This emulsification method is an important step in enabling the biological valorization of biomass-derived lignin.

Refs: 1. Johnson, C. W.; Beckham, G. T. *Metabolic Engineering* **2015,** 28, 240-247. **2.** Vardon, D. R.; Franden, M. A.; Johnson, C. W.; Karp, E. M.; Guarnieri, M. T.; Linger, J. G.; Salm, M. J.; Strathmann, T. J.; Beckham, G. T. *Energy & Environmental Science* **2015,** 8, (2), 617-628.

472414 Fractionated Lignin Rheology and Structure

Abstract:

Critical technological challenges remain in the development of commercially viable lignin extraction and utilization processes to yield high value commodity chemicals and novel materials such as aromatics, polymers, adhesives, and flavorings. Processing strategies need to be developed for chemical and enzymatic degradation methods, catalytic upgrading, and purification to make it feasible for commercialization. Designing strategies to process multiphase lignin slurries requires understanding the slurry characteristics, especially quantification of the rheological properties. Fractionating heterogeneous lignin structures yield relatively homogeneous sets that offer simpler processing strategies as compared to inherently heterogeneous whole lignin. The objective of this work was to characterize rheological measurements of fractionated lignin and correlate the rheology to fractionated lignin structures. Organosolv and Kraft lignins were fractionated into five fractions using solvents of increasing polarity. Rheology of lignin fractions dissolved in DMSO was characterized with a rheometer programmed to sweep a range of shear rates for non-Newtonian slurries. ^{13}C nuclear magnetic resonance (NMR) analysis was performed to determine changes in structure and functional group contents of each fractional set. Non-Newtonian behavior increased with increasing lignin concentration within any fractionated group. Although structure varied, there were no obvious rheological differences across solvents or lignin types.

473182 Selective Conversion of Biomass-Derived Lignin to Cyclic Hydrocarbons

Abstract:

A potential value-added stream produced in the DMR process from NREL is the â€œnative-likeâ€ and tractable lignin residues isolated from deacetylation and enzymatic hydrolysis. Nobel metal (i.e.Ru, Rh, Pt) and earth abundant metals (i.e. Cu, Fe, Ni, Zn) immobilized on an acidic carbonaceous material act as efficient catalysts in the hydrodeoxygenation of technical lignins under batch conditions. A synergistic interaction between the metal sites and acid groups within the bifunctional and bimetallic catalyst leads to enhanced catalytic activities for the overall transformation of technical lignins to cyclic hydrocarbons as compared to the individual steps catalyzed by the separate catalytic functionalities. In addition, the selectivity of catalysts will also be addressed in our presentation.

470224 The Evolution of Multi-Functional Nanoporous Metal Composite Electrocatalysts

Abstract:

The unique physicochemical properties of nanoscale materials and the continual development of synthesis and fabrication techniques has led to a renaissance in the production of complex catalytic materials with increasing degrees of multi-functionality. Nanoporous metals are an interesting subset of nanoscale electrocatalytic materials as they possess a high surface-to-volume ratio, are intrinsically conductive, are core-shell in nature, and their morphology, composition and compositional gradient are readily tuned through a manipulation of dealloying conditions. Here we present our mechanism for the formation and evolution of nanoporous metal morphology and their application to the catalysis of several elementary electrochemical reactions including the oxygen reduction reaction (ORR) and oxygen evolution reaction (OER)[1]. Nanoporous metals are inherently metastable. In addition to the standard mechanisms of catalyst deactivation and surface area loss including Ostwald ripening and coalescence, surface energy driven coarsening is a new pathway for the loss of electrochemically active surface area in nanoporous materials. With a better understanding of the interplay between nanoporous structure coarsening and transition metal corrosion, we can propose strategies to mitigate coarsening and improve operational catalyst stability. Additional utility of these porous materials lies in the ability to take advantage of the interior volume of the nanoporous metals, incorporating a material that can address the various kinetic limitations of the catalyst. Here we demonstrate the use of hydrophobic, protic ionic liquids on the ORR where the unique properties of the ionic liquid aid the optimization of the interaction between catalyst, reactive intermediates and water.

[1] Snyder, J.; Livi, K. & Erlebacher, J. *Adv. Funct. Mater.* **23,** 5494-5501 (2013).

472499 Effect of Doping on the Activity of Nickelate Oxides Toward Surface Oxygen Exchange and Oxygen Reduction

Abstract:

The first-order Ruddlesden-Popper (RP) series (A_2BO_4) oxides, known as nickelate oxides (e.g. $La_2NiO_{4+Î´}$, LNO), have recently attracted increasing interest due to their good mixed ionic and electrical conductivities suitable for the electrocatalysis. For instance, they have been reported as potential cathode materials for intermediate temperature (600-800 â„ƒ) solid oxide fuel cells (IT-SOFCs)[1]. We have previously demonstrated that the morphology and surface structure of lanthanum nickelate oxide can significantly affect its activity toward the surface oxygen exchange/reduction.[2] We found that LNO nanorods highly terminated by Ni oxide (001) surfaces exhibit superior performance to the traditional nanosphere LNO catalysts. To further enhance the activity of nickelate oxides, in this contribution we explore the effect of doping on the activity toward the surface oxygen exchange/reduction process using density functional theory (DFT) calculations combined with experimental studies. We focus on the substitution of the Ni B-site with Mn, Fe, Co, and Cu transition metals, and the La A-site with Ca, Sr, and Ba to tune the activity.

The surface lattice O diffusion into the bulk to form surface O vacancy is investigated first, and two possible diffusion pathways are proposed. The apical O assisted pathway is found to be more favorable. We find that the surface chemistry of La_2NiO_4 can indeed be tuned via modification of the Ni B-site and La A-site. With the B-site metal changing from Mn, Fe, Co, to Ni, the calculated binding energy of O_2 on the surface O vacancy neighboring the transition metal becomes gradually weaker. Moreover, the barrier for O_2 dissociation becomes linearly higher as the binding energy of O_2 becomes weaker, whereas, the surface O vacancy formation energy becomes lower. A volcano-type relationship between the calculated rates and the binding energies of O_2 is found, suggesting that the binding energy of O_2 might be a good descriptor to screen for nickelate oxides with optimal oxygen exchange activity.

References

1. A. Tarancon, M. Burriel, J. Santiso, S. J. Skinner, J. A. Kilner, J. Mater. Chem. 2010, 20, 3799.

2. X. F. Ma, J. S. A. Carneiro, X. K. Gu, H. Qin, H. L. Xin, K. Sun, E. Nikolla, ACS Catal. 2015, 5, 4013.

462588 Oxyhydroxides As a Platform for Confined Oxygen Electrochemistry

Abstract:

Oxyhydroxides as a Platform for Confined Oxygen Electrochemistry
Authors: Andrew D. Doyle, Michal Bajdich, Aleksandra Vojvodic

As we move towards fossil-free electrification a number of strong renewable energy technologies have emerged. However, some promising candidates (notably solar and wind power) provide only intermittent energy generation, which cannot immediately meet the total demand on a national scale. One proposed solution is to stabilize these renewable sources by coupling to external energy storage, including regenerative fuel cells which convert energy between electricity and the chemical bonds in water, oxygen, and hydrogen.

Unfortunately, researchers have shown that the catalytic efficiency of regenerative fuel cell technology is fundamentally limited by energetic scaling between reaction intermediates [1], [2]. However, our recent results have demonstrated that nanoscale channels may circumvent this scaling, and in the process allow for higher efficiency devices [3].

Additionally, experimental work has shown that doped NiOOH and similarly-structured materials have promise as possible next-generation catalysts for oxygen electrochemistry (water oxidation and oxygen reduction), possibly due to their unique crystal structure [4], [5].

In this work we will present results from density functional theory (DFT) calculations to explore NiOOH as a natural template for sub-nanometer channel formation with particular emphasis on the activity of bulk-like sites towards the oxygen evolution reaction. We will discuss the energetics of binding between hydroxide layers (relative to the surface), the impact of iron doping, and the possibility that subsurface activity contributes to the measured activity.

[1] I. C. Man, H.-Y. Su, F. Calle-Vallejo, H. a. Hansen, J. I. Martínez, N. G. Inoglu, J. Kitchin, T. F. Jaramillo, J. K. Nørskov, and J. Rossmeisl, "Universality in Oxygen Evolution Electrocatalysis on Oxide Surfaces," *ChemCatChem*, vol. 3, no. 7, pp. 1159–1165, Jul. 2011.

[2] V. Viswanathan, H. A. Hansen, J. Rossmeisl, and J. K. Nørskov, "Universality in Oxygen Reduction Electrocatalysis on Metal Surfaces," *ACS Catal.*, vol. 2, no. 8, pp. 1654–1660, Aug. 2012.

[3] A. D. Doyle, J. H. Montoya, and A. Vojvodic, "Improving Oxygen Electrochemistry through Nanoscopic Confinement," *ChemCatChem*, vol. 94025, 2015.

[4] M. S. Burke, S. Zou, L. J. Enman, J. E. Kellon, C. a Gabor, E. Pledger, and S. W. Boettcher, "Revised Oxygen Evolution Reaction Activity Trends for First-Row Transition-Metal (Oxy)hydroxides in Alkaline Media," 2015.

[5] K. Klingan, F. Ringleb, I. Zaharieva, J. Heidkamp, P. Chernev, D. Gonzalez-Flores, M. Risch, A. Fischer, and H. Dau, "Water oxidation by amorphous cobalt-based oxides: Volume activity and proton transfer to electrolyte bases," *ChemSusChem*, vol. 7, no. 5, pp. 1301–1310, 2014.

455764 Large-Size Graphene Tube Catalysts for Bifuncational Oxygen Reduction and Oxygen Evolution Electrocatalysis in Alkaline Media

Abstract:

The oxygen reduction reaction (ORR) and the oxygen evolution reaction (OER) are one pair of the most technologically important reactions for a variety of electrochemical energy storage and conversion technologies (*e.g.*, fuel cells, metal-air batteries, and water splitting). It is important to design and develop advanced nonprecious-metal catalysts capable of facilitating these sluggish reactions at sufficient rates for practical applications. In our recent efforts,[1-7] cost-effective nitrogen-doped graphene nanocomposites have been prepared *via* the graphitization of nitrogen-containing polymers or compounds, combined with transition metals (Co, Ni or Fe). Among others, a new type of N-doped carbon tubes with large diameters (up to 500 nm) and relatively thin walls (less than 10 layers), which we call graphene tubes (GTs), were prepared via a scalable graphitization process of inexpensive dicyandiamide. Due to the high ratio of diameter-to-wall thickness, much increased surface area was obtained compared to conventional multi-walled carbon nanotubes (MWNTs). An effective strategy for tuning the size of large-diameter nitrogen-doped graphene tubes (GT) from 50 to 200 nm was developed by varying the transition metals used to catalyze the graphitization. This effort demonstrates optimal manipulation of morphology and surface area, thereby providing an effective approach to further improving the performance of nonprecious metal catalysts.

Furthermore, aiming to improve the activity and stability of the state-of-the-art Pt catalysts, the ORR active GT is used as a matrix to disperse Pt nanoparticles in order to build a unique hybrid Pt cathode catalyst. This is the first demonstration of the integration of a nonprecious metal catalyst with Pt catalyst in a nanocomposite. Relative to traditional Pt/C catalyst, much improved activity and stability were achieved. This work provides a new hybrid concept to design and synthesis novel cathode catalysts for fuel cells.

Apart from dramatically enhanced ORR activity, the large-size graphene tubes demonstrated excellent OER activity comparable to state-of-the-art Ir black in alkaline media. Importantly, unlike traditional carbon blacks and 2D reduced graphene oxides, the graphene tubes can be remain stable in harsh electrochemical environments (potentials up to 1.9 V) in alkaline media. This provide a great opportunity to develop bifunctional oxygen catalysts for reversible electrochemical energy applications such as reversible alkaline media and metal-air batteries.

References

1. G. Wu, K.L. More, C.M. Johnston, P. Zelenay, *Science*, 332, 443, 2011.
2. G. Wu, P. Zelenay, *Acc. Chem. Res*., 46, 1878, 2013.
3. G. Wu, N.H. Mack, W. Gao, S. Ma, J. Han, J.K. Baldwin, P. Zelenay. *ACS Nano*. 6, 9764, 2012.
4. Q. Li, P. Xu, W. Gao, S. Ma, R. Cao, J. Cho, H.L. Wang, G Wu, *Adv. Mater*., 26, 1378,2014.
5. Q Li, H Pan, D Higgins, R Cao, G Zhang, H Lv, K Wu, J Cho, G Wu, *Small* 11, 1443, 2015.
6. Q. Li, R. Cao, J. Cho, G. Wu, *Adv. Energy Mater*., 4, 1301415,2014.
7. X Wang, Q Li, H Pan, Y Lin, Y Ke, H Sheng, MT Swihart, G Wu, *Nanoscale* 7, 20290, 2015.

464805 *Atomic Layer Deposition of $NiAl_xO_y$ Catalysts for Electrochemical Oxidation of Water*

Abstract:

The electrolysis of water to form hydrogen gas and oxygen gas has been highlighted as a promising method to store electrical/solar energy in the form of chemical bonds. Of the two half-reactions involved in electrolysis, the oxygen evolution reaction (OER), 4 OH^- â†'Â 2 H_2O +O_2 + 4e^- (in alkaline conditions), suffers from slow kinetics leading to losses in efficiency. Of earth-abundant materials, nickel-iron oxide based catalysts have been found to be among the most active for OER. Recently, it has been shown that the activity of these nickel-iron oxide catalyst can be further improved through the addition of aluminum.[1]

To study the effects of aluminum on the activity of nickel oxide based catalyst, atomic layer deposition (ALD) was used to synthesize ternary $NiAl_xO_y$ films. ALD allows for the deposition of conformal thin films with angstrom-level control over thickness.Â Depositing ternary materials with ALD is shown to provide good compositional control by switching between the individual ALD chemistries. Electrochemical characterization of these synthesized thin films shows that aluminum improves the onset potential for OER and the Tafel slope when compared to nickel oxide films.

Studies of the intrinsic activity of nickel aluminum oxide catalyst are complicated by two factors: (1) nickel oxide films are prone to incidental incorporation of iron from trace impurities found in the electrolyte, (2) aluminum oxide leaches in alkaline conditions leading to changes in composition and porosity of the catalytic films. This work addresses the first issue through additional purification of the electrolyte to remove trace iron. With an iron free environment, nickel aluminum oxide catalyst is shown to outperform nickel oxide catalyst. When $NiAl_xO_y$ films are allowed to age in electrolyte containing trace iron, their activity significantly improves, performing better than as-deposited $NiFe_xO_y$ films. The second complicating factor was further explored by comparing thin and thick films to examine the effect of porosity and its effect on performance. Though the as-deposited $NiAl_xO_y$ films are initially unstable under alkaline conditions, our studies show that the films retain their high activity after aluminum leaching equilibrates with the electrolyte.

1. James B. Gerken, Sarah E. Shaner, Robert C. Massé, Nicholas J. Porubsky, and Shannon S. Stahl. *Energy Environ. Sci.,2014, 7, 2376*

468156 The Role of Superoxide in the Non-Aqueous Oxygen Reduction Reaction in Li-O_2 Batteries

Abstract:

The non-aqueous Li-air (O_2) battery has attracted much research interest because it can potentially far exceed the gravimetric energy storage density than the current rechargeable battery technologies. The oxygen reduction reaction by Li (Li-ORR) is a key electrochemical process at a Li-O_2 cathode. Presently Li-O_2 batteries suffer from poor round-trip efficiency, low current density, and capacity fading,[1-3] To overcome these challenges, it is crucial to have a fundamental understanding of the reactive O_2 species involved in the ORR on different electrodes.

Recent studies have shown that the inclusion of metals in the cathode of Li-O_2 batteries can alter the cathode efficiency and possibly alter the nature of the discharge product.[1,2,4-8] We have performed a series of theoretical calculations of Li-ORR on metals and other materials. Our results show that the discharge potential is controlled by the formation of intermediates on the electrode surface. For instance, the initial O_2 reduction on Au occurs associatively with the formation of molecular lithium superoxide (LiO_2) at an equilibrium potential of ca. 2.0 V.[9] The *intrinsic* activity of Li-ORR on different metals (including Au, Ag, Pt, Pd, Ir, and Ru) forms a volcano-like trend with respect to the adsorption energy of oxygen.[9] Similar to the hydrogen-ORR, Pt and Pd are the most active of these metals, which has been borne out experimentally.10 This can be traced to the competition between the electrode on one hand and Li on the other hand for bonding with the oxygen.

The aprotic nature of typical Li-O_2 cells, however, makes the one-electron reduction of O_2 to the superoxide anion (O_2^-) possible (i.e., without concurrent Li^+ transfer), and allows O_2^- to have appreciable lifetime, particularly on a

relatively inert metal such as Au. O_2^- formation would occur at appreciably higher potential than molecular LiO_2 on many of the metals. DFT calculations, in conjunction with spectro-electrochemical studies,[11,12] therefore suggest that O_2^- formation needs to be taken into account when predicting the observed activity of ORR in Li-O_2 cells. On Au, for instance, O_2^- formation is the key process that corresponds to the discharge of a Li-O_2 cell at low overpotentials and determines how and where the solid discharge products are formed.[12] Our findings contribute to an improved understanding of the ORR in non-aqueous cells and to the further development of Li-O_2 batteries.

References

(1) Abraham, K. M.; Jiang, Z. *J. Electrochem. Soc.* **1996**, *143*, 1.

(2) Débart, A.; Bao, J.; Armstrong, G.; Bruce, P. G. *J. Power Sources* **2007**, *174*, 1177.

(3) Girishkumar, G.; McCloskey, B.; Luntz, A. C.; Swanson, S.; Wilcke, W. *J. Phys. Chem. Lett.* **2010**, *1*, 2193.

(4) Allen, C. J.; Mukerjee, S.; Plichta, E. J.; Hendrickson, M. A.; Abraham, K. M. *J. Phys. Chem. Lett.* **2011**, *2*, 2420.

(5) Lu, Y.-C.; Gasteiger, H. A.; Shao-Horn, Y. *J. Am. Chem. Soc.* **2011**, *133*, 19048.

(6) Peng, Z.; Freunberger, S. A.; Chen, Y.; Bruce, P. G. *Science* **2012**, *337*, 563.

(7) Lu, J.; Lei, Y.; Lau, K. C.; Luo, X. Y.; Du, P.; Wen, J. G.; Assary, R. S.; Das, U.; Miller, D. J.; Elam, J. W.*et al. Nat. Commun.* **2013**, *4*, 2383.

(8) Lu, J.; Jung Lee, Y.; Luo, X.; Chun Lau, K.; Asadi, M.; Wang, H.-H.; Brombosz, S.; Wen, J.; Zhai, D.; Chen, Z.*et al. Nature* **2016**, *529*, 377.

(9) Dathar, G. K. P.; Shelton, W. A.; Xu, Y. *J. Phys. Chem. Lett.* **2012**, *3*, 891.

(10) Lu, Y.-C.; Gallant, B. M.; Kwabi, D. G.; Harding, J. R.; Mitchell, R. R.; Whittingham, M. S.; Shao-Horn, Y. *Energ. Environ. Sci.* **2013**, *6*, 750.

(11) Peng, Z.; Chen, Y.; Bruce, P. G.; Xu, Y. *Angew. Chem. Int. Edit.* **2015**, *54*, 8165.

(12) Zhang, Y.; Zhang, X.; Wang, J.; McKee, W. C.; Xu, Y.; Peng, Z. *J. Phys. Chem. C* **2016**, *120*, 3690.

464819 Developing Structure-Property Relationships Between Reactant Structures, Ionic Liquids, and Reaction Rate Constant

Abstract:

Recent years have seen a burgeoning interest among researchers in the prediction of reaction rate constants using structure-property relationships. The rate constant has been modeled in terms of reactant structures (Chaudry and Popelier, 2003), solvent structures (Zhou et al., 2014) or a combination of reactant and solvent structures (Datta et al., 2016). The development of models for prediction of rate constants serve two primary purposes. First, such models can be utilized for the computer-aided molecular design (CAMD) of reactants and solvents such that the reactions are accelerated. Secondly, such models can help avoid conducting many cumbersome experiments in order to calculate the rate constant. With regards to CAMD, Zhou et al. (2014) recently designed solvents that enhance the rate of the Diels-Alder reaction. While much work has been carried out to model the rate constant in terms of molecular solvents and molecular descriptors that represent them, comparatively less work has been performed in modeling the rate constant in terms of the structures of ionic liquids. Ionic liquids offer a range of beneficial properties and hence have captured the interest of researchers in recent years. These include, negligible vapor pressure under standard conditions, high thermal stability and structural tunability (Priede et al., 2015). To the best of our knowledge, no previous studies have been published on rate constant modeling with respect to structures of both ionic liquids and reactants. Through our work, we wish to fill this research gap. We have utilized multiple linear regression (MLR), principal component analysis (PCA) and support vector regression in our work to model the reaction rate constant of the Knoevenagel condensation reaction. The rate constant has been modeled in terms of topological indices of reactants and ionic liquids. The model performance has been compared using ten-fold validation.

References:

U. A. Chaudry, P.L.A. Popelier, 2003, J. Chem. Phys. Chem. A, 107, 4578-4582

T. Zhou, K. Mcbride, X. Zhang, Z. Qi, K. Sundmacher, 2014, AIChE Journal, 61, 1, 147-158

468355 Reverse Design of Ionic Liquids for CO_2 Absorption

Abstract:

Chemical engineering design problems have evolved in recent years from large scale bulk processes to targeted small batch approaches. As a result, product design is reaching beyond the traditional solutions to classic chemical engineering problems such as extraction, purification, removal of sulfur from fuel, CO_2 capture and sequestration, energy storage and biomass pre-treatment. Ionic liquids can provide environmentally benign solutions that can be tailored to specific process requirements. Due to the unique structure of ionic liquids, the cation, anion and length of the alkyl chain can be varied to create a molecule with specific physical properties. It has been estimated that over 10^{14} unique molecules exist at room temperature, so experimental trial and error methods will not be effective. More advanced methods must be created determine the ideal combination for a specific set of process requirements.

The purpose of this work is to determine the ideal ionic liquid to capture CO_2, a question which has been the topic of many experimental laboratory studies. However, only a small percentage of the potential ionic liquids have been synthesized and tested for the solubility of CO_2. This innovative approach decouples the solution from time-consuming laboratory study and streamlines the search for the most promising candidates that should be synthesized.

A characterization based group contribution method is combined with density functional theory to determine the ionic liquid that can most effectively absorb CO_2. Infrared spectra data contains descriptor data that can be used to estimate properties of ionic liquids, but does not exist for all ionic liquids. Density functional theory is used to create IR data based on a training set of experimental data. Principal component analysis and partial least square techniques will be employed to reveal important features and patterns in the molecular architecture. A characterization based group contribution method is used to estimate the properties of the new structures. The reverse design of potential ionic liquid molecules will be completed by an exhaustive search of combinations with various cation, anions and lengths of alkyl chains until a candidate molecule is found that provides the highest solubility of CO_2.

461251 Application of Systematic Methodology for Design of Tailor-Made Blended Products: Lubricant Design

Abstract:

Application of systematic methodology for design of tailor-made blended products: Lubricant design.

Marina Fedorova[1], Laurent Grosset[2], Fang Zhou[3], Rafiqul Gani[1]

[1] Department of Chemical and Biochemical Engineering, Technical University of Denmark, Lyngby, Denmark

[2] TOTAL Marketing&Services, Solaize, France

[3] Tsinghua University, Beijung, China

Over past decades, chemical product engineering has received much attention among the chemical engineering community. This is due to the transformation of industries in manufacturing and selling chemical products based on the product performance rather than compositional specifications.

Traditional methods used in new product design/development combine a broad knowledge of existing products with scientific experimentation. The chemical product is developed based on scientific hypothesis, intuition or simple trial-and-error. However, this approach is costly, very time consuming and only a limited number of experiments can be performed. In order to efficiently design chemical products, a systematic methodology followed by focused experimental verification is a better approach. It can be implemented at the initial stage of the design, where it could minimize the required number of experiments, as well as to identify possible solutions that normally would lie outside the scope of traditional approaches, thus increasing the chances of finding better solutions with less resources.

The systematic methodology for solving mixture/blend design problem consists of four main tasks: 1) problem definition – identification of the final product needs and translations of them into target properties; 2) property models identification; 3) mixture/blend design; 4) experimental or model-based verification. Additional tools and methods needed to perform a specific task in the methodology, which are the property model library, the chemicals database and the mixture/blend design algorithm, also need to be available.

Application of the methodology is highlighted with several case studies related to lubricant design. Lubricants are required in almost all modern machines. Lubricating oil is a substance introduced between two moving surfaces to reduce the friction between them, improving efficiency and reducing wear. The functions of the lubricant depending on application may also include cooling, cleaning and suspending, protection and transfer power.

The application of lubricant is classified into two types – engine lubricants and non-engine lubricants. In this work, the lubricants are designed as engine oils, which is the largest application of lubricants. Typically lubricants are a mixture of base oil (65-98%) and performance additives to achieve required performance and end-user requirements. The target properties for the lubricant blend can be divided into performance properties, long life properties and environmental properties.

Focus of this work is performance properties, as they are central for any lubricant application. These properties include, for example, viscosity, thermal properties, low and high temperature properties, friction coefficient and film thickness that lubricant provides. The first part of the case study is to design the base oil, which is the main component in lubricant. The second part focuses on the development of the specific lubricant additive - its properties are thoroughly studied and the database of the chemicals is created to be used in the mixture design. Currently this database includes 1202 items. Novel blends with besigned functional properties are found in all cases. The method is generic and also be applied for other blended products.

468374 Physical Properties for Lipids Based Process and Product Design

Abstract:

Lipid processing covers several oil and fats technologies such as: edible oil production, biodiesel production, oleochemicals (e.g.: food additives, detergents) and pharmaceutical product manufacturing. New demands regarding design and development of better products and more sustainable processes related to lipids technology, emerge according to consumers demanding improved product manufacturing from sustainable resources and new legislation regarding environmental safety [1]. Physical and thermodynamic property data and models for prediction of pure compound properties and mixtures properties involving lipids represent the basic and most important requirements for process-product design, simulation and optimization. Experimentally measured values of involved compounds are desirable, but in most of the cases these are not available for all the compounds and properties needed. The lack of properties is even larger for mixtures properties. Therefore there is a highly need of predictive properties models.

Oleochemical industry, along with all other lipids processing technologies, presents many challenges in relation to thermodynamic and physical properties as a result of lack of experimental data for pure compounds and their mixtures. Due to these gaps, lipids involved industries are not able to exploit all the advances achieved within computer aided methods and tools, for which property data and models represent the basic requirements to perform process design, simulation and optimization, as well as product design. Another challenge comes from the availability of the data and models within commercial available software [2].

Recent developments within lipid property modelling, as well as new physical and thermodynamic data availability, allowed the development of our in-house SPEED Lipids Database. The database is directly connected with external computer aided tools, such as PROII and ICAS suite, and it is organised within two parts: one is dedicated to the pure compounds and their properties, while the other is dedicated to the mixtures phase equilibria. In product-process design, the chemicals database together with the lipids database is used.

For the development of the database â€“ pure compound part, available data for lipids in terms of their molecular description and a range of properties were collected in special attention to the needs for model-based design and analysis within oleochemical industry. Available experimental data was collected for the identified lipids and their corresponding properties. The database contains 330 pure compounds organised within 18 categories (e.g.: fatty acids, triacylglycerides, diacylglicerides, monoacylglycerides, methyl esters, ethyl esters, isopropyl esters, carotenoids, triterpenealcohols, phospholipids, fatty alcohols, sterols, vitamin E, etc.). The experimental data was collected for 7 primary properties (e.g.: critical temperature, pressure and volume, Gibbs free energy of formation, enthalpy of formation, melting point, and normal boiling point), and 6 temperature dependent properties (vapour pressure, liquid heat capacity, liquid density, liquid thermal conductivity, viscosity, and surface tension). To fill in the gaps within the database models for 14 primary properties (e.g.: molecular weight, critical properties, melting point, normal boiling point, enthalpy of formation, enthalpy of fusion, Gibbs fee enthalpy of formation, dipole moment, acentric factor, specific gravity, dipole moment, liquid volume) and 10 temperature dependent properties (e.g.: vapour pressure, liquid enthalpy, liquid viscosity, liquid thermal conductivity, ideal enthalpy, surface tension, latent heat, liquid density, vapour density, vapour thermal conductivity) were implemented. The chemicals database contains 45500 chemicals and more than 27 different properties.

The mixture phase equilibria component of the lipids database contains a collection of over 4500 measured data point corresponding 332 different phase equilibria data sets (92 VLE, 91 LLE, 70 SLE and 79 solubility data) for binary and multi component mixtures. Consistency test were performed for all the VLE and SLE data sets available in the database. From the available VLE data sets, original UNIFAC parameters were regressed, allowing better representation and prediction of the phase equilibria.

The following features are available within SPEED Lipids database: property data consistency test; optimization based data regression to estimate the model parameters; knowledge representation and search engine for database use knowledge representation and search engine for database; computer aided modelling tool to quickly and efficiently develop the necessary property prediction models. Also, a systematic work flow for database maintenance and development is provided. All databases are integrated with process and product simulators. The presentation will highlight the use of the database and property models in case studies involving process-product design.

Reference

[1] Diaz-Tovar C, Mustaffa A., Hukkerikar, A., Quaglia A, Sin G, Kontogeorgis G, Sarup B, Gani R. Lipid Processing Technology: Building a Multilevel Modeling Network. In: Pistikopoulos E. N., Georgiadis M. C., Kokossis A, editors. Proceedings of the 21st European Symposium on Computer Aided Process Engineering 2011, 29:256â€“260.

[2] Sarup B. Advances and Challenges in Modelling of Processing of Lipids. In: Gernaey K. V., Huusom J. K., Gani R., editors. Proceedings of the 25th European Symposium on Computer Aided Process Engineering 2015, 37:117-122.

471395 an Extension of COSMO-Based Methodologies for Computer-Aided Mixture Design

Abstract:

Group contribution methods like UNIFAC have been successful in capturing thermodynamics in mixture design problems while remaining amenable to a mathematical optimization framework. Despite being accurate in many domains, these methods have an inherent limitation: because they are often regressed over large datasets of neutral, ground state structures, they can only consider such structures when applied to predict mixture thermodynamics. This means that UNIFAC-like methods cannot be extended to arbitrary systems as they would fail to capture the electronic complexities of species like transition states, reactive intermediates, and ionic liquids. In this work, we propose the use of quantum chemistry calculations via the COSMO solvation model as a means to incorporate arbitrary systems into mixture design problems. Every component of these systems that is not in the design space can be modelled at an ***ab initio*** level of accuracy and then mapped to a charge density histogram called a sigma profile. In COSMO post-processing methodologies like COSMO-RS[1] and –SAC[2], these sigma profiles represent the main information required to estimate mixture thermodynamics properties. To incorporate these techniques into a mixture design problem, we develop group contribution methods to estimate the sigma profiles of molecules in the design space. That is, we only rely on group contribution estimates for every species that is not fixed in the mixture. We also include several extensions to our existing COSMO-based mixture design methodology[3] to improve the accuracy of the estimates.

We address the optimization of the mixture design problem using derivative-free optimization (DFO) methods. By projecting the problem onto a lower-dimensional space, we exploit the efficiency of DFO techniques at solving problems with few degrees of freedom. We apply these techniques to three case studies: (1) a reaction rates optimization problem, (2) control of a halogen-metal exchange reaction's selectivity, and (3) maximizing the selectivity of an intramolecular nucleophilic aromatic substitution reaction. It is important to note that all of these problems cannot be modeled directly with UNIFAC-like

approaches as they either involve species which do not have available UNIFAC groups or exhibit complex electronic properties. We consider all of these problems first from the perspective of unrestricted solvent design, meaning we consider mixtures of solvents which may not exist. Next, we consider all of these problems from a more practical point of view, designing an optimal blend of common laboratory and industrial solvents. Overall, these case studies demonstrate the ability of COSMO-based approaches to incorporate highly accurate quantum chemical information directly into mixture design applications.

[1] F. Eckert and A. Klamt. Fast solvent screening via quantum chemistry: COSMO RS approach.. ***AIChE Journal.*** **2002.** 48(2), 369-385.

[2] S. T. Lin and S. I. Sandler. A priori phase equilibrium prediction from a segment contribution solvation model. ***Industrial & engineering chemistry research.*** **2002.** 41(5), 899-913.

[3] N. D. Austin, N. V. Sahinidis, and D. W. Trahan . A COSMO-based approach to computer-aided mixture design. ***Chemical Engineering Science.*** **2016.** ***submitted***

465531 A Mixed-Integer Linear Programming Approach for the Design of Nanostructured Catalysts

Abstract:

Recent advances in nanotechnology have enabled the fabrication of materials with precision on the micro- or even nano-scale. These precisely fabricated materials have properties that are wildly different from their bulk counterparts, enhancing useful functionalities in a number of application contexts. For metallic nanoparticles, in particular, recent work has highlighted enhanced catalytic activities [1], unique optical properties [2], and the potential for use in drug delivery [3], among other exploitable properties. With a flourishing synthesis capability comes the need to answer the question: "Which design leads to the best catalytic performance?" Unfortunately, the answer to this question is not always obvious due to the combinatorial number of ways in which matter can arrange itself. However, such a material design problem can be formalized as a mathematical optimization problem like those traditionally cast and solved in many process systems engineering contexts. In essence, one seeks to identify a material design (i.e., a specific instantiation of the given class of materials of interest) that maximizes some metric of performance.

In this work, we focus on the design of nanostructured surfaces of transition metal crystals, a material class that has received a lot of attention due to its potential for catalyzing many important reactions [4,5]. We show how correlations that link catalytic activity to site descriptors, such as commonly available volcano plots [6,7], can be used to determine those surface designs that maximize the total reactivity. Volcano plots in practice tend to have steep sides, implying that only the "ideal" sites, i.e., those exhibiting features that lie very close to the peak of the volcano, contribute substantially to the overall reaction rate [7,8,9]. With this observation in mind, the problem of designing the surface of highest reactivity simplifies to that of designing the surface that packs the most ideal sites in the unit area, essentially neglecting the contributions from the remaining, non-ideal sites. Furthermore, when the material designs can be expressed via discrete variables, as is the case with materials conforming to well-defined crystalline lattices, and when appropriate site descriptors can be encoded via linear constraints and disjunctions, as is the case of descriptors based on site coordination numbers, we can cast mixed-integer linear programming (MILP) models and utilize well-developed MILP optimization software in order to determine the optimal surface packings. Besides the tractability advantages it offers, an ancillary benefit of using MILP solvers to address this design problem is that several optimal or near-optimal solutions can be collected at a marginal computational cost via a solution pool approach. In this manner, multiple designs can be identified for comparison against some secondary criteria or further consideration by experts.

The proposed MILP approach depends on a suitably defined graph, which serves as a "design canvas." The nodes of this graph constitute the lattice locations present in a periodic, space-filling tile of a thin layer on the surface of a crystal, while the edges of the graph signify those locations that are considered neighbors for determining a site's coordination number. Such a tile can be enforced to be periodic and replicated ad infinitum along a plane so as to form a full surface. The key decision variables in the model are a set of binary variables which indicate the presence or absence of metal atoms in the crystalline lattice locations. A variety of auxiliary site descriptors can then be encoded, as well as variables to indicate whether a location constitutes an ideal site.

Using our approach, we conducted extensive computational studies involving an array of crystallographic lattices. These design lattices were tested against a set of structure function relationships spanning a range of ideal site descriptors in order to identify optimal surface patterns relating to a range of catalytic systems. Our proposed approach validates certain surface patterns that were previously known to be optimal but also reveals a number of non-intuitive designs. We also demonstrate that the optimal surface patterns depend strongly on the target application, which strengthens the need to keep developing those fabrication capabilities that can yield higher levels of structural control. Additionally, it was found that the periodicity of the design canvas leads to considerable symmetry in our MILP model, which retards the progress of branch and bound search algorithms. To that end, we also investigate several symmetry breaking symmetry strategies and report the ones that appear to work best in practice.

[1] Narayanan R., El-Sayed M. A., "Shape-dependent catalytic activity of platinum nanoparticles in colloidal solution," *Nano Letters*, 4(7):1343–1348, 2004.

[2] Halas N. J., "Playing with Plasmons: Tuning the Optical Resonant Properties of Metallic Nanoshells," *MRS Bulletin*, 30(5):362–367, 2005.

[3] Caldorera-Moore M., Guimard N., Shi L., Roy K., "Designer nanoparticles: incorporating size, shape and triggered release into nanoscale drug carriers," *Expert opinion on drug delivery*, 7(4):479–495, 2010.

[4] Hammer B., Norskov J. K., "Theoretical Surface Science and Catalysis — Calculations and Concepts," *Advances in Catalysis*, 45:71–129, 2000.

[5] Stamenkovic V., Mun B. S., Mayrhofer K. J. J., Ross P. N., Markovic N. M., Rossmeisl J., Greeley J., Nørskov J. K., "Changing the activity of electrocatalysts for oxygen reduction by tuning the surface electronic structure," *Angewandte Chemie - International Edition*, 45(18):2897–2901, 2006.

[6] Nørskov J. K., Bligaard T., Rossmeisl J., Christensen C. H., "Towards the computational design of solid catalysts," *Nature Chemistry*, 1(1):37–46, 2009.

[7] Nørskov J. K., Bligaard T., Logadottir A., Bahn S., Hansen L. B., Bollinger M., Bengaard H., Hammer B., Sljivancanin Z., Mavrikakis M., Xu Y., Dahl S., Jacobsen C. J. H., "Universality in Heterogeneous Catalysis," *Journal of Catalysis*, 209(2):275–278, 2002.

[8] Calle-Vallejo F., Martínez J. I., García-Lastra J. M., Sautet P., Loffreda D., "Fast prediction of adsorption properties for platinum nanocatalysts with generalized coordination numbers," *Angewandte Chemie - International Edition*, 53(32):8316–8319, 2014.

[9] Calle-Vallejo F., Loffreda D., Koper M. T. M., Sautet P., "Introducing structural sensitivity into adsorption-energy scaling relations by means of coordination numbers," *Nature chemistry*, 7(5):403–410, 2015.

466287 First-Principles Rheological Modelling of Smart Drilling Nanofluids

Abstract:

Oil drilling campaigns need rely on reliable, high-performance drilling fluids which can facilitate in-situ control of fluid rheology, making their transport properties tunable. This benefits their application in different downhole environments, where a wide range of cumbersome technical and economic challenges arise. A nanofluid can perform several tasks in the fluid system, and facilitate drilling with a reduction in total solid and/or chemical content of the equivalent mud, whilst also reducing costs. Smart drilling fluids ensure greater filter cake (and thus well) stability in Low Pressure-Low Temperature (LPLT) conditions, thus enabling lower fluid losses, greater well stability and remarkably reduced capital expenditure (Mahmoud et al., 2016).

The development of first-principles models for the rheology of nanofluids is essential because it can effectively guide their tailored preparation: characterising the nanofluid behaviour as a function of shear rate, nanoparticle volume fraction and temperature is critical toward modelling, design and planning of cost-effective drilling campaigns. Upon successful validation, these models have general applicability to all function forms of nanoparticle-enhanced drilling fluids, eliminating the need for limited-value approximations and ad hoc parameterisations of fluid rheology.

First-principle rheological models describing the shear stress and viscosity of novel nanoparticle-based fluids have been derived on the basis of an isoelectric point assumption (Reilly et al., 2016). Model development relies on establishing a force balance between the Stokes drag force exerted upon the (idealised nanoparticle) sphere and the van der Waals interparticle attraction force described by the Hamaker equation. The interparticle distance variation is accounted for by means of an equation which relies on an equilibrium assumption between thickening and thinning rates (Pouyafar & Sadough, 2013). Nanoparticle interactions have been modelled employing an exclusive consideration of particle pairs, with a centre-to-centre separation distance of four radii at zero shear rate. Nevertheless, nanoparticles have been considered adequately dispersed at high shear rates, as a result of the attenuation of local energy gradients because of the interaction of vortices generated by particle rotation. The maximum separation distance can be computed on the basis of a homogeneous particle distribution assumption (Masoumi et al., 2009). The variation of interparticle distances between a minimum and a maximum estimated value has been approximated using a shear-dependant probability distribution, assuming that all particle pairs exist at one of these two states. These dispersion effects have been quantified as additional contributions to base fluid properties.

A univariate equation with a single fitting parameter characterised experimental data from two sources (Barry et al., 2015; Gerogiorgis et al., 2015) with a high coefficient of determination (0.999 in both cases). Coefficients of determination were more rigorously assessed using a logarithmic transformation (0.994 and 0.993, respectively), correcting for a highly asymmetric dataset (a skewness of 2.6 and 2.2 has been computed, respectively). A more elaborate trivariate equation characterised viscosity for variations in shear rate, concentration of nanoparticles, and temperature with high predictive potential (0.986). The coefficient of determination has been rigorously assessed using a logarithmic transformation (calculated to be 0.983). Temperature effects have been explicitly accounted for with an Arrhenius relationship: the activation energy of 11.24 kJ/mol computed is in accordance with values measured in traditional fluids (Khalil et al., 2011). Significant error attenuation improvements in viscosity estimation have been observed when comparing the novel correlation with the standard Herschel-Bulkley model, for a range of shear rates ($5\ s^{-1}$ to $1020\ s^{-1}$); in contrast to the Herschel-Bulkley model which failed to yield a coefficient of determination higher than 0.90, the first-principles model has achieved coefficients over 0.98, for both datasets and univariate as well as trivariate model instances (Reilly et al., 2016).

LITERATURE REFERENCES

1. Barry, M.M. et al., 2015. Fluid filtration and rheological properties of nanoparticle additive and intercalated clay hybrid bentonite drilling fluids. *J. Pet. Sci. Eng.*, **127**: 338–346.

2. Gerogiorgis, D.I., Clark, C., Vryzas, Z., Kelessidis, V.C., 2015. Development and parameter estimation for an enhanced multivariate Herschel-Bulkley rheological model of a nanoparticle-based smart drilling fluid. *Comput.-Aided Chem. Eng.* **37**: 2405-2410.

3. Khalil, M., Jan, B.M. & Raman, A.A.A., 2011. Rheological and statistical evaluation of nontraditional lightweight completion fluid and its dependence on temperature. *J. Pet. Sci. Eng.* **77**(1): 27–33.

4. Mahmoud, O. et al., 2016. Nanoparticle-based drilling fluids for minimizing formation damage in HP/HT applications, SPE Paper 178949, in: *SPE International Conference and Exhibition on Formation Damage Control.*

5. Masoumi, N., Sohrabi, N., Behzadmehr, A., 2009. A new model for calculating the effective viscosity of nanofluids. *Journal of Physics D: Applied Physics*, **42**(5): 055501 (6).

6. Pouyafar, V. & Sadough, S.A., 2013. An enhanced Herschel–Bulkley model for thixotropic flow behavior of semisolid steel alloys. *Metallurgical and Materials Transactions B*, **44**(5): 1304–1310.

7. Reilly, S.I., Vryzas, Z., Kelessidis, V.C., Gerogiorgis, D.I., 2016. First-principles rheological modelling and parameter estimation for nanoparticle-based smart drilling fluids, *Comput.-Aided Chem. Eng.* (in press).

469396 Modelling the Deposition of Actives on Cotton Fabrics during the Washing Process

Abstract:

Many organic textiles such as cotton absorb in the short UV-region of the spectrum extending into the blue range which promotes a yellowing shade of textiles. The extent of this shade depends on several properties of the polymer such as number of conjugated systems, degree of substitution and degradation products [1]. The removal of the yellowish colour is carried out by bleaching of textiles during its processing prior to conduct further finishing steps. However gradual yellowing and whiteness loss of used textile items occur over time due to several causes such as accumulation and aging of unremoved oily soils, being human sebum an important source [2 - 4].

Thus, during the washing process of textiles in a commercial washing machine, whiteness maintenance depends on many different transformations, namely (a) soil removal from fabrics' structure, (b) suspension and anti-re-deposition of removed soil, (c) bleaching of soils remaining in fabrics, (d) deposition of shading dye actives, which are used to shift the yellowness of fabrics to a more preferred blue hue and (e) deposition of Fluorescent Whitening Agents (FWAs), also known as optical brighteners.

Soil Release Polymers (SRPs) play a key role in detergency due to its use for soil release from fabrics and prevention of re-adsorption of soils. FWAs are widely used in the textile and paper industry for improvement and maintenance of whiteness. These molecules absorb light in the UV region (350 nm) and emit on the blue region of the spectrum compensating yellowing of materials [5, 6].

Understanding and modelling the chemical and physical processes driving deposition and retention of the actives driving each of the vectors mentioned above is key for optimising whiteness performance by an optimisation of the detergent formula.

In the present work the adsorption behaviour of two stilbene derivative FWAs typically used in detergent formulas onto flat cotton fabrics has been investigated and modelled as a function of temperature, water hardness, concentration of FWAs in solution, fabric: water ratio and detergent concentration. Furthermore, the impact of SRPs on soil release and soil anti- redeposition on cotton fabrics is investigated.

The adsorption of actives on textiles is tracked by a spectrophotometry based real – time methodology which allows obtaining high time resolution data of the active remaining in solution.

The resulting whiteness of textile garments is assessed by measuring their reflectance under different light conditions by spectrophotometry. Calibration curves that correlate the adsorbed mass of active per area of flat cotton fabrics with the resulting whiteness of textiles have been developed which allow predicting the final whiteness for any adsorbed concentration of actives. This correlation is typically non – linear where the whiteness benefit (delta b* in L*a*b* colorspace) plateaus as we reach high levels of adsorbed active.

Based on experimental data, a mechanistic model for molecular deposition has been developed which considers the following transformations in the system: (I) Dissolution of active in the washing liquor, (II) convective mass transfer flow into the fabrics, (III) Fickian diffusion of molecules to the fabric surface across boundary layer and stagnant layer of water in yarns and (IV) adsorption/desorption on fabrics' structure (adsorption/desorption isotherms).

The time to reach 95% of equilibrium as well as the equilibrium concentration, increase with increasing water to fabric ratio for a fixed active concentration since there is more mass of active per surface area to be adsorbed. Therefore, depending on the type of washing machine (Top Loader vs Front Loader Washing Machine) where the water to fabric ratio varies significantly (typically 0.02 m^3:kg vs 0.004 m^3:kg), washing time plays a key role on the final adsorbed mass of active on cotton fabrics and therefore on the resulting whiteness of textiles requiring almost twice the time for actives in the higher water:fabric ratio machine to reach equilibrium. The importance of these models is that experiments can be run under laboratory conditions with less variable high water to fabric ratios and results scale up to real wash conditions.

A lack of fit is observed for actives depositing on cotton fabrics when Fickian diffusion across a single boundary layer is considered. The hypothesis presented to explain this disagreement between experimental and predicted data is the presence of a complex dual porosity fabric structure [7]. Fabrics have two main porosities, inter-yarn porosity due to the spaces between the yarns of the textiles and intra – yarn porosity due to spaces between the fibres forming the yarns. As a consequence, molecules diffuse across different diffusing distances before they reach the fabric surface. Initially a rapid decay of the bulk concentration is experimentally observed which is thought to correspond to the adsorption of molecules on the outer surface of the fabrics, which progressively decreases until saturation is reached leading to diffusion to deeper areas of fabrics which would promote a slower decrease of the bulk concentration due to larger diffusion distances. The approach followed for modelling the entire process is based on the consideration of two different surfaces with different surface area and diffusion distance where molecules can diffuse and get adsorbed simultaneously in parallel.

References:

[1] Kramer, J.B. (1992), Fluorescent Whitening Agents, The Handbook of Environmental Chemistry, Volume 3, Part F, Ed. O. Hutzinger, Berlin.

[2] Rounds, M.A.J., Purchase, M. E., Smith, B.F. (1978), *Fatty Soil Retention and the Influence of Particulate and Fatty Soils on Yellowing of an Unfinished Dacron/Cotton Blend Fabric,* Textile Research Journal, 517 – 523.

[3] Chi Y-S., Obendorf, S.K., (1998), *Aging of Oily Soils on Textiles Materials: A literature Review.* Journal of Surfactants and Detergents, 1, (3), 407 – 418.

[4] Chi, Y-S., Obendorf, S.K, (2001), *Effect of fibre Substrates on Appearance and Removal of Aged Oily Soils,* Journal of Surfactants and Detergents, 4, (1), 35 – 41.

[5] XianNan, H., et al. (2008). *Study of adsorption kinetics for fluorescent whitening agent on fibre surfaces.* Sci. China Ser. B-Chem., 51 (5), 473 -478.

[6] Iamazaki, E.I. and Atvars, T., (2007). *Sorption of Fluorescent Whitening Agents (Tinopal CBS) onto Modified Cellulose Fibers in the Presence of Surfactants and Salt.* Langmuir, 23 (26), 12886 – 12892.

[7] Warmoeskerken, M.M.C.G, Van der Vlist, P., Moholkar, V., Nierstrasz, V., (2002). Laundry process intensification by ultrasound. Colloids Surf A: Physiochem. Eng. Aspects 210, 277-285

474797 Progress in Electro Acoustic Signal Analysis

Abstract:

We have recently determined a correlation between the acoustic time of flight and intensity and the structure, state of charge, and state of health of closed system electrochemical energy storage. We have shown that regardless of chemistry and geometry, an acoustic time of flight signal can be related to these properties of the cell.

Because a closed cell is effectively a mass redistribution reactor, and in a standard cell the volume is effectively fixed, the distribution of density within a battery must change as a function of state of charge and, along with density, the bulk moduli of the anode and cathode changes as well. The shifts in density and modulus also change the acoustic attenuation in a battery

In this presentation we will review the physical basis of our hypothesis and present progress in 1) correlating structural evolution and failure analysis with acoustic signal evolution, 2) present preliminary inverse models which describe the changes we see in the acoustic/state of charge response and 3) present â€œphysics freeâ€ machine learning results to examine the uniqueness of an acoustic signal for a given battery in a given state of charge and health.

463541 Model-Assisted Development of Microfabricated 3D $Ni(OH)_2$ Electrodes with Rapid Charging Capabilities

Abstract:

Model-assisted Development of Microfabricated 3D $Ni(OH)_2$ Electrodes with Rapid Charging Capabilities

Recently, several studies have focused on the development of materials that realize batteries with improved power densities, and thus, rapid charging and discharging capabilities. Yet, they are primarily based on non-deterministic fabrication techniques which limit the control over critical electrode dimensions affecting the power performance of the battery, such as the electrode surface area and the active material thickness. The study reported herein involves rationally designed and deterministically engineered 3D structures with nearly precise control over the optimized dimensions that address the aforementioned critical factors affecting the high-power capability and ultimately determine the performance of the electrochemical system.

A series of fabrication methods encompassing electrochemical techniques and microfabrication technologies have been utilized for the formation and characterization of the scalable, well-ordered, and high-surface-area 3D architectures that can be potentially used as high-power electrodes in a variety of applications, ranging from autonomous microsystems to macroscale portable electronics. These versatile structures were based on a large number of thin layers of two different materials, i.e., Cu and Ni, the former being sacrificial while the latter was structural, fabricated through a temporary photoresist mold in an alternating sequence via an automated robotic system. The final structure was obtained by the selective removal of the sacrificial Cu layers, leaving behind a large number of separate Ni layers to be coated with the electrochemically active material. Nickel hydroxide ($Ni(OH)_2$), one of the most well-studied active material in secondary (i.e., rechargeable) battery systems along with Li-ion chemistry, was selected as the active material to be deposited onto the 3D multilayer structure. Electrodes with areal capacities as high as 2.43 mAh cm^{-2}

were realized, which were able to deliver more than 50% of their capacities when charged at ultrafast rate of 150C. The resultant electrodes also featured remarkable cycling stability when charged and discharged at high rates for more than 80 cycles.

For the optimization of the characteristic dimensions of the electrodes, a two-dimensional mathematical model has been developed in COMSOL 5.2 by employing fundamental mass transport and reaction kinetics principles. The geometry of a repeating unit of the multilayer electrode structure was studied. Fickâ€™s second law, Nernst-Planck equation, and Butler-Volmerâ€™s equation were used to describe the solid phase diffusion, the electrolyte phase transport, and the electrode/electrolyte interfacial kinetics, respectively. The model was shown to successfully predict the energy/power tradeoff of the fabricated multilayer electrodes. Diffusion coefficient of protons within the $Ni(OH)_2$ active material, a critical parameter that determines the power performance of the $Ni(OH)_2$ electrode, was experimentally measured using cyclic voltammetry. The obtained value of $2\sim3*10^{-10}$ cm^2/s is within the range reported from existing literatures and close to the best-fit value ($3.4*10^{-10}$ cm^2/s) obtained from the model. Accordingly, effort was directed towards finding the optimal electrode dimensions that lead to enhanced power performance based on the model. Geometrical parameters that were taken into account included the active material thickness, the interlayer spacing between two individual layers, and the nickel current collector thickness. Projections showed that multilayer structures based on the optimized dimensions exhibited superior performance than the experimentally fabricated non-optimized structures. Depending on the energy and power needs, this model opened up the possibility for the realization of deterministically engineered electrodes for both micro- and macroscale applications.

464527 Electrode-Supported Macroporous Ceramic Membrane Separator for Lithium Ion Batteries

Abstract:

The emergence of lithium-ion batteries as the most popular energy storage system is mainly due to their high energy density and long cycle life. They find application in a wide variety of portable electronic devices as well as electric vehicles and smart grids. However, the relatively frequent occurrences of battery fires and explosions are a major cause of concern. A significant source of this flammability, along with the electrolyte, is the polymeric separator used widely in commercial lithium-ion batteries. Lithium ion batteries with inorganic separators offer advantages of safer operation. An inorganic macroporous membrane separator for lithium-ion battery was prepared by blade coating an α-Al_2O_3 slurry directly on the electrode followed by drying. The improved separator preparation involves a twice-coating process instead of coating the slurry all at once in order to obtain a thin (~40μm) and uniform coat. Unlike state-of-the-art polyolefin separators such as polypropylene (PP) which are selectively wettable with only certain electrolytes, the excellent wettability of α-Al_2O_3 allows the coated separator to function with different electrolytes. The coated α-Al_2O_3 separator has much higher resistance to temperature effects than polyolefin counterparts. The inorganic coated separator retains its dimensional integrity at temperatures as high as 200ºC, thus eliminating the possibility of a short circuit during thermal runaway. Lithium ion batteries half-cells and whole cells with coated α-Al_2O_3 separator exhibit electrochemical performance comparable with that with PP separator at room temperature. However, the coated separator shows better cycling performance than PP separator under extreme temperatures when tested over a wide temperature range of -30°C to +60°C. Therefore, the coated α-Al_2O_3 separator is very promising for uses in safe lithium-ion batteries for energy storage applications.

465548 First-Principles Density Functional Theory Modeling of Li Binding: Thermodynamics and Redox Properties of Quinone Derivatives for Lithium-Ion Batteries

Abstract:

First-Principles Density Functional Theory Modeling of Li Binding: Thermodynamics and Redox Properties of Quinone Derivatives for Lithium-Ion Batteries

Ki Chul Kim,[1] Tianyuan Liu,[2] Seung Woo Lee,[2] and Seung Soon Jang[1]

[1] *School of Material Science and Engineering, Georgia Institute of Technology, 771 Ferst Drive NW, Atlanta, GA 30332-0245, USA*

[2] *G. W. Woodruff School of Mechanical Engineering, Georgia Institute of Technology, 801 Ferst Drive NW, Atlanta, GA 30332-0405, USA*

Abstract

The Li-binding thermodynamics and redox potentials of seven different quinone derivatives are investigated to determine their suitability as positive electrode materials for lithium-ion batteries. First, using density functional theory (DFT) calculations on the interactions between the quinone derivatives and Li atoms, we find that the Li atoms primarily bind with the carbonyl groups in the test molecules. Next, we observed that the redox properties of the quinone derivatives can be tuned in the desired direction by systematically modifying their chemical structures using electron-withdrawing functional groups. Further, DFT-based investigations of the redox potentials of the Li-bound quinone derivatives provide insights regarding the changes induced in their redox properties during the discharging process. The redox potential decreases as the number of bound Li atoms is increased. However, we found that the functionalization of the quinone derivatives with carboxylic acids can improve their redox potential as well as their charge capacity. Through this study, we also determined that the cathodic activity of quinone derivatives during the discharging process relies strongly on the solvation effect as well as on the number of carbonyl groups available for further Li binding.

466882 Conductivity Degradation of Polyvinylidene Fluoride Binder during Cycling: Measurements and Simulations for Lithium-Ion Batteries

Abstract:

Battery electrodes are complex multiphase composites which must provide efficient bicontinuous networks for transport of electrons (through the particle phase) and positive lithium ions (through the electrolyte filled pores of the electrode). A crucial but often neglected element of battery electrodes is the binder, typically a mixture of polyvinylidene fluoride (PVDF) and carbon black. The binder has two primary roles â€" to provide mechanical integrity and to improve electrical conduction of the electrodes. Migration of the binder has also been implicated as a potential mechanism of capacity fade in rechargeable lithium ion batteries.

We will present experimental characterization of the polymer binder for battery applications. Mechanical properties of the composite binder will be shown for both dry films and also binder swollen with carbonate electrolytes used in rechargeable lithium batteries. The electrical properties are strongly dependent on the applied stress. The evolution of mechanical and electrical properties of the binder after repeated cycling will be shown.

Then we will examine the impact of cyclic mechanical stresses on the performance of lithium cobalt oxide cathodes. Experimental results will be presented on the evolution of electrical conductivity as a function of mechanical cycling. Mechanisms for electronic conductivity in these complex porous composite electrodes will be investigated using mesoscale simulations using experimentally determined three dimensional structures of battery cathodes. Implications for battery internal resistance and cycling stability will be discussed.

Sandia National Laboratories is a multi-program laboratory managed and operated by Sandia Corporation, a wholly owned subsidiary of Lockheed Martin Corporation, for the U.S. Department of Energyâ€™s National Nuclear Security Administration under contract DE-AC04-94AL85000. SAND2016-4371 A

474683 Polymeric Electrolyte Additives for Suppressing Zinc Dendrite Formation in Rechargeable Batteries with Zinc Anodes

Abstract:

Rechargeable zinc-air batteries are attractive because they utilize non-flammable aqueous electrolytes and offer high energy densities. However, a major roadblock in the development of rechargeable zinc-anodes is the dendritic morphology evolution during zinc electrodeposition (i.e., during battery charging). Zinc dendrites are a major source of capacity loss and early cell failures.

In the present talk, we will demonstrate the use of polymeric additives, such as polyethylenimine, as electrolyte additives for effectively suppressing dendrite formation during zinc electrodeposition. Dendrite suppression by additives will be characterized using in situ optical microscopy. Steady-state and transient electrochemical polarization measurements on a rotating disk electrode, combined with electrochemical quartz crystal microgravimetry will be used to unravel the mechanism by which additives suppresses zinc dendrite growth. Additive injection studies implemented in a rotating disc electrode configuration, coupled with diffusion-adsorption modeling, will be discussed as a reliable technique for precisely measuring the diffusion coefficient and adsorption rate constant of aforementioned polymeric additives. These additives' properties provide insights into the rate-controlling processes responsible for additives-assisted zinc dendrite suppression.

466874 Managing Trade-Offs Between Food, Renewable Energy and Ecosystem Services

Abstract:

Competition between renewable energy and food production, driven by increasing food and energy demands, creates incentives to increase agricultural productivity. In the short term, productivity can be increased through the use of intensive farming management practices that use extensive tillage and large quantities of chemical fertilizers. Such practices have been shown to contribute to ecological degradation of farmland via an increase in soil erosion and a decrease in soil carbon, among other impacts. The surrounding watershed is also affected negatively by the nitrate and phosphorus-contaminated runoff from fertilized fields, which degrades the water quality and contributes to the growth of harmful cyanobacteria and other microorganisms. In the long term, this ecological degradation results in reduced agricultural productivity, such that increasing amounts of chemical fertilizers and other technological inputs are required. Farming for maximum productivity without accounting for the ecological impacts thus creates a vicious cycle, in which productivity is maintained by inflicting ecological harm.

Experimental studies have demonstrated that conservation farming practices, with reduced tillage and in some cases reduced amounts of fertilizers, insecticides and herbicides, can mitigate the ecological damage of conventional agriculture. Farming with reduced or no tillage helps to increase and maintain the carbon content in soil, providing an offset for the emissions produced in the agricultural life cycle, and reduces the rate at which soil erodes. Reducing chemical fertilizers or partially replacing them with soil amendments such as agricultural residues reduces the concentration of nitrates and phosphates in contaminated runoff and cuts down on wasted fertilizer as more of the applied fertilizer is taken up by crops. Conservation practices offer myriad ecological benefits but do not necessarily offer similar productivity benefits: the rate at which biomass is produced can be decreased under conservation farming, creating trade-offs between ecological preservation, agricultural production, and farm economics. With demand for both food and energy likely to increase for the foreseeable future, the need for food and bioenergy crops must be balanced against the need for ecological preservation.

The objective of this work is to design a system for the co-production of food and renewable energy with respect to three conflicting criteria: food productivity, energy productivity and net ecosystem service (ESS) supply. Ecosystem services are the goods (food, fuel, materials) and services (climate regulation, water quality regulation, erosion control) provided by nature that are necessary to support human activities, including agriculture. Here we define net ESS supply as the difference between ESS supply, provided by nature, and ESS demand, required by human activity. Climate regulation is a simple example. The net climate regulation supply is the quantity of greenhouse gas emissions generated by human activities, less the quantity of emissions that can be sequestered or otherwise removed from the atmosphere by ecological processes such as biomass growth and sequestration in soil. Net ESS supply provides a quantitative way to measure ecological degradation: a negative net supply means that the demand for ESS is greater than can be supplied by nature, and ecological degradation is the likely consequence. In this work, we will identify options for food and renewable energy co-production that have a net ESS supply greater than or equal to zero, for key services.

The system we consider is comprised of two stages: land use and biomass-to-energy conversion. Land use includes several options for crops both grain and cellulosic, rotation systems and farming management practices, as well as the production of wind and/or solar electricity. Land can also be used to create additional ESS supply through reforestation; however, land used for ESS supply cannot also provide food or energy, thus there is competition between the land uses. Options for biomass-to-energy conversions cover a range of liquid fuels, both ethanol and hydrocarbon, as well as electricity. Grain crops can be sold directly as food products or converted to liquid fuels; the system thus has a variety of configurations in which both food and renewable energy is produced from the same farmland. The land use and biomass-to-energy conversions are modeled at the value chain and equipment scales, respectively, of the process to planet (P2P) multi-scale modeling framework. For this system, the value chain scale may be thought of as a regional scale. The third and largest scale of the P2P framework, the economy scale, is used to capture the life cycle of all inputs to the land use and biomass-to-energy conversion stages. Net supplies of two ESSs are quantified as decision criteria: climate regulation, quantified with carbon dioxide emissions and carbon sequestration, and air quality regulation, quantified with nitrogen oxide emissions and removal. Carbon dioxide emissions (climate regulation demand) is quantified at all scales, equipment, regional and life cycle. Carbon sequestration (climate regulation supply) is quantified at the regional scale only, as the land use options are the only components of the system that provide a variable supply of ESS. Air quality regulation is quantified only at the smaller two scales, with supply again provided only by the land use options and demand created by both the land use options and the biomass-to-energy conversion options. Air quality regulation is not considered at the life cycle scale because the spatial presence of criteria air pollutants and its effects, are either local or regional in nature depending on the pollutant molecule.

To analyze trade-offs between food, energy and ecosystem services, the co-production system is optimized under several constraints. Food and energy production are constrained relative to the maximum production possible, to ensure that all optimal system designs are truly co-production systems. Where possible, net ESS supply is also constrained to be greater than or equal to zero; this is not always achievable for air quality regulation, but net climate regulation supply can reach above zero for many system configurations. Within these constraints, the system is optimized for maximum food production, maximum energy production, and maximum net ESS supply for both services. With the design space thus defined, Pareto curves are found between food production and net ESS supply, and between energy production and net ESS supply, in order to identify sets of compromise system designs. Preliminary results indicate that maximizing either food or energy productivity leads to the worst possible performance for net ESS supply, as expected. However, by either sacrificing some amount of productivity or by using land for both food and energy production and ESS supply, designs can combine adequate net ESS supply with relatively high food and energy productivity.

461799 Efficient Solar Thermal Hydrogen, Electricity and Fresh Water Coproduction Process Synthesis

Abstract:

Global population growth, increase in per capita GDP and change in consumption habits create increasing pressure on food, energy and water (FEW) supply. For the past century, with lower FEW needs and abundant fossil resources, energy, chemical, fertilizer and even fresh water supply via desalination have been predominantly dependent on limited fossil resources. However, concerns regarding limited fossil fuel reserves and increasing greenhouse gas (GHG) emissions from fossil fuels accelerate advancements and implementation of renewable energy conversion processes. Among these alternatives, solar energy conversion processes are particularly prominent because of sunâ€™s tremendous potential [1]. Yet intermittencies and land availability constraints for solar energy collection are the grand challenges for solar energy conversion technologies and demand high conversion efficiency and synergistically integrated energy storage solutions [2-3]. Recently, it has been shown that hydricity concept, synergistic solar thermal hydrogen and electricity coproduction, presents a potential solution for continuous and efficient power supply and also an exciting opportunity to envision and create a sustainable economy to meet all the human needsâ€"namely, food, chemicals, transportation, heating, and electricity [1].

Here, we introduce an integration strategy for the coproduction of fresh water, hydrogen and electricity from concentrated solar thermal energy and present potential impacts of the proposed system on FEW nexus. The integration of solar thermal power production and solar thermal hydrogen production techniques reduces the exergy losses associated with each one of the systems and thus provide an efficient solution. Solar conversion processes are evaluated based on the process *exergy efficiency* that refers to the fraction of incident solar exergy that is directly recovered as the net exergy output, which is defined as the sum of electricity and the hydrogen exergy output.

The proposed efficient solar trigeneration process presents: i) a continuous power and fresh water supply solution that

can achieve high solar conversion efficiencies when combined with the proper hydrogen power cycle and ii) numerous opportunities in chemical industry, transportation sector and in particular in food production via continuous supply of fresh water for irrigation and fertilizer production from coproduced hydrogen. Coproduced products can supply all FEW needs especially in the case of production of fertilizers from the coproduced hydrogen. [1] Gençer E, Mallapragada DS, Marechal F, Tawarmalani M, Agrawal R. Round-the-clock power supply and a sustainable economy via synergistic integration of solar thermal power and hydrogen processes, Proceedings of the National Academy of Sciences (PNAS), 112(52), 15821-15826, 2015.

[2] Gençer E, Al-musleh E, Mallapragada D, Agrawal R. Uninterrupted Renewable Power through Chemical Storage Cycles. Current Opinion in Chemical Engineering, 5, 29-36, 2014.

[3] Gençer E, Agrawal R. A commentary on the US policies for efficient large scale renewable energy storage systems: Focus on carbon storage cycles, Energy Policy, 88, 477-484, 2016.

465212 Food-Energy-Water Nexus: Modeling Energy and GHG Emissions of Water Embodied in U.S. Domestic Food Transfers

Abstract:

Global population growth, climate change, and increasing urbanization are pressurizing already constrained essential food, energy, and water resources. Food, energy and water (FEW) systems are deeply interwoven, interconnected, and necessitate an integrated management approach. Central to understanding the complexity of the FEW nexus is quantitatively mapping and understanding their complex interdependencies. Existing studies have focused mostly on a single or two dimensions of the FEW nexus. There is a critical need for a holistic systems-level approach for studying FEW systems in a joint manner as opposed to studying them in isolation to avoid unintended consequences and encourage optimal outcomes. We integrate concepts of network theory with life cycle information to develop an understanding of the domestic FEW nexus for the United States.

Using publicly available disparate datasets, we develop a weighted and directed network model of interstate food transfers for the United States. The interstate food transfer model is translated into networks of embedded irrigation water, energy, and greenhouse gas (GHG) emission flows utilizing information on water footprints, irrigation datasets, and life cycle inventory databases. Specifically, we focus on four food commodity groups: livestock, cereal grains, meat, and milled grains. We utilize network theory tools and metrics to understand the structure, robustness, and environmental sustainability of these networks. Preliminary results indicate that over 600 million tons of food commodities were transferred across the U.S. along with 230 billion m^3 of virtual irrigation water. Additionally, 450 billion MegaJoules (MJs) of primary energy and 30 billion kg of CO_2-equivalent emissions were embedded in virtual irrigation water transferred across the U.S. It is observed that livestock and meat contributed only 13% by mass but accounted for 60% of embodied water, energy, and GHG transfers compared to grain based commodities. From a network perspective, the unweighted food transfer network is well-connected with majority of states participating in

high volume of trade. However, the weighted network structure reveals that the majority of the food flow and embodied resources and emissions are controlled by a few states. Furthermore, meat and milled grain networks are denser compared to livestock and cereal grains networks. A robustness analysis highlights that despite the presence of vulnerable key states, the network is fairly robust to both random and targeted disruptions. These results reveal that while the domestic food transfer network is robust against extreme disruptions on specific nodes, the dense food transfer patterns of resource intensive products have important implications for sustainability of the food network.

465536 Water Footprint of Hydrotreated Renewable Jet Fuel Produced through Rapeseed Rotation with Wheat and Other Crops in North Dakota

Abstract:

Hydrotreated Renewable Jet (HRJ) fuel from oilseed plants has increasingly become important for the aviation sector to address energy security and climate change mitigation. Rapeseed is considered a promising HRJ source for its high quality oil content and the potentially attractive agro-economic benefit to replace the fallow period in wheat/fallow rotations. However, concerns have been raised about the viability of this important change in land use, within which water supplies is one important concern as large scale biofuels production may place pressure on fresh water supplies and water quality. Despite these achievements on analyzing the water impact of production of selected biofuel crops, investigation of rapeseed production for biofuel has been limited. This study develops assessment informed by inputs from multiple models to evaluate the environmental impacts of water utilization and water quality of the different production options for HRJ production from rapeseed cultivation in North Dakota. The biogeochemical based EPIC model is incorporated to simulate the crop growth that influences the hydrological cycle. Output of the land-use management changes from the EPIC is used to run the blue water footprint analysis using the LCA software SimaPro 8.0. Evapotranspiration data from the EPIC model is incorporated to calculate green water footprint during cultivation. The UNESCO Grey Water Footprint Accounting Tier 1 Guideline is followed to estimate the grey water footprint caused by nutrient leaching and runoff. In addition, inputs for multi-modal transportation logistics were obtained from the Alternative Fuel Transportation Optimization Tool (AFTOT) models in collaboration with the US Department of Transportation Volpe National Transportation Systems Center. Co-product allocation methods were employed in scenario analyses, such as energy allocation and market value allocation, therefore producing a robust analysis of water footprint of the rapeseed HRJ life cycle with compiled life cycle inventory.

Our presentation details the methods, assumptions and initial water footprint results for different scenarios of importance in North Dakota, one of the 10 US states studied in this project, when considering the rapeseed HRJ life cycle. Preliminary results indicate that HRJ fuel derived from rapeseed in rotation with wheat has a much larger water footprint when comparing to the conventional jet fuel. In addition, although it is usually neglected, the indirect blue water footprint caused by the energy usage and the material inputs during HRJ production pathway is worth considering due to its important impact on the overall water footprint. Therefore, policy makers should consider highly the water factor when making decisions on transition to large share of biofuels.

468516 NexSym – a Local Nexus Simulation System

Abstract:

Food, energy and water are essential needs to sustain human life. The complex interactions and interdependencies among the infrastructures and decision making processes involved in the provisioning of such needs have recently generated great debate under the umbrella of the Food-Energy-Water (FEW) nexus (also referred to as Water-Energy-Food nexus). As society becomes increasingly aware that resources available to satisfy the needs of a growing population are finite, there is a call to look at the interactions and start creating more efficient solutions to manage the wide range of nexus scenarios, especially under conditions of climate change, increased urbanisation and waste generation. Furthermore, there is awareness of the importance of maintaining ecosystems to achieve sustainable production while expanding land use and using it more intensively to obtain more food, energy and clean water. The nexus may manifest in unique ways in different localities. Similarly, the global conditions will affect each location differently. Therefore, it is urgent to develop holistic analytical tools that could inform decision making for managing the interdependencies of the FEW nexus at the local scale. This is one of the ways chemical engineers are able to contribute with solutions by applying their skills of modelling processes and managing constraints associated with an interconnected system.

In this work, we present the **Local Nexus Simulation System (NexSym)** that integrates models for technological process components involved in the production of food, energy and clean water as well as waste treatment. Furthermore, the techno-ecological interactions can be simulated by integrating models for ecosystem processes, including agricultural production. The tool then allows the user to build a "flowsheet" of the local system with the component of interest (e.g. bioenergy production) and their interacting local system components (e.g. biomass cultivation, water provisioning). This enables examination of the whole picture of FEW nexus interactions at the local scale, which can then be investigated within a global context. Options for managing the nexus from an engineering perspective can then be devised and further evaluated under scenarios of interest to the user. The **NexSym** is expected to have wide application by engineers working across the nexus system components, planners for urban or rural development, policy makers and decision makers in general. The tool will be illustrated with sample case studies featuring interactions between food, energy, water provisioning and waste treatment to demonstrate the usefulness of such a tool in understanding, managing and creating sustainable solutions to the nexus.

472086 Gibbsian Game Theory for Tragedy of the Commons Problems in Food-Energy-Water Sustainability

Abstract:

Sustainability problems in the food-energy-water nexus often involve "tragedy of the commons" (TOC) problems. TOC challenges are often modeled in game theory using n-player Prisoner's Dilemma (NPD) games and other games. However, classical solutions to game theory problems – here called "Nashian games" – have a number of shortcomings, and do not always conform to experimental data. In this talk we treat strategic games as a set of chemical reactions, and we treat choices as types of reactive molecules. Thus, this model provides new solutions to classical games. By using the usual Gibbsian approach to chemical reaction equilibrium, we are able to employ entropic considerations and minimize a type of Gibbs free energy to find game theory equilibria, overcoming the usual shortcomings in game theory. We solve classical games, including the 2-player PD game and the NPD game, using the "Gibbsian game theory" approach, which accounts for "entropic choices". We then show how treating strategic games as a set of chemical reactions can be used to represent decision situations more accurately than typical games, and we solve for the equilibrium of sets of reactions. Then, we apply Gibbsian solutions to NPD games to problems in sustainability. By quantifying concepts like "trust", "power", and "corruption", we explore opportunities for nudging outcomes of contested decisions in sustainability decisions. When overlapping considerations occur, as in the food-energy-water nexus, the NPD games reveal fundamental political conditions that increase the probabilities for when genuine human solutions can advance. In this talk we describe Gibbsian game solutions for NPD problems, and show how they apply to sustainability problems in the food-energy-water nexus.

461657 Water and Energy Systems in Sustainable City Development: Agent-Based Modelling and Resource Technology Optimization

Abstract:

The grand challenges facing human societies are all closely interconnected with the sustainable provisioning of energy, water and material resources for constantly growing and developing populations, as well as the subsequent processing and management of wastes and pollutions [1]. Urban water and energy systems are expanding and require attention and careful planning to minimize their economic and environmental impact both from an operational and investment perspective. However, there is a lack of a sufficiently systematic approach to address these impacts and achieve the optimal strategies either in the context of research or real-world decision making [2].

Facing the challenges of energy and water systems in urban development, this work describes a decision-making platform combining a comprehensive database, agent-based simulation modelling and resource technology network optimization. The current regional demographics, land-use, infrastructure and economic information is used as input for the initial state of the focused urban system incorporating energy, water and other related resources. Detailed spatio-temporal demand data, obtained by simulating a synthetic population using agent-based models, is further used to plan capacity utilization and expansion by supply-side matching on a cost optimal basis, eventually aiming to explore the optimal design and operational strategies for residential, commercial, industrial, and other sectors in urban energy and water systems, and especially their nexus. Moreover, long-term socio-economic scenarios are addressed in the urban system development through a real-time feedback loop.

The outputs depict, with fine spatial and temporal resolutions, an overall energy-water landscape of the studied urban region (e.g., Greater Accra Metropolitan Area in Ghana as a prototype [3]). The material and energy balance of supply and demand from both macro and micro perspectives are obtained, and further used to propose environmental friendly and cost effective sustainable city investment and development strategies. Several representative use cases are built

under different scenarios to maximize utilization of resources, minimize net waste generation, and improve social welfare to engage local stakeholders. This work is to become a core component of the *resilience.io* platform as an open-source, data-driven, integrated systematic tool gathering social, environmental and economic data to inform urban planning, investment and policy-making for city-regions globally.

References

[1] Bringezu, S. and Bleischwitz, R. eds., 2009. *Sustainable resource management: global trends, visions and policies*. Greenleaf Publishing.

[2] Keirstead, J. and Shah, N., 2013. *Urban energy systems: an integrated approach*. Routledge.

[3] Koppelaar, R., Sule, M.N., Kis, Z., Mensah, F., Wang, X., Triantafyllidis, C., van Dam, K.H., Shah, N., 2016. Urban water and sanitation flows in the Greater Accra Metropolitan Area. *Urban Water Journal* (submitted).

469926 Biodiesel Produced from Soybean Oil with the Two Flow Reactors in Series

Abstract:

Viscous vegetable oils containing triglycerides of various fatty acids and free fatty acids are mostly converted to less viscous fatty acid methyl esters (FAME) through the methanol trans-esterification process, using homogeneous catalysts in batch reactors. FAME biofuels produced with homogeneous alkali catalysts through transesterifcation processes contain undesirable substances such as soap and homogeneous catalysts in addition to glycerol and water. Additional refining and purification processes are needed to remove these undesirable substances from FAME biofuels, generating wastewater. In our research, viscous soybean oil is converted to less viscous FAME biofuels with methanol and honeycomb monolithic catalysts in a two-flow reactor assembly in series at moderate temperatures and pressures to develop monolithic catalysts for efficient methanol transesterification processes.

Reaction mixtures from the first catalytic flow reactor consist of biodiesel (fatty methyl esters), glycerol, unreacted soybean oil, and excess methanol. Glycerol is removed from reaction mixtures from the first catalytic flow reactor to feed the second catalytic flow reactor. Reaction mixtures of biodiesel and unreacted soybean oil produced from the first catalytic flow reactor are fed into the second catalytic flow reactor to increase yields of biodiesel by shifting equilibrium conversion of soybean oil in favor of increasing conversion of soybean oil into biodiesel. Active and stable monolithic catalysts are formulated with various active metals and honeycomb monolithic catalyst supports to produce FAME biofuels from soybean oil.

459906 Biomass Conversion to Acrylonitrile Monomer-Precursor for Production of Carbon Fibers

Abstract:

Biomass Conversion to Acrylonitrile Monomer-Precursor for Production of Carbon Fibers

Jadid E. Samad, Lindsey Chatterton, Govedarica Zora and Amit Goyal

Energy and Environment Division

Southern Research, Durham, NC 27712

Over the last few years, the global demand for carbon fibers has been growing at an impressive rate. With newer applications in the horizon and the major consumers (e.g., aerospace, defense, automotive, construction) already gearing towards more use, this trend is projected to continue in the foreseeable future [1]. Of the several types of carbon fibers, Polyacrylonitrile (PAN) based fibers are of particular interest for its specific mechanical properties (250 ksi tensile strength and 25 Msi Youngâ€™s modulus) required for automotive applications.

Acrylonitrile (ACN), a monomer of PAN is commercially produced from fossil sources (e.g., propylene). This process has come under scrutiny due to volatile propylene price and non-sustainable nature of petroleum based feedstock. An alternative production route using biomass derived starting materials e.g., glycerol has gained steam of late. This route enables high selectivity to ACN with an array of valuable byproducts but is limited to purified glycerol obtained from biodiesel production. As a result, although promising, this route has yet to become economically attractive [2].

Southern Research is developing a biomass to ACN (B2ACN) process under a cooperative agreement with the

Department of Energy [3]. The goal of this work is to pursue a cost effective process for ACN production at mild condition using biomass derived sugars as raw materials. From sugar to ACN, the entire B2ACN process consists of multiple catalytic reaction steps. In the first reaction step, several novel multi-functional catalysts have been used to convert sugar to. The produced oxygenates are converted to a gas phase intermediate (acrolein) which will be converted to acrylonitrile in the subsequent step. Based on preliminary results significantly reduced cost (~15-22%) as well as greenhouse gas (GHG) emissions (~37%) compared to typical ACN processes have been forecasted. The work is currently at lab scale development/testing phase. A scaled up integrated bench-scale reactor system will be designed and demonstrated after successful completion of this phase.

In the previous paper, results of catalyst development and optimization from the first reaction (sugars to oxygenates) has been discussed. This paper will primarily focus on the results obtained from the second reaction (oxygenates to acrolein) involving oxygenate conversion. Several catalyst candidates have been identified and tested for optimum catalytic performance in this part of the study.

[1] Mark Holmes, Global carbon fiber market remains on upward trend, Reinforced Plastics, Volume 58, Issue 6, Novemberâ€"December 2014, Pages 38-45

[2] M. Olga Guerrero-PÃ©rez, Miguel A. BaÃ±ares, Metrics of acrylonitrile: From biomass vs. petrochemical route, Catalysis Today, Volume 239, 1 January 2015, Pages 25-30

[3] http://energy.gov/eere/articles/energy-department-announces-11-million-advance-renewable-carbon-fiber-production

470088 Catalytic Conversion of Polyols to Olefins

Abstract:

Catalytic Conversion of Polyols to Olefins

Bryan E. Sharkey and Friederike C. Jentoft*

Department of Chemical Engineering, University of Massachusetts, Amherst, MA 01003, USA

**Corresponding author: fcjentoft@umass.edu*

Hydroxyl groups are abundant in the cellulosic fraction of lignocellulosic biomass. To obtain fuels or a wider variety of chemicals from lignocellulose-derived feedstocks, the hydroxyl groups in these molecules must be removed or replaced by other, more useful functionalities. This conversion is best achieved after hydrolysis of biomass, which delivers a stream consisting of sugars or, after hydrogenation, sugar alcohols. Neighboring OH groups in sugars or sugar alcohols can be converted via deoxydehydration (DODH), that is, the simultaneous removal of one oxygen atom and a molecule of water (Eq. 1). The product is an olefin and thus a versatile chemical intermediate. The oxidation of the reductant, which is needed in stoichiometric quantities, and the elimination of a molecule of water make the reaction thermodynamically feasible.

$R\text{-}CHOH\text{-}CH_2OH + Red$ à $R\text{-}CH{=}CH_2 + RedO + H_2O$ (Eq. 1)

Original catalysts for this conversion were molecular (homogeneous) and include high-oxidation-state rhenium,[1,2] molybdenum[3], or vanadium complexes.[4,5] Heterogeneous catalysts have been reported only in recent years, and exclusively contain rhenium as the active transition metal species.[6,7] These catalysts suffer from leaching and contain expensive rhenium. In the present work, a second generation of rhenium-containing catalysts and a first generation of molybdenum-containing catalysts were developed and characterized.

Perrhenate or molybdate were deposited on various oxidic supports via incipient wetness impregnation or equilibrium adsorption of ammonium salts. The materials were optionally calcined. The conversion of diols was performed in benzene at temperatures of 135 to 200 °C, using various reductants including triphenylphosphine. The catalysts were characterized by various spectroscopic methods.

Catalysts consisting of oxide-supported ReO_x were highly selective in converting alkane diols to the respective terminal olefins. Decanediol could be converted to decenes with yields of more than 80% (at 100% diol conversion). Differences in conversion and yield could in part be explained by adsorption of both the diol and the olefin on the supports. The reaction progressed rapidly and selectively at 150 °C, whereas at higher temperature (175 °C), double bond shift occurred and internal olefins were produced in addition to 1-decene. The reductant efficiency was excellent with the ratio of decenes to triphenylphosphine oxide being close to one. Titania and silica were identified as the best oxide supports with respect to product yields, surpassing oxides of iron, zirconium, and aluminum.

Leach tests indicated some loss of rhenium, in particular for silica-supported catalysts. Consistently, the activity of recovered catalysts was generally lower than that of fresh catalysts, but there was always residual activity even after several uses. Spectroscopic characterization gave some indication that the structure of the supported oxo-rhenium species determines the proclivity for leaching.

A first generation of molybdenum-containing catalysts showed promise although they were less active than rhenium catalysts, and a higher reaction temperature was required to achieve DODH of decanediol to decene. More side products were observed, which included partially deoxygenated molecules and heavy species.

The activity of both oxo-rhenium and oxo-molybdenum species supports the hypothesis that an active catalyst must contain a metal that can form a dioxo unit on the surface and can cycle between two states by changing its oxidation state by ± 2. Deoxydehydration is confirmed as an attractive route to convert 1,2 diols into α-olefins.

References

1. G.K. Cook, M.A. Andrews, *J. Am. Chem. Soc.* **118**, 9448-9449 (1996).
2. J.E. Ziegler, M.J. Zdilla, A.J. Evans, M.M. Abu-Omar, *Inorg. Chem.* **48**, 9998-10000 (2009).
3. J.R. Dethlefsen, D. Lupp, B.-C. Oh, P. Fristrup, *ChemSusChem* **7**, 425-428 (2014).
4. G. Chapman, Jr., K.M. Nicholas, *Chem. Comm.* **49**, 8199-8201 (2013).
5. V. Tirupathi, K.M. Nicholas, *ACS Catalysis*, **6**, 1901-1904 (2016).
6. A.L. Denning, H. Dang, Z. Liu, K.M. Nicholas, F.C. Jentoft, *ChemCatChem* **5**, 3567-3570 (2013).
7. L. Sandbrink, E. Klindtworth, H.-U. Islam, A.M. Beale, *ACS Catal.* **6**, 677−680 (2016).

469634 Catalytic Oligomerization of Ethylene: Experimental Insight into the Effect of Supercritical Reaction Conditions

Abstract:

Ethylene oligomerization is a valuable chemical process for the production of carbon backbones for a variety of commodity chemicals ranging from fuels to surfactants and lubricants. Traditionally, this reaction has been covered extensively in homogeneous catalysis. The development of a heterogeneous catalytic process, however, is of significant academic and industrial relevance as it abides by green chemistry principles and is more easily scalable. However, coke deposition is a major issue for most heterogeneous systems. It has been determined previously that supercritical fluids have a unique ability to solvate coke. In this talk, we will explore the effect of supercritical reaction conditions on ethylene oligomerization. Ni-Hb catalysts were synthesized in the laboratory via a dropwise impregnation technique and used in a single-pass, high pressure packed bed flow reactor. Below its critical point, ethylene was a reactant for oligomerization reactions into C4-C12 olefins. As the partial pressure of ethylene exceeded its critical pressure, ethylene gas was observed to participate both as a reactant and as a solvent. This solvation effect was most evident as the amount of carbon deposition observed on the catalysts at supercritical conditions was found to be considerably lower than that of non-supercritical conditions.

466898 Mesoporous Catalysts for Conversion of 2,3-Butanediol to Butene

Abstract:

2,3-butanediol has significant potential as a platform chemical for production of renewable fuels and chemicals since it can be produced with high productivity via fermentation and provides a C4 building block for further synthesis. Previous work has shown that Cu/ZSM-5 can convert 2,3-butanediol to butene with high selecitivty (~70%) [1]. This study expands on these results by investigating the role of pore size in 2,3-butanediol conversion to butene. Three different types of mesoporous materials (Al-MCM-48, Al-SBA-15 and meso-ZSM-5) were loaded with ~20wt% CuO and tested in the conversion of 2,3-butanediol to butene. Mesoporous aluminosilicate molecular sieves Al-MCM-48 with different SiO_2/Al_2O_3 ratios of 23, 50, 100 and 200 were synthesized at room temperature by using N-Hexadecyltrimethylammonium bromide (CTAB) as the surfactant, tetraethyl orthosilicate (TEOS) and aluminum isopropoxide ($Al(^iProx)_3$) as Si and Al source [2]. Mesoporous aluminosilicate Al-SBA-15 with different SiO_2/Al_2O_3 ratios of 23, 50, 100 and 200 were prepared by the "pH-adjusting" method [3], using triblock copolymer Pluronic 123 surfactant ($EO_{20}PO_{70}EO_{20}$) as the structure-directing agent in a strong acidic HCl solution (2M), TEOS and $Al(^iProx)_3$ as Si and Al source. Meso-ZSM-5 was prepared by using a simple NaOH treatment of parent HZSM-5 zeolite [4]. Al-MCM-48 and Al-SBA-15 materials were characterized by small-angle X-ray scattering (SAXS), nitrogen adsorption and transmission electron microscopy (TEM), which clearly showed that the obtained Al-MCM-48 and Al-SBA-15 materials have highly ordered three-dimensional (3D) *Ia3d* cubic and two-dimensional (2D) *p6mm* hexagonal mesostructure, respectively. The pore (mesopore) size of Al-MCM-48, Al-SBA-15 and meso-ZSM-5 are about 2 nm, 10 nm and 20 nm, respectively, which is determined by the pore size distributions derived from the N_2adsorption-desorption isotherms using BJH method. It should be noted that, apart from the mesopores, micropores with diameter of 0.5 nm were observed on meso-ZSM-5 zeolite, which is typical for the conventional ZSM-5.

Reactions were performed over reduced catalysts at the same reaction conditions (feed rate of 2,3-butanediol of 3.0 mL/h, hydrogen to 2,3-butanediol molar ratio of 5, reaction temperature 250 °C). The catalytic results demonstrated that 20%CuO/Al-SBA-15(50) exhibited the highest initial selectivity of butenes, which is 76.6% at 10 min of reaction. In addition, the results showed that the existence of mesopores on the catalysts (Al-MCM-48 and Al-SBA-15 types) could decrease the selectivities of products from cracking reactions, especially $C_3^=$ and $C_5^=$- $C_7^=$, by comparison with the previous report on the catalyst 20%CuO/ZSM-5(280) [1]; meanwhile, the selectivity of $C_8^=$ was found to increase with increasing pore size of the catalyst. However, the activities of the catalysts with MCM-48 and SBA-15 types were decreased dramatically over time. With respect to CuO/meso-ZSM-5(280) catalyst, it can be seen that the catalyst has the composited performance of both CuO/ZSM-5(280) catalyst and mesoporous copper catalysts (CuO/Al-MCM-48 and CuO/Al-SBA-15). Cu/meso-ZSM-5(280) is displayed high activity on the cracking reaction ($C_3^=$, $C_5^=$~$C_7^=$), and the oligomerization ($C_8^=$) as well. Interestingly, Cu/meso-ZSM-5(280) showed an excellent catalytic stability; the selectivity of butenes dropped from 71% to 61% in 670 min of reaction, which is much better than the catalyst with Cu loaded on the conventional ZSM-5(280), the selectivity of butenes dropped from 71% to 50% in 550 min.

[1] Q. Zheng, M.D. Wales, M.G. Heidlage, M. Rezac, H. Wang, S.H. Bossmann, et al., Conversion of 2,3-butanediol to butenes over bifunctional catalysts in a single reactor, J. Catal. 330 (2015) 222–237. doi:10.1016/j.jcat.2015.07.004.

[2] K. Schumacher, C.D.F. Von Hohenesche, K.K. Unger, R. Ulrich, a. Du Chesne, U. Wiesner, et al., The Synthesis of Spherical Mesoporous Molecular Sieves MCM-48 with Heteroatoms Incorporated into the Silica Framework, Adv. Mater. 11 (1999) 1194–1198.

[3] S. Wu, Y. Han, Y.C. Zou, J.W. Song, L. Zhao, Y. Di, et al., Synthesis of Heteroatom Substituted SBA-15 by the "pH-Adjusting" Method, Chem. Mater. 16 (2004) 486–492. doi:10.1021/cm0343857.

[4] J.C. Groen, J. a. Moulijn, J. Perez-Ramirez, Desilication: on the controlled generation of mesoporosity in MFI zeolites, J. Mater. Chem. 16 (2006) 2121. doi:10.1039/b517510k.

465037 Polar Aprotic Solvent Effects on the Catalytic Conversion of Fructose to 5-(hydroxymethyl)Fulfural or Lactic Acid

Abstract:

The efficient production of primary renewable building blocks from biomass-derived carbohydrates must be achieved to develop economically viable biorefineries. Lactic acid (LA) and 5-(Hydroxymethyl)furfural (HMF) are two examples of primary renewable building blocks that can be obtained from fructose as starting feedstock. LA is currently used industrially for the production of green solvents and biodegradable polymers and is extensively used in the food industry. HMF is used as a building block for the manufacturing of solvents, polymers and fuel precursors. The heterogeneous catalytic conversion of fructose in water is limited, in most cases, by low selectivities, especially towards the production of these two platform molecules. Dumesic and coworkers [1] found that polar aprotic organic solvents such as g-valerolactone (GVL) and tetrahydrofural (THF) cause significant increases in reaction rates compared to water in addition to increased product selectivity for Brønsted acid-catalyzed reactions like conversion of xylose into furfural, dehydration of 1,2-propanediol to propanal and for the hydrolysis of cellobiose to glucose. Here we report a similar effect for the Lewis and Brønsted acid-catalyzed reaction of fructose to lactic acid or HMF, respectively. Sn-Beta catalyst was employed as catalyst for the Lewis acid-catalyzed reaction and a sulfonic acid-based CMK-3 catalyst was used for the Brønsted acid-catalyzed reactions. We demonstrate that the employment of aprotic polar solvents, especially GVL, in addition of the aforementioned catalysts is an effective combination for the selective catalytic conversion of fructose to LA or HMF.

The catalysts were characterized using nitrogen adsorption, X-ray Diffraction, X-ray Photoelectron Spectroscopy, Fourier Transform Infrared Spectroscopy and Inductively Coupled Plasma Atomic Emission Spectroscopy. Characterization of the modified materials shows that the desired physical and chemical properties of the catalysts were attained. Catalytic performance studies were performed in 10 mL batch reactors. The solvent effects on the catalytic performances were assessed at 160°C for the conversion of fructose to LA and at 130°C for the conversion of fructose to HMF. The results demonstrate that using aprotic polar solvents, especially GVL, as a reaction solvent significantly increases the fructose conversion turnover frequency to both, LA and HMF.

1. Mellmer, M. A., Sener, C., Gallo, J. M. R., Luterbacher, J. S., Alonso, D. M., and Dumesic, J. A. *Angew. Chem. Int.* Ed. 53, 11872 (2014).

475724 High-Efficiency Electrochemical Conversion of CO_2 to Value Added Fuels Current Status, Challenges, and Future Directions

Abstract:

The electrochemical reduction of carbon dioxide (CO_2) into liquid fuels especially coupling with the intermittent renewable electricity offers a promising means of storing electricity in chemical form, which reduces the dependence on fossil fuels and mitigates the negative impact of anthropogenic CO_2 emissions on the planet. Although converting CO_2 to fuels is not in itself a new concept, the field has not substantially advanced in the last 30 years primarily because of the challenge of discovery of structural electrocatalysts and the development of membrane architectures for efficient collection of reactants and separation of products. In this presentation, I would summarize recent advances in electrocatalytic conversion of CO_2 and presents the challenges and future directions in producing value-added fuels.

Currently, copper, silver and tin are the most extensively scrutinized electrocatalysts, which can convert CO_2 to CO, CH_4, C_2H_4, alcohols and carbolic acid in aqueous electrolytes. A long-lasting challenge in the electrosynthesis of fuels from CO_2 is to achieve a durable electrochemical performance, an equally important characteristic of an electrode as the activity and selectivity, but it has not yet been clearly addressed in the open literature. Hypotheses for the origin of the deactivation of CO_2 reduction electrode included the formation of intermediates like oxide, hydroxide and carbon species, as well as the accumulation of a soluble CO_2 reduction product in the electrolyte causing a remarkable decrease in the selectivity. Hori et al. examined the deactivation of Cu electrode and postulated that the degradation of the cell performance was originated from the deposition of iron and zinc contaminants in the electrolyte on the active surface of the catalyst. The impurity alone, however, cannot explain the degradation since the ultrapure electrolyte coupled with pre-electrolysis was used, in which the degradation was still observed. I will discuss

approaches to simultaneously achieve high efficiency and high stability of an electrocatalyst for the conversion to CO_2 to value-added fuels.

456332 Design and Economic Analysis of IGCC Power Plants with CO_2 Capture

Abstract:

Power producing industry has largely relied on coal as a fuel source and will continue to depend on it for coming decades under the current situation. However, in order to curtail CO_2 emissions while keeping high efficiency and economic feasibility of the processes is a key challenge. Post combustion coal-fired power plants usually have lower capital costs and cost of electricity without CO_2 capture. On the other hand, pre-combustion plants tend to be less expensive if the current CO_2capture systems are to be added. This analysis suggests that pre-combustion technology for the power generation can be an attractive option if carbon capture and storage (CCS) technology is implemented on a large scale. In this study IGCC process has been simulated in order to analyze the overall plant output with CO_2 capture using a physical solvent. Three design schemes have been developed to analyze the overall plant performance and economics. In order to ensure a fair evaluation of analysis, a consistent and transparent methodology has been followed for all the cases. First two cases use the water gas shift reactions scheme with sour shift catalysis process. The resulted syngas free of CO_2 can either be combusted by using air or O_2. The first case uses air as an oxidant for burning H_2 and the combustor temperature is controlled by air as well. In the second case, O_2 is used as an oxidizing agent in the H_2 combustion. However, combustion temperature control is achieved by recycling the captured CO_2. In the third model, WGS reactor has been removed. The syngas composed of CO and H_2 is sent directly for combustion which makes it similar to the oxy-fuel combustion process. In this way, captured CO_2 can be recycled back to control the combustor temperature in 2nd and 3^{rd}case. The results showed that the net power generation capacity and efficiencies calculated for case 1, case 2 and case 3 are 375Â MWe, 354 MWe, 388 MWe and 35.87%, 33.89%, 37.15% respectively. Finally, an economic analysis has been performed which shows thatÂ the total investment cost

(TIC) and operation and maintenance (O&M) required per unit (MW) of electricity generated for case 1, case 2 and case 3 is (2.97, 3.40, 2.64) and (0.2486, 0.2706, 0.2242) Mâ‚¬â€™s respectively. With the highest net power generation capacity and least TIC and O&M cost requirements, case 3 has been evaluated as the most feasible design. If the sole purpose of IGCC power plant is power generation with CO_2 capture, case 3 design offers a cost reduction of 8.17% and 6.59% in the total investment and O&M cost requirements respectively as compared to the base case design.

468074 Optimization of the CCS Network Under Harsh Geographical Conditions, Using GIS and MINLP Modeling

Abstract:

With the evermore increasing amount of greenhouse gas emissions every year, various efforts to mitigate its amount have earned worldwide attention. Carbon capture and storage, a practical technology to achieve this task, has been studied from different viewpoints, namely capture, transportation, and utilization of CO2. While the CO2 capture technology has been studied very deeply, studies for practical implementation of the CO2 transportation infrastructure has received relatively less attention. Although the CO2 capture technology is the most critical from an economic point of view, research on the topic of economic transportation of CO2 is essential for commercialization of the CCS technology, since mass amounts of it have to be transferred across long distances. For mass amounts of CO2 transport, pipelines are deemed as the most economical option. Previous studies concerning this problem have introduced the method of pipeline network optimization regarding the geographical conditions to minimize the cost of nation-wide transport. These studies used a combination of geographical information systems and mixed-integer program (MIP) modeling to optimize the network. However, geographical conditions vary greatly according to the region of pipeline installation, and linear pipeline models are inadequate in rigorously evaluating the cost of pipelines. In this study, a pipeline network for transporting mass amounts of CO2 in harsh geographical conditions was optimized, using mixed-integer nonlinear programming (MINLP). A commercial GIS software ArcGIS was used to obtain the least cost paths from the CO2 sources to the reservoirs, then the cost factor for penalizing the pipelines built on a certain plain was segmented into more rigorous levels. This data was transferred into GAMS, an optimization software and designed into a MINLP problem, using a weight-based cost model for evaluating the pipeline costs, for more accurate calculations. A case study in South Korea was conducted to test the effect of rigorous geographical penalization and cost model usage. The results of the study showed that a more rigorous network can be obtained by implementing more segmented penalty factors corresponding to the varying elevation. Also, it showed that rather than avoiding the rough landscape entirely, it may be more economical to gather the emitted CO2 from nearby sources into local junction nodes, then transport it through mainstream pipelines to the reservoir.

AMINE-BLENDS SCREENING AND CHARACTERIZATION FOR CO_2 POST-COMBUSTION CAPTURE

*Abdullah Al Hinai, Dang Viet Quang, and Mohammad R.M. Abu Zahra**
Department of Chemical and Environmental Engineering, Masdar Institute of Science and Technology, P.O. Box 54224, Masdar City, United Arab Emirates
** Tel.: +97128109181 - email: mabuzahra@masdar.ac.ae*

Abstract

This work studied the performance of novel amine blends in order to find a blend of alkanolamines, which possess all the required properties for efficient CO_2 capture. 2-(Butylamino)ethanol (2BAE), a secondary amine was selected to blend with three tertiary amines including 2-(dimethylamino)ethanol (2DMAE), N,N,N′,N′-Tetramethyl-1,3-propanediamine, (TMPDA), and N-methyldiethanolamine (MDEA). Different blends with a total amine concentration of 30 wt% were tested at 40°C using a flow micro reaction calorimeter to determine CO_2 loading and heat of absorption. Obtained results reveals that blending 2BAE with tertiary amines results in the increase of absorption heat compared to single tertiary amines. 2BAE/TMDAP blend has the highest CO_2 loading; 5, 10, and 15 wt% of 2BAE blending with 25, 20 and 15 wt% of TMDAP absorbed 1.09, 0.82 and 0.74 mole CO_2/mole amine, respectively. Results from CO_2 loading, absorption heat, and absorption kinetics revealed that blend containing 5wt% of 2BAE with TMDAP 25wt% or 2DMAE 25wt% provides the best CO_2 absorption performance.

Keywords: CO_2 capture; Absorption; CO_2 loading; Post-combustion; Amine blend;

Introduction

Concerns about global warming and climate change have urged scientists to find more effective methods to reduce anthropogenic CO_2 emission. Global CO_2 emission has increased significantly in recent years [1]. This causes great increase in atmospheric CO_2 concentration, which has reached 402 ppm in August 2016 [2]. Fossil fuel accounts for about 90% of total global CO_2 emission in which fossil fuel fired power plants are considered to be major sources of CO_2 emission [1]. Many efforts have been taken to reduce CO_2 emission in which carbon capture and storage (CCS) has been considered to be the most viable technology for CO_2 reduction. Three different CO_2 capture concepts have been introduced including pre-combustion capture, oxy-fuel combustion capture and post-combustion in which post-combustion capture is the feasible because it can be easily retrofit to the existing fleet of fossil fuel fired power plants [3]. The post combustion capture using amine solvents is the most major technology for CO_2 capture, which involves a chemical absorption of CO_2 into aqueous amine like ethanolamine [4-7]. Many works have been carried out to demonstrate the feasibility of this technology at pilot scale [8-11]. The application of this technology, however, is hindered by high energy consumption for the regeneration process, degradation, and corrosion. However, capturing CO_2 using amine solvent is still attractive due to its high absorption kinetics, which allow this technology to capture CO_2 from various sources including diluted CO_2 gas stream.

Various amine compounds including primary, secondary, and tertiary amines have been investigated for CO_2 capture application in which monoethanolamine (MEA), diethanolamine (DEA), and methyldiethanolamine (MDEA) are the most widely used amines for CO_2 absorption . The primary and secondary alkanolamines have a high reaction rate upon CO_2 absorption, which begins with the formation of a zwitterion (1) that then transfers a proton to an amine to form carbamate (2). Their maximum possible CO_2 loading is 0.5 mole of CO_2/mole of amine (two amine molecules react with one CO_2 molecule). At high pressure, the carbamate could be hydrolysed to form a free amine and bicarbonates, and then the free amine will again react with CO_2 (3).

$$R_1R_2NH + CO_2(aq) \leftrightarrow R_1R_2NH^+COO^- \quad (1)$$
$$R_1R_2NH + R_1R_2NH^+COO^- \leftrightarrow R_1R_2NH^+ + R_1R_2NCOO^- \quad (2)$$
$$R_1R_2NCOO^- + H_2O \leftrightarrow R_1R_2NH + HCO_3^- \quad (3)$$

For tertiary amines, the maximum CO_2 loading could be 1 mole of CO_2 per mole of amine (one molecule of amine reacts with one molecule of CO_2) with the formation of bicarbonates (4). Although the tertiary alkanolamines have higher CO_2 loading, they have low reactivity with CO_2 in comparison with the primary or secondary amines.

$$R_1R_2R_3 + CO_2(aq) + H_2O \leftrightarrow R_1R_2R_3NH^+ + HCO_3^- \tag{4}$$

Conventionally, aqueous MEA solution 30 wt% has been widely investigated and used for CO_2 capture [12]. However, a MEA based CO_2 capture process is very energy intensive due to the high energy requirement for MEA regeneration and this prevents this conventional post-combustion process from being used for industrial applications [13]. Recently, great effort has been spent to find alternative amines that may help increase CO_2 absorption capacity and reaction kinetics while lower regeneration energy. Amine system associated with high CO_2 absorption capacity, low regeneration energy, and high reaction kinetics results in lower cost of CO_2 capture. It is not easy to find a single amine that satisfies all above requirements. Our previous study did a survey on a large number of primary, secondary and tertiary amines bearing different functional groups and found that the primary and secondary amines have fast reaction kinetics but high absorption heat, whereas tertiary amines possess low absorption heat but slow reaction kinetics [14]. To take the advantage of both amine types, an idea to combine these amines in an amine blend system has been considered.

Following a previous experimental work which shows 2-(Butylamino)ethanol (2BAE) as a suitable amine for carbon capture as it is a secondary amine with a CO_2 loading of 0.73 mole CO_2/mole amine and a heat of absorption of -74.38 kJ/mol of CO_2. Tertiary amines including 2-(dimethylamino)ethanol (2DMAE), N,N,N′,N′-Tetramethyl-1,3-propanediamine, (TMPDA), and N-methyldiethanolamine (MDEA) have high CO_2 loading but slow reaction kinetics have been selected to blend with 2BAE. For a total of 30 wt% of aqueous amines, the 2BAE concentration is varied from 0 to 15 wt% while the respective tertiary amine contributes the remaining part to make a total of 30 wt%. CO_2 absorption performance and heat of absorption were investigated using a flow Micro Reaction Calorimeter (URC) to evaluate the effectiveness of the blends.

Experimental

Material

2-(Butylamino)ethanol, 98% (2BAE), 2-(dimethylamino)ethanol, 98% (2DMAE), N,N,N′,N′-Tetramethyl-1,3-propanediamine, 99% (TMPDA), N-methyldiethanolamine, 99% (MDEA) were purchased from Sigma Aldrich and used without further purification. CO_2 gas (99.9 %) was supplied by Gulf Industrial Gases CO. L.L.C.

Determination of CO2 loading and absorption heat

A description of the equipment can be found in the previous study [14]. A flow micro-reaction calorimeter that was supplied by Thermal Hazard Technology (UK) was used to evaluate CO_2 loading and the absorption heat for aqueous amine solutions (enthalpy of reaction/ΔH (kJ/mole of CO_2)) (Figure 1 (a)). The calorimeter is controlled by URC control software from Thermal Hazard Technology. In a typical experiment, a sample-containing cell was placed inside the calorimeter, and the temperature was set to 40 °C and atmospheric pressure. As the experiment starts, the variation in power (mW) over time (s) is recorded by the software. When the signal is stable, CO_2 gas is introduced into the cell at a rate of 0.5 ml/min (± 0.01 ml/min). The CO_2 flow is controlled by a mass flow controller, and the flow passes through a desiccant column to remove moisture before entering the sample cell. Because the CO_2 reaction with the aqueous amine solution is exothermic, the power signal initially increases, and then the signal decrease becomes constant when the reaction is complete (equilibrium) (Figure **1**1 (b)). The CO_2 loading (mole CO_2/mole amine) was calculated by dividing the mole of CO_2 absorbed per the mole of lean amine as described in equation 5.

$$CO_2 \text{ loading } = \frac{(m_1 - m_{0)/44}}{m_0 * (\frac{C1}{M1} + \frac{C2}{M2})} \tag{5}$$

Where, m_0 is the mass of lean amine and m_1 is the mass of CO_2 loaded amine that is the mass difference between the analysis cell containing amine after CO_2 adsorption and empty cell. C1 and M1 are concentration and molecular weight of amine 1 in the blend and C1 and M2 are concentration of amine 2 in the blend.

The integral heat Q (in kJ) was determined by using URC analysis software. The heat of absorption (ΔH - kJ/mole of CO_2) was calculated by dividing integral heat Q by the moles of absorbed CO_2. The absorption heat is proportional to amount of CO_2 absorption, therefore accumulative absorption heat evolved is used to monitor the CO_2 absorption progress and CO_2 absorption kinetics.

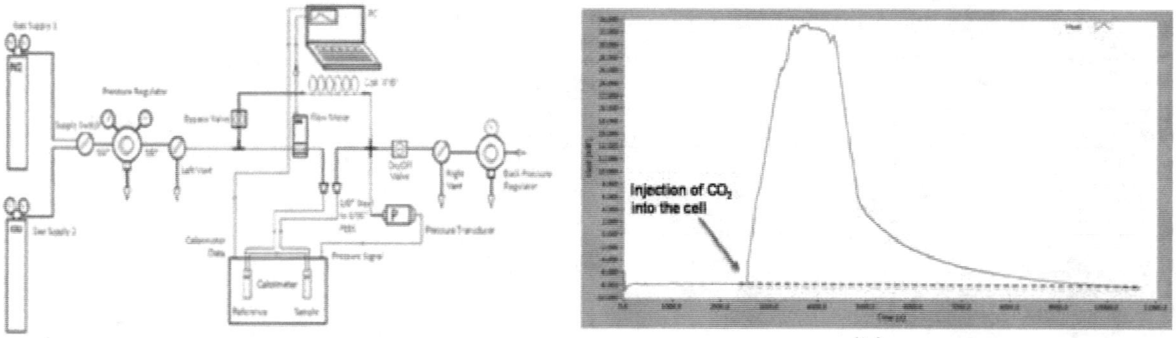

(a) (b)

Figure 1. Schematic of the flow Micro Reaction Calorimeter (a) and the data obtained from this equipment (b) [14].

Results and discussion

Previous study screened numerous amines and indicated that 2BAE is a promising candidate for CO_2 capture [14]. Aqueous 2BAE 30 wt% solution has CO_2 loading of 0.73 mole CO_2/mole amine and absorption heat of 74.4 kJ/mole CO_2. This is better than conventional MEA solution at the same concentration, which has CO_2 loading of 0.69 mole CO_2/mole amine and absorption heat of 85 kJ/mole CO_2. Three tertiary amines with different properties have been selected for blending with 2BAE including TMDAP, 2DMAE, and MDEA. These tertiary amines have lower absorption heat compared to 2BAE, however, their CO_2 loading ranges from lower to higher than that of 2BAE. TMDAP, 2DMAE, and MDEA have CO_2 loading of 1.32, 0.77, and 0.52 mole CO_2/mole, respectively. CO_2 loading of TMDAP is much higher, while CO_2 loading of MDEA is lower than that of 2BAE, meanwhile CO_2 loading of 2DMAE is similar to that of 2BAE. Experimental CO_2 loading and absorption heat of 2BAE blended with 3 tertiary amines are presented in Table 1.

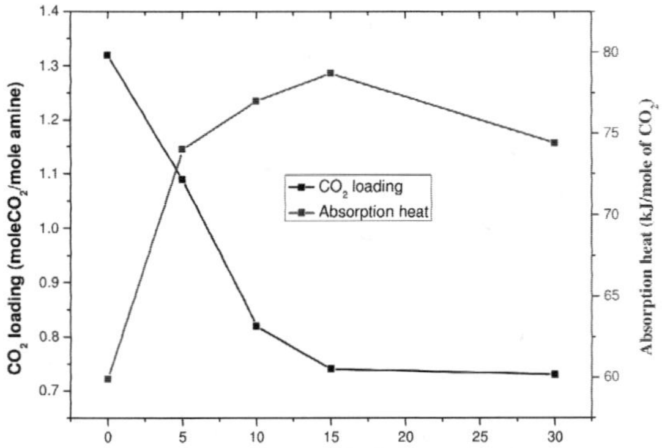

Figure 2. Variation of CO_2 loading and absorption heat as a function of 2BAE concentration in 2BAE/TMDAP blend

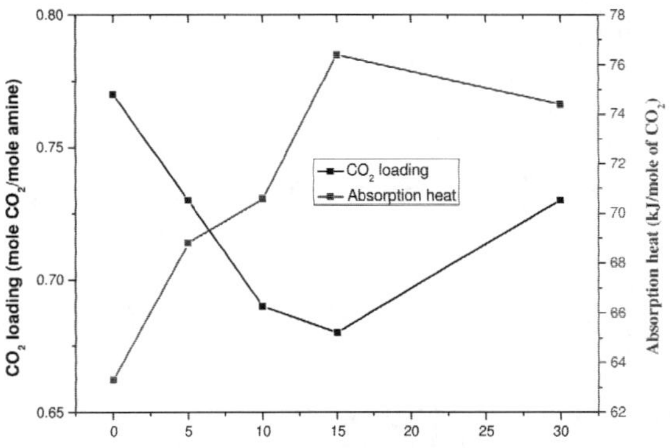

Figure 3. Variation of CO_2 loading and absorption heat as a function of 2BAE concentration in 2BAE/2DMAE blend

For 2BAE/TMDAP blend, the absorption heat of blend increased, whereas the CO_2 loading decreased compared with that of single TMDAP amine. CO_2 loading of TMDAP reduced from 1.32 to 1.09 mole CO_2/mole amine as 5 wt% of TMDAP was replaced by 2BAE and the CO_2 loading of blend is equal to that of single 2BAE as the 15 wt% of TMDAP was replaced with 2BAE (50:50 weight ratio). The result reveals that blend is efficient only when the 2BAE/TMDAP ratio is below 50%. Figure 2 shows variation of CO_2 loading and absorption heat as a function of 2BAE concentration in the blend and indicates that the best combination between 2BAE and TMDAP is 5 wt% of 2BAE and 25 wt% TMDAP in total 30 wt% amine in solution.

Table 1. CO_2 loading and absorption heat of amine blends. Tests were conducted at 40 °C using pure CO_2 at atmospheric pressure

	2BAE/TMDAP Blends				
2BAE concentration (wt%)	30	15	10	5	0
TMDAP concentration (wt%)	0	15	20	25	30
Heat of absorption (kJ/mole CO_2)	74.4	78.7	77.0	74.0	59.9
CO_2 loading (mole CO_2/mole amine)	0.73	0.74	0.82	1.09	1.32
	2BAE/2DMAE Blends				
2BAE concentration (wt%)	30	15	10	5	0
2DMAE concentration (wt%)	0	15	20	25	30
Heat of absorption (kJ/mole CO_2)	74.4	76.4	70.6	68.8	63.3
CO_2 loading (mole CO_2/mole amine)	0.73	0.68	0.69	0.73	0.77
	2BAE/MDEA Blends				
2BAE concentration (wt%)	30	15	10	5	0
MDEA concentration (wt%)	0	15	20	25	30
Heat of absorption (kJ/mole CO_2)	74.4	70.9	65.1	58.8	52.8
CO_2 loading (mole CO_2/mole amine)	0.73	0.59	0.57	0.55	0.52

Blending 2BAE with 2DMAE also results in increasing absorption heat and decreasing CO_2 loading of 2DMAE, but the change is not significant; 5 wt% of 2BAE in the blend caused reduction in CO_2 loading of 2DMAE from 0.77 to 0.73 mole CO_2/mole amine and raising absorption heat to 68.8 from 63.3 kJ/mole CO_2. Figure 3 reveals that replacing from 5 to 10 wt% of 2DMAE with 2BAE in 2DMAE 30 wt% solution will result in the best blend in term of CO_2 loading and heat of absorption. Contrast to TMDAP and 2DMAE, the addition of 2BAE in to MDEA result in increasing CO_2 loading of the blends at all 2BAE/MDEA ratios. The higher CO_2 loading of blend in comparison with single MDEA is contributed by 2BAE, whose CO_2 loading is higher than MDEA.

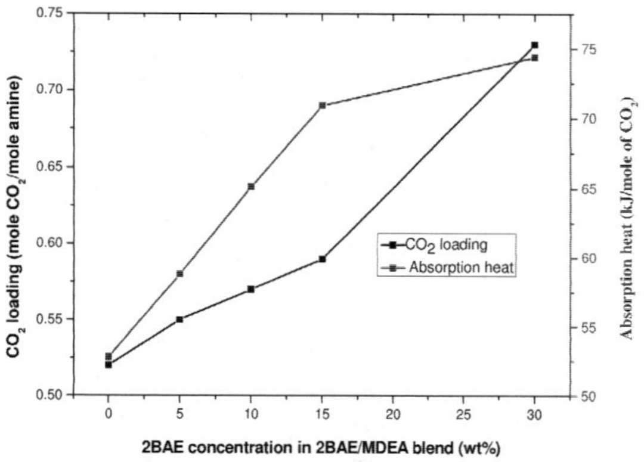

Figure 4. Variation of CO_2 loading and absorption heat as a function of 2BAE concentration in 2BAE/MDEA blend

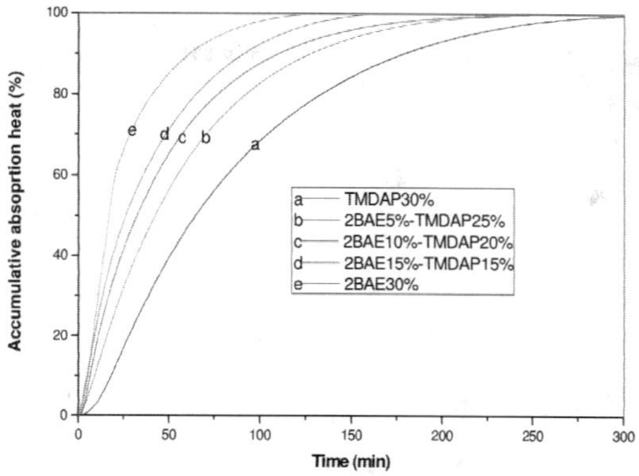

Figure 5. CO_2 absorption kinetics of 2BAE/TMDAP blends

Slow CO_2 absorption kinetics of tertiary amines is expected to be improved by blending with primary or secondary amine. In fact, a blend with low absorption heat and high CO_2 loading but slow absorption rate is still not realistic to CO_2 capture. Therefore, in addition to CO_2 loading and absorption heat, the CO_2 absorption kinetics should be investigated. In this study, heat evolved during CO_2 absorption recorded by flow micro calorimeter is proportional to amount of CO_2 absorbed. To interpret this data to kinetics, the percentage of accumulative absorption heat over absorption time was used. The CO_2 absorption kinetics of 2BAE, TMDAP, and their blends were exhibited in Figure 3. As can be seen in this figure, absorption rate of TMDAP 30 wt% solution is very slow, it took almost 178 min to complete 90% absorption, while 2BAE 30 wt% solution needed only 61 min. 2BAE5wt%/TMDAP25wt% blend required 127 min and 2BAE10wt%/TMDAP20wt% blend took 109 min to complete 90% absorption. This result indicated that replacing 5 wt% of TMDAP with 2BAE in blend helps reduced reaction time considerably. Similarly, blending with 2BAE helps improve CO_2 absorption kinetics of both 2DMAE and MDEA (Figure 6 and 7). Blending 5 wt% of 2BAE with MDEA do not significantly improve the absorption kinetics compared with single MDEA, however, the same blend concentration of 2BAE with TMDAP and 2DMAE results in considerable improvement of absorption kinetics in comparison with single counterpart tertiary amines. This further confirms that 5 wt% of 2BAE blending with TMDAP or 2DMAE was the best combination, which produced the best amine blend in terms of CO_2 loading, heat of absorption, and absorption rate.

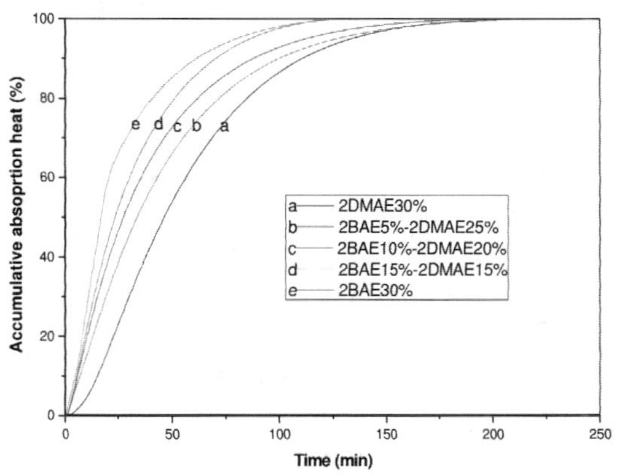

Figure 6. Absorption kinetics of 2BAE/2DMAE blend

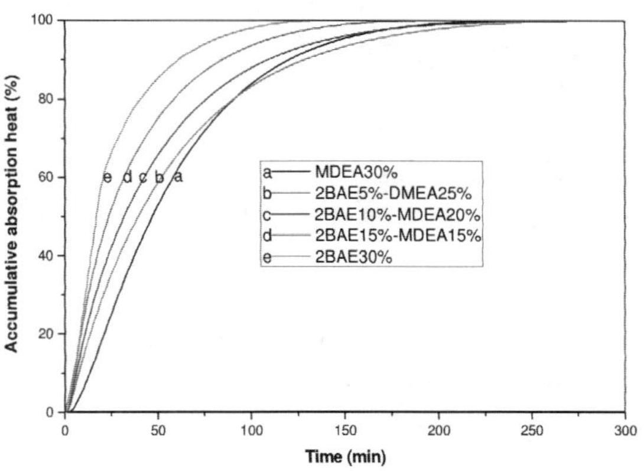

Figure 7. Absorption kinetics of 2BAE/2DMEA blend

Conclusion

Various blends of 2BAE (secondary amine) with three tertiary amines including TMDAP, 2DMAE, and MDEA were prepared and tested. Blending 2BAE with tertiary amine helps improve CO_2 absorption kinetics of tertiary amine, while remains reasonable CO_2 loading and absorption heat. Blends contain 5 wt% of 2BAE provide good CO_2 absorption performance in terms of CO_2 loading, heat of absorption, and absorption kinetics.

References

[1] J.G.J. Olivier, G. Janssens-Maenhout, J.A.H.W. Peters, Trends in global CO_2 emissions, Netherlands Environmental Assessment Agency, 2012.

[2] http://www.esrl.noaa.gov/gmd/ccgg/trends/#mlocited cited on 20/09/2016

[3] IPCC, Intergovernmental Panel on Climate Change (IPCC) Special Report on Carbon Dioxide Capture and Storage, Cambridge University press, Cambridge, UK, 2005.

[4] B. Lee, D. Kim, J. Cho, S. Park, A comparative study on the carbon dioxide capture power between 30 wt% 2-amino-2-methyl-1-propanol and 30 wt% methyldiethanol amine aqueous solutions, Korean J. Chem. Eng., 26 (2009) 818-823.

[5] M.R.M. Abu-Zahra, L.H.J. Schneiders, J.P.M. Niederer, P.H.M. Feron, G.F. Versteeg, CO_2 capture from power plants: Part I. A parametric study of the technical performance based on monoethanolamine, International journal of greenhouse gas control, 1 (2007) 37-46.

[6] M.R.M. Abu-Zahra, J.P.M. Niederer, P.H.M. Feron, G.F. Versteeg, CO_2 capture from power plants. Part II. A parametric study of the economical performance based on mono-ethanolamine, International journal of greenhouse gas control, 1 (2007) 135-142.

[7] A. Adeosun, N. El Hadri, E. Goetheer, M.R. Abu-Zahra, Absorption of CO_2 by Amine Blends Solution: An Experimental Evaluation, International Journal of Engineering and Science, 3 (2013) 12-23.

[8] P. Feron, M. Abu-Zahra, P. Alix, O. Biede, P. Broutin, H. de Jong, J. Kittel, J. Knudsen, L. Raynal, P. Vilhelmsen, Development of post-combustion capture of CO2 within the CASTOR Integrated Project: Results from the pilot plant operation using MEA, 3th International Conference on Clean Coal Technologies for our Futur, Cagliari, Italy, (2007).

[9] R. Idem, M. Wilson, P. Tontiwachwuthikul, A. Chakma, A. Veawab, A. Aroonwilas, D. Gelowitz, Pilot Plant Studies of the CO2 Capture Performance of Aqueous MEA and Mixed MEA/MDEA Solvents at the University of Regina CO2 Capture Technology Development Plant and the Boundary Dam CO2 Capture Demonstration Plant, Industrial & Engineering Chemistry Research, 45 (2005) 2414-2420.

[10] Y. Artanto, J. Jansen, P. Pearson, T. Do, A. Cottrell, E. Meuleman, P. Feron, Performance of MEA and amine-blends in the CSIRO PCC pilot plant at Loy Yang Power in Australia, Fuel, 101 (2012) 264-275.

[11] M.R. Abu-Zahra, Z. Abbas, P. Singh, P. Feron, Carbon Dioxide Post-Combustion Capture: Solvent Technologies Overview, Status and Future Directions, Materials and processes for energy: communicating current research and technological developments, Formatex Research Center2013, pp. 923-934.

[12] F.-Y. Jou, A.E. Mather, F.D. Otto, The solubility of CO2 in a 30 mass percent monoethanolamine solution, The Canadian Journal of Chemical Engineering, 73 (1995) 140-147.

[13] A. Chakma, CO2 capture processes — Opportunities for improved energy efficiencies, Energy Conversion and Management, 38, Supplement (1997) S51-S56.

[14] N. El Hadri, D.V. Quang, E.L. Goetheer, M.R.A. Zahra, Aqueous amine solution characterization for post-combustion CO 2 capture process, Applied Energy, (2016).

465184 Economic and Exergy Analysis of Integrated Purification Process with Carbon Capture Process (CCP) Using Selexol in a 500 MW Integrated Gasification Combined Cycle (IGCC) Power Plant

Abstract:

The enormous CO_2 emission from power plants raises the significant effects on global climate. In order to minimize greenhouse gas emissions from the utilization of fossils fuel, integrated coal gasification combined cycle (IGCC) with carbon capture process (CCP) takes a spotlight for high efficient and eco-friendly electricity generation. In this study, an integrated purification process with CCP for a 500MW IGCC power plant was simulated to evaluate energy efficiency and cost. First, only dual-stage Selexol processes with 80%, 85%, 90% and 95% carbon capture were preliminary simulated. According to preliminary results, energy consumption of Selexol process showed an optimum at more or less 90% carbon capture. Since each unit is integrated with the other units, the operating condition of one can affect the others. Then, as a second step, an integrated purification process with CCP simulation was carried out to evaluate energy analysis and cost estimation. It consists of contained water gas shift reactor (WGSR), dual-stage Selexol process, Claus process, scrubber unit, sour water stripper process, hydrogenation process, drying process, and compression. Finally, the exergy sankey diagram for whole integrated purification process with CCP, based on second law of thermodynamic, was performed to improve overall capture efficiency. Then, the development direction for reducing energy consumption was suggested from the exergy analysis.

470573 Influence of Alloying on CO_2 Electroreduction on Ag-Zn System

Abstract:

Electrochemical reduction of CO_2 is an attractive pathway for the sustainable production of fuels and chemicals, as it allows for the effective use of renewable energy sources while mitigating the effects of CO_2 in the atmosphere. While significant decrease in overpotentials and increase in reaction rates have been reported for the two-electron reduction products, reports on improved activity and selectivity for the products that require more than two electrons ($> 2\ e^-$ products) have been limited on monometallic transition metal catalysts. Recent theoretical work indicates that this difficulty of reducing CO_2 to $> 2\ e^-$ products is due to the linear scaling of the binding energies of CO_2 reduction intermediates on transition metal catalysts, and the necessity of catalyst systems that can break such scaling has been emphasized.(1)

Alloys, which can provide bifunctional sites that selectively stabilize intermediates, are a promising class of materials, and a large amount of effort has been focused on the investigation of their effectiveness as CO_2 reduction catalysts. Recent works have reported on improved CO_2 reduction activity and selectivity on alloy systems, which were attributed to the synergy between the two metal atoms at the surface. However, most improvements reported in these works are pertaining to two-electron products, and alloy systems that demonstrate improvements for $> 2\ e^-$ products remain elusive.

Here, we report on the investigation of polycrystalline AgZn foil, which resulted in the observation of enhanced activities and selectivities for methane and methanol productions compared to Ag and Zn foils. Although CO is the main CO_2 reduction product as expected of an alloy between two metals active for CO production, the selective enhancement observed for the $> 2\ e^-$ products is noteworthy, as it indicates that alloys are promising candidates in the development of effective CO_2 electroreduction catalysts. The production of CO as major product is explained by the binding energies of intermediates, and the importance of optimal binding energies are discussed.

References

1. A. A. Peterson and J. K. Nørskov, *The Journal of Physical Chemistry Letters*, **3**, 251 (2012).

470038 *In-Situ* Infrared Spectroscopic Investigation of Pyridine-Mediated CO_2 Reduction on Pt Electrocatalysts

Abstract:

The pyridine-mediated electrochemical reduction of CO_2 to methanol, formate, and formaldehyde was first demonstrated by Bocarsly et. al in 1994. Since then, many mechanistic investigations of this process have been conducted using reactivity, cyclic voltammetry, or computational methods; however, direct experimental mechanistic investigation has been lacking. We employ surface-enhanced infrared absorption spectroscopy, coupled with traditional electrochemical techniques in order to probe the electrochemical interface during the pyridine-mediated electrochemical reduction of CO_2 on Pt electrocatalysts. Previously proposed mechanisms are discussed and refined in light of new insights obtained in these studies.

471986 Investigations of Surface Chemistry for Pyridine-Catalyzed CO_2 Reduction on Gallium Phosphide

Abstract:

The surface chemistry of N-containing heteroaromatics, molecular co-catalysts that enable the selective electrochemical reduction of CO_2 to fuels, is discussed. The presented experimental results focus on elucidating the role of the electrode surface in CO_2 reduction reactions that are co-catalyzed by pyridine. For this catalysis, exceptionally high selectivity for reduced fuels has been reported when the reaction occurs at the surface of a GaP photocathode. For this reason, experimental emphasis is placed on assessing preferential adsorption sites and bonding interactions of adsorbates on surfaces of GaP. A surface science approach is used, whereby ultrahigh vacuum conditions facilitate the fabrication of highly characterizable electrode-adsorbate systems. The use of single crystal surfaces permits analysis of surface chemistry independent of complicating factors such as grain boundaries and morphology. Surface-sensitive core-level and vibrational spectroscopy techniques, including high-resolution X-ray photoelectron spectroscopy, synchrotron-based photoemission, and high-resolution electron energy loss spectroscopy, are used to probe adsorbate-substrate and adsorbate-adsorbate interactions for pyridine, water, hydrogen, and carbon dioxide on GaP. Scanning tunneling microscopy (STM) was used to obtain molecular orbital-resolved images of adsorbed molecules. Conclusions from experimental results on these model systems are supported by calculations using density functional theory. This work assists in generating a molecular-level understanding of the heterogeneous processes important to the reaction mechanisms involved in the efficient photoelectrocatalytic generation of carbon-containing fuels with high energy densities.

Two recent highlights from these studies will be discussed. First, we report on a combined investigation through experiment and theory on the surface-bound species on GaP(110) formed upon interaction with water. Experimentally, surface-bound species over 10 orders of magnitude of pressure were spectroscopically identified *in situ* using synchrotron-based ambient pressure photoelectron spectroscopy (APPES). Ga3d and O1s core-level spectra indicate that the interaction of GaP(110) with H_2O induces formation of a partially dissociated adlayer, characterized by the presence of both Gaâ"OH and molecular H_2O species. Measurements of the P2p core level indicate formation of a negatively charged hydride that irreversibly bonds to surface P in vacuum. The surface densities of the hydroxide and hydride species increase with increasing pressure (surface coverage) of water. Isobaric measurements at elevated pressures were used to probe the thermal stabilities of adsorbed species as well as the oxidation of surface Ga and P. The observation of stable surface hydride formation induced by interaction with water is especially notable given the critical role of hydride transfer to catalysts and CO_2 during chemical fuel synthesis reactions in aqueous environments. Secondly, we describe a scanning tunneling microscopy (STM) and density functional theory investigation of the orbital-resolved adsorption state defining the dative bonding interaction between a GaP(110) surface and a N-containing heteroaromatic (pyridine).Â By examining the distribution of unoccupied molecular orbitals, we show that STM images can be used to positively identify the sites on pyridine susceptible to nucleophilic attack, consistent with frontier orbital theory. This indicates that STM can be used to explore the local reaction centers of adsorbed ambidentate electrophiles and nucleophiles relevant to artificial photosynthesis, and more broadly to generate critical mechanistic information for various heterogeneous acidâ"base reactions.

461727 Implications of Surface Reconstructions Impacting Py-Catalyzed CO_2 Reduction on Semiconductor Photoelectrodes

Abstract:

The (photo)electro-catalytic reduction of CO_2 to useful fuels *via* energy from sunlight has received significant attention as a promising route for generating carbon-neutral fuels and value-added chemicals. Experimental evidence demonstrates that pyridine (Py) is an effective co-catalyst during CO_2 reduction over GaP, CdTe, and $CuInS_2$ semiconductor electrodes. Identifying the role of Py during catalytic reduction is essential for optimizing the design of such photocatalytic processes. Recent studies suggest that the semiconductor surfaces themselves can facilitate a heterogeneous reduction mechanism, and for this reason a detailed understanding of the interaction between Py and the various surfaces is required. Surface reconstructions occurring during operation alter the nature of adsorption sites available for interaction with the solution, therefore impacting the performance of the electrode. To address this issue, we employ density functional theory to assess the stability of reconstructed GaP, CdTe, and $CuInS_2$ surfaces, as well as to identify adsorption trends of Py-derived intermediates across sites created by such reconstructions. We also use many-body Greenâ€™s function theory combined with calculations of work functions to determine band edge positions of the explicitly solvated, reconstructed surfaces, which we compare to calculated reduction potentials involved in proposed elementary steps in the overall CO_2 reduction mechanism. This allows us to determine which reduction steps are thermodynamically feasible based on the energy of a photo-excited electron from the conduction band of the semiconductor. These results identify stable intermediate species along the CO_2 reaction path over reconstructed surfaces, thus lending insight into the Py-catalyzed reaction mechanism.

472650 Activity and Stability of Pure and Modified Coooh for Oxygen Evolution Reaction in Alkaline Medium

Abstract:

Improving the efficiency of the oxygen evolution reaction (OER) is critical for a number emerging energy conversion devices such as metal-air batteries, fuel cells and water-splitting cells. High overpotentials are required to increase the reaction kinetics of the OER in both alkaline and acidic conditions, which leads to significant efficiency lost. As a result, significant research efforts have been devoted to the development of active OER catalysts to improve device efficiency. To date, iridium and ruthenium oxides are considered the most active catalysts for OER, however their high cost and low abundance prevent large-scale application. Thus, active, stable, and affordable OER catalysts are highly sought after.

Transition metal oxides are very promising candidates for OER due to their high abundance, high OER activities, and availability in a wide range of structures and chemical compositions that allows optimization for OER activity. Recently, theoretical calculations as well as empirical evidence from Raman and X-ray absorption experiments have identified oxyhydroxides to be the OER active phase for a number of transition metal oxides in alkaline conditions under OER-relevant potentials. As a result, understanding the activity and stability of oxyhydroxides in alkaline conditions is critical and can aid in the design and optimization of OER catalysts. In this study, we wish to understand the correlation between OER activity and the structure and chemical composition of transition metal oxyhydroxides. To achieve this goal, we synthesized cobalt oxyhydroxide (CoOOH) with a well-controlled nanowire morphology and different dopant concentrations. The pure and modified CoOOH nanowires were synthesized directly onto a conductive substrate that offers excellent electrical contact between the nanowire catalysts and improves the area accessible to the electrolyteâ€"both of which are important considerations for practical OER electrodes. Pure and modified CoOOH catalysts were characterized using electron microscopy, X-ray diffraction, Raman spectroscopy, and X-ray photoelectron spectroscopy prior to and after electrochemical testing.

These electrodes showed improved OER activity in 0.1M KOH for CoOOH with nickel incorporation, whereas Mn incorporation did not affect the OER activity of CoOOH. Ni-modified CoOOH exhibited a lower Tafel slope compared to the pure and Mn-modified CoOOH. This suggests a difference in the reaction rate-determining step for Ni-modified CoOOH. Using electrochemical impedance spectroscopy, we determined that Ni modification of CoOOH lowers the charge transfer resistance and enhances the stability of surface species, whereas Mn modification did not alter the charger transfer resistance and decreases the stability of surface species. These results provide new insights regarding the effects of incorporating transition metals in CoOOH on OER activity and illustrate the stabilization of surface species as a strategy for optimization of CoOOH catalysts for the OER in alkaline conditions.

468323 Fe-Ni Core-Shell Hydroxide Nanoparticles As an Active Oxygen Evolution Reaction (OER) Catalyst

Abstract:

Hydrogen production through the electrolysis of water continues to be a key next-generation energy conversion and storage strategy with advantages including minimal greenhouse gas emissions, the lack of fossil fuels, and the potential as an alternative fuel source for fuel cells and reactant for industries such as ammonia production. Recent work by the Boettcher research group and others has demonstrated that iron incorporation into nickel hydroxide based catalysts can significantly enhance the activity of the catalyst for the oxygen evolution reaction (OER) [1-3], which is the half reaction of water electrolysis known to limit the overall process with high overpotential and slow kinetics. With these recent developments in understanding the role of iron incorporation into nickel hydroxide and the importance of iron as the catalytic site for OER within these bimetallic electrocatalysts [1], there is now an opportunity to further develop nanostructured bimetallic iron-nickel hydroxide catalysts for improved OER performance in an alkaline electrochemical environment.

In this presentation, our results on the synthesis, characterization, and electrochemical testing of an iron-nickel core-shell hydroxide nanoparticle catalyst will be presented. Bimetallic iron-nickel nanoparticles were synthesized using a multi-step procedure in water under ambient conditions. When compared to monometallic iron and nickel nanoparticles, the Fe-Ni nanoparticles show enhanced catalytic activity for OER under alkaline conditions (1 M NaOH). The bimetallic nanoparticles demonstrated an improvement in OER overpotential as well as a significant increase in maximum measured current density, as compared to the monometallic iron and nickel nanoparticles. At 1 mA/cm^2, the overpotential for the monometallic iron and nickel nanoparticles was 421 mV and 476 mV, respectively, while the bimetallic Fe-Ni nanoparticles had a greatly reduced overpotential of only 256 mV. At 10 mA/cm^2, bimetallic Fe-Ni nanoparticles had an overpotential of 311 mV. Electron microscopy and elemental analysis results

will be presented with a detailed discussion of unique aspects of the FeNi nanoparticle catalyst. Results suggest that while the nanoparticles are nominally in a core-shell morphology, there is significant migration of iron into the nickel shell as well as incorporation of some phosphorus into the nanoparticle shell, likely originating from the phosphonate-based stabilizer used during nanoparticle synthesis. X-ray photoelectron spectroscopy suggests that the primary phase of nickel is nickel hydroxide, and x-ray absorption spectroscopy characterization suggests that the primary phase of nickel is the more disordered alpha phase of nickel hydroxide. The presence of a small amount of nickel (oxy)hydroxide is also likely present, based on characterization results.

References

[1] M.S. Burke, L.J. Enman, A.S. Batchellor, S.H. Zou, S.W. Boettcher, Oxygen evolution reaction electrocatalysis on transition metal oxides and (oxy)hydroxides: Activity trends and design principles, Chemistry of Materials, 27 (2015) 7549-7558.

[2] M.S. Burke, S.H. Zou, L.J. Enman, J.E. Kellon, C.A. Gabor, E. Pledger, S.W. Boettcher, Revised oxygen evolution reaction activity trends for first-row transition-metal (oxy)hydroxides in alkaline media, Journal of Physical Chemistry Letters, 6 (2015) 3737-3742.

[3] L. Trotochaud, S.L. Young, J.K. Ranney, S.W. Boettcher, Nickel-iron oxyhydroxide oxygen-evolution electrocatalysts: The role of intentional and incidental iron incorporation, J. Am. Chem. Soc., 136 (2014) 6744-6753.

473700 Layered - Hyperthin Electrocatalysts for the Water Splitting Reactions

Abstract:

We report scalable, solution-processing methods for synthesizing two different layered hyperthin nanostructure electrocatalysts for the water splitting reactions. First, we show that 2D FeS_2 disc nanostructures are an efficient and stable hydrogen evolution electrocatalyst. By changing the Fe:S ratio in the precursor solution, we were able to preferentially synthesize either 1D wire or 2D disc nanostructures. The 2D FeS_2 disc structure has the highest electrocatalytic activity for the hydrogen evolution reaction, comparable to platinum in neutral pH conditions. The ability of the FeS_2 nanostructures to generate hydrogen was confirmed by scanning electrochemical microscopy (SECM), and the 2D disc nanostructures were able to generate hydrogen for over 125 h. Second, we report a novel solution-processing synthesis route for creating nickel-iron oxyhydroxides. Nickel-iron oxyhydroxides are non-precious metal electrocatalysts for the oxygen evolution reaction (OER) that have high efficiency in alkaline media. It has been proposed that the layered structure of the oxyhydroxide allows for electron hole transport from the bulk material to the solution interface, which creates very high active site densities. We compared the intrinsic electrocatalytic activity and active site densities for two $Ni_{0.8}$:$Fe_{0.2}$ oxyhydroxides with similar electrochemical surface areas (ECSA) â€" one synthesized via electrochemical conditioning of crystalline $Ni_{0.8}$:$Fe_{0.2}$ oxide and the second synthesized via our new solution-processing route. The overpotential for oxygen evolution on $Ni_{0.8}$:$Fe_{0.2}$ oxyhydroxide decreased from 366 mV on the crystal-derived structure to 268 mV on the new solution-processing structure at 1 mA cm^{-2} when normalized to the ECSA of the catalyst. We measured the active site density of both $Ni_{0.8}$:$Fe_{0.2}$ oxyhydroxides directly using the surface interrogation mode of scanning electrochemical microscopy (SI-SECM).

465065 Stabilizing Electrodeposited Nanoparticle Electrocatalysts on Si MIS Photocathodes for Solar Water Splitting

Abstract:

As the cost of solar energy continues to drop, the major hurdle limiting the widespread use of intermittent renewable solar energy is the lack of efficient and cost-effective energy storage. Photoelectrochemical cells (PECs) offer one solution through their ability to convert solar energy directly into storable solar fuels such as hydrogen. Hydrogen is a versatile energy carrier which can be used for a variety of applications (*e.g.*, chemical precursor for ammonia production, fuel for transportation, or converted back to electricity). The essential component of a PEC is the semiconducting photoelectrode that absorbs sunlight to drive non-spontaneous reactions such as water electrolysis. Commercialization of PEC technology is inhibited by challenges to identify and develop photoelectrodes that have long term stability, high efficiency, and low-cost materials.

One promising approach to achieving stable and efficient PEC water splitting is the composite photoelectrode architecture known as a metal-insulating-semiconductor (MIS) photoelectrode. [1,2,3,4] Within the MIS design, the semiconductor efficiently absorbs sunlight and a thin insulating layer (< 2 nm thick) protects the underlying semiconductor from corrosion and facilitates carrier transport between the semiconductor and metal through quantum mechanical tunneling. A metal catalyst is deposited on top of the insulator, to collect photo-generated carriers and catalyze the water splitting reaction. Through this composite design, the MIS architecture is able to decouple efficiency and stability tradeoff that typically limits conventional photoelectrodes, and has been applied to c-Si and InP photoelectrodes with great success in recent years. [1,2,3,4] In prior work on MIS photoelectrodes, the metal layer is typically deposited by physical vapor deposition in vacuum, which is an expensive and non-scalable method for deposition. If this technology is to become commercially viable, a scalable fabrication process is needed in addition to further improvements in MIS photoelectrode performance. In particular, improved photovoltage and maximized photo-current density generated by MIS photoelectrodes is essential for achieving DOE solar-to-hydrogen conversion

efficiency targets.

In this work, electrodeposition has been explored as a potentially low-cost and scalable means of depositing ultra-low loadings (1 to 20 Î¼g cm^{-2}) of discontinuous Pt nanoparticles onto SiO_2-covered p-Si photoelectrodes. Well-defined, nano-scale catalytic metal islands,[3,5] such as those achieve by electrodeposition, can facilitate high optical transmission of incident light into the semiconductor to improve photo-current density. However, as-deposited MIS photoelectrodes are found to exhibit very poor stability and performance. We report how this issue can be overcome through application of a thin transparent SiO_x overlayer. The SiO_x/Pt nanoparticle structure is similar in nature to core-shell nanoparticles that have previously been used for high temperature heterogeneous catalysis to enhance thermal stability [5], as well as for reducing undesired back-reactions in photoelectrochemical applications [6]. Surprisingly, it is found that the overlayer also results in a substantial improvement in performance in comparison to as-deposited MIS photoelectrodes of identical Pt loading. The combination of electrodeposition with the transparent overlayer thus offers an exciting opportunity for improving both the efficiency and durability of photoelectrochemical energy conversion. After describing a side-by-side comparison of MIS photoelectrodes with and without overlayer, we highlight reasons for the differences in their performance and stability.

References:

[1] Y.W. Chen, P. McIntyre, et al., Nat. Mater., 10 (2011)

[2] M.H. Lee, et al., Angew. Chem. Int. Ed., 51 (2012).

[3] J.C. Lee, A. Bard, J. Ekerdt, et al., Nature Nanotech., 10, 84-90 (2015)

[4] D.V. Esposito, I. Levin, T.P. Moffat, and A.A. Talin. Nature Materials, 12, 562-568 (2013).

[5] T. D. Gould, A. Izar, A. W. Weimer, J. L. Falconer, J. W. Medlin, ACS Catal. 2014, 4, 2714â€"2717.

[6] K. Maeda, K. Teramura, D. Lu, N. Saito, Y. Inoue, K. Domen, Angew. Chemie - Int. Ed. 2006, 45, 7806â€"7809.

468580 Two Distinct Types of Surface-Bound Atomic Hydrogen and Their Role in Dictating Product Selectivity in Artificial Photosynthesis Reactions

Abstract:

Understanding the surface reaction mechanisms, the nature of surface reaction sites, and the chemical physics behind differently-promoted hydrogen transfer stepsÂ in the photocatalytic reduction of CO_2 and H_2O is critical to enable the systematic design and improvement of catalysts for the production of higher hydrocarbons and oxygenates from solar energy.Â Our recent studies indicate that the product distribution, e.g., H_2 and CO vs. CH_4, in the photocatalytic reduction of CO_2 is connected to two distinct types of hydrogen transfer steps â€" electrochemically-coupled H-transfer and excited-electron mediated H-transfer.Â Presence of protonic hydrogen (H^+) on photocatalyst surface correlated with electrochemically-coupled H-transfer steps to preferentially produce H_2. The electrochemical steps for CO production in this case likely follow similar chemical physics. On the contrary, when metallic hydrogen (H^0) is encountered, results suggest that the transfer of H^0 to surface-bound organic fragments occur via excited electron induced vibrational excitation of the uncharged H^0, resulting in production of CH_4. A clear connection was observed between the presence of H^0 on surface and CH_4 evolution, and conversely, the presence of H^+ on surface and H_2 evolution. In our photoreaction experiments, we were able to modify the catalyst selectivity from CH_4 to H_2 evolution by transforming the adsorbed metallic H^0 into protonic H^+ by incorporating more electronegative reaction sites into the photocatalyst surface. Theoretical calculation results suggested a shift from excited electron-mediated H^0 transfer phenomena to electrochemical H-transfer phenomena resulting in H_2 evolution. Understanding the electronic nature of adsorbed hydrogen on photocatalyst surface and the nature of surface reaction sites will potentially lend insights into tuning the overall photocatalytic product distribution, which may lead to the sustainable production of higher

hydrocarbons and oxygenates from CO_2 and H_2O using solar radiation.

457102 Plasmon-Enhanced Photocatalytic CO_2 Reduction on Nanostructured Composite Electrodes

Abstract:

Using inputs of only sunlight, electricity, carbon dioxide, and water, photocatalytic CO_2 reduction could mitigate the effects of climate change caused by the burning of fossil fuels and produce valuable fuels or chemical feedstocks. However, current CO_2 reduction technologies suffer from high overpotentials and low selectivity, producing a mixture of carbon monoxide, methane, ethylene, formate, methanol, and other products.

Plasmon-assisted photocatalytic CO_2 reduction leads to greater selectivity and lower overpotentials by unlocking unique mechanistic pathways. It has been well-documented that the strongly localized near fields at the surface of plasmonic nanoparticles promote electron-hole pair generation in nearby semiconductors.[1–3] Plasmon decay generates an excited "hot" electron which can be transferred to a surface molecule for direct reduction or injected into an adjacent wide band gap catalyst, effectively limiting carrier recombination through charge separation and expanding the usable portion of the solar spectrum.[4–7] The electron dynamics in an irradiated plasmonic nanoparticle can alter the electronic coupling with surface adsorbed CO_2 and reaction intermediates, thereby changing the binding energy of these species and the catalytic properties of the plasmonic metals. This plasmon-enhanced effect was demonstrated when optically excited surface plasmons at the roughened surface of an Ag electrode in CO_2-containing solutions produced strong enhancement of the cathodic current.[8–9] The wide tunability of the plasmon resonance frequency with shape, size, and material confers fine control over these catalytic mechanisms, allowing for optimization of the photocatalytic performance.

To take advantage of plasmonic catalysis, nano-sized spheres, cubes, and wires of gold, silver, and copper

have been synthesized as these materials exhibit strong plasmonic and electrocatalytic behavior. For the semiconducting component of the photoelectrodes, nanostructured TiO_2 co-catalysts have also been prepared. Transmission electron microscopy (TEM) and scanning electron microscopy (SEM) were used to determine the shape and size distribution of the nanoparticles. X-ray diffraction (XRD) and optical spectroscopy were employed to characterize the electrodes. The CO_2 reduction ability of the plasmonic metal nanoparticles with and without a semiconductor co-catalyst has been tested in a custom glass photoelectrochemical cell with a quartz window that allows for front- and back-illumination of the electrode. The two electrode chambers are separated by a membrane and can achieve a gas-tight seal that allows for precise pressure measurements. Experiments were run both under isolation of the pressurized cell and with a constant CO
spectrometer (GC-MS) that allows detection and quantification of gaseous products. *Ex-situ* high performance liquid chromatography (HPLC) was used to quantify liquid products.

This material is based upon work performed by the Joint Center for Artificial Photosynthesis, a DOE Energy Innovation Hub, supported through the Office of Science of the U.S. Department of Energy under Award Number DE-SC0004993.

The work is supported by the National Science Foundation Graduate Research Fellowship under Grant No. 1106400.

Work at the Molecular Foundry was supported by the Office of Science, Office of Basic Energy Sciences, of the U.S. Department of Energy under Contract No. DE-AC02-05CH11231.

References:

(1) Thomann, I.; Pinaud, B. A.; Chen, Z.; Clemens, B. M.; Jaramillo, T. F.; Brongersma, M. L. *Nano Lett.* **2011**, *11*, 3440–3446.

(2) Ingram, D. B.; Linic, S. *J. Am. Chem. Soc.* **2011**, *133*, 5202–5205.

(3) Gao, H.; Liu, C.; Jeong, H. E.; Yang, P. *ACS Nano* **2012**, *6*, 234–240.

(4) Mukherjee, S.; Zhou, L.; Goodman, A. M.; Large, N.; Ayala-Orozco, C.; Zhang, Y.; Nordlander, P.; Halas, N. J. *J. Am. Chem. Soc.* **2014**, *136*, 64–67.

(5) Mubeen, S.; Lee, J.; Singh, N.; Krämer, S.; Stucky, G. D.; Moskovits, M. *Nat. Nanotechnol.* **2013**, *8*, 247–251.

(6) Naya, S.; Teranishi, M.; Isobe, T.; Tada, H. *Chem. Commun.* **2010**, *46*, 815–817.

(7) Lee, J.; Mubeen, S.; Ji, X.; Stucky, G. D.; Moskovits, M. *Nano Lett.* **2012**, *12*, 5014–5019.

(8) Kostecki, R.; Augustynski, J. *Chem. Phys. Lett.* **1992**, *194*, 386–390.

(9) Kostecki, R.; Augustynski, J. *J. Appl. Phys.* **1995**, *77*, 4701–4705.

459985 Photon, Electron, and Ion Management in Artificial Photosynthesis: Realizing Efficient Renewable Energy to Fuel Conversion

Abstract:

Direct production of fuels from sunlight, air (N_2 or CO_2), and water that can be stored, transported, and later converted into hydrogen or electricity to provide power for transportation and distributed energy generation, have received recent attentions worldwide. This technology could also provide chemicals as synthetic precursors or realize grid-level storage of intermittent solar energy. In artificial photosynthesis based on semiconductor photoelectrochemistry, production of chemical fuels generally requires the coupling of separated electrical charges with electrocatalysts for multi-electron chemical reactions. Meanwhile, the development of integrated, efficient and stable photoelectrochemical (PEC) systems requires the pairing of light-absorbing materials with an optimum bandgap combination. The development of such systems has been hindered in part by the lack of semiconducting materials that can simultaneously provide efficiency and stability in a corrosive aqueous electrolyte, typically strong acid or base. In this talk, I will present our recent progress in the development of efficient and stable PEC devices for H2 production from solar-driven water splitting. First, I will summarize some recent development of photon management in PEC devices and module designs from modeling to experiment. Then, I will discuss methods for heterogeneous interfacial energetic modification on covalence-bond semiconductors for efficient free carriers (electrons/holes) separation. Finally, I will present a novel approach to managing the ionic environment for creating a defect-tolerant condition, which leads to a record-setting solar water-splitting prototype.

462500 Tandem Core-Shell Si-Ta_3N_5 Photoanodes for Photoelectrochemical Water Oxidation

Abstract:

Tantalum nitride (Ta_3N_5
type doping, relatively small bandgap of 2.1 eV, and band positions that straddle the redox potentials of both the hydrogen and oxygen evolution reactions.[1] However, there are other challenges associated with Ta_3N_5 including poor charge transport properties, late photocurrent onset and photodegradation. To address all of these challenges, we fabricated efficient and stable core-shell heterostructured Si-Ta_3N_5 nanowire photoanodes coated with $CoTiO_x$ and NiO oxygen evolution co-catalysts.

Previous device architecture strategies for metal oxide semiconductors with charge transport limitations have included nanostructuring or scaffolding. In this work, a similar approach was applied to Ta_3N_5 to make a nanostructured core-shell Si-Ta_3N_5 device where the Si is a semiconducting, nanowire scaffold and the Ta_3N_5 is coated onto Si as a thin shell through which charge extraction should be efficient. With n-type Si as the scaffold, a heterojunction between Si and Ta_3N_5 is formed enabling a dual-absorber configuration resulting in a 200 mV cathodic shift for photocurrent onset due to the photovoltage gained from Si and therefore improves the overall performance of the Si-Ta_3N_5 device. The Si nanostructures were fabricated by a pseudo-Bosch deep reactive ion etching process, then tantalum oxide (Ta_2O_5) was coated onto Si by atomic layer deposition (ALD), and finally the Ta_2O_5 was nitrided in pure ammonia gas at 900 °C.

To gain a deeper understanding of charge transport within the thin Ta_3N_5 shells and to optimize the electrode

performance, we varied the thickness of the shell between 10 and 70 nm and tested for ferrocyanide oxidation as a facile photoelectrochemical reaction. We found that in a core-shell Si-Ta_3N_5 nanostructured device, photocurrent decreased with increasing Ta_3N_5 thickness, which indicates that charge extraction becomes problematic with thicker Ta_3N_5 films. To verify that decreasing photocurrent was due to poor charge transport in thicker films, we performed absorbed photon to current efficiency (APCE) measurements with Ta_3N_5 deposited on planar Si and quartz substrates. Again, we found that APCE decreased once the films were thicker than 30 nm, demonstrating that the minority carrier diffusion length is likely tens of nanometers.

Finally, to address the problem of Ta_3N_5 photodegradation and to demonstrate photoelectrochemical performance for water oxidation in basic electrolyte with the core-shell system, $CoTiO_x$ and NiO oxygen evolution catalysts were coated onto Si-Ta_3N_5 by dipcoating from a sol-gel and ALD, respectively. In terms of activity, we found that $CoTiO_x$ interfaces well with Ta_3N_5 and yields a high photocurrent density compared to NiO. With NiO we were able to achieve a photocurrent density close to that of $CoTiO_x$ after Fe incorporation into NiO (henceforth called Ni(Fe)O).[2] However, with $CoTiO_x$, the device is not stable; but with only ~1 nm of Ni(Fe)O we see a marked improvement in stability. The device loses 40-50% of its peak photocurrent over a 24 hour chronoamperometry test. Compared to the current Ta_3N_5 reports in the literature, the Ni(Fe)O catalyst affords noteworthy stability.[3-5] Therefore, in this work, we demonstrate a heterostructured Ta_3N_5 system with a tandem core-shell Si-Ta_3N_5 configuration that enabled earlier onset of photocurrent, efficient charge extraction from Ta_3N_5 and relatively stable long term water oxidation with a Ni(Fe)O co-catalyst.

1. Chun, W.-J.; Ishikawa, A.; Fujisawa, H.; Takata, T.; Kondo, J. N.; Hara, M.; Kawai, M.; Matsumoto, Y.; Domen, K. *The Journal of Physical Chemistry B* **2003,** 107, (8), 1798-1803.

2. Trotochaud, L.; Young, S. L.; Ranney, J. K.; Boettcher, S. W. *J Am Chem Soc* **2014,** 136, (18), 6744-6753.

3. Li, Y.; Zhang, L.; Torres-Pardo, A.; González-Calbet, J. M.; Ma, Y.; Oleynikov, P.; Terasaki, O.; Asahina, S.; Shima, M.; Cha, D.; Zhao, L.; Takanabe, K.; Kubota, J.; Domen, K. *Nat Commun* **2013,** 4.

4. Liao, M.; Feng, J.; Luo, W.; Wang, Z.; Zhang, J.; Li, Z.; Yu, T.; Zou, Z. *Advanced Functional Materials* **2012,** 22, (14), 3066-3074.

5. Liu, G.; Fu, P.; Zhou, L.; Yan, P.; Ding, C.; Shi, J.; Li, C. *Chemistry â€" A European Journal* **2015,** 21, (27), 9624-9628.

462505 Heterostructured c-Si/$BiVO_4$ Core-Shell Tandem Photoanode for Unassisted Photoelectrochemical Water Splitting

Abstract:

Photoelectrochemical (PEC) water splitting is a promising way of storing solar energy in hydrogen, which can be used as a carbon-free fuel or a chemical feed stock. In principle, a PEC electrode consists of a single active material capable of handling the key processes: absorption of solar photon energy, charge separation and transport, and electrocatalysis to make and break chemical bonds at the electrode surface. In practice, there is no existing material that can handle all of these functions simultaneously; thus, the multicomponent system that integrate different materials with more specialized functions is a more viable approach.

A tandem-junction PEC device offers a high theoretical solar-to-hydrogen (STH) efficiency due to the additive voltages across two photoabsorbers and the better utilization of the solar spectrum. Modeling STH efficiency as a function of band gap for a tandem device shows that over 20% STH can be achieved with band gaps of 1.2 and 1.8 eV.[1] One promising photoabsorber material combination is c-Si paired with a metal oxide, especially $BiVO_4$. In addition to the appropriate band structure alignment of both semiconductors, c-Si and $BiVO_4$ are low cost, earth abundant, and industrially scalable. However, the key challenges for these materials are the instability of c-Si in aqueous environment and the poor electronic properties of $BiVO_4$.

In this work, we present the fabrication of a heterostructured c-Si/$BiVO_4$ core-shell tandem based device capable of performing spontaneous water splitting without any precious metal. The wafer scaled device architecture was prepared by (1) nanostructuring of the c-Si substrate to serve as both a bottom absorber and a scaffold to improve the charge separation of $BiVO_4$; (2) tuning the c-Si p-n junction depth to maximize the device photovoltage; (3) engineering the SnO_2 interfacial layer to passivate the c-Si/$BiVO_4$ interface and protect the c-Si p-n junction; (4) conformally coating a $BiVO_4$ thin film on the nanostructured Si; and (5) decorating the c-Si and $BiVO_4$ surfaces with both oxygen and a hydrogen evolution catalysts.

(1) Seitz, L. C.; Chen, Z.; Forman, A. J.; Pinaud, B. A.; Benck, J. D.; Jaramillo, T. F. *ChemSusChem* **2014,** *7* (5), 1372-1385.

462617 The Fabrication of Cu_2o/g-C_3N_4/WS_2 Triple-Layered Photocathode for Photoelectrochemical Hydrogen Evolution

Abstract:

The Fabrication of Cu_2O/g-C_3N_4/WS_2 Triple-layered Photocathode for Photoelectrochemical Hydrogen Evolution

Xintian Xu, Yuanzhi Zhu, Xiaobin Fan, Guoliang Zhang, Wenchao Peng*

School of Chemical Engineering & Technology, Tianjin University, Tianjin 300072, China.

wenchao.peng@tju.edu.cn

Hydrogen is being vigorously pursued as a future energy carrier due to the fossil energy depletion. Photoelectrochemical (PEC) water splitting is an emerging hydrogen production technology by utilizing the solar light. This technology integrates in the same device both solar energy collection and water electrolysis. Both photo-active semiconductor and catalysts for hydrogen evolution reaction (HER) are necessary in this device. Cu_2O is a direct P-type semiconductor with a narrow band gap (~2.0 eV), which can be activated by the visible light. Moreover, Cu is a kind of earth abundant element, and the synthesis of Cu_2O is facile. Cu_2O is therefore one of the most promising low-cost photoactive semiconductor reported until now. However, the activity of bare Cu_2O is still poor, and it is not stable towards chemical oxidants and light corrosion. Combing with other semiconductors (e.g. TiO_2, ZnO) could improve the activity and stability of Cu_2O effectively.

As we know, Pt and other precious metals are the most efficient catalysts for hydrogen evolution reaction (HER).

However, these metals are rare and expensive to apply. Transition-metal dichalcogenides (TMDs), such as Molybdenum disulfide (MoS_2), tungsten disulphide (WS_2), are emerging HER catalysts that could be used as a substitute for noble metals in HER.

In this study, a novel Cu_2O/g-C_3N_4/WS_2 triple-layered photocathode was designed for the PEC hydrogen evolution. Cu_2O was electrodeposited on the FTO glass as the active layer for the light harvesting. A metal-free g-C_3N_4 was exfoliated and covered on the surface of Cu_2O as protection layer. G-C_3N_4 could also form p-n junction with Cu_2O to increase the activity. 1T-$WS_2$was obtained by lithium intercalation and used as HER catalyst for the further modification of the photocathode.

The obtained photocathode was then used for PEC HER test in the Na_2SO_4/H_2SO_4 solution (Î»â‰¥420nm, 100 mW/cm^2). With the irradiation, electron-hole pair will be generated in the p-type Cu_2O, and the excited electron could be then extracted through the g-C_3N_4 protective layer to produce hydrogen on the WS_2. The efficient extraction of excited electrons by this triple-layered photocathode leads to photocurrents of up to -6 mA/cm^{-2} at -0.4 V versus RHE (pH 6.0). The corresponding overpotential is -0.23 V (versus RHE) for the H_2 evolution. In sharp contrast, nearly no H_2 evolution happened under dark conditions. This photocathode is facile to synthesis and has great potential for developing low-cost PEC water-splitting device.

473001 *Sustainable System Synthesis and Analysis using a Novel Sustainability Concept*

Abstract:

Sustainable system analysis and synthesis

using a novel sustainability concept

One of the first definitions for sustainability has been introduced by United Nations in 1987, in a report entitled, Our Common Future (also known as the Brundtland report) which define sustainable developments as "...development that meets the needs of the present without compromising the ability of future generations to meet their own needs." Although this definition captured the essence of the word sustainability but it is very vague in order to be applicable for other organizations. Efforts have been made by researchers to quantify the abstract definitions of sustainability and give it an applicable view. Previous attempts include the ecological footprint (1992, 1996), the use of Fisher information theory (2002), the pressure-state-response model of environmental sustainability (2003), the environmental sustainability index (2005), the sustainability assessment by fuzzy evaluation (SAFE) model (2001-2009), and the AIChE Sustainability Index (2007).

In the current work a novel sustainability concept is put forward, which allows a rigorous quantification of sustainability for a variety of system descriptions. The new concept allows societal and personal input to be incorporated in the sustainability assessment process. Rigorous mathematical criteria are provided for its assessment, involving the global solution of finite dimensional optimization problems. In addition, the proposed concept lends itself to the synthesis of sustainable systems. Case studies for 2-dimensional and 3-dimensional dynamical systems are used to illustrate the introduced new concept, and its applicability to both sustainability analysis and synthesis.

473967 Systematic Approach Towards Establishing Thermodynamic Principles of Sustainable Coupled Industrial-Natural Systems (CINS)

Abstract:

Fundamentals of thermodynamics apply to wide variety of systems at multiple scales such as materials and biogeochemical cycles [1, 2] - ranging from quantum to planetary scale. It will not be an exaggeration to say that systems that violate physical laws of thermodynamics cannot exist as stable systems. This raises two very critical questions for the field of sustainability: a) What are the thermodynamic principles of sustainability science and b) How do we approach hypothesizing, establishing and validation of these principles in a scientifically rigorous way? For the first question, I will qualitatively present some plausible theories of thermodynamics in context of sustainability such as Maximum Entropy principle [3] and non-equilibrium thermodynamics [4]. For the second question, I will present a framework that can be adopted to study thermodynamic properties of coupled industrial-natural systems (CINS) and relating it to approaches for testing the applicability of thermodynamic principles for CINS. The quantitative study will focus on the thermodynamic properties of nutrient flows (Nitrogen) in an industrial ecosystems coupled with regional-ecological nitrogen cycle. Following the course of establishment of thermodynamic laws in other fields, it is suggested that quantitative modeling framework to observe thermodynamic properties over a long span may provide clues to answer question 1 whereas more validation approaches may need to be developed for drawing some general principles.

References

1. D Panagiotaras et al., 2015, "Biogeochemical Cycling of Nutrients and Thermodynamic Aspects", J. Thermodynamic Catalysis, 6:2
2. J.J. Vallino and C.K. Algar, 2016, "The Thermodynamics of Marine Biogeochemical Cycles : Lotka Revisited", Annu. Rev. Mar. Science , 8:333-356
3. R. E. Ulanowicz and B.M. Hannon, 1987, "Life and the production of entropy", Proc. R. Soc. Lond., B 232, 181-192
4. C.G. Chakrabarti and K. Ghosh, 2009, "Non-equilibrium thermodynamics of ecosystems : Entropic analysis of stability and diversity", Ecological Modelling, 220, 1950-1956

474368 Analysis of United Nations Clean Development Mechanism Carbon Emission Reduction Projects from a Life Cycle Assessment Perspective

Abstract:

We analyze, from a life cycle perspective, the effectiveness of the United Nations Clean Development Mechanism (CDM) emission reduction projects implemented in developing countries. Typically, the carbon offset credit allocated to these projects account for avoided direct emissions of GHGs but do not include indirect emissions that occur during upstream and downstream steps (for both avoided as well as the project emissions). We explore, through a case study approach, how these projects perform when life cycle calcualtions are used. Further, we also explore if these proejcts lead to lower (or higher) health and ecosystem impacts (co-benefits). A small but diverse set of renewable energy generation CDM projects are compared on an equivalent bassis to assess any potential problem shifting issues. Results from this study can help policy makers assess effectivenss of such projects in a more holistic manner and fine tune their emission calculation methodologies.

468174 Evolution and Robustness of the Global Agricultural-Phosphorus Trade Network

Abstract:

Containing one of the three essential nutrients for plant growth, phosphatic fertilizers are an integral input to modern agriculture, and will always have a global demand. As the world population continues to grow through mid-century, a significant increase in food production must follow, potentially furthering our current dependence on synthetic fertilizers. While the global resource-criticality of phosphate rock reserves – from which the majority of these fertilizers are derived – may be uncertain, the trade network upon which many countries depend for access to these fertilizers (and their precursors) is a well-documented complex system, subject to an array of physical, economic, and geopolitical constraints. If we are to collectively achieve global food security by 2030, as per the U.N. Sustainable Development Goals, the global security of agricultural nutrients must also be ensured. To formalize questions of access to these nutrients, network analysis provides an attractive quantitative framework with which to examine features of individual relationships and of the underlying structures of complex systems.

The present work focuses on characterizing the structure and understanding the systemic vulnerabilities of the global agricultural phosphorus trade network (GAPTN). We accomplish this by modeling a directed, weighted global network of annual phosphorus-containing commodity exchanges, including all synthetic phosphatic fertilizers and their chemical precursors. All commodities are further normalized to their weight content of elemental phosphorus. As such, nodes are taken to be countries participating in phosphorus trade, and edges are taken to be the aggregate of the directed trade flows of elemental phosphorus between each pair of countries, resulting in the construction of a global agricultural phosphorus trade network (GAPTN). The primary objectives of this study are to (1) characterize the network topology and evolutionary trends of the GAPTN. (2) Elucidate the sets of nodes and edges that are critical to the overall function of the network. (3) Identify those countries most dependent on the network's functioning, and outline engineering and policy measures such that they may ensure their consistent access to this essential resource.

Edges within the GAPTN display mild heterophily, preferentially connecting nodes of dissimilar degree. Based on statistical testing, the distribution of GAPTN node degrees follows a heavy-tail distribution, as does the distribution of in-, out-, and total-strength for 2014. Such distributions indicate that a small subset of all of the participating countries exchange most of the resources that comprise the system, leaving the network robust to random disruptions and highly vulnerable to inopportune ones. The potential for inopportune disruptions stems from both the price volatility of phosphatic fertilizers with respect to energy prices, and the geopolitical tension that surrounds Western Sahara & Morocco, which owns some 75% of the estimated global phosphate rock reserves. Further confirming the implications of the degree distributions, a robustness analysis of randomized versus targeted removal of nodes from the network indicates that some 80% of the network's overall activity halts when some 5% of the nodes do not participate. Results from node-level analyses reveal a small subset of countries to be particularly vulnerable to supply disruptions, even as the overall network has become more redundant and dense from 1990 to 2014. In light of these findings, both international cooperation on phosphorus management and country-level efforts to close local nutrient cycles can aid in improving food security while also minimizing the ecological impacts associated with fertilizer run-off, especially eutrophication.

471802 A Framework for Considering Synergies Between Nature and Engineering from Process to Planetary Scales

Abstract:

Two major requirements for designing sustainable systems include considering all the systems through the entire life cycle to prevent the shifting of impacts to different scales; and to account for the role played by nature while making decisions in order to stay within the carrying capacity of ecosystems to prevent unintended harm or ecological degradation.

The recently developed Process-to-Planet (P2P) modeling framework [1,2] integrates fundamental engineering models and key processes in the upstream supply chain with macroeconomic systems. It improves upon conventional Sustainable Process Design (SPD) methods by incorporating the entire upstream life cycle while solving a process design problem, thus reducing the chances of shifting of environmental impacts across narrow system boundaries. The Techno-Ecological Synergy (TES) framework [3], on the other hand takes into account the carrying capacity of ecological systems by quantifying the demand and supply of ecosystem services, accounting for all the interactions between technological and ecological systems at multiple scales. Including ecosystems within the boundary of traditional engineering design decisions is expected to identify design solutions that aim to keep the impact of engineering activities within the regenerative capacity of ecosystems, depending on the scale of relevance of ecological systems. While the TES framework is the first of its kind to account for interactions between engineered systems and ecological systems, the framework lacks connections with life cycle models and while the P2P considers a comprehensive analysis boundary, it does not consider ecological systems at any scale.

This work introduces the methodology of "Process-to-Planet Techno-Ecological Synergy" (P2P-TES) for designing sustainable systems. The objective behind developing this framework is to make design decisions at the local scale by capturing the entire upstream life cycle boundary. The Structural Path Analysis algorithm (SPA) embedded within the P2P framework helps to distribute the processes or sectors in the system among the different scales. Inclusion of ecosystem services through TES within the P2P framework will expand the solution space and help to discover optimal solutions which enhance both the environmental and economic profitability of the system. In other words, P2P-TES framework is expected to provide "win-win" solutions to the process design problem compared to one where ecosystem services are not included. In addition, accounting for the imbalance between the demand and supply of ecosystem services will enable making design decisions that keep human activities within the carrying capacity of ecological systems. Ecosystem services that will be included in the framework are carbon sequestration, within the national boundary and air quality regulation service within a regional serviceshed.

P2P-TES framework will be applied to two separate case studies which present their own different set of challenges. Successful application to both of them will prove the robustness of the framework as well expose existent deficiencies.

References

[1] Hanes, Rebecca J., and Bhavik R. Bakshi. "Process to planet: A multiscale modeling framework toward sustainable engineering." AIChE Journal 61.10 (2015): 3332-3352.

[2] Hanes, Rebecca J., and Bhavik R. Bakshi. "Sustainable process design by the process to planet framework." AIChE Journal 61.10 (2015): 3320-3331.

[3] Bakshi, Bhavik R., Guy Ziv, and Michael D. Lepech. "Techno-ecological synergy: A framework for sustainable engineering." Environmental science & technology 49.3 (2015): 1752-1760.

462411 Tungsten Oxide Decorated Zinc Telluride for the PhotoelectrochemicalÂ Water Splitting

Abstract:

Due to the increasing adverse effects on greener energy impacted by fossil fuel dependency and the resulting consequences on energy sustainability, a more succinct approach becomes inevitable for the continuous supply of clean and renewable energy. Tungsten Oxide (WO_3) has exhibited high photoactivities and electrochromic capabilities in the search for durable materials for hydrogen (H_2) production from photoelectrochemical water (H_2O) splitting. With a bandgap of 2.6 eV makes it an excellent candidate for harvesting the blue light of the solar spectrum with improved efficiency. Various experimental investigations have shown that electrochemically deposited WO_3 can be used to this effect. Sensitization with organic dyes that possess sunlight trapping capabilities have also been employed to enhance the performance of WO_3 as a photo catalyst for H_2O splitting. The successes in dye sensitization suggest that another metal compound, zinc telluride (ZnTe), which has an appropriate bandgap alignment for high utilization of the visible light of the solar spectrum can be used as a sensitizer as well. The fact that this material has not been studied in great detail in conjunction with WO_3 usage, makes it an interesting candidate of study in improving the performance of WO_3 as a photocathode for H_2 production. This material has a suitable bandgap of 2.26 eV for light harvesting and its conduction band edge positioned at -1.63 eV offers a large driving force for interfacial electron transfer from the semiconductor to the WO_3 acceptor. The objective of this work is to study the contribution of WO_3 decorated ZnTe in the splitting of H_2O. We will present the result of our investigation on cathodic electrodeposition of WO_3-ZnTe thin film on indium-tin oxide substrate, with its catalytic properties as an electrode for H_2 production.

468632 Molybdenum Silicide and Disulfide Protection Schemes for Silicon Photocathodes

Abstract:

Photoelectrochemical (PEC) water splitting offer a promising route to sustainable production of clean, renewable hydrogen.[1] In particular, silicon shows great potential as a small band gap photocathode in tandem photoelectrochemical (PEC) water splitting due to its appropriate band structure, energetic alignment for driving hydrogen evolution, and its band gap of 1.1 eV.[2] However, some challenges remain to ensure that silicon photoelectrodes possess the required catalytic activity and stability for use in highly effect PEC water splitting devices.

Molybdenum disulfide (MoS_2) is a low permeability material known for its excellent stability and HER activity in acid. We have previously demonstrated MoS_2 protected silicon photocathodes which operate for 100 hrs without significant loss of activity.[3] Here, we further this study to design strategies, such as the utilization of a molybdenum silicide interface layer, for improved activity and >800 hour stability with unchanged photoelectrochemical performance (onset potential, fill factor and light limited photocurrent density), which is among the best performance of any precious metal free photocathode.

Critical to developing future electrodes with enhanced stability, we probe the failure mechanism of the MoS_2 protected photocathodes. By XPS we observe that the MoS_2 is fully absent post stability testing. Detailed failure mechanisms of the photocathodes and electrolyte are uncovered by electrochemical methods, SEM, Raman microscopy and ICP. Furthermore, we implement increased light intensity, as well as light and dark cycling to probe accelerated durability testing protocols utilizing MoS_2 protected photocathodes. From such analyses we facilitate insightful design of future corrosion resistant protection schemes.

(1) Walter, M. G.; Warren, E. L.; McKone, J. R.; Boettcher, S. W.; Mi, Q. X.; Santori, E. A.; Lewis, N. S. Solar Water Splitting Cells. *Chem. Rev.* **2010**, *110*, 6446â€“6473.

(2) Seitz, L. C.; Chen, Z.; Forman, A. J.; Pinaud, B. A.; Benck, J. D.; Jaramillo, T. F. Modeling Practical Performance Limits of Photoelectrochemical Water Splitting Based on the Current State of Materials Research. *ChemSusChem* **2014**, *7* (5), 1372â€“1385.

(3) Benck, J. D.; Lee, S. C.; Fong, K. D.; Kibsgaard, J.; Sinclair, R.; Jaramillo, T. F. Designing Active and Stable Silicon Photocathodes for Solar Hydrogen Production Using Molybdenum Sulfide Nanomaterials. *Adv. Energy Mater.* **2014**, *4* (18), 1400739.

463462 Utility of Dual-Layer Photoanode for Photoelectrochemical Biomass Conversion

Abstract:

Photoelectrochemical (PEC) cells use solar energy to power desirable chemical reactions, such as splitting water into hydrogen, a fuel source, and oxygen. It has been shown that certain catalysts improve the efficiency of PEC water splitting, but less work has been done developing catalysts for chemical production in PECs. Here, we present the synthesis and characterization of electrocatalysts which enhance the efficiency of other oxidation reactions, where water oxidation is actually the competing reaction. Specifically, we have demonstrated improved performance of dual layer photoanodes on the oxidation of 5-hydroxymethylfurfural (HMF), a biomass derivative. Under 1.5 AM G illumination, we observe 21% conversion of HMF on a single layer photoanode while the addition of an electrodeposited catalyst increases HMF conversion to 33%. Both linear sweep voltammetry and chronoamperometry measurements indicate the catalyst positively affects the biomass oxidation reaction but not water oxidation. By adjusting reaction conditions and controlling the interfaces between the photoanode layers as well as the electrolyte, electrocatalysts may be fine-tuned to behave differently. This result opens new doors to the utility of PECs for applications beyond water splitting.

465069 Molybdenum Disulfide As a Protection Layer and Catalyst for Gallium Indium Phosphide Solar Water Splitting Photocathodes

Abstract:

Gallium indium phosphide ($GaInP_2$) is a semiconductor with a nearly ideal bandgap for solar water-splitting as the top absorber in a dual junction tandem absorber device. It has been used in conjunction with a gallium arsenide (GaAs) bottom absorber in an overall water splitting cell with 12.4% solar-to-hydrogen (STH) efficiency, one of the highest STH efficiencies for an integrated photoelectrochemical (PEC) water-splitting device reported to date.[1] However, $GaInP_2$ suffers from one of the biggest challenges facing the field: instability due to electrochemical corrosion in aqueous electrolytes. Molybdenum disulfide (MoS_2) nanomaterials can be used to both protect $GaInP_2$ and significantly improve its catalytic ability since it is resistant to corrosion and also possesses high activity for the hydrogen evolution reaction (HER). In this work, we demonstrate that $GaInP_2$ photocathodes coated with thin MoS_2 surface protecting layers exhibit excellent activity and stability for solar hydrogen production and we probe the details of failure mechanisms using a novel flow cell microscopy technique.

Our $GaInP_2$ photocathodes demonstrated no loss in performance (photocurrent onset potential, fill factor, and light limited current density) until 60 hours of operation which represents a five-hundred fold increase in stability compared to bare p-$GaInP_2$ samples tested in identical conditions.[2] We believe this to be one of the first successful attempts to stabilize $GaInP_2$ using a thin film protection layer scheme. Furthermore, as this protection scheme has previously been used successfully on silicon photocathodes[3], this work highlights the potential for MoS_2 to be used as a thin film protection layer for many different semiconductor water splitting devices that are unstable in acid. Using a custom-designed flow cell coupled with various microscopic techniques, we gained a greater understanding of the failure

mechanisms of MoS_2 as a thin-film protection layer. We discovered that pinhole formation in the MoS_2 layer exposes the $GaInP_2$ substrate, which readily corrodes in the acidic conditions, ultimately leading to device degradation. The flow cell further allowed us to capture the time scale of this pinhole formation. These insights represent a deeper understanding of MoS_2 as a protection layer and can be leveraged to improve the stability of thin film protected semiconductor water splitting devices.

References

(1) Khaselev, O.; Turner, J. A., A Monolithic PhotovoltaicPhotoelectrochemical Device for Hydrogen Production Via Water Splitting. Science 1998, 280, 425427.

(2) Britto R.J., Benck J.D., Young J.L., Hahn C., Deutsch, T.G., Jaramillo T.F., Molybdenum Disulfide as a Protection Layer and Catalyst for Gallium Indium Phosphide Solar Water Splitting Photocathodes. Journal of Physical Chemistry Letters. *Accepted (2016)*

(3) Benck, J. D.; Lee, S. C.; Fong, K. D.; Kibsgaard, J.; Sinclair, R.; Jaramillo, T. F., Designing Active and Stable Silicon Photocathodes for Solar Hydrogen Production Using Molybdenum Sulfide Nanomaterials. Advanced Energy Materials 2014, 4, 18

468243 Bioelectricity Generation from a Carbon Soot Electrode Using a Paper Based Microbial Fuel Cell

Abstract:

Disposable energy harvesting devices on a paper based platform in Microbial Fuel Cell (MFC) can provide low cost and affordable power supply in resource limited regions[1]. Electrodes play crucial role in these devices. The tedious and multiple steps involved in the fabrication of electrodes currently limit the application of MFC. Present research work deals with a readily available eyeliner as a novel conductive ink for the electrode preparation in paper based MFC. The inherent capillary action of the paper matrix allows a quick absorption of the ink when coated[2]. This results in the formation of a highly porous integrated paper electrode. The application of these coated electrodes was shown in a paper based air-breathing MFC where both anode and cathode were prepared using such coated conductive papers. An MFC employs exoelectrogenic-microbes as catalysts in the anode chamber to harness power from wastewater. When an air-cathode is used in the device configuration, it yields water as the only byproduct[3]. Using such device on a paper platform, a maximum power of 12.5 W/m^3 was obtained from a single device with *Shewanella putrefaciens* as the biocatalyst. Such an electrode can further find various applications in flexible electronics, electrochemical sensors and analytical devices.

References

[1] F Sharifi, S Ghobadian , F R Cavalcantiand NastaranHashemi, *Renew. Sust. Energ. Rev.*, 2015, 52, 1453–1472.

[2] L. Hu, J. W. Choi, Y. Yang, S. Jeong, F. La Mantia, L.-F. Cui and Y. Cui, *Proc. Natl. Acad. Sci. U. S. A.*, 2009, **106**, 21490–21494.

[3] R. Veerubhotla, A. Bandopadhyay, D. Das and S. Chakraborty, *Lab Chip*, 2015, **15**, 2580–2583.

459762 Sustainable Power Sources Based on High Efficiency Thermopower Wave Devices

Abstract:

There is a pressing need to find alternatives to portable power sources such as Li-ion batteries, which contain toxic metals, present recycling difficulties due to harmful inorganic components, and rely on elements in finite global supply. Thermopower wave (TPW) devices, which convert chemical to electrical energy by means of self-propagating reaction waves guided along nanostructured thermal conduits, have the potential to address this demand. Herein, we demonstrate orders of magnitude higher chemical-to-electrical conversion efficiency of thermopower wave devices, in excess of 1%, with sustainable fuels such as sucrose and NaN_3 for the first time, that produce energy densities on par with Li-ion batteries operating at 80% efficiency (0.2 MJ/L versus 0.8 MJ/L). We show that efficiency can be increased significantly by selecting fuels such as sodium azide or sucrose with potassium nitrate to offset the inherent penalty in chemical potential imposed by strongly p-doping fuels, a validation of the predictions of Excess Thermopower theory. Such TPW devices can be scaled to lengths greater than 10 cm and durations longer than 10s, an over 5-fold improvement over the highest reported values, and they are capable of powering a commercial LED device. Lastly, a mathematical model of wave propagation, coupling thermal and electron transport with energy losses, is presented to describe the dynamics of power generation, explaining why both unipolar and bipolar waveforms can be observed. These results represent a significant advancement toward realizing TPW devices as new portable, high power density energy sources that are metal-free.

468865 Nanostructuring, Oxygen Anion Diffusion Study and Electrochemical Performance of Double Perovskite Electrode for SOFC

Abstract:

The state of the art requires the synthesis of nanostructured materials to be used as electrodes, which will show improved electro-catalytic activity and bulk transport, thus in-turn providing better electrochemical performance. Chemical synthesis methods show limitations in obtaining reduced particle sizes. For example on adopting a chemical synthesis method an average particle size of 40 nm was obtained of $GaBaCo_2O_{5+d}$ (GBCO). Herein an attempt is made to reduce the particle size further by using a relatively green and novel top-down bio-milling process. The GBCO powder was subjected to a fungus, *Sacchromyces cerevisiae*, which was cultured in 250 mL YPD medium for 96h at 301K. We report the reduction in particle size of phase pure (fig1a) GBCO to a size lesser than 20 nm as observed using HR-TEM. Thin-film of $La_{0.6}Sr_{0.4}Co_{0.2}Fe_{0.8}O_{3-d}$ (LSCF) electrodes (thickness ~500 nm), were fabricated using a spray deposition method. In order to test the electro-catalytic performance of the GBCO nanoparticles, symmetric cells of GBCO on gadolin doped ceria (GDC) electrolyte were fabricated and electrochemical performance measured in area specific resistance was compared to the chemically synthesized GBCO (fig1b).Molecular dynamics simulations were performed to study the transport of oxygen anion [1] in anisotropic GBCO. The diffusivity of the GBCO was observed to be thermally activated with an activation energy (E_a) of 50.8 kJ/mol (fig1d).The diffusion coefficient was calculated to be of the value of 5 x 10^{-8} at 873K in *a-b* plan which was observed to be higher by an order of magnitude in comparison to the diffusion *c* axis (3.33 x 10^{-9} cm^2 s^{-1}) (fig1c). Similarly, the effect of Fe dopant on diffusivity at B site was measured in case of $GdBa_{0.5}Sr_{0.5}Co_{2-x}Fe_xO_{5.5}$ structure for x = 0 to 1. Where, it was concluded that the oxygen anion diffusion was maximum in case of 50% Fe substituted composition as shown in Table1

Table 1: Oxygen anion diffusion coefficient of double perovskite materials

Material	Diffusion (cm^2 s^{-1}) (923K)	E_a (KJ/mol)
$GdBa_{0.5}Sr_{0.5}Co_2O_{5.5}$	4 x 10^{-8}	38.4
$GdBa_{0.5}Sr_{0.5}Co_{1.5}Fe_{0.5}O_{5.5}$	5.13 x 10^{-8}	40.6
$GdBa_{0.5}Sr_{0.5}Co_{1.5}Fe_{0.5}O_{5.5}$	7.5 x 10^{-8}	42.8

Figure 1: a)XRD and HR-TEM image of biomilled nanoparticle, b) Nyquist plot of GBCO nanoparticle doped electrode, c) oxygen density profile where, blue represent Ba and yellow and green represent Co and Gd atoms respectively and d) Arrhenius plot of oxygen diffusion calculated from MD.

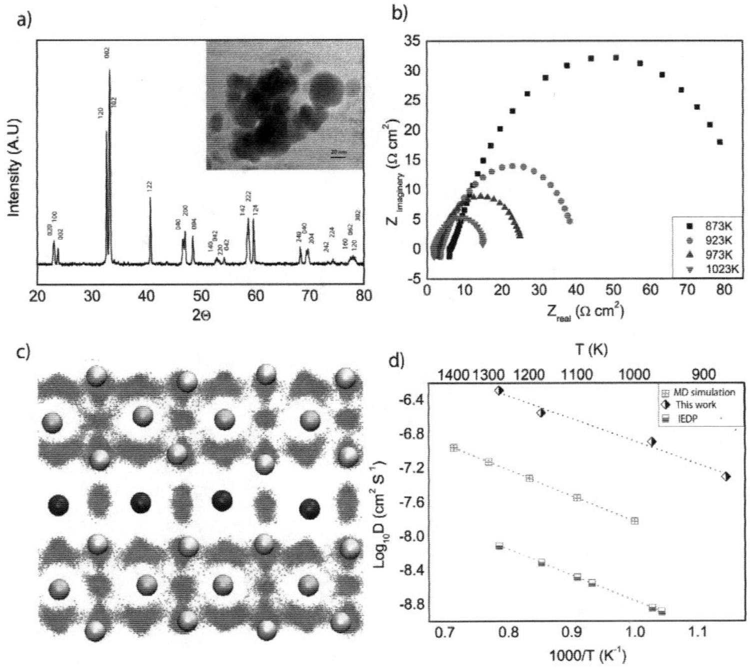

467700 Application of Photosystem I Multilayer Films for Photovoltage Enhancement in Aqueous Natural Dye Sensitized Solar Cells

Abstract:

Among the photosynthetic proteins, Photosystem I (PSI) is notable for achieving the most energetic reduction potential in nature and for its outstanding internal quantum efficiency. Stimulated by absorption of indigo, blue, and red photons, PSI facilitates electron excitation and sub-microsecond cross-membrane transport, all while retaining 1.1 eV of absorbed energy. As such, PSI is a model photodiode and the subject of broad scientific study and engineered emulation. In this study, we report photovoltage enhancement through application of PSI multilayer films atop TiO
glucoside—extracted from blackberries—and mediated by an aqueous $Fe(CN)_6^{4-/3-}$ redox shuttle. The enhancement phenomena is found to be dependent on both the amount of PSI deposited on the surface as well as the concentration of reduced redox group present in the electrolyte solution. We attribute the effect to PSI's photocatalytic reduction of photogenerated oxidized redox species in solution, which attenuates recombination rates at the dye/electrolyte interface. The blue and red absorption bands of PSI's light harvesting chlorophylls complement the strong green absorption of the sensitizing dye, allowing for better utilization of the visible spectrum.

468345 Compositions and Structures of High-N-Content Mesoporous Carbon Oxygen Reduction Electrocatalysts

Abstract:

Nitrogen- and transition-metal-containing carbon materials have been the subject of intense recent research interest, due to their desirable properties, including high oxidation stability, high conductivity, and high electrocatalytic activity, which enable their application as replacements for platinum-based oxygen reduction reaction (ORR) catalysts in fuel cells. Typically, these properties have been shown to be related to the N contents in the materials. However, N contents greater than ~4 wt% usually results in dramatically reduced electrical conductivity, and correspondingly reduced suitability for use as electrode materials. Nevertheless, a novel set of precursors have been identified which yield materials with order of magnitude greater N contents while retaining electrical conductivities up to 2.0 S/cm. When these materials are synthesized with high mesoporosities, surface areas in excess of 800 m^2/g, and the inclusion of iron-containing salts, they have been found to exhibit high ORR activities. Understanding the molecular compositions and structures of these functionalized mesoporous carbon materials aids the optimization of synthesis conditions, which yields improved performances of the electrochemical devices. Solid-state ^{15}N nuclear magnetic resonance (NMR) spectroscopy and X-ray photoelectron spectroscopy measurements demonstrate that the structures and compositions of these materials can be strongly influenced by choice of templating material, transition-metals included, and synthesis conditions, which result in physiochemical material differences that can be used to enhance electrocatalytic activities. Of particular interest are synthesis choices which can selectively alter surface chemistry without changing pore structure, and thus provide insight into the identity of the oxygen reduction active sites. Comparing and correlating the molecular compositions with the electrocatalytic activities crucially establishes that our materials with tunable compositions and cost-effective precursors exhibit ORR activities comparable to current Pt-based catalysts.

450281 Invited - Solar Light NO Removal By Flame-Made Sub-Nano Pd Clusters on Nano-TiO_2

Abstract:

Noble metal clusters deposited on metal oxide supports are broadly used. Size reduction of such noble metal particles is a classic way to enhance their activity with minimizing the use of expensive noble metals. Recently, loading of extremely small Pd clusters (~ 0.5 nm) on TiO_2 exhibits superior photocatalytic activity for NO removal[1] and its further size reduction from the clusters to the isolated atoms have a great potential to maximize the NO_x removal. Here, TiO_2 with different Pd loadings (0-3%) were prepared by scalable flame aerosol technology to attain atomically dispersed Pd atoms on TiO_2. The photocatalytic NO (1 ppm) removal efficiency of those Pd/TiO_2 under solar light (100 mW/cm^2) was evaluated based on standard ISO procedures.

Under solar light irradiation, these materials remove NO_x 3 or 7 times faster than commercial TiO_2 (P25, Evonik) with or without photodeposited Pd on it. X-ray photoelectron spectroscopy (XPS) reveals that such photodeposited Pd consists of metallic Pd along with several Pd oxidation states. In contrast, flame-made Pd subnano-clusters on TiO_2 dominantly consist of an intermediate Pd oxidation state between metallic Pd and PdO. In that intermediate state, the Pd subnano-clusters are stable up to, at least, 600 °C for 2 hours in air. However, a fraction of them is reduced into relatively large (> 1 nm) metallic Pd nanoparticles by annealing in N_2 at 400 °C for 2 hours, as elucidated by XPS and scanning transmission electron microscopy. The Pd subnano-clusters interact with oxygen defects on the TiO_2 surface as shown by Raman spectroscopy. This interaction suppresses CO adsorption on Pd as observed by diffuse reflectance infrared Fourier transform spectroscopy (DRIFTS), analogous to strong metal support interactions (SMSI) of nano-sized noble metals on TiO_2.

1. Fujiwara, K.; Müller, U.; Pratsinis, S. E., Pd Subnano-Clusters on TiO_2 for Solar-Light Removal of NO, *ACS Catalysis,* 6, 1887-1893 (2016).

470756 *Visible Light Photocatalytic Degradation of Organic Pollutant Using N Doped TiO_2 nanoparticles Synthesized By One Step Liquid Flame Aerosol Method (LFSP)*

Abstract:

Currently, there is a growing interest in the synthesis of nanomaterials. Nitrogen doped titanium dioxide is attracting a continuously increasing attention because of its potential as material for environmental photocatalysis. Conventional wet synthesis methods need multiple processing steps, high energy input and long processing times, In contrast, flame synthesis, especially liquid flame spray pyrolysis (LFSP), is a versatile one-step process for synthesizing a variety of anion doped nanoparticles. In the present work, we have developed a novel, inexpensive, green and highly reproducible method of the making the liquid precursor for the N incorporation in TiO_2 by using the rapid LFSP method. A combination of various physico-chemical characterizations were performed. The photocatalytic activity and stability of the N-doped TiO_2 samples were evaluated through using the decomposition of acetaldehyde as model reaction under visible light irradiation. Characterization results show that the nitrogen dopants have a significant effect on the crystallite size and optical absorption of TiO_2. It is found that the N-doped TiO_2 catalysts have enhanced absorption in the visible light region, and exhibit higher activity for photocatalytic degradation of VOCs. We have produced The UV-vis spectroscopy results of N incorporated TiO_2 showed enhancement of light absorption in the visible range by selected composite materials in the UV range (250-400 nm). We have found the Ti/N ratio of 13% for the materials synthesized from our XPS data. These interesting results will be discussed in the presentation.

452386 Invited - DNA Encapsulated within Silica Nanoparticles for Environmental Monitoring and Tracing

Abstract:

From an analytical chemistry perspective short DNA sequences are unique as they can not only be detected at extremely low concentrations (1 molecule by PCR), but that they can carry identifying information (barcodes). In order to make DNA more resilient to application in environmental monitoring and tracing, we have developed a method to encapsulate DNA into amorphous silica nanoparticles by a sol-gel method.(1,2) Within the particles the DNA is protected from hydrolysis and can withstand increased temperatures and prolonged exposure in oxidizing conditions.

Within this presentation we will discuss how such encapsulated DNA barcodes can be utilized to a) monitor the faith of silica nanoparticles in the environment, and b) to monitor the flow/distribution of liquids, foodstuff and commodity products within the environment and throughout a product life cycle. Examples include the use of the particles in understanding nanoparticle clearance in waste water treatment plants (3), the tracing of pesticides in the environment (4) and the tracking of milk and gasoline(5,6).

We will also show how the concept of DNA tracers can be utilized for the remote and distributed sensing of optical irradiation and temperature (7,8) and discuss future possibilities of combining bio analytics and inorganic chemistry for applications in the field of chemical engineering.

(1) D Paunescu, R Fuhrer, RN Grass, Angew. Chem. Int. Ed 52, 4269 (2013)

(2) D Paunescu, M Puddu, JOB Soellner, PR Stoessel, RN Grass, Nat. Protoc. 8 , 2440 (2013)

(3) RN Grass, J Schälchli, D Paunescu, JOB Soellner, R Kaegi, WJ Stark, ES&T Lett. 1, 484 (2014)

(4) CA Mora, HJ Scharer, T Oberhänsli, M Ludwig, R Stettler, PR Stoessel, RN Grass, WJ Stark, ES&T Lett. 3, 19 (2016)

(5) MS Bloch, D Paunescu, PR Stoessel, CA Mora, WJ Stark, RN Grass, J. Agr. Food Chem. 62 , 10615 (2014)

(6) M Puddu, D Paunescu, WJ Stark, RN Grass, ACS nano 8 , 2677 (2014)

(7) G Mikutis, CA Mora, M Puddu, D Paunescu, RN Grass, WJ Stark, Adv. Mater. 28, 2765 (2016)

(8) M Puddu, G Mikutis, WJ Stark, RN Grass, Small, 12, 452 (2016)

459882 Textile Dye Removal from Water Using Catalytic Nanoparticle-Carbon Composites

Abstract:

In recent years, increasing consumer spending and new emerging markets in the world trade have caused a boom in production in the textile industry. With this growth comes the contamination of the environment through substances called azo dyes. In the presence of digestive tracts, these azo dyes can cleave creating aromatic amines, which, in some cases have been proven to be carcinogenic. Also, the azo dyes often cannot be completely filtered out of water naturally due to their highly hydrophilic nature and very small size. Studies have shown that metal nanoparticles reduce the concentration of these azo dyes in aqueous solution to a certain extent, but many studies have focused on precious metals such as gold, copper, and palladium. Due to the cost of these metals, there is interest in developing reactive nanoparticles that are based entirely on non-precious metal combinations. For this study, we have chosen to focus on iron-nickel bimetallic nanoparticles.

In this study, our research focuses on demonstrating the effectiveness of bimetallic iron-nickel nanoparticles in the treatment of Orange G azo dye for different molar ratios of nickel to iron and for the case where these nanoparticles are deposited onto a high surface area carbon support material. If reactive nanoparticles are to be used in realistic treatment scenarios, the nanoparticles must be immobilized within a treatment system, and high surface area carbon materials are a potentially ideal choice for immobilization. Carbons, such as activated carbon, are often used in water treatment to adsorb and remove water contaminants and thus may serve not only as an immobilization strategy but as an additional available surface for adsorptive contaminant removal.

In our experimental work, nanoparticles were created by using an aqueous nanoparticle synthesis technique and using an alloyed nanoparticle morphology. The process uses an iron sulfate heptahydrate salt along with a stabilizer (ATMP) and a nickel chloride hexahydrate salt with its respective stabilizer (PVP). These metals are mixed under argon gas, then reduced using sodium borohydride. This procedure results in the formation of nanoparticles and hydrogen gas. The nanoparticles are then added to an aqueous solution of Orange G in a batch reactor setup and mixed while samples are taken at specific time intervals to evaluate Orange G removal. Our results thus far indicate a strong correlation between the decrease in Orange G concentration and the molar ratio of nickel to iron. Experiments performed with monometallic nanoparticles (10 g/L iron and 10 g/L nickel) showed little to no decrease in Orange G concentration while the bimetallic counterparts showed promising results in decreasing the concentration of Orange G. Results showed that at a lower molar ratio of nickel to iron increased the effectiveness of the bimetallic nanoparticles in decreasing the concentration of Orange G. In comparison to nanoparticles alone, nanoparticles were deposited onto several high surface area carbon support materials and tested in the same batch reactor setup for removal of Orange G. Preliminary results suggest that nanoparticle immobilization onto carbon results in an increase in the rate of removal of Orange G and an increase in the overall amount of Orange G removed.

In this talk, we will present results for immobilized iron-nickel nanoparticles on several key high surface area carbon support materials. In particular, two types of commercially-available activated carbons will be compared to several novel surface-functionalized carbons. The novel carbon materials are functionalized with specific heteropolyacids thought to have potential oxidizing capabilities in aqueous solution. Results for immobilized nanoparticles will be compared to results for non-immobilized nanoparticles in a batch reactor setup. The influence of nickel to iron molar ratio on Orange G removal for both immobilized and non-immobilized nanoparticles will be discussed. Characterization results for nanoparticle and nanoparticle-carbon composite morphology, composition and phase will be presented. Characterization methods include electron microscopy, power x-ray diffraction, and x-ray photoelectron spectroscopy. Finally, a comparison of the different tested carbon support materials will demonstrate that there are differences in Orange G dye removal depending on what carbon support is chosen for nanoparticle immobilization.

468651 Grafted Nanopyroxene for the Removal of Polar Organic Compounds from Produced Wastewater

Abstract:

A large amount of water is needed for bitumen extraction processes and, subsequently, a significant amount of Oil Sand Process-affected Water (OSPW) is produced and discharged into tailing ponds. The OSPW contains organic compounds such as Naphthenic Acid (NA), which is considered the main organic pollutant occurring naturally in tailing ponds. The removal of NA from OSPW has been of paramount importance in order to meet government environmental regulations and has been challenging due to its high solubility in water, which in turn makes it resistant to the conventional oil water separation processes. The current technologies used for the removal of NA present several shortcomings, such as high capital requirements and operating costs, and they are largely ineffective in meeting government regulations. Despite the fact that the adsorption process is promising for the removal of dissolved contaminants in the OSPW, classical adsorbents lack several properties that can be best established using nanoparticles.

Nanoparticles poses unique properties that make them desirable, such as large surface area, high efficiency, high reactivity, dispersibility, low cost and selectivity. Nanoparticle technology is becoming the new frontier in wastewater treatment due to its high efficiency, low cost and easy to implement processes.

In this study, silica-based nanoparticles were prepared in the lab for their proven advantages over classical adsorbents. These nanoparticles needed to be modified in order to attract polar organic compounds. After several studies, grafting was chosen, this entails the modifying and functionalizing of the surface of the nanoparticle with a naturally occurring and environmentally friendly polymer in order to achieve high affinity towards the polar organic compounds present in the OSPW. The preparation of the Grafted Nano Silica (GNIS) was accomplished and optimized using different polymer molecular weights as well as different preparation procedures.

A synthetic wastewater solution was prepared by dissolving NA in water. Microscopic batch adsorption took place to carefully study several factors influencing the adsorption process; including: contact time, initial concentration, GNIS amounts and the pH of the solution. The Sips, Langmuir and Freundlich adsorption isotherm models were employed using the experimental data to describe and understand the adsorption behavior.

This study confirmed that GNIS could be employed as efficient adsorbents for naphthenic acid removal from wastewater where classical adsorbents could be ineffective or costly.

456741 Invited-2D Materials for Packaging and As Nanoparticle-Surfactants

Abstract:

Our group has carried out systematic study on 2D materials, including liquid crystal phase formations, nanocomposite fabrication, and understanding of Janus platelets as nanoparticle surfactants for Pickering emulsification. I will discuss the discovery of discotic smectic phase, which can be used in more effective industrial sealants, improved food packaging, enhanced electronic displays, and improved fuel cell technology. We also developed method to mass produce Janus nanoplates, first using the model system of Zirconium Phosphate nanoplates, then using clay and graphene Quantum Dots with reduced cost. I will discuss the applications of these nanoplate surfactants to make stable Pickering emulsions and foams, to stable graphene suspensions, for enhanced oil recovery, and for nano-encapsulations.

1. "Hydrothermal synthesis of layered a-zirconium phosphate disks: control of aspect ratio and polydispersity for nano-architecture", M. Shuai, Andres F. Mejia, Y. Chang and Z. Cheng **Cryst. Eng. Comm. 15,** 2013, 1970-1977.

2. "Aspect Ratio and Polydispersity Dependence of Isotropic-nematic Transition in Discotic Suspensions" A.F. Mejia, Y. Chang, R. Ng(undergraduate), M. Shuai, M.S. Mannan, and Z. Cheng, **Physical Review E** 85, 2012, 061708. Figure selected as part of PRE kaleidoscope images, June 2012.

3. "Stable smectic phase in suspensions of polydisperse colloidal platelets with identical thickness" D. Sun, H-J. Sue, Z. Cheng, Y. Martinez-Ratonm and E. Velasco, **Physical Review E**, 80, 2009, 041704

4. "Gelation via Ion Exchange in Discotic Suspensions" Y. Chang, A.F. Mejia, Z. Cheng, D. Di, and G.B. McKenna, **Phys. Rev. Lett. 108,** 2012, 247802.

5. "Pickering Emulsions Stabilized by Amphiphilic Nano-sheets" A.F. Mejia, A. Diaz, S. Pullela, Y. Chang, M.S. Mannan, A. Clearfield, and Z. Cheng, **Soft Matter 8,** 2012, 10245-10253 (**Journal cover**).

6. "Stabilization of Pickering Foams by High-aspect-ratio Nano-sheets" J. S. Guevara (undergraduate), A.F. Mejia, Y. Chang, M. Shuai, M.S. Sam Mannan, Z. Cheng, **Soft Matter** 9, 2013, 1327-1336.

7. "Nano-encapsulated PCM via Pickering Emulsification", X. Wang, L. Zhang, Y. Yu, L. Jia, M. S. Mannan, Y. Chen & Z. Cheng, **Scientific Reports** 5, 2015, 13357.

465761 Ionophore-Decorated Magnetic Graphene Oxide As a Composite Adsorbent for Precious Metal Recovery

Abstract:

Recovery of lithium from aqueous streams has gained significant attention as it can contribute in resolving the anticipated challenges on shortage of lithium supply in the future. Given the variety and complexity in the composition of aqueous Li^+ sources, more advanced materials must be developed for selective Li^+ capture. It is also desirable to synthesize multi-functional materials which can be readily regenerated and reused for long term application.

Herein, a multifunctional adsorbent (CE-rGO-Fe_3O_4) was successfully synthesized and used as a selective lithium ion (Li^+) adsorbent. The adsorbent was decorated with crown ethers (CE) as Li^+-specific ionophores, magnetite (Fe_3O_4) for its easy separation and recyclability, and graphene oxide (GO) as a two-dimensional, high-aspect ratio support of the CE and Fe_3O_4.

The GO was prepared by oxidizing graphite powder through a modified Hummer's method. The Fe_3O_4 nanoparticles were then immobilized on the surface of the GO through a solvothermal technique which also partially reduced the GO nanosheets (rGO-Fe_3O_4). Alkyne moieties were grafted on the rGO-Fe_3O_4 surface using 4-ethynylaniline The alkyne terminals in 4-ethynylaniline served as "clickable" CE-specific attachment sites. Meanwhile, Li^+-selective CE (i.e. 2-hydroxymethyl-12-Crown-4 Ether) was prepared via a two-step mesylation-azidation reaction to obtain 2-(azidomethyl)-12-Crown-4 Ether. The azidated CE was then "clicked" on the alkyne- rGO-Fe_3O_4 at 60 °C via the 1,3 cycloaddition click chemistry reaction.

The synthesized composite adsorbent (12CE4-rGO-Fe_3O_4) was characterized using Transmission Electron Microscopy (TEM), Boehm Titration, X-ray Diffraction Spectrometry (XRD), Fourier-Transform Infrared Spectroscopy (FT-IR), Thermogravimetric Analysis (TGA), Raman Spectroscopy, and X-ray Photoelectron Spectroscopy (XPS). TEM was used to observe the morphology of the adsorbent, which revealed the presence of Fe_3O_4 NPs on the surface of the GO nanosheets. The XRD spectrum and FTIR conservatively confirmed the presence of different functional groups on the GO support, which was also consistent with the TGA results. Raman Spectroscopy also reflected the degree of functionalization of the GO as indicated by the changes in the D- and G-bands and their corresponding I_D/I_G ratios. The changes in the chemical states of the elements present in the adsorbent upon subsequent modification and functionalization was also confirmed via XPS. XPS results revealed the presence of the Fe_3O_4 in the adsorbent (12CE4-rGO-Fe_3O_4) as indicated by the Fe2p signals whereas the successful attachment of the CEs via click reaction was confirmed with the presence of the peak attributed to the triazole functional group found in the N1s spectra.

The adsorption results show the Langmuir-type Li^+ uptake of the 12CE4-rGO-Fe_3O_4, with the maximum Li^+ uptake at 7.33 mg g^{-1}. The results also show that the material has high selectivity towards Li^+ as compared to the other cations. The integration of magnetite allowed the easy separation and recyclability of the adsorbent. Overall results demonstrate the suitability of the 12CE4-rGO-Fe_3O_4 for long-term Li^+ adsorption application.

This research was supported by the Basic Science Research Program through the National Research Foundation of Korea (NRF) funded by the Ministry of Science, ICT and future Planning (2015R1A2A1A15055407) and by the Ministry of Education (No. 2009-0093816).

474957 Developing and Piloting Robust Downstream Processes to Utilize Variable Composition and Blended Biomass Feedstocks

Abstract:

Biomass variability can play a key role in end-product quality and downstream conversion process performance and economics. This is a key challenge to the biofuel and bio-based chemical industries in light of the differences in composition and even the identify and utility of biomass crops and residues in different seasons and geographic regions across the US. The Advanced Biofuels [and Bioproducts] Process Demonstration Unit (ABPDU) at the Lawrence Berkeley National Laboratory is working with several National Lab and industrial partners to address these issues. Bench-scale and pre-pilot prototyping of biomass conversion and upgrading processes will be presented using case studies from collaborative projects with the Idaho and Sandia National Laboratories focused on advanced biofuel and chemical product production processes. In addition, results from industrial collaborations and sponsored research at the ABPDU will be used to highlight the importance of robust and semi-universal processes for the conversion of a diverse set of feedstreams to fuels and other bio-based products. The learnings from these bench- and pilot studies will inform larger scale IBR demonstrations focused on multiple feedstocks.

474958 A Review of Minor Feedstock Characteristics that Have a Major Impact on Process Design and Commercial Success

Abstract:

Technology development in the field of lignocellulosics to fuels and chemicals continues, albeit at a reduced rate due to the recent collapse in the global crude price. Key physical and chemical properties of the selected lignocellulosic feedstocks that must be thoroughly understood for successful implementation of first-of-a-kind process technologies will be discussed. A thorough understanding of feedstock properties is not only critical to the design and selection of commercial scale systems but also must be considered during the technology development and scale-up phases. Though focus has historically been on issues related to sugar recovery and downstream conversion process efficiencies, the inorganic fraction composition and general physical properties often introduce unexpected challenges during scale-up and commercialization. When making capital allocation decisions failure to adequately account for feedstock variability and physical properties can result in less than desirable results, sometimes in unanticipated ways. Technology selection practices and investor-related constraints can be in conflict and may extend the project development phases resulting in unnecessary risks and escalated cost.

474955 Innovative Solutions to Feedstock Supply System Challenges Faced by Biorefineries: the U.S. Department of Energy's Feedstock Supply and Logistics Program

Abstract:

Biorefineries face many challenges associated with feedstock supply system establishment and maintenance as well as in-feed handling. Dr. Alison Goss Eng will provide an overview of these challenges as well as potential opportunities for further technology development within the context of the Bioenergy Technologies Office at the US Department of Energy, describing how the Department is working to overcome these challenges with its partners.

468420 Aquathermolysis of Waste Triglycerides in a Continuous Flow Reactor for Jet Fuels Production

Abstract:

The unavoidable dwindling fossil fuel and its attendant climate change in our time have given impetus to researchers in the energy sector to delve into different sustainable energy sources. One of such energy sources is the abundant waste triglycerides whose long chain hydrocarbons can be separated from their short oxygen chain via hydrodeoxygenation, decarbonylation, and decarboxylation in the presence of a heterogeneous catalyst and hydrogen. However, the use of gaseous hydrogen is limited by a diffusion process since it is not readily soluble in liquid (polar) components and requires an excess use of hydrogen. To address these challenges, a single-pot â€˜Formic Acid Assisted Autotheromolysis to Jet Fuel (FAT2JF)â€™ process is studied. The use of aqueous phase (formic acid and water mixture) serves as an *in situ* source of H_2 (hydrogen precursor and hydrogen transfer agent) and organic solvent where the reactions in aqueous media (aquathermolysis) helps in reducing mass transfer resistances that results in significant increase in reaction rates in this FAT2JF process. The phase diagram of water, formic acid and some typical triglycerides will be presented.

The present study on aquathermolysis is focused on developing a metal oxide-based low cost catalyst (e.g. Fe-Ni/activated carbon, Fe-Ni-Co/Al_2O_3) system that is stable in aqueous media. The study is conducted in a packed bed continuous flow tubular reactor in the range of 250 to 400áµ’C under hydrothermal conditions. In order to alleviate coke formation and evaluate the stability of the catalyst, the effects of reactants flowrates, feedstock composition, pressure, temperature, and formic acid concentration are studied. GC-MS and elemental analyzer are used to analyze the product compositions. Thermal gravimetric analysis (TGA), energy dispersive X-ray, SEM, BET method, FT-IR, and XRD are used to characterize both fresh and used catalyst. The proposed novel approach has certain advantages over conventional commercial hydrotreating processes including no handling of hydrogen gas, faster reaction times,

and use of renewable formic acid as hydrogen source.

469513 Rheology of Biomass Slurries to Determine Pumpability for Hydrothermal Liquefaction

Abstract:

The pursuit of renewable fuels and chemicals has led to the development of many different biomass conversion processes. One of these processes, hydrothermal liquefaction (HTL), has the advantage of high energy recovery, no drying step, and utilization of a green aqueous reaction media.[1] However, transport of biomass and water slurries during HTL has proven to be a challenge.[2,3] Typical reactors can only process biomass between 10 and 15 wt% solids because higher solids loadings plug reactor systems. To improve the efficiency of HTL, and other processes reliant on pumping biomass slurries, it is essential to increase solids loading. Increasing solids loading reduces the volume of waste streams and increases process throughput. In an effort to increase biomass solids loadings many people have studied the rheological properties of biomass slurries to improve pumpability.[4] Despite these efforts there has yet to be a strong correlation developed between typical rheological properties, like viscosity, and biomass pumpability.

This work takes a unique approach to studying biomass slurry rheology by combining traditional rheometric measurements, such as viscosity and storage/loss moduli, with a novel flow characterization system. Traditional rheometric measurements are made on a rheometer capable of analysis at temperatures up to 300 °C and 2000 psi, which mimics conditions found in some traditional HTL systems. Novel flow studies performed at Idaho National Laboratory are made on an in-house designed syringe attached to an Instron load frame. This system measures the force required to push biomass through a reducing orifice and can study plugging phenomena in biomass slurry transport systems. These studies have shown how plugging can be related to biomass dewatering, followed by bridging across the flow channel. Pumpability studies have been performed on a wide variety of materials from herbaceous to woody crops and residues. In addition to characterizing pumpability in biomass slurries this work has also focused on improving the pumpability of such slurries through preprocessing techniques including: size reduction, blending, and

torrefaction. This work will ultimately lead to improvements in feed handling related to hydrothermal liquefaction and other biomass slurry transport systems.

1 Toor, S. S., Rosendahl, L. & Rudolf, A. Hydrothermal liquefaction of biomass: A review of subcritical water technologies. *Energy* **36**, 2328-2342, doi:10.1016/j.energy.2011.03.013 (2011).

2 Elliott, D. C., Biller, P., Ross, A. B., Schmidt, A. J. & Jones, S. B. Hydrothermal liquefaction of biomass: Developments from batch to continuous process. *Bioresource technology* **178C**, 147-156, doi:10.1016/j.biortech.2014.09.132 (2015).

3 Berglin, E. J., Enderlin, C. W. & Schmidt, A. J. *Review and Assessment of Commercial Vendors/Options for Feeding and Pumping Biomass Slurries for Hydrothermal Liquefaction*. (Pacific Northwest National Laboratory, 2012).

4 Stickel, J. J. *et al.* Rheology measurements of a biomass slurry: an inter-laboratory study. *Rheologica Acta* **48**, 1005-1015, doi:10.1007/s00397-009-0382-8 (2009).

469372 The Upgrading of Biomass-Derived Dimethyl Ether to High-Octane Hydrocarbons: The Effect of Process Conditions on Catalyst Performance

Abstract:

Dimethyl ether (DME) and/or methanol can be converted into C_{5+} branched hydrocarbons at low temperatures and pressures over large-pore acidic zeolites. Recently, we have demonstrated that Cu-modification of the existing commercial catalyst (H-BEA) resulted in a 2-fold increase in total hydrocarbon productivity at 200 °C with co-fed H_2.[1] The Cu/H-BEA catalyst incorporated H_2 with DME into the desired branched hydrocarbons, while maintaining the high C_{5+} selectivity of the parent H-BEA. Here, we compare the performance of Cu/H-BEA and the parent H-BEA at 175, 200 and 225 °C, and at weight-hourly space velocities (WHSV) for DME ranging from 0.43-1.4 h^{-1}. The productivity of C_{5+} products increased with temperature over H-BEA, with no significant changes over Cu/H-BEA due to a decrease in C_{5+} selectivity at elevated temperature, which was observed for both catalysts. Operation at 175 °C resulted in a significant increase in C_{5+} selectivity (75-85% for Cu/H-BEA), but lower productivity than at 200 °C due to lower reactant conversion. Lower space velocities increased reactant conversion over both catalysts, but significantly decreased the C_{5+} productivity over Cu/H-BEA. The Cu/H-BEA catalyst was stable for 100 h at 200 °C after an initial 40 h break-in period. In previous research, we demonstrated that C_4-C_8 olefins produced in the DME homologation reaction can be oligomerized to C_{12}-C_{22} branched hydrocarbons with properties that align with a synthetic paraffinic kerosene product.[2] Following a simple vacuum distillation, the fuel properties of the distillate-range product align well with a jet fuel blend when compared with 5 ASTM methods. This DME-to-hydrocarbons reaction represents a promising route to transform renewable biomass sources into commercially viable, high-performance biofuels.[3]

[1]Schaidle, J. A.; Ruddy, D. A.; Habas, S. E.; Pan, M.; Zhang, G.; Miller, J. T.; Hensley, J. E. *ACS Catal.* **2015**, *5*, 1794.

[2]Behl, M.; Schaidle, J. A.; Christensen, E.; Hensley, J. E. *Energy & Fuels* **2015**, *29*, 6078.

[3]Tan, E. C.; Talmadge, M.; Dutta, A.; Hensley, J.; Snowden-Swan, L. J.; Humbird, D.; Schaidle, J.; Biddy, M. *Biofuels, Bioproducts and Biorefining* **2016**, *10*, 17.

462619 A System-Level Analysis on Biomass Thermal Fractionation and Catalytic Upgrading Processes

Abstract:

Biomass presents a promising source for sustainable liquid fuels production. There are three principle routes for converting biomass into liquid fuels, classified by the initial treatment of the biomass: gasification, hydrolysis, or pyrolysis and liquefaction. The pyrolysis route involves thermal decomposition of the biomass at high temperature (500-800°C), and it has demonstrated high yields. Unfortunately, the wide range of components in the resulting pyrolysis vapor (oil) makes it unstable, corrosive, and challenging to catalytically upgrade. Furthermore, the high content of light (<C6) oxygenates in the pyrolysis are unsuitable for liquid fuels and are therefore of limited utility. Multi-stage torrefaction of biomass with catalytic upgrading presents a promising method of overcoming these challenges. Importantly, the decomposition of the biomass over several stages can reduce the number of product species within each fraction, allowing one to mold the catalytic upgrading strategy to target a subset of the chemical functionalities.

Although staged thermal decomposition of biomass coupled with targeted upgrading has a number of advantages over single stage pyrolysis with hydro-treating, there are many unknowns regarding the best way to design such a system. Importantly, the optimization of the number and conditions of each thermal decomposition stage is a complex problem that cannot be solved without knowing the influence on the final yields and process complexity resulting from the subsequent catalytic upgrading. Furthermore, there are a wide variety of chemistries available to perform upgrading. In addition, the reaction sequence must also be configured to optimize the overall process. In order to improve the impact of such an approach, one must also determine if general practices from this strategy may be applicable to different types of biomass feeds.

Accordingly, the ultimate goal of this work is to design a broad roadmap that can be used to identify the key parameters and trade-offs in a bio-refinery process based on multi-stage torrefaction of biomass with catalytic upgrading. In order to investigate the effect of the number of torrefaction stages and the specific chemistries on the yield of fuel-range grade product and the cost, we first derived a series of process alternatives based on experimental torrefaction yields obtained for a 3-stage torrefaction process (270 °C, 360 °C, and 500 °C). In designing the alternatives, we carefully selected the chemistries that are used to upgrade the thermal decomposition fractions, to maximize the yield of fuel-grade product and to minimize the hydrogen consumption, considering the abundance of various chemical functionalities within the biomass fractionation product. We suggested twelve cases where there is 1, 2, or 3 thermal decomposition stages and where each fraction is upgraded 1, 2, 3, or 4 times. Through the analysis, we found that there exists a trade-off between fuel-range carbon yields versus the number of upgrading steps. In general, increasing the complexity of the fraction upgrading systems increases the ultimate yield of C6+ product, thought there are diminishing returns on the increase in product yield versus the complexity of the catalytic upgrading sequences. In contrast, however, the choice of the number of thermal decomposition stages is found to be not simple, requiring careful consideration of the chemistries available to upgrade different components and the relative abundances of these different components. More thermal decomposition stages allow for more fine-tuning of the upgrading processes but lead to more complex and potentially more expensive processes. Thus, the design of the decomposition conditions and number of stages must be done simultaneously with the design of the fraction upgrading system. Finally, we also suggest some hybrid processes by eliminating several upgrading steps that have small marginal effects on the fuel-range carbon yield and discuss about separations that could enhance the process performance via prevention of unwanted side reactions or protection a functionality from conversion.

461802 Performance of a Bench-Scale Continuous Hydrothermal Carbonization Reactor

Abstract:

Hydrothermal carbonization (HTC, thermal hydrolysis, or wet torrefaction) is a treatment process which converts moist feedstocks into homogenized, carbon rich, and energy dense solid fuel, called hydrochar. One of the main advantages of HTC compared to other thermochemical treatment processes is the use of moisture as reaction medium and reactant, thereby precluding the need for drying prior to treatment. Thermodynamic properties of water change greatly in the subcritical region from 180-280 °C, and as a result, subcritical water behaves as a non-polar solvent and mild acid and base catalyst simultaneously. When subjected to HTC biomass undergoes rapid hydrolysis and other reactions, releasing oxygen-containing volatiles and producing a highly hydrophobic hydrochar.

Although HTC offers a relatively simple and straightforward solution to process diverse biomass feedstocks, the requirements of high pressure and high temperature make the process complex and potentially costly to design and operate. The lab-scale batch process has already been demonstrated in various laboratories around the world, but batch process is not cost-effective for industrial-scale deployment. A continuous process offers a relatively small footprint, energy recovery and efficiency, and notably, economies of scale.

In this study, a bench-scale continuous HTC reactor system has been designed, constructed, commissioned, and operated with model compounds and with whole biomass feedstocks. The maximum temperature and pressure of the reactor are 250 °C and 40 bar. The throughput of the reactor system was maintained at 0.3 L/min with solids loading of ~5%, while the reaction time was maintained at 5 min. The basic reactor configuration is a section of straight pipe, and numerous engineering challenges were encountered during design and commissioning. The primary challenges are related to instrumentation durability, temperature control, and flow control.

Results are presented for conversion of glucose and microcrystalline cellulose. Both solid and liquid products were tested for their physico-chemical properties and compared with the corresponding products from batch process generated in a 2L batch Parr reactor. It was found that temperature and pressure were stable during operation and products are relatively similar to that of batch process.

474623 Spectroscopic and Thermal Characterization of Carbonized Food Wastes

Abstract:

Waste-to-energy conversion technologies may play an important role to improve the efficiency and sustainability of global food supply. In this worked, performed in collaboration with researchers at UNICAMP (Brazil), we used thermal and spectroscopic methods to characterize the carbonized residues that remain after flow-through hydrothermal extraction and hydrolysis. A recent energy analysis of flow-through hydrothermal hydrolysis finds that it may be energetically favorable compared to other pre-treatment and saccharification techniques. Three factory wastes produced by food industries were studied: sugarcane bagasse, green coffee powder, and defatted coffee cake. We foucs here on results obtained for the coffee wastes. Thermogravimetric analysis (TGA) was used to monitor extraction of semi-volatile components after treatment at 150 °C, followed by hydrolysis and carbonization as temperatures were increased to 250 °C. Infrared spectroscopy indicated that the extracted semi-volatile compounds were a combination of valuable lipids and organic acids. At higher temperatures, infrared spectroscopy identified preferential hydrolysis of hemicellulose components, with cellulose and lignin remaining. Raman microscopy gave further detail, identifying the specific compound chlorogenic acid, a valuable compound marketed as a nutraceutical, as one of the semi-volatile compounds to be extracted from the coffee. At higher temperatures, Raman microscopy provided evidence of carbonization of the coffee waste, as suggested by the appearance of graphitic and defect bands previously associated with hydrothermal char. As shown in the Figure, our results suggest temperature-programmed extraction, hydrolysis, and carbonization of coffee wastes to produce biorefinery streams consisting of valuable organic compounds, bioenergy precursors, and carbonized char.

474753 Drying and Pyrolysis of Solid Waste on Spacecraft for Water Recovery and Biochar

Abstract:

Manned spaceflight outside of low-Earth orbit will require significant advances in closing loops within life support systems, especially the recycling of solid and liquid wastes to produce oxygen, food, and fresh water. Pyrolysis is one waste conversion method that has received considerable attention due to its potential to enable improvements in waste reduction, material recovery, and energy production.

Here, we describe the development of a slow pyrolysis reaction system that can transform solid waste and brine from the water treatment system into a nutrient-rich crop growth medium, while recovering water and carbon dioxide. The feedstock for the process followed a standard s recipe used by NASA to represent the average daily solid waste produced by astronauts on manned missions. This recipe included packaging materials (polyethylene, nylon, aluminum foil), clothing and towels (mostly cotton), personal care products (i.e. shampoo, toothpaste, facial tissue, hand wipes, diapers, etc.), food waste, metabolic waste simulant, and miscellaneous materials such as nitrile gloves, duct tape, and cleaning wipes. Solid waste processing consisted of two steps: drying to recover water and pyrolysis at moderate temperatures (400-550°C). Process steps were optimized to recover the most water and produce the most suitable biochar for plant growth applications within the assumed 8 hour total operational time window. Recovered water and biochars were characterized using a variety of physical, chemical and soil nutrient analysis methods. We discuss the challenges of working with this waste feedstock compared to lignocellulosic materials, how this system fits into an overarching life support architecture, and future work needed to adapt this system to a microgravity environment.

474765 Evaluating Physical and Electronic Structure Developments of Loblolly Pine Derived Biochar and Activated Carbon

Abstract:

Biochar is a major byproduct of fast pyrolysis process. Until recently, biochar is considered as an inexpensive solid fuel that is re-consumed for the pyrolysis reactor heating. However, advances in nanotechnologies reveal the potential high-end applications of biochar as electrodes for electrochemical energy storage systems and conductive thin film materials for sensors and electronic devices. The high-end applications of biochar not only contribute to the profitability of overall fast pyrolysis process, but also contribute to the growth of sustainable material market. To attain the objective, it is imperative to understand the complex biomass derived carbon structural development during the carbonization and the activation. In this study, the physical and the electronic structures of loblolly pine (*Pinus Taeda*) derived biochar and activated carbon were systematically analyzed by interpreting X-ray diffraction patterns, electron energy loss spectra, and electrical conductivities. Loblolly pine was carbonized at nine different temperatures (300 , 350 , 400 , 500 , 600 , 700 , 800 , 900 , and 1,000 ,) using quartz tube furnace. Then, four biochar samples carbonized at 300 , 350 , 500 , and 700 were impregnated with NaOH to produce the activated carbon. Carbon sp^2 content was quantitatively determined by the electron energy loss spectroscopy. The sp^2 content of biochar produced at 300 is 58% and the sp^2 content gradually increases to 78% as the carbonization temperature is elevated to 1,000 . The sp^2 contents of activated carbon are 74%, 78%, 79%, and 85% where precursor biochar carbonization temperatures are 300 , 350 , 500 , and 700 . Polyaromatic cluster size and graphitic stacking thickness were determined by analyzing the X-ray diffraction patterns. Four stages of biochar physical structural development were determined. After major biomass components decompose, the average polyaromatic cluster size and the graphitic stacking thickness increase as the carbonization temperature increases. Carbon paste electrode

was produced by mixing biochar and activated carbon powder with polyvinylidene fluoride solution. Biochar electrodes features were closer to insulator that the electrical conductivity was not measurable. Activated carbon electrodes had good electrical conductivities which values are 1.14 S/cm, 1.41 S/cm, 1.3 S/cm, and 1.58 S/cm. The detailed model of physical and electronic structural developments will be discussed.

466364 Characterization of Free Radicals By Electron Spin Resonance Spectroscopy in Biochars from Pyrolysis at High Heating Rates and at High Temperatures

Abstract:

Understanding fast pyrolysis of biomass-derived materials is an important step in optimization of combustion processes. Similar to coal combustion, the fuel burn out is known to be influenced by the yield and reactivity of chars, produced during pyrolysis. The rapid heating of small biomass particles and the short residence time at high temperatures minimize the char yield and increase char reactivity. The differences in chemical composition of organic and inorganic matter between wood and herbaceous biomass affect the operational flexibility of power plants, and increase the complexity of mathematical models that can predict yields, composition and rates of product (char, tar, light gases) formation from fast pyrolysis. The modeling of cross-linking and polymerization reactions in biomass pyrolysis includes the formation of free radicals and their disappearance. Knowledge about these radical reactions is important in order to achieve the high fuel conversion at short residence times. However, little is known about the extent of free radical reactions in pulverized biomass at fast pyrolysis conditions.

The concentration and type of free radicals from the decay (termination stage) of pyrolysis at slow and fast heating rates and at high temperatures (above 1000°C) in biomass char have been studied. A room-temperature electron spin resonance spectroscopy study was conducted on original wood, herbaceous biomass, holocelluloses, lignin and their chars, prepared at high temperatures in a wire mesh reactor, an entrained-flow reactor, and a tubular reactor. The radical concentrations in the chars from the decay stage range up between $7x10^{18}$ and $1.5x10^{19}$ spins g^{-1}.

The results indicated that any differences in the biomass major constituents (cellulose, hemicellulose, lignin) had a minor effect on remaining radical concentrations compared to potassium and silica contents. The higher radical concentrations in the wheat straw chars from the decay stage of pyrolysis in the entrained flow reactor compared to the wood chars were related to the decreased mobility of potassium in the char matrix, leading to the less efficient catalytic effects of potassium on the bond-breaking and radical re-attachments. The high Si levels in the rice husk caused an increase in the char radical concentration compared to the wheat straw because the free radicals were trapped in a char consisting of a molten amorphous silica at heating rates of 10^3°C s^{-1}. The experimental electron spin resonance spectroscopy spectra were analyzed by fitting to simulated data in order to identify radical types, based on g-values and line widths. The results show that at high temperatures, mostly aliphatic radicals ($g = 2.0026$-2.0028) and PAH radicals ($g = 2.0027$-2.0031) were formed. Free radicals structure from the termination stage was modelled by density functional theory (DTF).

466172 Shalexenvironment: A Multi-Disciplinary Effort to Assess the Environmental Implications for Shale Gas Exploration Ad Production in Europe

Abstract:

In this presentation we will discuss the multi-disciplinary efforts implemented within the ShaleXenvironmenT consortium (http://shalexenvironment.org) to assess the potential impacts of shale gas exploration and production in Europe. The scientific activities implemented range from characterisation of shale core samples, quantification of the fluids contained in them, study of the fluid behaviour in nano- and meso-pores, design of fracturing fluids, optimisation of the water (both flowback and production) reclamation strategies, life-cycle assessment and comparison to other energy sources, as well as an analysis of socio-economical aspects related to the development of a shale gas industry in Europe.

463849 Study on Cold Energy Utilization System at Lcng Fueling Stations

Abstract:

Today the world energy market is in great transition and the demand of clean energy such as LNG increases rapidly. According to market prediction, the world demand for natural gas will increase about 2.1% yearly to 2030 and a higher proportion for Asian countries. In China, with the need of the environment protection and support from the government, the amount of natural gas vehicles is rapidly increasing, correspondingly with fueling stations. And it is known that the complete gasification of LNG (Liquefied Natural Gas) converting to CNG (Compressed Natural Gas) can return about 880kJ/kg of cold energy. If the cold energy recycle system is designed for station construction, it will bring not only economic but also environmental benefits. Nevertheless, comparing to the previous utilization of cold energy, key restrictions of utilization for LCNG fueling stations embody on three aspects. Firstly, discontinuous flow makes it hard to reuse the cold energy. It is necessary to convert it to a sustainable form such as reserved by PCM (phase change material). Secondly, CNG needs to be pressured as high as 25MPa where most heat exchangers in the market cannot sustain such pressure. A specific type of heat exchanger with matching flow pipe and measuring apparatus needs to be designed. Thirdly, considering the investment and maintenance cost, cold energy utilization system should be simple and compact. Besides, some other details are required to give sufficient consideration to make the whole system more economical and practical. Rare literature or patent relates to cold energy utilization in LCNG fueling stations. This paper will present an idea to make use of the cold energy and a set of apparatus to verify its feasibility.

Here are brief introduction of solutions to the cold energy utilization presented in this paper. Firstly, a suitable PCM to store the discontinuous cold energy of LCNG should guarantee safety, low toxicity and low flammability. At the same time, itâ€™s better to choose a common PCM to match downstream using cold energy apparatus. Thus, ice storage is a good choice satisfying the entire request above. On one hand, ice storage facilities are common in the market which can be bought and applied easily. On the other hand, ice/cold water can be used for air-conditioning in the LCNG stations. Secondly, traditional CNG fueling stations choose the vaporizer which exhausts cold energy directly to air. So a heat exchanger that could withhold high pressure up to 25MPa is quite important for the system. There are many types of heat exchanger such as plate, plate-fin, shell-and tube and so on. At present, according to the literature, only U-pipe and spiral wounded heat exchange can bear ultrahigh pressure. U-pipe ones are generally used in large scale system which needs more space. So spiral wounded heat exchanger with better exchange performance and less scale may be the suitable choice. Thirdly, besides the high pressure flow line and ice storage tank, a single-stage circulation with suitable intermediate refrigerating fluid would achieve the desired effect, which only involves one pump, one tank and pipe lines. Intermediate refrigerating fluid is divided into phase change material and non-phase change material and different material has different properties to affect the design of the size of heat exchanger and pipe diameter. In paper, investment cost, maintenance charge, and the cost payback period are discussed to make the whole system more attractive and practical.

Simulation cannot be avoided to make sure the feasibility of the system. HYSYS is a chemical engineering process dynamic simulation software, which allows designers to use the conceptual design and simplify the production process to complete the project work. Before the simulation, some fundamental parameters should be given combining references with experience. In consideration of safety in the lab, LNG is substituted by LN_2 (liquid nitrogen). EGS (ethylene glycol solution) which is chosen as the intermediate fluid should guarantee its fluidity when it is working at low temperature. The condensation point decreases while the mass concentration rises. When mass concentration reaches 70%, the condensation point could be the lowest point at -65â„ƒ. Considering the viscosity, 50% mass concentration of EGS is suitable. In addition, the values for the flow rate of LN_2, the efficiency of the pump, temperature, pressure and other parameters are explained in paper. By means of optimization by HYSYS, a series of process parameters of cold energy utilization system is given in details.

According to the design of cold energy utilization, a set of experimental apparatus is built to analyze the efficiency of the system. The whole equipment should approach to the real condition as much as possible which means the high pressure and discontinuous cold resource and limited working area. In the high pressure flow line, a made-to-order heat exchanger and pump make LN_2(liquid nitrogen) high up to 25MPa. And special venting pipe is designed to give off the N_2reducing the pressure and velocity without noise. Measuring transmitters connected to the data collecting computer are arranged in every inlet and outlet of the heat exchange device so that the amount of cold energy could be recorded. The equipment would achieve three different working conditions: ice-filling, ice-melting and ice-filling and melting at the same time. These conditions are based on the real demand in the air conditioning system in LCNG working space, which shows a good connection between cold energy recycle and utilization.

In the end, a discussion of the efficiency comparison between theory simulation and practical application demonstrates the feasibility of the cold energy utilization system and suggestions are made for further optimization and research.

key words: LCNG fueling stations, cold energy, ice storage, phase change material, experimental apparatus efficiency

463150 Techno-Economic Analysis of Shale Gas-to-Dimethyl Ether (DME) Process Via Direct Synthesis

Abstract:

Due to the volatile oil market and uncertainty in the crude oil supply in the future, it is necessary to develop cleaner and more carbon efficient non-petroleum based transportation fuels.[1] In the recent years, the use of dimethyl ether (DME) as an alternative transportation fuel is being strongly investigated. DME is used as a high quality fuel and has the potential to replace diesel due to its high cetane number (55-60).[2] DME can be produced by indirect and direct synthesis of syngas. The direct synthesis route has higher productivity compared to the indirect route but it has not been commercialized yet. Countries in the Asia-Pacific regions use coal to produce DME and are currently the largest consumers of DME. With low cost shale gas being abundantly available in the United States, DME can be produced from shale gas with less CO_2 emission and capital investment. Even though several studies have been conducted to developing catalysts and reactors for direct syngas-to-DME processes and plant-wide analysis of coal-to-DME processes, there is hardly any study in the open literature on a shale gas-to-DME process. With this motivation, we will present a techno-economic study of a shale gas-to-DME plant via direct synthesis using the process model developed in Aspen Plus® and economic model developed in Aspen Plus Economic Analyzer (APEA®).

In this work, a technically-feasible shale gas-to-DME process is synthesized first. Shale gas is first converted to syngas via pre-reformer and autothermal reformer and then converted to DME via direct synthesis. In the syngas production unit, a pre-reformer is required to convert the heavy hydrocarbons and prevent coking in the autothermal reformer. The raw syngas is sent to an acid gas removal unit to separate CO_2. The clean syngas is then sent to the DME production unit. The effluent of the DME reactor is then fed to a separation section where the unconverted syngas, DME, CO_2, water, and methanol are separated. A portion of the CO_2 recovered in the separation section is

utilized by recycling to the syngas production unit.

The presentation will focus on the following aspects: (1) development of kinetic models for pre-reformer, autothermal reformer, and DME synthesis reactor using the rate parameters regressed from experimental data, (2) development of a low cost novel separation system to separate unconverted syngas, DME, CO_2, water and methanol using DME as solvent, (3) development of the plant-wide model of a shale gas-to-DME process in Aspen Plus® and validation of plants section for which experimental data exist, (4) techno-economic analysis in APEA using the model developed in Aspen Plus®, (5) sensitivity study to analyze the effect of key design parameters on the main economic measures (i.e. net present value, internal rate of return, payback period), including design parameters such as steam to carbon ratio and oxygen to carbon ratio in the autothermal reformer, H_2/CO ratio in the DME synthesis reactor, technologies for acid gas removal (chemical or physical absorption), extent of CO_2 recycle, and utilization.

Reference

1. Azizi, Z.; Rezaeimanesh, M.; Tohidian, T.; Rahimpour, M. R. Dimethyl Ether: A Review of Technologies and Production Challenges. Chemical Engineering and Processing: Process Intensification. 2014, 82, 150â€“172.
2. Ogawa, T.; Inoue, N.; Shikada, T.; Inokoshi, O.; Ohno, Y. Direct Dimethyl Ether (DME) Synthesis from Natural Gas. Studies in Surface Science and Catalysis Natural Gas Conversion VII, Proceedings of the 7th Natural Gas Conversion Symposium. 2004, 379â€“384.

468659 Thermodynamic Modeling of Produced Water with Electrolyte NRTL Model: Aqueous Sr^{2+}- Na^{+} - $so_4{}^{2-}$- Cl^{-} Quaternary System

Abstract:

Oil and gas activities especially hydraulic fracturing lead to production of significant amount of highly saline flowback produced water. The environmental concerns related to disposing the produced water as well as fresh water conservation strategy, encourage the reuse of the flowback produced water especially the direct reuse for hydraulic fracturing. However, scaling of some minerals at fracturing temperature and pressure is a serious concern for using saline water for fracturing. One of the leading cations in sulfate type scaling is strontium (Sr^{2+}) which indicates the importance of comprehensive thermodynamic modeling of aqueous system including Na^{+}-Cl^{-} ions which cause more than 90% of salinity of brine solution, and Sr^{2+}- $SO_4{}^{2-}$ions.

In prior studies, Pitzer model has been mostly used for saline water modeling. However, the model results have been shown to be accurate only for dilute solutions with ionic strength up to 6 molal. Further, the need for ternary interaction parameters in addition to binary interaction parameters makes the application of this model for multi-component produced water cumbersome. In most cases, there are not enough available experimental data to correlate numerous Pitzer model parameters.

Therefore in this study symmetric eNRTL activity coefficient model is used to model the aqueous Sr^{2+}-Na^{+}- $SO_4{}^{2-}$-Cl^{-} quaternary system. The model requires only molecule-electrolyte and electrolyte-electrolyte binary interaction parameters. The temperature dependency of these interaction parameters are described via Gibbs-Helmholtz equation

including up to three temperature coefficients. Experimental data such as osmotic coefficient, mean ionic activity coefficient, vapor pressure, heat capacity, excess enthalpy, and salt solubility at different temperature up to 473.15 K and also at different electrolyte concentration up to saturation are used to identify the model parameters. The model provides accurate representation for all thermodynamic properties of the quaternary system.

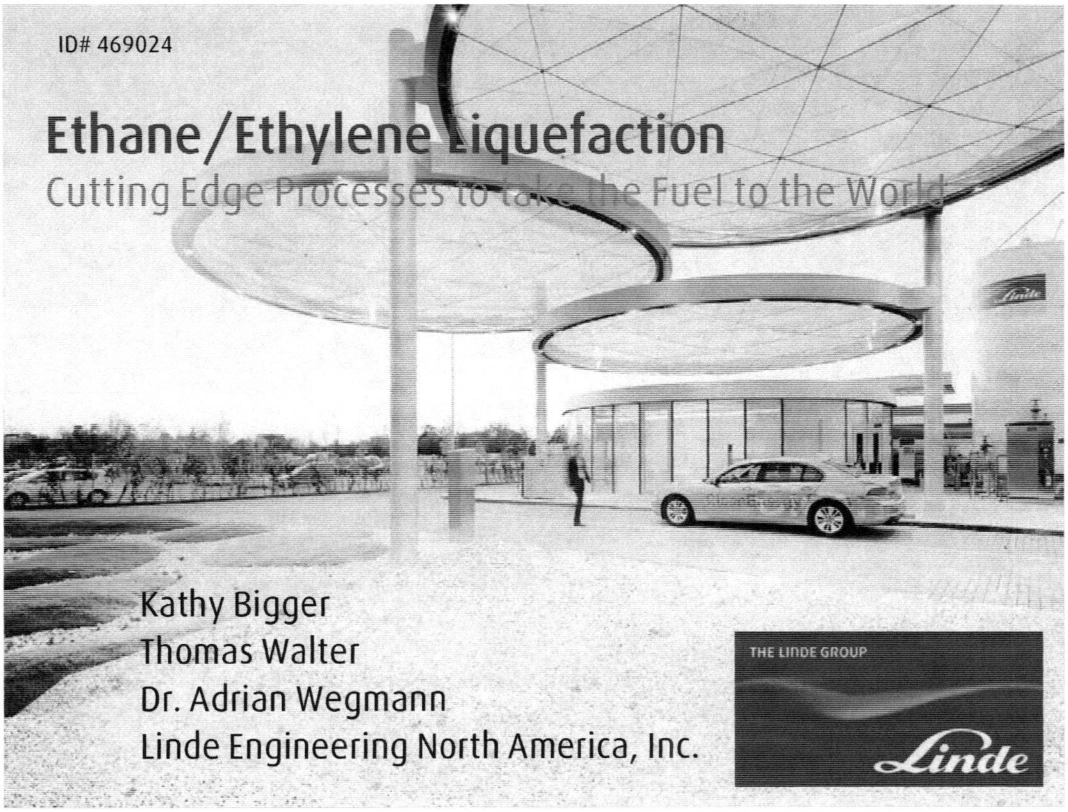

ID# 469024

Ethane/Ethylene Liquefaction

Cutting Edge Processes to take the Fuel to the World

Kathy Bigger
Thomas Walter
Dr. Adrian Wegmann
Linde Engineering North America, Inc.

THE LINDE GROUP
Linde

Energy in the world

THE LINDE GROUP
Linde

- Total World Energy Consumption expected to grow 48% between 2012 and 2040
- Non OECD energy consumption to rise 71%, OECD energy consumption to rise 18%
- Economic growth (GDP) is key determinant for energy demand

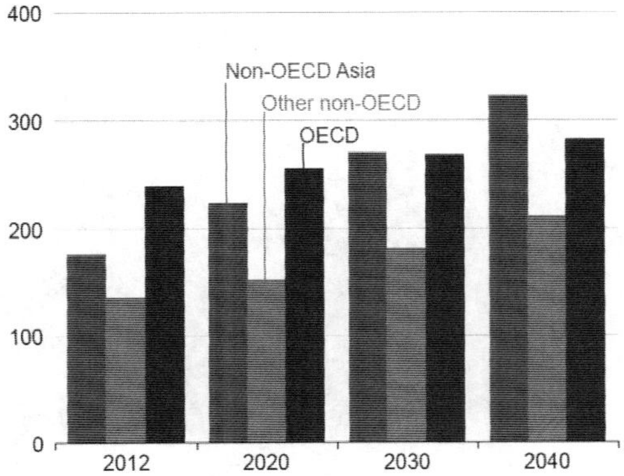

EIA: "International Energy Outlook 2016", Figure ES-1, page 1
World energy consumption by country grouping, 2012-40 (quadrillion BTU)

Energy in the world

THE LINDE GROUP
Linde

- Growth per category:
  - Renewables
  - Nuclear
  - Natural gas
  - Liquid fuels
  - Coal

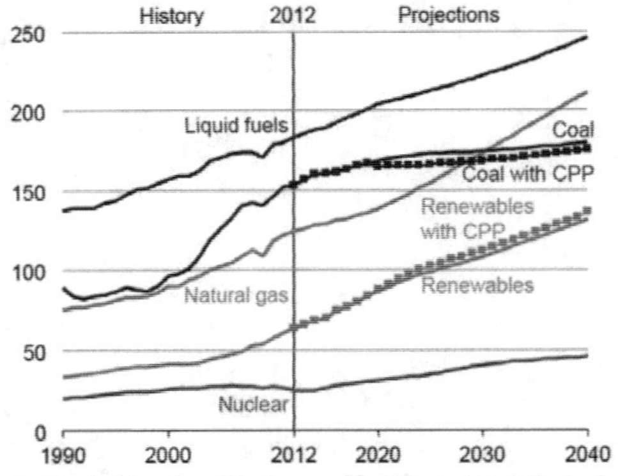

Note: Dotted lines for coal and renewables show projected effects of the U.S. Clean Power Plan.

EIA: "International Energy Outlook 2016", Figure ES-2, page 1
Total world energy consumption by energy source, 1990-2040 (quadrillion BTU)

Natural gas is projected to be the fastest-growing fossil fuel!

Natural gas – sources

THE LINDE GROUP
Linde

- Ethane is a byproduct of the natural gas production
  - Shale gas
  - Tight gas
  - Associated gas
  - Coalbed methane
  - Other natural gas

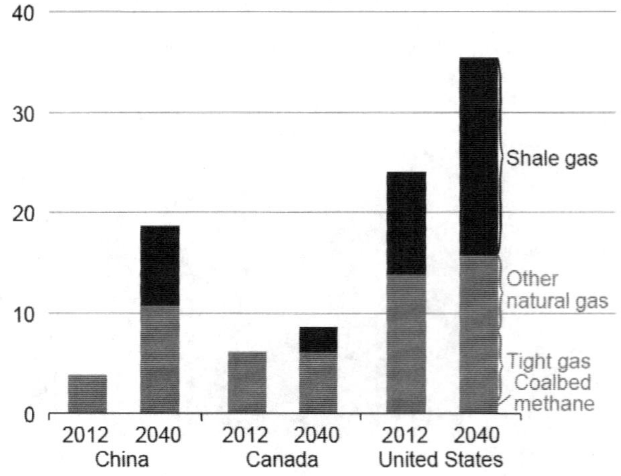

EIA: "International Energy Outlook 2016", Figure 3-3, page 38
Natural gas production by type, 2012 and 2040 (trillion cubic feet)

Investment in ethane/ethylene infrastructure

- Driven by abundant availability
  - Current surplus in the United States: ~500'000 bbl/day
  - Expected surplus in the United States in 2020: ~150'000 bbl/day
- Required by geographical mismatch between source and consumer
- Ethane Cracker capacity is being expanded due to cost advantage over naphtha crackers
  - Installed ethylene capacity in the USA in 2012: ~24 million tons per year
  - Anticipated capacity in 2020: ~35 million tons per year
- Challenges:
  - Lack of dedicated pipeline system
  - Physical properties require highly specialized tank equipment similar to LNG
- Solutions:
  - Liquefaction
  - Processing capabilities in unusual locations (Badlands NGL Polyethylene plant in North Dakota)

Ethane export

- Export
  - Pipeline to Canada since late 2013
  - Marcus Hook, Pennsylvania: 35'000 bbl/d capacity since March 2016
  - Morgan's Point, Texas: 200'000 bbl/d capacity since September 2016

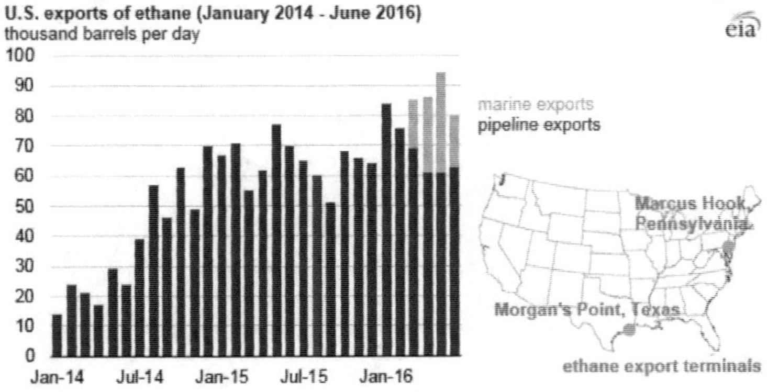

EIA: "Today in Energy, September 22, 2016 – First shipment of ethane from U.S. Gulf Coast arrives in Europe"

Drivers for finding the optimized solution

- Customer desires
  - OPEX – Compressor power, cost of refrigerant
  - CAPEX – Equipment count & size
  - Compliance implications, legal requirements
- Limiting factors
  - Physical size of plate fin heat exchangers (PFHX)
  - Refrigerant compressor power
- General dependencies
- Range of feasible solutions

Parameter study

THE LINDE GROUP
Linde

- Heat exchanger size – number of cores
- Piping arrangement
- Coldbox size
- Process conditions (pressure determines vaporization temperature)

- **30'000 variations of process setup**

Pareto Frontier

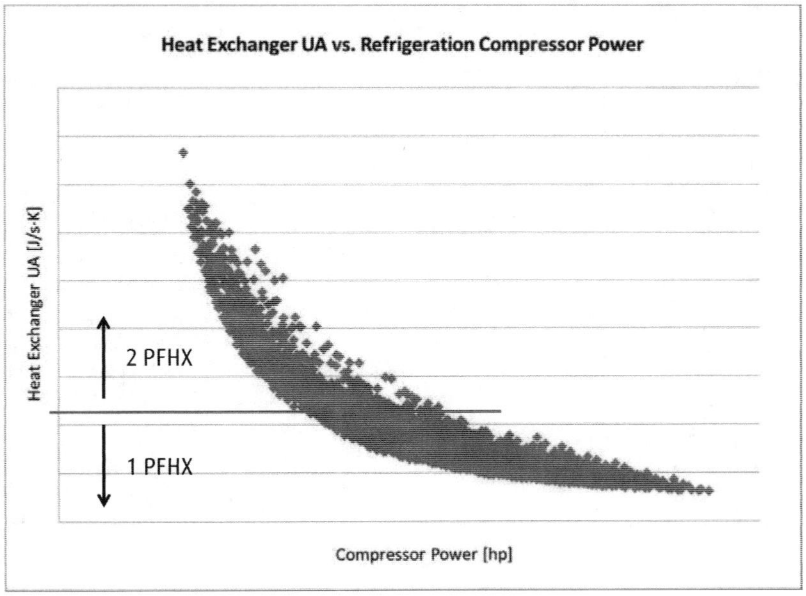

Conventional liquefaction process
Open Loop

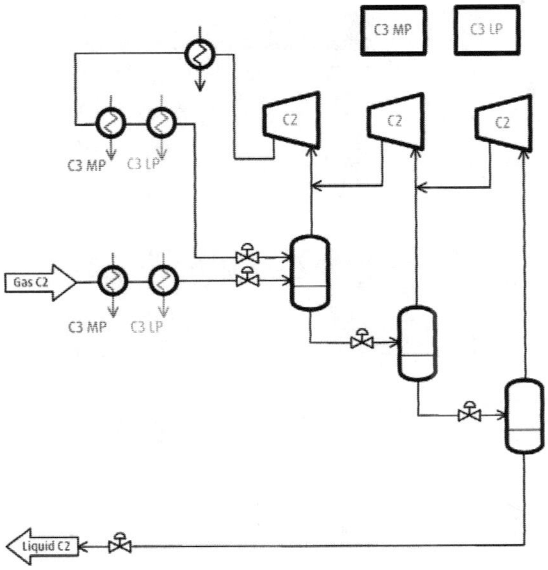

Open loop Process

- Conventional process utilizing STHE and block in shell to liquefy the C2
- Typical inlet pressure from pipeline >60 bar (900psi)
- MP and LP propane or propylene refrigeration needed to precool the C2
- Liquefaction by multi-stage flash and recompression of the flash gas
- Conventional process can also be designed as a closed loop resulting in increased power consumption

Mixed refrigerant liquefaction process
LIMUM®1

THE LINDE GROUP Linde

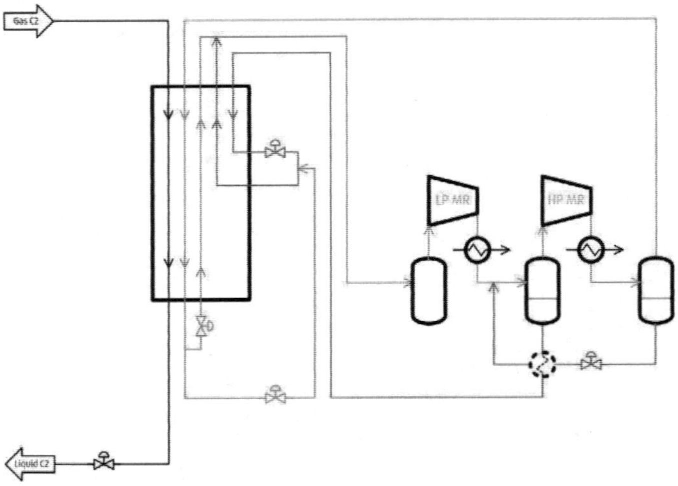

Single Cycle Process

- Typical for peak shaving or small scale LNG plants; process adapted to liquefaction of C2
- Simple process, low equipment count
- One refrigerant cycle providing refrigeration duty
- Pre-cooling, liquefaction and sub-cooling cycle realized with a plate-fin heat exchangers
- Lower specific energy consumption compared to an expander process

ID# 469024: Ethane/Ethylene Liquefaction - Cutting Edge Processes to take the Fuel to the World
Linde Engineering North America, Inc. - Kathy Bigger, Thomas Walter, Adrian Wegmann 11

StarLNG™
Liquefaction of CH_4 vs. C_2H_6 / C_2H_4

THE LINDE GROUP Linde

@1.013 bar/ 14.7 psi	CH_4	C_2H_6	C_2H_4
Liquefaction temperature	-161.6°C (-258.9°F)	-88.7°C (-127.7°F)	-104.1°C (-155.4°F)
Heat of vaporization	512.2 kJ/kg (220.2 Btu/lb)	487.5 kJ/kg (205.7 Btu/lb)	476.2 kJ/kg (204.7 Btu/lb)

@40 bar/ 580 psi	CH_4	C_2H_6	C_2H_4
Liquefaction temperature	-87.3°C (-125.2°F)	22.6°C (72.6°F)	-1.1°C (29.9°F)
Heat of vaporization	160.7 kJ/kg (69.1 Btu/lb)	171.6 kJ/kg (73.8 Btu/lb)	179.2 kJ/kg (77.0 Btu/lb)

The heat of vaporization is similar, but the liquefaction temperature is significantly different.
Rule of thumb: The specific energy consumption compared to LNG is approx. 5.6 times smaller for Ethane and 4.1 times smaller for Ethylene!

ID# 469024: Ethane/Ethylene Liquefaction - Cutting Edge Processes to take the Fuel to the World
Linde Engineering North America, Inc. - Kathy Bigger, Thomas Walter, Adrian Wegmann 12

Comparison energy consumption and equipment count
Mixed refrigerant vs. conventional

THE LINDE GROUP
Linde

	Mixed refrigerant LIMUM®1	Conventional Open loop
Relative power consumption	**1.0** 1), 2)	**0.97** 1), 2)
Heat exchanger	2x Air Cooler/Shell&Tube 1x Plate-fin 1x Shell&Tube (optional)	3x Air Cooler/Shell&Tube 3x Shell&Tube 1x Block in Shell
Compressor	2x Mixed Refrigerant stages	3x C2 Compressor stages 2x C3 Compressor stages
Separators	3x Suction/Discharge Vessel	5x Suction/Discharge Vessel
Make-up	2-3x Storage CH_4, C3(opt.) and C4	1x Storage C3
Total 3)	**10-12 Equipment**	**18 Equipment**

1) The power consumption for an Ethylene liquefaction unit was 94.4 vs. 91.3 kWh/t_{C2}
2) Power consumption strongly dependent on process conditions, cooling water vs. air cooler, etc.
3) Liquefaction unit only. Equipment for Dehydration, BOG and Storage is identical

Summary

THE LINDE GROUP
Linde

- Ethane is abundantly available in the US for the foreseeable future
- Ethylene production is forecast to substantially increase in the near term
- Both ethane and ethylene are considerably easier to liquefy than methane
- There is a large potential for the US to become a supplier of both ethane and ethylene to nations without any such resources readily available
- Modern, optimized technical concepts can provide sustainable solutions for the ever more challenging energy landscape on our planet!

Thanks for your attention.

463765 Absorption of Mixed Refrigerant in Lubricant Oil

Abstract:

Absorption of mixed refrigerant in lubricant oil

Yang,Song Wensheng,Lin

Institute of Refrigeration and Cryogenics, Shanghai Jiao Tong University,P.R.China

Abstract

Of all processes producing liquefied natural gas (LNG), mixed refrigerant cycle is considerably appealing for its high efficiency. In a small scale mixed refrigerant cycle, utilizing injected screw compressors makes the whole compression unit more compact. However mixed refrigerant may dissolve in lubricant oil existing in injected screw compressors. It is pretty essential to understand the detailed knowledge of mixed refrigerant absorption in refrigerant-lubricant oil mixtures, as the change in compositions of mixed refrigerant can directly influence some important operation parameters such as process energy consumption, heat exchangers performance, etc.

To determine the optimized components of mixed refrigerant in single mixed refrigerant (SMR) liquefaction process, without considering the absorption of refrigerant in oil, the Knowledge Based Optimization (KBO) method was adopted. During this optimization process, there were two limiting conditions for operation: the minimum heat transfer temperature difference in the LNG heat exchanger was set 3K and liquid entering compressors was not allowed. The P-R equation of state, which is suitable to calculate the hydrocarbon mixtures in particular, was adopted in the simulation. Typically, energy consumption of unite LNG product directly affects the running cost of the liquefaction system, so the minimum specific power consumption was set as objective function. Then refrigerant components in SMR LNG process were optimized for both four and five compositions by adjusting sub-flow of the components from low to high boiling point according to the KBO method. At the end of the optimization process, the optimal constitution was gained with methane 21%, ethylene 14%, propane 58%, nitrogen 7% in the four-composition mixed refrigerant and methane 19%, ethylene 30%, propane 4%, i-butane 39% and nitrogen 8% respectively. These optimal results were then used as basis of follow-up studies on the effect of single refrigerant component change on operation parameters as well as experimental investigation on refrigerant absorption in lubricant oil.

After the optimal constitution of mixed refrigerant was gained, composition change due to dissolving in lubricant oil was considered. In order to investigate the impact of the change of each component on the operation parameters referred to above, a variable-controlling approach based on HYSYS was adopted. More precisely, the influence of methane, ethylene, propane and isobutene was investigated respectively. Firstly, we investigated the effect of each component change on the specific power consumption. Different component reducing affect specific power consumption differently due to their thermophysical property difference. To study this effect more specifically, the content of each component was reduced gradually, each time 0.01kmol/h, while other components remained the same and the specific power consumption was recorded at the same time. Methane, ethylene, propane, and i-butane was individually operated in the same way until the minimum heat transfer temperature difference in the LNG heat exchanger was less than 3K. During this process, we can obtain that with the decrease of each kind of components, the specific power consumption declined in a linear trend. However, reducing the same amount of each component, the specific power consumption reduced in a different degreeâ€"reducing methane had the greatest impact, followed by ethylene, propane in the four-composition refrigerant process and the five-composition refrigerant shared the same trend with i-butane had the smallest impact. It is noteworthy that, nitrogen is hardly soluble in lubricant oil, so we did not consider its reduction. Secondly, the effect of each component change on the minimum temperature difference in LNG heat exchanger was investigated. Each component was reduce in the same way referred to above, during this process the minimum temperature difference was recorded. It can be obtained that in the four-composition refrigerant process, reducing propane impacted to the greatest extent, followed by methane and ethylene. However, the five-composition refrigerant process saw a different pattern with i-butane impacted most, followed by propane, ethylene and methane. This trend can be explained by temperature sensitive range of each component.

As well as single component analysis based on HYSYS, experiments focusing on overall absorption of mixed refrigerant in lubricant oil were also conducted.During this experiment process, we studied four-component gas, which was constituted by 21% methane, 14% ethylene, 58%propane, 7%nitrogen, entering lubricant oil. The model number and brand of lubricant oil were determined by actual LNG process. In this experiment, we tested three different kinds of lubricant oil (CP-1010-100, CP-1005-100, CP-1516-100) which were widely used in the SMR process compressors. In order to ensure the lubricant was sufficient, the tank was designed to be 10 liters.

Initially, the tank was evacuated, after which lubricant oil was added to the desired weight (10kg). Then turn on the heater until the oil reached 303K by regulating the regulator. After that, open the valve of the high pressure gas bottle, then the mixed gas was introduced into the lubricant oil. Regulating the valve to make sure the pressure in the tank was 0.5MPa initially. Finally, the component of the exhaust mixed gas was measured by gas chromatograph. When the component measured kept unchanged for two times, wrote down the specific percentage of each component and mixed gas mass flow of intake and exhaust. Apart from that, the instantaneous system pressure and temperatures in vapor and liquid regions were both recorded at temperatures of 303K, 313K, 323K, 333K, 343K and 353K. When the mixed gas in the bottle was used up, the lubricant oil should be discharged and the tank was cleaned for the next kind of lubricant oil. Because nitrogen is hardly soluble in lubricant oil, its mass flow kept unchanged, which could be used as a leak detection during the experiment process.

The mass of each component of mixed refrigerant absorbed in lubricant oil was calculated by the experiment data. When the calculation was completed, the results were then used to investigate the impact of mixed refrigerant absorbed in lubricant oil on operation parameters change in the LNG liquefied process through HYSYS.

474889 All Biomass Is Local: Paying Farmers More for Cellulosic Biomass Is a Game Changer

Abstract:

Current models of cellulosic biofuel systems require that the delivered price of the cellulosic

biomass feedstock be kept low. Thus the predicted biorefinery size is relatively small, limiting potential economies of scale. However, it is actually the ultimate selling price of the biofuel that largely determines market penetration. We relaxed the constraint of low delivered feedstock price and explored the resulting effects on biofuel price, biofuel volume produced, and global warming impact (GWI). Feedstock price greatly affects the feedstock supply chains that may develop. Increased feedstock price does not affect the final ethanol selling price very much, but higher feedstock prices greatly increase the amount of ethanol produced. Farmers will supply much more cellulosic biomass at higher feedstock prices, leading to shorter transportation distances with reduced transportation costs and enabling larger biorefineries with improved economies of scale, thereby reducing the ethanol selling price.

The cellulosic feedstock supply chain systems were studied as a function of feedstock prices by

determining potential feedstock supply clusters and the maximum capacity of cellulosic biorefineries across the United States. Supply clusters were determined by minimizing costs associated with ethanol production. The analysis is based on county-level cellulosic feedstock production data projected in the *US Billion-Ton Update* report. Each biomass supply cluster is unique in terms of local and regional characteristics (e.g. area, feedstock types), biorefinery capacity, ethanol selling price, and GWI. Very large-scale biorefineries (â‰¥20 000 dry Mg dayâ€"1) may be feasible in some regions.

471184 Impact of Alkaline Deacetylation and Pelleting of Corn Stover at the Depot

Abstract:

Currently, approximately 258 million tons of the Billion Tons (DOE goal) is considered accessible in the existing feedstock supply systems. Additionally, much of this biomass is highly variable in properties important to handling, densification and conversion. An important role of the distributed processing concept is to deliver feedstock to biorefineries with low variability and consistent conversion performance. Achieving this goal would decrease risk to the biorefinery and reduce operating expenses. This work examines the impact of deacetylation at the depot on both the feedstock supply chain and the conversion facility.

Single-pass corn stover was ground to 1 inch using a Wiley Mill and stored at room temperature and humidity. The corn stover was split and deacetylated in a stirred-batch reactor. The recovered solid residuals were collected and compared against raw corn stover using analyses that measured total ash, ash composition and proximate/ultimate. In addition, the chemistry of the deacetylation medium was determined along with the ash and CHNO composition of solids precipitated from the deacetylation medium. In this manner, data was collected to support a full mass balance of the process. The energy to grind and pellet the raw material and alkaline deacetylated material was measured. The alkaline deacetylation resulted in a 13.8% decrease in required grinding energy, and a 23.1% decrease in required pelleting energy. It also resulted in increased carbohydrate content and reduced ash for the biorefinery feedstock. In addition, it was observed that the deacetylated material could be ground at much higher moisture contents. This would allow corn stover to be pelleted, and then dried using a more efficient process. Finally, raw and deacetylated material was subjected to different pretreatment severities with and without disc refining. Key findings were that deacetylation and pelleting were found to: 1) negate the need for disc refining; 2) increase sugar yields by over two-fold with when no pretreatment was applied; 3) require significantly lower severity of pretreatment to achieve the same, or greater, sugar yield; and 4) increase maximum sugar yield to near 100%.

474787 Liquefaction of Lignocellulose

Abstract:

The liquefaction of biomass materials is a process where particulate wood, corn stover, sugarcane bagasse and other forms of lignocellulose feedstocks are converted to an aqueous suspension of solid materials that is suitable for further processing to fermentable sugars and biofuel materials. A number of reports have described the liquefaction of pretreated materials using either cellulose enzyme or an enzyme biomimetic (maleic acid). Our work has evolved to begin to examine the liquefaction of lignocellulosic substrates at 25% solids loading prior to pretreatment, since the ability to obtain pumpable slurries prior to pretreatment has significant benefits in equipment design and operating conditions for the pretreatment step itself. Challenges in obtaining pumpable slurries of biomass materials, either pre- or post-pretreatment include the shear thinning character of the slurry, mixing of the particulate biomass during the liquefaction process, and definition of operating conditions that make efficient use of the biocatalysts required for the liquefaction process. This paper presents an analysis of the insoluble particle characteristics that enable slurries to be obtained, and proposes a model for the liquefaction process that relates particle size, shape, and composition to the liquefaction process.

469070 Technoeconomic Analysis of Pyrolysis and Electrocatalysis Depots

Abstract:

Biomass upgrading depots are small-scale facilities that serve to densify raw feedstocks into transportable and salable commodities with increased energy and economic value. Decentralized systems seek to find the optimum balance of "economies of scale" and "economies of transportation," while creating a venue for value addition to rural communities. Fast pyrolysis is a thermochemical technology that can be deployed in small-scale depots to deconstruct and densify biomass by rapid heating (400-600°C) without oxygen to create bio-oil, biochar, and combustible gas. Bio-oil is a bulk dense and energy dense liquid that is less costly to transport than raw biomass. However, bio-oil is limited as a fuel intermediate because of its poor chemical properties, namely high oxygen content, corrosiveness, and potential to react and form sludge. As a remedy, electrocatalytic hydrogenation and deoxygenation (ECH) has been proposed to partially upgrade bio-oil to create a stable fuel intermediate. ECH works by saturating carbon-carbon and carbon-oxygen double bonds as well as cleaving certain ether linkages on a catalytic cathode. Though ECH is capable of upgrading bio-oil, the cost of doing so must not be prohibitive to its adoption. The economic viability of pyrolysis-ECH depots is a function of capital costs, operating costs, and product values. In this regard, a technoeconomic analysis of pyrolysis-ECH depots will be described to reveal the tradeoffs between depot capacity and the variable costs of transportation. Sensitivities to biomass purchase price, moisture content, electricity source, and electricity cost will be examined. ECH could be the keystone technology in biomass upgrading depots, leading to regionally produced commodities and revenue generation in rural communities.

474887 Electrochemical Bio-Oil Stabilization As an Alternative for Distributed Processing of Biomass

Abstract:

The conversion of biomass to value-added chemicals and liquid fuels is a developing alternative to fossil sources [1]. Bio-oil, which can be obtained from thermal breakdown of forest and agricultural biomass, can be produced near the material source. The distributed production of bio-oil allows the densification of the biomass energy and could reduce the transportation and production costs. However, compared with fossil hydrocarbon sources bio-oil is a chemically unstable micro emulsion that cannot be fractionated [1, 2]. The high concentration of oxygenated compounds (hydroxyaldehydes, hydroxyketones, sugars, carboxylic acids, and phenols) reduces the storage and temperature stability of the bio-oil and results in 50% lower calorific value when compared to hydrocarbon fuels [1, 3, 4]. The typical route to removing oxygen involves a process called hydrodeoxygenation (HDO). Typical HDO conditions (425â—¦C and 2000 psi H2 pressure) promote condensation reactions leading to the formation of coke. A multi-step hydrotreatment process, which include low temperature (140â—¦C) hydroprocessing stabilization under high H2 pressure (1200 psi H2 pressure), reduce the reactivity for HDO [5, 6]. In distributed production mild conditions are desired to reduce capital requirements at remote sites near biomass sources.
In this work an electrochemical process has been investigated as a low temperature alternative for the electrochemical reduction (ECR) and stabilization of bio-oil. The process attempts to reduce site requirements for hydrogen and achieve even milder conditions than used for existing catalytic stabilization. A polymer electrolyte membrane (PEM) configuration allows the processing of the complete bio-oil product without the addition of supporting electrolyte. While work to this point has focused on use of PEM cells employing cation exchange membranes, future work will attempt to use anion exchange membranes where current would be carried by anions of small acid compounds (acetate and formate). This configuration could simultaneously reduce reactivity of oil and also remove small acid molecules that contribute acidic content and consume hydrogen in downstream processes in production of low value gases. Description of the electro-catalytic systems and results obtained with different transition metal catalysts will be

presented.

References
[1] D. Mohan, C.U. Pittman, P.H. Steele, Pyrolysis of Wood/Biomass for Bio-oil: A Critical Review, Energy & Fuels, 20 (2006) 848-889.
[2] A.V. Bridgwater, Renewable fuels and chemicals by thermal processing of biomass, Chem. Eng. J., 91 (2003) 87-102.
[3] S. Czernik, A.V. Bridgwater, Overview of Applications of Biomass Fast Pyrolysis Oil, Energy & Fuels, 18 (2004) 590-598.
[4] A. Oasmaa, E. Kuoppala, Y. Solantausta, Fast Pyrolysis of Forestry Residue. 2. Physicochemical Composition of Product Liquid, Energy & Fuels, 17 (2003) 433-443.
[5] D.C. Elliott, Historical Developments in Hydroprocessing Bio-oils, Energy & Fuels, 21 (2007) 1792-1815.
[6] S. Jones, P. Meyer, L. Snowden-Swan, A. Padmaperuma, E. Tan, A. Dutta, J. Jacobson, K. Cafferty, Process design and economics for the conversion of lignocellulosic biomass to hydrocarbon fuels: fast pyrolysis and hydrotreating bio-oil pathway, in, National Renewable Energy Laboratory (NREL), Golden, CO., 2013.

462787 Chemical Looping Combustion for Carbon Capture Efficiency

Abstract:

Among the well-known state-of-the-art technologies for CO_2 capture, Chemical Looping Combustion (CLC) stands out for its potential to capture with high efficiency the CO_2 from a fuel power plant for electricity generation. CLC involves combustion of carbonaceous fuel such as coal-derived syngas or natural gas via a red-ox chemical reaction with a solid oxygen carrier circulating between two fluidised beds, air and fuel reactor, working at different hydrodynamic regimes. Avoided NO_x emissions, high CO_2 capture efficiency, low CO_2 capture energy penalties and high plant thermal efficiency are the key concepts making worthy the investigation of the CLC technology. One of the main issues about the CLC technology might concern the cost of the solid metal oxides and therefore the impact of the total solid inventory, solid make-up and lifetime of the solid particles on the cost of the electricity generated. In this work, macro scale models of fluidised beds (i.e. derived applying macroscopic equations) are developed and implemented in Aspen Plus software. Kinetic and hydrodynamic phenomena, as well as different operating conditions, are taken into account to evaluate their effect on the total solid inventory required to get full fuel conversion. Additionally, a 2D â€œmicroâ€ scale model of the fuel reactor (i.e. derived applying partial differential equations), making use of a CFD code, is also developed. The results, in terms of the effect of the different kinetic and hydrodynamic conditions on the outlet gas conversion, drive the improvement of the macro scale model to capture the main physics influencing the performance of the fuel reactor. The latter macro scale model is embedded into a natural gas fired power plant configuration where mass and energy balances are solved simultaneously. Two CFD platforms (namely MFIX and OpenFOAM) are implemented to devise the influence that the reactorâ€™s internals have on the hydrodynamics and then on the conversion. Finally, a thermal efficiency evaluation of the CLC power plant is carried out together with a detailed economic analysis. The effect of the fuel price and lifetime of the solid particles on the Levelised Cost Of Electricity (*LCOE*) is investigated and the resulting outcomes are critically discussed in comparison with post-combustion technology employing amines.

469869 Cold Flow Model Studies of a Counter-Current Moving Bed Syngas Chemical Looping Pilot Unit for High Purity Hydrogen and Electricity Co-Generation with Carbon Capture

Abstract:

The syngas chemical looping (SCL) process developed at the Ohio State University (OSU), which used Fe-based oxygen carrier and a counter-current moving bed reactor design, has the capability of achieving nearly 100% carbon capture and the flexibility of co-generating high purity hydrogen and electricity. A 250 kW_{th} â€“ 3 MW_{th} pilot unit operated under 10 atm pressure was designed and operated. The pilot unit consists of three major reactors, the reducer and the oxidizer operated as counter-current moving beds, and the combustor operated as a fluidized bed, to perform reductionâ€“oxidation reaction cycles for converting gaseous fuels to highly concentrated discrete streams of hydrogen (H_2) and carbon dioxide (CO_2). An L-valve is used to transport the solids between the moving bed reactors and the fluidized bed reactor with a controllable solids circulation rate. To investigate the hydrodynamic characteristics and optimize the system design, a one-to-one ratio cold flow model is built and tested. The cold flow model is operated under ambient temperature and pressure with both simulated Geldart Group D particles (glass beads) and actual oxygen carrier particles. The operational conditions of the cold flow model tested are to simulate the expected relative differential pressures and gas-solid flow regimes of the three reactors at operating temperature and pressure. Multiple solid circulation rates and gas processing capacities were tested to understand range of operating conditions achievable in the pilot unit while maintaining proper gas sealing between each reactor and solid circulation rate. The solids circulation rate with respect to the aeration gas flow rate was tested to determine a calibration curve. For solid flow measurement, an iso-kinetic device was designed and tested. To date, over 200 hours of operational results were completed with SCL cold flow model with over 19 test cases completed.

467632 Development of a Carbon Stripper Particle Separation System for Chemical Looping Applications

Abstract:

Chemical looping combustion (CLC) is a promising new technology for reducing carbon emissions from coal-based energy generation. The most significant limiting factor for implementation of this technology is the ability to remove unburnt carbon (char) from the oxygen carrier material as it is cycled between the fuel and air reactors within the CLC unit, thus preventing formation of CO_2 formation in the air reactor. In an effort to address this critical issue, a novel carbon stripper unit has been developed for the separation of unburnt char from oxygen carrier compounds for use with chemical looping combustion (CLC) systems. The carbon stripper unit consists of a conical spouted bed with a thin annular ring located between the spout cone and the bed wall. A sweep gas is injected into this annular region such that the sweep gas velocity within the annulus falls between the terminal velocities of the oxygen carrier and char particles. This facilitates separation of particles by weight, as the heavier oxygen carrier particles are allowed to fall down through the annular region, while the lighter char particles are prevented from entering the annulus. The char particles are ultimately entrained out of the dense bed region and into the freeboard region by the spouting gas, where they are carried out of the carbon stripper unit via elutriation. Preliminary results from experiments carried out in a cold flow prototype unit have demonstrated favorable separation characteristics.

Validation Attempts of Proposed Particle-level Growth Mechanism of Ash Agglomerates in Fluidized Bed Combustors

Aditi B. Khadilkar[1], *Peter L. Rozelle*[2], *Sarma V. Pisupati*[1]
[1] *John and Willie Leone Family Department of Energy and Mineral Engineering, and The EMS Energy Institute, The Pennsylvania State University, University Park PA 16802, USA*

[2] *Office of Fossil Energy, U.S. Department of Energy, FE-221/Germantown Building, 1000 Independence Avenue, S.W., Washington, DC 20585-1209 USA*
emails: abk165@psu.edu; peter.rozelle@hq.doe.gov; sxp17@psu.edu

Increase in particle size occurs during several processes that use fluidized beds. While it is desirable in processes such as granulation and pellet-making its occurrence can cause undesirable operational issues in combustion and gasification processes. Models that incorporate interdependencies and variations in both bed-hydrodynamics and binder chemistry are required for accurate predictions of agglomerate growth kinetics. Penn State has developed a unique two-particle collision-based model that combines these two aspects. The model is tested with poly-disperse fluidized bed combustion and gasification systems.

In fluidized bed gasification and combustion systems, agglomeration occurs by sticking of fuel ash particles that are wetted by slag-liquid. The non-uniform temperature, gaseous atmosphere and heterogeneity in ash chemical composition need to be accounted for in the prediction of the slag-liquid binder formation and chemistry. Additionally, the poly-dispersity of bed ash, particle velocities and collision frequencies are the important physics-based factors that affect particle growth rate. In the Penn State model, thermodynamic equilibrium calculations are used to estimate the amount of slag-liquid in the system, and the changes in particle collision frequencies are accounted for by continuously tracking the number density of the various particle sizes. The particle number density is also affected by the ash chemistry as solid particles melt to form slag. Computational fluid dynamics modeling is used in conjunction to obtain the initial granular physics inputs to the model. Thus, the heterogeneities in ash chemistry and granular physics are integrated in this model. The model development was presented previously and the present study focuses on validation of the modeling results and the methodology developed. For this purpose, ash agglomerates were generated in a laboratory-scale fluidized bed combustor at Penn State. Ash agglomerates were produced by operating the reactor at a superficial gas velocity close to the minimum fluidization velocity of the particles, using rejects from a bituminous coal with about 82 % ash content, under oxidizing conditions. Agglomerate samples were also obtained from a fluidized bed combustor at Canada. Polished cross-sections of these agglomerates were studied using scanning electron microscopy with energy disperse x-ray scattering (SEM-EDX) to relate the particle-level slag formation and sticking to the chemical composition. FactSage simulations of the slag-forming components were used to estimate the agglomeration temperatures.

The results of this study indicated that agglomerate growth in fluidized bed combustors (FBC) is initiated at the particle-level by low-melting components rich in iron- and calcium-based minerals. Although the bulk ash chemical composition does not indicate potential for agglomeration, study of particle-level heterogeneities revealed that agglomeration can begin at lower temperatures than the FBC operating temperatures of 850 °C. After initiation at the particle-level, more slag is observed to form from alumino-silicate components at slightly higher temperatures caused from changes in the system, and agglomerate growth propagates in the bed. A post-mortem study of ash agglomerates using SEM-EDX helped to identify stages of agglomerate growth.

Keywords- fluidization, agglomerate growth, SEM-EDX, slag, hydrodynamics

1. Introduction

Prediction of agglomeration is challenging since the heterogeneity in ash chemical composition, gaseous atmosphere and distributive granular physics properties affect the process simultaneously. Several instances of solids agglomeration have been reported, even at the relatively low operating temperatures in bubbling fluidized bed (BFB) and circulating fluidized bed (CFB) systems[1, 2]. Mineral matter in fuel ash is distributive and the bulk chemical composition is not representative. Based on mineral processing operations used, such as grinding, the mineral content is distributed amongst particle classes. Thus, the bulk fuel contains classes of particles that may be rich in specific minerals such as a class that contains heavy, iron-based minerals or calcium-based minerals. The particles of such a class can be assumed to be reasonably homogeneous in composition, while each particle class would have a composition distinct from another. The term "particle-level" refers to the differences that exist at the level of these ash particle classes. The term "bulk" composition refers to the overall composition of the whole coal that contains these specific mineral-rich particle classes. The particle-level differences in size and composition have been ignored in using only the composite fuel for analysis and prediction of agglomerate growth. A consideration of particle-level non-uniformities in the prediction of ash behavior and agglomeration is warranted. Khadilkar et al.[3, 4] showed that the bulk ash chemical composition did not indicate slag formation at low temperatures, while the study of particle-level heterogeneities revealed that it can begin at lower temperatures than the FBC operating temperatures of 850 °C.

2. Research objective

To develop a mechanistic understanding of agglomerate growth in fluidized bed combustors, based on the modeling results and to discuss attempts to validate the proposition.

3. Materials and Experimental Methods

The fluidized bed agglomeration modeling methodology that combines particle-level heterogeneities in ash chemistry and granular physics is based on two-particle collisions, tested using the Stokes' criterion. If the particles' kinetic energy is dissipated by the viscous slag, they remain stuck on collision. The kinetics of agglomerate growth depends on the collision frequency of the particles, which is calculated based on the kinetic theory of granular flow. The slag chemistry and surface properties determine the surface coverage and also affect the rate of agglomeration. The model development has been described in detail in previous work[5].

A study of the slag formation tendencies at the particle-level using density separated coal gravity fractions, was conducted in previous work, under oxidizing[4] as well as reducing[3] conditions. A lower slag formation onset temperature occurred for particles rich in iron- and calcium-based phases. The amount of slag formed at the onset temperatures was less than 10 %. The extent of agglomerate growth was estimated using the agglomeration model at different slag amounts to gain a better understanding of the conditions that promote agglomeration.

Additionally, agglomerate samples obtained from a laboratory-scale combustion facility in Canada were analyzed using SEM-EDX. Agglomerate samples were also generated in a laboratory-scale fluidized bed combustor at Penn State as described below using Pittsburgh No. 8 coal rejects for analysis. The chemical composition of these rejects is shown in Table 1.

Table 1: ICP-AES and sulfur analysis of Pittsburgh No.8 coal rejects

Species	SiO_2	Al_2O_3	Fe_2O_3	CaO	TiO_2	K_2O	MgO	Na_2O	BaO + SrO	MnO	S
Wt. %	62.57	24.82	5.74	2.17	1.26	2.5	0.83	0	0.09	0.02	1.61

3.1. Laboratory scale fluidized bed reactor

A laboratory-scale fluidized bed[6] was used to produce agglomerates from a fraction of Pittsburgh No. 8 seam coal. The rejects of this coal were used for this purpose as the rejects have a high ash content of 81.6 % and would reduce effects of the presence of carbon, as that is not accounted for in the model. The reactor was operated as a combustor by using 3% oxygen, 15% carbon dioxide, 10% water and 72% nitrogen as the fluidizing gas. The particle size distribution used consisted of about 48 % in the size range of 250-355 μm and 52% between 149-250 μm. The size and velocity distribution were chosen such that the superficial gas velocity would be greater than the minimum fluidization velocity and less than the terminal velocity of majority of the particles. A 200 g sample was used and the experiment was run for 2 h 15 min, using a gas flow rate of about 10 L/min.

3.2. Scanning electron microscopy coupled with energy dispersive x-ray spectroscopy (SEM-EDX)

The samples were tested in a FEI Quant 200 SEM coupled with a 10 mm^2 EDX detector (Oxford Instruments, Concord, MA). This technique was used to conduct post mortem analysis of agglomerate samples. They were given a conducting iridium coating to avoid excessive charging of the sample surface with electrons during the experiments. This helped to identify the elements in the regions that formed slag bridges as compared to the solid particles. The elemental analysis was performed using the software Aztec, version 2.4.

4. Results and Discussion

Slag formation onset temperature was as low as 850 °C for density fractions of Pittsburgh No. 8 and Illinois No. 6 coals and Skidmore anthracite. The amount of slag formed at initiation was less than 10 %. In order to determine the extent of growth at low slag amounts, a sensitivity analysis to the amount of slag was performed using the agglomeration modeling methodology. Particle agglomerate growth was limited by slag-liquid amounts up to 20 wt. % slag after which it plateaued as shown in Figure 1. When the slag is less than 15 wt. %, the diameter does not increase significantly since the slag viscosity is too high and amount too low. As the slag-liquid amount increases further, although the extent of particle wetness increases, the number of solid particles decreases as they melt to form the slag-liquid. So, it is believed that the collision frequency of particles decreases although the wetness of particles increases. However, the amount of slag that can undergo solidification into a larger agglomerate is higher, as explained in regime 3 described below.

Three regimes of agglomerate growth are proposed as follows-
Regime 1: Slag-liquid amount less than 10%- In this regime, the agglomerate growth is limited by the amount of slag-liquid available for sticking. Thus, the probability that the particles undergo a wet collision is low. The rate of growth is slow and the time required for the particles to reach the defluidization size would be very long.

Regime 2: Slag-liquid amount between 10-30 %- In this regime, the agglomerate growth rate is maximum. An adequate number of particles are available in the system and there is enough slag-liquid for particle sticking on collision. The particles in this regime are likely to be completely coated with slag-liquid layers and the probability of wet collisions will be high. The particle growth rate is high and significant defluidization would occur.

Regime 3: Slag-liquid amount greater than 50%- In this regime although a large amount of slag is available for wetting ash particles, since this slag is formed by the melting of solid particles, the number of particles in the system is low. The collision frequency of particles is likely to be low and although there is adequate amount of slag-liquid, the particle-particle interaction is limited. In

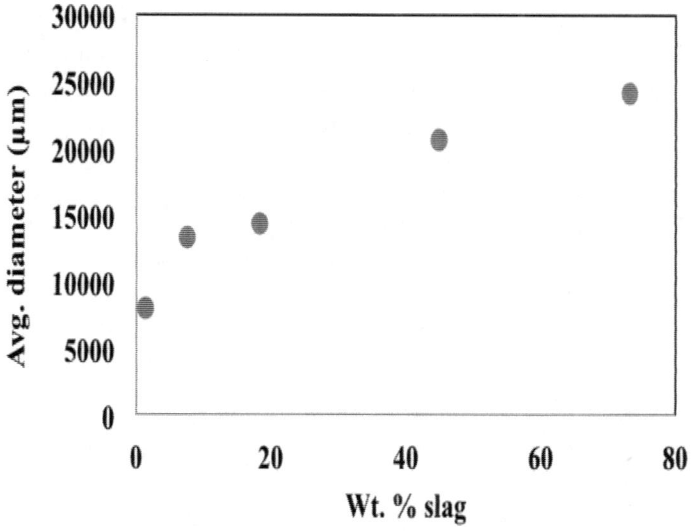

order to demonstrate the effect of the large slag amounts present in regime 3, the average agglomerate diameter was calculated assuming the extreme case that all the slag coalesces and deposits into a large solid agglomerate. Thus, Figure 1 shows an increase in average diameter with increasing slag amount. In the case of bituminous coals and anthracite, at the low operating temperatures in fluidized bed combustors, regime 1 and 2 are more likely to occur in the bed. Additionally, regime 1 and 2 are of interest in understanding the initiation of agglomerate growth and are the focus of the modeling study.

Figure 1: Effect of slag-liquid amount on agglomerate growth with solidification of slag formed

*4.1. **Proposed mechanism of the process of agglomeration in fluidized beds***

The heterogeneous behavior of bed ash helps to explain the occurrence of agglomeration at temperatures lower than operating temperatures in fluidized bed combustors. At low temperatures, certain low-melting components such as those in particle classes with high iron- and calcium-based components studied, can initiate the slag formation in regime 1. The agglomerate growth is likely to propagate in Regime 2.

Experimental studies by Ataköl et al.[7] showed that agglomeration occurs at 125-200 ˚C lower than characteristic IDT and about 300-400 ˚C lower than the flow temperature (FT) in a laboratory-scale fluidized bed combustor. They identified the temperature at which agglomeration was initiated based on the detection of bed temperature fluctuations and pressure drop measurements, as well as pictures of the active bed surface. Since the iron- and calcium-based mineral rich SG4 fraction has been identified as the initiator of agglomeration, in the present study, a comparison of IDT of bulk coal with the slag-liquid formation onset temperature obtained for the SG4 density fraction of Pittsburgh No. 8 coal and Illinois No. 6 coal is shown in Table 2. This comparison shows that the initiation of slag formation occurs at temperatures around 400-600 °C lower than the measured IDT temperature. This indicates that agglomerate growth is likely to be initiated around particles of certain compositions such as of the SG4 particle class, at temperatures significantly lower than AFT predictions that are based on bulk coal properties.

Table 2: IDTs of Bituminous coals from AFT measurements under reducing conditions[8]

	Pittsburgh No. 8	Illinois No.6
Initial deformation temperature (IDT) using whole coal (°C)	1,237	1,201
Particle-level Slag Formation Onset Temperature (°C)	840	650

The proposition that agglomeration is initiated around a few particles and then propagates throughout the fluidized bed reactor is supported by the study of agglomerates presented in a report by the U.S. Department of Energy (DOE)[9]. In this study undertaken by the U.S. DOE, agglomerates formed in thirteen fluidized bed combustion units were observed using SEM, which showed the presence of dark cores with rims. The agglomerates were also found to have a high calcium sulfate and iron oxide content. These observations further support the onset of conditions favorable for agglomeration by sticking of particles with composition similar to the SG4 fraction, which could have a dark iron core at locations of initiation of agglomerate formation with rims of alumino-silicate rich slag. Below 850 °C, the SG4 fraction in all the three coals studied began to form slag under reducing conditions[10]. The amount of slag is less than 10 weight % near the onset temperature. Agglomerate growth is likely to be initiated around these particles but as shown by the modeling results in Figure 2, it is not likely to lead to defluidization at these low slag-liquid amounts. At higher temperatures that can occur in bed, the SG4 fraction can form up to 90 % slag. Once initiated, agglomerate growth is likely to propagate in the bed due to one or more of the following reasons:

1) Temperature excursions due to changes in system hydrodynamics- Once particle size increases, and defluidization begins, degree of mixing will be reduced and temperature fluctuations increase. Local hot spots can form in a poorly mixed bed[11]. An empirical equation to obtain the critical velocity needed to avoid dead zone formation for particles of a certain size in fluidized beds was developed by Wen et al.[12]. As per their work, an increase in dead zone formation occurs with an increase in particle size. Similarly, a decrease in mixing at the onset of defluidization, resulted in temperature fluctuations and differences in temperatures measured by thermocouples at different heights in a fluidized bed combustor, in the study by Scala et al.[13]. A discussion of the effect of hydrodynamics on heat transfer in fluidized beds by Bock et al.[14] showed that heat transfer decreases with an increase in particle size and subsequent changes in bubble motion. Hence, with an increase in particle size with agglomerate growth and also dead zone formation at the onset of defluidization, heat transfer is likely to decrease causing an increase in particle temperatures. Bottom ash particles that are agglomerating are primarily of Geldart type A or B and as they grow in size, they will eventually move into type D. For BFBs, an increase in size of type A and B particles causes a decrease in the heat transfer co-efficient. As the particle size increases from 100 to 1,000 µm, the heat transfer co-efficient of heat transfer to heat exchanger surfaces decreases by about 4.5 times, mainly due to decrease in particle convection that causes decreased renewal of particles from heat transfer surfaces[15]. Additionally, after agglomeration is initiated, as particle size increases, the ratio of the particle velocity to the minimum fluidization velocity of the larger particles can decrease. This will increase the possibility of particle segregation. Hence it is possible that once agglomeration is initiated by particles melting at low temperatures, such as those rich in iron- and calcium-based minerals, the change in bed hydrodynamics would lead to further agglomeration due to decreasing particle velocities and decreasing bed voidage. As agglomeration proceeds, the volume of the expanded bed decreases due to the lesser fluidization of the heavier particles formed. Thus, with decreasing voidage and subsequent increasing collision frequency, agglomeration rate may increase initially. As it progresses, decrease in granular temperature and particle number density and the resulting collision frequency decrease would also decrease agglomeration rate. Thus particles grow quickly once the process of agglomeration is initiated until a stable size is reached as the agglomeration rate begins to decrease.

2) Temperature rise due to deposition on heat-transfer surfaces- One more plausible mechanism of the propagation of agglomeration in the system is due to decrease in heat transfer efficiency after initiation of agglomeration. Once particles of larger sizes form by sticking of mineral matter with low slag-formation temperatures, the particle velocity decreases and they can deposit on to heat transfer surfaces, thereby decreasing the efficiency of heat transfer. The decrease in heat transfer can cause a rise in bed temperature leading to melting of alumino-silicate rich components, which have higher slag-formation temperatures as shown from the analysis of

density fractions. The increase in slag amount would then cause an increased agglomeration rate. This mechanism is prone to occur when there are in bed heat exchanges, which are generally restricted to bubbling fluidized bed systems.

3) Presence of local pockets of reducing conditions in the bed- Besides a local increase in particle temperatures, agglomeration is also likely to occur due to changes in the local gaseous atmospheric conditions. The temperature required for the formation of a certain amount of slag-liquid is lower in a reducing gaseous environment than an oxidizing gaseous environment[4]. Local reducing conditions can exist in a fluidized bed reactor around regions in which char particles are concentrated. The literature shows that mineral matter within a carbonaceous particle experiences a high temperature and a reducing atmosphere[16]. Once a few particles such as those from the SG4 density fraction, rich in iron- and calcium-based minerals become sticky at low temperatures, char particles can stick to their surface. The concentration of char particles within mineral matter and slag can create these local reducing conditions. The included minerals in these char particles are also then subjected to the reducing environment and begin to form slag at lower temperatures.

In order to support the mechanism of agglomerate growth in fluidized bed reactors, microscopic analyses of agglomerate samples was performed as follows-

4.2. SEM-EDX analysis of agglomerates obtained from a laboratory scale FB combustion facility at Canada

Agglomerates that formed within two hours during combustion in a laboratory-scale combustion facility in Canada were used as samples to conduct a post-mortem study to understand the agglomerate growth mechanism. In order to avoid a biased analysis, microscopic imaging was performed in randomly selected regions across the entire sample surface. In this way, the entire sample surface was scanned to identify particles with different morphologies. Also, EDX of the whole sample cross section was done at a low magnification (50x to 200x) to identify regions with higher concentration of a particular type of element. Regions distinct in elemental composition were then magnified for detailed study with microscopic images and elemental maps, so as to account for all different types of particle behaviors. Primarily two distinct morphologies were observed across the sample surface as shown in Figure 2- small particles adhering to form a larger sphere and slag-coated particles with bridges. The latter showed that greater slag amounts were formed and coated the particles. Morphologies showing discrete adhered particles indicated lower slag amounts. The formation of higher amount of slag was believed to have occurred at higher temperatures. Hence, the slag-coated particles were likely to be a later stage (causing propagation at higher temperatures) in the agglomeration process, after initiation by the smaller adhered particles. In order to test this implication of the two distinct morphologies observed, the elemental composition of each of them was studied using EDX. EDX was performed on 8 to 10 spots of the two morphologies identified in Figure 2.

Figure 2: Morphologies of initiators and slag-coated particles observed by SEM of agglomerates

The EDX results are shown in Figure 3. The orange sphere in the figure is mapped to high concentrations of iron- and calcium-containing phases. Previous work using FactSage, TMA and XRD also indicated that density fractions rich in calcium- and iron-based minerals

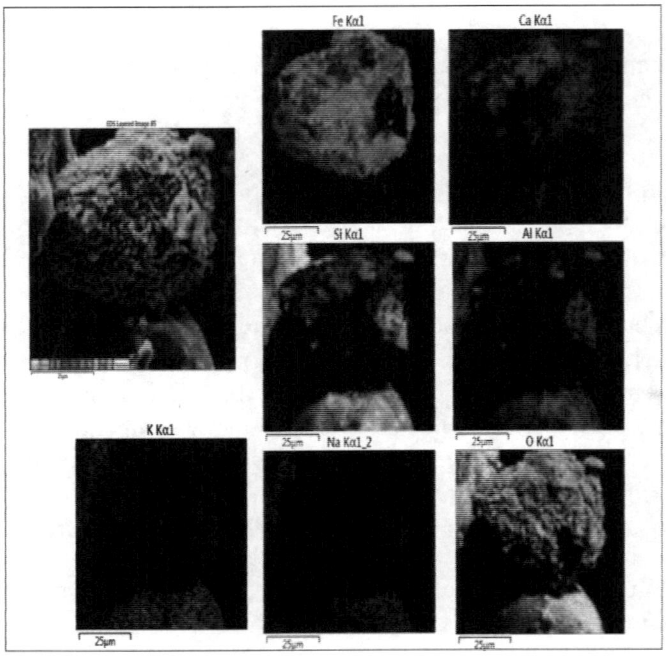

Figure 3: SEM-EDX of initiator and alumino-silicate slag coating

showed slag formation onset temperature less than 850 °C. The amount of slag obtained at initiation for these density fractions was less than 10 %. When the slag amount is less than 15%, the modeling results in Figure 1 showed that the agglomerate growth was restricted and particles did not reach their defluidization size in 10 hours. These classes rich in calcium- and iron-based minerals were thus believed to initiate agglomerate growth at the particle-level, which would then be followed by propagation. Further, molten slag (blue) seen to adhere to these initiators, is rich in silicon and aluminium with some sodium and potassium content as seen in Figure 3. In order to obtain a semi-quantitative estimate of the composition of the initiators and slag-coated particles, EDX was used and the average composition of the studied spots is reported in Table 3.

In order to verify this, FactSage thermodynamic simulations were performed to predict the initiation temperature of slag formation in these regions. It was found that the initiators, which are rich in iron phases, began slag formation at 820 °C as shown in Figure 4, while the alumino-silicate rich slag was formed at a higher temperature of 920 °C. Slag-coated agglomerates began to form bridges with this higher slag amount. The amount of slag formed from the components in the initiators was also less than 10% up to 950 °C. Since the particles do not melt completely, the SEM images show small discrete particles of these initiators adhering to form larger spheres.

Table 3: EDX compositional analysis of stage 1 and stage 4

	Small adhered particles (Initiators/Stage 1)		Slag melting at higher temperature (Slag on slag-coated agglomerates/stage 4)	
Element	Wt.%		Wt.%	
	Mean	Standard deviation	Mean	Standard deviation
O	36.9	8.4	57.5	4.4
Na	0.2	0.2	1.8	0.3
Mg	1.1	0.6	1.5	0.5
Al	1.9	0.5	10.7	1.5
Si	4.4	2.1	21.5	2.7
K	0	0.1	1.8	0.5
Ca	3.1	1.9	1.8	1.4
Mn	2.2	1.1	0	-
Fe	50.3	8.2	3.1	0.8

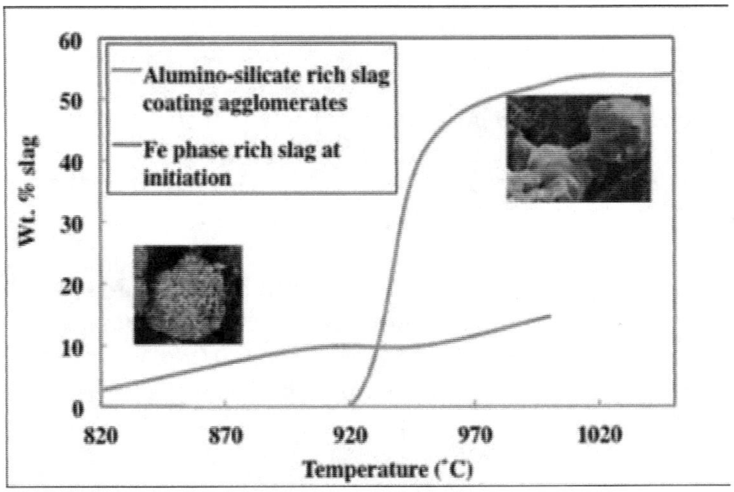

Figure 4: Slag formation tendency of components in stage 1 and stage 4 using FactSage simulations

Through this SEM-EDX study of agglomerates, four different stages of agglomeration were identified as shown in Figure 5. Stage 1 shows small particles rich in iron–based minerals adhered together to form a larger sphere. These particles begin to form slag at a lower temperature of 820 ˚C, as shown in Figure 6. The amount of slag formed from this composition is less than 20 % at temperatures up to 1,000 ˚C. Hence, this is adequate to initate agglomeration and lead to the formation of bigger spheres that are rich in iron, but not sufficient to make the particles grow up to their defluidization size. Stage 2 shows that slag formed at the higher temperatures, begins to coat these iron spheres. Stage 3 shows a completely coated agglomerate from stage 1. Finally, with the formation of up to 40 % slag at higher temperatures of above 920 °C, the coated agglomerates begin to stick to one another as shown in stage 4, wherein bridging is clearly seen between two slag-coated agglomerates.

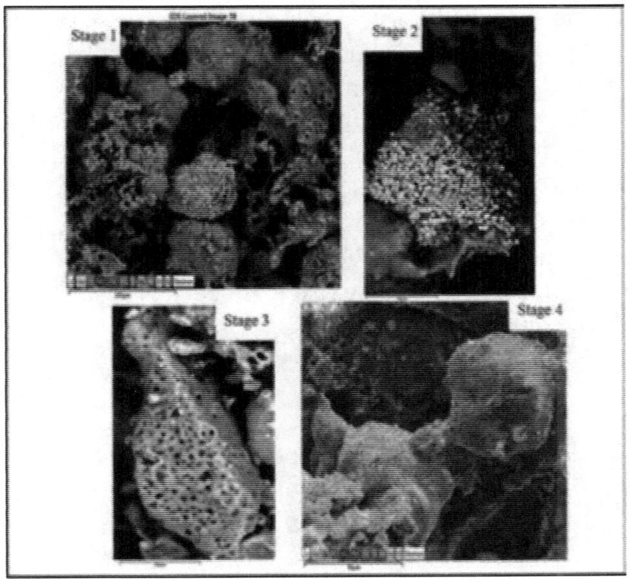

Figure 5: Stages of agglomerate growth observed through SEM-EDX of agglomerates

4.3. SEM-EDX analysis of agglomerates produced in laboratory fluidized bed reactor at PSU

Agglomerates of Pittsburgh No. 8 seam coal rejects (Figure 6) were produced in the laboratory-scale reactor at Penn State. The composition of this material is shown in Table 1.

Figure 6: Agglomerates of Pittsburgh No. 8 coal rejects produced in PSU reactor

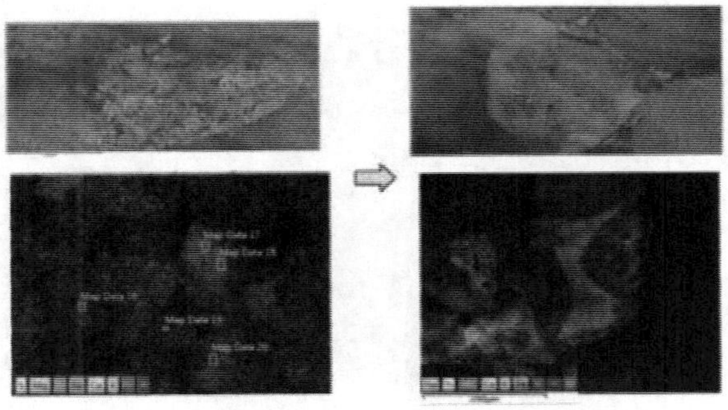

Figure 7: SEM-EDX images of agglomerate of Pittsburgh No. 8 coal rejects

The SEM-EDX analysis of the agglomerates shows two particles rich in siliceous minerals (violet) adhered by a slag bridge (orange) rich in iron and calcium containing minerals. SEM-EDX analysis of these agglomerates also showed regions with smaller, discrete particles adhering together as shown in Figure 7. Across the entire cross section of the sample, 25 regions including those with bridging and with smaller particles adhering were observed. 9 spots in regions with initiators (small, discrete particles adhering) were studied using EDX and found to be rich in calcium-containing compounds with the composition as shown in Table 4. Similarly, 15 spots in the regions that formed slag bridges as shown in Figure 7 were also analyzed for their elemental composition. The composition of the highly molten slag is shown in Table 4.

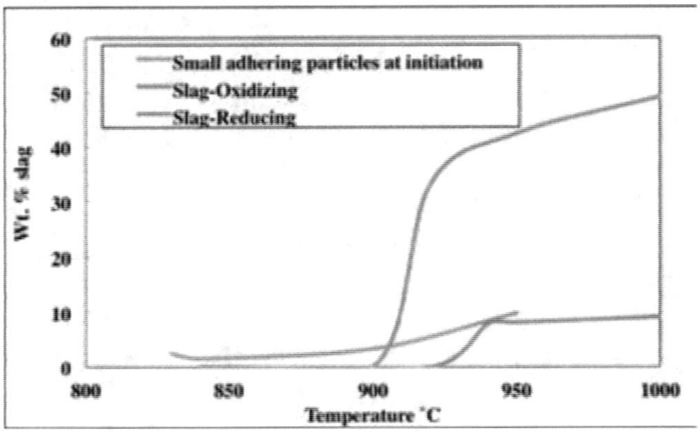

Figure 8: Slag formation tendency of components in small adhering particles and alumino-silicate slag-forming components using FactSage

FactSage simulations under oxidizing conditions were performed with the ash compositions in Table 3 to identify the temperature at which slag formation begins for the initiators rich in calcium-based minerals. The results shown in Figure 8 indicate that the agglomeration was initiated at 830 °C by the calcium-rich particles. The temperature of particles in bed is then likely to rise by one of the plausible mechanisms discussed, causing propagation of agglomeration with the formation of liquid bridges above 930 °C. This onset temperature of slag-formation from the alumino-silicate components that lead to bridging was calculated using FactSage simulations under oxidizing conditions.

Table 4: EDX of initiators and alumino-silicate, high temperature melting slag

Element	Initiator		Alumino-silicate Slag	
Wt.%	Mean	Standard Deviation	Mean	Standard deviation
C	2.1	0.94	1.3	0.39
O	40	0.64	40	9.5
Na	0.3	0.12	0.4	0.21
Mg	0.4	0.02	0.4	0.16
Al	4.4	0.88	9	1.8
Si	6.7	1.11	12.3	2.2
S	0.9	0.19	0.1	0.07
K	0.5	0.41	1.6	0.62
Ca	5.1	0.17	0.7	0.43
Ti	0.3	0.86	0.5	0.08
Fe	40	1.7	34	14

The slag formation onset temperature of the alumino-silicate components that have relatively higher melting temperatures was also computed under a reducing environment. Under this condition, the temperature of onset of slag formation was lowered to 910 °C and the amount of slag was 30-50 %, which was higher than under oxidizing conditions. Hence, in the presence of pockets of reducing gaseous conditions in the reactor, these components could cause propagation of agglomeration. FactSage analysis showed that the amount of slag-liquid formed from the alumino-silicate components was lower in case of Pittsburgh No. 8 coal rejects than the agglomerates obtained from Canada. Hence, the intermediate stage of the formation of a slag-coating on agglomerates, before bridging, which was seen for the agglomerates from Canada, was not seen in the Pittsburgh No. 8 rejects. Partially slag-covered agglomerates formed bridges in case of agglomerates from Pittsburgh No. 8 coal rejects.

5. Conclusions

In this work, a mechanism for agglomerate growth based on initiation at the particle-level followed by propagation in bed is proposed. SEM-EDX study of agglomerates from fluidized bed combustors substantiated this mechanism. It showed that iron-based minerals such as pyrite and also eutectics involving iron- and calcium-based minerals were identified as low-temperature initiators (≤ 850 °C) of agglomeration. The amount of slag formed at initiation is less than 10 %. As shown by the modeling work, this is enough only for sticking together of few smaller particles thereby initiating an increase in particle size. At that stage, the particles do not exceed the defluidization size. The alumino-silicates, which are present in larger amounts in bituminous coals, begin to form slag at higher temperatures (850 °C to 950 °C) and help propagation of the agglomeration process. Similarly, formation of a local reducing environment around char particles entrapped in agglomerates is likely to lead to larger amounts of slag at lower temperatures.

References

1. Tung, S.E., Williams, G.C., Atmospheric fluidized bed combustion—a technical source book, DE-AC21-80MC14536, U.S. Department of Energy, **1987,** 17-89–17-95.
2. Anthony, E.J., Talbot, R.E., Jia, L., Granatstein, D.L., Agglomeration and fouling in three industrial petroleum coke-fired CFBC boilers due to carbonation and sulfation, *Energy and Fuels,* 14, **2000,** 1021–1027.
3. Khadilkar, A. B.; Rozelle, P. L.; Pisupati, S. V., A study on initiation of ash agglomeration in fluidized bed gasification systems, *Fuel,* **2015**, 152, 48–57.
4. Khadilkar, A.B., Rozelle, P., Pisupati, S., Effect of heterogeneity in coal ash chemical composition on the onset of conditions favorable for agglomeration in fluid beds, *Energies* **2015**, *8* (11), 12530–12545.
5. Khadilkar, A.B., Rozelle, P., Pisupati, S., Incorporation of Hydrodynamics and Ash Chemistry into a Single Model for Prediction of Particle Size Increase in Fluidized Beds, *AIChE Annual Meeting,* **2015,** Utah, USA.
6. Khadilkar, A.B., Development of a fluidized bed agglomeration modeling methodology to include particle-level heterogeneities in ash chemistry and granular physics, Dissertation in Energy and Mineral Engineering, The Pennsylvania State University, 2016.
7. Ataküľ H., Hilmioğlu, B., Ekinci, E., The relationship between the tendency of lignites to agglomerate and their fusion characteristics in a fluidized bed combustor, *Fuel Processing Technology*, **2005,** 86, 12–13, 1369–1383.
8. Matuszewski, M, Detailed coal specifications, Final Report- DOE/NETL-401/012111, **2012**.
9. Brown, R.C.; Dawson, R.; Smeek, J. L. Bed material agglomeration during fluidized bed combustion, Final report for U. S. Department of Energy, No. DE-FG22-92PC92530, 1996.
10. Khadilkar, A., Rozelle, P. and Pisupati, S., Comparative study of fluidized bed agglomeration of high rank coal ash with consideration to particle-level heterogeneities, *Proceedings of the International Conference of Coal Science and Technology*, **2015,** Melbourne, Australia.
11. Visser, H. J. M., van Lith, S. C., Kiel, J. H. A., *Journal of Energy Resources Technology* **2008**, *130* (1), 011801-1- 011801-6.

12. Wen., C. Y., Krishnan, R., Kalyanaraman, R., Particle mixing near the grid region of fluidized beds, *Proceedings of the International Fluidization Conference* **1980**, 405.
13. Scala, F., Chirone, R., Characterization and Early Detection of Bed Agglomeration during the Fluidized Bed Combustion of Olive Husk, *Energy & Fuels* **2006**, 20, 120-132.
14. Bock, H.-J., Molerus, O., Influence of hydrodynamics on heat transfer in fluidized beds, *Proceedings of the International Fluidization Conference*, New Hampshire, U. S. A., **1980**, 217.
15. Basu, P., Combustion and Gasification in Fluidized beds, CRC Press, Florida, **2006.**
16. Scott, D. H., Ash behaviour during combustion and gasification, IEA Coal Research Report, **1999.**

453406 Numerical Simulation of the Effects of Size Change in a Coal Gasifier Using Method of Moments Approach

Abstract:

The population balance model (PBM) and the governing computational fluid dynamics (CFD) equations were used to simulate the effect of size change in a coal gasification process using the finite size domain complete set of trial functions method of moments (FCMOM) approach. The goal of this study was to demonstrate the effect of property (e.g. size and density) variation of coal char particles in the simulation, design, and scale-up of gasification processes. In this study, the process of coal char gasification in a fluidized bed at elevated temperature in the presence of hydrogen and steam as gasification media was simulated. The coupling of coal char size variations, hydrodynamics of the system, and heterogeneous gasification reaction rate were included in the simulation. The heterogeneous gasification reaction was modeled based on the available experimental data from Gas Technology Institute (GTI) using the shrinking core model. The simulation results showed that the effect of particle property variations (e.g., size) during a process such as gasification of coal char in a fluidized bed plays a very significant role in the modeling and design of such processes. Initially, as the coal char particles were gasified, the coal became smaller; therefore, the effect of mixing in the particulate phase became more pronounced (as a result of the formation of bubbles). Ultimately, the effects of mixing became less significant due to a reduction in the bed height, which, in turn, suppressed the formation of larger bubbles.

464063 Study of Structured Flow Pattern and Bubble Characteristics of a Pulsed Fluidized Bed Using Kinetic Theory and Turbulence Model

Abstract:

This work presentsÂ simulation results and analysis of the hydrodynamics and flow patterns of a pulsed fluidized bed using Eulerian-Eulerian two fluid modeling approach. The pulsed inflow of gas phase is in a form of sinusoidal function as $u_0/u_m = A + B \sin(2Ï€ft)$ where four pulsating frequencies of 3, 4, 5, 6 HZ are used. The kinetic theory and Johnson and Jackson Frictional Model for particles motions coupled with k-epsilon Turbulence Model for fluid phase, and Gidaspowâ€™s Drag Model for addressing interactions between two phases are used to close the system. The simulation successfully captured the structured and regular bubble patterns reported in other experimental papers. The hydrodynamics, bubble patterns, bubble sizes are obtained and analyzed for each frequency of inlet flow. The effects of inlet velocity, particle size and diameter on the hydrodynamics, flow patterns are also discussed. Discrete Fast Fourier Transform study is carried out on pressure drop, fluid velocities and volume fraction of a series of specific points in the bed to better understand the fluctuation and structure details of the fluid flow. A map of the structured pattern performance of particles with various density and size are marked on the Geldartâ€™s Chart for type B particles. Two Fluid Modeling results improve the understanding of the chaos and structured properties of the conventional and pulsed fluidization.

470285 Powder Flow Characterization at Low Consolidation: 1. Modeling and Experimental Values of Torque Estimation

Abstract:

The flow properties of fine powders at very low consolidation levels are relevant to small scale industrial application of powder flow, such as in small process hoppers and mixers of pharmaceutical powders. In addition, many process applications of powder technology involve gas-solids systems. The knowledge of flow properties of aerated powders is necessary for proper handling.

Characterization of powders and granular materials can be made with a variety of commercially available shear testing devices. However, in fact, sensitivity of these shear testers has been decreased greatly at low levels stress. Furthermore, materials which these testers find to be cohesion-less may have practical differences in flow behavior, especially in regions of low consolidation stress.

New testing devices, which use non-traditional methods (such as centrifugal force) to estimate cohesive forces at low levels of consolidation stress, are not well-established and their correlation to either traditional shear testers or practical problems is incomplete. The equipment used in this research project was the state of the art powder cell purposely developed by Anton Paar to fit on an Anton Paar rheometer and carry out torque measurements on both aerated and fluidized powders. In this instrument, the state of consolidation of the powder can be adjusted by changing the impeller depth and the aeration rate. Four free flowing powders, according to the Geldart classification, Sand, Glass beads, Alumina, and Calcium lactate powders were used in this research project. The theory developed by Bruni *et al.* 2007 and applied by Tomasetta *et al.* 2011 is used to evaluate powder flow properties of the powders. Two different models considering the state of stress, active and passive state of stress, during torque measurement were considered. The results indicated that the instrument allows to successfully exploring low consolidation ranges and well correlates with the expected powder flow properties.

REFERENCES

[1] Bruni, G., Barletta, D., Poletto, M., & Lettieri, P. 2007. A rheological model for the flowability of aerated fine powders. Chemical engineering science, 62(1), 397-407.

[2] Tomasetta, I., Barletta, D., Lettieri, P., & Poletto, M. 2012. The measurement of powder flow properties with a mechanically stirred aerated bed. Chemical Engineering Science, 69 (1), 373-381.

455145 CFD-DEM Simulation of Tube Erosion in a Bubbling Fluidized Bed with a Tube Bundle

Abstract:

CFD-DEM simulation of tube erosion in a bubbling fluidized bed with a tube bundle

Yongzhi Zhao (yzzhao@zju.edu.cn), Lei Xu

Department of Chemical and Biological Engineering, Zhejiang University, Hangzhou 310027, China

The erosion of the immersed tubes in a bubbling fluidized bed is studied numerically using CFD-DEM method and a particle-scale erosion model, SIEM (shear impact energy model), which is proposed based on the shear impact energy of particles in DEM simulations. By analyzing the simulation results, some characteristics of the tube erosion in the fluidized bed are obtained. The distributions of the erosion around the tubes located in different positions of the bed are different. For the tubes in the center of the fluidized bed, the profiles of erosion rate around the tubes have two peaks at the lower surface of the tubes. On the contrary, for the tubes near the sidewall of the fluidized bed, the maximum erosion usually occurs on the upper surface of the tubes. The values of the erosion rates on the tubes located in different positions of the bed are also different. With the increasing of the vertical position, the erosion rate of the tubes, including the maximum erosion rate and the average erosion rate, increases gradually. In the horizontal direction, the erosion rates of the tubes at the center of the fluidized bed are larger than that of the tubes on both sides of them in the same horizontal plane. For all tubes in the fluidized bed, the maximum erosion rates are about 1~3 times to the corresponding average erosion rates and most of the ratios are near to 2. The erosion model and the algorithm are proved to be effective for predicting the erosion of the tubes in the fluidized bed. The accuracy of the approach is good for predicting not only the distribution of the erosion around the tube, but also the absolute value of the erosion rate. The micro-scale behavior of particles around the tubes is also revealed and the linear relationship between the erosion and the shear impact energy is confirmed by the simulation results and experiment (Johansson et al., Chem. Eng. Sci., 2004, 59: 31-40). The agreement between simulation and experiment proves that the micro-scale approach proposed in this paper has high accuracy for predicting erosion of the tubes in the fluidized bed, and has potential to be applied to modeling the process in other chemical equipment facing solid particle erosion.

458377 Advances in Autothermal Reformer Development

Abstract:

Advances in autothermal reformer development

At the Forschungszentrum Jülich, at the Institute of Energy and Climate Research (IEK-3), intensive research and development in the field of reactors for autothermal reforming (ATR) of diesel fuel and kerosene was performed within the last 15 years. It is described in a number of scientific papers. Most reactors have in common that they are equipped with a pressure swirl nozzle for cold fuel injection. They all have a fuel evaporation chamber, in which fuel is vaporized and mixed with steam, as well as an air mixing area, in which air is injected and blended with steam and evaporated fuel. The core component of any of Jülich's autothermal reforming reactors is the catalyst. In most cases, it is a bimetallic RhPt species, which is coated on an Al_2O_3-CeO_2 washcoat, which in turn is deposited on a monolithic cordierite substrate. Steam is generated in an internal device for heat exchange using the waste heat from the product gas flow of the autothermal reforming reaction. Figure 1 illustrates these common characteristics of Jülich's ATRs.

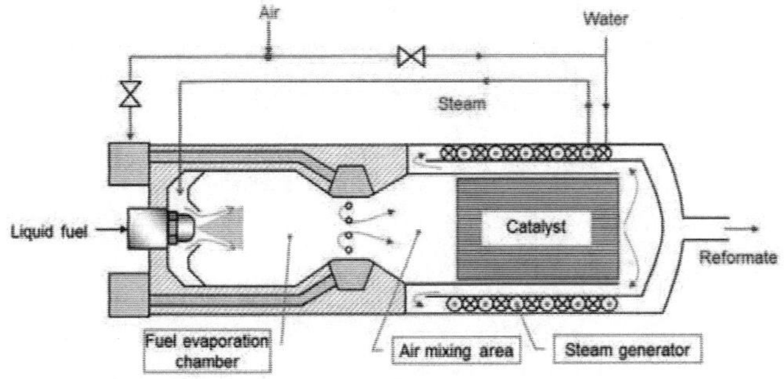

Figure 1 Basic layout of Jülich's reactors for autothermal reforming of diesel fuel and kerosene

This contribution deals with Jülich's recent technical and scientific advance in the field of reactors for autothermal reforming. This reactor generation is denoted as ATR 12.

ATR 12 is a consistent progression of ATR AH2, which was recently published in [1]. It is reported in this paper that the improvements of ATR AH2 in comparison to former ATR reactor generations from Jülich consisted of an additional pressure swirl nozzle for the injection of cold water and a steam generation chamber. As a result, no external process configuration for steam supply was necessary any more. Additionally, the cross-sectional area of the steam generator in ATR AH2, through which also a significant air flow is fed to the reformer, was increased, while in parallel its length was reduced. Thereby, the pressure drop of the steam generator drastically decreased. These modifications in the autothermal reformer design are not decisive for the stand-alone operation of the reformer, but very beneficial from the fuel cell system perspective since they lower the parasitic losses caused by the compressor for air supply and make the system layout simpler. For ATR 12, this way to consider important technical requirements for the design and construction being imposed by the fuel cell system was continued. For ATR 12, the technical execution of the internal steam generator was substantially changed and further improved. The coiled tubing for transferring the waste heat from the product gas flow of autothermal reforming to the stream of saturated steam coming from the catalytic burner of the fuel cell system was substituted by concentric shells, in which turbulence inserts were placed to improve the heat exchange properties of the new construction. This modification aims at an additional decrease in the pressure drop in this part of the reactor and a homogenization of the media streams, when they enter the fuel evaporation chamber. Apart from the above mentioned turbulence inserts, the concentric shells also offer enough space for the incorporation of an electric heating wire. This modification provides the option of fast and autonomous start-up of the autothermal reformer. In this respect, Figure 2 shows the specific design of Jülich's recent reactor type for autothermal reforming ATR 12.

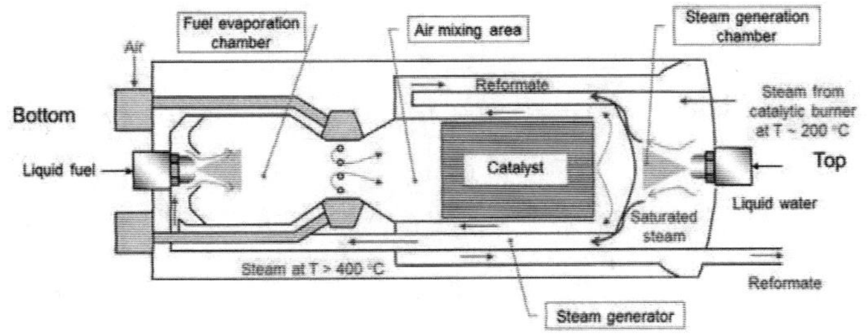

Figure 2 Specific design of Jülich's recent reactor type for autothermal reforming ATR 12

Three different experimental procedures for fast heat up of ATR 12 and additional steady-state experiments are presented in this contribution. Steady-state experiments aimed at validating the concept for heat management of ATR 12. Critical temperatures (internal steam, fuel evaporation chamber, air mixing area, to water-gas shift reactor) and product gas concentrations (H_2, CO, CO_2, CH_4, ethane, ethene, propene and benzene) of ATR 12 are presented as a function of the reformer load and the mass fraction of cold water to the nozzle on the top side of the reformer.

[1] J. Pasel, R.C. Samsun, A. Tschauder, R. Peters, D. Stolten, A novel reactor type for autothermal reforming of diesel fuel and kerosene, Applied Energy 150 (2015) 176-184.

453117 Stable Start-up Process for 1 kW_eÂ Diesel Autothermal Reformer for Auxiliary Power Unit Applications

Abstract:

Nowadays, Auxiliary Power Units (APUs) have a great potential to provide distributed power for remote regions. Also, APU can be used for mobile applications such as heavy duty trucks. Energy density is an important factor for fuels for the APU applications due to its limited space; volume and weight are restricted to ensure mobility. In that manner, diesel can be one of the good candidate fuels for the APU applications due to its high energy density.

However, typical diesel engine generator shows low electric conversion efficiency and makes high pollution. Therefore, a Solid Oxide Fuel Cell (SOFC) based APU is suggested to generate electricity, providing higher efficiency and lower pollutant emission. To utilize SOFC for the APU applications with diesel fuel, high quality hydrogen feed should be provided. With autothermal reforming (ATR) reaction of diesel fuel, hydrogen-rich gas can be produced with catalytic reactors.

During the operation of the diesel ATR, stable start-up process is a critical factor to determine its long-term stability. Initial degradation can severely occur if improper start-up process is applied, and the degradation affects to the durability of the diesel reformer. Also, start-up time must be reduced to minimum to provide fast electricity generation. Therefore, it is needed to develop stable and fast start-up process for the diesel autothermal reformer.

In this study, start-up strategies for diesel autothermal reformer are investigated. With initial heat-up process with electrical heater, partial oxidation mode boosts the start-up process. Temperature distributions within the reformer parts are monitored to manipulate inlet flow conditions. As the result, start-up strategy which enables less than 30 minutes of start-up time is obtained for the 1 kW_e diesel autothermal reformer.

467034 Production of Hydrogen Via Partial Dehydrogenation of Fuels

Abstract:

Production of Hydrogen Via Partial Dehydrogenation of Fuels, to Feed a Fuel Cell

Mélanie Taillades, Julien Belloc, Gilles Taillades, Deborah Jones, Jacques Rozière

Institut Charles Gerhardt UMR 5253, Agrégats, Interfaces et Matériaux pour l'Energie,

Université de Montpellier, Place Eugène Bataillon, 34095 Montpellier Cedex 5, France

One of the objectives of the industries is to provide more efficient and greener transport environment [1]. In this context a new generation of more electrified vehicles must be developed. One of the technologies on this roadmap is a fuel cell secondary power generation unit replacing the current Auxiliary Power Units (APU) [2-4] and proton exchange membrane fuel cells (PEMFC) are the best candidates for the role as an APU replacement. PEMFC operate on pure hydrogen and air between 50 to 100 °C (120 to 200 °C for high temperature PEMFC), and produce electricity and water, with heat and oxygen-depleted air as by-products

The two main issues for the actuation of such an energy system based on hydrogen are its production and storage. Hydrogen production mainly relies on fossil fuels and application to the transport sector would mean a big increase of the demand and hydrogen storage has the drawback of weight, overall storage size and security. This means a big challenge to find a safe and economic way to store it and make it available for applications of interest, in particular for transportation on-board application.

Among the different processes for the hydrogen delivery to fuel cells, a promising one is the generation of hydrogen by catalytic dehydrogenation of liquid hydrocarbons (HCs). This seems to be an appropriate solution as it can produce the hydrogen directly from vehicle fuel, but is only feasible if the process is able to produce sufficient pure hydrogen.

Ours works are based on the concept for hydrogen generation by partial dehydrogenation (PDh) of fuels, allowing the production of dehydrogenated hydrocarbons in liquid phase and hydrogen in gaz phase [5, 6, 7]. Hydrogen has high purity and is CO_2 free, allowing the direct alimentation of fuel cells for supply of electrical energy to auxiliary systems, without a purification step. The liquid phase, composed of partially dehydrogenated hydrocarbons, maintains its original fuel properties with the requisite specifications to be reused as fuel. This process allows the production of hydrogen either on board with the PDh of the vehicle fuel or in a generator connected to the existing fuel storage facilities. The modified fuel can be re-injected into the pool of the vehicle or in fuel storage existing facilities.

This work describes research works on the partial dehydrogenation of gasoline for the production of enough pure hydrogen to feed a fuel cell on board a car. The choice of the catalyst is crucial for the partial dehydrogenation process; it must produce H_2 without compromising the original fuel properties. An ideal catalyst must generate sufficient hydrogen of high purity, be selective to dehydrogenation and avoid cracking reactions responsible for coke deposition and catalyst deactivation. Advanced material, composed by an active phase of 1% Pt-1 % Sn impregnated on a new γ-alumina type support has been developed, characterized and evaluated as a catalyst in the reaction of partial dehydrogenation. The influence of catalyst composition on the activity, selectivity and stability as well as a regeneration mechanism were studied. This material allowed a hydrogen production of 2500 NL.h-1.kgcat^{-1}, with a purity of 85% vol. corresponding to 21.3 NL.L_{fuel}^{-1} which can generate an electrical power supplied by a fuel cell of 2.5 kW. This process allows a hydrogen production on-board or in distribution sites.

[1] ITF, I. T. F. *Reducing transport ghg emissions: opportunities and costs*; 2009

[2] Carter, D.; Wing, J. *The Fuel Cell Industry Review*; 2013

[3] U.S.DOE. *Hydrogen production and delivery: summary of annual merit review of the hydrogen production and delivery program*; 2013

[4] U.S.DOE. *Hydrogen , Fuel Cells & Infrastructure Technologies Program - multiyear research, development and demonstration plan*; 2010.

[5] Resini, C.; Lucarelli, C.; Taillades-Jacquin, M.; Liew, K.-E.; Gabellini, I.; Albonetti, S.; Wails, D.; Roziere, J.; Vaccari, A.; Jones, D. *Int. J. Hydrogen Energy* **2011**, *36*, 5972–5982

[6] Gianotti, E.; Reyes-Carmona, Á.; Taillades-Jacquin, M.; Taillades, G.; Rozière, J.; Jones, D. J. *Appl. Catal. B Environ.* **2014**, *160-161*, 574–581.

[7] Gianotti, E.; Taillades-Jacquin,M ; Reyes-Carmona, A; Taillades, G.; Rozière, J.; Jones, D. *J. Appl. Catal. B Environ. 2016, 185, 233-241*

459522 Pure Hydrogen Production Via Methane Catalytic Decomposition on a Ni-Based Catalyst

Abstract:

The CO concentration in the hydrogen stream produced from methane catalytic decomposition (MCD) reaction on Ni-based catalyst has been an interesting topic. In this work, the active and stable Ni based catalysts derived from hydrotalcite-like precursor prepared by coprecipitation method were employed for the MCD reaction. To avoid oxygen impurities in the feed gas, a high purity undiluted methane and two oxygen traps (purchased from Grace Davison Discovery Sciences) were used in the reaction. The effect of the reduction temperature, reaction temperature and metal loading amount on the CO content in the exhaust from MCD reaction were studied. The results show that the reduction temperature has almost no effect on the activities of Ni-Cu-Al catalysts, but the CO formation concentration decreases with the increase of the reduction temperature. The CO concentration and methane conversion of Ni-Cu-Al catalyst both increase with the increase of reaction temperature. The results of metal loading effect on CO formation show that the CO concentration decreases with the increase of Ni loading. Besides, it shows that the oxygen all comes from the support and the incomplete reduction of metal oxide via the long term reaction test performed on 15Ni-3Cu-2Al catalyst. The CO concentration over the 15Ni-3Cu-2Al catalyst at the optimized condition (reduced at 700 ^{o}C and react at 600 ^{o}C) is below the tolerance concentration for PEMFCs (10 ppm) after the first hour of the reaction. This work proves the possibility of supply the hydrogen produced from MCD reaction directly to PEMFC.

451924 Production of Hydrogen from Various Feedstock Options Using Plasma Reforming

Abstract:

Ceramatec has developed a method for reformation of various hydrocarbons using a non-thermal plasma system to catalyze the reformation process. The plasma provides the same type of catalytic activity as a solid catalyst but since the plasma is constantly renewed it is not sensitive to sulfur or various other contaminants that adversely affect traditional solid catalysts. The reformer can operate in a partial oxidation or steam reforming mode but these modes are usually combined in an autothermal process. Any hydrocarbon that can be vaporized can be reformed with the system. The reformer can operate with air, oxygen enriched air, or pure oxygen as an oxidant. The reformer has been deployed in the field in sizes as large as 1 MMSCF per day of natural gas. The paper will discuss the technology associated with plasma reforming, the results using various feedstock options (natural gas, high sulfur diesel fuels, condensed solvents, and residuals from biomass gasification), and current programs for further development.

471282 Computational Fluid Dynamic Modeling of a Microplasma Fuel Reformer

Abstract:

Batteries are present in everyday life as the primary means to provide portable power to most of our electrical devices. Due to the low energy densities of batteries, portable fuel cell technologies have been heavily considered as a potential replacement. Fuel cells have the beneficial advantages of higher overall efficiencies, no moving parts, and low temperature operation, but exhibit the limitation of not having a readily available stream of hydrogen. Microplasma reactors have been successfully designed, fabricated, and experimented as a novel fuel reformer. However experimental measurements from the devices presented low energy efficiencies making the dynamics of the fluid transport in the novel devices an important aspect of further investigation. To learn more about the reactive microplasma environment a computational fluid dynamic (CFD) model of the microplasma reactor was created. Additionally this model will ultimately be used to inform the ideal system conditions and improve microreactor designs in future generations of the experimental devices.

The low efficiencies are likely due to the large bulk volume allowing for a large portion of the feed to pass around the microplasma. The volume of the microreactor holder is 375 μL, where the reactive microchannel is at most 0.5 μL. The CFD model was designed in COMSOL Multiphysics version 5.2. The object of study is the channel and the bulk volume above the reactor. In the model only the channel volume is reactive. The model captures both momentum and mass transport. As reactants are fed into the holder in a gas phase they pass over and through the microchannel, the material in the microchannel is reacted and then products diffuse into the bulk volume, while more reactant diffuses into the channel. Based on the inconsistency of the microplasma size throughout experiments, three model variations were developed for each experiment that accounts for the 100% of the channel being reactive as well as 5% and 50%. Each model simulates experimented results to determine accuracy of the model. Due to reactor variation the model are also parameterizable with respect to width, depth, oxide thickness, inlet flow rate, reaction rate and electric current.

To date all models have been tested with a carbon dioxide decomposition reaction; forming oxygen and carbon monoxide. The carbon dioxide reaction study was performed to analyze the microreactors, where carbon dioxide was chosen due to its ease in developing a plasma and simplicity in products. Currently the CFD model has presented reaction rates from 125 to 18,153 mol/m^3-s (2.5E-8 to 9E-8 mol/s). This work aim is to not only determine unique reaction characteristics of the microplasma environment, but also to determine key parameters to study optimization of device performance. We plan to present the results of the model, and the ideal parameters to obtain the highest energy efficiency of the microplasma fuel reformer.

462329 Syngas Production Using Steam/Red Mud Gasification of Coal

Abstract:

Coal is one of the major sources of fossil fuels used for power generation in the USA. Being one of the most polluting fuels, its use significantly contributes to environmental pollution and greenhouse gas emissions. Coal gasification is an alternative to mitigate pollutant emissions. Syngas produced by coal gasification can be used for electricity generation, chemical feedstocks, or production of synthetic natural gas. Gasification of coal using steam and catalysts provides more flexibility in the control of the hydrogen and carbon monoxide ratio, which is critical in converting syngas intro transportation fuels such as gasoline and diesel using the Fischer-Tropsch process. Herein, in this work, sub-bituminous coal was gasified at atmospheric pressure in a bubbling fluidized bed reactor at 700, 800, and 900 °C on red mud (RM) under steam. During coal gasification on RM under steam the classic water-gas shift reaction (WGSR) occurred. This reaction, which consists of reaction of carbon monoxide and water vapor to form carbon dioxide and hydrogen was promoted by the RM. Thus, the syngas was significantly enriched with hydrogen. The RM also promoted the cracking of the tars at relatively low temperatures compared to those reported in literature. Gasification of coal using RM and carbon dioxide also increased the carbon monoxide content and increased the hydrogen content through the WGSR. Addition of biomass to the coal improved the hydrogen content of the syngas because of WGSR which converted the large fraction of CO produced by the biomass to hydrogen. Additionally, the use of red mud in the gasification process prevents the formation of clinker in fluidized bed reactor.

452426 Hydrogen Production from Lignocellulosic Biomass Residues Via Gasification in Supercritical Water Catalyst Activity and Process Optimization Study

Abstract:

Abstract

Hydrogen production from biomass is an attractive alternative for clean energy generation. Supercritical water gasification (SCWG) is a promising technology for effectively conversion of wet biomass into a pressurized, clean and hydrogen rich gas [1]. Catalysis is crucial for achieving promising conversion of biomass at lower temperature, which could bringing down the cost of this process [2].

In this study, firstly, characterization and evaluation for two groups of catalysts for hydrogen production via SCWG process are carried out. For Ni-Co/Mg-Al catalysts, it was found that higher hydrogen yield could be correlated to strong acidic sites in the range of 400 -600 °C. For catalyst preparation co-precipitation is better than impregnation in terms of hydrogen yield. Also, for Ni/TiO_2 catalysts, 5 wt% Ni loading is optimum for hydrogen production, however, adding promoters such as Ru, Co, Ce, Mg or change in the active metal to Ru at 5 wt% loading did not improve the hydrogen yield.

By catalysts screening, K_2CO_3 and 20Ni-0.36Ce/Al_2O_3 were identified as the best catalysts. Then, an optimization study relating the effects of different parameters based on Taguchi experimental design was conducted, and biomass including wheat straw, canola meal, and timothy grass were used as feedstock. The order of relative importance for process parameters for hydrogen production is: temperature catalyst loading catalyst type biomass type. High temperature (~650 °C), and high catalyst loading (~100%) are favorable for hydrogen production. Using different

waste biomass, the average hydrogen yield was in the order of: canola meal > wheat straw > timothy grass.

Key words

Supercritical water (SCW), hydrogen, biomass, catalysis

References

[1] R.M. Navarro, M.C. Sanchez-Sanchez, M.C. Alvarez-Galvan, F.d. Valle, J.L.G. Fierro. Hydrogen production from renewable sources: biomass and photocatalytic opportunities. Energy & Environmental Science. 2 (2009) 35-54.

[2] Y. Guo, S.Z. Wang, D.H. Xu, Y.M. Gong, H.H. Ma, X.Y. Tang. Review of catalytic supercritical water gasification for hydrogen production from biomass. Renewable and Sustainable Energy Reviews. 14 (2010) 334-43.

454502 *Featured:* Organic Aqueous Flow Batteries for Massive Electrical Energy Storage

Abstract:

The ability to store large amounts of electrical energy is of increasing importance with the growing fraction of electricity generation from intermittent renewable sources such as wind and solar. Flow batteries show promise because the designer can independently scale the power (electrode area) and energy (arbitrarily large storage volume) components of the system by maintaining all electro-active species in fluids. Wide-scale utilization of flow batteries is limited by the abundance and cost of these materials. We have developed an approach to electricity storage using the aqueous redox chemistry of small, inexpensive organic molecules such as quinones [1,2]. The absence of active metal components in both redox chemistry and catalysis represents a significant shift away from the direction modern battery R&D has been taking. This new approach may enable massive electrical energy storage at greatly reduced cost.

[1] B. Huskinson, M.P. Marshak, C. Suh, S. Er, M.R. Gerhardt, C.J. Galvin, X. Chen, A. Aspuru-Guzik, R.G. Gordon and M.J. Aziz, â€œA metal-free organic-inorganic aqueous flow batteryâ€, *Nature* **505**, 195 (2014).

[2] K. Lin, Q. Chen, M.R. Gerhardt, L. Tong, S.B. Kim, L. Eisenach, A.W. Valle, D. Hardee, R.G. Gordon, M.J. Aziz and M.P. Marshak, "Alkaline Quinone Flow Battery", *Science* **349**, 1529 (2015).

458957 *Featured:* Hydrogen-Based Redox FLOW Batteries

Abstract:

Demand for large-scale energy storage is increasing, especially as renewable energy resources, such as solar and wind, become more prevalent. These variable energy generators require that large-scale energy storage is used to arbitrage and minimize fluctuations into the grid. As a promising candidate for energy storage and load leveling, redox flow batteries (RFBs) have been considered. However, due to the challenging issues such as low cell performance, durability, and high electrolyte cost, their wide-spread adoption has not occurred.

An electrochemical RFB system capable of meeting these demands is the hydrogen-bromine system. It has highly reversible and kinetically favored electrochemical reactions, soluble chemical species, high current-density operation, and a relatively inexpensive electrolyte. We have shown that the optimization and understanding of the primary losses can result in power densities of 1.5 W/cm^2 and limiting current densities over 4 A/cm^2 for discharge at ambient conditions. However, a key attribute of any RFB system is robust and durable performance, especially upon cycling. In this talk, we will report on the cyclic performance of the RFB, and especially the effect of operating conditions such as electrolyte concentration, cut-off potential, and current on the cycling performance.

In addition, other hydrogen chemistries will be examined including inexpensive (H_2-Fe-ion) and high-voltage (H_2-Ce) systems. The various performances and tradeoffs of these systems will be detailed. Finally, the various results will be put into context using experiments and simulations to elucidate the most critical issues in the field and outline areas of needed research.

460720 Achievable Alcohol Concentrations and Membrane Requirements for Artificial Photosynthetic System

Abstract:

Artificial photosynthesis of liquid fuels is a potential source for clean energy. Alcohols are particularly attractive products because of their high energy density and market value per amount of energy input. The major challenges in photo/electrochemical synthesis of alcohols from sunlight, water and CO_2 are low product selectivity, high membrane fuel-crossover losses, and high cost of product separation from the electrolyte. The maximum achievable concentrations of various alcohols in the catholyte are of the order of 1 mol % at partial current densities of 10 mA cm^{-2}. However such concentrations can only be achieved about 100 h of batch operation and result in in 100% fuel-crossover if a Nafion 117 membrane is used to separate the catholyte and anolyte compartments. Reducing the alcohol permeability of the ion-exchange membrane by an order of magnitude only increases the liquid fuel concentration to 10 mol%, but also raises the batch operation time needed to achieve this concentration to 1000 h. To overcome these limitations, we propose an artificial photosynthesis scheme for direct synthesis and separation to almost pure ethanol with minimum product crossover using saturated salt electrolytes. The ethanol produced in the saturated salt electrolytes can be readily phase separated into a microemulsion, which can be collected as pure products in a liquid-liquid extractor. A novel design of an integrated artificial photosynthetic system is proposed that continuously produces >90 wt% pure ethanol using a polycrystalline copper cathode at a current density of 0.85 mA cm^{-2}. The annual production rate of > 90 wt% ethanol using such a photosynthesis system operating at 10 mA cm^{-2} (12% solar-to-fuel (STF) efficiency) can be 15.27 million gallons per year per square kilometer, which corresponds to 7% of the industrial ethanol production capacity of California.

454317 Integrated Solar Energy Conversion and Storage Using Solar Flow Batteries

Abstract:

Photovoltaics are rapidly approaching cost parity with conventional fossil fuel sources of electricity, but they are hampered by the intermittent availability of sunlight. Electrochemical devices like batteries and fuel cells can facilitate large scale solar energy storage, but the capital intensity and long-term costs of current technologies remain prohibitive in most markets. We have made considerable progress toward realization of efficient solar conversion, storage, and delivery as electricity using an integrated photo-electrochemical device we call the Solar Flow Battery (SFB). We believe that a mature SFB technology will provide considerable cost advantages relative to discrete solar+battery systems due to the intensification of the overall energy conversion and storage process into a single device. In this presentation, recent results on SFB device modeling and realization of laboratory-scale prototypes will be discussed.

466161 Dynamic Energy Supply By a Pilot Scale Liquid Organic Hydrogen Carrier Unit

Abstract:

Liquid Organic Hydrogen Carriers (LOHCs) can store hydrogen in a chemically bound form at ambient conditions. If there is a need for energy, a dehydrogenation reactor is used to release hydrogen at temperatures at around 580 K. To function as a storage for electrical energy, the chemical reactor has to be supplemented by further equipment. In terms of efficiency heat recovery is crucial, because of the endothermal nature of the reaction. Furthermore, depending on the reconversion technology, hydrogen has to fulfil certain purity demands. Using a PEM fuel cell as a hydrogen reconversion system, its dynamics are limited by an insufficient supply with hydrogen in times of increased electric demand. There are two major process parameters that can be used to increase the hydrogen production rate in the dehydrogenation reactor.

Volume flow of the reactants: increasing the pump flow rate is a fast and dynamic possibility, but results in lower LOHC discharge levels and efficiency due to reduced residence time.

Temperature: increased temperature enhances the hydrogen release, as the reaction responds quite fast to temperature changes. However, dynamics in practice suffer from the thermal inertia of the system.

If hydrogen release and electrical power production do not match, the hydrogen pressure in the LOHC system is changing. Since there is a free gas volume between the fuel cell and the chemical reactor, the hydrogen gas phase can serve as an interim energy buffer. This free volume is constituted not only by the tubing, but also by the heat exchangers and hydrogen purification units. Thus, the LOHC system is able to cover dynamic electrical energy demand, which would exceed the dynamic capability of the dehydrogenation reactor.

To characterize the dynamics of the whole LOHC process, a pilot scale reaction system based on dibenzyltoluene was combined with a heat integration, a hydrogen purification system, a PEM fuel cell and different electricity consumers. In this contribution the dynamic performance of such a system is discussed.

474228 Fabrication of $CoTiO_3$/g-C_3N_4 Hybrid Photocatalysts with Enhanced H_2 Evolution

Abstract:

Hydrogen (H_2) has gained much attention as a sustainable and carbon-neutral fuel carrier. Since Fujishima and Honda reported H_2 evolution from water splitting by using TiO_2 electrode under light irradiation in 1972, photocatalytic water splitting has become one of the most fascinating and promising technologies. Among numerous photocatalysts, graphitic carbon nitride (g-C_3N_4) has been rapidly developed due to its non-toxic, proper energy band level for H_2 production (2.7 eV), ability of absorbing visible light, good physicochemical stability and facile fabrication via one-step polymerization. Unfortunately, the pure g-C_3N_4 seriously suffers from poor photocatalytic efficiency because of a high recombination rate of photo-generated charge carriers. To settle this problem, g-C_3N_4-based heterojunctions were prepared. However, the redox ability of photoexcited electrons and holes would be declined as the typical heterojunction forms.

This work reports a novel direct Z-scheme $CoTiO_3$/g-C_3N_4 (CT-U) photocatalytic system with different weight percentage of $CoTiO_3$ using a facile in-situ growth method for H_2 evolution from water splitting. The as-prepared CT-U composites composed of 1D $CoTiO_3$ microrod and 2D g-C_3N_4 nanosheet were characterized by various techniques including XRD, SEM, TEM, XPS, FTIR and UV-vis. Results demonstrate that the CT-U composite photocatalysts were successfully fabricated, with intimate interfacial contact and heterojunction interaction between g-C_3N_4 and $CoTiO_3$ which can significantly boost the photocatalytic activity compared with prinstine g-C_3N_4 and $CoTiO_3$. The most enhanced H_2 evolution rate of 858 Î¼molÂ·h^{-1}Â·g^{-1} and high quantum efficiency (38.4% at 365 nm, 3.23% at 420 Â± 20 nm) are achieved at an optimal 0.15% CT-U. Meanwhile, the 0.15% CT-U sample exhibits good

photocatalytic stability in recycling H_2 evolution. Accordingly, direct Z-scheme mechanism capable of leading efficient charge carrier separation and strong reduction ability for enhanced H_2 production was proposed, and further evidenced by PL, Photoelectrochemical analysis and ESR assay.

464431 Combining Thermochemical with Sensible Heat Storage for Increasing the Stored Energy Density and Providing Stable Heat Transfer Fluid Outflow Temperature

Abstract:

Concentrated solar power (CSP) plants with integrated thermal energy storage (TES) are considered to be a promising option for the cost-effective electricity generation and dispatchability because they allow the temporal decoupling of electricity generation and solar insolation. TES systems can be classified as sensible heat, latent heat, or thermochemical storage (TCS) systems [1]. In a TCS, the heat is looped via reversible chemical reactions: it is stored as the enthalpy of the endothermic reaction and recovered as sensible heat through the exothermic recombination of the reactants.

A competitive advantage of TCS compared to sensible or latent TES is the ability to control the charging and discharging temperatures through an adjustment of the equilibrium temperature of the reaction [2]. This requires a physical separation of the chemical reactants from the heat transfer fluid (HTF), e.g., by placing the participants of a gas-solid reaction in metal tubes. By connecting the tubes to an external gas compression unit, the partial pressure of the gas reactant can be continuously adjusted to maintain the reaction at the desired rate and temperature and thereby heat the HTF passing the tubes to the desired temperature.

The ability to control the temperature of the HTF passing a TCS section is expected to provide several advantages compared to current state-of-the-art sensible and latent heat storage systems:

- By heating the HTF to a constant outflow temperature during discharging, the power block can be operated at its design point and therefore at a high thermal-to-electricity efficiency.
- By cooling the HTF to a constant outflow temperature during charging that is equal to the temperature of the HTF leaving the power block, these two HTF streams can be merged and recirculated to the solar field without creating exergy losses due to mixing.
- By distributing multiple TCS sections along the TES axis, the axial temperature gradient can be increased, leading to increased energy storage densities.

To reduce the overall material costs of the TES, the TCS sections may be combined with sections comprised of packed low-cost sensible-heat storage (SHS) materials such as rocks or steel slag. If the HTF and the storage and encapsulation materials in the TCS and SHS sections are selected accordingly, the operating temperature can exceed the limit of ~600°C that is typically encountered in state-of-the-art two-tank TES systems.

Whereas the adjustment of the gas reactant pressure in a TCS offers some distinct advantages compared to other TES concepts, it introduces an additional energy penalty. Future work will focus on (1) reducing the losses due to adjusting the gas-reactant pressure, (2) identifying suitable TCS materials, and (3) optimizing the combined storage performance through a variation of the TCS materials, storage geometry and/or operating conditions.

References

[1] S. Kuravi, J. Trahan, D. Y. Goswami, M. M. Rahman, and E. K. Stefanakos, "Thermal energy storage technologies and systems for concentrating solar power plants," *Prog. Energy Combust. Sci.*, vol. 39, no. 4, pp. 285–319, Aug. 2013.

[2] Y. Q. Yu, P. Zhang, J. Y. Wu, and R. Z. Wang, "Energy upgrading by solid-gas reaction heat transformer: A critical review," *Renewable Sustainable Energy Rev.*, vol. 12, no. 5, pp. 1302–1324, 2008.

476849 Next-Generation High-Efficiency Hybrid Solar to Hydrogen Conversion with Integrated Storage

Abstract:

By 2050 solar technologies are projected to deliver the majority of the world's electricity. Solar energy can be used to generate both heat and electrical power. Most solar panels are designed for only one of these purposes, so an electrical photovoltaic (PV) panel is typically less than 20% efficient.

It is well known that PV cells experience a deterioration in performance (efficiency) when they are operated at high temperatures, and that this leads to particularly high losses when the solar resource is at its highest. For example, a drop in PV cell efficiency in the range 10-20% can be expected, depending on the PV technology, when operating PV cells at temperatures of ~65 °C (or, 40 °C higher than room temperature), which is easily experienced in hot climates. Hybrid PV-thermal (PV/T) solar collector technology combines PV modules with the contacting flow of a cooling fluid (gas or liquid) in a number of configurations and geometries, and offers some advantages when space is at a premium and there is demand for both heat and power. Moreover, a major motivation in PV/T technology is to cool the PV cells and therefore to increase their electrical efficiency, while delivering a potentially useful thermal output (hot fluid stream). PV/T collectors are a highly efficient technology, capable of achieving system efficiencies (electrical plus thermal) in excess of 70%.

By far the most common use of the thermal-energy output from PV/T systems (in fact most solar-thermal collector technologies) is to provide hot water at 50-60 °C for households or commercial use, however, a wide range of opportunities arise at higher temperatures whence additional power-generation cycles (e.g., with organic Rankine cycles, thermoelectric generators, amongst other) or thermal-cooling technologies (e.g., with desiccant, ad/absorption refrigeration cycles, amongst other) can be combined with PV/T modules. These additional options become viable at temperatures typically above ~80 °C, and importantly, become increasingly efficient at progressively higher temperatures.

Operating PV/T panels efficiently at high temperatures for this purpose is a significant scientific and engineering challenge, since the two modes of heat-transfer loss that result directly in the drop in performance, namely convection and radiation, are both exacerbated at higher temperatures. The reduction of convective losses can be addressed by evacuation of the panel, while radiation losses can be decreased by using selective coatings. This talk will discuss recent advances in evacuated PV/T technology with optimized surfaces and selective coatings combined with a secondary power generation sub-system based on organic Rankine cycles (ORCs) for high-efficiency conversion of solar energy to electrical power and consequently hydrogen via electrolysis. Projections at this stage suggest an improvement of the order of 20-30% relative to PV-only systems, for the same area.

In this solar-to-electricity-to-hydrogen system, energy storage options include thermal storage, electrical storage and hydrogen storage. Of these, thermal storage is relatively low cost and can be performed over a range of scales with sensible heat storage (e.g., hot water tanks), phase-change materials (PCMs) covering a range of organic or inorganic substances (e.g., molten salt tanks), or via thermo-chemical energy storage which utilizes the energy alternately absorbed and released in reversible chemical reactions and has the highest theoretical storage capacity of these solutions. Electrical-energy storage on the other hand is highly dependent on scale. At small scales of application a variety of battery technologies exist or are currently under development, which are highly suitable solutions. At large-scales, on the other hand, the costs and potential environmental impact of the widespread use of batteries (arising from the use of hazardous or scarce substances) allow serious competition from mechanical or thermo-mechanical systems, which become the technology of choice. Relevant conventional electricity-storage solutions include pumped hydro (PHS) and compressed air energy storage (CAES).

A significant effort has been placed in recent years on the development of additional electricity storage technologies, driven by the increasing penetration into the grid of renewable energy technologies for electricity generation, many of which have an unpredictably intermittent nature. Of increasing interest are the so-called cryogenic 'liquid-air' (or, liquid-nitrogen) energy storage (LAES) system and a novel storage technology known as Pumped Thermal Electricity Storage (PTES). LAES uses excess electricity to operate a refrigeration cycle with which to liquefy air, which is then stored in a tank at atmospheric pressure. When the electricity is needed, the liquid air is evaporated and used to drive a turbine. PTES, on the other hand, uses excess electricity to operate a heat pump cycle with which to convert the electrical energy into thermal energy. The thermal energy, both hot and cold relative to atmosphere, is then stored in two large reservoirs (hot and cold). The reservoirs contain a material with a high heat capacity, and are able to store the energy much more compactly than PHS. When required, the thermal energy can be converted back into electricity by running the heat pump in reverse, as a heat engine. The projected round-trip efficiencies of both LAES and PTES are lower than PHS, but these systems have a number of potential benefits, including low capital cost, no geographical constraints (i.e., elevations, caverns, etc.) in terms of deployment, and no associated environmental impact. This talk will present recent advances in the development of these technologies, compare the various options with each other and with alternative thermal and hydrogen storage in the context of solar-energy conversion and storage, and finally discuss the next-steps that would enable their successful deployment. In isolation, thermal-energy storage emerges as the technological option with the lowest cost and highest technical maturity, whereas at larger scales electricity storage can benefit as part of a wider storage, transmission and distribution network that includes additional intermittent generation from renewables.

481381 Understanding and Eliminating Interaction of Cells in an Electrochemical Stack

Abstract:

Typically, both the fuel cells and electrolysis cells are stacked in series by means of impermeable graphite or metallic bipolar plates. The open-circuit voltage of most types of fuel cells is approximately 1 V. To support sufficient voltage and power required for commercial applications, cells are assembled in a stack. By connecting multiple cells in series, the voltage of a stack is multiplied, generating the required high power output. Similarly, electrolysis cells are stacked to uptake the high voltages and power, and reach a demanding hydrogen production; the hydrogen generation-rate of a single cell is limited to electrodesâ€™ active area.

In an ideal stack, the design and operating conditions for all cells should be identical throughout the stack. Overall stack performance would then be the summation of the equal stand-alone performances of cells. However, in a real stack, the cells are not perfectly identical and they might differ either in design or operating conditions or both for a variety of reasons: imperfections during manufacturing, a non-uniform inlet flow distribution, degradation of one or more cells and external boundary conditions, such as heat transfer with the surroundings. Though such non-uniformities are encountered at a single cell level, the performance differences are translated, very often not to a single cell only, but to a group of cells located in neighborhood. In other words, one poorly operating or perturbed cell could affect the performance of other cells in the stack. Such a coupling of the cells is the focus of investigation here as it could be detrimental for the performance and lifetime of the stack.

As described above, the individual cells in a stack are separated by impermeable bipolar plates. Thus, we can classify the transport phenomena in their stacks according to the region of influence. Mass, momentum and species transport are confined to individual cells, whereas charge and heat transfer take place globally across the cells such that the cells become coupled and influence each other. Since the cells are coupled via heat and charge transport, abnormal behavior

of one cell can propagate to the neighboring cells. Two adjacent cells can be said to be (1) electrically decoupled when their local current density distributions do not influence and perturb each other, and (2) thermally decoupled if there is negligible heat transfer between the two cells. We analyze charge and heat transport across the cells in a stack for the above two decoupling conditions. In short, electrical decoupling of the cells in a stack is found to depend on the geometrical and material properties of the bipolar plate and the operating parameters. Similarly, the thermal decoupling is governed by the design and operating parameters of coolant flow fields. We finally identify the design guidelines that ensure decoupling of the cells via both the transport modes. The guidelines could be of use for stack designers when selecting the bipolar plate material and dimensions and coolant flow field conditions, so as to minimize the spread of disturbances in a cell to neighboring cells.

Acknowledgements: The research is supported by the National Research Foundation, Singapore under its Competitive Research Program (Award No. NRF-CRP8-2011-04).

454078 Advancement of Hydrogen Storage Technologies

Abstract:

Hydrogen is seen as one of the major storage options for renewable energy. However, its storage still faces a number of challenges. Safe, dense and efficient hydrogen storage has to overcome a number of thermodynamic constraints. A big issue, for example, is the compromise between simple hydrogen release with a small energy demand on the one side and loss-free storage at ambient conditions with a high density on the other side. The interplay between enthalpy and entropy change during hydrogen uptake plays an important role in this respect.

In this contribution different steps in the development of hydrogen storage technologies, such as Liquid Organic Hydrogen Carrier (LOHC), will be presented. First, a study on the identification of potentials for the improvement of hydrogen storage technologies is conducted. This outlines the impact of different measures and gives insight into the question what should be improved.

Based on these findings, a systematic development of carrier materials and technologies will be demonstrated. A comparison of different hydrogen storage options is discussed.

The role of process simulation in this context will be outlined. In the last step a special emphasis will be put on the issue of reliability and uncertainty of the respective process models.

458291 Thermodynamic and Economic Evaluation of Hydrogen Transport Technologies

Abstract:

Due to its high gravimetric energy density, hydrogen is considered as an interesting energy carrying substance for a future energy system based on renewable resources. In this context, energy distribution and thereby efficient, cheap and secure transport of hydrogen is an important aspect.

Today, hydrogen transport is almost exclusively realized by compressed or liquefied gas. Transportation as compressed gas affords high pressures, generating a certain security risk while the amount of transported hydrogen is still low due to the low storage density. Liquefied gas can reach substantially higher storage densities and thereby capacities but suffers from large energy demands for liquefaction and boil off-losses during storage. Several alternative storage technologies came into the focus of research to overcome these limitations. In the context of hydrogen transport, especially carrier-based hydrogen storage technologies like metal hydrides or liquid organic hydrogen carriers (LOHCs) seem promising as these technologies enable a dense storage of hydrogen at ambient conditions.

In this contribution, a thermodynamic and economic evaluation of several hydrogen transport technologies is presented. The analysis commences with the determination of the energy demands for hydrogen conditioning (e.g. compression, liquefaction), the actual transport and the provision of hydrogen gas at the point of use (e.g. hydrogen liberation from carrier). For all technologies considered in the analysis the energy demand for the actual transport is comparatively low. The conditioning (for pressurized and liquefied gas) and provision step (for carrier-based transport) have a far higher energy demand. Options to reduce the energy demands are pointed out. The cost calculation includes investment, maintenance, staff, energy, and material costs and shows the potential advantage of chemical carrier technologies over the established transport technologies.

459518 State-of-the-Art of the Liquid Organic Hydrogen Carrier (LOHC) Technology

Abstract:

LOHC are a safe and dense solution of storing hydrogen at ambient conditions. The basis is a reversible chemical reaction of hydrogen with the LOHC. This technology is studied worldwide since 5 years.

Despite the fact that several component pairs (loaded and unloaded form) are able to be a LOHC, the basic process design is known.

The lecture highlights technical applications of this technology. This progress will push the development of a hydrogen-based energy system.

481365 On using CO_2 and Renewable Energy for Autonomous Cities

Abstract:

An innovative concept of district heating using carbon dioxide is proposed to supply energy services in cities. The principle is to use CO2 as a district heating fluid distributed in liquid and gas form at a temperature of 15-20 °C. Heating needs are then supplied by condensing CO2 while cooling needs are supplied by evaporation, the temperature levels being adjusted using heat pumps. The energy balance is closed by exchanging heat between the district heating fluid and the environment accessing therefore attractive sources. The advantage of such system is the supply of the heating/cooling services at the temperature it requires and the possible exchange between heating and cooling demands. The use of the district heating system allows indeed to connect the most attractive environmental sources (geothermal, lake, river, waste water, industrial waste heat,. . .) with the users in order to maximise the efficiency of the system. Using this type of technology, it has been demonstrated that heating and cooling the city center of a European city can be reduced by a factor 5 to 6 (i.e. 80% energy savings) while having a pay back period of around 7 years . The concept can be extended to integrate the renewable energy resources in particular solar photovoltaics (PV). One of the difficulty of the PV systems is the excess of electricity available in the summer that needs to be stored to supply services in the winter. The idea here is to electro-chemically convert excess of electricity into methane by combining electrolysis of water and the Sabatier reaction that converts CO2 and H2 into methane. This can be done in co-electrolysis concepts
using solid oxide electrochemical membranes. Methane can then be stored in liquid form. In the winter time, the methane is converted in decentralised solid oxide fuel cells into electricity and heat while separating the CO2. The CO2 is collected in the district CO2 network at 50 bar and stored in the liquid form for the summer season when it will again be converted into methane. The produced electricity is used to supply the needs of the households, to drive the heat-pumps of the CO2 based district network and even electrical cars. It is therefore possible to develop cities that are supplied only by renewable sources. Furthermore, the methanation concept can also be applied to convert the waste biomass produced in the city into methane.

471444 Invited - Preventing Nanoparticle Plate-out in Heat Exchangers during Carrier Fluid Phase Change and Implications for Power Generation

Abstract:

For over a decade, a great deal of research has been conducted on the enhancement of heat transfer using nanofluids. Most exclusively, these results are rooted in the measured increase in thermal conductivity of the nanofluid when compared to the base fluid. Far less research points to negative drawbacks such as degradation of specific heat of the carrier fluid and viscosity increase with the addition of nanoparticles. We focus on the use of metal organic frameworks (MOFs) to enhance the energy transfer of a working fluid as it changes phase. The benefit of MOFs as a nanoparticle manifests itself within the energy equation in the form of the enthalpy of adsorption and desorption. We detail the many challenges faced over several years in working with nanofluids that use MOFs. Ultimately, using MOFs in power cycles can enhance the overall efficiency of the cycle with short payback period. As with most energy conversion cycles, working fluid phase change is a necessity, and resulted in nanoparticle plate-out on the various surfaces where phase change occurs. Our team has looked at several techniques for preventing this plate-out. One technique that shows promise in solving the plate-out problem involves forming a dispersion of the nanoparticles in a non-volatile low viscosity oil that is then added to the working fluid. Optical and IR spectroscopic techniques show that working fluid adsorption-desorption occurs unimpeded in the oil dispersion.

461379 Invited - Particle ALD Cost/Performance Advantages

Abstract:

The ability to place uniform nanoscale films on primary non-agglomerated particles, including nanoparticles, is a platform technology with many applications in the areas of electrochemistry, lighting, catalysis, micro-electronics, personal care, biotechnology, defense, construction and other areas - many yet to be discovered. ALD on particles is being successfully demonstrated on numerous particle systems including metals, ceramics, and polymers. Particle ALD allows the cost effective synthesis of unique materials that cannot be synthesized by conventional coating processes such as chemical (CVD) and physical (PVD) vapor deposition, or sol-gel methods. The process provides for a pathway to efficiently use expensive precursors with almost no waste product, hence, the low cost option for many functionalized particle applications. Although invented in the late 1990s, it has taken over a decade for commercial applications to be taken seriously and developed. Much of this has been the stigma attached to the higher cost of coating flat low surface area materials used in the micro-electronics industry. This presentation will review the development of Particle ALD on primary particles, their processing, scalability, and the opportunities presented for advanced applications.

453177 Invited - Thermal and Plasma Enhanced Atomic Layer Deposition on Powders and Particles

Abstract:

Surface engineering of micro- and nanoparticles is of great importance in fields such as catalysis, energy and sensing. For many of these applications particles are required with different bulk and surface properties. Atomic layer deposition (ALD) is known as a reliable technique for surface functionalization of powders, by conformally coating the individual particles with ultrathin layers. However, to perform ALD on powders, fluidization or agitation of the particles is required. During the past decades, both batch and spatial ALD reactor concepts have been developed, including fluidized bed, rotary reactors and pneumatic transport.[1] Although fluidized bed reactors are most often used for ALD on particles, this reactor concept is not compatible with plasma enhanced ALD, which is advantageous for e.g. coating on temperature sensitive polymer particles or deposition of metals and metal nitrides.

In this work, we report on a rotary reactor concept that was developed to enable both thermal and plasma-enhanced ALD on powders.[2] Particles ranging from nanometer size to millimeter size were successfully coated with layers of Al_2O_3, TiO_2, SiO_2, AlN and TiN. The ALD processes were characterized in-situ by means of mass spectroscopy (MS) and optical emission spectroscopy (OES). The composition and conformality of the coatings were evaluated by X-ray fluorescence (XRF), X-ray photoelectron spectroscopy (XPS), energy-dispersive X-ray spectroscopy (EDX), and transmission electron microscopy (TEM). These ultrathin conformal coatings may, for instance, be applied for encapsulation, inducing or enhancing catalytic activity, or altering electrical conductivity.

Our results prove that the developed rotary reactor enables conformal deposition on nano- and micropowders by thermal and plasma enhanced ALD. In this way, surface engineering of such particles can be achieved.

---

[1] D. Longrie et al., Journal of Vacuum Science and Technology A 32, 010802 (2014).

[2] D. Longrie et al., Surface & Coating Technology 213, 183-191 (2012).

460846 Employing Synergetic Effect of Doping and Thin Film Coating to Boost the Performance of Lithium-Ion Battery Cathode Particles

Abstract:

Atomic layer deposition (ALD) has evolved as an important technique to coat conformal protective thin films on cathode and anode particles of lithium ion batteries to enhance their electrochemical performance. Coating a conformal, conductive and optimal ultrathin film on cathode particles has significantly increased the capacity retention and cycle life as demonstrated in our previous work. In this work, we have unearthed the synergetic effect of electrochemically active iron oxide films coating and partial doping of iron on $LiMn_{1.5}Ni_{0.5}O_4$ (LMNO) particles. The ionic Fe penetrates into the lattice structure of LMNO during the ALD process. After the structural defects are saturated, the iron starts participating in formation of ultrathin oxide films on LMNO particle surface. Owing to the conductive nature of iron oxide films, with an optimal film thickness of ~0.6 nm, the initial capacity improved by ~25% at room temperature and by ~26% at an elevated temperature of 55 °C at a 1C cycling rate. The synergy of doping of LMNO with iron combined with the conductive and protective nature of the optimal iron oxide film leads to a high capacity retention (~93% at room temperature and ~91% at 55 °C) even after 1,000 cycles at a 1C cycling rate.

452329 Particle Atomic Layer Deposition for Synthesis of Fuel Cell Catalytic Material

Abstract:

Fluidized bed Atomic Layer Deposition (ALD) is a scalable method for nanoscale catalyst design, and has potential to address many of the persistent material challenges associated with PEM fuel cells, namely utilization and durability of the platinum catalyst. ALD was carried out to deposit platinum nanoparticles on functionalized and unmodified carbon black with both combustion and hydrogenation chemistries. Reactions throughout the ALD cycle were monitored using mass spectrometry and IR spectroscopy to clarify the deposition chemistry. The synthesized materials were evaluated in hydrogen pumping tests and as oxygen reduction reaction (ORR) catalysts in proton exchange membrane fuel cell experiments. Hydrogenation made finer, more dispersed Pt nanoparticles that performed similarly to their commercial equivalent when pumping hydrogen; oxygenation made a coarser catalyst that underperformed commercial Pt/C. The hydrogenation catalysts deposited on functionalized carbon were found to be ineffective in ORR testing due to poor water management in the fuel cell, which was attributed to reduced hydrophobicity within the catalyst layer.

472102 Nanostructured Powders Made from Flame Spray Pyrolysis for Li-Ion Cathode

Abstract:

Flame processes are the most economic ones to produce fine powders and have been used to produce pigment titanium oxide and fused silica in large volumes. In this work a flame spray pyrolysis (FSP) process is used to make positive electrode materials for Li-ion battery, which are made of complex metal oxide powders, such as $Li(Ni_{1/3}Mn_{1/3}Co_{1/3})O_2$. In particular, the work was focused on making the NMC materials with stoichiometric compositions and desired morphologies. During the pyrolysis process, different morphologies can be formed, involving powders often with primary nanoparticles. Such nanostructures were found to have affected the battery performance. Detailed results of the synthesis, characterization, and battery testing of the NMC powder materials will be presented.

465170 Dry Coating of Electrode Particle with Model Particle of Sulfide Solid Electrolytes for All-Solid-State Secondary Battery

Abstract:

We here present a dry coating of electrode particle with a model material of sulfide solid electrolytes for all-solid-state secondary batteries. LiNi1/3Co1/3Mn1/3O2 (NCM) was used as a host particle. Sodium sulfate (Na2SO4) was selected as a model material of sulfide solid electrolytes and used as a guest particle. Three different processes, which were vibration mixing, mortar-and-pestle mixing, and dry impact-blending process known as â€œHybridizerâ€, were used and compared to each other. FESEM and EDX images showed that the dry impact-blending process produced composite particles in which NCM was more fully covered with Na2SO4 than using the vibration mixing and mortar-and-pestle mixing. Cross sectional observations of composite particles showed that using the dry impact-blending process continuous layer of Na2SO4 were formed on NCM particle. In a compacted pellet prepared from the composite particles produced by the dry impact-blending process, following features were observed: better dispersion of NCM particles in Na2SO4 matrix without NCM agglomerates; less void space; and intimate NCM-Na2SO4 interfacial contacts. These features suggest that performance of all-solid-state batteries are likely to remarkably improve using the composite particles prepared from the dry coating process [1].

[1] T. Kawaguchi, H. Nakamura, S. Watano. Powder Technology, in press.

462966 Mixed-Oxide Based Redox Catalysts for Hydrocarbon Oxidation, Water-Splitting, and CO_2 Utilization

Abstract:

Using transition metal oxides as the reaction intermediate, the chemical looping strategy seeks to intensify fossil fuel conversion processes via a cyclic redox scheme. In such a process, carbonaceous molecules are oxidized by active lattice oxygen (O^{2-}) in the transition metal oxide, which functions as an oxygen carrier, a.k.a. redox catalyst. In a subsequent step, the O^{2-} deprived oxygen carrier is replenished by a gaseous oxidant. Although conventional chemical looping processes seeks to fully oxidize fossil fuels for carbon capture, tailored redox catalysts can be applied for a variety of applications including selective oxidation, water-splitting, and CO_2 reduction.

The present study exemplifies novel applications of iron and manganese containing mixed oxides for methane partial oxidation, oxidative dehydrogenation (ODH) of ethane, water-splitting, and CO_2 reduction. A number of redox catalysts composed of iron and/or manganese oxides are synthesized and characterized. Rational strategies to improve the redox activity, product selectivity, and/or water-splitting and CO_2 reduction conversions are investigated through the assistance of spectroscopy and *ab-initio* calculation tools. Compared to traditional heterogeneous catalysts, the redox catalysts in the current study show the potential to be more selective. They also possess excellent redox stability. Close to 100% water and CO_2 reduction efficiencies are shown to be achievable through rational selection of mixed metal oxide based redox catalysts.

Catalytic Co-pyrolysis of Microalgae and Low Density Polyethylene Waste to Aromatic Hydrocarbons using Activated Carbon

Emmanuel Ansah[a], Lijun Wang[b,c*], John Eshun[a], Quazi Rahman[a], Abolghasem Shahbazi[bc]

[a] Department of Energy and Environmental Systems

[b] Department of Natural Resources and Environmental Design

[c] Department of Chemical, Biological and Bioengineering

North Carolina Agricultural and Technical State University, 1601 E Market Street

Greensboro, NC 27411, U.S.A.

Abstract

Catalytic and non-catalytic fast co-pyrolysis of micro algae (chlorella) and low density polyethylene (LDPE) waste were studied in multi-shot pyrolyzer using activated carbon as catalyst. Dried microalgae was co-pyrolyzed with LDPE in both the presence and absence of activated carbon to investigate the interaction between algae and LDPE during thermal decomposition and catalytic upgrade of the pyrolysis vapor. In the non-catalytic co-pyrolysis study in micro-pyrolyzer-GC/MS, the results indicated the interaction between microalgae and LDPE during co-pyrolysis had a pronounce effect on both microalgae derived oxygenated compounds and C_{10}-C_{20} aliphatic hydrocarbons. The total oxygenated compounds (mainly carbohydrate and lipid derived) increased by 18.7 % whiles olefins and alkanes saw an increase of 6.9% and 14.2% respectively. During activated carbon catalyzed co-pyrolysis of algae and LDPE, as temperature increased from 500°C to 700°C, the conversion of olefins and alkanes and the deoxygenation of oxygenated compounds from microalgae in the pores of the activated carbon produced high yield of aromatic hydrocarbon. The pyrolysis of algae produced relatively small aromatic hydrocarbons whilst pyrolysis of LDPE produced aromatic hydrocarbons with high yield at increasing temperatures. Addition of Algae to LDPE increased the yield of total aromatic hydrocarbons in Algae/LDPE/catalyst (an increase of 71.8% from 600°C to 700°C) compared to Algae/catalyst alone (an increase of 12.7% from 600°C to 700°C) and LDPE/catalyst alone (an increase of 6.1% from 600°C to 700°C) indicating high synergetic effect at rather higher temperature.

1. Introduction

A wide variety of biomass types such as grass, wood, agricultural crops and residues, animal waste, municipal solid waste and aquatic plants have been examined to determine their efficacy for the production of biofuels. Micro-algae are a hydrogen-deficient feedstock, thus leading to adverse impacts on the quality of liquid fuel especially low content of hydrocarbon content during pyrolysis. Liquid fuels

* Corresponding author, Tel: +1 336 285-3833, Fax: +1 336 3347270, Email: lwang@ncat.edu (L.J. Wang)

from microalgae were reported to have undesirable properties such as a high oxygen content and elevated acidity (i.e., high acid number). There is a growing awareness that the introduction of additional hydrogen-source is inevitable for improving the quality of bio-oil from microalgae feedstock. Waste commodity plastics, low and high-density polyethylene (LDPE and HDPE), polyethylene terephthalate (PET), polystyrene (PS), and polyvinyl chloride (PVC) have a high hydrogen to carbon ratio and molecular chain structures suitable for liquefaction and pyrolysis (Ballice, 2002). Polyethylene (PE) that is a common municipal solid waste and has a hydrogen to carbon ratio of 2 can potentially act as a cheap and available hydrogen source in a co-pyrolysis process (Zhang et al., 2015).

Pyrolysis vapors from the co-pyrolysis of microalgae and LDPE feedstocks can be converted to aromatic hydrocarbon by catalyst such as zeolites and activated carbon. Activated carbon structure is developed by imperfect aromatic sheets of carbon atoms, as well as incompletely saturated valences and unpaired electrons on the surface (Tsyntsarski et al., 2015). Few authors (Hu et al., 2014) and (Bu et al., 2013; Bu et al., 2014; Bu et al., 2015)) have reported methods to selectively produce syngas and phenolic-rich bio-oil respectively through catalytic pyrolysis of biomass using activated carbon as catalyst. Our study will seek to explore the non-catalytic and catalytic co-pyrolysis of microalgae and low density polyethylene in micro-pyrolyzer-GC/MS and determine how the thermal and catalytic interactions influence the yield of hydrocarbons.

2. Materials and methods

2.1 Materials

Microalgae (*Chlorella* spp) and Low density polyethylene (LDPE) were used in co-pyrolysis experiments. Proximate and ultimate analysis were performed on samples and shown in Table 1.

Table 1. Analysis of raw material

Description	Microalgae (chlorella spp)	Low Density Polyethylene
Ultimate analysis (dry basis)		
Carbon (wt. %)	42.60	84.90
Hydrogen (wt. %)	6.70	15.02
Nitrogen (wt. %)	6.71	0.01
Sulfur (wt. %)	1.25	0.02
Oxygen (wt. %)	42.7	-
Proximate analysis (dry basis)		
Moisture (wt. %)	1.73	-
Volatiles (wt. %)	63.63	99.98
Fixed carbon (wt. %)	32.38	-
Ash (wt. %)	2.26	.02

2.3 Micro-pyrolyzer-GC/MS (Py-GC/MS) experimental procedure

Fast pyrolysis was conducted in a multi shot pyrolyzer system (Frontier Laboratories Ltd, Japan) operated in a single-shot mode which is connected to GC/MS system (Agilent Technologies). The temperature of valve interface was set up at 300 ◦C and the temperature of the micro-pyrolyzer was set at temperatures of 400°C, 500 ◦C, 600 °C and 700 °C for the study. The GC front inlet temperature was set up at 300 ◦C to prevent the condensation of the products. Initial oven temperature was set at 40◦C and held for 2 min. The oven temperature was then ramped to 220°C at a rate of 5◦C/min and held at 200◦C for 15 min. Helium at a flow rate of 1 ml/min was used as carrier gas with a split ratio of 50:1.

Results and discussion

3.1 Effect of co-pyrolysis feedstock ratios and temperature on product yield

To investigate the influence of Microalgae/Plastic mass ratio on the product distribution from the co-pyrolysis of *Chlorella spp* and LDPE, experiments were performed at 400 , 500 and 600°C with Microalgae/LDPE mass ratios of 3:1, 1:1, 1:3 and the results shown in Fig 1. Fast pyrolysis of the pure forms of microalgae and LDPE at 400°C shows that LDPE only partially depolymerize at 400°C whilst microalgae almost completely devolatilize at same temperature. Microalgae showed optimum peak area at 500°C and decreased when temperature was increased to 600°C. LDPE on the other hand increased sharply when temperature was increased to 500°C and slightly increased further when increased to 600°C. During co-pyrolysis, it is seen that at 400°C, an increase in the ratio of LDPE to microalgae decreased the total peak area of volatiles due to the partial decomposition of LDPE at 400°C. However, at a co-pyrolysis feedstock ratio of equal mass and temperature of 500°C, the total peak area of volatiles lies between the total peak areas of microalgae and LDPE. Similar trend is observed when the temperature increased from 500°C to 600°C, however, the total peak area of feedstock ratio of equal mass was slightly lower than when microalgae was pyrolyzed alone. At temperature of 500°C and 600°C, an increase in the ratio of LDPE to microalgae increased the total peak area during pyrolysis. This indicates that the co-feeding of LDPE improved the total volatiles at the expense of bio-char due to the high percentage of volatile matter and low ash content in plastic (Xie et al., 2015).

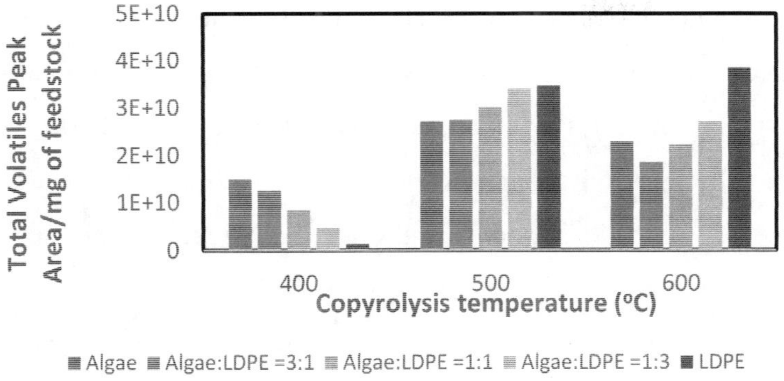

Fig. 1 Total volatiles yield at varying temperatures for co-pyrolysis of microalgae and LDPE at different feedstock ratios.

3.1.2 Product analysis

The chromatographic peaks were carefully selected and analyzed by means of the NIST MS library and literature (Kaewpengkrow et al., 2014; Lu et al., 2013; Zhang et al., 2013). For each chemical composition, a calibrated peak area% was achieved to represent its relative content. The total calibrated chromatographic peak areas could be used to reveal the changes of product yields since each peak area was normalized by the mass of feedstock during each experiment.

3.1.3 Non catalytic co-pyrolysis

Micro algae (chlorella spp) was co-pyrolyzed with LDPE to determine thermal interaction during co-pyrolysis of microalgae and PE. Typical pyrolysis products of micro algae and PE have been reported in literature (Aysu and Sanna, 2015; Chaiwong et al., 2013; Schwarzinger et al., 2015; Tan et al., 2014; Thangalazhy-Gopakumar et al., 2012; Xie et al., 2015; Xue et al.). The major pyrolysis products of microalgae (chlorella spp) obtained from NIST library in this study include CO_2, Toluene, 2-Furanmethanol, 2-hydroxy-3-methyl 2-Cyclopentene-1-one , P-cresol or 3-methyl phenol, Indole, 3-methyl 1-H-Indole, Phytol acetate, 3,7,11-trimethyl-1-Dodecanol, 3,7,11,15-Tetramethyl-2-hexadecen-1-ol, n-hexadecanoic acid. During LDPE pyrolysis, polyethylene is depolymerized to olefins and alkanes (together defined as aliphatic) with various chain lengths and therefore in the present study, we considered C_{10} to C_{20} since they were detected by GC/MS with greater certainty. The non-catalytic co-pyrolysis products used in this study were classified into seven groups namely CO_2, Toluene, Indoles, Oxygenated Compounds, Olefins, Alkanes and Aromatics. To assess synergetic thermal interaction, a ratio was calculated by dividing the GC/MS peak area of a pyrolysis product (which is proportional to the yield) during co-pyrolysis of microalgae and LDPE by the peak area of the same product when microalgae or LDPE was independently pyrolyzed. The ratios of oxygenates and aliphatic hydrocarbons for feedstock ratios of 1:1 and the optimum pyrolysis temperature at 500°C are reported and shown in fig. 2. In these graphs, a ratio higher than 1 indicates that co-pyrolysis promotes the formation of the product, where as a value lower than 1 indicates that co-pyrolysis suppresses the formation of the product. From the results, the interaction between microalgae and LDPE during co-pyrolysis had a pronounce effect on both microalgae derived oxygenated compounds and C_{10}-C_{20} aliphatic hydrocarbons from the depolymerization of LDPE. As seen from Fig 4, the co-pyrolysis resulted in an increment in yield (depicted as product ratios) of most of the $C_{10} – C_{20}$ aliphatic hydrocarbon ranging from an increment of 6.5% to 33%. Some oxygenated compounds (phytol, phytol acetate and 3, 7, 11-trimethyl-1-Dodecanol) mostly from decomposition of lipid content of microalgae showed an increase in yield ranging from 1.4% to 83%. This means that co-pyrolysis of LDPE with microalgae, enhanced the decomposition of both polyethylene and lipid content of microalgae leading to increase yield of aliphatic hydrocarbons and oxygenated compounds. From Fig 2(c), the total oxygenated compounds (carbohydrate and lipid derived) increased by 18.7 % whiles olefins and alkanes saw an increase of 6.9% and 14.2% respectively. It can be seen from same figure 2(c) that the yield of Indoles and Toluene were suppressed during co-pyrolysis as the product ratios were less than 1.

Fast pyrolysis of LDPE does not produce solid residue, however, the chain length of olefins and alkanes produced from depolymerization of LDPE can be as high as over C_{20+} during fast pyrolysis due to the short reaction time (Xue et al.). It has been reported that there is increasing probability that abstraction of hydrogen by biomass derived oxygenated compounds can facilitate the cleavage of LDPE polymer chain and its derivatives during co-pyrolysis. It is therefore probable that hydrogen transfer from LDPE to microalgae compounds likely reduces polymerization and cross-linking reactions and also suppresses decarboxylation and decarbonylation reactions (Vichaphund et al., 2014; Xue et al.).

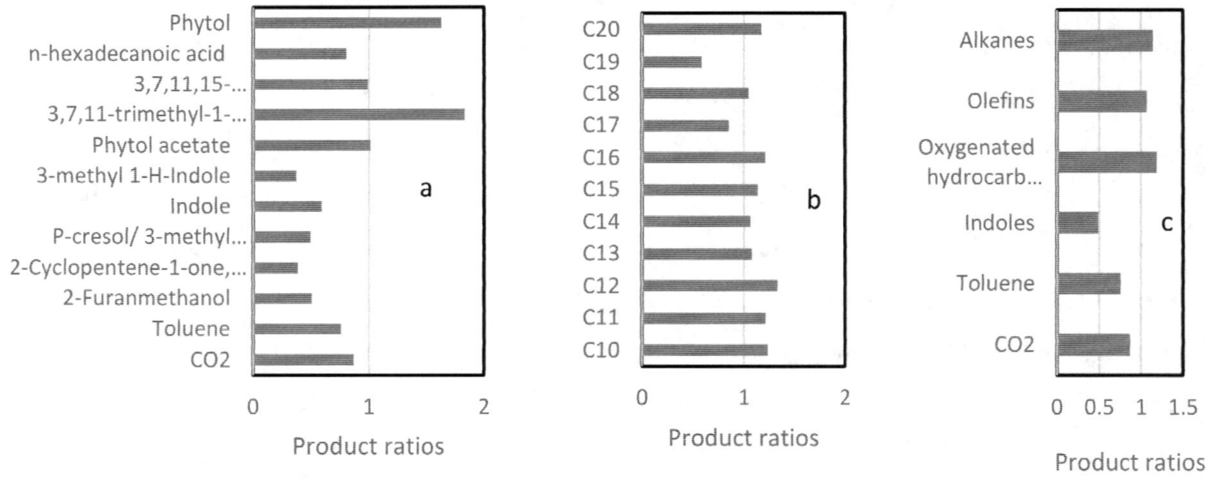

Fig 2. Ratios of product yields during co-pyrolysis of microalgae and LDPE in feedstock ratio of 1 to pyrolysis of individual feedstock determined by GC/MS peak area at pyrolysis temperature of 500°C.

3.1.4 Effect of catalytic co-pyrolysis on yield of aromatic hydrocarbons.

Pyrolysis of Algae, LDPE and Algae/LDPE mainly produced oxygenated compounds, nitrogen containing compounds and aliphatic hydrocarbons made up of olefins and alkanes. The catalytic pyrolysis of Algae/A.C, LDPE/A.C and Algae-LDPE/A.C on the other hand, mainly produced alkylated benzenes (toluene, ethylbenzene, p-xylene, propyl benzene, 1-ethyl-2-methylbenzene, 1-methyl-2-propylbenzene, butyl benzene and biphenyl), naphthalene and alkylated naphthalene (1-methylnaphthalene, 2-methylnaphthalene and 2-ethyl naphthalene) and aliphatic hydrocarbons (olefins and alkanes). These compounds from the activated carbon catalytic pyrolysis of algae, LDPE and mixture of algae-LDPE constitutes aromatic hydrocarbons and aliphatic hydrocarbons. The comparison of aromatic yield from individual algae and LDPE catalytic pyrolysis with catalytic co-pyrolysis of algae-LDPE (1:1) is presented in Figure 3. The catalytic pyrolysis of algae produced relatively small aromatic hydrocarbons whilst catalytic pyrolysis of LDPE produced aromatic hydrocarbons with high yield at increasing temperatures. Addition of Algae to LDPE increased the yield of total aromatic hydrocarbons in Algae/LDPE/catalyst (an increase of 71.8% from 600°C to 700°C) compared to Algae/catalyst alone (an increase of 12.7% from 600°C to 700°C) and LDPE/catalyst alone (an increase of 6.1% from 600°C to 700°C) indicating high synergetic effect at rather higher temperature. Similar results were reported by (Hassan et al., 2016) in pyrolysis of wood and polystyrene. During LDPE pyrolysis, free radical reactions with random chain scission would normally occur which results in the formation of olefins and alkanes. Activated carbon catalyst had a significant shape-selective effect on the products of aromatic hydrocarbons .When the primary product vapors from pyrolysis passed through the activated catalyst layer, they would diffuse into the internal pores of activated carbon catalyst and form aromatic hydrocarbons through a hydrocarbon pool mechanism in the activated carbon framework. The formation of aromatic hydrocarbons could come from two phenomena. The first one is the conversion of olefins and alkanes from LDPE pyrolysis to aromatic hydrocarbons. The second phenomenon is the deoxygenation of oxygenated organic compounds as it is processed through the catalyst layer.

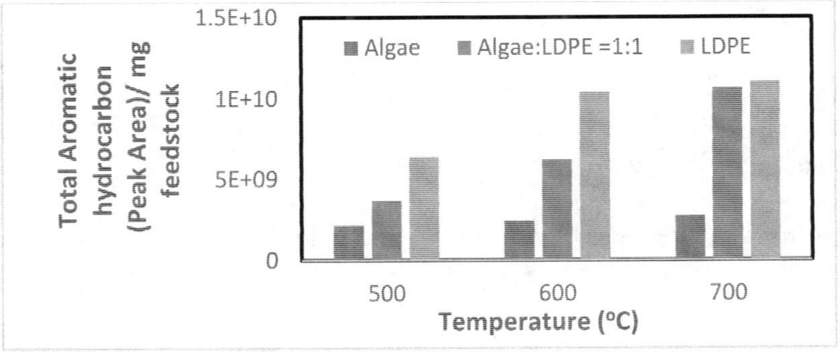

Fig 3. Yield of total aromatic hydrocarbon produced from individual catalytic pyrolysis of algae and LDPE and catalytic co-pyrolysis of algae-LDPE (1:1) at 500, 600 and 700°C.

4. Conclusion

In this study, catalytic and non-catalytic fast co-pyrolysis of microalgae and low density polyethylene were carried out using micro-pyrolyzer-GC/MS to investigate initial active pyrolysis temperature, effect of interaction of non-catalytic co-pyrolysis of microalgae and LDPE on product yields and the catalytic effect of activated carbon on the selectivity of aromatic hydrocarbons. The results show the non-catalytic interaction between microalgae and LDPE during co-pyrolysis had a pronounce effect on both microalgae derived oxygenated compounds and C_{10}-C_{20} aliphatic hydrocarbons from the depolymerization of LDPE. The pyrolysis of algae produced relatively small aromatic hydrocarbons whilst pyrolysis of LDPE produced aromatic hydrocarbons with high yield at increasing temperatures. Addition of Algae to LDPE increased the yield of total aromatic hydrocarbons in Algae/LDPE/catalyst (an increase of 71.8% from 600°C to 700°C) compared to Algae/catalyst alone (an increase of 12.7% from 600°C to 700°C) and LDPE/catalyst alone (an increase of 6.1% from 600°C to 700°C) indicating high catalytic effect at rather higher temperature.

References

Aysu, T., Sanna, A., 2015. Nannochloropsis algae pyrolysis with ceria-based catalysts for production of high-quality bio-oils. Bioresource Technology 194, 108-116.

Ballice, L., 2002. Classification of volatile products evolved during temperature-programmed co-pyrolysis of low-density polyethylene (LDPE) with polypropylene (PP). Fuel 81, 1233-1240.

Bu, Q., Lei, H., Wang, L., Wei, Y., Zhu, L., Liu, Y., Liang, J., Tang, J., 2013. Renewable phenols production by catalytic microwave pyrolysis of Douglas fir sawdust pellets with activated carbon catalysts. Bioresource Technology 142, 546-552.

Bu, Q., Lei, H., Wang, L., Wei, Y., Zhu, L., Zhang, X., Liu, Y., Yadavalli, G., Tang, J., 2014. Bio-based phenols and fuel production from catalytic microwave pyrolysis of lignin by activated carbons. Bioresource Technology 162, 142-147.

Bu, Q., Lei, H., Wang, L., Yadavalli, G., Wei, Y., Zhang, X., Zhu, L., Liu, Y., 2015. Biofuel production from catalytic microwave pyrolysis of Douglas fir pellets over ferrum-modified activated carbon catalyst. Journal of Analytical and Applied Pyrolysis 112, 74-79.

Chaiwong, K., Kiatsiriroat, T., Vorayos, N., Thararax, C., 2013. Study of bio-oil and bio-char production from algae by slow pyrolysis. Biomass and Bioenergy 56, 600-606.

Hassan, E.B., Elsayed, I., Eseyin, A., 2016. Production high yields of aromatic hydrocarbons through catalytic fast pyrolysis of torrefied wood and polystyrene. Fuel 174, 317-324.

Hu, Z., Ma, X., Li, L., Wu, J., 2014. The catalytic pyrolysis of microalgae to produce syngas. Energy Conversion and Management 85, 545-550.

Kaewpengkrow, P., Atong, D., Sricharoenchaikul, V., 2014. Catalytic upgrading of pyrolysis vapors from Jatropha wastes using alumina, zirconia and titania based catalysts. Bioresource Technology 163, 262-269.

Lu, Q., Zhang, Z.-b., Yang, X.-c., Dong, C.-q., Zhu, X.-f., 2013. Catalytic fast pyrolysis of biomass impregnated with K3PO4 to produce phenolic compounds: Analytical Py-GC/MS study. Journal of Analytical and Applied Pyrolysis 104, 139-145.

Schwarzinger, C., Hintersteiner, I., Schwarzinger, B., Buchberger, W., Moser, B., 2015. Analytical pyrolysis in the determination of the aging of polyethylene. Journal of Analytical and Applied Pyrolysis 113, 315-322.

Tan, X.-C., Zhu, C.-C., Liu, Q.-K., Ma, T.-Y., Yuan, P.-Q., Cheng, Z.-M., Yuan, W.-K., 2014. Co-pyrolysis of heavy oil and low density polyethylene in the presence of supercritical water: The suppression of coke formation. Fuel Processing Technology 118, 49-54.

Thangalazhy-Gopakumar, S., Adhikari, S., Chattanathan, S.A., Gupta, R.B., 2012. Catalytic pyrolysis of green algae for hydrocarbon production using H+ZSM-5 catalyst. Bioresource Technology 118, 150-157.

Tsyntsarski, B., Stoycheva, I., Tsoncheva, T., Genova, I., Dimitrov, M., Petrova, B., Paneva, D., Cherkezova-Zheleva, Z., Budinova, T., Kolev, H., Gomis-Berenguer, A., Ania, C.O., Mitov, I., Petrov, N., 2015. Activated carbons from waste biomass and low rank coals as catalyst supports for hydrogen production by methanol decomposition. Fuel Processing Technology 137, 139-147.

Vichaphund, S., Aht-ong, D., Sricharoenchaikul, V., Atong, D., 2014. Catalytic upgrading pyrolysis vapors of Jatropha waste using metal promoted ZSM-5 catalysts: An analytical PY-GC/MS. Renewable Energy 65, 70-77.

Xie, Q., Addy, M., Liu, S., Zhang, B., Cheng, Y., Wan, Y., Li, Y., Liu, Y., Lin, X., Chen, P., Ruan, R., 2015. Fast microwave-assisted catalytic co-pyrolysis of microalgae and scum for bio-oil production. Fuel 160, 577-582.

Xue, Y., Kelkar, A., Bai, X., Catalytic co-pyrolysis of biomass and polyethylene in a tandem micropyrolyzer. Fuel.

Zhang, B., Zhong, Z., Ding, K., Song, Z., 2015. Production of aromatic hydrocarbons from catalytic co-pyrolysis of biomass and high density polyethylene: Analytical Py–GC/MS study. Fuel 139, 622-628.

Zhang, H., Xiao, R., Jin, B., Shen, D., Chen, R., Xiao, G., 2013. Catalytic fast pyrolysis of straw biomass in an internally interconnected fluidized bed to produce aromatics and olefins: Effect of different catalysts. Bioresource Technology 137, 82-87.

458989 Modeling the Impact of Biomass Particle Residence Time on Fast Pyrolysis Yield and Composition

Abstract:

Bubbling fluidized bed and circulating fluidized bed reactors are the most common unit operations for carrying out fast pyrolysis of biomass. These reactors are designed to contact biomass particles with hot inert gas (such as nitrogen) in a bed of solids (sand) under turbulent multiphase flow conditions. In bubbling bed reactors, the gas flow is sufficient enough to cause mixing of the solids and gas without entraining most of the particles, except for char and ash generated from the biomass which can elutriate out of the sand bed then exit from the top of the reactor. In circulating bed reactors, the gas flow is much higher and the solids are typically smaller, consequently most of the solids are entrained out the top of the reactor vessel. The entrained solids are separated from the gas and most of them are recirculated back through the reactor with new gas, hence the name circulating fluidized beds.

Numerous studies have demonstrated that biomass fast pyrolysis yields depend on the residence time distribution (RTD) of the pyrolyzing particles, which is a function of the inherent feed properties as well as the reactor design and operating conditions. Experimental measurements indicate that it should be possible to represent the RTDs for biomass particles in both bubbling and circulating bed pyrolysis reactors with relatively simple functional approximations involving only two or three physically relevant parameters. We expect similar approaches can also be used to model the impact of catalyst particle residence time in vapor phase upgrading reactors.

In this presentation, we summarize literature on low-order reactor models for the particle RTDs in bubbling and circulating fluidized bed reactors and discuss the implications for the conversion of biomass particles undergoing fast pyrolysis. Based on this information, we demonstrate how particle RTDs can be combined with reaction kinetics and simplified reactor mass balances to create low-order fast pyrolysis models that can be used to summarize, interpret, and extrapolate observed pyrolysis trends in experiments and more detailed computational simulations. Our goal is to

provide flexible analytical tools for consolidating information from more detailed computational reactor simulations and improving high-level process simulations of bio-oil production needed for techno-economic analyses and reactor design.

472546 Ring-Locking Enables Selective Anhydrosugar Synthesis from Carbohydrate Pyrolysis

Abstract:

Anhydrosugars are common but low-yield products of carbohydrate pyrolysis. Their fixed conformation, protected stereoselectivity, and retention of most of their chemical functionality make these sugar derivatives highly desirable precursors for biologically active compounds, biodegradable surfactants, and polymers. Anhydrosugar production is very inefficient though, due to the nonselective nature of carbohydrate pyrolysis. Here we demonstrate how they form with excellent purity by ring-locking sugars prior to thermal treatment. We found that the selectivity to 1,6-anhydro-Î²-D-glucopyranose (levoglucosan, LGA) increased from 2% to greater than 90%, after alkoxy or phenoxy substituents were placed at the anomeric carbon of glucose, followed by fast pyrolysis of the resulting sugar at 600 Â°C. LGA formation becomes the dominant reaction pathway when the substituent group inhibits the pyranose ring from opening and fragmenting into non-anhydrosugar products. These findings introduce the ring-locking concept to sugar pyrolysis chemistry and offer a chemical-thermal treatment approach for upgrading simple and complex carbohydrates.

471419 Pyrolysis Kinetics to and from Levoglucosan

Abstract:

Levoglucosan is a major and characteristic product of cellulose pyrolysis, formed catalytically in the absence of metal ions by R-OH molecules like water and other hydroxylated sugar-pyrolysis products [Seshadri and Westmoreland, *J. Phys. Chem. A* **116**:49 (2012) 11997-20013]. No radicals are necessary, although radical chemistry must be important at high temperatures. Other pericyclic unimolecular reactions may also occur, often involving hydroxyl groups. Here, we extend the previous work through quantum-chemistry predictions of transition states for altered glucose structures. Specifically, the OH groups on C1 and C6 that are involved in levoglucosan formation are substituted by NH_2 and CH_3. Nitrogen has one lone pair of electrons, and the methyl carbon has none; the calculations thus reveal how oxygen's two lone pairs influence the kinetics of glucose pyrolysis, and the results are then applied to cellulose. Added insights result from examining solvation by explicit water molecules, solvations by continuum models, and the effect of ionic acids. Finally, the routes by which levoglucosan may decompose are examined.

464144 Molecular-Level Simulation of Thermogravimetric Analysis (TGA): A Cellulose Pyrolysis Example

Abstract:

Molecular-level Simulation of Thermogravimetric Analysis (TGA): A Cellulose Pyrolysis Example

Juan C. Lucio-Vega, Scott R. Horton, and Mike T. Klein

Department of Chemical and Biomolecular Engineering, University of Delaware

Thermogravimetric analysis (TGA) of cellulose pyrolysis was simulated using a molecular-level kinetics model. A temperature ramp between 600-1000°C was imposed in order to achieve the TGA simulation. The underlying celllose pyrolysis model accounted for polymer polydispersity whose decomposition was described using Flory-Stockmayer statistics and the degree of polymerization [1]. A recursive optimization method was utilized to calculate the degree of polymerization of the starting cellulose from literature bulk properties [1, 2]. The model's primary reaction pathways for the decomposition allowed for the clipping of the active cellulose by thermolysis in order to create levoglucasan (LGA) and glucose. The chemistry proposed by Lin et al [2] was then applied as secondary reactions where the LGA molecules reacted to produce intermediates and levoglucosenone (LGO). Further breakdown of the LGO molecules formed volatiles, furans, aldehyde, ketones, and char through a series of complex reactions. Linear free energy relationships were applied to minimize the number of model kinetic parameters. Executing the cellulose pyrolysis model in TGA mode provided a prediction of the mass loss, which in turn allowed the model parameters to be optimized to literature TGA data [2]. The optimized model showed good agreement with the experimental trends.

[1] Scott R. Horton, Rebecca J. Mohr, Yu Zhang, Francis P. Petrocelli, and Michael T. Klein. "Molecular-Level

Kinetic Modeling of Biomass Gasification." Energy & Fuels 30.3 (2016): 1647-1661.

[2] Yu-Chuan Lin, Joungmo Cho, Geoffrey A. Tompsett, Phillip R. Westmoreland, and George W. Huber, "Kinetics and Mechanism of Cellulose Pyrolysis" Journal of Physical Chemistry 113 (2009): 20097–20107.

464586 Impact of Carrier Gas and Temperature on Biomass Fast Pyrolysis Using a New Microsphere Microreactor Approach

Abstract:

Biomass fast pyrolysis is emerging as a viable alternative for carbon-based resources. Yields and selectivity for fast pyrolysis is a function of numerous factors including rate of heating and quenching (residence times for heterogeneous and homogeneous events) and gas phase conditions including the composition of the carrier gas. Carrier gas composition is a parameter that has not been explored as comprehensively as other factors. In this study four gases were used, He, Ar, CO_2 and N_2, for the pyrolysis of crystal cellulose, tall fescue and switchgrass at temperatures between 500 and 900 °C. Individual manufactured biomass microspheres weighing nominally 10 micrograms each were pyrolized in a micro-reactor. The rate of hydrocarbon gas production was measured using a fast flame ionization detector (fast-FID). The effect of carrier gas on shrinkage of the biomass at different temperatures and char properties was investigated.

454300 Study of Transmethylation As Primary Step on the Decomposition of Methoxy-Rich Lignin Model Compound over Zeolite Catalyst

Abstract:

In the context of the bio-based economy, lignin is a major source of aromatic compounds. Fast pyrolysis of lignin followed by bio-oil upgrading is a promising route for retrieving these aromatic chemical platforms and making them available for subsequent processing. Fast pyrolysis of lignin and upgrading of the derived bio-oil has been previously studied. However, the conversion pathway followed by each compound (or functional group) in the complex liquid product still needs to be accurately understood in order to tailor the process towards desired products.

Significant number of compounds containing methoxy functional group, such as anisole, guaiacol, syringol, and their derivatives, are found in primary liquids obtained from the thermochemical conversion of lignin. These compounds are further converted into phenolic and aromatic compounds via secondary pyrolysis and upgrading reactions. Anisole is normally used as model compound of methoxy-rich compounds present in the liquids from fast pyrolysis of lignocellulosic biomass because the only functionality in the molecule is the methoxy group. Research on anisole decomposition has been mainly focused on the deoxygenation process of the phenyl-oxygen bond. Study of the mechanism of transmethylation as reaction occurring prior to deoxygenation is less extensive, despite it is essential to understand the entire process of anisole decomposition.

In this work the transmethylation as the primary step on the catalytic decomposition of anisole is investigated. The decomposition of anisole was carried out in a fluidised bed reactor, and HZSM-5 zeolite was used as catalyst. The yield and product distribution in the liquid fraction, with particular focus on the phenolic compounds, was evaluated by gas chromatography. Non-catalytic anisole decomposition experiments were also conducted for comparison purposes.

Maximum yield of liquid products was observed at the minimum temperature at which complete conversion of anisole

is achieved (650°C for non-catalytic decomposition and 400°C for catalytic decomposition). Phenolic compounds were the primary products at these key temperatures and below. In the case of catalytic decomposition, phenol, and ortho-cresol and para-cresol were the main products. In addition, multi-methylphenols (xylenols and trimethyl-phenols) were abundantly produced over HZMS-5. Yields of aromatic compounds were larger than those of phenols at higher temperatures.

Experimental results suggest that transmethylation occurs as primary reaction when anisole conversion is not complete at low temperatures. A plausible mechanism for low-to-moderate temperatures is that, in the presence of the zeolite catalyst, the anisole is converted into phenol followed by the relocation of the methyl radical to form o-cresol and p-cresol. The addition of another methyl radical to the cresol molecule gives rise to the formation of xylenols. The rearrangement to trimethyl phenols is observed to a lesser extent. In all cases the major orientations for transmethylation are the ortho- and para-positions, and are favoured by the increase in temperature. As the temperature increases, anisole conversion is maintained but phenolic compounds yields decrease which points out to phenolics as precursors of aromatic compounds. Formation of aromatics occurs by means of deoxygenation as secondary step during anisole decomposition.

455165 Optimal, Flexible and Feasible Heat Exchanger Network Design Under Severe Uncertain Industrial Operation Conditions

Abstract:

[1]Engineering Department, Crop Science Division of Bayer, Institute, WV, 25112

[2]Dan F. Smith Department of Chemical Engineering, Lamar University, Beaumont, TX 77710

Abstract

Heat Exchanger Network (HEN) synthesis plays an important role in improving the economic performance and energy efficiency of industrial plants, and has been the subject of numerous investigations during the past decades. However, traditionally, most of the research focuses on the design based on determined condition. The operational flexibility and feasibility are less discussed. Chemical process and its production are complex and flexible. Production rate, utility stream quality, process stream temperature, ambient weather condition, and various other uncertain factors, will cause disturbances to the heat exchanger network operation. In this presentation, the authors propose a novel approach for optimal, flexible, and feasible HEN design considering the severe operation uncertainties. It utilizes the Probability Bounds Analysis (PBA) theory to model the uncertain industrial operation conditions, double loop method to sample the uncertainty, and Aspen Energy Analyzer for optimization design. The capability of this approach is demonstrated by a case study which generates a better solution with less cost and good flexibility in uncertain operations than the one generated at a deterministic condition.

461174 From Fossils to Bio-Based Economy: A Revolutionary Transformation in Process Industries

Abstract:

Increasing utilisation and dependence on fossil fuels are attributed to the growing global population and soaring energy demand, and led to the global warming and energy security issues. Employing renewable energy sources such as biomass in new or existing systems and improving the performance of the existing facilities through maximising resource utilisation, enhancing energy efficiency and lowering environmental impact are the keys in addressing these issues. Process industries such as energy, chemical, refinery and manufacturing industries are the major consumers of fossil resources and sources of greenhouse gases emission. The replacement of traditional process design approach by a more holistic approach is exigent for the development of sustainable industrial systems and for effectively reducing the energy and emission intensities in the process industries. The transition from fossil fuel technologies to mature commercialisation of biorefinery may require a couple of decades. Hence, the most strategic solution during this transition is to improve the existing facilities through deployment of decarbonisation system, enhancing the system performance through process integration approach, while carrying out intensive research and investigations into biorefinery technology. This paper provides high level discussion on the global issues on energy and the strategic approach during the transition from fossil era to bio-based economy era. Modern process design approaches using various state-of the-art concepts associated with integrated biorefinery systems, decarbonisation strategies such as CO
mitigating environmental impact and meeting the global demand, will be discussed.

OPTIMAL CO-PRODUCTION OF MARKET BASED POWER GRID SUPPORT AND RENEWABLE FUELS OR CHEMICALS

Robert Weiss, Lotta Kannari, Jari Pennanen, Teemu Sihvonen, Jouni Savolainen*
VTT Technical Research Centre of Finland Ltd, Finland
**Corresponding author: robert.weiss@vtt.fi*

Introduction

This paper addresses the operational profitability and benefits of co-production and dynamics in renewable Power-to-Fuel or Power-to-Chemical processes, when utilized to stabilize the power grid. To reach climate neutrality, the transport sector must switch to climate-neutral or climate-positive fuels or power. This can be done by a proper mix of electric vehicles where the power is based on renewable electricity, hydrogen vehicles where the hydrogen is based on side-product hydrogen or renewable power based processes, or by using in existing vehicle fleet renewable biofuels instead of fossil fuels. For hydrogen-based or hydrocarbon-based chemical industry, similar change needs are obvious. Today, renewable fuels can be produced with biomass based plants or in bio-refineries. If biomass cannot be utilized or there is a lack of some sustainable chemicals like hydrogen in the bio-process, new so called Power-to-X plants [1, 2 and 3] could be used to produce renewable fuels or chemicals of non-biomass origin (X refers to the end product of the plant). These processes are utilizing hydro, solar or wind power via electrolysis for production of the needed additional hydrogen fraction [1]. Optimized day- or week-ahead planning can increase considerably the value of renewable Power-to-X plant operation. Such optimal operation can enhance commercial viability of the Power-to-X plants, and in this way introduce these climate-positive, mid-sized and distributed units into a smart grid. In this paper, we show examples on the value of such optimal dynamic operation together with the value of the dynamics of key unit processes and intermediate storages, while at the same time reaching a high share of sustainable solar and wind power in the end product.

Methodology

Sustainable Power-to-X Concept

Power-to-X (PtX) plant's main products can be typically fuels (H_2, CH_4, CH_3OH, Fischer-Tropsch-products), chemicals (NH_3, light olefins C_2H_4, C_3H_6), or reduction agents (H_2 for metals). Side products are high purity oxygen and low grade waste heat from the electrolysis operation, as well as process steam from some PtX processing units that are highly exotermic, like catalytic methanation. These side products can generally only be economically utilized locally "over-the-fence", and were neglected in this analysis for generality reasons.

For sustainability and avoidance of fossil fuels, the PtX operations should avoid fossil fuel based grid power, and instead utilize high enough content of sustainable renewable power. In our case examples we target to utilize as much wind and solar power as possible which are intermittent. On the other hand, parts of the PtX process may not be as flexible as the intermittent renewable power. Or the demand for the end product itself might even be 24/7, stable, but completely inflexible industrial demand. Finally, for sustainability the PtX plant should participate in power grid stabilization. This operating environment, depicted in Fig. 1, render the production planning a challenging task, necessitating the optimization approach adopted in this paper.

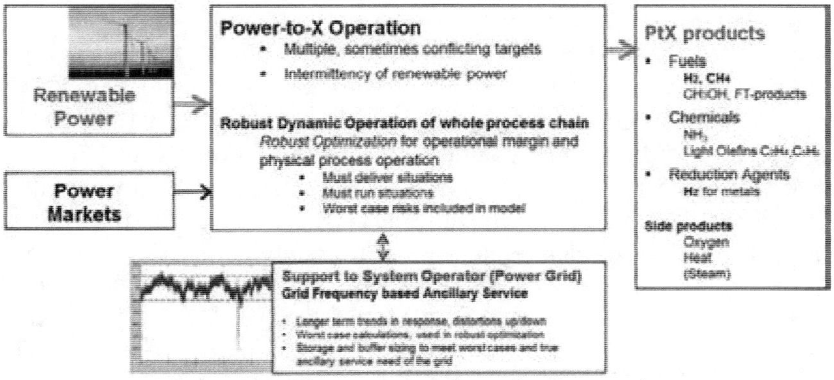

Figure 1 PtX concept and product examples.

Power-to-X Plant Operation and Dynamic Optimization Model

In our analysis, PtX power market contract handling as well as the physical production process is formulated as a dynamic, linear optimization model described in detail in [4], which minimizes the production cost. To our model, we added risk handling related to the power grid ancillary service as well as the utilization of available intermittent solar and wind power as described in the next sub-chapters.

For our analysis, we configured a 4x1.25 MW_e, state-of-the-art commercial PEM electrolyzer system utilizing confidential manufacturer data. As illustrative examples, we present the following example cases in our analysis: ***Case1*** describes on-site electrolytic production of H_2 as metal reduction agent, with a stable but unflexible H_2 demand. ***Case2*** describes Power-to-Gas production of CH_4 as green synthetic natural gas for sustainable mobility, with a flexibility range 70-100% in the catalytic methanation operation. In both cases, we investigate the impact of a 0-24h H_2-buffer storage. The optimization model schematic and case boundaries are depicted in Fig. 2.

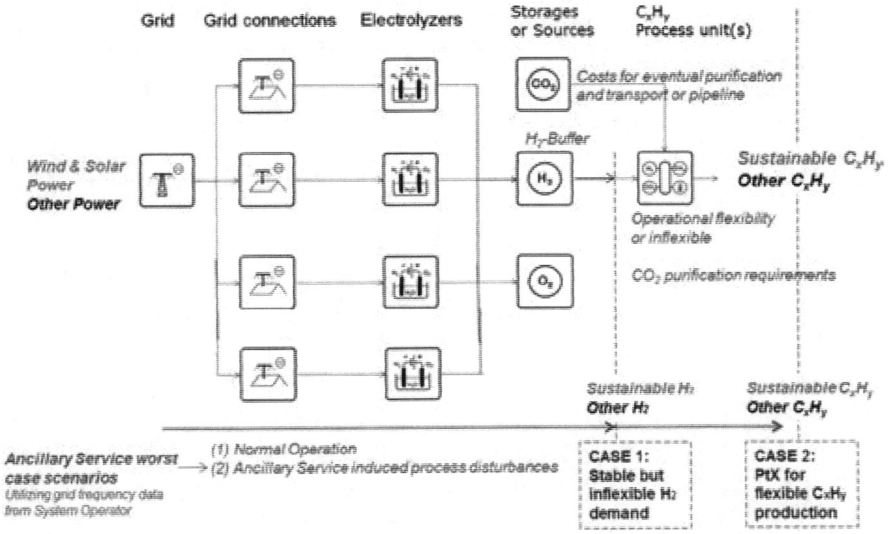

Figure 2 Schematic of PtX scheduling and dynamic optimization model, with the boundaries of example ***Case1*** and ***Case2***.

Markets for Wholesale Power and Ancillary Services

Deregulated electricity markets are generally divided into wholesale energy markets managed by power exchanges, and ancillary service markets managed by Transmission System Operators (TSO). Wholsesale energy is sold and purchased typically on international day-ahead markets for the 24 hours, see e.g. [5] for Nordic System, Baltic countries and Germany, or [6] for Germany, Austria, France and Switzerland. Concerning the ancillary service markets, situation of the Nordic countries compared to Netherlands, Germany and Poland (UCTE area) is well described in [7]. We list some examples on European ancillary service products in Table 1 below.

Table 1 Example on European Primary Frequency Control Schemes

Frequency Containment Reserve - Normal Operation **(FCR-N)** Also known as Primary Frequency Control or PRL	European 50Hz Grids: 49.9 – 50.1Hz or 49.8 – 50.2 Hz control band. Generally a symmetric reserve which is continuously triggered. Asymmetric bidding possible in some markets e.g. Denmark. Day-ahead markets in e.g. Finland, Denmark, and Norway. Week-ahead markets in Germany, Austria, Switzerland and The Netherlands.
Frequency Containment Reserve – Disturbance **(FCR-D)**	Scandinavian 50Hz Grids: 49.9 – 49.5 Hz (only disturbances). Asymmetric, fast. Not often triggered. Day-ahead markets in e.g. Finland, Denmark, and Norway.

In a power system with high penetration of intermittent renewable power production, grid balancing support and ancillary service prices tend to more and more volatile. As example on this effect, we utilized the hourly markets of the Finnish primary frequency control products FCR-N and FCR-D [8]. In this market, a large number of Combined Heat and Power plants (CHP) are on-line in the winter resulting in high power system intertia and a corresponding low need for additional FCR-N and FCR-D. In contrast, these plants are generally off-line in the summer, with nuclear power plants in maintenace, resulting in low intertia in the system and high prices for hourly additional FCR-N and FCR-D. Examples on this price volatility is shown in Fig. 3 below.

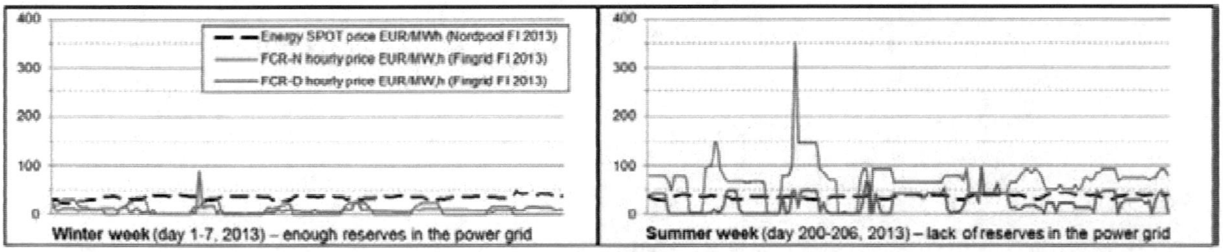

Figure 3 Hourly prices for energy [5] and FCR [8] during winter and summer example weeks.

Uncertainty in Disturbances and Ancillary Service Activations, Robust Optimization Solution

Most ancillary services, including primary frequency control like FCR-N and FCR-D, are capacity options with sometimes a strong uncertainty in the real need for their activation. The power system can have prolonged periods of under- or overfrequency, leading even in symmetric freqency control products (which theoretically should have 0 as net cumulative response) to long cumulative distortions towards large up or down regulation amounts, as depicted in Fig. 4.

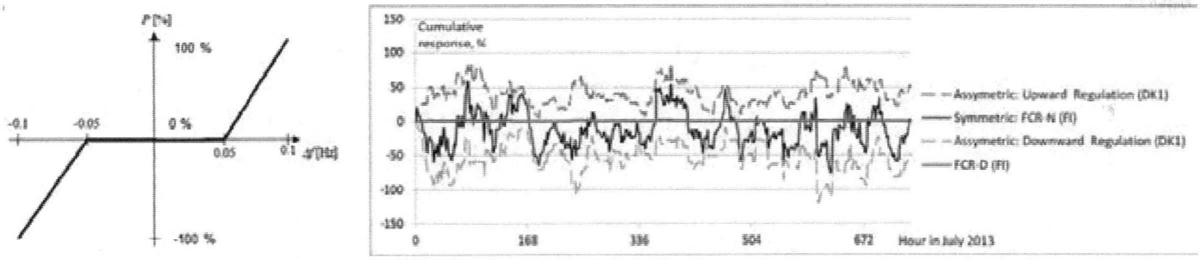

Figure 4 ***Left:*** FCR-N symmetric control scheme in Finland [8]. ***Right:*** Example on the 24 hour cumulative distortions up/down for the response of some primary frequency control schemes.

In our model, we handle such uncertainty with a robust optimization approach, where we solve the conflict between the need for stable unit process conditions versus the ancillary services' need for capacity reservation, fast reactions as well as prolonged activation worst case distortions up/down. Our ancillary service activation risk calculations utilize the *budget uncertainty set* methods from robust optimization theory [10, 11]. Uncertainty sets were defined for combined maximum and duration distributions of the ancillary service control response. These uncertainty set definitions were integrated into the optimization model for the electrolyzer power consumption and H_2 production, H_2 storage as well as H_2 and power balances. This resulted in a min-max bi-level optimization problem which was transformed into a LP model utilizing the duality of the problem. For the Finnish FCR-N and FCR-D schemes, budget uncertainty sets were statistically identified based on the primary frequency control scheme responses and cumulative worst cases identified from TSO frequency data, with a 0.1s resolution for the entire year 2013[8]. These identified uncertainty sets were utilized in our analysis.

Renewable Power and Dimensioning of a PtX Plant

Wind power from wind parks or larger areas can have short storm periods of very high power output still representing only a small energy amount of the entire yearly wind power production, as indicated in Fig 5. If a high annual wind energy share is desired for the PtX production, oversizing the nominal wind power capacity should be considered. Using additional solar power can be an advantage, since wind and solar power production are generally not strongly correlated. Also, in Northern countries like Finland, winds can generally be stronger in autumn and winter while solar production is concentrated to spring and summer. In our example, national wind power measurements from [8] and local solar PV measurements from [9] where re-dimensioned to 133% and 100% nominal capacity of the PtX plant, respectively, as shown in Fig. 5. Together, they resulted in a theoretical minimum curtailment energy loss below 2% of the yearly solar and wind energy available.

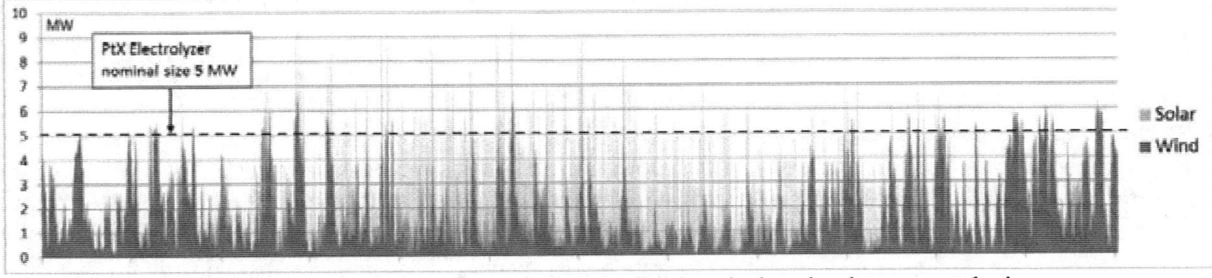

Figure 5 PtX plant nominal size compared to available wind and solar power during one year.

In our optimization study, we assumed wind and solar power to be directly and locally contracted on a yearly base, with only fixed costs but zero marginal costs when available. Forecasting errors of available wind and solar power were assumed to be balanced on average to zero costs on the balancing market. For hours when available wind or solar power where insufficient, the needed additional power could be purchased from the wholesale market, subject to transmission fees and taxes but no transmission congestions. For simplicity, we assumed that surplus wind and solar power not used had to be curtailed at zero prices, not able to reach the market for various congestion reasons.

Results and Discussion

In the analysis below we define the ***part load level sizing*** of the electrolyzer as the ratio of stable H_2 demand divided by the electrolyzer nominal H_2 capacity. ***Net power costs*** are defined as the wholesale power cost + wind and solar power costs – income from FCR-N and FCR-D hourly ancillary service contracts.

CASE 1: PtX with Stable but Inflexible H_2 Demand

Net power costs seem to be clearly lower if the electrolyzer is sized to 50-70% partial load (i.e. oversized) and larger H_2 buffer size decreased clearly the net power costs at all part load levels, as shown in Fig. 6 below. This finding applies both for the whole year analysis, as well as for the selected example weeks displayed with strongly differing pricing in FCR ancillary service (see Fig. 3). The FCR income was very high in the summer weeks, up to 3 US$/kg$H_2$, if large H_2-buffer space was available. For smaller H_2-buffers, our robust optimization restricted strongly the utilization of FCR according to the identified uncertainty sets for FCR control activation and PtX plant dynamics.

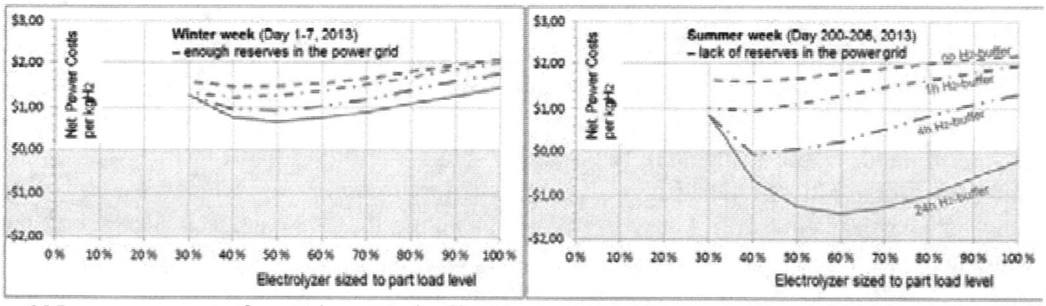

Figure 6 Net power costs for various H_2-buffer sizes during the example weeks. For a 24h H_2-buffer, the FCR income would have led to a net power *income* in the summer week.

Current state-of-the art commercial PEM and AEC electrolyzer technology can also be offered with an overload option, meaning that the electrolyzer may run for a short time substantially over its nominal capacity, but with less efficiency, and must thereafter cool down in the peripheral equipment like cooling circuits and transformers. In our analysis, we evaluated a theoretical +60% overload with a maximum utilization cycle of 0.5h overload + 0.5h cooling, which is quite close to current offerings on the PEM markets. The overload option showed no benefit during the winter example week due to low hourly FCR-N&D prices and absence of heavy peaks in wind power. However, during the summer example week, the overload option could bring substantial savings but only if a large enough H_2 buffer was available storing the overproduction (because of inflexible H_2-demand).

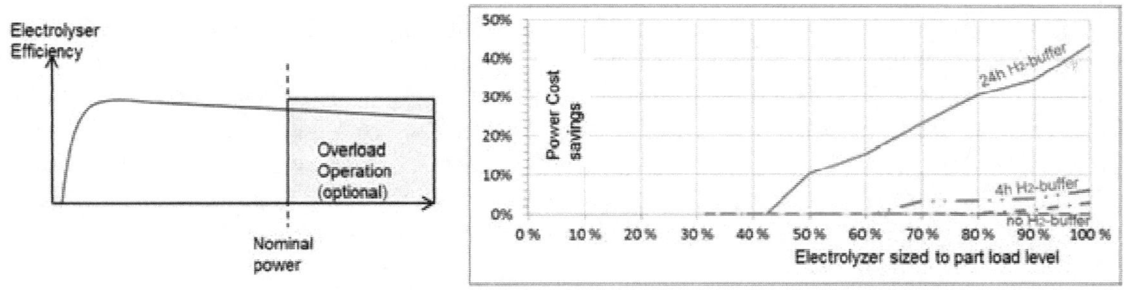

Figure 7 Electrolyzer overload option and power cost savings during the summer example week.

Calculating the hourly dynamic operation over an entire year, we analyzed the situation using a 24h H_2-buffer and 60% overload option, with results displayed in Fig. 8. Almost all available wind and solar power was used if the electrolyzer is sized to 100% i.e. full load. However, the wind and solar power usage decreases if the electrolyzer is sized to part load, which was accelerated below a 70% part load sizing. The reason for this utilization decrease was not the PtX plant flexibility, but the lack of H_2 demand at part load sizing. The share of sustainable power-based H_2 shows an opposite pattern: The share is as low as 50% for full load sizing, but can grow to close to 90% for a 30% part load sizing, being 65% at a 70% part load sizing. The reason for this utilization increase was the increased PtX plant flexibility at part load.

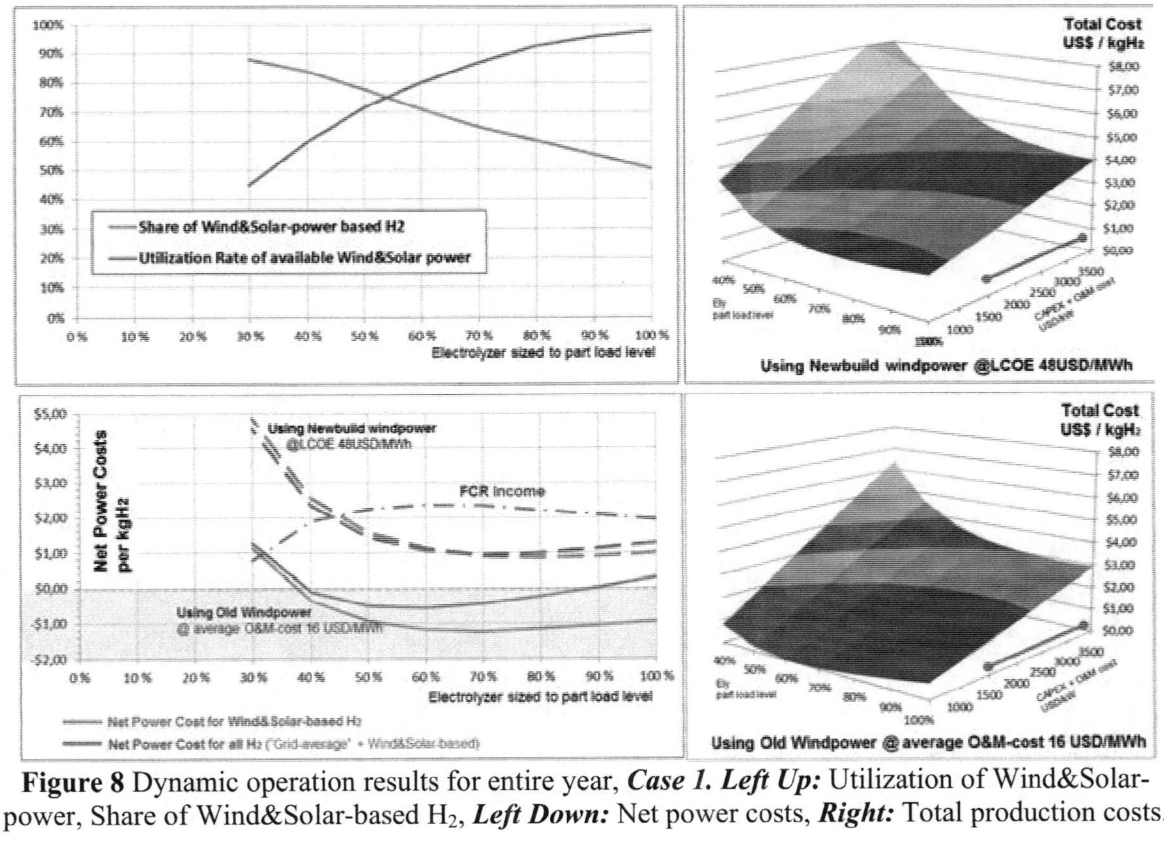

Figure 8 Dynamic operation results for entire year, ***Case 1. Left Up:*** Utilization of Wind&Solar-power, Share of Wind&Solar-based H_2, ***Left Down:*** Net power costs, ***Right:*** Total production costs.

Based on estimates in [3] and confidential information from manufacturers, for current commercial 1-1.5 MW_e AEC or PEM electrolyzer we estimate a fixed cost range between 1500 to 3500 US$/$kW_e$ for CAPEX and 10 year O&M + insurance costs, which is indicated with a red bar in Fig 8. If we use old "stranded" wind power at an O&M cost of 16 US$/MWh, the cost range would result in total H_2-prices 1.3-3.0 US$/kg$H_2$, which are close to or even competitive with steam reforming (avg. cost 1.7 US$/kg). When using new wind power at LCOE of 48US$/MWh, competitiveness can be reached only if electrolyzer CAPEX is clearly below 1000 US$/kWe.

CASE 2: PtX for flexible CH_4 production

Results were generally similar to Case1 and are displayed in Fig 9 below. Under-sizing the methanation unit towards the electrolyzer with a ratio 0.7:1 reduced the utilization of available solar and wind power, but increased the share of solar and wind based CH_4 in the end product gas mixture. The H_2 buffer can be kept at a minimum size of 1-4 hours, and larger buffer does not affect results at all, except if a longer term storage (week or larger) can be added. A more flexible methanation (operation range 10-100% instead of 70-100%) increased by 10% the share of solar and wind based CH_4 in the end product gas mixture, when the methanation unit was sized 1:1 towards the electrolyzer.

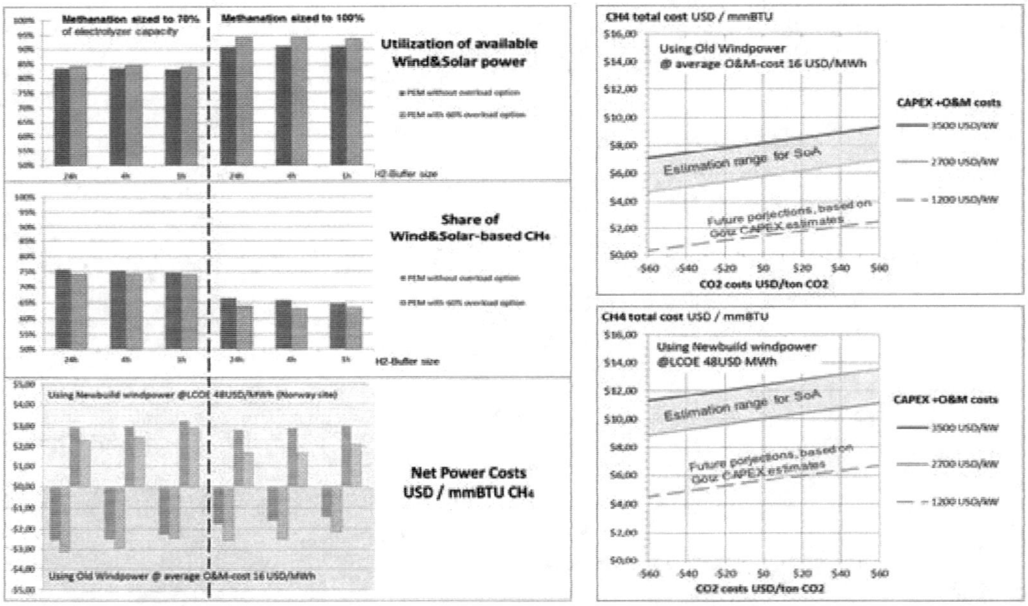

Figure 9 Dynamic operation results for entire year, ***Case 2. Left:*** Utilization of Wind&Solar-power, Share of Wind&Solar-based CH_4, and Net power costs, ***Right:*** Total production costs.

Methanation CAPEX estimation was challenging, since small- and medium scale methanation is not available on the market today. Using the fixed cost estimates for the electrolyzer as presented in previous sub-chapter, and estimates in [3] for methanation, we can see from the results in Fig. 9 that competitive CH_4-prices compared to current US SPOT prices (< 3 US$/mmBTU) seem challenging to reach. Higher CO_2-allowance prices and/or cost efficiency improvements in the electrolyzer and methanation manufacturing, and technology improvements could however change the picture.

Conclusions

PtX example cases for solar- and wind power-based H_2 and CH_4 production were shown for Finnish climate, power market and grid conditions for fast ancillary services. Evidently, concurrent sizing and operational planning is needed to find economical potential solutions and plant configurations. Participation in power grid ancillary services, e.g. primary frequency control, is essential both for plant economics and power system sustainability. Also, state-of-the-art PEM and AEC electrolyzer offer capacity to temporarily overload during primary frequency control price peaks, which is economically beneficial. We demonstrated the ability of robust optimization to handle the activation uncertainty of fast ancillary services, which proved to be an essential part of the optimization solution.

Competitive H_2-prices seem to be reachable for PtX compared to steam reforming, even for an inflexible end customer. In contrary, the sustainable CH_4 case seems to require technological advances or more benefits for the reuse CO_2 and avoidance of fossil fuels. Alternatively, dedicated markets have to be found or created where conventional natural gas cannot be offered, e.g. bio-methane as a sustainable transportation fuel.

References

1. Gahleitner, Gerda (2013), "Hydrogen from renewable electricity: An international review of power-to-gas pilot plants for stationary applications", *International Journal of Hydrogen Energy*, 38(5), February, pp. 2039-2061.
2. Rönsch, Stefan, Schneider, Jens, Matthischke, Steffi, Schlüter, Michael, Götz, Manuel, Lefebvre, Jonathan, Prabhakaran, Praseeth and Siegfried Bajohr (2015), "Review on methanation – From fundamentals to current projects", *Fuel* 166, pp. 276–296.
3. Götz, Manuel, Lefebvre, Jonathan, Mörs, Friedemann, McDaniel Koch, Amy, Graf, Frank, Bajohr, Siegfried, Reimert, Rainer and Thomas Kolb (2015), "Renewable Power-to-Gas: A technological and economic review", *Renewable Energy* 85, January, pp. 1371–1390.
4. Weiss, Robert, Savolainen, Jouni, Peltoniemi, Pasi and Eero Inkeri (2016), "Optimal scheduling of a P2G plant in dynamic Power, Regulation and Gas markets", in: *10th International Renewable Energy Storage* (IRES 2016), Düsseldorf.
5. Nordpool (2016), http://nordpoolspot.com/
6. EPEX (2016), https://www.eex.com/en/products/power/power-spot-market
7. Grande, Ove S., Doorman, Gerard and Bjorn H. Bakken (2008), "Exchange of balancing resources between the Nordic synchronous system and the Netherlands / Germany / Poland." *SINTEF Energy Research*, Technical Report.
8. Fingrid Oyj (2016), Finnish Transmission System Operator, Helsinki, www.fingrid.fi
9. HELEN Ltd (2016), https://www.helen.fi/sahko/kodit/aurinkosahko/suvilahti/
10. Ben-Tal, Aharon, El Ghaoui, Laurent and Arkadi Nemirovski (2009), "Robust Optimization", Princeton University Press.
11. Bertsimas, Dimitris, Brown David B. and Constantine Caramanis (2011), "Theory and Applications of Robust Optimization", *SIAM Review* 53(3):464-501.

453906 Improved Thermal Efficiency Analysis of Biomass Cookstoves

Abstract:

Biomass is an important sustainable energy source, especially in the developing world. One common use of biomass in the developing world is through biomass cookstoves. Biomass cookstoves are used by approximately 2 billion people worldwide on a daily basis. In order for this use to be as sustainable as possible, efforts to reduce wood usage are important to help minimize environmental and health effects associated with pollution and deforestation. Currently the effects of biomass cookstoves on deforestation have been shown to be relatively small, though minimizing wood usage can reduce health, safety and environmental risks.

Because of these risks, the safety and efficiency (both thermal and combustion) of biomass cookstoves has been considered and improved by many researchers through the design of cookstoves and associated auxiliary features such as pot skirts. Once a new cookstove or auxiliary feature has been designed, the design is assessed and optimized. The most common test for assessing cookstoves is the Water Boiling Test (WBT). The WBT consists of three tasks. The first task is to heat water from room temperature to boiling using a "cold" cookstove that hasn't been in use. The second task involves repeating the first task but starting with a "hot" cookstove that was recently used and still has thermal energy in the cookstove structure. The final task directly follows the second task by lowering the feed rate of fuel and keeping the water a few degrees below boiling for 45 minutes.

Measurements of emissions (CO , CO_2, particulate matter, etc.), fuel used, water boiled off, and the time to complete each task are performed during the WBT. Metrics, including thermal efficiency, time to boil, burning rate, firepower, turn-down ratio, and emissions per fuel burned are then calculated from the measurements. Unfortunately, some of these metrics are averages based on endpoint analysis (measurements taken at the beginning and end of the test). This work expands upon the use of the thermal efficiency metric by developing and validating a model that can be used to estimate the average thermal efficiency at any point during the boiling process. This transient average thermal efficiency can be used to gain insight into the attributes of tested cookstoves. Such insights can include the minimal testing time required to determine the endpoint thermal efficiency used for cookstove comparisons, the effects of testing perturbations on thermal efficiency, and the difference in the behavior of thermal efficiency between cold starts and hot starts.

The WBT protocol along with the developed method for estimating thermal efficiency with time was performed using a typical Peruvian cookstove made of two small parallel walls of bricks, modified with both a grate under the fire and pot skirts. The studies using this stove showed that the time required to reach steady-state thermal efficiency was significantly different for cold and hot starts. Additionally, the time and temperature difference required to reach steady state thermal efficiency was significantly lower than required to actually boil the water as required by the WBT. The experimental runs with time-based endpoints were also found to be in good agreement with the estimation shown in this work. This expansion of the WBT protocol should allow for more efficient analysis and optimization of biomass cookstoves.

474844 Pilot-Scale Catalytic Biomass Pyrolysis Studies

Abstract:

Catalytic fast pyrolysis is the direct liquefaction process used by RTI to produce a hydrocarbon-rich biocrude that can subsequently be upgraded by hydrotreating to make refinable intermediates and â€œdrop-inâ€ advanced biofuels. As part of the technology development, the process has been scaled up from bench-scale to 1 ton per day (1TPD) pilot-scale. Evaluation of the process at the pilot-scale is one of the critical steps being taken by RTI to address a number of technical challenges facing catalytic pyrolysis and to collect required processing and engineering data to support the demonstration and commercialization of the technology.

Up to date, several experimental campaigns have been conducted to optimize the catalytic biomass pyrolysis process to achieve high degree of deoxygenation while maximizing the bio-crude production and improving the biocrude thermal stability. A commercially available spray dried, non-zeolite, alumina-based catalyst has been exclusively studied so far. The quality and biocrude yield has been evaluated at different temperatures and other process conditions. Typically, partially deoxygenated biocrude with oxygen contents between 20 and 25 wt% is produced. Gas Chromatography Mass Spectrometry (GC-MS) analysis have shown high amount of oxygenated aromatics in the partially deoxygenated biocrude. Current results from RTIâ€™s 1 ton per day pilot-scale catalytic pyrolysis unit show a biocrude yield up to 19.2 wt% is achievable. This presentation will be used to discuss in detail the results that have been obtained from several studies and highlight the progress that has been made.

Catalytic Pyrolysis of Bio-Oil Model Compounds over La/Ce/Ni Modified HZSM-5 Zeolites

Fengwen Yu, Qiang SONG, Yong NIE, Zhichao XU, Jianbing JI
Zhejiang University of technology, Hangzhou, China

Abstract :

- With the contradiction between supply and demand of petroleum resources and environmental pollution, more and more attention has been paid to the study of clean and renewable energy, especially converting bio-oil into biofuel. Bio-oil catalytic pyrolysis has several advantages, such as low capital cost, easy handling and safe controlling. It has been widely used in improving the quality of bio-oil, for fossil fuels and chemicals production.
- The experimental present a catalytic strategy to upgrade a simulated bio-oil consisting of ethyl acetate, diacetone alcohol, furfural and guaiacol, aiming to study the effect of La/Ce/Ni modified HZSM-5 zeolites on catalytic pyrolysis. The catalysts properties are investigated by BET and NH_3-TPD and the major constituents of the pyrolytic products are carried out by GC/MS. The results have pointed out the yield of organic compounds is increased from 42.26% to 65.05% by Ce-HZSM-5, whilst keeping lower coke formation rates after adding metal cations under the condition of T=450℃, WHSV=4 h^{-1}. The major components of catalytic pyrolysis are aromatic hydrocarbons, ketones, phenols and so on. The optimal selectivity of aromatic reached 46.37% over Ce-HZSM-5. The La-HZSM-5 improved the selectivity of partial deoxidation products from 4.88% to 21.97% .

2016 AIChE Annual Meeting

AIChE

Catalytic pyrolysis of bio-oil model compounds over La/Ce/Ni modified HZSM-5 zeolites

Zhejiang Province Key Laboratory of Biofuel,
College of Chemical Engineering,
Zhejiang University of Technology,
Hangzhou, China

浙江工业大学
Zhejiang University of Technology

Fengwen YU
Qing SONG
Nie YONG
Zhichao XU
Jianbing JI

CONTENTS

Part 1

Background

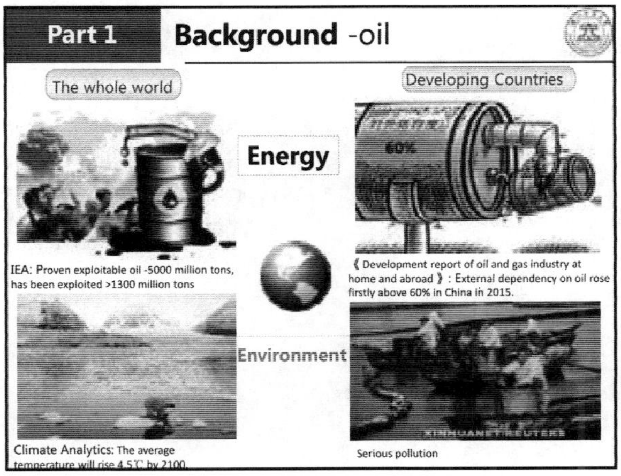
Part 1
Background -oil
The whole world
Developing Countries
Energy
IEA: Proven exploitable oil -5000 million tons, has been exploited >1300 million tons
《Development report of oil and gas industry at home and abroad》: External dependency on oil rose firstly above 60% in China in 2015.
Environment
Climate Analytics: The average temperature will rise 4.5℃ by 2100
Serious pollution

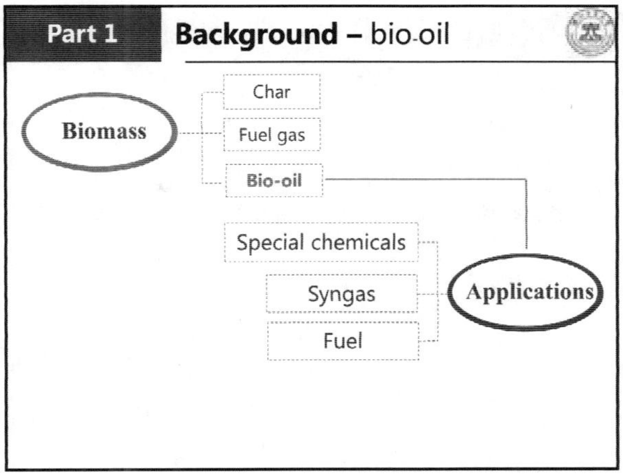
Part 1
Background – bio-oil
Biomass
Char
Fuel gas
Bio-oil
Special chemicals
Syngas
Fuel
Applications

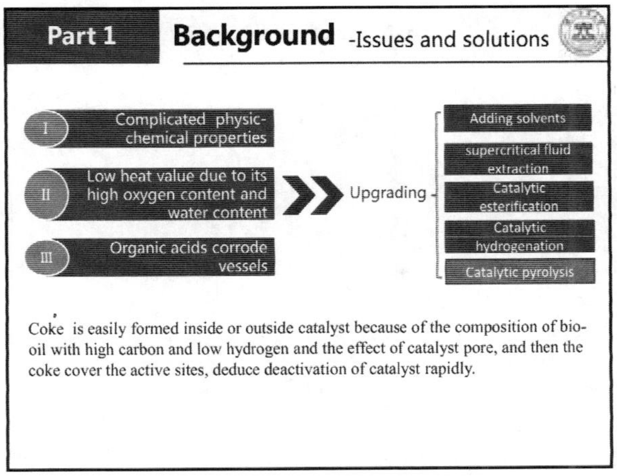
Part 1
Background -Issues and solutions
I Complicated physic-chemical properties
II Low heat value due to its high oxygen content and water content
III Organic acids corrode vessels
Upgrading
Adding solvents
supercritical fluid extraction
Catalytic esterification
Catalytic hydrogenation
Catalytic pyrolysis
Coke is easily formed inside or outside catalyst because of the composition of bio-oil with high carbon and low hydrogen and the effect of catalyst pore, and then the coke cover the active sites, deduce deactivation of catalyst rapidly.

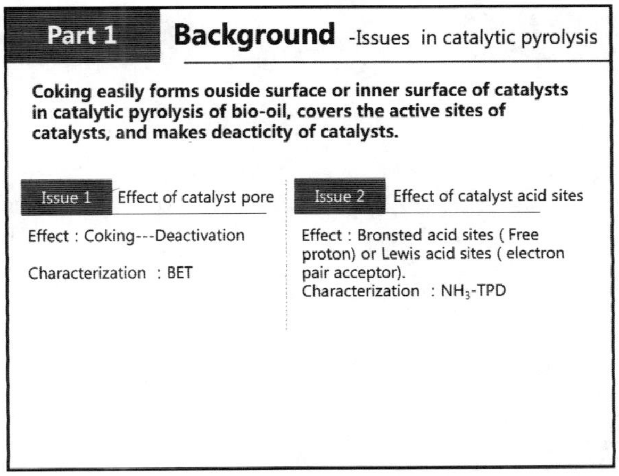
Part 1
Background -Issues in catalytic pyrolysis
Coking easily forms ouside surface or inner surface of catalysts in catalytic pyrolysis of bio-oil, covers the active sites of catalysts, and makes deacticity of catalysts.
Issue 1 Effect of catalyst pore
Effect : Coking---Deactivation
Characterization : BET
Issue 2 Effect of catalyst acid sites
Effect : Bronsted acid sites (Free proton) or Lewis acid sites (electron pair acceptor).
Characterization : NH3-TPD

Part 1 Background –purpose of the research

The key point lies in screening and modifying catalysts in order to obtain high degree of pyrolysis, high effect of deoxidation, and low carbon deposition and coking.

Part 2 Experimental

Part 2 Experimental

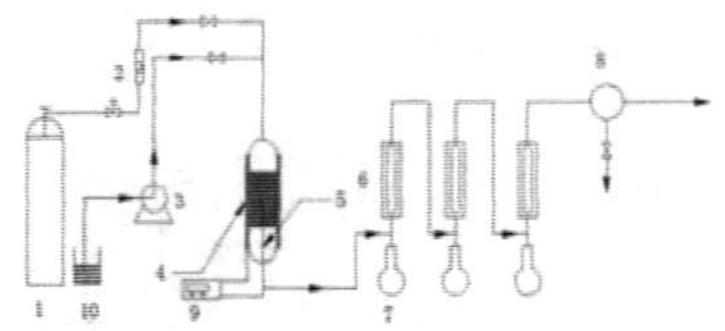

Fig. Experiment flowsheet of catalytic pyrolysis of bio-oil compounds
1-N_2 tank, 2-Rotameter, 3-Peristaltic pump, 4-Heating jacket, 5-Reactor, 6-Condensation pipe, 7-Liquid product receiving bottle,8-Gas buffer bottle, 9-Automatic temperature control unit, 10-Bio-oil (model compounds with Diacetone alcohol, Ethyl acetate, Furfural, Guaiacol)

- Continuous and convenient to control the temperature

Part 2 Preparation of modified HZSM-5 catalysts

Materials

HZSM-5 (Si/Al=50)
$La(NO_3)_3 \cdot 6H_2O$
$Ni(NO_3)_2 \cdot 6H_2O$
$(NH_4)_2HPO_4$

Catalysts

1.0 wt.%La-HZSM-5,
1.0 wt.%Ce-HZSM-5,
1.0 wt.%Ni-HZSM-5

Procedure

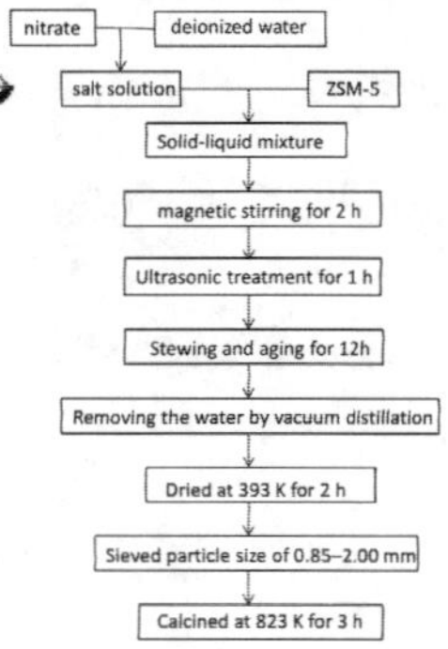

Part 3 Results and discussion

Part 3 Properties of modified HZSM-5 molecular sieve

Table Pore structure parameters of catalysts

Types of Catalysts	$S_{BET}/(m^2 \cdot g^{-1})$	Pore size/nm	Pore volume/ $(ml \cdot g^{-1})$
HZSM-5	308.61	0.77	0.15
La-HZSM-5	300.62	0.79	0.15
Ce-HZSM-5	297.72	0.83	0.15
Ni-HZSM-5	290.42	0.80	0.14

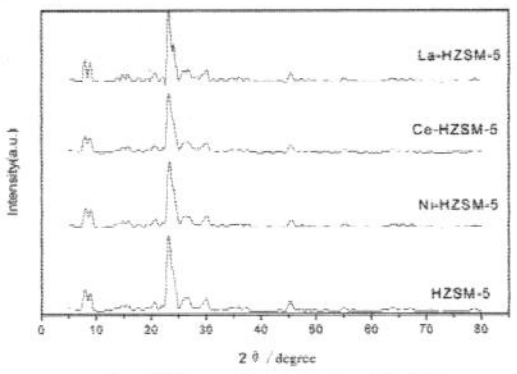

Fig. XRD patterns of La/Ce/Ni modified HZSM-5 zeolites

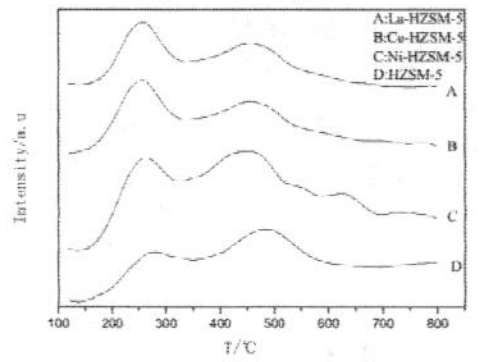

Fig. NH_3-TPD profiles of La/Ce/Ni modified HZSM-5

Part 3 Properties of modified HZSM-5 molecular sieve

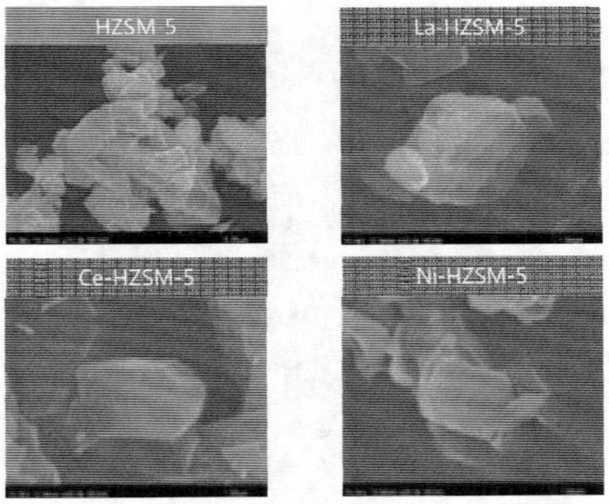

Fig. SEM images of La/Ce/Ni modified HZSM-5 zeolites

Part 3 Effect of temperature on model compounds pyrolysis

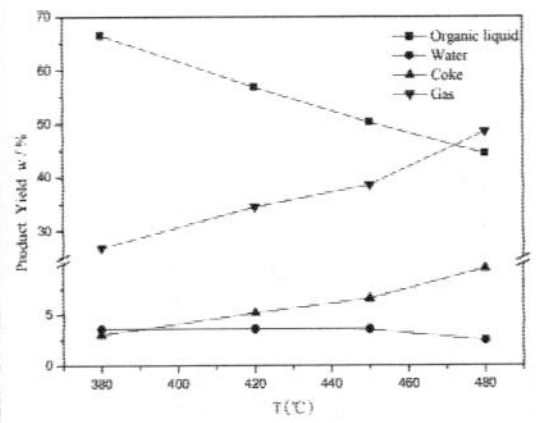

Fig.Products distribution at different temperatures (Feed rate=1.5 ml·min^{-1}, WHSV =4 h^{-1}, no catalyst)

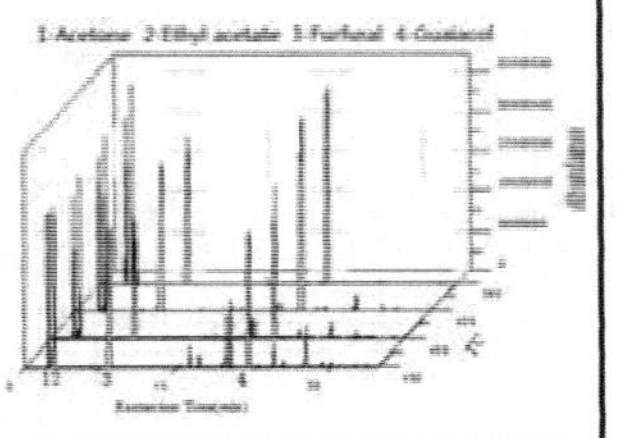

Fig. TIC of organic liquid obtained at different temperatures

- 450°C is optimal temperature due to higher pyrolysis yield and moderate product distribution.

Part 3 Effect of mdified catalysts on product distribution

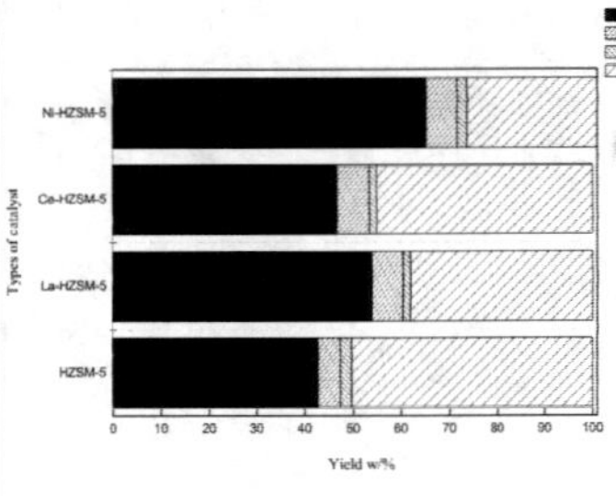

Fig. Products distribution on different catalysts
(At 450℃, feed rate=1.5 ml·min^{-1}, WHSV =4 h^{-1})

- Organic liquid yield increased
- Coking yield decreased
- Gas yield decreased

Part 3 Effect of mdified catalysts on conversions

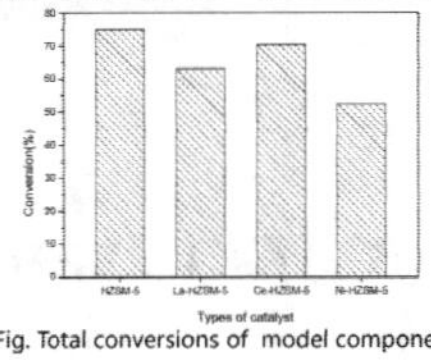

Fig. Total conversions of model components

- Total conversions decreased

- Single conversions decreased except EA over Ce-HZSM-5 and DAA

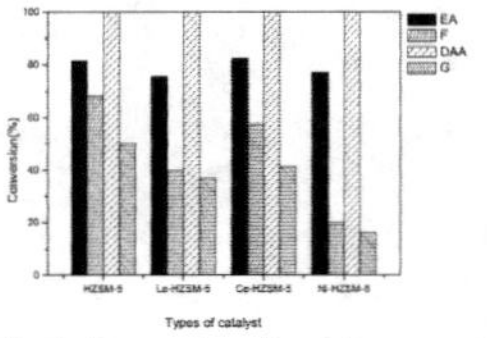

Fig. Single conversions of model component

Part 3 Deoxidizing products over HZSM-5

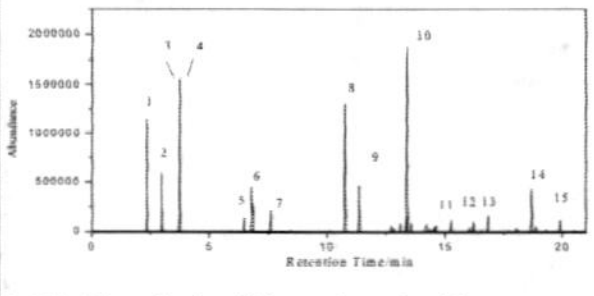

Fig. TIC of the major deoxidizing products of model components

Table Major deoxidizing products of model components

Number	Retention time/min	Compounds	m/z
1	2.343	Benzene	78
2	2.669	**2-Methyl Furan**	81
3	2.978	**2-Vinyl Furan**	94
4	3.737	Toluene	91
5	6.465	Ethyl benzene	91
6	6.828	M-Xylene	91
7	7.399	P-Xylene	91
8	10.755	**Phenol**	94
9	11.357	**Benzofuran**	118
10	13.374	**O-Phenyl phenol**	108
11	14.193	**M-Cresol**	108
12	15.250	**2-Methyl benzofuran**	131
13	16.817	**Dihydrobenzofuran**	120
14	18.043	Naphthalene	128
15	19.903	**2-Ethyl benzofuran**	131

Complete deoxidizing products

Partial deoxidizing products

Part 3 Effect of mdified catalysts on Deoxidizing degree

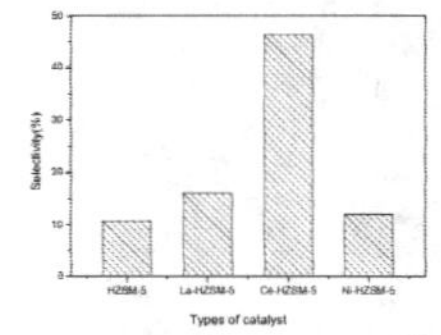

Fig. The selectivity of aromatic on catalysts

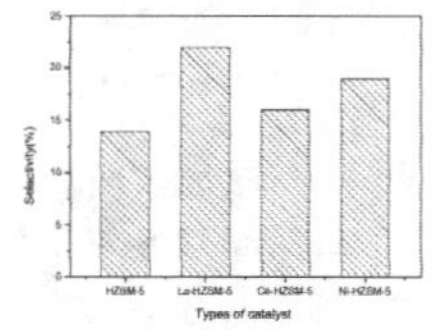

Fig. The selectivity of partial deoxidizing products on catalysts

- Selectivity of aromatic increased, Ce-HZSM-5 was the best.
- Selectivity of partial deoxididation increased, La-HZSM-5 was the best.

Part 4

Conclusion

Part 4

Conclusion

(1) Organic liquid yield increased, coking yield decreased, gas yield decreased over La/Ce/Ni modified HZSM-5 zeolites.

(2) Under the pyrolysis condition of 450℃, WHSV =4 h^{-1}, the selectivity of aromatic increased, Ce-HZSM-5 was the best, 46.37%; the selectivity of partial deoxididation increased, La-HZSM-5 was the best, 21.97%.

(3) Weak acid was increased and strong acid was decreased over La/Ce modified HZSM-5 zeolites, So the anti coking property and the ability of deoxididation were improved.

Part 5

Acknowledgments

Part 5

Acknowledgments

The authors thank the 863 Program (Grant No. 2014AA022103) and Doctoral Fund of Ministry of Education of China (Grant No. 20103317110001).

Part 6 References

Part 6 References

[1] Eoin Butler; Ger Devlin; Dietrich Meier; Kevin McDonnell , A review of recent laboratory research and commercial developments in fast pyrolysis and upgrading [J]. Renewable and Sustainable Energy Reviews, 2011, 15(8): 4171-4186.

[2] Fengwen Yu; Longchao Gao; Weijin Wang; Guodong Zhang; Jianbing Ji. Bio-fuel production from the catalytic pyrolysis of soybean oil over Me-Al-MCM-41 (Me-La, Ni, or Fe) mesoporous materials[J]. Journal of Analytical and Applied Pyrolysis, 2013, 104:325-329.

[3] W T Tsai, M K Lee, Y M Chang. Fast pyrolysis of rice husk: Product yields and compositions[J]. Bioresource Technology, 2007, 98(1): 22-28.

[4] Sathish K. Tanneru, Philip H. Steele. Direct hydrocracking of oxidized bio-oil to hydrocarbons[J]. Fuel, 2015,154: 268－274

[5] A.V. Bridgwater. Review of fast pyrolysis of biomass and product upgrading[J]. Biomass and Bioenergy, 2012, 38:68-94.

[6] YANG Wen-yan, ZENG Yan, LUO Jia, TONG Dong-mei, QING Ren-wei, FAN Yong, HU Chang-wei. Production of bio-oil by direct and catalytic pyrolysis of Nannochloropsis sp. [J]. Journal of Fuel Chemistry and Technology， 2011， 39(9)： 664－669.

[7] Supawan Vichaphund ; Duangdao Aht-ong;Viboon Sricharoenchaikul;Duangduen Atong. Production of aromatic compounds from catalytic fast pyrolysis of Jatropha residues using metal/HZSM-5 prepared by ion-exchange and impregnation methods [J]. Renewable Energy, 2015, 79:28-37.

[8] CHANG Sheng, ZHAO Zeng-li , ZHANG Wei, ZHENG An-qing, WU Wen-qiang, LI Hai-bin. Comparison of chemical composition and structure of different kinds of bio-oils[J]. Journal of Fuel Chemistry and Technology， 2011， 39(10)： 746－753.

[9] Feiyan Gong, Zhi Yang, Chenggui Hong, Weiwei Huang, Shen Ning, Zhaoxia Zhang, Yong Xu, Quanxin Li. Selective conversion of bio-oil to light olefins: Controlling catalytic cracking for maximum olefins [J]. Bioresource Technology, 2011,102: 9247–9254.

[10] Ji Hongqiang , Zhang Qiang , Dong Xue, Shan Honghong. PERFORMANCE OF METHANOL AROMATIZATION ON NICKEL MODIFIED HZSM-5 CATALYSTS[J]. Petroleum Processing and Petrochemicals, 2015,46(2):58-63.

[11] Shanling Li;Suping Zhang;Zhanyuan Feng;Yongjie Yan. Coke formation in the catalytic cracking of bio-oil model compounds [J]. Environmental Progress and Sustainable Energy , 2015,34(1): 240-247.

[12] S L González-Cortés, S Rugmini, T Xiao, M.L.H. Green, S.M. Rodulfo-Baechler, F.E. Imbert. Deep hydrotreating of different feedstocks over a highly active Al_2O_3-supported NiMoW sulfide catalyst[J]. Applied Catalysis A:General, 2014, 475:270-281.

[13] Minghao Zhou, Yuan Wang, Yanbin Wang, Guomin Xiao. Catalytic conversion of guaiacol to alcohols for bio-oil upgrading[J]. Journal of Energy Chemistry, 2015,24 (4): 425-431.

[14] Wen Chen, Zhongyang Luo, Chunjiang Yu, Yi Yang, Guoxiang Li, Jixiang Zhang. Catalytic conversion of guaiacol in ethanol for bio-oil upgrading to stable oxygenated organics[J]. Fuel Processing Technology, 2014,126: 420-428.

[15] CHEN Jiao-jiao , CHEN Guan-yi , MA Wen-chao , MA Long-long, WANG Tie-jun, ZHANG Qi, LU Wei. Experimental study of aromatics production from catalytic cracking of bio-oil model compounds. Journal of Fuel Chemistry and Technology， 2013， 41(2)： 183-188.

[16] Arakshita Majhi;Y.K. Sharma;Rajaram Bal, Babita Behera, Jagdish Kumar. Upgrading of bio-oils over PdO/Al2O3 catalyst and fractionation[J]. Fuel, 2013,107: 131-137.

浙江工業大學
ZHEJIANG UNIVERSITY OF TECHNOLOGY

THANKS!

464724 Process Modeling of Fluidized Bed Biomass-CO_2 Gasification Using Aspen Plus

Abstract:

Biomass can be converted to biofuels using various pathways which are broadly classified as biochemical or thermochemical. Gasification is a common thermochemical process that converts biomass to a gaseous product (syngas) which can be used for power generation, chemical production and as a fuel. The use of air, steam and oxygen as oxidizing agents in this process has been widely studied and reported. Improved CO concentrations have been observed when CO_2 is added as a co-reactive agent in the gasification of biomass.

In this work, a comprehensive model simulating the biomass gasification process using CO_2 in a fluidized-bed gasifier is developed using ASPEN Plus. The model utilizes additional reactors and mechanisms simulating various steps in the gasification process to predict the effect of operating parameters such as temperature and CO_2 flowrate on biomass conversion and syngas composition. A sensitivity analysis of product gas composition and a comprehensive sustainability assessment is conducted to evaluate the economic, environmental, and social performance of the biomass-CO_2 gasification process.

The results from this model are compared to published gasification experimental data obtained on a bench scale fluidized bed gasifier.

451759 Simulation, Heat Integration and Rectisol-Based Decarbonisation for the Production of Synthetic Natural Gas from Biomass Gasification and Landfill Gases

Abstract:

The outline of this research is to develop a systematic heat-integrated gas purification process, followed by methanation for the production of synthetic natural gas or substitute natural gas (SNG). SNG production from biomass is currently considered as an alternative to reduce dependency on high-priced natural gas import and an opportunity to reduce greenhouse gases by applying CO_2 Capture and Sequestration (CCS) technologies. Biomass is thermally converted (e.g. gasification) to produce CO_2, CO, H_2O and H_2 gases that are cleaned to remove CO_2 before being converted to SNG (methane) via a methanation process.

The focus of this project is to develop a comprehensive simulation framework of heat-integrated decarbonized process suitable for the removal of contaminants from biomass-based produced gases to meet SNG production specifications.

In this study, a detailed heat-integrated steady-state simulation framework for biomass thermal conversion, water gas shift (WGS) reaction, gas purification by Rectisol process and SNG production via methanation process has been developed. The main objective was to explore heat recovery opportunities, cleaner operation by reducing contaminants and optimised selectivity with minimal cost. The simulation framework was constructed using Aspen HYSYS (V.2006) process simulator, which is capable to predict the optimised operating conditions of gasification, methanation and Rectisol process at which containments and solvent loss are minimized. Optimization software (SPRINT) is used for data extraction and analysis of heat integration opportunities across the entire process model.

In this project, a biomass (straw) was initially pyrolyzed to mainly produce a gas stream consisting of CO_2, CO, H_2O and H_2 along with tar and char. These streams were then mixed with landfill gases that comprise of methane (CH_4) (40-60%), CO_2 and H_2S as major components that also considered as major greenhouse gases (GHG). Furthermore, these gases are contaminated with sulphur, nitrogen, oxygen, siloxanes, water, halogenated compounds and many other chemicals which require complex gas clean-up process. This mixture was gasified to produce a gas stream consisting mainly of CO_2, CO, H_2O, H_2 and H_2S. The gas stream was then fed to a water gas shift reactor to increase H_2/CO ratio to 3. This gas mixture was cleaned up by a Rectisol process to remove CO_2, H_2S and other containments. The cleaned gas (CO, H_2O, H_2) is finally fed to a methanation process to convert it virtually to pure methane (CH_4), which is the desired final product (i.e. SNG).

The Rectisol process uses refrigerated methanol as a solvent, where 99% removal of CO_2 and H_2S that are particularly present in landfill gases and product gas from biomass gasification was achieved. Two-level refrigeration cycle was developed to maintain the operating temperature of the overall gas purification process. The methanation route was selected for the production of SNG. Methanation is an exothermic reaction and hence, a comprehensive sensitivity analysis was carried out to optimize the process by effective removal of heat of reaction and to achieve maximum methane (CH_4) formation. This analysis showed that four reactors were used to achieve maximum conversion CO to CH_4 which corresponded to CO conversion of 97.8%. After an in-depth heat integration study of the overall process model, 2.5 MW of steam at 320 °C was generated and 0.45 MW process heat recovery was achieved, which resulted in energy savings.

In summary, this project establishes process simulation framework and heat-integrated design of biomass gasification, gas purification and methanation technology. Extensive heat integration strategies have been established based on the following criteria; maximum heat recovery, cleaner and economical production of SNG. The research focused on determining optimal operating conditions for WGS reactor to maintain H_2/CO ratio of 3, for Rectisol Process to maintain 99% removal of CO_2 and H_2S and for methanation process to maximise the conversion rate to SNG. The following conclusions have been drawn from this research:

1. The overall efficiency of SNG production depends significantly on the energy content of the biomass feedstock.
2. Optimizing the refrigeration system for the Rectisol Process plays a vital role in minimizing the total energy requirement of the whole process.
3. Environmental and sustainability factors should always be taken into account prior to selecting the energy conversion process.
4. To meet the SNG specification regarding composition, relative density, calorific value and Wobbe Index; the methanation process must be optimised for maximum conversion of CO to CH_4.
5. Systematic heat integration of the overall process is essential to reduce the utility requirements and provides opportunity for heat and power generation.
6. The cost of electricity from steam generation and cost of carbon capture were estimated to be a function of biomass feedstock cost. It is concluded from this study that biomass feedstock cost is absolutely critical in determining the overall operating cost.

461719 Production of Synthetic Gas through Agricultural Waste Using Fixed BED Gasifier

Abstract:

Gasification is a process that converts organic fuels or fossil fuels (carbonaceous materials) into synthetic gas. It is referred as a producer gas from agricultural waste using gasifier. Biomass can be obtained from living organisms like wood, waste materials and agricultural waste were to be gasified and convert into synthetic gas by gasification process. The results and experiments of solid waste by gasifier is operating based on stoichiometric calculations referred as three different types of agricultural waste like cotton crop waste, tree barks and tree leafs. The material with higher content of the moisture was tree barks waste that was noted as 13.3% whereas cotton crop waste had 12.23% of moisture content and tree leafs waste had 12.81% moisture content. The study showed that cotton crop waste had almost same production of synthetic gas as of tree barks but the productivity was achieved higher in tree barks which had greater quantity of fixed carbon which was noted as 14.13%. The tree leafs also had higher amount of fixed carbon was noted as 12.85% but the ash content in tree leaf was noted as 18.92 %. The composition of synthetic gas from cotton crop waste was 27.52% CO and 32.17% H_2. The composition of synthetic gas from Tree barks was 28.72% CO and 31.88% H_2. The composition of synthetic gas from tree leafs were 26.07% CO and 28.12% H_2. As examined the increasing fuelâ€™s demand and consumption in a wide country to complete and fulfill the requirement of fuel and energy to achieve the desired targets and goals in Pakistan through the production of alternative fuels with inexpensive way. An innovative process of synthetic gas from solid waste and agricultural biomass is gasification process to generate SYN gas that can be used as replace of natural gas, LPG, power generation and also used for the production of DME (dimethyl ether) to run the mechanical vehicles in the great performance where compared to other fuel resources.

466240 The Steam Iron Process for Pressurized Hydrogen Production

Abstract:

The commitment to fuel cell powered electric vehicles offer a carbon emission free mobility in the near future. A key requirement to achieve this goal is the supply of low cost hydrogen produced out of renewable resources. Currently more than 90% of the hydrogen production involves the utilization of fossil fuels in large scale centralized plants. The supply of cost-competitive hydrogen by renewable resources requires a different production infrastructure. Transportation distances of resources and of hydrogen itself have to be minimized and the production has to be realized in decentralized small scale production facilities. At Graz University of Technology the reformer steam iron process is developed as a scalable, on-site process, to emphasize on the requirements of natural resources. The process combines the steam reforming with the steam iron process (a chemical looping water splitting process) in one single unit.

In the first step of the process a syngas is generated by catalytic steam reforming of hydrocarbons according to eq.1. $C_xH_y + xH_2O$ -> $(x+y/2)H_2 + xCO$ (1) This syngas is directly used for the reduction of an oxygen carrier (eq. 2) which is a mixture of hematite (Fe_2O_3) and supporting metal oxides, prepared by mechanical mixing. Fe_2O_3 + $3H_2/CO$ -> $2Fe + 3H_2O/CO_2$ (2) The consecutive oxidation of iron with steam leads to the formation of magnetite (Fe_3O_4) and pure hydrogen (eq. 3), an additional oxidation step with air regenerates the magnetite back to its initial hematite state (eq. 4). $3Fe + 4H_2O$ -> $Fe_3O_4 + 4H_2$ (3)

$2Fe_3O_4 + 1/2O_2$ -> $3Fe_2O_3$ (4) In the reformer steam iron process the reduction is performed at ambient pressure to utilize a high fuel gas conversion efficiency due to the favorable reaction equilibrium. The oxidation on the other hand can be executed at an increased pressure to yield compressed pure hydrogen without additional gas compressors, which improves the efficiency of the system.

Isothermal cyclic pressure experiments at 750 °C were performed in a lab scale test rig as a proof of concept for pure pressurized hydrogen production. The goal of the experiments was to investigate (i) possible draw-backs of the production of compressed hydrogen by stability constrains of the oxygen carrier or operational limitations, (ii) influences of an increased system pressure on the hydrogen purity regarding carbon contaminations, (iii) the characterization and quantification of the contaminations and (iv) the analysis of the oxygen carrier conversion regarding efficiency and stability.

The reduction reactions were performed at ambient pressure with hydrogen or a synthesis gas mixture of $H_2/CO/CO_2$. The re-oxidation of the oxygen carrier at an increased pressure was achieved by pumping liquid water into the system which is evaporated in the heated inlet. The steam oxidizes the oxygen carrier by releasing hydrogen. The pressure build-up is realized by closing the out-going valve after the cooling system. In this case the unreacted steam is condensed and the pressure increase can be directly related to the amount of produced hydrogen.

Â A successful hydrogen production up to 55 bar was achieved in the experiments with a hydrogen purity to fit the requirements of low temperature fuel cells. The oxygen carrier showed signs of structural degradation with no evident impact of the different pressure conditions in the prior oxidations. The linear pressure increase in the ï¬rst part of each oxidation maintained at a stable level throughout the entire investigation thus the oxidation rate was not reduced by the solid degradation. Sintering was identified as the source for the loss of cycle stability which is mainly inducted by the high reaction temperatures of 750 Â°C. The oxidations indicated no visible influence on the purity by the different system pressures. Thus the process is suitable for the direct production of pressurized hydrogen using hydrocarbons as feed.

Acknowledgements

This work is funded by the IEA-Research Cooperation on behalf of the Austrian Federal Ministry of Transport, Innovation and Technology. The authors gratefully acknowledge support from NAWI Graz.

473113 Integrated Thermochemical Water-Splitting Reactor Fuel Cell System for Electricity Generation

Abstract:

A continuous thermochemical water-splitting reactor for hydrogen generation has been developed and integrated with a fuel cell for electricity generation. To perform a water-splitting reaction, ferrite materials were synthesized inside a porous yttria-stabilized zirconia (YSZ) substrate using a sol-gel technique. These materials were analyzed by scanning electron microscopy, x-ray diffraction and BET surface area analyzer. In particular, Li-based ferrite materials were immobilized onto a porous YSZ substrate and loaded inside a continuous water-splitting reactor where both regeneration and water-splitting steps were performed at lower temperatures. Hydrogen and oxygen were continuously monitored and recorded using the online sensors interfaced with a computer. The outlet from a continuous thermochemical water-splitting reactor is integrated with a hydrogen fuel cell module. The fuel cell current was continuously recorded and analyzed at different load conditions. The power output and fuel cell or stack current were analyzed at different hydrogen production rates. The results obtained on the characterization of ferrite materials, hydrogen volume generation and the electricity generation will be presented.

466622 The Development of Microchannel Reactors for Hydrogen Production: Experimentation and CFD Modelling

Abstract:

A microchannel reactor for hydrogen production from ammonia is simulated with a three-dimensional (3D) CFD model using Comsol Multiphysics, in order to obtain a mathematical framework to understand the reaction-coupled transport phenomena associated with the microchannel reactor. The catalysts used consisted of commercial Ni-Pt/Al_2O_3 and Ru/Al_2O_3 catalysts in the form of a porous wash-coat deposited on the walls of the micro channels. The transport processes and reactor performance are elucidated in terms of velocity, temperature, and species concentration distributions, as well as local reaction rate and ammonia conversion profiles. The baseline case is first investigated to determine the behaviour of the microchannel reactor, then microstructural design and operating parameters are methodically altered around the baseline conditions to explore the optimum values. The simulation results show that an optimum ammonia space velocity (GHSV) yields 99.1% ammonia conversion and a power density of 32 $kW_e\ L^{-1}$ at the highest operating temperature of 973 K for the Ni-Pt based catalyst. It is shown that a 40-Î¼m-thick porous wash coat is most desirable at these optimum conditions. Also a small channel hydraulic diameter (225 Âµm) is observed to contribute to high ammonia conversion. It is reported that the Ru/Al_2O_3 gives a superior yield than the Ni-Pt/Al_2O_3 at optimal conditions and that experimental and model results agree very well. The results obtained confirm that a catalytic microchannel reactor can be most desirable for small scale hydrogen production from ammonia, required especially for the operation of fuel cells.

458104 Thermodynamic Modeling of the Hybrid Sulfur (HyS) Cycle for Hydrogen Production

Abstract:

Abstract

The hybrid sulfur (HyS) cycle has received much attention over the past few decades as a possible means of efficient large-scale production of clean hydrogen from high-temperature nuclear or solar heat sources [1]. A typical HyS cycle consists of a thermochemical step and an electrochemical step, the net result of which is the splitting of water into hydrogen and oxygen. In order to fully understand the limiting HyS heat and electric power requirements and to optimize the operating and design parameters, it is important to perform a thermodynamic analysis of the highly nonideal aqueous SO_2-H_2SO_4 system. Therefore, we develop a comprehensive thermodynamic model for the sulfuric acid-sulfur dioxide-water ternary system based on the symmetric electrolyte NRTL (eNRTL) [2] activity coefficient model framework in conjunction with the PC-SAFT [3] equation-of-state. This model is an extension of the previously published thermodynamic model of the sulfuric acid-water [4] binary system and additionally considers one step hydration of sulfur dioxide molecules. The eNRTL binary interaction parameters and the chemical equilibrium constants are regressed from the experimental SO_2solubility data in aqueous sulfuric acid solutions and the ternary model is validated over a wide temperature range from 283.15 to 383.15 K, and acid concentrations from pure water to pure acid. The model should be very useful in supporting process research, development, and design of the advanced direct solar water splitting process based on the HyS cycle.

Keywords:

Hybrid-Sulfur (HyS) cycle, Electrolyte NRTL, Hydrogen, SO_2 solubility, Sulfuric Acid

References:

[1] Gorensek MB, Summers WA, "Hybrid sulfur flowsheets using PEM electrolysis and a bayonet decomposition reactor," *Int. J. Hydrogen Energy*(2009) 34, 4097-4114.

[2] Song Y, Chen C-C, "Symmetric electrolyte nonrandom two-liquid activity coefficient model," *Industrial & Engineering Chemistry Research*(2009) 48, 7788–7797.

[3] Gross J, Sadowski, G, "Application of the Perturbed-Chain SAFT Equation of State to Associating Systems," *Ind. Eng. Chem. Res.* (2002) 41, 5510-5515.

[4] Que H, Song Y, Chen C-C, "Thermodynamic Modeling of the Sulfuric Acid-Water-Sulfur Trioxide System with the Symmetric Electrolyte NRTL Model," *J Chem Eng Data* (2011) 56, 963-977.

471842 Utility Perspectives on the Hydrogen Economy

Abstract:

Hydrogen is poised to become a preferred energy carrier in a deeply decarbonized energy future. While today's utilities make, move, or sell today's energy carriers of electricity and natural gas, future utilities may include hydrogen. This talk will focus on potential roles for today's utilities as they contemplate energy carriers of the future. It includes a survey of hydrogen production, not only from conventional steam methane reforming or coal gasification, but also technologies such as advanced electrolysis, methane thermolysis, water photolysis, and thermochemical hydrogen production cycles. It will discuss hydrogen delivery, a potential future of spatial and transient energy storage using hydrogen, the emergence of a hydrogen economy, and roles for hydrogen to provide clean energy in the near-term. Finally, the hydrogen economy will be contrasted against alternative future scenarios.

461264 Convective Heat Supply Reactor Concepts for the High Temperature Pyrolysis of Methane

Abstract:

Contrary to the popular belief, fossil energy consumption is not limited by its availability but is constrained by environmental considerations. Over the past decades, the world's natural gas proven reserves have been increasing steadily, from 2592 trillion cubic feet in 1980 to 6972 trillion cubic feet in 2014 [1], and if non-conventional sources like methane hydrates or shale gas are included, this increase is even steeper. Many scenarios for the future production of energy without CO_2-emissions have been investigated. The consensus seems to be directed towards renewable energy; however, a complete transition to renewable energy is not foreseeable in the near future [2]. It is thus necessary to search for bridging technologies. Among these, hydrogen presents itself as a very attractive alternative energy carrier due to the possibility of using existing natural gas infrastructure as well as for its facile integration in fuel cells and other clean energy technologies.

Hydrogen can be generated with low to zero CO_2-emissions by pyrolysing hydrocarbons at high temperatures. The pyrolysis of methane can be described by the reaction:

$$CH_4 \rightleftarrows C(s) + 2H_2$$

$$\Delta H_{R,1000°C} = 91.52\ kJmol^{-1}$$

$$\Delta G_{R,1000°C} = -49.28\ kJmol^{-1}$$

CH4⇄Cs+2H2
DHR,1000°C=91.52 kJmol-1
DGR,1000°C=-49.28 kJmol-1

This reaction is endothermic and becomes thermodynamically spontaneous above 820 K and kinetically feasible at temperatures around 1073 K. The reaction can be catalysed at lower temperatures, but the carbon deposition on the catalyst and the limited equilibrium conversions leading to a separation problem and recycle streams are still issues that have to be addressed. The high temperature pyrolysis remains an alternative, but is not without challenges: carbon deposition is still a major hurdle, which can lead to blockage of the reactor, and the endothermicity of the reaction necessitates the introduction of heat at higher temperatures.

Convective heat supply reactor concepts offer advantages that can help circumvent both problems. In this work, three concepts are presented:

The first two concepts are based upon the use of liquid molten media, such as molten metals. On the one hand, the liquid medium acts as a barrier preventing carbon deposition at the heat transfer surface, as well as a heat transfer material due to its excellent transport properties. The use of molten metal bubble columns was first proposed by Steinberg [3]. A modified version was explored by Serban et al. [4] exhibiting a relatively high conversion (57%) for low temperature operation (750°C). In this publication, carbon deposition occurred at both the reactor surface and in the sparger used to feed the methane. Currently a bubble column reactor with a fixed-bed is being investigated by the group of Prof. Abanades in the IASS in Postdam in cooperation with the Karlsruhe Institute of Technology amongst others [4]. Their results show a hydrogen yield of 30% at 1000°C with little carbon deposition on the reactor surface. However, the lack of control over the residence times in bubble columns inherently leads to a decrease in conversion. To overcome this difficulty, the use of a capillary reactor using molten metals in slug flow regime (Fig. 1) was studied at temperatures up to 1100°C [5], showing promising results, although only relatively low conversions (30% methane) have been achieved so far.

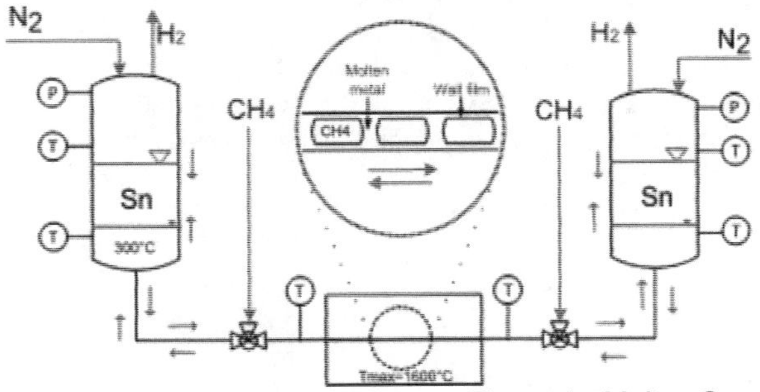

Figure 1. Molten Sn capillary reactor in slug flow regime

To increase conversions, working at still higher temperatures is the logical approach when the reactor length is limited. However, initial experimental results (Fig. 2) indicate increased carbon deposition, which can be explained by the increased rates of carbon formation, coupled with insufficient wetting of the heat transfer area by the molten metal as well as morphological differences in the carbon formed. To resolve this problem, two techniques can be used: the first one is to tweak the wetting properties of the molten metal, to avoid carbon deposition. A second approach would be to modify the reactor design appropriately to guarantee film formation while maintaining the advantages of slug flow operation.

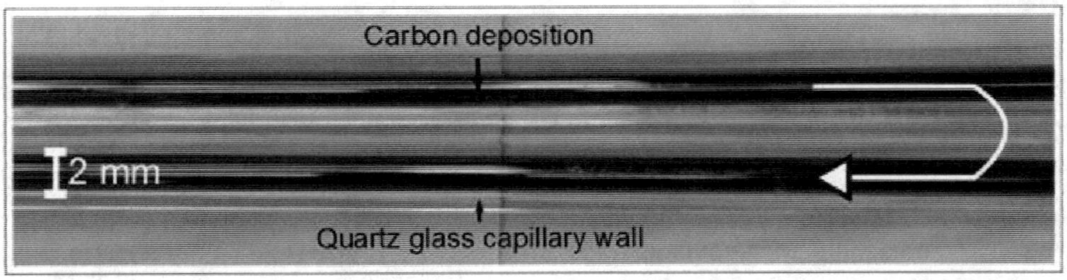

Figure 2. Carbon deposition in a capillary reactor at 1300°C

In this work, the experimental results for the first approach are presented, where the wetting properties of the melt were modified using a different alloy (Ga-In-Sn) rather than just tin. For the second approach, a theoretical study on a countercurrent falling-film reactor (Fig. 3) is presented. The falling-film reactor uses a similar concept to the capillary reactor, employing the molten metal as both a heat transfer medium and to protect the heat transfer surface, thus retaining the advantages of the slug flow regime. In addition, countercurrent operation inherently prevents the occurrence of the back reaction which could potentially take place in the slug flow regime. Furthermore, the residence time of the gas although not as well-defined as for the slug flow regime, can be still considered narrow enough to assure plug flow. The mathematical modelling and the simulation results are presented together with preliminary experimental results regarding film stability.

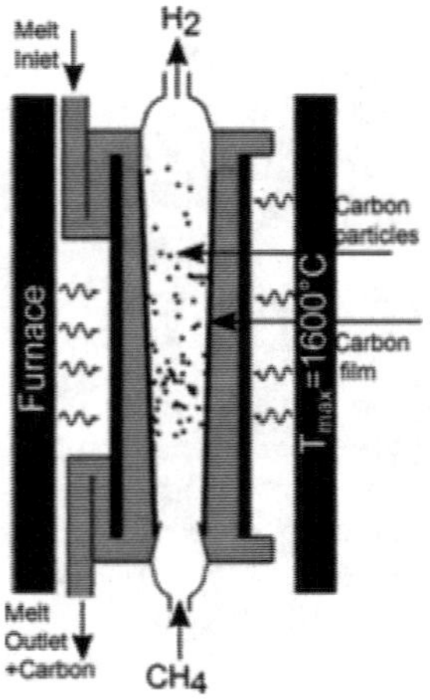

Figure 3. Falling molten Sn film reactor.

Finally, the third concept uses gas as a convective heat carrier. Previous work at our laboratory used a porous-wall reactor [6], where the carrier gas flowed through a porous wall into the reaction zone, avoiding the carbon deposition on the outer walls. However, methane diffusion through the membrane proved to be higher than anticipated, and as a result carbon deposits were still found in the outer region. To avoid this diffusion, a side-jet reactor (Fig. 4) concept was developed, where the carrier gas flows through small holes in the inner tube generating a greater gas velocity in comparison to the porous wall. This minimizes the rate of diffusion of methane into the outer zone and provides a better mixing with the methane and thus an improved heat transfer. In this work, the modelling of the side-jet reactor is presented as are the preliminary experimental results.

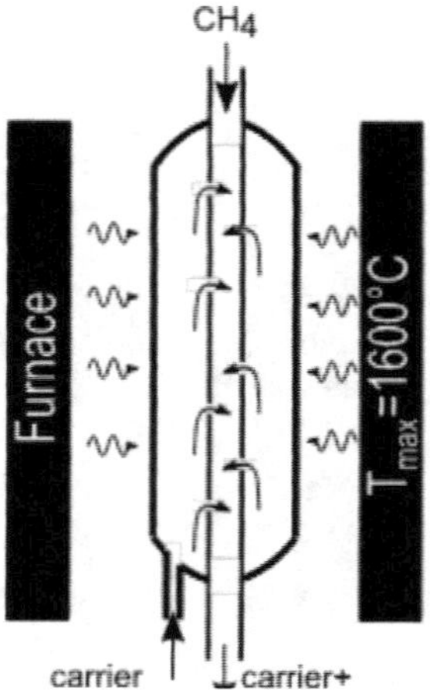

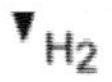

Figure 4. Side-jet reactor concept.

Acknowledgements

Financial support from the German Research Council DFG (Grant No. AG 26/15-1) is gratefully acknowledged.

References

[1] U.S. Energy Information Administration. International Energy Statistics. Retrieved from: http://www.eia.gov/cfapps /ipdbproject/iedindex3.cfm?tid=3&pid=3&aid=6&cid=regions&syid=1980&eyid=2015&unit=TCF on October 2015.

[2] G. Kreysa, DGMK Conference "The Future Role of Hydrogen in Petrochemistry and Energy Supply", Berlin, Germany, October, 2010.

[3] M. Steinberg, Int. J. Hydrogen Energy. 24(8) (1999) 771-777.

[4] M. Serban, M. A. Lewis, C. L. Marshall, R. D. Doctor, Energy Fuels. 17 (3) (2003) 705-713

[5] T. Geißler , M. Plevan , A. Abanades, A. Heinzel, K. Mehravaran, R.K. Rathnam, C. Rubbia, D. Salmieri, L. Stoppel, S. Stückrad, A. Weisenburger, H. Wenninger, T. Wetzel, Int. J. Hydrogen Energy. 40 (41) (2015) 14134-14146.

[6] I. Schulz, D.W. Agar, Int. J. Hydrogen Energy. 40(21) (2015) 11422-11427

[7] Fabian Goebel, Bachelor Thesis, Technische Universität Dortmund, 2015.

464729 Production of Levulinic Acid in Integrated Biorefineries from Municipal Solid Waste

Abstract:

Demand for material goods is expected to increase due to rising human population and managing municipal solid waste (MSW) resulting after consumption will be challenging. This will require holistic integration of waste processing for efficient resource use, and minimisation of environmental impacts, through pursuit of a circular economy [1]. Integrated waste biorefinery systems are a promising solution for profitable and sustainable waste valorisation [2]. This paper reports an integrated conceptual mechanical biological chemical treatment (MBCT) system in the form of an advanced biorefinery system, for the production of levulinic acid to increase economic margin of municipal solid waste processing by 110-150% [3]. After mechanical separation recovering recyclables, metals (iron, aluminium, copper) and refuse derived fuel (RDF), lignocelluloses from remaining municipal solid waste (MSW) are extracted by supercritical-water. This lignocellulosic stream is then valorised by a chemical process comprising hydrolysis in dilute H_2SO_4 catalyst producing LA as main product and an effluent stream containing some furfural and formic acid (FA). This effluent and the pulping effluent are anaerobically digested into biogas and fertiliser. Produced biogas (6.4 MWh/t), RDF (5.4 MWh/t), char (4.5 MWh/t) are combusted, heat recovered into steam generation in boiler (efficiency: 80%); on-site heat/steam demand is met; balance of steam is expanded into electricity in steam turbines (efficiency: 35%). The economic margin (product prices – annual cost – MSW cost) of the system is 118 Euro/t MSW, with margin of various products, decreasing in the order chemical > metals > fuel > energy > composting, respectively. Value analysis shows diversion of compost like output into chemical eliminates needs for gate fees. This conceptual system demonstrates how the judicious application of process integration combined with economic value and life cycle analyses allows generating competitive and environmentally friendly municipal solid waste biorefinery systems and can contribute to the development of a circular economy.

[1] Satchatippavarn S, Martinez-Hernandez E, Hang MYLP, Leach M, Yang A. Urban biorefinery for waste processing. Chemical Engineering Research and Design 2015; 107, 81-90.

[2] Sadhukhan J, Ng KS, Martinez-Hernandez E, Biorefineries and Chemical Processes: Design, Integration and Sustainability Analysis. Wiley, Chichester, UK, 2014.

[3] Sadhukhan J, Ng KS, Martinez-Hernandez E, Novel integrated mechanical biological chemical treatment (MBCT) systems for the production of levulinic acid from fraction of municipal solid waste: A comprehensive techno-economic analysis. Bioresource Technology 2016. *In press.*

464135 Synthesis of Processing Paths Fot the Valorization of Specialty Chemicals in Microalgae Biorefineries

Abstract:

Over the last years, biorefineries are gaining ground and numerous studies have been developed with main focus on the exploitation of lignocellulosic feedstock for biofuels and bioproducts. Advances on synthesis, design and integration of such biorefineries are well established and continue to evolve. Third generation biorefineries are currently in the spotlight, especially due to the plethora of chemicals that can be incorporated in the portfolio of products. Microalgae constitute a very promising feedstock with the ability to utilize CO_2 and convert it to specialty chemicals. Thus, challenges arise as to the probe of such an incentive, which does not only include the synthesis and design of a microalga biorefinery, but also its integration with other types of biorefineries.

The study addresses the challenges in the synthesis of microalgae biorefineries by developing a combinatorial approach, in order to simultaneously screen for potential processing paths and formulate the desired product portfolio. The approach enables high-throughput screening with focus on securing the feasibility and profitability of the incentive. The selection of processes and the formulation of the product portfolio are treated as degrees of freedom. The model takes into consideration feedstocks, processes, intermediate fractions and products in a conceptual form and employs economic criteria for the selection of the optimal path. The study takes into account the stages after cultivation and harvesting of the microalgae, where screening paths of candidate technologies and product portfolios is a matter of necessity due to the numerous available options. Competitive chemistries characterizing microalgae as well as the lack of thermodynamic properties regarding involved chemicals add to the complexity of such designs. Extensive knowledge about the use of suitable technologies and required specifications is imperative, thus the need for experimental data becomes indispensable. The methodology deals with these limitations and practicalities, which derive mainly from the lack of industrial applications to draw knowledge from. Therefore, extensive collaboration with practitioners is of vital importance.

For the purpose of this study, the methodology is applied to a real-life case of a microalga biorefinery undertaken by the D-Factory, an FP7 collaborative project, which is relative to the valorization of high-value chemicals from the microalga Dunaliella. This particular microalga is considered a great candidate for mass outdoor cultivation especially because of its extraordinary tolerance to salt stress and potential environmental contaminants. The addressed methodology deals with encountered limitations and constitutes a stepping stone to the development of a systematic approach for the design of integrated microalgae biorefineries. It provides flexibility due to the high profit of specialty chemicals and guidelines of vital importance for future design analysis.

470586 Solvolytic Conversion of Biorefinery Lignin to Value-Added Products

Abstract:

Next generation biorefineries will produce a significant amount of lignin-enriched residues, which are currently just burnt or underutilized. Therefore, new lignin valorization strategies targeting biorefinery lignin streams are important from technoeconomic standpoint. Herein, we show the solvolytic depolymerization of lignin obtained from ionic liquid (IL) pretreatment/saccharification process into low molecular weight compounds. Joint Bioenergy Institute (JBEI) has recently developed one-pot IL pretreatment / saccharification process using biocompatible and bio-derived ILs, which was found to be effective for biomass pretreatment. Residual lignin recovered after one-pot process was subjected to a depolymerization reaction using a Parr reaction with hydrogen donor solvent. The effects of reaction parameters, including reaction temperature (250 ~ 300 °C), time (1 ~ 4 hrs) and catalyst (Ru/C) on depolymerization behavior of lignin were explored. Mass balance analysis shows that approximately 70 % liquid is obtainable from biorefinery lignin. In-depth analysis will be performed to characterize the final products, which will include GC, GPC and NMR. This study will help to provide insights for developing effective depolymerization pathway of biorefinery lignin.

AUTHOR INDEX

AUTHOR INDEX

AUTHOR INDEX

AUTHOR INDEX

AUTHOR INDEX

AUTHOR INDEX

AUTHOR INDEX

AUTHOR INDEX

AUTHOR INDEX

AUTHOR INDEX

AUTHOR INDEX

AUTHOR INDEX

AUTHOR INDEX

AUTHOR INDEX

AUTHOR INDEX

AUTHOR INDEX

AUTHOR INDEX

AUTHOR INDEX

AUTHOR INDEX

AUTHOR INDEX

AUTHOR INDEX

AUTHOR INDEX

AUTHOR INDEX

AUTHOR INDEX

AUTHOR INDEX

AUTHOR INDEX

AUTHOR INDEX

AIChE
120 Wall Street, FL 23
New York, NY 10005-4020

ISBN 978-1-5108-3420-0

CURRAN ASSOCIATES INC.
proceedings.com

International Congress on Energy 2016 (ICE)

Topical Conference at the 2016 AIChE Annual Meeting

San Francisco, California, USA
13 - 18 November 2016

Volume 2 of 2

AIChE
The Global Home of Chemical Engineers

International Congress on Energy 2016 (ICE)

Topical Conference at the 2016 AIChE Annual Meeting

San Francisco, California, USA
13 - 18 November 2016

Volume 2 of 2

ISBN: 978-1-5108-3420-0

Printed from e-media with permission by:

Curran Associates, Inc.
57 Morehouse Lane
Red Hook, NY 12571

CURRAN ASSOCIATES INC.
proceedings
.com

Some format issues inherent in the e-media version may also appear in this print version.

Printed by Curran Associates, Inc. (2017)

AIChE
120 Wall Street, FL 23
New York, NY 10005-4020

Phone: (800) 242-4363
Fax: (203) 775-5177

www.aiche.org

Additional copies of this publication are available from:

Curran Associates, Inc.
57 Morehouse Lane
Red Hook, NY 12571 USA
Phone: 845-758-0400
Fax: 845-758-2633
Email: curran@proceedings.com
Web: www.proceedings.com

TABLE OF CONTENTS

VOLUME 1

VOLUME 2

459504 Depolymerization of Kraft Lignin to Value-Added Chemicals over a MoC_{1-X} /Cu-$MgAlO_x$ Catalyst

Abstract:

Different contents of Mo were impregnated on $CuMgAlO_x$ prepared with a co-precipitation method, and the as prepared material was reduced in CH_4/H_2 flow to obtain the MoC_{1-x}/Cu-$MgAlO_x$ catalyst. Lignin conversion experiments were carried out in ethanol with MoC_{1-x}/Cu-$MgAlO_x$ catalyst of different Mo loading at 300 °C for 6 h. The liquid products were identified with a GCâ€"MS instrument and quantified with a GC-FID instrument. Fig. 1 shows the gas chromatography of the liquid product of lignin depolymerization obtained with a 30 % Mo loading catalyst. About 50 kinds of compounds were detected and the products of ethanol self-reactions were excluded by comparison to a blank reaction. 25 compounds from lignin are categorized into five groups, i.e. alcohols, esters, arenes, phenols and benzyl alcohols, of which the detailed yields are given in Fig. 2. When $CuMgAlO_x$ (Mo loading=0) was used as the catalyst, the yield was relatively low with a 584 mg/g lignin total yield. When MoC_{1-x}/Cu-$MgAlO_x$ was used as catalyst, almost all yields increased with the increase of Mo loading. When Mo loading reaches 30 %, all yields reach the maximum values and the yields of alcohols, esters, phenols, arenes and benzyl alcohols are 319, 768, 2, 207 and 125 mg/g lignin, respectively, with the total yield of 1421 mg/g lignin. This yield is lower than that obtained with MoC_{1-x}/AC as a catalyst as we previously reported. However, the selectivity of MoC_{1-x}/Cu-$MgAlO_x$ is different from that of MoC_{1-x}/AC, which obtained higher yields of arenes and benzyl alcohols and lower yields of alcohols, esters and phenols. The difference in selectivity may be caused by the basicity of the $CuMgAlO_x$.

467188 Glycerol Oligomers Synthesized from Etherification of Glycerol with Zeolite-Supported Catalysts

Abstract:

Glycerol oligomers synthesized from etherification of glycerol with zeolite-supported catalysts

Chin-Kai Chen, Bing-Hung Chen*

Department of Chemical Engineering, National Cheng Kung University,

1 University Road, Tainan 70101, Taiwan

e-mail: bkchen@mail.ncku.edu.tw (BHC)

Abstract

In this work, zeolite-supported Ca/Mg/Sr catalysts were synthesized to produce glycerol oligomers from the catalyzed etherification of glycerin which is the abundant byproduct from transesterification of oils and fats for biofuel, using. Both zeolite ZSM-5 and zeolite LTA were chosen as supports impregnated with Ca, Mg, or Sr, to catalyze the etherification reaction of glycerol to glycerol oligomers, mainly the diglycerol and triglycerol, in the Dean-Stark apparatus as a semi-batch reactor. The properties of these zeolite bases catalysts were characterized with XRD, FT-IR, SEM, BET, TGA and TPD. In addition to the heterogeneous zeolite based catalysts, homogeneous catalysts, both acid (sulfuric acid) and base (NaOH), were tested to provide the benchmark. The chemical structure of the product from etherification of glycerin were analyzed with GC-MS, GC-FID and ^{13}C-NMR.It It indicated that diglycerol and triglycerol, either in linear or cyclic form, were dominant products. Moreover, the yield of polyglycerols (diglycerol + triglycerol) could exceed 85% after 6 h of reaction at 260¢XC using Ca-impregnated zeolite ZSM-5. The activation energy of 107.71 kJ/mol of the pseudo-first order reaction confirmed a high energy barrier in glycerin etherification.

Keywords: *Glycerin, Diglycerol, etherification, Zeolite ZSM-5, impregnation.*

Figure 1: The XRD patterns of Zeolite ZSM-5 impregnated with $Ca(NO_3)_2$ solutions

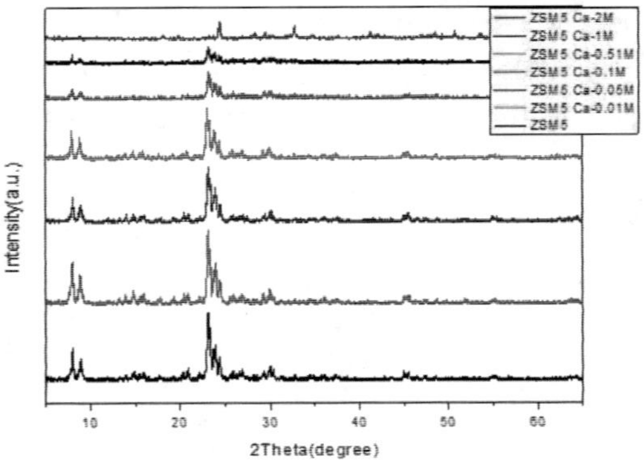

Figure 2: Etherification of glycerin catalyzed by zeolite ZSM-5 impregnated with alkaline earth metals. (Reaction conditions: (1) catalyst loading = 2 wt% of glycerin, (2) reaction time = 8h, (3) reaction temperature = 260¢XC, (4) N_2 flow = 50 sccm).

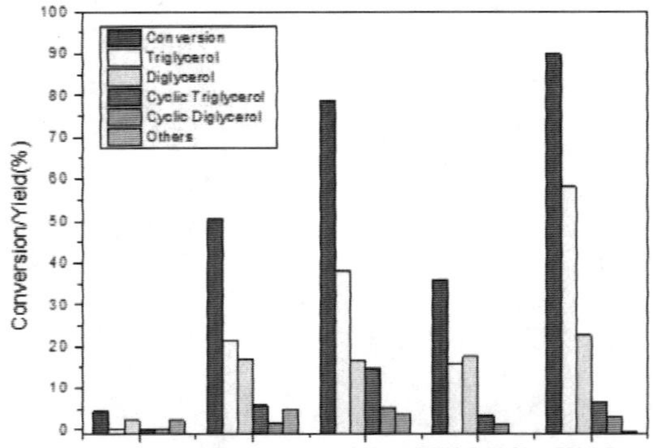

465940 Influence of Pyrolysis Temperature on Characteristics and Aromatics Adsorption Capability of Magnetic Biochars Derived from Rice Straw Pyrolysis Oil Distillation Residue

Abstract:

Influence of pyrolysis temperature on characteristics and aromatics adsorption capability of magnetic biochars derived from rice straw pyrolysis oil distillation residue

Hao Li, Shuqian Xia*, Peisheng Ma

Key Laboratory for Green Chemical Technology of State Education Ministry, Collaborative Innovation Center of Chemical Science and Engineering (Tianjin), School of Chemical Engineering and Technology, Tianjin University, Tianjin, PR China

The bio-oil, derived from rice straw fast pyrolysis, deservedly offers an opportunity as alternatives to replace fossil fuels due to its renewable characteristics. Unfortunately, pyrolysis oil is an unstable product and the physical and chemical properties are changing during storage, which limit its application. Distillation as an effective method for upgrading biomass fast pyrolysis oil has been gained great attention. Although the high-value chemicals or high-quality fuel were obtained from the pyrolysis oil distillation, the distillate yield did not reach 100% and a fraction of solid residue was formed after pyrolysis oil distillation. In this work, the magnetic biochars were synthesized by pyrolyzing the ditillation residue derived from rice straw pyrolysis oil at different temperature of 400, 600 and 800 °C. The effects of pyrolysis temperature on the physicochemical and morphological properties of magnetic biochars were investigated. The experimental results of TGA, FTIR, SEM and XRD indicated that the increase in pyrolysis temperature led to the formation of magnetic biochar with a stable aromatic carbon structure. When the pyrolysis temperature was increased up to 600 °C, the surface area and pore volume of magnetic biochar were enhanced, while a further increase in the pyrolysis temperature lowered the value of surface area and pore volume due to pores widening or blocked pores in the magnetic biochar during pyrolysis. Experimental studies related to the adsorption behaviors of the different magnetic biochars toward various aromatic contaminants (i.e., anisole, phenol and guaiacol), key affecting factors and the underlying mechanisms proposed to explain the adsorption behaviors, have been investigated. The Freundlich adsorption model could describe successfully the adsorption performance and the used magnetic biochars could be easily separated from the aqueous solutions by an outer magnet. An analysis of adsorption mechanisms suggests that the specific surface area and pore volume were the dominant mechanism for adsorption of anisole and phenol on the magnetic biochars while the adsorption of guaiacol on the magnetic biochars were controlled by the ¦Ð-¦Ð electron donor¨Cacceptor (EDA) interaction. Furthermore, compared with phenol and guaiacol, the higher adsorption capacity of anisole can be attributed to the high hydrophobicity (or low polarity) of magnetic biochars.

473748 Process to Planet Framework for Sustainable Design: Systematic Approach for Developing a Multiscale Model and for Multiobjective Optimization

Abstract:

Over the last decade, sustainable process design (SPD) research has utilized life cycle assessment(LCA) methods to incorporate an environmental dimension in process systems engineering(PSE). Conventional SPD involves a â €œbottom up approachâ€ of modeling the system of interest at a smaller scale with detailed process specific fundamental engineering models, while the upstream life cycle of the product is modeled at a larger scale using process based LCA. System boundary selection is a significant challenge for process LCA due toÂ infinite nature of the upstream life cycle network for any process [1]. Process based LCA can lead to significant truncation errors in life cycle resource consumption and emissions, leaving a wide margin for shifting of impacts across narrow system boundaries and leading to illusive optimal design solutions that may in fact be inferior due to incomplete life cycle information [2]. Hybrid LCA(HLCA) methods aim to address the boundary selection problem by combining process level life cycle inventories with environmentally extended input-output analysis(EEIO), which allows for a significant part of the product life cycle to be included in the assessment. The Process to Planet (P2P) framework [3] was developed by introducing the concept of integrated hybrid LCA [4] to SPD. Detailed engineering models of the primary process being designed and other key processes in the supply chain are integrated within a macroeconomic system represented by economic input-output (EIO) model. This ensures solution of process design problems while reducing chances of environmental impacts being neglected due to incomplete life cycle inventories. While bridging the knowledge gap between PSE and LCA, the P2P framework not only improves upon the life cycle assessment component of SPD, but also addresses a major weakness in Hybrid LCA. HLCA is limited to the impact assessment stage of life cycle assessment. Very few studies have dealt with the fourth and final stage of LCA, improvement analysis. This situation is lamentable because inventory scope and impact analysis provide little benefit if not followed by recommendations and decisions that should be adopted to reduce environmental burden. The P2P framework can be visualized as an extension of HLCA method where it enables the incorporation of design variables through process design and allows optimization of the system to improve its environmental footprint. It was successfully demonstrated by application to a corn based ethanol production plant and highly insightful results were obtained that proved the efficacy of the framework [2].

This presentation proposes extensions and modifications that address some of the inadequacies that are inherent in the original P2P framework. While strictly applied to P2P, these extensions can be extrapolated to answer challenges that exist in broad areas of literature, like HLCA. The selection of significant processes that are modeled at the process based life cycle scale and the insignificant processes that are studied at the economy scale is arbitrary. In other words, delineation of the boundary between the different scales of the P2P framework was a subjective task. Previously, this was done based on data availability, accuracy and detail requirements and the complexity of the overall problem which is proportional to the solving time [5,6]. The problem was first mentioned and some methods were proposed for addressing it in a paper by Suh [7]. Lenzen proposed Structural Path Analysis (SPA) involving the decomposition of the Leontief Inverse for addressing this problem [8]. Treloar also used this method to study energy use of individual residential buildings in Australia [9,10]. Lave et al. [11] and Lenzen [12] used SPA to quantify the truncation errors in process-based LCA. A path exchange method using SPA was also developed as an alternative to the method of combining process LCA and IO LCA within integrated hybrid LCA as proposed by Suh. We intend to address this knowledge gap by proposing a path analysis algorithm where the P2P framework superstructure model generation is guided by a hierarchical structural path decomposition method which determines the coverage of the different scales. It is one of the first applications of SPA for solving the boundary selection problem involving the distribution of different upstream processes between multiple scales.

A large number of studies have dealt with multi objective optimization in process design and energy systems involving environmental and economic impacts. A comprehensive list of such literature can be found in Diwekar et al.[13] Reviewing these articles, it becomes apparent that it is necessary to simultaneously work towards both economic profitability and environmental sustainability of the system. Previously, the P2P framework focused only on the environmental dimension of process design. In this work, we propose a multi objective optimization problem, involving the trade-off between an economic objective and an environmental objective function. The novelty in this addendum is that it makes the P2P framework more fruitful for designing any generic process while exercising control over both the economic and environmental scope.

Preliminary results have provided insightful views on the application of SPA algorithm for model generation as well as exposed existing deficiencies to work on. While SPA is successful in mapping the upstream networks of the process to be designed and calculate their contributions to the entire life cycle, the quantification of the quality of the generated model still needs to be addressed. For example, considering two scenarios where an upstream sector is modeled at the IO model scale and process LCA scale respectively, we need to develop parameters that can quantitatively compare between the two cases and rank them objectively. As of now, subjective decisions are being used to build the model, but we intend to develop an algorithm based on error analysis or uncertainty information to determine the model quality. Multiobjective optimization solutions demonstrated that, highly erroneous optimal solutions and counter intuitive results are obtained from practicing conventional SPD approach rather than P2P. In addition to this, it was observed that there is a 59% chance of choosing an inferior solution over the optimal solution when solving the design problem using conventional SPD.

References

[1] Sangwon Suh. Functions, commodities and environmental impacts in an ecological economic model. 48:451-467, 2004.

[2] Rebecca J Hanes and Bhavik R Bakshi. Sustainable process design by the process to planet framework. AIChE Journal, 61(10):3320-3331, 2015.

[3] Rebecca J Hanes and Bhavik R Bakshi. Process to planet: A multiscale modeling framework toward sustainable engineering. AIChE Journal, 61(10):3332-3352, 2015.

[4] Sangwon Suh and Shinichiro Nakamura. Five years in the area of input-output and hybrid lca. The international journal of life cycle assessment, 12(6):351-352, 2007.

[5] H Hondo and S Sakai. Preliminary life cycle inventory analysis (pre-lci) using an economic input-output table. In The Fourth International Conference on EcoBalance, Tsukuba, Japan, pages 181-4, 2000.

[6] G Norris. Selecting and evaluating boundaries for life cycle assessment using economic input-output analysis. In Proceedings from the SETAC 17th Annual Meeting, Washington, DC, pages 17-21, 1996.

[7] Sangwon Suh, Manfred Lenzen, Graham J Treloar, Hiroki Hondo, Arpad Horvath, Gjalt Huppes, Olivier Jolliet, Uwe Klann, Wolfram Krewitt, Yuichi Moriguchi, et al. System boundary selection in life-cycle inventories using hybrid approaches. Environmental Science & Technology, 38(3):657-664, 2004.

[8] Manfred Lenzen. A guide for compiling inventories in hybrid life-cycle assessments: some australian results. Journal of Cleaner Production, 10(6):545-572, 2002.

[9] Graham J Treloar. Extracting embodied energy paths from input-output tables: to-wards an input-output-based hybrid energy analysis method. Economic Systems Re-search, 9(4):375-391, 1997.

[10] GJ Treloar, PED Love, OO Faniran, and U Iyer-Raniga. A hybrid life cycle assessment method for construction. Construction Management & Economics, 18(1):5-9, 2000.

[11] Lester B Lave. Using input-output analysis to estimate economy-wide discharges. Environmental Science & Technology, 29(9):420A-426A, 1995.

[12] Manfred Lenzen. Errors in conventional and input-output based lifecycle inventories. Journal of Industrial Ecology, 4(4):127-148, 2000.

[13] Urmila Diwekar and Yogendra Shastri. Design for environment: a state-of-the-art re-view. Clean Technologies and Environmental Policy, 13(2):227-240, 2011.

462158 Application of a Shale Environmental Footprint Optimization Tool to Enhance Operational Excellence– a Case Study of Selecting Process Fuels with Lower Cost and Less Potential for Impacts to Power Chevron Operations

Abstract:

Unconventional resources represent significant business opportunities; however, their development is subject to much public and environmental scrutiny. Currently there are over 400 local bans and 2 state bans on hydraulic fracturing. Many countries besides the U.S. have placed moratoriums on hydraulic fracturing. The challenge and opportunity that the oil and gas industry is facing is: how to unlock the potential in unconventionals, achieve better economic and operational performance, while still fulfilling our commitment to communities and the environment.

The Shale Environmental Footprint Optimization Tool is a tool developed at Chevron to enable the integration of environmental considerations in process design and decision making related to shale development. It simulates shale operations at the play level, where material and energy flows of in-scope processes are aggregated together and converted to quantitative environmental indicators (i.e. energy consumption, water consumption, and air emissions). It is built with a commercial life cycle assessment (LCA) software (GaBi® Professional) that contains >4000 datasets, and adopts international standards (ISO 14040-10444). The tool also allows sensitivity and uncertainty analysis to better assess and interpret results under varying reservoir and operational conditions.

This tool has been used by business units to better characterize their operations, assess potential environmental risks associated with different decision alternatives, and understand the trade-offs. It also provides them a more holistic view across the value chain, and help them pinpoint major contributors to the total footprint. These insights help business partners prioritize R&D efforts, make informed decisions on mitigation options, and ultimately lead to reduced environmental impact. Externally, the tool can be used address stakeholder concerns and for benchmarking our performance.

In the presentation, we'll show how this tool has been used to assess real world situations (using field data and business scenarios) and generate information to support a critical decision – evaluating and identifying the preferred alternative fuel to power drilling rigs. Optimizing the selection of process fuels could result in significant environmental, social, and economic benefits. One of the scenarios we'll include is to switch from LNG to gas produced in the field for power generation, which may reduce truck traffic on the roads. Reducing trucks is beneficial to safety (MVC, avoid backing/maneuvering incident on well pad), cost (hauling and road repair), land disturbance (less roads and processing units), and emissions (air pollutants & spills). In some locations, it also means improved reliability and fewer burdens on business continuity.

GHG Life Cycle Assessment for the United Arab Emirates Electricity Sector Combining Optimization and Simulation Tools

Paper ID #: 466062

*Alberto Betancourt-Torcat, Mohammed Alkatheri, Ali Almansoori**

Department of Chemical Engineering, The Petroleum Institute, Sas Al Nakhl Campus, Abu Dhabi, P.O. Box 2533, United Arab Emirates.

*Corresponding Author's e-mail: aalmansoori@pi.ac.ae

Keywords: Greenhouse Gas, Life cycle assessment, Electricity sector, Economics, Optimization, Simulation.

Abstract

The United Arab Emirates (UAE) relies heavily on natural gas for power generation, and the demand for gas in a country with one of the highest per capita power consumption rates in the world position the electricity sector as the leading CO_2 emission source. Although the country is very rich in energy resources it is currently facing challenges to meet its domestic gas demand. Thus, the country has been forced to develop its large, but highly sour (high-sulfur content) gas reservoirs. Given the gas high-sulfur content, the sweetening process is particularly very energy and greenhouse gas (GHG) emission intensive compared with conventional gas resources. Additionally, the volume of sweet gas yield is low due to the high proportion of acid gases in the feed gas (e.g., over 35% of H_2S and CO_2 combined). This work quantifies the well-to-grid (WTG) GHG emissions per kilowatt hour (kWh) of domestic gas-based electricity generated in the UAE using mathematical models, simulation tools, and engineering principles. WTG emissions include emissions from sour gas extracting, sour gas processing, sweet gas product transportation, electricity generation, and electricity transmission in the UAE. Two types of gas resources were examined: conventional and ultra-sour gas. The results show that the WTG GHG emissions of domestic gas-based power generation in UAE ranges between 444 to 501 g CO_2 eq./kWh; placing the country in the low-mid range of values compared with the literature 413-587 g CO_2 eq./kWh.

Introduction

Over the last decade, the rapid economic and demographic growth of the United Arab Emirates (UAE) has pushed the country's electricity grid to its limits. Most of the UAE's generation capacity is natural gas-based [1]. Although natural gas is considered to be a relatively clean primary energy source, it is still a fossil fuel source that generates significant greenhouse gas (GHG) emissions. The UAE electricity demand reached a domestic record of 105 billion kWh in 2013, placing the country as one of the highest per capita power consumers in the world [2]. Nuclear and renewable power capacities are currently being added to increase the country's energy mix and reduce GHG emissions [3]. As a result of the steadily growing electricity demand, the UAE plans to increase natural gas production to meet the domestic demand. The majority of the growth will come from the country's largely sour (high-sulfur content) gas reservoirs. The UAE's gas reservoirs have a relatively high sulfur content, which makes it corrosive and difficult to process [4]. The process is particularly very energy intensive and environmentally challenging. Thus, the process is highly GHG-emission intensive, as well vast amounts of sulfur are generated as by-product [5].

There is currently a need to assess the environmental footprints of the UAE electricity sector in terms of its GHG emission life cycle. This could allow benchmarking the UAE's electricity sector against other nations, while finding suitable niches for GHG mitigation strategies. Moreover, since over 98% of the electricity generated in the UAE is gas-based, an analysis on the UAE's natural gas life cycle represents a key factor on determining the overall environmental impact of the electricity sector. The present analysis considers a well-to-grid (WTG) GHG lifecycle assessment; which includes the upstream and downstream operations. The upstream operations includes all activities before fuel is combusted at power plants. Such activities comprise: sour natural gas extraction, sour gas processing, and sweet gas transport to the power plants gate. On the other hand, the downstream operations include the electricity generation and transmission. For the present study, fundamental engineering principles were used to mathematically estimate process specific energy consumption and related GHG emissions; namely for sour gas extraction, sweet gas product transportation, and electricity transmission. Additionally, ProMax® v3.2 [6] was used to simulate the sour gas processing, while performing a parametric analysis to improve the performance of these processes. Also, an optimization mathematical model developed in the General Algebraic Modeling System (GAMS®) [3] was used to illustrate the operation of the UAE power sector, and obtain the average GHG emissions of the country's gas-based power fleet.

Methodology

In the UAE the well-to-grid life cycle starts with the extraction of natural gas from the reservoirs and ends at the electricity transmission. Natural gas is extracted using wells and sent to processing plants where: water, carbon dioxide, sulfur, and heavier hydrocarbons are removed. Then, the resulted sweet gas product is transported through pipelines to the power plants to generate electricity, and transmitted through the electricity grid. The UAE's WTG GHG emissions life cycle can be divided into two inventories: upstream and downstream. The upstream inventory includes all activities before fuel is combusted at a power plant. Such activities comprise: sour natural gas extraction, sour gas processing, and sweet gas product transportation to the power plants gate. On the other hand, the downstream inventory includes electricity generation and transmission. In this study, the GHG emissions associated to gas extraction, sweet gas product transportation, and electricity transmission were estimated based on typical average emission values found in the literature [7]. On the other hand, the GHG emissions generated by the natural gas processing and combustion in power plants were calculated using ProMax® v3.2 (simulation software) [6] and a mathematical model developed in GAMS® (modeling system) [3], respectively. There are two main types of natural gas resources in UAE: 1) ultra/highly-sour natural gas, and 2) conventional natural gas. Therefore, for the present analysis the WTG GHG emissions life cycle was estimated taking into account both natural gas resources. Typical compositions for both types of gas resources are shown in Table 1.

Table 1. Typical compositions for ultra-sour and conventional natural gas resources in UAE.

Main Components	Ultra-sour gas [4]	Conventional gas [8]
H_2S	22	5
CO_2	10	4
C1	54	76
C2	4	8
C3+	5	3
H_2O	5	4

The first stage in the WTG GHG emissions life cycle is the natural gas extraction using a gas well. This operation encompasses various processes such as: rig construction, drilling, well completion, and gas production. Energy is consumed during the drilling process generating GHG emissions. The emissions of natural gas during the extraction stage are considered episodic emissions because they are not part of the daily, steady state of the well operations. For instance, it was assumed that around 20 g CO_2 eq./kWh is emitted to the atmosphere in gas extraction (e.g., mainly point source and fugitive losses) [7]. Moreover, gas production starts after the well is completed. Gas from the wellhead is sent to surface processing, where it is sweetened, dehydrated, and compressed for pipeline transmission. The raw gas from the wellhead is first sent to a three phase gravity separator to separate the gas, light liquids (e.g., mostly sour hydrocarbon liquids C3+), and heavy liquids (mostly sour water). Since no fuel is consumed during this process, it is not considered a significant source of GHG emissions.

After the phase separator, the gas is treated in a gas sweetening unit. In this work, di-glycol amine (DGA) is used as the gas sweetening solvent for ultra-sour natural gas; while methyl-di-ethanol amine (MDEA) is used as sweetening solvent for conventional gas. The units that consume energy in the gas sweetening process are: feed pre-coolers, amine reboilers, circulation pumps, solvent coolers, and acid gas condensers. The amine re-boiler is assumed to be powered by low pressure steam (LPS) (i.e. saturated steam at 3.5 barg), the circulation pump and solvent air cooler are powered by electricity, the feed pre-cooler and acid gas condenser employ cooling water. Later, the wet sweet gas is sent to the dehydration unit; whereas the acid gases (i.e. H_2S and CO_2) are sent to the sulfur recovery unit. In the dehydration unit, water is removed from the sweet natural gas since it promotes corrosion. Tri-ethylene glycol (TEG) is used as liquid desiccant to dehydrate the sweet gas. The glycol reboiler, sweet gas pre-cooler, and pumps are the major energy-consuming units in this process. In the dehydration unit, high pressure steam (HPS) (i.e., saturated steam at 41 barg) is used to power the glycol reboiler. Moreover, electricity is used to drive the pumps, whereas cooling water is used to pre-cool the wet sweet gas. The dry sweet gas from the dehydration unit can be combusted at an onsite power plant to produce electricity. Sometimes the dry sweet gas from the dehydration unit is sent to a natural gas liquid recovery unit to recover ethane and ethylene. However, for simplicity the natural gas liquid recovery unit was not considered in this study.

The liquid hydrocarbons are treated in a liquid sweetening unit using DGA as solvent; while the sour water is treated using a sour water stripper. The sole purpose of both aforementioned treatment processes is the removal of sour components from the hydrocarbon liquids and water. The liquid hydrocarbons sweetening process is similar to the gas sweetening process; thus, involving equivalent energy consuming units. Conversely, mainly two types of units consume energy in the sour water stripper: the water stripper reboiler and pumps.

All acid gases effluent from the gas sweetening unit, liquid sweetening unit, and sour water stripper are collected and sent to the sulfur recovery units (SRU). In these units, the H_2S in the acid gas stream is converted into elemental sulfur. A variety of configurations are available for sulfur recovery. In this study, it was used the "straight-through" Claus process. This process consists of an initial combustion step at high temperature (e.g., above 980 °C), followed by a series of reactions (carried out in a Claus Bed) and condensation steps. Generally two or three such steps are used. Steam is generated in the combustion step; thereby, contributing to meet the steam requirements of the process. The units consuming energy in sulfur recovery are condensers and re-heaters. Condensers and re-heaters are powered by cooling water and HPS, respectively.

The ProMax® v3.2 simulation software was used to calculate all forms of energy required to process conventional and ultra–sour natural gas. However, fundamental engineering principles were employed to calculate the associated GHG emissions. Furthermore, the amount of fuel gas (i.e., dry sweet natural gas) required to produce process steam (e.g., high and low pressure steam needed to process natural gas) was also estimated using the simulator. It is assumed that the fuel gas first combines with air, and then is combusted in a burner. The heat generated from the combustion process is used to boil water and generate high and low pressure steam. From this analysis, it was found that the average emission per slandered of fuel gas is around 66.7 g CO_2 eq./scf. This emission rate is in agreement with values reported by the US Environmental Protection Agency (EPA) [9].

The electricity requirements for the cooling water loop can be estimated using the cooling water flow rate. This can be done calculating the power requirements in pumps and cooling tower fans. For instance, the power requirements in a cooling water pump with a volumetric flow rate (V (m^3/s)) and pressure drop (ΔP (kPa)) can be calculated as follows:

$$Power = \frac{1}{\text{efficiency}} * V * \Delta P \tag{1}$$

For the cooling water loop, it was assumed an overall efficiency of 75% [10], the pressure drop around the water loop was estimated to be 266.7 kPa [10]. Accordingly, the power requirements can be calculated as follows [11]:

$$Fan\ Power = \left(\mathrm{V}\frac{m^3}{s}\right)\left(\frac{60s}{min}\right)\left(\frac{264.172\text{gallon}}{m^3}\right)\left(\frac{0.5ft^2}{\frac{gallon}{min}}\right)\left(\frac{0.041\text{hp}}{\text{ft}^2}\right)\left(\frac{0.7457\text{kw}}{\text{hp}}\right) \tag{2}$$

The emissions of transporting the sweet gas product from process facilities to power plants was assumed to be approximately 30 g CO_2 eq./kWh according to information reported in the literature [7]. Furthermore, a mathematical model developed in GAMS on the UAE power infrastructure [3] was used to calculate the average GHG emissions generated by the gas-based power plants fleet in the country. The simulation results show that in average 370 g CO_2 eq. are generated per kWh of power produced. This value was used to calculate the GHG emissions associated to the electricity requirements involved in the analyzed life cycle. Finally, the GHG emissions associated to electricity transmission was also taken from the literature (i.e., 3.4 g CO_2 eq./kWh) [7].

Results and Discussions

The ProMax® v3.2 simulator was used to calculate the utilities (e.g., high/low pressure steam, cooling water, and electricity) consumed in the processing of conventional and ultra–sour natural gas. Table 2 shows the utility requirements for both types of gas resources per standard cubic feet (scf) of dry sweet gas produced.

Table 2. Utility requirements for conventional and ultra–sour natural processing.

Utility Type	Units	Ultra-sour gas	Conventional gas
Electricity	(Wh/scf)	20.8	4.0
High Pressure Steam (HPS)	(g/scf)	6.1	1.2
Low Pressure Steam (LPS)	(g/scf)	41.5	11.0
Low Pressure Steam (LPS) generated from SRU	(g/scf)	23.7	3.7
Net Required Low Pressure Steam (LPS)	(g/scf)	17.8	7.3
Cooling water	(liter/scf)	1.0	0.3

As shown in the table, the processing of ultra-sour gas requires much more energy than conventional gas per unit of sweet gas yielded. For example, the energy requirements to process ultra-sour gas are roughly four times higher than that for conventional gas. This can be explained by the fact that as the amount of H_2S in the feed gas increases so does the absorption plat utilities demand. This because more acid gases must be removed by the absorber leading to higher recirculating solvent flow rates. Therefore, the power consumed to cool and pump the regenerated solvent back into the absorber, as well as the steam used to regenerate the alkanolamines increases. Both energy forms (i.e., steam and electricity) account for the majority of the utilities consumed in the absorption plant. Likewise, more sulfur must be processed by the sulfur recovery units. Moreover, the sweet gas product recovery rate is much lower for the ultra-sour gas feed. Table 3 shows the recovery percentage for ultra-sour and conventional gas resources.

Table 3. Recovery percentages for ultra-sour and conventional gas.

Feed type	**Ultra-sour gas**	**Conventional gas**
Recovery rate	60 %	90 %

As shown in the table, almost 40 % of the ultra-sour gas feed volume is loss due to the high acid gases content; which are removed to obtain the sweet gas product. Conversely, for conventional gas only 10% of the volume is loss. Consequently, more energy is required per unit of sweet gas produced from ultra-sour gas. The electricity requirements for the cooling water loop were estimated using Equations (1)-(2). Accordingly, the GHG emissions corresponding to the electricity requirements were estimated using the gas power generation emission factor. Moreover, the process steam requirements associated to sour gas processing were obtained simulating the gas sweetening plants with ProMax® v3.2 [6]. The GHG emissions values related to HPS and LPS usage, and electricity for processing the natural gas are reported in Table 4. These values are given per standard cubic feet (scf) of dry sweet gas produced.

Table 4. GHG emissions related to conventional and ultra–sour gas processing.

Type of Utility Consumed	**Units**	**Ultra-sour gas**	**Conventional gas**
Electricity	(g CO_2 eq./scf)	7.7	1.5
High Pressure Steam (HPS)	(g CO_2 eq./scf)	0.7	0.4
Light Pressure Steam (LPS)	(g CO_2 eq./scf)	2.7	1.1
Total GHG emissions	(g CO_2 eq./scf)	11.1	3.0

From the table it can be clearly noticed that the GHG emissions associated with ultra-sour gas processing are much higher than those for conventional gas. Additionally, Table 4 shows that more than 50% of the GHG emissions are due to electricity consumption. Using the GAMS model [3], the average amount of natural gas needed to produce one-kilowatt hour of electricity in UAE was estimated. This value was found to be approximately 6.97 scf of gas/kWh. Additionally, all GHG emissions values from natural gas processing were also expressed in terms of per unit of kWh of electricity produced. The GHG emissions involved in natural gas extraction and final product transportation were estimated using factors reported in the literature [7]. The WTG GHG emissions life cycle for UAE domestic gas-based power generation can be estimated as the summation of the GHG emissions from: gas extraction, sour gas processing, sweet gas transportation, electricity generation (i.e., combustion of natural gas in power plants), and electricity transmission. Table 5 shows the single components and total GHG emissions life cycle for electricity produced from natural gas for both type of gas resources.

Table 5. Single emissions and total GHG emissions life cycle for domestic gas-based power generation in the UAE.

Life Cycle Stage	**Ultra-sour gas (g CO_2 eq./kWh)**	**Conventional gas (g CO_2 eq./kWh)**
Gas Extraction	20	20
Gas Processing	77.6	20.6
Product Transportation	30	30
Gas Combustion (power plants)	370	370
Electricity Transmission	3.4	3.4
Total GHG	501	444

As shown in the table, ultra-sour gas involves higher GHG emissions than conventional gas; which results from greater energy requirements. It is worth pointing out that most of the GHG emissions originate from the combustion of the sweet natural gas product in the power plants. Additionally, the WTG GHG life cycle analysis shows that for ultra-sour gas the emissions are approximately 13% greater than those involved using conventional gas. Figure 1 shows the percentage share of each stage/activity in the GHG emissions life cycle.

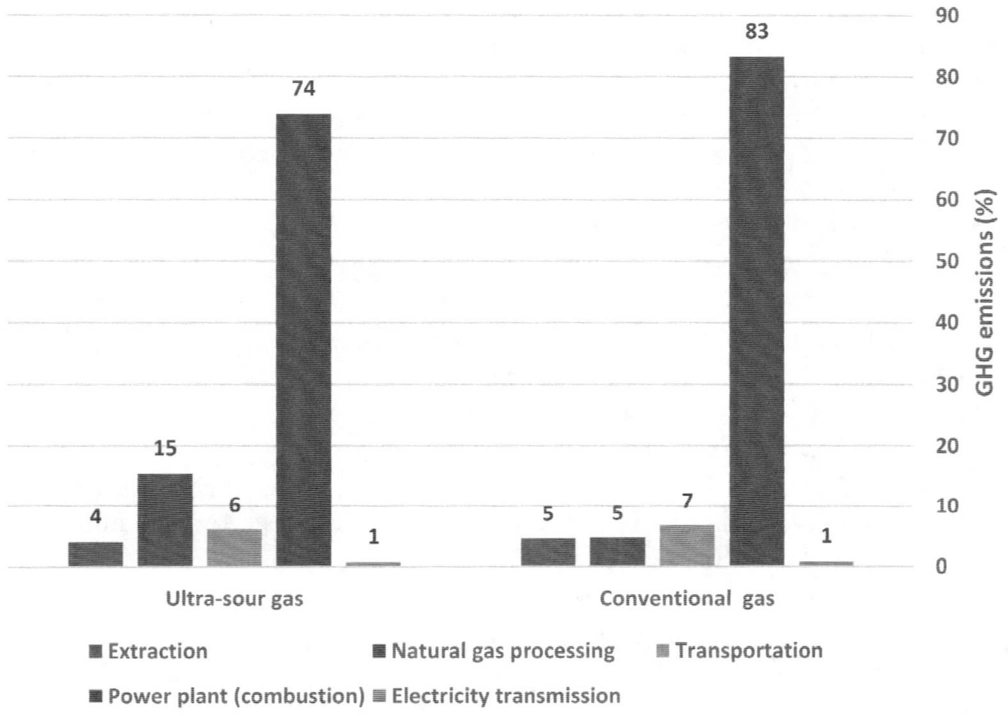

Figure 1. Percentage share per stage/activity in the GHG emissions life cycle for ultra–sour and conventional gas.

As shown in the Figure, sour gas processing and electricity generation (at the power plants) account for more than 88% of the total GHG emissions in the life cycle. Moreover, electricity generation is clearly the leading GHG emission source in the analyzed life cycle. Nevertheless, the GHG emissions generated by the electricity generation operations are rather low compared with other countries. This since the majority of the UAE's current power fleet is composed of modern gas-fired combined cycle plants; which are comparatively more efficient than other gas-based plants. Also, the total WTG GHG emissions of domestic gas-based power generation in UAE places the country in the low range (for conventional gas) and middle range (for ultra-sour gas) of values reported in the literature (e.g., 413-587 g CO_2 eq./kWh) for similar systems worldwide [7, 12, 13].

Conclusion

In this study, the well-to-grid (WTG) GHG emissions lifecycle for domestic gas-base power generation in the UAE is analyzed. The analysis results show that gas-based power generation using ultra-sour gas is more emission intensive than using conventional gas as the power plants' feed. This because much more energy is consumed in ultra-gas processing. Additionally, the yield of sweet gas product is much lower when processing ultra-sour gas compared with conventional gas. Regarding the total emissions of the life cycle, sour gas processing and electricity production account for almost 90 % of the total GHG emissions. Electricity generation is clearly the leading GHG emission source in the analyzed WTG lifecycle.

References

[1] A. Betancourt-Torcat and A. Almansoori, "Design multiperiod optimization model for the electricity sector under uncertainty – A case study of the Emirate of Abu Dhabi," *Energy Conversion and Management,* vol. 100, pp. 177-190, 8// 2015.

[2] Energy Information Administration, "United Arab Emirates " U.S. Energy Information Administration2015.

[3] A. Almansoori and A. Betancourt-Torcat, "Design optimization model for the integration of renewable and nuclear energy in the United Arab Emirates' power system," *Applied Energy,* vol. 148, pp. 234-251, 6/15/ 2015.

[4] Al Hosn Gas. (2015, 22/5/2016). *Abu Dhabi Gas Development Company Ltd. (Al Hosn Gas).* Available: : http://www.alhosngas.com/.

[5] J. Slagle, "Highly Sour Amine Sweetening Process Study for Conditions Common in the Middle East," presented at the The Abu Dhabi International Petroleum Exhibition and Conference, Abu Dhabi, UAE, 2013.

[6] Bryan Research & Engineering Inc., "ProMax," ed. Bryan, Texas, U.S.: Bryan Research & Engineering, 2013.

[7] T. Skone, J. Marriot, J. Littlefield, and B. A. Hamilton, "Life Cycle Assessment of Natural Gas Extraction, Delivery and Electricity Production," National Energy Technology Laboratory, United States of America2012.

[8] P. Pal, A. AbuKashabeh, S. Al-Asheh, and F. Banat, "Role of aqueous methyldiethanolamine (MDEA) as solvent in natural gas sweetening unit and process contaminants with probable reaction pathway," *Journal of Natural Gas Science and Engineering,* vol. 24, pp. 124-131, 5// 2015.

[9] Environmental Protection Agency, "U.S. annual non-baseload CO2 output emission rate," US Environmental Protection Agency, Washington, DC, US2015.

[10] R. Turton, R. C. Bailie, W. B. Whiting, and J. A. Shaeiwitz, *Analysis, Synthesis and Design of Chemical Process*, 3rd ed. Boston, MA, U.S.: Prentice Hall PTR International, 2008.

[11] R. H. Perry and C. H. Chilton, *Chemical Engineers' Handbook*, 5 ed. New York, US: McGraw-Hill, 1973.

[12] L. S. Abrahams, C. Samaras, W. M. Griffin, and H. S. Matthews, "Life Cycle Greenhouse Gas Emissions From U.S. Liquefied Natural Gas Exports: Implications for End Uses," *Environmental Science & Technology,* vol. 49, pp. 3237-3245, 2015/03/03 2015.

[13] World Nuclear Association, "Comparison of Lifecycle Greenhouse Gas Emissions of Various Electricity Generation Sources," World Nuclear Association London, UK2011.

468547 Data Envelopment Analysis Coupled with Thermodynamic and Life Cycle Assessment Metrics for Solvent Screening: Application to CO_2 Capture

Abstract:

The growing trend towards more sustainable products and processes calls for advanced decision-making tools for the assessment and optimisation of the economic, environmental and social impact of chemicals and chemical processes. The pressure for more sustainable processes is particularly strong in the chemical industry, in which thousands of products are developed and launched at a very fast pace, raising the question (before scaling them up) of whether they are more sustainable than the existing ones or not.

Particularly, in green chemistry applications, engineers and scientists are confronted with the task of screening chemicals (e.g. mainly solvents, additives, etc.) according to sustainability principles. In practice, it is quite hard to find a single chemical/process performing best in the whole range of sustainability metrics, as inherent trade-offs typically arise between conflictive thermodynamic, environmental and safety metrics. There has been a substantial amount of research in the field of multi-criteria assessment of chemical processes.[1-3] In contrast, the problem of screening chemicals according to several criteria has received much less attention to date. In the absence of any rigorous and systematic screening method, practitioners assess them by defining, in either an implicit or explicit manner, subjective weights that are attached to every category of interest (e.g. viscosity, density, solubility, etc.). These weights enable the calculation of aggregated indicators on the basis of which the chemicals can be ranked.[3-5]While it is straightforward to implement, this approach is highly sensitive to the weights defined and provides in turn no insight into the underlying multi-criteria problem.

In this work we propose a new methodology to screen and select chemicals according to the extent to which they adhere to sustainability principles that is based on Data Envelopment Analysis (DEA), a technique developed in economics for efficiency assessment. Essentially, DEA is a non-parametric analytical tool for quantifying the relative efficiency of a set of decision making units (DMUs) that makes use of linear programming techniques.[6,7]DEA serves two purposes: i) it identifies, among a set of DMUs - each producing multiple outputs from multiple inputs - the efficient ones (i.e. those with the best ratio desired outputs to required inputs); and ii) it establishes efficient targets for the inefficient units that if achieved would make them efficient. DEA has recently found many applications in environmental studies, but to the best of our knowledge has never been applied to assess chemicals in green chemistry applications. In essence, we combine here DEA with a set of thermodynamic and life cycle assessment indicators so as to identify the chemicals that perform best according to technical (economic), environmental and social aspects simultaneously.

The capabilities of the DEA-based approach in the context of green chemistry are demonstrated through the screening of 125 conventional amine-based solvents for CO2 capture according to 10 performance indicators. Our approach eliminates 36% of the solvents for being inefficient, establishing quantitative targets and clear guidelines on how to improve them. The DEA models are capable of identifying the practical downsides of amines as main sources of inefficiency through the calculation of clear improvement targets for properties such as vapour pressure, acute toxicity and several life cycle impacts. The final aim of the DEA-based analysis is to facilitate the selection of more sustainable chemicals in the transition towards a more sustainable chemical industry.

References

1. C. Capello, U. Fischer and K. Hungerbühler, *Green Chem.*, 2007, **9**, 927.
2. G. Wernet, S. Papadokonstantakis, S. Hellweg and K. Hungerbühler, *Green Chem.*, 2009, **11**, 1826.
3. S. Hellweg, U. Fischer, M. Scheringer and K. Hungerbühler, *Green Chem.*, 2004, **6**, 418.
4. M. Tobiszewski, S. Tsakovski, V. Simeonov, J. Namieśnik and F. Pena-Pereira, *Green Chem.*, **17(10)**, 4773-4785.
5. A. I. Papadopoulos, S. Badr, A. Chremos, E. Forte, T. Zarogiannis, P. Seferlis, S. Papadokonstantakis, C. S. Adjiman, A. Galindo and G. Jackson, *Chem. Eng. Trans.*, 2014, **39**, 211–216.
6. W. D. Cook and L. M. Seiford, *Eur. J. Oper. Res.*, 2009, **192**, 1–17.
7. P. Zhou, B. W. Ang and K. L. Poh, *Eur. J. Oper. Res.*, 2008, **189**, 1–18.

470934 Using Multiobjective Optimization and Life Cycle Assessment for the Design of More Sustainable National and International Energy Systems

Abstract:

Using multiobjective optimization and life cycle assessment for the design of more sustainable national and international energy systems

Sabio, N[a*]., Volkart, K[b]., Densing, M[b]., Strachan, N[a].

***a* UCL Energy Institute, University College London, 14 Upper Woburn Place, WC1H 0NN**

***b* Paul Scherrer Institut, Energy Economics Group, 5232 Villigen, Switzerland**

Nowadays, an increasing energy demand driven by population and economic growth, and the need to reduce green-house gas emissions are the two major axis along which energy systems need to evolve. Not in vain, the design of sustainable energy systems has also been classified as a major challenge in the area of process systems engineering [1,2].

Energy systems models were developed after the first oil crisis using the methods of linear programming for informing policy makers concerned with the importance of energy systems planning and design at the time [3]. These models are still nowadays the main decision-support tools used for informing energy policy.

Although they were initially conceived for analysing energy security and costs, climate regulation has turned environmental factors in the major focus of energy systems analysis. For instance, after the UK committed to reduce its greenhouse gas emissions by 80% in 2050 [4], the Committee of Climate Change produced a set of carbon budgets based on energy systems analysis that outline the pathways and expected changes in the energy system needed to achieve the target [5].

Conflicting objectives lie at the heart of any design problem. Multi-objective optimization (MOO) frameworks were pervasive in PSE since the early days, the combination of multi-objective optimization frameworks combined and life-cycle analysis [6] has gained more recent attention [1,7,8]. Nevertheless, to the extent of our knowledge, this framework has not been applied to energy systems models.

One reason for this might be that multi-criteria analysis is not seen as a popular method in energy systems modelling community [9]. Although the advantages of multi-objective optimization have already been more widely recognised when applied to distributed generation problems involving a variety of stakeholders [10]. However, the challenges for the implementation of life cycle assessment in a whole energy systems model has been subject of more extensive attention in the literature [11].

In the present work we propose a framework for simultaneously optimizing several life cycle environmental impact indicators and the economic performance in energy systems models. For that purpose we use the classical ε-constraint deterministic multi-objective method that a side from being suitable for convex Pareto-sets, it allows to exploit the problem structure showing the different objective spaces without the need to articulate a priori preferences (i.e., weights), and avoiding the numerical problems arising from normalization.

Our results evaluate the effect of simultaneously optimizing several life-cycle environmental impact metrics against the more traditional total system cost optimisation with CO_2 emissions constrained for a newly developed national energy systems model of the UK (UKTM-UCL) [12] and for an international multi-regional energy systems model (GMM) [13].

***Corresponding author: n.sabio@ucl.ac.uk**

References

[1] Grossmann, I.E., Guillén-Gosálbez, G. (2010) Scope for the application of mathematical programming techniques in the synthesis and planning of sustainable processes, ***Computers and chemical engineering***, 34: 1365-1376

[2] Grossmann, I.E.(2012) Advances in mathematical programming models for enterprise-wide optimization, ***Computers and Chemical Engineering,*** 47: 2-18

[3] Stefan Pfenninger, Adam Hawkes, James Keirstead, Energy systems modeling for twenty-first century energy challenges, Renewable and Sustainable Energy Reviews, Volume 33, May 2014, Pages 74-86

[4] Acts of Parliament. ***Climate Change Act 2008***. The Stationary Office Limited, UK

[5] The Committee of Climate Change (CCC) The 5th Carbon Budget: The next step towards a low carbon economy, November 2015

[6] Guinee JB, Goree M, Heijungs R, Huppes G, Kleijn R, de Koning A, van Oers L, Wegener Sleeswijk A, Suh S, Udo de Haes HA, de Bruijn H, van Duin R, Huijbregts MAJ. Handbook on Life Cycle Assessment. Operational Guide to the ISO Standards. Dordrecht: Kluwer Academic Publishers, 2002.

[7] Azapagic A, Clift R. (1999) The application of life cycle assessment to process optimisation. ***Computers and Chemical Engineering***, 10:1509–1526.

[8] N Sabio, C Pozo, G Guillén Gosálbez, L Jiménez, R Karuppiah, Vasudevan, V., Sawaya, N., Farrell, J.T. Multiobjective optimization under uncertainty of the economic and life cycle environmental performance of industrial processes, ***AIChE Journal,*** 60 (6): 2098-2121

[9] Hall, M.H. L., Buckley, A.R. (2016) A review of energy systems models in the UK: Prevalent usage and

categorisation. ***Applied Energy,*** 169: 607-628

[10] Alarcon-Rodriguez, A., Ault, G., Galloway, S. (2010) Multi-objective planning of distributed energy resources: A review of the state-of-the-art. ***Renewable and Sustainable Energy Reviews,*** 14 (5): 1353-1366

[11] Volkart K. (2014).Development of a New Methodology for the Integration of LCA and Energy-Economic System Modelling, ***LCA XIV***, San Francisco, USA

[12]UCL Energy Institute Models Website, Available at: https://www.ucl.ac.uk/energy-models/models/uktm-ucl [10 May 2016]

[13] Turton H., Panos E., Densing M., Volkart K. (2013). Global Multi-regional MARKAL (GMM) model update: Disaggregation to 15 regions and 2010 recalibration, PSI Bericht 13-03

472120 Life Cycle Assessment and Multiobjective Optimization in a Natural Gas Based Petrochemical Complex

Abstract:

In this work, we address the optimization of a natural gas based petrochemical complex using economic and environmental criteria. The mathematical model includes linear and non-linear models for single plants, taking into account main operating variables, intermittent deliveries and inventory variable profiles. The environmental objective is measured with the global warming potential (GWP) and ReCiPe metrics according to the life cycle assessment procedures using indicators at two levels: eighteen midpoint indicators and three endpoint indicators realised in SimaPro (Pre-Consultants, 2015).The resulting mixed integer nonlinear programming (MINLP) models are implemented in GAMS. The bi-criteria MINLP model is solved with the Îµ-constraint method. Mid-point indicators are presented for current conditions. Numerical results show that the petrochemical complex can satisfy product demands while improving the environmental performance by decreasing greenhouse gases emissions in 33 %, from 1,042 to 695.33 kt CO_2-eq/year in the case of GWP minimization.

457916 Comparison Between the OLI-MSE and eNRTL Models in Predicting Thermodynamic Properties of the $NaNO_3$-HNO_3-H_2O Ternary System

Abstract:

Chemical engineers engaged in spent nuclear fuel and nuclear waste processing have pursued high fidelity process modeling and simulation as have their counterparts in the conventional chemical industries. Although current modeling and simulation is capable of representing the physical and chemical processes in nuclear waste treatment at a certain level of sophistication, accurate thermodynamic models for the systems involved are essential not only to predict their chemical and physical properties correctly, but also to support high fidelity process simulation. Since reprocessing usually involves extraction with concentrated nitric acid followed by neutralization with caustic soda (e.g., the PUREX process), large amounts of sodium nitrate are formed. Consequently, three compounds – nitric acid, sodium nitrate, and water – are typically the most abundant species in nuclear waste liquids. The OLI-MSE (mixed solvent electrolyte) and eNRTL (electrolyte non-random two-liquid) are two of the activity coefficient models most widely used in the nuclear industry for aqueous nitric acid solutions. In the present work, the two models are compared side-by-side for the first time by having their results checked against literature experimental data on vapor-liquid equilibrium, calorimetric properties (i.e., heat capacity and liquid molar enthalpy), and speciation for the $NaNO_3$-HNO_3-H_2O ternary system and its subsystems.

Formatting Issue - Best Available Paper/Graphic

462300 Multi-Agent Optimization Framework (MAOP) for Synthesizing Optimal Radioactive Waste Blends

Abstract:

Multi-agent Optimization Framework (MAOP) for synthesizing optimal radioactive waste blends

Berhane H. Gebreslassie and Urmila M. Diwekar

Center for Uncertain Systems, Tools for Optimization and Management (CUSTOM):Vishwamitra Research Institute, Crystal Lake, IL 60012 â€" USA

berhane@vri-custom.org; urmila@vri-custom.org

Abstract

The multi-agent optimization (MAOP) framework provides a way of combining various algorithms in one platform and exploits the strength possessed by each algorithm. Most of large scale process system engineering problems include nonlinear and non-convex problems. In this work, we propose a multi-agent optimization framework for solving complex large scale process system engineering problems. The framework uses a variety of different algorithmic agents which include the gradient based local optimizers and metaheuristic algorithms (efficient simulated annealing, efficient genetic algorithm and efficient ant colony optimization algorithms). Each agent encapsulates a particular problem-solving procedure. We investigate the effect of cooperation among agents of the multi-agent system combined into a framework designed to solve large scale combinatorial optimization problems. Computational experiments are carried out using benchmark problems and real world case study. The proposed methodology enables to improve the quality of solutions and the computational efficiency in comparison with non-cooperative multi-agent framework and standalone agents. The MAOP framework which includes the five major parts of the algorithm (figure 1) is given in the figure 2.

Fig. 1. The major parts of the MAOP algorithm and the information flow direction.

Fig. 2. The basic flow diagram of the heterogeneous MAOP algorithm

Case Study:

A real-world problem of synthesizing optimal waste blends is analyzed to test the applicability of this novel approach in addressing a general synthesis problem. The Hanford Site in southeastern Washington produced nuclear materials using various processes for nearly fifty years. Radioactive waste was produced as by-products of the processes. This waste can be retrieved and separated into high-level and low-level wastes and subsequently immobilized for future disposal. The high-level waste is converted into a glass form before disposal. However, the glass must meet both process-ability and durability restrictions. The process-ability conditions ensure that during processing, the glass melt has properties such as viscosity, electrical conductivity and liquidâ€™s temperature such that they lie within the range of known to be acceptable for the vitrification process. Durability restrictions ensure that the resultant glass meets the quantitative criteria for disposal in a repository. There are also bounds on the composition of the various components in the glass. The radioactive waste and frit are mixed and heated to form a glass that satisfies the constraints related to the glass property and configuration.

Hanford has 177 tanks (50,000 to 1 million gallons) containing radioactive waste. These wastes vary widely in composition, and the glasses produced from these wastes will be limited by a variety of components. The minimum amount of frit would be used if all the high-level wastes were combined to form a single feed to the vitrification process. Because of the volume of waste involved and the time span over which it will be processed, this is logistically impossible. However, much of the similar benefit can be obtained by forming blends from sets of tanks. The problem is how to divide all the tanks into sets to be blended together such that the minimal amount of frit is used. Therefore, the goal of this work is to find the optimal blend configuration that minimizes the amount of frit added to convert the radioactive waste into glass. Minimizing the amount of frit added to a minimum amount has two explicit reasons.

Fig. 3. Processing of radioactive wastes to glass

References

1. Siirola JD, Steinar Hauan S, and Westerberg AW, 2003. Toward agent-based process systems engineering: proposed framework and application to non-convex optimization. *Computers and Chemical Engineering* 27: 1801-1811.
2. Gebreslassie BH, and Diwekar UM, 2015. Efficient ant colony optimization for computer aided molecular design: Case study solvent selection problem. *Computers and Chemical Engineering* 78: 1-9.
3. Diwekar UM, Xu W, 2005. Improved genetic algorithms for deterministic optimization and optimization under uncertainty. part I. algorithms development. *Industrial and Engineering Chemistry Research* 44(18) 7132-7137.
4. Venkatesh Narayan V, Diwekar UM, Hoza M. Synthesizing Optimal Waste Blends. *Computers chem. Engng* Vol. 20, Suppl., pp. SI443-SI448, 1996.
5. Chaudhuri P, Diwekar UM, 1999. Synthesis Approach to the Determination of Optimal Waste Blends under Uncertainty. *AIChE Journal*. 45: 1671-1687.

462734 Thermodynamics Analysis and Optimization of a Stirling Cycle for Lunar Surface Nuclear Power System

Abstract:

A nuclear powered stirling engine system in lunar surface composed of a reactor core, a hot heat exchanger, a stirling engine, a cold heat exchanger and a radiator. A mathematical model for the overall thermal efficiency of the lunar surface nuclear powered stirling engine with the heat loss of the hot heat exchanger, finite rate heat transfer, regenerative heat loss, finite regeneration process time and conductive thermal bridging losses is developed. The optimal efficiency of the system was determined by the finite time thermodynamics. The effect of some system parameters such as the absorbed temperature, the heat rejection temperature, the convection heat transfer coefficients of the hot and cold heat exchangers on the optimal efficiency were analyzed in details. The results showed that the overall thermal efficiency increased and then decreased with the absorbed temperature increase. The highest thermal efficiency was about 28%, which was far away from the corresponding carnot efficiency. Higher rejection temperature had bad effect on the thermal efficiency but could improve the waste heat rejection to the outer space. Enhancement of the heat transfer could increase the thermal efficiency. The results obtained here may provide a new ideal to design lunar surface nuclear powered stirling cycle.

468396 Simulation of Helium Transport Near Prismatic Dislocation Loops in Tungsten

Abstract:

We analyze the effect of sub-surface prismatic dislocation loops on the surface morphology and helium clustering behavior of plasma-facing tungsten through the use of molecular dynamics simulations. These simulations are moderately large, 25 nm x 25 nm x 24 nm, consisting of approximately 830,000 atoms and simulated to times on the order of one microsecond. This approach eliminates some finite-size effects common in smaller simulations and reduces the flux to approximately 5.5 x 10^26/m^2/s, including ions that reflect back into the plasma. This flux is approximately a factor of 15 lower than is typically used in smaller simulations. These results indicate that prismatic loops with radii of approximately 3 nm that are centered 10 nm below the surface with Burgers vectors parallel to the surface initially cause helium atom clusters to accumulate at the edges of the dislocation core relatively quickly (within 100-150 ns of the onset of plasma exposure). Subsequent growth of these clusters, however, is relatively minimal. This is partially explained by the relatively high helium implantation flux, which causes bubbles to accumulate 0-7 nm below the surface and "block" the region of the metal containing the dislocation. Another effect results from the strain field around the loop itself: the compressive regions along the direction of the Burgers vector repel helium, but the tensile region initially attracts helium and traps it. However, we think that the attractive tensile stress region is effectively shielded by the formation of helium clusters above it, and these bubbles subsequently experience relatively slow growth.

468183 Atomistic and Continuum Drift-Diffusion Simulations of Helium Transport in Tungsten

Abstract:

Helium from linear and tokamak-like plasma devices has been known to cause "fuzz" and other nanometer-scale features on the surface of tungsten and other metal surfaces after a few hours of plasma exposure. The mechanisms involved occur on multiple length and time scales, and understanding all of those details is the subject of significant multi-scale modeling efforts. In particular, the spatial and temporal scales necessary to simulate even a small portion of the tungsten divertor in a real fusion device make atomistic simulations intractable, so we have sought a continuum-based description that is informed by atomistic simulations of more accessible systems. This effort comes in four parts: (1) large-scale atomistic simulations of helium diffusion in tungsten, which aims to discover mechanisms and provide a benchmark against which more coarse-grained models can be tested; (2) small, targeted molecular dynamics and statics calculations that quantify the rates of specific mechanisms, such as drift forces near surfaces and grain boundaries, and (3) coarse-grained, drift-reaction-diffusion models that incorporate information from small-scale molecular dynamics simulations and use large-scale simulations as a benchmark for still-larger, more realistically-sized simulations of plasma-facing tungsten. The orientation of the surfaces, the presence of grain boundaries, and the presence of other bubbles all have a strong impact on the rate of local helium transport, resulting in a spatially-dependent rate of helium transport in the divertor. These tools aim to provide a tractable and reasonably accurate discription of helium bubble distributions over time, an important step in the process of understanding tritium retention and material fatigue in nuclear fusion reactors.

469392 Engineering *Escherichia coli* strains for Optimal Performance in Co-Culture with *Trichoderma Reesei* for Consolidated Bioprocessing

Abstract:

Consolidated bioprocessing, in which enzyme production, hydrolysis, and fermentation occur in a single step, is predicted to offer the lowest cost configuration for cellulosic biofuel production given a sufficiently efficient process. Use of synthetic microbial consortia can help to address some of the limitations of monoculture bioprocessing, including high metabolic burden on the cells and the need to optimize multiple pathways or functions in the same species. Design of synthetic microbial consortia, however, brings new challenges including identifying or engineering strains that can perform optimally within a mixed population.

We are working on a synthetic co-culture of cellulase hyper-producer *Trichoderma reesei* strain RUT-C30 and *Escherichia coli* carrying a pathway for the drop-in biofuel isobutanol from glucose. This co-culture can produce isobutanol directly from cellulosic substrate in one consolidated bioprocess [1]. The *E. coli* strain NV3 [2] initially employed in this system performs suboptimally, achieving lower yield and titer, higher byproduct formation, and notably higher plasmid loss compared to monoculture conditions. Here we present our work to generate *E. coli* strains with superior co-culture performance. We have taken a directed evolution approach, selecting for *E. coli* strains with improved growth on valine analog norvaline in co-culture with *T. reesei*. Additionally we have integrated and optimized the copy number of the isobutanol production pathway genes in the *E. coli* genome to achieve a more stable process.

[1] Minty JJ, Singer ME, Scholz SA, Bae C-H, Ahn J-H, Foster CE, Liao JC, & Lin XN (2013) Design and characterization of synthetic fungal-bacterial consortia for direct production of isobutanol from cellulosic biomass. *Proc Natl Acad Sci USA* **110**(36): 14592-14597.

[2] Smith KM & Liao JC (2011). An evolutionary strategy for isobutanol production strain development in *Eschericia coli*. *Metab Eng* **13**(6): 674-81.

472567 PHA Production Using Yarrowia Lipolytica and Alternative Feedstocks

Abstract:

Polyhydroxyalkanoates (PHAs) are naturally-occurring polyesters produced by a number of different bacterial species as a means of carbon and energy storage. PHAs have been suggested as possible replacements and supplements to petroleum-derived plastics due to PHAsâ€™ tunable mechanical properties and high potential for feedstock flexibility. Feedstock flexibility allows production of a desired product even under conditions of environmental or economically-induced scarcity. Natural producers of PHAs are often highly sensitive to the substrates and media on which they are grown. This variability precludes feedstock flexibility. In order to address this issue, we have engineered a far more metabolically-robust organism, the oleaginous yeast *Yarrowia lipolytica*, to produce PHAs from a number of different feedstocks.

To accomplish this, the PHA-producing enzymes, called PHA Synthases, from the natural PHA producers *Pseudomonas aeruginosa* and *Ralstonia eutropha* were incorporated into the yeast. Flux to this enzyme was increased by overexpression of the native acetyl-CoA carboxylase (ACC1) and modification of Multi-Function Enzyme (MFE1). ACC1 facilitates the synthesis of longer carbon chains, which is desirable when building a specific larger product from smaller substrates. The modification to MFE1 prevents this double-acting enzyme from degrading a highly desired PHA precursor. *Yarrowia lipolyticaâ€™s* native catabolic pathways allow it to be an effective metabolizer of fats, sugars, and proteins. Our engineered organisms demonstrated PHA production using pure and waste fats, crude glycerol, and hexose and pentose sugar feedstocks. The final PHA products were also characterized for their structure and composition: necessary information for the tuning of the mechanical properties.

473357 Rewiring Native Metabolism of Yarrowia Lipolytica for Enhanced Assimilation of Complex Biomass-Derived Sugars and High-Yield Production of Organic Acids

Abstract:

While the model oleaginous yeast *Yarrowia lipolytica* is well known for decades to grow mainly on C6 sugars, its native metabolic capability to assimilate xylose and cellobiose, dominant sugars derived from lignocellulosic biomass, is poorly understood. In this study, we elucidated and activated native xylose and cellobiose pathways of *Y. lipolytica* through comprehensive metabolic and transcriptomic analyses. We identified 7 glucose-specific transporters, 16 xylose-specific transporters, and 4 cellobiose-specific transporters that were transcriptionally upregulated for growth on respective single sugars. While *Y. lipolytica* is capable of using xylose as a carbon source, xylitol dehydrogenase (XYL2) is the key bottleneck of xylose assimilation. *Y. lipolytica* has a set of 5 extracellular and 6 intracellular beta-glucosidases (BGLs) to assimilate cellobiose via both extra- and intra-cellular mechanisms, but the latter being dominant for growth on cellobiose as a sole carbon source. By tuning the expression of native metabolic enzymes and transporter genes, we engineered a *Y. lipolytica* mutant with enhanced assimilation of xylose and cellobiose as well as their mixtures with glucose. Further, we demonstrated the performance of *Y. lipolytica* mutant for high-yield production of high-value organic acids (e.g., alpha-ketoglutaric acid) using ionic liquid-pretreated lignocellulosic biomass in a simultaneous saccharification and fermentation process.

467715 Using Protein Design to Evaluate the Relationship Between Protein Surface Potential and Protein-Lignin Binding for the Eventual of Low Lignin Binding Cellulases

Abstract:

Increasing total turnover number or recycling cellulases used in deconstruction is a key step in making microbial biofuel production from lignocellulosic feedstocks economically viable. Nonspecific binding interactions between cellulases and lignin denature cellulases and hinder cellulase recycling. Previous literature supports that this binding event is influenced by the surface potential characteristics of the cellulases themselves, namely surface hydrophobicity and electrostatic surface potential. To further investigate the relationship between surface potential and nonspecific protein-lignin binding, we used a hybrid computational/experimental approach to (i.) computationally redesign surface potentials spanning a large range of net charge, total charge density, and surface hydrophobicity on a model protein; and (ii.) experimentally characterize the resultant designs for function, stability, and protein-lignin binding. In all, we tested sixteen designs spanning the physiological ranges of the above potentials. We found that designs were, on average, better expressed and more stable that the parental protein, showing that the computational design process is robust. We characterized variant binding to AFEX pretreated, dioxane extracted cornstover lignin with two different experimental methods â€“ quartz crystal microbalance with dissipation (QCM-D) and fluorescence based subtractive mass balance assay. Comparing the binding capacity of variants between characterization methods highlights that both methods isolate the same low, medium, or high lignin binders. Analysis of determined binding capacity with known surface potential properties suggests a weak relationship between protein net charge and protein:lignin binding capacity, while other parameters were not statistically significant. We then applied this method to redesign surfaces of different catalytic domains on endocellulases. Implications for improving the deconstruction step in the next generation of biorefineries will be discussed.

454097 Towards Development of *Rhodococcus Opacus* As a Microbial Cell Factory: Conversion of Lignin-Derived Aromatic Compounds into Lipids

Abstract:

Lignocellulosic biomass is a renewable feedstock that can be converted into biofuels and biochemicals using microorganisms. However, pretreatment of lignocellulosic biomass releases toxic compounds that inhibit microbial growth. Additionally, lignin has been an untapped carbon source due to its recalcitrance and toxic monomeric units (e.g., aromatic compounds). To achieve sustainable and economic utilization of lignocellulosic biomass, we must develop technologies that utilize the complete biomass, including lignin, which is currently considered as waste. To this end, our work focuses on developing *Rhodococcus opacus* as a new chassis for conversion of aromatic compounds into triacylglycerols (TAGs), biodiesel precursors. *R. opacus* is a promising host due to its high tolerance to and utilization of aromatics as a sole carbon source. Importantly, *R. opacus* can accumulate TAGs up to ~80% of cell dry weight under nitrogen-limiting conditions. To enhance its innate aromatic-degrading capacity, we applied adaptive evolution, a growth-based strain selection method, by sequentially sub-culturing cells in diverse combinations of lignin-derived aromatic compounds as sole carbon sources. Our adapted strains demonstrated higher growth rates and higher lipid accumulation compared to the wild type strain. Whole genome sequencing, RNA-seq, and ^{13}C-fingerprinting analysis have identified possible aromatic tolerance and utilization mechanisms such as upregulation of degradation pathways and putative transporters for aromatic compounds. In addition, synthetic biology tools have been developed to enable engineering of the identified tolerance genes, regulators, and pathways. We will present progress towards development of *R. opacus* as a microbial cell factory.

471737 Defining the Minimal Set of Microbial Genes Required for Valorization of Lignin Biomass

Abstract:

Lignin is the second-most abundant biopolymer in the world, but substantial previous efforts to break it down into useful carbon-based platform chemicals in a reliable and scalable manner have been largely unsuccessful, rendering lignin substantially underutilized. By contrast, numerous fungi and bacteria readily degrade this exceptionally recalcitrant biopolymer using a milieu of specialized enzymes, and are thus able to metabolize lignocellulose as an energy source. Reconstituting the lignin-degrading machinery of these specialized microbes in a genetically-tractable model organism may therefore enable a route to efficient valorization of lignin biomass. To date, however, no engineered heterologous host has been shown to have degradation activity toward whole lignin biomass. Identification of the minimum necessary set of enzymes required for the conversion of lignin biomass into valuable feedstock chemicals has likewise not yet been realized. This study investigates the co-expression of putative lignin-degrading enzymes in the model host organism *Saccharomyces cerevisiae.* Specifically, recombinant strains were engineered to drive efficient export of peroxide-generating enzymes as well as ligninase candidates identified through transcriptomic and secretomic data from lignin-degrading fungi. A combinatorial approach was employed to develop an enhanced understanding of the interconnected roles of putative ligninases in degrading lignin biomass.

474992 Invited Presentation: Lignin Conversion By Biological Funneling and Chemical Catalysis

Abstract:

Lignin is an underutilized feedstock for producing renewable fuels and chemicals, where it is instead burned for process heat because its inherent heterogeneity and recalcitrance make it difficult to selectively valorize. Despite many decades of lignin depolymerization research, most catalytic strategies to break down lignin yield a highly heterogeneous slate of aromatic compounds. In Nature, some organisms have evolved metabolic pathways that enable the utilization of lignin-derived aromatic molecules as carbon sources. Aromatic catabolism typically occurs via upper pathways that act as a â€œbiological funnelâ€ to convert heterogeneous substrates to central intermediates, such as protocatechuate or catechol, which undergo ring cleavage and are further converted via the Î²-ketoadipate pathway to central carbon metabolism. We are using a natural aromatic-catabolizing organism, *Pseudomonas putida* KT2440, to demonstrate that these metabolic pathways can be harnessed and engineered to convert both aromatic model compounds and heterogeneous, lignin-enriched streams into value-added compounds. To make this concept a reality will require tailoring the lignin chemistry of the feedstock to the catalytic deconstruction strategy and the biological pathways for upgrading. Overall, this work demonstrates that the use of microbial aromatic catabolism may one day enable an approach to valorize lignin by overcoming its inherent heterogeneity to produce fuels, chemicals, and materials.

475260 Next-Generation Testing Platform for Strain Optimization, Screening, and Fermentation Testing

Abstract:

Amyris is a renewable products company that engineers yeast to convert sugar into valuable chemicals. We have successfully produced four isoprenoid molecules at commercial scale, replacing petroleum, animal or plant-sourced chemicals with reliable, sustainable alternatives. Underlying this commercial success is an automated pipeline for the construction, testing, and deployment of our yeast strains that has made each strain development timeline shorter with each successive molecule. We are now in the midst of expanding our footprint on the metabolic map into chemical targets outside isoprenoids, and consequently, expanding the capabilities of our strain construction pipeline. This talk will focus on our strain testing technologies, and describe the next-generation testing tools we are developing to enable testing of strains targeting diverse chemical space and multiple global markets.

475263 HarnessingÂ Biological Complexity and Diversity through Metabolic Engineering: The Manus Bio Approach

Abstract:

Specialized metabolites represent a class of compounds with unique and complex chemistries and which possess a myriad of commercial applications, from pharmaceuticals and food products to industrial chemicals. Despite their intrinsic value, however, sourcing remains a bottleneck to more widespread use due to their low abundance in nature and the inherent difficulties in applying metabolic engineering and industrial biotechnology approaches to produce these complex natural chemistries in microbial hosts. At Manus Bio, we focus on addressing the challenges of microbial fermentation of plant secondary metabolites through the combined application of our proprietary Multivariate Modular Metabolic Engineering (MMME) and Pathway Integrated Protein Engineering (PIPE) technologies for optimizing heterologous enzymes and pathways. With this approach, we have successfully demonstrated the potential of using a single microbe to catalyze several complex chemical reactions, including polymerization, cyclization, multiple oxidations, and up to six glycosylations, all while producing the target metabolite at greater than 95% purity. In this presentation, we will highlight some of the important insights and guiding principles used to engineer complex biochemical pathways in simple bacterial systems, such as *Escherichia coli*.

475265 The Emerging Organism Engineering Industry

Abstract:

There is an emerging demand for sourcing plant-derived extracts (nutraceuticals, flavors, fragrances, sweeteners, etc.) from engineered microbes. While recent advances in synthetic biology and metabolic engineering provide feasible approaches to engineering such organisms, commercial success for developing these â€œculturedâ€ ingredients presents specific challenges. Unlike biofuels, where efforts can be focused on one particular molecule given the enormous market size, cultured ingredients require developing different organism lines in a rapid and low cost fashion. This requires a scalable solution for bio-manufacturing of organisms, which is provided by our state of the art foundry that continues to grow. I will describe how organism development at Ginkgo leverages our foundry to accelerate the design/build/test using specific examples. In particular, I will highlight the value of combining computer-aided engineering software tools, cheap gene synthesis and high resolution-accurate mass LCMS to develop engineered microbes. Finally, Iâ€™ll touch on how our improvements in manufacturing organisms lend to opportunities outside of cultured ingredients.

475269 General Purposing Synthetic Biology:Â Â a Biofoundry in a Box

Abstract:

What could you build with unlimited DNA writing? Productivity of DNA synthesis continues to increase at exponential rates. The oligonucleotide synthesis market alone is forecast to reach nearly $2B USD by 2020 with an annual growth rate in excess of 10%. This increase in oligo synthesis capacity and the associated reduction in cost is transforming synthetic biology; cheap oligo synthesis will remake the impossible to the new routine. Muse's foundational technology is a proprietary synthetic biology platform that scales directly with multiplex oligo synthesis, enabling designer DNA construction at 1000X lower cost than the current state of the art. To date, Muse has demonstrated the ability to generate and track over 50,000 designer mutants simultaneously. Applications of this technology include sequence to activity mapping for site saturation mutagenesis, protein and pathway engineering, and strain reconstruction of complex evolved strains to name a few. Gene editing capabilities have been demonstrated in various industrially relevant organisms, including applications across biotechnology. Further, this technology is being automated in the form of a bench-top device that allows a single user to perform a broad range of molecular biology endeavors at rates 1000X beyond the current state of the art.

475259 Making Silk without Silkworms: Using Industrial Biotechnology to Make Performance Protein-Based Fibers

Abstract:

Bolt Threads is a Bay Area startup that believes that the combination of molecular biology and materials science can revolutionize how people think about and use natural materials. Bolt's materials can dramatically improve the performance and environmental impact of a broad range of consumer and industrial materials. Specifically, Bolt is developing the next generation of sustainable performance apparel by leveraging the latest advances in industrial biotechnology, automation, and materials science. Derived from our knowledge of spider silk protein, the company has developed a high-throughput platform to generate thousands of different silk proteins, all from a scalable microorganism host. Using industrial biotechnology processes and fiber spinning, Bolt is able to make commercial volumes of protein and fiber. Precision control from primary protein sequence through fiber extrusion allows the company to produce materials that combine the benefits of both natural and synthetic fibers, and to go beyond even what is currently available in nature and the marketplace. Leveraging the unique properties of its fibers, Bolt Threads will be delivering its own line of consumer apparel made with its Engineered Silk(TM) as well as working with partners across industry verticals.

475548 The Development of Platform-Based Technologies for the Optimization of Sustainably Produced Chemicals

Abstract:

Genomatica has established an integrated computational/experimental bioengineering platform to design, create, and optimize novel high-producing organisms and bioprocesses. Here we present the use of our platform to develop *E. coli* strains for the production of the industrial chemical 1,4-butanediol (BDO) from carbohydrate feedstocks. These efforts have culminated in commercial scale production of over 5M lbs of BDO from dextrose. BDO is a chemical intermediate produced commercially through energy-intensive petrochemical processes from hydrocarbon feedstocks and is used in a variety of products including automotive, electronics and apparel. Therefore, this product represents an opportunity to make a significant impact on the replacement of traditional petrochemical processes with benign bioprocesses using renewable feedstocks.

We are continuing to enhance our technology platform, enabling faster development of new processes with increased chance of success. Here we describe the application of cell-free transcription-translation (TX-TL), towards improving the design-build- test-learn cycle for the construction of a high-performing microorganism capable of producing BDO and other industrial chemicals from carbohydrates. We used systems biology approaches including 13 C-flux analysis, proteomics, and metabolomics to characterize TX-TL applied to different engineered metabolic pathways, including BDO, and compared the results to the analogous in vivo processes. These studies provide critical insight into the active metabolism of TX-TL and mechanisms of redox and energy regeneration. Learnings from these studies were applied towards improving the TX-TL platform for the production of intermediate and specialty chemicals. The presentation will highlight the successful integration of the TX-TL technology into our rational and high-throughput strain engineering platform, and will show the potential advantages for using cell-free systems towards the optimization of sustainably produced chemicals.

478302 Imagine, Design, Create - Cloud Based Software for Biological Design

Abstract:

Over the past 10 years the ability to sequence, edit, and build life has grown exponentially. Software in this space has not kept pace with the rapid pace of biological innovation resulting in a tremendous gap between the modeling and manufacturing of biological systems. The Autodesk BioNano team works to bring design solutions to biology, making the interaction of scientists with software more seamless and efficient. Our goal is to allow scientists to focus on what matters, the science. Autodesk has experience in designing and displaying complex, man-made mechanical structures in the cloud. It is time to apply this knowledge to biology to show how proteins and molecules interact to create life. Autodeskâ€™s BioNano group hopes to empower scientists to Imagine, Design, and Create a better world.

463022 Applications of Omics Technologies for Enhanced Process Robustness in a Chemically Defined Fed-Batch Industrially Relevant CHO Cell Culture Process

Abstract:

We have previously reported on an undesirable high lactate phenotype, caused by overfeeding a mAb-producing CHO cell line, in a chemically-defined process. To better understand this phenomenon, we utilized metabolic flux analysis (MFA) using isotopic labeling with multiple tracers (glucose and glutamate) and untargeted metabolomics. The results from these analyses provided new insights into our cell culture process. We first identified two supplements capable of delaying the onset of the high lactate phenotype by ~3 days. The addition of the first supplement increased the peak viable cell density by 15% and harvest titer by 25%. The second supplement increased both peak viable cell density and harvest titer by 25%. Following this observation, untargeted metabolomics was performed to investigate the mechanism of action for these supplements by comprehensively characterizing the impact of the supplements on the levels of intracellular and extracellular metabolites. Focusing on central carbon metabolism, we conducted isotopic labeling experiments with glucose and glutamate tracers, and analyzed the results using MFA to characterize the fluxes in and around the TCA cycle. We identified overflow of malate and citrate from the mitochondria as a potential mechanism underlying growth inefficiency, which was attenuated by the two medium supplements. In addition, we assessed the redox potential in our cell culture, identified supplements capable of preventing a redox imbalance, and confirmed their ability to delay the high lactate phenotype in our production culture. Utilizing the insight gained through our systems biology approach, we evaluated the proposed mechanism of action by reformulating our nutrient feed to implement the inhibition of mitochondrial metabolite overflow independent of the two supplements.

471844 Scale Related Variables and Their Potential Impact on Process and Product Quality in the Production of Biologics

Abstract:

Cell culture technologies for the production of recombinant protein therapeutics have evolved over the years and span from batch to fed batch, perfusion and other production methods. Process intensification has resulted in new challenges during process scale-up. We assessed dissolved oxygen and pH control strategies at 2-L, 100-L and 2000-L scale to evaluate process and product consistency across the three scales. Differences in mass transfer characteristics and offline sample handling across the scales caused up to 20% differences in integral viable cell density. The changes in the cell growth profiles resulted in shifts in product glycan profiles and associated effector function attributes. Understanding the cell culture operating space based on the selected production method and associated control variables is key to successful process scale-up.

458796 Design and Simulation of a Continuous Rotating Annular Bioreactor with an Internal Spiroid

Abstract:

This research introduces a continuous bioreactor designed to increase cell oxygenation, growth rate and overall high productivity through its novel geometry. The bioreactor consists of an outer shell cylinder and inner core cylinder, each with adjustable rotational velocities to control the mixing in the reactor and provide a variety of shear environments. The reactor may be operated with or without headspace. Inlet flows are controlled to provide medium and oxygen to the reactor volume. A spiroid tube is attached to the inside of the bioreactor shell wall to increase the gas-liquid contact area and thus improve oxygen transfer. The complex flow patterns were successfully analyzed by using computational fluid dynamics (CFD) simulations. Simulations were conducted for multiphase fluids with various operating conditions to predict the actual reaction environment in the bioreactor prior to experimental studies. Various experiments along with mathematical models were used to calculate the volumetric mass transfer coefficients at different operating conditions for bioreactor models with or without the spiroid tube. The comparison of the computational with experimental results was used to support the improved oxygen transfer performance and to predict the higher cell cultivation performance of this bioreactor due to the addition of the spiroid tube.

468411 Cell Culture of the Termite Gut Microbiome Using a 3D-Printed Synthetic Microhabitat

Abstract:

A community of microbes living in association with a host organism is sometimes called a microbiome. The interactions among members of a microbiome can give rise to important communal functionality that cannot be achieved by its constituent organisms alone. Examples of microbiome function include normal human digestion, productive and resilient plant root systems, and the degradation of lignocellulose in lower termites such as *Reticulitermes flavipes*. A microbiome lives in a complex and often dynamically changing microhabitat. No single set of cell culture conditions is suitable for the entire microbiome. As a result, microbiomes are often studied in ways that destroy the community such as the use of metagenomic analysis. However, emerging technology in microfabrication is leading to novel cell culture systems that better emulate complex features of microhabitats. Here, the steep oxygen gradient that naturally exists in the lower termite hindgut was emulated in a microfluidic device using a rapid prototyping and iterative design approach. 3D-printed devices with a 200-µm resolution were used as casting molds for oxygen-permeable, polydimethylsiloxane (PDMS) cell culture chambers. The microbiota from *Reticulitermes flavipes* hindguts were inserted into synthetic microhabitats under anaerobic conditions before allowing diffusive flux from an 8 ppm dissolved oxygen reservoir to reach one side of the culture chamber, thereby closely emulating oxygen gradients in the real hindgut. Initial results indicate spatial reorganizations of protists within this synthetic microbiome as micro-oxygen gradients evolve. Several enabling technologies are under development that will further enhance complex microhabitat cell culture systems. One such technology is semipermeable membranes incorporated directly into disposable and easy to use 3D printed cell culture chambers for the delivery of oxygen and other nutrients. Our work contributes to a better understanding of the *Reticulitermes flavipes* hindgut microbiome by enabling hypothesis-driven research of how culture conditions and perturbations influence community structure and microbiome function. Eventually, gradient bioengineering technology may enable efficient, adaptive conversion of waste biomass into valuable chemicals and fuels.

460147 Single High-EDR Exposure of Cultured Human Cells Reveal Effects of Prolonged Doublings and Microcarrier Attachment on Cell Shear Susceptibility

Abstract:

Background: Human Mesenchymal Stem Cells (hMSCs) are very important for cell therapy use, and are able to avoid immune responses when used in allogeneic therapies [1-3]. hMSCs are cultured at centralized manufacturing centers before being shipped to where they are needed and then differentiated. Currently there are no standardized protocols for the isolation, expansion and harvesting of the hMSCs[4] Additionally there is concern about the effects of shear stress on hMSCs, as damage during scale up and manufacturing have been reported. Based on these reports, we expected that hMSCs would be more sensitive to shear stress than cells typically used in large scale cell culture, specifically compared to Vero cells. Additionally, scale-up issues are present with the large scale production that would be needed for therapeutic use, where up to a trillion cells could be necessary. Microcarriers are one method that could be used to produce cells, but different microcarriers can have different effects on growth [5], and it is important to determine if the microcarrier type has an impact on the shear sensitivity of cells grown on that carrier.

Materials and Methods: Beginning with cells grown in T-flaks, a contractional flow device, called a "torture chamber" was used, flowing hMSC and Vero cells through with flow rates, subjecting them to previously defined hydrodynamic forces[6]. The Vero and hMSC cells were suspended in their respective media and then were flowed at various flow rates through the torture chamber. Following this the cells were collected and centrifuged. A cytotoxicity assay (Promega) was used to quantify the cell damage due to the shear stress. T-flasks were reseeded using cells that were not passed through the torture chamber. An approximate count of the cells that attached one day after seeding was gathered to determine the generation count. The number of cell divisions was estimated for each hMSC batch run through the torture chamber. These were placed into bins of generation counts for comparison to Vero cells using ANOVA, each bin representing what is traditionally considered a "passage". Following this microcarrier effects on shear sensitivity were studied. Human Embryonic Kidney cells (HEK-293) were grown on two different types of microcarriers, collogen coated (Solohill) and dextran matrix (GE cytodex). The HEK-293 cells were grown on the carriers in spinner flasks under protocols typically used for growth on the microcarrier, and then the carriers were flowed through the torture chamber at varying concentrations. These different concentrations were formed by collecting different amounts of media with carriers from the spinner flaks and diluting them to a constant volume, with the highest concentration run being 4 times the lowest concentration, and intermediate concentrations run at 2 and 3 times the lowest concentration. The cells and carriers were then collected, centrifuged, and tested using the same cytotoxicity assay to assess the cell damage due to the shear stress. After taking a sample of the supernatant for the cytotoxicity assay, the cells were then lysed, putting the LDH level on a per cell basis.

Results and Discussion: For suspended cells, the amount of LDH released after shear stress exposure for the hMSC cells after 20.5-24.5 and 24.5-28.5 were not significantly different than the vero cells. However, the hMSCs that had generation counts over 28.5 released significantly ($p<0.05$) more LDH than vero cells at energy dissipation rates above $6.0*10^6$ W/m^3. Specifically, the vero cells and the hMSC cells at generation counts below 28.5 show less than 10 percent LDH, compared to around 20 for the higher generation hMSC cells. At EDR rates of around $4.0*10^7$ W/m^3, the hMSC cells with a generation count of 32.5 to 36.5 exhibit about 60% LDH, indicating a high amount of cell damage and death. The generational effect could be due to changes in the cytoskeleton that have been previously found, possibly contributing to tears and LDH release[7]. The work with the micorcarriers is currently ongoing, but preliminary results show that the percent LDH increases with increasing micorcarrier concentration. The percent LDH increases sharply at higher concentrations, indicating a concentration where the shear effects on the cells become a significant factor. In our preliminary results, the LDH levels increase dramatically between the 3X concentration trial 4x concentration trial.

Conclusions: Our results indicate that the hMSCs might be just as tolerant to sheer stress as other cells currently used for manufacturing, as long as the cells are used when they are still young and aren't entering the process of senescence. The results also indicate that the torture chamber can be used to screen conditions that would allow for optimal growth of cells. Computational modeling has been previously done on the chamber to determine EDR rates for the different flow rates used, and data is available for EDR in spinner flaks and bioreactors based on agitation speed, so the torture chamber could be used to determine the agitation speed that allows for minimal sheer stress effects. The torture chamber can also be used to determine the concentration of carriers that can be used with minimal cell death. More carriers provide more surface area for the cells to grow on, but too many carriers introduces cell damage due to collisions. The use of the chamber allows for the determination of the concentration that would have the most surface area without introducing excessive cell damage. The torture chamber can be used to screen conditions for production of various products without needing to waste production runs to try out conditions to determine viability.

1. Wei, X., Yang, X., Han, Z., Qu, F., Shao, L., Shi. Y., 2013, "Mesenchymal stem cells: a new trend for cell therapy", Acta Pharmacologica Sinica, Vol 34, pp. 747-754.
2. Kim, N., Cho, S., 2013, "Clinical applications of mesenchymal stem cells", Korean Journal of Internal Medicine, Vol. 28, pp. 387-402.
3. Ankrum, J., Ong, J., Karp, J., 2014, "Mesenchymal stem cells: immune evasive, not immune privileged", Nature Biotechnology, Vol. 32, pp. 252-260.
4. Ikebe, C., Suzuki, K., 2014, "Mesenchymal Stem Cells for Regenerative Therapy: Optimization of Cell Preparation Protocols", BioMed Research International, http://dx.doi.org/10.1155/2014/951512
5. Rowley J., Abraham, E. Campbell, A., Brandwein, H., Oh S., 2012, "Meeting Lot-Sized Challenges of Manufacturing Adherent Cells for Therapy", Bioprocess International, Volume 10, pp. 16-22.
6. Shenkman, R., Godoy-Silva, R., Papas, K., Chalmers, J., 2009, "Effects of Energy Dissipation Rate on Islets of Langerhans: Implications for Isolation and Transplantation", Biotechnology and Bioengineering, Vol. 103, pp. 413-423.
7. Maloney, J., Nikova, N., Lautenschlager, F., Clarke, E., Langer, R., Guck, J., Van Vliet, K., 2010, "Mesenchymal stem cell mechanics from the attached to the suspended state", Biophysical Journal, Vol 99, pp. 2479-2487.

463688 Metabolite Pathway Analysis for Analyzing Consumption of Amino Acids in Hybridoma Cell in Different Types of Growth Media and Cell Culture Process Conditions

Abstract:

In hybridoma cell cultures, amino acid and glucose consumption, cell growth and antibody production are key considerations during upstream process optimization, particularly media optimization. Gaining knowledge on their interrelations and the relevant cellular physiology could provide good insight for higher and more consistent antibody production. The amino acid consumption pattern was analyzed in hybridoma cell cultures producing monoclonal antibody (mAb). The amino acid concentrations, product concentrations, and cell growth profiles were measured in three different cell cultures subject to an amino acid supplementation strategy. In this research, we examine the amino acid metabolism and its impact on cell growth and mAb production during the different type of cell culture processes. We found that a group of key amino acids and metabolic pathways, the highest mass fraction of the product, and these could become limiting for product expression. In this study, metabolic pathway analysis was applied to show that certain metabolic pathways are significantly different between the culture process and media composition. Metabolic pathway analysis was helped to better evaluate the impact of amino acid supplementation on the most relevant pathways in the culture process.

474761 Confessions of a Cell Culture Engineer: Process Impact on Protein Products in a QbD World

Abstract:

In producing therapeutic proteins from cell culture processes, it is necessary to develop processes capable of delivering consistent protein characteristics, particularly those that could affect the structure or function of the biomolecule and potentially impact the product safety or efficacy. Over the past thirty years, cell culture scientists and engineers have navigated the waters of process understanding in tackling protein product quality challenges. In many cases, the root cause of protein product attribute variation is related to cell culture process conditions.

The impacts of process conditions on protein attributes may be due to biosynthetic (biological origin) or degradative (chemical or physical origin) effects, and these effects may be driven by intracellular or extracellular factors. This presentation will review industrially relevant experiences, and discuss approaches toward better understanding and control of recombinant protein product characteristics through cell culture manipulations. Specific protein attribute variation addressed may include amino acid sequence, glycosylation (micro and macroheterogeneity), and charge variant distributions, and their potential for impact on various aspects of protein quality. The linkage between this process knowledge and todayâ€™s Quality by Design (QbD) approaches to process and product development will be discussed.

461824 Systematic Process Design Strategies for Efficient and Synergistic Integration of Solar Thermal Hydrogen, Electricity and Fresh Water Production Processes

Abstract:

There is an ever-increasing demand for food, energy and water (FEW) due to population growth and change in consumption habits. For the past century, energy, chemicals, fertilizers and even fresh water via desalination or water treatment have been supplied by using fossil fuel resources. However, with decaying fossil fuel reserves and increasing greenhouse gas (GHG) emissions, the grand challenge is to sustainably meet FEW needs without compromising the current level of civilization. Development and implementation of renewable energy conversion processes by systems approach are essential to ensure a smooth transition to a sustainable economy. Among these alternatives, solar energy is particularly prominent because of its tremendous potential [1]. High conversion efficiency and integrated energy storage systems are indispensable to overcome intermittent and dilute nature of harnessing solar energy [2-3]. Recently proposed hydricity concept, synergistic solar thermal hydrogen and electricity coproduction, presents a potential solution for continuous and efficient power supply and also an exciting opportunity to envision and create a sustainable economy to meet all the human needsâ€"namely, food, chemicals, transportation, heating, and electricity [1].

Here, we introduce an integrated process design strategy based on the hydricity concept for the coproduction of fresh water, hydrogen and electricity from concentrated solar thermal energy. Process simulations for the proposed integration are performed in an integrated Matlab and Aspen Plus platform [4]. The operating conditions of various units and topological structure of the integrated process are determined via sensitivity analysis and optimization in Matlab. Detailed process simulations of thermal desalination processes are validated by comparison with performance data from the current operating desalination plants. The resultant desalination processes are further integrated with

solar thermal electricity and hydrogen processes to coproduce electricity, hydrogen and fresh water. Standalone and integrated solar conversion processes are evaluated based on the process energy efficiency and *exergy efficiency*that refers to the fraction of incident solar exergy that is directly recovered as the net exergy output, which is defined as the sum of electricity and the hydrogen exergy output. We have identified exergetically inefficient steps and implemented new process designs to improve the thermal efficiency of standalone solar desalination processes.

The integration of solar thermal power generation, solar thermal hydrogen production and thermal desalination techniques reduces exergy losses associated with each one of the standalone processes and thus provide a synergistic strategy. The implementation of the proposed process synthesis strategy results in efficient solar thermal process designs, which coproduce i) baseload power, ii) fresh water, and iii) hydrogen. Depending on the targeted end-use, hydrogen can be used as an energy career, transportation fuel, feedstock for the production of chemicals, fuels or fertilizers for food production. Coproduced products can supply all FEW needs unveiling a sustainability roadmap.

[1] Gençer E, Mallapragada DS, Marechal F, Tawarmalani M, Agrawal R. Round-the-clock power supply and a sustainable economy via synergistic integration of solar thermal power and hydrogen processes, Proceedings of the National Academy of Sciences (PNAS), 112(52), 15821-15826, 2015.

[2] Gençer E, Al-musleh E, Mallapragada D, Agrawal R. Uninterrupted Renewable Power through Chemical Storage Cycles. Current Opinion in Chemical Engineering, 5, 29-36, 2014.

[3] Gençer E, Agrawal R. A commentary on the US policies for efficient large scale renewable energy storage systems: Focus on carbon storage cycles, Energy Policy, 88, 477-484, 2016.

[4] Gençer E, Tawarmalani M, Agrawal R, Integrated solar thermal hydrogen and power coproduction process for continuous power supply and production of chemicals, Computer Aided Chemical Engineering, 37, 2291-2296, 2015.

472761 Ideas As a Process Intensification Tool with Application to Natural Gas Reforming Based Hydrogen Production

Abstract:

In this work, process intensification is identified as a process synthesis activity aiming at significant improvements over traditional process designs. The infinite-dimensional state-space (IDEAS) conceptual framework is put forward as a systematic tool for process intensification, which can identify fundamental limitations to the attainable performance by networks of technologies under consideration. The employed approach allows for a broad search of the state-space for process intensification opportunities, even by low dimensional IDEAS linear programming approximations that always retain design feasibility. The proposed process intensification methodology is applied to a case study of natural gas reforming based hydrogen production. The obtained results identify intensified process designs featuring hot/cold utility costs that in certain cases can be lower by more than an order of magnitude to those of traditional designs.

464734 Optimal Design of Integrated Upgrading Plant and Utility System for the Oil Sands Industry

Abstract:

Profitability in the oil sands industry historically has been quite unstable and it is significantly affected by the fluctuating crude oil price. Specially, the margin between bitumen production costs and the synthetic crude oil market price is insignificant these days. However, despite the unstable market and low profitability, oil sands production is predicted to be tripled by the year 2020, when taking into account existing projects, projects under development, and announced investments. Hence, it is imperative to reduce the operating costs for both existing and developing upgrading plants to allow for a more sustainable industry with a larger profit margin.

Limited number of works studied the optimization of oil sands upgrading and utility system. Ordorica-Garcia et al. [1] provided a model for the estimation of the energy requirements associated with producing crude oil products from oil sands according to the current commercially used production schemes. In their later work, Ordorica-Garcia et al. [2] presented a model for optimizing energy production for oil sands operations to reach the minimum of the total annual cost of energy consumptions in the oil sands industry. The model determined the optimal combinations of power and hydrogen plants to meet the given energy demands. Betancourt-Torcat et al. [3] incorporated the two previous studies into a more complex model simultaneously searching for the optimal set of oil production schemes and the corresponding energy infrastructures that meet the total production demands. The proposed superstructure involves multiple extraction and upgrading technologies and it is useful for production costs estimation, evaluating future production schemes and energy demand scenarios, and identifying the key parameters that affect the oil sands industry. In a more recent work [4], an energy optimization model for the oil sands upgrading operations was presented to reach the minimum downstream operating costs at the upgrading part in the oil sands industry, and the proposed superstructure of this work only focused on upgrading plants.

In this work, optimal design of integrated upgrading plants and utility systems in the oil sands industry is investigated. For the upgrading plant part, several state-of-the-art applicable upgrading schemes are considered, which include atmospheric and vacuum distillation, LC-fining, delayed coking and hydrotreatment. For the utility system part, a detailed superstructure model is proposed in this work. Compared to previous works [1-4] where the utility systems were just taken account as energy producers and very rough approximations were used for their operational and economic features, this work considers detailed features including process capacity, heating rate, capital cost, and operation and maintenance cost. A more complex polygeneration superstructure is developed in this study for the utility systems which can produce power, hydrogen and steam of varying quality. It includes gasifier (to produce syngas from coke or residue withdrawing from the upgrading plants), syngas cleaning, air separation unit, gas turbine, heat recovery steam generator, gas-fired boiler, and steam turbine. Results of various case studies demonstrate that the proposed work can efficiently address the optimal design of the integrated upgrading plant and utility system, which can lead to significant cost reduction of the oil sands industry operations.

References:

[1] G. Ordorica-Garcia, E. Croiset, P. Douglas, A. Elkamel, and M. Gupta, "Modeling the Energy Demands and Greenhouse Gas Emissions of the Canadian Oil Sands Industry," *Energy & Fuels,* vol. 21, pp. 2098-2111, 2007.

[2] G. Ordorica-Garcia, A. Elkamel, P. L. Douglas, E. Croiset, and M. Gupta, "Energy optimization model with CO2-emission constraints for the Canadian oil sands industry," *Energy and Fuels,* vol. 22, pp. 2660-2670, 2008.

[3] A. Betancourt-Torcat, G. Gutierrez, A. Elkamel, and L. Ricardez-Sandoval, "Integrated Energy Optimization Model for Oil Sands Operations," *Industrial & Engineering Chemistry Research,* vol. 50, pp. 12641-12663, 2011.

[4] J. Charry-Sanchez, A. Betancourt-Torcat, and L. Ricardez-Sandoval, "An optimization energy model for the upgrading processes of Canadian unconventional oil," *Energy,* vol. 68, pp. 629-643, 2014.

452326 Design and Operation of a 10 MWe Supercritical CO_2 Recompression Brayton Power Cycle

Abstract:

Supercritical carbon dioxide (sCO_2) Brayton power cycles are gaining increasing attention as an attractive alternative to conventional Brayton and Rankine cycles using steam. Brayton cycles with sCO_2 as the working fluid have the potential for more compact design, higher thermal efficiencies, and lower cost of electricity. In view of these promising advantages, the U.S. Department of Energy under its Supercritical Transformational Electric Power Program is in the process of designing, with the intent to build, a nominal 10 MWe (net) sCO_2 Brayton cycle test facility to accelerate scale-up, development, and commercial deployment. In this work, steady-state and dynamic process models of a 10 MWe indirect recuperated sCO_2 recompression Brayton cycle have been developed to analyze cycle design and operation for use with power generation applications. The process models are capable of predicting the nominal design point, off-design, and part-load performance.

In this presentation, the 10 MWe sCO_2 recompression Brayton process model, assumptions, and design parameters will be described and the key results for the nominal steady-state operating point will be presented. The additional specifications and methods applied to provide increased model fidelity for generating the rigorous pressure-driven dynamic model will also be discussed, including the use of performance curves for the turbomachinery and flow-dependent pressure drop correlations, volume specifications, and heat capacity calculations for the heat exchangers. The dynamic model predicts time-dependent profiles of key outputs such as power generated and thermal efficiency as a function of compressor inlet temperature and pressure, turbine inlet temperature, bypass recompression fraction, and other key variables for the purpose of designing effective control strategies.

From the dynamic perspective, this presentation will highlight the transient responses of the sCO_2 cycle to reductions in heat input, representing typical off-design and part-load scenarios arising from integration with a load-following energy plant. Based on the open-loop cycle results without controllers, several operational strategies will be analyzed for maximizing cycle efficiency by maintaining turbine inlet temperature during heat input turndown subject to process constraints.

464793 Integrated Thermochemical Process for Optimal Co-Production of Liquid Fuels and Chemicals

Abstract:

Although the current crude oil price is quite low, it is still necessary to develop reliable biorenewable conversion technologies from a long-term perspective due to diminishing petroleum reserves and increasing market requirements. As a very versatile energy resource, biomass can not only be converted to liquid transportation fuels for vehicular use but can also be transformed to aromatics, olefins, lubricants, industrial chemicals, and many other products currently derived from petroleum or natural gas. Indeed, co-producing these value-added chemicals in an integrated biorefinery can incentivize the de-risking of the â€œfront-endâ€ processes that are necessary for liquid fuels production.

An integrated thermochemical conversion based biorefinery plant for coproducing premium quality liquid transportation fuels, olefins, and aromatics is systematically designed and synthesized in this presentation. The central elements contain gasification, fast pyrolysis, hydrocarbon upgrading, methanol synthesis, methanol-to-olefins, methanol-to-propylene, and methanol-to-aromatics. In general, the pretreated biomass is directed to the gasifier for syngas production and to the fast pyrolyzer for crude bio-oil production, respectively. The major portion of the hot syngas from the gasifier is used to provide heat for the fast pyrolyzer and to promote the fluidization. The crude bio-oil is converted to liquid fuels through sequential hydrotreating and downstream separation. The fluidizing gas along with the syngas from the fast pyrolyzer will finally serve as the feedstocks for methanol production. The produced methanol will be converted to olefins and aromatics through alternative routes. In the developed process superstructure, which considers multiple catalysts and reactor types for certain processing units, is formulated as a generalized disjunctive programming model. To reduce the computational complexity of optimization model, a symbolic regression based approach is proposed for generating surrogate models to describe those complex processing units. The bilinear, trilinear, quadrilinear, and concave terms are reformulated for enhancing the optimization model to be solved to global optimality. Water, power and heat integration studies are simultaneously carried out to reduce the utility cost. Trade-offs between process economics and hydrogen source selection are made. The top adjustable factors on overall economic performance are identified through ranking relevant candidates. Through this analysis, the designed integrated process exhibits high flexibility to cope with the uncertainties on market demands and feedstock sources.

465824 Design and Optimization of Integrated Carbon Capture and Conversion with Natural Gas to Produce Syngas

Abstract:

Direct chemical recycling of CO_2 to fuels and chemicals can significantly reduce fossil fuels-related greenhouse gas emission, and create great economic opportunities for energy and chemical sectors worldwide. As much as 2 Gt/yr of fuels and 200 Mt/yr of chemicals can be produced from CO_2 utilization [1]. In this work, we develop and optimize a novel process which takes power plant flue gas and cheap methane sources (e.g., renewable biogas, landfill gas, coalbed methane, CO_2-contaminated natural gas, shale gas) as raw materials and produces syngas (CO and H_2) - the universal precursor for hydrocarbon-based fuels and chemicals. The process is integrated such that it can simultaneously separate CO_2 from power plant flue gas as a source of carbon, and convert CO_2 with methane as a source of hydrogen to produce syngas. The process also utilizes renewable energy such as concentrated solar energy to partially supply the process energy demand.

Specifically, we combine a CO_2-selective zeolite [2] and a dry-reforming catalyst in a sequential configuration to capture and store CO_2 from dilute flue gas, and convert it to syngas using renewables. A special instance of this is when both the zeolite and catalyst are packed inside a single column in a layered configuration. The process is dynamic and operates in a cyclic manner, where each cycle consists of at least two steps. In the first step, the zeolite section separates and intermittently stores CO_2 from flue gas at ambient temperature. In the second step, the zeolite section automatically supplies CO_2 and methane toward the catalyst section for conversion, due to desorption triggered by solar heating and/or the introduction of methane-rich feed. The high temperature for conversion is partially achieved by using concentrated solar energy. The simultaneous desorption and conversion of CO_2 completely eliminates a

separate regeneration step, which is typically the most energy-intensive operation in current capture technologies.

The integration of CO_2 adsorption, intermittent storage, desorption using reactor feed, and conversion with methane is largely controlled by the trade-offs between these competing phenomena. Hence, a predictive model of the overall process has been difficult to create – largely because of the time-dependent, distributed and cyclic nature of the process and the complex interaction between the adsorption and conversion sections. In this work, we use a nonlinear, algebraic and partial differential equation (NAPDE)-based model that captures the equilibrium, transport and kinetic behavior of the gases for the dynamic adsorption, desorption and conversion operations. The optimization of the NAPDE model is also challenging. To this end, we have recently developed and applied a data-driven platform for the simulation-based optimization of complex process models, which treats the NAPDE-based model as a black-box system and use it to generate the required data for optimization. We optimize the process for different feed conditions and syngas specifications (H_2:CO ratio). Once optimized, the integrated technology shows significant promise for reducing the cost of CO_2 capture by eliminating the need for a material regeneration step, and reducing the cost of CO_2 conversion by storing solar energy as chemical energy while mitigating GHG emission.

References:

[1] Quadrelli, E. A.; Centi, G.; Duplan, J.-L.; Perathoner, S. Carbon Dioxide Recycling: Emerging Large-Scale Technologies with Industrial Potential. *ChemSusChem.*2011, 4:1194–1215.

[2] Hasan, M. M. F.; First, E. L.; Floudas, C. A. Cost-effective CO_2 capture based on *in silico* screening of zeolites and process optimization. *Physical Chemistry Chemical Physics* 2013, 15 (40):17601–17618.

464736 Optimal Design and Operation of a Semi-Closed Oxy-Combustion Combined Cycle Power Plant

Abstract:

Optimal design and operation of a semi-closed oxy-combustion combined cycle power plant

Holger Teichgraeber[1] Adam Brandt[1]

[1]Department of Energy Resources Engineering, Stanford University, Stanford, California 94305-2220, USA

As gas turbines provide flexibility to energy systems today, in a low carbon future energy system, natural-gas-based power systems with CO_2 capture may take over that role. Hence, we analyse the design and operations of an oxyfuel-based CCS system that is designed to operate flexibly on timescales of order one hour. Such a system can provide an intermediate level of supply flexibility beyond that provided by short response batteries. Such a system could also fill gaps in seasons of low renewable energy availability. (e.g., winter).

Specifically, we present optimal design and hour-to-hour operation of a flexible semi-closed oxy-combustion combined cycle power system. Our method includes co-optimization of the required air separation unit used to produce O_2. Computational optimization is used to maximize net present value, including both operational and design optimization. Operating profit is increased through running the air separation unit at full capacity at times of low electricity prices, storing oxygen and running the oxyfuel-combustion combined cycle at full load in times of high electricity prices. This avoids energy-intensive air separation when the economic penalty is the greatest.

The system consists of a cryogenic air separation unit, oxygen storage facilities, a natural gas combustion turbine, a heat-recovery steam generator, a steam turbine and a CO_2 compression unit. In this integrated system, all components are represented in a modular fashion using energy and mass balances. The air separation unit is modelled for operation at full capacity and partial load based on industrial data. The gas turbine's, steam turbine's and CO_2 compression unit's partial load operation are modelled based on existing literature. In order to model the oxyfuel-combustion gas turbine, the partial load gas turbine model is expanded with pure oxygen combustion and a recirculation fluid cycle. The heat recovery steam generator model is based on previous work [1]. System configuration and the sizing of various components have a substantial impact on operations and system economics.

The system in this work is able to operate flexibly and its overall operation including time-varying electricity prices is optimized. It also includes the air separation unit and an oxygen storage facility within the system boundary. Amongst all system components, the air separation unit is the least flexible and gaseous or liquid oxygen storage would be beneficial in order to provide fast ramping in case of changing electricity prices.

We maximize net present value (NPV) which is a function of design and operational decision variables. A similar optimization approach has been taken and described in previous work [2]. The design optimization is formulated as a mixed integer nonlinear programming problem. It uses the operational optimization as an inner loop in order to find the maximum NPV. This problem is solved using an algorithm that consists of Particle Swarm Optimization (PSO) and Mesh Adaptive Direct Search (MADS). The optimal operation maximizes annual profit using a Sparse Nonlinear OPTimizer (SNOPT) algorithm with multiple starting points due to the nonconvexity of the problem.

[1] Kang, Charles A., Adam R. Brandt, and Louis J. Durlofsky. "Optimizing heat integration in a flexible coal–natural gas power station with CO 2 capture." *International Journal of Greenhouse Gas Control* 31 (2014): 138-152.

[2] Brodrick, Philip G., Charles A. Kang, Adam R. Brandt, and Louis J. Durlofsky. "Optimization of carbon-capture-enabled coal-gas-solar power generation." *Energy* 79 (2015): 149-162.

462461 Natural Gas to Liquids, Olefins, and Aromatics Under Uncertainty in Feedstock and Product Prices

Abstract:

Although natural gas to liquid transportation fuels processes have been investigated for several decades, they were considered economically unfavorable due to high feedstock costs. [1,2] Recent studies have demonstrated the profitability of such refineries can be significantly increased by coproducing valuable chemicals, such as olefins and aromatics, in tandem with liquid fuels. [3,4] Even more recently, new studies have investigated the potential of stand-alone natural gas based chemical refineries, where the main products output are mainly olefins and aromatics, and assessed the technoeconomic and environmental impacts of such processes. [5,6]

The economics of such natural gas facilities is intricately dependent on the complexity of the refinery configuration and the type, quantity, and quality of products produced (i.e., the product slate). The large chemical space available for the manufacture of liquid fuels and chemicals has been expanded even further in recent years due to novel and emerging technologies that can produce liquid fuels, olefins, and aromatics. [3-6] The types of technologies to utilize and the types of products to produce is a tremendously difficult challenge to address, one that is further complicated with the uncertainty surrounding the costs and prices of feedstocks and products, respectively. In this talk, we address these challenges through a robust optimization framework that is incorporated in a large-scale process superstructure that contains several alternatives for the production of fuels, olefins, and aromatics from natural gas. [6-10] We introduce an important production parameter that allows us to investigate the economic effects of natural gas to liquids, olefins, and aromatics refineries. The resulting large-scale, nonconvex, mixed-integer, nonlinear optimization (MINLP) model is solved using our novel global optimization algorithm. [11]

Our comprehensive optimization-based process synthesis framework contains several direct and indirect natural gas conversion technologies and numerous hydrocarbon production alternatives, including Fischer-Tropsch refining and methanol synthesis; each of which is compared to determine the optimal processing pathway. [2-6] Multiple commercial and novel technologies for the production of the olefins and aromatics, such as methanol-to-olefins and toluene alkylation with methanol, are rigorously modeled. [5,6] The key production parameter is introduced as a function of the important products output from the refinery, namely, gasoline, diesel, kerosene, aromatics, and olefins. Robust counterparts are formulated in order to address the uncertainty associated with feedstock and product prices.

Several case studies are presented to investigate the effect of plant capacity and production ratios on the overall profit of the refinery and incorporate uncertainty in the optimal solutions obtained. A discussion on the robust refinery designs that incorporate the effect of uncertainty will follow with emphasis on the major topological decisions as a function of the production parameter. The economic and environmental tradeoffs of the refineries are presented.

[1] Floudas, C. A.; Niziolek, A. M.; Onel, O.; Matthews, L. R. Multi-Scale Systems Engineering for Energy and the Environment: Challenges and Opportunities. AIChE Journal 2016, 62, 602-623.

[2] Baliban, R. C.; Elia, J. A.; Floudas, C. A. Novel natural gas to liquids (GTL) technologies: Process synthesis and global optimization strategies. AIChE Journal 2013, 59, 505–531.

[3] Onel, O.; Niziolek, A. M.; Elia, J. A.; Baliban, R. C.; Floudas, C. A. Biomass and natural gas to liquid transportation fuels and olefins (BGTL+C2_C4): Process Synthesis and Global Optimization. Industrial & Engineering Chemistry Research 2014, 54, 359-385.

[4] Niziolek, A. M.; Onel, O.; Elia, J. A.; Baliban R. C.; Floudas, C. A. Coproduction of Liquid Transportation Fuels and C6_C8 Aromatics from Biomass and Natural Gas. AIChE Journal 2015, 61, 831-856.

[5] Onel, O.; Niziolek, A. M.; Floudas, C. A. Optimal Production of Light Olefins from Natural Gas via the Methanol Intermediate. Industrial & Engineering Chemistry Research 2016, 55, 3043-3063.

[6] Niziolek, A. M.; Onel, O.; Floudas, C. A. Production of Benzene, Toluene, and Xylenes from Natural Gas via Methanol: Process Synthesis and Global Optimization. AIChE Journal 2016, 62 (5), 1531 – 1556.

[7] Li, Z.; Ding, R.; Floudas, C. A. A comparative theoretical and computational study on robust counterpart optimization: I. Robust linear optimization and robust mixed integer linear optimization. Industrial & Engineering Chemistry Research 2011, 50, 10567–10603.

[8] Li, Z.; Tang, Q.; Floudas, C. A. A comparative theoretical and computational study on robust counterpart optimization: II. probabilistic guarantees on constraint satisfaction. Industrial & Engineering Chemistry Research 2012, 51, 6769–6788.

[9] Li, Z.; Floudas, C. A. A comparative theoretical and computational study on robust counterpart optimization: III. improving the quality of robust solutions. Industrial & Engineering Chemistry Research 2014, 33, 13112–13124.

[10] Guzman, Y. A; Matthews, L. R; Floudas, C. A New a priori and a posteriori probabilistic bounds for robust counterpart optimization: I. Unknown probability distributions. Computers & Chemical Engineering 2016, 84, 568-598.

[11] Baliban, R. C.; Elia, J. A.; Misener, R.; Floudas, C. A. Global Optimization of a MINLP Process Synthesis Model for Thermochemical Based Conversion of Hybrid Coal, Biomass, and Natural Gas to Liquid Fuels. Computers and Chemical Engineering 2012, 42, 64-86.

471927 Examining the Partitioning Coefficients of Butanol and Other Fermentation Metabolites Produced By *Clostridium Pasteurianum* into a Lipid Vesicle Solvent

Abstract:

Continuous extractive fermentation is critical to increasing biofuel production. Extracting the fermentation products reduces product inhibition, increases the metabolic activity and drives solvent production, and can increase solvent selectivity. Liquid-liquid extraction using organic solvents has shown to be effective for extractive fermentation, but the solvents employed are often toxic to the microbes and can limit productivity. Self-assemblies such as vesicles and micelles provide an alternative to organic solvents for the extractive fermentation of biofuels. We are currently developing a process where lipid vesicles are used as the solvent phase to recover butanol from fermentation broths and remove toxic impurities present in substrate feedstocks. This approach is based on the high thermodynamic partitioning coefficients observed for butanol from water to lipid membranes, which is one of the main causes of product inhibition as butanol partitions into and fluidizes microbial membranes. This presentation will report on the partitioning behavior of butanol, as well as other metabolites produced by *Clostridium* pasteurianum (acetate, ethanol, 1,3-propanediol), to lipid vesicles as a function of lipid composition, pH, and temperature. We have shown that butanol partitioning increases with the amount of unsaturated lipid present within the vesicles due to an increase in the size of the â€˜binding pocketsâ€™ for amphiphilic butanol to partition into at the vesicle/water interface.

458249 Intensified Liquid-Liquid Extraction with Chemical Reaction By Emulsification

Abstract:

Biobased effluents are commonly defined by undesirable characteristics, such as a low concentration of the target constituent in aqueous solution or separation barriers like azeotropes. The isolation of the constituents from such effluents challenges standard unit operations to their limits. To accomplish this task liquid-liquid extraction has been widely used, but for an increase of efficiency and operability further improvement is necessary. Improvement can be obtained through intensification of the extraction step by combination with chemical conversion of the target constituent. Thus, the separability of the constituent as well as economics can be significantly increased.

Effluents from biorefinery processes like pulping serve as representative separation task. Such effluents often contain a significant amount of carboxylic acids like acetic acid or formic acid. Through spiking an appropriate solvent with an aliphatic alcohol, the extraction rate of the carboxylic acids can be increased by chemical conversion via esterification which yields a favorable change of the physio-chemical properties of the constituents. Admixture of homogeneous catalysts to the solvent provides an additional benefit by increasing the reaction rate of the esterification step. *4-*dodecylbenzenesulfonic acid (4DBSA) is a suitable catalyst. This highly potent catalyst, which additionally acts as a strong anionic surfactant, offers access to two different process routes.
The first route focuses on the prevention of emulsification by adjusting the hydrophilic-lipophilic balance value (HLB value) through adding a second surfactant to the system. Admixture of nonionic surfactants (e.g. SPAN®85) significantly enhances primary phase separation.
The second process route makes use of emulsification in order to increase the interaction between reactants and catalyst, thus increasing the reaction rate although splitting of the emulsion must be applied prior to product isolation. This technology is a promising alternative to state of the art isolation of carboxylic acids from aqueous effluents by

liquid-liquid extraction.

447293 Hybrid Shortcut-Modell for Estimating Partition Coefficients in Aqueous Two-Phase Extraction of Therapeutic Proteins

Abstract:

The development of industrial scale bioprocesses has come into high scientific and industrial interest over the last decades. Especially in the red and white biotechnology, process costs are often dominated by 50 - 80 % by the downstream processing. First investigated in the 1896, aqueous two-phase systems (ATPS) were shown to have an enormous potential for the extraction of these biomolecules. In general ATPS show a high biocompatibility and are cheap in comparison to chromatographic methods. The most common ATPS consist of a hydrophilic polymer and a salt serving as phase-forming components. The distribution of the target protein within these systems can either be influenced by the choice of the phase forming components, and additionally by a choice of a displacement salt. The displacement salt influences both the solubility and the distribution of the protein in the two phases. Thus, knowledge of the influence of the different salts potentially useful to enhance the distribution coefficient is necessary.

One promising and fast measure accounting for complex interactions of biological macromolecules is the second osmotic virial coefficient (B_{22}) characterizing the interactions between the biological macromolecules in solution and the cross virial coefficient (B_{23}) characterizing interactions of the biological macromolecules with the salts / impurities. It combines the different influence factors such as pH, Temperature, solvent, salt and salt concentration within one measure.

B_{22} as well as B_{23} data was measured using composition gradient multi angle light scattering (GC-MALS) for different proteins and salts in aqueous solution. The results were then used to characterize the distribution behavior of the target protein in the ATPS dependent on the displacement salt. The results point to the fact, that based on B_{22} and B_{23} the suitability of a salt as displacer for proteins in aqueous two-phase systems can be investigated. One major advantage

using this approach is the fact, that the partition coefficient can be estimated using a hybrid ShortCut Modell based on a minimal set of experimental data. Furthermore the influence of the displacer (salt) on the solubility of the protein can be judged qualitatively serving as an ideal first measure in the development of aqueous two-phase extraction systems. The results will help to improve the development and optimization of ATPS used for efficient separation of processes.

461385 Novel Purification Process for Bio-Derived Citric Acid Using Solvent Extraction and CO_2 Antisolvent Purification

Abstract:

Citric acid (CA) is a chemical commodity directly used in food and confectionery, as well as in the pharmaceutical and chemical industry. CA is used as raw material for calcium supplements to prevent osteoporosis, for the manufacture of biobased plasticizers, and to produce citrate salts used as food preservatives, buffer and chelating agents. Generally, CA is produced from a fungal broth fermentation with *Aspergillus niger* using sucrose as the main substrate. As the microorganism is able to withstand acidic condition, CA is obtained in the acidic form at the end of the fermentation process. Traditional downstream processing involves removal of biomass and suspended solids by filtration, followed by CA precipitation with calcium hydroxide. Impure calcium citrate is acidified with sulfuric acid to release CA in aqueous solution, and precipitating gypsum.

In a typical purification process, CA solution is subjected to ion exchange to remove metal and remaining sucrose, adsorption with activated carbon, followed by evaporation and crystallization. Sucrose traces removal is of major concern as it heavily affects CA crystallization, and prevents from obtaining high CA purity (i. e. higher market value). This process is energy and mass intensive, and generates considerable amounts of liquid and solid waste.

Taking the aforementioned into account, this work describes a novel CA purification process using direct solvent extraction from the clarified fermentation broth and compressed gas (CO_2) as an antisolvent treatment to remove impurities. After the CA is extracted with a medium polarity solvent mixture, precipitation of impurities was carried out by compressing the liquid under CO2 pressure between 10 and 45 Bar, and different contact times (from 2 to 30 h). Under these conditions the impurities, mostly containing sucrose, precipitate out from the solution, allowing high purity CA recovery after solvent evaporation.

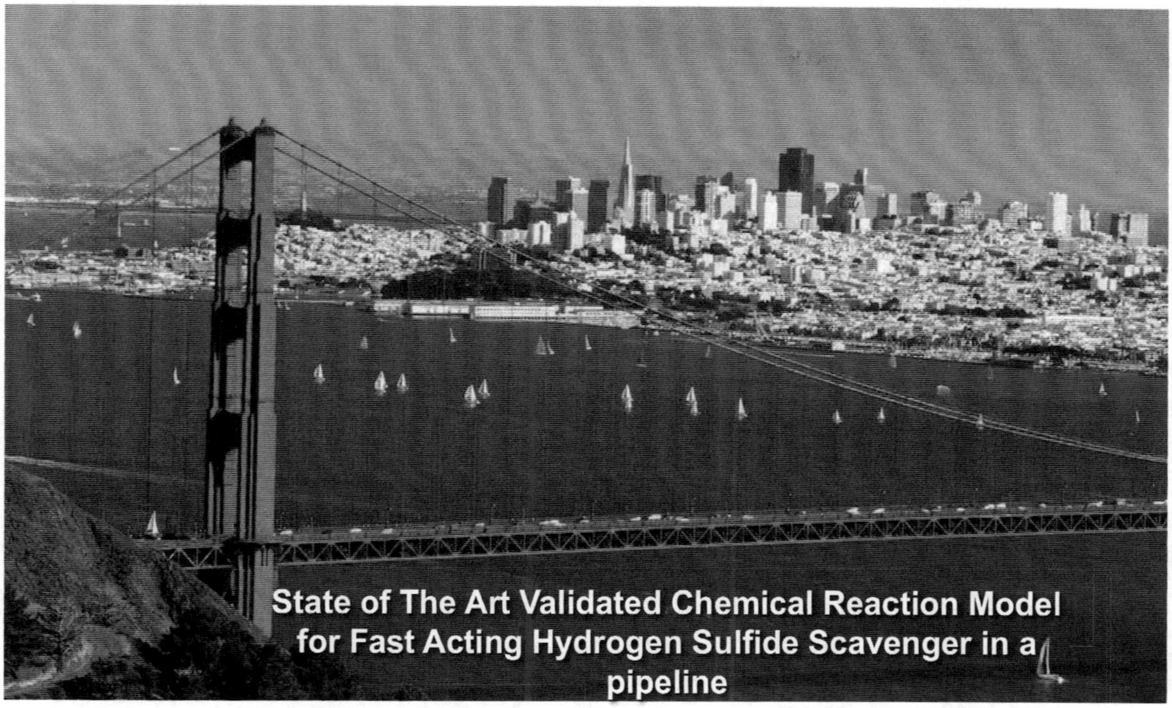

Emanuel Marsis, PhD.; Tudor Ionescu, PhD.; Paul Stead; Scott Lehrer; Sunder Ramachandran, PhD.

Outline

- Motivation
- Introduction
  - H_2S Scavenger
  - Computational Fluid Dynamics
- Test Results
- CFD Model
- Discussion & Next Steps
- Conclusion

Motivation

- Accurate prediction of H_2S concentration during the scavenging process is important to highlight any risk associated with high localized H_2S quantities
- Factors affecting the dynamics of H_2S scavenging:
  - Fluid phases
  - Fluid dynamics
  - Injection rate
  - Scavenger particle diameter
  - Reaction rates
  - Mass transfer

Introduction (H_2S Scavenger)

- Risk of having H_2S in a production system
- It can be present in either gas or liquid
- H_2S Scavengers are application sensitive
- An accurate model of H_2S scavengers is needed to minimize the risk on structure integrity and environmental impact due to H_2S production
- Triazine is a widely used H_2S scavenger, but it can be slow acting
- A fast acting H_2S scavenger was developed, tested and modeled

Introduction (Computational Fluid Dynamics)

- CFD Applications
  - Fluid Dynamics
  - Multiphase Flow
  - Chemical Reactions
  - Moving Geometry (impellers)
  - Thermodynamics
  - 2D & 3D
  - Steady State & Transient

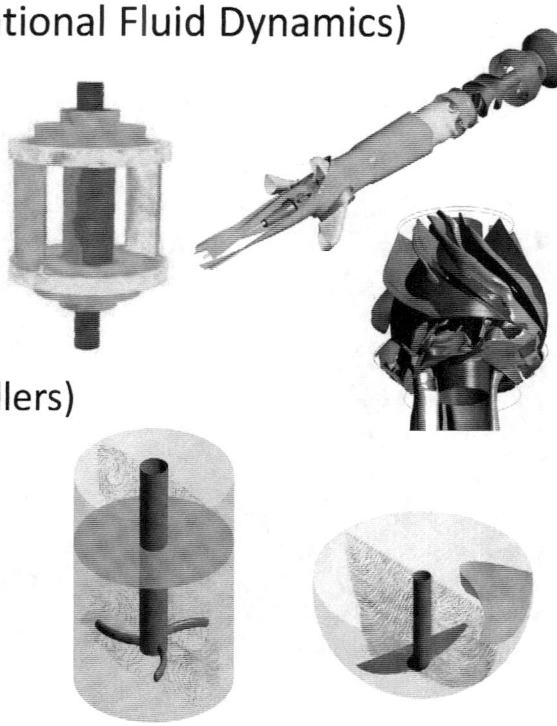

Test Setup

- **Features**
  - Oil / brine/ mixed phase
  - Continuous mixed gas flow
    - N_2 (dilution gas), CO_2, H_2S
  - Stirred
  - Continuous H_2S measurement
  - Heated (atmospheric pressure)
- **Benefits**
  - Relatively time efficient
  - Measures scavenger efficiency & kinetics

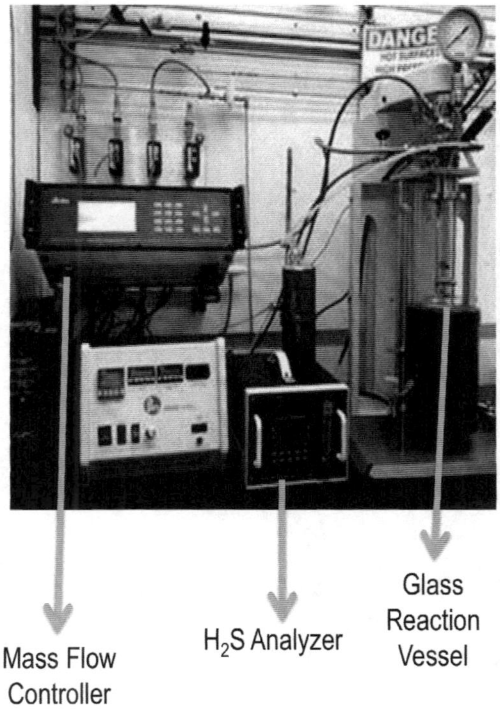

Test Setup

Test Setup

Reactor

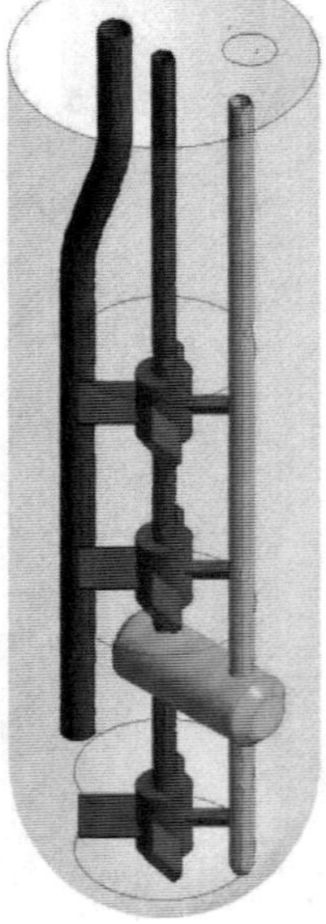

3D Computer Model

Test Results

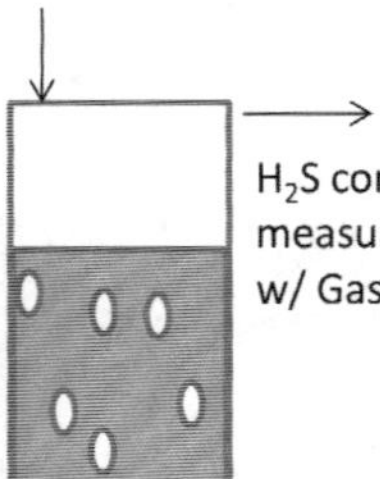

Gas Composition
(5% CO_2; 3200 ppm H_2S)

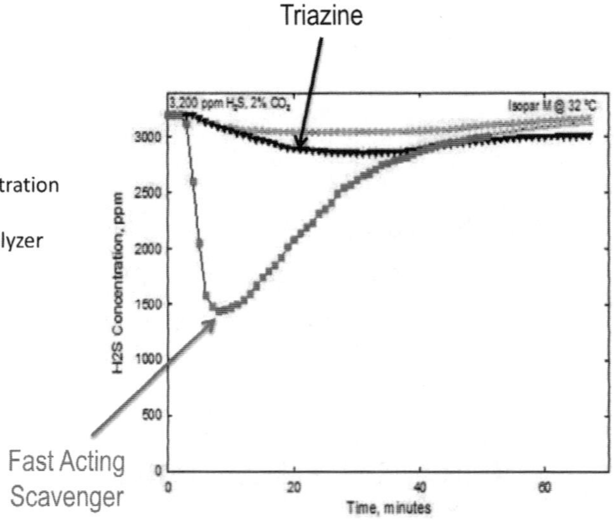

CFD Model

- Full 3D Transient
- Moving Mesh (Rotating Impellers)
- Eulerian Multiphase
- Species Model
- Mass transfer of H_2S from liquid to gas phase
- Chemical reaction takes place in the liquid phase only
- Polyhedral mesh

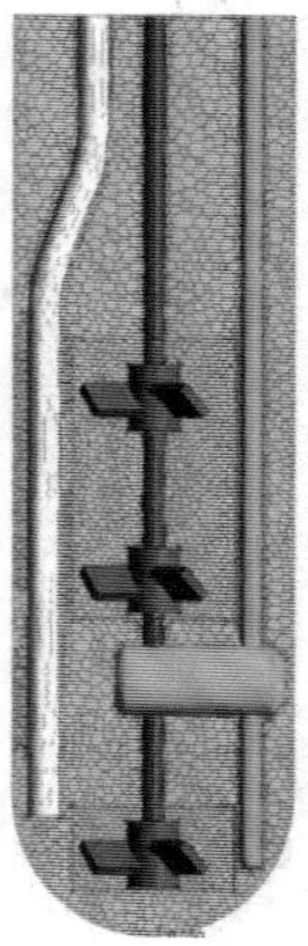

CFD Model

- All the species were defined separately for each phase
- Mass transfer rate of H_2S was obtained from the literature
- The slope of H_2S Concentration with time was obtained from the CFD model
- The reaction rate was determined in the CFD model by matching experimental data with CFD data

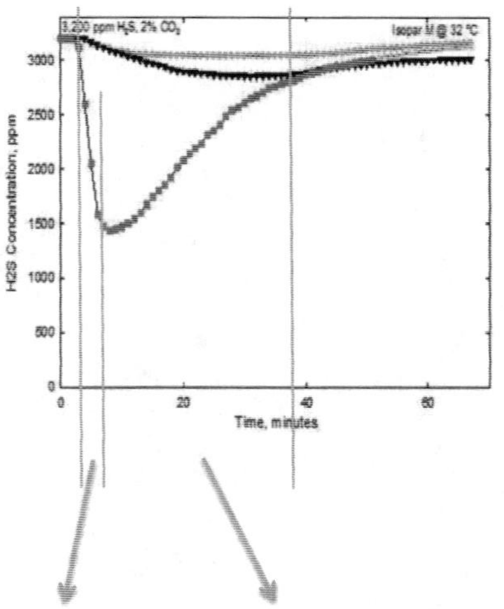

CFD Model

- Gas Volume Fraction

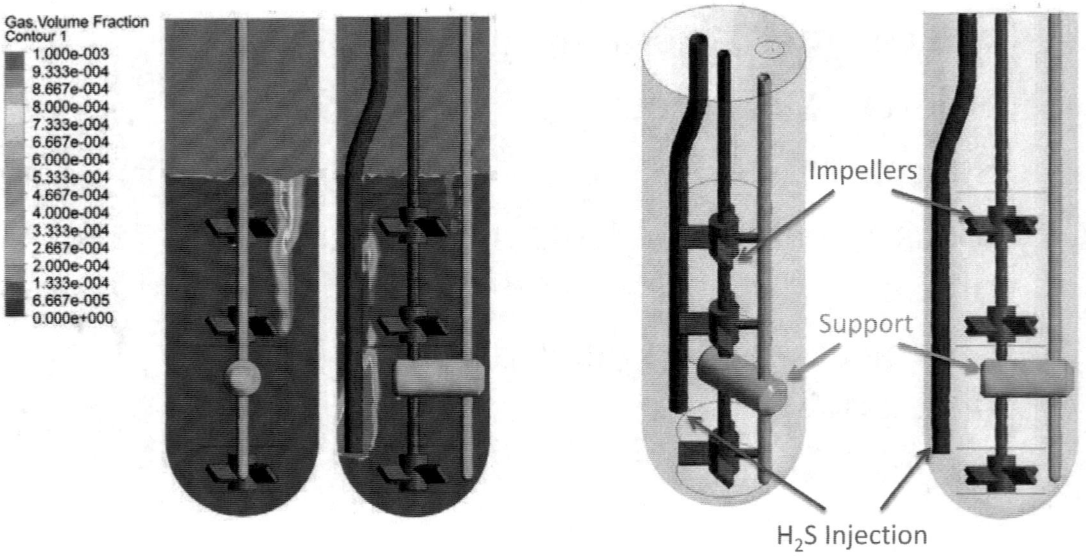

CFD Model

- H_2S Injection

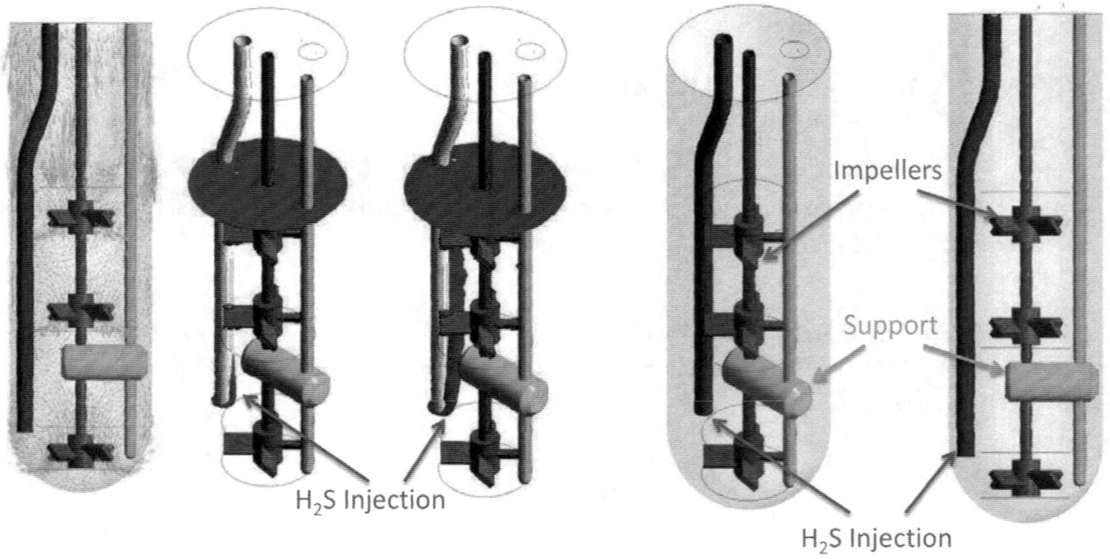

CFD Model

- Injecting the Fast Acting H_2S Scavenger

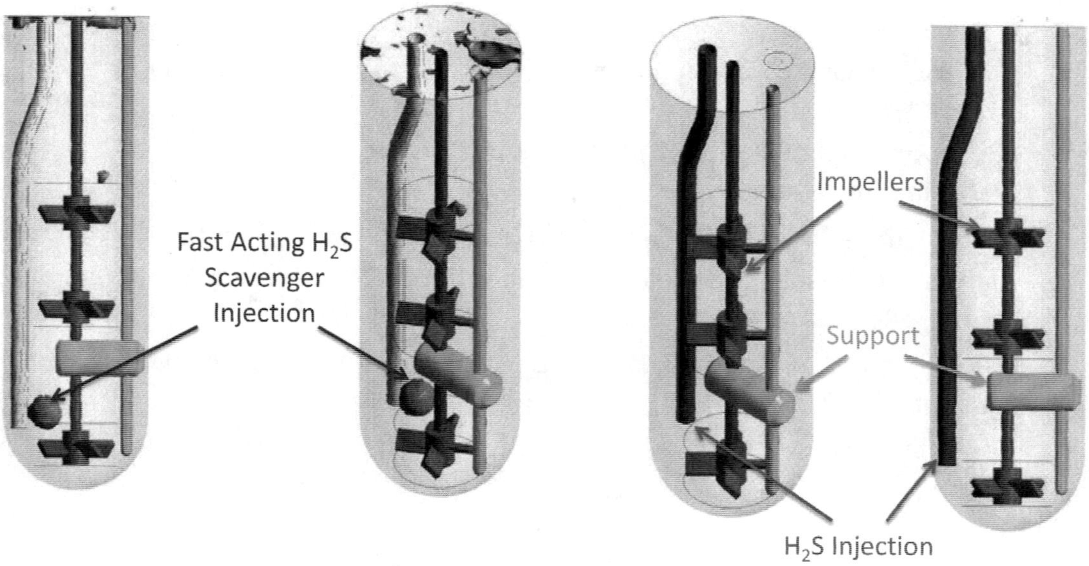

CFD Model

- Monitoring the H_2S Concentration vs Time
- Different CFD models represent different kinetic rates
- Kinetic rate in CFD Model #3, shown in the graph, matches the experimental results
- Limitations on the model
  - Time step size

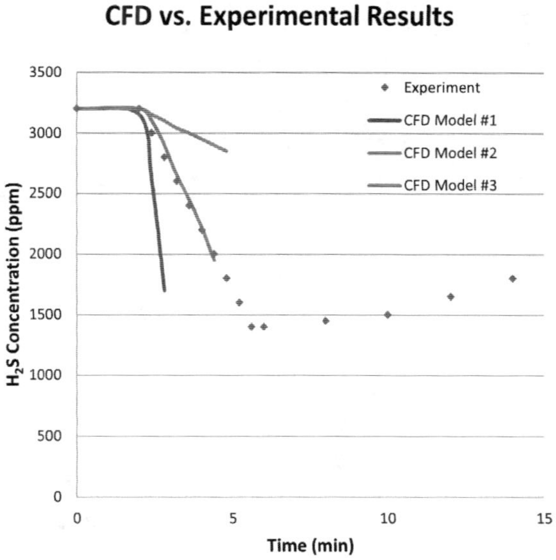

Discussion & Next Steps

- H_2S Concentration is dependent on:
  - Reaction rates
  - Fluid dynamics (rpm of the impeller, gas injection rate, mass transfer between the phases)
- By modeling the whole setup, an accurate prediction of the kinetic rate was carried out
- The validated model can be used in complicated field applications; like valves, wellheads, flowlines, or pipelines to predict the concentrations of H_2S at different locations
- Next Step is to use the validated model to predict H_2S Scavenging in a gas line

Discussion & Next Steps

- A production tubing was modeled using a species model
- The reaction rates computed earlier using the lab test were used to show the localized concentration of hydrogen sulfide in the production tubing

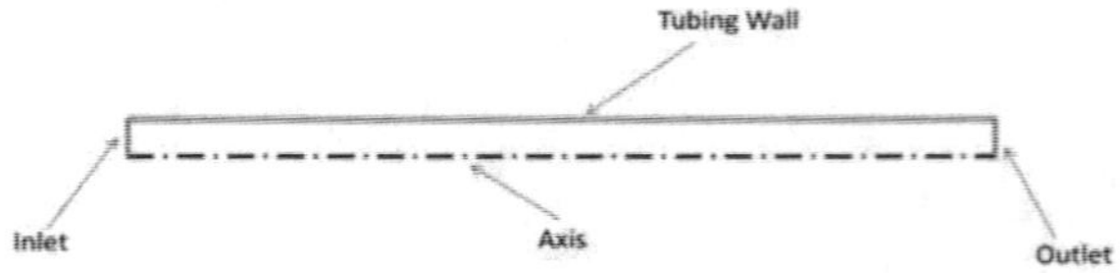

Discussion & Next Steps

- CFD allows the use of Discrete Phase Models (DPM), which is a Lagrangian particle tracking model to simulate H_2S Scavenger particles flowing in a gas line
- The mass transfer from the gas to the liquid particle was identified
- The reaction is taking place only in the liquid phase

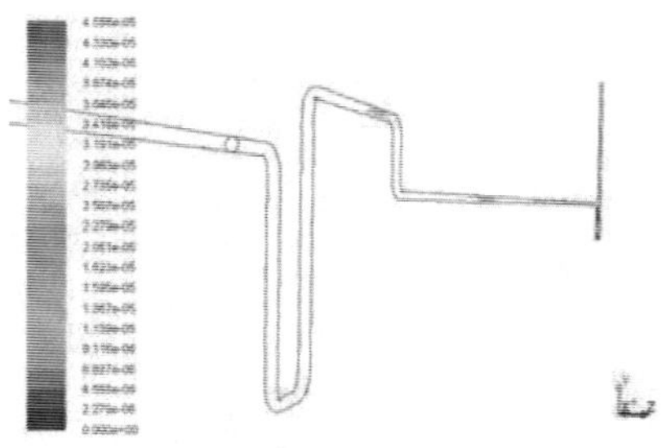

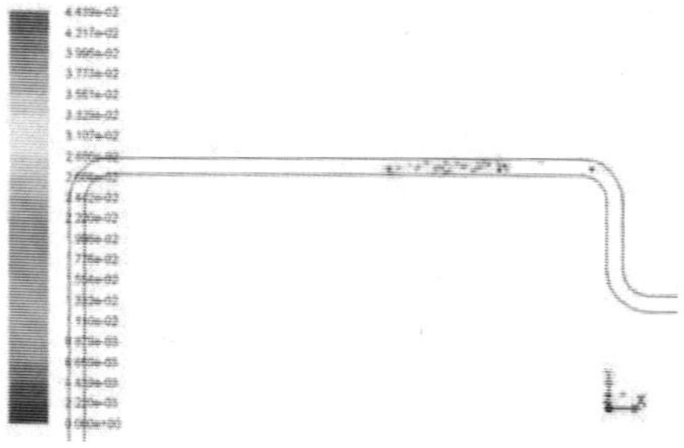

Conclusion

- Fast Acting H_2S Scavenger was tested in the lab and compared to Triazine Scavengers at the same concentrations
- The Fast Acting H_2S Scavenger has much higher kinetic rate compared to Triazine
- CFD was used to build a 3D model of the test setup and predict the reaction rates including the effect of fluid dynamics
- CFD is capable of modeling multiphase flow with chemical reactions and multiple species including mass transfer between the phases
- CFD modeling can be used to simulate complicated production problems and predict H_2S concentration in a given geometry using the fast acting H_2S Scavenger

QUESTIONS

Computational study of the bubbling-to-slugging transition in a laboratory-scale fluidized bed

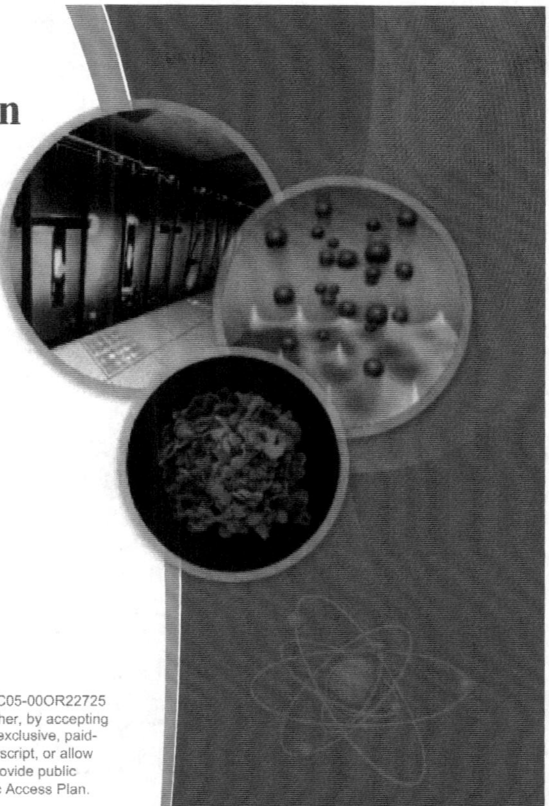

Emilio Ramirez[1,2], C. Stuart Daw[1], Charles E. A. Finney[1], Sreekanth Pannala[3], Jack Halow[4], Qingang Xiong[5]

[1] Oak Ridge National Laboratory, Oak Ridge TN 37831 USA
[2] University of Tennessee, Knoxville TN 37996 USA
[3] SABIC Americas, Sugar Land TX 77478 USA
[4] Separation Design, Garards Fort, PA 15334 USA
[5] FCA US LLC, Auburn Hills, MI 48326 USA

Presented at the 2016—AIChE Meeting
Group/Topical: Particle Technology Forum

San Francisco, California
November 16, 2016

OAK RIDGE National Laboratory | NATIONAL TRANSPORTATION RESEARCH CENTER

Acknowledgements

This research was supported by the U.S. Department of Energy Bioenergy Technologies Office (BETO). The authors would like to thank BETO sponsors Jeremy Leong, Cynthia Tyler, and Kevin Craig for their guidance and support.

NETL – Tingwen Li, Rupen Panday, William Rogers

ORNL - Gavin Wiggins, Maggie Connatser, Mike Kass, Jae Soon Choi, John Turner

MIT - Akhilesh Bakshi, Christos Atlantzis

NREL reactor experimentalists
Rick French, Danny Carpenter, Kristiina Ilsa, Mark Jarvis, Esther Wilcox, Kristin Smith, Katherine Gaston

University of Tennessee - Bredesen Center for Interdisciplinary Research and Graduate Education

Why are we interested in slugging?

"The arrival time of a space probe traveling to Saturn can be predicted more accurately than the behavior of a fluidized bed reactor!"
- Derek Geldart

"The **art** of engineering science is the art of seeing which steps should be taken to erase from real life (hopefully) irrelevant details which make it so complex, to look at the essential aspects of it, or, in other words, to decide what is the appropriate meta-problem to look at: to conceive the appropriate chain of idealization."
- Gianni Astarita and Julio Ottino

Are there challenges with slugging correlations?

Literature slugging correlations vary between studies
Possibly due to:
- qualitative methods used to identify slugging

Quantitative methods can provide a robust method to determine:
- onset of bubbling-to-slugging transition
- fully developed slugging

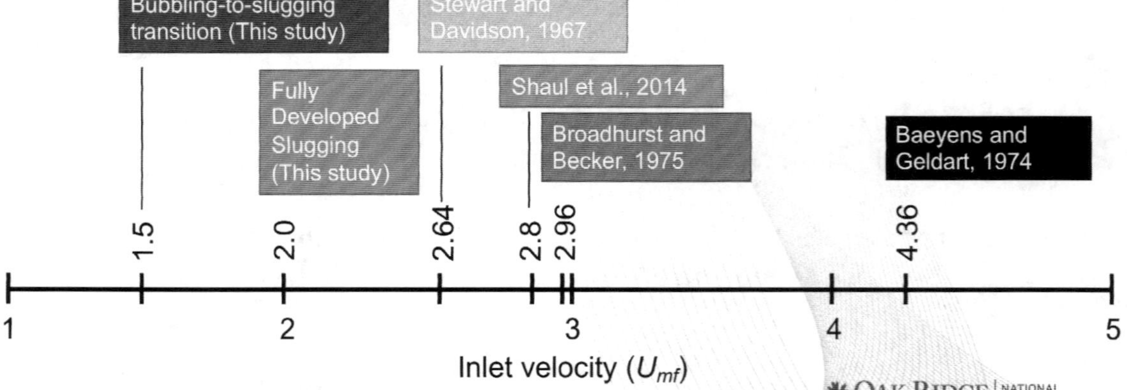

How does slugging affect biomass pyrolysis oil yield & composition?

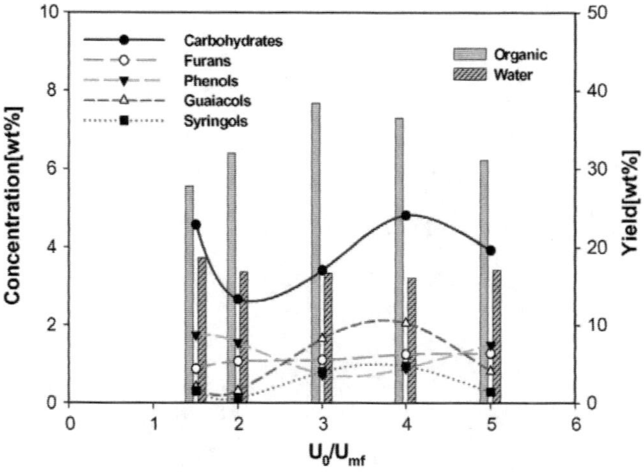

Figure: S.-H. Lee, M.-S. Eom, K.-S. Yoo, N.-C. Kim, J.-K. Jeon, Y.-K. Park, B.-H. Song, S.-H. Lee, The yields and composition of bio-oil produced from *quercus acutissima* in a bubbling fluidized bed pyrolyzer, J. Anal. Appl. Pyrolysis 83 (2008) 110-114. http://dx.doi.org/10.1016/j.jaap.2008.06.006

Hydrodynamics directly impact:

1. Particle residence time
2. Gas residence time
3. Particle heating rate
4. Particle attrition/fragmentation
5. Particle and ash elutriation
6. Particle segregation

All the above significantly impact raw oil yield and composition.

Particle heating rate

Affects rate of chemical kinetic yield and composition

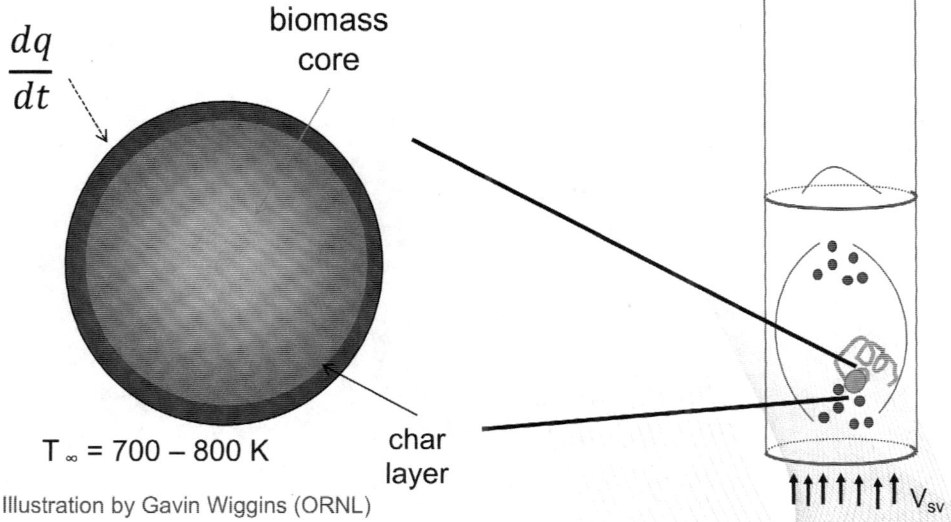

Illustration by Gavin Wiggins (ORNL)

Residence time, elutriation, and segregation

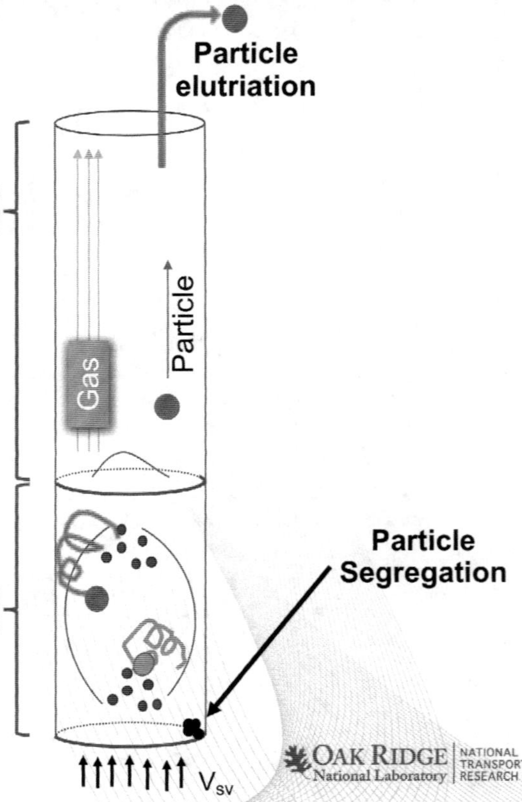

Particle and gas residence time, $\boldsymbol{\tau}$, in the bed and freeboard

$$\tau_{freeboard} = \frac{H_{reactor} - H_{exp\,bed}}{V}$$

$$\tau_{\exp bed} = \frac{H_{\exp bed}}{V / \varepsilon_{\exp bed}}$$

Particle attrition/fragmentation

Caused though mechanical impact, agitation, shearing, and internal pore structure

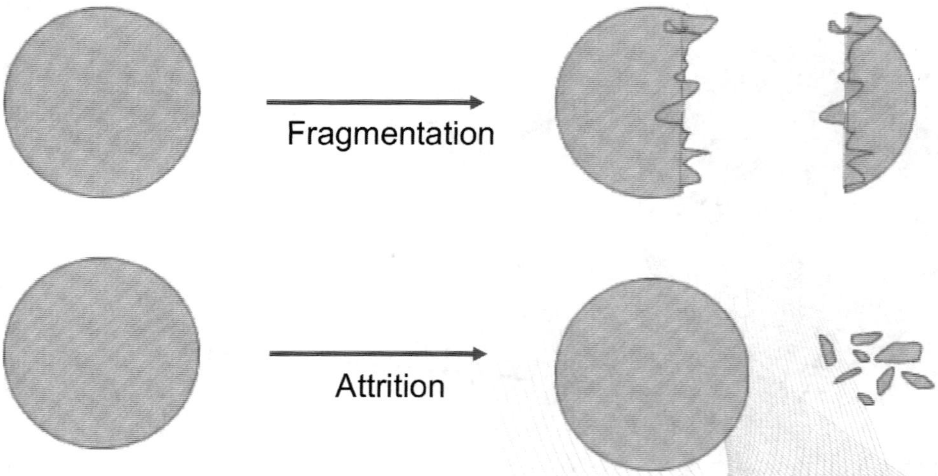

Objectives

Develop clearer understanding of slugging physics and develop possible on-line diagnostic
- What measurement can be used?
- Can we be quantitative?

Can a numerical model be used as a tool to understand the physics involved in slugging beds?
- How can an MFiX simulation be used to validate slugging studies?
- What quantitative numerical tools can we use?

- Scope of this study:
  - Bubbling/slugging beds
  - Bench scale reactor ~2 inch
  - Geldart type B particles
  - Hydrodynamics
  - Bubble analysis
  - Pressure analysis

Motivation: Slugging transition appears to be associated with a critical performance limit

Heat up time of biomass particles depends on particle size, shape, thermal properties, and reactor conditions.

Low tar yield and incomplete particle conversion if $\tau_{particle} < \tau_{tar}$

Tar yield decreases due to secondary reactions if $\tau_{particle} > \tau_{tar}$

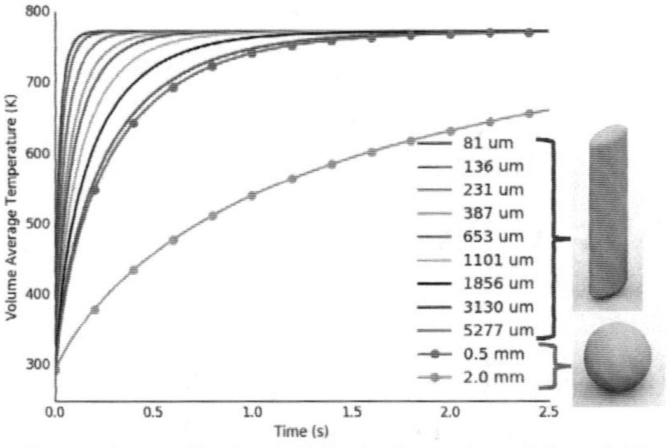

Temperature profiles from loblolly pine low-order particle model for D_F = 81 – 5277 um and single sphere with D = 0.5 and 2 mm.

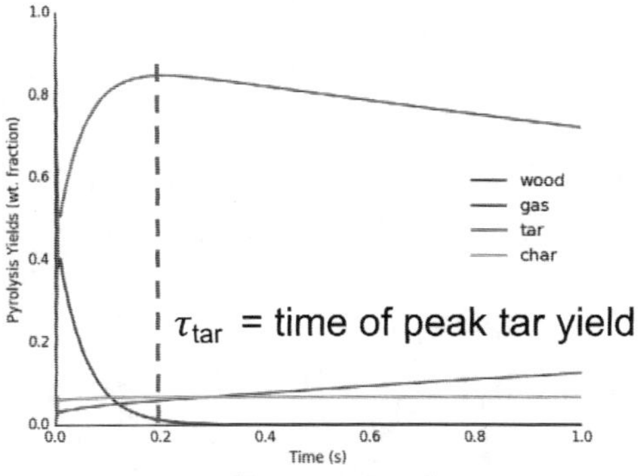

Estimated pyrolysis yields from Miller and Bellan 1997 kinetic scheme for beech wood at 500°C.

See the following AIChE 2016 session for details of this work:
458989 Modeling the Impact of Biomass Particle Residence Time on Fast Pyrolysis Yield and Composition

Background: Experiment of interest

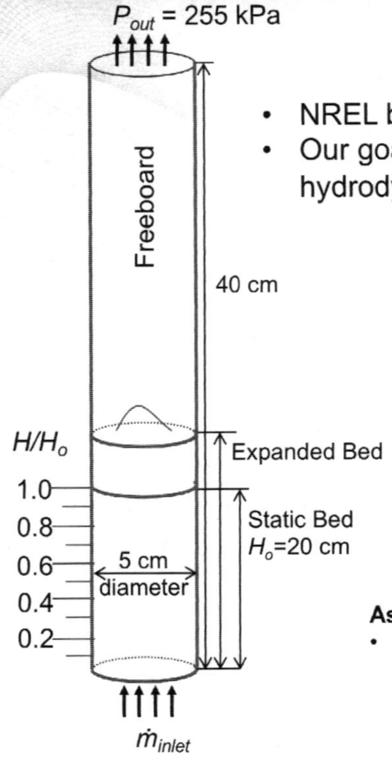

- NREL bubbling bed reactor used to study biomass fast pyrolysis
- Our goal is to help improve control and interpretation of the hydrodynamic aspects of this experiment

definition	Units	experiment	simulation
Particle diameter	m	250×10^{-6}	250×10^{-6}
Particle density	kg/m^3	2484	2484
Bulk density	kg/m^3	1552	
Temperature	K	773	773
Pressure	kPa	289	289
Fluidizing N_2 (range)	kg/s	$8.46\text{-}18.6 \times 10^{-5}$	$7.70\text{—}23.1 \times 10^{-5}$
Minimum fluidization	kg/s	-	7.70×10^{-5}
Minimum fluidization	m/s	-	0.03
coefficient of restitution	-	-	0.8
angle of repose	°	-	30
friction coefficient	-	-	0.1

Assumptions:
- Minimum fluidization – Richardson correlation
  - N_2 density – ideal gas temperature and pressure corrected
  - N_2 viscosity – NASA 7-coefficient polynomial, temperature corrected

Technical approach: computational

MFiX TFM

MFiX simulation parameters:
- Eulerian-Eulerian (Two-Fluid Model)
- Syamlal-O'Brien drag-model
- Kinetic theory of granular flow
  - Schaeffer frictional stress tensor formulation
  - Sigmoidal blending stress function
- Second order discretization
- Modified SIMPLE with variable time stepping
- Jackson and Johnson partial-slip wall boundary condition
- Cylindrical Mesh

Technical approach: Identify spatio-temporal transition metrics in MFIX simulation results

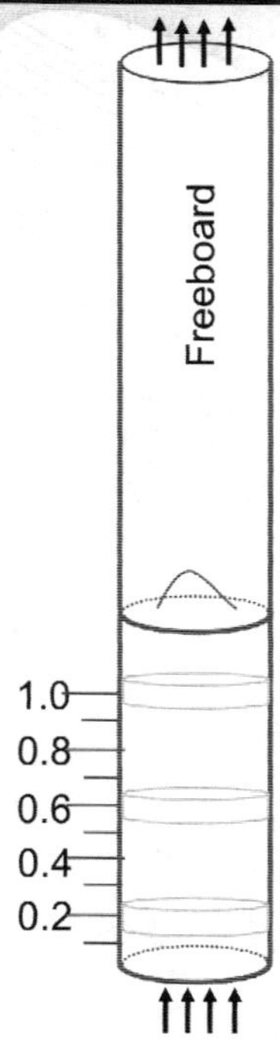

MFiX simulated pressure and bubble probes utilized data from locations relative to static bed height (H/H_o)

Pressure Statistics

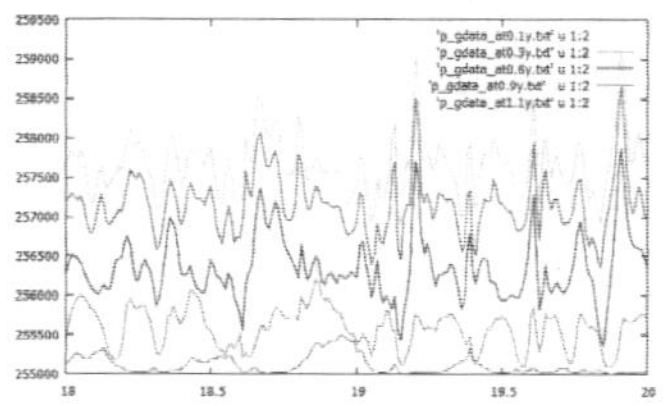

Analysis of PostMFiX pressure time series
- fluctuation distributions and statistics
- time-scale metrics

This work

Bubble Statistics

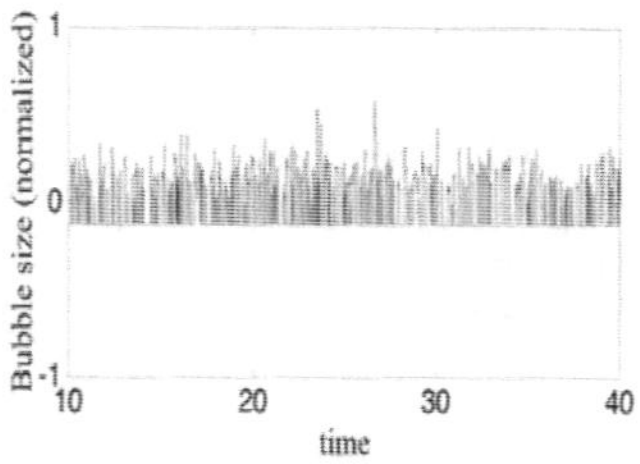

Analysis of PostMFIX void fraction time series processed with MS3DATA
- bubble size distributions and statistics
- time-scale metrics

A. Bakshi, C. Altantzis, R.B. Bates, A.F. Ghoniem, Multiphase-flow Statistics using 3D Detection and Tracking Algorithm (MS3DATA): Methodology and application to large-scale fluidized beds, CEJ, Volume 293, 1 June 2016, Pages 355-364.

OAK RIDGE National Laboratory | NATIONAL TRANSPORTATION RESEARCH CENTER

13 Computational study of the bubbling-to-slugging transition in a laboratory-scale fluidized bed

Results (1): Bubble statistics indicate the slugging transition occurs gradually, not abruptly

Max and min bubble frequency indicates beginning and end of bubbling to slugging transition in upper part of bed

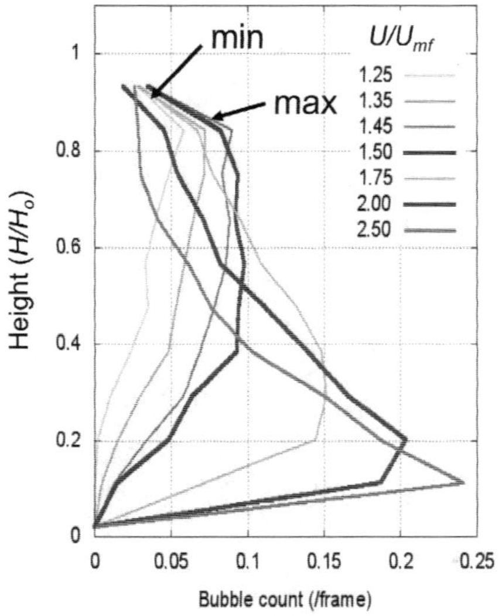

Bimodal distribution appears at bubbling to slugging transition which varies with inlet velocity and height

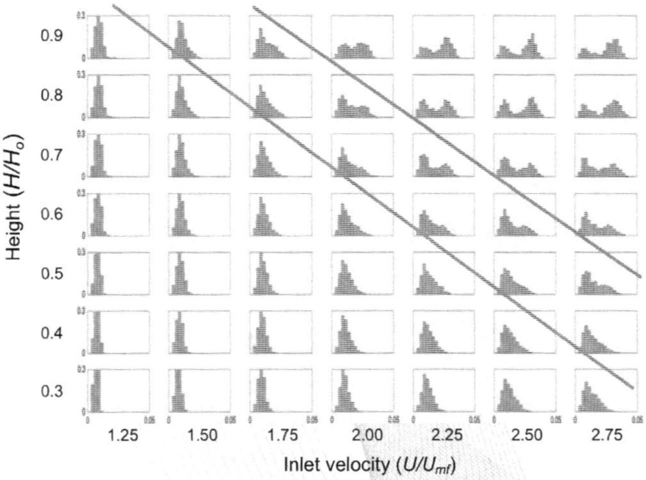

OAK RIDGE National Laboratory | NATIONAL TRANSPORTATION RESEARCH CENTER

14 Computational study of the bubbling-to-slugging transition in a laboratory-scale fluidized bed

Results (2): Pressure variations appear to track with bubble statistics but the relationship is complex

Pressure distribution tails reveal different asymmetries at different heights

1.25 U_{mf} – Free Bubbling

Pressure variations at different heights look similar

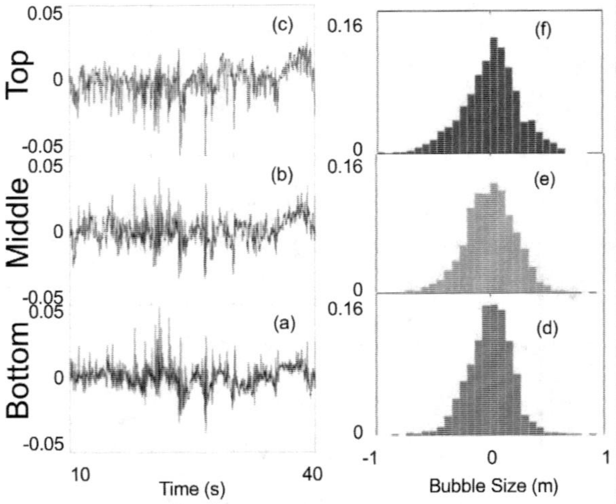

2.75 U_{mf} – Fully developed slugging

Bimodal patterns emerge

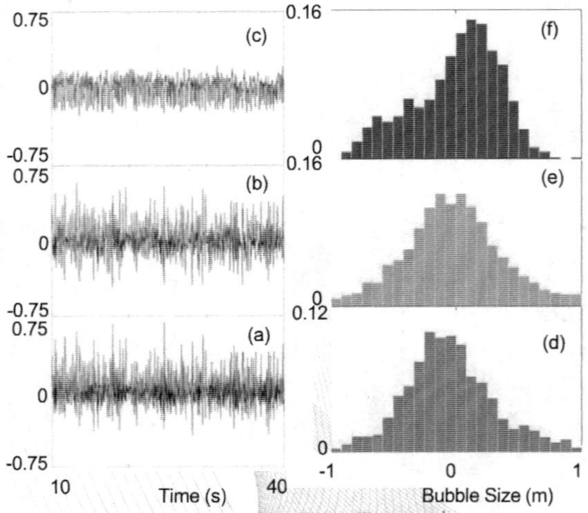

Results (3): Shifts in statistical metrics for pressure fluctuations change with location

1.25 U_{mf} – Free Bubbling

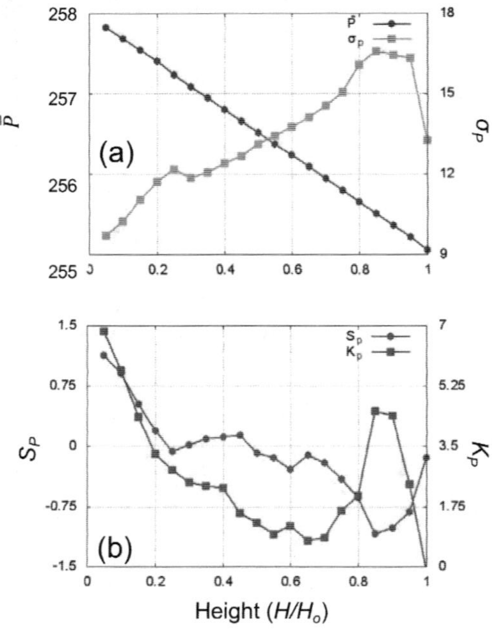

2.75 U_{mf} – Fully developed slugging

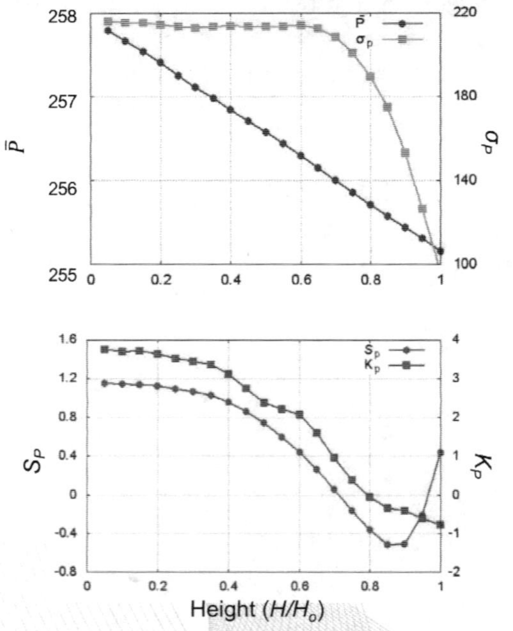

Results (4): MFIX predicted pressure trends appear consistent with experimental studies

Example: Predicted P fluctuation trends at 0.7-0.9 H/H_o appear to agree with Lee et al. 2002 experimental results in upper bed region (0.75 H/H_o).

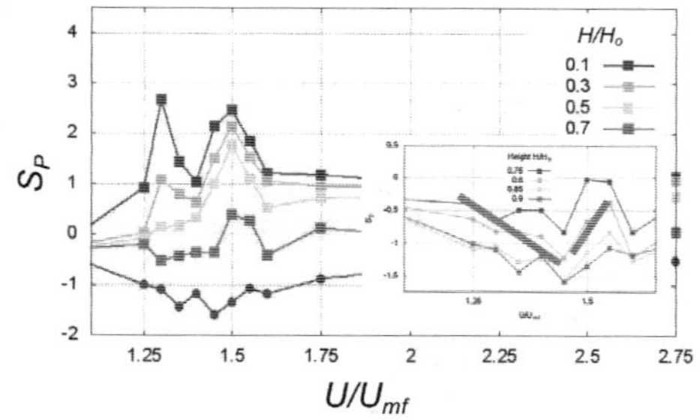

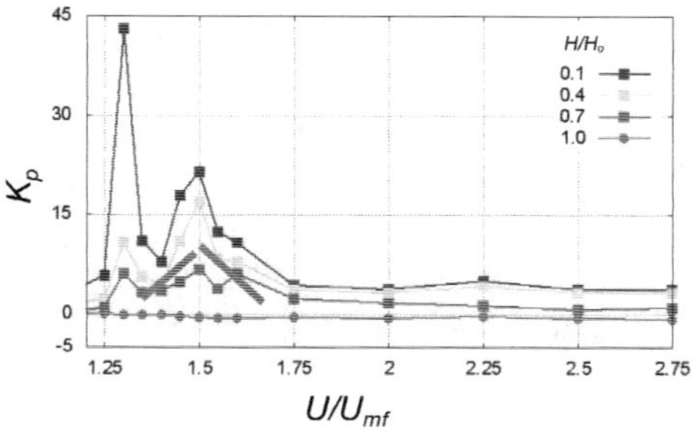

Skewness at 0.75 H/H_o has a downward spike

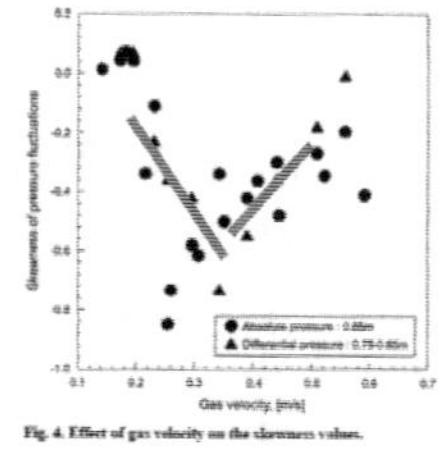

Kurtosis at 0.75 H/H_o has a upward spike

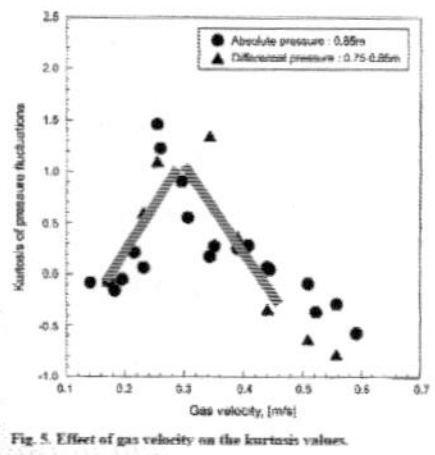

Statistical and Deterministic Chaos Analysis of Pressure Fluctuations in a Fluidized Bed of Polymer Powders, 2002
See Hoon Lee, Sang Done Kim† and Soung Hee Park

OAK RIDGE National Laboratory | NATIONAL TRANSPORTATION RESEARCH CENTER

17 Computational study of the bubbling-to-slugging transition in a laboratory-scale fluidized bed

Results (5): MFIX results imply temporal metrics can be used to relate bubbles and pressure measurements

- Pressure and void fraction signal appear related
- Cross correlation magnitude increases with inlet velocity
- Upper bed section appears the most useful for slugging diagnostic

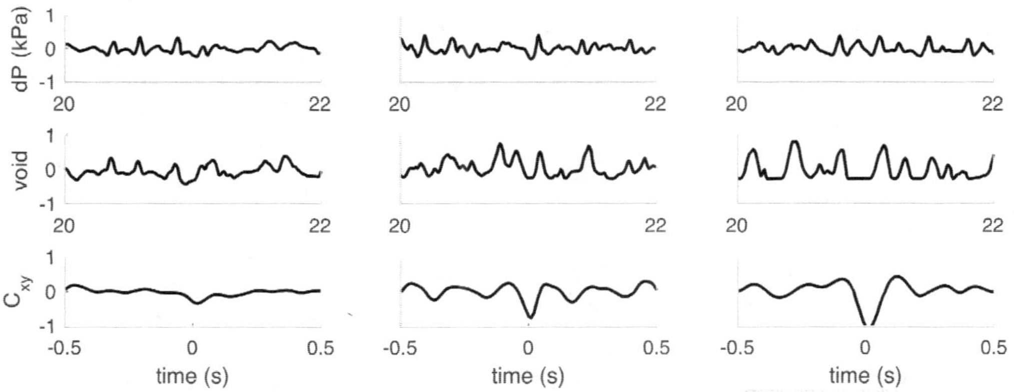

1.25 1.75 2.50
Inlet velocity (U_{mf})

OAK RIDGE National Laboratory | NATIONAL TRANSPORTATION RESEARCH CENTER

18 Computational study of the bubbling-to-slugging transition in a laboratory-scale fluidized bed

Summary and Conclusions to Date

- The bubbling-to-slugging transition is a gradual process, not an abrupt event
  - The bubbling-to-slugging transition occurs over a range of flows
  - Bubble coalescence begins near top of the bed and progresses downward as gas flow increases
  - Slugging becomes fully developed when the first major bubble coalescence occurs neat the distributor
- It may be easiest to detect the initial transition toward slugging at higher levels in the bed
- Statistical and temporal analyses of local pressure and void fluctuations can be used to develop quantitative metrics to indicate/diagnose the slugging transition
- MFIX and MS3DATA are useful tools for understanding the complex relationship between dynamic pressure and void fluctuations (bubble behavior)
- The optimal degree of slugging for fast pyrolysis of biomass has not yet been resolved

Future work (1): Relate simulated pyrolysis yields and bubble patterns

Preliminary 2D simulation

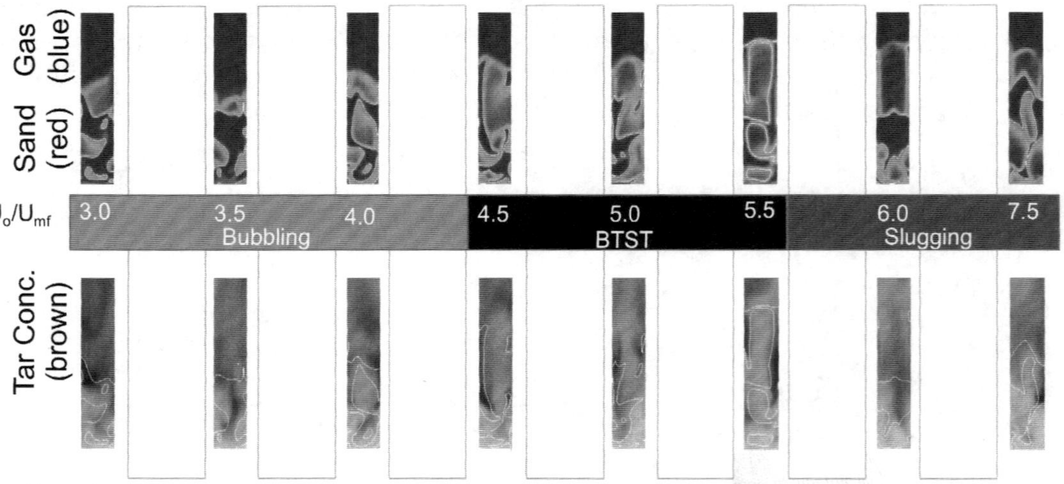

Bubble contours at 0.7 void cut off

Gas/sand emulsion and tar (oil) generation are highly correlated

Future work (2): Improve MFIX kinetic and elutriation parameters

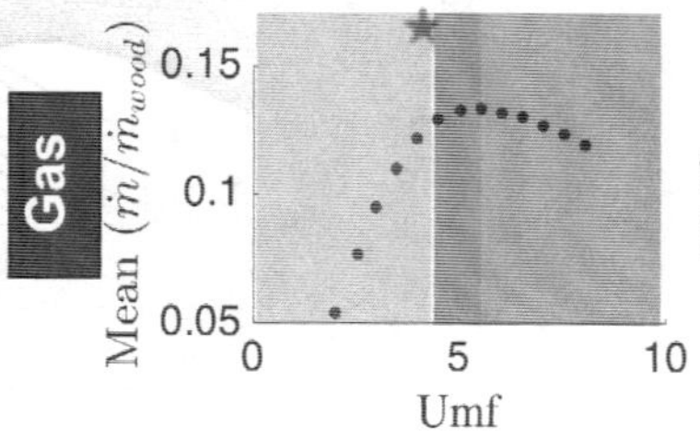

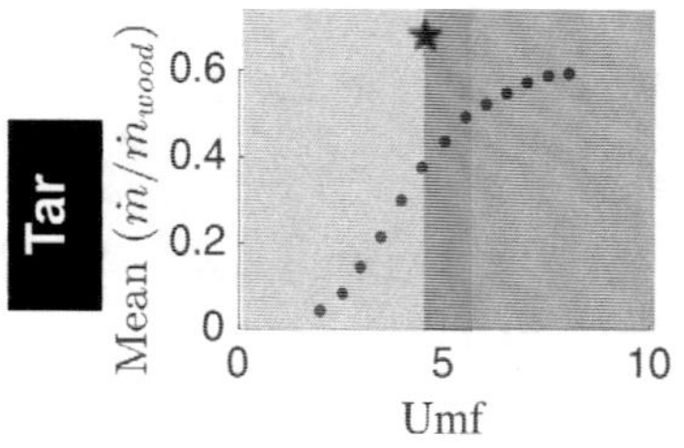

Bubbling = 3.50 U_o/U_{mf}
Transition (BTST) = 5.00 U_o/U_{mf}
Slugging = 7.50 U_o/U_{mf}

Preliminary Kinetic yield trends appear to be affected by the bubbling-to-slugging transition.

Table: Yield at ~4.4 U_{mf}

	Experiment	Simulation
Tar	0.661	0.377
Gas	0.162	0.129
Char	0.119	0.494

Mean yields did not match experiment

Uncertainty not accounted in results

2D simulation does not capture the correct physics in the fluidized bed

Future work (3): Do exit pyrolyzer fluctuations impact downstream processes?

Amplitude of tar oscillations increase with increased superficial velocity

Oscillations have potential to affect down stream equipment operation, such as VPU or LPU reactors.

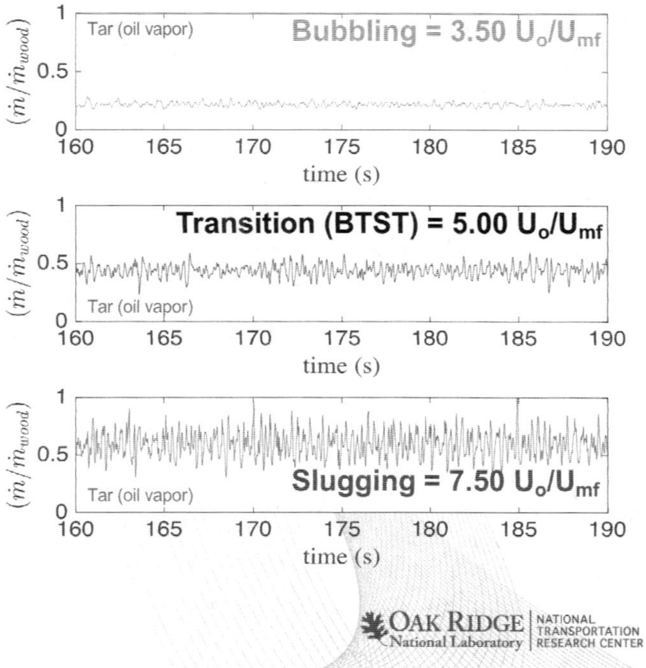

OAK RIDGE National Laboratory | NATIONAL TRANSPORTATION RESEARCH CENTER

Future work (4): Experimental

Experimental validation

NETL
– ambient reactor experimental trends vs. MFIX predictions

NREL
– Analysis of existing reactor data
– Added instrumentation and diagnostics on the reactor

Questions ?

Emilio Ramirez – eramire2@vols.utk.edu

E. Ramirez, C.E.A. Finney, S. Pannala, C.S. Daw, J. Halow, Q. Xiong, Computational study of the bubbling-to-slugging transition in a laboratory-scale fluidized bed, Chemical Engineering Journal 308 (2017) 544-556.http://dx.doi.org/10.1016/j.cej.2016.08.113

Computational Pyrolysis Consortium
http://cpcbiomass.org

Transitioning to Consortium for Computational Physics and Chemistry (CCPC)

469537 Particle and Gas Velocities in a Vortex Driven Externally Recirculated Fluidized Bed

Abstract:

The ever demanding need for lower capital costs requires the development of new contactors that can potentially utilize some hydrodynamic feature to enhance heat and mass transfer and thereby achieve process intensification. One such concept being explored at NETL is for a vortexing circulating fluidized bed. In this reactor concept, the gas enters tangentially, the rotating gas flow establishes a pressure gradient between the wall and center of the reactor. The geometry of the riser design exploits this gradient to generate a recirculating flow of solids from the top outer wall of the reactor to the bottom center. Barracuda computational findings and experimental results from a test reactor will be compared to assess the capabilities, and potential of the system. Specifically, gas and solids velocities will be compared between the simulation and the experimental results as well as comparing the pressure profiles.

CFD DEM COUPLING VALIDATION FOR PNEUMATIC SEPARATION OF SUGARCANE BAGASSE

Eduardo de Almeida, Nicolas Spogis and Maria A. Silva, University of Campinas, Campinas/SP, Brazil

Introduction

Sugarcane bagasse can be defined as the byproduct of the sugarcane milling process consisting of the sugarcane's fibrous portion after the extraction of its juice. Currently many countries ferment and then distill the sugarcane juice to produce ethanol; an important renewable fuel used as fossil fuel substitute aiming, among other factors, to reduce the global warming caused by fossil fuel burning.

The bagasse is nowadays used as a burning fuel in boilers in sugarcane mills, as animal feed, and in paper production. However, if the cellulose present in this biomass was converted to ethanol, by hydrolysis process, called second generation ethanol the yield per hectare of planted sugarcane could rise from current 5500 L ethanol/ha to around 8000 L ethanol/ha without increase the planted area in Brazil's case (1,2).

However, the hydrolysis process still presents some important bottlenecks that need to be overcome like high cost for example. The high cost is affected by many factors highlighting one important step of the process called physical pretreatment responsible for homogenize the material and decrease the bagasse particle's size allowing good hydrolysis yields but consuming high energy because the bagasse should be submitted to milling, grinding and chipping or a combination of these.

In order to propose alternatives to the physical pretreatment, consuming less energy than other ways of obtaining particles of different sizes it was analyzed the performance of a device, called pneumatic classifier equipment, developed at the School of Agricultural Engineering (Feagri / Unicamp). Based its operation principle on fluid dynamic theories, the pneumatic classifier equipment makes the separation of bagasse in an air column by dragging the small particles to the top and collecting the coarse particles at the bottom. The results from experiments showed the pneumatic classifier equipment allowed the continuous separation of sugarcane bagasse particles wherein the small particles, around 37% of total bagasse at dry basis, reached excellent yields when submitted to enzymatic hydrolysis without any other extra pretreatment thus avoiding energy consumption costs (3).

Inspired on previous results, it has started a theoretical study for bagasse separation process where the scope was to obtain a simulated model validated by experimental findings in order to be applied for example in different equipment structures aiming to evaluate the best one for separation of particles with a desired range of sizes or to help the simulation of other studies involving pneumatic transport of bagasse like drying, pyrolysis, gasification and others.

Pneumatic conveying and particle separation process are a complex phenomenum that requires a tighter coupling between air/fluid flow and particle-particle interaction. In order to better characterize these interactions, this work was performed coupling ANSYS FLUENT (16.2) and ROCKY (3.11). ANSYS FLUENT is a computational fluid dynamics (CFD) software used to help analyses of systems involving fluid flow, heat transfer, combustion, etc., by means of computer simulation based on finite volume to solve the continuum equations of motion. ROCKY uses a numerical technique called discrete element method (DEM) for predicting the behavior of polydispersed solids. The coupling between the simulated air flow at FLUENT and the particles motion at ROCKY allows a realistic approach of the separation process.

To validate the simulation results it was used a PhD thesis entitled “Characterization of sugarcane bagasse for energy generation” from School of Mechanical Engineering (FEM/Unicamp) where the author used a pneumatic separator device to evaluate bagasse particle’s fluid dynamics characteristics besides to determine bagasse’s physical properties from experimental data. As these results were not published in scientific journals, they are described in the next session (4).

Materials and Methods

The next paragraphs describe the equipment used to obtain the bagasse experimental data by Paulo Lenço in a PhD thesis entitled “Characterization of sugarcane bagasse for energy generation” and the models used by the softwares ANSYS FLUENT and ROCKY.

Operation Principle of the Separator Equipment

In Lenço’s (4) work, the first step to start the bagasse fractions separation into the equipment is to turn on the centrifugal fan to cause a pipeline pressure differential, forcing a flow of ambient air through the opening at the bottom of the separation column to balance this difference. After the ascending airflow stabilization is reached in the air column, the bagasse is fed by an endless worm and falls by gravity into an opening near to the outlet of the column [see Figure 1].

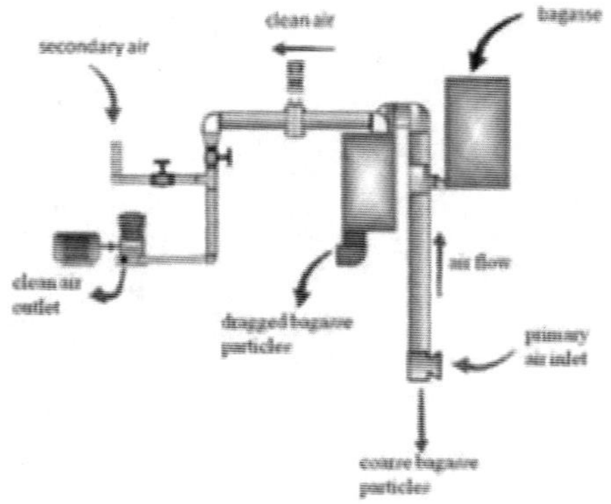

Figure 1. Equipment scheme used to separate bagasse particles (4)

Thus, the ambient air in contact with the bagasse, in a countercurrent flow, promotes the entrainment of the fine particles in the ascending flow until their collection in a particle precipitator at the top of the equipment. The coarse particles heavy enough to not be either dragged or put into lifting, fall by gravity action into an opening at the bottom of the air column. These coarse particles collected at the bottom are fed again and a higher velocity was used and a new fraction was separated. This process was repeated until all the remaining particles are dragged. In such study 7 fractions were found.The properties of the bagasse fractions as mass separated, particle size and apparent density were evaluated according the different velocities of separation. The basic dimensions of the riser are height 1.5m and internal diameter 0.1m.

Physical Properties of the Air and Bagasse

The air physical properties followed the default FLUENT properties for air, dynamic viscosity 1.7894×10^{-5} $kg.m^{-1}s^{-1}$ and density 1.225 $kg.m^{-3}$.The bagasse properties were that obtained by Lenços' [9] studying the biomass fluid dynamics characteristics according to the experiments in the separator equipment. Table 1 summarizes all the properties applied on simulations [mass, d_c [particle diameter], L [particle length] and density] and the calculated terminal velocity using the Ganser's drag coefficient (5) that was a parameter to define the velocities simulated according the selected group to be dragged: 1.5 m/s; 1.7 m/s; 2.0 m/s; 3.0 m/s; 3.5 m/s and 4.0 m/s.

Table 1. Values Obtained by Practical Experiments (4) and calculated terminal velocity

Particle group	**Air velocity experimenal (m/s)**	**Mass (kg)**	**d_c (m)**	**L (m)**	**Density (kg/m^3)**	**Terminal velocity calculated (m/s)**
1	0.501	0.00924	0.00028	0.00061	545.3	0.74
2	0.710	0.00579	0.00091	0.00286	143.2	0.99
3	1.062	0.01091	0.00164	0.00587	164.3	1.64
4	1.578	0.01301	0.00191	0.00736	287.0	2.39
5	1.963	0.00609	0.00317	0.01279	248.0	2.81
6	2.277	0.00337	0.00318	0.01317	273.7	2.94
7	>2.277	0.01174	0.00329	0.01318	471.5	3.90

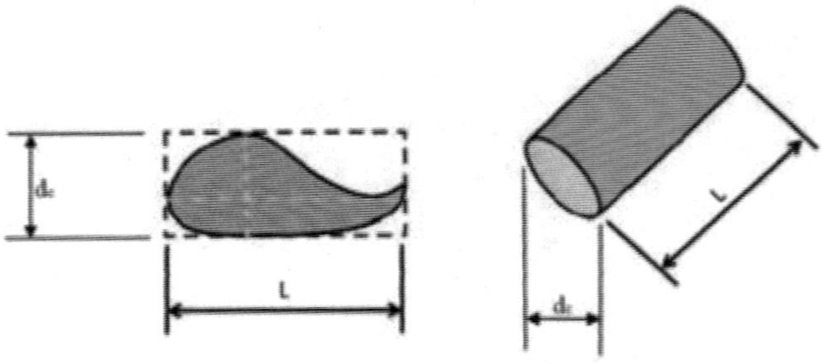

Figure 2. Particle dimensions (4)

Computational Fluid Dynamics (CFD)

The differential equations that describe the air flow into the riser are determined from Navier-Stokes' Equations of mass and momentum conservation.

Mass conservation equation (6):

$$\nabla.(f_g\rho_g V_g) = \frac{\partial(f_g\rho_g v_{g,x})}{\partial x} + \frac{\partial(f_g\rho_g v_{g,y})}{\partial y} + \frac{\partial(f_g\rho_g v_{g,z})}{\partial z} = S_g^{\rho} \quad [01]$$

Where, f_g represents the volumetric fraction of air

ρ is the air density [$kg.m^{-3}$];

V is the velocity vector that can be decomposed into $v\hat{x}$, $v\hat{y}$ e $v\hat{z}$ [m/s];

S_g^p is the source term, in this case, $S_g^p = 0$

Momentum conservation equations to directions x, y and z:

$$\sum_{\xi=x,y,z} \frac{\partial}{\partial \xi}\left[f_g \rho_g v_{g,x} v_{g,\xi} - f_g \mu_g \left(\frac{\partial v_x}{\partial \xi} + \frac{\partial v_\xi}{\partial x}\right)\right] = S_{g,x}^m \quad [02]$$

$$\sum_{\xi=x,y,z} \frac{\partial}{\partial \xi}\left[f_g \rho_g v_{g,y} v_{g,\xi} - f_g \mu_g \left(\frac{\partial v_y}{\partial \xi} + \frac{\partial v_\xi}{\partial y}\right)\right] = S_{g,y}^m \quad [03]$$

$$\sum_{\xi=x,y,z} \frac{\partial}{\partial \xi}\left[f_g \rho_g v_{g,z} v_{g,\xi} - f_g \mu_g \left(\frac{\partial v_z}{\partial \xi} + \frac{\partial v_\xi}{\partial z}\right)\right] = S_{g,z}^m \quad [04]$$

Where, μ_g is the air dynamic viscosity [kg.m^{-1}.s^{-1}];
S_g^m represents the momentum transformation in the air.

These differential equations are used to describe the fluid motion, establishing that changes in momentum and acceleration of a fluid particle are resulting from changes in the pressure and viscous dissipative forces, originated from the molecular interaction, acting within the fluid.

The air turbulence was taken into account in the model using *k*-ε model. This model deal with two separated transport equations obtaining the turbulent kinetic energy, *k*, and the kinetic energy dissipation, ε, whereas the Reynolds Stress Models (RSM) are proportional to the average velocity gradients.

The turbulence kinetic energy, *k*, and its rate of dissipation, ε, are obtained from the following transport equations (6):

$$\nabla \cdot \left(f_g \rho_g \bar{v}_g k\right) = \nabla \cdot \left[\left(\mu_{g+} \frac{\mu_{g,t}}{\sigma_k}\right) \nabla k\right] + P_k - f_g \rho_g \varepsilon \quad [05]$$

$$\nabla \cdot \left(f_g \rho_g \bar{v}_g \varepsilon\right) = \nabla \cdot \left[\left(\mu_{g+} \frac{\mu_{g,t}}{\sigma_\varepsilon}\right) \nabla \varepsilon\right] + \frac{\varepsilon}{k}\left(C_{\varepsilon 1} P_k - C_{\varepsilon 2} f_g \rho_g \varepsilon\right) \quad [06]$$

In which the constants take the values: $C_{\varepsilon 1} = 1{,}44$; $C_{\varepsilon 2} = 1{,}92$; $\sigma_k = 1{,}0$ e $\sigma_\varepsilon = 1{,}3$.

P_k is the turbulent product due to viscous and buoyancy forces, expressed by equation (4):

$$P_k = \mu_{g,t} \nabla v_g \cdot \left(\nabla \bar{v}_g + \nabla \bar{v}_g^T\right) - \frac{2}{3} \nabla \cdot \bar{v}_g \left(3\mu_{g,t} \nabla \cdot \bar{v}_g + f_g \rho_g k\right) + P_{kb} \quad [07]$$

Considering the buoyancy production term (P_{kb}) null because there is no turbulence generation by it.

Discrete Element Method (DEM)

Discrete Element Method (DEM) models the total force experienced by individual particles or grains of a granular system and subsequent lists the accelerations, velocities and positions in a time period. The total force is the sum of the contact forces (particle-particle and particle-boundary) and non contact forces (such as gravity, magnetism and electrostatic forces). DEM is a mesh-free method and does not solve the continuum equations of motion instead of it solves the equations of motion for every individual particle that are numerically integrated with time.

DEM can be coupled with CFD to study fluid-particulate systems in which both solid and fluid phases contributes for dynamic flow, such as pneumatic transport and fluidized beds. In this approach, commonly referred as Lagrangian-Eulerian method, the solid phase is first modeled discretely using DEM, and these results are coupled with the CFD model.

The accomplished work by ROCKY DEM and ANSYS allows ANSYS FLUENT to work with Rocky in parallel CPU (Central Processing Unit) and GPU (Graphics Processing Unit) capabilities providing very fast DEM-CFD results. This processing methodology also allows to simulate large multiphase flow problems with millions of particles presenting different sizes and shapes.

For this study it was used the one-way coupling simulation because the particle loading is low and coupling parameters are almost null, thus, CFD solution provides information to the DEM solution, i.e., the fluid flow affects the particle trajectory but the particles do not affect the fluid flow.

The contact forces used by DEM, including in ROCKY software, consist of two parts, normal-to-contact plane and tangential-to-contact plane. The contact plane is a plane perpendicular to the line connecting closest points of the particles and boundaries. So, the algorithm for determining the contact plane for non-round particles, like those simulated in this study, takes account the calculations for one of any the following options: the closest points of two particles; or the closest points of a particle and triangle; or the two points with the maximum overlap distance in the case of a physical contact, depending on the contact type for each pair of particles.

The model that has been chosen to predict the normal contact forces in this work was an elastic-plastic model. The elastic-plastic models follow two major requirements that are: the force has to be a repulsive one; the normal contact force model has to allow significant energy dissipation since a granular medium is an extremely dissipative system. For the tangential contact prediction it was used an elastic-frictional force model.

One of the elastic-plastic models, the linear hysteresis model was adopted by this study because it presents many advantages as: to avoid the overhead of long simulation time when simulating the plastic energy dissipation on a contact; to make the energy dissipation insensitive to other contacts, since no viscous damping term is used and the energy dissipation is not dependent on the relative velocities of neighboring particles; and to allow modeling compressible materials accurately due to the fact that the contact forces can be almost zero even at residual overlaps.

The following equations describe the linear hysteresis model used by ROCKY (7):

$$F_n^t = min(F_n^{t-dt} + K_{nu} \cdot ds_n, K_{nl} \cdot S_n) \quad \text{if } ds_n \geq 0 \qquad [08]$$

$$F_n^t = min(F_n^{t-dt} + K_{nu} \cdot ds_n, 0.001 \cdot K_{nl} \cdot S_n) \quad \text{if } ds_n < 0 \qquad [09]$$

$$ds_n = s_n^t - s_n^{t-dt} \qquad [10]$$

Where, F_n^t is the normal elastic-plastic contact force at the current time, *t*

F_n^{t-dt} is the force at the previous time, *t-dt*, where *dt* is the timestep

ds_n is the change in the contact normal overlap during the current time (it is assumed to be positive as particles approach each other and negative when they separate)

s_n^t and s_n^{t-dt} are the normal overlap values at the previous and the current time
K_{nl} and K_{nu} are the values of loading and unloading contact stiffness respectively

The particle's loading stiffness is calculated in the following manner (7):

$$K_{npl} = E_p \cdot L \quad [11]$$

$$K_{nbl} = E_b \cdot L \quad [12]$$

Where, E_p is the particle material bulk Young's Modulus or Elastic Modulus
E_b is the boundary material Young's Modulus
L is the particle size.

For the contact of two particles, or of a particle with a boundary, the loading and unloading stiffness are defined respectively as (7):

$$K_{nl} = \frac{K_{npl\_1} \cdot K_{npl\_2}}{(K_{npl\_1} + K_{npl\_2})} \quad [13]$$

$$K_{nl} = \frac{K_{npl} \cdot K_{nbl}}{(K_{npl} + K_{nbl})} \quad [14]$$

$$K_{nu} = \frac{K_{nl}}{\varepsilon^2} \quad [15]$$

Where, ε is the coefficient of restitution, in ROCKY it is a measure of energy dissipation for the contacting pair of materials; and subscripts *_1* and *_2* refer to particle 1 and particle 2 of two contacting particles.

Considering the tangential force model it is an elastic-frictional force model described by the equations (7):

$F_t^t = min(F_t^{t-dt} + K_{nl} \cdot ds_t, \mu_s \cdot F_n^t)$, if no sliding is taking place for the contact [16]

$F_t^t = min(F_t^{t-dt} + K_{nl} \cdot ds_t, \mu_d \cdot F_n^t)$, if sliding is taking place for the contact [17]

Where, F_t^t is the tangential contact force at the current time;
F_t^{t-dt} is the same force at the previous time;
ds_t is the tangential relative particles' displacement during the current time;
μ_s and μ_d are the static and dynamic friction coefficients;

If the tangential force exceeds the limit of $\mu_s \cdot F_n^t$ the sliding is considered to be taking place for the contact, otherwise if the tangential force falls below value of $\mu_d \cdot F_n^t$, the contact is considered non-sliding.

Results

The group 1 of particles, around 450,000, weren't simulated because they would all certainly be dragged by the air due the chosen velocities and the simulation time would increase a lot considering the computational power available making the simulations infeasible. Nevertheless, the simulation with the other groups of particles, representing 82% of the true bagasse mass, proved to be very predictive as shown hereafter.

Figures

The next figures show the visual results obtained by Rocky software with zoom of particles near the bottom and top of the equipment.

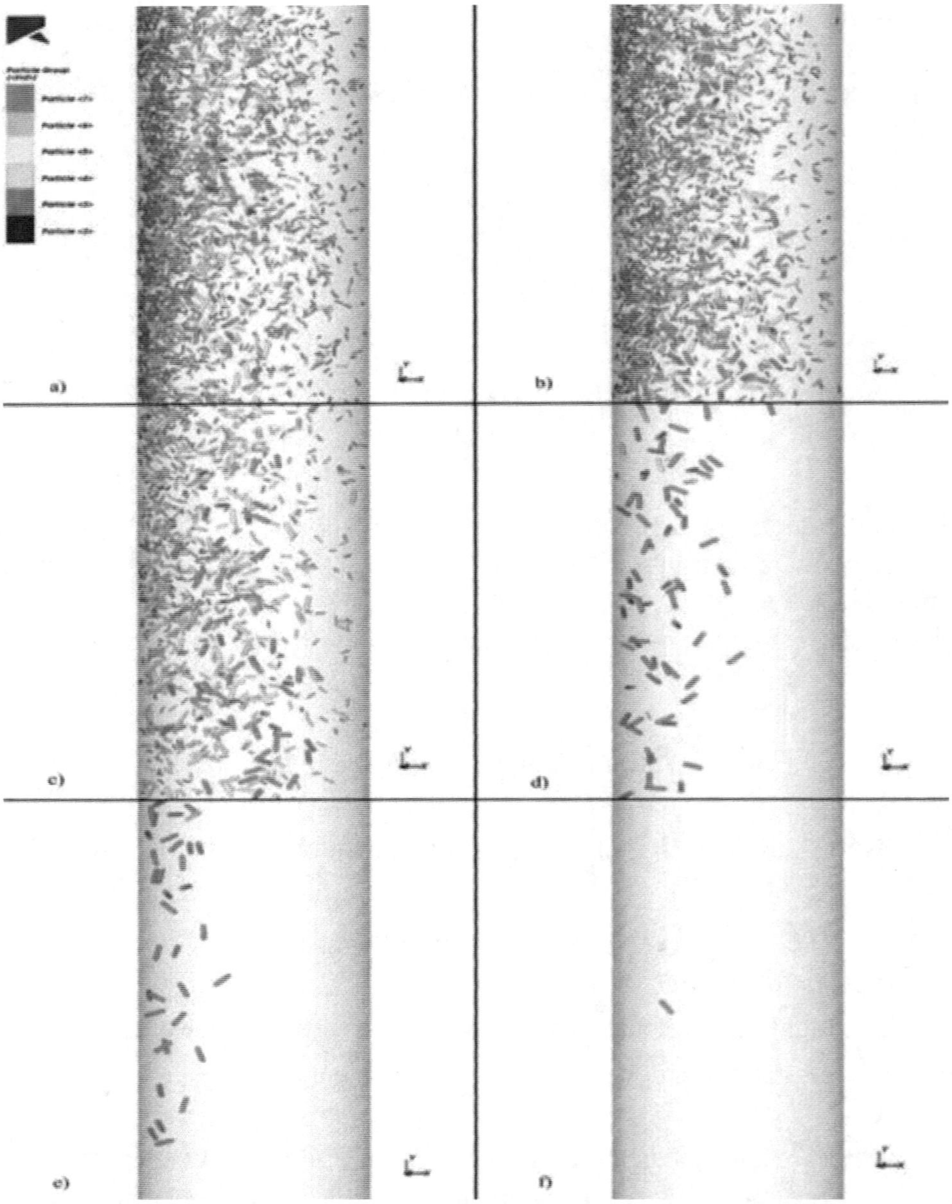

Figure 3. Zoom in bagasse separation simulation near the bottom at same time using air average velocities of a) 1.5m/s b) 1.7 m/s c) 2.0 m/s d) 3.0 m/s e) 3.5 m/s and f) 4.0 m/s

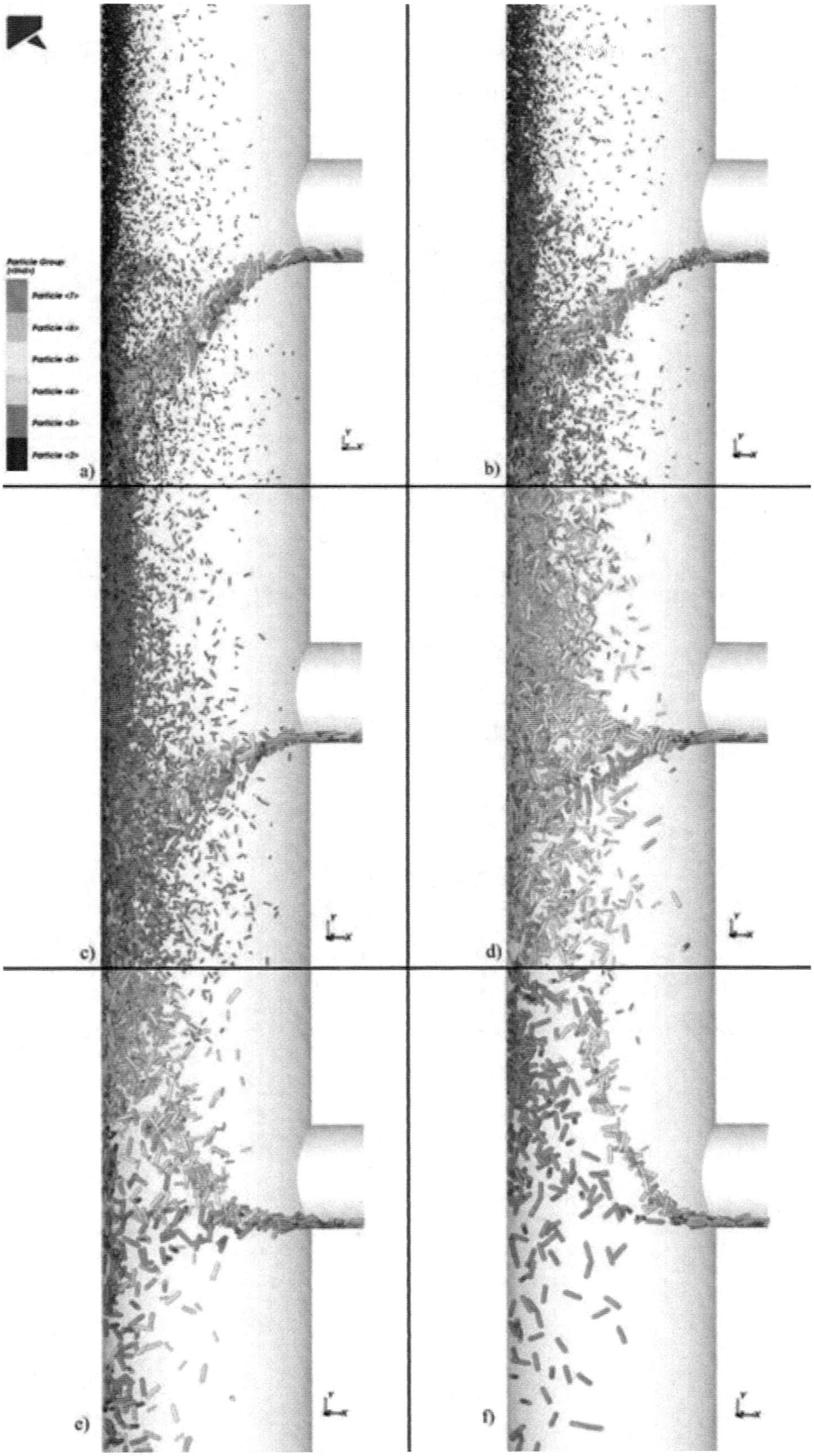

Figure 4. Zoom in bagasse separation simulation near the top at same time using air average velocities of a) 1.5m/s b) 1.7 m/s c) 2.0 m/s d) 3.0 m/s e) 3.5 m/s and f) 4.0 m/s

Graphics

The graphics show de particle distribution of groups collected on the top and bottom of the equipment varying the air velocity of separation.

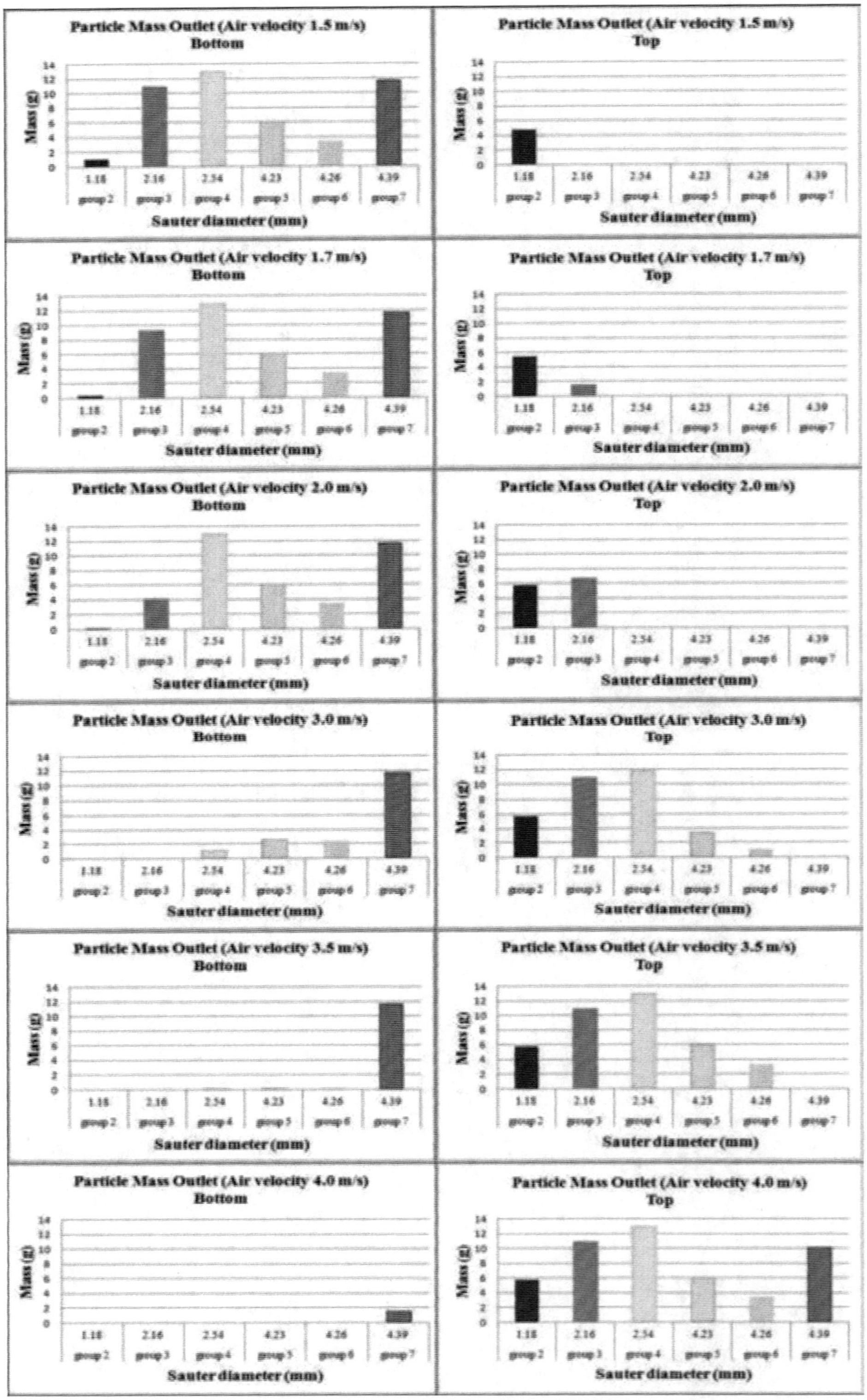

Figure 5. Mass distribution of particles on top and bottom of the equipment varying the simulated air velocities

Analyzing the results both by images and mass distribution was possible to understand that at lower velocities even particles that should be dragged to top are collected by bottom possibly because with these velocities the contacts particle/particle and particle/wall are higher. Thus, the particle contact with the riser wall decrease the drag force over it, prevailing the weight force on the forces balance, causing the particle fall by gravity action and besides, the falling larger particles can also drag the smaller to the bottom due to contact during separation.On the other hand, using the air average velocity of 3.5 m/s into the riser, almost all the particles were dragged to the top just separating group 7, as expected, when compared the air velocity with the particles` terminal velocities in Table 1. Simulating the air velocity 4.0 m/s almost all the particles were completely dragged separating just few grams of particles from group 7 at the bottom outlet.

Conclusions

CFD/DEM techniques were used in this paper to simulate the bagasse particles` separation that showed to be a powerful tool since the adopted model described with good accuracy the particles movements and their interactions with other particles and boundaries. When the air velocity was enough to separate intermediate group of particles it was observed that the particle/particle and particle/wall contacts were larger than when high air velocities were adopted, thus the separation was more efficient when the objective was to separate the coarsest particles.The parameters of simulation studied can be extended to other processes that use sugarcane bagasse fluidization as drying, pyrolysis, gasification and others, allowing the study and development of equipment close to an ideal configuration saving time and money for it.

Acknowledgments:

CNPq - National Council for Scientific and Technological Development (Brazil)
ESSS - Engineering Simulation and Scientific Software

References

1. Companhia Nacional de Abastecimento: Acompanhamento da safra brasileira: cana-de-açúcar. http://www.conab.gov.br/OlalaCMS/uploads/arquivos/12_04_10_09_19_04_boletim_de_cana.pdf (2012) Accessed 28 April 2016
2. Leite, R.C.C., Leal, M.R.L.V., Cortez, L.A.B., Griffin, W.M., Scandiffio, M.I.G. (2009), "Can Brazil replace 5% of the 2025 gasoline world demand with ethanol?", *Energy 34*, pp. 655-661.
3. Almeida, E., Cortez, L.A.B., Silva, M.A.(2013), "Sugarcane bagasse pneumatic classification as a technology for reducing costs on enzymatic hydrolysis process. " *In: International Society of Sugar Cane Technologists*, Brazil, Abstract, Brazil: STAB & The XXVIIIth ISSCT Organising Committee, p.358.
4. Lenço, Paulo A. (2010), "Characterization of sugarcane bagasse for energy generation," PhD Thesis, University of Campinas, p.181.
5. Ganser, Gary H.(1993),"A rational approach to drag prediction of spherical and non-spherical particles," *Powder Technology 77*, p.p. 143-152.
6. Ansys Fluent 16.2 (2011), *Theory Guide*
7. Rocky 3.11 (2015), "A brief introduction to DEM and an Overview of Contact Force models in Rocky," *Theory Guide*

464140 CFD-DEM Simulation of Hydrodynamics in Wet Gas-Solid Fluidized Beds

Abstract:

It is common in many industries, including energy, pharmaceuticals and polymer production, for small amounts of liquid to be injected into gas-solid fluidized beds. This liquid can serve as a reactant or facilitate agglomeration or heat transfer. The liquid often covers the surface of particles and proceeds to form cohesive liquid bridges between particles, ultimately leading to the agglomeration. The cohesive force caused by small amounts of liquid can alter hydrodynamics dramatically, since agglomerates, rather than individual particles, need to be fluidized. In extreme cases, liquid bridging and agglomeration can lead to local or device-scale defluidization. Despite this industrial importance, the effect of wetness on hydrodynamics is not fully understood. A limited number of experimental studies show differences in bed fluidity [1,2], particle velocities [3–6], bed height [1], bubble size [6,7] and minimum fluidization velocity [6] caused by liquid bridging in gas-solid fluidized beds.

Various efforts have been undertaken to model the effect of wetness on fluidization computationally. Most work has involved an Euler-Lagrange approach often referred to as the Computational Fluid Dynamics – Discrete Element Method (CFD-DEM) [8] in which each individual particle is modeled, gas flow is simulated on a CFD grid and the two phases are linked via a drag law. In this approach, contacts between particles are modeled, often with a coefficient of restitution dictating the level of inelasticity in the collisions. A number of studies have accounted for wetness by lowering this coefficient of restitution [9,10], since the cohesive force provided by liquid bridges will act to slow the particles from rebounding off of one another after a collision. A fundamental issue with approach is that liquid bridging does not merely make collisions more inelastic; liquid bridges provide cohesion necessary to keep particles together for long periods of time and form agglomerates. To address this issue with a modified coefficient of restitution approach, a few more detailed studies have directly modeled forces arising from liquid bridges [11,12]. One issue with these studies is that they have assumed liquid bridges to form instantaneously. In fact, the surface tension and viscosity of the liquid will affect the rate at which liquid pours from the surface of a particle into a liquid bridge [13], and in certain cases, particles can rebound off of one another before a significant liquid bridge has had time to form.

In this study, we model the effects of liquid on fluidization hydrodynamics using a CFD-DEM model in which liquid bridges are modeled directly and liquid bridges form at a finite rate. We simulate laboratory-sized fluidized beds and investigate the effects of liquid content, viscosity and surface tension on bed height, particle velocity, bubble behavior and minimum fluidization velocity for comparison with effects seen in the literature.

References:

[1] S. L. McDougall, M. Saberian, C. Briens, F. Berruti, and E. W. Chan, Int. J. Chem. React. Eng. **2**, (2004).

[2] S. McDougall, M. Saberian, C. Briens, F. Berruti, and E. Chan, Chem. Eng. Process. Process Intensif. **44**, 701 (2005).

[3] V. S. Sutkar, N. G. Deen, A. V. Patil, E. A. J. F. Peters, J. a. m. Kuipers, V. Salikov, S. Antonyuk, and S. Heinrich, AIChE J. **61**, 1146 (2015).

[4] Q. Zhang, Y. Zhou, J. Wang, B. Jiang, Y. Yang, S. Stapf, C. Mattea, and Q. Gong, Chem. Eng. Technol. n/a (2015).

[5] M. L. Passos and A. S. Mujumdar, Powder Technol. **110**, 222 (2000).

[6] Y. Zhou, C. Ren, J. Wang, and Y. Yang, AIChE J. **59**, 1056 (2013).

[7] J. Ma, D. Liu, and X. Chen, Ind. Eng. Chem. Res. **55**, 624 (2016).

[8] Y. Tsuji, T. Kawaguchi, and T. Tanaka, Powder Technol. **77**, 79 (1993).

[9] P. Darabi, K. Pougatch, M. Salcudean, and D. Grecov, Powder Technol. **214**, 365 (2011).

[10] V. S. Sutkar, N. G. Deen, J. T. Padding, J. a. m. Kuipers, V. Salikov, B. Crüger, S. Antonyuk, and S. Heinrich, AIChE J. **61**, 769 (2015).

[11] T. Mikami, H. Kamiya, and M. Horio, Chem. Eng. Sci. **53**, 1927 (1998).

[12] M. Girardi, S. Radl, and S. Sundaresan, Chem. Eng. Sci. **144**, 224 (2016).

[13] M. Wu, J. G. Khinast, and S. Radl, in (Barcelona, 2014).

CIRCULATION AND SEPARATION OF BINARY SOLIDS IN CONNECTED FLUIDIZED BEDS

[1]Yusumi Nagahashi, [1]Hideki Takeuchi, [1]Tomohiro Kawamura
[1]National Institute of Technology, Kochi College, Nankoku, JAPAN
[2]John R. Grace
[2]University of British Columbia, Vancouver, CANADA
[3]Yuataka Asako
[3]MJIIT, University Technology Malaysia, Kuala Lumpur, MALAYSIA

Introduction

In many fluidization and particle processing systems, mixing and segregation of multi-particle mixtures occur, sometimes being problematic. On the other hand, segregation is useful in some applications for separating one type of particle from other particles. An understanding of the mixing and segregation of particles is therefore of considerable importance in the operation of fluidized bed processes, e.g. in fluidized bed reactors for fast pyrolysis producing bio-oil, both for pyrolysis and char removal.

Although a fluidized bed pyrolyzer has some advantages for processing capacity and fast residence time compared with other types of pyrolysis reactor, it also has disadvantages such as the difficulty of char removal from the bed material heated medium. Because wood-char acts as a vapor cracking catalyst, rapid and effective separation and removal of char particle from the bed material is needed for high quality and favorable oil-yields [(1)], and it is also essential for continuous operation of bio-oil production.

There have been many investigations related to particle mixing and segregation in binary solids fluidized beds, e.g. those by Rowe et. al [(2), (3)]. Mixing and segregation depend on particle density, size and gas flow rate in binary solids mixture gas-fluidized bed. In general, large low-density particles such as wood char gather in the upper region as flotsam in gentle large-density particle beds, whereas high-density small particles of bed material like silica sand behave as jetsam. In this work we utilize Rowe's segregation terminology (char: flotsam, bed material: jetsam).

Working on char removal from the fluidized bed pyrolysis, Berruti's ICFAR group has reported unique results of char removal using movable partition plates which divide the bed into two sections for separation and mixing [(4), (5)]. Their results indicate that separation efficiency depends on biomass feed rate. The driving force transporting particles into the separation section in their configuration could only be hydraulic head difference (bed pressure difference) between two sections. Therefore it needs a high feed rate for faster separation, but excessive feed rates lead to choking when transporting char particles to the separation section.

In this work, particle circulation and separation of binary solids have been investigated in a novel fluidized bed system as a fast pyrolysis reactor for bio-oil production. The system was constructed with two sections connected through a slit installed on the partition wall immersed in the bed. This configuration is similar to ICFAR's facility, but with a unique feature, an inclined gas distributor [(6)] which assists particle transportation into the separate section. One side contains a vigorously fluidized bed with an inclined distributor for biomass pyrolysis, while the other provides gentle fluidization to facilitate particle separation.

Particle exchange through the slit is observed and analyzed by PIV (Particle imaging velocimetry) to explain the driving force transporting particles between sections. Influences of particle properties of modeled chars and slit locations and also gas velocities of each section on separation performance are investigated in the bed with a conventional flat gas distributor and an inclined gas distributor. The system with the inclined distributor is shown to perform well for char-particle separation and removal.

Experimental Apparatus and Procedure

The experimental apparatus is shown in Fig. 1. Experiments were carried out using an acrylic two-dimensional fluidization column of rectangular cross-section 23 mm × 160 mm (D_1 = 56 mm for the particle separation section, D_1 = 100 mm for the mixing section), and height 270 mm. The inclination angle of the inclined distributor was set at 20° with respect to the horizontal. Silica sand of mean diameter 0.11 mm (minimum fluidization velocity U_{mf} = 0.00295 m/s) was the bed material and six types of particles represented char particles. Table 1 shows the properties of the modeled char (d_p: particle diameter, ρ_0: particle density, ρ_e: particle bulk density). Most experiments were carried out with L_0/D = 1.0, where L_0 is the static bed height and D is the column width of the mixing section.

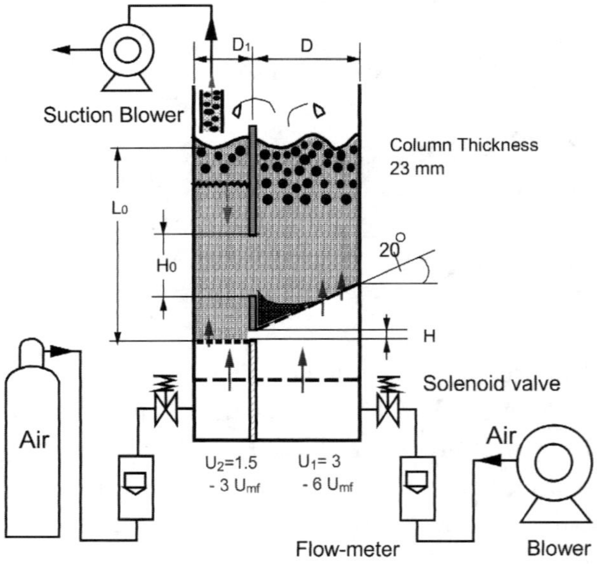

Fig. 1. Experimental Set-up

Table 1. Particle Properties of Modeled Char in Air at Atmospheric Temperature and Pressure

	d_p [mm]	ρ_0 [kg/m^3]	ρ_e [kg/m^3]	U_{mf} [m/s]
Glass Beads	1.0	2450	1410	0.42
Actv. Carbon 1 (sphere)	0.78	850	560	0.19
Actv. Carbon 2 (broken block)	0.80	-	450	0.25
Actv. Carbon 3 (broken block)	1.1	-	450	0.39
Actv. Carbon 4 (broken block)	1.45	-	450	0.48
Polystyrene	3.2	1010	510	0.70

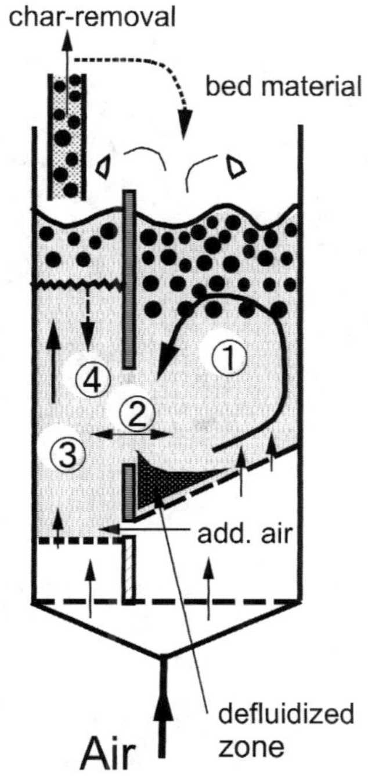

Fig. 2. Pattern and Mechanisms of Particle Circulation and Particle Exchange

Figure 2 shows the pattern and mechanisms of particle circulation and particle exchange. Based on our observations, the following mechanism is presented. (1) Particle circulation occurs along the distributor plate. (2) Particles in the separation section are entrained into the mixing (pyrolysis) section. (3) Mixed char-inert particles transfer into the separation section via the slit. (4) Char is removed as segregated flotsam from the top of the bed, as shown by Rowe [(2), (3)], in the gentle fluidized bed.

Results and Discussion

Observation of Particle Exchange

Figures 3 and 4 indicate particle transportation and return between the two compartments. Enhanced serial pictures (Fig. 3), when model char particles were placed the bed surface initially to the same level as the bed surface, illustrate particle movement due to surface-level-difference (hydraulic head difference). For batch operation, the same phenomena occur, but movement was less severe. Particle exchange occurs by entrainment and local pressure differences, even if the levels are the same in the two compartments, as shown in Fig. 4, which shows particle return from the separation section to the mixing section.

Particle Exchange (transfer) through a Slit

In order to confirm particle exchange through a slit, PIV analysis was carried out at the center of the slit. Figure 5 shows the particle oscillatory movement (velocity vector chart) at the center of slit. In this figure, the positive velocity means that there was transportation from the separation section to the mixing section. The negative value of the velocity means that particles from the mixing section moved to the separation section. The results indicate high-frequency oscillating movement, with a frequency of a few Hz. From these results, it can be confirmed that the particle exchange between two sections is possible, even when they have the same bed level.

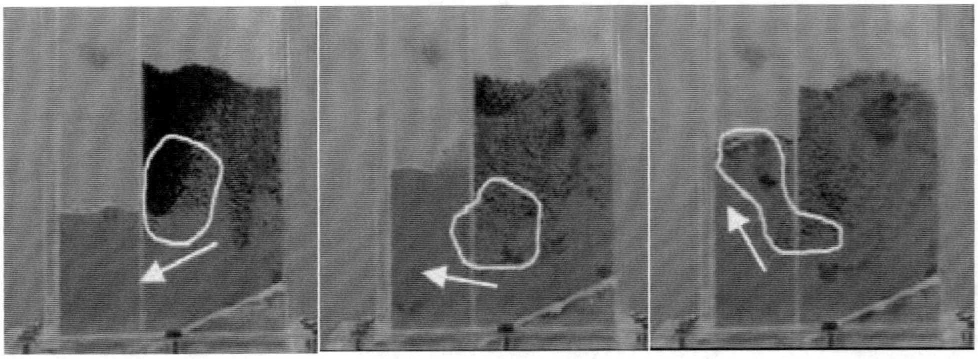

Fig. 3. Particle Transportation to Separation Section

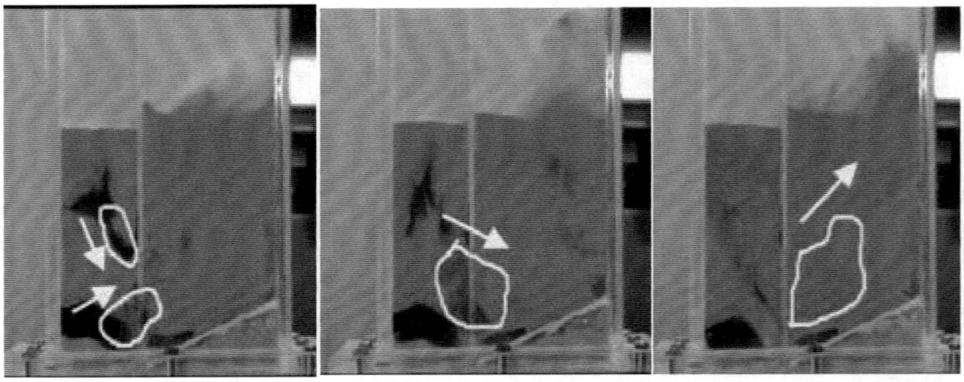

Fig. 4. Particles Return to Mixing (pyrolysis) Section

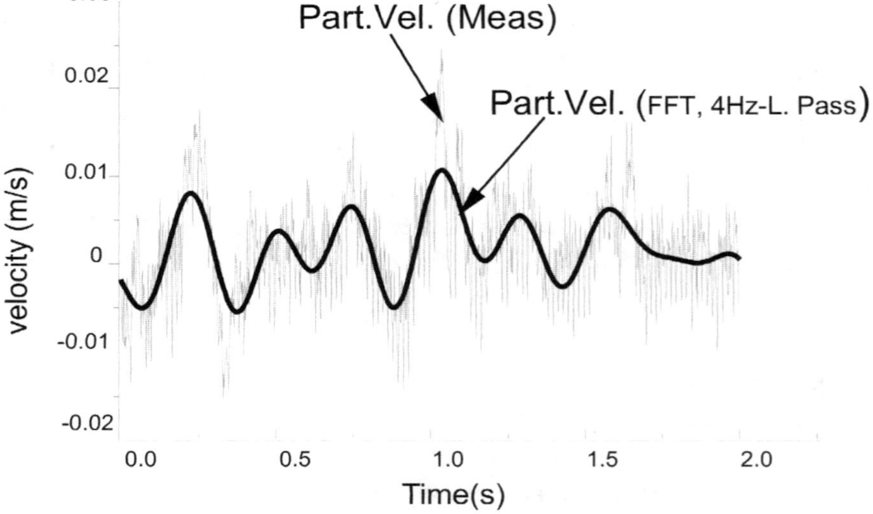

Fig. 5. Particle Oscillatory Movement at Center of Slit (vector chart)
(+) Separation Section → Mixing Section, (−) Separation Sect ← Mixing Sect

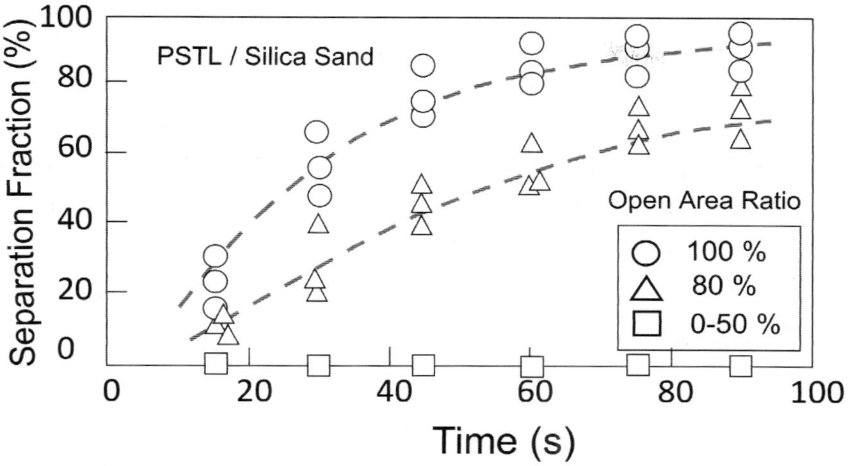

Fig. 6. Effect of Added Air on Separation Fraction
(Air flow rate depends on open ratio: open height 0-3 mm)

Effect of Added-Air on Particle Separation

Fluidization with the inclined distributor was found to become poor because of the high pressure drop at the bottom of the inclined distributor. Earlier work (7) on reducing a defluidized zone with an inclined gas distributor, indicates that particle circulation can be improved by controlling the local gas velocities, with an optimum ratio of surrounding gas velocity to core area gas velocity of $u_{0s}/u_{0c} = 2$. In this work added air is also supplied to improve the mobilization of particles around the defluidized zone as shown in Fig. 2. Figure 6 shows the effect of added air on the separation of polystyrene particles. The particle separation fraction is defined by m_s/m_0, where m_s is the mass of particles suctioned from the separation section and m_0 is the initial mass of the input particles. It was found that the particle separation fraction increased with increasing open area ratio which results in changing the air flow rate. As the particle separation fraction for the open area ratio of 100% (3 mm / 3 mm) was > 90% after 90 s, the 10%-slit height (= 3 mm, slit height is 30 mm as shown in Fig. 1) for the open height gives separation performance comparable to a flat plate distributor. This means about 10% height clearance of the slit height gives sufficient air flow to improve the mobilization around the low end of the inclined distributor plate.

Particle Separation Fraction and Particle Transport Fraction

Figure 7 indicates the particle separation fractions for all five types of particle for modeled char particles (Actv. Carbon 1-4 and Polystyrene in Table 1) obtained with batch operation. The suction and separation were carried out for 15 s, and then measurements were taken, and only bed material was returned to the mixing bed. In general, the separation fraction for binary solid fluidized beds strongly depends on the density ratio of these particles. Larger particles give better separation performance for particles of the same material (density). This tendency is also seen in Fig. 7, with the separation fraction improving with increasing particle diameter. The separation performance was worst for the smallest spherical particles.

Figure 8 shows the particle separation fraction for the flat and inclined distributors. These results are also from repetition of 15 s batch operation. After 90 s of operation, the results for the inclined and flat distributors were almost the same, but the inclined distributor performed better in short-time operation (up to 75 s). Since faster char particle removal resulted in better oil quality and much larger yields, the inclined system is desirable for reactor design. These results suggest that the particle circulation forced by the inclined gas distributor enables rapid transportation into the separation bed.

The particle transport fractions for the flat and inclined distributors are shown in Fig. 9. The particle transport fraction is defined by m_t/m_0, where m_t is the mass of particles transported from the mixing section to the separation section. For convenience and to obtain the results under the same bed level conditions, the particle transport fraction (strongly correlated with separation fraction) was also measured. In these experiments, the mass of char particles in the whole separation section was measured during 60 s of continuous operation. It is seen that a higher slit location gave a larger transport fraction. The inclined distributor system also resulted in a good transport fraction for the whole slit location. The transport fraction (obtained in this continuous operation) was less than the separation fraction obtained in a batch system because of no-head-difference.

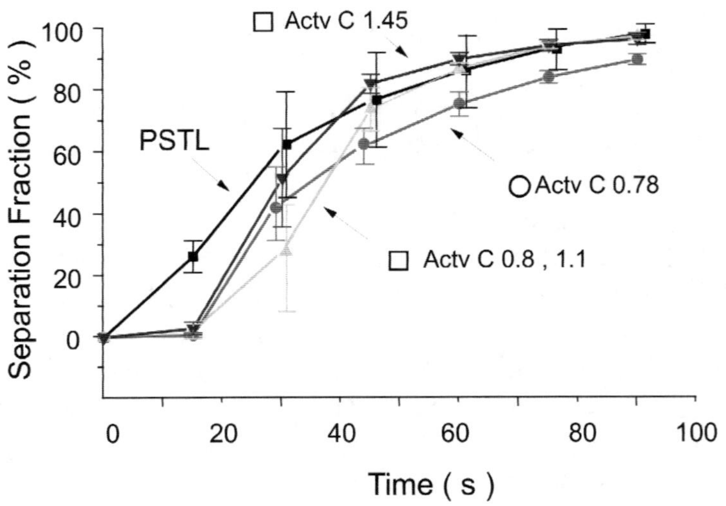

Fig. 7. Particle Separation Fraction (5 char models)

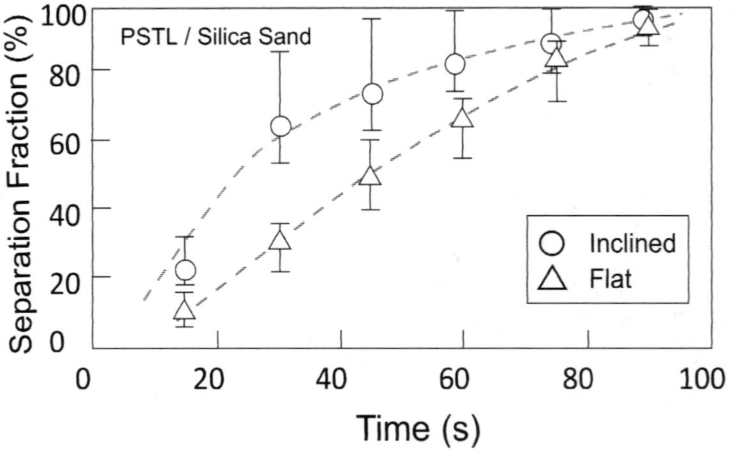

Fig. 8. Particle Separation Fraction for Flat and Inclined Distributors

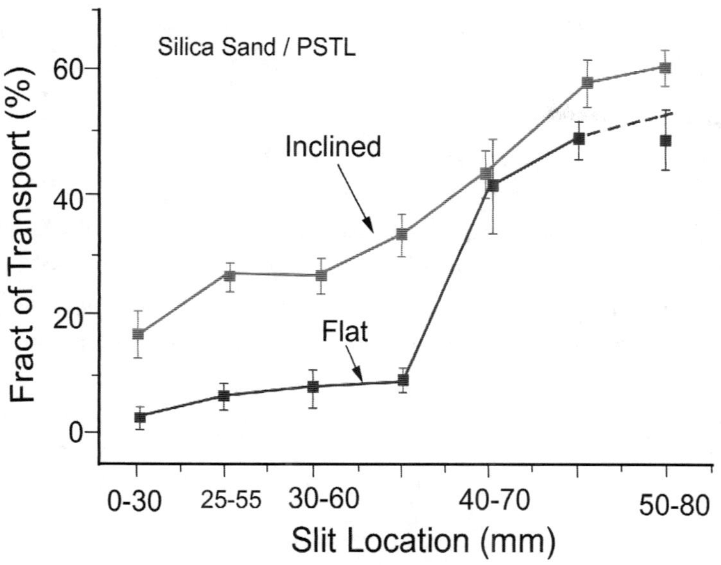

Fig. 9. Particle Transport Fraction for Flat and Inclined Distributor

Conclusions

A fast pyrolysis reactor, with char separation and char removal from bed material, was investigated in a novel connected fluidized bed system under intermittent operation. This investigation led to the following conclusions:

1. Particle exchange through a slit occurs, probably due to pressure difference around the slit zone. It is enhanced by pressure difference when char particles are continuously withdrawn from the bed.
2. The inclined gas distributor performed well in transporting particles from the bed to the other connected bed.
3. The system with the inclined distributor for particle separation and removal worked well, with its hydrodynamics elucidated by PIV analysis around the slit region.

These results are based on intermittent operation. Continuous operation (biomass feed and char removal) will be studied in the near future.

Acknowledgments

The authors are grateful for support from both the Natural Sciences and Engineering Research Council of Canada (RGPIN 7111-11) and JSPS KAKENHI Grant Number JP24560248.

References

1. V. Bridgewater (2003), "Renewable Fuels and Chemicals by Thermal Processing of Biomass", *Chemical Journal*, **91**, pp.87-102.
2. N. Rowe, A.W. Nienow and A.J. Agbim (1972), "The Mechanisms by which Particles Segregate in Gas-Fluidized Beds - Binary Systems of Near Spherical Particles", *Trans. Inst. Chem. Eng.*,**50**, pp.310-323.
3. N. Rowe and A.W. Nienow (1976), "Particle Mixing and Segregation in Gas Fluidized Beds. A Review", *Powder Technol.*, **15**, pp.141-147.

4. Mara, L. Ferrante, F. Berruti and C. Briens (2011), "New System for the Continuous Removal of Biochar from Bubbling Bed Pyrolyzers", *Bioenergy 3*, Lanzarote, Canary Islands, Spain.
5. M. Adegboye (2014), Master Thesis of Eng. Sci., Univ. Western Ontario, (Electronic Thesis and Dissertation Repository).
6. Y. Nagahashi, J.R. Grace, N. Epstein, Y. Asako and A. Yokogawa (2007), "Flow Characteristics of Fully Wetted Binary Solid Mixtures in Gas Fluidized Beds with Inclined Gas Distributors", *Fluidization 12*, ed. X, Bi, F. Berruti and T. Pugsley, ECI, Brooklyn , NY, pp.337-344.
7. T. Kita, H. Kamiya and M. Horio (1998), "Mixing and Segregation of Solid Wastes in a Fluidized Bed with an Inclined Distributor", *Kagaku Kogaku Ronbunshu*, **24-6**, pp. 901-906.

463998 Numerical Study on the Hydrodynamics Behaviour in Bubbling Fluidized Bed for Sewage Sludge Gasification

Abstract:

Sewage sludge is high moisture bio-solid produced during the sewage treatment. Gasifying sewage sludge based on the fluidization technology becomes a promising method of waste management to achieve energy recovery for sustainability and environmental protection. The hydrodynamics behaviour of sewage sludge in the fluidized bed is crucial to obtain an effective gasification process. In this numerical study, sewage sludge and saw dust are adopted as the feedstock for gasification, and air is continuously injected from the bottom of the fluidized bed. The multiphase flows of sewage sludge, saw dust and air are studied and compared under different operating conditions based on the Euler-Euler model. The results are capable of serving as reference for the design and optimisation of the sewage sludge gasification to improve the effectiveness.

460783 Realization, Control and Stability Analysis of Multiple Temperature Zones in Liquid-Containing Gas-Solid Fluidized Bed Reactor

Abstract:

Multiple temperature zones fluidized bed reactors (MTZFBRs) can overcome the limitations of uniform temperature in traditional gas-solid fluidized beds, and thus are widely used in numerous chemical engineering processes, such as fluid catalytic cracking, coal gasification, drying and granulation to mention a few. Investigations of the common scientific problems from the perspective of multiple temperature zones and multiple fluidization patterns coexistence, will greatly enhance the understanding of the fluidization characteristics and control mechanism of MTZFBR. Various mature technologies have shown that spraying liquid into a fluidized bed is one of the most effective means to realize MTZFBR. Therefore, this thesis proposes a novel MTZFBR by means of liquid-spraying within a fluidized bed reactor based on gas-phase fluidized bed polyethylene condensed mode operation process. Due to its optimization and improvement in the fluidized bed reactor process, high performance polyethylene products can be expected, which provides a new research direction and realization for product optimization and reactor process development. However, liquid-containing operation significantly affects hydrodynamic behavior and fluidization stability of fluidized beds. Realization and control of multiple temperature zones in liquid-containing gas-solid fluidized beds on the premise of stable fluidization states, is an extremely challenging research task which possesses great theoretical significance and industrial value. This thesis focuses on realization, control and stability of the MTZFBR and thus the research work has been carried out from the following five aspects:

1. A new research methodology, with combination of time scale analysis and force balance analysis, has been proposed in this thesis and thus the stability analysis model for liquid-bridge induced particle agglomerations is acquired. Firstly, the judgment of the effective particle agglomerations among liquid-coating particles is made based on time scale analysis of liquid-related key processes (such as droplet evaporation, droplet-particle collision and liquid-coating particles collision). To be specific, the particle agglomerations are induced by the liquid bridge when the droplet evaporation time scale is greater than the collision time scale among liquid-coating particles. Furthermore, the accurate judgment of fluidization instability caused by particle agglomerations is made according to the proposed force balance analysis and important criterion. Finally, based on results of stability analysis model, the spectrums of gas-liquid-solid (G-L-S) three phases fluidization patterns under different liquid-solid contact states are obtained as well as the particle agglomeration behaviors and the dominant mechanisms during the liquid-containing fluidization processes.
2. Multi-scale characterization method has been established with multiple measurement techniques including acoustic emission, pressure fluctuation and camera, which has been used for the first time to reveal the variation law of particle, bubble behavior and overall fluidization states simultaneously in the liquid-containing gas-solid fluidized bed reactor. Results demonstrate that when particle size reflected by acoustic signal varies significantly, the bubble scale (from pressure fluctuation) and overall fluidization state scale (from camera) behaviors show regular variation trends, such as formation of bubble shrinkage and gas channeling. Moreover, through Hurst and V-statistics analysis of the acoustic signal, the cyclic behavior characteristics induced by increased liquid in the fluidized bed are distinguished for the first time. The frequency of the characteristic cycle component is 400 Hz and thus the cycle time is 2.5 ms, which indicates the newly formed characteristic behavior is correlated to motion behavior of the meso-scale particle agglomeration. There are verification and complementarity among three kinds of measurements results, and all the three techniques can reflect the unstable fluidization states such as agglomerations and gas channeling during the liquid addition process.
3. With a self-designed hot mode fluidized bed apparatus, systematic experiments have been performed to realize MTZFBR. Results demonstrate that compared with the bottom liquid-spraying scheme, the upper liquid-spraying scheme shows larger temperature differences between upper zone and bottom zone, as well as higher fluidization stability, and thus the upper liquid-spraying scheme is preferable in the realization of MTZFBR. Based on upper liquid-spraying scheme, liquid flow-rate, bed height and gas inlet temperature are found to be the most significant operating parameters to affect multiple temperature zones. Moreover, studies have shown that liquid evaporation and liquid bridge are the two competitive factors during liquid-containing fluidization and thus the relative intensity of these two factors will significantly affect the temperature differences and fluidization stability. Furthermore, based on the stability analysis method proposed in the Chapter 4, two different three phases fluidization patterns are found to coexist in the stable MTZFBR. One is G-L-S three phases fluidization pattern with dynamic particle agglomerations mechanism dominated by a balanced action between liquid evaporation and liquid bridge, and the other is G-L-S three phases fluidization pattern with agglomeration breakup mechanism dominated by liquid evaporation action.
4. In the MTZFBR, the action law of particle, bubble and particle circulation pattern induced by increased liquid flow-rate have been revealed. The study shows that both increases in liquid flow-rate and decreases in gas velocity cause lower gas-liquid relative action intensity, and thus the liquid-containing fluidization process dominant mechanism shifts from gas flow (or liquid evaporation) controlling mechanism to liquid bridge controlling mechanism. As a result, the particle circulation pattern changes from one mode (particles move upward in the center and particles move downward in the wall) to another mode (downward motion of agglomerations has been enhanced significantly), meanwhile the particle and bubble motion behaviors also change with regular trends. Besides, the minimum fluidization velocity of particles is found to increase with liquid content increase. Based on temperature profile measurements, one feasible method is proposed to reflect particle circulation modes in the fluidized beds, which helps to provide a new means to study particle circulation modes.
5. For industrial multiple temperature zones fluidized bed polymerization reactor, based on the concept of G-L-S and G-S fluidization patterns coexistence, an improved emulsion-bubble two phase fluidized bed model has been proposed with help of the liquid evaporation model. The model can simulate accurately the temperature profile of industrial reactor, which proved that the industrial MTZFBR can be studied by means of two fluidization patterns coexistence. Finally, the process simulation results show that high performance products can be realized by using the reactor with composite fluidization patterns.

457454 Graphene Nanoribbons As Conductive Pathways in Directly Deposited Silicon Nanofiber Anodes for High Performance Lithium-Ion Batteries

Abstract:

Silicon is the best candidate for the next generation of Li-ion battery anodes due to its high specific capacity of 4200 mAh/g .However, it expands to 300% of its original size upon insertion of lithium ions[1], which results in crack formation and pulverization of active material with cycling and thus capacity fading. Crack formation also results in constant solid electrolyte interface formation on the surface of the newly exposed silicon, which brings about unstable performance. Several strategies such as use of nanoparticles[2] and composites[3] have been employed by research groups to overcome aforementioned issues. However, all the modifications to silicon has to be done in an economical manner for the product to become an industrial realization.

In the present work we are engineering the nanostructure of a silicon composite with graphene nanoribbons (GNRs) and polyvinyl alcohol (PVA) to create a reliable anode for lithium ion batteries. Silicon nanoparticles and GNRs are dispersed along PVA nanofibers via a water-based electrospinning process to directly form functional fiber mats on top of the copper current collector. This method is compatible with a role to role production, and bypasses several long-winded steps in the conventional paste-based electrode making. Physical and chemical properties of the nanoparticles and their dispersion affect the fiber morphology and ultimately electrochemical performance of the composite material. Graphene nanoribbons, made by unzipping carbon nanotubes, offer unique electrochemical and structural properties when inside the composite fibers. It was observed that GNRs outperform their precursor CNTs in terms of overall capacity and retention. The Silicon-GNR-PVA fiber anode exhibited an outstanding 2000 mAh/g at a rate of C/15 and 1200 at C/5 over 200 cycles. Inclusion of GNRs improved the structural stability and conductivity of the fibers, while protection of nanoparticles from the electrolyte offered by the polymer resulted in stable cycling of the directly deposited electrodes.

[1] C.-M. Wang, X. Li, Z. Wang, W. Xu, J. Liu, F. Gao, L. Kovarik, J.-G. Zhang, J. Howe, D. J. Burton, Z. Liu, X. Xiao, S. Thevuthasan, D. R. Baer, *Nano Lett.* **2012**, *12*, 1624.

[2] M. M., Ryu I, Wu H, Liu N, Hu L, Nix WD, Cui Y Yao Y, *Nano Lett.* **2011**, *11*, 2949.

[3] Y.-S. Hu, R. Demir-Cakan, M.-M. Titirici, J.-O. Müller, R. Schlögl, M. Antonietti, J. Maier, *Angew. Chem. Int. Ed.* **2008**, *47*, 1645.

467753 Si-Based Nanofiber Anodes for Li-Ion Batteries Prepared Using Particle/Polymer Electrospinning

Abstract:

The demand for rechargeable batteries with high energy and power densities has never been greater. The functionality of many portable electronic devices is limited by battery size and lifetime. Similarly, electric vehicle propulsion systems require significant improvements to satisfy consumer expectations for long drive distances and short recharge times. To meet the demands of energy-hungry consumers, new battery electrodes with high capacities and long cycle life at fast charging rates must be developed.[1-2] These requirements have motivated research efforts to develop next-generation Li-ion battery materials, such as nano-structured silicon anodes. There is also a need for new electrode architectures to replace conventional slurry cast materials. One approach for an electrode design is the use electrospun electrodes containing engineered void spaces to enhance Li^+ transport rates for high areal capacities at fast charge/discharge rates.

Silicon has received considerable attention as an active material for Li-ion battery anodes due to its high theoretical capacity (3,579 mAh/g vs. 372 mAh/g for graphite). However, silicon based anodes typically suffer from rapid capacity fade due to particle cracking and electronic isolation caused by large volumetric changes between the lithiated and delithiated states. While particle fracture can be eliminated by using nanosized Si, electronic isolation remains problematic. Another significant obstacle for commercialization of Si in Li-ion batteries is the irreversible consumption of Li during formation of the solid-electrolyte interphase (SEI) layer.

The present work is the latest installment by Pintauro and co-workers on the use of particle/polymer electrospun nanofiber mats for fuel cell and Li-ion battery electrodes.[1-5] Such electrodes are a viable and attractive alternative to electrospun mats prepared by carbonizing polymer fiber precursors at high temperature. In particle/polymer electrospinning, there is no fiber pyrolysis; the electrospinning and post-processing are performed at room temperature which simplifies the process and preserves the beneficial characteristics of the polymer binder. Furthermore, the structure of the fibers can be easily altered for improved battery performance (e.g., by using core-shell or hollow-bore fibers or fibers with a high internal porosity). Due to these characteristics, high gravimetric, areal, and volumetric storage capacities can be achieved through the use of thick, densely packed particle/polymer nanofiber mats.

In this presentation, recent results will be presented on the fabrication and use of electrospun particle/polymer nanofiber anodes containing Si nanoparticles, carbon powder, and poly(acrylic acid) binder (Si/C/PAA). Charge/discharge properties of the nanofiber electrodes will be presented, and electrode capacities for different fiber mat thicknesses and mat porosities will be discussed. Notably Si/C/PAA nanofiber anodes have been prepared with high gravimetric, areal, and volumetric capacities of 1,100 mAh/g, 3.1 mAh/cm^2, and 560 mAh/cm^3, respectively, which were stable for 50 cycles at 0.1C. These results demonstrate that proper micron-scale design of electrodes can enhance Li-ion battery performance.

References

1. E. C. Self, R. Wycisk and P. N. Pintauro, *Journal of Power Sources*, **282**, 187 (2015).

2. E. C. Self, E. C. McRen, and P. N. Pintauro, *ChemSusChem*, **9**, 208 (2016).

3. W. Zhang and P. N. Pintauro, *ChemSusChem*, **4**, 1753 (2011).

4. M. Brodt, R. Wycisk and P. N. Pintauro, *Journal of The Electrochemical Society*, **160**, F744 (2013).

5. M. Brodt, T. Han, N. Dale, E. Niangar, R. Wycisk and P. Pintauro, *Journal of The Electrochemical Society*, **162**, F84 (2015).

463852 *In Situ* Activation of Nitrogen-Doped Graphene-Based Materials Anchored on Graphite Foam for High-Performance Energy Storage

Abstract:

Contact interface between the active materials and the current collector is essential for electron transfer rate and the mechanical properties of an electrode. With a traditional drop-casting electrode preparation method, the nano-sized active materials should mix with the nonactive and nonconductive binder, thus inevitably affect the total electrode resistance. Activation of graphene-based materials (e.g. graphene, carbon nanotube) by KOH can rebuild the crystalline structure to obtain continuous interface contact to facilitate electron transfer, while the molten KOH can act as the filler to enlarge the interlayer space to increase the electrolyte ion transfer rate.

Here, three-dimensional free-standing nitrogen-doped porous graphene-based materials/graphite foam composites (aNG/GF) were fabricated *via in situ* activation of nitrogen-doped graphene-based materials on highly conductive graphite foam (GF). During activation, the molten KOH reacts with the carbon to generate nano-holes as well as to rearrange the carbon, yielding a distribution of meso- and micro-pore channels and increases the specific surface area. After *in situ* activation, intimate 'sheet contact' was observed between the graphene sheets and the GF. The 'sheet contact' produced by *in situ* activation is found to be superior to the 'point contact' obtained by the traditional drop-casting method, and facilitates electron transfer. Due to the intimate contact as well as the abundant nano-holes and channels, the aNG/GF electrodes can exhibit high electron and electrolyte ion transfer rate, which is of great important for high-performance energy storage devices. The aNG/GF electrodes can deliver high gravimetric capacity and volumetric capacity with respect to the whole electrode mass and volume (including the active materials and the GF current collector). Furthermore, the tailored contact interface and the unique micro-structure could guarantee a high rate performance of the electrode even under ultrahigh current density, indicating its promising use as

a high-rate LIB anode.

470624 Solution Combustion Synthesis for High Performance $ZnCo_2O_4$ Anode in Lithium-Ion Batteries

Abstract:

Nanostructured, tailored $ZnCo_2O_4$ is synthesized by solution combustion synthesis using two fuels, glycine and citric acid, to explore how reaction conditions impact electrochemical performance as a Li-ion anode. Owing to the unique combustion reaction mechanism, the novel strategy of using mixed fuels and ammonium nitrate creates anode material with improved cycling stability and capacity. Characterization shows a mesoporous spinel product, composed of nanoparticles assembled into micron-sized secondary structures, with properties varied by the fuel ratio and type. A stable capacity of 1,200 mAh g^{-1} is achieved at C/10 and 1000 mAh g^{-1} at 1C, showing high energy and power capabilities as compared to the bulk material. Based on an understanding of the reaction mechanisms and influence of reaction parameters, solution combustion synthesis is a powerful and flexible technique to produce superior electrode materials.

473388 a General and Mild Approach to Controllable Preparation of Manganese-Based Micro/Nanostructured Bars for High Performance Lithium-Ion Batteries

Abstract:

To meet the requirements for large-scale applications in electric vehicles, hybrid electric vehicles and stationary energy storage, growing interest has been directed at the economical, sustainable and environmental friendly methods to produce electrode materials for rechargeable lithium ion batteries (LIBs) with high energy and power densities.

Recently, manganese-based electrode materials, such as layered LiNixCo1-x-yMnyO2, spinel LiNi0.5Mn1.5O4 and layered Li-rich xLi2MnO3â€¢(1-x)LiNi1/3Mn1/3Co1/3O2 (0 â‰¤ x â‰¤ 1) have received a flurry of research activities due to their advantages of high energy and power densities, low cost and high cycling performance.

In respect to thermal stability, larger particles with a higher tap-density would be less reactive in a highly oxidized state. On the other hand, an electrode consisting of uniform particles with an appropriate size could ensure a homogeneous intercalation reversibility from particle to particle over repeated cycles. Therefore, the control of particle size and morphology for tailoring lithium-ion battery electrode materials has been paid more and more attention in recent years. Till now, it has remained a challenge to develop a general strategy that allows the large-scale preparation of a wide array of 1D micro/nanostructured manganese-based electrode materials with a controllable phase, composition and shape, as well as a narrow size distribution.

Herein we have established a Chimie Douce approach --- by mixed solvent co-precipitation --- to selectively prepare an extensive series of 1D micro/nanostructured manganese-based electrode materials, both cathode and anode for LIBs with high energy and power densities, attesting to its generality and effectiveness. Based on the kinetic elaboration of crystallite nucleation and growth through mixed solvent mediation, the new route is easily controllable, repeatable, and generalizable to the fabrication of a large series of some highly promising manganese-based electrode materials with tunable 1D micro/nanostructures for enhanced performances in lithium ion batteries. As a significant difference from the conventional methods, the present method is conducted in ethanol/water mixed solutions at room temperature, without having recourse to long-chain surfactants or hydrothermal condition. In addition, it can be employed to produce different aspect ratios of microbars with mixed transition metal oxides by tuning the ethanol/water volume ratio in the mixed solvent. It is anticipated that the mixed solvent mediated co-precipitation strategy will open opportunities to heuristically and systematically design mixed transition metal oxides with controllable phase and composition, including not only manganese-based but also other metal-based materials, with desirable micro-/nano-structures for developing high performance lithium-ion batteries.

460764 Advanced Energy Materials Derived from Bijels

Abstract:

Pressing energy demands for national security, distributed power, and portable technologies require the development of next-generation batteries and supercapacitors that can simultaneously deliver high power and energy densities. In order to achieve this goal, the kinetics of ion and electron transport within the electrodes must be enhanced while maintaining a large capacity for energy storage. Therefore, the idealized electrode microstructure has been envisioned as a three-dimensional, co-continuous arrangement of percolating passages for efficient ion and electron transport that would also enable a large volumetric density of active materials. We report a novel method to produce this distinctive microstructure through the use of soft matter templates derived from bicontinuous interfacially jammed emulsion gels (bijels), which are a new class of soft materials with a co-continuous microstructure where two interpenetrating fluid domains are kinetically arrested through jamming of particles at the fluid-fluid interface during spinodal decomposition. Bijels can be converted to electrolytically active composites with co-continuous internal microstructure through simple chemical post processing steps. In this talk, we will present microstructured composite electrodes that have been synthesized by this technique and demonstrate how their electrochemical characteristics can be tuned over a wide range, allowing for concurrent delivery of high energy and power. The broader applications of this new class of soft materials in electrochemical energy conversion and storage, sensing, catalysis, and tissue engineering will also be discussed.

457746 Sulfur Self-Doped Micro/Mesoporous Carbon Derived from Lignin and Its Application on Supercapacitor and Oxygen Reduction Reaction

Abstract:

The growing global concerns about energy needs, fossil fuels consumption and the related environmental issues have motived scientists to find new, green and sustainable energy resources and technologies. In this work, byproduct lignin was successfully converted into self-sulfur-doped graphitic carbons via in situ hydrothermal carbonization (HTC) by following the post-annealing treatment at elevated temperature. The surface, porosity, textural, structural and morphological properties of synthesized materials were investigated by SEM, XPS, Raman, FTIR and surface area analyzer. Electrochemical properties of the prepared materials were analyzed by cyclic voltammetry (CV), charge-discharge (CD), electrochemical impedance spectroscopy (EIS), linear sweep voltammetry (LSE) in a KOH electrolyte solution. Up to 3.5 wt. % sulfur intercalating on carbon molecules was successfully achieved. The resulting materials electrode displays high surface area (up to 550 m^2 g^{-1}) with mesoporous structure and partially graphitic/amorphous carbon structure. The obtained original electrode materials result in high electrochemical activity for supercapacitor applications. The SGC-850 electrode exhibited a specific capacitance value of 225 g^{-1} at a scan rate of 1 A g^{-1}. The SGC-850 electrode revealed a very consistent response over 10000 cycles at a harsh condition of 5 A g^{-1} current density. In addition to supercapacitor application, the obtained material was tested for oxygen reduction reaction (ORR) application. The ORR CV curve of the SGC-850 presents well-defined cathodic peak in the O_2 electrolyte. The SGC-850 goes under 4 electron pathway for the O_2 reduction. To sum up, metal-free SGC-850 is promising candidate electrode materials for supercapacitors and ORR applications.

468442 The U.S. Department of Energy's R&D Program for Carbon Use and Reuse

Abstract:

The U.S. Department of Energy Office of Fossil Energyâ€™s National Energy Technology Laboratory has an active carbon capture and storage (CCS) program. Â The goal of the Carbon Use and Reuse Focus Area is to identify and develop a suite of technologies to beneficially use CO_2 to produce useful products that can generate revenue to offset capture costs associated with CCS implementation. Identification of candidate chemistries and economically feasible approaches that utilize large amounts of CO_2 as a feedstock for high-demand product is very challenging while also ensuring that the overall process has a smaller greenhouse gas footprint that traditional pathways.

462065 Economic Production of Syngas from CO_2 and H_2O: New Routes to Synthetic Fuels

Abstract:

Economic production of syngas from CO_2 and H_2O could produce new routes to synthetic fuels. In this paper we will describe recent work on the electrolysis of CO_2 in a 5 cm^2 alkaline electrolyzer using standard Fuel Cell Technologies fuel cell hardware. Previous work had demonstrated CO_2 conversion with steady state currents of 50 mA/cm^2 or less at 3 V. In this paper we find that by optimizing the composition of the membrane and ionomer we can raise the current to 280 mA/cm^2 at 3V and 600 mA/cm^2 at 3.25-3.35 V.

Table 1 shows how the performance of the electrolyzer varies with membrane composition for several different membranes using bare silver cathode catalysts, ruthenium anode catalysts and a variety of membranes. In all cases the measurements were done by sandwiching the membrane between two pieces of carbon cloth that were coated with membrane, mounting the resultant MEA in 5 cm^2 fuel cell hardware, supplying humidified CO_2 to the cathode, applying 3 V, waiting 20 minutes to 1 hr for the current to stabilize and recording the current.

Membrane	Faradaic Efficiency	Current at 3V (mA/cm^2)	Current for carbon containing products (mA/cm^2)
Nafion 117	0%	72	0
Neosepta	34%	24	8
Acid doped CMI-7000	0.02%	35	0.007
Alkali doped AMI-7001	25%	10	2.5
SPEEK	2.5%	5	0.13
Alkali doped PVA	52%	15	7.5
Alkali doped PEI/PVA	16%	10	1.6
H_3PO_4 doped PBI	14.7%	8	1.2
Sustainion™ X1	95%	80	75
Sustainion™ X2	93%	80	72
Sustainion™ X3	83%	25	20.8

Notice that the faradaic efficiency for the conversion of CO_2 to CO varies from 0 to 95%. While the current varies from 5 to 80 mA/cm^2. Sustainion™ membranes give higher currents and faradaic efficiency than the others. These results demonstrate that the membrane composition makes a significant difference to the electrolyzer performance.

We have also determined shows how the polarization curve of the cell varies as we add different amounts of Sustainion™-X5 to the cathode layer in a cell with a Sustainion™-X4 membrane. Notice that one can obtain currents of 1.4 amps (280 mA/cm^2) at 3 volts with a Sustainion™-X4 membrane and and Sustainion™-X5 ionomer.

Steady state runs were done at 200 and 600 mA/cm^2. The cell is able to sustain 200 mA/cm^2 a 2.9 V and 600 mA/cm2 at between 3.25 and 3.35 V. By comparison, the closest competitor needed almost 7 V to obtain the same current.

These results demonstrate that one can use electrolyzers with Sustanion™ membranes to convert CO to CO_2 at industrially important rates

Acknowledgement

Parts of this work were supported by ARPA-E under contract DE-AR0000345 and by 3M. The opinion are those of the authors and may not reflect the opinions of ARPA-E.

472626 Electrochemical Carbon Dioxide Reduction As an AlternativeÂ Source ofÂ Fuels and Chemicals

Abstract:

Cost-effective industrial CO_2 recycling using renewable energy sources (ECO2R) could form the basis of an artificial carbon cycle that replaces a wide range of products that are currently derived from fossil resources. ECO2R combines CO_2, water, and electricity, and converts them into useful products using engineered metal catalysts. At full scale, this technology could eliminate our dependence on fossil resources by providing an alternative source of carbon-based compounds for fuels and commodity chemicals. However, commercial fuel and chemical production via ECO2R is challenging, because the current state of the technology is not cost-effective enough to compete with conventionally manufactured fuels and chemicals already on the market.

The key cost-drivers of ECO2R are the energy efficiency, product selectivity, and current density. Based on these metrics, we have developed a model for the cost of ECO2R-derived compounds. We will use the model to compare the cost of ECO2R derived compounds to fuels and chemicals derived from traditional sources. In doing so, we will identify the technical barriers that need to be overcome to reduce the cost, including enhanced electrocatalysis [1-3] and achieving high current densities needed for a cost-competitive process because of limited carbon dioxide solubility.

Opus 12 has also developed a prototype ECO2R reactor. The Opus 12â€™s prototype contains improved catalysts and a reactor design with high energy efficiency, high product selectivity and high current densities. We will present our technical progress in terms of these key performance metrics. Using these metrics, we can estimate the current cost of ECO2R and present a roadmap of improvements needed to further reduce costs to compete with traditional chemicals and fuels derived from fossil resources.

[1] K.P. Kuhl, E.R. Cave, D.N. Abram, T.F. Jaramillo, New insights into the electrochemical reduction of carbon dioxide on metallic copper surfaces, Energy Env. Sci., 5 (2012) 7050-7059.

[2] K.P. Kuhl, T. Hatsukade, E.R. Cave, D.N. Abram, J. Kibsgaard, T.F. Jaramillo, Electrocatalytic Conversion of Carbon Dioxide to Methane and Methanol on Transition Metal Surfaces, Journal of the American Chemical Society, 136 (2014) 14107-14113.

[3] F.S. Roberts, K.P. Kuhl, A. Nilsson, High Selectivity for Ethylene from Carbon Dioxide Reduction over Copper Nanocube Electrocatalysts, Angewandte Chemie International Edition, 54 (2015) 5179-5182.

461439 Enhanced Asphaltene Precipitation in Paraffinic Froth Treatment Using Carbon Dioxide

Abstract:

Paraffinic froth treatment is an essential step in oil sand surface mining operations for the removal of water and mineral solids from bitumen froth. In this process some of the asphaltenes present in the bitumen precipitate out and form aggregates with emulsified water droplets and dispersed mineral solids, accelerating the separation of solids and water from the bitumen and producing a clean bitumen product. In order to achieve the required asphaltene separation, a large volume of paraffinic solvent such as pentane or hexane is added to bitumen froth.

It has been reported that CO_2 dissolved in heavy oil results in asphaltene precipitation during enhanced oil recovery. Because asphaltene precipitation is considered a key step of paraffinic froth treatment, we conducted a lab-scale investigation to explore whether adding CO_2 to paraffinic froth treatment solvent could enhance asphaltene precipitation.

Froth treatment tests were conducted using both supercritical fluid CO_2 and liquid CO_2 mixed with pentane-diluted bitumen froth in an autoclave. In these tests both pentane and CO_2 concentrations were varied. The results indicated that when mixing CO_2 with n-pentane-diluted bitumen froth significant asphaltene precipitation was observed. At a pentane-to-bitumen ratio of 1.0 and lower, which is below the onset asphaltene precipitation point with solvent alone, adding CO_2 to the diluted bitumen produced significant asphaltene precipitation and resulted in a much reduced asphaltene content in the bitumen product. The combined solvent-to-bitumen ratio (CO_2 plus solvent) was much lower than that using n-pentane alone when precipitating the same amount of asphaltenes. For example, at the combined solvent-to-bitumen ratio of about 1.4, the asphaltene content in the product bitumen decreased to about 5-7 wt% from an initial 17 wt%. To reach these levels of asphaltene content using pentane alone, pentane-to-bitumen ratios of 2.2-2.5 are needed. Low water content (below 0.3 wt%) in the diluted bitumen phase was also observed. These results suggest that CO_2, when dissolved into diluted bitumen, may enhance the paraffinic froth treatment process.

450850 CO_2 Conversion to Novel Solid Materials

Abstract:

In the recent years, a great effort was made for CO_2 conversion by my group, leading to novel approaches. In this presentation, advances in CO_2 conversion will be briefly reviewed, our important results for CO_2 conversion to carbon nitrides and 3D graphene will be discussed, and the energy application of those solid materials from CO_2 will be demonstrated.

460814 Characterization of TFA Regenerated Cellulose with a Two-Step Treatment to Understand the Impact of Cellulose Structure on Hydrolysis

Abstract:

Concentrated trifluoroacetic acid (TFA) is known to depolymerize cellulose. In this paper we report the effect of temperature and treatment time on physiochemical and morphological changes to cellulose. We present x-ray diffraction, SEM, degree of polymerization, and FTIR characterizations and compare them to acid-catalyzed hydrolysis of the decrystalized cellulose and conversion of the resultant glucose to HMF and levulinic acid by maleic acid-$AlCl_3$ catalysis. Treatment with concentrated TFA for 2 hrs at 0°C resulted in a 28% reduction in crystallinity index (CrI) of the cellulose. Additional treatment for 5 hrs at 50°C resulted in a CrI of 0. Both changes of crystallinity index (CrI) resulted in about 40% improvement in overall yield of HMF and levulinic acid. Enzymatic hydrolysis study shows that TFA treatment for 2 hrs at 0°C results a 52% glucose conversion yield, which had higher conversion yield than the additional treatment for 5 hrs at 50°C with 45% glucose conversion yield. Results showed the additional reduction in CrI through heating step does not further improve yields from enzymatic hydrolysis.

462550 The Role of Atmosphere, Solvents and Acids on the Production of 5-Hydroxymethylfurfural from Biomass Derived Fructose

Abstract:

The Role of Atmosphere, Solvents and Acids on the Production of 5-Hydroxymethylfurfural from Biomass Derived Fructose

As global energy demands and atmospheric CO_2 levels continue to rise, sustainable green energy sources will become necessary solutions. One such alternative energy source is corn Stover, a domestically produced, cellulosic biomass which is low enough in cost and of sufficient abundance for production of liquid fuels on a large scale to be competitive with petroleum. Sugars, such as fructose, can be derived from biomass through pretreatment of biomass to glucan enriched solids, these solids can undergo saccharification and isomerization to produce fructose-enriched streams. This fructose stream can be processed to produce platform chemical feedstocks, like 5-hydroxymethylfurfural (5-HMF). 5-HMF is known as a platform chemical, as it is easily converted to biofuels or high valued commodity chemicals.

Past studies have demonstrated that it is difficult to improve the selective and/or conversion of biomass sugars to 5-HMF in aqueous solutions. The majority of these studies rely on increasing the catalyst loadings or using high-temperature processes. Furthermore, numerous studies have explored the use of organic solvents to improve the selective conversion of biomass sugars to furanic compounds. However, these studies are often times inconsistent and have not conclusively demonstrated the combined effects of acids, solvents, and atmosphere on the conversion of fructose to 5-HMF.

Presently, the highest reported yields of HMF obtained, arises from acid-catalyzed dehydration of fructose in dimethyl sulfoxide (DMSO). A combination of Brønsted acids and DMSO has been proposed to provide an effective catalytic

pathway and intermediate stabilization that maximizes HMF yields. However, significant contradictions exist in literature regarding the roles of DMSO, O_2, and H_2O in the mechanism of fructose conversion. It is also unclear whether the fructose conversion mechanisms are radically different between: 1) various acids in DMSO or 2) different aprotic solvents vs. DMSO. To address these questions we utilized a 96 well plate high-throughput batch reactor system loaded and sealed under inert or atmospheric conditions. The reactor was loaded into a high pressure vessel and a Fulton electric steam boiler was utilized to rapidly heat the vessel to systematically and accurately explore the influence of pH, O_2, H_2O, solvent, time and temperature on fructose conversion and 5-HMF selectivity.

We report three primary results: 1) DMSO decomposes in O_2 at elevated temperatures to produce Brønsted acids that drive fructose conversion. (2) In O_2 free environments, increasing H_2O content does not negatively influence the HMF yield. 3) Lastly, 5-HMF production follows an almost identical, negative linear relationship between conversion and selectivity, which universally captures the behavior regardless of solvent, reaction time, reaction temperature and acid loading. These systematic studies have allowed us to identify realistic conditions where fructose conversion to 5-HMF can be executed with high yields, >80% at reasonable pH (~2), water contents, temperatures, and times. Furthermore, we utilize these findings to propose a universal mechanism for the fructose conversion pathways driven by homogeneous Brønsted acids and present insights into reactor design and conditions to optimize 5-HMF yields at minimal cost.

469433 Lignin Depolymerization into Aromatic Monomers over Acidic Mesoporous Silicates

Abstract:

Lignin Depolymerization into Aromatic Monomers over Acidic Mesoporous Silicates

Kakasaheb Y. Nandiwale[a,b], Andrew M. Danby[a], Anand Ramanathan[a], Raghunath V.Chaudhari[a,b], and Bala Subramaniam[a, b]

[a]Center for Environmentally Beneficial Catalysis, The University of Kansas, Lawrence, KS 66047, USA.

[b]Department of Chemical and Petroleum Engineering, The University of Kansas, Lawrence, KS 66045, USA.

Keywords: Biomass, Lignin depolymerization, mesoporous silicates

We report on novel applications of Zr-incorporated mesoporous silicates for the one-pot conversion of lignin into value added aromatic monomers. Quantitative analysis of the depolymerized products along with mass balance based on the overall solid and liquid products are reported.

Cubic mesoporous silicates, KIT-6 and KIT-5, modified with different Zr loadings (1-6 wt%) display dominant Lewis acidity. These materials display superior lignin depolymerization activity compared to commercial zeolites that are typically Brønsted acids. Briefly, batch depolymerization of dealkaline lignin (TCI chemicals) was carried out in a 300 mL Parr reactor at 250°C, 100 psi N_2, 1000 RPM for 30 min in a methanol + water solvent mixture. The performances of 12 commercially available zeolite catalysts were evaluated. H-ZSM-5 (Si/Al = 11.5) was found to be the best catalyst among the commercial zeolites evaluated, providing 92% lignin conversion. Approximately 54 wt.% of the products are soluble in THF while the lignin substrate is insoluble in THF. The identified aromatic monomers in the THF-soluble fraction represent 4.4 wt. % of the initial lignin. These results with H-ZSM-5 are consistent with those reported in the literature. Commercial H-ZSM-5 catalyst was therefore selected for further comparative study with Zr-KIT-5 and Zr-KIT-6 catalysts. Zr-KIT-5(Si/Zr = 20) provided the best catalytic performance with 97% lignin conversion, yielding 65% of THF soluble products containing 5.5% identified monomers. Gel permeation chromatography (GPC) analysis of depolymerized product from both runs indicates a decrease in the molar mass when compared to the lignin substrate, clearly implying lignin deconstruction. Approximately 25 monomers were identified (by GC-FID, GC-MS) in the THF-soluble product mixture and 15 monomers were identified using external standards. Most of the identified products are typically expected from lignin depolymerization as reported in the literature. These results, along with fundamental insights into the correlation between catalyst activity and total acidity of catalyst are presented.

472651 Characteristics and Activity of Steamed Zeolites for Butanol Upgrading to Higher Olefins

Abstract:

The increasing prevalence of renewable alcohols derived from biomass feedstocks make them an ideal platform for developing into transportation fuel. Butanol is a promising bioproduct that can be generated either from direct fermentation or ethanol condensation chemistry, and can be further upgraded to fuels and chemicals in a selective manner. Development in the methanol-to-gasoline (MTG) process has shown zeolite catalysts to be particularly effective for the production of higher olefins, capable of catalyzing both the dehydration and oligomerization reactions in a single step and directly converting butanol to jet or diesel range hydrocarbons. However, changes to the catalyst in the initial startup phase and over time are reflected in the dynamic product distribution, shrinking the optimal reaction window. Dealumination through steaming has been shown to be an effective modification method, impacting the distribution and coordination of aluminum as well as the overall zeolite structure. Tests showed that presteaming ZSM-5 zeolites prior to the reaction of butanol resulted in the immediate stabilization of the reaction product distribution. Catalyst characterization revealed significant differences in the zeolite lattice structure and acid sites from steaming and after reaction. In this presentation, the specific effects of steaming on the catalyst structure and reaction sites will be discussed along with their role in stabilizing the reaction product distribution.

466460 Tandem Catalysis in a New, Solid-Phase Route to Renewable Aromatic Chemicals from Biobased Furanics

Abstract:

Efficient production routes for 'drop-in', renewable aromatics are currently bring highly sought after due to diminishing reserves and changes in the composition of fossil fuel resources. In this light, sugar-derived furanics offer attractive alternative resources for the production of aromatic chemicals. Recently, we reported on a new three-step procedure to the synthesis of (substituted) phthalic anhydrides from biobased furanics by introducing an intermediate hydrogenation step in the classical Diels-Alder (DA) aromatization route, typically run at elevated temperatures, to address the general issue of the reversible nature of the DA addition step [1]. The hydrogenated DA adduct of, e.g. 2-methylfuran and maleic anhydride (MFMA[H]), is thermally stable and can be efficiently aromatized to give 3-methylphthalic anhydride and *ortho-*/*meta-*toluic acid in a tandem catalytic reaction using a physical mixture of acidic zeolites and Pd/C in toluene. Variation in acidity and textural properties of, e.g. H-Y zeolites, in the combined dehydration/dehydrogenation catalyst provides a handle on selectivity toward aromatic products [2]. A linear correlation exists between the yield of 3-methylphthalic anhydride and the product of the strong acid/total acidity ratio and mesopore volume of H-Y, indicating the need for a balanced catalyst acidity and porosity of H-Y zeolite.

Here, we show that the hydrogenated DA adducts can also be conveniently converted to renewable aromatics in a solid-phase reaction (i.e., without any solvent), using only H-Y zeolites [3]. The remarkable performance of these catalysts in the solid-phase can still be further enhanced upon the addition of a small amount of a dehydrogenation catalyst (e.g., 1 wt. % Pd/C or Pt/C) or by the use of bifunctional catalysts (e.g., 1 wt. % Pd/H-Y or Pt/H-Y). Monitoring the evolution of H_2, H_2O, and CO_2 as a function of temperature and time provided valuable insights into the mechanism of aromatization and decarboxylation. The effects of various experimental parameters, different concentrations of H-Y zeolites and dehydrogenation catalysts, on the reactivity of MFMA[H] were studied to effectively control the reaction rate

and product selectivity. The use of Pt/H-Y (Si/Al=30) resulted in an excellent yield of 94% total aromatics, which is the highest reported to date. The bifunctional catalysts were also found highly selective for the aromatization of other hydrogenated DA furan derivatives. The development of new heterogeneous catalysts for the production of aromatic chemicals from biomass-derived furanics can therefore be guided by taking into account the influence of the acidic and textural properties of H-Y zeolites on selectivity.

[1] S. Thiyagarajan, H. C. Genuino, M. Śliwa, J. C. van der Waal, E. de Jong, J. van Haveren, B. M. Weckhuysen, P. C. A. Bruijnincx, D. S. van Es. Substituted phthalic anhydrides from bio-based furanics: A new approach to renewable aromatics, *ChemSusChem* **2015**, *8*, 3052-3056.

[2] H. C. Genuino, S. Thiyagarajan, J. C. van der Waal, E. de Jong, J. van Haveren, D. S. van Es, B. M. Weckhuysen, P. C. A. Bruijnincx. Selectivity control in solid acid and Pd-catalyzed tandem aromatization of bio-based furanics, *Manuscript in preparation.*

[3] S. Thiyagarajan, H. C. Genuino, J. C. van der Waal, E. de Jong, B. M. Weckhuysen, J. van Haveren, P. C. A. Bruijnincx, D. S. van Es. A facile solid-phase route to renewable aromatic chemicals from biobased furanics, *Angew. Chem. Int. Ed.* **2016**, *55*, 1368-1371.

Reaction Pathway and Microkinetic Model for Levulinic Acid Hydrodeoxygenation over Sulfided $NiMo/Al_2O_3$

Miha Grilc[a], Blaž Likozar[a], Janez Levec[a]

[a] Department of Catalysis and Chemical Reaction Engineering, National Institute of Chemistry, Hajdrihova 19, 1000 Ljubljana, Slovenia, miha.grilc@ki.si

Keywords: Platform chemicals, levulinic acid, specialty chemicals;

Summary

Hydrodeoxygenation of levulinic acid was studied in a batch three-phase slurry reactor over sulphided $NiMo/\gamma\text{-}Al_2O_3$ catalyst in order to find optimal reaction conditions for high selectivity towards γ-valerolactone production and high hydrodeoxygenation rates Since two different reaction pathways are dominant, the aim of this study was to find appropriate window of parameters to catalyse the hydrodeoxygenation reactions and inhibit the thermal decarboxylation reaction. The kinetic study included the variation of temperature between 200 and 300 °C, hydrogen pressure between 25 and 75 bar, stirring speed, amount of catalyst addition, particle size etc. As expected, the temperature and pressure had the highest influence on the reaction rates of hydrodeoxygenation of decarboxylation. To obtain high selectivity of hydrodeoxygenation, temperatures must be kept below 250 °C, since above this temperature the homogeneous C-C bond cleavage occurs and results in excessive decarboxylation of carboxylic group and 2-butanone formation.

1. Introduction

In the process of acidic hydrolysis or solvolysis the cellulose, hemicellulose and sugars are converted into 5-hydroxymethylfurfural (HMF), that can be further deformylated to levulinic acid [1]. Levulinic acid (LA) is therefore the main product of lignocellulosic biomass liquefaction by acidic solvolysis or hydrolysis, where up to 83 mol% yields were reported from cellulose [2]. Levulinic acid was reported to be an ideal platform chemical, as it is a member of the gamma-keto acids and can be therefore utilized to produce various targeted bio-chemicals or just a liquid fuel.

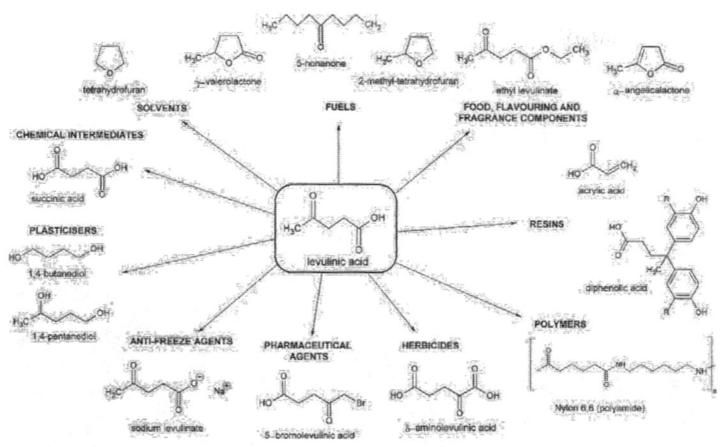

Figure 1: Levulinic acid as a platform chemical for specialty chemicals and liquid fuel [2].

2. Methods and Procedures

Catalytic hydrodeoxygenation (HDO) step is introduced to reduce the presence of chemically bonded oxygen from the feedstock. Levulinic acid can be hydrodeoxygenated by the cyclisation, resulting in formation of γ-valerolactone and water. Reaction takes place at temperatures above 200 °C in the presence of hydrogen and hydrodeoxygenation catalysts either based on transition metals (Mo, W, Ni, Co) in their sulphide, metal, oxide, nitride or carbide form or noble metals (Pt, Ru, Pd) on different supports [3]. In this study industrial sulphided NiMo/γ-Al_2O_3 catalyst was used. Figure 3 shows most important reactions for oxygen removal, where oxygen can be removed in the form of water, carbon monoxide and carbon dioxide.

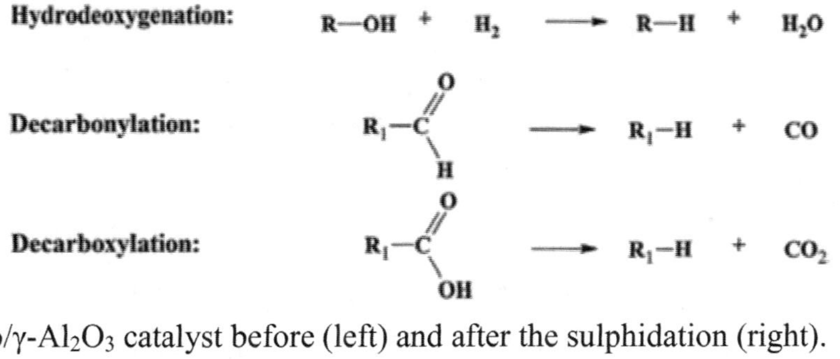

Figure 2: NiMo/γ-Al_2O_3 catalyst before (left) and after the sulphidation (right).

Decarbonylation and decarboxylation are homogeneous reactions caused by thermal decomposition, so presence of hydrogen or catalyst is not strictly necessary. HDO is most desired deoxygenation reaction, as no carbon atom is removed, which results in a higher yield of the final product.

3. Experimental

Since NiMo/γ-Al_2O_3 was obtained in its oxide form, it had to be sulphided first. The catalyst was sulphided using DMDS at 25 bar of hydrogen at temperature of 350 °C for 60 minutes in a stainless steel autoclave.

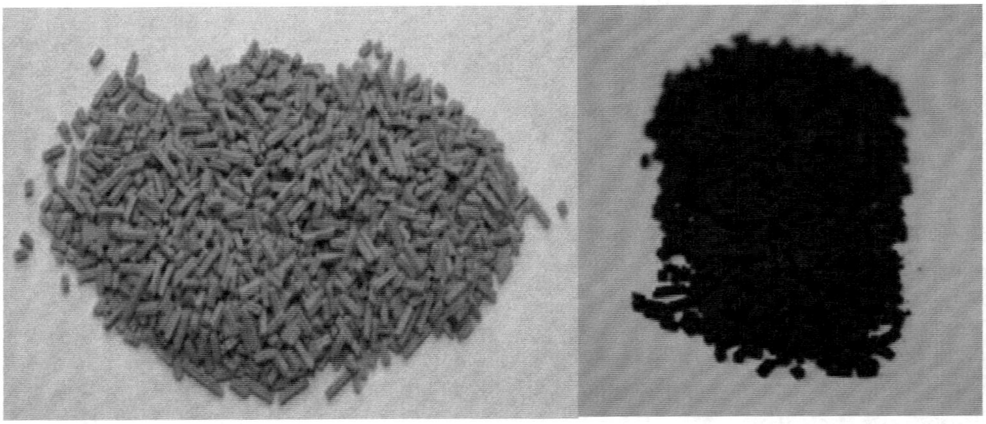

Figure 3: NiMo/γ-Al_2O_3 catalyst before (left) and after the sulphidation (right).

$$MoO_3 + 2\ H_2S + H_2 \rightarrow MoS_2 + 3\ H_2O$$

$$3\ NiO + 2\ H_2S + H_2 \rightarrow Ni_3S_2 + 3\ H_2O$$

Hydrodeoxygenation was investigated in the same 300 mL autoclave equipped with a magnetically driven turbine stirrer and electric heating jacket. Hydrodeoxygenation took place in solventless conditions, so only solid catalyst, liquid levulinic acid and gaseous hydrogen were in contact in the initial stages of the experiment.

4. Results

The analysis of gaseous (GC) and liquid phase (GC, FTIR) reveals the presence of two deoxygenation mechanisms: hydrodeoxygenation and direct decarboxylation. Hydrodeoxygenation reaction only took place in the coexistent catalyst presence and elevated hydrogen pressure. If catalyst particles were removed or nitrogen gas was used instead of hydrogen, the hydrodeoxygenation reaction was negligible. In the presence of sulphided NiMo/γ-Al_2O_3 the main product is γ-valerolactone, while other products of hydrogenation or dehydroation were rare. Few percent of valeric (pentanoic) acid was detected, while 4-hydroxypentanoic acid is obviously very reactive, since its concentration was very low during the runs. Products of decarboxylation and subsequent hydrogenation were also followed. At higher temperatures where decarboxylation takes place, 2-butanone is being formed, while its subsequent hydrogenation product (2-butanol) was never detected in the reaction mixture. Keton group is obviously resistant to hydrogenation at reaction conditions that were used in this study.

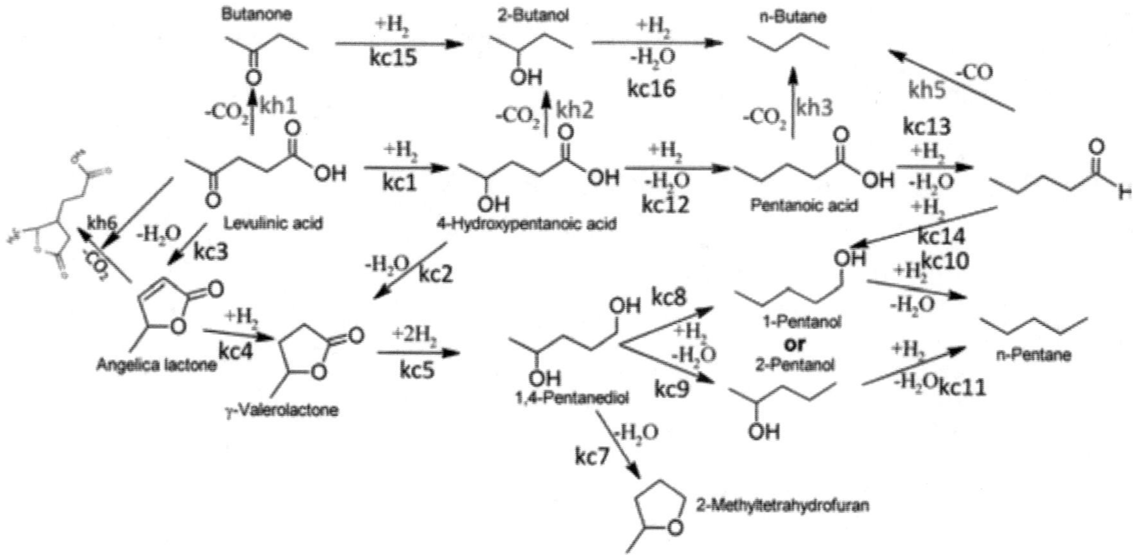

Figure 3: Proposed reaction pathway network of competitive hydrodeoxygenation and decarboxylation reaction of levulinic acid and its products.

5. Conclusions

- Levulinic acid is an important biomass-derived platform chemical
- γ-valerolactone is the main product of LA hydrodeoxygenation over NiMo/γ-Al_2O_3
- Hydrodeoxygenation and decarboxylation are competitive reactions
- Selectivity towards the γ-valerolactone can be regulated by temperature and pressure and catalyst loading
- Undesired decarboxylation is difficult to inhibit, especially at higher temperatures

Acknowledgements

The authors gratefully acknowledge the financial support of the Ministry of Education, Science and Technology of the Republic of Slovenia through the Research Program No. P2-0152.

Literature

[1] B. Girisuta, et al., Chemical Engineering Journal 217 (2013) 61-70.

[2] D.W. Rackemann, W.O.S. Doherty, Biofuels, Bioproducts and Biorefining 5 (2011) 198-214.

[3] M. Grilc, B. Likozar, J. Levec, Applied Catalysis B: Environmental 150–151 (2014) 275-287.

455098 The Sunshot Inititiative: New Opportunities for R&D in Concentrating Solar Power

Abstract:

The SunShot Initiative is a collaborative, national endeavor by the United States Department of Energy's Solar Energy Technologies Office to make electricity generated from the sun cost competitive – on an unsubsidized basis – with traditional forms of electricity generation by the year 2020. Concentrating Solar Power (CSP) constitutes one technological approach to achieving this ambitious goal, generally identified as a levelized cost of electricity (LCOE) ≤ $\$0.06/kWh_e$. CSP focuses the sun's rays onto a receiver that converts the photons into thermal energy. Part of this thermal energy can then be transported to storage for use when the sun does not shine, while the remainder is transported to a conventional turbine to generate electricity. The turbine, or heat engine, can realize increased thermodynamic efficiency as the input temperatures increase; however, high temperature CSP systems come with significant materials and cost challenges. Evaluating the interfaces found in the solar receiver, thermal energy storage technologies, heat exchangers, and power blocks will help define the technical challenges and their requisite success metrics. The presentation concludes with remarks, links, and information on DOE's SunShot Initiative so that the audience may stay aware of future Funding Opportunity Announcements.

469740 Non-Stoichiometric Perovskite Oxides As High-Temperature Energy Storage Media and Their Application to Concentrating Solar Power Generation and Hydrogen Production

Abstract:

Although there is a significant abundance of solar resource, the intermittency of the resource compromises its deep integration into the U.S. energy portfolio. However, solar-thermal processes such as concentrating solar power (CSP) are unique in that thermal energy can be efficiently stored in engineered heat-absorbing media. The quantity of heat stored and subsequent power-generation efficiency generally increase as temperature increases, motivating solid particulate storage media for high-temperature systems. However, the storage of only sensible energy limits the potential of both the quantity of stored energy as well as the possible applications of the stored energy. For this reason, energy storage media have been engineered to store both sensible heat and reaction heat, a practice known as thermochemical energy storage (TCES).

Here we present the concepts behind the engineering, synthesis, and characterization of novel doped calcium manganese perovskites ($CaAl_yMn_{1-y}O_{3-\delta}$). These materials exhibit rapid and reversible reduction/oxidation behavior through the introduction of oxygen vacancies in the material structure: $ABO_3 + \Delta H \rightarrow ABO_{3-\delta} + \delta/2\ O_2$. This approach allows the storage of chemical energy through thermal reduction without the kinetic limitations inherent in decomposition reactions. The dopant concentrations of this materials family have been optimized in our laboratories to produce the largest energy storage capacity of a high-temperature perovskite system to date.

Additionally, two applications of the material are presented. First, these materials are applied to a CSP system for electricity generation. In this system, the material is first heated and thermally reduced in the incident solar beam of a solar receiver. The material is then stored in large thermally-insulated bins, after which both the sensible and chemical energy can be utilized by a reaction and heat-transfer process with a working fluid. This high-enthalpy working fluid is passed to a turbine to create power on demand, even when the solar resource is unavailable. Secondly, these materials can operate as thermal-to-chemical energy converters, where the thermal reduction results in a chemically energy-rich species which can be used either fully or in part to facilitate subsequent reactions, such as water-splitting for hydrogen generation. In this embodiment, only the chemical energy is useful, and the sensible energy must be recovered to maximize process efficiencies.

Sandia National Laboratories is a multi-program laboratory managed and operated by Sandia Corporation, a wholly owned subsidiary of Lockheed Martin Corporation, for the U.S. Department of Energy's National Nuclear Security Administration under contract DE-AC04-94AL85000. This work is supported by the U.S. Department of Energy, SunShot Initiative, under Award Number DE-FOA-0000805.

471159 Solar Thermochemical Energy Storage Based on Strontium Carbonate Chemistry

Abstract:

The $SrO/SrCO_3$ chemistry offers the potential for high temperature heat release (~1100-1200°C) upon the carbonation of SrO, which could be a pathway to high efficiency, combined cycle power generation. Building upon previous work on this cycle, this talk will propose a kinetic model for the carbonation reaction derived on experiments using both TGA and packed bed systems. In addition, preliminary technoeconomic and exergy analyses are performed for the cases of CO_2 storage as 1) a compressed gas and 2) ab/adsorbed species in a liquid or solid matrix. Probabilistic methods are used to elucidate the cost drivers and areas for future consideration in evaluation of the cycle feasibility.

452449 Design of Manganese Oxide-Based Particles for High-Temperature Thermochemical Energy Storage

Abstract:

High-temperature thermochemical energy storage shows promise in aiding concentrating solar power plants in meeting variable, grid-scale electricity demand. In this work, manganese oxide-based mixed metal oxide particles have been designed and tested for thermochemical energy storage. Particles are designed for high energy storage capacity, flowability, and physical and chemical stability. We evaluate the effects of Al_2O_3, Fe_2O_3, and ZrO_2 in Mn_2O_3-based spray-dried particles in a TGA between 650°C and 1,200°C over six consecutive redox cycles. Results are compared with thermodynamic predictions from 400-1,400°C under oxidizing and reducing atmospheres. A mixture of 2:1 Fe_2O_3:Mn_2O_3 formed iron manganese oxide spinel (Fe_2MnO_4) on calcination, and demonstrated the highest thermochemical activity despite particle agglomeration and deformation. Conversely, zirconia was an inert support that does not react with manganese oxide. Differences in redox performance between materials with different Fe to Mn ratios have been attributed to ion diffusion and secondary phase formation.

452774 Grid-Scale Thermochemical Energy Storage Using Mixed Metal Oxide Redox Cycles

Abstract:

Thermochemical energy storage (TCES) systems are of great interest in concentrated solar power (CSP) applications. Storing sunlight as chemical energy during the day can enable power generation at night or during cloudy periods, effectively alleviating the inherent intermittency of solar sourced electricity. Metal oxides are among the most attractive TCES materials because they possess high-energy density and high reduction/oxidation temperatures (>1000°C) suitable for driving high-efficiency thermodynamic power cycles. However, limitations involving high-temperature heat transfer, particle handling, sintering, and chemical performance of the metal oxide materials are still problematic in practical application.

A novel, grid-scale CSP/TCES system design that addresses many of the traditional problems associated with metal oxide TCES has been developed at the ANU. This presentation serves as a broad overview of recent work surrounding the designed CSP process, including improvements to the metal oxide active materials, solar reactor design, optical field development, and preliminary techno-economic analyses. Improved chemical performance of the active material is being explored through metal co-doping and the intentional formation of solid solutions. The incorporation of fluidized bed reactors for solar reduction and off-sun oxidation allows for improved heat transfer and enhanced chemical kinetics, and also enables high temperature pneumatic transport of gas-solid reacting flows. The initial solar reactor concept consists of one or several vertical fluidized bed reaction tubes in a beam-up solar cavity receiver, which allows for well-controlled fluidization, high optical efficiency, and minimized convective heat losses. The techno-economic analysis of a 100 MW_{Th} grid-scale facility is employed to predict the levelized cost of electricity.

465618 Kinetic Analysis of High Temperature Thermochemical Energy Storage Based on Calcination–Carbonation Chemical Looping Reactions

Abstract:

This work aims to improve the performance of carbonate storage systems, a form of thermochemical energy storage (TCES) for concentrated solar power (CSP) technologies. TCES has great potential due to its high energy storage densities and inexpensive materials, making it highly competitive with commercial molten salt storage systems. The storage system is based on metal oxide calcination–carbonation chemical looping, a cyclic process for which a broad range of metal oxide candidate materials are considered. Metal oxide produced in the solar-driven endothermic calcination reaction is cycled to the exothermic carbonation reaction, from which thermal energy is released to drive a power cycle.

This work details a comparative thermodynamic and kinetic study of calcination and carbonation reactions, using commercial samples and synthesized mixtures of carbonates such as $CaCO_3$ and $SrCO_3$. Experiments are carried out under a CO_2 atmosphere using thermogravimetric analysis (TGA) and differential scanning calorimetry (DSC) and the materials are selected to attain the desired storage temperature (typically above 565°C), energy density and charging and discharging rates. Chemical and structural cyclability of candidate materials are evaluated as a function of the calcination–carbonation cycle number using several characterization techniques such as scanning electron microscopy (SEM), transmission electron microscopy (TEM), N_2 adsorption-desorption and X-ray diffraction crystallography. We show how different mixtures and synthesis precursors can improve the cyclability of metal oxide–carbonate systems.

465613 Thermodynamic Analysis of $CaCO_3$/CaO Looping for High Temperature Thermochemical Energy Storage

Abstract:

High temperature power cycles, such as the supercritical carbon dioxide (CO_2) cycle, are promising technologies currently being investigated by the international research community for concentrated solar thermal power generation (CSP). These power cycles have the potential to increase the overall efficiency of the power plant compared to state-of-the-art steam Rankine cycles and thereby reduce the levelized cost of electricity (LCOE) of CSP.Â Therefore, novel thermal energy storage materials and processes are required to successfully couple concentrated solar power collectors with high temperature power cycles.

In this context, the $CaCO_3$/CaO thermochemical cycle is an attractive process for thermal energy storage that can provide dispatchable concentrated solar thermal energy at temperatures between 650â€“980Â°C, appropriate to use with high temperature power cycles. Current research efforts focus on material development and solar reactors. However, there is a need for investigation into the field of system integration to support and verify the viability of this concept.

In this work, we evaluate the performance of a $CaCO_3$/CaO system coupled to a supercritical CO_2 power cycle for continuous electricity production. This investigation has identified high-energy demanding subsystems that require minimization to improve the overall efficiency of the process. The effect of the operational conditions (temperature and pressure) in the calcination receiver and carbonation reactor on overall efficiency are also investigated.

471254 Study of Bio-Inspired Nanostructured Materials for Solar Energy Conversion

Abstract:

Solar energy conversion has been drawn a great interest in our scientific community due to the demand for renewable energy. The necessary steps to improve solar energy conversion, including photovoltaics and water splitting, are solar energy harnessing, charge separations, and catalytic reactions. Nature always inspires us to mimic natural photosynthetic systems such as, Purple photosynthetic bacteria, revealing a new light harvesting protein with efficient channels for energy and electron transport.

This study focuses on developing artificial photosynthetic systems for understanding the fundamentals of energy transfer, and electron transport in both water splitting, and photovoltaic devices with uniquely designed nanostructures. In the artificial design systems, dye-sensitized solar cell systems were a base for investigating the above fundamental problems. For the efficiency of solar energy harnessing and better electron transport systems, uniquely assembled nanostructures of ZnO , TiO_2 , and an acceptor/donor energy system were fabricated using electrochemical approaches. The ZnO nanostructures including nanorods, nanosheets and nanowries were then decorated with small band gap semiconductors including, Fe_2O_3, CuO and Cu_2O by spin coating techniques. These structures were then investigated for electron transfer using characterization methods such as: SEM, TEM, XRD, and electrochemical potentiostat methods. Spectroscopic methods such as: UV-vis spectrometer, and time-resolved photoluminscence spectroscopy are being investigated now. This fundamental study will lead us to a better design for artificial photosystems.

470640 Direct Glycerol Fuel Cell with Polytetrafluoroethylene (PTFE) Thin Film Separator

Abstract:

Anion-exchange membrane direct glycerol fuel cells can yield a high power density, but challenges exist in developing chemically stable ion exchange membranes. In this work, we demonstrate a porous PTFE thin film, a well-known chemical, electro-chemical, and thermal robust material can work as a separator, and achieved high performance of direct glycerol fuel cells (DGFCs). DGFC using PTFE thin film with PdAg/CNT anode catalyst exhibit a remarkable peak power density of 214.7 mW cm^{-2}, which is about 22.6% lower than state-of-art AEM fuel cell with A901 membrane. The durability test was performed in order to evaluate the degradation rate of cell voltage of both PTFE thin film and A901 membrane. We report a 5.81% and 11.08% decrease in cell voltage for PTFE thin film and A901 anion-exchange membrane. Transmission electron microscopy (TEM) was used to evaluate the particle size of the anode catalyst after long-term discharge, and the results suggest that the PdAg particle size growth may be the main reason to lead performance drop. Our work shows, development of PTFE thin films could potentially alleviate the expense associated with anion exchange membranes for wide-spread fuel cell applications.

468826 Catalytic Behavior in Internal Steam Reforming of Methane in SOFC Conditions over $La_XSr_{2-X}MnO_4$ Materials

Abstract:

CATALYTIC BEHAVIOR IN INTERNAL STEAM REFORMING OF METHANE IN SOFC CONDITIONS OVER $La_XSr_{2-X}MnO_4$ MATERIALS

Karen Silva[1], Sebastian Vecino[1*], Monica Sandoval[1], Patrick Gèlin[2], Paola Gauthier-Maradei[1], Gilles Gauthier[1]

1: INTERFASE, Escuela de Ingeniería Química, Universidad Industrial de Santander, Carrera 27, Calle 9 Ciudadela Universitaria, Bucaramanga, COLOMBIA

2: Université Lyon 1, CNRS, UMR 5256, IRCELYON, Institut de recherches sur la catalyse et l'environnement de Lyon, 2 avenue Albert Einstein, F-69626 Villeurbanne, France.

*e-mail: jusevema13@hotmail.com

ABSTRACT

Nowadays, solid oxide fuel cells (SOFC) are one of the most attracting and interesting energy conversion devices. They can convert the chemical energy directly to electrical energy with high efficiency (>60%) and low environmental pollution, employing several types of fuels such as hydrogen, coal gas, natural gas (methane) or another hydrocarbons fuels. These electrochemical systems operate at high temperatures (600-1000°C) using oxygen from the air as oxidant, and hydrogen directly or by internal steam reforming (ISR) of some compounds present in the fuels to produce heat as well as water and electricity. The ISR is a highly endothermic process that takes place, catalytically, at the anode of the

SOFC. However, during this process and employing state of art anode material Ni/YSZ, the hydrocarbons decompose into hydrogen and additionally in an undesired compound: solid carbon (coke). This coke tends to deposit on the anode surface causing a rapid deterioration of the cell due to the deactivation of catalytic activity and the inhibition of hydrogen diffusion into the anode to continue with the electrochemical reaction. In order to overcome this limitation, it is necessary to develop new anode materials that catalyse the ISR and reduce significantly the problem described above. According to literature, single-phase oxide materials have been studied with excellent results. In this work, La0.5Sr1.5MnO4 was studied as potential new anodic material for SOFC directly fed with methane, operating at high temperature and under water deficient conditions. The material was synthesized by Pechini method and characterized by XRD. Measurements of catalytic activity in steam reforming of methane were carried out using 40 mg of catalyst at 900°C during 40h in a U-shaped quartz reactor. Different methane/water ratios (0.66≤CH4/H2O≤10) were employed to determine kinetic parameters. Coke formation during catalytic testing was measured by temperature programmed oxidation (TPO). The catalytic activity was compared with a reference material CGO (Ce0.9Gd0.1O2-d). Finally, the material was characterized again by XRD to evidence any damage or changes in the catalyst structure. This catalyst was found very active (high hydrogen production 9000ppmv) and resistant to coke formation (<69µmol/gcat), close to CGO catalyst (13000 ppmv hydrogen and 50.2µmol/gcat). The materials present an excellent stability during the test run; no structural change was found after the reaction. The small increase in the lattice parameters was attributed to the manganese reduction during the catalytic reaction (Mn+4/Mn+3 to Mn+3/Mn+2).

Key Words: *Solid oxide fuel cell, Internal steam reforming, Methane, Manganite, Energy.*

469639 Direct Fast Pyrolysis Bio-Oil Fuel Cell

Abstract:

Bio-oil derived from the pyrolysis of lignocellulosic biomass shows a great promise, however, needs further upgrading to potentially serve as an alternative to fossil fuels. Herein, we demonstrate that crude fast pyrolysis bio-oil can be directly used as a fuel for anion exchange membrane fuel cells (AEMFCs) to generate high power density electrical energy at low temperature. Direct fast pyrolysis bio-oil AEMFCs with the Pd/CNT anode catalyst and a commercial Fe-based cathode catalyst exhibit a remarkable peak power density of 42.7 mW cm^{-2}. Levoglucosan was identified as the major sugar compound with 11.1 wt% of the bio-oil composition, along with disaccharides, pyrolytic lignin, and oligomer of lignin-derived phenolic compounds. Cyclic voltammetry (CV) studies investigated the electrocatalytic oxidation of high purity levoglucosan over the four noble metal catalysts in half cell, as levoglucosan is the dominant sugar component in bio-oil. Pd/CNT, compared to other catalysts, displayed the highest activity and lowest onset potential of electrocatalytic oxidation of levoglucosan. AEMFC with high purity sugars shows ~ 1.2 to 3 times higher power density than that with fast pyrolysis bio-oil fuel.

461692 Nature-Inspired Fractal Flow Field for PEM Fuel Cells

Abstract:

Nature-Inspired Fractal Flow Field for PEM Fuel Cells

Jason In Sung Cho[1,2], Toby Neville[1,2], Jeffrey Marquis[3], Panagiotis Trogadas[1], Dan Brett[1,2] and Marc-Olivier Coppens[1]

[1] *Department of Chemical Engineering, University College London, London WC1E 7JE, United Kingdom, m.coppens@ucl.ac.uk*

[2] *Electrochemical Innovation Lab, Department of Chemical Engineering, University College London, London WC1E 7JE, United Kingdom*

[3] *Isermann Department of Chemical & Biological Engineering, Rensselaer Polytechnic Institute, 110 8th St., Troy New York, United States*

One of the ongoing challenges faced by polymer electrolyte membrane (PEM) fuel cells for broader commercialisation is the accomplishment of uniform gas distribution across the catalyst layer. Ensuring reactant homogeneity is essential to improve fuel cell efficiency and to forestall adverse side reactions that are detrimental to fuel cell longevity [1]. Within the confines of the conventional flow fields, such as serpentine channels, this is an unavoidable consequence, due to reactant depletion across the channel and over the active area as a whole [2].

This work aims to address the issue by implementing a novel 3D flow field inspired by the fractal geometry of the upper respiratory tract of the human lung. The human lung has been shown to have an optimal architecture that uniformly distributes oxygen from the trachea to the alveoli. Furthermore, the human lung transitions between two flow regimes: 14-16 upper generations of branches dominated by convection, and 7-9 lower generations of space-filling acini dominated by diffusion[3, 4]. The upper generations of branches are designed to slow gas flow to a rate compatible with the rate in the diffusional regime (Pe≈1) [5] and the entropy production was shown to be uniformly distributed in both regimes [3, 4]. Furthermore, an attractive aspect is the scalability of fractal networks. They can bridge multiple length scales by the addition of further generations, while preserving the building units and microscopic function of the system [6]. These characteristics that make the lung an optimal transport structure are all desirable for a PEM fuel cell. However, rather than uniformly distributing oxygen in a given volume, we wish to distribute the gases uniformly over a plane to ultimately deliver a uniform concentration of reactant throughout the catalyst layer [7].

The fractal pattern consists of repeating "H" shapes and, by employing a fractal dimension of 2, where daughter "H's" are half the parent's size, a plane-filling geometry is attained. Furthermore, the three-dimensional branching structure allows only the outlets of the fractal inlet channel to be exposed to the MEA, eliminating the reactant depletion along the channel, and thus providing uniform local conditions on the surface of the catalyst layer.

In this study, the reactant flow and cell performance were modelled for different numbers of fractal branching generation. Optimal performance was reached for branching levels N ^{3} 5. Based on the simulation results, three fractal flow field prototypes were produced via selective laser sintering (SLS), a «3D printing» method, each representing a branching generation level of N = 3, 4 and 5, and validated against a standard serpentine flow field.

The results suggest improved mass transport for the fractal flow fields at high current density, due to the homogeneous gas distribution. The fractal prototype with N = 4 generations demonstrated improved performance over the serpentine at all operating conditions investigated in this study and a good correlation with the results predicted by the model. The worse-than-expected performance of for N = 5 generations at higher humidity level reveals insufficient convective liquid water removal due to the slow air velocity from each fractal outlet hampering effective diffusion within the porous medium. The issues with flooding were alleviated by operating the fuel cell at lower humidity levels, which led to improved mass transport for the prototype with N = 5 generations. However, fuel cells should ideally be operated under fully humidified condition to minimise local membrane drying and subsequent Ohmic losses from reduced proton transport.

The findings elucidate the need for implementation of active water management systems to alleviate flooding for the fractal flow fields, especially at higher generation levels where flooding is projected to be more pronounced.

References

[1] Liu, Z., Yang, L., Mao, Z., Zhuge, W., Zhang, Y., and Wang, L., *Behavior of PEMFC in starvation.* Journal of Power Sources, **2006**. 157(1): p. 166-176.

[2] Tüber, K., Oedegaard, A., Hermann, M., and Hebling, C., *Investigation of fractal flow-fields in portable proton exchange membrane and direct methanol fuel cells.* Journal of Power Sources, **2004**. 131(1–2): p. 175-181.

[3] Kjelstrup, S., Coppens, M.-O., Pharoah, J., and Pfeifer, P., *Nature-inspired energy-and material-efficient design of a polymer electrolyte membrane fuel cell.* Energy & Fuels, **2010**. 24(9): p. 5097-5108.

[4] Trogadas, P., Ramani, V., Strasser, P., Fuller, T.F., and Coppens, M.-O., *Hierarchically Structured Nanomaterials for Electrochemical Energy Conversion.* Angewandte Chemie International Edition, **2016**. 55(1): p. 122-148.

[5] Coppens, M.-O., *A nature-inspired approach to reactor and catalysis engineering.* Current Opinion in Chemical Engineering, **2012**. 1(3): p. 281-289.

[6] Trogadas, P., Nigra, M.M., and Coppens, M.-O., *Nature-inspired optimization of hierarchical porous media for catalytic and separation processes.* New Journal of Chemistry, **2016**.

[7] Marquis, J., Nature-inspired hierarchically structured high-efficiency PEM fuel cell, PhD Thesis, **2013**.Rensselaer Polytechnic Institute, Troy, NY.

Acknowledgements

The authors wish to thank the National Science Foundation Fuel Cell IGERT at Rensselaer, DGE-0504361 for supporting initial work. Financial support from an EPSRC "Frontier Engineering" Award (EP/K038656/1) and a UCL Faculty of Engineering Sciences Dean's Scholarship are also gratefully acknowledged.

466171 Membraneless Fuel Cells Modified with Gold Nanoparticles and Laccase for Direct Conversion of Recalcitrant Carbohydrates into Electricity

Abstract:

The need for clean and viable sources of alternative energies has prompted the development of biofuel cells (BFCs), which are characterized by renewability and high power density. In particular, enzymatic BFCs are a special type of fuel cell in which enzymes are used as catalysts to oxidize specific substrates on anodes and reduce O_2 on cathodes. Candidates for substrates include simple sugar like glucose to complex ones such as starch and cellulosic biomass. Since most readily available biomass is in the raw, unprocessed cellulosic form, there is a challenge for BFCs to utilize these complex and recalcitrant substrates. Furthermore, more complex feedstocks such as hydrolysates from lignocellulosic biomass, however, have rarely been used as substrates in BFCs. Typically, some pretreatment or saccharifications are required for these complex feedstocks before they can be used as substrates to generate a high current density. Thus, there is motivation to design, construct and modify BFCs such that they are able to efficiently and effectively utilize various carbohydrate substrates.

In this study, we aim to address problems of using complex substrates by leveraging the catalytic properties of gold nanoparticles (AuNPs) on cathodes and laccase on anodes in a membraneless hybrid BFC. AuNPs are versatile in catalyzing a wide variety of substrates and present a viable alternative to substrate-specific enzyme catalysts that are traditionally used. This is beneficial especially when mixed substrates derived from biomass are used as fuel. We first fabricated and modified cathodes with AuNPs and characterized the effect of the presence of AuNPs (as compared to unmodified electrodes and that modified with bulk gold), AuNP sizes and loadings on the catalytic action of the BFC. Next, the harnessing of stable and useful currents from a variety of carbohydrate substrates was studied: In addition to glucose, other carbohydrates including recalcitrant disaccharides and polysaccharides such as fructose, sucrose and carboxymethylcellulose were also found to successfully generate significant current densities. To further demonstrate the versatility of the BFC, complex feedstocks substrates such as sugarcane bagasse hydrolysate was also shown to deliver a stable current, which was further validated with the lighting up of devices such as light emitting diodes.

476420 Role of Hydrogen in Future Clean Energy Landscape

Abstract:

Given the future energy demand growth, sustainable oil constraints and global mandates against GHG induced climate change; hydrogen **is** poised to be a strong contender in the future energy landscape. Strategic scenarios and **hydrogen roadmaps** have been laid out for addressing the future energy needs against the backdrop of environmental targets and clean energy strife. Hydrogen's appreciable leverage and share is put on the transport and power sector with hydrogen pathways rendering a strong impetus in the longer run.

Hydrogen's role as a unique carrier of ultra clean energy is imminent despite any barriers or roadblocks put by skeptics debating its claimed viability. Based on its unparalleled leverage and source flexibility, whether fossil, renewable or nuclear into 'one clean energy platform', the **hydrogen economy** is destined to be in the offing.

Extensive deliberations for transitioning from direct hydrogen substitution as fuel (like H-CNG or HICE applications) and **fuel-cells** based stationary distributed power to transport vehicles starting with fleets and marine applications, is now expanding to a sizeable deployment for personal (HFC) cars. Marking the onset of the hydrogen movement, Toyota had recently launched their personal HFC car "Mirai" last year in California, thus projecting an exponential expansion in coming years.

Furthermore, in view of the setback and limitations on nuclear power and also renewable energy options in terms of large scale deployment and overall economics, potential future trajectories are being forecasted on **hydrogen** as a dominant energy carrier in the longer term.

For realization of such demand, the fossil-derived hydrogen is still the strong contender and anticipated to cater for a substantial portion of future hydrogen generation, especially large centralized facilities and several medium-sized units, with carbon capture provisions.

It has been further boosted given the drive to exploit the abundant low-cost unconventional gas as a cost-effective low-carbon feedstock for generating large scale hydrogen using state-of-the-art technology. Additional drivers are ultimate well-to-wheel efficiency, economy of scale and qualification for CO_2 capture and sequestration. Eventual options of using renewable power to generate "**green hydrogen**" are also on the horizon.

The talk will provide an overview on these interesting facets on the role of hydrogen in the future energy landscape.

470455 Population Ensemble Modeling of Cellulose Dissolution: Insights for Efficient Biomass Processing

Abstract:

A major barrier to the efficient utilization of biomass is the recalcitrance to dissolution of semicrystalline cellulose. The present study addresses the kinetics of swelling and dissolution of cellulose particles in conditions emulating large-scale processing where the particles exhibit a distribution of size and degree of crystallinity. To this end, we have developed a model in which the behavior of a population of particles is obtained from an ensemble of individual cellulose particle dissolution models. The dissolution of individual solid cellulose particles is based on the relevant transport phenomena and kinetics and reveals decrystallization and disentanglement as two important and potentially rate-determinant steps in the process. The average value or the number distribution of any intra-particle property captured by the individual particle model can be determined by simulation of a sufficient number of individual particles such that ensemble averages are independent of the particle numbers and the computed particle distributions are acceptably smooth. Using this population model, various cellulose particle size and crystallinity distributions are analyzed for different dissolution parameters. The findings from this study would be useful for the rational design and optimization of pretreatment processes to reduce the particle size and degree of crystallinity, leading to enhanced biomass utilization.

475618 Renewable Aromatic Hydrocarbon Production from Furfural By a Continuous Dual-Stage Hydrogenation-Cocracking Process

Abstract:

Aromatic hydrocarbon production from furfural by catalytic cracking over zeolite catalysts is significant for realizing biobased aromatics generation, because furfural can be produced from biobased sugars through different transformation routes such as liquid phase conversion and fast pyrolysis. However, due to the high unsaturation degree and high oxygen content, direct cracking of furfural faces low aromatic hydrocarbon yield and coking problems. Therefore, in this work, two improving methods namely hydrogenation pretreatment and methanol-assisted cocracking over HZSM-5 catalyst were combined to convert furfural into aromatic hydrocarbons, and the influence of hydrogenation catalysts was investigated. It was found that in the hydrogenation stage, although the furfural was nearly completely converted over $10Cu/SiO_2$, the furan ring was not efficiently hydrogenated. $5Ni\text{-}5Cu/SiO_2$ and $10Ni/SiO_2$ achieved the further hydrogenation of some furan rings, but over $10Ni/SiO_2$ the primary conversion of furfural was lower and more saturated gaseous hydrocarbons were generated. The different hydrogenation behaviors of furfural affected the subsequent cracking processes. The hydrogenation-cocracking of furfural and methanol over $5Ni\text{-}5Cu/SiO_2$ generated the highest oil phase yield of 26.4%, which was mainly composed of mono-aromatic hydrocarbons such as benzene, toluene, xylenes and trimethylbenzenes. However, because of the insufficient hydrogenation of furan ring over $10Cu/SiO_2$ and the low conversion of aldehyde group as well as some excessive hydrodeoxygenation over $10Ni/SiO_2$ in the hydrogenation stage, the yield and quality of the oil phases in the corresponding following cracking stage both decreased. The comparison of hydrogenation-cocracking process using the optimum hydrogenation catalyst $5Ni\text{-}5Cu/SiO_2$ with other traditional cracking processes including single-stage cocracking of furfural and methanol, hydrogenation-cracking of pure furfural using $5Ni\text{-}5Cu/SiO_2$ as the hydrogenation catalyst and direct cracking of furfural indicated that the hydrogenation-cocracking process was the most favorable technique for the generation of liquid aromatic hydrocarbons and the maintaining of catalyst activity. Finally, based on the experimental results, the hydrogenation-cocracking mechanism of furfural and methanol was proposed to explain the different hydrogenation behaviors of furfural over three catalysts and also the subsequent different cocracking performances.

481269 Biomass Anaeorobic Convertion Technology Development

Abstract:

Base on the demands for environmental protection, energy security and modern agriculture development, with drawing on experiences of vigorously developing anaerobic biomass conversion from developed countries, innovative research on high efficient anaerobic conversion technologies has been conducted by Biomass Bio-chemical Conversion Laboratory, Chinese Academy of Sciences Guangzhou Institute of Energy Conversion.

In recent years, our Laboratory has undertaken more than 30 projects funded by the National Science & Technology Program, National High Technology Research and Development Program, National Natural Science Foundation, National Science & Technology Pillar Program, Provincial and Ministerial Science & Technology Programs and also programs supported by Chinese Academy of Sciences and corporations, focusing on bio-chemical pretreatments of lignocellulosic materials, nutrient balancing regulation base on complementary characteristics of substrates, high concentration co-digestion of multi-feedstock, two-stage continuous desulfuration, treatments of industrial waste (Cassava starch Dreg, Molasses Alcohol Wastewater etc.). Technologies and reactors for breeding/cultivation/pretreatment/anaerobic digestion of energy grassesï¼Œfermentable organics extraction from municipal solid wastes with multiple impurities, two-phase& three-stage anaerobic digestion, combined oil-gas-fertilizer production from food waste have been developed base on our research; integrated technological system for high-efficient, continuous and stable production of biogas from organic wastes has been established. Maximum organic loading rates of 8-10 kgVS/(m^3Â·d), highest solid concentration of 20% as well as stable operation can be achieved within the system; the best performance results of 75% methane content, 89% volatile solid removal rate, 70% energy conversion rate and 45% lignin removal rate have be obtained; the biogas production cost in this system can be lower than 0.5 yuan/m^3. The international leading position of this achievement has been appraised based on the aggregate technology and economic indicators, and it has won the First Award of the Science and Technology Progress of Guangdong Province and Third Award of Energy Science and Technology Progress.
The aim of this report is to systematically introduce our achievement, and we are looking forward to cooperation on relevant research and application with domestic and foreign organizations. This work is funded by national high-tech development programs(No.KFJ-EW-STS-138, KGZD-EW-304-1, 2012AA101802 etc..)

474806 Oxidative Desulfurization of DBT with H_2O_2 Catalysed By TiO_2/Porous Glass

Abstract:

Oxidative desulfurization of DBT with H_2O_2 catalysed by TiO_2/porous glass

Chun Shen[†], Yujun Wang*[,‡], Jianhong Xu[‡], Tianwei Tan[†], Guangsheng Luo*[,‡]

[†]Beijing Key Laboratory of Bioprocess, College of Life Science and Technology, Beijing University of Chemical Technology, Beijing 100029, China

[‡]The State Key Laboratory of Chemical Engineering, Department of Chemical Engineering, Tsinghua University, Beijing 100084, China

Corresponding author: Tel: 86-010-62788568, Fax: 86-010-62788568

Email address: wangyujun@tsinghua.edu.cn and gsluo@tsinghua.edu.cn

Abstract

Aimed at ultra-deep oxidative desulfurization (ODS) of dibenzothiophene (DBT) and 4,6-dimethyldibenzothiophene (4,6-DMDBT) to control air pollution, we specially designed and prepared porous glass supported with TiO_2 nanoparticles acting as an amphiphilic catalyst. Hydrogen peroxide which is considered as the "green" oxidant was used, and for the extreme liquid–liquid phase ratio (usually larger than 1500) reaction system, the pore volume of 0.19 mL/g of the catalyst provides enough space for storage of hydrogen peroxide. The as-prepared catalyst offers a high interfacial surface area of 116.9 m^2/g and enhances the reaction by facilitating the mass transfer. The mono-dispersed TiO_2 exhibited good crystallinity. The mean diameter varied from 2.1 to 7.8 nm with the loading amount increased from 1.27 wt.% to 9.85 wt.%. The catalyst showed high activity and good stability for producing ultra-clean fuels: 100% conversion was obtained within 2 min and the conversion just decreased from 100.0±1.0 % to 94.3±0.6 % after 5 cycles. Overall, this new reusable catalyst provided an alternative for highly efficient ultra-deep desulfurization in a green way.

475573 Hydrolysis of corncob using a modified carbon-based solid acid catalyst

Abstract:

Lignocellulosic biomass is a very important renewable resource that could be used to produce biofuels and various platform chemicals. The conversion of biomass to reducing sugar is an essential intermediate step that can impact the economic benefits of the producing process. The modified carbon-based solid acid is a novel catalyst used in the hydrolysis process. A catalyst material was prepared through the incomplete carbonization and sulfonation of the mixture of glucose and *p*-toluenesulfonic acid, followed by oxidation using 30% H_2O_2 to promote the acidity. The hydrolysis efficiency of the solid acid was significantly increased. The effects of the carbonization temperature, time, and the glucose / *p*-toluenesulfonic acid ratio on catalytic efficiency were investigated. Under the appropriate preparation condition, the catalyst could hydrolyze corncob to produce 78% xylose and 34% glucose. The activity of the catalyst slightly declined after five-cycles application.

474464 Simultaneous Catalytic Conversion of Bio-Oil and Bio-Char for Hydrogen Production

Abstract:

Based on the integrated processes for hydrogen production from biomass, the simultaneous catalytic conversion of bio-oil and bio-char for hydrogen production was investigated in this paper. The potential hydrogen (H_2+CO) yield was increased from 45.70 to 82.19 mmol/g bio-oil with the presence of metals in the bio-char. The effects of metals (K, Ca, Mg, Fe, Zn, and Al) in bio-char on the catalytic conversion of bio-char and bio-oil were also investigated. The catalytic activities of the metals for the conversion of bio-char and bio-oil were followed by the sequence K>Ca>Mg>Fe>Zn>Al. Meanwhile, the alkali and alkaline earth metals (AAEM), including K, Ca, and Mg, were found to play the dominating roles among the metals in bio-char, and the catalytic effect of K was the most obvious. The potential hydrogen yield and tar yields were 143.50 mmol/g bio-oil and 4.96 mg/g bio-oil when adding the metal of 5 wt.% K.

The effects of metals on the biochar and biomass pyrolytic vapor (bio-oil) were investigated. Al had an inhibition on the catalytic activity of biochar, as a result, the soot yield was increased. Naphthalene was the primary tar in reaction temperature was 900 °C, also had a thermal stable in 900 °C. Fe was deactivated by coke and sintering easily.

The integrated process for hydrogen production from biomass was developed

Basing on the gas-solid simultaneous gasification process of the integrated processes of biomass, the effects of metals (K, Ca, Mg, Fe, Zn, and Al) in biochar on the catalytic gasification and catalytic reforming processes were investigated in this paper.

The catalytic reforming of biomass pyrolytic vapor (MPV) over biochar in a fixed bed was investigated. The carbon conversion and potential hydrogen yield of biomass pyrolytic vapor were increased respectively from 68.46% to 86.88% and 42.16% to 82.76% in the presence of biochar. The catalytic reforming activities of different metals were found with the following the sequence: K>Ca>Mg>Fe>Zn>Al. Meanwhile, the alkali and alkaline earth metals (AAEM) like K, Ca, and Mg were found to play the dominating roles in the conversion of MPV, and especially, the catalytic effect of K was the most obvious among the metals in bio-char. The carbon conversion and potential hydrogen yield of pyrolytic vapor were increased by increasing temperature and ratio of steam to pyrolytic vapor.

The effects of metals on the biochar catalytic gasification were also investigated. The catalytic activities of metals on biochar catalytic gasification were in the sequences: K>Ca>Mg>Zn>Fe>Al. AAEM (K, Ca and Mg) and Fe can promote water-gas shift reaction, while Al had an inhibition effect on water-gas shift reaction. The increasing of temperature and steam were benefit for biochar catalytic gasification. When the carbon conversion was more than 90% (S = 20 g/h and T =900 °C), biochar and demineralized biochar need 40 min and 115min respectively.

The effects of metals on the biochar and biomass pyrolytic vapor (bio-oil) simultaneous catalytic conversion were investigated. The potential hydrogen (H_2+CO) yield was increased from 45.70 to 82.19 mmol/g bio-oil with the presence of metals in the biochar. And the tar yield was decreased from 24.88 to 13.84 mg/g bio-oil. The catalytic conversion of biochar and bio-oil were followed by the sequence K>Ca>Mg>Fe>Zn>Al. Meanwhile, the potential of hydrogen and tar yields were 143.50 mmol/g bio-oil and 4.96 mg/g bio-oil when adding the metal of K. The performance of AAEM (K, Ca and Mg) was comparable with Ni in catalytic reforming bio-oil. AAEM can promote water-gas shift and methane reforming reactions, also promote the elimination of soot. Al had an inhibition on the catalytic activity of biochar, as a result, the soot yield was increased. Naphthalene was the primary tar in reaction temperature was 900 °C, also had a thermal stable in 900 °C. Fe was deactivated by coke and sintering easily.

465897 The Pretreatment of Pennisetum Hybrid with Composite Solid Base Catalyst

Abstract:

The Pretreatment of Pennisetum hybrid with Composite Solid Base Catalyst

Zhuang Xinshu, Tan Xuesong, Yu Qiang, Qi Wei, Wang Wen, Wang Qiong, Zhou Gui Xiong,

Yuan Zhenhong*

(Key Laboratory of Renewable Energy, Guangzhou Institute of Energy Conversion, Chinese Academy of Sciences, Guangzhou 510640, China)

Abstract: Pretreatment is a critical step for the downstream enzymatic hydrolysis to producing cellulosic ethanol. Solid base catalyst can be a potential way for lignin removing, xylose releasing and phenols producing. In this paper the effects of the reaction conditions on enzymatic digestibility and total sugar yield of Pennisetum hybrid pretreated in low temperature aqueous environment with composite solid alkaliâ€"CaO/MgO as the catalyst are investigated. The results indicated that 20% catalyst content, 110â„ƒ,80min and 1:20 ratio of solid are the optimal pretreatment conditions, the lignin removal rate is up to 54.18%.The enzymatic digestibility increases from 41.23% to 81.04% with the enzyme loading of 40FPU/g after72h, substrates can be fully hydrolyzed. After pretreatment, the compact structure of the feedstock is damaged, more fibers exposing, the chemical groups such as hydrogen bond in or between are destroyed.

Keywords: Composite Solid Base; Pennisetum hybrid; Enzymolysis; Pretreatment

This work is financially supported by the National Natural Science Foundation of China (51476179) and the Sino-Thai project (51561145015).

Corresponding author: Yuan Zhenhong, 86-20-87057735; yuanzh@ms.giec.ac.cn.

450795 Realization of Sustainable and Cost-Effective Cellulosic Ethanol Production By Deploying Enzyme and Advanced Yeast Technologies in an Integrated Approach

Abstract:

Ethanol is considered the most potential next generation automotive fuel because it is carbon-neutral and could be produced from renewable resources like lignocellulosic biomass. Key to economic viability of cellulosic ethanol is cheap production and application of different enzymes hydrolyzing lignocellulosic carbohydrates into fermentable sugars and efficient conversion of the derived monomeric sugars into ethanol at high productivities and yields using advanced yeast strains. DSM is a frontrunner in the development of bioconversion and fermentation technologies for the production of cellulosic ethanol, enabling the industry to diversify from starch crops to lignocellulosic agricultural residues, unlocking the full industrial potential of this cheap, abundantly available and sustainable type of feedstock.

Within DSM, research projects aim at developing both commercially competitive cellulolytic enzymes for the liquefaction and saccharification processes, and genetically engineered yeast strains that ferment hexoses and pentoses to ethanol. DSM's enzyme cocktail is uniquely active at high temperature and low pH, offering many benefits and product differentiation. Further strain and process development, enzyme cocktail ratio optimization, protein and genetic engineering have led to significant improvements in productivity, efficiency and costs. Fermentation of hexoses and pentoses from lignocellulosic hydrolysates by DSM's yeast strains has been significantly improved by applying both rational and evolutionary engineering strategies, resulting in C6- and C5-sugars fermenting *Saccharomyces cerevisiae* strains being commercially available. Additionally, next generation advanced yeast strains, which have accelerated pentose utilization, improved robustness, as well as the ability to convert inhibitors and specific biorefinery waste products into additional ethanol, are currently finding their way to commercial implementation.

In this context, it is of great importance to fully integrate the design, construction and operation of the aforementioned technologies into a biorefinery for biofuels production. In this presentation, we will discuss the latest developments and how those fit within the platform of on-site manufacturing of enzymes and yeast fermentation.

468107 Alternative Transport Fuels and Their Production Using Surplus Renewable Electricity, Water and CO_2

Abstract:

Electrical power generated from wind energy fluctuates strongly. As put forward by the EU Commission, in 2050 80 % of the electricity demand should be covered by renewables. Therefore, the integration of fluctuating electricity generation from renewables plays a fundamental role in Europe's and especially Germany's future energy system. Additionally for 2050, the EU Commission postulates a reduction of anthropogenic CO_2 emissions of about 80-95 %, compared to 1990. In Germany, the transport sector was accountable for approximately 20 % of the domestic CO_2 emissions in 2014, while contributing 30 % to the domestic energy demand. One reason for this high share is that 90 % of the transport sector is run on fossil fuels. Hence, to achieve the goals for 2050, non-fossil alternative fuels need to be found. A popular option is the further integration of fuel cell cars. However, especially for heavy duty vehicles and planes, there is long term need for liquid fuels such as diesel and kerosene. Another promising option is the idea of Power-to-Fuel (PTF): the synthesis of fuels based on sustainably produced hydrogen and usage of CO_2 as a feedstock. The energy required to run those processes can be provided by surplus electricity from renewable sources. This is energy which cannot be fed into the grid when there is no demand on the consumer side.

If an alternative renewable fuel is able to penetrate the market, two main requirements need to be met: its potential for an implementation strategy into an existing market as well as a technically feasible and economic production. Put differently, a renewable fuel should not require a significant alteration of the fuel infrastructure, thereby allowing a gradual integration in the existing infrastructure. Additionally, a range of different fuels for different usages need to be offered. Just as the fossil fuel market, the renewable fuel market needs diversification.

At the Institute of Energy and Climate Research (IEK-3) of Forschungszentrum Jülich, intensive research is being carried out on a comparable overview of the several products and process routes of PTF. In this contribution, investigations on two different routes from synthesis gas to an alternative renewable fuel will be outlined. The respective products of the two outlined processes are Fischer-Tropsch (FT)-diesel and polyoxymethylene dimethylene (POMDME). Both products can be blended with conventional diesel and with each other because they have appropriate compatibility. Therefore, an implementation strategy into the existing fuel market is practicable.

Despite an equal product application, the comparison of both technologies is widely diversified. At the molecular level, FT-diesel is closely related to conventional diesel; this applies less to POMDME. POMDME also has a considerably lower heating value; however, it has great potential as a fuel due to its beneficial combustion properties.

Although nowadays FT is not used for PTF in large scale, state-of-the-art FT-processes have been used industrially for many decades. In contrast to FT, owing to low product demand, processes for POMDME have not yet been developed on an industrial scale. Both processes have various options for the interconnection of system components and for complete process control. The system's interconnections as well as the process parameters have influence on raw material and energy demand and also product quality, co- and byproducts. Just as the product specifications, the process analysis results will play an important role in the fuel implementation strategy.

462845 Co-Products from the Catalytic Fast Pyrolysis of Biomass

Abstract:

Catalytic fast pyrolysis (CFP) of biomass is a potentially viable process for producing drop-in, renewable hydrocarbon transportation fuels and the economics may be improved by providing alternative pathways to valuable co-products. This provides flexibility to a biorefinery to adjust conditions to improve economics or to valorize waste low value streams. As an example, CFP using zeolite catalysts such as HZSM-5 produces aromatic molecules with 1-4 rings. The single ring products are excellent additives for gasoline because of their high octane numbers and their high heats of combustion. However, the larger aromatics have boiling points too high for gasoline and low cetane numbers, eliminating them as potential diesel additives. Hydrotreating these molecules is required to convert them to a fuel, but they may have value as co-products. Likewise oxygenated intermediates and products, such as phenols, formed during CFP may be high value products that can also help the economics of hydrocarbon fuels. These oxygen functionalities are often difficult to add to hydrocarbon molecules extracted from petroleum and producing these molecules from biomass may be intrinsically more environmentally benign. In addition, leaving oxygen in the molecules may improve the yields of biofuels and bioproducts, which would again improve the economics. We will discuss studies to improve selectivity of co-products and efforts to separate them and prepare polymers.
Catalytic fast pyrolysis (CFP) of biomass is a potentially viable process for producing drop-in, renewable hydrocarbon transportation fuels and the economics may be improved by providing alternative pathways to valuable co-products. This provides flexibility to a biorefinery to adjust conditions to improve economics or to valorize waste low value streams. As an example, CFP using zeolite catalysts such as HZSM-5 produces aromatic molecules with 1-4 rings. The single ring products are excellent additives for gasoline because of their high octane numbers and their high heats of combustion. However, the larger aromatics have boiling points too high for gasoline and low cetane numbers, eliminating them as potential diesel additives. Hydrotreating these molecules is required to convert them to a fuel, but they may have value as co-products. Likewise oxygenated intermediates and products, such as phenols, formed during CFP may be high value products that can also help the economics of hydrocarbon fuels. These oxygen functionalities

are often difficult to add to hydrocarbon molecules extracted from petroleum and producing these molecules from biomass may be intrinsically more environmentally benign. In addition, leaving oxygen in the molecules may improve the yields of biofuels and bioproducts, which would again improve the economics. We will discuss studies to improve selectivity of co-products and efforts to separate them and prepare polymers.

480817 Application of Molecular Characterization to Bituminous Crude Oil to Study Asphaltene Precipitation

Abstract:

Application of molecular characterization to bituminous crude oil to study asphaltene precipitation

Md Rashedul Islam, Yifan Hao, Meng Wang, Toni Kirkes, and Chau-Chyun Chen

Department of Chemical Engineering, Texas Tech University

Abstract

Bituminous crude oils are vulnerable to asphaltene precipitation when blended with paraffinic samples. Precipitation behavior depends on the nature and content of the paraffins in the blend. A thermodynamic modeling of asphaltene precipitation from crude oil is important as part of the larger effort to understand the compatibility of bitumen or heavy crude oils with the paraffinic ones. Recently, asphaltene precipitations are predicted from wide spectrum of solvents by aggregation thermodynamic framework [1] developed from Yen-Mullins model [2]. However, precipitation prediction from crude oils is still very difficult because of lack of reliable techniques to represent petroleum fluids with real molecules. In this work, a molecule based characterization approach [3] is utilized to represents Cold Lake bitumen with real hydrocarbon molecules. Additionally, asphaltene precipitations are predicted form mixing n-alkanes with Cold Lake bitumen.

References

[1] M. Wang, Y. Hao, M. R. Islam, and C.-C. Chen, "Aggregation thermodynamics for asphaltene precipitation," *AIChE Journal,* vol. 62, pp. 1254-1264, 2016.

[2] O. C. Mullins, "The Modified Yen Model†," *Energy & Fuels,* vol. 24, pp. 2179-2207, 2010.

[3] C.-C. Chen and H. Que, "Method of Characterizing Chemical Composition Of Crude Oil For Petroleum Processing," US Patent Application Number: 2013/0185044, July 18,2013.

464593 Rapid Analysis of Syngas Using Micro GC Fusion

Abstract:

Rapid Analysis of Syngas using Micro GC Fusion

In 2012, the World Bank estimated that 140 billion cubic meters (bcm) of the 3370 bcm of natural gas produced worldwide was flared. Flaring burns surplus gas and is detrimental to the environment, contributing over 350 million tons of CO2 in annual emissions. An alternative to flaring is to convert the associated natural gas into synthesis gas (syngas), which is an intermediate gas used to produce synthetic liquid fuel. Converting syngas to liquid chemicals through the Fischer-Tropsch (FT) process presents an economical and greener alternative to flaring. Syngas is composed predominantly of hydrogen and carbon monoxide, though it may also contain carbon dioxide, nitrogen, methane, ethane, ethylene, or H2S in the high ppm to percent level. After catalytic conversion, determining the syngas composition is a critical step before using the gas in any downstream process. To ensure efficiency, the analysis of syngas must be analyzed quickly and accurately. Building on proven microelectromechanical systems (MEMS) technology, Micro GC Fusion is capable of providing rapid analysis of typical syngas components in a compact instrument. A method equipped with temperature programming provides an area %RSD of less than 1.0% repeatability. This accurate and repeatable analysis of syngas is performed in less than 50 seconds.

470239 Discovery and Manipulation of a Native Polyketides Influencing ABE Production in *Clostridium Acetobutylicum.*

Abstract:

The production of ABE (acetone, butanol, ethanol) by the anaerobic bacteria *Clostridium* is a bioprocess capable of converting various renewable substrates into high value solvents and fuels. *Clostridium acetobutylicum* ATCC 824 has long served as a model organism for clostridial ABE production, with an extensive body of literature ranging from fundamental cellular biology to industrial-scale fermentation. Despite the abundance of knowledge concerning this organism, the series of cellular events responsible for the dramatic and somewhat mysterious transition from acid- to solvent- production (deemed the â€œmetabolic switchâ€) remain unclear. Towards addressing this phenomena, we investigated a polyketide synthase (PKS) gene from *C. acetobutylicum* ATCC 824, which has previously been shown to be upregulated directly preceding the metabolic switch in ABE fermentation of this organism. Through targeted gene knockouts, HRMS-guided comparative metabolomics, and extensive structural characterization, we identified a previously undiscovered family of three polyketides synthesized by the PKS gene product. Batch fermentations of PKS-modified strains revealed that while the polyketides are not essential for the metabolic switch (and hence, ABE production), the presence or absence of these compounds could influence both the solvent titers and productivities of *C. acetobutylicum*. Through this work, we hope to highlight the potentially significant role of secondary metabolites in clostridial fermentations, and provide a relatively unexplored strategy for manipulating these industrially-relevant organisms.

463439 Machine Learning-Driven Pathway Optimization: Application of Automation in Industrial Biotechnology

Abstract:

Microbial biosynthesis of fuels and chemicals often requires heterologous expression of metabolic pathways. To achieve the highest efficiency, the pathway flux must be carefully tuned by changing the expression level of each enzyme in the pathway coupled with a high-throughput screening. However, the high-throughput screening is unavailable for most targets. Here, we present a versatile platform for pathway optimization where a machine learning algorithm directs the experimentation to find the highest producing combination with a minimum number of experiments. Compared with other methods like Exterior Derivative Estimation (EDE) used by Lee et al., [1], our machine learning algorithm was able to find the maximum of the test data with less than half of evaluations required by EDE model. As proof of concept, we sought to optimize the lycopene biosynthesis pathway with 24 different expression levels for each gene resulting in more than 13,000 different combinations. The expression level gradient was achieved by mutating both the T7 promoter and ribosome binding sites (RBS). Although we have developed the iBioFAB platform [2], assembly and characterization of all these combinations is very difficult and expensive, if not impossible. By adding the machine learning component and developing an automation friendly lycopene extraction assay, we can achieve an unprecedented level of automation in pathway optimization. This machine learning driven pathway optimization strategy can be applied to optimize any pathway of interest, especially when a high throughput screening method is not available.

1. Lee, M.E., et al., *Expression-level optimization of a multi-enzyme pathway in the absence of a high-throughput assay.* Nucleic Acids Res, 2013. **41**(22): p. 10668-78.

2. Chao, R., Y. Yuan, and H. Zhao, *Building biological foundries for next-generation synthetic biology.* Sci China Life Sci, 2015. **58**(7): p. 658-65.

461757 Engineering Methanogenesis Pathway of Methanogenic Archaea for Biofuel Production

Abstract:

Engineering methanogenesis pathway of methanogenic archaea for biofuel production

Maryam Raeeszadeh-Sarmazdeh, Jacqueline Gonzalez, Wilfred Chen

Department of Chemical and Biomolecular Engineering, University of Delaware, Newark, DE 19716

Methane gas is the second most prevalent greenhouse gas and an abundant source of energy in the United States. Methane is both produced and consumed in marine sediments through anaerobic methanogenesis. Methanogenic archaea primarily use one-carbon substrates such as CO_2, acetate, or methanol for growth and methanogenesis through CO_2 reduction, aceticlastic, or methylotrophic pathways, respectively. The key step in the methanogenesis pathway, reducing methyl-coenzyme M (CH_3-S-CoM) and coenzyme B (HS-CoB) to methane, is catalyzed by Methyl-coenzyme M reductase (MCR). This reaction, which leads to the formation of a heterodisulfide substrate (CoM-S-S-CoB), is exergonic. Therefore, for reverse methanogenesis to occur, the methane oxidation reaction needs to be coupled with an endergonic reaction that provides a thermodynamically favorable situation.

The genetic toolbox in methanogenic archaea is limited. Therefore, we have investigated Nanoluciferase expression, as a sensitive reporter, through episomal expression and integration into the genome in *Methanosarcina acetivorans* . In order to engineer anaerobic Methanogenesis pathway to produce biofuel, we have expressed functional Methyl-coenzyme M reductase (MCR) from *Methanothermobacter marburgensis* in *Methanosarcinals*. The recombinant strains showed improvement in methane formation yield compared to the wild-type strain. Furthermore, the effect of soluble electron acceptors such as Fe(III), and the regeneration of a heterodisulfide substrate on methane oxidation were investigated.

459970 Production of Wax Esters As High-Value Compounds in Yeast *Yarrowia Lipolytica*

Abstract:

Wax esters have a lot of applications in lubricant, cosmetics, inking, and coating industries. Currently the main biological source for wax esters is the seeds of jojoba. Mainly due to the limited supply of jojoba, wax esters are currently produced at a high cost. To alleviate this issue, we introduced the wax ester biosynthetic pathway into the oleaginous yeast, *Yarrowia lipolytica*, which is well known for its superior ability to accumulate triacylglycerol (TAG) under appropriate culturing condition and genetics combinations. To divert the metabolic flux from TAG biosynthesis to wax ester biosynthesis, the genes involved in the wax synthetic pathway from difference sources with unique substrate specificities were tested, and several metabolic engineering strategies have been applied to increase precursor availability and remove substrate competition. The resulting strain produced wax esters at a titer 5-fold higher than the previous record in literature. With a comprehensive optimization on medium recipe, we anticipate a further significant improvement in wax ester titer.

471627 A Microbial Factory for Sustainable Production of Polymer Precursors: Chain Length-Specified ω Hydroxy Fatty Acids and α,ω-Dicarboxylic Acids

Abstract:

In the hope of advancing sustainable plastic production, great progress has been made toward efficient microbial conversion of renewable feedstock to short-chain (C2-C8), bifunctional polymer precursors (e.g. succinic acid, cadaverine, 1,4-butanediol). Relatively little attention has been given to medium-chain (C12-C14) monomers such as ω-hydroxy fatty acids (ω-OHFAs) and α,ω-dicarboxylic acids (α,ω-DCAs), which nonetheless are precursors to high performance polyesters and polyamides. In our work, we are engineering complete microbial conversions of glucose to various medium chain ω-OHFAs and α,ω-DCAs, with precise control of product chain length. We combine bioinformatics and *in vivo* production to screen a wide range of enzymes across phyla and identify combinations that yield complete conversion of intermediates to product α,ω- DCAs. This approach, coupled with optimization of culture conditions has allowed us to achieve near g/L titers of specific medium chain α,ω- DCAs. In particular, our results indicate potential for this microbial factory to enable commercially relevant, renewable production of C12 α,ω- DCA – a valuable precursor to the high-performance plastic nylon-6,12.

472557 Model-Guided Metabolic Engineering of Increased 2-Phenylethanol Production in Plants

Abstract:

2-Phenylethanol (2-PE) is a naturally occurring aromatic with properties that make it a candidate oxygenate for petroleum-derived gasoline. In plants, biosynthesis of 2-PE competes with the phenylpropanoid pathway for the common precursor phenylalanine. The phenylpropanoid pathway in plants directs approximately 30% of carbon flux towards the biosynthesis of lignin, a major constituent of plant cell walls that impedes the process of biofuel production. Therefore, we propose a genetic engineering strategy at the phenylalanine branch point, whereby a portion of the carbon flux towards lignin biosynthesis is diverted towards the production of an economically valuable product, 2-PE. Transgenic *Arabidopsis thaliana* were generated that overexpress aromatic aldehyde synthase (AAS) in tandem with phenylacetaldehyde reductase (PAR) from tomato introducing a pathway for the production of 2-PE. The primary stem was selected as the experimental system for analyzing competition of Phe between phenylpropanoid metabolism and the engineered 2-PE pathway. Excised 5-week-old stems were exogenously fed different concentrations of $^{13}C_6$-ring labeled Phe. Both the amount and isotopic enrichment of downstream intermediates was quantified using LC-MS/MS at multiple time points after feeding. A kinetic model of the phenylpropanoid network was constructed, and the parameters were identified through non-linear optimization with training datasets, and validated with data from an independent experiment. *In silico* analysis of the results from our model predicted that the endogenous cytosolic Phe pools limit the 2-PE production in these transgenic plants. This prediction was tested by combining overexpression of PAR and PAAS with overexpression of a feedback-insensitive 3-deoxy-D-arabino-heptulosonate 7-phosphate (DAHP) synthase, the latter of which has been previously shown to have hyper-induced phenylalanine biosynthesis in Arabidopsis. This transformation led to a significantly increased accumulation of 2-PE in transgenic *Arabidopsis*. The use of kinetic modeling combined with time-course *in vivo* metabolite profiling is shown to be a promising approach to rationally engineer plants that accumulate high-value commodity chemicals.

474734 The Emerging Organism Engineering Industry (Featured Presentation)

Abstract:

There is an emerging demand for sourcing plant-derived extracts (nutraceuticals, flavors, fragrances, sweeteners, etc.) from engineered microbes. While recent advances in synthetic biology and metabolic engineering provide feasible approaches to engineering such organisms, commercial success for developing these "cultured" ingredients presents specific challenges. Unlike biofuels, where efforts can be focused on one particular molecule given the enormous market size, cultured ingredients require developing different organism lines in a rapid and low cost fashion. This requires a scalable solution for bio-manufacturing of organisms, which is provided by our state of the art foundry that continues to grow. I will describe how organism development at Ginkgo leverages our foundry to accelerate the design/build/test using specific examples. In particular, I will highlight the value of combining computer-aided engineering software tools, cheap gene synthesis and high resolution-accurate mass LCMS to develop engineered microbes. Finally, I'll touch on how our improvements in manufacturing organisms lend to opportunities outside of cultured ingredients.

470510 *Smartwaste*- Biomass Conversion into Low Oxygenated Hydrocarbons in Subcritical Water

Abstract:

Sustainable management of food wastes as well as animal manure and other agricultural wastes is a major challenge facing food production worldwide. The dairy and food industries generate approximately 19 and 36 million tons of waste per year, respectively, only in the U.S. The majority of these residues, rich in organic carbon and valuable nutrients, are nowadays disposed in landfills. The current management strategies for manure and food waste are mainly based on biological processes, converting organic wastes into biogas (anaerobic digestion). Limited biogas conversion yield (40-50%) generates a large volume of liquid effluent, containing an aqueous mixture of carbohydrates, proteins, lipids, minerals and nutrients that should be further processed to exploit its economic value. The *SmartWaste* approach is a process integration strategy that uses a series of thermochemical processes to recycle multiple waste streams, producing chemicals, fuels, and nutrients. Furthermore, these fast, selective, and efficient waste-to-fuel processes can be easily integrated into current bio-refineries flow-chart.

Among recycling approaches for high water content waste feedstocks, *SmartWaste* proposed the hydrothermal biomass conversion technology, using both sub- and supercritical water. The suggested approach is based on three-stage processing of waste. First, the pretreatment step focuses on blending different waste streams. Achieving and appropriate feed composition is key to increase the bio-oil production yield in the second stage. In the second step, hydrothermal liquefaction (HTL) of the blended waste stream with subcritical water significantly reduces the oxygen content of the feed to approximately 10% by dewatering and condensation reactions. In the third step, the hydrogenation process, the produced bio-crude oil is further upgraded, by reducing the oxygen content.

The *SmartWaste* process depolymerizes organic waste into smaller structural units. The HTL step is based on fast hydrolysis reactions, followed by dehydration and condensation of sugar, lipids, proteins, and their degradation products, in a single step reaction using subcritical water. This process benefits from the modification in chemical behavior of supercritical water while approaching to the critical point conditions (374 °C and 22.1MPa). For instance, the increase of ionic product (K_W) favors ionic, and acid-base reactions without the addition of a catalyst. Therefore, the optimum processing temperature for this stage is expected to be between 250 and 350 °C. In terms of reaction time, these reactions are usually completed within 20-60 minutes. This represents a big step towards process intensification (reduction of required reactor volume) compared to biochemical processes (enzymes) that require few days.

The main challenges facing residual biomass recycling are the highly complex polymeric composition, together with the high oxygen content, making this feed low in energy content. The flexibility of the hydrothermal technology allows the manipulation of process conditions (reaction time, temperature, pressure) to adapt to different feed compositions and achieve higher bio-oil yields (around 70% on a carbon basis), with higher energy values (oxygen content of ~10% w/w). This oxygen content represents a reduction of 70% with respect to the feed waste. As a consequence of the dehydration and condensation reactions of waste compounds for the generation of bio-oil, an aqueous-soluble carbon phase is also produced. In addition, the solubilized-hydrolyzed amount of solids is not complete, being bio-char being one the products of HTL. To complete the product distribution, a gaseous phase is also generated. The relative amount of any of these four phases depends on reaction conditions, apart from the presence of catalyst.

In this presentation, we will present experimental data for the HTL compartment of the *SmartWaste* project. Carbon mass balances for a wide range of biomass feedstocks will be discussed. This included modeling compounds (polysaccharides, monosaccharides, proteins and long-chain fatty acids) as well as real waste biomass (diary manure digestate collected from anaerobic digestion in upstate NY, and carbohydrate rich food waste from Cornell's dining halls). An example for carbon mass balances is presented in Fig. 1, showing the highest production of bio-crude oil from a mixture of polysaccharides, proteins and lipids at 300 °C with negligible char production. The quality of the bio-crude oil is showed in the Van-Krevelen diagram (Fig. 2), representing atomic rations H/C vs. O/C. HTL significantly reduces the O/C and H/C ratios for all thermal conditions studied. During the HTL process, the quality of the oil intermediate can be highly improved by choosing the adequate biomass feed. For example, the alcoholic groups used for condensation can be synthesized directly in the reactor by lignin hydrolysis, and high pH catalytic media can be achieved by proteins hydrolysis and ammonia release.

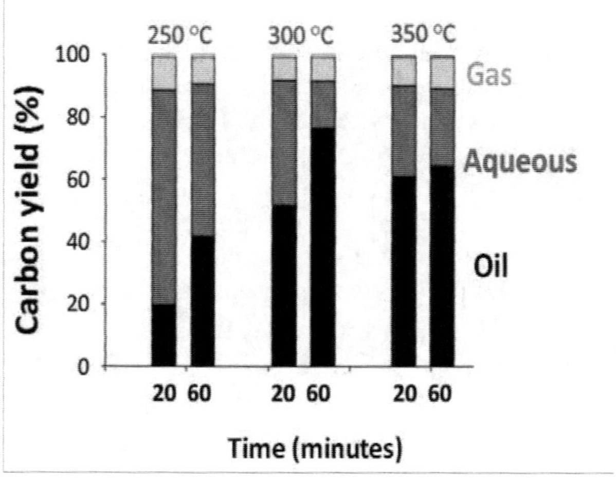

Fig. 1. Carbon mass balance for hydrothermal conversion of ternary modeling mixtures containing polyssacharides, proteins and lipids. The amount of gas phase was estimated as 100-Oil-Aqueous-Solid %.

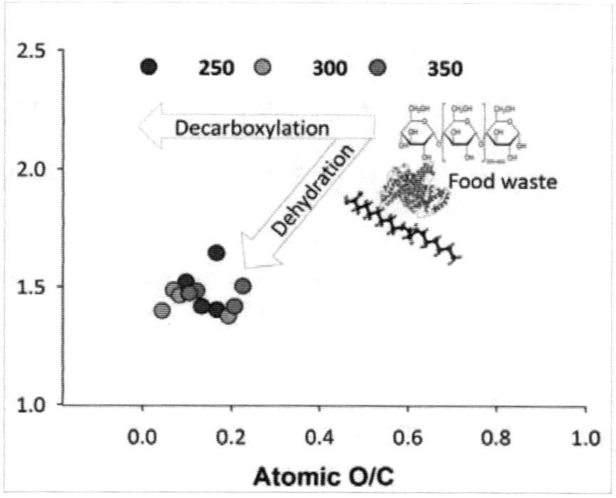

Fig. 2. Van-Krevelen diagram of the produced bio-crude oil upon hydrothermal conversion of ternary modeling mixtures containing polysaccharides, proteins and lipids.

The target product of the *SmartWaste* approach is a stream of alkanes with minimal oxygen content. Further reduction of the bio-oil oxygen content by hydrogenation on supported noble metal catalysts will maximize the energy density of the final products. The production of low polar stream of light straight-chain alkanes (4 to 10 carbons) is achieved by functional reaction routes including hydrolysis, hydrogenation, dehydration, and hydro-deoxygenation steps. Thus, bifunctional acid/redox catalysts are required for the hydrogenation step, where hydrolysis and dehydration processes are carried out on solid acid sites and hydrogenation on active metal sites of the catalyst. The *SmartWaste* produces carbon streams that have similar characteristics to the streams out of an oil refinery. Therefore, future production of biofuels, using small and selective hydrothermal reactors and taking advantage of the available infrastructure for oil refining, without large capital investments, is a reasonable and promising application.

In this research project it has been demonstrated that the hydrothermal liquefaction of waste biomass is an efficient process to convert low value material into enriched fuel and chemicals precursors. The HTL process was able to reduce the oxygen content of the waste streams in 70%. This represents the major oxygen reduction in the whole process. It was also demonstrated that the composition of the biomass reacting is a key factor to consider when designing the process. Adapting the correct combination of sugars, proteins and lipids could increase the carbon oil yields from 20-30% w/w up to 70% w/w.

468527 Catalytic Hydrothermal Gasification of Post Hydrothermal Liquefaction Waste Water for Biogas Production

Abstract:

Hydrothermal liquefaction (HTL) of complex organic materials yields products including bio crude oil, post HTL wastewaster (PHWW). The PHWW contains approximately 100,000 mg/L COD of organics, which are mostly water soluble organic acids and nitrogen-containing aromatics. In order to further assimilate the residual energy and nutrients in PHWW, a sequential catalytic hydrothermal gasification (CHG) was used in this study to produce valuable biogas to maximize the energy recovery from HTL process, reuse the nutrients, and at the same time clean the PHWW. The effect of catalyst selection was investigated using ruthenium (Ru), sodium hydroxide (NaOH), and zeolite, which were used individually and in combination (e.g. Ru plus NaOH) to determine the optimal organic removal rate, gas yield, chemical compositions of gas products, and nutrient fate. Preliminary results demonstrated that a high gasification efficiency of 98% could be achieved in terms of COD removal. Hydrogen production was able to reach up to 28 mmol per g of organics in the PHWW, and methane (CH_4) production up to 21 mmol per g organics in the same condition. The concentration of water soluble short chain organic acids (C1-C4) increased, compared to PHWW feedstock, at lower temperature of 400°C but then decreased and was almost completely removed at higher temperature of 600°C. The concentration of ammonia-nitrogen was raised by 68% in the PHWW after CHG treatments, indicating that a significant denitrogenation of the nitrogen-containing aromatic compounds took place, thus better nutrient recovery was achieved.

463234 Development of an Integrated Process for Sustainable Production of Biodiesel and Plant Protein Isolates from Underutilized Feedstock

Abstract:

â€œFood or Fuelâ€ is an international concern about the use of agricultural land for biofuel production, which results in higher food prices and adversely affects the living standards of poor people. Consequently, it is essential to find compromises that meet the demand for both food and green transportation fuels. Yellow mustard is a Canadian food crop that is a viable source of both oil and protein. At present neither of these valuable components are utilized as most of the seed is now only used as condiment. The proteins have a well-balanced amino acid distribution making them potentially useful as a food ingredient, and the inedible oil, which is high in erucic acid, is desirable in biodiesel due to its superior lubricating properties. In this study, an integrated process was developed to produce food-grade protein isolates, free of solvent contact/residues, and fuel-grade biodiesel from dehulled yellow mustard flour. This process makes novel contributions to the link between energy, food and the environment.

The aqueous extraction of proteins form mustard flour was started by contacting the flour with water at room temperature and alkaline pH 11 followed by centrifugation. The protein was recovered in the form of protein-rich skim fraction with minimal protein damage, while the oil was released as a separate fraction and tied up in a very stable oil-in-water emulsion. The protein-rich skim fraction was further concentrated and purified by membrane processing followed by isoelectric precipitation to produce two high quality food-grade protein isolates: isoelectric precipitated protein isolate (PPI) and acid soluble protein isolate (SPI). The PPI and SPI with desired functionalities had 96.0 and 83.5% protein content on a moisture and oil free basis, respectively.

The underutilized oil-rich emulsion fraction was characterized where its remarkable stability was due to the presence of protein emulsifiers of high molecular weight along with the mixed phospholipid-oleosin layer. The emulsion was successfully destabilized by a novel chemical approach where the organic solvents including isopropyl alcohol (IPA), tetrahydrofuran (THF) or dioxane were used to solubilize the emulsion to produce a single-phase oil-solvent-water miscella. The ternary phase diagrams of THF/oil/water, dioxane/oil/water, and IPA/oil/water were prepared, and the solubility of the oil in these organic solvents in the presence of water was determined. The application of ternary phase diagrams resulted in the development of three-stage destabilization process at which all of the oil was successfully recovered from the emulsion into the miscella. The oil-solvent-water miscella was dehydrated in the next phase by adsorption over zeolite 4A using either batch or continuous fixed-bed systems. The dehydrated miscella was reacted with methanol in a single-phase base-catalyzed transmethylation process with high yields (99.4%) to fatty acid methyl esters (FAME). The resulting FAME satisfied the international standards for use as biodiesel fuel.

Approximately 69% of the protein originally in the flour was recovered as high-quality protein products and over 62% of the oil in the flour was recovered as standard biodiesel through this integrated process. The preliminary economic analysis showed an overall positive margin. The application of this processing approach could help increase Canadian production of both food ingredients and biofuel, without impacting the cost of edible oils, as very large areas of Canada can be readily converted to mustard production.

COMBUSTION OF PYROLYSIS CHAR WITH A FLUIDIZED BED REACTOR FOR GENERATING PROCESS HEAT

Nicholas, R. Schwartz, Paul E. Yelvington*

Mainstream Engineering Corporation, Rockledge, FL

Corresponding Author: nschwartz@mainstream-engr.com

Abstract

Portable biomass fast pyrolysis processes require combustion of byproducts to provide heat to the reactor and to dry feedstocks. Heat integration and byproduct utilization are especially important for small-scale, distributed processes where heat losses become more problematic. Pyrolysis char produced with a 1 ton per day (tpd) pilot reactor from pine woodchips was combusted with a pilot-scale fluidized bed reactor (FBR) to demonstrate the feasibility of process heat generation. The char product was characterized prior to combustion to determine carbon content, moisture, ash content, and heating value. Particle size distributions (PSD) of the char feed and collected ash material were measured. The fluidizing gas flow rate, gas inlet temperature, excess air percentage, and reactor bed temperature were varied to determine the effects on emissions and combustion efficiency. Gaseous emissions were continuously monitored for a range of fluidizing gas velocities and reactor bed temperatures. The effects of excess air on gaseous emissions were determined along with the inlet temperature of the fluidizing gas. The effects of reactor conditions on CO, SO_2, and NO_x were assessed using the FBR with a variable feed rate of 0.3 to 0.4 kg/h. Based on the experimental results, increasing the excess air and reactor bed temperature reduced CO emissions. Increasing the fluidizing gas flow rate decreases the residence time resulting in increased CO. Particulate emissions were determined after each run by measuring the ash blowout from the cyclone separator. Total fine particulates were collected within a baghouse filter after 16 hours of operation. Reactor performance data using the pyrolysis char fuel will be used to scale-up the FBR design to a 7 kg/h system, right-sized for a nominally 1-tpd distributed pyrolysis process.

Introduction

Fast pyrolysis is a process whereby organic material is converted to a liquid bio-oil, char, and combustible non-condensable process gas. The fast pyrolysis process depolymerizes organic material at high temperatures (500°C) in an oxygen-free environment with short vapor residence times, which favors bio-oil production.[1-4] The bio-oil is similar in refinement to a fossil crude oil (higher oxygen and water content), which can be burned directly or further upgraded to higher value products. Pyrolysis is the ideal platform for conversion of organic materials and waste to bio-oil, char, and process gas and has very low emissions compared to incineration. Pyrolysis offers the advantage of processing a wide range of materials at a single set reactor conditions. Byproducts of pyrolysis can be burned for process heat for a near self-sustaining process. Burners tailored to pyrolysis byproducts offer cleaner combustion and lower emissions compared to direct combustion of solid waste material. Bio-oil and char can be stored for later heat or power production with a significant reduction in waste volume.

Previously, a continuous bench-scale unit was optimized over a range of process conditions to maximize the bio-oil yield and quality. The heating values and typical percent yields of the bio-oil, char, and gas for fast pyrolysis are listed in Table 1.

Table 1. Heating Value and Yields for Bench-scale Pyrolysis

Component	Lower Heating Value (MJ/kg)	Yield (wt%), dry
Bio-oil	17.5	65
Char	29.7	15
Non-condensable Gas	7.3	20

Methods

Pyrolysis char collected from the 1-tpd reactor was characterized prior to combustion within a fluidized bed reactor (FBR). Chemical and physical properties of char and ash were characterized by PSD, heating value, moisture content, carbon content, and elemental composition. The PSD was completed following ASTM standard D6913 for the pyrolysis char (shown below in Figure 1).[5] The char product was fine and uniform with 85% by weight is smaller than 0.149 mm and an average particle size of 0.124 mm.

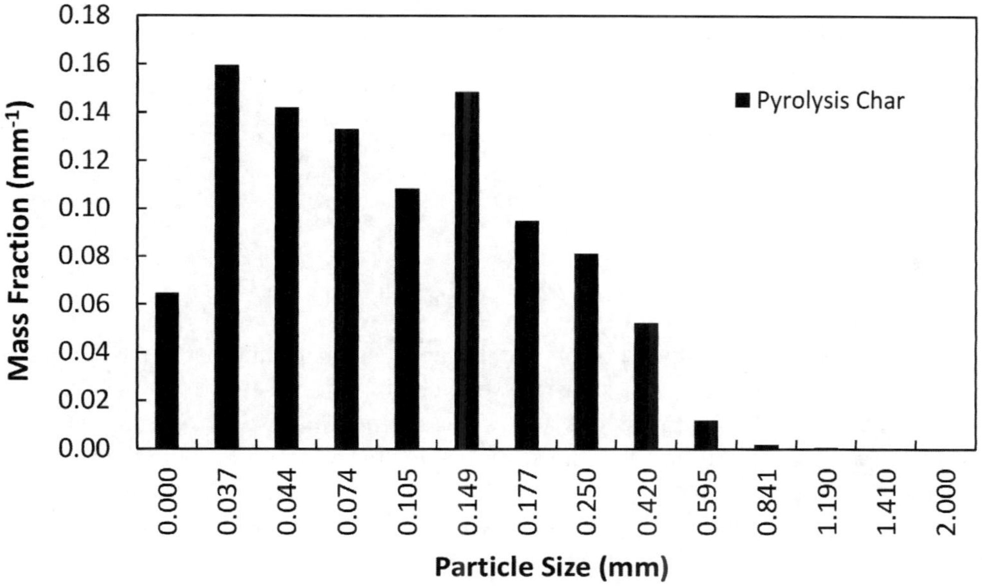

Figure 1. Particle size distribution for pyrolysis char produced by pyrolysis of pine

Heating value, density, PSD, and moisture content are important variables for designing the char combustion FBR. The complete analysis assisted with predicting reactor performance and combustion byproducts. Heating value, moisture, ash, and fixed carbon content for each char material were also determined (results listed in Table 2).

Table 2. Heating Value, Moisture, Ash, and Fixed Carbon Content of Pyrolysis Char

Property	Pyrolysis Char
Heating Value,$_{dry}$ (MJ/kg)	29.7
Moisture (wt%)	6.03
Ash Content (wt%)	5.26
Fixed Carbon (wt%)	82.70

The elemental analysis of the char product determined all detectable constituents of the char and ash. Major constituents of the ash for the char are listed in Table 3. The ash was comprised of mostly CaO, K_2O, Al_2O_3, MgO, and SiO_2, which are beneficial soil additives.[6] One study measured volatile elements produced during pyrolysis and determined most elements are retained within the char.[7]

Table 3. Elemental Composition of Major Ash Constituents

Composition (wt%)	Pyrolysis Char
C	82.70
CaO	7.43
K_2O	3.18
Al_2O_3	2.31
MgO	2.22
SiO_2	1.86
P_2O_5	0.90
Fe_2O_3	0.39
Na_2O	0.26
SO_3	0.25
MnO	0.13

Pyrolysis char produced by the 1-tpd reactor was burned using an automated FBR at the Energy and Environmental Research Center (EERC), which simulates bed chemistry, ash interactions, and emissions under controlled conditions (shown in Figure 2). Emissions and process temperatures were continuously monitored for one hour at steady-state at each set of reactor conditions. Pyrolysis char was fed into the reactor through a single screw feeder. Preheated fluidized gas entered the bottom of the reactor and fluidized the sand and char particles fed into the combustion chamber. External ceramic heaters were used to maintain the combustor and cyclone temperatures. Sand and ash leaving the FBR were collected using an ash cyclone, where the remaining gas exited the system to the gas analyzers.

Figure 2. Fluidized bed char combustor tested at EERC with pyrolysis char

The fluidizing gas temperature, gas velocity, percentage of excess air, and reactor bed temperature were adjusted to determine the effects of process conditions. While burning the pyrolysis char within the FBR, the CO, CO_2, SO_2 and PM emissions were continuously monitored over time for each reactor condition listed in Table 4.

Table 4. Test Matrix for Pyrolysis Char Burning with Fluidized Bed Reactor

Test	Preheat Temp. (°F)	Gas Velocity (ft/s)	Excess Air (%)	Avg. Bed Temp. (°F)
1	400	2.5	25	1,550
2	675	2.5	25	1,550
3	675	2.5	35	1,550
4	675	2.5	15	1,550
5	675	2.5	25	1,650
6	675	2.5	25	1,450
7	675	1.5	25	1,550
8	675	3.5	25	1,550

Results

Flue gas emissions, process temperatures, and flow rates were continuously measured for the FBR for all evaluated reactor conditions. The NO_x, CO, and SO_2 emissions were corrected to a common O_2 content (3% O_2). Average emissions were also reported in terms of a mass-energy basis (lb/MMBtu) for comparison to other fuels. Reactor temperatures were varied from 1,450 °F to 1,650 °F, gas velocity from 1.5 ft/s to 3.5 ft/s, and the excess air from 15% to 35%. Due to the low sulfur content in the pyrolysis char fuel, SO_2 emissions were extremely low and averaged across all tested reactor conditions.

NO_x Emissions

Reactor bed temperature, fluidizing gas velocity, and excess air level all have a significant impact on NO_x emissions. NO_x emissions typically increase with increasing temperature (Figure 3), but some fuels result in decreasing NO_x emissions with increasing temperature. NO_x was highest for the low temperature at 1,450 °F (198 ppm) and lowest for the high temperature at 1,650 °F (178 ppm).

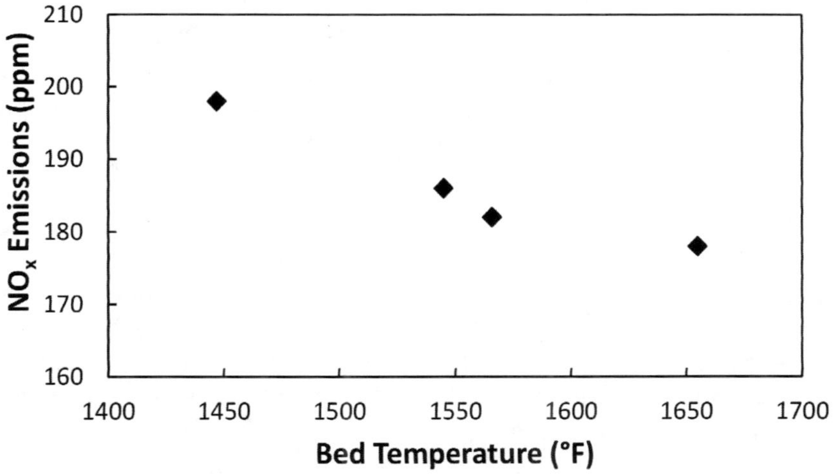

Figure 3. Effect of bed temperature on NO_x emissions

Fluidizing gas bed velocity, which is inversely proportional to gas residence time, has an even greater impact on NO_x emissions (Figure 4). As velocity increased from nominally 1.5–3.5 ft/s, the residence time decreased from approximately 3 to 1 seconds, NO_x increased from 163 to 205 ppm.

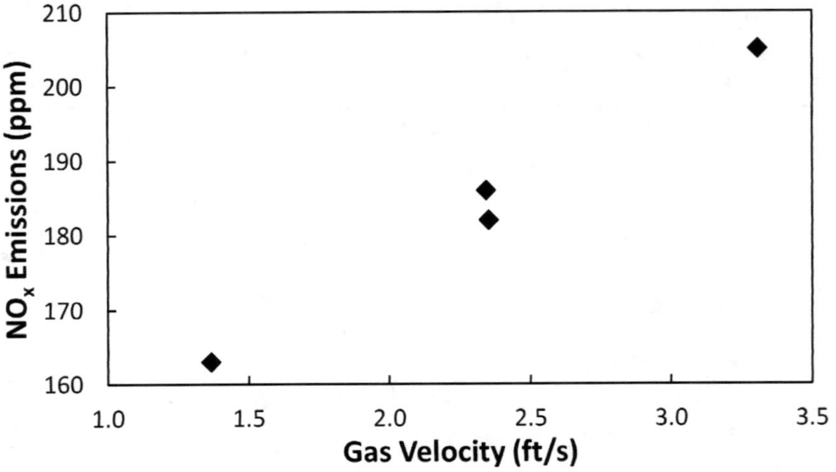

Figure 4. Effect of superficial gas velocity on NO_x emissions

NO_x emissions increased with higher levels of excess air, as expected, increasing from 175 ppm at an excess air level of 11% to 194 ppm at 36% excess air (Figure 5). The NO_x emissions were on average 184 ppm at 26% excess air. As lower excess air improves NO_x emissions, determining optimal reactor conditions will be a trade-off with CO emissions. Average NO_x emissions over all test conditions was 0.331 lb NO_x/MMBtu.

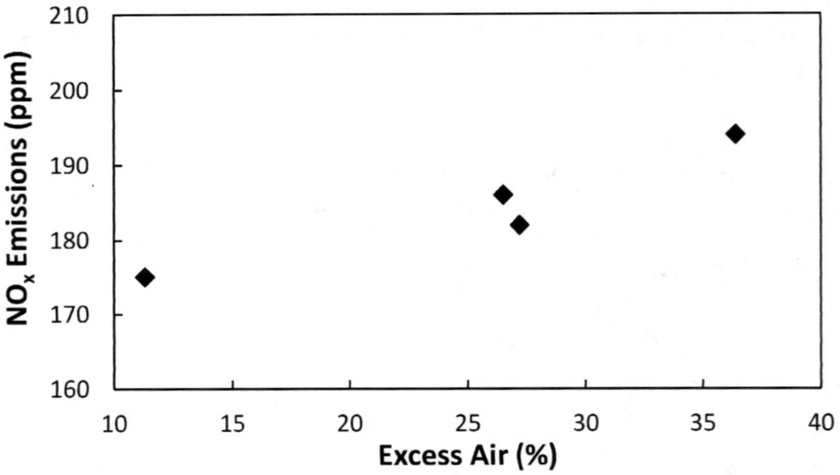

Figure 5. Effect of excess air on NOx emissions

CO Emissions

CO emissions are an indicator of combustion efficiency, with lower CO levels indicating high carbon burnout. As with NO_x, CO emissions can be affected by reactor bed temperature, excess air, and fluidizing gas velocity. Increased reactor temperature typically results in better carbon burnout, which was demonstrated with pyrolysis char (Figure 6). At a high temperature of 1,650 °F, the CO emission was 48 ppm; and at 1,450 °F, the CO emissions averaged 605 ppm.

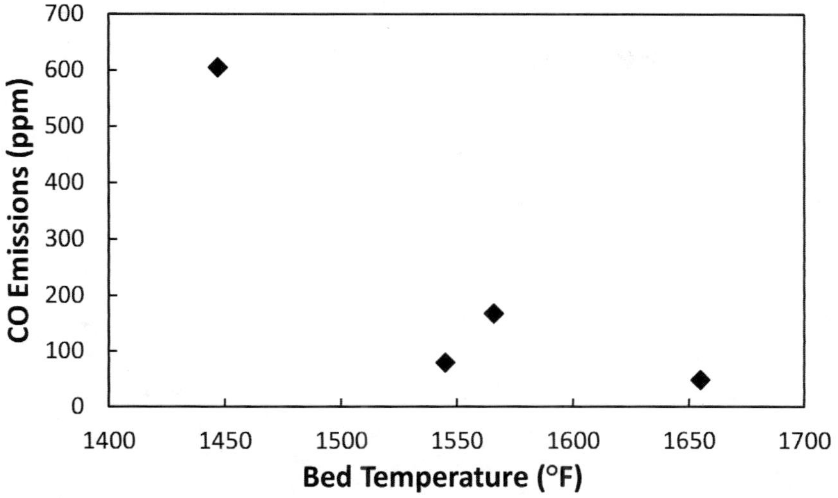

Figure 6. The effect of reactor temperature on CO emissions

The effect of fluidizing gas velocity on CO emissions was more pronounced in comparison to NO_x emissions, as shown in Figure 7. CO averaged 758 ppm at the highest velocity tested (3.3 ft/s), which is significantly higher compared to 83 ppm at 1.4 ft/s and 168 ppm at 2.3 ft/s. There appears to be a specific point at which CO increases sharply with increased gas velocity and the combustion efficiency decreases.

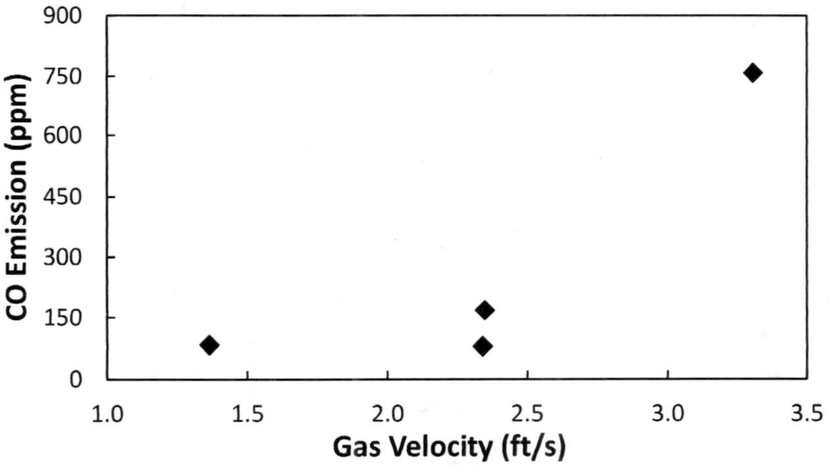

Figure 7. The effect of gas velocity on CO emissions

Excess air had much less of an effect than either reactor temperature or velocity. The CO emissions measured at varying levels of excess air are shown in Figure 8. CO emissions were slightly higher at low excess air (131 ppm) than at high excess air (96 ppm). When the pyrolysis char was combusted with 26% excess air, the CO emissions were 79 ppm with char collected from the reactor bed and 168 ppm with fuel collected after the reactor. Another test that had a lower temperature fluidizing gas velocity with the same level of excess air produced CO emissions of 134 ppm. Over all test conditions, the average CO emissions were 0.224 lb CO/MMbtu.

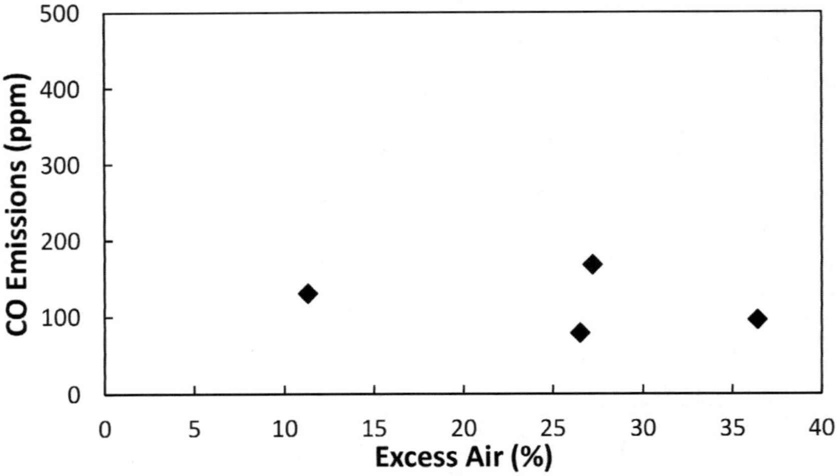

Figure 8. The effect of excess air on CO emissions

Particulate Matter & SO_2 Emissions

The sulfur content of the char was very low, which resulted near-zero SO_2 emissions. Average SO_2 emissions across all reactor conditions was 3.4 ppm (0.020 lb SO_2/MMBtu). Particulate matter (PM) emissions were determined by measuring the total solids captured by the baghouse filter downstream of the FBR over all reactor conditions. The mass balance across the FBR, cyclone, and baghouse filter for ash and sand bed material is detailed in Figure 5. Overall mass balance closure for ash and sand bed material was 85.1%.

Table 5. Mass Balance for Char over FBR, Cyclone, and Baghouse Filter

Total Material In (g)		Total Material Out (g)	
Fuel Feed	6557 g	Cyclone	1200 g
Ash Content	1809 g	Baghouse	17 g
Sand Material	1600 g	Fluidized Bed	1685 g
	3409 g		2902 g

The total amount of exhaust gas was totaled over the 16 hour operating period for an average volumetric flow rate of 1.5 ft^3/min (0.0425 m^3/min). Only as the solids collected by the baghouse filter (total of 17 g) were assumed to contribute to the PM emissions, which was estimated to be 417 mg/dscm. Implementing a baghouse filter into a pyrolysis char burner would capture and eliminate any PM emissions.

Conclusions

Pyrolysis char produced from fast pyrolysis can be combusted for heat generation within a FBR. The effects of reactor temperature, fluidizing gas flow rate, and excess air on NO_x and CO emissions were determined. Low NO_x and CO emissions and high combustion efficiency can be achieved by combusting pyrolysis char at an operating temperature of 1,550 °F or greater within the FBR. Lower fluidizing gas velocities (1.5 ft/s) and excess air (11%) resulted in lower NO_x emissions and improved carbon burnout. CO emissions were strongly dependent on reactor temperature and gas velocity, where higher bed temperatures and lower gas velocities improved combustion efficiency. Changes in excess air had minimal influence on CO emissions across the tested reactor conditions. No agglomeration or fouling was observed during the test, which indicated the pyrolysis char burned cleanly. The fluidized bed pressure drop slowly increased over time, which indicates that intermediate bed draining may be required to maintain pyrolysis char burner conditions. Measured emissions over the tested reactor temperature, flow rates, and excess air provides a starting point for large-scale pyrolysis char FBR operation. Bed depth, diameter, and vapor space height will influence control of reactor conditions and fluidization characteristics. Effects of reactor geometry must be considered along with reactor conditions for the scale-up design for the pyrolysis char burner.

References

1. Bridgwater, A. V., *Fast Pyrolysis of Biomass: A Handbook*. CPL Press: Newbury, UK, 2002; Vol. 2, p 426.
2. Bridgwater, A. V., *Fast Pyrolysis of Biomass: A Handbook*. CPL Press: Newbury, UK, 2005; Vol. 3, p 221.
3. Bridgwater, A. V.; Meier, D.; Radlein, D., Overview of fast pyrolysis of biomass. *Organic Geochemistry* **1999,** *30* (12), 1479-1493.
4. Bridgwater, A. V.; Peacocke, G. V. C., Fast pyrolysis processes for biomass. *Renewable & sustainable energy reviews* **2000,** *4* (1), 1-73.
5. ASTM, Standard Test Methods for Particle-Size Distribution (Gradation) of Soils Using Sieve Analysis. 2009; Vol. ASTM D6913.
6. Heitkotter, J.; Marschner, B., Interactive effects of biochar ageing in soils related to feedstock, pyrolysis temperature, and historic charcoal production. *Geoderma* **2015,** *245-246*, 56-64.
7. Liaw, S. B.; Rahim, M. U.; Wu, H., Trace Elements Release and Particulate Matter Emission during the Combustion of Char and Volatiles from in Situ Biosolid Fast Pyrolysis. *Energy and Fuels* **2016,** *30* (7), 5766-5771.

472362 Small Scale Distributed Ammonia Produced: Analysis of Pilot Plant Runs and Routes to Improve the Economics of Scale

Abstract:

Small Scale Distributed Ammonia Produced: Analysis of Pilot Plant Runs and Routes to Improve the Economics of Scale

Mahdi Malmali[‡], Michael Reese[†], Cory Marquart[†], Kevin Wagner[‡], Eric Buchanan[†]

Alon McCormick[‡] and Edward L. Cussler[‡]

[†] West Central Research and Outreach Center

[‡] Department of Chemical Engineering and Materials Science

University of Minnesota, Minneapolis, MN 55455

Ammonia is one of the most important chemical commodities in the U.S. and will be a key component in helping the world meet the rising demand for food and energy. Ammonia is needed in distributed locations for agriculture (as fertilizer for small grain and corn production), for indirect hydrogen storage[1] (transported as a liquid at moderate pressure to hydrogen stations), or as a liquid fuel[2] (for internal combustion engines or solid oxide fuel cells). Currently, ammonia is manufactured by the Haber-Bosch process, a highly capital- and energy-intensive process requiring large quantities of fossil fuels. Ammonia synthesis is only feasible in massive centralized locations; ammonia consumption thus relies on moving anhydrous ammonia from distant locations to regions where ammonia is needed.

Recently, there has been significant effort to develop scalable technologies for conversion of intermittent energies (e.g., solar, wind) into energy dense carbon-neutral liquid fuels, and ammonia is considered to be a promising option. For the synthesis of carbon-neutral ammonia, hydrogen and nitrogen are delivered from electrolysis of water and pressure swing adsorption of air. Once produced, nitrogen and hydrogen are fed to a catalytic reactor to produce anhydrous ammonia. All these processes are carbon-free and powered by wind turbines or solar arrays.

In this talk, we are going to present a benchmark for the performance of a small-scale ammonia synthesis pilot plant powered by wind energy[3]. The Renewable Hydrogen and Ammonia Pilot Plant located in Morris, MN, is the first local farm-to-coop scale system for ammonia production. The energy used for this plant is stranded, located in areas far from urban centers where ammonia is primarily used as a fertilizer. This plant targets production of small amounts of ammonia for local demand, thus representing distributed production. We will present the analysis of small plant runs, along with some laboratory-scale kinetic studies, to assess the performance of different units, including reaction, phase separation and recycle. A simple but insightful model is developed, in order to understand the performance of the small-scale Haber-Bosch pilot plant and to determine the optimal conditions for operating. This model is successfully employed to predict the performance of each unit, and gives us insight into selecting appropriate unit operating conditions. Such a straightforward model may help us to understand the design of other small-scale processes in the future.

(1) Sørensen, R. Z.; Hummelshøj, J. S.; Klerke, A.; Reves, J. B.; Vegge, T.; Nørskov, J. K.; Christensen, C. H. Indirect, Reversible High-Density Hydrogen Storage in Compact Metal Ammine Salts. *J. Am. Chem. Soc.* **2008**, *130* (27), 8660.

(2) Wojcik, A.; Middleton, H.; Damopoulos, I.; Van herle, J. Ammonia as a Fuel in Solid Oxide Fuel Cells. *J. Power Sources* **2003**, *118* (1-2), 342.

(3) Reese, M.; Marquart, C.; Malmali, M.; Wagner, K.; Buchanan, E.; McCormick, A.; Cussler, E. L. Performance of a Small-Scale Haber Process. *Ind. Eng. Chem. Res.* **2016**, *55* (13), 3742.

468147 A Supply Chain Optimization Framework for Distributed Renewable Ammonia Production

Abstract:

Ammonia is a critically important chemical for global food production which can be either used directly as or processed further into fertilizer. The Haber-Bosch process revolutionized global food production and greatly increased mankind's ability to support a larger population. However, the way ammonia is produced and transported currently is unsustainable. Ammonia is produced using hydrogen obtained from fossil fuel sources, ultimately polluting the environment. It is then transported over long distances by train, pipeline, or even ship: over one-third of ammonia consumed in the United States is imported. It would be more desirable from a sustainability standpoint to have a cleaner, smaller scale, and more local ammonia supply chain. Motivated by this goal, a first of its kind plant has been built in Morris, MN, using the wind-powered electrolysis of water to produce hydrogen without the carbon emissions inherent in fossil fuel gasification. This plant has been demonstrated to produce renewable ammonia with minimal carbon emissions at a scale of 26 t/y. Using data from this plant, it is reasonable to ask how, and at what scale, renewable ammonia plants would be economically feasible with respect to the overall ammonia supply chain.

To address this question, this work aims to create a framework for the analysis of renewable plants in a general region's ammonia supply chain. Candidate sites for renewable plants are placed in towns with nearby wind turbines with maximum plant capacity determined by wind turbine energy output. A supply chain optimization problem is fomulated to decide where and how large to build new renewable plants, as well as how much ammonia should be transported between conventional plants, renewable plants, distribution centers, and demand centers to meet local demand. The problem is a nonlinear program due to economies of scale when scaling up the Haber-Bosch reactor and related separation units. The problem can be set up to minimize economic or environmental objectives. When doing so, a clear tradeoff emerges: a minimal cost supply chain obtains most of its ammonia from a few large centralized plants, while a minimal emissions supply chain obtains its ammonia from many small distributed plants. A multiobjective optimization problem is formulated to analyze these tradeoffs via a Pareto curve of annual cost vs. annual carbon dioxide emissions.

The developed framework is then applied in two separate case studies to two upper midwestern states, Minnesota and Iowa, respectively. Despite the fact that these states are neighbors, they have various key differences that make separate analysis instructive. For example, Iowa has more distribution centers, more pipeline connections, more overall demand, more wind potential, and a more geospatially homogeneous ammonia demand distribution than does Minnesota. Despite these differences, we establish that for both states, a renewable plant scaled up to the order of 100,000 t/y ammonia production is selected for inclusion in the base case optimal supply chain. We then further explore how the differences between the two states affects the solution sensitivity, the impact of a carbon tax on the supply chain, and the improvements required for smaller scale plants to be economically feasible.

NH3 – The Optimal Transportation Fuel
2016 AIChE Annual Meeting

Norm Olson and John Holbrook
Iowa Energy Center
NH3 Fuel Association (https://nh3fuelassociation.org/)

November 8, 2016

Any analysis to help identify an optimal alternative fuel should begin with a logical, comprehensive set of desired characteristics. The following criteria were used to help identify the fuel closest to an "ideal" alternative fuel.

Characteristics of an Ideal Fuel

1. Production flexibility – i.e. the ability to be produced from all primary energy sources (e.g. solar, nuclear, wind, biomass, OTEC, natural gas with CCS, coal with CCS, etc,). Another highly desirable characteristic is that the production of the fuel be a relatively simple process with a wide range of scalability.
2. Affordability. This involves infrastructure costs, production CapX and OpX costs, efficiency and other considerations. The fuel should be cost competitive with gasoline and diesel fuel.
3. Existing storage and delivery infrastructure
4. End use flexibility – suitable for use in spark ignition engines, compression ignition engines, combustion turbines, fuel cells, etc.
5. High engine/fuel cell efficiency potential - the single most important characteristic for high efficiency SI engines is octane rating. A high octane rating allows the use of high engine compression ratios and therefore highly efficient engines. A closely related issue in SI engines is resistance to pre-ignition. High efficiency fuel cell operation is also important.
6. Excellent environmental and health performance
7. Sustainability – renewable resources and widely available, low-cost raw materials
8. Acceptable safety characteristics
9. The ability for any country to produce the fuel and become both power and fuel independent
10. Having additional non-fuel uses of significant benefit to society. E.g.– fertilizer, renewable energy storage, chemical precursor, refrigerant, hydrogen source, household cleaner, etc.

Alternative Fuel Candidates

There are many alternative fuel candidates that partially meet the ideal fuel criteria listed above including but not limited to hydrogen, NH3, drop-in fuels, propane, DME, ethanol, methanol, and butanol. Only hydrogen and NH3 can be produced without using biomass (or a somewhat convoluted, inefficient path involving CO2 capture plus hydrogen). The International Energy Agency (IEA) and the Energy Information Administration (EIA) both estimate the upper limit for biomass-derived alternative fuels at 30% for transportation fuels. This is a very significant issue and would result in continued addiction to petroleum indefinitely.

Algae-based fuels are currently receiving attention as a source of biomass for fuels. In the 1950's algae was viewed as the ultimate food source due to its excellent nutritional value and efficient growth characteristics.

Food vs fuel issues are even more relevant for algae than for corn. Negligible amounts of the type of corn used to produce ethanol can be used for human consumption, this type of corn is used nearly exclusively for livestock feed. If algae can be grown cost-effectively in large quantities, it would make more sense that it be used as a highly efficient, high-quality source of food than as a fuel source. It may prove to be problematic if the future liquid transportation fuel is made exclusively from biomass (although biomass-based fuel additives could be acceptable). In a world where ending the continuing addiction to petroleum is desired, biomass should be used to produce food first, chemicals (as replacements for petrochemicals) second and thirdly as fuel additives.

NH3 (aka "The Other Hydrogen") rises above other alternative fuel candidates based on the criteria listed above. Simply stated, NH3 is the most cost-effective, energy dense means of delivering hydrogen. It has excellent environmental performance, extremely high engine efficiency potential, production flexibility characteristics second to none, universal end use potential, and a decades-long, proven, acceptable safety record. The fact that NH3 also serves as the most cost-effective means of delivering nitrogen fertilizer provides enormous additional benefits, especially to developing countries. NH3 is an efficient, environmentally friendly refrigerant, an excellent household cleaner, is an efficient medium for long-term storage of intermittent renewable energy and a versatile chemical precursor. No other alternative fuel can deliver the comprehensive benefits associated with NH3.

NH3 Production Flexibility

NH3 production is a relatively simple industrial process compared to most petroleum refinery processes. NH3 production facilities are highly scalable and NH3 has multiple, high-value uses including use as a transportation fuel, power fuel, fertilizer, refrigerant, energy storage medium for renewable energy, and a versatile chemical precursor.

NH3 can be competitively priced when produced from natural gas today and has a very small carbon footprint using CO2 sequestration or via thermal cracking co-producing carbon black. It can be produced via nuclear energy and all renewable energy sources when these routes become more cost competitive.

NH3 can be produced using natural gas, coal, biomass, wind, solar, hydro, OTEC, wind, wave, nuclear energy and all other primary energy sources. One of the most promising methods of storing and transporting stranded renewable energy utilizes NH3 as the storage medium.

The two most commonly discussed methods for NH3 production are solid state ammonia synthesis (SSAS) and the Haber-Bosch process both of which are very scalable. Numerous innovative improvements on the Haber-Bosch are currently being developed.

Affordability

NH3 is primarily produced from coal and natural gas at the present time and is normally cost-competitive with gasoline and diesel fuel on a gallon of gasoline equivalent (GGE) basis. The chart below provides characteristic costs for NH3 for a range of natural gas prices and compares these costs to gasoline.

NH3 Cost (GGE) vs Natural Gas Cost

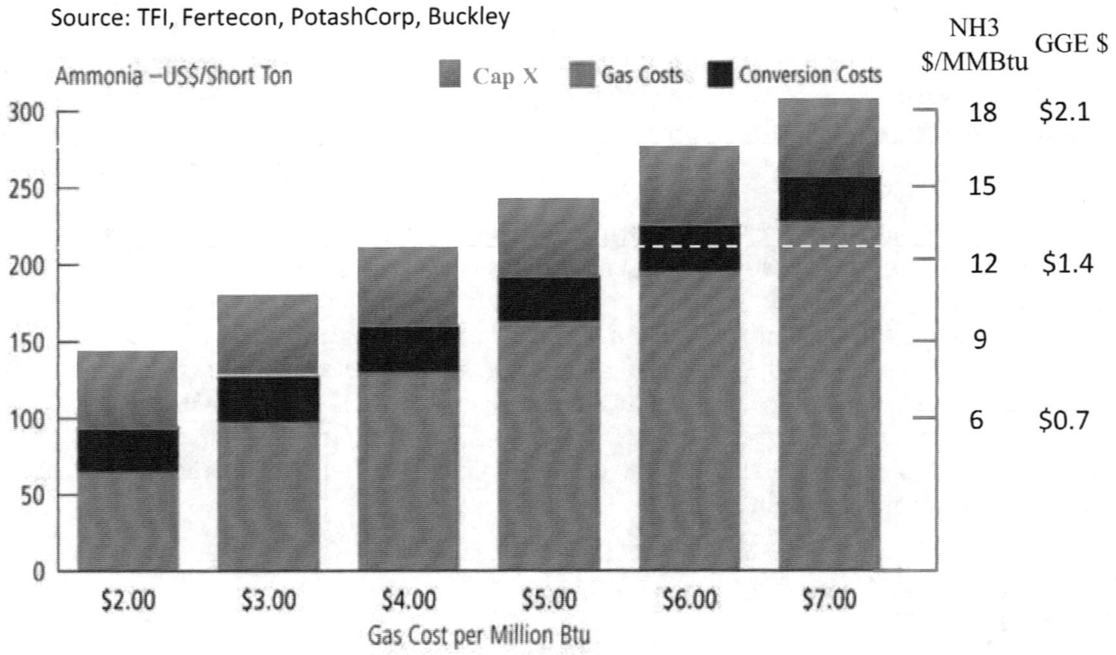

NH3 production facilities, much like ethanol, biodiesel and methanol are very scalable and low-cost relative to petroleum refineries. This makes NH3 "refineries" affordable in most areas of the world. In countries where NH3 fertilizer is crucial to increased food production, a relatively simple NH3 refinery could be used for multiple purposes. NH3 could be used as a transportation fuel, an electrical generation fuel, a refrigerant, a household cleaner, and a chemical precursor in addition to its use as a fertilizer.

Infrastructure costs are low and are similar to propane infrastructure costs. NH3 has an extensive, existing, worldwide infrastructure already in place which will help lower implementation costs.

When NH3 is produced from renewable electricity such as wind or solar PV, assuming an electrical energy cost of 5 cents/kWh (including CapX and OpX costs) and an electrical consumption of 7000 to 9000 kWh/ton, the cost per ton of ammonia produced would be $350 to $450/ton. This corresponds to a gallon of gasoline equivalent (GGE) cost of **$2.45 - $3.15/GGE.**

Storage and Delivery Infrastructure

NH3 is the second most transported chemical in the world and has an extensive, world-wide storage and delivery infrastructure already in place. Infrastructure costs are relatively low and very similar to propane at low storage volumes (<4 ton or ~1500 gallon). At larger storage volumes (30 - 30,000 ton), ammonia can be stored at atmospheric pressure as a refrigerated liquid. It is transported as a refrigerated liquid in ocean-going ships as well. The existing, low-cost storage and delivery infrastructure provides a very valuable advantage to NH3 fuel.

Annual production of NH3 is over 160,000 tonne compared to <100,000 tonne of methanol produced annually. NH3 is the 2nd most transported chemical. Mild steel is the material of choice for pipeline and storage applications. Over 3000 miles of NH3 pipelines have been in place for decades in the U.S. Natural gas pipelines could be modified by replacing compressors with pumps to carry NH3. Atmospheric pressure storage at relatively mild conditions (-28F) compared to LNG (-265F) and liquid hydrogen (-423F). NH3 is less energy intensive to produce, deliver and store per pound of hydrogen delivered than 10kpsi or liquid hydrogen. The volumetric energy density of NH3 is 30% higher than liquid hydrogen. Storage infrastructure capital costs for NH3 can be one to two orders of magnitude lower than hydrogen infrastructure costs.

End Use Flexibility

The use of NH3 in gas turbines, spark ignition engines, compression ignition engines, gas burners and fuel cells has been demonstrated. Although NH3 has a relatively slow flame-spread-speed and high ignition energy, it can easily be partially cracked to produce a mixture of hydrogen and ammonia, which then combusts with characteristics very similar to natural gas (but carbon free!).

NH3 use in diesel engines normally requires the conversion of the engine to spark ignition or dual-fuel use, very similar in concept to dual-fuel diesel/natural gas engine conversions. Advanced compression ignition engines such as those produced by Sturman Industries have demonstrated successful operation using 100% NH3.

Solid oxide fuel cells are very compatible with NH3 as are alkaline fuel cells. Although PEM fuel cells cannot tolerate NH3, there is a commercially available PEM fuel cell that operates on cracked NH3 (i.e. cracked to nitrogen and hydrogen).

Gas burners and gas turbines have both been successfully operated on partially cracked NH3. A federally funded program in Japan has successfully demonstrated the use of NH3 in a gas turbine.

Engine and Fuel Cell Efficiency

High efficiency, affordable cost and low emissions are key metrics associated with optimized internal combustion engines, combustion turbines and fuel cells. The high octane rating and tunable resistance to pre-ignition allow NH3 to be used in the highest compression (and therefore highest efficiency) engines capable of being designed. Just as in real estate the three most important characteristics are location, location, location, the three most important characteristics of a transportation fuel are high octane, high octane, high octane. The highest engine efficiencies will be attained using high octane fuels which also have a high resistance to pre-ignition. Low/zero carbon fuels are extremely beneficial in attaining low emissions and longer lasting engines, due to low carbon deposits.

The most important characteristic of an ideal fuel for spark ignition engines from an efficiency standpoint is a high octane rating. A high octane rating allows the use of high compression ratios and therefore high efficiencies. A related issue is resistance to pre-ignition.

Fuel Octane Rating	RON	MON	(R+M)/2
Gasoline			87
Ammonia	>130[1]		
Ethanol	109	90	99
Methanol	109	89	99
n-Butanol	92	71	83
Hydrogen	>130[2]		
Methane	120	120	120

[1]The minimum ignition energy for ammonia is very high. It is approximately 500 times greater than the energy required for igniting hydrocarbons and 1000 to 10,000 times greater than that required for hydrogen. This makes NH3 much less susceptible to undesirable pre-ignition than hydrogen and other fuels.

[2]Hydrogen does not fit well into the normal definitions of octane number. It has a very high RON and a low MON, so that it has low knock resistance in practice due to its low ignition energy and extremely high flame spread speed.

NH3 is a "tunable" fuel, meaning that depending on the percentage cracked, it can be tuned to perform like propane, natural gas or even hydrogen. This is a very beneficial characteristic allowing NH3 engines to start very easily in cold weather conditions (similar to hydrogen cold start performance) and then be tuned to perform as a very high octane fuel with a very high resistance to pre-ignition in high compression/high temperature conditions, i.e. high efficiency engines. The tunable nature of NH3 also allows continuous performance adjustments for seasonal weathe8r conditions. Gasoline refineries are required to make significant seasonal adjustments to their operations to change from summer to winter blends. In fact, these required seasonal adjustments are often given as a reason for gasoline price increases in the spring and fall. The numerous gasoline specifications required to meet individual state standards has a large impact on the cost of gasoline. NH3 can be tuned to perform optimally for each engine and seasonal condition with firmware onboard the vehicle, providing another huge advantage.

NH3 is an excellent fuel for SOFC's. Intermediate temperature direct ammonia fuel cells have been demonstrated by several research groups.

Environmental and Health Characteristics

NH3 is commonly a naturally occurring chemical and also a common man-made chemical. It is not a known carcinogen, not a greenhouse gas, has an ozone depletion number of zero. NH3 is moderately caustic but is not a toxic chemical. It can be hazardous at fairly low concentrations and can produce base burns (akin to acid burns) when inhaled into lung tissue or when it comes into contact with eye tissue.

With conventional catalytic control systems, NH3 can be used in internal combustion engines with zero measurable criteria pollutants. NH3 fuel cell vehicles would also be zero emission vehicles.

Sustainability, Multiple End Uses, Developing Country Economic Development

NH3 production requires only two widely available raw materials: air (nitrogen) and hydrogen. An iron-based catalyst is used making catalyst availability a non-issue. Any and all primary energy source(s) (e.g. wind, solar, biomass, nuclear energy, OTEC, wave, coal, natural gas, etc.) can be used to produce NH3 and providing an opportunity for unlimited, sustainable production.

Making NH3 the fuel of choice could be a decision that provides developing countries with an opportunity to create a robust, sustainable local economy and greatly enhance the likelihood of developing an export economy as well. Oil refineries and gas-to-liquids facilities normally require multi-billion dollar investments and do not scale down well. NH3 production facilities can be very small and still be economically competitive. The smallest operational NH3 production facility in the world is located in Creston, Iowa. The Creston plant capacity is 100 ton per day which might cost approximately $65 million (~600,000 MMBtu or 5.2 million GGE or 13,000 vehicles at 12,000 miles per year and 30 mpg). For countries without significant fossil fuel infrastructure in place (i.e. natural gas pipelines, utility grid, oil pipelines, etc.), this is the most practical/economical option.

Local energy production creates jobs and energy security for every country in the world, reduces shipping costs. One relatively simple industrial facility producing NH3 can provide fuel for transportation and dispatchable electrical generation, fertilizer for local food production, a high performance long-term alternative energy storage medium (see chart below), an environmentally friendly refrigerant for space cooling and food storage freezers, an environmentally friendly disinfectant/cleaner, and a precursor for many other chemicals.

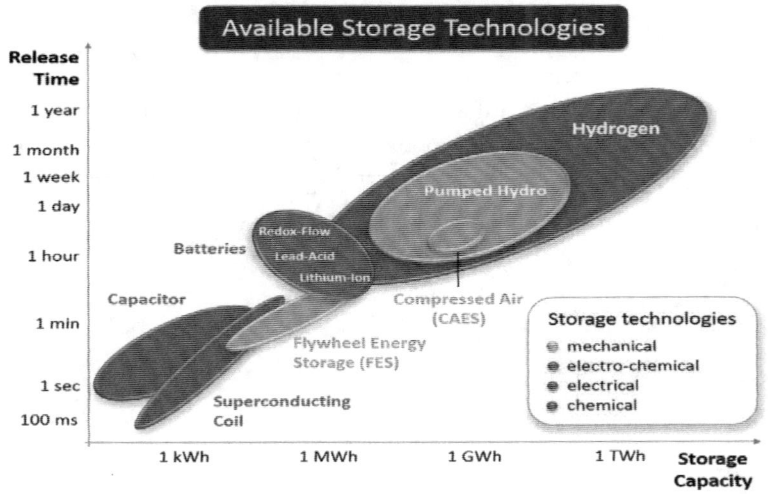

The investment required for a NH3 refinery is relatively low, providing an attractive option for developing countries to drastically improve their standard of living and to spur opportunities for additional economic development. Developing countries open the door to tremendous economic development opportunities by choosing to build their economy around NH3 and the biorefinery model (depicted below) where biomass is used to sustainably produce food, chemicals, fuels and co-products. These countries have an unprecedented opportunity to leap-frog the fossil-fuel addicted OEC's, weighed down by huge existing investments in fossil fuel infrastructure and associated lobbying interests, and to move forward with an affordable, robust, sustainable economy based on renewable NH3.

Biorefinery: Using NH3 to Catalyze the Creation of a Robust Economy via Increased Agricultural Production

Safety

Two highly credible safety studies have been completed to establish the comparative risks of NH3 verses propane and gasoline in transportation fuel applications. The studies: RISO (Europe) and Quest (U.S.) The conclusion of both reports was that NH3 would not present any unacceptable safety risks and in fact would be safer than propane in transportation fuel applications.

An entry-level product design engineer should be capable of designing a NH3 refueling and fuel tank system capable of meeting the world's most stringent safety standards. It is rather a trivial design problem compared to developing an affordable hydrogen infrastructure or an EV with performance equal to liquid fueled vehicles. NH3 fuel systems can be designed to be as safe as they need to be to meet public expectations.

NH3 fertilizer been used in large volumes in an acceptably safe fashion for decades in the Midwest of the United States. There are 800 retail outlets in the state of Iowa alone, many located near residential areas.

NH3 fueling stations be designed to be even safer than the existing fertilizer outlets. Chilled, atmospheric-pressure tanks located in underground containment vaults would provide a very safe design for NH3 fuel. Interlocking refueling hoses where no flow would be allowed until proper connection is verified would provide a re-fueling system that would be safer than existing gasoline re-fueling systems. A typical NH3 fueling station is depicted below.

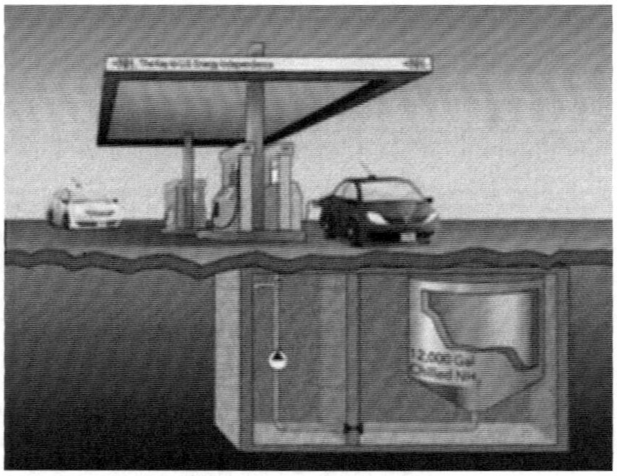

Summary

Unlike some technology areas where "all of the above" has significant advantages, there are tremendous advantages associated with choosing a single, optimized, liquid transportation fuel. The advantages of a logically chosen "ideal fuel" include: 1) low-cost, standardized production facilities; 2) affordable, standardized storage and delivery infrastructure; and 3) optimized, high-efficiency, low-cost, low/zero-emission engines and fuel cells. While electric vehicles will contribute to a reduction in the need for liquid transportation fuels, there will be a substantial need for liquid transportation fuels for the foreseeable future. The cost, efficiency and environmental benefits associated with choosing an optimized liquid transportation fuel are enormous and merit serious consideration.

NH3 most closely meets the criteria for an ideal liquid transportation fuel. It is the most efficient and cost-effective means of delivering hydrogen and has an extensive world-wide delivery system already in place. Additional infrastructure could be added at costs similar to propane infrastructure costs, one of the lowest-cost liquid transportation fuel infrastructure options. Using currently available catalytic controls, NH3 emissions can be even cleaner than hydrogen engine emissions. While NH3 can effectively be produced using biomass, it can also be produced from all other primary energy sources including wind, solar, nuclear, natural gas, OTEC, etc. NH3 can be used in fuel cells, gasoline engines, diesel engines, gas turbines and gas burners. Gasoline engines would require conversions similar to those required for propane. Currently available diesel engines could be modified in a similar fashion to dual-fuel natural gas/diesel fuel engines but NH3 substitution rates could be higher than natural gas substitution rates. Next generation diesel engines have demonstrated the ability to use 100% NH3. The tunable characteristics of NH3 allow its optimal use in almost any combustion application.

Developing countries could create extensive, robust agricultural and energy economies based on NH3 production facilities in a sustainable, environmentally friendly fashion. Competitive energy and agricultural economies would provide the basis for nearly limitless additional economic development opportunities.

Impressive federal programs in Japan and numerous patents field by Toyota combined with significant R&D efforts in the U.S., Canada and Europe are providing exciting momentum for the NH3 fuel movement. It is time to acknowledge the show-stopping deficiencies associated with the usual alternative fuel "suspects" and move forward with the closest thing to an ideal fuel, NH3.

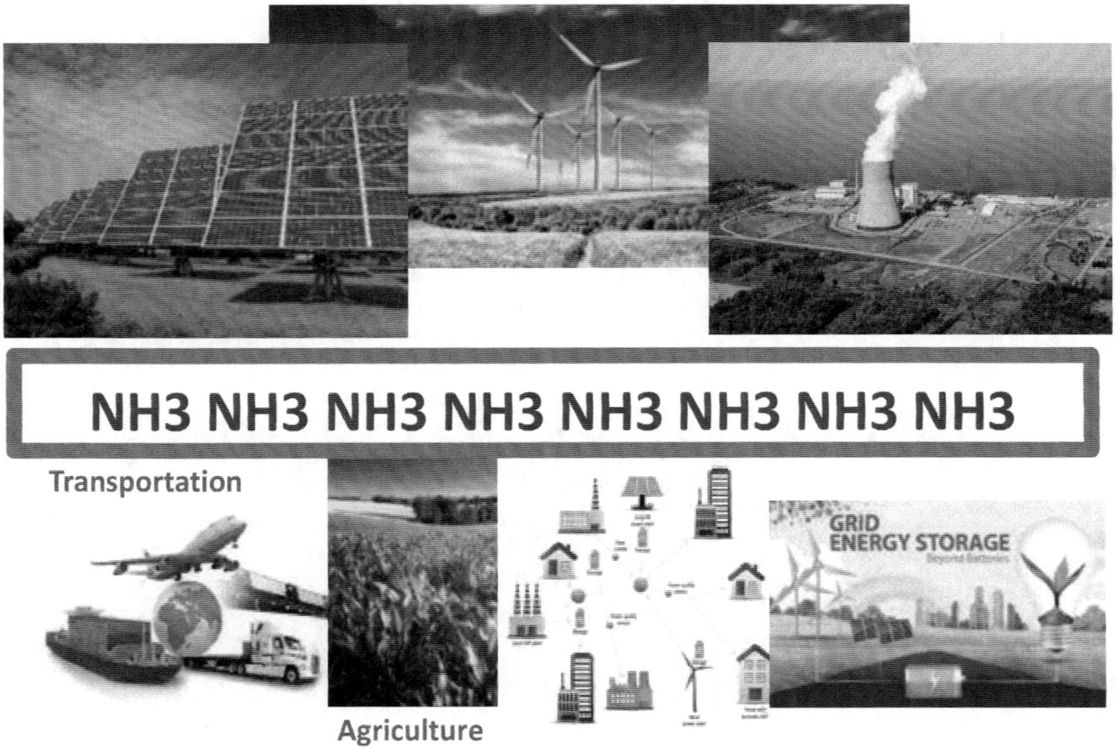

458643 Implementing Robust Vapor-Liquid Equilibrium Calculations in Nonsmooth Multi-Stream Heat Exchanger Models

Abstract:

Optimization of the liquefaction process for natural gas has proven to be a challenging task in particular for two reasons. First, the refrigeration cycles introduce computational loops that cause convergence problems. Second, the refrigerants are mostly multicomponent mixtures with strong interactions in the search for optimal heating profiles to match the cooling profile of the natural gas. In addition, the driving forces in the heat exchangers are very tight, asking for rigorous modeling to make sure the simulation mimics the true behavior of the process.

Both gradient based methods and stochastic search methods have been used to try to optimize processes for Liquefied Natural Gas (LNG) production, where commercial simulators are used to provide the function values in the optimization, however, with limited success beyond the most simple liquefaction processes. Our research groups have been involved in the development of a new model for these processes with the inherent features required to be able to utilize the latest developments in the field of Global Optimization [1].

As an important part of this new approach to LNG optimization, Barton and coworkers at MIT [2] has developed a new non-smooth multi-stream heat exchanger model accounting for phase changes in the process. The model was successfully used to simulate the Poly Refrigerant Integrated Cycle Operations (PRICO) process for liquefied natural gas production. However, it suffered from convergence difficulties when studying more complex single mixed refrigerant (SMR) processes. The primary challenge was flash calculations, which frequently failed even for initial guesses close to the solution. These problems will of course increase even further when moving to Dual Mixed Refrigerant (DMR) processes and the Mixed Fluid Cascade (MFC) process.

In this project, an alternative model structure is studied, replacing the equation-oriented framework with a hybrid solution. Rather than converging the vapor-liquid equilibrium calculations simultaneously with the other model equations, they are included as nested subroutines and solved sequentially.
Nonsmooth Newton-type methods and the implicit function theorem for lexicographically smooth functions are used for computing analytical derivatives in each subroutine [3, 4]. In addition, more robust flash calculations are attempted by developing a nonsmooth version of the "inside-out" algorithm by Boston and Britt [5] that can be incorporated in the existing model framework. The same SMR processes are then simulated in order to investigate whether the alternative framework is more robust.

[1] Floudas, C.A. and Gounaris, C.E., A review of recent advances in global optimization, *Journal of Global Optimization*, vol. 45(1), pp. 3-38, 2009.

[2] Watson, H.A.J., Khan, K.A. and Barton, P.I., Multistream heat exchanger modeling and design, *AIChE Jl.*, vol. 61(10), pp. 3390-3403, 2015.

[3] Facchinei, F., and Pang, J.S., *Finite-dimensional variational inequalities and complementarity problems,* vol. 2. Springer Science & Business Media, 2007.

[4] Khan, K. A., and Barton, P.I., Generalized derivatives for hybrid systems, *Submitted*, 2015.

[5] Boston, J.F. and Britt, H.I., A radically different formulation and solution of the single-stage flash problem, *Computers & Chemical Engineering*, vol. 2(2), pp. 109-122,

465812 Shared and Practical Approach to Conserve Utilities in Eco-Industrial Parks

Abstract:

Eco-Industrial Park (EIP) is a community of manufacturing and service facilities in close proximity. These enterprises collaborate with each other for improving the social, environmental, and economic performance of the whole complex in a sustainable manner by utilizing each other's resources efficiently, and sharing facilities, services, and utilities (Boix et al., 2015; Côté & Cohen-Rosenthal, 1998; Lowe et al., 1996). Examples of such successful industrial symbiosis are Jurong Island (Singapore), Kalundborg (Denmark), and Rotterdam (Netherlands). EIPs such as Jurong Island in Singapore are interested in assessing the potential for inter-plant heat integration among its inhabitants. The aim of such an enterprise wide heat exchanger network system is to exploit the synergistic heating/cooling needs and reduce total utility costs. However, there are many practical challenges in the implementation, construction, and operation of such an enterprise wide network (Chew et al., 2013) that need to be addressed while synthesizing inter-plant heat exchange network (HEN).

(1) The inter-plant heat exchange network synthesis (HENS) methodology should involve all the capital and operating expenses. Since, the exchanger and pipeline costs are comparable in case of long distance transport, inter-plant piping and transport costs cannot be ignored in configuring an EIP-wide HEN.

(2) The independent and diverse profit-making enterprises should be willing to participate and collaborate in such an EIP-wide HEN. Hence, there should be attractive economic benefits and fairness guaranteed to motivate the enterprises for such collaboration.

(3) The exchangers in an EIP-wide direct heat integration must have real physical locations. Furthermore, if a plant hosts a heat exchanger, it must bear the added responsibilities of maintaining and operating that exchanger, and guaranteeing the safety, security, and confidentiality of any guest streams.

(4) Operational disturbances in one plant shouldn't propagate to other plants, as the plants should be able to operate independently.

In this work, we present a novel strategy for EIP-wide HENS that addresses most of the issues highlighted above. We propose a central, shared, mutually agreed location for the HEN. The participating enterprises can appoint a third party logistics provider to manage and operate such a facility. We have considered pumping costs in the operating costs and the pipeline, pump and heat exchanger costs in the capital investment. By optimizing the HEN location, we can minimize the transport and pipeline costs. In our proposed model, we allowed the cooling (heating) of a hot (cold) stream both before entering and after leaving the HEN. To give operational flexibility and control to the parent company, existing in-plant utility heaters/coolers will be retained. Such an arrangement offers several advantages in terms of flexibility, control, cost, safety, security, confidentiality, accountability, operation, and management. We then propose a more practical and rational cost-benefit sharing arrangement. The contribution of an enterprise to capital expenses (CAPEX) and operating expenses (OPEX) is proportional to the utility cost savings it enjoys through the shared HEN. This ensures fairness among the enterprises as each enterprise enjoys an identical rate of return.

The EIP-wide HEN must trade off capital and operating costs versus utility savings. Therefore, we develop a mixed-integer non-linear programming (MINLP) model to maximize the net present value (NPV) of an EIP-wide HEN. The superstructure-based formulations allow sub-streams and non-isothermal mixing in a stage (Huang & Karimi, 2013; Yee & Grossmann, 1990). Since we wish to maximize the net present value of the shared HEN, discounted flows and rate of return has been included in the cost equation. Our model allows arbitrary cost expressions for exchangers with both fixed and variable components and cost multipliers for material and installation considerations. The single HEN location allows the full flexibility of multiple heat exchanges among streams from different plants. We have also improved the efficiency of formulation by adding additional constraints for eliminating redundant or complicated network designs. We have limited the number of sub-streams in a stage, avoided repeat identical matches, and included lower limits on exchanger areas and temperature drops to avoid exchangers with small-duty or small-area. If the HEN looks profitable (NPV > 0), each enterprise will enjoy annual savings. In return, it must share the various costs of the HEN in proportion to its annual savings.

Our proposed model for the case study in Hiete et al. (2012) offers an NPV of $5.71 million with a utility savings of $1.74 million/a for the EIP. The piping and pumping costs are major costs in this shared HEN, which restates their importance in EIP-integration. Enterprise E1 contributes 67.4 % to the CAPEX and OPEX, as it enjoys the most savings. In the game theory based formulation in Cheng et al. (2014), enterprise E2 trades streams to E1 and E3, but pays nothing for the exchangers. In contrast, our optimal HEN apportions costs on a more rational and practical basis of utility savings. Since E1 gains the most from the EIP-integration with a utility cost savings of $126 456/a ($29 411 /a for E2 and $71 183 /a) for E3, it contributes the largest share of the costs (E1:55.7 %, E2:13.0 % and E3:31.3 %). Like the case study of direct heat integration in Wang et al. (2015), our HEN also achieves a total utility savings of 95 %, which is the thermodynamic limit. Wang et al. (2015) did not clearly indicate the locations of their heat exchangers (or equivalently the streams being transported). E2 contributes 90.5 % of investment costs, as it supplies the cold streams that would otherwise require expensive hot utilities. Hence, solutions are fair, as companies saving expensive heating utilities will have to contribute more in the investment costs.

Heat integration has been employed over the years in several chemical processes, in particular refineries and petrochemical. As discussed, EIP-wide heat integration poses several unique economic, design, logistics, operations, and safety challenges. But, the utility costs for refineries and petrochemical plants can be huge every year, which is an incentive to reduce these costs in spite of the various challenges. In fact, modern EIPs like Jurong Island in Singapore and Jubail in Saudi Arabia are very much interested in assessing the potential for inter-plant heat integration. The methodology presented here paves the way for assessing and implementing heat integration for such "futuristic" systems.

Boix, M., Montastruc, L., Azzaro-Pantel, C., & Domenech, S. (2015). Optimization methods applied to the design of eco-industrial parks: a literature review. Journal of cleaner production, 87, 303-317.

Cheng, S.-L., Chang, C.-T., & Jiang, D. (2014). A game-theory based optimization strategy to configure inter-plant heat integration schemes. Chemical Engineering Science, 118, 60-73.

Chew, K. H., Klemeš, J. J., Wan Alwi, S. R., & Abdul Manan, Z. (2013). Industrial implementation issues of Total Site Heat Integration. Applied thermal engineering, 61, 17-25.

Côté, R. P., & Cohen-Rosenthal, E. (1998). Designing eco-industrial parks: a synthesis of some experiences. Journal of cleaner production, 6, 181-188.

Hiete, M., Ludwig, J., & Schultmann, F. (2012). Intercompany Energy Integration. Journal of industrial Ecology, 16, 689-698.

Huang, K. F., & Karimi, I. A. (2013). Simultaneous synthesis approaches for cost-effective heat exchanger networks. Chemical Engineering Science, 98, 231-245.

Lowe, E. A., Moran, S. R., Holmes, D. B., & Martin, S. A. (1996). Fieldbook for the Development of Eco-Industrial Parks: Final Report: Indigo Development.

Wang, Y., Chang, C., & Feng, X. (2015). A systematic framework for multi-plants Heat Integration combining Direct and Indirect Heat Integration methods. Energy, 90, Part 1, 56-67.

Yee, T. F., & Grossmann, I. E. (1990). Simultaneous optimization models for heat integration--II. Heat exchanger network synthesis. Computers & chemical engineering, 14, 1165-1184.

462355 Optimal Design and Operation of Integrated Multi-Vector Energy Networks

Abstract:

Integrated multi-vector energy networks are becoming more important because we have a much greater variety of primary resources of energy such as wind, solar, biomass etc. Whereas existing energy systems are dominated by fossil fuels as the primary resource, with electricity and natural gas as the energy vectors, increasing the penetration of renewable energy will change the energy mix and other energy vectors, such as hydrogen, methanol, syngas, ammonia etc. may also play important roles. There are many different pathways to satisfy energy service demands through different energy vectors: for example, hydrogen, biofuels or electricity are considered promising alternatives to petroleum for the transport sector.

The aim of this work is to determine the best design and operation of integrated energy network to obtain the most value from limited available resources while minimising its negative socio-economic and environmental impacts. This endeavour requires the use of mathematical models to provide an accurate representation of the potential technologies and transport infrastructures that may become part of the network, as well as optimisation techniques to determine the most promising designs among the many possible alternatives. This involves selecting the right combinations of conversion and storage technologies, where to locate them, what their capacities should be and how to interconnect them via transport infrastructures. Moreover, the mathematical models need to be sufficiently detailed to account for all of the operational issues at different time scales (e.g. hours, days, seasons, years) and interactions between the different energy vectors. In this conference, we will present a general mixed integer linear programming model that can simultaneously determine the design and operation of any integrated energy networks comprising technologies for conversion and storage and infrastructures for transport of resources (both material and energy). The model takes into account the spatial and temporal distribution of system properties and simultaneously determines the spatial structure, i.e. the location, number, capacity of the technologies and the structure of the transport infrastructures that connect them, as well as the hourly operation of the entire network. It is a dynamic model in order to properly account for the intermittency of renewables and dynamics of energy storage. The model includes detailed and coherent storage inventory balances, thus it is able to track the inventory level of each resource in storage at every hour over the entire planning horizon (e.g. out to 2050).

This presentation will describe the model and discuss a decomposition method that enables this very large, and otherwise intractable, optimisation model to be solved. We will also present case studies wherein the design and operation of an integrated electricity-hydrogen-syngas-natural gas network are determined in order to maximise the value (or minimise the cost or environmental impacts) of meeting the demands for heat, electricity and transport fuels in Great Britain.

464617 Simultaneous Optimization of Design and Operation Strategies for CHP Systems

Abstract:

Combined heat and power (CHP) facilities are a very promising path to reducing CO_2emissions and increasing efficiency in the power generation sector, especially when combined with residential solar photovoltaic (PV) power generation. CHP facilities rely on natural gas, a cleaner fuel than coal, to generate electricity, as well as hot and chilled water to provide heating and cooling. The ability to supply these essential residential utilities in an efficient way on a medium to large scale opens the path for combining district cooling, heating and power generation, and suggests that CHP plants are an appealing choice as an integrated utility supplier for the neighborhood of the future. Yet, there are currently no CHP plants that serve exclusively residential neighborhoods, and published works exploring this possibility are scarce.

In our previous work [1], we demonstrated that a CHP plant with PV integration can in principle meet the demands of a residential neighborhood during the summer months. However, a CHP plant operating in island (i.e., grid-disconnected) mode must be optimally sized to maximize efficiency and to lower the capital and marginal costs, while producing enough electricity, heating, and cooling to meet the neighborhood demands at all times. A majority of the approaches described in literature for the optimal CHP plant sizing tend to select equipment by analyzing the various utility demands (i.e. electricity, steam, and chilled water) individually. Also, they only minimize the capital and O&M costs, overlooking the transitional costs of equipment turning on, by either not allowing the equipment to turn off or ignoring the monetary value of the wear and tear on the equipment. Lastly, many micro-grid papers assume that the main grid will be there to support the system at all times, if the CHP plant is not able to produce the needed energy.

In the paper, we describe a novel simultaneous optimization of design and operation strategies for a CHP plant as a utility producer for a residential neighborhood operating in island (i.e., grid-disconnected) mode, incorporating residential photovoltaics, implemented in a centralized fashion. In order to more accurately predict the operational costs of the CHP plant over a long period of time, the integrated design optimization is combined with an operational-oriented scheduling, forming a two-level problem. The first level consists of the presented integrated design and scheduling optimization, which selects equipment for a plant that must meet hourly residential neighborhood utility demands while minimizing the overall cost of the plant. The second level uses the optimal equipment sizes determined from the first level, and schedules the plant for a year, giving a better result of the operation economics of the CHP plant over a longer period of time. Utilizing data collected by Pecan Street Research Inc., a non-profit smart grid demonstration project headquartered at The University of Texas at Austin, a year of hourly residential heating, cooling, and electricity demand are analyzed and evaluated. These data are then used to create a time-resolved energy demand profile describing residential energy use, which is subsequently serves as a basis for optimizing the size and operation of the CHP plant so that all utility demands are met, while minimizing the capital and marginal costs, including transition penalties.

References:

[1] Ondeck, A.D., Baldea, M., Edgar, T.F. (2015). *Data-Driven Modeling and Optimal Operation of District-level Combined Heat and Power and Photovoltaic Power Generation System*. Applied Energy, 156, pp. 593-606.

462056 Exploiting Dynamic Flexibility to Enable Participation in Multi-Scale Electricity Markets

Abstract:

There is significant interest in re-designing the control architectures for a broad spectrum of energy and manufacturing systems to exploit time-varying electricity prices. For example, the Alcoa Point Comfort Power Plant, which is a utility plant that provides electricity and steam to the adjacent aluminum manufacturing facility, re-optimizes its operations every 15 minutes in response to electricity and natural gas price. These new energy-oriented architectures provide load flexibility to the power grid in exchange for monetary payments. Most academic literature on industrial demand response optimization focuses on scheduling and planning decision layers and considers coarse (e.g., 1 hour) time discretizations with multi-day planning horizons. Emphasis has been placed on optimal planning under time-vary electricity prices, although recent work considers providing reserve capacity and interruptible loads. This emphasis missing flexibility at fast timescales (e.g., minutes to seconds) that may be monetized through participating in real-time energy markets and/or providing regulation capacity and other ancillary services.

We develop a mathematical model for multi-scale electricity markets run by the California Independent System Operator (CAISO). The model incorporates prices from both the day-ahead (1 hour intervals) and real-time markets (5 â€“ 15 minute intervals), and considers both energy and ancillary service (regulation, reserves, etc.) sales. Market rules are expressed as linear algebraic expressions with continuous and optionally discrete variables. The models are general and applicable to both conventional thermal generators and storage devices (e.g., batteries).

Using the market model, we optimize the operation policy of a combined heat and power utility system using real price data for all of 2015. We observe participating in energy markets run at multi-times decreases net operating costs (fuel minus market proceeds) by 14 â€“ 54%, depending on fuel price, without degrading satisfaction of onsite steam and electricity demand. The savings increase if the onsite steam and/or electricity demands are made flexible.

Approximately one-quarter to one-half of the anticipated savings are from regulation capacity sales to the market. In CAISO, the automatic generator control (AGC) system sends new electricity production setpoints to generators providing regulation every 5 seconds that are within a specified flexibility band (units: MW_e). Capacity payments are based on the size of this band. Thus regulation offers a way to monetize the inherent flexibility in manufacturing/energy systems at faster timescales (seconds compared to electricity price variations that occur at slower timescales (minutes to hours). In practice exploiting this flexibility and realizing revenues from electricity markets requires sophisticated controls strategies that manage multiple time-scales and consider uncertainty in market prices.

468112 A Rolling Horizon Scenario-Based Approach for Smart House Management Under Uncertainty

Abstract:

The development of sustainable, efficient and environmental friendly energy supply chains has led to advances in the area of Energy Systems Engineering. In this context, the growing interest in the exploitation of renewable energy sources has led to an increasing use of microgrids, which are decentralized networks able to integrate several heat/power generation systems of low/medium capacity (Hodge et al., 2011; Zhang et al., 2013), with the main purpose of distributing the generated heat/energy locally so as to reduce energy losses and improve the responsiveness to energy demand fluctuations (Kopanos et al., 2013; Tong et al., 2015). These kind of networks usually involve the presence of renewable generators. So, it is imperative that the variations in the behaviour of natural energy sources requires the consideration of uncertainty in microgrids. This variability must be considered in order to ensure the operability and the reliability of the system, since the optimal solution obtained under deterministic conditions can be sub-optimal or even infeasible. Particularly, the operations management of microgrids can be affected by several types of uncertainty, such as energy production and energy demand variations.

This work addresses the presence of uncertainty through a discrete-time two-stage stochastic programming approach, which is combined with a rolling horizon approach (Kopanos et al., 2014) for the simultaneous management of energy supply and demand in microgrids. The proposed Mixed Integer Linear Programming (MILP) formulation contains inequality constraints corresponding to the available heat and energy generation technologies, as well as equality constraints corresponding to heat and energy balance equations that describe flows, production, storage and consumption levels. This mathematical formulation uses a scenario-based stochastic programming approach where the scenarios are associated to internal variations in the duration of the energy consumptions as well as in the overall heat demand. However, the high complexity related to the estimated weather forecast, makes the consideration of all possible external scenarios, computationally intensive. The variability in weather conditions may affect the availability and production capacity of renewable energy generators. Thereby, updating input data (i.e., wind profile, heat and energy demand) is needed to ensure the adequate quality in the obtained results. Hence, an MILP two-stage stochastic programming approach was incorporated within a rolling horizon scheme that periodically updates input data information as the uncertain input parameters are revealed or considered to be known with certainty. The main decisions to be made to maximise the profit of the microgrid includes the execution of consumptions, the amount of heat and energy to be produced or purchased, the heat and energy storage levels and the amount of energy to export to the power grid. The methodology proposed above was tested on a case study and promising results demonstrating the applicability of the methodology for decision making were obtained. This work also forms a basis for future work to address more complex problems.

References

Hodge, B.M., Huang, S., Siirola, J.D., Pekny, J.F. & Reklaitis, G.V. (2011). A multi-paradigm modeling framework for energy systems simulation and analysis. Computers & Chemical Engineering, 35, 1725-1737.

Kopanos, G.M., Georgiadis, M.C. & Pistikopoulos, E.N. (2013). Energy production planning of a network of micro combined heat and power generators. Applied Energy, 102, 1522-1534.

Kopanos, G.M. & Pistikopoulos, E.N. (2014). Reactive scheduling by a multiparametric programming rolling horizon framework: A case of a network of combined heat and power units. Industrial & Engineering Chemistry Research, 53, 4366-4386.

Tong, C.D., Palazoglu, A., El-Farra, N.H. & Yan, X.F. (2015). Energy demand management for process systems through production scheduling and control. AIChE Journal, 61, 3756-3769.

Zhang, D., Shah, N. & Papageorgiou, L.G. (2013). Efficient energy consumption and operation management in a smart building with microgrid. Energy Conversion and Management, 74, 209-222.

450742 Optimization Models for Shale Gas Well Refracture Treatments

Abstract:

Optimization Models for Shale Gas Well

Refracture Treatments

Markus G. Drouven[1], Diego C. Cafaro[2] and Ignacio E. Grossmann[3]

[1,3]Department of Chemical Engineering

Carnegie Mellon University

Pittsburgh, PA 15213, USA

[2]INTEC (UNL – CONICET)

Güemes 3450, 3000 Santa Fe, Argentina

[1]mdrouven@cmu.edu, [2]dcafaro@fiq.unl.edu.ar [3]grossmann@cmu.edu

Abstract

Shale gas wells are well-known for their characteristically steep decline curves. In fact, some shale wells produce more than half of their total estimated ultimate recovery (EUR) within the first year of operation. The initial production peak after well completion is caused by the sudden release of previously trapped hydrocarbons in the pores of the reservoir. The subsequent decline is ultimately driven by pressure depletion and the ultra-low permeability of the shale play. Upstream operators oftentimes struggle with the production decline curves. For one, they are contractually obligated to providing steady gas deliveries to midstream distributors over time – which is difficult to accomplish given the characteristic, dramatic production decline immediately after wells are turned in line. Moreover, as Cafaro and Grossmann [1] suggest and Drouven and Grossmann [2] confirm, operators need to maximize the utilization of production and gathering equipment – such as pipelines and compressors – in order to stay profitable. In reality, however, production and gathering equipment is usually sized based on the initially high production rates. This means that within a matter of months shale gas wells feed into oversized pipelines and compressor stations, and equipment utilization drops. Worse even, in order to satisfy contractual gas delivery agreements, operators see themselves forced to open up new wells continuously to honor their obligations, and hence the process is repeated over and over again.

Refracturing presents a promising strategy for addressing the characteristically steep decline rates of shale gas wells [3]. The core idea behind refracturing is to restimulate the reservoir such that it yields previously untapped hydrocarbons and improves the overall production profile of a well. Whether or not a refracture treatment will reinvigorate a shale gas well depends on a number of factors including the characteristics of the reservoir and the initial completion design. Historically, refracture treatments have been applied predominantly to shale gas wells suffering from low production rates due to known suboptimal initial stimulations and completions. However, Dozier et al. [4] argue that even wells with effective initial treatments have shown significant production improvements when restimulated after an initial period of production and partial reservoir depletion.

The problem addressed in this paper can be stated as follows. We assume that a candidate shale gas well has been identified for refracturing. For this well, a long-term production forecast as well as the production profile after additional refracture treatments at any point in time over the planning horizon is given. A gas price forecast along with expenses for drilling, fracturing, completions and refracturing operations are also available as problem data. The problem is to determine: (a) whether or not the well should be refractured, (b) how often the well should be refractured over its entire lifespan, and (c) when exactly the refracture treatments should be performed. The objective is to maximize either: (a) the estimated ultimate recovery (EUR) of the well, or (b) the net present value (NPV) of the well development project.

First, we present a continuous-time nonlinear programming (NLP) model to determine whether or not a shale gas well should be refractured, and when to schedule the refracture treatment. The proposed NLP model relies on the assumption that the well productivity profile follows a decreasing power function of time. To predict well performance prior to and after a refracture treatment, we propose an effective forecast function that mimics real-life curves. Next, we extend the proposed framework to allow for multiple refracture treatments and present a discrete-time mixed-integer linear programming (MILP). In an attempt to reduce solution times to a minimum, we compare three alternative formulations against each other (big-M formulation, disjunctive formulation using Standard and Compact Hull-Reformulations) and find that the disjunctive models yield the best computational performance. Finally, we apply the proposed MILP model to two case studies to demonstrate how refracturing can increase the expected recovery of a well and improve its profitability by several hundred thousand USD.

References

[1] Cafaro, D. C.; Grossmann, I. E. Strategic Planning, Design, and Development of the Shale Gas Supply Chain Network. *AIChE J.* 2014. doi: 10.1002/aic.14405.

[2] Drouven, M. G.; Grossmann, I. E. Multi-Period Planning, Design and Strategic Models for Long-Term Quality-Sensitive Shale Gas Development. *AIChE J.* 2016 (accepted for publication).

[3] Jacobs T. Renewing Mature Shale Wells Through Refracturing. *SPE News*. 2014, May Issue.

[4] Dozier, G., Elbel, J., Fielder, E., Hoover, R., Lemp, S., Reeves, S., Siebrits, E., Wisler, D., Wolhart, S., Refracturing works. *Oilfield Review*. 2003; 10(8):38-53.

460724 A Composite-Curve-Based Biomass Procurement Planning Approach

Abstract:

A Composite-Curve-Based Biomass Procurement Planning Approach

Wenzhao Wu, Daniel Kurniawan, WenBo Zhu, Christos T. Maravelias*

Dept. of Chemical and Biological Engineering, University of Wisconsin-Madison, Madison, WI 53706

The production of fuels and chemicals from biomass has received considerable attention recently due to environmental concerns[1]. Since the production of biofuels involves relatively expensive feedstock and energy-intensive biomass transportation, any biomass-to-fuels strategy should include an efficient, both in terms of cost and environmental impact, biomass (feedstock) supply chain. Unlike fossil fuels, biomass, as a low-energy density resource, is sparsely distributed. The efficient biomass transportation thus requires biomass procurement planning methods[2]. In many studies, the farms are treated as points without shape or area[3]. This is a reasonable assumption when the transportation distance is so large that the shape and size of the farms can be neglected. In this case, the transportation problem is modeled as a point-to-point (farm-to-refinery) problem. However, the shape and size of the farms cannot be neglected when the refinery is close to the farm, which means that the size of the farm is not significantly smaller than the transportation distances, which in turn means that the error in approximating the real transportation distance with the distance between the center of the farm and the bio-refinery can be quite large. In this case, transportation should be treated as a region-to-point problem. To this end, we discuss a novel approach to biomass procurement planning on a region-to-point basis.

In terms of transportation, we propose a region-to-point modeling approach based on mathematical integration (in a polar coordinate system) over the sourcing region that has unique characteristics such as shape, location, and productivity. The transportation cost is correlated with the amount of biomass ("mass") procured from each farm. The final result is a function $C = f(M)$, where M is the mass, and C is the transportation cost. In other words, the proposed methods generate a function that returns the total cost of procuring M mass, which is graphically represented by a "procurement curve". Both algebraic and numerical solution methods are discussed and demonstrated with examples.

In terms of system-level procurement planning, we develop a composite-curve-based approach that incorporates the regional transportation modeling method, and aims at identifying the biomass procurement plan that minimizes the total procurement cost (including biomass purchasing, harvesting and transportation). The specific steps for the generation of the composite curve using the individual procurement curves, as well as insights into the procurement planning problem are discussed. An analogy to the cold/hot composite curves in the pinch design method for heat exchanger networks[4] is also presented. A case study involving 12 farms (which are approximated as polytopes) and one refinery is presented (see Figure 1A). The corresponding composite curve is shown in Figure 2, and the final procurement strategy is graphically represented in Figure 1B.

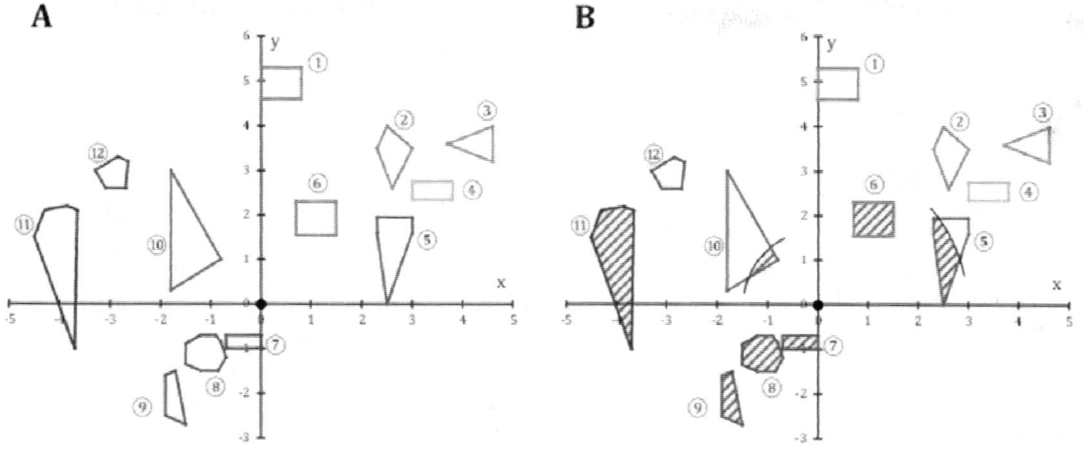

Figure 1. (**A**) Map of farms (the polytopes) surrounding a bio-refinery (the black dot at the origin) for the case study; (**B**) procurement strategy (represented by the green dashed areas) for a 11800 T/year demand. The farm numbers are labeled accordingly.

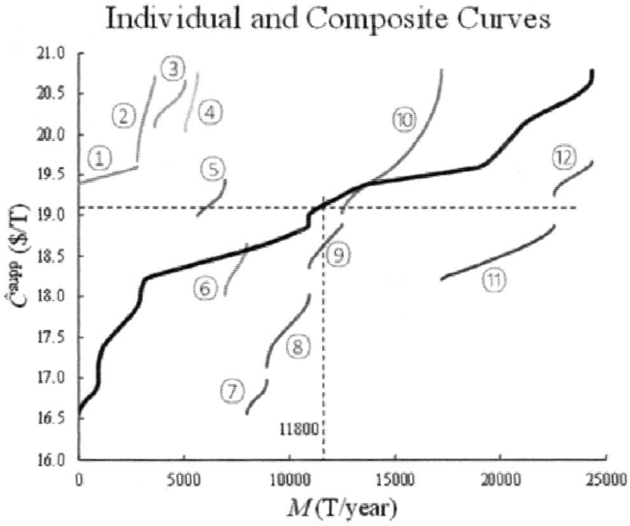

Figure 2. Individual procurement curves and the composite curve for the case study. The composite curve is marked thick. The demand of 11800 T/year and the corresponding per-mass supply cost ($\hat{C}^{supp}$) on the y-axis are marked with dashed lines. The farm numbers are labeled accordingly.

The proposed composite-curve-based method allows us to more accurately calculate transportation distance, and thus transportation costs and GHG emissions due to transportation. It also provides some key insights into the design of biofuel supply chains. In addition, the methods proposed in this work are integrated with mathematical programming to address complicated problems involving, for example, multiple feedstocks, multiple refineries, and multiple periods. We can solve such problems by either generating multiple composite curves, or directly incorporating the $C = f(M)$

function for each farm into a general supply chain optimization model.

References

[1] DOE Bioenergy Technologies Office, 2014. Multi-year program plan, Washington DC, USA: DOE.

[2] DOE/EERE, 2013c. Feedstock supply and logistics: biomass as a commodity, US: Department of Energy, Office of Energy Efficiency & Renewable Energy.

[3] You, F., Tao, L., Graziano, D. & Snyder, S. W., 2012. Optimal Design of Sustainable Cellulosic Biofuel Supply Chains: Multiobjective Optimization Coupled with Life Cycle Assessment and Input-Output Analysis. AIChE Journal, Volume 58, pp. 1157-1180.

[4] Linnhoff, B. & Hindmarsh, E., 1983. The pinch design method for heat exchanger networks. Chemical Engineering Science, 38(5), pp. 745-763.

453168 Enhanced Ethanol Production Via Activated Carbon Addition during Syngas Fermentation

Abstract:

Syngas made by gasification of biomass, coal, pet-coke and waste materials, or present in industrial gas waste streams can be used for production of fuels and chemicals. The biological conversion of syngas components, CO, CO_2 and H_2, to fuels and chemicals at commercial feasibility requires high conversion efficiency, product specificity and operation stability. Acetogenic bacteria convert CO, CO_2 and H_2 via the Wood-Ljungdahl pathway to acetic acid and ethanol as main products. The production of these products relies on efficient transfer of CO, CO_2 and H_2 to the cells at rates that match the bacterium's kinetic capability to process the gas. Cells are inhibited when too much CO accumulates in the fermentation broth, which also decreases H_2 utilization. Batch syngas fermentations using *Clostridium ragsdalei* in a 3-L CSTR with and without activated carbon were compared. Fine powdered activated carbon particles were added to the fermentation broth to alter the gas mass transfer to the bacterium and provide additional surface for growth. Modeled syngas containing 38.0:28.5:28.5:5.0 of $CO:H_2:CO_2:N_2$ was used. Results showed that the addition of activated carbon sustained activity of *C. ragsdalei* in batch fermentations increasing total CO and H_2 uptake six fold compared to no carbon. The increased gas uptake with activated carbon produced 19 g/L ethanol with less than 1 g/L total acetic acid. However, only about 1 g/L ethanol and 5 g/L acetic acid were produced without activated carbon and cell concentrations sharply decreased after 158 h of fermentation with slight syngas consumption measured after 200 h. Alternatively, syngas consumption and cell activity were sustained during the fermentation with activated carbon for over 450 h. The maximum CO and H_2 conversion efficiencies in the fermenter with activated carbon were 88% and 86%, respectively. The operation of syngas fermentation with activated carbon exhibits higher stability, selectivity and energy conservation than any previously reported results, indicating potential for commercial biofuels production and other biological gas conversion processes.

469891 A Discussion of Realistic Feedstocks: Can We Use 2-Year-Old Short Rotation Hybrid Poplar Coppice for Fuels and Chemicals Production Via Biochemical and Thermochemical Conversion?

Abstract:

The economic success of a potential biorefinery is directly related to the use of low cost biomass. Due to the high cost, it is unlikely that a wood-based biorefinery would use only whitewood chips from mature trees in conversion. Using a more heterogeneous raw material is less costly, but efficacy of the conversion using whole short ration coppice (including leaves, bark, branches and white wood) has not been investigated. This study investigated the influence of using whole 2-year-old hybrid poplar coppice on the overall sugar yield via biochemical conversion and the bio-oil production via thermochemical conversion.

By harvesting all aboveground parts of 2-year-old hybrid poplar coppice, the biomass used in this study was comprised of 37% leaves, 9% bark, 12% branches, and 42% whitewood chips (dry weight basis). From this, four biomass samples were prepared and studied: 1) 2-year-old poplar mixture, 2) 2-year-old poplar mixture without leaves, 3) 2-year-old poplar leaves, and 4) 12-year-old poplar wood chips. In biochemical conversion, all poplar samples were steam pretreated at 195ËšC for 5 minutes with 3% $SO_2$and enzymatic hydrolyzed at 5FPU/cellulose enzyme loadings. The sugar recovery and estimated ethanol yield were calculated. In thermochemical conversion, samples were fast pyrolyzed at 500 ËšC, for 1 second in a fluidized bed reactor. The final bio-oil yield and composition were measured.

Results show that converting the 2-year-old hybrid poplar coppice may be challenging, with up to 50% lower sugar yield and close to 30% lower bio-oil yield than 12-year-old poplar wood chips. However, the removal of leaves significantly increased the overall sugar recovery and bio-oil production by 20% and 15%, respectively. Unsurprisingly, the substrate derived from isolated leaves performed poorly in both biochemical and thermochemical conversion. These results show that leaf removal from 2-year-old hybrid poplar coppice prior to harvest may enhance the biorefinery production. Given the yields of both processes, we will evaluate the economics of converting 2-year-old hybrid poplar coppice to sugar and bio-oil via biochemical and thermochemical conversions.

467443 Pyrolytic and Electrochemical Upgrading of Lignins Extracted from Pretreated Biomass to Produce Liquid Fuel Intermediates and Value-Added Products

Abstract:

Pretreatment enhances the effectiveness of enzymatic strategies for depolymerizing the cellulose and hemicellulose fractions of plant biomass. Many different pretreatment strategies have been investigated, including the use of dilute acid, steam explosion, organosolv, alkaline hydrogen peroxide (AHP), and ammonia fiber expansion (AFEX). Extractive ammonia (EA) pretreatment, a more severe version of AFEX, recovers a separate lignin stream from biomass. Upgrading this lignin stream to create fuels and chemicals potentially adds value to decentralized processing depots or centralized biorefineries that practice EA pretreatment. One approach to valorizing this lignin stream is to deconstruct EA lignin using fast pyrolysis to create a lignin-derived bio-oil that is rich in phenolics. Fast pyrolysis involves rapid heating (400-600°C) in the absence of oxygen to create bio-oil, biochar and combustible gas. Upgrading of lignin bio-oil proceeds by electrocatalytic hydrogenation (ECH), preferably using electricity from wind and solar sources. In this study, model compounds that represent lignin bio-oil components were subjected to ECH under mild conditions using ruthenium on activated carbon as a catalytic cathode. To date, several such compounds have been successfully reduced to more stable forms, for example, phenol, guaiacol and syringol have been converted into cyclohexanol at appreciable yields and selectivities. In parallel to model compound studies, catalyst reusability studies have enhanced our understanding of electrocatalyst deactivation and methods for extending catalyst lifetime. By coupling fast pyrolysis and electrocatalysis for lignin deconstruction and upgrading, respectively, a major byproduct of alcohol fermentation can be valorized into fuels and chemical products. Integration of thermochemical, electrochemical and biochemical transformative pathways is paramount for realizing the full potential of plant biomass for displacing petroleum.

466853 Combined Biochemical and Thermochemical Processing of Lignocellulosic Biomass: Techno-Economic Evaluation

Abstract:

The need for environmentally friendly fuels and commodity chemicals has driven research in the production of biofuels and bio-derived chemicals. There is now a broad product portfolio and wide variety of stand-alone technological pathways for biomass conversion systems. In the biochemical pathway of ethanol production alone, there exists multiple conversion technologies available. One such method, the consolidated bioprocessing proposed by Brethauer and Studer [1], seems to offer a viable alternative, as high conversion rates are achieved while combining into one reactor the hydrolysis and fermentation steps. This method of using a microbial consortium for conversion would likely open pathways to other liquid fuels and commodity chemicals as well, as similar work has been done in the production of higher carbon alcohols [2]. But, analogous to an oil refinery, a stand-alone biorefinery must look at coupling heat integration, fuel production, and higher value co-product production. However, work has not yet been widely published to determine feasibility or economic potential for this method of ethanol production. Similarly, scale-up potential and feasibility studies are limited for many of these novel processes.

Moreover, many of these concepts are still in the development stage, which requires the use of process modeling techniques. Modelling of lab-scale based production processes allows for techno-economic analysis and optimization, as estimates of energetic efficiency and scale-up potential of novel conversion processes are lacking. Production feasibility will be highly dependent on the energetic efficiency, environmental sustainability, and economic potential of these bio-based fuels and chemicals, especially in comparison to their fossil-fuel derived counterparts.

This work presents techno-economic process models, using flowsheeting software, of co-production of value added products, liquid fuels, and heat from lignocellulosic feedstock to analyze and compare the annual costs and efficiencies of the different combined biochemical and thermochemical conversion pathways. Furthermore, to contribute to the development and implementation of sustainable biorefineries, a systematic methodology which combines process modeling and optimization approaches is applied and a comparison of the pathways based on economic and energetic efficiencies is then proposed. The results provide a powerful support for the design of biorefinery systems with improved process economics and reduced environmental impact.

468144 Total Site Analysis As a Synthesis Model to Select, Optimize and Integrate Processes in Biorefineries

Abstract:

The paper presents a new approach to apply energy integration in multiple-product plants. The incentives of this work originate in biorefineries. Biorefineries constitute the most promising route for the valorisation of bio-renewables and the replacement of fossil-oil in any application of chemicals and fuels production. Biorefineries are capable to utilize a wide range of biomass sources including agricultural and forestry residues, industrial and municipal wastes as well as aquatic biomass (algae). Several biorefineries have already taken off in USA, EU, China and Brazil; the majority of applications focus on the production of bio-fuels. However, a big number of conventional and novel bio-products â€“ bulk chemicals and specialties, polymers, coatings, resins, pharmaceuticals, enzymes, proteins, fuels etc â€“ can be generated via thermo-processing and bio-processing of biomass. Bio-chemicals production is projected to reach up to 10 % of $1.5 trillion worldwide annual chemicals market [1], while biorefineries are capable to consume up to 84 % less non-renewable sources increasing expectations for GHG emissions reduction from 60 % to 100 % [2].

Biorefineries can be benefited by the integration of multiple products and production lines through the exchange of materials and energy among them. The development of appropriate synergies among the different processes may result in significant reduction up to 90 % in CAPEX [2], 84 % in energy consumption [3] and 58 % in water demands [4]. Provided that numerous products and technologies can get integrated in biorefineries, critical questions rise about which processes portfolio maximizes efficiencies, if collocated and integrated together. Given the high impact of energy cost in sustainability of biorefining, this work focuses on energy integration techniques in upcoming applications.

Energy integration is a powerful tool that estimates the maximum feasible margins of energy savings by exchanging heat between hot (need cooling) and cold (need heating) streams involved in the process (direct integration). In the scope of integrating multiple-process plants (like biorefineries), energy integration takes the form of Total Site Analysis (TSA) [5]. TSA estimates steam utility savings by using excess heat from the individually (directly) integrated processes to cover steam demands of others via steam production-reuse (indirect integration). The graphical tool of Site Sources and Sinks Profiles (SSSPs) is used to estimate steam savings in Total Sites. However, the graphical tool appears impractical and ineffective to address the upcoming industry, since assumes fixed and known processes. In the case of biorefineries, processes are not essentially known and solutions are not profound given the range of candidate technologies that may enter the biorefinery. The biorefinery problem constitutes a synthesis problem, where processes are significant degrees of freedom in integration. Each process features a different energy profile and energy savings will change as far as processes are integrated across different process combinations. In contrast to conventional TSA, heat exchange options in biorefinery cases do not refer to a fixed cluster of processes, but actually expand among all candidates. A potential solution would be the exhaustive use of the graphical tool (SSSPs) to estimate steam savings for all possible combinations and detect the one with the lowest energy cost. In fact, this approach is impractical for large problems (number of combinations explodes) and appears to be inefficient and unsafe to solve the biorefinery problem, as it is proven in this work.

The candidate biorefinery routes from raw materials (biomass) to intermediate and final products are represented in the form of value chain paths. Each path denotes a candidate process, while a set of consecutive paths composes a potential biorefinery plant (Total Site). Alternative biorefinery structures can be developed by choosing each time different processes among competitive paths (they valorise the same chemical), which are branched along the value chain. In the scope of energy integration, the challenge is to systematically address all alternative structures and detect the biorefinery that maximizes steam savings and minimizes energy cost. This means that screening of process combinations and integration (direct and indirect) should simultaneously occur. To address the complexity and multiplicity of the synthesis problem, Total Site Analysis should be upgraded into a synthesis tool by combining thermodynamics with mathematical programming.

The use of SSSPs requires fixed and known processes and, thus, known energy contents of streams that will get integrated. This implies that processes from the value chain should be previously selected and get integrated using SSSPs; accordingly, one process among competitive can be selected and integrated with others each time. Nevertheless, higher savings can be succeeded by sharing upstream chemicals among competitive processes in any way, subsequently changing the capacity and the energy content of involved hot/cold streams in order to improve the heat exchange conditions and secure better synergies. In the courses of a real-life biorefinery problem, this approach resulted up to 8 % higher steam savings compared with savings obtained by using SSSPs. Provided that the graphical tool is not capable to meet with these needs, new comprehensive representations are proposed to aid with the mathematical formulation of Total Site integration and address process capacities as additional continuous variables. The proposed concept could be interpreted as if SSSPs involve moving (variable) curves.

Two new representations, which communicate each other, are introduced to describe the combinatorial synthesis and integration problem. The first, a Biomass Bipartite graph Representation (BBR) is used to translate value chain paths into representations that enable the development of mass balances along the value chain. The second, a Total Site Representation (TSR) uses a properly modified transshipment model that incorporates all heat exchange options (energy balances) related with direct and indirect integration and enable the mathematical formulation of Total Site integration. The mass and energy balances are constructed as an optimization model that is solved using mixed integer linear programming and yields the processes portfolio that minimizes energy cost of the under-construction plant.

The BBR uses three building blocks: (i) product nodes for chemicals (raw, intermediate and final), (ii) process units for paths and (iii) interconnections that connect nodes with units and describe the transportation of chemicals from path to path. The flowrate of each chemical is assigned to the relative interconnection and, thus, mass balances can be recorded for each product node and process unit, given the capacity of raw materials at the beginning of value chain and the conversion rate of each path (given by experimental groups). Accordingly, the flowrates of process inlet interconnections are used to denote the capacity of the associated process. Each path is previously simulated to estimate the normalized heat demands of involved streams per unit mass of inlet chemical. Assuming that mass and energy balances within a process are linearly depended on its capacity, the actual heat content of hot/cold streams can be linearly expressed by the normalized heat demands and the variable flowrates involved in mass balances. This expression secures the connectivity between BBR (selection of paths) and TSR (integration of selected paths).

The TSR incorporates a super-interval cascade, called Total Site Cascade (TSC), which is composed by the temperature intervals of all candidate processes. The process streams contribute to TSC according to flowrates selected by BBR and the modified cascade proceeds with the direct integration of all selected processes simultaneously rather than each separately. As a result, the utility needs of individually integrated processes are estimated by TSC. At the same time, a parallel set of blocks, called utility operators, gathers the utility needs from TSC and proceeds with indirect integration. Specifically, the utility operators (i) translate excess heat from TSC into steam production, (ii) estimate the overlap between generated and demanded steam and (iii) call for external (Total Site) utilities to cover the remaining demands. In fact, the utility operators identify which process are capable to work as steam-source or sink and detect (in combination with BBR) energy complementary process combinations. The energy balances at each internal of TSC and the utility operators are constructed as an optimization model that minimizes energy cost of Total Sites. The combined system of TSC and the utility operators composes the TSR. The model composed by the energy balances of TSR can be also individually used (without BBR) for Total Site integration of a fixed cluster of processes avoiding the graphical tool of SSSPs.

The proposed methodology was applied for the development of a real-life lignocellulosic biorefinery with 9 candidate paths and 11 bio-products yielding in the 5 best processes portfolio that consume 9 % and 14 % lower hot and cold utilities, respectively, compared with the best solution that obtained by exhaustively using SSSPs.

Literature

1. Austin A., Biomass Magazine, 2012, http://biomassmagazine.com/articles/2778/bio-roundtable-feedstock-flexibility-key-in-biochemical-industry-
2. EuropaBio, 2011, http://www.europabio.org/biorefinery-feasibility-study
3. Kokossis A.C., Tsakalova M., Pyrgakis K., 2015, Design of Integrated Biorefineries, Computers & Chemical Engineering, 81, p.p. 40â€"56
4. Nikolakopoulos A., Karagiannakis P., Galanis A., Kokossis A., 2012, A water saving methodology for the efficient development of biorefineries, Computer Aided Chemical Engineering, 30, p.p. 7-10
5. Dhole V.R., Linnhoff B., 1983, Total Site Targets For Fuel, Co-Generation, Emissions, And Cooling, Computers & Chemical Engineering, 1, S101-S109

472568 Multianalyte Electrochemical Sensors on a Monolith Electrode By Optically Probing the Electrical Double Layer

Abstract:

Electrochemical sensors are arguably the most successful chemical sensing devices primarily because of their high sensitivity at low cost. The key advantage of electrochemical devices is the active signal due to chemical reaction allowing specific recognition of the chemical with a low incidence of false positives. However, only one redox reaction per electrode is possible to measure. An opto-electrochemical method, called Scanning Electrometer for Electrical Double-layer (SEED), to quantitatively detect multiple, individual redox reactions on a monolith electrode will be described. SEED optically measure the charging/discharging of the electrical double layer to quantify the local redox current. By scanning a laser beam, local redox current density distribution can be quantitatively mapped to detect multiple analytes on a monolith (working) electrode. SEED is responsive to (redox) reactions of 0.1 atto-mole of molecules. The local redox current is measured as the change in the charge in the electrical double-layer which significantly changes due to redox reaction. SEED signal enhances as the sensing area reduces in size. Characteristics of SEED and application to chemical and biochemical sensing, and genomics will be described.

473889 Role of Defects in Enhancing the Electrochemical Properties of Transition Metal Oxide

Abstract:

Transition metal oxides are fascinating class of materials that are valued for their catalytic, ferroelectric, superconducting, and optical properties, and hence they find uses in diverse applications such as heterogeneous catalysts, sensors, solar cells, and high temperature superconductors. Such rich properties of metal oxide are partly enabled by their complex surface chemistry and the presence of lattice vacancies that significantly perturb the electronic structure of the material. Hence, an in-depth understanding of the origin and nature of these defect states is important for further expanding the scope of oxide-based systems for various applications. The focus of this presentation is on elucidating the nature of interaction of various chemicals with defects in metal oxide. The chemical-structure-property correlations developed in this work can enable rational engineering of a new type of nanoscale, low temperature catalysts having tailored functionality and selectivity that can significantly reduce the energy cost and carbon footprint of many important chemical reactions.

452861 The Degradation of Additives during the Electrodeposition of Copper for Metamaterial Fabrication

Abstract:

Metamaterials, a class of artificial optically-active materials with exceptional properties based on their patterned structures, typically use metallic and dielectric repeating features to provide a negative index of refraction. In the fabrication of metamaterials utilized in this study, copper is selected for the metallic part of metamaterial due to its high thermal and electrical conductivity, excellent mechanical properties and relative ease of acquisition.

Electrodeposition from a highly-acidic copper electrolyte is the most attractive approach for fabricating the conductive component of the metamaterial in this study due to its ease of control and high deposition rate. However, since the presence of special high-aspect ratio patterns in the metamaterial requires bottom-up super-filling and leveling of micron and sub-micron scale volumes; additives such as chloride ions, 3-mercapto-1-propanesulfonate (MPSA), polyethylene glycol (PEG), and polyvinylpyrrolidone (PVP) are introduced into the electrodeposition process in order to provide excellent conformality and adhesion. Most of the present study focuses on investigating the optimal parameters of electrodeposition of copper such as the effects of additive aging, the degradation of MPSA and the electrodeposition parameters of deposition voltage and its corresponding current density.

Electrochemical methods such as cyclic voltammetry, chronopotentiometry and controlled potential coulometry probe the mechanism of both the specific effects of additives and the interplay between additives. In parallel, COMSOL finite element modeling (FEM) software is used to simulate the 2D electrodeposition of copper to investigate the copper super-filling, mass transfer and the electrochemical behavior of the electrodeposition system.

Atomic force microscopy (AFM), four-point probe sheet-resistivity measurements, optical microscopy, optical profilometry and X-ray diffraction (XRD) are used as characterization tools to map the surface topography, sheet resistances, surface roughness and crystal structure of the deposited films obtained under different experimental conditions. Nuclear magnetic resonance (NMR) is conducted to analyze the degradation of the additives during the electrodeposition process. Combined with the electrochemical methods, optimized parameters for electrodeposition of copper are obtained and desired fabrication methods for metamaterials are achieved in this study.

470927 Field Effect Control of Electrochemical Reaction Kinetics at Back-Gated, UltrathinÂ Semiconductor Electrodes

Abstract:

An electrochemical reaction refers to a process of chemical changes caused by the electrical current, or a process of generating electricity by chemical reactions, in which direction and rate of the reaction can be controlled by electrode potential (*i.e.* energy of holes and electrons in the electrode). Because an electrochemical reaction involves with charge transfer across an electrode/electrolyte interface (though it may not be the only process), the reaction kinetics is determined to a large extent by the electronic structures (*i.e.* energy level distribution and electronic occupation) of the electrode material and the redox species in the electrolyte. In this talk, we propose a new approach exploiting the *field effect*, in addition to electrode potential, to modulate the electronic structures and electrochemical reaction kinetics at electrode/electrolyte interfaces.

As a model system, we prepared working electrodes with â€˜gate-insulator-semiconductorâ€™ structure, which is similar to that of field effect transistors (FETs). Herein, we employed ultrathin semiconductor layers so that electrochemistry at the semiconductor surface are effectively modulated by a voltage bias applied to the gate. On those *gate-tunable* electrodes, we observed continuous, *in-situ* modulation of outer-sphere electron transfer kinetics at the semiconductor/electrolyte interface with voltage biases applied to the gate. For example, the reduction potential of 2,3,5,6-tetrabromo-1,4-benzoquinone (TBBQ) on a 5 nm thick ZnO electrode could be shifted by ~0.4 V. With further control experiments, we found that the observed gate-controlled redox reaction kinetics is essentially attributed to band alignment shift at the electrode/electrolyte interface.

462103 Electrochemical Synthesis of Organic Nanorods on Gold Nanoparticles Seeds

Abstract:

This talk is focused on nanoparticle-directed synthesis of tetrathiafulvalene charge-transfer salt (TTFBr) and potassium tetracyanoplatinate (KCP) organic nanorods using electrochemical deposition. Gold nanoparticles (AuNPs) acting here as nucleation seeds are deposited on highly oriented pyrolytic graphite (HOPG). The nanorods, either TTFBr or KCP, are deposited subsequently on the AuNPs-decorated HOPG for the study of seed-mediated crystallization process at the nanoscale. Both the seed and the organic crystal are deposited using electrochemical chemistry. The deposition processes are monitored by cyclic voltammetry and in-situ atomic force microscope. The field emission scanning electron microscope is also employed to characterize the seed and crystal nanostructures. The TTFBr and KCP crystals deposited on AuNPs-decorated HOPG show preferential nucleation on the AuNP in comparison with crystals electrochemically deposited on the planar HOPG. The size of the crystals is controlled predominantly by the solution concentration and the seed size. Nanorods as small as 7 nm in height are nucleated on AuNPs of 20 nm in height. We hypothesize that the local curvature of the seed particle, that limits the cross sectional dimension and introduces an interfacial strain, contributes to the nanoconfinement effect observed. Currently we are also applying this seed-directed synthesis method to fabricate organic nanowire sensors on gold lithographical patterns. The prototype sensor shows potential to compete with commercial sensors in detecting toxic and flammable gases. This research contributes a better understanding of crystallization at the nanoscale and a possible low-cost technology for chemical sensing applications.

462273 Interfacial Structure and Capacitance of Li-Doped Ionic Liquid Electrolytes from Molecular Simulations

Abstract:

Ionic liquids have been proposed as candidate electrolytes for high-energy density, rechargeable batteries, supercapacitors, and hybrid energy storage devices. Though Li-salt is often present in these systems, its influence on interfacial properties is largely uncharacterized. We, thereby, present an extensive computational analysis, supported by experimental comparisons, of the properties of a representative set of these electrolytes at an ideal carbon interface as a function of Li-salt doping and voltage. We have performed polarizable molecular (MD) dynamics simulations, using the APPLE&P force field, to evaluate electric double layer (EDL) capacitance and distribution of Li^+ in the EDL. Differential capacitance exhibits the characteristic â€œcamelâ€ profile and is insensitive to Li-doping. Li^+ localizes in the second molecular layer of the EDL, which is a result of confinement from free energy barriers associated with ion layering. Joint MD/electronic structure computations show the electrochemical window of the electrolytes to be a weak function of Li-doping. Estimates of supercapacitor specific energy are made using the computed window and capacitance. The magnitude and trends in specific energy are in good agreement with experiment.

465711 Reactivity of an Alucone Thin Film Coating on a Lithiated Si Anode of Li-Ion Batteries

Abstract:

Lithium-ion batteries are one of the most common types of rechargeable batteries to power small, portable electronic devices. Current research efforts are focused on achieving the energy density required for electric cars and high-end portable electronics. Silicon anodes in Li-ion batteries yield higher theoretical capacities than graphite, the most common anode material; nevertheless, the significant volume changes occurring in the silicon anode during lithiation and de-lithiation limits its ability to retain capacity for an extended number of charge/discharge cycles. A solid electrolyte interphase (SEI) layer that naturally forms on the silicon surface from the reduction of the electrolyte components can improve the mechanical stability of the anode at the expense of hindering Li transport. In contrast, surface modification of the silicon anode may be a viable strategy to accommodate volumetric changes, preventing cracking and pulverization of the electrode, while maintaining or improving ionic and electric conductivity. Aluminum alcoxide (alucone) films to improve the stability of the anode have been successfully deposited and characterized on lithiated silicon surfaces, and reportedly become electronically conductive after Li atoms saturate the favorable sites of the film. In the present work, we use density functional theory and ab initio molecular dynamics simulations to study the reactivity of the film and the anode surface in contact with a typical organic solvent, ethylene carbonate. The solvent molecules are found to be able to penetrate the film and undergo reduction inside the film and at the film-anode and film-electrolyte interfaces via one- and two-electron mechanisms. Such reaction mechanisms are examined along with the structure, stability and suitability of the film to improve the anode, as well as the implications for the design of enhanced electrode materials for Li-ion batteries.

471645 Microstructure of Room-Temperature Ionic Liquids at Charged Surfaces Revealed By Integrated Modeling and Experimental Approaches

Abstract:

The interfaces of room-temperature ionic liquids (RTILs) at charged surfaces are of great importance for many applications, such as electrical energy storage and lubrication. Many nanoscale techniques (e.g., atomic force microscopy, AFM) are developed to reveal the microstructure of electrical double layers (EDLs) formed by RTILs at charged solid surface, and with the help of molecular dynamics (MD) simulation one can obtain detailed information of EDLs at molecular level and have a better understanding of the interfacial phenomena.

In this talk, firstly we report direct observation of structure and properties of the topological defects formed in ordered ionic liquid layers at carbon interfaces. The observation of layer structure at structural defects such as step edges reinforced by MD simulations define spatial resolution of the method. We have observed the internal structure of topological and structural defects on the ionic-liquid surface interfaces using 3D force mapping. The structural defects were found to result only in short range changes in ionic liquid ordering, with the changes in 3D force fields consistent with rigid shift of the layers with narrow disordered region between the two, consistent with MD simulations. At the same time, formation of serendipitous dislocation-type topological defects were observed on planar surfaces. The decrease in the ordering was observed within the defect, consistent with behavior expected for classical liquid crystals.

Secondly, further MD work was done to investigate the influence of the thickness of the electrode surface step on the microstructure of interfacial ILs. A strong correlation was observed between the interfacial IL structure and the step thickness in electrode surface as well as the ion size. Specifically, when the step thickness is commensurate with ion size, the interfacial layering of cation/anion is more evident; whereas, the layering tends to be less defined when the step thickness is close to the half of ion size. Furthermore, two-dimensional microstructure of ion layers exhibits different patterns and alignments of counterion/co-ion lattice at neutral and charged electrodes.

Thirdly, beyond carbon surfaces, we then explore the ILs at charged mica surfaces. By comparing the measured force profiles from AFM to ion density profiles from MD simulations, we were able to determine which ion of the ionic liquid the AFM tips are sensitive to during force measurements, and this is determined by ion larger ion volume and/or weight. By varying the anion or cation of the ionic liquids we are able to move from a regime where we detect the position of the anions (e.g. [emim][Tf_2N]) to one where we detect the position of the cations (e.g. [emim][BF_4]). Using AFM data alone we cannot identify which ion we are sensitive to, and through a direct comparison of experiment data with molecular dynamics simulation, we are able to gain a better understanding of the mechanisms leading to the measured force profiles, therefore making future interpretation of force-curves at ionic-liquid solid interfaces easier. We will also show the interfacial structure of humid ILs at charged mica surfaces, in particular, the water accumulation will differ from that at carbon surfaces.

470229 Mathematical Models for the Impedance Response of Subcutaneous Glucose Sensors

Abstract:

Since 1974, the artificial pancreas has been acknowledged by the greater scientific community to be a possible way to manage type 1 diabetes [1]. In 2008, a case study showed that a subcutaneous continuous glucose monitor sensor could be implanted for 28 days in a patient with type 1 diabetes [2]. However, commercially available subcutaneous continuous glucose monitors are generally changed every seven days to avoid unreliable results. The object of this work is to explore the potential of impedance spectroscopy to monitor the state of health of subcutaneous continuous glucose monitor sensors.

A mathematical model has been developed for the impedance response of a glucose oxidase enzyme-based electrochemical biosensor. The model accounts for a glucose limiting membrane GLM, which controls the amount of glucose participating in the enzymatic reaction. The glucose oxidase was assumed to be immobilized within a thin film adjacent to the electrode. In the glucose oxidase layer, a process of enzymatic catalysis transforms the glucose into peroxide, which can be detected electrochemically. This system may be considered to be a special case of the coupled homogeneous and heterogeneous reactions addressed by Levich [3].

The model development required two steps. The nonlinear coupled differential equations governing this system were solved under the assumption of a steady state. The steady-state concentrations resulting from the steady-state simulation were used in the solution of the linearized set of differential equations describing the sinusoidal steady state. The enzymatic catalysis was treated in terms of two homogeneous reactions, one consuming the glucose oxidase and forming gluconic acid, and the other regenerating the glucose oxidase and forming the peroxide.

References

1. A. M. Albisser, B. S. Leibel, T. G. Ewart, Z. Davidovac, C. K. Botz, and W. Zingg, Diabetes 23 (1974), 389â€“396.
2. J. J. Chamberlain and D. Small, Clin. Diabetes, 26 (2008), 138â€“139.
3. V. G. Levich, Physicochemical Hydrodynamics, Prentice Hall, Englewood Cliffs, NJ, (1962), p. 345-357.

470435 The Effect of Alkali and Quaternary Ammonium Cations on the Electrochemical Interface Studied By *in-Situ* Infrared Spectroscopy

Abstract:

Recent computational studies have suggested that the adsorption of alkali metal cations may have significant impact on electrochemical processes, particularly those which occur at low potentials such as the hydrogen evolution (HER) and oxidation (HOR) reactions. It has also been demonstrated that the reactivity and selectivity of other important electrochemical processes such as CO_2 reduction are influenced by the presence of alkali cations in the electrolyte. In this study we employ surface-enhanced infrared absorption spectroscopy to study the adsorption of alkali and various quaternary ammonium cations using CO as a probe molecule on Pt electrodes. Linearly adsorbed CO is displaced at low electrode potentials (below the potential of zero charge) by electrostatically bound cations and forced to a bridge-bonded configuration. It is found that larger alkali cations such as K^+ have stronger surface interactions with adsorbed CO than smaller alkali cations. Quaternary ammonium cations show the least pronounced effect.

459309 CFD Modeling and Computation for an Industrial Steam Methane Reforming Furnace

Abstract:

Hydrogen is one of the most important raw materials for petroleum refineries that convert crude oil into a variety of products with higher economic value, and its unavailability can limit the production rates of petroleum products. Hydrogen is produced on commercial scale most commonly and economically by the steam methane reforming (SMR) process. Specifically, the steam reforming process is an overall endothermic process in which raw natural gas, e.g., methane, reacts with high-pressure, high-temperature steam (superheated steam) in the presence of a nickel-based catalyst to produce hydrogen, carbon dioxide and carbon monoxide. Due to large production rates of hydrogen (e.g., the largest North American plant in 2004 could produce up to 300, 000 Nm^3/h of hydrogen [1]), the cost of raw materials can be on the order of millions of dollars [1]. Therefore, a small improvement in the process efficiency results in a great gain in profit margin of a plant. However, determining the optimal process parameters may be a time-consuming or risky process if performed experimentally, because it may require increasing the outer reforming tube wall temperature to improve efficiency, which can pose issues for the reactor material [1] [2]. To determine optimal processing parameters without experimental techniques, computational fluid dynamics (CFD) modeling is attractive because it can precisely capture all geometry characteristics of a given reformer through computer-aided design software, which in turn allows CFD models to generate simulation results that can be expected to serve as reasonable substitutes for experimental data, and the parametric effect can be visualized and quantified through CFD simulations.

This work develops a computational fluid dynamics (CFD) model of an industrial-scale steam methane reformer comprised of 336 industrial-scale reforming reactors, 96 industrial-scale burners and 8 industrial-scale flue gas tunnels. The industrial-scale reformer CFD model is developed by analyzing well-established physical phenomena, i.e., the transport of momentum, material, energy and turbulence, and chemical reactions, i.e., combustion and the SMR process, that take place inside the steam methane reformer.

Specifically, the P − 1 radiation model, standard k − turbulence model, compressible ideal gas equation of state and finite rate/eddy dissipation (FR/ED) turbulence-chemistry interaction model are adopted to simulate the macroscopic and microscopic events in the reformer. The simulation results generated by the industrial-scale reformer CFD model are verified to be in agreement with the available typical plant data, and also data reported in literature. Furthermore, the CFD model of a reformer can provide insights into the system which cannot be captured by experimental data generated by on-site parametric study or by solution of a complete reformer mathematical model (e.g., the species distributions inside the combustion chamber.) Through this work, a converging strategy that allows one to quickly acquire the converged solution of complex CFD models is developed.

[1] Latham D. Masters Thesis: Mathematical Modeling of an Industrial Steam Methane Reformer. Queen's University, 2008.

[2] Pantoleontos G, Kikkinides ES, Georgiadis MC. A heterogeneous dynamic model for the simulation and optimisation of the steam methane reforming reactor. International Journal of Hydrogen Energy. 2012;37:16346-16358.

[3] Lao L, Aguirre A, Tran A, Wu Z, Durand H, Christofides PD. CFD modeling and control of a steam methane reforming reactor. Chemical Engineering Science. 2016;148:78-92.

[4] Aguirre A, Tran A, Lao L, Durand H, Crose M, Christofides PD. CFD Modeling of a Pilot-Scale Steam Methane Reforming Furnace. Advances in Energy Systems Engineering, Kopanos G, Liu P and Georgiadis M (Eds.), Springer, in press.

462415 Toward Optimal NGL Conversion to Olefins: Advances in Steam Cracking Optimization

Abstract:

Recent advances in the shale gas industry resulted in a significant increase in natural gas liquids (NGL) production [1]. NGL production increased by more than 100% since 2008 and is expected to grow by another 27% until 2040 [2,3]. In some wet shale plays, the combined ethane and propane composition can exceed 30% [4]. These recent trends motivate industrial efforts to build several new ethane crackers with a combined ethylene production capacity of 12.5 million tonnes/year in the United States [5]. Similar to ethane, other natural gas liquids can undergo steam cracking to produce olefins [6,7] with different underlying kinetics and mechanisms. This suggests that each hydrocarbon has a unique set of optimal operating conditions for the steam cracking process.

Although the steam cracking kinetics are well-established [6,7], several side reactions exist and the underlying reactor model contains a complex set of ordinary differential equations. In a stiff system such as this one, there is a need to develop a mathematical optimization framework to simultaneously address all trade-offs for optimal sizing operation. Therefore, the differential equations regarding mass, energy, and momentum balances are transformed using orthogonal collocation on finite elements [8]. In this work, the proposed reactor model also incorporates the kinetics of coke formation as well as decoking downtime. Finally, the reactor length is optimized to minimize the reactor volume and the associated investment costs. The resulting mixed integer nonlinear optimization model (MINLP) is solved to optimize the inlet flowrates, external heat flux, and reactor length (volume) to maximize the profit. The optimal reactor topologies and configurations are compared with the industrial cases, since the overall profit increased significantly. Finally, novel reactor designs are proposed that input a mixture of hydrocarbon feeds as opposed to a pure feed (ethane, propane, or butane). These optimal designs unravel the untapped opportunity of optimal allocation of NGLs as petrochemical feedstocks, while establishing a systematic framework for kinetic reactor design and optimization.

In order to illustrate the practical benefits of our systematic framework over a complete shale gas to olefins process, the optimal cracking designs are investigated in an integrated process superstructure [9]. Through the use of optimal steam cracking reactors that we developed, the net present values of these systems are significantly improved (>10%).

[1]: Floudas, C. A.; Niziolek, A. M.; Onel, O.; Matthews, L. R. Multi-scale systems engineering for energy and the environment: Challenges and opportunities. AIChE Journal 2016, 62 (3), 602-623.

[2]: Energy Information Administration, Annual Energy Outlook, 2015

[3]: Energy Information Administration, Petroleum and Other Liquids, 2015

[4]: Hill, R. J.; Jarvie, D. M.; Zumberge, J.; Henry, M.; Pollastro, R. M.; Oil and gas geochemistry and petroleum systems of the Fort Worth Basin, 2007, AAPG Bulletin, 91(4), 445-473.

[5]: Chang, J.; New projects may raise US ethylene capacity by 52%, PE by 47%, 2014, ICIS Petrochemicals.

[6]: Sundaram, K. M.; Froment, G. F.; Modeling of thermal cracking kinetics â€" I: Thermal cracking of ethane, propane and their mixtures, 1977, Chemical Engineering Science, 32(6), 601-608.

[7]: Sundaram, K. M.; Froment, G. F.; Modeling of thermal cracking kinetics â€" II: Cracking of iso-butane, of n-butane, and of mixtures ethane-propane-n-butane, 1977, Chemical Engineering Science, 32(6), 609-617.

[8]: Cuthrell, J. E.; Biegler, L. T.; On the optimization of differential-algebraic process systems, 1987, AIChE Journal, 33(8), 1257-1270.

[9]: Onel, O.; Niziolek, A. M; Floudas, C. A Optimal Production of Light Olefins from Natural Gas via the Methanol Intermediate. Industrial & Engineering Chemistry Research 2016, 55 (11), 3043-3063.

467043 Decarbonization of Electricity Grids: A Multi-Scale Challenge

Abstract:

Decarbonization of Electricity Grids: A multi-scale challenge

Sharifzadeh Mahdi[1] , Shah Nilay.

Centre for process Systems Engineering (CPSE), Chemical Engineering Department, South Kensington Campus, Imperial College London, SW7 2AZ, UK.

Abstract - Electricity power consumption in a country has a direct relation to the quality of life and economic growth. Nevertheless, about a quarter of global greenhouse gas emissions is associated with electricity and heat production, [1]. Therefore, decarbonisation of electricity grids is one of the major environmental impact pathways. The main methods for the desired decarbonization include carbon capture from exiting fossil-fired power generation stations and substituting fossil fuels with renewable or cleaner resources. This presentation covers a part of our recent research activities for design and operation of low carbon electricity grids.

In the first part of the presentation, we review our research for integrated design and operation of electricity grids. We proposed a comprehensive framework in which design and operation of electricity grid is considered simultaneously using stochastic mixed integer optimization programming. The formulated mathematical model is comprehensive and include conventional members of the electricity grid, the potentials for including renewable generators such as wind and solar energies, energy storage and the transmission network. The proposed method is demonstrated for the case of retrofitting the UK electricity grid to include 50% penetration of the renewables energies. Five years of historical data was applied for generating the representative scenarios regarding availability of wind and solar power and fluctuations in demand. The features of the interest include the configuration of the electricity grid after retrofit, investment in renewable power generators and the operational scheduling for power generation under various stochastic scenarios, [2].

Electricity grids are subject to high degrees of uncertainties. Examples of uncertainties in electricity grid include the stochastic behaviours of the solar and wind energies and fluctuation in electricity demand. The key observation is that future electricity grids will consist of a heterogeneous portfolio of energy conversion technologies and will be associated with a large degree of uncertainties. Such operational uncertainties can be accommodated using standby power generation capacitates, often provided by "fast-acting" grid members such as gas-fired power plants or by investing in energy storage. However, there is a much cheaper third option and that is to enhance the "smartness" of the grid. In this part of our research, we investigate the application of artificial intelligence (AI) for enhancing the predictability of electricity demand, in addition to wind speed and solar radiance. Various AI methods are incorporated and compared and the benefits of grid intelligence are quantified in economic terms, [3].

Currently, 80% of the household energy consumption is associated with heating and is supplied by fossil fuels. While electrification of residential heating systems can deliver an important environmental impact, electricity grids, unlike natural gas grids, are not resilient enough to accommodate the uncertainties in the heat demand. The third and last part of this presentation will study the potentials for decarbonisation of urban energy systems using combined heat and power systems integrated with carbon capture and storage. The feature of interest is the overall network flexibility and is demonstrated on the case of the urban energy system in London, [4].

Keywords

Smart grid, renewable energy, optimization, Big-data, Uncertainty, Combined Heat and Power (CHP) generation

References

[1] EPA, Global Emissions by Economic Sector, accessed online:

https://www3.epa.gov/climatechange/ghgemissions/global.html

[2] Sharifzadeh M, Lubiano H, Shah N. Optimal design and operation of integrated renewable energy systems. Under review.

[3] Sharifzadeh M, Sikinioti-Lock, A., Lubiano H, Shah N. The economic implications of enhance electricity grid smartness. In preparation.

[4] Sharifzadeh M, Shah N. Decarbonization of urbane energy using combined heat and energy cycles. In preparation.

1 Corresponding author, Email: mahdi@imperial.ac.uk , **Address:** Room C603, Roderic Hill Building, Centre for Process Systems Engineering (CPSE), Department of Chemical Engineering, Imperial College London, South Kensington, London SW7 2AZ, UK.

472604 Efficient Global Optimization for a Mixed AC-DC Power Distribution System

Abstract:

This presentation is concerned with global optimization for integrated design and operation of a mixed AC-DC power distribution system. The system under consideration is a benchmark electricity distribution system of medium office buildings. Frank and Rebennack [1] developed an integrated design and operation problem for a subpart of the system. The goal is to determine the network architecture (i.e., AC or DC) to which each branch or bus is assigned, in order to minimize the total energy supplied from the AC electric utility while satisfying physical restrictions, for a certain number of operating time periods. This problem is a large-scale nonconvex mixed-integer nonlinear programming (MINLP) problem. The integer variables are binary design decisions on network architectures. The nonconvex constraints come from the operational subproblems over the different time periods, each of which is essentially a mixed alternating current optimal power flow (ACOPF) and direct current optimal power flow (DCOPF) problem. While the DCOPF part of the problem is linear, the ACOPF part of the problem is nonconvex quadratic.

ACOPF problem plays a central role in optimal design and operation of electrical power networks, and the formal study of this problem can date back to 1962 [2]. The vast majority of ACOPF literature focuses on operational ACOPF problems that are nonconvex continuous optimization problems, and only a few of the existing methods (e.g., [3]) can guarantee finding a global optimal solution. This is because operational ACOPF problems usually need to be solved many times a day and none of the existing global optimization methods can solve the problem fast enough. However, global optimization seems to be more attractive for ACOPF design problems, as it is reasonable to make a design decision in a day or a week. Frank and Rebennack [1] successfully applied a decomposition-based global optimization method, called nonconvex generalized Benders decomposition (NGBD [4]), to some mixed AC-DC power distribution systems, and demonstrated the computational advantage of NGBD over state-of-the-art global optimization solvers.

The performance of NGBD is highly dependent on tightness of the convex lower bounding problem used. When the size of the distribution system becomes larger and the convex lower bounding problem becomes less tight, the performance of NGBD degrades quickly, as identified by Frank and Rebennack [1]. Here we propose to integrate domain reduction techniques into NGBD, in order to enhance the tightness of the convex lower bounding problem progressively during the decomposition procedure. Two types of domain reduction techniques are considered for NGBD. One is to solve customized convex bound contraction problems whose sizes are independent of the number of time periods. These bound contraction problems are also strengthened with valid linear cuts derived from the previously solved convex NGBD subproblems. The other is to perform range reduction calculations [5] for nonconvex NGBD subproblems, according to the primal and dual solutions of the previously solved convex NGBD subproblems. The domain reduction techniques not only can reduce the number of NGBD iterations, but also can speed up the solution of nonconvex NGBD subproblems. It will be shown in the case study that, the proposed method is significantly faster than the NGBD used by Frank and Rebennack [1] for the mixed AC-DC power distribution system.

In the case study, the convex lower bounding problem used for NGBD is obtained through linear relaxation, which is known to be weaker than a second-order cone programing (SOCP) relaxation or semi-definite programming (SDP) relaxation for ACOPF problems (e.g., [6]). Therefore, our method can be further improved via the use of a tighter nonlinear convex lower bounding problem.

References

[1] S. M. Frank and S. Rebennack, "Optimal design of mixed AC-DC distribution systems for commercial buildings: A nonconvex generalized Benders decomposition approach," *European Journal of Operational Research*, vol. 242, no. 3, pp. 710 – 729, 2015.

[2] J. Carpentier, "Contribution to the economic dispatch problem," Bulletin de la Socit Franaise des Electriciens, vol. 8, no. 3, p. 431447, 1962.

[3] C. Chen, A. Atamturk, and S. S. Oren, "Bound tightening for the alternating current optimal power flow problem," IEEE Transactions on Power Systems, vol. PP, no. 99, pp. 1–8, 2015.

[4] X. Li, A. Tomasgard, and P. I. Barton, "Nonconvex generalized Benders decomposition for stochastic separable mixed-integer nonlinear programs," *Journal of optimization theory and applications*, vol. 151, no. 3, pp. 425–454, 2011.

[5] Ryoo, H. S., Sahinidis, N. V., "A branch-and-reduce approach to global optimization," *Journal of Global Optimization*, vol. 8, no. 2, pp. 107-138, 1996.

[6] X. Bai, H. Wei, K. Fujisawa, and Y. Wang, "Semidefinite programming for optimal power flow problems," International Journal of Electrical Power and Energy Systems, vol. 30, no. 67, pp. 383–392, 2008.

469324 Data-Driven Modeling of Gas Leakage from Shale Natural Gas Wells

Abstract:

Availability of large amount of data recorded during natural gas well construction and completion stages can be put to use for studying underlying reasons for some of the critical problems which occur during the production phase. One such problem is natural gas leakage from the cemented annulus of the gas wells. The gas leakage may severely pollute the environment and underground water aquifers. Therefore, it is important to understand the reasons responsible for gas leaks and subsequently seek solutions to limit such leakage from wells drilled and completed in the future.

A number of factors may cause poor cementing, which may lead to development of cracks and micro-channels in the cement, providing a continuous path for gas to migrate from the production zone to the well head (Nelson, 2006). The extent of gas leakage from the cemented annulus is measured by a parameter called *Sustained Casing Pressure* (SCP) which is simply the pressure exerted by gas in the cemented annulus at the well-head. In a previous case study that we conducted on a set of gas wells data, we developed linear models to identify factors causing high SCP (Wehrens, 2011). Although the linear models provided a reasonable approximation of the underlying cause-and-effect relationships, the model accuracy was somewhat restricted by nonlinearities existing in the process (Hastie et al., 2009). In this paper we present a nonlinear modeling approach to study underlying causes of high gas leakage. A number of data-driven nonlinear modeling approaches, revolving around dimensionality reduction, as well as model specifics will be included in the final presentation.

References:

1. Hastie, T., Tibshirani, R., & Friedman, J. (2009). The Elements of Statistical Learning: Data Mining, Inference, and Prediction (2nd ed.).

2. Nelson, E. B. (2006). Well Cementing (2nd ed.).
3. Wehrens, R. (2011). Chemometrics with R: Multivariate Data Analysis in the Natural Sciences and Life Sciences. New York: Springer.

467393 The Simulation of an Industrial Wet Flue Gas Desulfurization Absorber

Abstract:

A comprehensive computational fluid dynamics (CFD) model was developed to study the process of desulfurization in the absorption tower of a wet flue gas desulfurization (WFGD) unit at a coal fired power station. This study was motivated by the need for South African coal-fired power stations to meet internationally aligned compliance levels w.r.t. SO_2 emissions, and the fact that current WFGD designs have been largely tailored for European power stations. Currently, the design of WFGD absorbers are based on empirical and semi-empirical correlations which address a limited set of design parameters; therefore, CFD modeling is considered to be essential for further enhancement of WFGD design to ensure compliance with the stringent environmental requirements w.r.t. SO_2 emissions. The main objective was to develop a comprehensive CFD model for an industrial (plant size) WFGD to provide an in-depth analyses of (i) the slurry injection dynamics and the droplet behavior with the inclusion of droplet properties such as size distribution, droplet distortion, wall-droplet interaction and evaporation, (ii) the hydrodynamics of the respective slurry and the gas phases to analyze the velocity distribution, the L/G ratio and temperature profiles, and (iii) the SO_2 absorption efficiency by the limestone slurry. The model was subsequently validated against plant test data of an operational WFGD absorber, with respect to the flue gas velocity, temperature and desulfurization efficiency.

The modelling of the injectors was successfully achieved with a statistical model based on an optimized number of parcel streams to ensure converged results. The detailed modelling of the injector assembly enabled accurate description of the gas and droplet behaviour with particular reference to the velocity fields, the L/G ratio, and the droplet size distribution. The gas and droplet velocities were found to be within the ranges suitable for controlling conditions to avoid carry-over of the smaller droplets in the outlet gas stream, and thus to ensure effective operation of the mist eliminator. The distortion of droplets was found to be very small with the largest effect occurring at the gas inlet to the tower. The inclusion of evaporation in the model proved to be important to accurately describe the temperature and moisture concentration profiles within the column. Consequently, the model predicted that saturation is achieved relatively close to the flue gas inlet, and that saturated conditions are prevalent over most of the column. The dependence of the desulfurization efficiency on the slurry pH as predicted by the model were found to correlate well with the plant test results.

463432 Modeling Chemotactic Bacterial Transport in Physically Homogeneous Groundwater Systems Containing Distributed Contaminant Sources

Abstract:

Groundwater pollution is of major concern as contaminants are continuously released into the subsurface via oil spills, chemical waste streams, and other means. Often times these contaminants become trapped within regions of the soil matrix that are characterized by low hydraulic conductivity making them difficult to remediate. In bioremediation strategies, chemotaxis â€“ a phenomenon in which pollutant-degrading bacteria have the ability to detect chemical concentration gradients and swim toward the contaminant source â€“ may enhance the transport of bacteria to sources of contamination and concomitantly lead to increased pollutant accessibility and biodegradation, even in regions with low water permeability. The influence of chemotaxis on bacterial transport in aquifers containing a distribution of contaminants with localized concentration gradients is not yet fully understood. This study aimed to utilize computational techniques to predict and better understand the migratory response of chemotactic bacteria to contaminant microniches within a model subsurface environment. Focus was placed on a sand-packed column containing randomly distributed pollutant sources. The bacterial population was introduced as a pulse input at the column inlet. A modified advection-dispersion equation containing an additional advection-like term to describe chemotaxis was used to model bacterial transport. The chemotactic velocity is a function of the chemotactic sensitivity coefficient, which describes the strength of chemotaxis, and the chemotactic receptor constant, which is the concentration at which the chemotactic velocity of the bacterial population is equal to one-half of its maximum value. Sensitivity analyses were performed on the chemotactic sensitivity coefficient and receptor constant to determine their influence on bacterial transport. It was equally important to predict the pollutant concentration profile within the porous media, via a standard advection-dispersion equation, because the chemotactic velocity is also a function of the pollutant concentration and concentration gradient. Our results showed that when chemotaxis was occurring, bacterial transport in the direction of groundwater flow was retarded and the bacterial population was retained within the contaminated porous media for an extended period of time, compared to when chemotaxis was not occurring. Predictions from our numerical simulations were consistent with experimental observations from a sand column system analogous to our model set-up. Modeling chemotactic bacterial transport in porous media systems containing a distribution of contaminant sources is important for assessing the significance of chemotaxis in enhancing the retention of bacteria in sites of contamination.

472925 Analytical Calculation of Laminar Flow Reactor Effluent Concentrations without the Necessity of a CFD Approach

Abstract:

This work seeks to develop an analytical model capable of calculating effluent concentrations for segregated laminar flow reactors (SLFRs) based on the segregated flow model (SFM). The distinguishing element of the presented model is its applicability not only to isothermal, but also non-isothermal SLFRs. This essentially enables the accurate determination/calculation of effluent concentration for any type of laminar flow reactor, regardless of size or geometry. The sole requirement is the knowledge of the reactorâ€™s residence time distribution function (RTD), which is easily determined through simple pulse experiment. The accuracy of the developed model is demonstrated on two case studies featuring a Trambouze reaction scheme for both the isothermal and non-isothermal cases respectively. The obtained results are compared with outlet concentrations obtained from CFD simulations conducted for the same cases. The comparison clearly indicates the high accuracy of the analytical model consequently making a CFD approach superfluous.

466759 New Bimetallic Catalysts for CO_2 Hydrogenation to Chemicals and Fuels for Sustainable Energy Development

Abstract:

Abstract for AIChE CO2 Symp – Nov 2016

New Bimetallic Catalysts for CO_2 Hydrogenation to Chemicals and Fuels for Sustainable Energy Development

Professor Chunshan SONG*

EMS Energy Institute, Dept of Energy & Mineral Engineering, and Dept of Chemical Engineering, Pennsylvania State University, University Park, PA 16802, USA

*Email: csong@psu.edu; Tel: 814-863-4466; Fax: 814-865-3573

Capturing CO_2 and converting it into chemicals, materials, and fuels using H_2 produced with renewable energy, is an important path for sustainable development and a major challenge in 21st century. CO_2 has become a plentiful source of carbon, and there are long-term merits in using CO_2 as a feed for chemicals and materials and as a carrier for energy. The concentrated CO_2 can be used for manufacturing chemicals (lower olefins such as ethylene and propylene, methanol, and carbonates), and fuels (such as liquid transportation fuels including gasoline, diesel and jet fuels, or synthetic natural gas). Effective use of catalysis and adsorption over nano-porous materials can decouple the traditional link between energy utilization and negative environmental impacts. As an example, the hydrogenation of

CO_2 (using H_2 produced from H_2O using renewable energy) over novel bimetallic catalysts can selectively produce either C_2-$C_4^{=}$ olefins (and C5+ liquid hydrocarbons) using Fe-Co catalysts, or methanol over Pd-Cu catalysts. Nano-structured catalysts based on hollow zeolite-encapsulated Ni nanoparticles can create a highly stable and carbon-resistant catalysts for co-conversion of CO_2 and CH_4 in bio-gas to syngas for chemicals and ultra-clean fuels.

468853 Combined Magnesium OXIDE/Water Gas Shift-Based CO2 Capture Process

Abstract:

COMBINED MAGNESIUM OXIDE/WATER GAS SHIFT-BASED CO2 CAPTURE PROCESS

Andrew Lucero, Jiajia Meng, Shen Zhao, Santosh Gangwal

Southern Research, 5201 International Drive, Durham, NC 27712

alucero@southernresearch.org; (919) 282-1054

Southern Research is developing a combined magnesium oxide (MgO) CO2 sorbent/water gas shift (WGS)-catalyst-based CO2 capture process for integrated gasification combined cycle (IGCC) power plants. A heat-exchanger reactor with a highly efficient heat management capability is being developed for simultaneous CO2 capture and WGS to convert CO to CO2 and H2 and capture more than 90 % or more of the carbon from coal gasifier syngas. Simultaneous heat management and reaction allows the maintenance of thermodynamically favorable reaction temperatures for both steps. Also, by simultaneously converting the CO to CO2 over the shift catalyst, the overall reaction thermodynamics is shifted allowing greater carbon capture by the MgO sorbent than if the two steps were carried out separately. The ultimate project goal is to capture > 90 % CO2 and to reduce the cost of electricity by 30% over IGCC plants employing conventional methods of CO2 capture.

Extensive thermodynamic analyses from multiple sources of thermodynamic data have been completed for simultaneous CO2 capture and water-gas shift reactions. Based on these analyses, optimum temperature and CO2 partial pressure windows have been established for reactor design. A thorough analysis of the literature has been conducted for preparation of MgO sorbents with excellent multi-cycle durability and high CO2 capture capacity. An appropriate commercial water-gas shift catalyst has been obtained from a reputable catalyst manufacturer. Several MgO-based sorbents with and without promoters have been synthesized and have been tested using a thermo-gravimetric analyzer (TGA) at various temperatures and CO2 partial pressures determined from the thermodynamic analysis. The best performing sorbents have shown CO2 capture capacities greater than 20 wt %.

Present sorbent development/synthesis efforts are concentrating on (i) achieving excellent multi-cycle durability for these high capacity sorbents over 100s of cycles and (ii) mechanical forming of the sorbent in appropriate form for use in the heat-exchange reactor along with the WGS catalyst. A laboratory scale CO2 capture/ WGS process skid has been designed and constructed to test sorbents under relevant gasifier conditions in a ½ inch diameter, 18 inch long reactor. Following sorbent development and completion of laboratory scale experiments, the sorbent and process will be scaled up to bench-scale. Simultaneous heat management, CO2 capture and WGS will be demonstrated with high capacity and durability over 1000 cycles. A techno-economic evaluation will be conducted to demonstrate the potential of the process to achieve the required reduction in cost of electricity.

461222 Oxidative Dehydrogenation of Ethane and Propane By CO_2

Abstract:

A series of vanadium catalysts (V loading in the 0.7 ~ 16 V/nm^2 range) supported on silica have been prepared by incipient wetness impregnation method. The activity of ODH was studied in ethane and propane oxidative dehydrogenation in the presence of O_2, CO_2 and their mixture from 350 to 600°C and at atmospheric pressure. The products of oxidative dehydrogenation of paraffins were corresponding olefins, CO and CO_2, in the propane reaction a small part was cracked to ethylene and methane.

The vanadia silica catalysts at sub-monolayer VO_x coverage were found to be active and selective in propane ODH by O_2 and CO_2, while VO_x multilayers and nano-sized clusters showed lower activity and selectivity. Moreover, due to its acidic character of silica, it showed undesirable carbon deposition (coking) in propane ODH by CO_2 in a short period of reaction time and as a result catalyst deactivated quickly. Around 20 % weight loss observed for the used catalyst after 30 hours propane ODH by CO_2 reaction under conditions of TGA in air. On the other hand, the same catalyst employed to propane ODH by O_2, almost no carbon deposition was observed. Our experimental observations are in agreement with recent theoretical calculations indicating that in this catalytic system, carbon deposition is thermodynamically favorable, unless the surface character is modified by basic additives, such as Na^+. Moreover, basic additives were recently reported to enhance the monolayer coverage of VO_x species on silica which also improved their activity and selectivity in propane ODH by CO_2.

The catalytic activity of these species may be optimized by varying the surface coverage of VO_x species and the nature of the support and promoters.

451420 Conversion of Coal and Biomass to Liquid Hydrocarbons Using Gasification and a Hybrid Fischer Tropsch Catalyst

Abstract:

Ceramatec is leading a team to develop a two (2) BPD (barrel per day) Fischer Tropsch system to produce jet fuel. Funding is provided by the United States Air Force, through the US Department of Energy - National Energy Technology Laboratory. The feedstock for the system will be pulverized coal with some biomass to meet carbon emission limitations. The coal - biomass will be gasified and the resultant synthesis gas compressed to pressures suitable for Fischer Tropsch (FT). The synthesis gas stream will be cleaned using a sulfur removal system provided by IntraMicron, a company located in Auburn, AL. The clean synthesis gas will be fed to a 2 BPD modular FT reactor designed and assembled by Ceramatec.

Ceramatec is currently operating a small GTL facility at its Salt Lake City laboratory and portions of this facility will be used for this program also. The sulfur removal system and the 2 BPD FT reactor will be tested at this existing facility prior to shipment to the gasifier location for testing with the coal - biomass derived synthesis gas. A hybrid FT catalyst capable of terminating carbon chain growth in the liquids range will be used in the FT reactor.

The objectives and challenges of the 2 BPD CTL (coal-to-liquids) project will be described in the presentation. The presentation will also describe the capabilities of the team to address the issues associated with development of that integrated 2 BPD CTL system. The test data generated to date will be presented to demonstrate how the issues are being addressed. This data will include the carbon distribution on various test runs, catalyst performance, and synthesis gas utilization. The design for the integrated CTL facility will be described and individual components highlighted. Other relevant technical data from associated projects will be presented.

460840

The Effects of Carbon Dioxide on the Adsorption Capacity of Methane in Shale

Abstract:

- Abstract Text

The effects of carbon dioxide on the adsorption capacity of methane in shale

Bing Cao, Lu Wang, Lu Xia, Qingchun Yu

With population increases and economic development, energy consumption has increased worldwide. Consequently, balancing energy supply and demand is a priority. Shale gas is a clean and efficient non-conventional energy source that has attracted increasing attention in many countries. Generally, shale reservoirs contain carbon dioxide (CO_2). Under reservoir conditions, shale has a larger adsorption capacity for CO_2 than for methane (CH_4), especially in primary pores. Recent research showed that five CO_2 molecules could displace one CH_4 molecule. Therefore, the influence of the CO_2 content on the adsorption of CH_4 in shale is not negligible. There have been a number of national and international studies conducted on the adsorption of CH_4 and CO_2 mixtures by shale to evaluate the effect of CO_2 on CH_4 adsorption.In the present study, two shale samples were obtained from depths of about 838.5 m and 1035.4 m. To reduce the equilibration time of the sorption, the shale samples were pulverized to powders and screened through an 80 mesh sieve. Both of the samples were analyzed by scanning electron microscopy, and for their vitrinite reflectance and total organic carbon content. The experiments were conducted with a gas-solid two-phase adsorption instrument, which included a gas pressurization system, gas adsorption device, data acquisition system, vacuum pumping system, and gas composition analysis system, all designed by our research group. The gases injected into the pressurization system and used for adsorption were CH_4, a binary mixture of 95.06 % CH_4 and 4.95 % CO_2, and another binary mixture of 80.01 % CH_4 and 19.99 % CO_2. The gas adsorption measurements were performed at 303.15 K, 313.15 K, or 323.15 K and at a pressure between 0 and 10 MPa. For both CH_4 and the binary mixtures of CH_4 and CO_2, the absorption capacity decreased with increasing temperature but increased with increasing pressure. In addition, absorption of the CH_4/CO_2 mixture on the shale surface was affected by the total organic carbon content. As the total organic carbon content increased, the absorption was enhanced by the increased interfacial area between the gas and shale surface. The difference between the results for pure CH_4 and the mixture containing a small amount of CO_2 was not pronounced when the experiments were conducted at the same temperature and low pressure.Therefore, in an actual reservoir, the presence of trace amounts of CO_2 will likely have little effect on the adsorption of CH_4. However, the difference increased if either the amount or pressure of CO_2 was increased. Although the CO_2 content is low, the adsorption capacity for CO_2 is stronger than that for CH_4. This could occur because CO_2 molecules may be better accommodated in the pore structure of shale than CH_4 molecules. These results indicate that CO_2-enhanced shale gas recovery could be feasible for shales in the Qaidam Basin (China). Further tests and analysis on the adsorption behavior of each gas will be conducted at higher CO2 fraction. Results from this study may help in understanding the CH4 adsorption behavior and the effect of CO2 at low fraction (<10%), which is representative for realistic reservoir conditions.

464667
Flare Minimization for Simultaneous Turnaround Operations of Two Olefin Plants

Abstract:

Flare minimization for olefin plant turnaround operations such as start-up and shut-
down is a double-
win practice, which brings not only industrial economic benefits, but also local environm
ental credits. By far, almost all of flare minimization studies focus on single units or singl
e plant normal or abnormal operations. No systematic studies have been reported on flare
minimization coupling abnormal operations of two chemical plants. Conceivably, if mate
rial and energy exchange between two chemical plants could be effectively integrated and
scheduled, flare emissions from one plant might be adopted by another plant, so that flar
e minimization could be achieved to its maximum extent. In this paper, a systematic meth
od has been developed to study the flare minimization for simultaneous turnaround operat
ions of two olefin plants: one plant is experiencing start-
up while another plant is conducting shut-down operations.

When one olefin plant undergoes its shut-
down operation, its flare minimization strategy is conducted as: (i) keep the minium num
ber of furnace running by shuting down extra furnaces; meanwhile, reduce liquid invento
ries in tanks, drums, and columns; (ii) shut down the remaining furnaces and keep CGC (
crack gas compressor) running with vapor recycles; meanwhile, vaporize liquid remains i
n tanks, drums, and columns; (iii) shut down CGC after recovering liquids and vapor inve
ntory as much as possible; (iv) perform decommissioning for each facilities including left
liquid discharge, vapor discharge, and nitrogen purge. For another plant performing its st
art-
up operation, the general operating procedure includes: (i) perform nitrogen replacement
with hydrocarbons; (ii) input liquids to columns to build up lthe vapor -
liquid balance in various vessels; meanwhile columns are engaged in the full-
reflux mode; (iii) start CGC with vapor recycles; meanwhile, precooling the chilling train
and warm up columns; (iv) start up furances gradually; (v) gear up each column operatio
n to their normal status.

In this paper, flare minimization opportunities coupling shutdown operating steps (ii) thro
ugh (iv) together with start-
up operating steps (ii) and (iii) are exploried. To study the feasibility of this coupling turn
around operation and calculate flare emission sources, plant-
wide dynamic simulations for these two plants have been conducted. The developed duo-
plant flare minimization strategy has been virtually tested and demonstrated. The obtaine
d flare minimization strategies and simulation results provide an indepth understanding o
n synergies of two plant abnormal operations for flare minimization.

469916 Consequence Assessment for Event Tree Analysis in Transitions Using Dynamic Simulation

Abstract:

Incidents that occur in industrial facilities can lead to the release of hazardous materials, uncontrolled fire and explosions, economic losses and environmental contamination. Consequence assessment is an examination of the undesirable impacts of an accident. Fuzzy logic combined with Event Tree Analysis (ETA) is used for quantitative risk assessment (QRA).

In this work dynamic simulation was used to assess the consequences of event tree branches for two examples of equipment switching transitions in chemical plants. The first example was switching between two solid desiccant dryers in an ethylene plant. The other example was the switching of a reflux pump with its spare in a reformate stabilizer column. Both examples have standard operating procedures (SOP) commonly accomplished manually by operators. Numerous operator induced fault cases were considered. These SOPs and fault cases were simulated in Aspen HYSYS V8.0 using the event scheduler feature. The consequences were identified by monitoring variables during dynamic simulation of each case. Thus, the fuzzy ETA algorithm was used to evaluate the risk of equipment switching transition faults while dynamic simulation was used to assess the consequence.

AIChE Annual Conference 2016

Division: Process Development

Session: Process Research and Development for Industrial Sustainability

Title: Effect of chemical price uncertainty on project viability: an economic evaluation

Authors: Gaurav Arora, Mohammad Shafiei, Mike Huckman

Abstract: A case study is presented for evaluating economic viability of a chemical process. The discussion includes process design choices, cost estimates, and economic evaluation of the process. The economics are found to be very sensitive to the feed and product prices. Three specific scenarios for process are analyzed in detail – at current chemical price, low feed-high product price and high feed-low product price. While these scenarios are helpful in understanding the cost, revenue breakdown, and economics of the project, a better insight is obtained by analyzing the effect of chemical prices over the last 15 years on the economic viability of the project. Margin required to meet an IRR of 15% is calculated followed by a market price analysis to calculate the probability of this occurring in the last 15 years.

Keywords: Economic Analysis

455746 Automated Reaction Mechanism Generation Including Nitrogen As a Heteroatom

Abstract:

The open-source Reaction Mechanism Generator (RMG) software and its thermochemical and reaction kinetics databases have been extended to include nitrogen as a heteroatom. Specific changes to RMG and the mining of thermochemistry and reaction kinetics data will be discussed. This new version of RMG has been tested by generating a reaction mechanism for pyrolysis and oxidation of ethylamine ($CH_3CH_2NH_2$), and the generated model was benchmarked to recent shock tube studies. Ethylamine was selected as a model nitrogen-containing organic species since it is the simplest primary amine containing important characteristic chemical bonds (i.e., C–C, C–H, C–N, and N–H), and it is also a model molecule for nitrogen-containing species found in biofuels. Validation of the reaction network with experimental data showed that the generated model successfully reproduces measured radical levels (NH_2 and OH) as well as ignition delay times. This work presents the potential to automatically generate detailed chemical models for nitrogen-containing species, and specifically unveils the pyrolysis and oxidation mechanisms of ethylamine.

465018 Modeling the Combustion of Otto Fuel

Abstract:

Otto Fuel II is a monopropellant that is used in undersea propulsion. It consists of three components: propylene glycol dinitrate, (PGDN), dibutyl sebacate (DBS), and 2-nitrodiphenylamine (NDPA). PGDN is the oxidizer, DBS is added as a plasticizer, and NDPA is serves as a stabilizer. Our modeling results suggest that there are some interesting similarities and differences between monopropellant combustion and large molecular weight transportation fuels. At high temperatures, the decomposition kinetics of the oxidizer is effectively decoupled from the oxidation of the plasticizer, and DBS effectively undergoes pyrolysis and partial oxidation in a high-pressure $CH_2O/NO_2/NO$ bath gas. The stabilizer undergoes a nitrate to nitro conversion, followed by O-NO bond fission to yield a highly stable oxy-diphenylamine radical. This radical recombines to form a large number of polycyclic aromatic amines, which are believed to be responsible for the large amount of carbon deposits that are formed during operation.

455944 Molecular Weight Growth Kinetics of Olefin Pyrolysis Under Low to Intermediate Temperatures

Abstract:

In many conversion processes, such as solid oxide fuel cell operation or hot gas mixing upstream of steam reformers, hydrocarbon fuels might be exposed for sufficiently long times at temperatures between 500°C and 1000°C to allow gas-phase reactions to occur. This chemistry not only leads to a change in gas phase composition that enters the process, but it might also produce molecular weight growth species (MWG) large enough to precipitate from the gas phase on the surface to form deposits. Deposit formation can reduce the efficiency of the technical application. For example it may deactivate the reforming catalyst or fouling in the solid oxide fuel cell, or in severe cases lead to a catastrophic failure of the unit. Other examples include formation of gas-phase deposit precursors within steam crackers. These limit the time between decoking cycles. MWG species contained in exhaust gas emissions from IC engines also constitute an environmental and adverse health problem.

There have been amount of work focusing on understanding the reaction kinetics under high temperature environment, e.g., combustion. However, few studies focused on low to intermediate temperatures, in particularly the MWG kinetics for unsaturated hydrocarbons, such as alkene, is poor. In this work, propene and the three butene isomers are selected as the model compounds. Pyrolysis experiment were performed in a tabular flow reactor at pressure ~0.82atm in temperatures of ~500-800°C, with nominal residence time ~2.4s. These fuels showed substantially different reactivities, as well as product distributions. The pyrolysis data showed that the three linear alkenes (e.g., propene, 1- and 2-butene) generated similar amount of MWG species under the same fuel conversions, while the branched isobutene tended to form as much as twice amount. A detailed kinetic model was developed to understand the impact of fuel structures on the fuel conversion, product formation and especially MWG kinetics during olefin conversion. The knowledge obtained from this combined experimental and theoretical work could be applied to real-world

practices including reduce deposit formation in steam cracker and understand ignition kinetics of motor engines.

452102 Detailed Kinetic Modeling of Gas Evolution and Energy Recovery during Chemical Quenching of Acetylene Production in Thermal Plasma

Abstract:

Detailed Kinetic Modeling of Gas Evolution and Energy Recovery during Chemical Quenching of Acetylene Production in Thermal Plasma

Yan Cheng, Tianyang Li, Yi Cheng [*]

Department of Chemical Engineering, Beijing Key Laboratory of Green Chemical Reaction Engineering and Technology, Tsinghua University, Beijing 100084, PR China

Thermal plasma pyrolysis provides a convenient way to realize the one-step conversion from coal to acetylene. Taking advantage of the ultra-high temperature field provided by thermal plasma, the coal feedstock is heated up in milliseconds to give out volatiles. The volatiles undergo gas phase reactions afterwards to form acetylene owing to the fact that acetylene is thermodynamically stable at above 1500 K among small-molecule-hydrocarbons. However, acetylene will decompose into hydrogen and carbon black during cooling process. So efficient quenching is necessary to prevent acetylene decomposition so as to maintain the target products. It is noticed, on the other hand, that the high-temperature gas products still contain about 40% of the total energy input. So it is vital to make use of the latent energy during the quenching process.

In addition to the common physical quenching by water spray, chemical quenching method is an alternative solution to the process design. In chemical quenching, hydrocarbons are used as quenching media instead of water, which brings about co-products (e.g., ethylene) during the decomposition of hydrocarbons by the afterheat in gas products. The process profit can be accordingly improved as well.

In this work, we carried out modelling and simulation on the above-mentioned chemical quenching in thermal plasma technology. A gaseous detailed kinetics, coupled with a soot formation mechanism, was proposed to investigate the species evolutions during quenching. The model consisted of C1-C5 gas phase reactions, PAH (polycyclic aromatic hydrocarbons) growth and a simplified soot formation mechanism. The predicted acetylene decomposition and soot yield were well validated with published data. Then, numerical simulations were performed to reveal the process features, including acetylene decomposition evolution, chemical quenching behavior of different C1-C3 alkanes as well as the energy balance.

Finally, the model was used to theoretically propose an optimized chemical quenching design in a pilot-plant thermal plasma process. The results showed that under optimized operating conditions, propane quenching could recovery 35.4% of the total gas phase afterheat. At the same time, propane decomposed into smaller molecules in millisecond(s), which realized the co-production of ethylene at a ratio of 0.75 t C_2H_4/t C_2H_2. The optimal energy consumption was reduced by 43% per kilogram of C_2H_2+C_2H_4, which verified the feasibility of energy efficiency improvement via chemical quenching.

Keywords: Thermal plasma pyrolysis; Chemical quenching; Acetylene; Detailed kinetics

[*] Corresponding author. Address: Department of Chemical Engineering, Tsinghua University, Beijing 100084, P.R. China. Tel: +86 10 62794468, Fax: +86 10 62772051. *E-mail address*: yicheng@tsinghua.edu.cn.

474760 Predictive Modeling of Thermal Decomposition of Hydrocarbon Mixtures

Abstract:

Thermal decomposition of hydrocarbon mixtures is important in a wide range of applica- tions, including geochemical conversion of oil to gas, thermal coking and cracking processes in refineries, and steam-cracking to form olefins. Most detailed studies focus on the cracking mechanism of a single molecule, to reduce the complexity. Here we present recent advances in methods for modeling mixtures, and so seeing composition effects on the chemistry. Also we present experimental data measured on thermal decomposition of various mixtures at several temperatures using the gold tube technique to maintain the system in a single phase, and discuss the development of detailed a priori kinetic models for these experiments using the open source software Reaction Mechanism Generator (RMG). Recent improvements to RMG, including a method for more rapidly and accurately estimating the thermochemistry of fused polycyclics, are presented.

464972 Nondimensional Specific Energy Index for Methane Pyrolysis in Thermal Plasma

Abstract:

Acetylene is the raw material for several important chemical products including polyvinyl chloride (PVC), vinyl acetate, 1,4-Butylene glycol (BDO) and so on. Partial oxidation of natural gas is the main access to acetylene in Europe and America, yet the introduction of oxygen reduces the carbon selectivity to hydrocarbons. Methane pyrolysis in thermal plasma is an efficient way to produce acetylene. In this process, ultra-high temperature in reductive plasma allows for more preferred reactor performance such as methane conversion at above 90% and selectivity to acetylene at above 70%. A large amount of researches have been done on methane pyrolysis on various lab-scale plasma devices. However, due to the diversity of experimental conditions, it is hard to summarize the conclusions from different researches.

In this work, we introduced a nondimensional specific energy index, defined as the ratio of supplied power and enthalpy flow needed for plasma weighted with inert gas and methane content, to better evaluate the influences of supplied power and methane content in feed on the methane pyrolysis. This nondimensional value stands for the excess of energy needed for plasma and includes both the influences of energy provided for per unit of gas and methane content. It is worth noted that compared with enthalpy change of gas in plasma, reaction heat of methane to acetylene is negligible. In an ideal process, the index value should be 1, which means that the supplied power is precisely sufficient to maintain the plasma and the reaction. A series of experiments in a 2 kW DC plasma reactor were carried out to testify the applicability of the index. The experimental results showed that the conversion of methane increased from around 10% when the index was 0.5~0.6, and reached 90% when the index was 1.0~1.2, then slowly increased to a maximum around 95%~97% when the index kept increasing. This means that this nondimensional index positively correlates to the conversion and implies the energy excess state of the system as expected.

465112 Long Term Stability and Operational Limits of Lithium Sulphate for Thermal Energy Storage

Abstract:

This work explores the experimental feasibility of lithium sulphate as a thermal energy storage material for concentrated solar power (CSP) technologies. This salt has a crystalline phase change between cubic and monoclinic at temperatures around 576°C. By operating with a minimum cycle temperature, both sensible and latent thermal energy can be utilised. The operating temperatures of the proposed system are appropriate for current CSP technologies based on subcritical steam Rankine or supercritical CO_2 power cycles.

Current phase change materials (PCMs) under investigation utilize solid-liquid transformations which can achieve a higher storage performance than sensible storage materials within a low temperature difference. However, these PCMs suffer from some technical problems such as supercooling, corrosion and volume expansion. Such disadvantages can be overcome when solid-solid PCM is used as the storage material as it is less corrosive and undergoes a much smaller volume change.

In this work, the long term performance and stability of lithium sulphate has been evaluated. This involved long term experiments using 55 g of salt in an air furnace, with characterization of material properties before and after cycling using TGA-DSC, XRD and SEM. From experiment, it was determined that the highest temperature at which the thermal properties and volume of the salt were unchanged after repeated cycles was equivalent to its melting point (859°C). At temperatures lower than 130°C, the material is prone to hydration reducing the performance through significant changes to volume, density and consequently thermal conductivity.

474794 Thermal Properties and Storage Analysis of Nano Enhanced Organic Phase Change Material (NEOPCM) for Thermal Energy Storage Application

Abstract:

The objective of the present work is to model the thermal energy storage unit using Nano Enhance Organic Phase Change Materials (NEOPCM). NEOPCM is capable of absorbing or releasing the large quantity of latent heat during phase change. This paper investigatesNEOPCM with Lauric acid as a base material and nanoparticles (Copper Oxide, Aluminum Oxide)as supporting materials for the preparation of NEOPCM. In this paper, it was studied experimentally the effect of combinations of nanoparticles CuO and Al_2O_3 with different mass fractions to the base PCM, Lauric acid. To ensure the uniform dispersion a field emission scanning electron microscope (FESEM) has been used. In order to determine the changes in the compositional/functional groups during the cycle test, Fourier Transform Infrared (FT-IR) spectroscope is used. Crystalloid phase of NEOPCMs is investigated with an x-ray test conducted by an x-ray diffractometer (XRD). Thermal reliability is carried out by accelerated thermal cycle tester, phase change temperature and latent heat of fusion are measure by Differential Scanning Calorimeter (DSC).Thermal conductivity of NEOPCM is measure by using a KD-2 pro, which is based on hot wire technique.It was also verified experimentally that there is improvement in thermal conductivity and storage capacity of NEOPCM.

Keywords: Thermal energy storage, NEOPCM, Lauric acid, differential scanning calorimeter

474795 Thermal properties and heat transfer analysis of Al_2O_3 as a nano particle dispersed in the base mixture of Therminol VP -1 and Water

Abstract:

This paper presents multi-parameter investigation of Nano HTF with Al_2O_3 as a nano particles dispersed in the base material, Therminol VP-1 and water mixture (50/50 vol.%). In this paper, it was studied the effect of different volume fraction of nanoparticle dispersed in to the base HTF, Therminol VP-1 and water mixture. Nano-fluid sample is characterized by Transmission Electron Microscope (TEM) imaging, particle surface charge (zeta potential), particle aggregate size, light absorbance and PH. Dispersed particle aggregate size measure by particle analyzer based on the dynamic light scattering technique. The results show multiple perspective correlations between nano fluid characteristics and thermal properties. Al_2O_3 properties are very close to desired properties of heat transfer fluid properties and it is cheaper. Its thermal conductivity also higher than those of the conventional heat transfer fluids. It was found experimentally that there is significant improvement in thermal conductivity and considerable lowering of viscosity of Nano HTF which is the main requirement of heat transfer fluid for solar thermal applications.

Key words: Therminol VP-1, NEOHTF, Transmission Electron Microscope (TEM), Al_2O_3.

474519 Conversion of Cheese Whey to Lactobionic Acid

Abstract:

Lactobionic acid (LBA) is a versatile, high-value, low-volume chemical with numerous applications in the food, pharmaceutical, cosmetic, and chemical industries. Microbial production of LBA has the advantages of high selectivity and the ability to use cheese whey, a waste stream from the dairy industry, which is rich in lactose. However, microbial LBA production has not yet been developed as successfully as other organic acids, despite attempts being made to use wild type bacteria, fungi and yeast strains. There is no commercial implementation of microbial LBA production technologies due to the following factors: 1) The most robust strains are pathogenic, which prevents their industrial implementation. 2) The enzyme (cellobiose dehydrogenase or lactose dehydrogenase) which oxidizes lactose to LBA is produced only in the stationary phase. The highest level production of LBA was only achievable under resting cell conditions, which are not usually used in industrial fermentations. 3) The microorganism consumed LBA at the later stage of the fermentation. We report engineering a Neurospora crass strain for cellobionic acid production from cheese whey and solving the problems associated with LBA fermentation using metabolic engineering efforts

468385 Predictive Model and Bioconversion of Mixed Feedstocks

Abstract:

Currently, biorefineries are utilizing single feedstock inputs but in order to reduce feedstock dependency and the related risks, researchers are exploring the option of mixing multiple feedstocks. To understand the implication of mixing feedstocks on sugar yield, we relied on statistical design of experiments to build a predictive model for sugar yield based on varying feedstocks fractions, pretreatment types, and pretreatment reaction parameters. We performed pretreatments at low biomass loading, but for a biorefinery to be economically viable biomass deconstruction has to be performed at higher solid loading. Also, lignocellulosic sugars should be fermentable to produce high energy density molecules such as hydrocarbons.

In this study, three biomass feedstocks: corn stover, switchgrass, and energy cane were chosen based on the least cost formulation model developed by Idaho National Laboratory. These single and mixed feedstocks were pretreated with dilute alkali (DL), dilute acid (DA), and ionic liquid (IL) at severities predetermined by the statistical experimental design and were subsequently, enzymatically hydrolyzed with CTec2® at a loading of 10 mg protein/g glucan. The resulting sugar yields from all treatments, which varied from 25 to 100% w/w untreated biomass, were then fed into SAS JMP® to develop a predictive model that generated a continuous envelope of optimal feedstock mixtures. In order to achieve 90% sugar yield, we have to perform IL pretreatment at 140°C for 120 mins with 12%, 13%, and 75% (w/w) corn stover, switchgrass, and energy cane. More predictions of such kind were generated from this model that can potentially overcome the dependence on a single feedstock while maintaining the overall product yields.

These feedstock/treatment combinations, selected from the model, were tested at higher biomass loading to study the impact on downstream fermentation process to produce ethanol and hydrocarbon, such as bisabolene. At higher biomass loading, mixed feedstock conversion to ethanol performed very similar to single feedstock conversion to ethanol. Rheological characterization was performed on high solid pretreated biomass mixtures to better understand mixing, material handling, and mass transfer issues. This information will be applied in techno-economic analysis of this end-to-end study of deconstruction of optimal biomass mixtures at high solid loading for ethanol and hydrocarbon production.

459583 Two-Phase Kinetics of Enzymatic Depolymerization of Hemicellulose to Soluble Sugars for Liquid Biofuel Production

Abstract:

Hemicellulose is the second most abundant polysaccharide in nature, available not only in its pure form but also as a constituent of all lignocelluloses. The conversion of hemicelluoses to soluble sugars in parallel to its cellulosic counterpart would make liquid biofuels competitive in the market. Of the three key conversion steps (pretreatment of biomass, enzymatic hydrolysis of pretreated biomass to soluble sugars, ethanol from sugars by fermentation), hydrolysis is the slowest and the rate limiting step [1]. Adsorption of the liquid phase enzymes (endo- and exo-xylanases) to the insoluble solid substrate (xylan) is the first step in hemicellulose hydrolysis, followed by enzyme-catalyzed xylan depolymerization reactions in the solid and liquid phases. This work uses a coupled theoretical and experimental framework [2] to quantify the kinetics of product-inhibited solid and liquid phase depolymerization reactions that produce xylose and other reducing (soluble) sugars. We experimentally determine the type of adsorption of the enzyme on the solid substrate, the type of product inhibition in the solid and liquid phases, the transition of two-phase hydrolysis to single in batch reactors. Model-experiment comparisons allow us to quantify the xylobiose inhibition constants in the two phases. Our experiments show that the hydrolysis system transitions from a two-phase (solid-liquid) to a single (liquid) phase process at 11 hours. The enzymatic hydrolysis in both liquid and solid phases is found to be non-competitively inhibited by the product (xylose), with the inhibition being more potent in the liquid phase. Our multi-step reaction kinetic model is simulated using the liquid and solid phase kinetic data and validated with our experimentally measured temporal dynamics of xylose and reducing sugar yields in the liquid phase, and of the solid phase hemicellulose concentrations. We show that the product inhibition in the solid phase increases with the substrate loading, thus lowering the yield of reducing sugar, while the liquid phase product inhibition decreases with substrate loading, thus increasing the xylose yield. Our substrate inhibition studies allow us to recommend 5 mg/ml as the optimum xylan loading, above which substrate inhibition is observed to set in.

References:

[1] Chakraborty, S., Raju, S., and Pal, R., "A multiscale three-zone reactive mixing model for engineering a scale separation in enzymatic hydrolysis of cellulose," *Bioresource Technology*, **173, pp 140-147** (2014)

[2] Dutta, S.K., and Chakraborty, S., "Kinetic analysis of two-phase enzymatic hydrolysis of hemicellulose of xylan type," *Bioresource Technology*, **198, pp 642-650** (2015)

474465 Selective Fermentation of Six Carbons Sugars from Hot-Water Hardwood Extract Hydrolysates

Abstract:

Hot water wood extract hydrolysate contains a mixture of sugars: xylose, glucose, galactose, mannose, arabinose and rhamnose. This sugar mixture presents a challenge in fermentation to ethanol and other chemicals / biofuel. The fermentation efficiency is especially low for xylose. When genetically modified *E. coli* fbr5 was employed to ferment the sugar mixture to ethanol, glucose was the first sugar consumed, and then galactose, arabinose, mannose and rhamnose, xylose was the last to be consumed. While *E. coli* fbr 5 was genetically engineered to ferment Xylose, this outcome is disappointing. Most microorganisms prefer glucose, *E coli* is no exception. Xylose being the dominant sugar species in hardwood hot-water extract hydrolysate and the least preferred sugar for ethanol production present a challenge to the biomass energy industry development. On the other hand, xylose can be employed to produce xylitol, furfural, and a lot more products efficiently. The value of xylose is higher than glucose in the market, and yet it is limiting the efficiency in biomass conversion. Hot water wood extract hydrolysate was subject to fermentation with *S. cerevisiae* and xylose was left unconsumed. Xylose can be recovered from the fermentation broth, while ethanol was produced from glucose.

465914 Effect Temperature and Flow Rate on Lhw Pretreatment SB: Kinetic Study

Abstract:

Assuming xylan hydrolysis to xylose in sugarcane bagasse (SB) by liquid hot water (LHW) and xylose further degradation to furfural were two consecutive first-order reactions, a kinetic model was developed. SB was pretreated by LHW at different temperatures and flow rates, and xylan hydrolysis behavior in SB pretreatment was investigated using the developed kinetic model. Results showed the kinetic model could fit the data well (R^2>0.85). k_S (the rate constant of xylose production) and k_X (the rate constant of xylose degradation) at different temperatures was in accordance with Arrhenius equation. Activation energy (E_a) for xylose production and xylose degradation was 50.39 and 104.66 kJ/mol, respectively, which indicated that xylose degradation was relatively more difficult that xylose production during LHW hydrolysis. The relationship between k_S (or k_X) and flow rate was almost linear. The application of the model in xylan hydrolysis kinetics was successful.

468268 Molecular-Level Kinetic Model Informed Life Cycle Analysis of Green Diesel Production

Abstract:

The drive to reduce our reliance on petrodiesel coupled with the advent of new environmental policies and economics has prompted the production of green diesel as an alternative, sustainable fuel. In this study, a molecular-level kinetic model was used to investigate the impact of changing feedstock composition and reaction conditions on the life cycle analysis of green diesel. The kinetic model simulates a fixed bed hydrotreating unit that includes the detailed triglyceride and free fatty acid hydroprocessing chemistry to produce paraffin-rich diesel range fuels. Linear free-energy relationships and literature data were used to reduce and tune the parameters of the LHHW rate laws allowing for product cloud point and cetane number property calculations. The product green diesel has excellent combustion properties but poor cold-flow properties, so hydrocracking severity was used to regulate end use and value of the green diesel. The process inputs and outputs were used in Argonne National Laboratoryâ€™s Greenhouse Gases, Regulated Emissions, and Energy Use in Transportation (GREET) model to evaluate greenhouse gas emissions of green diesel compared to petrodiesel. Scenario testing was performed to study the relative emission reduction for the different reaction conditions.

470164 Application of the Advanced Distillation Curve Method to Characeterize Two Alternative Transportation Fuels Prepared from the Pyrolysis of Waste Plastic

Abstract:

The properties of alternative fuels must be evaluated to determine their suitability for use in existing engines, either as blends with conventional petroleum products or as stand-alone substitutions (that is, drop in replacements). Alternative fuels made from waste feedstocks are especially attractive since they address the dual purpose of fuel production and mitigation of environmental pollution environmental concerns. In this study, two alternative liquid fuels generated by pyrolysis of waste polyethylene plastic bags were assessed using a composition-explicit distillation approach, which provides a rich data matrix that includes including fuel volatility, composition, hydrocarbon classification, and energy content in each distillate volume fraction. The data collected using this approach provide valuable information for the development of equations of state to describe the thermodynamic behavior of these complex fluids. The analysis showed that, despite significant differences in composition, the distillation curves and associated data of the alternative fuels show promising similarity to their conventional counterparts.

469215 Application of Real-Component Based Molecular Characterization on Petroleum Refinery Simulation

Abstract:

Abstract

As light to medium sweet crude oils keep depleting while demand of oil products are increasing, heavier and sourer crude has become a growing component to global oil market. Accompany with that, extreme bulk properties (including high molecular weights, high boiling points, large densities, high viscosities, etc.) are expected. However, current experimental approaches are still limited in characterizing heavy crudes in two aspects: 1) chemical structures of hydrocarbons with carbon number up to merely 12 can be accurately analyzed; 2) distillation cuts can only be obtained up to 750 K of normal boiling point, even with deep vacuum fractionation apparatus[1]. With limited experimental data available, less than 50 wt. % of individual heavy oil or bitumen can be analyzed, which renders simple model extrapolation over the whole crude highly unreliable.

The present work compares performance of petroleum refining simulation between two different methods: 1) the traditional API correlations commonly used in industry, 2) a real-component based molecular characterization method[2]derived from thermodynamically-consistent correlation of available crude oil data on chemical and physical properties. Compared with the former method, the later one not only gives reasonable predictions for properties of crude oil including heavy cuts, but also provides chemical structures and compositions of model constituent components over the whole crude oil, which serves as a cornerstone of high fidelity refining simulation.

Reference

1. Sánchez-Lemus, M. C., et al. (2016). "Physical properties of heavy oil distillation cuts." Fuel 180: 457-472.

2. Chen, C.-C. and H. Que (2013). "Method of characterizing chemical composition of crude oil for petroleum refinery processing". US Patent Application 20130185044 A1.

460128 Economic Design of Bioconversion Process: Natural Gas to Isobutanol

Abstract:

The low cost and abundant supply of natural gas have been motivators for the development of many technologies that convert this low cost carbon feedstock into compounds of greater value. Natural gas bioconversion is a developmental technology which applies synthetic biology to a naturally methane consuming bacteria, a methanotroph, to program it to produce higher value materials of interest. Intrexon has formed joint ventures to commercialize methanotrophic bioprocesses for biofuels and high value chemical intermediates.

Using a methanotrophic bioprocess to produce fuels and chemicals has several advantages other than the low cost and minimal processing of feedstock. Natural gas bioconversion offers scaling opportunities beyond traditional gas-to-liquid (GTL) facilities, as well as biomass feedstock facilities. Theoretical yields of methanotroph/methane exceed that of yeast/sugar approaches by a wide margin. Biocatalytic processes that achieve high yields under milder processing conditions allow for reduced capital and operation expenditures, offering greater economic viability.

Intrexon's flagship molecule of interest in the methanotroph platform is isobutanol, a fuel for gasoline blending which has several positive attributes. Isobutanol is a clean burning "drop-in" fuel which can be blended with gasoline and is compatible with the existing petroleum infrastructure. Currently, provisions allow for bio-butanol blending up to 12.5% with gasoline and ASTM specifications exist for sale as fuel. Combustion of isobutanol has cleaner emissions as compared to gasoline by reducing hydrocarbon, carbon monoxide, nitrous oxide and sulfur dioxide production. Isobutanol also has advantages over first generation biofuels, such as ethanol, because of its higher energy density.

Currently, the production of both isobutanol and farnesene has been demonstrated in lab scale fermentation. To successfully bring this technology to commercial scale, consideration must be given not only to optimizing the metabolic pathway engineered into the methanotroph, but to key issues in the fermentor design, downstream processing and overall plant economics. Traditional engineering workflows, such as a front-end loading strategy, were used to shape the commercial scale process design. The biologic, technical and economic commercial feasibilities of the end-to-end natural gas to isobutanol process were critical to Intrexon's plant design. Laboratory scale learnings were applied to select technology that was commercially feasible at demonstration scale. This demonstration scale design was scaled down to pilot scale, where the operation will yield learnings into refinement of at-scale design, equipment selection, empirical data on heating and cooling, and controls strategies.

463172 Techno-Economic Analysis of Indirect, Direct, and Hybrid Coal-Biomass to Liquids (CBTL) Plants with CO_2 Capture and Storage (CCS)

Abstract:

Indirect coal liquefaction (ICL) and direct coal liquefaction (DCL) technologies are commercially available for producing alternative transportation fuels. However, both processes are plagued with high CO_2 emission. Because the syncrude produced in the ICL and DCL reactors cannot satisfy the current specification of transportation fuels, additional upgrading technologies are required. For example, syncrude from the ICL process has low octane number for naphtha cut but high cetane number for diesel cut, while syncrude from the DCL process has high octane number for naphtha cut but low cetane number for diesel. Therefore the hybrid coal liquefaction (HCL), a combination of ICL and DCL technologies, can reduce the penalty of downstream syncrude upgrading unit by optimal blending. Addition of biomass and application of carbon capture and storage (CCS) technologies are two possible solutions to reduce the carbon footprint, but would lead to higher operating cost and capital investment. Several studies have been conducted for indirect coal-biomass to liquids (CBTL) plants with CCS, while few studies have been conducted for direct and hybrid CBTL processes at the systems level. With this motivation, we will present a techno-economic study conducted in Aspen Process Economic Analyzer® (APEA®) environment for indirect, direct, and hybrid CBTL plants with CCS using high fidelity process models developed in Aspen Plus® and Excel.

In the indirect CBTL plant, coal and biomass are first gasified to syngas. Then the syngas is converted to syncrude via Fischer-Tropsch (FT) synthesis. CO_2 is captured from both raw syngas and FT vapor product. In the direct CBTL plant, coal and biomass are directly converted into syncrude in the catalytic two-stage liquefaction (CTSL) unit by adding hydrogen produced from gasification of coal/biomass/liquefaction residue or reforming of shale gas. Significant amount of CO_2 that is generated in the hydrogen production unit(s) is captured to satisfy the target extent of CO_2

capture. In the hybrid CBTL plant, pre-processed coal and biomass are sent to either syngas production unit or the CTSL unit. Produced syngas is sent either to FT unit or hydrogen production unit. Naphtha and diesel products from the FT unit and the CTSL unit are blended to reduce the syncrude upgrading penalty. Different CCS technologies are considered and optimized for the indirect, direct and hybrid CBTL plant depending on the sources of CO_2 containing stream and corresponding CO_2 partial pressure.

The focus of this presentation will be on optimal synthesis of the hybrid plant. In particular, following aspects will be presented: (1) development of plant-wide process model and rigorous economic model in Aspen Plus®, Aspen Exchanger Design and Rating®, Excel, and APEA®, (2) sensitivity studies to analyze the impact of key design parameters (i.e. biomass/coal ratio, extent of CCS, CCS technologies, blending ratio of indirect and direct syncrude in the hybrid route) and investment parameters (i.e. price of coal and biomass, project life, plant contingency and plant capacity) on key economic measures, such as net present value (NPV), internal rate of return (IRR) and break-even oil price (BEOP), (3) analysis of the potential environmental credits due to use of biomass and CCS for all liquefaction approaches, (4) comparisons and analysis of trade-offs of indirect, direct, and hybrid CBTL plants.

465496 Finished Fuel Blending Models for Assessing Integration of Biomass-Derived Products with Petroleum Refinery Products

Abstract:

As the bioenergy industry focuses on the near-term development of conversion pathways from biomass feedstocks to infrastructure-compatible fuel blendstocks, analysis capabilities and models are essential for assessing the values of biomass-derived blendstocks based on its fuel properties and the constraining parameters for blending them with petroleum-derived products. Funding from DOEâ€™s Bioenergy Technologies Office (BETO) has enabled the analysis team to develop both linear programming models (Aspen PIMS) and process models (Aspen Plus and Aspen HYSYS) to meet the new analysis needs. In addition to developing these analysis tools, the team has also utilized published experimental blending data to validate the results of the models and to improve blend property predictions. These models are already playing a vital role at NREL in understanding the potential to integrate fuel products from biomass conversion pathways under BETOâ€™s research and development portfolio.

This presentation will focus on (1) an overview of the analysis approaches and process models developed under this effort, (2) the benefits for utilizing the tools for assessing biomass-derived blendstocks, (3) examples of blending analysis results, (4) comparison of results from blending models relative to experimental data and (5) future work for this effort in assessing blending of oxygenated compounds.

474466 R-HB Production from Glucose By B. Cepacia

Abstract:

Polyhydroxylbutyrate (PHB) is one of the important Polyhydroxyalkanoates (PHAs) that are a promising alternative to conventional, fossil fuel derived plastics. Plastic is placing a burden on the environment by polluting ecosystems, overwhelming landfills, using finite oil resources, and contributing to greenhouse gas emissions. PHAs are polyesters created by microorganisms intracellularly under limited nutrient growth conditions as carbon and energy storage compounds. PHAs are not yet widely accepted and produced because of their high production costs, specifically the carbon source and recovery method, compared to those of petroleum based plastics. In addition, present technology in recovery PHA is not environmentally friendly, leading to whether PHA is a sustainable bioproduct. This presentation shows some of our research efforts in utilizing renewable carbon source. Glucose was employed to produce droxybutyric acid. Butyric acid can be polymerized to produce PHB. The production of butyric acid and polymerization to PHB enables the desired biodegradable plastics to be recyclable.

467000 Process Strategies for High Titers of Lipid Production By Oleaginous Yeasts in Undetoxified Hydrolyzates of Lignocellulosic Biomass

Abstract:

Oleaginous yeasts can accumulate up to 70% of cell biomass as lipid, predominantly as triacylglycerols. Yeast lipid fatty acid profiles have been reported to be similar to that of vegetable oils and consist primarily of oleic, palmitic, stearic and linoleic acids. This capability provides the opportunity to use yeasts to produce bio-based fuels and chemicals from agricultural residues and bioenergy crops, such as corn stover and switchgrass, respectively. Successful cultivation of these yeasts on inhibitory lignocellulosic hydrolyzates is uncertain because little has been published regarding their application to bioconversion of plant biomass. Several novel yeasts producing promising lipid concentrations (>10 g/L) from enzyme hydrolyzates of acid and base pre-treated lignocellulosic biomass were discovered in a prior screening of the ARS Culture Collection (NCAUR, Peoria, IL). Three of the top strains have not previously been reported for the bioconversion of lignocellulose to lipids. Process strategies for maximizing lipid production in one of the more benign hydrolyzates, AFEX-pretreated corn stover, and one of the more toxic hydrolyzates, dilute acid-pretreated switchgrass, respectively, were employed. Hydrolyzates were prepared at up to 20% solids loading to provide over 100 g/L sugars (at ~56:39:5 mass ratio glucose:xylose:arabinose). A two stage process boosting the molar C:N ratio from 60 to well above 400 in undetoxified switchgrass hydrolyzate was optimized with respect to nitrogen source, C:N, carbon loading and pH. Using this process three strains were able to consume acetic acid and nearly all available sugars to accumulate 50-65% of cell biomass as lipid (w/w), to produce 25-30 g/L lipid at 0.12-0.22 g/L/h and 0.13-0.15 g/g (or 39-45% of the theoretical yield) at pH 6 and 7, a performance unprecedented in lignocellulosic hydrolyzates. Depending upon plant biomass yields, we estimated that oleaginous yeasts have the capability of producing ~48 and 190 gal oil per acre from corn stover and switchgrass, respectively. As a frame of reference, ~48 gallons of oil per acre are produced from processing soybeans. The successful application of top-performing lipid-producing yeast in undetoxified lignocellulose hydrolyzates is expected to advance the economic feasibility of high quality biodiesel and jet fuels from renewable biomass, expanding the market potential for lignocellulose-derived fuels beyond ethanol for automobiles to the entire U.S. transportation market.

471952 Evaluation of Alcohol-to-Jet (ATJ) Conversion Technology for Renewable Jet Fuel

Abstract:

Abstract Submission to Be Considered As An Oral Presentation at AIChE 2016 Annual Meeting in the Biochemical Conversion Processes in Forest/Plant Biomass Biorefineries Session

Evaluation of Alcohol-to-Jet (ATJ) Conversion Technology for Renewable Jet Fuel

Scott Geleynse, Manuel Garcia-Perez, Michael Wolcott, Xiao Zhang

Washington State University

The FAA is working to enable the U.S. use of one billion gallons per year of "drop-in" sustainable alternative jet fuels by 2018. Airlines and aerospace manufacturers around the world have also expressed continuously increasing interest in the replacement of jet fuel from fossil fuels to renewable fuels from biomass. Despite the considerable investment in the research of several different conversion strategies, establishing the successful commercial production of biorenewable jet fuel remains a challenge. In this presentation, we will discuss technical aspects of strategies to use alcohols as a platform molecule to generate jet fuel and other products. By using process modeling and techno-economic analysis, we aim to assess the feasibility and readiness of the technology for biorefinery implementation. We hope to provide insight into the various political and economic challenges faced in making and meeting renewable fuel goals.

Presenting author:

Scott Geleynse

Bioproducts Sciences and Engineering Laboratory

School of Chemical Engineering and Bioengineering, Washington State University
2710 Crimson Way, Richland, WA 99354

Contact: scott.geleynse@email.wsu.edu, x.zhang@wsu.edu

473497 Kinetics of the Enzymatic Hydrolysis of Paper Pulp Fibers

Abstract:

Papermaking fibers, particularly waste fibers from recycling operations can be a valuable resource for bioenergy and products such as ethanol, bioplastics or biochemicals. The amount of lignin in such fibers is much lower than native wood or other plant based lignocellulosic resources eliminating the need for pretreatment steps. Moreover, waste fibers are readily available in the waste streams rejected from the papermaking operations to improve paper strength and runnability of the paper machines.

The present work develops a model for the kinetics of the enzymatic hydrolysis of cellulosic pulp fines based on the physico-chemical interactions. The cellulolytic enzymes adsorb to the outer surfaces of the fines after which they diffuse into the interior pores. The fine particles are assumed to be spherical, containing two distinct kinds of pores: one at the macro-scale and another at a micro-scale allowing slow progression of the enzymes into the solid cellulosic matrix. Enzymatic hydrolysis reactions are assumed to occur at the micro-scale within the cellulosic matrix. Convective mass transfer boundary layers reduce the enzyme transport across the external particle surfaces. The main utility of this model is in estimating the optimal reaction conditions and assist in scaling studies of these enzymatic hydrolysis processes.

Keywords: Enzymatic hydrolysis, cellulose, pulp fibers, fines, rejects, paper mill waste solids

472678 Comparing Sugar Titers and Ethanol Yields from Fed Batch Fermentation of Celf and DA Pretreated Poplar Solids at High Glucan Loadings

Abstract:

Biological conversion of lignocellulosic biomass to liquid fuels reduces greenhouse gas emissions and addresses other economic and strategic challenges. However, costs are high due to high enzyme loadings and capital costs, particularly at glucan loadings over 10% needed to realize economically attractive ethanol concentrations. Because solids concentrations above about 10% cannot be readily mixed in batch fermentations, a fed-batch strategy was applied to establish maximum carbohydrate conversions from enzymatic hydrolysis of THF Co-Solvent Enhanced Lignocellulosic Fractionation (CELF) and dilute acid pretreated poplar wood. At 15, 18 and 20% glucan loadings, glucose concentrations of 63, 70, and 75g/l were achieved in only 2 days from CELF pretreated solids while concentrations from hydrolysis of dilute acid solids were only 45, 53, and 60g/l, respectively.

458379 Co-Production of Bioethanol and Bio-Lactic Acid in a Biorefinery Concept: A Comprehensive Study

Abstract:

Biorefineries are seen as a key component for the transition toward a sustainable and low carbon bio-economy. In this research, alternative biorefineries annexed to a typical South African sugar mill are investigated to co-produce ethanol (EtOH), lactic acid (LA) and/or electricity, utilizing bagasse and a component of harvesting residues (brown leaves) as feedstock. Studied scenarios are EtOH production (scenario 1), EtOH form glucose and LA from xylose (scenario 2), EtOH form xylose and LA form glucose (scenario 3) and LA production (scenario 4), all of which are associated with some level of export electricity production. Aspen simulation of the scenarios were developed considering all supplementary units such as evaporation, water treatment, boiler and steam/power generation in addition to the process units. Economic evaluation, energy assessment and Life Cycle Analysis (LCA) of the developed simulations were carried out to determine the viability of each scenario. In order to supply the combined energy demands of both sugar mill and biorefinery, a 35 to 40% bypass of lignocellulose directly to the boiler section was required, to achieve integrated scenarios that were bio-energy self-sufficient, with no supplementation with fossil fuels. Based on the economic analysis, scenario 4 achieved the highest Internal Rate of Return (IRR) of 20.9%, while scenario 1 had the lowest IRR of 10.7%. Since EtOH, LA and enzyme prices strongly affect the profitability of biorefinery scenarios, sensitivity analysis was carried out, which identified scenario 2 as the most economically robust, having the least sensitivity to these price variations.

474486 Thermophiles and Their Thermostable Enzymes in Biofuel Synthesis

Abstract:

Second-generation feedstock, especially nonfood lignocellulosic biomass, is a potential source for biofuel production. Cost-intensive physical, chemical, biological pretreatment operations and slow enzymatic hydrolysis carried out at â‰¤50Â°C make the overall process of lignocellulosic conversion into biofuels less economical than available fossil fuels. Thermophiles and their enzymes can play important roles in many kinds of bioprocessing including in conversion of non-food biomass into biofuels. The Homestake gold mine (8000 ft. deep, Lead, SD) offers a unique opportunity for direct exploration of the deep biosphere environment. Using soil/biofilm samples of deep biosphere of the Homestake Gold Mine, compost facility (Rapid City, SD, USA), and Hot Springs State Park (Thermopolis WY, USA), we have isolated several thermophilic cellulose- and xylan-degrading and -fermenting pure cultures belonging to the genera *Brevibacillus*, *Paenibacillus*, *Clostridium*, *Bacillus*, and *Geobacillus*. Unique characteristics of lignocellulose-deconstructing enzymes produced by mine and compost thermophiles include optimum temperatures of >70°C, pH ranges from 4 - 8, and high thermostability (e.g., at 60°C, 50% of cellulases and xylanases activities were retained in 35 and 23 days of incubation, respectively). Hot Springs State Park thermophiles can grow on various inexpensive regional carbon and energy sources (e.g., prairie cord grass and corn stover), and produce biohydrogen in a single step bioprocessing of biomass. This talk will describe the limitations in exiting lignocellulose conversion technologies and possible ways to overcome those limitations using thermophiles and their enzymes. The influence of high temperatures on various existing lignocellulose conversion processes and those that are under development, including separate hydrolysis and fermentation, simultaneous saccharification and fermentation, and extremophilic consolidated bioprocess will discussed. Integrated decentralized thermophilic biofuel production employing cellulolytic- and fermentative-thermophiles in a single step consolidated process will also be discussed.

463989 Improving the Energy Efficiency of Carbon Capture Processes – Combining Enzyme Accelerated Solvent Systems and Improved Contacting Equipment

Abstract:

In order to achieve the ambitious targets specified by the United Nations Framework Convention on Climate Change in 2011 and reduce GHG emission in the EU by 2020 at least 20% below those in 1990 carbon dioxide (CO_2) emissions have to be reduced significantly. Fossil-fired power plants are identified as major targets for carbon capture processes, since they are responsible for 40 % of the total global CO_2 emissions (Mondal et al., 2012). Efficient post-combustion capture of $CO_2$is one of the most promising options to face this crucial challenge and to achieve at least the specified short-term targets (Rochelle, 2009).

As a result, a wide range of different fluid separation processes, including (reactive) absorption, adsorption, cryogenic distillation and gas permeation, have been intensively studied in recent years (Mathias et al., 2013). Nevertheless, the current state-of-the-art process for CO_2 capture is the reactive absorption in packed columns using amines. Especially primary amines, like Monoethanolamine (MEA), are often used because of the covalent bonding of the CO_2, resulting in high absorption rates (Kenig & Gorak, 2005). However, the recovery of the solvent, which is required for a continuous operation, becomes more challenging the stronger the $CO_2$is bound (Yildirim et al., 2015).

Therefore, the best absorption process will represent an optimal trade-off between fast absorption rates and energy efficient solvent recovery. The latter causes up to 80 % of the total capture costs for the MEA process and results in a significant loss in overall power plant efficiency (Notz et al., 2011). Consequently, solvent recovery represents the bottleneck of the reactive absorption/desorption process, which is practically not considered for large scale industrial application (Mondal et al., 2012). In order to overcome this restriction a lot of research effort is carried out with a focus on more energy efficient solvents. While tertiary amines or potassium carbonate require less energy in the regeneration step, application of these solvents is restricted by their relatively low absorption rate (Kunze et al., 2015).

Process intensification (PI) can help to overcome this restriction in different ways, which are investigated in this study. In order to compensate for the low absorption rates of tertiary amines enzymes are integrated as efficient biocatalysts to the solvent system. Here, especially the enzyme Carbonic Anhydrase (CA) is considered as the fastest catalyst for the hydration of CO_2, which is the prevailing CO_2consuming reaction in tertiary amine systems (Gundersen et al., 2014). A significant enhancement of the absorption performance was observed in previous studies with a wetted wall column set-up, validating the direct influence on the reaction rate (Kunze et al., 2015). However, besides the possible improvement, the application of carbonic anhydrase as a biocatalyst also introduces additional restrictions to the process due to sensitivity towards temperature and pH (Gundersen et al., 2014).

In addition to an intensification of the solvent system, combining high capacity solvents with the efficient biocatalyst, the mass transfer can be further improved by means of intensified contacting devices (ICD), other than making use of conventional columns. The use of a membrane contactor provides a well-defined interfacial area, with packing densities being a magnitude higher as in conventional equipment (Zhao et al., 2016). Due to the fully separated liquid and vapour flow, membrane contactors also allow operational modes different from conventional equipment, which are limited by flooding or loading. The modular nature and operational flexibility make membrane contactors a promising alternative to conventional equipment. However, mass transfer resistance might be increased due to the membrane and the operational window of the membrane contactor is limited as well, such that especially transmembrane pressure differences have to be controlled in order to avoid breakthrough of one of the two phases, which will lead to undesired phase dispersion.

Another ICD that can improve mass transfer in ab- and desorption is the rotating packed bed (RPB) technology. In RPB processes centrifugal forces are exploited to increase acceleration of the liquid. In a counter-currently operated RPB, gas is pushed through the torus-shaped rotating packing from the outside to the middle of the rotor. Liquid is sprayed into the eye of the rotor and flows counter-currently to the gas outwards, driven by the centrifugal forces. RPBs offer mass transfer improvements due to high turbulences and extend the operating window, having the rotational speed as an additional degree of freedom. These advantages result in high capacity combined with compactness (Sudhoff et al., 2015). Yet, application of RPBs is still restricted due to lack of reliable data on the performance efficiency.

While both, ICD and the use of an improved solvent system with CA as catalyst, have the potential to significantly intensify the CO_2capture process, literature studies and research on the combination of both is scarce. In order to investigate if the advantages of the different PI techniques can be exploited and if a combination is feasible, first investigations are performed in order to characterize the operating windows of such intensified processes and to evaluate the potential benefits for an actual implementation. In preliminary work (Kunze et al., 2015), the tertiary amine Methyldiethanolamine (MDEA) has shown promising results to be more energy-efficient compared to MEA and is therefore chosen as solvent (30 wt.-%) for this work. Absorption experiments in a column and the different ICD are performed to determine the operating windows and compare separation efficiency. Subsequently CA is added to the solvent system in order to investigate the potential synergetic intensification of ICD and CA versus the acceleration by CA in a conventional packed column set-up. The results enable the discrimination between the different means for process intensification depending on the determined operating windows and separation performances. Finally, the obtained results are evaluated and benchmarked against the current state-of-the-art process on a theoretical basis. Future work will also investigate the application of the different means for process intensification to the desorption step in order to provide a complete picture for a potential industrial application.

REFERENCES

Gundersen, M.T., Von Solms, N., Woodley, J.M., 2014, Energy Procedia, 63, pp. 624-632.

Kenig, E.Y., Górak, A., 2005, Integrated Chemical Processes: Synthesis, Operation, Analysis, and Control, pp. 265-311.

Kunze, A.-K., Dojchinov, G., Haritos, V.S., Lutze, P., 2015, Applied Energy, 156, pp. 676-685.

Mathias, P.M., Reddy, S., Smith, A., Afshar, K., 2013, Energy Procedia, 37, pp. 1863-1870.

Mondal, M.K., Balsora, H.K., Varshney, P., 2012, Energy, 46 (1), pp. 431-441.

R. Notz, I.Tönnies, H.P. Mangalapally, S. Hoch, H. Hasse, 2011, Int. J. Greenh. Gas Control, 5, 3, 413-421

Rochelle, G.T., 2009, Science, 325 (5948), pp. 1652-1654.

Sudhoff, D., Leimbrink, M., Schleinitz, M., Górak, A., Lutze, P., 2015, Chemical Engineering Research and Design, 94, pp. 72-89.

Zhao, S., Feron, P.H.M., Deng, L., Favre, E., Chabanon, E., Yan, S., Hou, J., Chen, V., Qi, H, 2016, Journal of Membrane Science, 511, pp. 180-206.

Yildirim Ö., Kiss A. A., Hüser N., Leßmann K., Kenig, E.Y., 2012, Chemical Engineering Journal, 213, pp. 371–391.

ACKNOWLEDGEMENT

The research leading to these results has received funding from the European Union Seventh Framework Programme FP7/2007-2013 under grant agreement n° 608535.

464321 Boosting the Multienzymatic Conversion of CO2 through Spatially Separated Immobilization Strategy

Abstract:

Multienzymatic catalysis is considered to be the next generation of biocatalysis. As a central part, rational design and construction of multeinzyme systems have attracted great attention and become a hot research topic nowadays. When designing an efficient multienzyme system, three crucial issues should be taken into consideration: 1) synergistic catalysis among multienzymes; 2) efficient mass transfer of substrates and products; and 3) easy recyclability of multienyzmes. Inspired by the existence form and synergistic catalytic behavior of multienzymes in organisms, we put forward to the strategy of constructing spatially isolated multienzyme systems. In our recent study, a series of spatially isolated multienzyme systems (containing formate dehydrogenase and formaldehyde dehydrogenase) enabled by nanospheres/microcapsules are constructed by combining biomimetic mineralization with other biomimetic platform techniques (*e.g.*, biomimetic adhesion, Pickering emulsion, surface segregation and layer-by-layer (LbL) self-assembly). When utilized for the conversion of CO_2, these systems exhibit desirable catalytic performance and easy recyclability.

Firstly, a spatially isolated multienzyme system enabled by active nanospheres was successfully constructed through the synergy between biomimetic mineralization and biomimetic adhesion, achieving the efficient process intensification. The first enzyme (formate dehydrogenase) was entrapped accompanying the formation of nanospheres through biomimetic titanification. After *in situ* surface functionalization of nanospheres with oligodopa, the second enzyme (formaldehyde dehydrogenase) was immobilized on the surface of active nanospheres through amine-catechol adduct reaction. When utilized for the conversion of CO_2, the as-constructed multienzyme system exhibited a formaldehyde yield of higher than 60.0%. Besides, after evaluating the performance of different-sized multienzyme systems, it was found that smaller sized multienzyme system possessed higher enzymatic activity but lower recyclability.

Secondly, a spatially isolated multienzyme system enabled by nanosphere-stabilized capsules (NSSCs) was constructed through the synergy of biomimetic mineralization and Pickering emulsion, achieving the efficient process intensification. The first enzyme-containing nanospheres were monodispersely spread on the surface of oil droplets through Pickeirng emulsion, maintaining their structural characteristics and integrity. Then, the gel titania, that was produced through sol-gel process of tetrabutyl titanate, was in charge of crosslinking the nanospheres, forming a stable and continuous capsule wall. The second enzyme was immobilized on the surface of capsule wall through amine-catechol adduct reaction between oligodopa and enzyme. During the construction process, the released butanol as a result of the sol-gel process of butyl titanate could be adsorbed on the surface of nanospheres, further inducing the self-assembly of the nanospheres at the oil/water interface. This process played key role on the formation of Pickering emulsions. When utilized for the conversion of CO_2, the as-constructed multienzyme system acquired a formaldehyde yield of higher than 50.0%. Meanwhile, the lower density of the oil core than that of water endowed the multienzyme system with excellent recyclability, especially the formaldehyde yield kept almost unaltered after 10 times recycling.

Thirdly, a spatially isolated multienzyme system enabled by mesoporous hybrid microcapsules was constructed through the synergy of biomimetic mineralization and surface segregation, achieving the efficient process intensification. The first enzyme was entrapped accompanying the formation of the PAH-segregated $CaCO_3$ template. Then, the PAH polymeric network was formed through crosslinking between GA and PAH. Afterwards, this polymeric network was utilized for implementing biomimetic mineralization with *in situ* entrapping the second enzyme. After removing the template, a spatially isolated multienzyme system was acquired. During the construction process, the $CaCO_3$ microsphere served as the dual templates for formation of both capsule lumen and mesopores on the capsule wall, whereas PAH played the following two crucial roles: (1) forming the bulk polymer network to render confined space for biomimetic mineralization and (2) triggering the mineralization of Ti-BALDH to produce titania nanoparticles. When utilized for the conversion of CO_2, the as-constructed multienzyme system exhibited a formaldehyde yield of higher than 70.0%. Meanwhile, the reaction rate was nearly two times higher than the average level. And no catalytic activity was lost after 10 times recycling. Besides, the mesoporous hybrid microcapsules can be utilized in various other applications and exhibited desired performance.

Fourthly, a spatially isolated multienzyme system enabled by mitochondria-like microcapsules was constructed through the synergy of biomimetic mineralization and LbL self-assembly, achieving the efficient process intensification. Inspired by the double membrane structure of mitochondria, we proposed to utilize process and structure biomimicism strategy instead of just process biomimicism strategy for constructing multienzyme systems. The first enzyme was entrapped accompanying the formation of the PSS-doped $CaCO_3$ template. The organic inner membrane was acquired *via* LbL self-assembly of oppositely charged polymers/proteins on the template. The silica template layer was then formed onto the inner membrane through biomimetic silicification with *in situ* entrapping the second enzyme, while the organic-inorganic hybrid outer membrane was acquired *via* biomimetic titanification subsequently. After removing the $CaCO_3$ template and the silica template, a spatially isolated multienzyme system was established. When utilized for the conversion of CO_2, the as-constructed multienzyme system exhibited a formaldehyde yield of higher than 82.0%, with a formaldehyde selectivity of 90.0%. Meanwhile, 70.0% of the initial activity was retained after 10 times recycling.

By using of a series of cases as the evidences, it can be conjectured the spatially separated immobilization strategy can boost the multienzymatic conversion of CO2. This study will shed light on the new strategies for the design and preparation of multicomponent catalysts.

466064 Evaluation of Consumption Energy in CO_2 Absorption Processes with Two-Liquid Phase Separation Type

Abstract:

The global issue of climate change is receiving ever-increasing attention. One of the main focuses of this attention is the reduction of CO_2 emission from activities such as power generation, iron and steels making process. The technology that has the potential to make a significant impact on emission is CO2 capture and storage.

The aqueous of amine CO2 capture proses can remove CO2 from exhaust gas in some industrial process. The major aqueous amine absorbent which is applied in industrial process are primary amines, the secondary amines and the tertiary amines. The aqueous of amine CO2 capture proses are consumed thermal energy for CO2 desorption. A previous study has revealed that a lot of energy for CO2 desorption are CO2 desorption heat, sensible heat of the aqueous for desorption temperature and the heat of vaporization the aqueous.

We suggest new CO_2 absorption processes with two-liquid phase separation type for reducing consumption energy. Absorbing liquid in new process consist of H2O, amines and ether or alcohol additive. In CO2 absorption section, the CO2 react to amine in absorber column. The CO2-rich solution liquid is transferred to the decanter. In the decanter, the CO2-rich solution liquid was separated hydrophilic liquid and hydrophobic liquid. Carbamate and bicarbonate ion exist in hydrophilic liquid. Hydrophobic liquid is only additive. The hydrophilic liquid is transferred to the stripper where comparatively high temperature is used to desorb the CO2. The hydrophobic liquid is transferred to mixer. It is mixed with hydrophobic liquid returned to stripper. The new CO_2 absorption processes can be reduced energy consumption for sensible heat of the aqueous for desorption temperature and the heat of vaporization the aqueous. In this work, we select 2-(Ethyl amino) ethanol, water and diethylene glycol diethyl ether to Absorbing liquid. We examine the CO2 absorption characteristics. We study the effect of the new Absorbing liquid on energy consumption for process. We determine the comparison of absorbing liquid. (H2O, EAE, diethylene glycol diethyl ether = 10, 30, 60 wt%) We measure the solubility of CO2 on this absorbing liquid. In experiment, we obtained amount of CO2 are measured to be 0.3 (mol-CO2/mol-amine) at absorber temperature 40 degree C. The new absorbing liquid has the enough loading amount of CO2. We find out that the enough performance is obtained as absorbing solution.

Consumption energy of the new process is calculated by process simulator. The consumption energy is calculated to be 2.5 (GJ/t-CO2). This value is lower than that obtained for MEA absorption process. Thus, CO_2 absorption processes with two-liquid phase separation type can be valuable in CO2 capture and separation technology.

466110 The Influence of DissociationÂ Constants on the Kinetics of Carbon Dioxide Absorption in Aqueous Tertiary Amines Solutions Containing Carbonic Anhydrase

Abstract:

American Institute of Chemical Engineers (AIChE) Annual Meeting

The influence of dissociation constants on the kinetics of carbon dioxide absorption in aqueous tertiary amines solutions containing carbonic anhydrase

Bin Liu[a], Xiao Luo[a*], Zhiwu Liang[a,b*], Paitoon Tontiwachwuthikul[a,b*]

[a]Joint International Center for CO_2 Capture and Storage (iCCS), Hunan Provincial Key Laboratory for Cost-effective Utilization of Fossil Fuel Aimed at Reducing CO_2 Emissions, College of Chemistry and Chemical Engineering, Hunan University, Changsha 410082, PR China

[b]Clean Energy Technologies Research Institute (CETRI), Faculty of Engineering and Applied Science, University of Regina, Regina, Saskatchewan, S4S 0A2, Canada

Keywords: Tertiary amines; CO_2 absorption; carbonic anhydrase

* Author for correspondence: Dr. Zhiwu Liang E-mail: zwliang@hnu.edu.cn

Tel: +86-13618481627 and +86-73188573033

ABSTRACT

The chemical absorption of CO_2 into primary amine and secondary amine, such as monoethanolamine (MEA), diethanolamine (DEA), is currently the most widely accepted commercial approach to carbon capture.[1] This solvent is extremely effective in forming stable carbamates when reacted with CO_2, leading to a very efficient removal of CO_2 with a 2:1 reaction stoichiometry.[2,3] However, regeneration of the carbamate species requires a large parasitic supply of energy to release the CO_2 during stripping.[1] This high energy requirement accounts for up to 80% of the operating costs for CO_2 capture and recovery.[4] It is also a highly corrosive substance that readily undergoes degradation in the presence of oxygen, emitting harmful volatile organic compounds.[5]

Considering the high absorption heat and high energy consumption during CO_2 capture process, there are a number of tertiary amines such as methyldiethanolamine (MDEA), dimethylaminoethanol (DMEA), in place of primary and secondary amine for carbon capture. While many of tertiary amines solvents are able to reduce the stripping energy requirement, the reaction rate for CO_2 sorption in the absorber is often slow. This means that it is not possible to recover high purity carbon dioxide within a reasonable column height. Carbonic anhydrase (CA), a naturally occurring enzyme and very efficient catalyst that enhances the reversible reaction of CO_2 to HCO_3^- , was first identified in 1933 in red blood cells and known to catalyze the conversion of CO_2 into bicarbonate (HCO_3^-) at extremely high turnover rates, [6-7] which may be useful for promoting the absorption rates of CO_2 from gas streams when these alternative solvents are used. There is great potential for using CA to hydrate CO_2, especially when it is known that the hydration step is the rate determining step in the CO_2 absorption process. Carbonic anhydrase is not just a single enzyme form, but a broad group of zinc metallo-proteins (enzymes) that exists in three genetically unrelated families of isoforms (¦Á, ¦Â and ¦Ã). Its catalyzed mechanism of CO_2 hydration has been introduced by Lindskog et al. [6-7] and simplified as shown in Figure 1. Penders et al. [8] has been working on CO_2 absorption into amine solutions catalyzed by human carbonic anhydrase in string cell reactor since 2012. This work focuses on the study of effect of dissociation constants (pKa) on the kinetics of carbon dioxide absorption in aqueous tertiary amines solutions containing carbonic anhydrase at various temperature using stopped-flow techniques. The examined alkanolamines were (Figure 2): Triethanolamine (TEA), Methyldiethanolamine (MDEA), Dimethylethanolamine (DMEA), 2-(Diethylamino)ethanol (DEEA), 1-Diethylamino-2-propanol (1DEA-2P), 3-Diethylamino-1-propanol (3DEA-1P) and 2-(Dimethylamino)isobutanol (2DEA-2M-1P). The values of pKa of different amine was determined using conventional pH method as shown in Figure 3 and its final results was presented a function of temperature in Figure 4.

It can be seen from the results (Figure 5) that DMEA with a higher pKa value than MDEA and TEA hence showed a higher reactivity with carbon dioxide in the absence of enzyme. It can also been found that the reaction rate constant (k_0) in amine-CO_2-H_2O system was significantly enhanced in the present of BCA, in which the enzyme activity is higher in TEA solutions. This is because Enzymes often display a pH-dependent changes in activity, in characterizing the pH dependence of the enzyme activity, the experimenter often observes a bell-shaped curve in plot of activity versus pH or s-shaped curves in plots of activity versus pH (either falling from optimal activity or rising to optimal activity) with an inflection point at some pH value. This result would provide a useful information to understand the reaction activity of CO_2 absorption into various amines system.

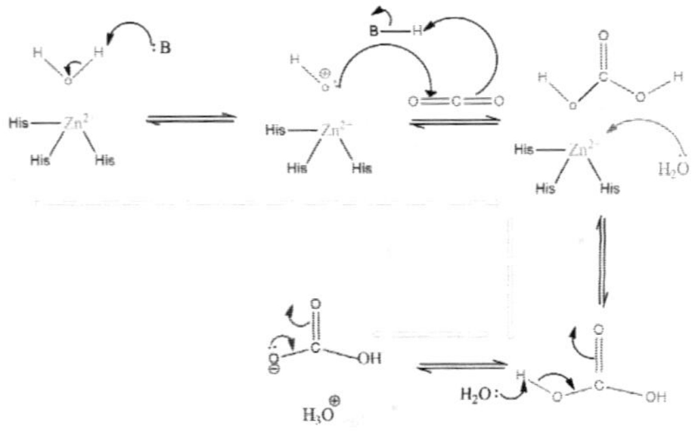

Figure 1. The catalytic Mechanism of enzyme in aqueous solutions

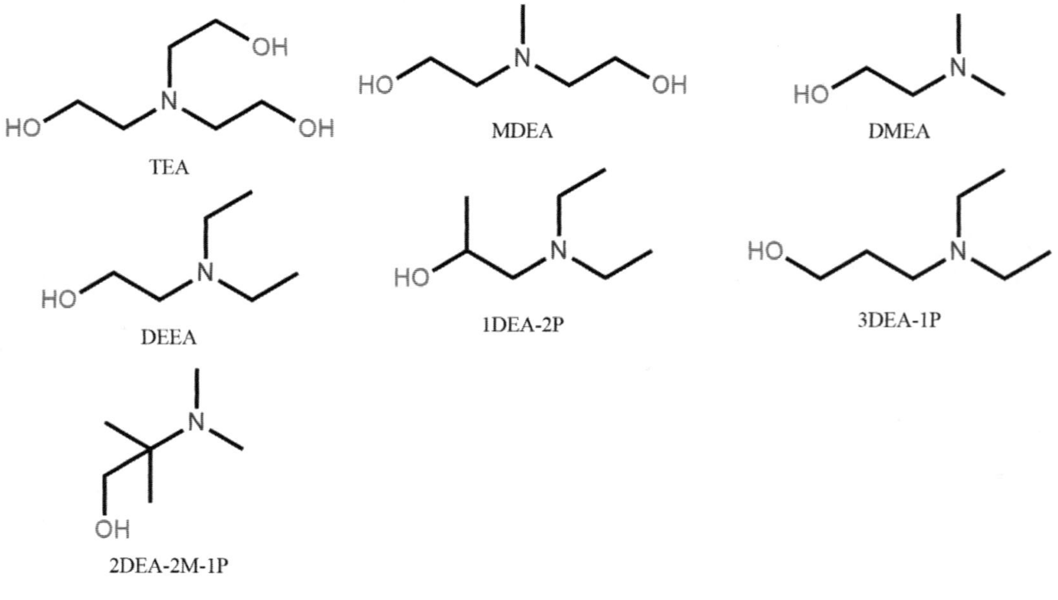

Figure 2. The chemical structure of amines studied in this work

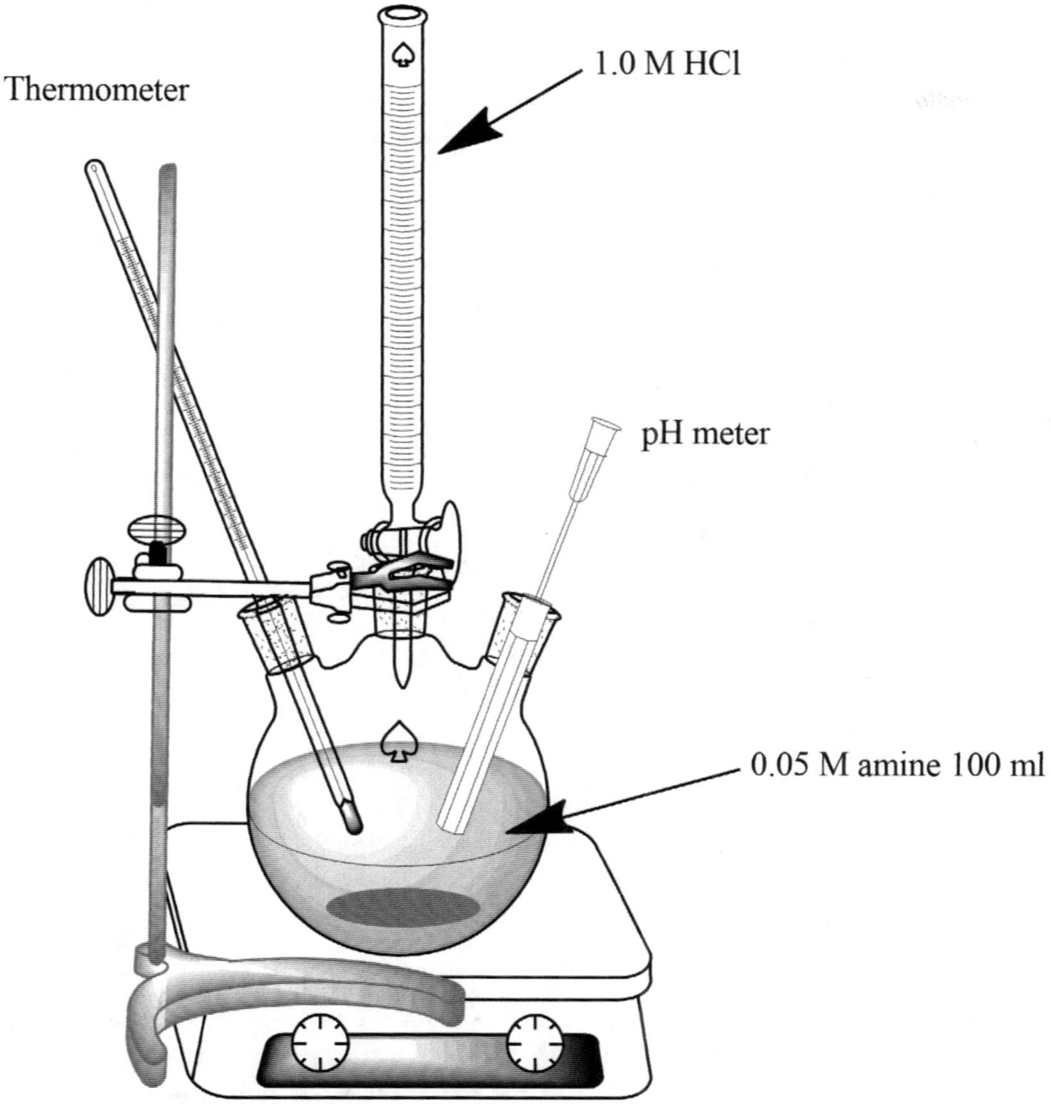

Figure 3. String cell for determination of amine pKa

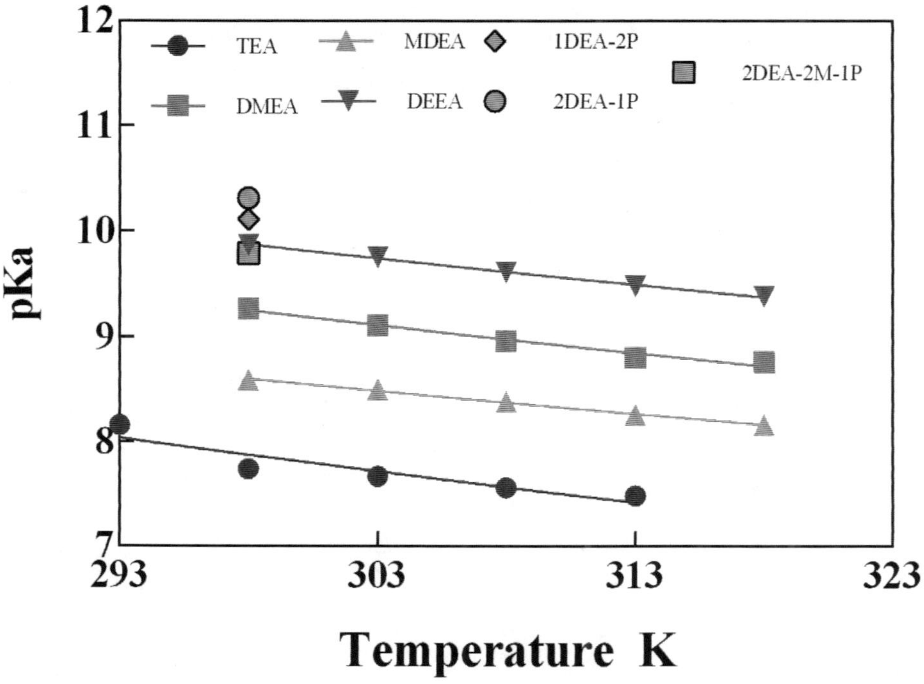

Figure 4. The dissociation constants show as a function of temperature

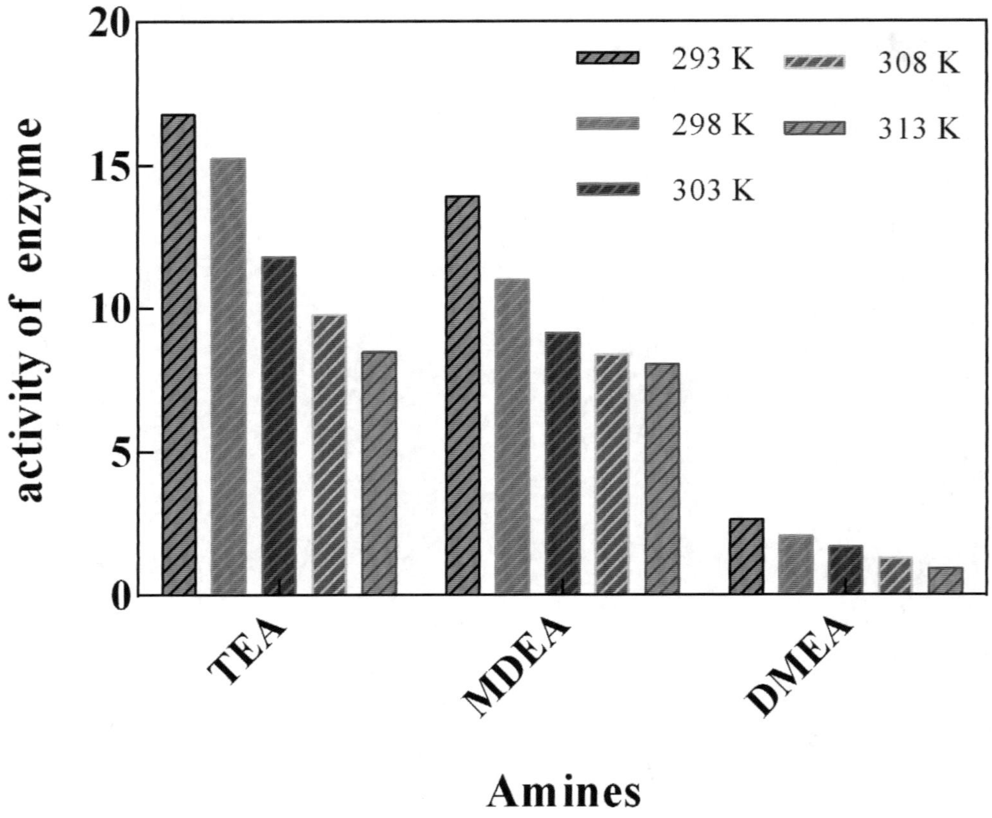

Figure 5. Comparison of enzymes activity in different aqueous amine solutions

Acknowledgment: The financial supports from the National Natural Science Foundation of China (Nos. 21476064 and 21406057), National Key Technology R&D Program (Nos. 2012BAC26B01 and2014BAC18B04), Innovative Research Team Development Plan-Ministry of Education of China (No. IRT1238), and China¡¯s State ¡°Project 985¡± in Hunan University-Novel Technology Research and Development for CO_2 Capture as well as Hunan University to the Joint International Center for CO_2 Capture and Storage (ICCS) is gratefully acknowledged.

References

1. Tontiwachwuthikul P, Idem R, Gelowitz D, Liang ZH, Supap T, Chan CW, et al. Recent progress and new development of post-combustion carbon-capture technology using reactive solvents. Carbon Management. 2011; 2: 261-3.

2. McCann N, Phan D, Wang X, Conway W, Burns R, Attalla M, et al. Kinetics and mechanism of carbamate formation from CO2 (aq), carbonate species, and monoethanolamine in aqueous solution. The Journal of Physical Chemistry A. 2009; 113: 5022-9.

3. Yu C-H, Huang C-H, Tan C-S. A Review of CO2 Capture by Absorption and Adsorption. Aerosol and Air Quality Research. 2012; 12: 745-69.

4. Liang ZH, Sanpasertparnich T, Tontiwachwuthikul P, Gelowitz D, Idem R. Part 1: Design, modeling and simulation of post-combustion CO2 capture systems using reactive solvents. Carbon Management. 2011; 2:265-88.

5. Reynolds AJ, Verheyen TV, Adeloju SB, Meuleman E, Feron P. Towards commercial scale postcombustion capture of CO2 with monoethanolamine solvent: key considerations for solvent management and

6. Lindskog S. Structure and mechanism of carbonic anhydrase. Pharmacology & therapeutics. 1997; 74: 1-20.

7. Lindskog S, Silverman DN. The catalytic mechanism of mammalian carbonic anhydrases. The Carbonic Anhydrases: Springer; 2000. p. 175-95.

8. Penders-van Elk NJMC, Derks PWJ, Fradette S, Versteeg GF. Kinetics of absorption of carbon dioxide in aqueous MDEA solutions with carbonic anhydrase at 298 K. International Journal of Greenhouse Gas Control. 2012; 9: 385-92.

A study of porous support amine-RTILs binary system for CO_2 capture

Min Xiao [a], Helei Liu [a,b], Raphael Idem [a,b], Paitoon Tontiwachwuthikul [a,b], and Zhiwu Liang [a,b]**

[a] Joint International Center for CO_2 Capture and Storage (iCCS), Provincial Hunan Key Laboratory for Cost-effective Utilization of Fossil Fuel Aimed at Reducing CO_2 Emissions, College of Chemistry and Chemical Engineering, Hunan University, Changsha 410082, PR China

[b] Clean Energy Technologies Research Institute (CETRI), Faculty of Engineering and Applied Science

University of Regina, Regina, Saskatchewan, S4S 0A2, Canada

Abstract

In this work, room temperature ionic liquids (RTILs) 1-butyl-3-methylimidazolium tetrafluoroborate ([BMIM]BF_4) and 1-butyl-3-ethylimidazolium tetrafluoroborate ([BEIM]BF_4) are synthesized. The cation structure of ionic liquids is confirmed by NMR spectroscopy. The thermal decomposition temperature of synthesized ionic liquids is measured and both ionic liquids show good thermal stability. The viscosity of ionic liquids is determined over the temperature range of 298 K to 353 K. A mixed system made by blending amine and ionic liquid is proposed for CO_2 capture. Investigation of the performance of this binary system on a porous support medium for CO_2 absorption shows that amine is able to greatly intensify the ionic liquids' ability to absorb CO_2. Furthermore, when blending tertiary amine and ionic liquids to absorb CO_2, it was found that only physical absorption takes place and that the viscosity of the absorbent increases slightly after CO_2 absorption. Based on the results of this work, the binary system of tertiary amine and RTILs is proven to be an alternative absorbent for CO_2 capture.

Keywords: *amine-RTILs; CO_2 capture; low viscosity; physical absorption*

* Author for correspondence:
Email: zwliang@hnu.edu.cn (Z. Liang) & lhl0925@hotmail.com (H. Liu)

1. Introduction

Global warming, which brings extreme weather, has become a serious issue and is more and more drawing people's attention. As one of many greenhouse gases, carbon dioxide is mainly blamed for the rise in global temperatures. Large amounts of CO_2 are directly released to the atmosphere through the burning of fossil fuels to meet the global demand for energy. There is an urgent need to reduce CO_2 emissions to protect the environment. Meanwhile, CO_2 is also a potential resource which can be applied to Enhanced Oil Recovery (EOR) or reformed with methane to produce synthesis gas etc. Thus, much research has been done on separating CO_2 from combustion exhaust gases. Although absorption of CO_2 with alcohol amine has been widely studied for years and proven to be able to capture CO_2 efficiently, its wide application is limited by a number of remaining problems such as corrosivity, volatility, degradation, and especially high energy consumption. The amount of energy required for regenerating rich amine solvents accounts for as much as 70-80% of the total operating cost of the system. It is thus of primary importance to reduce this energy consumption in order to bring down the cost of CO_2 capture.

A new kind of compound named room temperature ionic liquids (RTILs) has aroused people's interest because of their many unique properties. RTILs usually exist as liquid at room temperature with low vapor pressure, something that commends them as environmental friendly. Besides, RTILs show good thermal/chemical stability and tunable physicochemical characteristics, properties that have allowed them to be successfully used as green solvents in reactions and as catalysts for organic reactions [1]. There is also some work focusing on using RTILs for CO_2 capture. However, using pure RTILs for CO_2 capture as an answer to the world's CO_2 problem still has a very long way to go. When physical absorption dominates CO_2 capture with RTILs, the procedure is usually performed under extreme conditions, such as a very high pressure (90 bar or higher), and with a very long time being required to reach equilibrium (up to 24 h) [2]. To intensify CO_2 capture performance, a kind of ionic liquids named task-specific ionic liquids was developed. From the work of Bates et al. [3], the molar uptake of task-specific ionic liquids, which was modified with amino group, approached 0.5 mol CO_2•mol amine^{-1} during 3 h. However, high viscosity appears to be a new problem making those particular ionic liquids challenging for application in CO_2 capture [4]. The viscosity of ionic liquids is commonly high and will dramatically increase after CO_2 absorption, especially for task-specific ionic liquids [2].

To address these problems, a kind of low viscosity amine-IL mixed system is proposed for CO_2 capture in this work. Imidazole based ILs 1-butyl-3-methylimidazolium tetrafluoroborate ([BMIM]BF_4) and 1-butyl-3-ethylimidazolium tetrafluoroborate ([BEIM]BF_4) were chosen for their relatively low viscosity compared with other ILs. Conventional amines, Monoethanolamine (MEA) and Methyldiethanolamine (MDEA) were selected as the amines for blending with the ILs. The structures and abbreviation of amines and cation of ILs are shown in **Figure 1**.

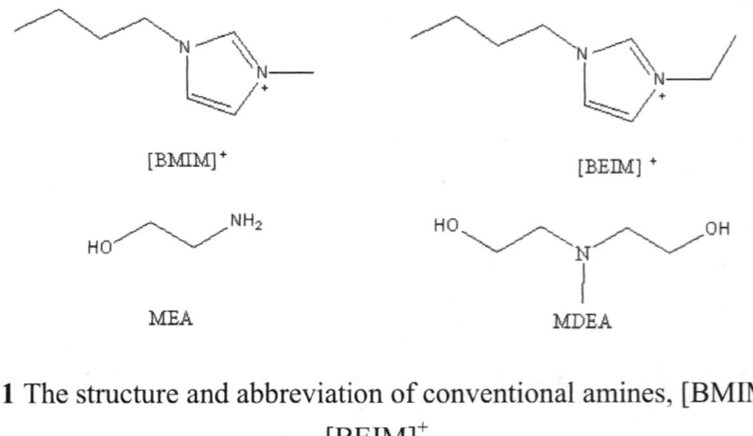

Figure 1 The structure and abbreviation of conventional amines, $[BMIM]^+$ and $[BEIM]^+$

2. Materials and experiment

2.1. Materials

CO_2 with purity of 99.9% was obtained from Changsha RiZhen Gas Co., Ltd. MDEA with purity of 98% was from Aladdin Industrial Corporation. MEA with purity of 99% and n-Butyl bromide (≥ 98%) was obtained from Sinopharm Chemical Reagent Co., Ltd. 1-Methylimidazole with purity of 99% and 1-Ethylimidazole with purity of 98% were supplied by Shanghai Macklin Biochemical Co. Ltd.

2.2. Experiment

$[BMIM]BF_4$ and $[BEIM]BF_4$ were synthesized on basis of imidazole. The general preparation was similar to the work of Harris et al [5]. Complete anion exchange was confirmed when no bromide could be detected with silver nitrate. After evaporating any unreacted components and water, the RTILs were stored in a vacuum drier. A ^{13}C NMR Spectrometer (Bruker 400MHZ Advance) was employed to characterize the cation structure of the RTILs and the absorbents after CO_2 absorption. Thermogravimetric Analysis (Shimadzu DTG-60) was introduced to evaluate the water content and thermal stability of the RTILs. Then the dynamic viscosities (Brookfield Viscometer DV2T) of pure RTILs and the amine-RTIL binary systems were measured.

The CO_2 absorption experiment process is briefly described below. Approximate 0.8 mL absorbent was stabilized into a porous filter candle which was then placed into a gas-washing bottle. The gas-washing bottle was immersed into a water bath to control the temperature. The pure CO_2 flow rate was fixed at 50 $mL{\bullet}min^{-1}$ by a mass flow controller, and fed into a conical flask with concentrated sulfuric acid to get rid of all trace amounts of water. Then, the CO_2 was fed to the gas washing bottle continuously. The amount of absorbed CO_2 was determined by detecting weight change using an electronic analytical balance with accuracy of 1 mg. When the weight of the absorbent stopped increasing, equilibrium was considered to be achieved.

3. Results and discussion

3.1. Thermal stability of RTILs

To evaluate the thermal stability of $[BMIM]BF_4$ and $[BEIM]BF_4$, Thermogravimetric Analysis (TGA) was introduced. The results for pure $[BMIM]BF_4$ and $[BEIM]BF_4$ are shown in **Figure 2**. Water content of the RTILs was confirmed to be less than 1% and thermal decomposition temperatures (10% mass loss) of $[BMIM]BF_4$ and $[BEIM]BF_4$ were found to be 635.65 K and 646.96 K, respectively. Results showed that both RTILs have good thermal stability. Because the side chain of $[BEIM]^+$ was one carbon longer than $[BMIM]^+$, it showed better thermal stability.

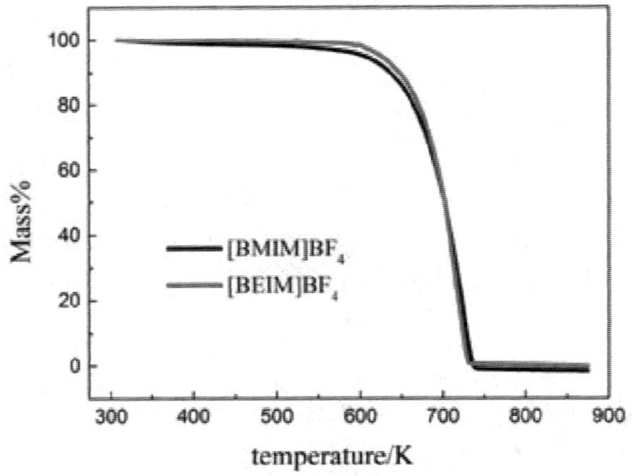

Figure 2 Thermogravimetric Analysis on $[BMIM]BF_4$ and $[BEIM]BF_4$

3.2. Viscosity of RTILs

The viscosities of $[BMIM]BF_4$ and $[BEIM]BF_4$ were measured over temperature ranges from 298 K to 353 K. It was found that $[BMIM]BF_4$ shows viscosity of 101.2 cP at 293 K, which agreed well with Harris's work with a deviation of 2.6% [5]. The viscosity of $[BEIM]BF_4$ is slightly higher than $[BMIM]BF_4$ due to its longer carbon chain. An extended Arrhenius equation (1) is applied to correct viscosity with temperature:

$$\ln\eta = \frac{E_a}{RT} + \ln\eta_\infty \cdots\cdots\cdots\cdots(1)$$

The R-square of viscosity fitting for $[BMIM]BF_4$ and $[BEIM]BF_4$ is 0.997 and 0.998, respectively. The equation is able to correct the experiment data well. The summaries of E_a and η_0 are listed in **Table 1**.

Table 1 Summaries of E_a and η_0 for pure $[BMIM]BF_4$ and $[BEIM]BF_4$

	characteristic parameter	
	E_a/(kJ•mol^{-1})	η_0/(•10^{2}cP)
$[BMIM]BF_4$, Harris's work [5]	33.53	0.14
$[BMIM]BF_4$, this work	34.15	0.0079
$[BEIM]BF_4$, this work	32.62	0.015

3.3. CO_2 absorption

In this work, the CO_2 absorption by RTILs and some binary systems was studied at the temperature of 298 K under atmospheric pressure. The absorption curves of the pure RTILs are shown in **Figure 3**. CO_2 equilibrium solubility of $[BMIM]BF_4$ and $[BEIM]BF_4$ at 298 K was 0.039 and 0.049 mol CO_2·mol ILs^{-1}, respectively. Equilibrium was reached within 10 minutes and $[BEIM]BF_4$ showed higher equilibrium CO_2 solubility than $[BMIM]BF_4$.

However, both RTILs revealed low physical CO_2 absorption without the amines. To improve CO_2 absorption, 30% mass fraction of MEA and MDEA was blended with the RTILs, respectively. It was found that CO_2 absorption could be greatly improved by these added amines and the CO_2 equilibrium could be achieved within 120 minutes. Because the amines dominate the absorption of CO_2, the result was expressed using the units of mol CO_2•mol amine^{-1}. Four CO_2 absorption curves are shown in **Figure 4**. The equilibrium CO_2 loadings of MEA-[BMIM]BF_4 and MEA-[BEIM]BF_4 at 298 K were 0.494 and 0.515 mol CO_2•mol amine^{-1}, respectively. Chemical reaction between MEA and CO_2 was confirmed by the detection of a new resonance at 165.08 ppm using ^{13}C NMR technology. Two MEA molecules combined with one molecule of CO_2 to form carbamate without the presence of water. The CO_2 absorption of MDEA-[BMIM]BF_4 and MDEA-[BEIM]BF_4 was accomplished within 80 minutes and the equilibrium CO_2 loadings at 298 K were 0.214 and 0.334 mol CO_2•mol amine^{-1}, respectively, which was relatively lower than the MEA-RTILs systems. It has to be pointed out that for the MDEA-RTILs systems, no additional peak could be detected after CO_2 absorption. This means that only physical absorption exists in this mixed system. The CO_2 molecule is stabilized in the absorbent by van der Waals force, which is much weaker than the force of chemical bonding. Consequently, regenerating the absorbent requires less heat compared with the system where chemical reaction happens. Furthermore, amine-[BEIM]BF_4 performs better than amine-[BMIM]BF_4 on CO_2 absorption. It can be concluded that RTILs with higher CO_2 capacity are able to enhance the amine absorption of CO_2 as well.

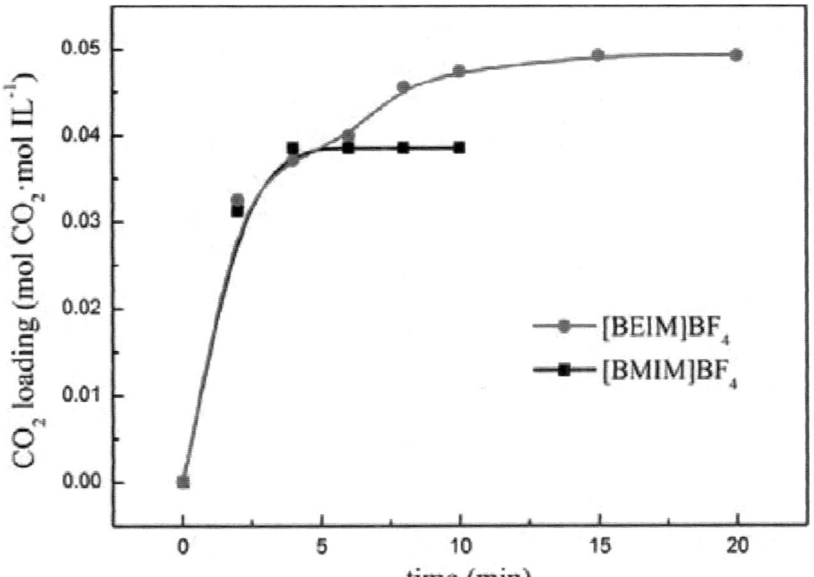

Figure 3 The CO_2 loading of pure RTILs versus time at 298 K

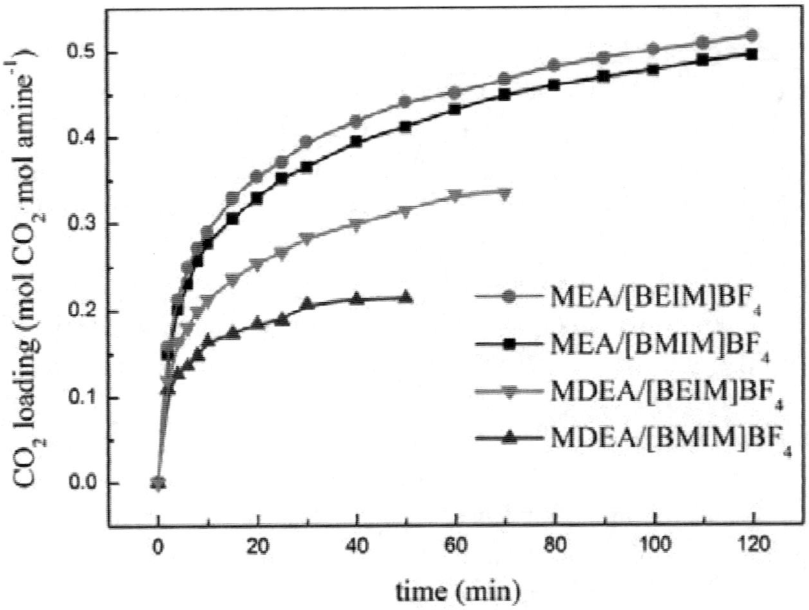

Figure 4 The CO_2 loading of mixed system with amine mass fraction of 30% at 298 K

3.4. Viscosity of mixed system before and after CO_2 absorption

To further study the viscosities of the mixed MDEA-RTILs systems, approximately 1 mL of each absorbent was exposed to a pure CO_2 atmosphere for 180 minutes. It has to be pointed out that during the CO_2 absorption using amine-RTILs mixed system, MEA was found to precipitate out in the form of carbamate after reacting with CO_2. The viscosity of MEA-RTILs system was thus unable to be measured. MDEA was fully dissolved into the RTILs before and after CO_2 absorption. Thus the viscosities of the MDEA-RTIL systems at 298 K were measured and are summarized in **Table 2**. Results show that the viscosity of MDEA-[BMIM]BF_4 increased from 49.46 cP to 63.40 cP after CO_2 absorption while for MDEA-[BEIM]BF_4, it increased from 49.79 cP to 77.77 cP. It can be concluded that the viscosity of both absorbents increased slightly after CO_2 absorption. In the case of [BEIM]BF_4, it showed higher viscosity than [BMIM]BF_4, the MDEA-[BEIM]BF_4 system revealed higher viscosity than MDEA-[BMIM]BF_4 before and after CO_2 absorption as well.

Table 2 The viscosity of mixed MDEA-RTILs system before and after CO_2 absorption at 298 K

	Before CO_2 absorption	After CO_2 absorption
	η(cP)	η(cP)
MDEA/[BMIM]BF_4	49.46 ± 1.02	63.40 ± 1.23
MDEA/[BEIM]BF_4	49.79 ± 1.02	77.77 ± 1.61

4. Conclusion

In this work, the thermal decomposition temperature and viscosity of ionic liquids [BMIM]BF_4 and [BEIM]BF_4 was measured. Both ionic liquids showed good thermal stability. [BEIM]BF_4 was found to be more stable than [BMIM]BF_4 due to its longer carbon chain. Meanwhile, the viscosity of [BEIM]BF_4 was slightly higher than that of [BMIM]BF_4. The absorption of CO_2 by amine-RITLs mixed systems was investigated as well. The viscosity of MDEA-RTILs was measured before and after CO_2 absorption. Results show that MDEA-[BEIM]BF_4 proves to be a potential binary system for CO_2 capture. The mixed system is thermally stable, able to capture CO_2 quickly and shows relatively low viscosity even after CO_2 absorption. Although MEA-RTILs show higher CO_2 capacity than MDEA-RTILs, it has to be pointed out that only physical absorption occurs in MDEA-RTILs systems, which means the mixed system can readily be regenerated by heating to strip out CO_2. Consequently, MDEA-[BEIM]BF_4 is considered to be an energy saving absorbent.

Acknowledgements

Financial support from the National Natural Science Foundation of China (Nos. 21476064, 21376067 and U1362112), National Key Technology R&D Program (Nos. 2012BAC26B01 and2014BAC18B04), Innovative Research Team Development Plan-Ministry of Education of China (No. IRT1238), Shanxi Yanchang Petroleum (Group) Co., LTD, Specialized Research Fund for the Doctoral Program of Higher Education (No. 20130161110025),and China's State 'Project 985' in Hunan University-Novel Technology Research and Development for CO_2 Capture as well as Hunan University to the Joint International Center for CO_2 Capture and Storage (iCCS) is gratefully acknowledged.

References

1. Joan F. Brennecke, Maginn EJ. Ionic Liquids: Innovative Fluids for Chemical Processing. *AIChE Journal.* 2001;47(11):2384-2389.
2. Zhang Y, Zhang S, Lu X, Zhou Q, Fan W, Zhang X. Dual amino-functionalised phosphonium ionic liquids for CO2 capture. *Chemistry.* 2009;15(12):3003-3011.
3. Eleanor D. Bates, Rebecca D. Mayton, Ioanna Ntai, James H. Davis. CO_2 Capture by a Task-Specific Ionic Liquid. *J. Am. Chem. Soc.* 2002;124(6):926-927.
4. Seo S, Quiroz-Guzman M, DeSilva MA, et al. Chemically tunable ionic liquids with aprotic heterocyclic anion (AHA) for CO_2 capture. *The journal of physical chemistry. B.* May 29 2014;118(21):5740-5751.
5. Harris KR, Kanakubo M, Woolf LA. Temperature and Pressure Dependence of the Viscosity of the Ionic Liquid 1-Butyl-3-methylimidazolium Tetrafluoroborate: Viscosity and Density Relationships in Ionic Liquids. *J. Chem. Eng. Data.* 2007;52:2425-2430.

COMPARATIVE STUDY ON LOW AND HIGH PRESSURE CO2 ADSORPTION CAPACITY OF ORGANIC MATERIALS

Mert Atilhan and Ruh Ullah*
Chemical Engineering Department, Qatar University, Doha

Santiago Aparicio
Department of Chemistry, University of Burgos, Burgos, Spain

Hasmukh Patel
Northwestern University, Evanston, USA

*Cafer T. Yavuz**
Department of Chemistry, KAIST, Daejeon, Republic of Korea

Abstract

Application of solid sorbents for CO_2 capture and separation has been under consideration since last decades owing to the promising feature of repeatable use, low cost of regeneration and thermo physical stability. However, the capturing capacity, selectivity and the overall performance of adsorbents at different temperatures and pressures still require further investigation to make these materials capable of replacing the mono-ethanol system. Here in we report synthesis, characterization and application of covalent organic polymers. Capturing capacity of COPs at low temperature and pressure has been experimented for two COPs materials (namely COP-9, COP-10) for CO_2 and N_2 adsorption. Materials were also characterized with BET and TGA.

1. INTRODUCTION

Till date cost of mono-ethanol amine regeneration and reduction in the electricity production are still among the major challenges of the plants using fossil fuels as an ultimate source for power generation. The demand of more power production due to the technological development for the improvement of life style is increasing world-wide which in turn enhances the usage of coal and subsequently rising CO_2 emission and concentration in the environment [1]. The renewal energy sources such as solar power, wind power, nuclear power and bio fuel power are still in the research and initial stages and consequently the world will have to depend mostly on fossil fuels for next few decades. It has been estimated that on average a single power plant using fossil fuels as power production source emit at least 3.5 million tons of CO_2 per year, whereas the number of such power generation industries increases globally. Cement industries and transportation, which are the second and third major sources of CO_2 emission and environment pollution are also multiplying every year and accordingly enhancing the concentration of CO_2 in atmosphere. Research has shown that the content of CO_2 in atmosphere may exceed 500 ppm by 2020 if not handle properly [2, 3]. This alarming increase in CO_2 emission has severely threatened life on the earth and issues like global warming, severe weather conditions spread of very dangerous disease such as respiratory illnesses and asthma are all due to the excessive and uncontrollable emission of CO_2 [4-9]. The new energy production plants using fossil fuels however now changes to oxy-combustion fuel system and integrated gasification combine cycle system where CO_2 can be effectively captured using advanced technologies as well as MEA[4, 10]. Solid sorbents possess various advantages such as simple regeneration, repeated applications, low manufacturing cost, no extra-cost on electricity production, environmental and economic viability, tunable capturing capacity and changeable selectivity. Solid sorbents mainly have applications in pre-combustion system like IGCC, where, CO_2 can be separated from syngas after water shift in the process[10]. However, this requires the adsorbents to be physically strong which can hold high pressure and high temperature unlike the post-combustion process where CO_2 can be separated at atmospheric pressure. In addition to the thermal and physical stability, solid sorbent needs to be extremely selective for separation of CO_2 from the stream, which have H_2 as fuels, should have high surface area and large pore volume and better affinity for CO_2. Various solid sorbents such as activated carbon, metal organic framework, zeolites, SBA-15, nanoclays, and many organic polymers have been investigated both for low and high-pressure CO_2 adsorption and separation. Attempts have been made to increase the CO_2 capturing capacity of solid materials by inserting various amine function groups within the porous structure of adsorbents[11]. As an example SBA-15 has been impregnated with different function groups and tested for low and high pressure CO_2 adsorption. Results indicated that amine modification have very limited effect on the capturing capacity at high pressure. Here in we report synthesis and characterization of covalent organic polymers, which possess internal nitrogen functionality and can be assumed to be the more favorable compounds as long as the affinity of CO_2 [12-15]. We prepare two COPs materials using different building block monomers and linked with two different linkers to investigate the effect of different functionality on the CO_2 and N_2 adsorption capacity at low and high pressure under different temperatures.

2. EXPERIMENTAL

2.1. Materials and Synthesis procedure

N,N-Diisopropylethylamine, 1,3,5-benzene tricarbonyltrichloride, 2,5-Dimethylpiperazine, cyanuric chloride, 2-imidazolidinone, dimethylacetamide, and tetrahydrofuran (THF), purchased from Sigma-Aldric, however, dioxane and ethanol were obtained from TCI, Japan. All the chemicals were used without any further modification and/or dilution. COP-9 and COP-10 were prepared through the following procedures. In a typical procedure N,N-Diisopropylethylamine (1.9 mL, 10.8 mmol) was added to trans-2,5-Dimethylpiperazine (0.48 g, 4.2 mmol) dissolved in THF at 0-5 ^{0}C. Cyanuric chloride (0.5 g, 2.7 mmol) dissolved in THF was added dropwise to the above solution with continuous stirring at 0-5 °C in N_2 environment. The solution was stirred at 0-5 °C for 1 h. Subsequently, the temperature of the reaction mixture increased to 25-30 °C and stirred for 2 h, followed by heating at 68-70 °C for 21 h. The precipitate was washed with THF, dioxane and ethanol (three times each). Finally, the precipitate was dried at room temperature for 2 h and 80-90 ^{0}C for 5 h under vacuum and labeled as COP-9. N,N-Diisopropylethylamine (2.1 mL, 12 mmol) was added to 2-imidazolidinone (0.4 g, 4.65 mmol) dissolved in Dimethylacetamide (DMAc, 35 mL) at 25 ^{0}C. 1,3,5-Benzenetricarbonyl trichloride (0.8 g, 3 mmol) dissolved in DMAc (25 mL) was added dropwise to the above solution with continuous stirring at 25 ^{0}C. The solution was stirred at 25 °C for 2 h. Subsequently, the temperature of the reaction mixture increased to 80 ^{0}C and stirred for 16 h. The reaction mixture cooled to room temperature and 250 mL of de-ionized water added which resulted precipitate. The precipitate was washed with water and ethanol (three times each) and it was dried at room temperature for 2 h and 80-90 ^{0}C for 5 h under vacuum and labeled as COP-10.

2.2. Characterization and Adsorption Setup

Thermal stability of COP-9 and COP-10 was conducted with Perkin Elmer Pyris 6 (Thermogravimetric Analysis) TGA model 4000 system. TGA system was properly balanced and calibrated according to the purging gas (N_2), crucible and sample holder stem. TGA machine was equipped with chiller from Huber model Pilot One to automatically maintain the temperature of the system and N_2 was used as the heating environment, where the materials were heated directly from 30 °C to 800 °C at the rate of 1° per minute. Physical parameters such as pore volume and surface area of materials were estimated using BET analyzer Model micrometrics ASAP 2420. The solid materials were degased for at least 12 hours prior to the BET measurement to avoid any blockage of pore due to the water.

PCT-Pro was used for the low-pressure adsorption process, where each pressure step was 0.1 bars up to the maximum of 1 bar. Here only CO_2 and N_2 were calculated at three different temperatures of 298 K, 308 K and 323 K [16]. Two gas cylinders (high purity grade CO_2, and N_2) were connected to gas dosing system (GDS) via external pump to pressurize the required gas into the measuring cell. The maximum set pressure (1 bar) was applied stepwise by increasing the pressure gradually from 0 bar up to 1 bar using each step of 0.1 bar. After reaching the maximum set pressure points, pressure of the system was gradually reduced exactly with the same measurement points starting from the high pressure and reaching to the last evacuation point to complete the desorption process. The system is fully automated and the pressure goes to next higher point after completing the previous measurement point.

3. RESULTS AND DISCUSSIONS

Thermal gravimetric analysis of both covalent organic polymers COP-9 and COP-10 (shown in figurer 1) reveals that COP-9 is thermally more stable than COP-10, since; the former is almost stable up to 400 °C. TGA further indicates that COP-10 is thermally degraded in three steps, while COP-9 decomposes in two steps. The initial degradation of COP-10 starts at around 75 °C which can be associated to the evaporation of stored water within the structure, 2nd degradation of COP-10 starts at around 200 °C, while the third one starts at 400 °C. Contrary, COP-9 is about completely stable up to 400 °C however, beyond this point it decomposed quickly and in another 100 °C increase in temperature, almost 50 % of the material losses. TGA analysis confirmed that COP-9 is much stable than COP-10, however, it degrades gradually while a sharp degradation in COP-9 can be observed after 400 °C.

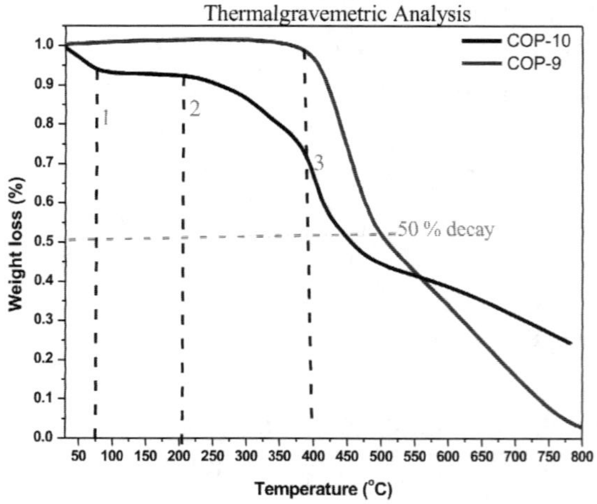

Figure 1. Thermal-gravimetric analysis of COP-9 and COP-10

Physical properties and proposed chemical structure of COPs are given in table 1 below which shows that COP-9 has smaller pore volume than COP-10, however, the former has larger surface area and lower tapped bulk density. It has been assumed that low pressure gases uptake by solid materials is mostly governed by the physical parameters such as pore volume and surface area[17]. On the other hand adsorption capacity at high pressure does not necessarily depend upon the porosity of the structure, but most often varies with affinity of the materials. Thus the material with lower porosity and lower surface area may have higher affinity at high pressure and therefore will capture more CO_2 at elevated pressure as compared to those adsorbents which have larger physical parameters [18, 19].

Table 1. Physical Parameters and structure of covalent organic polymers (COPs)

Sample	**Structure**	**Surface area (m^2/g)**		**Pore volume (cm^3/g)**	**Density (cm^3/g)**	
		BET	**Langmuir**		**Bulk**	**Tapped**
COP-9	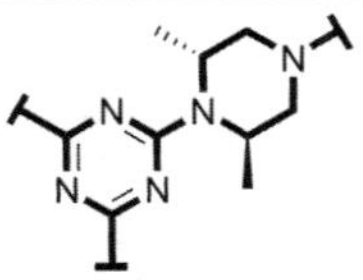	146.4	204.3	0.244	0.181	0.202
COP-10	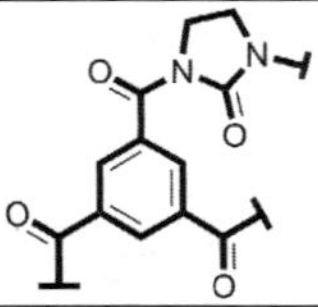	84.5	114.6	0.634	0.129	0.258

Figure 2, shows the low pressure CO_2 and N_2 adsorption of COP-9 (c, d) and COP-10 (a, b), which clearly indicates that due to the larger pore volume of COP-10 it captures comparatively more CO_2 at lower pressure of 1 bar. Xiang et al reported that COP-1 possess the highest surface area captured more N_2 and CH_4; however, COP-2 with larger pore volume captured more CO_2 and H_2 than all other organic polymers[20]. On the other hand COP-10 uptakes very low quantity of N_2 at 1 bar and three different temperatures as compared to COP-9. It must be noted that selectivity of COP-10 for CO_2:N_2 at low pressure is much higher than that of COP-9. Additionally as given in table 2 selectivity of COP-10 for CO_2:N_2:H_2 is also better than that of COP-9 at 308 K, while at all other temperatures selectivity of both polymers is similar.

Additionally, the selectivity of COP-10 at atmospheric pressure is increasing with rising temperature and becomes infinity at temperature above than 308 K. As shown in figure 2b COP-10 is totally inert for N_2 at higher temperatures, although it captures a bit smaller quantity of N_2 at standard temperature and pressure. Figure 2c and 2d indicates that COP-9 uptake less CO_2 than COP-10, but its N_2 uptake is comparatively larger than that of COP-10. Table 2 indicates that the overall performance of COP-10 at atmospheric pressure and different temperatures is comparatively better than COP-9, since the former adsorb more CO_2 and less N_2 which makes this material favorable for the pre-combustion CO_2 capture and separation.

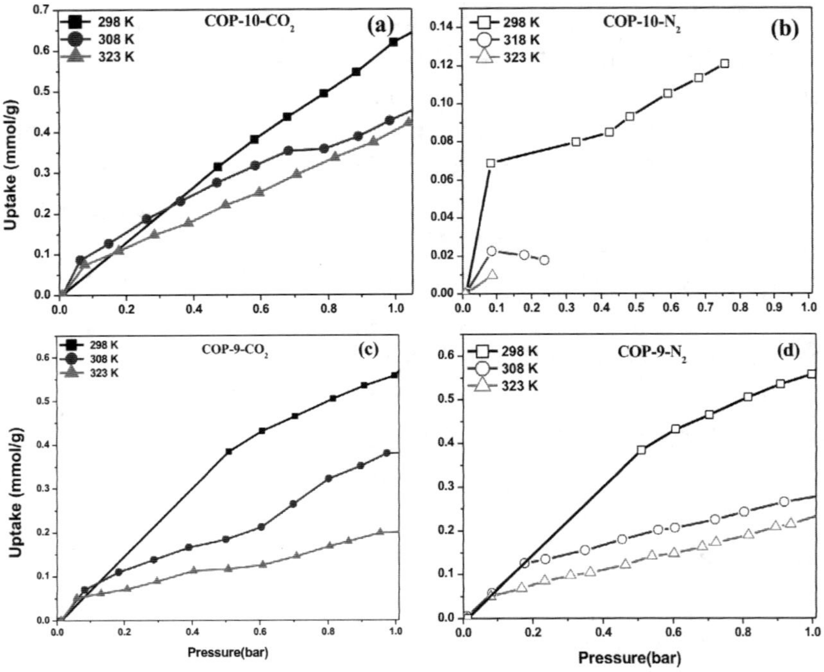

Figure 2. CO_2 adsorption of COP-9 (c & d) and COP-10 (a & b) at maximum of 1 bar and three different temperatures obtained with PCT-Pro.

It can be concluded that two covalently bonded organic polymers COP-9 and COP-10 were prepared from two different core building block monomers and linked with different linking agents. Both COPs material were tested for low and high pressure CO_2 and N_2 adsorption at three different temperatures. It was found that COP-10 has almost three times larger pore volume than COP-9 and therefore captures slightly larger CO_2 than COP-9 at atmospheric pressure and room temperature. Additionally, COP-10 uptakes significantly lower quantity of N_2 at atmospheric pressure and room temperature while it becomes almost N_2 phobic at high temperature and lower pressure, thus resulting in very good selectivity under these conditions.

In addition to the effect of pore volume and surface area on the gases uptake capacity of solid sorbents, affinity of the gas molecules toward the attractive sites also contribute to the adsorption capacity of materials. It can be assumed that, COP-9 may have more interacting sites than COP-10 since the core monomer of COP-9 has three nitrogen nodes whereas the core monomer of COP-10 has less nitrogen species. On the other hand COP-10 may have more oxygen OH functionality that may have less affinity for CO_2 than the N functionality. Thus it can be argued that, although COP-9 has lower pore volume but it has more and stronger attractive sites for CO_2 than COP-10, therefore, it captures comparatively more CO_2.

ACKNOWLEDGMENTS

This paper was made possible by the support of Qatar National Research Fund, National Priorities Research Program grant (NPRP# 5-499-1-088) and (NPRP# 6-330-2-140). The statements made herein are solely the responsibility of the authors.

REFERENCES

1. Aaron, D. and C. Tsouris, *Separation of CO2 from Flue Gas: A Review.* Separation Science and Technology, 2005. **40**(1-3): p. 321-348.
2. Boot-Handford, M.E., et al., *Carbon capture and storage update.* Energy & Environmental Science, 2014. **7**(1): p. 130-189.
3. Chu, S. and A. Majumdar, *Opportunities and challenges for a sustainable energy future.* Nature, 2012. **488**(7411): p. 294-303.
4. Ahmed, U., et al., *IGCC process intensification for simultaneous power generation and CO2 capture.* Chemical Engineering and Processing: Process Intensification, 2016. **101**: p. 72-86.
5. Awadallah-F, A. and S. Al-Muhtaseb, *Carbon dioxide sequestration and methane removal from exhaust gases using resorcinol–formaldehyde activated carbon xerogel.* Adsorption, 2013. **19**(5): p. 967-977.
6. Berger, A.H. and A.S. Bhown, *Selection of Optimal Solid Sorbents for CO2 Capture Based on Gas Phase CO2 composition.* Energy Procedia, 2014. **63**(0): p. 2092-2099.
7. Cavka, J.H., et al., *High Pressure Adsorption of CO2 and CH4 on Zr-MOFs.* Industrial & Engineering Chemistry Research, 2014. **53**(40): p. 15500-15507.
8. Chen, J., L.S. Loo, and K. Wang, *High-Pressure CO2 Adsorption on a Polymer-Derived Carbon Molecular Sieve.* Journal of Chemical & Engineering Data, 2008. **53**(1): p. 2-4.
9. Gonzalez-Salazar, M.A., et al., *Comparison of Current and Advanced Post-Combustion CO2 Capture Technologies for Power Plant Applications.* Energy Procedia, 2012. **23**: p. 3-14.
10. Ekström, C., et al., *Techno-Economic Evaluations and Benchmarking of Pre-combustion CO2 Capture and Oxy-fuel Processes Developed in the European ENCAP Project.* Energy Procedia, 2009. **1**(1): p. 4233-4240.
11. Hedin, N., et al., *Adsorbents for the post-combustion capture of CO2 using rapid temperature swing or vacuum swing adsorption.* Applied Energy, 2013. **104**(0): p. 418-433.
12. Ullah, R., et al., *Insights of CO2 adsorption performance of amine impregnated mesoporous silica (SBA-15) at wide range pressure and temperature conditions.* International Journal of Greenhouse Gas Control, 2015. **43**: p. 22-32.
13. McKeown, N.B. and P.M. Budd, *Polymers of intrinsic microporosity (PIMs): organic materials for membrane separations, heterogeneous catalysis and hydrogen storage.* Chemical Society Reviews, 2006. **35**(8): p. 675-683.
14. Modak, A. and A. Bhaumik, *Porous carbon derived via KOH activation of a hypercrosslinked porous organic polymer for efficient CO2, CH4, H2 adsorptions and high CO2/N2 selectivity.* Journal of Solid State Chemistry, 2015. **232**: p. 157-162.
15. Plaza, M.G., et al., *CO2 capture by adsorption with nitrogen enriched carbons.* Fuel, 2007. **86**(14): p. 2204-2212.
16. Deniz, E., et al., *A combined computational and experimental study of high pressure and supercritical CO2 adsorption on Basolite MOFs.* Microporous and Mesoporous Materials, 2013. **175**: p. 34-42.
17. Belmabkhout, Y., V. Guillerm, and M. Eddaoudi, *Low concentration CO2 capture using physical adsorbents: Are metal–organic frameworks becoming the new benchmark materials?* Chemical Engineering Journal, 2016. **296**: p. 386-397.
18. Veneman, R., et al., *Adsorption of CO2 and H2O on supported amine sorbents.* Energy Procedia, 2014. **63**(0): p. 2336-2345.
19. Zulfiqar, S., M.I. Sarwar, and C.T. Yavuz, *Melamine based porous organic amide polymers for CO2 capture.* RSC Advances, 2014. **4**(94): p. 52263-52269.

20. Xiang, Z., et al., *Covalent-organic polymers for carbon dioxide capture.* Journal of Materials Chemistry, 2012. **22**(42): p. 22663-22669.
21. Mason, C.R., et al., *Enhancement of CO2 Affinity in a Polymer of Intrinsic Microporosity by Amine Modification.* Macromolecules, 2014. **47**(3): p. 1021-1029.
22. McKeown, N.B. and P.M. Budd, *Exploitation of Intrinsic Microporosity in Polymer-Based Materials.* Macromolecules, 2010. **43**(12): p. 5163-5176.
23. Ullah, R., et al., *Synthesis, characterization and evaluation of porous polybenzimidazole materials for CO2 adsorption at high pressures.* Adsorption, 2016: p. 1-14.
24. Patel, H.A., et al., *Highly Stable Nanoporous Sulfur-Bridged Covalent Organic Polymers for Carbon Dioxide Removal.* Advanced Functional Materials, 2013. **23**(18): p. 2270-2276.
25. Patel, H.A., et al., *High capacity carbon dioxide adsorption by inexpensive covalent organic polymers.* Journal of Materials Chemistry, 2012. **22**(17): p. 8431-8437.

466434 Carbonic Anhydrase Enhanced Carbon Capture: Kinetic Measurements and Pilot Plant Trials

Abstract:

Carbonic Anhydrase Enhanced Carbon Capture: Kinetic Measurements and Pilot Plant Trials

Arne Gladis[1*], *Maria T Gundersen*[2], *Philip L Fosbøl*[1], *John M Woodley*[2], *Nicolas von Solms*[1]

1) Center for Energy Resources Engineering, Department of Chemical and Biochemical Engineering, Technical University Denmark, Kgs. Lynby 2800

2) CAPEC-PROCESS Center, Department of Chemical and Biochemical Engineering, Technical University Denmark, Kgs. Lyngby 2800

Keywords:

Wetted wall column, carbonic anhydrase, carbon capture, enzyme kinetics, amine, carbonate, absorption, carbon dioxide

Abstract:

In this study the effect of carbonic anhydrase addition on the absorption of CO_2 was investigated in a wetted wall column apparatus. Four different solvents: MEA (a primary amine), AMP (a sterically hindered primary amine), MDEA (a tertiary amine) and K_2CO_3 a carbonate salt solution were tested in concentrations from 5 to 50 wt%. Necessary mass transfer parameters such as liquid side mass transfer coefficient and solvent and enzyme reaction rates

were determined in a temperature range from 298 to 328 K and benchmarked to a 30 wt% MEA solution.

The study reveals that the addition of the enzyme carbonic anhydrase (CA) dramatically increases the liquid side mass transfer coefficient for 30 wt% MDEA and 15 wt% K_2CO_3. 30 wt% AMP has a moderate increase whereas 30 wt% MEA was unchanged. The results confirm that bicarbonate forming solvent which do not produce carbamate benefit from CA. The results reveal the impact of temperature in relation to CA. A temperature increase resulted in lower liquid side mass transfer rate for 30 wt% MDEA and 15 wt% K_2CO_3 but in higher rate for 30 wt% AMP. The overall first order enzyme reaction rate (s^{-1}) was linearly dependent on enzyme concentration for 30 wt% MDEA and 15 wt% K_2CO_3 at 313 K. The derived enzymatic reaction rate constant k_{enz} ($m^3\ kg^{-1}\ s^{-1}$) for 15 wt% K_2CO_3 at 313 K was about 9 times higher than for 30 wt% MDEA and 10 times higher than for 30 wt% AMP. Temperature and concentration did not observably influence the enzymatic rate constant in the concentration range of 5 to 15 wt% K_2CO_3. The higher solvent concentration only led to a slightly higher reaction rate. A solution with 20 wt% K_2CO_3 had almost 3 times higher enzyme reaction rate compared to 15 wt% at 298 K and increased with temperature to almost 5 times faster at 328 K. The enzymatic reaction rate for MDEA decreased with both temperature and solvent concentration from 15 to 30 wt%. An increase to 50 wt% resulted in a decrease in reaction rate due to less water present.

Pilot plant campaigns were carried out for different solvents and conditions and the results were successfully modelled using intrinsic data obtained from the wetted-wall column experiments

470117 Production of Terephthalate Esters Via Diels-Alder Cycloaddition-Dehydration Reactions of 1,2-Furandicarboxylate Esters and Ethylene

Abstract:

Diels-Alder cycloaddition-dehydration (DA) reactions of dienophiles and biomass-derived furans are promising routes to renewable aromatics. Catalysis of such reactions with ethylene as the dienophile has been the topic of numerous experimental and computational studies. DA reactions of furans with reduced side groups (e.g., dimethyl furan) have been shown to be catalyzed with high selectivity by microporous materials with Brønsted or Lewis acid sites.[1] However, furans functionalized with oxygenated side groups (e.g., partially oxidized products of 5-hydroxymethyl furfural) were shown to be more prone to side-reactions.[2,3] For these reations, Lewis acidic Sn- or Zr-containing molecular sieves with the zeolite beta topology were found to be more selective catalysts than the Brønsted acidic Al-Beta. Zr-Beta yielded the highest selectivities (ca. 80%), with methyl (5-methoxymethyl) furan-2-carboxylate (MMFC) as the diene. However, the further oxidized 1,2-furandicarboxylic acid (FDCA), or esters thereof, were shown to be unreactive with such catalysts.

Recently, we have reported on the synthesis and characterization of Zn Lewis acid sites in CIT-6 (Zn-Beta).[4Â] We showed that, in addition to enabling DA reactions of MMFC, CIT-6 is able to react the dimethyl ester of FDCA (DMFDC). However, in the latter case, catalysis of decarboxylation reactions of the furan compromised selectivity towards the diester aromatic products, with the highest selectivity of 36%.

Here, we report a set of microporous catalysts for such DA reactions that do not appear to promote decarboxylation of DMFDC, and the optimization of reaction conditions that enables terephthalate diester selectivities upwards of 88%.

References:

(1) Chang, C.-C.; Je Cho, H.; Yu, J.; Gorte, R. J.; Gulbinski, J.; Dauenhauer, P.; Fan, W. *Green Chem.* **2015**.

(2) Pacheco, J. J.; Davis, M. E. *Proc. Natl. Acad. Sci. U. S. A.* **2014**, *111*(23), 8363â€"8367.

(3) Pacheco, J. J. New Catalysts for the Renewable Production of Monomers for Bioplastics, PhD Dissertation, California Institute of Technology, **2015**.

(4) Orazov, M.; Davis, M. E. *Chem. Sci.* **2016**, *7*(3), 2264â€"2274.

*Corresponding author. E-mail: mdavis@cheme.caltech.edu

458119 Quantifying Lewis Acid Sites in Zeolites That Catalyze Glucose Isomerization

Abstract:

Quantifying Lewis Acid Sites in Zeolites that Catalyze Glucose Isomerization

James W. Harris, Michael J. Cordon, Juan C. Vega-Vila, Fabio H. Ribeiro, Rajamani Gounder*

School of Chemical Engineering, Purdue University, 480 Stadium Mall Drive, West Lafayette, IN 47907

*rgounder@purdue.edu

Pure-silica molecular sieves with tetravalent framework heteroatoms (M = Hf^{4+}, Sn^{4+}, Ti^{4+}, Zr^{4+}) catalyze intermolecular and intramolecular Meerwein-Ponndorf-Verley alcohol oxidation and Oppenauer aldehyde reduction (MPVO) cycles, which mediate stereoselective glucose isomerizations. Precise assessment of the reactivity of Lewis acid zeolites of different provenance, which invariably contain different densities of silanol defects and of open ((HO)-M-$(OSi)_3$) and closed (M-$(OSi)_4$) sites, require methods to accurately quantify these structures. Integrated molar extinction coefficients were determined for infrared (IR) peaks of deuterated acetonitrile (CD_3CN, 303 K) bound to open (2316 cm^{-1}) and closed (2308 cm^{-1}) Sn sites, and of pyridine (423 K) bound to Lewis acidic Sn sites (1450 cm^{-1}, 1610 cm^{-1}) in Sn-Beta zeolites [1]. Identical numbers of Lewis acidic Sn sites were titrated by CD_3CN and pyridine among Sn-Beta zeolites (>10 samples) of varying synthetic origin and composition, and by ammonia and n-propylamine only among samples with high Sn content (Si/Sn ≤ 150) and low-defect density. Pyridine and CD_3CN did not distinguish isolated Ti sites of different coordination in IR spectra. We will discuss our latest results using

carbon monoxide (CO, 100 K) to distinguish open and closed Sn and Ti sites in IR spectra, by extending similar reports on Zr-Beta [2].

The number of open Sn sites counted *ex situ* (CD_3CN IR) agreed quantitatively with the number of active sites in Sn-Beta zeolites that catalyze aqueous-phase glucose-fructose isomerization (373 K, 1% w/w glucose) at *initial* reaction times [1], consistent with experimental [3] and theoretical [4] evidence that open Sn sites are the dominant active sites for this reaction. Isomerization rate constants were measured on Sn-Beta and Ti-Beta in kinetic regimes that are first-order or zero-order in glucose concentration, and reflect free energy differences between the same kinetically-relevant 1,2-hydride shift isomerization transition state relative to two bound water molecules or to one bound glucose molecule at open Sn sites, respectively. Rate constants in both regimes (per active M, 373 K) were ~10-50x higher on hydrothermally-synthesized, low-defect M-Beta-F zeolites (Si/M = 100-220) than on post-synthetically prepared, high-defect M-Beta-OH zeolites (Si/M = 30-580), suggesting that isomerization rate enhancements within hydrophobic voids are caused by preferential stabilization of 1,2-hydride shift transition states. The number of silanol defects quantified from CD_3CN IR spectra (2275 cm^{-1}) correlated linearly with the molar water uptake determined from adsorption isotherms (293 K, $P/P_0 = 0.2$) for low-defect M-Beta-F samples, providing a quantifiable descriptor of hydrophobicity. These findings provide guidance to prepare more reactive M-Beta zeolites by maximizing the number of open sites while minimizing residual silanol defects.

[1] J. W. Harris et al. J. Catal., 335 (2016) 141-154.

[2] V. L. Sushkevich et al., J. Phys. Chem. C, 119 (2015) 17633-17639

[3] R. Bermejo-Deval et al. ACS Catal., 4 (2014) 2288-2297.

[4] N. Rai et al. ACS Catal., 3 (2013) 2294-2298; Y.P. Li et al. ACS Catal., 4 (2014) 1537-1545.

471562 Designing Cooperative Interactions to Tune Catalytic Activity and Selectivity for Biomass Conversion

Abstract:

Catalytic sites incorporating multiple catalytic units can act cooperatively to achieve higher rates than isolated units. It remains an open question how to design these cooperative interactions. Specific design concepts can be gleaned from enzymes that contain multiple functional groups positioned in close proximity to provide tuned interactions with the substrates. The tuned interactions and precise spatial positioning enable the enzymes to achieve high catalytic rates and high selectivity. Understanding how to incorporate these design concepts into heterogeneous materials would improve catalyst performance.

Previously, we and others have focused on understanding the different aspects of how to tune interactions to achieve cooperative catalysis.[1â€"5]Â This work has primarily been for pharmaceutically relevant reactions such as the aldol, nitroaldol, and Knoevenagel reactions. In this presentation, we extend our synthetic approach to control catalytic activity and selectivity for directly relevant biomass production reactions that produce 5-hydroxymethylfurfural. We exploit different elements of cooperative interactions to create a novel design of the catalytic active site. These novel materials are synthesized, characterized, and tested in catalytic reactions. In particular, we utilize advanced spectroscopic techniques to demonstrate formation of the desirable catalytic site. The resulting materials achieve increased catalytic selectivity and activity for these reactions relative to current state-of-the-art materials. Overall, the increased yields and selectivities increase the potential to produce many important commodity chemicals in a sustainable manner.

References

[1] N.A. Brunelli, C.W. Jones, J. Catal. 308 (2013) 60.

[2] N.A. Brunelli, K. Venkatasubbaiah, C.W. Jones, Chem. Mater. 24 (2012) 2433.

[3] N.A. Brunelli, S.A. Didas, K. Venkatasubbaiah, C.W. Jones, J. Am. Chem. Soc. 134 (2012) 13950.

[4] R.K. Zeidan, M.E. Davis, J. Catal. 247 (2007) 379.

[5] J.D. Bass, A. Solovyov, A.J. Pascall, A. Katz, J. Am. Chem. Soc. 128 (2006) 3737.

471727 Lewis Acid Zeolite Catalysts for Tandem Diels-Alder Cycloaddition and Dehydration of Dimethylfuran and Ethylene to Renewable p-Xylene

Abstract:

Recently, increasing demand for energy and commodity chemicals has led to accelerated research efforts in the conversion of renewable resource into chemicals and fuels for a sustainable economy. The processing of lignocellulosic biomass, an inexpensive, abundant and sustainable source of carbon, offers the promise of sustainable chemicals and carbon-neutral liquid transportation fuels. Lewis acid catalysts play an important role for selectively activating functional groups of the organic molecules involved in biomass conversion.

In this study,[1] Lewis acid zeolites including Zr-, Sn-, and Ti-BEA were examined for tandem [4 + 2] Dielsâ€"Alder cycloaddition of 2,5-dimethylfuran (DMF) and ethylene to form cycloadduct, which is further dehydrated to produce biorenewable *p*-xylene. *p*-Xylene is currently produced from petroleum processing, which serves as a precursor for polyethylene terephthalate (PET) polymers (e.g. PET bottle). Zr-BEA (Si/Zr = 168) offered superior performance with improved recalcitrance to deactivation, because of its low activity for the DMF oligomerization.[2] Zr-BEA also achieved the highest selectivity to *p*-xylene of 80% at 99% conversion of DMF. In the case of low catalyst loading, the reaction rate of *p*-xylene production was linearly proportional to the number of Lewis acid sites, whereas high catalyst loading was independent of Lewis acid sites. These findings were similar to the behaviors of Brønsted acid zeolite catalysts.

References

1. C.-C. Chang, H. J. Cho, J. Yu, R. J. Gorte, J. Gulbinski, P. Dauenhauer and W. Fan, *Green Chem.*, **2016**, 18,

   1368

2. J. Yu, S. Zhu, P. Dauenhauer, H. J. Cho, W. Fan and R. J. Gorte, *Catal. Sci. Technol.*, **2016**, **DOI:** 10.1039/C6CY00501B

459019 Comparative Studies of Grafted Fe(III) and Ti(IV) Active Sites on Silaceous Supports: Crystalline Delaminated Zeolite Vs. Amorphous Silica

Abstract:

Silica has been widely used as a catalyst support with tunable surface area, porosity, and microscopic structure[1], making it ideal for supporting metal catalysts with high dispersion and high active site loading, while avoiding internal diffusion limitations. In contrast, zeotype materials have limited use as catalyst supports in many applications, since the majority of their surface area resides in micropores, making it inaccessible to precursors necessary for dispersing metal catalysts and to reactants and products relevant to their catalytic applications. Delamination of zeolite precursors can significantly improve accessible external surface area of zeotype materials and provides a strategy for their use as a support for dispersed metal catalysts[2]. In particular, removal of framework boron or aluminum and substitution for a different catalytically active cation has been used as a successful approach for synthesizing heterogeneous catalysts which, in contrast to amorphous silica supported metals, contain active sites embedded into well defined, crystalline, zeolitic framework positions[3].

The consequences of altering the coordination environment of a grafted cation from an amorphous support such as silica, to a crystalline framework site with well-defined structure, such as that of a zeolite, remain poorly understood. In this work, two case studies will be explored, involving Fe(III) and Ti(IV) cations grafted onto zeolite framework positions, which are compared to their grafting on amorphous silica. While grafting on amorphous silica results in cations covalently attached onto the support surface, grafting on delaminated zeolite supports results in cations that are incorporated into the crystalline framework. This incorporation is achieved by removal of framework B from delaminated zeolite supports, generating vacant tetrahedral framework positions – silanol nests. These vacancies are then repopulated with other cations, resulting in heteroatoms that are incorporated into the crystallographic position previously occupied by B, a

coordination environment different from that of amorphous silica.

During catalyst synthesis, our data demonstrate significantly higher uptake of aqueous-phase Fe(III) precursors by the delaminated zeolite framework relative to amorphous silica. DR-UV spectra demonstrate the synthesis of two different types of highly dispersed Fe(III) sites in both of these catalysts, which are both active for liquid-phase oxidation of adamantane by H_2O_2. However, after reaction, zeolitic Fe(III) active sites remain nearly unchanged, whereas Fe(III) sites on amorphous silica aggregate, as evidenced by DR-UV spectroscopy. These data support the hypothesis that zeolitic Fe(III) sites are more robust compared to their amorphous silica counterparts, presumably because grafted Fe(III) is stabilized when embedded into the zeolite framework, as opposed to simply grafted onto the silica surface.

We also investigated isolated Ti(IV) cations grafted on the same two supports, as relevant models of industrial catalysts for the epoxidation of olefins with organic hydroperoxides. In order to further facilitate a direct comparison between the two supports, we defined the coordination environment from the top of the Ti(IV) cation with a calix[4]arene macrocycle and from the bottom with either a crystalline zeolitic or amorphous silica support. We investigate both catalysts for the epoxidation of demanding terminal olefins with *tert*-btuyl hydroperoxide as oxidant, and compare with calcined variants. Regardless of whether there is a calix[4]arene ligand present or not, our results demonstrate significant catalytic consequences that favor the zeolitic support over the amorphous silica support in terms of rate. This in turn suggests that the degree of connectivity to the support is less important in determining rate than the actual support type.

Altogether, these two examples provide insights into the importance of the support crystallinity, which dictates the molecular coordination environment of a grafted metal cations, in establishing the catalytic properties of atomically dispersed supported catalyst.

(1) Soled, S. *Science* **2015**, *350* (6265), 1171–1172.

(2) Corma, A. *J. Catal.* **2003**, *216* (1-2), 298–312.

(3) Ouyang, X.; Hwang, S.-J.; Xie, D.; Rea, T.; Zones, S. I.; Katz, A. *ACS Catal.* **2015**, *5* (5), 3108–3119.

456728 A Liquid Phase Spectroscopic Study of the Effect of Liquid Water on Acid Sites of NaY

Abstract:

Liquid phase heterogeneously catalyzed reactions are ubiquitous, especially in biomass upgrade processes. However, establishing structure-activity relations for liquid phase reactions is challenging because the effect of solvent is poorly understood. Spectroscopy of adsorbed probe molecules like pyridine have been widely used to characterize catalyst acidity, and these experiments have almost exclusively been conducted in vacuum. This motivates a need for liquid phase probe molecule characterization, both to more accurately reflect reaction conditions, and to study the effect of solvent. Attenuated Total Reflection Fourier Transform Infrared spectroscopy (ATR-FTIR) was used to detect pyridine adsorbed on sodium faujasite (NaY) in water. The slow accumulation of pyridinium on NaY in water was observed in the ATR cell, suggesting water promotes Brønsted acidity on NaY. The accumulation of protonated pyridine in NaY pores led to the question of whether NaY can support Brønsted acid catalyzed pathways in protic solvent. The Brønsted catalyzed ring opening of 2,5-dimethylfuran (DMF) to 2,5-hexanedione is not active on NaY in water, which is attributed to the lack of basicity of DMF. A weaker base than pyridine, deuterated acetonitrile, adsorbed on NaY in water demonstrated no Brønsted character.

460384 Gasification Reactivity of Model Refuse Derived Fuel (RDF) Char and Its Components in CO_2 and Steam

Abstract:

The discarded portion of Municipal Solid Waste (MSW) is subjected to pre-processing steps (e.g., glass and metal separation) to produce Refuse Derived Fuel (RDF). For this study, a model RDF composition was established based on the EPA data and pyrolysis results of individual components in different categories, e.g., plastics, paper and food waste. Pyrolysis studies were carried out in a thermogravimetric analyzer (TGA) at atmospheric pressure and at temperatures between 25 â€" 800 Â°C with a heating rate of 20 K/min. Gasification of pyrolyzed feed was conducted at 800 Â°C using pure CO_2 and steam as the gasification reagents.

Inorganic materials inherently present in the feed have an effect on the gasification reactivity. Alkali and alkaline earth metals were catalytically beneficial, while other inorganics, such as silicon and phosphorous, had an inhibiting effect. The three major inorganics affecting the gasification reactivity in the model RDF feed were potassium from orange peel, calcium from paper components and silicon from paperboard and rubber. Gasification behavior of the model RDF char in CO_2 showed a synergistic effect, which is attributed to the redistribution of inorganic species during the gasification process. Potassium present in the orange peel char catalyzes the gasification of other RDF char components due to its mobility at high temperatures. This observation was verified by comparing the sum of individual gasification reactivities with mixed RDF char composite reactivity. Additionally, EDX of the chars confirmed redistribution of potassium during gasification.

The effect of the inorganics was also evident when steam was used as the gasifying agent. However, the transient reactivity profiles in steam were different as compared to those during CO_2 gasification. Tissue paper (calcium-rich), orange peel (potassium-rich), rubber (silicon-rich) and model RDF were gasified in steam and the reactivity trends were compared to those observed during CO_2 gasification. Both the type and amount of inorganics present had an effect on the gasification reactivity in different gasifying mediums. In addition to the effect of the inorganics, steam gasification reactivity profiles were affected by hydrogen inhibition. Various RDF components rich in different inorganic species thus exhibited varying reactivity trends during steam gasification.

460629 Catalytic Hydrotreatment of Pyrolysis Oil Model Compounds in a Batch Reactor

Abstract:

Bio-oil generated from the pyrolysis of lignocellulosic feedstocks is a complex mixture of highly oxygenated hydrocarbons (>30wt% oxygen), and is poorly miscible with conventional fossil fuels. Therefore, catalytic hydrodeoxygenation (HDO) is necessary to lower its oxygen content and to produce a compatible â€œdrop-inâ€ transportation fuel. We conducted batch HDO reactor studies using representative phenolic compounds found in pyrolysis oil including anisole, m-cresol and phenol, both individually and in mixtures. These tests were conducted at pressures (50 bar) and temperatures (250 to 350°C) close to industrial conditions to obtain high conversion of the initial reactants.Â Experiments were conducted with finely powdered catalysts in a well-mixed Parr reactor to reduce mass transfer effects, and samples were collected regularly until completion. Instead of traditional sulfided HDO catalysts, the studies employed platinum on alumina (Pt/Al_2O_3) and palladium over carbon (Pd/C). Utilizing reaction mechanisms and pathways proposed in our previous studies, the data from the batch reactor experiments was used to fit the kinetic rate constants and Arrhenius parameters for the primary reactants and intermediate compounds. The analytical results show that many of the reactants proceeded through similar intermediates; therefore, a hybrid, lumped-parameter model was created and compared to the data. Higher temperatures led to shorter intermediate lifetimes, and more secondary reactions. While platinum on alumina was primarily focus, tests run with palladium on carbon exhibited less rearrangement of methyl groups. The reaction model developed may then be used to predict product yields, hydrogen consumption, energy requirements, and greenhouse gas (GHG) emissions. This capability is necessary for a more accurate Life Cycle Assessment (LCA) of this biofuel route, and to better assess the economic viability and sustainability of the process. Catalyst coking at longer test times was an issue, and thermal desorption tests followed by mass spectroscopy were performed to determine the rate and extent of the coking process.

467202 Supercritical Fluid Processing of Biomass-Derived Pyrolysis Oil

Abstract:

Liquid hydrocarbon fuels are essential to our current modes of transportation due to their high energy density. However, new technologies are needed to satisfy the demand for these fuels using sustainable resources. Biofuels are an attractive renewable substitute for petroleum-based fuels. The challenge is that woody biomass, although abundant enough to supply a meaningful fraction of the nationâ€™s fuel, is recalcitrant to many chemical processes. Fast pyrolysis to convert biomass into a crude bio-oil has gained recent attention as the basis of a distributed biofuels production capability. Despite promise, the disadvantages of fast pyrolysis are that the resulting bio-oil has low energy density, high water content, acidic pH, and poor storage stability. Vigorous research efforts have been mounted in the past few years to upgrade and stabilize pyrolysis bio-oil. Unfortunately, bio-oil upgrading has been plagued by unacceptable hydrogen consumption, catalyst failure, and inability to process whole bio-oil.

A particularly attractive scenario is to efficiently separate out multiple fractions of bio-oil so that each portion can be treated with a process specifically tailored to the functionalities present in that phase. Bio-oil can be selectively condensed to create multiple fractions, or separated during post-processing. Efficient extractive post-processing is an appealing approach due to the potential for tailoring an extraction for specific components, and facilitating in-situ upgrading where hydrotreating/cracking can be performed during the extraction. Supercritical fluid processing is especially promising for this application due to the ease of solvent recovery, good solvency for organic chemicals and the flexibility of tailoring the solvent for desired chemical fractions. Results of supercritical fluid processing of biomass-derived pyrolysis oil will be presented. This includes characterization of the fractions obtained during various processing conditions as well as conversion strategies tailored to these specific fractions.

462128 Commodity Chemicals from Lignocellulosic Biomass: Conversion of Furfural into 1,5-Pentanediol

Abstract:

Lignocellulosic biomass is a low cost, abundant, and renewable feedstock that can be used to produce commodity chemicals. α,ω-diols like 1,4-butanediol (BDO) and 1,6-hexanediol (HDO), are commodity chemicals produced at high volume (annual market of 2,500,000 ton and 138,000 ton, respectively) and high value ($2,200/ton and $4,400/ton, respectively). In this presentation we will discuss the production of 1,5-pentanediol (PDO) from furfural.

Furfural, is one of the most important platform chemicals produced from biomass. It is commercially produced from the hemicellulose portion of biomass by hydration to xylose and subsequent dehydration (Xing et al., 2011). Furfural can then undergo hydrodeoxygenation to produce PDO with tetrahydrofurfuryl alcohol (THFA) as a precursor. The furfural is first hydrogenated to THFA (Nakagawa and Tomishige, 2010; Nakagawa et al., 2012). The resulting THFA can then be converted into PDO by a Rh-ReOx/C catalyst at high selectivity (above 97.2% selectivity). However, the bottlenecks for this process are: (1) the production of PDO at low conversion; (2) the high catalyst cost resulting from using an expensive noble bimetallic RhRe catalyst; and (3) the high capital cost of reaction and separation systems resulting from low feedstock loading in the liquid phase.

In this work, we propose a novel multi-step catalytic process to convert furfural into PDO at high feedstock loading (up to 50 wt% in water) using non-precious metal catalysts. We also design a separation subsystem for the recovery of the final PDO product from process water. We have developed an experimentally based process simulation model (mass and energy balances) to determine the economic potential of PDO production. It is shown that this strategy leads to much lower catalyst costs and a higher overall PDO yield (over 80%) comparing to one-step THFA hydrogenolysis (Chia et al., 2011) using a noble metal catalyst. If we consider a plant processing 135 ton per day furfural (equivalent to 1,000 dry metric tonne per day white birch) feedstock and producing 38,520 ton per year PDO, the integrated strategy leads to a production cost of $1,600 per ton of PDO. We estimate that this cost is over 35% lower than the cost of production of PDO via a noble metal catalyst. Finally, we perform sensitivity analysis to identify the major economic drivers of the process and suggest future research directions.

References

1. Chia, M., Pagán-Torres, Y. J., Hibbitts, D., Tan, Q., Pham, H. N., Datye, A. K., Neurock, M., Davis, R. J., Dumesic, J. A. (2011). Selective Hydrogenolysis of Polyols and Cyclic Ethers over Bifunctional Surface Sites on Rhodium–Rhenium Catalysts. J. Am. Chem. Soc., 133 (32), 12675.
2. Nakagawa, Y., Nakazawa, H., Watanabe, H., Tomishige, K. (2012), Total Hydrogenation of Furfural over a Silica-Supported Nickel Catalyst Prepared by the Reduction of a Nickel Nitrate Precursor. ChemCatChem, 4: 1791–1797.
3. Nakagawa, Y., Tomishige, K. (2010). Total hydrogenation of furan derivatives over silica-supported Ni–Pd alloy catalyst. Catalysis Communications, 12 (3), 154-156.
4. Xing R., Qi W., Huber G. W., (2011). Production of furfural and carboxylic acids from waste aqueous hemicellulose solutions from the pulp and paper and cellulosic ethanol industries, Energy Environ. Sci., 4, 2193.

466868 Cyclization and Dehydration of Aldaric Acids to 2,5-Furandicarboxylic Acid

Abstract:

Aldaric acids (2,3,4,5-tetrahydroxy-1,6-hexanedioic acids) are formed by catalytic oxidation of glucose or by oxidation of uronic acids, the monomers of pectin and alginate biopolymers. Aldaric acids are emerging as important biorenewable intermediates for use in consumer goods and as platforms for further conversion. The direct cyclization and dehydration of glucaric acid and galactaric acid and their esters to form 2,5-furandicarboxylic acid (FDCA) has been carried out. The production of FDCA is attracting a great deal of attention, as it is a monomer of polyethylenefurandicarboxylate (PEF) polymers, which exhibit superior barrier properties to PETE packaging materials. Aldaric acid cyclization to FDCA takes place in an anhydrous environment in the presence of strong acids; we have found that under some conditions the presence of an ionic liquid solvent (1-butyl-3-methylimidazolium with various anions) enhances the yield of FDCA. An optimized set of reagent compositions and reaction conditions have been identified to give FDCA yields approaching 60% of theoretical from galactaric acid. Undesired reaction pathways include lactonization of the aldaric acids under anhydrous conditions, and total dehydration to form carbonaceous residues. A method to recover pure FDCA directly from the reaction mixture has been developed, along with recycling of the ionic liquid and acids. This new route to FDCA avoids the use of difficult-to-handle intermediates such as hydroxymethylfurfural (HMF), and appears promising given the recent advancements in commercialization of the aldaric acid precursors.

472905 Thermo-Catalytic Conversion of Biomass Derived Lipids to Fuels and Chemicals over HZSM-5

Abstract:

In this study, thermo-catalytic cracking of triglyceride was carried out in a continuous pyrolysis system with feed injected through an atomizer. This allowed introduction of micron-sized droplets of oil that could be rapidly vaporized inside the hot reactor. With this novel design, we were able to achieve short vapor residence times (optimum $Ï„_{vap}$ was 20s in order to achieve high liquid product yeilds and triglyceride conversion) without use of carrier gas, which would significantly reduce the overall cost of pyrolysis. Effects of reaction temperature ($450<T_{rxn}<500$ Â°C) and catalyst loading (HZSM-5; 0-5g) on conversion, product yields and composition were investigated. At T_{rxn} =500 Â°C and without catalyst, nearly quantitative conversion of feed was achieved and the yield of pyrolysis liquids was as high as 89% (relative to feed mass). Under these conditions, the identified products consisted of 37% hydrocarbons (C_5-C_{17}), 35% long-chain fatty acids (C_{16}-C_{18}, but primarily oleic acid) and 10% short-chain fatty acids (C_6-C_{12}). At T_{rxn} =500 °C using 5g of HZSM-5, the yield of liquid products was 65%. The identified liquid products contained 73% aromatics (C_6-C_{12}, but primarily benzene, toluene and xylene), 5% aliphatic and only 4% long-chain fatty acids. Overall, our results demonstrate the feasibility of producing liquid products at high yields, including a wide range of fuels (gasoline and jet) and enriched oleic acid (for oleochemicals production) with integration of conventional downstream separations, such as distillation.

469768 An Ontology Supported Integration Framework for Models and Data in Biorefining

Abstract:

Process modelling and simulation are vital tools to plan, evaluate, assess and develop different alternatives in the biorefinery design for production of value added chemicals, materials, fuels, and energy from biomass. However, modelling for biorefining processes poses a number of technical challenges due to the complexity of the characteristics of biomass feedstock, and associated processing technologies evolving multiple pathways, including thermochemical-, catalytic-, biochemical- and hybrid-conversion routes. Due to the complexity and variety of these processes, numerous software tools have been developed to obtain the best-of-breed software for each application area, to fulfil the requirements of particular engineering tasks. During the process of developing a new model, the best suited modelling tools for different parts of the process are employed. However, to understand process design as a continuous work process from an integrated perspective requires the use of tools from a diverse range of sources and disciplines simultaneously. With the current increasing number of Computer Aided Chemical Engineering (CAPE) tools, the emergence of interoperability between heterogeneous modelling tools and process simulators was demonstrated by Braunschweig et al. (2000), which has ultimately led to the CAPE-OPEN initiative.

The CAPE-OPEN has conceptualised and developed a standard to define a set of interface specification as a method pertaining to the interoperability of the mainstream flowsheet applications (Pons 2010). The interoperability is facilitated by a standard middleware service, which enables the use of a single set of common interface as a communication tool (Braunschweig et al. 2004; Morales-Rodríguez et al. 2008). Having the common interface increases the potential reusability of existing models without having to invest in the software development, which further reduce the developing time, efforts, and the need for expertise (van Baten & Pons 2014). However, to fully exploit the potential reusability of existing models, the task to become aware of and have access to all of those models from a variety of sources is a challenge, yet required in order to identify the most suitable model to achieve the best solution for a particular engineering problem. As a consequence, Yang et al. (2008) indicated the potential misuse of the models that can further lead to inefficient or even wrong solution to the problem.

This paper is intended to address the challenges that previous research, CAPE-OPEN framework has faced and hence to minimise user intervention in selecting a model for the purpose of model integration. To enable automated discovery process in order to reduce the reliance of user intervention, the semantic representation is used to describe the models and data in terms of their input and output parameters, preconditions, and functionality (Koo & Cecelja 2015). Additionally, the Semantic Web Services (SWS) are employed to create a semantic profile of models and data representing biorefining processes. To this end, the domain ontology assures unified description of models by guiding the model registration and hence instantiation of ontology through ontology parsing, whereas the SWS assures the process of model discovery, its invocation and consequently integration. To demonstrate proposed approach, the framework is built for the models and data residing in web-based semantic repository in the domain of biorefining. In the domain ontology, the models are characterised in terms of functionality, modelling methods, model granularities, required software environment, etc. and their input and output parameters are characterised in terms of type (i.e. material, utility, economic and emissions), compositions and associated parameters, etc. The classification and characterisation of parameters used in matching process are not limited to numerical values (i.e. value for the flow rate), but also includes descriptive (i.e. type of material, energy) or even composite (i.e. range of flow rate with minimum and maximum values) types. The options of the discovered models to the requesting model, the input-output (I/O) matching technique is employed to identify the models that matches to a certain degree of request, which permits for a high degree of flexibility by allowing a partial matching rather than a full (exact) matching between I/O parameters. The discovered models from matching process are then ranked according to a set of preferences and the best option that satisfies the matching criteria is proposed to the user. To verify the performance of model discovery and selection, models representing biochemical- and thermochemical conversion processes are used to demonstrate as a case study with an aim to evaluate the best suited model whilst avoiding wrong design solution.

References:

Braunschweig, B.L. et al., 2000. Process modeling : The promise of open software architectures. Chemical engineering progress, 96(9), pp.65–76.

Braunschweig, B. et al., 2004. CAPE web services: The COGents way. Computer Aided Chemical Engineering, 18(C), pp.1021–1026.

L. Koo, F. Cecelja, "Model Integration Using Ontology Input-Output Matching," Computer Aided Chemical Engineering, 2015, volume 37, pp. 2567-2572.

Morales-Rodríguez, R. et al., 2008. Use of CAPE-OPEN standards in the interoperability between modelling tools (MoT) and process simulators (Simulis® Thermodynamics and ProSimPlus). Chemical Engineering Research and Design, 86(7), pp.823–833.

Pons, M., 2010. How to make use of CAPE-OPEN? In 2010 AIChE Annual Meeting, 10AIChE.

van Baten, J. & Pons, M., 2014. CAPE-OPEN: Interoperability in Industrial Flowsheet Simulation Software. Chemie Ingenieur Technik, 86(7), pp.1052–1064. Available at: http://doi.wiley.com/10.1002/cite.201400009.

Yang, a. et al., 2008. A multi-agent system to facilitate component-based process modeling and design. Computers and Chemical Engineering, 32(10), pp.2290–2305.

466202 A Biorefinery Boutique Made in Switzerland

Abstract:

Many challenges pave the road from an oil-based to a renewable economy. In the chemical industry, a substantial amount of work needs to be undertaken for renewable-based processes to compete with their petrochemical counterparts. Nowadays, the biggest hurdle is to evaluate which biorefinery concepts are environmental friendly and economic competitive at an early stage. Therefore, an innovative exploration was undertaken to create a biorefinery boutique for Switzerland and Europe for different future industrial scenarios.

Traditionally, there are two approaches to find biorefinery concepts. The first is to use overly simplified models, which can be analyzed according to their production portfolio and evaluated based on their profitability. In the second, detailed process simulations are used, which give detailed technological insight into each technology, but cannot be connected with each other due to their high complexity. This work employs a novel technique which connects detailed model biorefinery technologies in a network and finds tailor made biorefinery concepts for multiple future business scenarios.

Besides state of the art bioprocessing technologies, the biorefinery boutique incorporates life cycle assessment to indicate environmental issues and an economic assessment to project future profitability.

460997 New Short-Cut Tools for Early-Stage Investment Evaluation of Biorefineries

Abstract:

New short-cut tools for early-stage investment evaluation of biorefineries

Biorefineries offer a promising alternative to fossil-based processing industries and have met rapid development during the last years. Biorefinery processes employ state-of-the-art technologies and thus, represent high risk business decisions. Researchers undertaking the development of a biorefinery, often have to estimate capital and manufacturing costs based on minimum information, in order to decide wisely on project continuity and justify further funding for their research. Most R&D engineers rely on literature information to estimate the costs as historical cost data are proprietary information and are seldom announced. They also draw on various costing techniques and heuristics, which were developed for the needs of the petrochemical and the Oil & Gas Industry and require high level of process detail, which is not available at the early-stages of process conception.

We have evaluated some of these methods with means of both deterministic and statistical comparison from reported literature estimates and commercial biorefineries' cost data. As the majority were published during the 1970's-80's and were derived from petrochemical processes, they do not account for the technological progress often met within the state-of-the-art biorefinery processes and report discrepancies in their results. Fig. 1 illustrates the aforementioned discrepancies for dry corn mill ethanol biorefineries. We have been collecting capital cost data for commercial corn-to-ethanol plants (marked as circles). Only three out of the six cost methods fall within the 90% Confidence Intervals of the best regression line (marked as triangles), while the capital estimation reported in literature reference greatly underestimates the plant's final cost (marked as x). Our work reports similar figures for chemical and thermochemical biorefineries, as well CITATION Tsa16 \l 2057 [1].

Therefore, we decided to attempt new capital and manufacturing cost estimation methods to meet the needs of non-experienced cost estimators which undertake the design of first-of-a-kind biorefineries: the methods require information available at the start of the process conception. The first capital cost estimation method proposes new investment factors that draw on historical investment costs along with probabilistic estimation of the total investment. The second capital cost estimation method proposes cost estimation relationships for chemical, biochemical and thermochemical biorefineries, as well as simple techniques to define the uncertainty of the estimate. Finally, we have developed a new rapid manufacturing cost estimation methodology with uncertainty: it relies on factors requiring minimal user input, i.e. raw materials cost calculation, for determining the net production cost of the process, employing probabilistic means of determination.

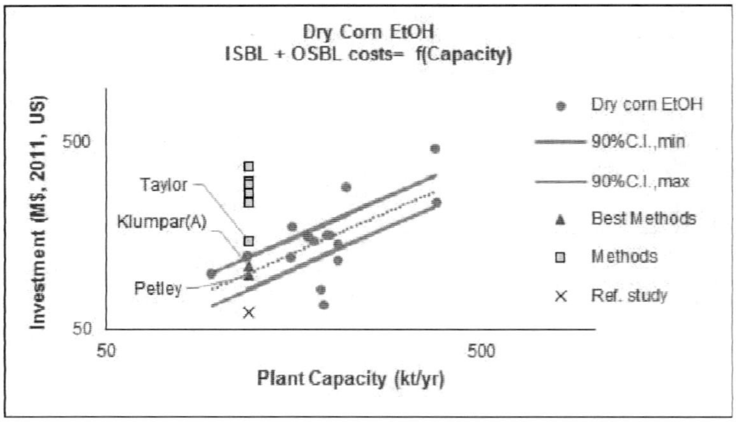

Fig. 1. Dry corn mill bioethanol production plants Investment Costs (M$, 2011,US) vs Plant Capacity (kt/yr)

Acknowledgements

The European Commission is gratefully acknowledged for funding the Renewable Systems Engineering project (RENESENG - FP7 Marie Curie project).

BIBLIOGRAPHY

[1] M. Tsagkari, J.-L. Couturier, A. Kokossis and J.-L. Dubois, "Early-stage capital cost estimation of biorefinery processes: a comparative study of heuristic techniques," *ChemSusChem*, 2016 *(under review)*.

466913 Global Sensitivity Analysis of Economic Assessment of Early Stage Process Design: The Case of the Glycerol Biorefinery

Abstract:

Nowadays it has become progressively important to achieve a sustainable process performance in order to keep a viable and competitive advantage in the international markets. Thus, the development of all-inclusive/holistic/comprehensive and systematic methods to accomplish this goal is the topic of this study. To this end, a multi-level framework for early-stage design and screening of process options is proposed for techno-economic and environmental sustainability analysis through risk assessment. The alternatives composing the design space are assessed by following the framework's steps, targeting to (i) quantify the economic risk; (ii) quantify the potential environmental risk under uncertainties by using monetary valuation of environmental impacts techniques; (iii) measure the concepts' eco-efficiency; and, finally (iv) facilitate the decision-making procedure by interpreting the trade-offs through a qualitative and quantitative sustainability risk assessment matrix. The framework application is highlighted through the critical assessment and screening of early stage glycerol-based biorefinery design concepts. It was found that the best potential options for glycerol valorisation is are obtained through the production of either: (i) lactic acid; (ii) succinic acid; or, (iii) 1,2-propanediol. To reduce the potential economic risk, a multi-product biorefinery is suggested, whose strategy is based on its capability of switching between the production of lactic acid and succinic acid.

The risk assessment tool described above provide valuable information on how uncertainties (technical versus external such as feed/product prices) impact the decision of optimal process concept and how to better mitigate these impacts by improving technology development. In this contribution, we take one step further and ask the question, a given a certain economic risk associated with a conceptual process alternative, what is the contribution of input uncertainties((e.g. feed composition, yield, feedstock prices,

product prices, etc)) to the risk? If risk is defined as variance (deviation from an expected value), what is the individual variance attributed to each input uncertainty? And how we can use this information to further minimize economic risk and improve technology advancement? To answer these questions, we use sensitivity analysis methods including local methods such as one-factor-at-a-time (OAT) as well as global methods such as variance decomposition to compute first order as well as total sensitivity indices.
We implemented algorithms in Matlab (R2015a) based on Monte Carlo sampling using Latin Hypercube Sampling with Iman Conover correlation control (Sin et al 2009) and sampling based on conditional probability of dependent variables proposed by Kucherenko et al 2012 to take into account the correlation structure of input uncertainties (especially price correlation matrix). The economic model used in this work is the discounted cash-flow rate of return (DCFR). The OAT sensitivity analysis has shown that deviations in the product's and feedstock prices, total production cost, fixed capital investment as well as discount rate, among others, have a high impact on the project's profitability (quantified with NPV and MSPA global sensitivity analysis of NPV indicated two major factors contributing to its variance (risk): uncertainty in fixed capital cost estimation (82%), and, uncertainty in product prices (14%). The NPV meta-model showed that improving the accuracy of fixed capital cost estimation (5% *std*) leads to a 95% reduction in the NPV's uncertainty. This global sensitivity analysis confirms the process engineering expectation that more detailed capital cost estimation supported through pilot scale studies or more rigorous model analysis is expected to decrease the economic risk and better judge the transferability of technology to the market. Global sensitivity analysis is an important and complementary tool to study and decompose impact of uncertainties to economic assessment of conceptual process design studies.

Keywords: techno-economic assessment, uncertainty analysis, glycerol-based biorefinery concepts, risk-based decision making, global sensitivity analysis.

References:

Kucherenko, S., Tarantola, S., & Annoni, P. (2012). Estimation of global sensitivity indices for models with dependent variables. *Computer Physics Communications*, *183*(4), 937-946.

Sin, G., Gernaey, K. V., & Lantz, A. E. (2009). Good modeling practice for PAT applications: Propagation of input uncertainty and sensitivity analysis. *Biotechnology progress*, *25*(4), 1043-1053.

Cheali, P., Gargalo, C., Gernaey, K. V., & Sin, G. (2015). A Framework for Sustainable Design of Algal Biorefineries: Economic Aspects and Life Cycle Analysis. In *Algal Biorefineries* (pp. 511-535). Springer International Publishing.

460678 General Bio-Separation Superstructure Optimization Framework

Abstract:

General Bio-Separation Superstructure Optimization Framework

Wenzhao Wu, Kirti Yenkie, Christos T. Maravelias*

Dept. of Chemical and Biological Engineering, University of Wisconsin-Madison, Madison, WI 53706

Recent advances in synthetic biology and metabolic engineering have enabled the production of a range of chemicals using engineered microorganisms[1,2]. However, despite the intriguing potential, the biological production of high-value chemicals is challenging because it is likely to have low titer and substantial amounts of excreted byproducts, and the purity specifications for high-value chemicals, as opposed to fuels, are rather strict, which means that separation costs are likely to represent a large fraction of the total production cost (more than 70% of total cost[3]). Thus, the efficient synthesis of bio-separation processes becomes a critical task. Although this synthesis problem has been studied for various chemicals in the past, these studies were mostly performed on a case-by-case basis. There has been limited research towards the development of systematic methods for bio-separations, applicable to all chemical targets. Accordingly, the goal of the present work is to develop a general bio-separation superstructure optimization framework, aiming to provide guidance on the preliminary synthesis of separation networks.

Based on general bio-separation principles and insights obtained from industrial separation processes for specific products, we first identify four separation stages: (1) cell treatment, (2) product phase isolation, (3) concentration and purification, and (4) final refinement. Specifically, cell treatment starts with the harvesting of cells (if the product is intracellular), which can be performed using solid-liquid separation technologies. Cell harvesting is followed by cell disruption to release the intracellular products. The first stage is skipped if the product is extracellular, and the separation starts from the second stage - product phase isolation, where the product-rich phase is isolated. In the concentration and purification stage, we remove large amount of water along with other impurities, by utilizing the differences between the product and the other components on volatility, molecular size, diffusivity, solubility in solvents, etc. In the final refinement stage, we further remove trace impurities and perform refined operations to satisfy special product specifications, such as colorlessness, complete dryness, and crystal form.

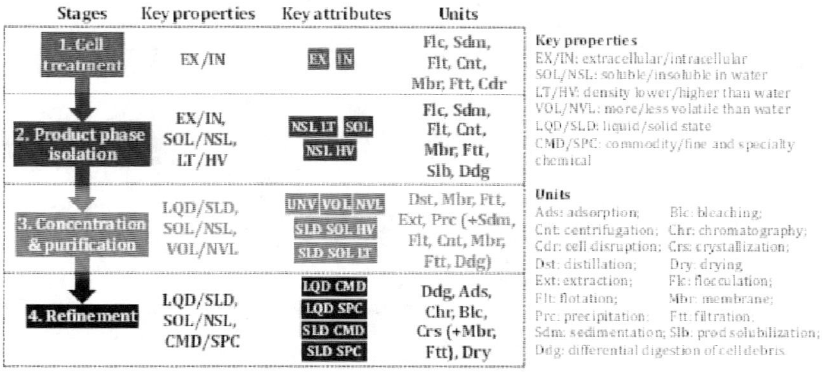

Figure 1. Stage-wise analysis of bio-separation processes, including key product properties, attributes, and units for each stage.

Next, we identify key product properties and key attributes that affect the separation processes in each stage, as shown in Figure 1. For example, in Stage 2, the different attributes imply different locations of the product (e.g., "NSL LT" indicates that the product is located at the top, while "SOL" indicates that the product is dissolved in water and evenly distributed in the system), and hence different separation processes. Major units in each stage are also identified.

Then for each stage, we systematically implement a set of connectivity rules[4] to develop attribute-specific superstructures. They are subsequently combined to generate a stage-specific superstructure. Finally, all the four stage-specific superstructures are integrated into a general superstructure (shown in Figure 2) that accounts for all types of bio-chemical products.

We further develop a superstructure reduction method to solve product-specific instances, based on product type, unit availability and suitability, case-specific considerations, and final product specifications. The reduced superstructure for an example case is shown in Figure 2, where an EX NSL LT NVL LQD CMD product (see abbreviations in Figure 1) is required to be completely colorless in its final product form, and all units in the general superstructure are available except for filtration.

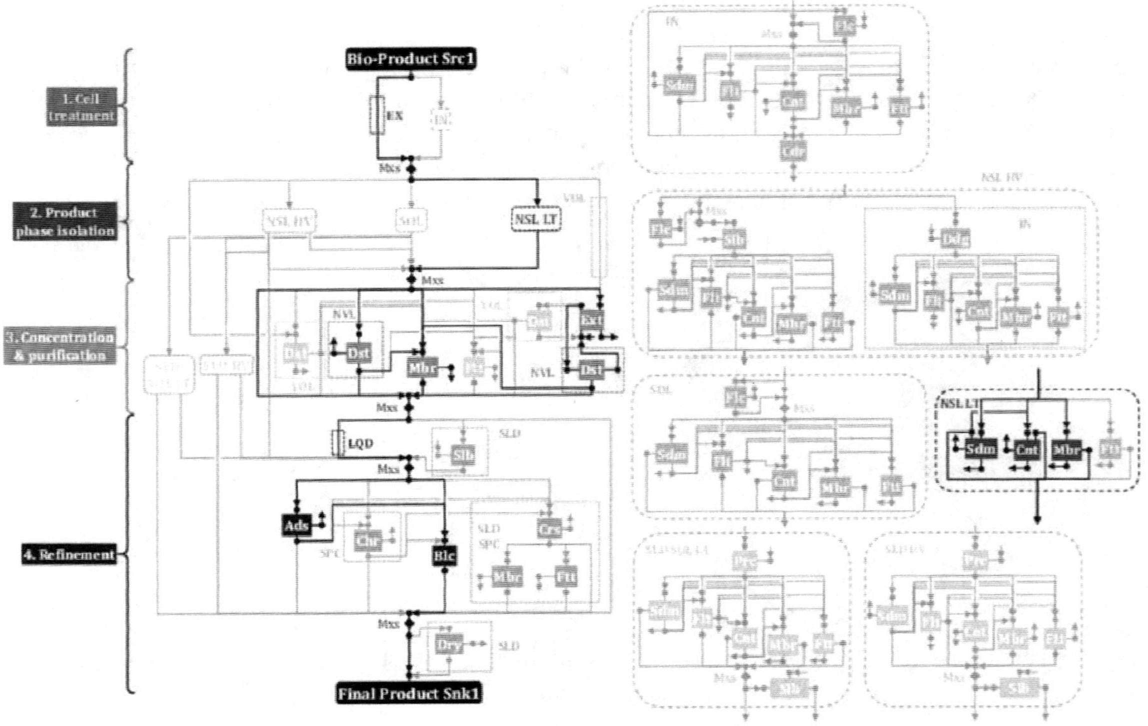

Figure 2. The general bio-separation superstructure (including the "dimmed" parts), and the reduced superstructure (excluding the dimmed parts) for an example case, where an EX NSL LT NVL LQD CMD product is required to be completely colorless in its final product form, and all units in the general superstructure are available except for filtration.

Finally, a general MINLP optimization model, including short-cut unit models (e.g., Fenske-Underwood equations for distillation unit models) for all types of separation units considered in the framework, is formulated. In the modeling of a specific reduced superstructure, only models of relevant units (which exist in the reduced superstructure) are included.

The proposed framework is applied to three case studies. Each case considers different set of product types, unit availability and suitability, case-specific considerations, and final product specifications. The common objective is to minimize total cost of purifying a dilute product stream (5 wt% concentration) to a product stream with at least 90 wt% purity.

References

[1] Gavrilescu, M. & Chisti, Y., 2005. Biotechnology- a sustainable alternative for chemical industry. *Biotechnology advances,* Volume 23, pp. 471-499.

[2] Bornscheuer, U. T. & Nielsen, A. T., 2015. Editorial overview: chemical biotechnology: interdisciplinary concepts for modern biotechnological production of biochemicals and biofuels. *Current Opinion in Biotechnology,* Volume 35, pp. 133-134.

[3] Kiss, A. A., Grievink, J. & Rito-Palomares, M., 2015. A systems engineering perspective on process integration in industrial biotechnology. *J Chem Technol Biotechnol,* Volume 90, pp. 349-355.

[4] Wu, W., Henao, C. & Maravelias, C., 2016. A Superstructure Representation, Generation and Modeling Framework for Chemical Process Synthesis. *AIChE J.*

459851 Technoeconomic Study of AB Synthesis for Biobutanol Production

Abstract:

Over the last 15 years there has been an important effort towards climate change mitigation. Among the different actions, one of the more challenging is reducing emissions in the transport sector. Ethanol and biodiesel, from different biomass sources, represent the typical alternatives to substitute crude based gasoline and diesel nowadays. However, there are other options such as FT fuels (Martín & Grossmann, 2011), glycerol ethers (Martín & Grossmann, 2014) or biobutanol. Biobutanol is considered an advanced fuel. It has similar properties to gasoline, it can be directly used in gasoline engines without modification, and it has higher energy density and boiling point than ethanol reducing evaporation losses. Moreover, it is less hydroscopic than ethanol (Durre, 2007).

The ethical issues involving first generation biofuels have led to focus on obtaining biofuels from agricultural residues. Corn being the major crop in the world, results in the fact that corn stover is widely available. However, in the case of biobutanol, the yield has been low due to the fact that together with biobutanol, ethanol and acetone are typically obtained. Thus, this work is divided in two parts: the first part is devoted to the experimental evaluation of the lignocellulosic biomass degradation. In the second one, a technoeconomic study of a new production process is described. The novelty of this work relies on the scale up of AB fermentations, that does not produce ethanol, (Qureshi et al., 1999). In addition, in-situ recovery techniques such as evaporation are also included at the process to make it economically competitive with conventional fuels and bioethanol.

In the first stage of the work, comprehensive evaluation of the experimental conditions to break down switchgrass and corn stover from Castille and Leon region (Spain) has been performed. The study considers both pretreatment and hydrolysis. With regards to pretreatment, several physio-chemical techniques such as acid (H2SO4), alkali (using KOH or NH4OH), ultrasounds processing, surfactants or organic solvents are

tested. With regard to hydrolysis, various enzyme cocktails were used. It turns out that corn stover reaches the highest yield, 677.00 mg of sugars by gram of biomass, when it is pretreated using a diluted sulfuric acid solution 0.5% v/v, autoclaving the mixture during 90 min at 121 ºC with no sonication stage. The hydrolysis method is set up at 50 ºC and 150 rpm with a solid loading of 2.5 % w/v and a ratio of enzymes/biomass of 400 and 440 µl/g of NS50013 (cellulose complex) and NS50010 (β-glucosidase cocktail) respectively. The sugar recovery obtained is competitive with current best available technologies i.e. (Sundaram and Muthykumarappan, 2016).

In a second stage, we synthesize a process for the production of biobutanol from corn stover including the use of novel technologies such as pervaporation to improve the yield from biomass and reducing the by-products by means of AB fermentation. The processes consists of four sections, pretreatment, hydrolysis, fermentation-pervaporation system and product purification. We use the experimental results to determine the yield of the pretreatment and hydrolysis. A hybrid simulation using EXCEL to develop black box kind of models for the experimentally validated units and ChemCAD has been performed to evaluate the technoeconomic feasibility of this process.

The results show an investment of 186 MM€ and a butanol production cost of 1.09 €/kg for a 27 kt/y plant. The yield to butanol is 0.21 kg/kg of biomass resulting, altogether, in a fuel closer to become competitive versus second generation bioethanol and fossil based fuels in terms of yield and cost.

References

Durre, Biobutanol: an attractive biofuel Biotech. J., 2 (2007), pp. 1525–15234

M.S. Madihah, A.B. Ariff, K.M. Sahaid, A.A. Suraini, M.I.A. Karim (2001) Direct fermentation of gelatinized sago starch to acetone–butanol–ethanol by Clostridium acetobutylicum. World J. Microbiol. Biotechnol. 17 (2001), pp. 567–576

Martín, M., Grossmann, I.E. (2011) Process optimization of FT- Diesel production from biomass. Accepted Ind. Eng. Chem Res, 50 (23),13485–13499

Martín, M.; Grossmann, I.E. (2014) Simultaneous dynamic optimization and heat integration for the co-production of diesel substitutes: Biodiesel (FAME & FAEE) and glycerol ethers from algae oil. Ind. Eng. Chem. Res. 53, 11371-11383

Qureshi, N., Hans P. Blaschek. Production of Acetone Butanol Ethanol (ABE) by a Hyper-Producing mutant strain of Clostridium beijerinckii BA101 and recovery by Pervaporation. Biotechnol.Prog. 1999, 15, 594-602.

Sundaram, V., Muthukumarappan. K., Influence of AFEXTM pretreated corn stover and switchgrass blending on the compaction characteristics and sugar yields of the pellets. Industrial Crops and Products 83 (2016) 537–544.

465757 Effect of Market and Technical Parameter Uncertainties on the Optimal Design of Integrated Biorefineries

Abstract:

Effect of Market and Technical Parameter Uncertainties on the Optimal Design of Integrated Biorefineries

As research is exploring possibilities for developing renewable bio-based fuels and chemicals, there arises a need for development of strategies which can design sustainable value chains that can be scaled up efficiently and provide tangible net environmental benefits. Recent ventures in biofuel production have been fraught with corporate failures. A driving reason for these unsuccessful ventures, in part is governed by the lack of proper planning in designing plants and supply networks. Modeling strategic and operational level decision processes of an enterprise which produces renewable-based fuels and chemicals, using engineering and financial tools, can provide valuable insight into the inter-play of technology and process interactions for different product streams.

In this study we present the development of a comprehensive decision support tool, within which the business value of a multiproduct biorefinery is optimized by considering all types of uncertainties including uncertainties at strategic and operational level. A stochastic linear model is first developed to optimize production capacity of the plant for the desired planning horizon and then process simulation coupled with stochastic optimization algorithm is employed to optimize the operating condition of the plant under uncertainty. The proposed methodology is based on an iterative framework that utilizes systems-based optimization in conjunction with detailed mechanistic modelling and simulation of the process. Market uncertainties are taken into account at the strategic planning level, and uncertainties related to parameters characterizing the processing technologies are addressed in operational level optimization. Monte-Carlo based simulation and global sensitivity analysis are utilized to identify the most critical parameters and optimize the operating conditions of the plant accordingly. Additionaly, risk management strategies are introduced to the framework for explicit treatment of strategic and operational risks. The other key feature of our approach is the development of dynamic data exchange strategy embedded as part of the process simulation to incorporate the complex kinetics of bio-reactions in the simulation model. The final results of the proposed framework include strategic capacity plan, optimal expected NPV, and optimal operating conditions of the process.

461131 Resource Recovery from Municipal Solid Wastes: A Case Study in the UK

Abstract:

Increasing waste generation in the UK has called for a sustainable municipal waste management strategy. It is imminent to reduce the volume of wastes going into landfill to avoid permanent resource loss, health impacts and environmental pollution. Prevention of waste generation is the most preferred option, however this is not easy to achieve in practice. Resource recovery from wastes by numerous treatment technologies is the second preferred option for not only to eliminate landfilling, but also to close the loop for a circular economy. Systematic techno-economic analysis and environmental impact assessments of various mechanical biological treatment (MBT) configurations for recovering metals, fuel and energy as resources from municipal solid wastes (MSW) are essential for selecting the most sustainable option. The systems under consideration include MBT systems with aerobic and anaerobic digestion (non-thermal) processes as well as incineration (thermal) processes. In the present study, MBT plant with aerobic digestion (windrows and in-vessel compositing); MBT plant with aerobic digestion and combined heat and power (CHP), with different degree of RDF and metal separation, with and without RDF incineration; and MSW incineration and CHP have been investigated. The MBT system with anaerobic digestion coupled with biogas and RDF CHP has shown promising economic performance giving an additional benefit of 113.5 Euro/t from the electricity generation from RDF incineration and CHP, that is almost 18 times increase compared to the case without electricity generation from RDF incineration and CHP, at an additional capital investment cost of 18 Euro/t. MSW valorisation through recovering metals such as aluminium, zinc and copper has the potential of generating additional revenue of 152 Euro/t. It has been identified that recovered aluminium is the main contributor to the economic value while recovered copper has the highest avoided environmental impact. Metals containing in the residues from mechanical separation worth 1.8 Euro per tonne of MSW could be lost if they are not recovered. Higher economic incentives and environmental impact savings for MSW treatment can thus be realised through effective resource recovery strategies, demonstrated in this work.

471007 The Case of Making Fuel Oil from Sanitary Sewage

Abstract:

What if we make fuel oil from our sanitary sewage? The ramifications of this question posed a decade ago turned out in various research areas such as resource recovery, wastewater treatment, chemical reaction engineering, catalysis, bioprocessing, separation processes, modeling-simulation, systems integration, and supply chain. The enhancement and extraction of oil from microbial consortia of activated sludge (AS) generated by wastewater treatment facilities have been studied with the potential as viable platform for organic waste-to-fuels and organic waste-to-chemicals transformation. The fundamental principles of this technology emulate the principles of single cell oil technology and resource recovery from wastewater treatment. The technology was established in 2005 through a project on biodiesel and value-added chemicals extraction from wastewater sludges [1]. Significant amounts of AS microbial oil were extracted and trans-esterified to biodiesel [2], but the enhancement of oil content is needed to make the technology economically feasible [3]. The oil content of AS was then enhanced using the principles of single cell oil technology in which high carbon to nitrogen ratio (mole C/mole N) triggers generation of structural lipids (phospholipids, sphingolipids, etc.) and reserve lipids (mainly TAGs) [4]. With sugars such as glucose and xylose as model carbon sources, researchers established that activated sludge oil can be enhanced and that the microbial process has similar microbial kinetics to that of a single cell oil accumulation [5, 6]. The commercialization of the aerobic lipid accumulation by AS, however, is challenged by the relatively high cost of the typical sugar substrates glucose and xylose. Several studies explored alternative carbon sources for microbial lipid accumulation and suggested the use of short chain fatty acids (SCFAs) such as acetic acid, propionic acid and butyric acid as substrates [7, 8]. These SCFAs are considered cost-effective because they are intermediate compounds in anaerobic digestion, a common wastewater treatment process [9]. Recent studies indicated that the microbial consortia of AS feeding on acetic acid (an SCFA) can accumulate lipids [10, 11]. This talk assesses the maturity of the technology in terms of the findings and the pertinent questions yet to be answered. On-going studies are also presented in order to set the immediate directions of some active research paths.

[1] Zappi ME, French WT, Hernandez R, Dufreche ST, Sparks DJ. Production of biodiesel and other valuable chemicals from wastewater treatment plant sludges Mississippi State University, US Patent, 2009.

[2] Dufreche S, Hernandez R, French T, Sparks D, Zappi M, Alley E. Journal of the American Oil Chemists' Society 2007;84(2):181-187.

[3] Mondala A, Liang K, Toghiani H, Hernandez R, French T. Bioresource Technol. 2009;100(3):1203-1210.

[4] Papanikolaou S, Aggelis G. European Journal of Lipid Science and Technology 2011;113(8):1031-1051.

[5] Mondala A, Hernandez R, Holmes W, French T, McFarland L, Sparks D, Haque M. AICHE J. 2013;59(11):4036-4044.

[6] Mondala AH, Hernandez R, French T, McFarland L, Santo Domingo JW, Meckes M, Ryu H, Iker B. AICHE J. 2012;58(4):1279-1290.

[7] Fei Q, Chang HN, Shang L, Choi J-d-r, Kim N, Kang J. Bioresource Technol. 2011;102(3):2695-2701.

[8] Fontanille P, Kumar V, Christophe G, Nouaille R, Larroche C. Bioresource Technol. 2012;114:443-449.

[9] Chang HN, Kim N-J, Kang J, Jeong CM. Biotechnol. Bioprocess Eng. 2010;15(1):1-10.

[10] Fortela DL, Hernandez R, French WT, Zappi M, Revellame E, Holmes W, Mondala A. Renew. Energ. 2016;96, Part A:11-19.

[11] Fortela DL, Hernandez R, Zappi M, WT French, Bajpai R, Chistoserdov A, Revellame E, Holmes W. J. Bioprocess. Biotech. 2016;6(4):275.

473515 Synthetic Diesel Production through Catalytic Pyrolysis of Biomass-Waste Tire Mixtures

Abstract:

The continuance of the modern world luxurious life-style, threatened by imminent depletion of cheap crude oil reserves, demands both the identification of alternative, sustainable, dense, cheap energy resources and means for effective waste reuse and disposal. This is of particular importance in the context of Jet fuels, for which energy demands are significant. The pyrolysis of waste tire rubber (WTR) results in the production of a pyrolytic oil (PO) in many ways similar to the crude exploited from fossil resources. PO has become a promising product since it addresses all of the aforementioned demands simultaneously. The yields and composition of this PO are strong functions of the production conditions and feedstock. In this work, the reaction conditions for PO synthesis using pulverized WTR alone and a mixture of WTR with sugarcane bagasse (30 and 70% w/w respectively) were experimentally assessed. Slow heating rates of approximately 5.5ºC/min were applied to 500 g of the feedstock in a batch reactor at a nitrogen flow rate of 2 L min^{-1} with and without MgO derived catalysts. Liquid yields were obtained in the range from 50 to 56% for WTR alone and 47 to 50% for WTR+SCB, with catalyst having a negligible effect on the conversion to PO. Cetane number of PO was estimated using NIR spectra, with average values of 43 for bulk oils of both studied feedstock. For WTR pyrolysis, gas yield was observed to be close to 6% for experiments run without catalyst with a two-fold yield increase if catalyst was present. In contrast, WTR+SCB yields approximately 20% gas, with no definitive catalyst effect on gaseous phase. Pyrolytic processes require energy, which may partly come from the fuel gasses obtained during the process itself; therefore, a characterization of the gas phase comes convenient. In the experiments run within the context of this work, the obtained gas of WTR pyrolysis was composed mainly of CH_4, with a peak measured composition of over 96% CH_4. SCB addition resulted in a 75% reduction of CH_4 and a 70% increase in CO_2 concentration.

471932 Conversion of Municipal Solid Waste to Methyl Ketone Using Ionic Liquid Based Process

Abstract:

Municipal solid waste (MSW) has the potential to serve as a feedstock for the production of advanced biofuels and biochemicals but has not been extensively studied to date in terms of conversion efficiency and hydrolysate quality. In this study, we evaluated the conversion of MSW blends to methyl ketones (MKs) using ionic liquid (IL)-based hydrolysates followed by fermentation using an engineered E. coli strain. The hydrolysates were produced from a one-pot IL process (which included both pretreatment and enzymatic saccharification) using different types of MSW blends and renewable, choline-based ILs. For the range of MSW blends and ILs tested, similar trends of OD600, sugar consumption, and MK production of hydrolysates were observed. Comparable or even greater amounts of medium-chain (C11-C17) MKs were produced using IL-based hydrolysates compared to the simple sugar controls with the same amount of sugars at the starting point. The results showed that this engineered E. coli strain can tolerate IL-based hydrolysates and that the neutralized ILs are acting as the extra carbon source for MK production.

462491 Upgrading of Hydrothermal Liquefaction Biocrude Oil Converted from Wet Biowaste

Abstract:

Wet biowaste is an increasingly studied feedstock for producing renewable bioenergy, reusing nutrients and improving the environment. Each year, the United States produces 79 million dry tons of wet biowastes from food processing and livestock manure and could be recovered as renewable energy. One of the limitations for wet biowaste to energy conversion is the high water content (80-99%).

Hydrothermal liquefaction (HTL) is a technology to reduce complex organic materials into biocrude oil and an aqueous fraction containing original nutrients and wastewater. Research has shown that HTL can convert 30-70% of different volatile solids into biocrude oil with heating values between 32-38 MJ/kg. HTL biocrude oil converted from different types of wet biowastes contains impurities: 10-20% oxygen, 3-7% nitrogen, and up to 20% of moisture. It is necessary to separate and/or upgrade the HTL biocrude oil for transportation fuel application. Distillation can reduce the oxygen content in the biocrude oil to 5% and increase the heating value to 41-45 MJ/kg. We hypothesize that with appropriate separation techniques, such as distillation, the HTL biocrude oil can be separated into different fractions, and some fractions can be used as transportation fuel including aviation fuel and diesel.

Our initial results suggest that distillation of HTL biocrude oil converted from swine manure (SW) and food processing waste (FPW) is very promising. Analyses demonstrate that the distillates from SW- and FPW-derived biocrude, respectively, contain 15 wt.% and 45-50 wt.% fractions that have heating values similar to those of petroleum aviation fuel and diesel without any further upgrading. In addition, an aviation biofuel blend prepared with 10 wt.% HTL distillates with the highest heating values (46-47 MJ/kg) and 90 wt.% petroleum jet fuel (or bio-kerosene BK10), exhibited the same properties including density, viscosity (6.3% lower), total sulfur content, copper corrosion performance, and net heat of combustion (1.30% lower) as those of commercial petroleum jet fuel. Although the distillation significantly improved the fuel's properties, more improvement is needed for use in current engines. For instance, the BK 10 blend had four times the acidity and a 20% lower flash point than those of petroleum jet fuel. We also found that oxygenates and nitrogen-containing compounds in the distillates from SW-derived biocrude led to excessive gum in BK10 blend. Thus, additional upgradation is needed.

Two strategies were investigated to make the HTL distillates converted from SW and FPW more compatible with petroleum transportation fuels. First, esterification was conducted to reduce the acidity in FPW-derived distillates. An orthogonal experimental design was operated to examine the effect of reaction temperature, reaction time, the catalyst loading (H_2SO_4), and the molar ratio of methanol on the esterification yield and acidity of the distillates. After esterification, the acidity of the distillates was reduced from 27.5 mg KOH/g to 0.8-7.3 with a yield of about 88 wt.%. The highest esterification yield was achieved at 70 °C for a 0.5 h reaction time with 2 wt.% catalyst loading under a 1:9 molar ratio of methanol; and the lowest acidity was realized at 50 °C for a 2h reaction time with 2 wt.% catalyst under a 1:15 molar ratio of methanol. The orthogonal test also indicates that the reaction time impacted the esterification yield the most, while the reaction temperature affected the acidity the most.

On the other hand, catalytic deoxygenation and denitrogenation under sub-/supercritical water are investigated to reduce the oxygen and nitrogen contents in the SW-derived distillates to reduce gum contents for BK10 blend. Preliminary data showed that supercritical water is a promising reaction media to inhibit the char formation and promote deoxygenation/denitrogenation for SW-derived distillates. Different types of catalysts (*e.g.,* Raney-Ni), reaction temperature, reaction time, and reaction media (e.g., supercritical water) are examined. The physicochemical properties of the upgraded distillates are analyzed to elucidate the upgrading mechanism. Elemental compositions, chemical compositions, heating values, flash point, viscosity, gum, and ash contents of the upgraded distillates are characterized and compared to those without upgrading.

471372 Evaluation of Optimum Cell Disruption for Microbial Lipid Extraction Using High Pressure CO2

Abstract:

The economy of lipid extraction from non-food sources of triglycerides (e.g. microbial oil) depends on lipid extraction and feedstock cost. Lipid can be extracted either from dried biomass (needs drying of the biomass up to 95%) or from wet biomass (consists mostly moisture). The microbial lipid extraction from wet biomass has advantages over dried biomass due to the elimination of energy intensive drying process. To improve lipid recovery from wet biomass, the cell wall of the microbes needs to be disrupted prior to lipid extraction for releasing the intracellular lipid from microbes that is usually difficult to extract using traditional solvent extraction process. The cell disruption is found to be affected by applied gas pressure, temperature, exposure time, depressurization rate, agitation, etc. This research will investigate the effect of different factors that affect the microbial cell disruption using high pressure CO_2. A fractional factorial design of experiment will be used to evaluate the controlling factors in microbial cell disruption. The number of cells before and after the treatment will be used as a response variable. After evaluating the main factors, a central composite design will be used to find the optimum cell disruption using a response surface methodology.

462954 Porous Structure Based Electrocatalysts for High Performance Fuel Cells

Abstract:

In this work, we systematically studied the effect of pore size and doping site position of metal-free catalysts in MEAs using well-defined ordered mesoporous carbon system to achieve highly active metal-free catalysts in MEA levels. At first, we synthesized mesosporous phosphorus doped ordered mesoporous carbon with large pore (~38 nm) by block copolymer assisted co-assembly. Then, addition nitrogen doping was conducted with 2 different conditions using different sized nitrogen precursors. Finally, we fabricated and tested MEAs that employ metal-free catalysts as a cathode catalyst. As a result, the MEA performance of synthesized catalyst with precisely controlled doping sites considerably exceeds that of previously reported metal-free catalysts.

465313 Design of Heterostructure Alloy Nanoparticles for Photocatalysis of CO_2 Reduction

Abstract:

Semiconductor nanocrystals of size in the strong quantum confinement regime, or quantum dots, exhibit size-dependent energy gap. The size-tunable shift of band-edge states can be a powerful tool in the control of exciton dynamics. For example, heterostructure nanocrystals with mere changes in geometric dimensions can lead to the band alignment from type I to type II, which could dramatically increase the exciton recombination lifetime.[1] In this presentation, I describe the design of heterostructure nanocrystals showing long exciton recombination lifetime. For the study, we synthesized various core/shell composite particles (e.g., PbSe/CdSe or CdSe/TiO2) and introduced CdS outer-most shell with varying thickness and morphology. [2] These heterostructure nanoparticles with prolonged exciton lifetime result in efficient photocatalysis of methylene blue reduction and hydrogen evolution from water.

In the way of addressing the heterostructures, we stumbled upon the observation that metal cocatalysts and surface can be engineered with relatively simple chemical approaches, rendering new photocatalytic systems suited in photocatalytic reduction of CO2 in gas phase. [3]

References

[1] Lee D C, Robel I, Pietryga J M and Klimov V I 2010 *J. Am. Chem. Soc.* **132** 9960

[2] Lee S, Lee K, Shin D J, Kim W D, Lee S and Lee D C 2014 *J. Phys. Chem. C* **118** 23612

[3] Jeong S, Lee S, Kim W D, Lee K, Lee S and Lee D C 2016 *ChemCatChem* in print

[4] Lee S, Jeong S, Kim W D, Lee K, Lee S and Lee D C 2016 *Nanoscale* in print

470971 Nanostructured Composite Intermediate-Temperature Solid Acid Fuel Cells Fabricated By Needleless Electrospinning

Abstract:

Fuel cells are the most efficient technology available today for the conversion of chemical energy into electricity. A new and promising fuel cell technology is the solid acid fuel cells (SAFCs), offering unique possibilities as a result of their intermediate-temperature region of fuel cells (100–300 °C). Their state of the art solid electrolyte, CsH_2PO_4 (CDP) is a solid acid compound with a superprotonic phase transition around 240 °C [1]. Under a dry condition the dehydration of CDP proceeds to form polymerization at temperatures above 200 °C, and CDP transforms into $CsPO_3$ [2]. Development of the catalytic performance can enable this type of fuel cell to be among the leading fuel cells on today's market.

Due to the pure ionic conducting nature of the electrolyte, the electrochemical reactions occur at the so-called triple phase boundaries, between the catalyst, electrolyte and the gas phase. Because of the triple phase, decreasing just the size of the catalyst does not increase the electrochemical reaction rates. However, in case the size of the electrolyte is also decreased, the electrochemical reaction rate is higher at the triple phase boundary resulting in increased power density of the fuel cell [1].

Needleless electrospinning technique was employed to fabricate nanostructured SAFC electrodes from CDP/$CsPO_3$-polymer solutions. Nanostructured CDP composite has the advantage of increased catalytic surface due to the large surface area. Needleless electrospinning is an inexpensive technique for producing nanostructured fuel cells at industrial scales. Instead of the commonly used nozzle, the needleless design employs a rotating drum immersed in a solution bath. Multiple Taylor cones form on the drum electrode surface resulting in rapid deposition of nanofibers over large areas. By applying $CsPO_3$ solutions, the phosphate groups form a chain, increasing the viscosity of the solution. This enables electrospinning with negligible amounts of polymers. The fibers produced from the solutions have diameters between 30 nm – 300 nm depending on the initial solution concentration, substrate temperature, and applied voltage. The deposition area is over 100 cm^2 and the deposition rate is up to 2 g/hr in the laboratory scale setup. Electrochemical characterization of the nano-composite electrodes, performed by A.C. impedance spectroscopy of symmetric cells under humidified hydrogen, confirmed the high activity of the electrospun nanocomposite electrodes. Fuel cell results using electrospun nanofibers from $CsPO_3$ solutions demonstrate that a 0.86 V cell voltage can be obtained at a current density of 50 mA/cm^2, which is ~10% higher than what can obtained using state-of-the-art techniques.

[1] C. R. Chisholm, D. A. Boysen, A. B. Papandrew, S. K. Zecevic, S. Cha, S., K. A. Sasaki, Á. Varga, K. P. Giapis, S. M. Haile, ***Interface***, 2009, ***18***, 53-59.

[2] Y. Taninouchi, T. Uda, Y. Awakura, A. Ikeda, S.M. Haile ***J. Mater. Chem.***, 2007, 17, 3182.

463690 Optimization of Pt/C Particles Electrocatalytic Activity By the Control of Carbon Nanostructures *Via* a Hybrid Aerosol-Colloidal Process

Abstract:

An optimum electrocatalytic performance of carbon-supported platinum (Pt/C) catalyst is determined by many factors, including the structure of the carbon support. In the study herein, we evaluated the effect of carbon nanostructures on the electrocatalytic performance of Pt/C catalyst and determined the most suitable nanostructure to deliver the best performance. Submicron carbon particles were prepared *via* spray pyrolysis of phenolic resin prior to being deposited with Pt nanoparticles using a microwave-irradiation process. Nanostructuration of the carbon particles was performed *in situ* owing to the colloidal interaction between phenolic resin and polystyrene latex (PSL) template, and was controlled by changing the properties of PSL. According to the results of the electrochemical analyses, Pt/C catalysts with porous carbon structure delivered the best performance compared to those with dense and hollow carbon structure. An evaluation on the pores size revealed the importance of a good balance between meso- and macropores for an effective electrocatalysis, which can be explained from the hydrodynamic profile in the catalyst layer.

469355 Highly Active Robust F Doped Transition Metal Oxide Based Solid Solution Electro-Catalyst for Acidic Medium Oxygen Evolution Reaction in PEM Based Water Electrolysis

Abstract:

The vast consumption of fossil fuels for meeting the energy demand has led to major unsustainable environmental concerns due to excessive greenhouse gas emissions warranting the identification and development of clean non-carbonaceous energy sources.[1-5] In this regard, hydrogen has been identified as a clean, non-carbonaceous and potential energy source with superior energy density than currently adopted carbonaceous energy sources for meeting the global energy demand. However, efficient and economic production along with cost-effective storage and distribution of hydrogen utilizing the clean non-carbonaceous approach is of paramount importance before universal adoption of hydrogen as a clean energy source in the impending non-carbonaceous fuel economy. Along these lines, electricity driven water splitting (water electrolysis) is a frontrunner among other clean pollution free approaches for hydrogen production (having low carbon footprint). The commercial development of water electrolysis is thwarted due to the need of expensive and precious noble metals based electro-catalysts (Pt, IrO_2) which exhibit excellent electro-catalytic activity and stability for oxygen evolution reaction (OER) in acid assisted proton exchange membrane (PEM) based water electrolysis. Hence, the identification and development of novel reduced noble metal containing electro-catalysts exhibiting excellent electro-catalytic activity and superior long term electrochemical stability similar/superior to state of the art OER electro-catalyst IrO_2 in harsh and highly acidic operating conditions of OER, will aid in the reduction of capital cost of water electrolysis cells and thus, its progression towards commercialization.[6]

With this aim in mind, exploiting theoretical first principles calculations of the total energies and electronic structures, fluorine doped transition metal (TM) oxide based solid solution electro-catalyst containing significantly lower noble metal content has been identified exhibiting surface electronic structure and electro-catalytic activity similar to that IrO_2.The thin films of 10 wt. % F doped $(TM_{1-x}Z_x)O_2$ (x=0.2, 0.3, 0.4; Z is a noble metal) coated on Ti foil were synthesized and studied as electro-catalyst for OER in acidic media. **Fig. 1** shows the SEM micrograph showing the nanoscale architecture of the representative composition $(TM_{0.8}Z_{0.2})O_2$:10F. Elemental x-ray maps show homogeneous distribution of elements throughout the material without any segregation at specific site.

Electrochemical characterization of the synthesized electro-catalysts has been carried out in a three-electrode configuration using 0.5 M sulfuric acid (H_2SO_4) solution as a proton source as well as the electrolyte, Pt wire as counter electrode and Hg/Hg_2SO_4 as the reference electrode (+0.65 V with respect to normal hydrogen electrode, NHE), with a scan rate of 10 mV/sec and at temperature of 40^0C. $(TM_{1-x}Z_x)O_2$:10F (for x = 0.2, 0.3, 0.4) electro-catalysts (total loading = 0.3 mg/cm^2) exhibit excellent electrochemical activity for OER with significantly reduced onset potential (~1.35 *vs* RHE) compared to that of IrO_2 (~1.43 *vs* RHE) (total loading=0.3 mg/cm^2). In addition, $(TM_{1-x}Z_x)O_2$:10F displayed outstanding ~8, ~14 and ~15 fold improved electro-catalytic activity at ~1.45 V (*vs* RHE) compared to IrO_2. Following chronoamperometry test conducted in 0.5 M H_2SO_4 solution at ~1.55 V (*vs* RHE) for 24 h, minimal loss in current density as well as excellent long term electrochemical stability was observed for $(TM_{0.8}Z_{0.2})O_2$:10F, similar to that of IrO_2.

These results show the promise of $(TM_{1-x}Z_x)O_2$:10F electro-catalyst portending ~80% reduction in noble metal content displaying significantly much higher electro-catalytic activity than IrO_2 and superior long term stability. In addition, these results show the promise of the formation F-doped solid solution in improving the electronic, physical, chemical and electro-catalytic properties of the doped TM oxide structures in acidic media. This will aid in significant reduction in the overall capital cost of PEM based water electrolysis for efficient and economic production of hydrogen. The results of these studies will be presented and discussed.

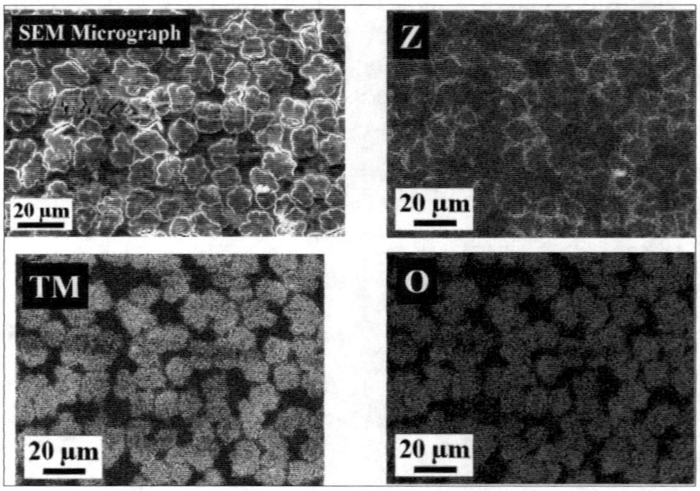

SEM micrograph with elemental x-ray maps of $(TM_{0.8}Z_{0.2})O_2$:10F electro-catalyst

References:

1 Patel, P. P. *et al.* Nanostructured robust cobalt metal alloy based anode electro-catalysts exhibiting remarkably high performance and durability for proton exchange membrane fuel cells. *Journal of Materials Chemistry A* **3**, 14015-14032 (2015).

2 Patel, P. P. *et al.* WO_3 based solid solution oxide–promising proton exchange membrane fuel cell anode electro-catalyst. *Journal of Materials Chemistry A* **3**, 18296-18309 (2015).

3 Patel, P. P. *et al.* High performance and durable nanostructured TiN supported Pt 50–Ru 50 anode catalyst

for direct methanol fuel cell (DMFC). *Journal of Power Sources* **293**, 437-446 (2015).

4 Patel, P. P. *et al.* Vertically aligned nitrogen doped $(Sn,Nb)O_2$ nanotubes – Robust photoanodes for hydrogen generation by photoelectrochemical water splitting. *Materials Science and Engineering: B* **208**, 1-14, (2016).

5 Patel, P. P. *et al.* Nitrogen and cobalt co-doped zinc oxide nanowires – Viable photoanodes for hydrogen generation via photoelectrochemical water splitting. *Journal of Power Sources* **299**, 11-24, (2015).

6 Datta, M. K. *et al.* High performance robust F-doped tin oxide based oxygen evolution electro-catalysts for PEM based water electrolysis. *Journal of Materials Chemistry A* **1**, 4026-4037, (2013).

Acknowledgements:

The authors gratefully acknowledge the financial support of NSF-CBET grant# 1511390. The authors also acknowledge the Edward R. Weidlein Chair Professorship funds and the Center for Complex Engineered Multifunctional Materials (CCEMM) for partial support of this research.

467668 Superwetting Nanoarray Electrodes for Gas-Involved Electrocatalysis

Abstract:

Superwetting Nanoarray Electrodes for Gas-Involved Electrocatalysis

Yingjie Li, Wenwen Xu, Zhiyi Lu, Xiaoming Sun*

Beijing University of Chemical Technology, Beijing 100029, China, Email: ~~sunxm@mail.buct.edu.cn~~

Electrochemical gas-involved reactions can be divided into two types: gas-evolution reaction and gas-consumption reaction, both of which are crucial for a variety of energy conversion processes and industries (e.g. HER and ORR). For gas-evolution reaciton, if generated gas bubbles pin at the electrode surface and cannot escape from the surface in time, the accumulated bubbles will reduce the effective electrode surface area, increase diffusion resistance and enlarge polarization effect, resulting in more energy consumption. How to construct a novel electrode to promote gas bubble release is critical for improving the electrochemical efficiency besides activity improvement. Inspired from bio-inspired superwetting surfaces, we found that the interface behavior of electrode could be tuned by surface architecture construction, for example, transferring from aerophobic to superaerophobic by engineering a series of nanoarray electrode, eg. MoS_2 (Fig. 1), pine-shaped Pt, NiFe LDHs and Cu films [1-7]. This kind of superaerophobic electrodes could decrease the critical size of gas overflowing from the surface by cutting the three phase contact lines into discontinue dots, and thus reduce the diffused impedance and maintain the integrity of the solid−liquid interface that is necessary for electrocatalysis (e.g. water splitting and hydrazine fuel cells). On the other hand, for gas-consumption reaction, the construction of nanoarray ¡°superaerophilic¡± electrode could accelerate gas diffusion to reaction zone via gas-phase to solve the issue of low solubility and slow diffusion of gases in traditional electrocatalysiss reaction system with limited current density (eg., ORR). Therefore, the construction of superwetting nanoarray electrodes is imperative to improve gas transport at electrodes surface and to enhance activity and stability of electrodes.

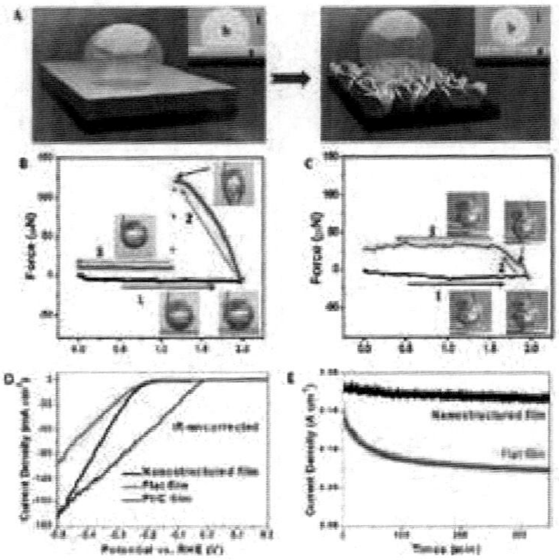

Fig. 1 (A) Schematic illustration of adhesion behaviors of gas bubbles on flat film (left) and nanostructured film (right), (B) and (C) Adhesive forces measurements of the gas bubbles on flat and nanostructured MoS_2 films, (D) Polarization curves (without IR-correction) of MoS_2 and Pt/C catalysts (E) Stability testing of flat and nanostructured MoS_2 electrodes

References

[1] Z. Lu, W. Zhu, X. Yu, H. Zhang, Y. Li, X. Sun, X. Wang, H. Wang, J. Wang, J. Luo, X. Lei, L. Jiang. Adv. Mater. 26 (2014), 2683-2687.

[2] Y. Li, H. Zhang, T. Xu, Z. Lu, X. Wu, P. Wan, X. Sun, L. Jiang. Adv. Funct. Mater. 25 (2015), 1737-1744.

[3] Z. Lu, M. Sun, T. Xu, Y. Li, W. Xu, Z. Chang, Y. Ding, X. Sun, L. Jiang. Adv. Mater. 27 (2015), 2361-2366.

[4] X. Liu, Z. Chang, L. Luo, T. Xu, X. Lei, J. Liu, X. Sun. Chem. Mater., 26 (2014), 1889-1895.

[5] M. Sun , Z. Lu , L. Luo, Z. Chang , X. Sun. Nanoscale, 8 (2016), 1479-1484

[6] W. Xu, Z. Lu, P. Wan, Y. Kuang, X. Sun. Small, 12 (2016), 2492¨C2498

[7] M. Jiang, Y. Li, Z. Lu, X. Sun, X. Duan. Inorg. Chem. Front. 2016, DOI: 10.1039/C5QI00232J.

Biography

Professor Xiaoming Sun gained his B.S. degree and Ph.D. in Department of Chemistry, Tsinghua University in 2000 and 2005, respectively. After postdoctoral work at Stanford University, he joined State Key Laboratory of Chemical Resource Engineering, Beijing University of Chemical Technology at 2008. His main research interests focus on separation and assembly of inorganic nanostructures, synthesis and separation of carbon nanomaterials and their composites, and structure control, opto-/electro-property investigations of nanoarrays and superwetting nanoarray electrodes for energy sciences. He has authored 63 journal articles (eg. ***J. Am. Chem. Soc., Angew. Chem. Int. Ed., Adv. Mater., ACS Nano.***), which have been cited >4600 times.

460470 Nanocatalysts for Water Splitting

Abstract:

Developing a clean and sustainable energy system is essential for the earth's sustainability. The hydrogen/water energy cycle, a clean and sustainable system, receives of great interest recently. Water splitting is one of the key reactions in the hydrogen/water energy cycle. Developing active and cost-effective nanocatalysts for water splitting is important to water splitting for the production of clean H_2 energy sources. This work will present various nanocatalysts for solar-driven and electric-driven water splitting reactions, especially for hydrogen evolution reaction (HER). The co-catalysts decorated Cu_2O/CuO photocathodes and MO/M heterostructured electrocatalysts were prepared by various interfacial engineering techniques. Combining advanced characterization and DFT computational techniques is capable of providing the insight of the heterostructured nanocatalysts for HER. It was found that the heterostructured interface plays a vital role and provides a highly active catalyst for HER. The design strategy for the HER catalysts can be developed via the understanding of the interfacial properties and reactions.

470723 Hydrogen Generation from Hydrous Hydrazine over Nickel-Doped Ceria Catalysts Prepared By Solution Combustion Synthesis

Abstract:

Hydrous hydrazine is considered to be a promising hydrogen storage material due to its high hydrogen content (8.0 wt.% for hydrazine monohydrate) and simple byproduct (only nitrogen) via complete decomposition. The key to exploit hydrous hydrazine in practical hydrogen generation applications is to develop efficient, selective, and economic catalysts for the decomposition. In this work, nickel-doped cerium oxides as new non-noble metal catalysts were prepared by solution combustion synthesis (SCS) of CeO_2, followed by impregnation of Ni or by one-step SCS of Ni/CeO_2. The effect of catalyst preparation conditions, including metal weight percentage (10%-80%), type of fuels (glycine, hydrazine, ethylene glycol, etc), ratio of fuel to oxidizer (0.5-3), and some additives on the properties of catalysts were investigated to optimize synthesis conditions and then their catalytic performance were tested. The experimental results demonstrate that the Ni/CeO_2 catalysts via SCS possess high catalytic activity and selectivity to generate hydrogen from hydrous hydrazine.

464096 Network Structure in Hybrid Solid Polymer Electrolytes (Invited Talk)

Abstract:

Batteries are of utter importance for energy storage and reducing carbon footprint to combat global warming and foster a sustainable future of our planet. To fabricate safer and more powerful energy devices including electric vehicles, one of the grand challenges is to fabricate mechanically robust and ionic conductive solid polymer electrolytes (SPEs) that form stable interface with electrodes. We herein report the design, synthesis and characterization of hybrid electrolytes with inorganic polyhedral oligomeric silsesquioxanes (POSS) as the crosslinker. The SPEs were prepared by a facile one-pot reactions between octakis[3-(glycidyloxy)propyldimethylsiloxy]silsesquioxane and bis(3-aminopropyl) terminated poly(ethylene glycol) (PEG) in the present of bis(trifluoromethane)sulfonimide lithium salt (LiTFSI). Uniform distribution of POSS and LiTFSI in the SPEs were confirmed by SEM-BSD and SEM-EDX. These hybrid network SPEs show promising ionic conductivity, mechanical properties, and lithium dendrite growth resistance.

455759 Design of High Transference Number Battery Electrolytes

Abstract:

High transference number electrolytes, those in which the electrochemically active lithium cation diffuses faster than its electrochemically inert counterion, have the potential to improve battery performance by limiting concentration polarization. High cation transference electrolytes are typically polymers with anions fixed to their backbone (ionomers), thereby only allowing cation diffusion over large length scales. In their dry state, such polymer electrolytes generally have substantially lower ion conductivity, particularly at room temperature, than liquid electrolytes, negating the advantages that polymers provide in terms of safety or Li^+ transference. Therefore, our research has focused on identifying routes to improve ion conductivity of ionomers through small molecule plasticization or improve Li^+ transference numbers of liquid electrolytes by using ionomers as the salt. Here, we have designed a series of poly(ethylene oxide) (PEO) based ionomers containing lithium sulfonate to explore these concepts. Our initial results on improving transference numbers of liquid electrolytes using these polymers as salts will be presented. By tuning the molecular weight of the polymer, solvent donor number, and concentration of polymer in solution, we begin to understand the relationship between electrolyte characteristics and performance. Preliminary results show that in the case of poorly dissociated ions, the ion and counterion move together, with a transference number of close to 0.5. However, in the case where ions are well dissociated, very high cation transference numbers (0.94) can be achieved.

461836 Effect of Transporting Enhancer in Fe_3O_4 Li-Ion Battery Anodes

Abstract:

Battery electrodes are complex mesoscale systems comprised of active material, conductive agents, current collector and binders [1]. For the improved design of a composite electrode, both electron and ion transport are critical factors to improve electrochemical performance. However, research rarely focuses on both factors.

The present work focuses on enhancing electron and ion transport in high capacity anode systems, by introducing a poly[3-(potassium-4-butanoate) thiophene] (PPBT) as a binder component and a polyethylene glycol (PEG) surface coating on Fe_3O_4 nanoparticles. PPBT is a water soluble carboxylate functionalized polythiophene, which seems favorable for forming porous structures. The PEG coating reduced the aggregate size and improved dispersion. Together the PPBT water system and PEG coating helped improve capacity retention by effectively decreasing charge-transfer resistance. PPBT and PPBT carbon composites also exhibited high electronic conductivity compared to PVDF, which was used as a reference due to its common use as a binder in Li-ion battery electrodes. SEM images illustrated that the PEG coating reduced the size of aggregates as compared to a Fe_3O_4/carbon/PPBT electrode from ~5mm to ~0.5-1mm. This improved dispersion, which positively contributed to battery performance. Furthermore, FT-IR and XPS provided evidence of chemical interaction between PPBT and Fe_3O_4 nanoparticles. Strong interactions between a binder and surface of active materials have been reported to improve stability [2]. The use of a PPBT binder and PEG surface coating improved both ion and electron transport, which provides a new approach for high-capacity energy materials [3].

An ongoing portion of this work takes a closer look at the active material, Fe_3O_4, and further studies the difference in particle sizes and PEG coating on battery performance. The PEG coating has already been shown to reduce aggregate size and improve dispersion [3]. Further testing is ongoing to study how crystallite size (10 nm vs. 20 nm) of nanoparticles affects battery properties. Preliminary SEM results indicate that at higher nanoparticle size (20nm), the aggregates from PEG are closer in size to the nanoparticles, which begs the question of how this impacts battery performance. Future testing is still needed to confirm this hypothesis and determine the "critical" nanoparticle size for greatest battery performance by enhancing the active material of the electrode.

[1] Y. H. Kwon, M. M. Huie, D. Choi, M. Chang, A. C. Marschilok, K. J. Takeuchi, E. S. Takeuchi, E. Reichmanis, *ACS Appl. Mater. Interfaces, 2016, 8* (5), 3452–3463.

[2] D. Mazouzi, B. Lestriez, L. Roué, D. Guyomard, *Electrochem. Solid-State Lett.* **2009**, *12*, A215.

[3] Y. H. Kwon, K. Minnici, M. M. Huie, A. C. Marschilok, K. J. Takeuchi, E. S. Takeuchi, E. Reichmanis. Electron/Ion Transport Enhancer in High Capacity Li-ion Battery Anodes, *submitted.*

466862 Integration of Ultrathin Polyaniline Films into Carbide Derived Carbon Supercapacitors Via Oxidative Chemical Vapor Deposition

Abstract:

The use of conducting polymers as the active materials for supercapacitors is a promising approach to improve the performance of supercapacitors[1, 2]. Conducting polymers enhance charge storage capacity through Faradaic redox reactions, and have the ability to be p- or n-doped. They are typically deposited via solvent-based techniques such as chemical bath deposition, electrodeposition, and casting from suspension. Nevertheless, challenges exist with making uniform conformal coatings via wet chemistry routes, especially with poor solubility of many conducting polymers and poor accessibility of the highly tortuous pore space within nanostructured electrodes. Oxidative chemical vapor deposition (oCVD) allows one to conveniently bypass these challenges as it is a solvent-free method where the reagents for oxidative polymerization, the monomer and oxidizer, are heated to vapor form that can easily penetrate into pores and polymerize (Figure 1). By understanding the oCVD polyaniline (PANI) synthesis parameters (reagent flow rates, reactor pressure, substrate temperature, saturation pressure), the reaction and diffusion rates can be carefully controlled to optimally coat the carbide derived carbon (CDC) electrode. By flowing vapors of aniline and antimony pentachloride (a strong oxidizer), a solid thin PANI film can be directly polymerized on a substrate. FTIR, UV-vis, and XPS confirm that oCVD produces an electrochemically active emeraldine state of PANI. The pore distribution of the CDC electrode, as determined by N_2 sorption, remains mostly unchanged before and after coating because the fine oCVD control over film thickness allows the pores to be coated without filling them, thereby maintaining the electrochemical double layer of the intrinsic carbon electrode, while imparting additional energy density through Faradaic redox reactions from PANI.

In this work, we are realizing a strategy proposed by Simon and Gogosti[3] to improve both the energy and power densities of electrochemical capacitors by integrating conducting polymers into nanostructures electrodes. In particular,

we demonstrate that charge storage capacity can be improved significantly through the integration of ultrathin (<30 nm) PANI films into mesoporous Mo2C-derived carbide derived carbon (CDC, 800 °C synthesis temperature) electrodes using a single-step oCVD process. The ultrathin coating preserves to a large extent the native electrode surface area and pore distribution while simultaneously improving capacitance. The oCVD process allows PANI to be integrated into pores as small as 1.7 nm, and the oCVD PANI integrated supercapacitors have a specific capacitance more than 100% higher than that of bare CDC ones (136 F/g for 11 wt% of PANI in the CDC electrode vs. 60 F/g for bare CDC at 10 mV/s). This yields a PANI-only capacitance of ~690 F/g, which is close to the theoretical value of 750 F/g [4]. Even after high scan rates of over 100 mV/s, the added pseudocapacitance from PANI remains evident (Figure 1). The composite device exhibits excellent cyclability, which decreases only by 10% of the initially stabilized value (~100 F/g) after 10,000 cycles. To our knowledge, this work is the first reported synthesis of PANI via oCVD and demonstrates oCVD's advantages compared to other approaches in the integration of ultrathin PANI films into carbon supercapacitors.

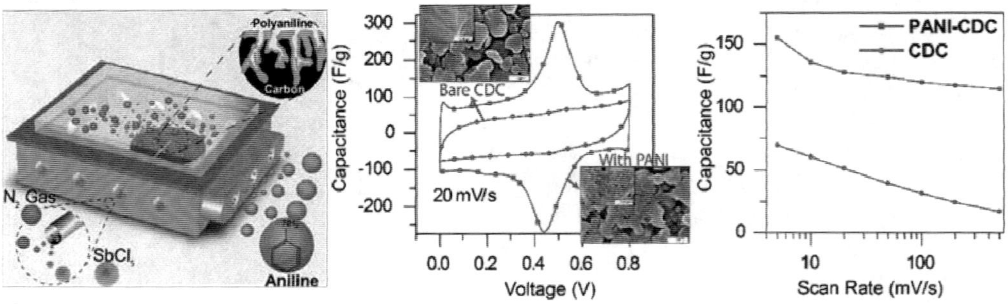

Figure 1. Left: oCVD process where the oxidant (SbCl5, orange) and monomer (aniline, green) vapors enter the reaction chamber and surface polymerize onto the electrode. Middle: A comparison of the cyclic voltammograms for both bare (red) and polyaniline (PANI) (green) electrodes at 20 mV/s. Inset shows SEMS of bare and PANI-coated CDC electrodes. Right: Rate performance of both devices based on cyclic voltammetry.

1. Laforgue, A.; Simon, P.; Sarrazin, C.; Fauvarque, J.-F. Journal of Power Sources 1999, 80, (1–2), 142-148.

2. Liu, D. Y.; Reynolds, J. R. ACS Applied Materials & Interfaces 2010, 2, (12), 3586-3593.

3. Simon, P.; Gogotsi, Y. Nature Materials 2008, 7, (11), 845-854.

4. Snook, G. A.; Kao, P.; Best, A. S. Journal of Power Sources 2011, 196, (1), 1-12.

472346 Controlling the Li-Air (O_2) Discharge Process with a Gel Polymer Electrolyte

Abstract:

The 2015 climate change conference in Paris espoused the need to stem the rise in global temperatures to below 2 °C. Therefore, there is a need to shift from fossil fuel use in electricity generation and transportation – industries responsible for the bulk of carbon dioxide (CO_2) emissions. Alternative energy conversion sources such as solar and wind are promising, but their intermittent nature affects long-term adoption. To complement these new conversion technologies, energy storage media such as batteries and fuel cells among others have been explored. The compact nature of batteries make them attractive over other energy storage media. Of commercially available batteries, lithium-ion batteries are the most energy-dense, and are used in electric vehicles and portable electronic devices. However, Li-ion batteries suffer from high cost limiting their use for stationary grid applications, and their energy densities are not high enough to allow for electric vehicles to compete with gasoline-powered cars in terms of mileage and cost. To address these limitations of Li-ion, other battery chemistries such as metal-air and metal-sulfur are being investigated. Lithium-air (O_2) batteries have the highest theoretical gravimetric energy density and have garnered tremendous research interest.

Li–O_2 chemistry is governed by the reduction of oxygen during discharge and the oxidation of the reduced oxygen discharge product during charge. In conventional Li–O_2 cells, non-aqueous liquid electrolytes are often used. The oxygen reduction chemistry is dominated by a 2 mol e^-/mol O_2 (peroxide) process that some have attributed to be partially responsible for the sluggish reduction and oxidation kinetics, limited current rate, and poor capacity retention of Li–O_2 batteries. Battery chemistries such as Na–O_2 and K–O_2 that utilize a 1 mol e^-/mol O_2 chemistry have been shown to support higher current rates and better energy efficiencies.

In this work, we incorporate ionic liquids in a polymeric matrix and show that controlling the lithium/ionic liquid molar ratio in the gel polymer electrolyte can allow for a 1 mol e^-/mol O_2 reduction process in a Li–O_2 battery. Ionic liquid cations has been shown to support a 1 mol e^-/mol O_2 process using cyclic voltammetry, but not in actual Li–O_2 cells, where a 2 mol e^-/mol O_2 process (and Li_2O_2) is observed. Furthermore, we use multiple spectroscopic tools to confirm for the first time the formation of a solid ionic liquid-superoxide discharge product. Knowledge gained from this work should spur development of newer and more stable ionic liquids and polymers that can allow for better long-term Li–O_2 cycling. In addition, the mechanism observed here could prove vital for other battery chemistries such as metal-air and metal-sulfur where controlling intermediate solubility is paramount.

460841 Water-Polymer Mobility and Distribution in Hydrated Aromatic Ionomer Thin Films

Abstract:

The confined state properties of ionomers are surprisingly different from bulk membranes and are still not well understood. Hydration of fuel cell ionomers in supported thin films leads to complex multimodal interactions among water, polymer chains and substrate and these interactions may lead to interesting mechanical and transport properties. In this work, thin films (~25-250 nm) of sulfonated aromatic ionomer, S-Radel were investigated to understand thickness and hydration effects on density, water-polymer distribution and mechanical properties. The density values obtained from quartz crystal microbalance and spectroscopic ellipsometry showed a similar trend with thickness to those obtained from neutron reflectometry. Thicker films were always more dense and less water-rich at the interface compared to thinner samples. A fluorescent rotor probe was incorporated into the polymer samples to predict the stiffness of the films using time resolved fluorescnece. The presence of voids rationalized greater mobility and lower stiffness in the thinner films in the dry state. When the same sample was hydrated, film density and interfacial water volume fraction significantly increased with the disappearance of voids. In addition, the water distribution near substrate interface in the aromatic s-Radel films was distinctly different from that in fluorocarbon based Nafion films seen in previous work. Antiplasticization, or stiffening of the films with the addition of water, was observed from fluorescence lifetime of rotor probes. The plasticization properties appeared to be controlled by film thickness, film density, and water mobility.

457801 Highly Proton-Conductive Polyelectrolyte Membranes with Supramolecularly Suppressed Water Swelling

Abstract:

Highly Proton-conductive Polyelectrolyte Membranes with Supramolecularly Suppressed Water Swelling

Joseph Aboki (jaboki@nd.edu), Shuangjiang Luo (sluo2@nd.edu), Ruilan Guo* (rguo@nd.edu)

Department of Chemical and Biomolecular Engineering, University of Notre Dame, Notre Dame, IN 46556-5637

Polyelectrolyte membranes based on sulfonated multiblock copolymers with alternating hydrophilic and hydrophobic blocks hold great potential as alternatives to benchmark Nafion® for fuel cell applications, due to their much better proton conductivity at low relative humidity (RH) levels and thermal stability. Previous studies suggest that long hydrophilic sequences (>10,000 g/mol) are necessary to effect sharp nanophase separation and form continuous proton conducting nanochannels enabling high proton conductivity at low RH, which, however, is always accompanied with excessive water swelling resulting in deterioration of dimensional stability and mechanical robustness of the membranes.

In this study, we report an innovative supramolecular strategy to suppress excessive water swelling by introducing triptycene structure units into the hydrophobic sequences of multiblock copolymers, which induce strong supramolecular interactions with the hydrophilic sequences via the so-called chain threading and interlocking effectively constraining the polyelectrolyte membranes from excessive swelling. Specifically, a series of triptycene-containing poly(arylene ether sulfone) multiblock copolymers were synthesized via a coupling reaction between phenoxide-terminated hydrophilic oligomers (BPS100) and fluorine terminated triptycene-containing hydrophobic oligomers (TRP0). The molecular weight (ranging from 5,000 to 20,000 g/mol) and end group functionality of oligomers were precisely controlled by adjusting the stoichiometric feeding ratio of monomers. Additionally, the ion exchange capacity (IEC) of the multiblock copolymers was controlled by coupling oligomers with even and uneven block lengths. The resulting multiblock copolymers yielded tough and ductile membranes via solution casting. In general, the multiblock copolymer membranes showed increasing water uptakes and proton conductivities with increasing block length. Anisotropic swelling behavior was observed for all the copolymer membranes suggesting nanophase separated lamellae-like membrane morphology. Most importantly, for copolymers with long hydrophilic sequences (> 10,000 g/mol), despite of their expectedly high water uptake, they all showed greatly reduced swelling ratios as compared to Nafion® and reported nontriptycene-containing multiblock copolymers. This observation indicates that the supramolecular chain threading and interlocking interactions induced by the triptycene units in the hydrophobic sequences are instrumental in suppressing membrane swelling and improving dimensional stability. For example, the multiblock copolymer with hydrophilic block length of 15,000 g/mol had a water uptake of 107%, an excellent proton conductivity of 0.155 S/cm in R.T. liquid water (almost 3 times that of a Nafion® 212) and a volume swelling ratio of just 27% (more than 65% reduction when compared to Nafion® 212 and previously reported multiblock copolymer systems). In addition, a super high liquid water proton conductivity (R.T.) of 0.224 S/cm was achieved by a multiblock copolymer with uneven block length (i.e., 10,000 g/mol of BPSH100 and 7,000 g/mol of TRP0). These multiblock copolymer membranes hold tremendous potential for applications in proton exchange membrane fuel cells and water desalination membranes.

468854 Role of Oligomeric Additives on P3HT/PCBM Domain Interfaces and Photovoltaic Performance

Abstract:

Controlling the nanophase separation of donor-type polymers and fullerene-based acceptors in bulk heterojunction (BHJ) thin films is crucial to optimizing the performance and enhancing the stability of organic photovoltaics (OPV) devices. A variety of strategies, including thermal or solvent annealing or incorporating small molecule processing aids or block copolymer compatibilizers, have been used as a way to mediate phase segregation and enhance OPV performance. In this presentation, we describe our efforts to characterize the influence of oligomeric donor-type materials based on phase segregation and performance of BHJ thin films comprising poly(3-hexylthiophene) and PCBM fullerenes. A combination of electron microscopy and X-ray and neutron scattering methods, augmented with insights from multiscale modeling, are used to develop a picture of how the size of low molecular weight P3HT additives and their loading level influences the distribution of the additives and their molecular orientation at donor-acceptor interfaces. These features of nanoscale structure are correlated with results from device performance testing, providing useful insights into the links between additive design, BHJ structure and device performance. While results are specific to P3HT/PCBM systems, we believe that the findings are generally applicable to systems where miscibility, crystallization, and packing at interfaces conspire to affect structure and performance of OPV devices based on the BHJ concept.

467773 Elucidating the Charge Transfer Mechanism in Conjugated Radical Polymers

Abstract:

Conjugated polymers bearing pendant nitroxide radicals have been an intriguing research topic in energy storage due to their dual electroactive nature with different redox potentials within the same polymer backbone. Recently, we demonstrated polythiophenes bearing pendant nitroxide radical groups such as the TEMPO radical (2, 2, 6, 6-tetramethyl-1-piperidinyloxy). Polythiophenes possess a high oxidation potential with tunable electronic conductivity, and the organic radicals possess a very high electron transfer rate. However these polymers exhibited worse battery performance than that of either PTMA (the polymeric form of TEMPO radical) or polythiophenes alone. One reason for this phenomenon is that conjugated polymers and nitroxide radicals are both redox active and that intra and inter-charge transfer is possible. As such, it is relevant to investigate the charge transfer pathways in such a system. In this study, we report the systematic study of polythiophenes carrying pendant TEMPO radicals. By engineering the alkyl chain spacer length between the thiophene ring and TEMPO radical, we tune the electronic and electrochemical properties of these electroactive polymers. In addition, the charge storage mechanism during the galvanostatic charge/discharge process is also interrogated and the respective contributions from each redox moiety are also discussed. Evidence of internal charge transfer is observed using a potentiostatic pulse technique. These results provide a path forward in the design of conjugated radical polymers for energy storage.

471685 The Solar Fuels Research Program within the Australian Solar Thermal Research Initiative (ASTRI)

Abstract:

The advanced concentrated solar thermal (CST) technologies being developed in the Australian Solar Thermal Research Initiative (ASTRI) can be applied for electricity generation and production of chemical fuels. ASTRI is an eight-year research program supported by the Australian Government, through the Australian Renewable Energy Agency (ARENA). The ASTRI program is a consortium of leading Australian universities and CSIRO, and in close partnership with several international collaborators. In particular, this solar fuels research program within ASTRI aims at demonstrating production of liquid fuels to increase the share of CST in Australiaâ€™s energy supply and lower greenhouse gases emissions.

With a vast array of possible processes and parameters involved in the conversion of solar heat to liquid fuels, a key part of this project has been to identify and prioritise potential pathways. A number chemical pathways, using various feedstock, are being developed and analysed. These include low cost fossil fuels, biomass, CO_2 and H_2O. Technologies being developed to support these processes are:

- A solar hybridised dual fluidised bed gasifier for processing of dry solid feedstocks, such as lignite, biomass or their mixture;
- A solarised supercritical water gasifier for processing of wet solid feedstocks, such as micro-algae;
- A solar redox reactor for syngas production from H_2O and CO_2 by redox cycling;

The proposed solar fuels technologies is complimented by with the development of processes required to convert the syngas, which produced by the above processes into high-value liquid hydrocarbon fuels. Technologies being investigated to support the advancement of solar thermal production of liquid fuels are:

- Catalytic materials that reduce the sensitivity of the Fischer-Tropsch (FT) reactor to variability of the quality and quantity of the feedstocks and/or lowering cost for the process
- Advanced Sabatier reaction to utilise the waste stream of CO2 emitted from the proposed solar fuels technologies by targeting enhanced production of higher value products than its conventional use for methane, such as propane and methanol.

The overall aim of this paper is to estimate the cost of the production of a drop-in solar fuel (interchangeable with convention fuels) that can be chosen as realistic target with a goal to produce liquid fuels at a cost well below AUD1.50/L (excise-free at the gate of the plant) in Australia.

459558 Solar Thermochemical CO_2 Splitting Via Redox Cycling with Low-Cost Metal Oxide Nanostructures

Abstract:

Solar Thermochemical CO_2 Splitting via Redox Cycling with Low-cost Metal Oxide Nanostructures

Xiang Gao[1], Peter Kreider[2], Alicia Bayon[3], Thomas Gengenbach[4], Jim Hinkley[3], Wojciech LipiÅ„ski[2], Antonio Tricoli[1]*

[1]Nanotechnology Research Laboratory, Research School of Engineering, The Australian National University, Canberra, ACT 2601, Australia

[2]Solar Thermal Group, Research School of Engineering, The Australian National University, Canberra, ACT 2601, Australia

[3]CSIRO Energy, P.O. Box 330, Newcastle, NSW 2300, Australia

[4]CSIRO Manufacturing, Bayview Avenue, Melbourne, VIC 3168, Australia

michael.gao@anu.edu.au

Solar thermochemical two-step metal oxide redox cycles are viewed as a viable pathway to produce syngas via water

splitting (WS) and carbon dioxide splitting (CDS). Syngas is an important intermediate in the creation of commercial products such as hydrocarbon fuels and ammonia. Using a carbothermal reduction step such as methane partial oxidation (MPO) can alleviate some of the process requirements by decreasing the reduction temperature and increasing the fraction of the oxygen carrier undergoing reaction. Recent studies on metal oxides for MPOâ€"CDS/WS solar thermochemical redox cycles have demonstrated potential improvements to the steady-state solar-to-fuel efficiency. A major challenge is engineering redox materials that can maintain reactivity and stability over many cycles, achieve high non-stoichiometry in the reduction step, and promise large-scale production at low cost.

Here, we present a novel low-cost earth abundant metal oxide nanostructure with as much as ten times higher redox capacity and up to 60% faster reaction kinetics compared to state-of-the-art ceria nano-powders. This composite shows promising long-term stability and maintains a significant fraction of its initially high syngas production rates over multiple MPOâ€"CDS cycles. Structural properties such as crystal phase composition and size, specific surface area, pore volume and particle size distribution were characterised before and after thermochemical cycling. These initial results demonstrate a promising pathway towards development of stable, reactive and low-cost redox material for two-step solar thermochemical fuel production.

460593 Evaluating the Effect of Modeling Variables and Experimental Conditions on Material Development for Solar Thermochemical Water Splitting

Abstract:

While solar energy is the most abundant renewable energy resource, the capture, storage, and distribution of it remains a challenge. Solar thermochemical water splitting (STWS) provides a promising route for efficient utilization of this disperse resource since it allows for use of the entire solar spectrum to convert water to an energy dense fuel, H_2. However, despite a significant number of materials having been examined, an optimal redox material to drive this process has yet to be developed. In this work we focus on developing standardized procedures, both computational and experimental, for assessing new STWS active materials. First, we apply density functional theory (DFT) to assess the effects of a number of variables, including crystal structure, cation disorder, and magnetic ordering, on the predicted solar thermal water splitting behavior of spinel and perovskite metal oxides. Accurate modeling of materials using DFT is particularly challenging for high temperature applications, but we show that all of the variables evaluated can critically impact the predicted STWS behavior and must be modeled accurately in order to predict high-temperature material properties. Experimental results from thermogravimetric analysis and redox cycling in a stagnation flow reactor are presented to validate DFT models. Second, direct comparison of experimental results for new active materials is extremely challenging given the wide variety of water splitting conditions and apparatus configurations used. We evaluate a number of known water splitting metal oxides, including ceria, hercynite (spinel), and SLMA (perovskite), at various experimental conditions in order to allow for a direct comparison and to aid in the effort of developing standard experimental procedures for STWS material development.

466776 Continuous Splitting of CO_2 into CO and O_2 Using an Isothermal Redox Membrane Reactor

Abstract:

Renewable liquid hydrocarbon fuels can be produced with two-step thermochemical redox cycles that convert CO_2 and H_2O with concentrated solar energy into renewable CO, H_2 (syngas) and O_2. Syngas can be further processed into liquid hydrocarbon fuels using commercially available Fischer-Tropsch technology. Typically, solar-driven splitting of CO_2 and H_2O are temperature-swing processes with cyclic production of syngas at lower temperatures and O_2 at higher temperatures. Alternatively, this work develops an isothermal redox membrane reactor that continuously produces O_2 and oxygen vacancies at the outside of a metal oxide membrane. The formed oxygen vacancies migrate along an oxygen vacancy gradient across the membrane. At the inside of the membrane the oxygen vacancies are consumed via oxidation of the membrane with CO_2 or H_2O, yielding continuously CO or H_2, respectively. This work employs thermochemical equilibrium computations to determine the thermodynamic limitations of this concept as a function of process parameters and membrane material. The concept is demonstrated experimentally for CO_2 splitting into CO and O_2 using tubular ceria redox membranes and inert alumina membranes as reference. Steady-state gas production rates are analyzed in-process using gas-chromatography. The membranes are characterized via scanning electron microscopy, X-ray diffraction, and energy dispersive X-ray spectroscopy. Thermodynamic limitations are confirmed experimentally via parametric investigation of the effect of operating temperature, partial pressure of O_2, and metal doping of the ceria membranes on the CO and O_2 production rates.

468650 Experimental Investigation of Methane Reformation Via the Nonstoichiometric Ceria Redox Cycle

Abstract:

Dry reformation of methane via the nonstoichiometric ceria (CeO_2 â€“ $CeO_{2-Î´}$) redox cycle was theoretically and experimentally examined for converting high-temperature solar process heat to syngas. The aforementioned cycle is conducted through two phases: (1) endothermic reduction of ceria in methane and (2) exothermic oxidation of the reduced ceria with carbon dioxide. In both regimes, chemical equilibrium calculations suggest total selectivity and favorability towards syngas production above 827 Â°C, enabling isothermal operation and providing motivation for the following experimental evaluation. An array of seven xenon short arc lamps provided solar-simulated, high-flux radiation to indirectly irradiate a packed bed of ceria situated within a cavity receiver. The endothermic reduction of ceria in methane was demonstrated at three discrete temperatures: 950, 1035, and 1120 Â°C. Upon completion of reduction, nonstoichiometry (Î´) equaled 0.07, 0.21, and 0.24, yielding methane conversions of 9, 41, and 51 % respectively. Subsequent oxidation was first performed with oxygen to indirectly quantify carbon deposition. This proved to be minimal, but its overall long term effect on cycling stability is still under investigation. Finally, isothermal operation was executed at 1120 °C and the solar-to-fuel conversion efficiency, defined as the ratio of calorific value of syngas (H_2 and CO) produced to the combined input solar radiative energy and the caloric value of methane converted, was measured as 3.71 % average and 4.56 % peak. Results indicate that a packed bed configuration is not optimal for this process and further reactor optimization efforts are underway.

464685 Membrane-Free Photo-Electrochemical Cell Design Involving Flow Field Hydrogen Separation

Abstract:

In order for solar fuel research to make an impact on the future of energy, it needs to transition from materials to devices. In the absence of a breakthrough in photo-absorber materials that fit all the requirements of water splitting, reactor designs that can operate efficiently with sub-optimal (but inexpensive materials) are crucial to realise this.

By applying research into membrane-free electrolysers to photo-electrochemical cells (PEC), the performance of such reactors can, theoretically, surpass many designs in literature. The hydrodynamics of the flowing electrolyte between two parallel plates can be controlled through judicious choice of Reynolds and Pechlet numbers for each of the dissolved product species so that convective forces dominate. This would allow product separation of the dissolved hydrogen and oxygen by fluid flow alone.

If the photo-absorber substrate is transparent (e.g. thin film on FTO, Fluorine-doped Tin Oxide), then PEC flow cells could be stacked perpendicular to the light source. This simple, compact, design would improve the overall solar to hydrogen efficiency as the light that is not captured by the photo-absorber on the first pass, is captured on the second or third thus creating a large photo-electrode surface area per unit volume.

A multi-physics model of the proposed cell arrangement has been developed and solved using MATLAB to implement a finite-difference numerical scheme. Using the model, it is possible to optimise fuel production, overall solar-to-hydrogen efficiency and pumping cost using a cost analysis of the various processes.

The results of the model show that there are multiple benefits of such a reactor design, not least in terms of reduced capital cost due to membrane free operation. It also crucially relaxes two key performance constraints of photo-absorbers, namely a) the photo-voltage required is reduced due to lowered ohmic losses in the electrode gap and b) the removal of the classical compromise between the optical absorption and excitation path length. This reactor design could thus help to realise the potential of sub-optimal semiconductor material such as ultra-thin hematite, which have a very high quantum efficiency but low optical absorbance.

We shall further touch on issues of bubble formation and high pressure operation to highlight elements for potential process intensification, also with respect to concurrent solar energy capture below the 1.23 eV threshold.

Experimental verification of the theoretical/model results will be demonstrated with a single membrane-free flow cell employing stable and inexpensive Fe based materials.

474681 Solar High-Temperature Processes for Solar Fuels Production and Solar Energy Storage Applications

Abstract:

When solar energy is employed as the energy source for the production of the raw materials for the synthesis of fuels, the latter are characterized with the term “solar fuels”. In the broad sense this term can contain in addition to “solar hydrogen”, synthetic liquid hydrocarbons and alcohols that are produced from reactions between H_2 and CO (syngas) that have originated from solar-aided dissociation processes as well as metal powders obtained by solar thermal reduction of metal oxides. Among the various solar energy variances that can be employed to produce such fuels, the thermochemical route uses solar heat at high temperatures supplied by Concentrated Solar Power (CSP) systems – i.e. special mirror assemblies that track the sun - for performing various high-temperature reactions that produce hydrogen or syngas from transformation of various fossil and non-fossil fuels. Such chemical reactions are steam methane (natural gas) reforming (SMR) or CO_2 (or dry) methane reforming (DMR), (reactions (1) and (2) respectively), gasification of solid carbonaceous materials like coal or biomass, or Water Splitting (WS) to hydrogen and oxygen; this last reaction often in combination with Carbon Dioxide Splitting (CDS) to CO and O_2, to produce syngas. In addition, CSP systems can be employed alternatively to photovoltaics for hydrogen/syngas synthesis indirectly, e.g. by supplying the (solar thermal) electricity for high-temperature electrolysis of steam or of steam/CO_2 mixtures.

The so-called “closed-loop” WS thermochemical cycles (TCs) are a series of consecutive chemical reactions (≥ 2), their "net" sum being the splitting of H_2O to H_2 and O_2, where the maximum-temperature (endothermic) step takes place at a temperature lower than that of the single-step water decomposition chemical reaction. Nevertheless, since they involve a highly endothermic step, they need the input of external energy. The most prominent examples of such solar-aided thermochemical cycles are cycles from the so-called sulphur family (the hybrid sulphur cycle described in reactions (3)-(5) and the sulphur-iodine process) and two-step redox cycles via multivalent metal oxide pairs, as in reactions (6) and (7). Each particular route exhibits advantages as well as drawbacks. An advantage of the latter route vs. the Sulphur-based ones is that the same thermochemical principle can be exploited for CDS as well for the production of CO (reaction (8)) that can be further combined with the H_2 produced via reaction (7) to syngas. On the other hand the high-temperature endothermic step of the sulphur cycles, SO_3 splitting to SO_2 and O_2 requires much lower temperatures than WS/CDS.

$$CH_4 + H_2O + (\Delta H) \quad 3\ H_2 + CO \qquad ...(1)$$

$$CH_4 + CO_2 + (\Delta H) \quad 2\ H_2 + 2CO \qquad ...(2)$$

$$2\ H_2SO_4\ (l) + (\Delta H) \rightarrow 2\ SO_3\ (g) + 2\ H_2O\ (g) \qquad ...(3)$$

$$2\ SO_3\ (g) + (\Delta H) \leftrightarrow 2\ SO_2\ (g) + O_2\ (g) \qquad ...(4)$$

$$2\ SO_2\ (g) + 4\ H_2O\ (l) \rightarrow 2\ H_2SO_4\ (l) + 2\ H_2\ (g)\ \text{(electrolysis)} \qquad ...(5)$$

$$MeO_{oxidized} + (\Delta H) \rightarrow MeO_{reduced} + \tfrac{1}{2}\ O_2\ (g) \qquad ...(6)$$

$$MeO_{reduced} + H_2O\ (g) \rightarrow MeO_{oxidized} + H_2\ (g) + (\Delta H) \qquad ...(7)$$

$$MeO_{reduced} + CO_2\ (g) \rightarrow MeO_{oxidised} + CO\ (g) + (\Delta H) \qquad ...(8)$$

Similar chemistry schemes can be employed in the same solar thermal power plant for the so-called thermochemical storage (TCS) of solar energy. For instance, Sulphur-based cycles can be modified to produce solid Sulphur in next steps of the cycle (disproportionation reaction (9)), which can be then combusted in air to produce high-temperature heat and SO_2 (reaction (10)). Similarly, the heat produced by e.g. a solar receiver during on-sun operation can be used to power an endothermic redox oxide reduction like (6); should this reaction be completely reversible the thermal energy can be recovered completely by the reverse reaction (11) taking place during off-sun operation. The "net" result of such cycles (reactions (3), (4), (5), (9), (10) or (6), (11)) is not the production of a particular chemical but the exploitation of the heat effects of the reactions for the "storage" of solar heat.

$$2H_2O(l) + 3SO_2(g) \rightarrow 2H_2SO_4(aq) + S(s)\ \text{(disproportionation)} \qquad ...(9)$$

$$S(l) + O_2(g) \rightarrow SO_2(g) \qquad ...(10)$$

$$MeO_{reduced} + O_2\ (g) + (N_2\ (g)) \rightarrow MeO_{ox} + (N_2\ (g)) + (\Delta H) \qquad ...(11)$$

Large scale production of solar fuels will require the economies of scale offered by heliostat fields with central tower receivers that can comfortably generate solar thermal fluxes in the MW capacity, although the deployment of such fields to date has been limited to electricity generation. For the efficient design and operation of solar aided-reactors for such processes, concepts from "traditional" chemical reactor engineering should be combined with ways to achieve efficient heating of the reactor via concentrated solar irradiation. Among the endothermic steps of the reaction schemes above, reforming and SO_3 splitting are catalytic reactions of gases on a solid catalyst surface. In contrast redox oxide-based cycles involve gas-solid reactions where the solid reactant (metal oxide) is not a "catalyst" present in much smaller quantities than those of the gaseous reactants, but a reactant itself, with non-negligible mass, that not only has to be heated to the reaction temperature but gets progressively depleted during the course of the reaction, having to be replenished. Therefore it either has to be fed constantly into the reactor if the reactions are to be performed in a continuous-mode, or alternatively, practical ways have to be "invented" for its' in-situ regeneration imposing rather challenging reactor operation conditions.

All processes described above, reforming, splitting and TCS, involve not only gas-solid chemical contact and reaction but heat exchange between either the solar receiver and the heat transfer fluid (SMR/DMR, WS/CDS) or the heat transfer fluid and the storage medium (TCS). The major technical challenge for these applications lies in the proper design and operation of reforming/splitting/storage reactors that will operate simultaneously and efficiently as heat exchangers. The two main research tasks are the improvement of solar interfaces and integrated heat recovery schemes in the one hand and solving the main materials-related issues and providing the right functional materials at reasonable costs in the other hand. The present analysis gives an overview on the recent developments and state-of-the art of those concepts at DLR as well as at international level, emphasizing on the most important performance parameters involved and the commonalities that can lead to convergence of reactor concepts employed up until now in different applications. Finally, the current outlook on further potential and necessary developments in the areas of solar fuels production and the use of chemical reactions to store solar heat is discussed.

448616 CO_2/Brine/Rock Interactions in Lower Tuscaloosa Formation

Abstract:

Saline aquifers are the largest potential continental geologic CO_2 sequestration resource. Understanding of potential geochemically-induced changes to the porosity and permeability of host CO_2 storage and sealing formation rock will improve our ability to predict CO_2 plume dynamics, storage capacity, and long-term reservoir behavior.

Experiments exploring geochemical interactions of CO_2/brine/rock on saline formations under CO_2 sequestration conditions were conducted in a static system. Chemical interactions in cores samples from the Lower Tuscaloosa formation from Jackson County, Mississippi with exposure CO_2 saturated brine under sequestration conditions were studied through six months of batch exposure. The experimental conditions to which the core samples of Lower Tuscaloosa Massive Sand and Selma chalk were exposed to was a temperature of 85 °C, pressure of 23.8 MPa (3,500 psig), while immersed in a model brine representative of Tuscaloosa Basin that was mixed with CO_2. Computed tomography (CT), X-Ray diffraction (XRD), Scanning Electron Microscopy (SEM), brine chemistry, and petrography analyses were performed before and after the exposure. Permeability measurements from the sandstone core sample before and after exposure showed a permeability reduction. No significant change of the permeability measurements was noticed for the core sample obtained from Selma chalk after it was exposed to CO_2/brine for six months. These results have implications for performance of the storage interval, and the integrity of the seal in a CO_2 storage setting.

In addition, a numerical core scale model was also developed to simulate reactive transport with porosity and permeability change of the sandstone from the Lower Tuscaloosa formation and Selma Chalk rock from the sealing formation above the Lower Tuscaloosa formation. The model predicted a 5.2% permeability decrease in the sandstone after 180 days of exposure to CO_2-saturated brine, which was close to laboratory-measured permeability results. For the Selma Chalk sample, there was a very small porosity decrease in the interior of the sample after 180 days of exposure. The model predicted no significant change of permeability for the Selma Chalk sample after 180 days of exposure, which was consistent with laboratory-measured permeability results.

455391 Biofuel Production Via Thermochemical Conversion of Algae and Optimizing Coproduct Utilization

Abstract:

Hydrothermal liquefaction (HTL) is a method for thermochemical conversion of wet algae that can produce energy dense biocrude oil as well as gaseous, aqueous and solid by products. In this study, three different algal species such as Galdieria, Kirchneriella and Micractenium were used as feedstock for the biofuel production. Process parameters like reaction temperature, time and solids loading were optimized to achieve maximum biocrude oil yield. A bench top 250 mL batch reactor was used for the HTL experiments and dichloromethane was used as a solvent for extraction. The maximum biocrude oil yield of 45.5% was obtained for Kirchneriella species at reaction conditions of 300 C, 90 bar, 30 min and 10% solids loading. Since the maximum biocrude oil yield achieved was less than 50% and to make this process economically viable, other co-products need to be utilized which adds value to the algal biorefinery. The aqueous phase will be rich in carbohydrates and soluble proteins and can be recycled to the cultivation facility. This nutrient rich aqueous phase upon required dilution was successfully used to cultivate algae. The biochar produced during the hydrothermal liquefaction has significant amount of phosphorous and nitrogen. These components were extracted as phosphates and ammoniacal nitrogen and were used as macronutrient supplement to grow algae. The growth experiments were investigated using a tissue Culture Roller drum apparatus with 16 mm borosilicate glass tube photo reactor and confirmed using a Microplate Assay at 40â°C and 2-3 % COâ,, in an incubator. The remaining biochar was dried and used to produce porous carbon. Various analytical instruments such as gas chromatography mass spectrometry (GC-MS), CHNS/O analyzer, spectrophotometer, semi micro calorimeter and inductively coupled plasma optical emission spectrometry (ICP-OES) were used to analyze algal biomass and different products. In conclusion, HTL along with coproduct development has the potential to make algal biofuels sustainable.

456357 Hydrothermal Pretreatment of Rice Husk with Biodiesel Waste As the Reaction Medium

Abstract:

Agricultural waste is becoming more popular as cellulosic biomass for bioethanol production because it is not competing with human food supply. Rice is the largest agriculture product of Japan including in the small district in Oasa, Kita Hiroshima, Hiroshima Prefecture. As the consequence, rice husk is produced annually in large amount in this region. Rice husk is mostly remains unutilized and it contains cellulose. Therefore, it has the potential to be converted into bioethanol. Hydrothermal process is known as environmental friendly pretreatment technique to convert cellulosic biomass into various products including bioethanol since it is using hot compressed water solely as the medium. However, studies shown that catalyst addition improve the yield of products. Alkali catalyst like sodium hydroxide is often used in hydrothermal pretreatment since it is low cost chemical and easily obtained. Nonetheless, the reduction of catalyst cost will be beneficial to make bioethanol more economically competitive. Interestingly, in district of Oasa, Kita Hiroshima Japan, biodiesel has been produced from vegetable cooking oil waste and being utilized for diesel engine vehicles. This biodiesel production generates biodiesel waste that has very high pH (alkaline) and treatment is required to neutralized it before discarding it into the environment. In this research, the possibility of using the alkali characteristic in biodiesel waste to replace commercial alkali catalyst for hydrothermal pretreatment of rice husk is studied. The study was conducted using autoclave reactor made from stainless steel. The experiments were carried out at reaction temperatures of 150, 200 and 250 °C, reaction times of 30 min, and the biodiesel waste addition varied 0, 10, 20, 50 % v/v. To measure the effectiveness of pretreatment, enzymatic hydrolysis was employed after pretreatment to obtain glucose yield. The effect of biodiesel waste addition on rice husk conversion to glucose by hydrothermal pretreatment at various temperature was elaborated.

Thermal Resistance Network Model for Heat Pipe – PCM Based Cool Storage System

Sean Hoenig[1*], CJ Pan[2], Chien-Hua Chen[1], and Richard Bonner III[1]
[1]Advanced Cooling Technologies, Inc., Lancaster, PA 17601, USA

Concept Overview

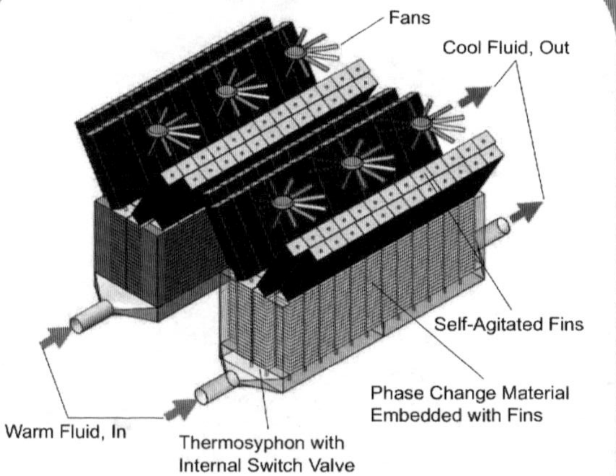

Figure 1. Schematic of proposed cool storage system

Motivation

- For the ARPA-e ARID Program, ACT proposed a supplemental cool storage system that utilizes an integrated heat pipe-phase change material (PCM) heat exchanger to transfer heat from a low-grade steam loop into a salt-hydrate PCM reservoir. During an operational cycle, the stored heat is rejected at night to the cooler ambient air, which solidifies the PCM for repeated use in the cooling system.
- To model the thermal performance and obtain a practical capital cost analysis of the system (<$150/kW$_{th}$), a transient numerical based thermal resistance network (TRN) was developed.
- A CFD based model is impractical to complete for a detailed parametric study and cost analysis. An Excel-based analytical design tool can demonstrate the optimization of design components, thermal capabilities, and overall capital cost to represent the technology's value proposition.

Model Configuration

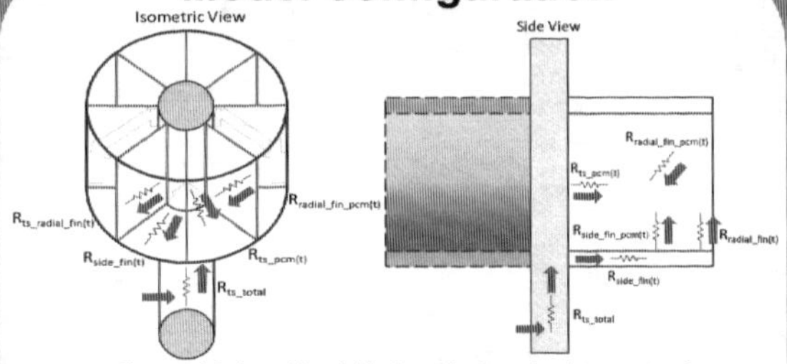

Figures 2 & 3. Isometric and side view of the thermal resistance network

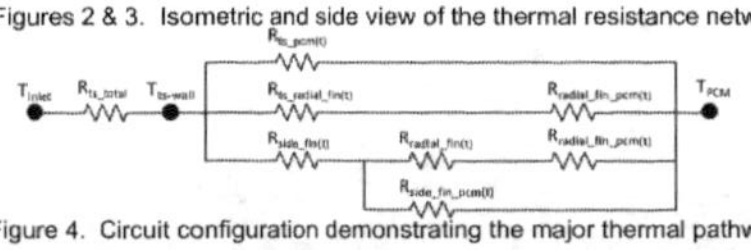

Figure 4. Circuit configuration demonstrating the major thermal pathways

Calculation Methods

- Based on Figure 4, a combined series-parallel resistance calculation is used to derive the changing melt front propagation with each time step. An energy balance based on the accumulated latent heat of PCM and heat into the system through the overall network is used to accomplish this task.

$$q = \underbrace{\frac{d(H\rho V)}{dt} = \frac{H\rho A_s \Delta L_c}{\Delta t}}_{\text{Accumulated}} = \underbrace{\frac{T_i - T_m}{R_{eq}}}_{\text{In}}$$

$$q_{total} = q_1 + q_2 + q_3$$

$$\Delta L_{c,1} = \frac{\frac{T_i - T_m}{R_{eq,1}}\Delta t}{H\rho A_{s,1}} \quad \Delta L_{c,2} = \frac{\frac{T_i - T_m}{R_{eq,2}}\Delta t}{H\rho A_{s,2}} \quad \Delta L_{c,3} = \frac{\frac{T_i - T_m}{R_{eq,3}}\Delta t}{H\rho A_{s,3}}$$

- As the PCM melts, a liquid PCM boundary layer develops, which is the major transient resistance pathway to be analyzed.
- A fin efficiency analytical method is used to account for the changing temperature gradients along the various fin structures as the PCM melts.

Results

- To validate the use of this model, comparisons with ANSYS Fluent were done.
- Figure 5 comparison was completed at experimental conditions and the melt time results match well.

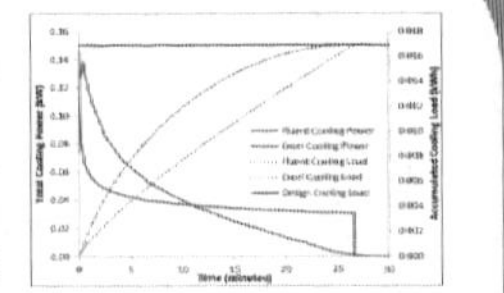

Figure 5. Model comparison with ANSYS Fluent

- Parametric studies determine the optimum trends in design (e.g. thermosyphon diameter, fin spacing).
- Larger diameter thermosyphons are beneficial due to the insignificant price increase and larger achievable heat flux, resulting in lower melt times.

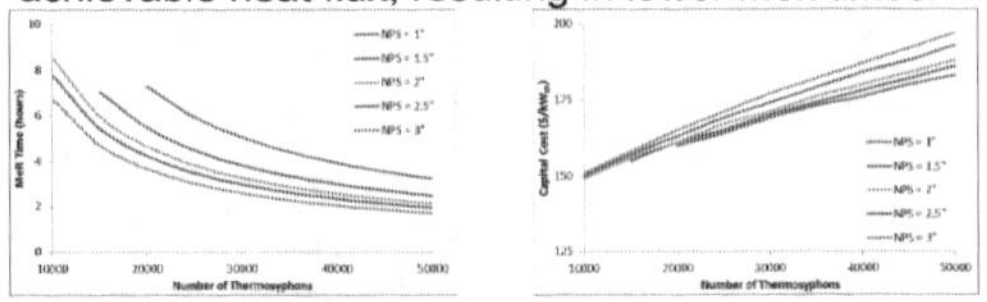

Figures 6 & 7. Initial parametric study to demonstrate thermosyphon sizing optimization

Future Work

- A numerical Excel-based thermal resistance network can demonstrate the thermal performance and capital cost of this technology.
- Further progression of the aforementioned high fidelity methods will ensure precise simulation of the PCM melting/solidifying process.
- Extensive parametric studies will demonstrate a holistic understanding of optimum system design.

Acknowledgements

This project is supported by ARPA-e Award DE-AR0000582.

*Corresponding author: Sean Hoenig, R&D Engineer
E-mail: Sean.Hoenig@1-ACT.com

The author would like to thank his coworkers at [1]ACT and the [2]Lehigh University team for their efforts in support of this work.

459853 Life Cycle Carbon Footprint of Renewable Electricity Generation from Aspen Forest Intensive Harvest in Wisconsin, USA

Abstract:

The Renewable Portfolio Standard mandates increased production of electricity from renewable sources. There are large quantities of timber net growth in excess of removals in the Great Lakes region. The excess timber wood can be a promising feedstock to address the forest products industry while also providing renewable feedstock for energy production. In this study, we aim to evaluate the life cycle greenhouse gas (GHG) emission of renewable electricity produced from cogeneration of aspen biomass in Wisconsin. The typical harvest age for aspen is 40 years. We propose that one way to provide biomass for electricity generation is to intensively harvest aspen by shortening rotation cycles. First, soil carbon (0-60cm depth) was measured from 4 age classes of aspen, namely, 10yr, 20yr, 30yr, and 40-60yr. CO_2 emissions due to intensive aspen harvest were estimated based on differences in soil C levels of various age classes. Next, the optimal biomass supply chain will be determined by minimizing the economic, environmental and social cost of shipping biomass from forest to the cogeneration facility. Finally, a life cycle assessment model is used to quantify the cradle-to-grave GHG emission of renewable electricity, including soil C change, wood harvest, trans-loading and transportation, and combustion in fluidized bed boiler. The life cycle GHG emissions of renewable electricity will be compared with that of Wisconsin grid mix.

460758 Modeling Differential Catalytic Hydrotreatment of Surrogate Compounds for Pyrolysis Oil

Abstract:

The final product distribution of catalytic hydrotreatment of pyrolysis oil depends strongly on processing parameters, especially temperature, hydrogen to oil ratio, and catalyst choice. In this study, we have investigated the hydrotreatment of individual surrogate compounds (anisole, m-cresol, guaiacol, and furfural) in a continuous reactor maintained at 7 bar ranging from 250-350°C, over 3 different catalyst types: acidic (ZSM-5), hydrogenation (Pd/C), and bifunctional (Pt/Al_2O_3). Reaction pathways were studied for each temperature and catalyst; higher temperatures led to greater conversions and more cracking, cracking and rearrangement of methyl groups was seen with ZSM-5, ring saturation dominated with Pd/C, and Pt/Al_2O_3 showed a combination of these functionalities. These tests aim for low conversions to study differential rate kinetics that can be used to model the process. This model uses a lumped parameter approach that groups reaction pathways together and sorts compounds by functional groups.

461080 Pretreatment of Black Liquor to Improve the Recovery Boiler Operation

Abstract:

This work presents a possible method to decrease the alkali related problems in a recovery boiler. This will improve the performance and make a better utilization of the renewable feedstock. The new process design is done by a thermal pretreatment in order to minimize entrained and evaporated alkali.

Black liquor is an intermediate stream that is formed during the production of kraft paper pulp where wood is used as raw material. It contains most of the lignin from the wood but also some parts of the hemicelluloses and celluloses. In addition, it also contains the ingredients of the cooking liquor used to deliberate the different wood components. This makes the black liquor highly rich in alkali and sulfur. It is not uncommon with a portion of 20 wt% for sodium and 7 wt% for sulfur on dry material basis. In order for the pulp mill to close its material balances and to produce necessary steam and electricity, the black liquor is combusted in a recovery boiler. Before the combustion, the black liquor is concentrated to about 80% before it is fed into the boiler through the liquor guns.

During combustion, there are three distinct phases for each individual droplet. These are: drying, pyrolysis and combustion. Due to the high concentration of alkali metals (sodium and potassium), the amount of these that are entrained in the gas flow or just evaporated due to the high temperature is rather large. This creates problems in the boiler due to deposition of material on the heat exchangers, which might cause plugging and thus decrease the heat transfer but also by blocking the passage and create an unwanted pressure drop. Another problem is the corrosive combination between alkali and chlorine (chlorine compounds are present in the original wood material and ends up in the black liquor) which are corrosive and thus detrimental for especially the super heater tubes inside the boiler. Most of the sodium reacts rapidly and forms sodium carbonate or sodium sulfur oxides such as sodium sulfates or sodium thiosulfates. Thereafter, the pyrolysed droplet starts to combust but as a design feature of the recovery boiler, the droplet should contain some carbon and not be completely burned out when it reaches lower part of the boiler in order to reduce the sulfur compounds to sulfides.

The thermal pretreatment has been performed in our laboratory on black liquor from two Swedish kraft pulp mills at different temperatures and heating rates. In combination to laboratory test for the amount of sodium leaving the black liquor during pyrolysis, we have also performed thermodynamic modelling of the lower part of the recovery boiler to investigate if all sodium would stay in the melt or if it is evaporated in this last stage. It has been previously showed that the sulfur oxide compounds in the black liquor will mainly form hydrogen sulfide in oxygen lean environment at temperatures below 600°C which is the condition for mild pyrolysis but the equilibrium also shows that it will completely form sodium sulfide at high temperatures (above 800°C) which is the condition in the smelt bed.

Within this project we have showed that the thermal pretreatment is a possible way to minimize the sodium loss due to entrainment and evaporation during the free fall period and that the equilibrium calculations indicates that this higher amount will stay in the molten phase and thus increase the direct recovery of sodium from the boiler. This could give higher efficiencies and a lower cost for boiler operation due to lower need for soot blowing steam, possible higher temperatures in the steam cycle and less maintenance.

461121 Characterization of Physical Properties of Macroalgal Pyrolysis Products for Process Simulation

Abstract:

Pyrolysis is a thermochemical conversion of carbon based material to lower molecular products. This conversion is performed at elevated temperatures ranging from 450°C to 500°C in the absence of oxygen and the products are solid carbon-rich char, organic liquids called bio-crude, and hydrocarbon gasses. A major goal of pyrolysis in the field of biofuel production is the maximization of bio-crude yield through appropriate optimization of process parameters. Since the number of components in macroalgal bio-crude is vast and mechanisms of product generation and distribution from macroalgal pyrolysis are not very well understood, accurate and representative simulation of such process is challenging.

Nevertheless, this study presents an approach to characterize the liquid product mixture, bio-crude, for simulation of macroalgae pyrolysis using Aspen Plus. Primarily, bio-crude was modeled in order to simulate the conversion accurately and in time efficient manner. Original liquid product mixture contained around 200 compounds (detected experimentally by GC/MS analysis) in water soluble and water non-soluble organic phases. In this study this number was reduced to 21 representative compounds. In its databases Aspen Plus contains information about physical properties of vast number of compounds. Often, simulations of thermochemical conversions of biomaterials require definition of properties for components which are not found in the databases. This study demonstrates the methodology of reducing compositional experimental data based on chemical structure and boiling point range similarities and translating into components which are already contained in the Aspen databases. This is, in essence, a crucial step of simulating a complex thermochemical conversion.

462358 Experimental Investigation on Expanded Granular Sludge Bed Reactor to Maximize the Methane Gas Production in the Biogas

Abstract:

Experimental Investigation on Expanded Granular Sludge Bed Reactor to Maximize the Methane Gas share in The Biogas Production

Haider Al-Rubaye, Joseph D. Smith, Ph.D.

Chemical & Biochemical Engineering Department, Missouri University of Science and Technology, Rolla, USA

Abstract

The increase in the global temperature, the greenhouse gas emissions, air pollution and the instability in the fossil fuel prices played an important role in making the world more aware of the renewable energy developments, By most definitions alternative energy(have low carbon emissions, a distinction which separates it from renewable energy which may or may not have significant environmental impact.

The Anaerobic Digestion Process, is one of the most contributed renewable energy sources as it generates biogas that could be used for heating and powering purposes, it could be used also in all equipment designed for the natural gas with minimal adjustment to account for the lower BTU content of the methane gas, upgrading the biogas sometimes is preferred in order to increase the composition of the methane gas in it. In general, the biogas consists of methane gas about 50-70% and carbon dioxide 30-50%.

There are many ways to upgrade the biogas and increase the methane share in it. One of the methods is a biochemical process, which basically introduce the hydrogen gas into the process in order to react with the carbon dioxide. In this way, methane increases about 20-30% while the carbon dioxide would decrease about 10-25%.

The experiments would be conducted in an expanded granular sludge bed reactor which treats the industrial and municipal waste water to generate biogas, sensitivity analysis will be performed to maximize the composition of the methane in the biogas. Moreover, the effect of the hydraulic retention time and the substrate concentrations will also be investigated.

462833 Modelling of Exponential and Stationary Phases in Microalgae Growth Using a Population Balance Equation

Abstract:

Being the fastest growing biomass on earth [1], makes microalgae an interesting bio-product [2] source and, depending on the environmental conditions, it is possible to further enhance the productivity. In the past few years, numerous works have focused on developing models able to predict the growth [3, 4], taking under consideration parameters such as temperature, light intensity, nutrients concentration, etc. However, all of these models do not consider the growth as a consequence of the cell life cycle. In other words, the reactor is considered as black box, that under certain conditions (input) gives a correspondent growth (output). Differently, aim of this work is to consider the whole life cycle of the cells, which means that the cell will first grow, then get mature and ready for division and, finally, give birth to daughter cells.

Looking at the downstream process, we know that the harvesting is one of the most energy consuming steps. Knowing that some microalgae may have a wide size range [5], we can use the model to predict which is the most suitable moment to collect the cells. On one side we have that larger cells will require less energy for their collection (assuming that the cell density doesnâ€™t change). Also, the moment in which harvesting is actuated, may affect the next batch. For example, if the collected population is old, the initial growth (in the subsequent batch) will be particularly good because the cells will be ready for reproduction and, at the same time, the amount of useful components (e.g. lipids) inside the cell will be much higher than the one present in younger cells.

The population balance equation has been used in numerous and diverse applications in science and engineering (dispersed phases and microbial populations to mention some). Also, applications related to microalgae can be found in literature [6]. The equation can allow to monitor the cell population over the time in different conditions. In this work, the objective is to simulate the exponential and the stationary phases of microalgae growth and evaluate the size

distribution.

The work has been experimentally validated. Inspired to open ponds, which are among the most common type of photobioreactors, the experiment was run using some open roof tanks (approximately 10 L) and the turbulence was maintained using some water pumps. Temperature was kept about constant using a temperature controller, connected to a heater placed inside the tank. Nutrients were introduced into the system at the beginning of the experiment, while lights were turned on during all the experiment The concentration of cells as well as the size distribution have been monitored daily. In few words, data related to mass concentration, number concentration and size distribution over the time were collected. The developed model can predict all of the mentioned parameters dynamically in the same experimental conditions.

References

[1] R. P. John, G. Anisha, K. M. Nampoothiri and A. Pandey, "Micro and macroalgal biomass: a renewable source for bioethanol," *Bioresour. Technol.*, vol. 102, pp. 186-193, 2011.

[2] M. A. Borowitzka, "High-value products from microalgaeâ€"their development and commercialisation," *J. Appl. Phycol.*, vol. 25, pp. 743-756, 2013.

[3] Q. BÃ©chet, A. Shilton and B. Guieysse, "Modeling the effects of light and temperature on algae growth: State of the art and critical assessment for productivity prediction during outdoor cultivation," *Biotechnol. Adv.*, vol. 31, pp. 1648-1663, 2013.

[4] H. Chang, Y. Huang, Q. Fu, Q. Liao and X. Zhu, "Kinetic characteristics and modeling of microalgae Chlorella vulgaris growth and CO 2 biofixation considering the coupled effects of light intensity and dissolved inorganic carbon," *Bioresour. Technol.*, vol. 206, pp. 231-238, 2016.

[5] K. Miklasz, *Physical Constraints on the Size and Shape of Microalgae,* 2012.

[6] E. Pahija, Y. Zhang, M. Wang, Y. Zhu and C. W. Hui, "Microalgae growth determination using modified breakage equation model," *Computer Aided Chemical Engineering,* vol. 37, pp. 389-394, 2015.

463767 Novel Process of Biofuel Production from Acid Oil and Hydrous Bioethanol Using Ion-Exchange Resin; II. Introducing Life Cycle Perspectives in Optimization of Operating Conditions

Abstract:

This research is separately reported in two presentations (Parts I and II) in sessions that best suit for the content. The two presentations therefore share parts of introduction section.

Liquid biofuels offer attractive advantages over other forms of biomass energy resources in terms of high energy density and ease of storage and transport. Bioethanol is useful as a gasoline additive (up to 10%), with production worldwide in 2014 of 94 million kL. Biodiesel, or fatty acid methyl ester (FAME), is also used as a fuel alternative (100%) to petroleum diesel, with worldwide production in 2014 at 30 million kL. However, there has been no significant increase in the production of these fuels in the past few years. The reason is that there is no profitability for producers owing to high production cost and no benefits for consumers because of unstable product quality and lower fuel economy and power. Moreover, emergence of next-generation automobiles (e.g., battery-powered and fuel cell electric vehicles) will lower the demand for gasoline. The water content of ethanol used as the gasoline additive should be less than 0.5 wt% and the dewatering process requires a large energy input. Thus, the necessity of anhydrous ethanol as gasoline additive fades and new technology which can use hydrous ethanol is desired. On the other hand, the diesel fuel continues to be used for heavy duty on-road commercial transportation, non-road construction, agricultural machines, railroads, and various types of ocean vessels. If new technology to economically produce high quality biodiesel can be developed, the necessity of biodiesel utilization will increase.

In Part I of this research[1)], biodiesel is produced in the form of fatty acid ethyl ester (FAEE) using hydrous ethanol by esterification of free fatty acid (FFA). The FAEE has higher cetane number and fuel economy and power than the

traditional fatty acid methyl ester (FAME) and FAEEâ€™s low temperature fluidity is also improved[2)]. However, the conventional technologies for esterification of FFA never use hydrous ethanol because the water in the reaction solution must be completely removed to obtain a high conversion. We have reported to produce FAME at more than 95% conversion by esterification of acid oil using a cation-exchange resin catalyst without dewatering and adding excess alcohol[3)]. This means that the catalytic activity of the resin is not inhibited by water. Thus, using the cation-exchange resin as solid catalyst the batch and continuous esterification experiments with hydrous ethanol are performed. The maximum water content in hydrous ethanol to obtain high conversion is clarified.

The experimental study exhibited how the rate of esterification reaction slows down as the water content becomes higher. Lower reaction rate leads to an increase in energy consumption, thus contributes negatively from a life cycle perspective by inducing more environmental impacts associated with energy. It will also lead to higher frequency of regeneration of resin, which requires additional loading. On the other hand, increased water content means lower duty for distillation of ethanol. To allow a system-wide optimization, in Part II of this study, modeling of the system with a scope that contains all the relevant product life cycles is performed. Using the data acquired from the experiments with continuous configuration, the abovementioned trade-off is highlighted using a case study with a Japanese example. The systems are modeled so that is possible to conduct a comparative LCA. For the bioethanol, the base case is set at the situation where fermented liquor is distilled and dehydrated to 99.5% to replace gasoline, whereas in the evaluated case, it is used to produce FAEE. For biodiesel, in the base case, FFA and methanol are subject to esterification reaction whereas in the evaluated case, FFA and hydrous ethanol at concentration of 90%-99.5% are assumed to be used for esterification reaction. The inventory data for the continuous production are obtained from the pilot scale experiments performed in this study. The details of the experiments are described in Part I[1)]of this report. Electricity consumption is continuously measured by built-in Watt-Hour meter. Molasses from cane sugar mills are assumed as the source of bioethanol. Yeast is used for fermentation to convert sugars to ethanol. The yeasts are recycled after the fermentation. Economic allocation is applied to the surplus yeast after fermentation as it is valuable as a fertilizer. In Japan, most of the cane sugar mills produce only the sugar as product, and the molasses are collected by a private company as raw material for other product. Heat and power required for sugar extraction is usually entirely supplied from bagasse boiler attached to the sugar mill. The bagasse is usually not enough if ethanol fermentation and distillation is required. Therefore, combustion of fuel oil in the bagasse boiler is assumed as the source of heat and power for bioethanol production, in the case study. The factors that influence the trade-off in life cycle greenhouse gas emissions and cost of operation over the water content in the bioethanol used for FAEE production is discussed.

Reference

1) N. Shibasaki-Kitakawa et al., Submitted for presentation at 2016 AIChE Annual Meeting

2) N. Shibasaki-Kitakawa et al., Kagaku Kogaku Ronbunshu, 42 (2016), 30-36.

3) N. Shibasaki-Kitakawa et al., Fuel, 139 (2015), 11-17.

464779 Advances in the Pyrolysis of Municipal Solid Waste Mixtures

Abstract:

Municipal solid waste (MSW) generation is one of the most critical tasks in environmental management and assuring public health of cities all over the world. Its storage in landfills is currently leading to many concerns, not only due to the scarcity of land available but mostly for the environmental issues related to it. According to the U.S. Environmental Protection Agency, landfill emissions contributed 18% to total methane emissions from 1990 to 2013 in the US. Therefore, alternative solutions for MSW disposal should be investigated. Thermochemical conversion offers an interesting option to address this issue, by providing a useful product (biofuel) with the possibility of reduce greenhouse gases (GHG) emissions. In a recent study by our research group we found that the production of alternative liquid transportation fuels using MSW can generate less GHG emissions compared to landfill storage even when considering mixtures of organic and plastic waste. However, little research has been reported about the effects of important operation conditions for this process.

In the present work a series of different operation conditions will be tested on a micro-pyrolysis scale coupled to a gas chromatograph-mass spectrophotometer. The objective will be to obtain processing conditions that increase bio-oil production. First, different mixture compositions of organic and polyolefin plastic waste will be tested, over the range of plastic waste blends up to 20% (wt.) in organic MSW (paper, cardboard, demolition wood). Pyrolysis temperatures from 500-600ËšC will be investigated to determine its effect on pyrolysis bio-oil composition and quality (oxygen content, molecular weight distribution). These findings will provide useful data to aid in the design and simulate economic and environmental effects of waste-to-fuels facilities, which appear as a promising alternative to solve waste management tasks by more sustainable means.

462697 Development of Methanol Synthesis Process Using Self-Heat Recuperation

Abstract:

Many methanol synthesis processes have reactant recycle system with product separation to achieve high conversion ratio of the whole process, as well as other chemical processes. For this product separation from reactant recycle stream, a gas-liquid separator has often been used and distillation column has often been used for product purification. However, this gas-liquid separator and distillation column are well known as energy consuming processes. In fact, a huge amount of coolant is required for the cooler located between reactor and the gas-liquid separator in the methanol synthesis process. Furthermore, there are two distillation columns are normally located in the methanol synthesis process to purify the crude methanol and these columns consumed a large amount of energy. In addition, some of the exothermic reaction heat are utilized for the preheating of the reactant stream, leading to reduction of recovered heat amount by steam generation.

Recently, authors have developed self-heat recuperation technology. In this technology, each process unit is divided into functions to analyze the process energy balance and not only the latent heat but also the sensible heat of the process stream can be circulated without any heat addition, leading to huge reduction of energy requirements in several chemical processes.

In this research, we investigated the feasibility of applying self-heat recuperation technology to a methanol synthesis process using industrial conditions and data and developed an innovative process for methanol synthesis process from the energy saving point of view. By installing the self-heat recuperation technology to the methanol synthesis process, the energy consumption of the process can be greatly reduced. In addition, all of the exothermic reaction heat of the methanol synthesis can be recovered by steam generation.

464428 The Impact of Fugitive Particulate Matter on the Environmental Health and Sustainability of Dry and Arid Regions

Abstract:

The developing arid and dry regions (i.e. North Africa, Middle East, Central Asia, and Australia) are experiencing a large wave of urbanization and industrialization, adversely affecting the area's air quality. In 2008, outdoor air pollution contributed primarily to 7.3% of the total deaths in the UAE (Gibson *et al.*, 2013). Lately, particulate matter (PM) pollution in these regions had become a topic of increasing concern (Nasser *et al.*, 2015; Tsiouri *et al.*, 2014), giving the recorded health hazards associated with its exposure. According to the air quality guidelines set by the World Health Organization, annual mean concentrations of $PM_{2.5}$ and PM_{10} should not exceed 10 µg m^{-3} and 20 µg m^{-3}, respectively (WHO, 2005). Nevertheless, annual PM levels estimated in several Middle Eastern cities between the years 2010-2012 have greatly exceeded these limits (WHO, 2014). Fugitive particulate matter (fPM) are unrestrained particles that escape to the atmosphere when applying a mechanical force on an exposed surface (e.g. emissions caused by wind shear, material transfer process, or other mechanical forces from non-point sources). fPM is also a substantial source of airborne pollution in these regions. However, accurate determination of fPM emissions and their impact on environmental health, humans' well-being, and overall sustainability of these regions has been an ongoing challenge for the air quality research community because of the induced health effects and the large uncertainty on their determination. Despite the efforts invested in developing emission inventories for standard air pollutants, it has been observed that (i) comprehensive spatially and temporally resolved emission inventories for fugitive PM sources in the developing arid regions are non-existent, (ii) the knowledge on fugitive (non-point) sources in the existing inventories is sparse (Pouliot *et al.*, 2012; Winiwarter *et al.*, 2009), and requires extensive attention.

In this study we present a holistic approach towards the assessment of the fPM impact to health and sustainability. We have selected the fPM emissions from loose Calcisols (i.e. soils with substantial accumulation of secondary carbonates) which are a characteristic soil for developing areas; usually very close to, or within residential areas. Multiple campaigns have been conducted to monitor fPM a) emissions from anthropogenic activities e.g. construction, b) building infiltration, and c) composition and size. We developed new emission factors for fPM from anthropogenic activities, assembled high resolution emission inventories, and finally assessed the impact of fPM on local air quality, environmental health, and sustainability of the involved society.

Acknowledgements

This publication was made possible by the support of the Qatar National Research Fund (a member of The Qatar Foundation) [NPRP 7 - 674 - 2 - 252] and the Ministry of Public Health [Contract No SCH/PA/18/2015] State of Qatar. We would like to extend our thanks to our research teams, owe to technical limitations not all contributing researchers have been included here. The statements made herein are solely the responsibility of the authors.

References

Gibson, J. M., J. Thomsen, F. Launay, E. Harder and N. DeFelice, 2013: Deaths and medical visits attributable to environmental pollution in the United Arab Emirates. *PLoS ONE*, **8**

Nasser, Z., P. Salameh, W. Nasser, L. A. Abbas, E. Elias and A. Leveque, 2015: Outdoor particulate matter (PM) and associated cardiovascular diseases in the middle east. *International Journal of Occupational Medicine and Environmental Health*, **28**:641-661.

Pouliot, G., T. Pierce, H. Denier van der Gon, M. Schaap, M. Moran and U. Nopmongcol, 2012: Comparing emission inventories and model-ready emission datasets between Europe and North America for the AQMEII project. *Atmospheric Environment*, **53**:4-14.

Tsiouri, V., K. E. Kakosimos and P. Kumar, 2014: Concentrations, sources and exposure risks associated with particulate matter in the Middle East area-a review. *Air Qual Atmos Health*:1-14.

WHO. (2005). WHO Air Quality Guidelines for Particulate Matter, Ozone, Nitrogen Dioxide and Sulfur Dioxide - Global Update 2005. In W. H. Organization (Ed.). http://www.who.int/phe/health_topics/outdoorair/outdoorair_aqg/en/.

WHO. (2014). Ambient (outdoor) air pollution in cities database 2014. http://www.who.int/phe/health_topics/outdoorair/databases/cities/en/.

Winiwarter, W., T. A. J. Kuhlbusch, M. Viana and R. Hitzenberger, 2009: Quality considerations of European PM emission inventories. *Atmospheric Environment*, **43**:3819-3828.

464865 Viologen-Catalyzed Carbohydrate Oxidation for Biofuel Cell Applications

Abstract:

Biomass has potential as a renewable source of electricity, but currently faces a number of challenges that inhibit its suitability for widespread use. One method of converting carbohydrates to electricity is with a biofuel cell. However, such fuel cells have been plagued by low reaction rates and incomplete conversion of the carbohydrate. Viologens are stable, non-biological catalysts that appear to enable nearly complete carbohydrate conversion. The process consists of a homogeneous reaction between viologen catalyst and the carbohydrate in solution, coupled with a second reaction (electrochemical or with oxygen) for re-oxidation of the catalyst. The purpose of this study is to investigate and quantitatively describe the reaction between the catalyst and the carbohydrate. To do so, experiments were performed as a function of temperature for a variety of carbohydrates obtainable from cellulose and hemicellulose at different molar ratios of viologen to carbohydrate. The rate of the reaction was tracked by spectroscopic measurement of the viologen concentration as a function of time. In addition, the extent of carbohydrate conversion was determined from the reaction products with use of 13C NMR. The reaction was observed to proceed rapidly for all of the biofuels tested; the most rapid rates measured were for the carbohydrates with fewer carbons. Of particular note is the observation of carbohydrate conversion >80% complete (i.e., ~20 of 24 electrons released from glucose). Finally, a comprehensive kinetic model was developed for the oxidation of carbohydrates by a viologen catalyst under anaerobic conditions. The results of this study support the feasibility of using viologen-catalyzed carbohydrate oxidation in a biofuel cell for power generation, and provide quantitative data needed for initial reactor design.

465067 Multi-Objective Optimization of Integrated Aspen Plus Unsteady-State Batch and Fed-Batch Fermentation and *in Situ* Gas Stripping Simulations

Abstract:

The batch Acetone-Butanol-Ethanol (ABE) fermentation process produces low final ABE concentrations, yields and productivity as a result of substrate and product inhibition to the microorganisms. Approaches to reduce product toxicity to the microorganisms include integrated fermentation and *in situ* product recovery techniques, such as gas stripping, adsorption, pervaporation, etc. Fed-batch fermentation reduces substrate inhibition by supplying substrate as it is consumed to keep the substrate concentration in the reactor below the inhibitory levels to the microorganisms.

Process simulation can predict the most informative experiments to conduct for process design, control and optimization. Previous Aspen Plus simulations and analyses of the ABE fermentation represented the batch and fed-batch processes with steady-state stoichiometric reactors with fixed product yields relative to key inputs, such as substrate (glucose) concentration. These steady-state representations of the batch and fed-batch fermentation processes are inadequate as these processes are inherently unsteady-state; the fermentation environment changes due time-dependent concentrations of cells, substrate, intermediates and products. This simulation approach results in Aspen Plus steady-state simulations that decouple the ABE fermentation kinetics from the fermentation environment and may not be representative of the integrated batch and fed-batch ABE fermentation and *in situ*gas stripping. The operating conditions of the fermentation must be directly linked to the fermentation kinetics in order to optimize productivity and control product profiles in the fermentation process.

To this end, this study will use a multi-objective optimization strategy to maximize the ABE productivity, yield and concentration by manipulating the operation conditions of integrated batch and fed-batch fermentations and *in situ* gas stripping processes. The integrated batch and fed-batch ABE fermentation process will be simulated as unsteady-state processes using a Fortran user kinetics subroutine linked to the batch reactor (RBatch) in Aspen Plus. This research can serve as a basis to evaluate the economic viability and technical feasibility of bioprocesses as a decision-support tool for researchers, investors and policy makers in the long run.

468409 Characterization Based Design of Ionic Liquids for CO_2 Absorption

Abstract:

Chemical engineering design problems have evolved in recent years from large scale bulk processes to targeted small batch approaches. As a result, product design is reaching beyond the traditional solutions to classic chemical engineering problems such as extraction, purification, removal of sulfur from fuel, CO_2 capture and sequestration, energy storage and biomass pre-treatment. Ionic liquids can provide environmentally benign solutions that can be tailored to specific process requirements. Due to the unique structure of ionic liquids, the cation, anion and length of the alkyl chain can be varied to create a molecule with specific physical properties. It has been estimated that over 10^{14} unique molecules exist at room temperature, so experimental trial and error methods will not be effective. More advanced methods must be created determine the ideal combination for a specific set of process requirements.

The purpose of this work is to determine the ideal ionic liquid to capture CO_2, a question which has been the topic of many experimental laboratory studies. However, only a small percentage of the potential ionic liquids have been synthesized and tested for the solubility of CO_2. This innovative approach decouples the solution from time-consuming laboratory study and streamlines the search for the most promising candidates that should be synthesized.

A characterization based group contribution method is combined with density functional theory to determine the ionic liquid that can most effectively absorb CO_2. Infrared spectra data contains descriptor data that can be used to estimate properties of ionic liquids, but does not exist for all ionic liquids. Density functional theory is used to create IR data based on a training set of experimental data. Principal component analysis and partial least square techniques will be employed to reveal important features and patterns in the molecular architecture. A characterization based group contribution method is used to estimate the properties of the new structures. The reverse design of potential ionic liquid molecules will be completed by an exhaustive search of combinations with various cation, anions and lengths of alkyl chains until a candidate molecule is found that provides the highest solubility of CO_2.

467962 Simultaneous Optimization for Heat-Exchanger Network Synthesis of Natural Gas Combined Cycle Power Plant Integrated with Post-Combustion CO_2 Capture and Compression

Abstract

In this work the thermal integration of natural gas combined cycle (NGCC) power plant integrated with post-combustion carbon capture (PPC) and CO_2 compression is proposed. With the objective of found a heat exchanger network (HEN) that decrease the energy impact of the PCC process, which decreases the power of the turbines as it requires a large amount of steam for its operation and also it uses part of the power generated by the NGCC power plant. This study consists of a 453 MW_e NGCC power plant with an MEA-based PCC process which are simulated in ASPEN PLUS® to get all the process inventory that includes stream flows, composition, temperature, pressure, equipment heat, work, and also the identification of hot and cold process streams. Using the model of simultaneous optimization of thermal integration SYNHEAT scheduled in GAMS™ it was obtained as a result the decrease in the use of steam in the striper reboiler to 65.43 Kg/s with a HEN that decrease also the use of hot water to heat the input of natural gas producing a 126.571 MW_e power output of the steam turbines increasing the thermal efficiency to 50.6996%, due to the use of the heat of CO_2 compressor train streams in the HEN.

- Abstract Text

468973 Overview on Evaluation of Sustainable Biofuel Conversion Processes

Abstract:

This paper reviews the existing methodologies on how to assess the overall sustainability of a biofuel production process, including approaches on how to quantify and compare the key indicators. It is always highly desirable to develop a more sustainable biofuel production process, for better economy, environment, and society. A holistic sustainability analysis for the production of biofuel plays a critical role on the improvement of the overall life cycle sustainability. Once a conceptual conversion process is developed, the next step is to evaluate its overall sustainability, based on a process sustainability evaluation methodology using various indicators in the areas of environment, efficiency, energy, and economics. One outcome of the sustainability evaluation is to allow researchers to identify process areas in need of sustainability improvement, as well as the challenges and opportunities for achieving the best possible sustainability target.

468418 The Role of ‘Free Radical Based Mechanisms’ in Wood-Feeding Termites for the Deconstruction of Lignocellulosic Biomass

Abstract:

Wood-feeding termites are very efficient in deconstructing the plant cell wall, leading to more than ~90% of cellulose and hemicellulose solubilization in a span of 24 hours. The key to this efficient natural process is the *mechanism of selective lignin modification* in the termite gut as shown by previous studies, which characterized lignin subunits in termite chewed wood particles and feces. These studies have led to an important scientific question- what is the mechanism that helps termite to cleave selective lignin bonds, enough to avoid lignin recalcitrance for saccharifying enzymes?

In the present research, we are answering this question by designing a speculated *free-radical driven system* that can participate in lignin bond cleavages in such a manner that it would avoid lignin recalcitrance and subsequently allow bigger molecules like-enzymes to enter the matrix and catalyze the saccharifying reactions. The free radicals are small, highly reactive and unstable molecules like hydroxyl, superoxide etc., which play an important role in termite based pretreatment. These radicals can be generated and utilized in many enzymatic and chemical reactions. Therefore, this proposed free-radical system could be targeted as a mimic for designing a novel pretreatment process.

469935 Life-Cycle Assessment of Biofuels and Chemicals from Unconventional Methane Feedstocks

Abstract:

This is the result of an ongoing collaboration with a team of researchers in an ARPA-E project to investigate the feasibility of utilizing stranded, small-scale sources of natural gas to produce liquid transportation fuels and value added chemicals. In addition to technical and economic considerations, environmental impacts are being studied through a cradle-to-gate life-cycle assessment approach. Here, we present preliminary findings for the fuel life cycle, involving methane procurement, conversion to lipids in a bioreactor, and further downstream processing. We have defined several potential scenarios for using methane derived from stranded gas, associated gas, and landfill gas. The requirements for procuring each gas stream are taken into account, along with fugitive emissions and considerations of alternate usage for each gas stream. Bioreactor input requirements, along with outputs of products and other material streams, were evaluated in consultation with LanzaTech process engineers. Environmental impacts being considered are global warming potential and fossil energy demand. Initial results reveal that major differences emerge between potential gas feedstock scenarios based upon alternate fate of the gas (flaring vs. utility generation, for instance). Estimated fugitive emissions rates also play a large role in determining the embodied global warming potential of feedstock gases and will need to be thoroughly documented in each commercial setting to verify optimal performance. Large improvements to the overall environmental performance of the system are gained when next-generation LanzaTech bioreactor technology is utilized, in comparison to conventional loop or bubble column reactors, due to improved mass transfer properties and lower gas compression requirements to facilitate the methane fermentation reaction inside the bioreactor.

469948 Climate Change Impacts of Bioethanol Production from Willow

Abstract:

Abstract:

The use of cellulosic biofuels is projected to increase in order to offset greenhouse gas (GHG) emissions from the combustion of non-renewable fuels. Hot water extraction (HWE) is a potential pretreatment step for multi-pathways biorefinery to yield natural, wood-based chemicals and materials. A cradle-to-grave life cycle assessment (LCA) was performed to assess the GHG emissions associated with the production of 1 MJ of ethanol by fermentation of sugar derived from HWE of willow biomass. The system boundary includes cultivation and harvesting of willow, transportation from the field to the biorefinery, hot-water extraction, fermentation and separation of ethanol, as well as the transportation of ethanol and its combustion. The system boundary was expanded to include co-products and underground carbon sequestered. Two conversion scenarios were considered. In the first scenario, electricity and heat required for the processes are cogenerated on site by the combustion of a fraction of the extracted willow biomass. In the second, electricity for the processes is supplied exclusively by the national grid. A sensitivity analysis and Monte Carlo analyses were performed for each scenario. Preliminary results show that GHG emissions for scenario one are -36.9 g CO_{2eq} MJ^{-1}. For scenario two, GHG emissions increase to 117.9 g CO_{2eq} MJ^{-1}. The amount of yeast used and the yield of willow were the most sensitive variable parameters. The fermentation and distillation of ethanol accounts for up to 70% of the total greenhouse gas emissions. The source of heat and electricity is the main driver in reducing greenhouse gas emissions when gasoline is replaced by ethanol produced from sugars that are extracted by HWE of willow.

470418 CO_2 and Water Adsorption on Seeded Growth of Mg-MOF-74 on Carbon-Nanotubes

Abstract:

Increased atmospheric CO_2 concentrations result in increased demand for novel and efficient CO_2 control technologies. Metal Organic Frameworks (MOFs) have recently attracted research interests due to their tailorable porous structure and potential application as novel adsorbents. The presence of open metal sites in some MOFs fortifies their adsorption sites, which is favorable for CO_2 capture. In this research, the adsorption performance of Mg-MOF-74 is enhanced by incorporating MOF on a membrane surface. Mg-MOF-74 was synthesized by solvothermal method and seeded on the surface of plasma treated carbon nanotube bucky paper which has demonstrated a superior CO_2 adsorption. The plasma treatment was conducted in either O_2/Ar or NH_3 gas flows. All samples were characterized using XRD, XPS, SEM, and CO_2 volumetric adsorption. XRD results show identical pattern for Mg-MOF-74 grown on CNT and Mg-MOF-74 alone, confirming the intact crystalline structure of MOF. XPS measurements illustrates that plasma treatment with O_2/Ar can yield samples with higher wettability which is favorable for further surface modification and increases affinity towards water vapor by anchoring hydroxyl functional group on the surface. The incorporation of MOF on CNT increased the CO_2 adsorption performance of MOF. The highest CO_2 adsorption capacity of 5.6, and 34.7 mmol/g was obtained for Mg-MOF-74, and Mg-MOF-74-CNT, respectively. Maximum water adsorption capacity of 9.0 and 20.6 mmol H_2O/g were obtained for Mg-MOF-74 and Mg-MOF-74-CNT, respectively. Comparing CO_2 and H_2O isotherms for prepared samples demonstrates that generally the seeded growth Mg-MOF-74 on CNT-BP performed better than Mg-MOF-74.

470328 Optimization of Biomass Processing through Physical Fractionation

Abstract:

The nascent cellulosic biofuels industry is struggling due to challenging economics derived from high cost of cellulosic sugars relative to starch and sucrose (and recently the substantial drop in the price of petroleum) with contributions arising from the high capital and operating costs associated with large centralized lignocellulosic biorefineries. Well-recognized, critical challenges include the low bulk density represents an important challenge for the logistics of feedstock transport and storage, the substantial within- and between-feedstock heterogeneity of herbaceous feedstocks such as corn stover and switchgrass that can contribute significant challenges to processing and conversion, and the need for high-value co-products or co-products targeting diverse market sectors to both add value and to improve the economic robustness to fluctuations in both global and local market forces.

Biomass heterogeneity has important impacts on a number of important processing variables. These include differing responses of the biomass to comminution that result in wide particle size distributions, substantial differences in particle hygroscopicity and drainability, a diversity of compositions (including lignin and ash), and a wide range of responses to chemical pretreatments and hydrolysis. This work will, work will assess physical fractionation approaches for corn stover and energy sorghum that mimic industrially relevant fractionations utilizing either manual classification that mimics in-field classification that could be achieved during harvest and combinations of comminution and sieving that could be performed either during harvest or at decentralized processing depots. Employing these approaches in this work, we demonstrate that physical fractionation can yield biomass fractions with diverse properties that include wide ranges for lignin content and hygroscopicity that impact feedstock processability (*i.e.*, response to pretreatment and enzymatic hydrolysis).

471527 Kinetics Modeling of Fed-Batch Lipid Accumulation of Activated Sludge

Abstract:

Recent studies have demonstrated the capability of activated sludge (AS) microbial consortia to accumulate lipid droplets under fed-batch cultivation on acetic acid as carbon source. The experiments were conducted in laboratory-scale bioreactors. The prediction of system performance at a large-scale implementation of the technology requires the mathematical modeling of the laboratory-scale bioprocess, so kinetic models must be fitted to the experimental data. This study models the kinetics of the AS microbial culture during the fed-batch cultivation.

Monod-type kinetic models were regressed to the experimental data using various numerical methods implemented in MATLAB. Initial data-fitting via unconstrained nonlinear optimization of least-square-error suffered from instability of solution and long computation time. Then the Markov Chain Monte Carlo (MCMC) method was implemented resulting to improved regression. That is, the model parametric constants converged while minimizing the least-square-error at a faster rate. This work describes the various components of the problem-solving strategy that enabled the kinetics modeling.

472988 Pyrolytic Fractionation of Algal Biomass: Catalytic Conversion of the Resulting Vapors to Fuels and Chemicals

Abstract:

Our group recently developed a novel pyrolytic fractionation method for converting oleaginous biomass (microalgae and/ or oil seeds) to fuels and products. We have observed that the biopolymer components (protein, starch and lipids) volatilize over narrow and distinct temperature regions. This allows recovery of oil and gas separately from each fraction. Through pyrolytic fractionation, volatile N-containing compounds and carbohydrate products can be removed from the biomass by first heating to ~320 °C and holding temperature steady until most protein and starch are thermally degraded or stabilized via polymerization to char. The biomass remaining after thermal removal of protein and starch (first pyrolysis fraction) can be further processed (at temperatures > 450°C) to recover low-N bio-oil from lipids, again as a separate fraction. We are investigating the production of higher value aromatics and nitriles from each of these fractions (320 °C and 450°C) by upgrading the generated vapors using an H-ZSM-5 catalyst bed. Our recent findings on this project will be presented and discussed

471255 Biomethane Potential Determinants of Different Agro Industrial Substrates

Abstract:

Anaerobic digestion is a promising technology offering the opportunity for conversion of biomass into biogas, a renewable fuel that can be used for energy generation in internal combustion engines, microturbines and boilers. Throughout the years this technology has been applied for the valorization of waste streams including but not limited to manure, olive mill waste and municipal sludge. While the method is offering a promising pathway for generation of biogas, it may suffer by low efficiency, inhibition by product or substrate accumulation and lack of available enzymes. The literature is rich with studies describing the biomethane potential of comparable (according to the presented analysis) substrates where the efficiency and the recovered yields showed significant variation. For example Pitk et al. (2013) studied the biomethane potential of different slaughterhouse wastes. The sterilized mass (i.e. bone and meat meal) used in this study achieved a production of 721 mLCH4gVSadded. In contrast Wu et al. (2009) reported a yield of only 381 mLCH4gVSadded while assessing bone and meat meal. Furthermore, Ebner et al. (studied the digestion of cattle manure and reported a biomethane potential of 238 mLCH4gVSadded which is at least 30% higher to this described by Amon et al. (2007). Bayr et al. (2012) assessed the biomethane production of fat and reported yields of upto 403 mLCH4gVS FM (FM = 935g/Kg, 100%VS) corresponding to approximately 430 mLCH4/gVSadded. In contrast to this, Martínez EJ et al. (2016) reported yields higher than 800 mLCH4/gVSadded. Despite the fact that all these values are well within the theoretical limits, the reason of the variability is not easily distinguished. While the variations can be the outcome of differences in the influent substrate and the presence of inhibitors, other technical issues and lack of standardization cannot be excluded.

In this work we investigated the anaerobic digestion of four substrates (animal fat, wheat straw, sterilized slaughterhouse wastes and glucose). The different substrates selected were representative of the most common feedstock used in industrial digesters.

The substrates were assessed in batch vials with the reaction time set to 30 days under mesophilic conditions. The loading rates applied were controlled to range between 10 and 20 kgVS/m3, while the inoculum to substrate ratio between 2 and 0.5. Furthermore, the experiments were divided into two stages with two complete sets of vials inoculated with different inoculums sourced from reactor operating under different loading rates and feed composition regimes. The last experiment targeted the recovery of data from where information related to the importance of the initial inoculum can be extracted.

According to the results the source of inoculum appears as the most important factor for the AD process. The inoculum from the reactors treating the agro industrial waste (Inoc A), was able to offer a robust process when fat was used as substrate with only the 0.5 inoculum to substrate ratio failing. On the other hand, when the inoculum recovered from the reactors treating manure (Inoc B) was used only digestion where the lowest loading rate applied was successful. In this case the ultimate yield was at least 25% lower compared to this achieved by Inoc A. The same is true for the case of the sterilized mass where the daily productivities of the batches with Inoc A were up to 100% higher than these offered by Inoc B. On the other hand when sugars were digested Inoc B trsulted in higher yields and productivities compared to Inoc A. The second important parameter is the initial inoculum to substrate (I/S) ratio. A low I/S ratio can delay the biomethane production and in some cases inhibit the overall process which can be easily described by the limited number of available cells in the inoculum and the detrimental effect that an inhibitor can have in an unbalanced biological process. The least important factor among the ones assessed is the initial organic load per batch. Loadings up to the 20 kgVS/m3 can be sustained by AD systems treating slowly hydrolyzed substrates. On the other hand, addition of 20kgVS/m3 of simple sugars into AD systems can inhibit the process through intermediate product accumulation, manifesting the inherent problem of the process and the different rates of the biochemical reactions involved.

Our analysis underlined the critical role that inoculum (both in terms of its source and in terms of its relative concentration in the reactor) plays on the determination of the overall methane production efficiency in anaerobic digestion systems. Further work will investigate the microbial and biological profile of the system to determine optimal operational regimes as a function of the performance determinants analyzed above.

Amon, T., Amon, B., Kryvoruchko, V., Zollitsch, W., Mayer, K., Gruber, L. 2007. Biogas production from maize and dairy cattle manureâ€"Influence of biomass composition on the methane yield. Agriculture, Ecosystems & Environment, 118(1â€"4), 173-182.

Bayr, S., Rantanen, M., Kaparaju, P., Rintala, J. 2012. Mesophilic and thermophilic anaerobic co-digestion of rendering plant and slaughterhouse wastes. Bioresource Technology, 104(0), 28-36.

Ebner, J.H., Labatut, R.A., Lodge, J.S., Williamson, A.A., Trabold, T.A. Anaerobic co-digestion of commercial food waste and dairy manure: Characterizing biochemical parameters and synergistic effects. Waste Management.

Martínez EJ, Gil MV2, ernandez C1, Rosas JG, X., G. 2016. Anaerobic Codigestion of Sludge: Addition of Butcher's Fat Waste as a Cosubstrate for Increasing Biogas Production. PLoS One(http://dx.doi.org/10.1371/journal.pone.0153139).

Pitk, P., Kaparaju, P., Palatsi, J., Affes, R., Vilu, R. 2013. Co-digestion of sewage sludge and sterilized solid slaughterhouse waste: Methane production efficiency and process limitations. Bioresource Technology, 134(0), 227-232.

Wu, G., Healy, M.G., Zhan, X. 2009. Effect of the solid content on anaerobic digestion of meat and bone meal. Bioresource Technology, 100(19), 4326-4331.

472961 Flash Pyrolysis and Fractional Pyrolysis of Oleaginous Biomass in a Fluidized-Bed Reactor

Abstract:

In this study, flash pyrolysis of oleaginous biomass was investigated to produce liquid fuels. Milled soy flakes were used and can be considered a surrogate for algal biomass due to similar compositional, proximate, and ultimate characteristics. A fluidized-bed flash pyrolysis was designed and constructed to facilitate rapid heat transfer to the biomass along with short vapor residence times to minimize secondary degradation of bio-oils and achieve high yields of bio-oil. In addition, a dry-ice packed bed condenser was used to recover the generated vapors. The novel condenser provides a large area of cold surfaces that allow dilute vapors to impinge, condense and coalesce effectively. Single-step pyrolysis experiments were performed at several temperatures over the range 250-610°C with a vapor residence time between 0.3-0.6s. Biooil yields were highest at 550°C (~70% g/g-feed) with only a 12% loss as uncondensed vapors. Moreover, 90% of the feedstock lipids were recovered in the bio-oil fraction. In addition, three-step fractional flash pyrolysis experiments were performed to assess the possibility of producing higher quality bio-oils since a large fraction of protein and carbohydrates degrade at lower temperatures (310-350°C). A low temperature pyrolysis step (250ºC) was first performed and was followed by pyrolysis of the residues at higher temperatures (310 and 530ºC). The bio-oil from last step (530ºC) resulted in higher quality bio-oil with low water- and nitrogen- content, and high heating value of 33 MJ/kg.

473026 Life Cycle Assessment of Renewable Hydrotreated Fuels from Castor

Abstract:

Hydrotreated renewable transportation fuels (HRTF), i.e., green diesel and bio-jet fuel, have become promising options for reduction of fosil fuels consumption and mitigation of climate change [1]. These biofuels have to be produced, however, in a sustainable way. The life-cycle assessment (LCA) have shown to be an useful and accurate tool to evaluate CO_2production and energy and water consumption in the Biojet Fuel Production [2].

LCA allows the evaluation of the environmental impact of the production of hydroprocessed renewable fuel against conventional fuels, the main objective of this alternative is to reduce the environmental impact. The limits and objectives of the LCA were defined in order to measure CO_2emissions during cultivation, transport and industrial production, everything applied at the conditions of Oaxaca State. This research is part of the research to demonstrate that hydrotreated renewable fuels are a real option to substitute conventional hydrocarbon-based fuels in Mexico.

Thus this work evaluates the life cycle assessment of green diesel and bio jet fuel from castor plant, which has a large growth in rural areas along Mexico. The LCA methodology has been used in this work to quantify GHG emissions. No land change has been considered, due the nature of the crop, which is a wild and not eatable plant. The presentation details the methods, assumptions and LCA results for the production of renewable fuels in Oaxaca State. Preliminary results indicate that HRTF are capable of being producedin large scale to supply a significant fraction of energy demand of the country, reducing at the same time the GHG emissions. These results contributed to a favorable carbon footprint result for castor HRTF, yielding savings of greenhouse gas emissions compared to fossil fuels.

References

[1] J.Han et al, "Life-cycle analysis of bio-based aviation fuels", Bioresource Technology, pp 1-10, 2013

[2] D. B. Agusdinata et al, "Life Cycle Assessment of Potential Biojet Fuel Production in the United States", Enviromental Science & Technology, pp 1-11, 2011.

473262 Enhanced Design of Hybrid Distillation Processes for Separation of Biomass Derivatives

Abstract:

Global warming is the current major environmental issue that modern society is facing due to the world's overdependence on the use of mineral resources as raw materials. Therefore, extensive studies of renewable resources as feedstock to substitute fossils have been carried out. Therefore, the interest of biomass, which is the only renewable resource of fixed carbon, has increased considerably in recent times. Biomass can not only produce conventional hydrocarbon liquid transportation, but also petrochemical products. Several value-added platform chemicals with a range of potential uses from biomass were investigated based on the potential markets, derivatives and the complexity of the synthetic pathways. On the other hand, biomass production on an industrial scale is limited due to the expensive feedstock, low synthetic yield and lack of detailed process design. Several promising hybrid separation processes for recovering valuable biomass derivatives from inexpensive raw materials have been investigated. This work proposes a systematic procedure of enhanced design of hybrid distillation process for separation of biomass derivatives. Initially, enhanced separation process configurations were assessed based on a literature survey, equilibrium distribution and a separation feasibility study. Several configurations were chosen for detailed design and optimization on an industrial scale and compared in terms of the energy requirement, total annual cost (TAC) and total annual CO_2 emission (TAE). All distillation columns were designed with a load of approximately 85% of the load at the flooding point to prevent flooding in the columns. Some preliminary simulations were conducted to determine the initial structures for the extractor and all distillation columns. A rigorous simulation using the simulator, Aspen Plus V8.6 was then conducted to identify the optimal design in terms of minimizing the TAC. The capital cost, total annual operating cost (TOC) and TAC were calculated based on Guthrie's modular method. The results showed that the innovative hybrid distillation configuration could save up the energy requirement and TAC significantly compared to the conventional sequence.

474714 A Chemical Engineer's Guide to Contemporary Problems of Climate, Energy and Environment

Abstract:

From artists to engineers, how does an instructor best prepare his or her students for issues they will face related to climate change, energy supply, and the nexus of these two issues? Future public policy decisions may very well be in the hands of those without a strong scientific background. Meanwhile, as many engineering undergraduates are developing the technological background to tackle these challenges, they may miss out on many aspects of a broader education that will put future solutions in context. Topics regarding energy and the environment are much broader and more complex than can be covered in the traditional one hour lecture period. Over the past year, the authors have been working on a general education book entitled, "How Science is Done: Contemporary Problems of Climate, Energy and Environment". Although targeted to a general (i.e. non-scientific) audience, the book is relevant to providing the future chemical engineering workforce a broader understanding of how technology will impact society and the environment. In the book, we have also worked to incorporate a number of detailed discussions on the physics of how the earth's climate is determined and regulated, the thermodynamics and mechanics of engines, energy conversion and conservation, and the mass flows of carbon throughout the environment. A number of these examples are relevant to the basic chemical engineering curriculum. Even more of these examples can be used to engage first-year engineering students in the impact engineering has on the society and the environment. Several classroom examples will be shared in this session, with a focus on how they might be utilized within the existing chemical engineering curriculum. A longer term "green engineering" project that incorporates hands-on campus-wide sustainability initiatives into an advanced undergraduate elective will also be discussed. To "engineer a sustainable future", our undergraduates will need a broad contextual understanding of energy usage, energy supply, carbon emissions, and environmental ramifications of technology.

475634 Residential Wood Pellet Heating System Optimization Using Process Dynamic Simulation

Abstract:

As a growing part in biomass energy utilization, the use of wood pellet boilers for residential space heating have been increased over the past decades in EU and North America. A typical wood pellet heating system has three major components: the boiler, thermal storage tank, and heat consumption unit. Considerable effort has been applied to the design of the boiler to build high efficiency and low emissions (HELE) boilers. However, currently there is no systematic process optimization in terms of the system efficiency, especially the optimization of the operation of the thermal storage tank with the heating system.

In this study, a 25 kW nominal output wood pellet boiler was installed and monitored through the past two heating seasons (from February 2015 to May 2016). Two 120 gallon tanks were connected in parallel with the boiler as the thermal storage unit. Field data were collected for boiler thermal efficiency calculation. Meanwhile, several runs of CTM-039 stack sampling were performed to evaluate the emissions from the boiler. In the end, a process dynamic simulation using VMGSim^{TM} was built for the residential wood pellet heating system. The boiler simulation is based on wood combustion modeling of the boiler and tuned with field data.

The whole system was optimized for a better combustion and a higher system efficiency from the following aspects. Firstly, optimize the combustion efficiency through control of excess air ratio, pellet quality, etc. Secondly, optimize boiler thermal efficiency by running the boiler at maximum load and employing more efficient cleaning mechanisms. Thirdly, optimize of the tank sizing with respect to boiler nominal output and heat consumption from the building for a higher tank efficiency and better use of water stratification in the tank. And lastly, a sensitivity analysis was performed for the validation of the model.

This work can substantially improve future residential space heating system designs based on wood pellet heating technology.

475766 Introduction and Study on Electron-Beam Assisted Stabilization Process for Cost-Effective Manufacture of Carbon Fibers: Process to Bring about Wide Application of Lightweight Structural Composites for Sustainable Environment

Abstract:

To lower energy consumption and carbon emission in transportation is essential choice for future sustainable environment. Employing a lightweight structural composite with high mechanical properties in the structure of transportations can be thought as one of the approaches. As a reinforcing material in the composite, a carbon fiber (CF) is a promising candidate due to its high strength and modulus per unit weight. Even though the CFs have been commercialized and available in the market, the application of CFs is still limited to specific transportation such as aircraft parts due to their high manufacture cost. The wide application to common transportations such as personal vehicles will be realized by decreasing the manufacture cost of CFs.

The manufacturing process of CFs derived from polyacrylonitrile (PAN, the most common precursor polymer of CF) can generally be divided into three processing steps: 1) spinning and stretching, 2) stabilization (200~300 Â°C) and 3) carbonization (â‰¥1000 Â°C). A number of experimental parameters associated in each step have been extensively optimized for the energy and time effective manufacture of the CFs. Among these three steps, the stabilization is the most crucial step as a determinant of not only high performance but also of processing cost due to lengthy heat treatment. The conventional stabilization is carried out by passing the fibers through a series of convection ovens with gradual temperature gradient (200~300 °C) for two to three hours, which is the most time and energy consuming step. As one of the methods to shorten such lengthy step with expensive energy consumption, electron beam irradiation on PAN precursor fibers has been employed prior to short thermal treatment. In conclusion, with short treatment of electron beam (200 sec), the irradiated fibers were successfully stabilized with short time of thermal treatment (30 min). In the presentation, we briefly overview the environmental impacts if the CF composites are commonly applied to transportation parts and approaches towards low manufacture cost of CF for each of manufacturing step. In particular, as one of the specific suggestions to the problem and study, the electron beam treatment for energy-effective stabilization will be explained in detail by presenting several characterizations of irradiated PAN fibers.

475416 Life Cycle Assessment of Pharmaceutical Manufacturing Processes

Abstract:

Continuous processing, as a form of process intensification, is one of the keys of green engineering research and development in the pharmaceutical industry. It has the potential to reduce solvent use and cost of production as well as increase production quality and operational safety. The goal of this work was to identify environmental hotspots, quantify and compare the differences in environmental burdens between batch and continuous manufacturing processes. A comparative cradle-to-gate life cycle assessment between the batch and continuous manufacturing processes of 4-D-Erythronolactone at pilot plant scale is presented as a case study. The assessment begins with defining the scope, the system boundaries, and functional unit. The system boundaries include from raw material extraction, transportation, the manufacturing process, equipment cleaning, facility requirements (lighting and heat, ventilation and air-conditioning) and waste incineration. Two other waste handling options (wastewater treatment and solvent recovery) are investigated as they are common operations employed by the pharmaceutical industry to treat waste. As part of a modular approach to obtain life cycle inventory data, each manufacturing system was part of a theoretical production campaign to fulfil the functional unit. A variety of methods such as kinetic and reactor modelling, input-allocation inventory models is used to fill data gaps.

For the case study, environmental hotspots of the batch system were centered on the production of inputs, washing activities and waste disposal. The adoption of continuous processing from batch manufacturing produced an overall reduction in impacts. With continuous processing, less solvents were used in washing and it translated to reduced impacts from washing and waste disposal. However, more energy was required for continuous processing due to the use of less concentrated starting materials. This case study provided insights into the life cycle impacts of batch and continuous manufacturing processes and helped develop a modular approach to compare the two manufacturing systems.

475872 Advancing Building Materials Product Transparency with the Living Product Challenge

Abstract:

General Concept, Background, Motivation

Product transparency has become a mega-trend in the building materials industry. Consumers now demand that increased scrutiny be placed on the materials they interact with on a daily basis. To that end, ingredients once considered proprietary to product manufacturers must now be disclosed.

Method

The Living Product Challenge is a comprehensive 3rd party certification that goes beyond single attribute programs to encompass all the life cycle stages of a product as well as the social impacts, in addition to environmental and financial impact. The International Living Future Institute's Living Product Challenge challenges the status quo; it encourages manufacturers, architects and builders to consciously increase the level of disclosure they offer the public through their structured product certification process. The work described in this project outlines Owens Corning's journey as a pilot participant in the Challenge to successfully gain certification for their unbounded loosefill fiberglass insulation as the world's first Living Product. The process caused Owens Corning to thoroughly examine the way it conducts business, manufactures products and quantify the positive impact its operations have the environment.

Results and Interpretation

The successful facility and document audit lead to imperative certification for Owens Corning's leading fiberglass insulation products. This test-and-learn project served as a case study for the Owens Corning internally for use as a model for documenting the sustainable impact of other products within the company's portfolio. The intent was to encourage other manufacturers to join Owens Corning in increasing their commitment to product sustainability.

The process, product and lifecycle are all integral players in the success of a sustainable edit to a manufacturing regime. The Living Product Challenge directly addresses marketplace demand for comprehensive product transparency while leveraging single attribute certifications to document the sustainability gains of industry leading companies like Owens Corning.

475935 Process Design, Modification, and Sustainability Assessment of Coal and Biomass Co-Fired Plants for Generation of Transportation Fuel: A Case Study in Kentucky

Abstract:

As demand for in-country generated transportation fuel increases, methods that utilize region-based resources to produce biofuels are sought. Such is the case in Kentucky, United States, where the biofuels blend rate in transportation fuels is likely to increase from 10% to 25% by 2022. Consequently, a total of 775 million tons of transportation fuel are demanded to keep the state from importing 90% of its fuel in 2022. This challenge can be met by investing the implementation of coal and biomass co-fired plants in Kentucky to produce various fuels, including diesel, kerosene, and gasoline. To analyze this possibility, process design and modification of coal and biomass co-fired plants was conducted for energy and material efficiency improvement and waste reduction. Process simulation was performed, with the evaluation of process sustainability using the IChemE Sustainability Metrix System. The results of this case study will be presented.

475948 Synthesis, Properties, and Processing of Melt-Spinnable Acrylonitrile Copolymers

Abstract:

Carbon fiber (CF) which is one of outstanding structural materials have received much attention for application such as composite materials, textiles, microelectrodes, and catalyst because of its superior mechanical properties, heat resistance, low thermal expansion, chemical resistance, and so on. CF has been developed from several kind of precursors such as polyacrylonitrile (PAN), pitch, and rayon. In particular, because CF derived from PAN shows especially high mechanical properties and its manufacturing process has been well studied, CF commercial makers has mainly used PAN as a CF precursor. Until now, CF and its composites have been only applied in high tech fields such as aerospace because of the cost problem. Recently, those application to automobile is a rising issue because those properties, light weight and high strength, can make increase of performance and fuel-efficiency. However, reducing the cost of CF is a precondition for these application. In addition, because the automobile application is not required for very high mechanical properties, a little decrease of the properties is allowed. In this regards, many research institutes have found the method for reducing manufacturing cost. There have been several approaches such as using low cost precursors and changing manufacturing procedure. In particular, traditional PAN precursors is prepared by the wet spinning which required expensive organic solvent recovery process and waste water treatment process. In this research, we have synthesized acrylonitrile copolymers containing functional components for reducing manufacturing cost by using melt spinning process which can spin fibers with a high speed than wet spinning and do not require solvent and water handling. In conclusion, we could prepare acrylonitrile fibers by melt spinning and stabilize the fibers by using other treatment with thermal method. In the presentation, the thermal and rheological properties of the synthesized polymers will be mainly discussed in detail.

470203 Self-Assembly of Antibody-Polymer Conjugates into Novel Sensing Materials

Abstract:

Affinity biosensors based on antibody binding represent one of the most successful and widely investigated categories of protein materials, having yielded technologies as ubiquitous in the US today as home pregnancy tests and often forming the basis for efforts at the development of low cost diagnostics. The sensitivity of these materials is fundamentally limited by the density of binding sites on the material surface, with the possibility of an order of magnitude boost in sensitivity of a transformative increase in surface binding sites could be achieved.

Here, we show that such an increase in binding sites can be achieved through antibody self-assembly into three-dimensional nanostructured materials. Using a site-selective chemistry, bioconjguates are prepared of IgG antibody with the synthetic polymer poly(N-isopropylacrylamide) (PNIPAM). Lyotropic self-assembly of the materials leads to the formation of large lamellar nanostructures that can be observed by both X-ray scattering and transmission electron microscopy. Studies of the self-assembly as a function of bioconjguate concentration and polymer length provide insight into the phase behavior of the system. This is the first such demonstration of antibody self-assembly into large structured arrays.

Using a simple flow coating and crosslinking process, the antibody bioconjguates can be immobilized in nanostructured thin films on different substrates. These films show a 3-4 times higher capacity for protein G binding than adsorbed antibody monolayers. As the size of the binding protein is increased, however, the difference is reduced, presumably because the large binder has trouble penetrating the polymer channels within the protein nanostructure, indicating a need for additional engineering of nanostructure size and transport rates. In addition to this, ongoing work is focusing on developing protein materials from smaller engineered protein binders such as nanobodies.

458100 Structural Studies to Determine the Mechanisms of Biobased Nanoparticle Synthesis

Abstract:

Protein and peptide directed synthesis of inorganic nanomaterials offers several advantages over traditional synthesis methods; however, despite being used to great effect, the fundamental mechanisms of how biomolecules control and drive nanomaterial synthesis is not well understood. We are using high-resolution structural biology techniques on a model gold binding and mineralizing peptide system to investigate how biomolecules are able to influence inorganic nanomaterial nucleation and growth. In this presentation we will present our current results, as well as discuss how this model gold mineralizing system is being used to develop an improved colorimetric ionizing radiation sensor for therapeutic use in cancer treatment.

460477 Biopolymers from a Thermophile: Production, Characterization, and Application

Abstract:

Exopolysaccharides (EPSs) are high molecular weight carbohydrate biopolymers that are composed of sugar residues linked together through glycosidic bonds, and secreted by microorganisms into the surrounding environment with certain properties and functions. The EPSs are one of the adaptations developed by various kinds of extremophiles in order to survive in the environments with extreme conditions, such as extreme temperature, salinity, or pH. Extremophiles are considered as potential producers of EPSs with novel and unusual characteristics which may hardly be found in more traditional polymers. Recently some extremophilic EPSs have already been characterized, and the properties and functions of EPSs from related extremophiles were also partly elucidated. However the accumulated knowledge concerning the structural and functional properties of EPSs secreted by extremophiles is still limited. In this research, a thermophilic strain, *Geobacillus* sp. (WSUCF-1) which produced unique type of EPSs has been studied. WSUCF-1 could utilize glucose as optimum carbon and energy source to produce significant amount of EPSs (382 mg/L). Two purified EPSs were obtained through ion-exchange and gel filtration columns, and both of them demonstrated higher degradation temperature (319.20 and 314.85 °C) than the thermophilic EPSs recently reported. Monosaccharide composition analysis showed that these two EPSs were composed by mannose and glucose with various relative molar ratios (1/0.2141 and 1/0.0270). Spectrum analysis also showed several unique features of the EPSs. EPSs from the relatively unexplored WSUCF-1 strain will likely provide a valuable resource for exploitation in novel biotechnological processes and opportunities in various industrial fields including bio-based polymeric materials.

471023 Genetically Encodable Acoustomagnetic Reporters for Background-Free Molecular and Cellular MRI

Abstract:

Technologies for non-invasive whole-organ imaging of molecular targets and cellular processes such as gene expression are essential to support biological studies, medical diagnostics, and the development of cellular therapeutics. While optical methods provide high-resolution images, they are fundamentally limited by the penetration depth of light in opaque tissues. In comparison, magnetic resonance imaging (MRI) and ultrasound are powerful and complementary non-invasive imaging modalities and have been widely used in clinics. Yet they have never been paired for molecular imaging. Such a pairing, especially in a way that one modality can control the image content from the other modality, could take advantage of the unique features of each modality to provide information not available via either technology alone. **Here, we describe the first contrast agents for acoustomagnetic imaging. These agents are based on gas vesicles (GVs), a unique class of genetically encoded gas-filled protein-only nanostructures formed by photosynthetic microbes as a means to regulate buoyancy.** These agents are genetically encodable, robustly detectable using susceptibility-dependent MRI sequences and "erasable" at specific ultrasound pressures. Thus they provide, for the first time, a direct method to eliminate the endogenous background, a major drawback for susceptibility imaging including functional MRI and superparamagnetic iron oxide (SPIO) nanoparticle imaging. In addition, clustering of these agents based on the presence of a specific analyte produces changes in MRI contrast that could enable the use of these agents as functional sensors.

We reasoned that the gaseous interior of GVs would have a different magnetic susceptibility from the surrounding aqueous media, and would thus induce a local magnetic field gradient that would alter the precession of nearby aqueous ^1H nuclear spins. Indeed, we show that GVs can be imaged at nanomolar concentration using gradient echo MRI and quantitative susceptibility mapping. Furthermore, GVs are "erasable" MRI contrast agents, since ultrasound pressure applied in a spatially selective manner can collapse these nanostructures, eliminating their gas compartment and thus the MRI contrast. This is very useful because, many endogenous tissue properties, such as the presence of blood vessels and the interface between different types of tissue, can produce hypointense contrast in gradient echo images, and the complication of endogenous contrast has been a major drawback to the design of susceptibility-dependent T2* MRI contrast agents. Due to their ability to be imaged before and after acoustic collapse, GVs enable, for the first time, the acquisition of background-free T2* contrast. We demonstrate the concept of acoustically erasable background-free imaging using GVs injected into mouse brains and collapsed using ultrasound.

Furthermore, acoustic collapse occurs at different specific pressures for GVs from different organisms, enabling multiplexed "acoustomagnetic" imaging using a pressure-scan paradigm. The paradigm could enable the tracking of multiple GV-labeled cell types using MRI. Finally, GVs can potentially be engineered as sensors of specific biological signals because the clustering of GVs produces a marked change in gradient echo images, as demonstrated ***in vitro*** by streptavidin-biotin cross-linking.

Overall, these results support the introduction of GVs as the first acoustomagentic reporters, which use the combination of ultrasound and MRI to enable background-free and multiplexed imaging. The ability of GVs to be genetically encoded and engineered opens the possibility of using this new form of contrast in a wide range of biological applications, especially in synthetic biology and cellular therapeutics.

474994 How Bolt Threads Is Using Chemical and Process Engineering to Grow Its Business of Engineering Silk (Featured Presentation)

Abstract:

Bolt Threads is a startup that believes that the combination of molecular biology, chemical engineering, and materials science can revolutionize how people think about and use natural materials. Bolt is developing the next generation of sustainable apparel by leveraging the latest advances in industrial biotechnology, automation, and materials science. The company is producing spider silk, a notoriously strong and lightweight material that scientists have tried to commercialize for decades. Boltâ€™s secret weapon is chemical and process engineering, which allows it to do what its predecessors couldnâ€™t: scale up its technology. From bench scale to industrial scale, process engineering is the foundation of building this high growth startup business. Leveraging the unique properties of its fibers, Bolt Threads will be delivering new products to consumers and brands worldwide.

474277 Catalytic Process for Levoglucosan Conversion into Distillate Range Molecules

Abstract:

Pyrolysis oils are low quality fuels which cannot be used in conventional gasoline or diesel engines largely due to their high oxygen contents. Upgrading strategies for deoxygenation of pyrolysis oils are imperative for use in transportation fuels. In this presentation, we describe a multi-step catalytic process for conversion of levoglucosan (LGA), a major anhydrosugar present in pyrolysis oil, to monofunctional compounds, which can subsequently be coupled to obtain distillate range molecules.

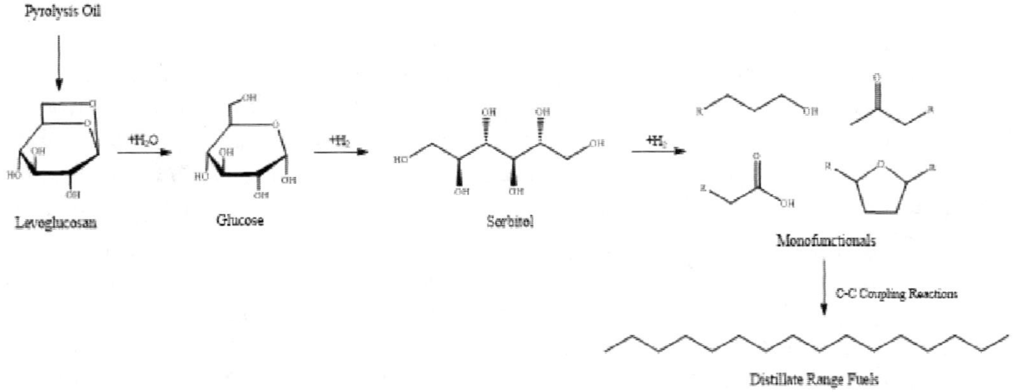

Figure 1. Pathway to convert levoglucosan to distillate range fuels

Sugar syrups produced via the pyrolysis of red oak pyrolysis oil are rich in LGA. Firstly, LGA in this sugar syrup is hydrolyzed to glucose. Following hydrolysis, glucose is converted to sorbitol with high yield by hydrogenation over Ru/C. Monofunctional intermediates are produced in the aqueous-phase hydrodeoxygenation (APHDO) of sorbitol with carbon yields up to 55%. These monofunctional species consist mainly of alcohols (71%) and heterocyclic species (24%). The majority of these species (52%) are C_5 or C_6 compounds, and only small amounts of ketones and acids are formed. The monofunctional intermediates are then converted into distillate range molecules by dehydration-oligomerization. The sorbitol APHDO effluent is fed to a flow-through reactor packed with a solid acid catalyst, SiO_2/Al_2O_3, to convert the monofunctional stream to olefins. Nearly 75% yield to mono-alkenes and dienes is obtained, based on the total carbon in the monofunctionals. Dienes originate from ring opening of heterocyclic compounds. C_2-C_6 olefins thus obtained can be oligomerized over acid catalysts to produce distillate range

hydrocarbons.

456831 Hydrodeoxygenation of Pinyon Juniper Catalytic Pyrolysis Oil to Hydrocarbon Fuels

Abstract:

As a renewable source, biomass is an essential option for diminishing dependence on conventional fossil fuel energy sources. Pyrolysis is a promising technology for the conversion of biomass into liquid fuels. However, several challenges associated with using pyrolysis oils such as their high acidity and low energy content inhibit their direct use as transportation fuels. We conducted a batch hydrodeoxygenation of pinyon juniper catalytic pyrolysis oil using Ni/SiO_2-Al_2O_3 catalyst to improve the following properties of the oil: heating value, acidity, oxygen content, water content, and viscosity. During the hydrogenation process, the influence of four experimental factors; temperature, catalyst loading, residence time, and hydrogen pressure was investigated. Once hydrogenation was completed, gas, coke, and a liquid product of two immiscible phases (aqueous and organic), were obtained. Maximum hydrogenation was obtained at a reaction temperature of 450 °C with catalyst loading of 20% (wt. % of total bio-oil feedstock), an initial hydrogen cold pressure of 1000 psi and a residence time of 30 minutes. At these conditions, bio-oil was deoxygenated by 96.17%. After hydrodeoxygenation, the higher heating value of the organic liquid product was 45.68 MJ/kg compared to 27.64 MJ/kg of the bio-oil. The water content of the organic liquid was zero compared to 1.63% of the bio-oil using Karl Fischer titration method. The aqueous fraction of the liquid product was 99.61% water. Furthermore, pH of the organic liquid was 6.87 compared to 3.46 of the starting material. The viscosity of bio oil was119.37 cP while it was 1.27 cP for the organic liquid product. The hydrogenation process provided a means of producing upgraded bio-oil, which possessed properties similar to that of gasoline.

458036 A Synthetic Biology Chassis for Lignin Valorization into Muconic Acid and Other Value-Added Chemicals

Abstract:

Lignin represents an untapped portion of biomass for the production of fuels and chemicals. In nature, many aerobic organisms degrade lignin-derived aromatics through conserved intermediates including protocatechuate and catechol. Employing this microbial approach offers a potential solution for the bioconversion of lignin into value-added chemicals for a future biorefinery. In this study, two hybrid biochemical routes combining chemical catalysis, lignin bioengineering, and metabolic pathway reconstruction were demonstrated for the conversion of lignin into value-added products. In the biochemical route 1, alkali lignin was chemically catalyzed for the release of vanillin and syringate as major products, which were further bio-converted into *cis, cis*-muconic acid and pyrogallol through engineered *E. coli* strains. The highest yields achieved were at 0.63 g ccMA/g vanillin in the fermentation broth and 0.69 g ccMA/g vanillin in whole cells mixture. While the highest yield of pyrogallol in the fermentation broth was 7.3 mg pyrogallol/g syringate. In the second biochemical route, the lignin biosynthesis pathway of plant *Arabidopsis* was engineered to accumulate protocatechuate (PCA) as an intermediate compound. The PCA extracted from the engineered *Arabidopsis* was further converted into ccMA using the engineered *E. coli* strains. In this biochemical route, the extraction yield of PCA was 1.45 mg PCA/ g plant biomass. The highest yield of ccMA was achieved at 0.31 g ccMA/ g PCA or 0.45 mg ccMA/ g plant biomass. In addition to ccMA, catechol was produced at a higher yield of up to 0.54 g catechol/ g PCA or 0.79 mg catechol/ g plant biomass. This study reports for the first time a direct process for converting lignin into ccMA and pyrogallol as value-added chemicals. It provides a great alternative pathway for the production of fuels and valuable from lignin.

468091 Porous Carbonaceous Solid Acids Derived from Farm Animal Waste and Their Use in Catalyzing Biomass Transformation

Abstract:

We demonstrate here the transformation of waste cow manure into N rich porous carbons (NPC) with large amount of micropores and large Brunauer–Emmett–Teller (BET) surface areas. The transformation process was achieved *via* carbonization of cow manure in presence of $ZnCl_2$ and $FeCl_3$ additives, without using additional templates. The resultant NPC could be converted into microporous carbonaceous solid acids (NPC-[C_4N][X]) by quaternary ammonization of samples with 1,4-butane sultone, and ion exchanging with strong acids such as HSO_3CF_3 and H_2SO_4. X-ray diffraction patterns and Raman spectra showed that NPC-[C_4N][X] exhibit amorphous carbon network. N_2isotherms and SEM images showed that NPC-[C_4N][X] have abundant nanopores with large BET surface areas. XPS, FT-IR and elemental analysis results showed the N rich network in NPC support, has been successfully and homogeneously functionalized with strongly acidic ionic liquid and sulfonic groups. ^{31}P solid NMR spectra confirmed that NPC-[C_4N][X] have typical Brønsted acid centers with strong and controllable acidity. Catalytic tests showed that NPC-[C_4N][X] exhibits excellent activity and reusability in production of biofuels and fine chemicals through esterifications, transesterifications, and depolymerization of microcrystalline cellulose. This approach may provide a novel and cost effective way to recycle the waste cow manure into efficiently porous carbon based solid acids for heterogeneous acid catalysis.

471686 Analysis of Carboxylic Acids in the Biocrude Derived from Hydrothermal Liquefaction of Lignocellulosic Biomass

Abstract:

The authors report (a) hydrothermal liquefaction (HTL) of lignocellulosic biomass using homogeneous nitrate salt catalyst: $Ni(NO_3)_2$, $Ca(NO_3)_2$, $Co(NO_3)_2$ and $Fe(NO_3)_3$ for the production of carboxylic acid enriched biocrude, and (b) detailed investigation of biocrude generated from a biomass emphasizing on the selectivity of lactic acid in particular and (c) study of the possible reaction pathways for production of carboxylic acids, biocrude and bio-oil with their energy content from a biomass. A biomass was loaded in the high temperature high pressure SS316 reactor and HTL of biomass was performed at 200 ^{o}- 300 ^{o}C. During HTL reaction, liquid samples were withdrawn at regular time intervals and the processed slurry was centrifuged to recover biocrude and residue. The biocrude was further analyzed by TOC analyzer, GC-MS and HPLC. The biocrude and residue obtained were further extracted in different solvents followed by evaporation on Rota evaporator to recover bio-oil and the bio-oil was analyzed using GC-MS. The HPLC analysis of biocrude revealed the presence of sugars, C_1-C_3 carboxylic acids such as lactic, propionic, acetic and formic, HMF/furfural and other oxygenated compounds. Among different C_1-C_3 acids, lactic acid was observed in the major amount of 83.92 wt% using pinewood: solvent of 1:75 at 250°C. The results obtained on carboxylic acid enriched biocrude, bio-oil composition, selectivity of organic acids such as lactic acid obtained during HTL of biomass substrates and possible reaction pathways will be presented in detail.

472041 Catalytic Pyrolysis of Sugar Cane Bagasse and Rice Straw

Abstract:

Proper utilization of agricultural wastes is particularly important not only for prevention of environmental pollution but also creation of value added by-products from them. In rural areas, the agricultural wastes is a good resource for animal feed if the transportation cost are low. If there is no local demand for such wastes burning in the fields is a common practice; which although is an easy solution for the farmers create air pollution for the close and distant localities. Sugar cane bagasse (SCB) and rice straw (RS) are potential environmental thread in countries where huge amount of such waste have no industrial utilization. Biochemical or thermochemical conversion technologies can be used to convert agricultural waste streams into biobased chemicals and biofuels. In this research, SCB and RS supplied from Egypt were pyrolyzed in a 2-inch fast pyrolysis reactor at 450°C. Commercial HZSM-5 and granulated red mud were used as the catalyst and nitrogen was used for fluidizing the system. The yields of SCB and RS pyrolysis liquids ranged from 35% and 28% over HZSM-5 catalyst and 36% and 24% over red mud catalyst respectively. The pH of the SCB and RS pyrolysis oils was 2.0-3.0 . The yields of char were 16% and 42% respectively for SCB and RS over HZSM-5 catalyst and 31% and 40% respectively for SCB and RS over red mud catalyst. The gas products were 30% (SCB) and 40% (RS) over HZSM-5 catalyst and 32% (SCB) and 36%(RS) over red mud catalyst. The majority of the gas were CO_2, CO and CH_4 together with H_2 and C_2-C_4 hydrocarbons at lower concentrations.

465741 Engineering Highly Active Brookite Titania Nanorods for Sustainable Hydrogen Production

Abstract:

Titanium dioxide (TiO_2, titania) nanomaterials are the most studied photocatalysts because of their abundancy, non-toxicity, stability and activity. Anatase and rutile polymorphs have been deeply investigated because they are the energetically favored forms at the nanoscale and in the bulk, respectively. Brookite, instead, has been rarely studied, despite theoretical and experimental data support its higher activity in some photocatalyzed transformations. By taking advantage of methods to produce 1-D phase-pure brookite nanorods, we show not only that brookite is a much more active photocatalyst than anatase, but also that its activity is length-dependent. In particular, we attribute the high activity to both the 1-D nature of the nanostructure, which favors electron-hole separation, and to the defective structure of the rods, which are subject to strain that causes favorable changes in electron-hole recombination processes. By tuning length and strain we are able to prepare samples that show among the highest rates of hydrogen production from biomass-derived compounds under solar simulated irradiation reported to date.

460638 Hydrogen Production Via Photocatalytic Water Splitting over Mesoporous Pt/TiO_2

Abstract:

Mesoporous TiO_2 semiconductors were synthesized by sol-gel technique, using a triblock-copolymer-templating, named Pluronic F-127. In order to minimize the electron-hole pair recombination, Pt was added as an electron trap in different percentages (*i.e.* 1.00 wt. %, 2.50 wt. %) during the sol-gel synthesis of the TiO_2 semiconductors. This was followed by calcination at 500 °C and 550 °C. Semiconductors' activity was evaluated in a bench-scale Photo-CREC-Water II (PCWII) photoreactor. In this sense, water splitting was carried out under near-UV light for hydrogen molecule production in the PCWII Reactor using ethanol at 2 % v/v as a renewable organic scavenger (electron donor). Ethanol acts as a good hole scavenger. The PCWII consisted of a 2.65 L internally irradiated slurry annular reactor with a 15 W T8 Black-Light-Blue (F15T8/BLB) lamp. The optimal photocatalyst loading was 0.15 g L^{-1} in acid medium (pH = 4.0). The volume of the aqueous suspension was 6.0 L. Thereby the present study shows that mesoporous TiO_2 with 2.50 wt. % of Pt exhibits the best activity for hydrogen production under the same photoreaction conditions. This semiconductor material showed: a) An anatase phase with XRD, b) A reduced band gap at 2.34 eV was calculated, employing UV-Vis spectroscopy and a Tauc plot, c) 10-30 nm TiO_2 particles with SEM-EDX, d) A 140 m^2/g BET surface area as well as 10 nm and 20 nm bimodal pore size distributions. Furthermore, the absorbed photons on the photocatalysts have been studied by employing macroscopic balances. These macroscopic energy balances considered for incident, scattered and transmitted near-UV photons. In this respect, evaluated Quantum Yields (QY) displayed a promising 0.27 (27%) when utilizing mesoporous 2.50 wt. % Pt/TiO_2-550°C.

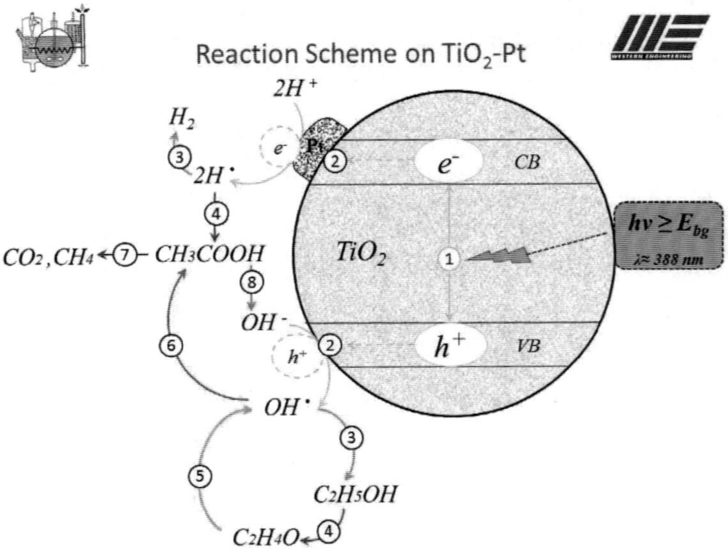

467466 How Ceria Redox Materials Split H_2o and CO_2 with Concentrated Solar Energy into H_2, CO, and O_2

Abstract:

Heterogeneous catalyst design has been revolutionized in the past several years by the discovery of adsorption-energy scaling relations on transition-metal catalysts. These and correlated activation energies of chemical reactions are understood to limit the catalytic activity of a surface for a number of interesting reactions, such as production of H_2 and CO via splitting of H_2O and CO_2. Metal oxide redox materials exchange oxygen with a gas phase and store it in the solid state. This allows circumventing scaling relations that limit catalytic reactions, as a chemical reaction may be split into reduction and oxidation steps that are conducted at different process conditions. This work employs electronic structure calculations to investigate the reaction mechanism of CO_2 and H_2O splitting with ceria. Conceptually, non-stoichiometric ceria is reduced at elevated temperature with concentrated solar energy, liberating O_2. In a second step, the lattice oxygen is replenished via oxidation with H_2O and CO_2, yielding renewable H_2 and CO (syngas). Syngas is the precursor for synthetic liquid hydrocarbon fuel, such as diesel, gasoline and kerosene. The presented computations outline the effect of reduction and oxidation temperatures, partial pressure of O_2, and metal doping at the ceria surface on the free energy landscape of the two studied chemical reactions. Limits of isothermal processing as well redox trade-offs when controlling the reduction potential of the oxide surface are discussed.

472043 Methane Conversion to Syngas Using Dopant Modified Metal Oxide Composites in Chemical Looping Reforming

Abstract:

Metal oxide composites are used as enabling materials for operation in many energy conversion systems including shale gas and coal gas reforming using chemical looping systems. When metal oxide composites are involved in redox reactions, ionic diffusion takes place which is accompanied by the creation and diffusion of defects that decide such their morphology and chemical properties. Rare earth element dopants can affect the redox reactions by providing alternative reaction sites, thus change the ionic diffusivity and reaction reactivity of metal oxides. In this study, we examine at nanoscale to delineate the characteristics of undoped and doped iron oxide that undergo methane reforming reaction. 0-2% lanthanum doped iron oxides are fabricated by modified wet chemical synthesis. XRD and SEM/EDS characterization implies that the dopant changes lattice parameters of metal oxides whose surface energy is changed. The atomistic thermodynamics methods and ab initio calculations are carried out to investigate the interaction between methane molecules and undoped/doped iron oxide. It is found that dissociation of C-H bond in the CH-La species is the key step for CH_4 partial oxidation reaction, and low-fold O atoms neighboring dopants in metal oxide composites could more readily dissociate from external surface to create oxygen vacancies. Moreover, the presence of rare earth element dopants can significantly decrease the barrier of CH_4 disociation, hence the reaction between CH_4 and oxygen carrier is promoted, which is in good agreement with our experimental results. The fundamental insight into the mechanism of cyclic redox reactions of La dopant modified metal oxide composites at nanoscale revealed in this study can be applied to other rare earth element dopant modified metal oxide systems and to guide future efforts to improve oxygen carriers for methane conversion.

471655 A New Thin Layered Structural Coating on a Metal Substrate for Enhanced Hydrogen Production from Steam Methane Reforming

Abstract:

The recent increase in the availability of hydrocarbons, especially methane, from shale gas exploration has led to renewed interest in converting or upgrading methane to higher value chemicals. Steam reforming is still the most developed and widely practiced industrial process for converting methane to hydrogen/syngas and its derivatives. It has no oxygen requirement and the most favorable H2/CO ratio for methanol and ammonia production of all available technologies. However, opportunities still remain to lower the energy requirement of steam reforming, including development of more efficient catalyst systems in terms of higher activity and effectiveness factors. Recently, Alloy Surfaces Co. Inc. developed an intrinsically bound thin-layered skeletal catalyst adhered on a metal substrate. These structures having relatively uniform coating thickness allow the physical manipulation into a highly active complex monolith structure that greatly enhances the catalyst effectiveness and interfacial heat and mass transport without applying an intermediate washcoat. A collaboration among City University of New York, City College, Alloy Surfaces, Co. Inc., Innoveering, and Université catholique de Louvain has been established to develop novel catalyst systems to more efficiently convert methane. Recently the team studied and analyzed this catalyst developed for hydrogen generation via steam methane reforming. Results have demonstrated that this catalyst is intrinsically more active than the conventional Ni catalyst and shows an increased selectivity toward CO2 and H2 and higher WGS rates. Methane-to-steam ratios ranging from 2.0 to 6.0 were tested over a temperature range between 400°C to 900°C at pressures of 1.0, 1.5, 3.0 bar with GHSV from 50,000 to 350,000 h-1. Mass balances were closed to within 2% and methane conversions were maintained below 20% of the potential equilibrium conversions possible at the particular test conditions. Results demonstrate the catalyst is capable of producing H2/CO ratios up to 425 from a reactant steam to carbon ratio of 3 at 515oC with no measureable change in activity over the time tested. The WGS reaction is about an order of magnitude faster than on the conventional nickel catalyst with nearly a factor of 2 enhancement of the CH4 + 2 H2O = CO2 + 4 H2 reaction. This presentation will show the results obtained over a wide range of conditions. A

kinetic model with rate parameters was derived from the experiments with a specially designed microreactor and used to quantify the enhance selectivity toward hydrogen and carbon dioxide under typical commercial operating conditions. Finally some details will be given on how the enhanced performance is achieved.

460805 Promotion Mechanisms of Iron Oxide-Based High Temperature-Water Gas Shift (HT-WGS) Catalysts By Chromium and Copper

Abstract:

The Cr and Cu promotion mechanisms of high temperature water-gas shift (HT-WGS) iron oxide catalysts, synthesized by co-precipitation, were investigated as a function of reaction conditions. XRD and *in situ* Raman characterization showed that the initial calcined catalysts consisted of the Fe_2O_3 (hematite) bulk phase and transformed to the Fe_3O_4 (magnetite) phase during the HT-WGS reactions. *In situ* NAP-XPS and HS-LEIS surface analysis revealed that Cr was surface enriched as Cr^{+6} for the initial catalyst and reduced to Cr^{+3} during the HT-WGS reactions, with the Cr^{+3} dissolving into the bulk iron oxide lattice forming a solid solution. *In situ* NAP-XPS, XANES and HS-LEIS characterization indicated that Cu was initially present as Cu^{+2} cations dissolved in the Fe_2O_3 bulk lattice and reduced to metallic Cu^0 nanoparticles (~3nm) on the external surface of the iron oxide support during the HT-WGS reactions. *In situ* HS-LEIS surface analysis also suggests that ~1/3 of the surface of the Cu nanoparticles was covered by a FeO_x overlayer. The CO-TPR probe demonstrated that Cr does not chemically promote the iron oxide catalyst and that only Cu is a chemical promoter for the iron oxide HT-WGS catalysts. The Cu promoter introduces new catalytic active sites that enhance the reaction rates of the WGS reactions.

472372 Catalytic Ammonia Decomposition to Hydrogen and Nitrogen on Supported Nanosized and Single-Site Cobalt Catalysts: Effects of Atomicity, Ligand Environment and Support

Abstract:

Hydrogen is one of the most widely used chemicals in the world. Petroleum refining and ammonia production are the major fields with the largest demand for hydrogen whereas automotive fuel is an emerging sector with the hydrogen demand growing incredibly fast. The total expected value of the whole hydrogen market is 152 billion USD by 2021. The main disadvantage of using pure hydrogen is the difficulty associated with its storage and transport.

Ammonia, on the other hand, is a hydrogen rich fuel (17.65 wt% hydrogen) that can be liquefied, stored and transported easily. As a liquid it has a volumetric hydrogen density about 45% higher than liquid hydrogen. To use this hydrogen, ammonia has to be decomposed over a catalyst to produce the desired clean fuel - hydrogen (H_2) along with nitrogen (N_2) a non-toxic, non-greenhouse gas that has no contribution to carbon emissions.

The most active catalysts used nowadays for ammonia decomposition are carbon nanotube-supported ruthenium nanoparticles. They are expensive and not always straightforward to produce. Therefore, finding new catalysts for ammonia decomposition that contain Earth-abundant metals utilized with atomic efficiency could be a viable alternative to expensive ruthenium catalysts. Moreover, gaining fundamental knowledge about metal-catalyzed ammonia decomposition, starting from the single-atom presents an important challenge that would allow us to establish more precise structure-catalytic property relationships, as well as create a pathway to produce more active/stable catalysts.

Herein, we demonstrate for the first time that the single-site silica-supported cobalt catalyst is an effective and stable catalyst for ammonia decomposition to nitrogen and hydrogen with activity starting at 480 °C and high TOF at elevated temperatures. The catalyst retains its single-site nature up until 630 °C, as evidenced by XPS, Raman and Diffuse-Reflectance UV-VIS spectroscopic measurements. Reduction of the catalyst takes place above 630 °C (with accompanying color change from blue to black) producing a reduced supported cobalt or cobalt nitride Co_xN_y phase, which is even more active in catalytic NH_3 decomposition, with activity above 390 °C. We discuss the proposed mechanism of NH_3 decomposition by single-site Co(II)/SiO_2 catalyst: even though the catalyst itself is very robust and it cannot be reduced or oxidized below 630 °C, it has unique activity towards nitrogen-containing molecules like ammonia and pyridine; those molecules chemisorb on Co(II) with the concomitant change of distorted tetrahedral environment of CoO4 unit to square pyramidal and octahedral as evidenced by Diffuse Reflectance UV-VIS. The catalytic activity of Co(II)/silica catalyst is assessed considering the previous report by A. Hock et al which indicates Co(II)/SiO_2 is an active and selective catalyst for propane dehydrogenation at temperatures close to 600 °C. Based on the similarity of N-H and C-H bond strengths, the ability of Co(II)/SiO2 to break strong N-H bonds therefore signifies a general trend of this catalyst to break strong X-H bonds. However, the single-site cobalt catalyst is less active than extended metal or metal nitride surface, exemplifying the important fundamental and mechanistic difference between the regimes of operation for the single-site and nanosized supported Co catalysts for NH_3 decomposition under given conditions.

Moreover, we show that the ligand environment around cobalt center can be modified on a single-atom scale: from CoO_4 to CoN_4 by incorporating single Co(II) atoms into mesoporous carbon nitride C_3N_4. This change in the ligand environment is not favourable for ammonia decomposition, with activity starting at much higher temperature compared to analogous Co(II)/SiO_2 catalyst.

Additionally, we demonstrate that the same dehydrogenation principle can be applied to ethylamine $C_2H_5NH_2$, with the Co(II)/SiO_2 catalyst capable of selectively dehydrogenating the mixed RCH_2-NH_2 bond of ethylamine to produce acetonitrile CH_3CN.

To get a more general and fundamental understanding of the catalytic behavior of mononuclear Co(II) species embedded into silica, we tested it in two bench mark reactions for which nanosized supported cobalt and cobalt oxide particles are predominantly used: CO and CO_2 hydrogenation. CO hydrogenation activity starts at temperatures above 380-400 °C with the main product being methane (~90%) and minor C_2 products at 2 MPa CO/H_2 pressure; CO_2 hydrogenation of CO_2/H_2 mixture at 1 bar leads to reverse water-gas shift reaction and formation of CO as the main CO_2 reduction product, detectable at temperatures above 300 °C.

In conclusion, Co(II)/SiO_2 was shown to be catalytically active for ammonia decomposition as well as carbon monoxide/carbon dioxide hydrogenation. The direct comparison of two catalysts with atomically dispersed and nanodispersed supported metal atoms/particles once again exemplifies the fundamental difference between catalysis by single metal-atoms (cationic species) and traditional nanosized supported catalysts.

473927 Refineries in a Carbon Contrained Future - the Potential for Facilty-Wide CO2 Capture at Plant Utilities

Abstract:

What can refineries do in a carbon constrained future? Obviously refineries will produce products to satisfy the product demand slates â€“ primarily liquid transportation fuels. However, refineries can also remove the CO2 emissions associated with their operations. We will explore a CO2 capture concept comprised of a utilities plant that and produces hydrogen for combustion in the refinery furnaces and CO2 to send to storage outside the refinery boundary. This option could be the least-cost pathway for refineries to remove their CO2 emissions.

CO_2-Utilization in Chemicals Production Based on Evidence from Start-Ups and Literature

Abstract:

Greenhouse gas emissions and the resulting anthropogenic climate change create tremendous economic, ecologic and social challenges for societies today and in the future. To mitigate these effects, world leaders agreed to limit global warming to two-degrees compared to preindustrial times, which translates to an atmospheric CO_2-concentration of 450 ppm [1]. To reach this goal, global annual greenhouse gas emissions have to decrease by 40 to 70 percent until 2050 and to near-zero until 2100. Major reductions are suggested to be generated from energy efficiency measures, renewable energies and carbon capture and storage [2]. One additional emissions reduction pathway is CO_2 utilization, which takes CO_2 from point sources or from air and utilizes it in large-scale applications. This way CO_2 becomes a process agent or even a carbon source, which leads to the substitution of fossil fuels and lower overall greenhouse gas emissions [3].

Whether or not emission reductions from CO_2 utilization can be realized, largely depends on the rapid implementation in large-scale industrial processes and markets. While the political pressure and support is high, most current CO_2 consuming reaction pathways have not yet passed lab-stage technology levels and industrial feasibility is mostly not yet proven [4]. Many projects described in literature are in the stage of catalyst and reaction optimisation. In many cases evaluations of early-stage base chemicals, intermediates, polymer or fuel reaction pathways have been carried out in literature (e.g. [6], [7], [8], [9]), often described as "techno-economic" assessment. However it remains unclear what indicators such a techno-economic assessment should include or which procedure it should follow. While critical success factors for large-scale implementation lie within the fields of technologic, economic but also environmental feasibility, very few studies manage to evaluate technologies in all three fields. Additionally assessments in high level of detail remain challenging, since many processing parameters are not defined at an early stage of technology. A multidimensional evaluation with short-cut indicators seems to be beneficial to clarify potentials and challenges with proposed CO_2 utilization technologies compared to current market benchmarks.

In this contribution we first outline the approaches of techno-economic assessments performed in chemistry-related start-ups, gathered by qualitative interviews. Furthermore we analyse current techno-economic assessment approaches in the literature. Based on this evidence we present an integrated approach for rapid and systematic evaluation of chemical CO_2 utilization technologies in the three fields of technologic, economic and environmental feasibility. The method leads to recommendations for further catalysis research and process development in the prior domains and facilitates decision making for start-ups, funding agencies and strategic investors. Additionally we show first results of its application in scientific literature of selected CO_2 utilization reactions.

This contribution is funded by EIT Climate KIC's project "enabling CO_2 reuse" EnCO_2re.

[1] International Energy Agency (2012): World Energy Outlook Special Report. Redrawing the Energy Map. Paris. Available online http://www.worldenergyoutlook.org/media/weowebsite/2013/energyclimatemap/RedrawingEnergyClimateMap.pdf.

[2] IPCC (2013): Climate Change 2014 Mitigation of Climate Change. Geneva. Available online http://www.ipcc.ch/report/ar5/wg3/.

[3] P. Styring, E.A. Quadrelli, K. Amstrong (2015): Carbon Dioxide Utilisation: Closing the Carbon Cycle. 1. edition. Amsterdam, Oxford: Elsevier.

[4] A. Boxin-Dumitriu, M. Perez Fortes, E. Tzimas, T. Sveen (2013): Carbon Capture and Utilisation Workshop Background and proceedings. Luxembourg. Available online http://publications.jrc.ec.europa.eu/repository/handle/111111111/30232.

[5] H. Audus, H. Oonk (1997): An assessment procedure for chemical utilisation schemes intended to reduce CO2 emissions to atmosphere. Energy Conversion and Management, 38, S409-S414.

[6] K. Mueller, W. Arlt (2014): Shortcut Evaluation of Chemical Carbon Dioxide Utilization Processes. Chemical Engineering & Technology, 37 (9), 1612-1615.

[7] M. Lehner, M. Ellersdorfer, R. Treimer, P. Moser, V. Theodoridou, H. Biedermann (2012): Carbon Capture and Utilization (CCU) – Verfahrenswege und deren Bewertung. BHM Berg- und Hüttenmännische Monatshefte, 157 (2), 63-69.

[8] A. Otto, T. Grube, S. Schiebahn, D. Stolten (2015): Closing the loop: captured CO2 as a feedstock in the chemical industry. Energy & Environmental Science, 8 (11), 3283-3297.l

[9] I. Dimitriou, P. Garcia-Gutierrez, R. H. Elder, R. M. Cuellar-Franca, A. Azapagic, R. W. K. Allen (2015): Carbon dioxide utilisation for production of transport fuels: process and economic analysis. Energy & Environmental Science, 8 (6), 1775-1789.

460305 Assessing the Potential of CO_2 Utilization with an Integrated Framework for Producing Power and Chemicals

Abstract:

Energy is an indispensable part of our daily lives. The last decade of development has resulted in a sharp rise in energy demand. Reducing CO_2 emissions urgently is a global issue, and the stakes are increasing with the passage of time. The major focus so far has been on sequestering the captured CO_2, but this has its own problems and concerns. Thus, the idea of "using" or "converting" CO_2 rather than "disposing" seems more attractive! The direct use of CO_2 being limited, we need ways to convert CO_2 into something useful such as valuable chemicals and fuels. Carbon dioxide utilization (CDU) is seen as a strategic measure to mitigate CO_2 emissions. In CDU, CO_2 is viewed as a carbon source for producing chemicals and fuels instead of treating CO_2 as a mere waste. However, a quantitative assessment of the global impact of CDU is a crucial yet unaddressed issue.

In this work, we propose the concept of CO_2 neutral chemicals and power production. We propose a network of reactions that begins with oxy-combustion of methane (the energy source) and ends with production of chemicals that have significant global demands. Dimethyl ether and methanol are considered as potential alternatives for conventional fuels. The reactions in the network are categorized as combustion reactions, hydrogen (H_2) producing reactions, CO_2 consuming reactions and secondary reactions. CO_2 consuming reactions use any CO_2 produced within the system. Secondary reactions produce any other reactants required to achieve CO_2 neutrality. The superstructure of all possible reactions is solved as an optimization problem to obtain the reaction network that achieves CO_2 neutral chemicals and electricity production. The process is entirely self-sufficient in its thermal and electrical energy needs with a net surplus of electricity, and completely utilizes all CO_2 produced.

The only caveat in achieving CO_2 neutrality mentioned above is the availability of hydrogen, as hydrogen produced internally is not sufficient to achieve both CO_2 neutrality and meet global demands for the chosen chemicals. Hydrogen can either be produced within the proposed framework (i.e. fossil based), or obtained from electrolysis of water using renewable energy resources. Thus, based on the source of hydrogen, we present three different scenarios. Firstly, we assumed availability of unlimited renewable energy to produce hydrogen by electrolysis of water. In this scenario, the estimated CO_2 avoidance was 59% as compared to the 2013 emission level. In the second scenario, we assumed the availability of renewable energy at current global levels. CO_2 avoidance dropped dramatically to 6% in this scenario. Finally, in the third scenario, we assumed zero external hydrogen and the avoidance of CO_2 emissions reduced to a mere 1%. Our analysis quantitatively establishes the fact that availability of renewable hydrogen will play a pivotal role in determining the future of CDU. Note that our framework does not advocate centralized production of bulk chemicals in one place. We also tested the idea of producing individual chemicals within our framework, but it obviously results in lower CO_2 avoidance.

Our work has taken a process systems viewpoint to show that utilizing CO_2 to produce alternate fuels along with bulk chemicals has the potential to combat CO_2 emissions at the global scale. Our framework allows us to obtain an upper bound for the global impact of CO_2 utilization by taking into account the maximum current and projected global demands for all chosen chemicals. Market demand of chemicals derived from CO_2 will have a significant impact on CDU. With the rising concerns about CCS, CDU may well be the next strategy for consideration in the long run. However, availability of abundant renewable hydrogen will be necessary to realize the full potential of CDU.

463458 CO2 Utilization for Methanol Production As an Intermediate for Formaldehyde: Technoeconomic and Sustainability Assessment of the Value Chain in the Case of Sweden

Abstract:

To enable CO_2 re-utilization, a study in West Sweden aimed to identify partnerships between companies producing CO_2 and companies that may reuse CO_2 in their industrial processes. The study was initiated by a methodology for location, quantification and characterization of CO_2sources providing the first test-bed with large replication potential to other regions in Europe. The test-bed currently focuses on technologies mature enough to be implemented in a short-term timescale (5 to 10 years). On this basis methanol production has been prioritized as one of the technologies where emphasis should be given (e.g., some of the others being algae production, horticulture production, power to gas, concrete curing, mineral carbonation, etc.).

Methanol is one of the most common platform chemicals (i.e., a precursor of MTBE, DME, formaldehyde, acetic acid, ethanol, olefins, etc.) and is traditionally produced from synthesis gas (syngas). Syngas is mainly produced by the reforming of natural gas; however, gasification of solid (e.g., wood residues, coal) and liquid (e.g., black liquor from pulp and paper industries, refinery by-products) fuels is also gaining attention. The global annual use of fossil-derived syngas is approximately 6 EJ, 10% of which is used for methanol production. Currently, the methanol synthesis plants operate in a stable, high-throughput production mode at low CO_2/CO ratios in feed (e.g., preferred by copper catalysts up to concentrations of 10-12% CO_2). It is an issue of on-going research to develop catalysts and processes that can utilize high-CO_2content feeds in a techno-economically viable operating mode (generally high pressures, low temperatures are preferred for increasing the methanol yield).

Depending on the source of CO_2, hydrogen, the other methanol synthesis reactant, may be already present in the gas mixture (e.g., syngas from gasification, steel work off-gases), may be available from a different industrial source or provided from water hydrolysis using external, cheap energy sources (e.g., surplus energy from the grid). Hybrid solutions have also been proposed, e.g., where methanol is produced by combining biomass based synthesis gas with steel plant off-gases typically having very diverse composition (e.g., 3-65% H_2, 5-60% CO, 0-20% CH_4, 2-25% CO_2, 3-55% N_2) and LHV values (e.g., 3-18 MJ/Nm^3).

This study focuses on the perspective of utilizing available industrial CO_2 sources to produce methanol as an intermediate for the subsequence synthesis of formaldehyde, which is of particular industrial interest for chemical production companies in West Sweden. It compares 6 process alternatives of different technology readiness level (TRL), including synthesis routes through the reversed-water-gas-shift reaction, formic acid, organo-carbamates, methyl formate, dimethyl carbonate and an upcoming indium oxide based technology. Rigorous process modelling is used to provide the necessary data for the technoeconomic (i.e., considering CAPEX and TRL) and sustainability (i.e., considering LCA and hazard assessment). The potential impact of impurities is discussed for the whole value chain from the CO_2 and H_2 sources to methanol and the subsequent formaldehyde production. The assessment considers uncertainties in the information for the CO_2 and H_2sources, the process modelling, and various future scenarios in the development of the energy technology sector.

474798 Life Cycle Emissions Assessment of a Renewable Fuel Process: Impact of Catalyst Performance on the Net GHG Emissions of Methanol Production By Direct Electrocatalytic Reduction of CO2

Abstract:

The ability to synthesize liquid hydrocarbon fuels from renewable energy and CO2 would provide a stable, energy-dense storage medium for solar energy, and provide a non-fossil, low-emissions route to liquid fuels. Research is now underway to address technical challenges such as reducing the energy cost of CO2 capture and increasing the efficiency of CO2 reduction catalysts. However, the full system-wide emissions balance of such a process has received little attention. Although the CO2 that is incorporated into a product is diverted from emission to the atmosphere, other aspects of the utilization process may produce greenhouse gas (GHG) emissions that negate some or all of this benefit. To determine whether a CO2-to-fuel process would provide a net emissions benefit requires a life cycle assessment (LCA) that extends across the full life cycle of the utilization process.

We present a GHG LCA for a CO2-to-fuel process to assess the net emissions benefit of such a process, and to evaluate the importance of catalyst performance and other system parameters in achieving net emissions reductions. As an example of a CO2-to-fuel process, we examine the low-temperature, six-electron reduction of CO2 to methanol ("deCO2rr"). We consider a scenario in which CO2 is supplied from a concentrated point source (coal plant flue gas), and process energy for the electrocatalytic reduction is provided by a low-emissions source (wind generated electricity). We evaluate the total GHG emissions from producing both the methanol fuel and the electricity output from the power plant, and compare to the emissions when producing these from conventional processes.

The analysis projects that the deCO2rr pathway could produce methanol and electricity with lower GHG intensity than conventional processes, with a sufficiently efficient CO2 reduction catalyst (e.g. 40% faradaic efficiency, 40% selectivity for methanol). The maximum emissions reduction potential of the deCO2rr pathway, using a near-ideal catalyst, is a 60% reduction relative to the conventional processes for methanol and electricity production. When the faradaic efficiency of the CO2 reduction catalyst is above 30%, further improvements in faradaic efficiency have little impact on the full system emissions of the deCO2rr route. This analysis demonstrates the opportunity for full system LCA to complement fundamental research and guide technology strategy for sustainable outcomes.

470127 Integration of Sustainability Assessment in the Early Design Phases of Amine Based Post-Combustion CO_2 Capture

Abstract:

Climate change is one of the biggest environmental and engineering challenges of the 21st century. Post-combustion CO_2 capture (PCC) is one of the most mature technologies to be deployed in an attempt to maintain atmospheric CO_2 levels below 450 ppmv. Controlling CO_2 levels would be necessary to keep the average global temperature rise below 2 °C compared to pre-industrial levels to avoid irreversible changes as a result of climate change. Given the environmental concerns driving process development, it is important to fully study the impacts of deployment of such technologies beforehand to avoid replacing one problem with potentially more severe ones.

Monoethanolamine (MEA) is currently the industry standard for amine solvents in PCC applications. However, there is increasing pressure to develop alternative solvents that avoid some of the drawbacks encountered in the MEA process without posing additional health and environmental risks. A new solvent would have to reduce the energy penalty associated mainly with solvent regeneration as well as circumvent problems caused by solvent degradation while maintaining a high reactivity and capacity for CO_2 absorption. Progress in development of more effective solvents and processes has been slow due to the need for extensive experimental research and pilot plant campaigns. There is a vast number of combinations of potential solvent candidates, including solvent blends, and process configurations; the experimental evaluation of each option is expensive and time-consuming.

To help speed up this effort, computer-aided design methods are used to guide the search for improved solvents and process schemes as well as to focus the experimental effort. The early integration of sustainability assessment in the process and solvent design is crucial to avoid costly choices and impacts that will be difficult to mitigate in later design and operation stages. Sustainability assessment is defined here as a combined hazard and life cycle assessment. The hazard assessment evaluates potential damage in accidental scenarios, while the life cycle assessment presents the environmental impact of normal process operation. A rigorous hazard and life cycle assessment is first conducted for the well-studied MEA process and compared to amine-based CO_2 capture processes that have been proposed in literature. The findings of this assessment are used to guide the development of a robust sustainability assessment tool to be integrated into computer aided molecular design (CAMD) applications for initial solvent screening.

Typical solvent design applications use properties such as solubility parameters, pKa, vapour pressure, density and viscosity among others as proxy indicators for the technical feasibility of a solvent and its performance in a capture process. Such properties could represent a solventâ€™s capacity for CO_2 capture, expected fugitive losses, pump work and heating loads for solvent regeneration. All these factors are also important as indicators for the life cycle assessment. QSAR models for estimating physical properties are largely available. The hazard assessment requires the estimation of additional properties such as solvent flammability, toxicity and persistence in the environment. In a smaller extent, the use of shortcut group contribution methods also allows for the integration of the impacts of solvent production to the life cycle assessment. Severe data gaps however exist for the prediction of a solventâ€™s potential to degrade and the impact of solvent degradation. Solvent degradation can lead to many operational problems such as reduced heat transfer efficiency, foaming, corrosion and the formation of hazardous degradation products in addition to loss of CO_2 absorption capacity. Therefore, degradation potential is an important indicator for a solventâ€™s process, hazard and life cycle performance and should be included as early as possible.

A database of hazard and environmentally related properties is created comprising 80,000 chemical structures. Physical and chemical properties are collected mainly from available material safety datasheets. This database facilitates the automation of the online connection of the CAMD optimization routine and the complimentary sustainability assessment module. The connection can be established where an algorithm combines the proposed molecular groups into all potential structure formations and runs the proposed structures through the internal database for hazard related properties. If all proposed molecules are found the mean of the properties is taken, otherwise QSAR models are used wherever possible to predict the missing properties Experimental results are used in combination with literature data to develop a QSAR model for predicting solvent degradation rates. The model is developed using Principal Component Regression (PCR) and Multi Non-Linear Regression (MNLR) to correlate degradation rates with molecular structure, thermodynamic properties and operating conditions.

In conclusion, this work introduces a framework for the integration of the sustainability assessment at early stages of solvent and process design for post combustion CO_2 capture. It shows the translation of rigorous hazard and life cycle assessment into indicators that can be easily integrated in a CAMD routine. It presents a database of hazard and environmental properties, which facilitates the online connection to an optimization routine. Finally, it proposes a QSAR model for the determination of solvent degradation rates under the PCC conditions, which enables the addition of an important and currently missing solvent feature at molecular design stages for a more comprehensive assessment.

472144 Optimal Design and Techno-Economic Analysis of CO_2 Mineralization Process Using Aqueous NaOH

Abstract:

CO_2 mineralization is one of the prospective CO_2 utilization methods, which can achieve both economic benefit and CO_2 reduction by converting CO_2 to more stable and valuable mineral form. As a potential CO_2 mineralization process, sodium bicarbonate ($NaHCO_3$) production using aqueous sodium hydroxide (NaOH) was suggested earlier by Skyonic Corporation. The overall process is composed of chlor-alkali process, CO_2 absorption process, and product recovery process. One feature of this mineralization process is that a variety of product choices are possible. In addition to $NaHCO_3$, one can produce hydrochloric acid, sodium hypochlorite, and hypochlorous acid from NaOH, Cl_2 and H_2 which are intermediate products of the chlor-alkali process. In this study, optimal design of the CO_2 mineralization process using NaOH is mainly examined. A process design achieving maximum CO_2 utilization is considered as a base case in which all of NaOH produced is consumed to convert CO_2. The base case is designed as a reference to the Skyonicâ€™s process using AspenPlus^{TM}. Additionally, techno-economic analysis is carried out to discuss the feasibilities of the base case in terms of economics and CO_2 reduction in comparison with a conventional $NaHCO_3$ production process (â€˜Solvayâ€™ process). Finally, optimal process designs leading to maximum profit are suggested among several possible process designs with different production portfolios. Particularly, various scenarios of the market demand and product prices are considered in finding the optimal design.

463256 New Conceptual Design Models for Carbon Capture and Utilization Via Carbonate-Based Construction Materials

Abstract:

Carbon capture and sequestration by amine-based absorption technology is too costly for widespread deployment and also suffers from safety and environmental concerns about the ultimate fate of the sequestered fluid. We prefer carbon capture and utilization as carbonate-based cementitious material [1]. This CO_2 utilization strategy exhibits two major advantages; it turns a waste into a useful product and obtains the collateral benefit of reducing CO_2 emissions from cement kilns. The overall cost to produce carbonate containing structural materials will be highly sensitive to process design, so the problem of analyzing the technical and commercial feasibility of carbonate-based supplementary cementitious material becomes an exercise in creative conceptual design and process intensification. Point sources of CO_2 exist at power generation facilities as low pressure, relatively dilute gaseous mixtures. Geologic brines are a reasonable candidate for calcium and magnesium sources to minimize limestone calcination. Therefore, rational conceptual design in this system requires a framework for collectively considering three-phase contact, rate-based chemical reaction, and multicomponent, electrolytic mass transport. There are methods in the literature for dealing with these effects in binary combinations, but the combined effect of concentrated multicomponent reaction and diffusion among electrolytes occurring in conjunction with both absorption and precipitation processes does not find a complete treatment in the literature. Here, we present a framework for considering these effects collectively and consistently using the Maxwell-Stefan description of diffusion [2]. We present some preliminary results and address the remaining challenges that exist in applying these methods to the production of carbonate containing cementitious material to capture and utilize CO_2.

References

[1] Constantz, Brent R., Andrew Youngs, and Terence Hollands. Reduced-carbon Footprint Concrete Compositions. Calera Corporation, assignee. Patent WO 2010048457 A1. 29 Apr. 2010.

[2] Taylor, R., & Krishna, R. (1993). *Multicomponent mass transfer* (Vol. 2). John Wiley & Sons.

472620 Renewable N-Containing Chemicals Production By Ammonization of Biomass Derived Lipid over Heterogeneous Catalyst

Abstract:

N-containing chemicals are widely used in the pharmaceutical and polymer industries. Nitrilation of fatty acids by direct reaction with ammonia is mostly carried out in a batch reactor, in which the acid reactant is in liquid phase and the nitrile remains in the reactor. The process needs a few hours of reaction at high temperature, during which undesirable side reactions are likely to occur, especially with the unsaturated fatty acids (e.g. oleic acid). Alternatively, ammonization of fatty acid can be carried out in vapor phase and at a significantly shorter residence time to decrease the side reactions and improve the nitrile yield.

We are investigating the reaction of fatty acids with ammonia in the vapor phase and presence of a heterogeneous solid catalyst bed. The feed is continuously introduced into the reactor through an atomizer. Our previous work has shown that this approach allows the introduction of micron-sized droplets of feed that are rapidly vaporized inside the reactor. The fatty acid vapors can then easily react with NH_3 in the presence of metal oxide catalysts to produce fatty amide as intermediate products. Subsequently, the fatty amide dehydrates and produces fatty nitrile. We will present the effect of reaction temperature, vapor residence time and catalyst on conversion, yield and selectivity of nitriles formation.

467828 Catalytic Conversion of 5-Hydroxymethylfurfural into Tetrahydrofuran Dimethanol in Aqueous Media Using a Continuous-Flow Reactor

Abstract:

AIChE meeting 2016

Catalytic Conversion of 5-Hydroxymethylfurfural into tetrahydrofuran–dimethanol in aqueous media using a continuous-flow reactor

Sérgio Lima, Klaus Hellgardt, David Chadwick

Chemical Engineering Department, Imperial College London , London SW7 2AZ, United Kingdom

During the coming decades, the world is being challenged to reach its goals to replace a major part of its use of oil based feedstock for the production of new commodity chemicals and fuels. In this context, a particular attention is now given to the introduction of renewable carbon (e.g. biomass) in the chemical industry. Among the large variety of biomass available on earth, lignocellulosic biomass, a non-edible resource, has the potential to provide a direct access to valuable chemicals, ideally with similar or improved properties as compared to fossil-derived chemicals. Its high oxygen content, low accessibility to (bio)catalysts and its polyfunctionality makes the transposition of current catalytic processes to lignocellulosic biomass rather challenging. Tetrahydrofuran–dimethanol (THFDM) is a key derivative for the production of a wide range of non-petroleum-derived chemicals. In particular it can be used for the production of 1,6-hexanediol (1,6-HD; Scheme 1). 1,6-HD can be used to produce caprolactam, the monomer for nylon-6, a widely used synthetic polymer and is extensively used in the production of polyesters for polyurethane elastomers, coatings, adhesives, and polymeric plasticizers.[1,2]

In this work we wish to report on the heterogeneous catalysed continuous synthesis of tetrahydrofuran–dimethanol (THFDM) from 5-hydroxymethylfurfural (5-HMF) in water using a continuous-flow reactor. Raney-Cu and Raney-Ni were found to be effective catalysts for this reaction, using water as solvent in a two-steps process. The highest yield of THFDM (higher than 90 %) was achieved by first hydrogenation of 5-HMF to 2,5-furandimethanol (FURDM) in the presence of Raney-Cu follow by hydrogenation of FURDM to THFDM under 90 bar H_2 at 90 °C. Results indicate that large-scale production of low-cost platform and commodity chemicals for the polymer industries from carbohydrate biomass can be achieved in continuous flow processes.

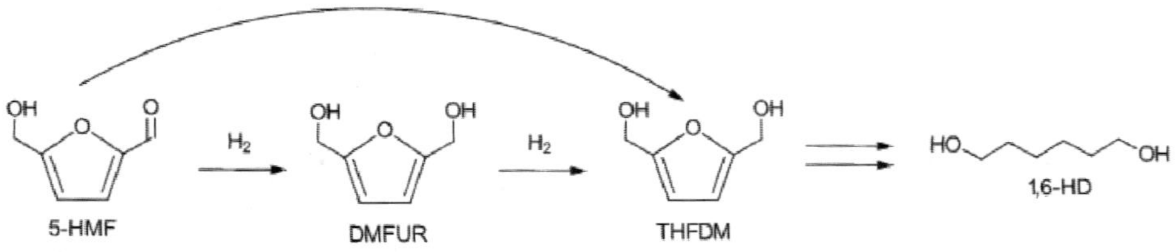

Scheme 1

1. Buntara T., Noel S., Phua P. H., Cabrera I. M., de Vries J. G., Heeres H. J., *Angew. Chem. Int. Ed.* **2011**, 50, 7083-7087.

2. Figueiredo F. C. A., Jorda E., Carvalho W. A., *Appl. Catal. A* **2008**, 351, 259-266.

472847 Catalytic Co-Hydropyrolysis of Biomass and Plastics for Improved Product Quality

Abstract:

Catalytic co-hydropyrolysis of biomass and plastics for improved product quality

Yuan Xue[a], Xianglan Bai[a]

[a] Department of Mechanical Engineering, Iowa State University

Every year, a tremendous amount of Municipal Solid Waste (MSW) is generated in the US. In 2013, Americans produced 254 million tons of trash and only recycled 1/3 of it. The environmental pollution and land usage from the improper disposal of MSW has become an increasingly urgent issue. Among MSW, the major compositions of waste plastics are polyethylene (PE), polypropylene (PP), polystyrene (PS) and polyethylene terephthalate (PET), which are petroleum-derived synthetic polymers, rich in carbon and hydrogen. In the present work, catalytic co-pyrolysis of the plastics and biomass is investigated by co-pyrolyzing red oak with PE, PP, PS or PS using HZSM-5 zeolite catalysts. The work is conducted in a tandem micropyrolysis both in-situ and ex-situ under a helium and hydrogen environment. It was found that in-situ catalytic co-hydropyrolysis significantly improved the product distribution for all cases compared to the helium environment, by dramatically increasing the yield of aromatic hydrocarbons and reducing the formation of solid residues. The yield of alkanes also increased at the expense of the decreasing yield of alkenes for co-hydropyrolysis of red oak with PE and PP. For the co-pyrolysis of red oak with PS and PET, the yield of CO increased, whereas the yield of CO_2 decreased in hydropyrolysis compared to pyrolysis under a helium environment.

For the ex-situ catalytic co-pyrolysis, hydropyrolysis was less effective as there were fewer changes in the product distribution compared to pyrolysis under a helium environment. The study suggests that the external hydrogen plays an important role in promoting a series of reactions involving the biomass and plastics over the zeolite catalyst, which include cracking, deoxygenation and aromatization.

464230 Pt Catalysts for Efficient Aerobic Oxidation of Glucose to Glucaric Acid in Water

Abstract:

Pt Catalysts for Efficient Aerobic Oxidation of Glucose to Glucaric Acid in Water

Jechan Lee, **Basudeb Saha**, and Dionisios G. Vlachos

Catalysis Center for Energy Innovation & Department of Chemical & Biomolecular Engineering, University of Delaware

Non-food biomass is an appealing and sustainable starting material for renewable liquid fuels and chemicals that can minimize our dependence on petroleum and mitigate concerns on greenhouse gas emissions. One such chemical is glucaric acid which has use in detergents and polymeric products. Glucaric acid can be upgraded to adipic acid (AA) for the production of nylon 6,6, polyurethanes and adipic esters. The current manufacturing process of AA (>2.3 million tons per year) from petroleum-based KA oil[1-3] employs corrosive nitric acid and generates greenhouse gas, N_2O.[4,5] Therefore, development of sustainable manufacturing processes for glucaric acid for use in detergents and for upgrading to AA is a viable strategy.[6] However, efficient aerobic oxidation of glucose to glucaric acid has proven to be challenge because of slow and difficult oxidation of gluconic acid, an intermediate oxidation product of glucose.[7]

We studied aerobic oxidation of glucose to glucaric acid using activated carbon, SiO_2 and Al_2O_3 supported Pt catalysts in water. The Pt/C catalyst exhibits the highest activity of these in glucose conversion and glucaric acid yield. Experiments are conducted systemically in a wide range of pH, O_2 pressure, temperatures and glucose/Pt to optimize the reaction conditions. The reactions in base-free and mild basic conditions give best yield to glucaric acid as compared to acidic and highly basic (pH 13.2) media. A maximum 74% glucaric acid yield at initial pH of 7.2, 80 °C, 13.2 bar O_2 and glucose/Pt molar ratio of 54 was achieved, which is the highest reported to date. C–C bond scission of glucaric acid in basic pH via retro-aldol results in low carbon chain carboxylic acids. Higher temperatures and higher Pt loadings cause degradation of glucaric acid, resulting in lower yields. Importantly, the Pt/C catalyst is stable after five consecutive catalytic cycles. Recyclability and characterization studies reveal that the catalyst is stable after five cycles with no sign of Pt leaching into the solution.

Mechanistic studies of the oxidation process entails rapid conversion of the aldehyde group of glucose and much slower oxidation of the $-CH_2OH$ group of gluconic acid, limiting the overall process efficiency. In order to rationalize the observed phenomenon, we have estimated the free energy of adsorption of various compounds. Kinetically, the transformation of an alcohol group $-CH_2OH$ to an acid requires first dehydrogenation to –CHO followed by oxidation to –COOH.[8-10] Upon dehydrogenation, the chemistry becomes facile. These fundamental findings are consistent with the general knowledge that aldehydes are much more reactive than alcohols. We propose that the oxidation of gluconic acid is slow due to the (oxidative) dehydrogenation of $-CH_2OH$ to –CHO being slow; this dehydrogenation step, which is important for activating the molecule, is unnecessary in oxidation of glucose to gluconic acid. We believe high pressure of O_2 facilitates oxidation of gluconic acid to glucaric acid in our study. This explanation is in line with previous work where the formation of glucaric acid is reported at high pressure of O_2.[11]

References:

1. K. Weissermel and H.-J. Arpe, *Industrial Organic Chemistry*, Wiley-VCH, Weinheim, 2003.

2. S. V. d. Vyver and Y. Román-Leshkov, *Catal. Sci. Technol.*, 2013, **3**, 1465-1479.

3. R. Beerthuis, G. Rothenberg and N. R. Shiju, *Green Chem.*, 2015, **17**, 1341-1361.

4. R. A. Reimer, C. S. Slaten, M. Seapan, M. W. Lower and P. E. Tomlinson, *Environ. Prog.*, 1994, **13**, 134-137.

5. C. Aellig, C. Girard and I. Hermans, *Angew. Chem. Int. Ed.*, 2011, **50**, 12355-12360.

6. T. R. Boussie, E. L. Dias, Z. M. Fresco, V. J. Murphy, J. Shoemaker, R. Archer and H. Jiang, *US Pat.,*2010/0317823 A1, 2010.

7. X. Jin, M. Zhao, J. Shen, W. Yan, L. He, P. S. Thapa, S. Ren, B. Subramaniam and R. V. Chaudhari, *J. Catal.*, 2015, **330**, 323-329.

8. B. N. Zope, D. D. Hibbitts, M. Neurock and R. J. Davis, *Science*, 2010, **330**, 74-78.

9. S. E. Davis, B. N. Zope and R. J. Davis, *Green Chem.*, 2012, **14**, 143-147.

10. S. E. Davis, A. D. Benavidez, R. W. Gosselink, J. H. Bitter, K. P. d. Jong, A. K. Datye and R. J. Davis, *J. Mol. Catal. A-Chem.*, 2014, **388-389**, 123-132.

11. V. J. Murphy, J. Shoemaker, G. Zhu, R. Archer, F. Salem and E. L. Dias, *US Pat.*, 2011/0306790 A1, 2011.

454177 Synthesis of Methyl Levulinate from Glucose Using Solid Acid Catalyst

Abstract:

1. Introduction Levulinic acid or alkyl levulinate is one of the important chemicals, which can be an intermediate of many types of chemical products. The production method of levulinic acid is, for example, treating saccharide with mineral acids. However, the recovery of levulinic acid after using homogeneous catalyst is generally difficult, and there are few reports about recovery or further purification of produced levulinic acid. In this study, the production method of methyl levulinate using solid acid catalyst was examined. Furthermore, the purification of methyl levulinate and the reusability of recovered catalyst and methanol was examined as well.

2. Experimental method Glucose was used as a saccharide material. 4 types of zeolites consist of different SiO_2/Al_2O_3 ratio (Name: Cation type: SiO_2/Al_2O_3 = HS-320: NH_4^+: 6, HS-341: H^+: 7, HS-500: K^+: 5.5, HS-690: H^+: 200) were used as solid acid catalysts after the calcination under air at 550°C for 6 h. A certain amount of glucose, catalyst, and methanol were sealed in a batch reactor, and then plunged into preheated oil bath. After the reaction, a reactor was cooled in a water bath, and then products were recovered by vacuum filtration. A recovered catalyst was again calcined under the same condition for reuse. A filtrate was distilled at 80°C and then at 120°C for the recovery of methanol and the purification of the product.

3. Results and discussion

3.1 Conversion of glucose into methyl levulinate using zeolite 0.2 g of glucose, 0.1-1.0 g of zeolite, and 10 mL of methanol were reacted at 80-160°C for 1 h. The conversion pathways of glucose into methyl levulinate consist of the isomerization of glucose to fructose, dehydration of fructose to 5-hydroxymethylfurfural, hydration of 5-hydroxymethylfurfural to levulinic acid, and at any time intermediates form dehydration products of methyl-intermediates. When 0.1 g of HS-320 was used at 120°C, fructose was obtained at the yield of 42mol% and the selectivity of 78%. This yield was close to glucose-fructose equilibrium which indicated that HS-320 was an effective catalyst for glucose isomerization. Lewis acidity of zeolite is expressed by the cleavage of Si-Al bonds, therefore, Lewis acidity of zeolite is considered to be stronger as its SiO_2/Al_2O_3 ratio is closer to 1. Isomerization of glucose to fructose is generally promoted by Lewis acid, therefore, HS-320 whose SiO_2/Al_2O_3 ratio was the closest to 1 among examined catalysts resulted in the best yield of fructose. The best yield of methyl levulinate among the examined condition was 54mol% which was presented by 0.5 g of HS-341 via 4 h of the reaction at 160°C.

3.2 Conversion of glucose into methyl levulinate using several types of zeolites To increase the yield of methyl levulinate, the use of several types of zeolites was examined. Firstly, sequential reaction of isomerization using HS-320 and conversion using HS-341 was examined ("2-steps synthesis"). As the 1st step, 0.2 g of glucose was treated with 0.1 g of HS-320 and 10 mL of methanol at 120°C for 1 h, and then as the 2nd step, HS-320 was exchanged with 0.5 g of HS-341 and treated at 160°C for 0.5-4 h. Secondly, simultaneous reaction of isomerization using HS-320 and conversion using HS-341 was examined ("one-pot synthesis"). 0.2 g of glucose was treated with 0.5 g of blended catalyst and 10 mL of methanol at 160°C for 0.5-4 h. The blended catalyst consisted of 0.1 g of HS-320 and 0.4 g of HS-341. Through "2-step synthesis", the yield of methyl levulinate was lower than that obtained using only HS-341 at any reaction time. This is probably because of the material loss at the 1st step as byproducts. On the other hand, the yield was slightly increased by "one-pot synthesis", which resulted in the yield of 58mol% after 4 h of the reaction.

3.3 Recovery and reuse of catalyst

During the reaction at 160°C for 4 h, the catalyst HS-341 changed its color from white into brown. This was because undesirable byproduct of humin was adsorbed on the surface of the catalyst. The brown catalyst turned back into white after re-calcination, however, the yield of methyl levulinate changed from 54mol% to 45, 41, 34mol% after 1st, 2nd, 3rd time of reuse, respectively. For the reuse of the catalyst, therefore, it was implied that the further examination of the calcination condition which could only remove humins was indispensable.

3.4 Purification of methyl levulinate and reusability of recovered methanol Filtrate obtained after the reaction at 160°C for 4 h with HS-341 contained 1wt% of methyl levulinate. Through the simple distillations at 80°C and 120°C, colorless transparent liquid which was considered as methanol was obtained, and the concentration of methyl levulinate increased to 33wt%. The obtained clear liquid through the distillation at 80°C was then used as a reaction medium. 0.2 g of glucose was treated with 0.5 g of HS-341 and 10 mL of obtained liquid at 160°C for 4 h. As the result, 51mol% of glucose was converted into methyl levulinate. The yield decreased by 3mol% comparing with the yield using pure methanol, however, the decrease was not considerable. Therefore, it could be concluded that some amounts of methanol was reusable in this process.

4. Conclusions Glucose was converted into methyl levulinate using solid acid catalyst at the yield of 58mol%. The possibility of reuse of the catalyst was indicated. The purification of the product was indicated to be possible through the simple distillation, and the reusability of methanol was confirmed.

465690 Processing Carbonaceous Feedstock Using Ionic Liquids for Sustainable Carbon Engineering

Abstract:

Energy security is an important issue for the sustainable development of the global community. The United States faces major energy security challenges due to the high energy demand from residential, commercial, industrial, and transportation sectors, which spur the development of low energy and clean processes that produce fuels and chemicals from domestically available carbon-based feedstocks. Using lignocellulosic biomass to produce biofuels for transportation has emerged as a focus of interest in recent years. Biomass provides a renewable carbon-based source to reduce GHG emissions through the capture of CO_2 during photosynthesis, and decrease the reliance on petroleum for the production of both fuels and chemicals. However, biomass conversions to fuels and chemicals also suffer from its' own set of problems: the distributed nature of biomass resources, low energy density, high cost of transporting sufficient biomass to allow refineries to take advantage of economies of scale.

One promising alternative is to develop a hybrid feedstock system. This hybrid energy system by combining biomass with other carbonaceous feedstocks in a single conversion platform would reduce supply fluctuations, mitigate GHG emissions, and be more politically palatable. However, current research has mostly focused on thermochemical conversion of single type of feedstock at high temperature and elevated pressure, which requires high energy consumption. Developing low energy, environmental benign, feedstock agnostic technologies, is critically important to address concerns of energy security and process economics. Ionic Liquids (ILs) as a class of novel environmental benign solvents are receiving increased attentions. Many researches have focused on ILs' applications in lignocellulosic biomass dissolution for biofuels production, while the work on blending biomass with other feedstock sources to reduce the supply cost and improve the conversion efficiency is limited. This works looks into the feasibility of processing woody biomass mixture with other carbonaceous feedstocks using IL solvents for production of clean energy and high valuable materials. Various types of ILs and feedstocks, and process conditions are investigated. The IL processed carbonaceous materials are further characterized for their compositional, physiochemical and rheological properties changes to advance process optimization and development towards an economic feasible carbon engineering technology by employing wide range of regional feedstocks.

469354 A New Lignin Based Polymer

Abstract:

Lignin, the most abundant renewable aromatic polymer on earth, is usually treated as a waste stream in the biorefinery industry. Developing new applications for lignin as valuable products or materials is a well-recognized challenge. Here, we present a template free method for synthesizing lignin-resorcinolâ€“formaldehyde porous gel with high lignin loading, by utilizing lignin with low molecular weight and less condensation. Up to 75% lignin loading was obtained by blending lignin.

471770 Process Modeling and Techno-Economic Assessment of the Conversion of Biorefinery Lignin to Dicarboxylic Acids

Abstract:

The development of commercially viable, integrated biorefineries remains a difficult challenge. A key barrier is the lack of a strong value proposition for the lignin waste stream. Most approaches towards lignin valorization have focused on the depolymerization of lignin to phenolic compounds, but a class of chemicals â€“dicarboxylic acidsâ€“ have largely been overlooked as a potential lignin product. Dicarboxylic acids are important and highly valuable industrial chemicals and intermediates used across the biopolymer, pharmaceutical, and food additives industries. Though the oxidative aromatic cleavage of lignin is a well-known reaction during the delignification of wood pulp, it is only recently that the selective production of dicarboxylic acids from biorefinery lignin has been demonstrated as a mild, environmentally friendly process. Development of this new pathway and its potential impact on biorefineries remains to be explored.

In this study, we will present several process models for the oxidative conversion of biorefinery lignin to dicarboxylic acid products, developed in Aspen Plus. We will evaluate and compare their technical and economic feasibilities, and present sensitivity analysis on key process parameters. Process and cost bottlenecks will be identified and discussed as potential areas for future research and improvement.

471856 Is Lignin an Ideal Source for Renewable Hydrocarbon Fuel?

Abstract:

We will compare reductive and oxidative chemistries for the conversion of lignin to renewable hydrocarbon fuel.

472014 Integration of Renewable Jet Fuel Production with the Pulp Industry through Alcohol Conversion

Abstract:

Abstract Submission to Be Considered As An Oral Presentation at AIChE 2016 Annual Meeting in the Chemical Conversion Processes in Forest/Plant Biorefineries Session

Integration of Renewable Jet Fuel Production with the Pulp Industry through Alcohol Conversion

Scott Geleynse, Senthil Subramaniam, Xiao Zhang

Washington State University

Among increasing interest in the production of alternative fuels from biorenewable resources, considerable research has provided numerous routes to renewable jet fuels. Along with these new technologies, there is need for design concepts to integrate them with existing industries in order to enable new facilities and ease the transition toward the new renewable biorefinery industry. In this presentation, we explore one such idea in which the alcohols-to-jet conversion technology is implemented into an established kraft pulp mill. By utilizing a large portion of a pulp mill's processes, a biorefinery can be constructed on already established supply lines and facilities with significantly reduced capital investment. This may not only be a possibility to enable to generation of new renewable jet fuels on a large scale, but an option for pulp mills to benefit by reinventing their process and enter entirely new markets.

Presenting author:

Scott Geleynse

Bioproducts Sciences and Engineering Laboratory

School of Chemical Engineering and Bioengineering, Washington State University
2710 Crimson Way, Richland, WA 99354

Contact: scott.geleynse@email.wsu.edu, x.zhang@wsu.edu

472054 Novel Solvent Deconstruction of Woody Biomass and Isolation of High Purity Lignin

Abstract:

Abstract Submission to Be Considered As An Oral Presentation at AIChE 2016 Annual Meeting in the Chemical Conversion Processes in Forest/Plant Biomass Biorefineries Session

Novel Solvent Deconstruction of Woody Biomass and Isolation of High Purity Lignin

Karissa Garcia, Ruoshui Ma, Scott Geleynse, Xiao Zhang
Washington State University

Lignin is the most abundant renewable aromatic polymer on territorial earth present in all high plants. Emerging interests in commercialization of lignocellulosic biofuel process injected new enthusiasm to lignin valorization. Due to the main objective of optimizing cellulosic fiber and/or carbohydrate recovery, most of current lignin isolated from either paper making or biorefineries are often difficult to purify and/or has been indiscriminately modified depending on the process conditions. It appears that one strategy has been overlooked is to devise a new method and/or identify a new solvent to extract lignin from plant biomass with a high purity and minimized structural heterogeneity.

Herein, we will report a new method of applying Deep Eutectic Solvents (DES) for extracting lignin from woody biomass with high yield and high purity. Four different DES were evaluated for treatment of hardwood and softwood. It was found that these DES treatments can selectively extract a significant amount of lignin from wood with high yields: 78% from hardwood and 58% from softwood. The extracted lignin has high purity (95%) with unique structural properties. The mechanism of DES cleavage of ether bonds between phenylpropane units was investigated.

The results from this study demonstrate that DES is a promising solvent for wood delignification and the production of a new source of lignin with promising potential applications.

Presenting author:

Karissa Garcia

School of Chemical Engineering and Bioengineering

Washington State University

Contact: karissa.garcia@wsu.edu, x.zhang@wsu.edu

466262 On the Systematic Integration of Different Generation Biorefineries

Abstract:

As the world strives to create a more sustainable environment, the biorefinery community fights to minimize the environmental impact of bioprocesses and build a green economy. The use of biomass as raw material for production of energy carriers and chemicals has emerged as a promising alternative to fossil resources for mitigating climate change and enhancing energy security. Second generation biorefineries (2G) came up as the natural evolution of the first generation (1G) bioprocesses to preserve food resources, while, recently, seaweed and microalgae processes have also started to hatch as the third generation (3G) biorefieneries. As different generation biorefineries treat different feedstock, their integration can reduce energy consumption and result in a more economical design with lower environmental impacts than the stand-alone.

To date, first and second generation processes have been integrated by the adoption of grassroots design methodologies [1-4], while, in reality, their coupling is far more complex. The upgrade of first generation plants constitutes a retrofit problem, constrained to make the best use of the invested equipment of the existing plant. Second and third generation upgrades account for new technology and investment very flexible to adjust so that the upgrades are best, but the portfolios of products need also to be investigated. Apart from integrating energy, there is also significant potential to exchange (by-) products. For example, bagasse can be used as feedstock in the second generation process, and, similarly, lignin can be used as fuel for the co-generation unit, while wastewater could be used for algae cultivation. Moreover, the processes may share common sections, like the downstream processes. Finally, the co-generation (CHP) unit, after modifications, can support both plants. Thus, the integration of first and second generation biorefineries is a combined grassroots–retrofit design problem, much larger and complex than most conventional applications, that hides inner trade-offs and proves the deficiency of present-day practices in proposing realistic. The size and the complexity of the problem impose the need for a systematic approach

in the form of mathematical models.

This work studies the different symbiotic options in the integration of different generation biorefineries, scoping for the development of a new systematic methodology, which can combine different process systems engineering tools for the integration of grassroots-retrofitted problems. Detailed designs for sugarcane (1G), bagasse (2G), and seaweed (3G) ethanol production are developed in Aspen Plus®, validated by the industrial partner (CIMV Process™) and laboratory experiments, and the potential of cultivating microalgae in wastewater is also investigated. The detailed models feed the mathematical models, developed in GAMS®. The processes are integrated following a mixed mathematical model. The model applies the methodology of retrofit integration on first generation [5] and grassroots integration on second and third generation technology [6, 7], following the transhipment analysis. Preliminary results indicate a significant potential in internal trade-offs savings, both in operating and capital cost.

Acknowledgment: The financial support from Marie Curie European Research Program, RENESENG (FP-607415) is gratefully acknowledged.

References

[1] R. Palacios-Bereche, A. Ensinas, M.Modesto, and S. Nebra, 2012, Ethanol production by enzymatic hydrolysis from sugarcane biomass–the integration with the conventional process. In Proceedings of the Ecos 2012–25th international conference on efficiency, cost, optimization, simulation and environmental impact of energy systems (pp. 26-29).

[2] M. Dias, O. Cavalett, R. Maciel Filho, and A. Bonomi, 2014, Integrated first and second generation ethanol production from sugarcane, Chemical Engineering Transactions, 37, 445-450.

[3] P.R. Lennartsson, P. Erlandsson, and M.J. Taherzadeh, 2014, Integration of the first and second generation bioethanol processes and the importance of by-products, Bioresource technology, 165, 3-8.

[4] J. Moncada, J. A.Tamayo, and C.A. Cardona, (2014). Integrating first, second, and third generation biorefineries: Incorporating microalgae into the sugarcane biorefinery. Chemical Engineering Science, 118, 126-140.

[5] R. Smith, M. Jobson, and L. Chen, 2010, Recent development in the retrofit of heat exchanger networks, Applied Thermal Engineering, 30(16), 2281-2289.

[6] A. Kokossis and A. Yang, 2010, On the use of systems technologies and a systematic approach for the synthesis and the design of future biorefineries, Computers & Chemical Engineering, 34, 1397-1405.

[7] A. Kokossis, M. Tsakalova, and K. Pyrgakis, 2015, Design of integrated biorefineries, Computers and Chemical Engineering, 81, 40–56.

470769 Use of Multi-Objective Optimization for Selecting Optimally Integrated Biorefinery Processes

Abstract:

Biorefineries are integrated processing facilities that convert biomass into value added products and energy after many physical, chemical and thermal conversion steps, aiming the maximization of the economic value of biomass while minimizing the waste streams. Recently, the attention has been focusing on developing feasible integrated biorefinery concepts as well as on the identification of the most promising bio-based chemicals. This is due to rising needs for renewable feedstocks as alternatives to petroleum and high environmental impacts of petroleum-derived products. The design of a biorefinery brings a significant number of challenges due to the large number of processes and products available. Hence, it requires the use of advanced process synthesis and optimization methods to ensure that these systems are economically viable, energy efficient, have minimum environmental impacts and realize the best use of the biomass resource.

In this study, a systematic process design methodology which incorporates process optimization and energy integration is applied to assess the thermo-economic performance, the environmental impact and the energy requirement for several processing pathway options. A superstructure of process models in a lignocellulosic biorefinery converting the feedstock in sugars (biochemical conversion) and syngas (thermochemical conversion) platforms for different bio-based fuels and valuable chemicals is developed. The heat recovery is represented by characterizing the process units energy requirement using pinch analysis.

Multi-objective optimization is implemented, using different objective functions such as economic and environmental criteria that are simultaneously considered to show the trade-offs between these conflicting objectives. Different performance targets are established to compare the alternative designs, to increase the utilization of biogenic carbon and to understand the best combination of products and the synergies between them. The overall design problem is formulated as mixed integer linear programming (MILP) optimization problem and epsilon-constraint method is combined with integer cut constraints (ICC) algorithm to systematically generate the list of competing options in a Pareto front. The results of the proposed approach provide a set of non-dominated solutions (i.e., Pareto front), and each solution shows different biorefinery configuration. Finally, these trade-off solutions can be used as a decision support for engineers and decision-makers to identify the best biorefinery pathways.

Keywords: Integrated Biorefinery Systems, Process Design, Process Integration, Multi-Objective Optimization

465681 Process Models to Assess the Co-Processing of Gasoil and Bio-Based Feedstocks in Hydroprocessing Unit Operations

Abstract:

As the bioenergy industry focuses on the near-term development of conversion pathways from biomass feedstocks to infrastructure-compatible fuel blendstocks, a major area of interest is the integration of partially upgraded, biomass-derived intermediates into existing refinery upgrading unit operations. Development of analysis capabilities and process models are needed to assess these new ideas for refinery integration strategies on the basis of potential economic viability, impact on quality of refinery fuel products and benefits versus ill effects to reliability of petroleum refinery operations. One major area of focus for the modeling effort, enabled by funding from DOEâ€™s Bioenergy Technologies Office (BETO), is the development of hydroprocessing models in Aspen Plus and Aspen HYSYS to assess the co-processing of fossil-derived gasoils and bio-based feedstocks. One of the case studies explored through this initiative is the co-hydrotreating of gasoil and triglycerides.

This presentation will focus on (1) an overview of the analysis approaches and process models developed under this effort, (2) the benefits for utilizing the tools for assessing co-processing scenarios for biomass-derived intermediates, (3) insights gathered from model development and analysis for gasoil and triglyceride co-processing and (4) future work for this effort in assessing blending of oxygenated compounds.

469498 Design and Economic Analysis of a Macroalgae-to-Butanol Process Via a Thermochemical Route

Abstract:

Design and Economic Analysis of a Macroalgae-to-Butanol Process via a Thermochemical Route

Chinedu O. Okoli[1], Thomas A. Adams II[1][*], Boris Brigljević[2] and Jay J. Liu[2]

[1]Department of Chemical Engineering, McMaster University, 1280 Main Street West, Hamilton, Ontario, L8S 4L7

[2]Department of Chemical Engineering, Pukyong National University, 365 Sinseon-ro, Nam-gu, Busan 608-739,Republic of Korea

-

Biofuels produced from macroalgae are termed "third generation" biofuels and are becoming of increasing interest in the renewable fuels community. The advantages macroalgae have over first and second generation biofuel feedstocks include fast growth rates, which enable up to 4 - 6 harvest cycles per year, and the ability to be grown in the sea thus eliminating issues relating to land use and irrigation water [1]. Despite these advantages, question marks around their economic and environmental potential remain because of uncertainties around optimal macroalgae cultivation and harvest methods as well as the technologies to convert them to fuel. The objective of this presentation is to address the question of macroalgae conversion to fuel.

One such fuel receiving increased attention is biobutanol, as it is a better alternative to bioethanol as a gasoline replacement in current automobile engines and fuel pipeline networks [2]. Two major routes exist for biobutanol production; the biochemical route which primarily proceeds via fermentation of biomass feedstock to butanol, and the thermochemical route which proceeds via gasification of biomass feedstock to syngas and the conversion of syngas over catalysts to butanol. This work assesses a first-of-its-kind process for butanol production from macroalgae through a thermochemical pathway.

The design and simulation of different configurations was performed in Aspen Plus. In addition, the potential of the different configurations were assessed using economic and environmental metrics such as the minimum butanol selling price (MBSP), and cost of CO_2 equivalent emissions (CO_2e) avoided under different market scenarios. The resulting values of the metrics were then compared amongst the different configurations and also against standard literature references of similar processes. Finally, the impact that variations in key parameters have on the metrics were assessed using sensitivity analysis.

The results showed that the lowest MBSP was obtained with configurations which import natural gas and electricity as utility sources alongside the macroalgae feedstock. However these options perform poorly when the cost of CO_2e avoided is factored in. On the other hand, macroalgae-only configurations provide the best potential for cost of CO_2e avoided but perform poorly for the MBSP metric. Finally, evaluation of the sensitivity analyses results show that gasoline price changes have a high impact on the South Korean based plant configurations as shown in Fig. 1 [3].

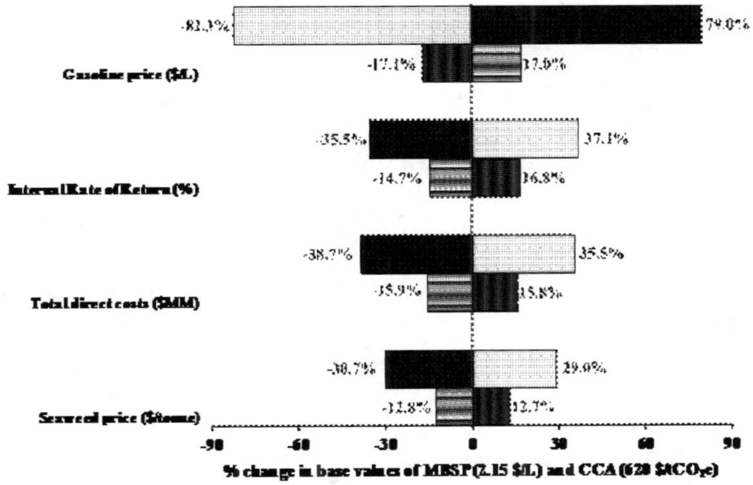

Figure 1: Sensitivity of key parameters from their base case values on MBSP (bottom bars with hatched fill) and CCA (top bars with solid fill) of the South Korea - self-sufficient plant scenario.

REFERENCES:

[1] G. Roesijadi, S. B. Jones, L. J. Snowden-Swan, and Y. Zhu, “Macroalgae as a biomass feedstock: a preliminary analysis,” Pacific Northwest National Laboratory, Richland, 2010.

[2] M. Kumar and K. Gayen, “Developments in biobutanol production: New insights,” *Appl. Energy*, vol. 88, no. 6, pp. 1999–2012, Jun. 2011.

[3] C. Okoli, T. A. Adams, 2016. "Design and economic analysis of a macroalgae-to-butanol process via a thermochemical route". Manuscript submitted for publication, 2016.

[*] Corresponding author: Email: tadams@mcmaster.ca; Ph: (905) 525-9140 x24782

462493 Robust Optimization of Biomass and Natural Gas to Liquid Transportation Fuel Refineries: Process Synthesis Under Uncertainty in Feedstock and Product Prices

Abstract:

Biomass and natural gas are two important feedstocks for the future production of liquid transportation fuels in the United States [1]. Biomass, as a renewable feedstock, can provide significant greenhouse gas (GHG) reductions for refineries, and the increased abundance of natural gas has significantly lowered its purchase price. Through process synthesis techniques and utilization of deterministic global optimization algorithms, optimum refinery topologies can be found that utilize both biomass and natural gas as feedstocks [2-7]. There are significant benefits to hybrid refineries in which both biomass and natural gas are converted to fuels, often through synthesis gas as an intermediate before upgrading through methanol or Fischer-Tropsch synthesis routes. These hybrid systems can offset the potentially high cost of biomass through low-cost natural gas, while providing significant GHG emission reductions based on the utilization of biomass.

In most process synthesis approaches, model parameters are assumed to be known and retain static parameter values. In reality, this is often not the case for parameters such as prices for feedstocks and products, among others; uncertain parameter realizations can have drastic impacts on the objective function values of the optimal solution, or even on overall model feasibility. Robust optimization is a framework for incorporating uncertainty which allows the optimization of large-scale process synthesis models which have uncertain parameters participating linearly in model constraints [8,9]. By imposing uncertainty sets for the parameters onto the model constraints, optimal solutions are found that ensure feasibility with known probabilities of constraint violation. Recent advances in robust optimization theory have greatly reduced the conservatism of robust solutions and allow the determination of competitive refinery topologies at known levels of risk through an iterative method utilizing *a priori* and *a posteriori* probabilistic bounds

[10-13].

The process synthesis superstructure for converting biomass, specifically hardwood or switchgrass, and natural gas to liquid transportation fuels is a non-convex, nonlinear mixed-integer optimization problem, and is solved to global optimality using a rigorous branch-and-bound algorithm [14]. Robust solutions will be presented when robust optimization model counterparts are incorporated into the process synthesis superstructure for a refinery with uncertain price parameters. These parameters appear in the objective function, allowing probabilistic guarantees on the level of profit to be provided. The impact of uncertainty on the objective function value, feedstock utilization, product distribution, and investment costs will be discussed. Case studies will be conducted using the box, interval + ellipsoidal, and interval + polyhedral uncertainty sets, and an iterative method will be utilized in order to provide high quality robust solutions at low probabilities of constraint violation.

[1] Floudas, C. A.; Niziolek, A. M.; Onel, O.; Matthews, L. R. Multi-scale systems engineering for energy and the environment: Challenges and opportunities. AIChE Journal 2016, 62 (3), 602-623.

[2] Baliban, R. C.; Elia, J. A.; Floudas, C. A. Biomass to liquid transportation fuels (BTL) systems: process synthesis and global optimization framework. Energy Environ. Sci. 2013, 6 (1), 267-287.

[3] Baliban, R. C.; Elia, J. A.; Floudas, C. A. Biomass and Natural Gas to Liquid Transportation Fuels: Process Synthesis, Global Optimization, and Topology Analysis. Industrial & Engineering Chemistry Research 2013, 52 (9), 3381-3406.

[4] Baliban, R. C.; Elia, J. A.; Floudas, C. A.; Gurau, B.; Weingarten, M. B.; Klotz, S. D. Hardwood Biomass to Gasoline, Diesel, and Jet Fuel: 1. Process Synthesis and Global Optimization of a Thermochemical Refinery. Energy & Fuels 2013, 27 (8), 4302-4324.

[5] Onel, O.; Niziolek, A. M.; Elia, J. A.; Baliban, R. C.; Floudas, C. A. Biomass and Natural Gas to Liquid Transportation Fuels and Olefins (BGTL+C2_C4): Process Synthesis and Global Optimization. Industrial & Engineering Chemistry Research 2015, 54 (1), 359-385.

[6] Niziolek, A. M; Onel, O.; Hasan, M. M. F.; Floudas, C. A Municipal solid waste to liquid transportation fuels - Part II: Process synthesis and global optimization strategies. Computers & Chemical Engineering 2015, 74 (0), 184-203.

[7] Matthews, L. R; Niziolek, A. M; Onel, O.; Pinnaduwage, N.; Floudas, C. A Biomass to Liquid Transportation Fuels via Biological and Thermochemical Conversion: Process Synthesis and Global Optimization Strategies. Industrial & Engineering Chemistry Research 2016, 55 (12), 3203-3225.

[8] Li, Z.; Ding, R.; Floudas, C. A. A Comparative Theoretical and Computational Study on Robust Counterpart Optimization: I. Robust Linear Optimization and Robust Mixed Integer Linear Optimization. Industrial & Engineering Chemistry Research 2011, 50, 10567-10603.

[9] Li, Z.; Tang, Q.; Floudas, C. A. A Comparative Theoretical and Computational Study on Robust Counterpart Optimization: II. Probabilistic Guarantees on Constraint Satisfaction. Industrial & Engineering Chemistry Research 2012, 51 (19), 6769-6788.

[10] Li, Z.; Floudas, C. A. A Comparative Theoretical and Computational Study on Robust Counterpart Optimization: III. Improving the Quality of Robust Solutions. Industrial & Engineering Chemistry Research 2014, 53 (33), 13112-13124.

[11] Guzman, Y. A; Matthews, L. R; Floudas, C. A New a priori and a posteriori probabilistic bounds for robust counterpart optimization: I. Unknown probability distributions. Computers & Chemical Engineering 2016, 84, 568-598.

[12] Guzman, Y. A; Matthews, L. R; Floudas, C. A New a priori and a posteriori probabilistic bounds for robust

counterpart optimization: II. A priori bounds for known symmetric and asymmetric probability distributions. 2016, In Preparation.

[13] Guzman, Y. A; Matthews, L. R; Floudas, C. A New a priori and a posteriori probabilistic bounds for robust counterpart optimization: III. A posteriori bounds for known probability distributions. 2016, In Preparation.

[14] Baliban, R. C; Elia, J. A; Misener, R.; Floudas, C. A. Global optimization of a MINLP process synthesis model for thermochemical based conversion of hybrid coal, biomass, and natural gas to liquid fuels. Computers & Chemical Engineering 2012, 42, 64-86.

465132 Dynamic Modeling of a Continuous Acetone-Butanol-Ethanol (ABE) Fermentation Process

Abstract:

Continuous acetone-butanol-ethanol (ABE) fermentation has drawn interest for large-scale production of biobutanol due to biobutanolâ€™s superior fuel properties over bio-ethanol and biodiesel [1]. Compared to batch fermentation, continuous ABE fermentation enables better control over process parameters and the possibility of achieving higher biobutanol productivity owing to the continuous product removal [1,2]. Dynamic modeling and analysis of integrated ABE fermentation and separation processes offer opportunities for optimal process design and operation. Dynamic models have been reported for fermentation processes integrated with product recovery units such as hollow fiber membrane, liquid-liquid extraction, and pervaporation [3]. In this work, we present a dynamic model for an integrated continuous ABE fermentation process that consists of a continuous fermentation unit [4], a tangential filtration unit, a membrane vapor extraction (MVE) unit [5], and a recycle stream. MVE is a novel product recovery technique that separates the vaporized solvent products from fermentation via solvent transport through a microporous, omniphobic membrane followed by isothermal absorption into a nonvolatile solvent [5]. The MVE unit enables continuous product removal with minimal energy consumption. An extensive dynamic sensitivity analysis has been performed to systematically investigate the effects of various kinetic and process parameters on the butanol productivity. The analysis provides guidelines for future process integration and optimization.

[1] P. Dürre. Biobutanol: An attractive biofuel. *Biotechnology Journal*. 2007; 2:1525.

[2] E. M. Green. Fermentative production of butanol- The industrial perspective. *Current Opinion in Biotechnology*.2011; 22:337.

[3] R. Mayank, A. Ranjan, and V. S. Moholkar. Mathematical models of ABE fermentation: Review and analysis.

Critical Reviews in Biotechnology. 2013; 33: 419.

[4] E. A. Buehler and A. Mesbah. Kinetic study of acetone-butanol-ethanol fermentation in continuous culture. *PLoS One*. 2016; Submitted.

[5] D. E. Liu, C. Cerretani, R. Tellez, A. P. Scheer, S. Sciamanna, P. F. Bryan, C. J. Radke, and J. M. Prausnitz. Analysis of countercurrent membrane vapor extraction of a dilute aqueous biosolute. *AIChE Journal*. 2015; 61: 2795.

457852 An Overview of the U.S. Doe Carbon Storage Program's Approach to Developing Safe, Effective Commercial Carbon Storage Technologies

Abstract:

The Carbon Storage Program implemented by the U.S. Department of Energy's (DOE) Office of Fossil Energy and managed by the National Energy Technology Laboratory (NETL) is helping to develop technologies that safely and permanently store carbon dioxide (CO_2) without adversely impacting natural resources or hindering economic growth. Specifically, the Carbon Storage program is developing and advancing carbon capture and storage (CCS) technologies both onshore and offshore that will significantly improve the effectiveness of the technology, reduce the cost of implementation, and be ready for widespread commercial deployment. Since 1997, DOE's Carbon Storage program has significantly expanded the CCS knowledge base in selected technology areas through a diverse portfolio of applied research projects and field demonstrations. The Storage program is comprised of three primary technology components: (1) Advanced Storage R&D, (2) Storage Infrastructure, and (3) Sub-Disciplinary R&D. These areas work together to address significant technical challenges in order to meet program goals that support the scale-up and widespread deployment of CCS.

The Carbon Storage Program pursues the advancement of fossil fuel power-generation technologies by supporting the development of technologies which follow a sequential progression of scales (stages) toward making the technology available for commercial deployment. The scale of a technology is the size of the system relative to the final scale of the application, which in this case is a full-scale commercial power production or industrial facility. As RD&D progresses, the scale of the tests increases incrementally from lab/bench scale, to pilot scale, to pre-commercial scale, and to full-commercial scale. The acceptable degree of system integration at each scale is based on the scope of the technology under development within a particular research effort. As the test scale increases, the duration and cost of the projects increase, as does the probability of technical success. In order to assess the progress towards commercialization of a specific technology, a Technology Readiness Level (TRL) concept has been adopted by the Carbon Storage Program to help guide and evaluate the RD&D process. TRLs provide an assessment of the technology development progress on the path to meet final performance specifications. The TRL for a technology is established based upon the scale, degree of system integration, and test environment in which the technology has been successfully demonstrated.

CCS and other clean coal technologies are expected to play a critical role in mitigating CO_2 emissions while supporting energy security throughout the world. DOE's Carbon Storage program has positioned the United States on a path toward ensuring that the enabling technologies will be available to address the demands of new regulations and impact CCS projects in the 2020–2030 timeframe. Continued U.S. leadership in technology development and future deployment is important to the cultivation of economic rewards and new business opportunities, both domestically and abroad.

456707 Impacts of Realistic and Varying Rates of CO_2 Injection on *Safe* CO_2 Storage in the Bunter Sandstone of the Southern North Sea

Abstract:

Carbon capture and storage (CCS) has been recognized as a key technology to reduce CO_2 emissions and mitigate global climate change. Recently, CCS was included in several countries' emissions reduction target as part of the United Nations Framework Convention on Climate Change Paris agreement. In the UK, CCS will enable us to meet an 80% target reduction in CO_2 emissions from 1990 levels by 2050 as set by the Climate Change Act 2008.

In order to maximize the potential for CCS in any region, it is vital to develop a thorough understanding of a country's potential storage reservoirs and their dynamic behavior in response to CO_2 injection. This is particularly key as the pore volume of a reservoir does not equate in a simple way to storage capacity. This is because pressure constraints limit the rates of injection. This in turn has complex dependencies on reservoir geometry, and nature of connectivity and the bounding seals.

Carbon storage in the UK is recognized as an important transition technology for mitigating emissions of fossil fuel power plants while the capacity factor of renewable energy sources increase over time. For that reason, we investigate scenarios of gradual CCS deployment, with plants starting capture at different times and eventually shutting down progressively. Accordingly CO_2 capture rates and injection into storage reservoirs vary over time.

The Bunter Sandstone Saline Aquifer in the UK Southern North Sea is believed to provide one of the largest storage potentials for the UK. In this work we have simulated dynamic injection of CO_2 into the Bunter Sandstone – using a geological model and the commercial reservoir simulator Eclipse. The study assumes that the field has closed boundaries and contains 12 injection sites. We do not consider reservoir management techniques such as brine production. Limiting the bottomhole pressure at each injection site to 75% of the lithostatic pressure imposes a conservative constraint minimizing risk to fracturing of the sealing caprocks.

Four different geographical and market-based scenarios of the UK's CCS deployment and CO_2 capture rates are fed to a pipeline network towards the Bunter Sandstone field. We have found that there may be a significant start up time delay before CO_2 injection can be achieved at the target rates provided by the plants. This is caused by limited early-time CO_2 injectivity which evolves as more CO_2 is injected into the reservoir. The effect is sensitive to reservoir permeability, depth of the injection site and the bottomhole pressure limit at the injection point. Our results show that limited CO_2 injectivity may persist for the first few years of injection and play a significant role in planning for large scale deployment of CO_2 storage.

The key outcome of this work is demonstrating the ability to use reservoir simulation to reflect prospective consequences and responses to realistic CCS deployment and reaction to CO_2 supply. Hence, the reservoir generates a feedback to the CO_2 transport network system - controlling the injection rates delivered to the storage site - the capture system - feeding into the network - and the market demand for fossil fuel products that in turn controls the CO_2 supply. Understanding this link between CO_2 supply and reservoir behavior allows us to inform decision makers on how best to develop CO_2 storage in line with capture and transport system deployment.

461703 Geologic CO_2 Storage Leakage Detection with Statistical Analysis of Controlled-Release Field Data

Abstract:

For geologic CCUS to be accepted, leak detection capability must be guaranteed. Furthermore, it is ideal for leak detection algorithms be very practical and applicable at any geologic CCUS site. The primary objective of this analysis is to develop and test algorithms for effective detection of leakage from geologic carbon capture and sequestration (CCUS) sites. Other leak detection techniques such as monitoring eddy flux and atmospheric modeling have proven effective at detecting simulated leaks CCUS sites. These methods require knowledge of turbulent flow patterns and complex boundary layer models. Monitoring eddy flux is difficult due to the significant impacts that variable terrain, buildings, seasonal cycles, diurnal cycles, and biological processes have on the flux measurement. It is evident that we need a more straightforward, probabilistic method to leak detection and identification that could be employed at any monitoring site. The techniques developed in this study have filled that need.

As part of the monitoring, verification, and accounting (MVA) efforts of the Southwest Regional Partnership, this study developed and tested a leak detection methodology using single and multiple greenhouse gas detection towers on the University of Utah campus. Controlled releases of CO2 and CH4 were used to simulate various point sources by changing duration, location, and flow rate of the point source emissions. High frequency background data (on the order of 10 Hz) and the controlled release experiment data were collected. Point source emission concentration thresholds for both CO2 and CH4 were determined by probabilistic analysis of the data. Data was also used to construct a visual "probability density rose" (PDR) diagram. A PDR is a polar diagram that represents the probability of a leak coming from a certain direction with respect to the detection tower. PDR's were constructed using CO2 and CH4 concentration data and meteorological data collected at the detection tower over a specified time span, which was typically at least 24 hours to account for varying wind directions throughout the day. The wind direction was split into a designated bin size, which for this study was 5°. The number of measurements above the concentration threshold of a leak for each bin was divided by the total number of measurements in each bin to give a "probability" of the concentration exceeding the leakage concentration threshold for each bin.

The resulting statistical analysis (both descriptive and circular statistical analyses) on the data collected indicated that there were significant differences in wind direction, wind speed and background concentration data throughout the day as well as from day to day. This confirmed the importance of collecting long periods of background data for successful leak detection methodology development. Furthermore, in most instances the direction from which readings representative of a point source emission were significantly different than the dominant wind direction. When compared to traditional wind rose diagrams and pollutant concentration rose diagrams, PDR diagrams were shown to be far more useful in determining the direction from which leakage signals were originating from. This was attributed to the probabilistic nature of the diagram and that by dividing the number of measurements above the concentration threshold by the total number of measurements in each bin, the diagram has accounted for the likelihood of readings coming from that direction. It was concluded that PDR diagrams were not only effective visual tools, but also effective at determining the leakage source location with respect to the detection tower.

Ongoing and planned future work includes conducting an uncertainty analysis on the leak detection methodology developed on the University of Utah campus. This analysis will focus on quantifying uncertainty associated with variable monitoring scenarios such as the number of detection towers, location of detection towers, and various site characteristics associated with the monitoring effort. Furthermore, the data collected during the controlled release experiments described in the above paragraphs will be used to determine whether or not an inverse model can be built and used to effectively locate and quantify leaks based on data collected at the detection tower and machine-learning algorithms.

470601 A Numerical Simulation Update of the Aquistore CO_2 Storage Project

Abstract:

Aquistore serves as the storage site for the world's first commercial postcombustion carbon capture, utilization, and storage (CCUS) project from a coal-fired power generation facility. The operation site is located near SaskPower's Boundary Dam Power Station, just outside of Estevan, Saskatchewan, Canada. CO_2 injection and monitoring began in April 2015.The Energy & Environmental Research Center (EERC), through its Plains CO_2 Reduction (PCOR) Partnership, is collaborating with the Petroleum Technology Research Centre (PTRC) and SaskPower to continue PCOR Partnership efforts to numerically model and interpret the performance of the project's injection well and observation well. The results of performance interpretation, modeling and simulation will improve MVA (monitoring, verification, and accounting) strategies based on history matching and predictive simulation results.

The deep saline system targeted for the Aquistore project comprises the Deadwood and Black Island Formations, the deepest sedimentary units in the Williston Basin. This paper describes the performance and interpretation of the injection and observation wells since the start of CO_2 injection, utilizing the dynamic simulation model and GEM – a compositional reservoir simulator from Computer Modelling Group Ltd. As of the end of April 2016, approximately 65,000 tonnes of CO_2 has been injected into the reservoir. The injection rate and length of the operating periods have been variable, but injection rate was increased near the end of 2015 to a level of 350 to 550 tonnes a day.

The injection and observation wells have been closely monitored, and history matching was performed while reconciling rate, pressure, temperature, variations in injectivity, injection flow profiles, and pulsed-neutron logging profiles. The current data set is well replicated in the simulation model by applying skin factor adjustments and a local permeability reduction near the injection wellbore, possibly caused by a previous water injection test. However, variations in bottomhole pressure response remain considerable, and injection rate has experienced sudden large changes. Near the end of year 2015, injection rate increased along with an improved injectivity index, from a previous index value of 0.02 to 0.06 tonnes/day per kPa before November, to 0.08 to 0.16 tonnes per kPa after November, an increase of several hundred percent. This improvement is associated with a lower downhole temperature operating range.

CO_2 breakthrough at the observation well has been observed using pulsed-neutron logging, while the injection well flow profile indicates that channeling may have occurred and that CO_2 passes the assumed damage zone that limited the injectivity during early operations. A recent geophysical imaging attempt to observe the CO_2 plume will also be used for performance confirmations and to update the model correspondingly in order to reduce the uncertainty, thus providing more reliable performance prediction. Additional reporting of project performance may be appropriate in the future.

ACKNOWLEDGMENTS

This material is based on work supported by the U.S. Department of Energy National Energy Technology Laboratory under Award No. DE-FC26-05NT42592.

471125 Lessons Learned in Near-Surface Monitoring for Large-Scale CO_2 Storage

Abstract:

A 5-year near-surface monitoring program was developed and conducted by the Energy & Environmental Research Center as part of the Plains CO_2 Reduction Partnershipâ€™s large-scale (>1,000,000 tons CO_2/yr injection) CO_2 storage demonstration site. The purpose of the program was to establish baseline (preinjection) conditions and variability of surface water, soil, and shallow groundwater aquifer chemistries in the vicinity of geologic CO_2 injection that can be used in conjunction with continued assurance monitoring to 1) provide a scientifically defensible source of data to show that near-surface environments remain unaffected by fluid or gas migration and 2) identify and evaluate anomalies that could be indicative of an out-of-zone migration event should they occur during assurance monitoring.

The near-surface monitoring program implemented consists primarily of acquisition and analysis of soil gas and water chemistry data. This program comprised three parts: sampling of surface water features, sampling of shallow groundwater aquifers, and sampling of soil gas in the shallow vadose zone. Assessment of the analysis data indicated that a high degree of natural variability within these systems is possible, especially at select sampling locations. However, the techniques proved to generate sufficient data to both detect and attribute anomalies to natural processes within these environments.

The lessons learned from execution of the program encompass proper establishment of baseline site conditions, effective and efficient assurance monitoring following the start of CO_2 injection operations, and critical data management of monitoring results. Baseline activities included site characterization and reconnaissance, establishment of sampling locations and techniques, and evaluation of results to determine natural variation. Assurance monitoring allowed for protocols to be developed to address any potential anomaly detection and for prioritizing near-surface

sampling frequency, types of samples and analyses conducted, and client/landowner relations. Data management required strict organization in preserving raw data, processing data during evaluation, and database development for housing and accessing results. Extensive quality assurance/quality control practices were also developed and integrated throughout all aspects of the monitoring program. The methodology developed and lessons learned will allow future carbon capture, utilization, and storage operators to make informed decisions regarding site-specific monitoring programs at other commercial-scale injection sites throughout the region.

465072 Economic Modeling of Carbon Dioxide for Geologic Sequestration and Enhanced Oil Recovery in Jacksonburg-Stringtown Oil Field, West Virginia, USA

Abstract:

Continued global climate change caused by the use of fossil fuels may require effective control or mitigation of greenhouse gases. Geological carbon dioxide (CO_2) capture, utilization and storage (CCUS) is one of the three options to limit CO_2 emissions into the atmosphere. Injection of CO_2 into depleted oil and gas reservoir coupled with enhanced oil/gas recovery (EOR/EGR) is an economic-feasible and short-term methodology. CCUS costs encompass the full chain of CCUS process (i.e., capture, transport, storage and monitoring, including any utilization of CO_2 that results in its long-term sequestration). In order to harmonize the costs of CCUS, a common cost-benefit model is developed to evaluate the economic feasibility of geologic CO_2 storage implemented in conjunction with enhanced oil recovery (EOR) in Jacksonburg-Stringtown oil field in West Virginia. The cost-benefit analysis framework developed in this study includes a reservoir performance model integrated with a cost model and a discounted cash flow economic model. Realization of the economic incentives is associated with CO_2-EOR storage operations in Jacksonburg-Stringtown oil field. This model will improve the clarity and consistency of cost estimation for CCUS and may prove the impetus for commercial scale deployment of CCUS systems from the numerous coal-fired electric power plants in the Ohio River region of the Appalachian basin.

455372 Carbon Dioxide-Induced Liberation of Methane from Laboratory-Formed Methane Hydrates: A Pathway to Economical Carbon Sequestration

Abstract:

The presence of vast methane (CH_4) hydrate reserves around the globe are known for decades. Gas production by external stimulation techniques such as steam injection, gas depressurization or hydrate-inhibitor injection is well-established but an economical production method remains a challenge. In parallel, there is a sense of urgency to minimize carbon dioxide (CO_2) levels in the atmosphere that have exceeded 400 ppm. Both CH_4 and CO_2 gases form structure sI hydrates and at temperatures below 10ºC, CO_2 hydrates are more thermodynamically stable at lower pressures than CH_4 hydrates. Considering that burying CO_2 in deep formations is one of the carbon sequestration options being considered, the CH_4-CO_2 exchange in natural methane hydrate reservoirs may be a way to address both issues. The CH_4-CO_2 exchange option was attractive enough to warrant a commercial test by ConocoPhilips in 2012 in which over 24,210 m^3 of CH_4 was released when a 23 mol% CO_2 in N_2 mixture was injected in a hydrate reservoir in Alaska. Carbon dioxide was preferred over N_2 in the hydrate phase, as about 70% of the N_2 gas injected was recovered, while only 40% of the 1376 m^3 of CO_2 injected was retrieved. Thus, the CH_4-CO_2 exchange reaction merits further research.

In this talk, we will describe several laboratory-scale experiments to understand the CH_4-CO_2 hydrate exchange kinetics. Five hydrate formation-decomposition runs focused on CH_4-CO_2 exchange, two baselines and three with host sediments, were performed in a 200 mL high-pressure Jerguson cell fitted with two glass windows that allowed visualization of the time-resolved hydrate phenomenon. The data show that the induction time for hydrate appearance was the largest at 96 h with CH_4 while with CO_2, the time shortened by a factor of four. However, when the secondary gas (CO_2 or CH_4) was injected into the system containing preformed hydrates, the entering gas formed the hydrate phase instantly (within minutes) and no lag was observed. In a system containing host Ottawa sand (104 g) and artificial seawater (38 mL), the induction period reduced to 24 h. The CO_2 hydrate formation in a system that already contained CH_4 hydrates was facile and they remained stable, whereas CH_4 hydrate formation in a system consisting of CO_2 hydrates as hosts were initially stable, but CH_4 gas in hydrates quickly exchanged with free CO_2 gas to form more stable CO_2 hydrates. In all five runs, even though the system was depressurized, left for over a week at room temperature, and flushed with nitrogen gas in between runs, hydrates formed instantaneously, irrespective of the gas used. The â€œmemory effectâ€ was exhibited with either gas, a result in contradiction with that reported previously in the literature. The facile CH_4-CO_2 exchange observed under temperature-pressure conditions that mimic naturally-occurring CH_4 hydrates show promise to develop a commercial carbon sequestration system.

474016 Long Life Cycle Lithium–Oxygen Battery Using Molybdenum Disulfide Nanoflakes

Abstract:

Long Life Cycle Lithium–Oxygen Battery Using Molybdenum Disulfide Nanoflakes

Mohammad Asadi, Baharak Sayahpour, Amin Salehi-Khojin

Department of Mechanical and Industrial Engineering, University of Illinois at Chicago, Chicago, IL, 60607, USA.

Abstract:

Today, the world transportation system largely relies on liquid hydrocarbons (e.g., gasoline) which are limited in source and cause severe environmental issues. Although electrical motors are very clean and efficient compared with internal combustion engines, their utilization in transportation units is still pending on the development of electricity storage technologies with comparable power densities to those of liquid hydrocarbons[1–4]. Recently, limited numbers of electric vehicles (EVs) have been manufactured using existing rechargeable battery systems i.e., Li-ion battery. However, this battery technology suffers from low energy density (250 Wh/kg) that needs to be improved by a factor of five to make long distance (at least 500 miles) drive of EVs feasible on a single charge[4].

Here, we report a rechargeable Lithium-Oxygen (Li-O_2) battery system with an energy density of 1800-

2100 Watt-hour per kilogram of catalyst resulting in a driving range of more than 500 miles based on the efficiency of commercial EVs. The Li-O_2 battery cell integrated with molybdenum disulfide (MoS_2) nanoflakes as a catalyst and ionic liquid (IL) as an electrolyte preforms remarkably well evidenced by its high round trip efficiency (85%), small discharge/charge polarization gap (~0.8) as well as excellent stability and cyclability tested up to more than 700 cycles. Various in-situ (e.g., Differential electrochemical mass spectroscopy) and ex-situ (e.g., X-ray diffraction and Raman spectroscopy) characterization methods were used to elucidate the cell performance. We believe that this system with excellent reversibility without clogging issue can provide a unique platform to extend the Li-O_2 concept to the real air system (with impurities) and find a solution for EVs' energy storage systems that can operate at much higher efficiency and cyclability than available battery systems.

References:

(1) Lee, J. S.; Kim, S. T.; Cao, R.; Choi, N. S.; Liu, M.; Lee, K. T.; Cho, J. Metal-Air Batteries with High Energy Density: Li-Air versus Zn-Air. *Adv. Energy Mater.* **2011**, *1*, 34–50.

(2) Armand, M.; Tarascon, J.-M. Building Better Batteries. *Nature* **2008**, *451*, 652–657.

(3) Noorden, R. Van. The Rechargeable Revolution: A Better Battery. *Nature* **2014**, *507*, 26–28.

(4) Bruce, P. G.; Freunberger, S. A.; Hardwick, L. J.; Tarascon, J.-M. Li–O2 and Li–S Batteries with High Energy Storage. *Nat. Mater.* **2011**, *11*, 19–29.

457405 The Next Generation High Power, $LiFePO_4$ Cathode Material

Abstract:

<span" roman"="Roman"" new="New">The use of lithium-ion batteries for transportation applications can be broken up into two segments, low and high voltage. Low voltage systems encompass applications such as 12 V starter batteries and start-stop platforms, while high voltage systems are generally associated with PHEVs and EVs. Thus, when considering the use of rechargeable lithium-ion batteries in the transportation industry, and the development of active materials for this industry, it is important to consider the application to be served. For example, low voltage systems require high power densities with the ability to sustain high current densities at low temperatures. Conversely, high voltage applications main technology driver is energy density while enhancing cell safety and reducing cost. The focus of this presentation is on the development of active materials well suited for use in lithium-ion cells for low voltage applications. This is achieved by engineering electrode materials with irregular surfaces which result in high interfacial surface areas and short characteristic diffusion lengths. This can be accomplished through a porous secondary particle structure and/or nanoscale primary particle size. A study using X-ray Diffraction (XRD) has been utilized to measure the primary particle size and investigate the plane orientation in the crystalline structure and its correlation to performance. Knowing that Li diffusion is more favorable through specific crystalline planes may assist in material engineering development for better and faster Li-ion transport. These attributes are expected to provide lithium-ion batteries with high power densities. Additionally, this presentation will explain how the above key points were approached and resulted in high powder cathode materials with a Direct Current Resistance (DCR) < 9 ohm at -20°C; a key target set by USABC that enables Li-ion starter battery applications.

<span" roman"="Roman"" new="New">

<span" roman"="Roman"" new="New">

471084 Solid State Thermal Reaction of NaOH and Mn3O4 Drives the Formation of Sodium-Manganese Oxide Birnessite for Aqueous Electrochemical Energy Storage

Abstract:

Sodium-ion aqueous electrochemical energy storage devices have been widely studied recently due to their low-cost and environmental-friendliness compared with commonly used non-aqueous Li-ion batteries. Here, sodium-manganese oxide materials were synthesized *via* a solid state reaction by thermal treatment of Mn_3O_4 and NaOH at various molar ratios. Energy-dispersive x-ray spectroscopy (EDS) measurements confirmed that the atomic ratio of sodium to manganese varied from 0.02 to 0.29. The atomic structures of sodium-manganese oxide were studied with synchrotron X-ray and neutron scattering using pair distribution function analysis and Rietveld refinement. The results showed the structural evolution from monoclinic Mn_5O_8 nanoparticles to layered MnO_2 birnessite with the increasing sodium content. Cyclic Voltammetry (CV) measurements showed $Na_{0.29}MnO_2$ had a specific capacitance of 211 F g^{-1} at a scan rate of 5 mV s^{-1} in a 0.1 M Na_2SO_4 electrolyte in a three-electrode half-cell, 60% higher than that of pure manganese oxide (132 F g^{-1}). *In-situ* XRD measurements showed crystalline structure change and the *d*-spacing of (001) diffraction peak of sodium-manganese oxide varied 4% during charge and discharge processes, demonstrating its ability to achieve high capacity. Electro-kinetics analysis showed that more capacitive contribution was observed in sodium-manganese oxide compared with that of pure manganese oxide, indicating its high-rate power performance. Our synthesis methods as well as structure and electrochemical studies of sodium-manganese oxide materials provided an insight into the development of innovative electrode materials with high capacity and high power performance for aqueous electrochemical energy storage.

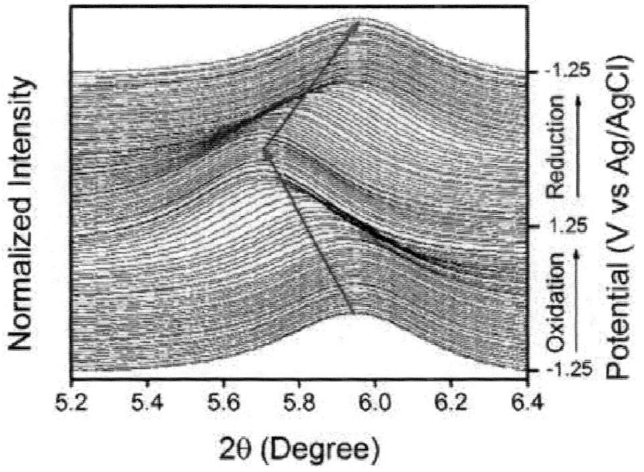

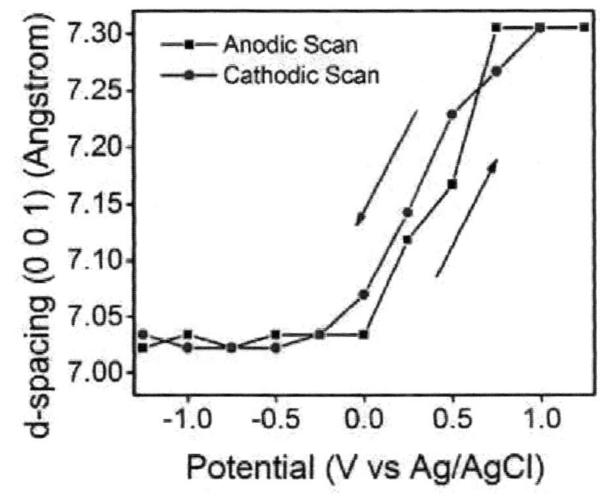

470774 Capacitive Behavior of Natural Biomaterials in High-Performance Renewable Supercapacitor

Abstract:

Supercapatior, as a promising high performance energy-storage device, has drawn great attention of researchers due to its outstanding advances, like high power, rapid charging/discharging rate and reliable runnability. Theoretically, the capacitive performance of a supercapacitor is dominated by two factors, electrochemical double-layer and redox reaction. Recently, an increasing critical requirement from environmental viewpoints has motivated the development of sustainable energy-storage materials. Many biomaterials and their derivatives own unique insights to be applied as supercapacitor. In this work, we successfully prepared biomass-based supercapacitor with good performance and investigated the effects of biomass in detail. Our preliminary results demonstrated the importance of biomass in renewable supercapacitor and indicated that it contributed to the capacitive behavior with more than one mechanism.

459708 Bipolar Plates for Redox Flow Batteries: Relating Conductivity to Morphology of Carbon-Based Polymer Composites

Abstract:

The increasing expansion of renewable sources of electricity stimulates the need for a large-scale storage of energy. A very promising candidate in this respect is the family of redox flow batteries (RFB) due to their good scalability and the possibility to decouple the capacity and power. For the broader commercialization, the price per kilowatt-hour of the energy stored by these devices need to be competitive with other technologies, such as pumped storage hydroelectricity, conventional rechargeable batteries or hydrogen-based storage. The relatively high price of the state-of-art RFB technologies is caused by their expensive components. Therefore, there is a large motivation for the development of cheaper materials with high performance and durability for the application in RFB. This work is focused on the composite bipolar plates used in RFB and their development and optimization using the combination of experiments and mathematical modelling.
Carbon-based polymer composites represent highly conductive materials with corrosion stability (even in acidic environment) and good mechanical properties at a reasonable price. Due to these properties they are widely used as bipolar plates in various electrochemical energy convertors, such as RFB and polymer electrolyte fuel cells. These composites consist of a polymer matrix and carbon fillers. Carbon fillers are responsible for the transport of electrons through the material. The resulting conductivity of the composite is strongly influenced by the nature of the carbon-based fillers and their content in the material. In order to obtain good mechanical properties and processability of the material it is crucial to minimize the total content of the fillers, while preserving high conductivity of the composite. The composites of polypropylene with various fillers (graphite, black, fibers) were prepared using laboratory compounder Plasticorder Brabender PL2000. The effect of the filler content on the electrical conductivity of the composite plate was observed by the measurements of in-plane and through-plane conductivity. The morphology of the composites was observed using AFM, SEM and X-ray micro tomography. The mathematical model is based on the Poissonâ€™s equation and computes the effective conductivity of the composite plate from different conductivities of the matrix and fillers and from the morphology of the plate. This morphology can be either obtained from 3D-reconstructed images of the experimental samples or generated artificially using algorithms for the sphere or fiber deposition.
Our results indicate that a significant increase of the plate conductivity can be achieved by the careful tuning of its composition. For example, the addition of the carbon black into polymer-graphite composites resulted in a dramatic decrease of the percolation threshold due to synergetic effect of the fillers. Moreover, even with the same total filler content, different morphologies of the composite can further increase the conductivity, an effect observed in both our experimental and simulation results. Namely, the shape of the filler particles largely determines the effective conductivity. Anisotropic particles were found to be more efficient in achieving high effective conductivities, but at the same time their orientation becomes an important issue leading to the anisotropy of the whole composite.
The basic idea of this work is to improve the properties of a complex composite material through the understanding and subsequent tuning of its morphology. The developed combination of the experimental and theoretical approach offers a promising tool for the further optimization of carbon-based composite materials.

458505 Fabricating of High-Performance Functional Graphene Fibers for Micro-Capacitive Energy Storage

Abstract:

Although graphene is a typical two dimentional material, it was converted to multi-dimentional stuff to enhance special properties. The one dimensional graphene fiber is an example. Here we fabricated graphene fibers by utilizing ionic liquid as coagulation and functional diamines as cross-linkers to connect graphene oxide layers. The fibers showed excellent mechanical properties. The tensile strength of the resulant fibers reaches ~729 MPa after annealing at 2800 °C. All-solid-state flexible micro-capacitors are fabricated as well with promising energy storage performance. The fibers show a specific volumetric capacity as high as ~225 F/ cm^3 (measured at 103.5 mA cm^{-3} in a three-electrode cell), as well as a long cycle life of 2000 times. The initial results have confirmed that these fibers can be complement or even replacement of batteries in miniaturized portable electronics systems.

450678 Solid Dispersion Flow Battery Material Synthesis and Battery Characterization

Abstract:

Flow batteries are energy storage systems that store reversible electrochemical energy within active material species contained within the liquid electrolyte. Conventional flow battery systems are limited in their energy density by the solubility of the active materials within the electrolyte because above the solubility limit electrochemically inactive solid particles will precipitate. In addition, most conventional flow battery systems involve aqueous electrolytes, which limit the operating voltage due to the electrochemical stability window of water. Recently, researchers have explored using solid particles as the active material in flow batteries because starting with solid electroactive materials bypasses solubility limitations and results in substantial increases in the energy density of the flow battery system. In addition, often these semisolid flow batteries are comprised of lithium-ion battery materials dispersed in organic electrolytes. Using organic electrolytes and lithium-ion battery electrode materials increases the operating voltage of the battery system and further improves the energy density of the flow battery. One of the major challenges within this system, however, is that carbon additives are used to form interconnected particle structures within the fluid. This structured solvent results in high electrical connectivity and thus high power density in the semisolid flow batteries, however, the tradeoff is extremely high viscosities in the system making the electroactive fluids difficult and energy-intensive to pump.

Within our research group, we have developed a flow battery that consists of a solid dispersion of particles in the electrolyte. This solid dispersion flow battery consists of solid particles, and thus has the improved energy density that comes with bypassing the solubility limitations of conventional flow batteries. However, in contrast to many other semisolid flow batteries reported in the literature, by avoiding the use of carbon additives the viscosity is kept much lower in these fluids. The tradeoff of having a lower viscosity is that the power density is lower because electrochemical reactions proceed via collisions with the current collector, rather than throughout and interconnected

carbon structure.

In this talk, recent efforts in our group to synthesize and characterize materials for the solid dispersion flow battery will be described. In particular, we have developed a scalable templated synthesis of active material nanoparticles of lithium-ion battery cathode materials that are electrochemically active and small enough to remain stably dispersed within the flow battery electrolytes. Recent efforts to characterize and understand the electrorheological coupling within this system, and the limits of the power capability within the flow battery architecture, will also be discussed.

465564 Highly Cyclable and Energy Dense Manganese Dioxide Cathodes for Advanced Alkaline Batteries

Abstract:

Manganese dioxide (MnO_2) and zinc (Zn) are one of the most abundant, safest and cheapest materials available. Together, they are found in common household batteries like Duracell, Energizer, etc. as small cylindrical alkaline cells. These cells or batteries are used as primary batteries, i.e., as single use batteries, where the entire capacity of the battery is delivered once and then discarded. The disadvantage of primary batteries is that it takes a lot of energy to produce the battery than the energy that can be actually obtained from it, and also, it creates environmental waste. However, the manufacturing of primary cells has still been rampant due to the cost of manufacturing MnO_2-Zn cells being very cheap. In terms of improving the overall energy efficiency, reducing waste and maintaining its cost advantage, it makes good sense, economically and environmentally, to make MnO_2-Zn cells rechargeable. However, the main deterrent to this direction has been the fundamental material and chemical problems of the main raw components, i.e., MnO_2 and Zn.

Manganese dioxide can theoretically deliver a capacity of approximately 617mAh/g. It delivers this capacity through a 2 electron electrochemical reaction (each electron providing around 308mAh/g). MnO_2 has been found to be rechargeable when the capacity has been limited to around 5-10% of the 617mAh/g. It suffers a crystal structure breakdown as more of the capacity is accessed, and it inherently forms electrochemical irreversible phases. If the entire 2 electron capacity can be accessed then theoretically it can reach energy density numbers near lithium-ion batteries. Similar problems are associated with the zinc electrode, where higher utilization of its capacity causes dendrite formation, shape change and formation of inactive zinc oxides that ultimately lead to electrode failure. These are the main deterrent to a cheap and safe battery that could be a disruptive technology in the energy storage field.

At the City College of New York, we have made a breakthrough in accessing the second electron capacity by altering the crystal structure of manganese dioxide through dopants and novel and cheap synthesis routes. We have cycled the MnO2 electrode to well over 6000 cycles at rates that are of interest in the battery community. Also, in this talk I will go through the various breakthroughs that have been made at the City College of New York with regards to solving other fundamental problems and the scope of these batteries in the current market.

474742 Interfacial Assembly and Engineering of Ordered Functional Mesoporous Materials for Applications

Abstract:

With recent progresses made in modern nanoscience and nanotechnology, ordered mesoporous materials have been one of the hottest research topics in scientific community spanned chemistry, materials science, physics and biology. The construction of mesoporous materials is mainly concerned with building monodispersed mesosized (2-50 nm) pore voids and arranging them in a long-range ordered array. Generally, two kinds of templates are used to produce the mesopores: supramolecular aggregates such as surfactant micelle arrays, and rigid preformed solids such as ordered mesoporous silica, carbon, and colloidal crystals. Noticeably, besides the templates, the interface also plays a central role in the synthetic process, because it provides a rich and crucial space for the assembly and construction of mesostructures. Generally, two kinds of interfaces involve in the synthetic system. The first one is at between surfactant templates and guest species, which has been extensively investigated. Another important interface is the two-phase (solid, liquid and gas) one, including liquid-solid, gas-liquid, liquid-liquid, gas-solid, and solid-solid interface, which has been well developed for the synthesis of ordered mesoporous materials. Compared with the one phase synthesis referring to homogeneous nucleation and growth, the introduction of a two-phase interface in the system can change the growth behaviors of mesoporous materials and lead to the formation of molding or multifunctional mesoporous materials. For example, mesoporous thin films or membranes have been widely fabricated on a substrate *via* an evaporation-induced self-assembly (EISA) method. Multifunctional core-shell structured mesoporous materials can be obtained by rationally depositing mesoporous shells on well-designed cores at the interface. Recently we have developed a novel facile approach i.e. a solvent evaporation-induced aggregating assembly (EIAA) to synthesize large pore mesoporous silica materials. In addition, the well-known hard-templating method for mesoporous materials is also a typical interface reaction.

448965 High-Energy Density Metal-Free Biobatteries Powered By Soft Drinks

Abstract:

More and more portable electronics urgently require better batteries featuring high-energy density, good safety, and low environmental impacts. Enzymatic fuel cells are a promising micro-power source for small electronics. However, incomplete oxidation of sugars mediated by few redox enzymes and low faraday efficiency result in low energy storage densities. Here we demonstrate the complete oxidization of sucrose, glucose, and fructose at the anodic compartment of biobatteries via a synthetic 15-enzyme pathway. The biobatteries generated maximum power densities from 0.65 to 0.98 mW cm^{-2}. Also, a nitrogen-doped activated carbon cloth was used as a flexible cathode to achieve oxygen reduction reaction faster than that catalyzed by platinum-coated membrane electrode assembly. Biobatteries comprised of 15 enzymes oxidize sucrose, fructose and glucose simultaneously, generated nearly 24 electrons per hexose with faraday efficiencies of approximately 95%. Metal-free biobatteries powered by commercial soft drinks whose energy-storage density was ~400 Ah kg^{-1} of drinks exhibited maximum power densities of up to 0.55 mW cm^{-2}, implying that soft drinks could be perfect fuels for bioelectricity generation.

474830 Lego-like Micropillar/Microwell Chip for High-Throughput Functional Analysis of Genes Encoding Pathogen-Specific Antimicrobial Enzymes

Abstract:

The overuse and misuse of antibiotics have resulted in the emergence of antibiotic-resistant pathogens along with the perturbation of healthy human microbiota via killing putatively beneficial commensal bacteria. Hence, targeting infectious bacterial pathogens is important for reducing the evolution of antibiotic-resistant bacteria and preserving the endogenous human microbiome. Cell lytic enzymes including bacteriophage endolysins, autolysins, and peptidoglycan targeting enzymes are useful antibiotic alternatives, and genetic information of numerous cell lytic enzymes is currently available. However, the identification of their antimicrobial function and specificity has been limited due to time-intensive protocols to identify their specific targets. Here, we developed Lego-like sandwich microchip for rapidly accessing the function of diverse genes that are suggestive of encoding cell lytic enzymes. This approach can be used to quantify antimicrobial activity in a high-throughput manner by detecting bacterial cell growth in 40 nL alginate gel spots on micropillar supports. In addition, cell-free protein synthesis could be achieved in a complementary microwell chip. The combination of the chip-based antimicrobial assay with parallel cell-free protein synthesis enabled the rapid identification and evaluation of species-specific antimicrobial function of 11 genes encoding known and putative cell lytic enzymes.

466405 Chaperonin-Inspired Enzyme Protection By Mesoporous Silica

Abstract:

Enzymes can catalyse bioorganic transformations with much higher specificity than traditional chemical catalysts, which makes them valuable for fine chemical and pharmaceutical manufacturing. However, enzymes have evolved to operate within their host cells: most often in neutral, aqueous solutions and at moderate temperatures. Therefore, they tend to have poor stability in chemical reactors that involve high temperatures, acidic or basic conditions, or organic solvents.

In vivo, enzymes are stabilised from high temperature or unfavourable solution conditions by chaperonins. Chaperonins, such as the GroEl/ES complex, help refold partially unfolded proteins within a narrow, cylindrical pore that becomes negatively charged. While chaperonins' complete mechanisms are complex, we are inspired by some of their fundamental properties when developing materials for enzyme immobilisation.

Mesoporous silica SBA-15 is used as a synthetic chaperonin analogue because of its controlled mesopore diameter and its negatively charged interior surface. Enzymes are immobilised on SBA-15 of different pore diameters, and their stabilities and biocatalytic activities are tested. SBA-15 is shown to be a valuable candidate for enzyme immobilisation, because it can protect enzymes from denaturing conditions. Selected findings include: immobilised myoglobin's protection from acidic reaction conditions, with activity improvement of up to 350%; immobilised myoglobin's complete protection from the protease pepsin; immobilised lysozyme's protection from basic reaction conditions, with activity improvement of up to 200%; and the improvement of immobilised lysozyme's activity by altering SBA-15's pore diameter.

More fundamentally, SBA-15 is a useful material for the systematic investigation of enzyme immobilisation parameters because of its highly tuneable pore diameter and its easily functionalised surface. The relative importance of hydropathy, electrostatic attraction, and steric confinement are studied in order to improve the future design of enzyme immobilisation materials.

464833 Intermediate Channeling Via Nanoscale Confinement

Abstract:

Multi-step, enzyme-catalyzed reaction cascades occur frequently in metabolic and catabolic pathways of living cells with high efficiency, selectivity, and yield. Understanding these metabolic pathways can lead to engineered synthetic pathways for the production energy and high-value molecules.

Transfer of intermediates between the active sites without diffusing into the bulk is known as "substrate channeling." Confinement by molecular tunnels, electrically charged pathways, spatial organization of catalysts and covalent bonding of intermediates are examples of substrate channeling mechanisms. Under most relevant reaction conditions, the rate of diffusion is faster than the rate of reaction.[1] Hence, confining the active sites and intermediate by creating a physical tunnel between active sites enables controlled transport without bulk diffusion.

In the present study, computational modeling has been performed to study the effect of geometric, kinetic, and transport parameters on intermediate channeling via confinement, using a continuum model. Efficiency of transport is quantified by reactant yield and flux control coefficients (FCC).[2] Retention of intermediates within the tunnel, with minimal access to bulk solution, is shown to be the key to efficient channeling.

Acknowledgement

We gratefully acknowledge support from Army Research Office MURI (#W911NF1410263) via The University of Utah.

References

1. I. Wheeldon, S. D. Minteer, S. Banta, S. Calabrese Barton, P. Atanassov and M. Sigman, "Substrate channelling as an approach to cascade reactions", *Nat. Chem.*, **8**, 299–309 (2016). doi:10.1038/nchem.2459.

2. D. A. Fell, "Metabolic control analysis: a survey of its theoretical and experimental development", *Biochem. J.*, **330**, 313–330 (1992). doi:10.1042/bj2860313.

474743 Engineering Ultrastable Protein Scaffolds for the Controlled Assembly of Multifunctional Nanobiomaterials

Abstract:

Self-assembling protein templates are of increasing interest in the field of nanoscale fabrication of biomaterials, where precise patterning of functional biomolecules, such as enzymes, is often desired. In particular, protein building blocks can be strategically chosen to exhibit desired functionality, while engineering their assembly allows for the controllable positioning of the subunits. The filamentous protein g-prefoldin (g-PFD) from the hyperthermophilic archaeon *Methanocaldococcus jannaschii* is an excellent candidate for such a tunable scaffold. Its remarkable stability, unique modularity, and self-assembly into filaments with chaperone activity render it an ideal candidate for the bottom-up construction of novel protein nanostructures.

We have recently reported the assembly of g-PFD monomers into geometrically defined templates of tunable dimensions and symmetry, as well as the controlled positioning of nanoparticles along the filaments. We aim to achieve more complex patterning, while expanding the applicability of g-PFD templates to diverse enzymatic systems. Fusing different enzymes to each subunit should enable periodic positioning of multiple enzymes along the filament to catalyze sequential reactions and metabolic pathways. In addition, we have demonstrated the ability to position cytochrome C proteins at high density along the filament as a first step to creating novel conductive nanowires. Finally, we are studying g-PFDâ€™s ability to form 3D crosslinked multifunctional protein hydrogels. Ultimately, the ability to position functional molecules in a controlled manner will considerably enhance our ability to fabricate advanced multifunctional nanobiomaterials.

460929 Concanavalin A Enabled High Performance Enzyme Cascades on Magnetic Nanoparticles

Abstract:

Concanavalin A (ConA) is a lectin that specifically binds to carbohydrates containing non-reducing terminal α-D-mannosyl or α-D-glucosyl moieties. Here we present a facile and aqueous method preparing multiple glycolenzyme catalysts using ConA coated magnetic nanoparticles (ConA-MNP). The first step is to from a ConA layer on the surface of MNP (ConA-MNP) by crosslinking with glutaraldehyde. Then glycoenzymes were agglutinated on the surface of ConA-MNP spontaneously, followed by crosslinking with glutaraldehyde.

The resulted GOx-HRP@ConA-MNP showed an over 80% residual activity compared to their native counterparts in solution. In contrast, merely no activity was observed when the enzymes were directly immobilized on MNP via crosslinking with glutaraldehyde. The GOx-HRP@ConA-MNP retained over 90% of the initial activity after 10 cycles. Above mentioned procedure was applied to form GOx-Cat@ConA-MNP for catalyzing the conversion of glucose to gluconic acid. It was evidenced by both structural characterization and reaction kinetics that the co-localization of the cascade enzymes enabled a substrate looping between two enzymes, i.e., oxygen produced by Cat was confined to GOx producing H_2O_2 as a substrate for Cat. As a result GOx-Cat@ConA-MNP achieves 91% glucose conversion in 75 minutes whereas the aqueous GOx and Cat reached a conversion of 86%. The GOx-Cat@ConA-MNP retained 99% after 8 cycles with a glucose conversions over 90% in each run.

The high catalytic performance and the ease of operation makes the multiple enzymes @ConA-MNP appealing for enzyme cascades for various applications.

467438 Design of a Heterogeneous Biocatalyst for Cofactor Regeneration and Improved Catalytic Characteristics

Abstract:

Enzyme immobilization generally allows for improved lifetime and stability of catalytically active protein. Typically, immobilization systems involve adsorption to surfaces, cross-linked enzyme aggregates or entrapment of enzyme in a polymer network. However, these systems often can introduce unwanted limiting effects, such as high resistance to mass transfer or distortion in the enzyme structure due to adsorption or crosslinking. To overcome these shortcomings, it is possible to use more porous materials which are imbued with immobilization domains. Supraparticles formed by a hierarchically structured self-assembly of protein-inorganic nanoflowers provide a porous and stable support for enzyme capture with a high surface area and low mass transfer resistance. In this work, we used hybrid protein-inorganic supraparticles made of calcium phosphate and protein with leucine zipper binding domains for high-affinity immobilization of fusion proteins containing the complementary binding domain and enzymes. Particularly, we showed the modularity of the supraparticle system by immobilizing two distinct enzymes on the particle in tandem.

As a model system, we used the amine dehydrogenase, cFL1AmDH, which is coupled for cofactor regeneration with formate dehydrogenase, cbFDH. The cFL1AmDH oxidizes NADH to convert 4-fluorophenylacetone and ammonia to 4-fluoroamphetamine, an important chiral intermediate. However, the enzyme lacks desirable kinetic characteristics and reaction condition, given a current optimum at 5M ammonium chloride and a pH of 9.6. The cbFDH regenerates NADH from NAD+ by oxidizing formate to carbon dioxide, reducing the cost associated with replacing an expensive cofactor. By co-immobilizing these enzymes, we seek to improve the activity of cFL1AmDH by supplying a constant source of NADH. To account for the effects of alkalinity and ionic strength from the optimal cFL1AmDH reaction, the supraparticles were synthesized with several concentrations of ammonia and at neutral and alkaline pHs. The particles synthesized under the reaction conditions proved to be comparable to the neutral, ammonia-free synthesis in size and number. We then generated a set of fusion proteins with four different linker peptides connecting the catalytic domain to the immobilization domain and tested their activity. After selecting the enzymes with the highest specific activity, we incubated them with the particles and quantified the enzyme immobilization and specific activity. The particle porosity, effectiveness factor, and Thiele modulus were quantified to determine the effects of the particles on mass transfer. This work demonstrates our ability to create an industrially relevant heterogeneous biocatalyst which incorporates cofactor regeneration, is stable in enzymatically relevant conditions, and allows for enzyme recycling.

1. Sheldon, R. A. Enzyme Immobilization: The Quest For Optimum Performance.ChemInform. 2007, 38, 1289–1307.
2. Park, W. M.; Yee, C. M.; Champion, J. A. Self-Assembled Hybrid Supraparticles That Proteolytically Degrade Tumor Necrosis Factor-α. J. Mater. Chem. B. 2015, 4, 1633–1639.
3. Bommarius, B. R.; Schürmann, M.; Bommarius, A. S. A Novel Chimeric Amine Dehydrogenase Shows Altered Substrate Specificity Compared to Its Parent Enzymes. Chem. Commun. 2014, 50, 14953–14955.

461059 Concanavalin a Coated Activated Carbon for Enzyme Immobilization

Abstract:

While the large pore size, good mechanical property, non-toxicity, as well as the availability of chemical or physical modification makes activated carbon (AC) an ideal support for enzyme immobilization, direct adsorption of enzymes onto AC often yields poor apparent activity and reusability. This might be attributed to the strong interaction that leads to undesired conformational changes of the adsorbed enzymes. Here we present a novel and facile method, as featured by coating AC with Concanavalin A (ConA), a lectin that specifically binds to glyosidic molecules, to prepare AC immobilized enzymes.

During a run, AC was loaded into ConA solution of appropriate concentration and incubated at room temperature for 12 hours. Then glycoenzyme such as laccase and HRP was loaded into the suspension of ConA coated AC and incubated for 12 hours. BSA, which was frequently used as a stabilizer of aqueous protein or during free drying was also used as a control to evaluate the stabilization effect of ConA.

It is shown that HRP immobilized on ConA coated AC remained 50% initial activity after 25 cycles. In contrast, the directly adsorbed HRP on AC lost all activity after 12 cycles. Replace of ConA with BSA resulted in 18% apparent activity remain after 25 cycles. The laccase immobilized on ConA coated AC appeared 96% of initial activity after 11 cycles. In contrast, the direct adsorption of laccase on AC led to a reduction of apparent activity by 74%. Moreover, the immobilized enzymes showed a 2.5-fold increase in activity compared to direct adsorption and an improved stability over a wider range of pH and at a higher temperature. The high yield of activity, enhanced reusability and stability of immobilized enzyme, in combination with the high adsorption ability of AC, make the enzyme immobilized on ConA coated AC promising for various applications of enzymatic catalysis.

475491 R&D Progress in Air Pollution Control Technologies in China (Featured talk)

Abstract:

For recent years, people in most northern and eastern parts of China have suffered from the heavily-polluted haze. Besides natural climate variations, emissions from transportation and industrial sources are the main contributors. In this presentation, current situation of industrial air pollution emissions in China will be summarized and applications of pollution control technologies will be critically reviewed with emphasis on the catalytic oxidation of volatile organic compounds (VOCs) and potential opportunities for novel materials.

461082 Removal of Toxic Chemicals Using Metal-Organic Frameworks

Abstract:

Metal-organic frameworks (MOFs) have been widely investigated for a wide range of applications, most notably gas storage, separations, and sensing. Only recently have advances been made in the area of toxic chemical remediation focusing on potentially catalytic chemistries for removal. Here we discuss the use of several metal-organic frameworks to remove chemical warfare agent as well as other less-toxic, yet still problematic, chemicals such as chlorine and alkyl halides. In particular, we focus on multiple removal mechanisms, including hydrolysis as well as nucleophilic attack, to mitigate these chemicals.

469056 Adsorption Energies of Light Hydrocarbons in Functionalized Uio-66(Zr)

Abstract:

Differential adsorption energies of ethane, ethylene, and acetylene on the metal-organic framework (MOF) UiO-66(Zr) and UiO-66(Zr) functionalized with â€"NH_2 and â€"NO_2 were measured by adsorption calorimetry at 293K using a Tian-Calvet calorimeter. Measurements were taken at low coverage ranging from 0.05 mmol/g to 1.2 mmol/g. In all cases the MOFs were heated under vacuum to dehydroxylate the MOFs prior to testing. The adsorption energy of ethylene was highest for UiO-66(Zr)-NO_2 and was lowest for UiO-66(Zr). For UiO-66(Zr)â€"NO_2, the adsorption energy of ethane was higher than ethylene and only slightly lower than acetylene. Given that the overall molecular polarity of ethane is lower than ethylene and that the kinetic diameter of ethane is larger than ethylene, confinement effects may be a dominating factor for ethane adsorption in the -NO_2 case. For adsorption of both ethane and acetylene on UiO-66(Zr)-NO_2, the adsorption energies decreased by ~5 kJ/mol as the coverage increased from 0.05 mmol/g to ~0.6 mmol/g. For ethylene, the adsorption energies were essentially constant over the entire range of coverage investigated. The change in adsorption energy for ethane and ethylene at low coverage may indicate that UiO-66(Zr)-NO_2 has multiple types of adsorption sites. We are currently investigating low-temperature calorimetry measurements for this system, which will allow one to more clearly isolate the energies of stronger and weaker adsorption sites.

473309 Selective Gas Adsorption on Functionalized H3btc and Di-Isophthalate Based Metal Organic Frameworks

Selective Gas Adsorption on Functionalized H_3BTC and Di-Isophthalate Based Metal Organic Frameworks

Prudhviraj Medikonda[1], Sastri Chivukula[2] and Sasidhar Gumma[1*]

1 Department of Chemical Engineering, Indian Institute of Technology Guwahati, India

2 Department of Chemistry, Indian Institute of Technology Guwahati, India

E-mail: *s.gumma@iitg.ernet.in*

Functionalization of organic linkers towards metal organic frameworks (MOFs) containing open metal sites having more attention and interest provides predictable pore size, surface area and more adsorption sites (Collins, 2009). The Cu-BTC MOF is a widely studied adsorbent for gases separation and storage, but limited work has been done to modify the H_3BTC ligand or functionalize Cu-BTC to optimize its properties for gas adsorption. On the other hand, organic linkers containing double-bond functionalities have rarely been studied for gas adsorption using MOFs and N containing MOFs are more selective for CO_2 over N_2, CH_4 and CO (Liu et al. 2007, Belof et al. 2007). In this work we report the effect of functionalization by comparing the gas adsorption properties of 1 with 2 and those of 3 with 4. The role of the functional group and the polarity of the gas will be discussed.

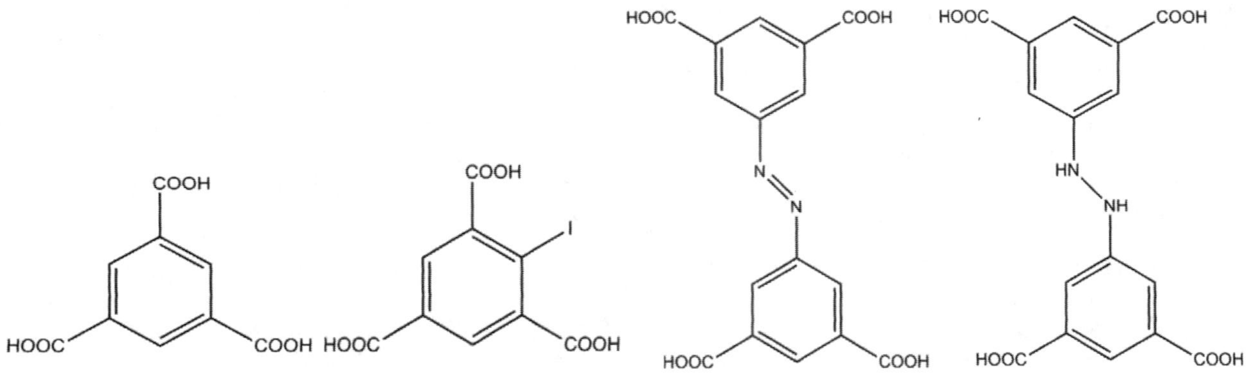

1 **2** **3** **4**

Scheme 1: Structures of organic linkers H_3BTC (**1**), H_3IBTC (**2**), $H_4AzoBTC$ (**3**) and H_4 reduced AzoBTC (**4**) used in this work

References

D. J. Collins, S. Ma, H. C. Zhou. *Hydrogen and methane storage in Metal-Organic Frameworks: Design and Application,* Wiley-VCH, Weinheim 2009.

Y. Liu, J. F. Eubank, A. J. Cairns, J. Eckert, V. Ch. Kravtsov, R. Luebke, M. Eddaoudi. *Angew. Chem. Int. Ed. 46*, 3278, 2007.

J. L. Belof, A. C. Stern, M. Eddaoudi, B. Space. *J. Am. Chem. Soc., 129*, 15202, 2007.

Y. Cai, Y. Zhang, Y. Huang, S. R. Marder, K. S. Walton. *Cryst. Growth Des.,* 12, 3709–3713, 2012.

454249 A One-Step Ion Exchange Method to Functionalized Zif-67 for Enhanced CO2 Capture

Abstract:

A one-step ion exchange method to functionalized ZiF-67 for enhanced CO_2 capture

Fujiao Song, Qin Zhong*

School of Chemical Engineering, Nanjing University of Science and Technology, Nanjing 210094, PR China. E-mail: Song_fj2006@126.com (Fujiao Song), zq304@mail.njust.edu.cn (Qin Zhong).

ZiFs have shown great potential applications in gas separation [1]. Ion exchange could improve the surface area, pore volume as well as surface charge of the adsorbents, further enhancing the CO_2 adsorption capacities. However, it is notable that the standard ion-exchange process needs to be repeated for several times, which brings a series of disadvantages, such as complicated in synthesis, time consuming and solvent consuming [2, 3].

In this work, we develop a new one-step ion exchange method (donated as ion as-exchange technique) to overcome the disadvantages mentioned above. Since the cations for exchange (Li^+ and Na^+) was added before the hydrothermal preparation of ZiF-67, time consuming and solvent consuming in the standard ion-exchange procedure was eliminated. The product was designated as M-as, where ¡°M¡± represents the alkali metal ion (Li and Na). By comparison, the standard ion-exchange procedure was also carried out to synthesize the adsorbents, and the product was designated as M-post. The structure, surface charge and CO_2 adsorption properties of the adsorbents prepared via the two methods were investigated and compared.

Fig. 1(A) shows the XRD patterns of ZiF-67 exchanged with Li^+ and Na^+. all the ion exchanged ZiF-67 maintains crystallinity of the original phase [4]. And almost no changes in peak intensity and position could be observed. Fig. 1(B) presents the particle size distributions of as-synthesized and ion exchanged ZiF-67 samples, which shows the Gaussian type distribution with different distribution ranges.

The CO_2 adsorption isotherms of parent and these ion-exchanged ZiFs at 0 ¡ãC are shown in Fig. 1(C). The CO_2 uptake of parent ZiF-67 at 1 bar is 35.4 cm^3/g. Li/Na-post exhibited decreased CO_2 adsorption capacity (30.4 and 31.2 cm^3/g) compared to the parent, while Li/Na-as showed improved CO_2 uptake values (48.7 and 45.1 cm^3/g). The highest CO_2 capacity is achieved by Li-as, which is 37.6 % higher than that of the parent ZiF-67. An interesting phenomenon is found that all of the isotherms show linear patterns with stable adsorption rate from 0 bar to 1 bar, which reveals the materials could get excellent adsorption performance at higher pressure range (> 1 bar).

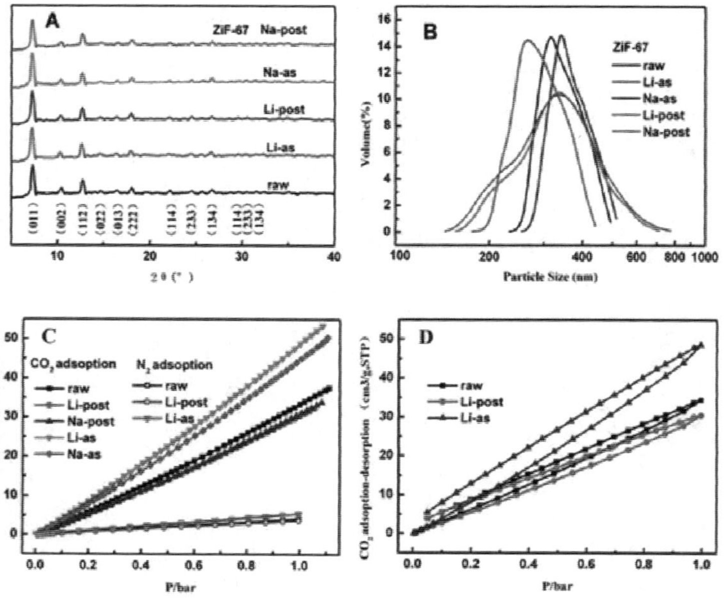

Fig. 1. (A) XRD patterns of parent ZiF-67 and ion exchanged with Li^+ and Na^+, (B) Particle size distribution, (C) CO_2/N_2 adsorption isotherms and (D) CO_2 adsorption-desorption isotherms

The CO_2 adsorption mechanism of the adsorbents prepared via the two methods was investigated. The surface area and pore volume (Table 1) of Li^+/Na^+ as-exchanged ZiF-67 increase significantly because the radius of Li^+ and Na^+ is smaller than that of Co^{2+}. However, the surface area of post-exchanged materials decreases contrarily due to the agglomerate and pore block. PZC characterization (Table 2) indicates that parent ZiF-67 has a neutral character, whereas ion-exchanged ZiFs has basic surfaces. And pH_{PZC} values of Li^+ and Na^+ as-exchanged ZiFs is higher than that of post-exchanged samples, which could improve CO_2 adsorption due to the coordination interaction. All of the CO_2 adsorption isotherms show linear patterns with stable adsorption rate from 0 bar to 1 bar, which reveals the materials could get excellent adsorption performance at higher pressure range (> 1 bar). Van der Waals interaction determined by the surface area and coordination interaction resulting from electrostatic interaction [5] work in synergy to enhance CO_2 adsorption performance of Li^+/Na^+ as-exchanged ZiF-67 adsorbents.

Table 1. Physical properties of parent ZiF-67 and ion exchanged ZiF-67.

ZiFs	BET (m^2/g)	Pore volume (m^3/g)	SF Median pore width (nm)
parent	1424.59	0.84	0.868
Li-as	1450.71	0.87	0.875
Na-as	1435.15	0.86	0.871
Li-post	1366.14	0.76	0.851
Na-post	1359.50	0.77	0.864

Table 2. The pH_{PZC} values of the parent ZiF-67 and ion exchanged ZiF-67 before and after CO_2 adsorption

Sample	pH_{PZC} values	
	before CO_2 adsorption	after CO_2 adsorption
parent	6.92	5.36
Li-post	7.79	5.10
Na-post	7.51	5.63
Li-as	9.26	5.27
Na-as	8.83	5.49

References

[1] J. An, N.L. Rosi, J. Am. Chem. Soc. 132 (2010) 5578-5579.

[2] M. Eddaoudi, J. Eubank, Y. Liu, V.Ch. Kravtsov, R. Larsen, J. Brant, Stud. Surf. Sci. Catal. 170 (2007) 2021-2029.

[3] F. Nouar, J. Eckert, J. Eubank, P. Forster, M. Eddaoudi, J. Am. Chem. Soc. 131 (2009) 2864- 2870.

[4] J. Qian, F. Sun, L. Qin, Mater. Lett. 82 (2012) 220-223.

3601-3606.

[5] H. Zhuo, Q. Li, W. Li, J. Cheng, New J. Chem. 39 (2015) 2067-2074.

*Corresponding author. Tel.:+86 025-84315517; fax: +86 025-84315517.

E-mail: zq304@mail.njust.edu.cn (Qin Zhong).

453025 CO_2 Capture in Uio-66-Type Metal-Organic Frameworks (MOFs)

Abstract:

CO_2 Capture in UiO-66-Type Metal-Organic Frameworks (MOFs)

Dan Zhao*

Department of Chemical and Biomolecular Engineering,

National University of Singapore, 117585 Singapore

E-mail: chezhao@nus.edu.sg

The large-scale anthropogenic carbon dioxide (CO_2) emission has aroused world-wild concerns. One of the key steps in carbon mitigation is CO_2 capture from stationary CO_2 emission resources. Metal-organic frameworks (MOFs) are coordination polymers consisting of metal ions or clusters as nodes and organic ligands as spacers. The study of CO_2 capture using MOFs as adsorbents is booming because of their ultra-high surface area, uniform yet tunable pore size, and versatile chemical compositions. However, several issues still exist preventing their large scale applications, such as selectivity, stability, and material cost. In our research, we aim at circumventing the above problems through rational ligand design and crystal engineering based on UiO-66-type MOFs, with the emphases on a) mitigating moisture interference, b) increasing CO_2 selectivity, c) strengthening materials stability, and d) reducing material cost. The synthesized MOF candidates have underwent various tests for their eligibility in CO_2 capture, such as stability, gas sorption, and breakthrough tests. We have identified several promising MOF candidates with the CO_2 capture performance equivalent or even better than the industrial benchmark material zeolite 13X. These materials will be integrated into the next-level process and system research.

References

1. Hu, Z. G.; Castano, I.; Wang, S. N.; Wang, Y. X.; Peng, Y. W.; Qian, Y. H.; Chi, C. L.; Wang, X. R.; Zhao, D.* Modulator effects on the water-based synthesis of Zr/Hf MOFs: quantitative relationship studies between modulator, synthetic condition, and performance. *Cryst. Growth Des.* **2016**, DOI: 10.1021/acs.cgd.6b00076.
2. Hu, Z. G.; Nalaparaju, A.; Peng, Y. W.; Jiang, J. W.; Zhao, D.* Modulated hydrothermal synthesis of UiO-66(Hf)-type metal-organic frameworks for optimal carbon dioxide separation. *Inorg. Chem.* **2016**, *55*, 1134-1141.
3. Hu, Z. G.; Faucher, S.; Zhuo, Y. Y.; Sun, Y.; Wang, S. N.; Zhao, D.* Combination of optimization and metalated-ligand exchange: an effective approach to functionalize UiO-66(Zr) MOFs for CO_2 separation. *Chem.-Eur. J.* **2015**, *21*, 17246-17255.
4. Hu, Z. G.; Peng, Y. W.; Kang, Z. X.; Qian, Y. H.; Zhao, D.* A modulated hydrothermal (MHT) approach for the facile synthesis of UiO-66-type MOFs. *Inorg. Chem.* **2015**, *54*, 4862-4868.
5. Hu, Z. G.; Khurana, M.; Seah, Y. H.; Zhang, M.; Guo, Z. G.; Zhao, D.* Ionized Zr-MOFs for highly efficient post-combustion CO_2 capture. *Chem. Eng. Sci.* **2015**, *124*, 61-69.
6. Hu, Z. G.; Zhang, K.; Zhang, M.; Guo, Z. G.; Jiang, J. W.; Zhao, D.* A combinatorial approach towards water-stable metal-organic frameworks for highly efficient carbon dioxide separation. *ChemSusChem* **2014**, *7*, 2791-2795.

461320 Tuning the Kinetic Water Stability and Adsorption Interactions of Mg-MOF-74 By Partial Substitution with Co or Ni

Abstract:

Varying amounts of Co and Ni were substituted into the metal-organic framework Mg-MOF-74 via a one-pot solvothermal reaction, and the effects of these substitutions on CO_2 adsorption and kinetic water stability properties were examined. Based on elemental analyses, Co and Ni are more favorably incorporated into the MOF-74 framework from solution than Mg. In addition, reaction temperature more strongly impacts the final metal composition in these mixed-metal (MM) MOF-74 structures than does the reaction solvent composition. Single-component CO_2 adsorption isotherms were measured for the MM-MOF-74 systems at 5, 25 and 45 °C and isosteric heats of adsorption were calculated. These results suggest that CO_2 sorption properties can be adjusted by partial metal substitution. Water adsorption isotherms were also measured for the MM-MOF-74 samples, with powder X-ray diffraction patterns and Brunauer-Emmett-Teller surface areas measured both before and after water exposure. Results show that Mg-MOF-74 can gain partial kinetic water stability by the incorporation of Ni^{2+} or Co^{2+} metal ions that are less vulnerable to hydrolysis than Mg^{2+}. Of particular note, Mg-Ni-MM-MOF-74 shows a significant increase in water stability when incorporating as little as 16 mol% Ni into the Mg-MOF-74 structure.

475638 Redox Active Metal Oxide-Based Thermochemical Processes for Producing Solar Fuels and Storing Thermal Energy (Invited Talk)

Abstract:

Reversing combustion and recycling carbon dioxide and water back to liquid hydrocarbons is an attractive option for storing solar energy and mitigating the growth of atmospheric CO_2 concentrations. For any such process, high solar-to-fuel efficiency and material availability is critical for large scale viability and favorable economics. Redox active metal oxide-based thermochemical approaches for solar-to-fuel have the potential to be highly efficient as they avoid inherent limitations of photosynthesis and sidestep the solar-to-electric conversion necessary to drive electrolytic reactions. This presentation highlights the challenges and progress from multidisciplinary and international efforts that has been progressing down a technical path for systems, novel reactors, and materials design and discovery for making hydrogen, liquid hydrocarbon fuels, and storing thermal energy from concentrated sunlight based on a two-step metal oxide redox cycle motivated to address the dual challenges posed by the strategic and economic importance of petroleum and the increasing concentration of atmospheric carbon dioxide.

454582 Doped Ceria for Solar Thermal Water Splitting: What Works, What Doesn't, and How to Improve It

Abstract:

Solar thermal water splitting (STWS) has the potential to achieve high solar to hydrogen efficiencies because of its use of the entire solar spectrum to split water. Two-step STWS uses a metal oxide oxygen carrier to couple two reduction/oxidation half reactions, one of which generates O_2 and the other which splits water to generate H_2. However, an optimal material which facilitates these reactions has not been identified. Among the candidate materials that are known to be capable of driving STWS, CeO_2 is considered the most promising. However it requires very high reduction temperatures (≥1500°C) to achieve acceptable H_2 production capacities. Many researchers have tried to dope ceria to bring down the reduction temperature, to varying degrees of success. In this work we use *ab initio* calculations based on density functional theory to develop a fundamental understand and explanation for why only Hf, and Zr have "improved" the STWS abilities of ceria. Based on this understanding we propose a novel set of dopant combinations and modifications strategies that, based on DFT results, improve the hydrogen production capacity of ceria at lower temperatures and/or result in higher H_2O conversion extents during the oxidation reaction.

454960 Stabilizing SiC for Solar Thermal Water Splitting Applications

Abstract:

Stabilizing SiC for Solar Thermal Water Splitting Applications

Amanda Hoskins[1], Aidan Coffey[1], Charles B. Musgrave[1,2], Alan W. Weimer[1]

[1]Department of Chemical and Biological Engineering, University of Colorado, Boulder, CO 80303, USA.

[2]Department of Chemistry and Biochemistry, University of Colorado, Boulder, CO 80309, USA.

Solar thermal water splitting (STWS) has been proposed as a means for hydrogen production which has the benefit that both the energy source and the fuel are renewable and readily available. However, considerable technological challenges must first be addressed in regards to hydrogen production, storage and transport in order for hydrogen to become an economically viable fuel on a large enough scale to significantly contribute to future energy demands. A major limiting factor to the viability of this technology is the lack of high temperature materials able to withstand the extreme chemical and physical environment. Specifically, materials that can be used to fabricate efficient heat exchangers for steam/steam heat recuperation will significantly impact the progress of this technology.

Silicon Carbide (SiC) is an ideal material for high-temperature heat exchange due to its stability at much higher temperatures than the current state-of-the-art material (Inconel) which cannot be used above ~1000°C. Using SiC for heat exchangers involves challenges related to the lifetime of SiC in an oxidative environment. Silicon carbide degrades in water-rich environments through the formation and subsequent degradation of a silica (SiO_2) layer at the exposed surface. In a dry environment, the silica has a low permeability to oxygen, however the presence of water, especially at high temperatures, increases the rate of oxidation, causing cracking and volatilization of the silicon from the structure. We propose that the application of nanostructured films grown with atomic layer deposition (ALD) will significantly improve the oxidation resistance of SiC in extreme environments. We have targeted a variety of high temperature coating materials chosen based on thermal properties, structural characteristics, and stability in oxidative environments. We have grown ALD layers of our desired materials on SiC particles using a fluidized bed particle ALD reactor. We have characterized all films using energy dispersive X-ray spectroscopy (EDS) mapping, scanning electron microscopy (SEM), transmission electron microscopy (TEM), and inductively coupled plasma mass spectrometry (ICP). Using thermogravimetric analysis, it has been shown that these coatings improve the oxidation resistance of SiC.

We have applied density functional theory (DFT) to model the performance of these coatings in an oxidative environment. Computational modeling of the chosen coating material properties, as well as oxygen diffusion through these materials can identify promising coatings to further extend the lifetime of SiC. We expect to use this understanding to develop more advanced stabilizing coatings for SiC and the myriad of other applications that require high-temperature oxidation resistant materials. This presentation will focus on both the fundamental DFT calculations and the experimental results for thin ALD films being developed to protect SiC surfaces from water oxidation at temperatures above 1000°C.

460329 Development and Test Operation of a Demonstration Plant for Sulfuric Acid Splitting at the Dlr Concentrating Solar Power Tower Facility

Abstract:

Sulfuric acid splitting is a key step of the hybrid sulfur cycle (HyS) for solar thermochemical hydrogen production. This exothermal reaction can be divided into two steps: firstly, the evaporation of liquid sulfuric acid (H_2SO_4) at about 400 Â°C forming sulfur trioxide (SO_3), and secondly, the decomposition of SO_3 to sulfur dioxide (SO_2) and oxygen (O_2) at 800 â€“ 1000 Â°C. While the first sub-reaction has fast kinetics, the second one is rather slow and requires the introduction of catalysts to achieve sufficient conversion. Since 2004 the concept of a solar receiver-reactor for sulfuric acid splitting has been developed by DLR and operated in its solar furnace during the European projects HYTHEC and HycycleS. In the follow-up European project SOL2HY2, a scale-up of this concept has been designed, developing a solar tower demonstration plant. This demonstrator has a design flow rate of 1Â l/min of sulfuric acid (50Â w%) and consist of four main components arranged in series and connected by Joule heated piping: a 60 kW electrical evaporator for vaporization of the liquid acid, a solar receiver for superheating the SO_3 to about 1000 °C, an adiabatic reactor with a fixed bed of an iron(III) oxide catalyst and a scrubber. The Joule heated evaporator consists of six vertical steel tubes with a siliconized silicon carbide (SiSiC) inner tubes filled with porous SiSiC foam structures for enhanced heat transfer. Liquid acid is injected at the bottom and vaporizes while passing upwards through the pipes. The vapors are collected in a steel manifold and, subsequently, conveyed to the solar receiver with an outer shell also made of steel. Concentrated radiation from the solar field passes through a quartz glass window closing the receiver and heats up a sectioned absorber composed of porous SiSiC foam structures. After superheating in the receiver, the process gas passed through the catalyst bed for an adiabatic reaction forming SO_2. Before neutralization with sodium hydroxide solution in the scrubber, the SO_2 concentration is measured by a customized gas analysis

system via UV/Vis spectroscopy. The pilot plant was constructed and assembled on the research platform of the DLR concentrating solar power tower facility in Juelich, Germany. The layout of the plant was accompanied and supported by thermo-mechanical modelling of the most important components like the evaporator and solar receiver. Initial operation of the demonstrator was performed with air and water as process fluids. During water operation, the solar receiver reached the predicted design parameters achieving a gas outlet temperature of 1000 °C at an absorber front temperature of 1200 °C and a solar power on aperture of 50 kW. In the adiabatic reactor, however, temperatures of only about 400 °C were measured which are too low for SO_3 decomposition. Therefore, the system was modified placing the catalytic fixed bed directly behind the solar absorber in the outlet section of the receiver. In the following test runs, the temperatures of this adiabatic reaction zone were sufficiently high with a minimum temperature in excess of 750 °C below which the catalyst would be deactivated due to sulfate formation. As a result, testing could proceed with sulfuric acid as the feed successfully demonstrating decomposition of SO_3. A detailed analysis of all results of the systematic on-sun test series is given in the present paper.

466891 Modelling of a Continuous Solar Tubular Aerogel Reactor for the Thermochemical Reduction of CeO_2: The Effect of Residence Time, Particle Diameter and Particle Loading

Abstract:

A simulation is developed and implemented to describe a continuously operated and indirectly irradiated solar reactor for the thermochemical reduction of CeO_2 to Ce_2O_3. An aerogel, consisting of N_2 and CeO_2 (particle diameter between 1-20 microns), enters the reactor at 900 K. The aerogel flows upward through a 30 cm long by 2.54 cm diameter tube, made of Tungsten coated with Silicon Carbide at 2500 K. The model aims at calculating the concentration and temperature profiles within the reaction tube at steady state, the conversion of CeO_2 particles and the Ce_2O_3 particles mass flow exiting the reactor.

The radiative transfer model within the reactor is implemented by considering a gray, non-isothermal, absorbing, emitting and scattering media. Scattering is considered to be anisotropic, as described by the Henyey-Greenstein phase function. Optical parameters where calculated from Mie Scattering theory, weighted according the blackbody spectra at 2500 K. The thermochemical reduction of CeO_2 is modeled by considering an endothermic, first order, source term. Temperature dependence follows the Arrhenius equation. Convective heat and mass transport mechanisms disregard any thermo-hydrodynamic effects, and assume instead plug flow. Finally, gas conduction is accounted and CeO_2 particle diffusivity is estimated from the Einstein-Smoluchowsky equation for brownian motion of spherical particles.

Simulations consider axisymmetric fields, i.e. dependent on the axial and radial position within the tube. First, the radiation heat transfer mechanisms where implemented using an implicit hybrid Monte Carlo-Finite Volume scheme, linearizing the fourth order emission term using a Taylor series expansion. On the other hand, the heat and mass convection, diffusion and reaction where implemented using an explicit iterative scheme. The steady state distributions of all combined physical phenomena are obtained by using the false transient-state approximation, while considering an iterative-splitting scheme.

The parameters considered within this study where the average flow velocity, the CeO_2 particle loading and the particle diameter. Simulations are carried out considering particle loadings of 10 and 100 mg/l, particle diameter of 1,5,10 and 20 microns and average flow velocities of 1,5 and 10 mm/s. Obtained results suggest that high particle loadings tend to favour temperature gradients, which hinder the overall reactor performance. Higher gas velocity lowers the conversion, but slightly increases the mass Ce_2O_3 flow rate. Finally, it is interesting to point out, that higher particle diameter, presents low specific extinction coefficients, which allow relatively high radiation heat transfer rates throughout all the reactor, minimizing temperature gradients and yielding overall higher CeO_2 conversion and mass Ce_2O_3 flow rate.

468882 Conceptual Analysis of Process Alternatives for Solar Thermochemical Methanol Production: The Role of Chemical Storage

Abstract:

Conceptual Analysis of Process Alternatives for Solar Thermochemical Methanol Production:

The Role of Chemical Storage

Bruno A. Calfa[1] and Christos T. Maravelias NOTEREF OLE_LINK1 \h * MERGEFORMAT 1

08D0C9EA79F9BACE118C8200AA004BA90B02000000080000000A0000004F004C0045005F004C0049004E004B0031000000

In this talk, we study the production of fuels and chemicals from a renewable source, namely solar energy. In a solar refinery, solar energy is utilized to convert raw materials, such as CO_2 emitted from fossil fuel power plants (and sometimes with H_2O), into value-added products; e.g., methanol and Fischer-Tropsch (F-T) fuels (Herron et al., 2015). While different solar-based technologies have been investigated to carry out the conversion, in this paper, we consider the solar thermochemical splitting of CO_2 and H_2O (see Meier and Steinfeld (2013) and references therein), which has some similarities with Concentrated Solar Power (CSP) technology in the electric power industry.

The viability, based on systems-level analyses, of a solar-driven (thermochemical cycle technology) process to produce methanol and/or F-T fuels from syngas produced from CO_2 and H_2O splitting has recently been evaluated in the literature (Kim et al., 2011; Kim et al., 2012; Kim et al., 2013; Rytter et al., 2016; Falter, Batteiger, and Sizmann, 2016). However, all aforementioned works considered a continuously operated process. However, one fundamental aspect of the production of solar fuels is the inherent *intermittency* of solar energy. Therefore, since the splitting reactions cannot take place when there is not sufficient solar irradiance (i.e., cloudy days or at night), there is a need to consider *chemical storage*.

In this work, we provide a systems-level analysis of three process configurations, a base case and two alternatives, which employ solar-thermal technology for the conversion of CO_2 and H_2O to syngas, which is further upgraded to methanol. We propose and assess the use of chemical storage to cope with the intermittency of solar-based chemical technologies. The base case requires intermediate gas storage, since all streams are treated as continuous from a high-level perspective (typically assumed in the literature).

First we determine the number of compression stages that minimizes the total annualized cost of compression and storage of CO_2/CO and H_2 streams. Material and energy balances followed by an economic analysis indicate that gas compression and storage significantly contribute to the production cost in the base case, which makes it less economically favorable than its alternatives. We perform sensitivity analysis for key process parameters (e.g., chemical conversion of the solar reactors, cost of the solar field and solar reactors, and solar refinery capacity) to investigate their effect on the production cost of methanol.

The two alternative process configurations do not require gas storage, since the syngas production step is operated entirely intermittently. One process alternative uses a reverse water-gas-shift reactor to further convert CO_2 to CO and correct the stoichiometric number (H_2:CO ratio) required for the methanol synthesis step, whereas the other process alternative has a membrane-based CO_2/CO separation that is operated intermittently. Our analyses show that the process alternatives have overall lower total capital investment and operating costs than the base case, whose compression, storage, and amine-based CO_2/CO separation costs represent major components of the production cost.

References

Falter, C.; Batteiger, V.; and Sizmann, A. 2016. Climate Impact and Economic Feasibility of Solar Thermochemical Jet Fuel Production. Environmental Science & Technology. 50(1):470-477.

Herron, J. A.; Kim, J.; Upadhye, A. A.; Huber, G. W.; and Maravelias, C. T. 2015. A General Framework for the Assessment of Solar Fuel Technologies. Energy & Environmental Science. 8(1):126-157.

Kim, J.; Henao, C. A.; Johnson, T. A.; Dedrick, D. E.; Miller, J. E.; Stechel, E. B.; and Maravelias, C. T. 2011. Methanol Production from CO2 Using Solar-Thermal Energy: Process Development and Techno-Economic Analysis. Energy & Environmental Science. 4(9):3122-3132.

Kim, J.; Johnson, T. A.; Miller, J. E.; Stechel, E. B.; and Maravelias, C. T. 2012. Fuel Production from CO_2 Using Solar-Thermal Energy: System Level Analysis. Energy & Environmental Science. 5(9):8417-8429.

Kim, J.; Miller, J. E.; Maravelias, C. T.; and Stechel, E. B. 2013. Comparative Analysis of Environmental Impact of S2P (Sunshine to Petrol) System for Transportation Fuel Production. Applied Energy. 111:1089-1098.

Meier, A., and Steinfeld, A. 2013. Solar Energy. New York, NY: Springer New York. chapter Solar Energy in Thermochemical Processing, 521-552.

Rytter, E.; Sou__kov‡. F; Lundgren, M. K.; Ge, W.; Nannestad, . D.; Venvik, H. J.; and Hillestad, M. 2016. Process Concepts to Produce Syngas for Fischer-Tropsch Fuels by Solar Thermochemical Splitting of Water And/Or CO_2. Fuel Processing Technology. 145:1-8.

---

[1] Department of Chemical and Biological Engineering. University of Wisconsin-Madison. Madison, WI 53706. USA.

471163 Hercynite Reduction and ITM Membrane Oxygen Removal System for Solar Thermal Water Splitting

Abstract:

Solar Thermochemical Water Splitting Processing (STWS) is a promising technology for producing renewable H_2[1]. A metal oxide, such as hercynite ($FeAl_2O_4$), is first reduced with high temperature concentrated sunlight to produce O_2[2]. Then, the reduced material is exposed to steam to re-oxidize it and produce H_2. Low O_2 partial pressure is required to drive the reduction reaction. An intriguing potentially highly efficient process is the recycle of inert sweep gas at near atmospheric pressure using a high temperature O_2 ionic transport membrane (ITM)[3], since the recycled gas is already at the required high temperature. In this work, we present preliminary experimental results for hercynite reduction using a TGA and separately the removal of O_2 from 1% O_2/inert down to ppm levels using a laboratory high temperature ITM. The impact of separation energy requirements is evaluated for the overall solar to H_2 cycle.

1. Muhich, C.L., et al., *A review and perspective of efficient hydrogen generation via solar thermal water splitting.* WIREs Energy Environ 2016, **5**:p. 261â€"287. doi: 10.1002/wene.174

2. Muhich, C.L., et al., *Predicting the solar thermochemical water splitting ability and reaction mechanism of metal oxides: a case study of the hercynite family of water splitting cycles.* Energy & Environmental Science, 2015. **8**(12): p. 3687-3699.

3. Ehrhart, B.D., et al., *System efficiency for two-step metal oxide solar thermochemical hydrogen production â€" Part 3: Various methods for achieving low oxygen partial pressures in the reduction reaction.* Int. J Hydrogen Energy (online) http://dx.doi.org/10.1016/j.ijhydene.2016.07.106.

474467 Monte Carlo method to calculate the lifetime efficiency of a solar reactor for reduction of zinc oxyde

Abstract:

Monte Carlo method to calculate

the lifetime efficiency of a solar reactor for reduction of zinc oxyde

O. Farges, J.J. B′ezian, and M. El Hafi

Universit′e F′ed′erale de Toulouse Midi-Pyr′en′ees, Mines Albi, UMR CNRS 5302, Centre Rapsodee, Campus Jarlard,

f-81013 Albi ct cedex 09, France

Concentrated solar plant are ommonly used to produce electricity through ther- modynamic cycles (Gemasolar Power Plant, Ivanpah Solar Power Plant, ...) and can also be seen as an alternative to fossil fuel-based methods for H2 generation [5]. High temperature solar thermochemical processes make use of concentrated solar energy to realize metal oxydes reduction. These processes can be considered as a solar energy storage : products may be stored and transported (H2 in this case). Among all metal oxydes reduction, the zinc oxyde to zinc (ZnO/Zn) redox reactions are particularly attractive as the first part of a two step water splitting cycle (eqs. (1a) and (1b)).

As presented on fig. 1, a thermochemical reactor *R*, placed at the top of a solar tower receives concentrated solar energy emited from the sun *S* and reflected by heliostats *H*. Thus, the reactor, feeded with zinc oxyde, reaches high temperature, *ie* above 1400 K, to achieve reduction of zinc oxyde and to procude zinc Zn.

$$ZnO \rightarrow Zn + \frac{1}{2}O_2 \quad (1a)$$
$$Zn + H_2O \rightarrow ZnO + H_2 \quad (1b)$$

In order to evaluate the performance of a thermochemical solar plant and to allow an optimally design of both the reactor and the whole facility (heliostat field,

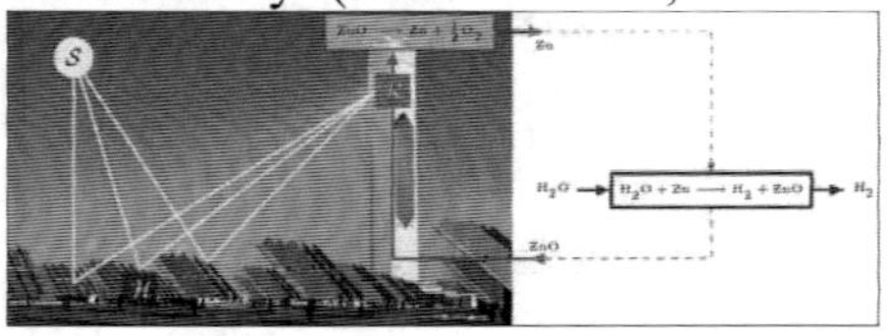

Figure 1: H2 solar production process

tower), numerical modelling of this process is required. We present here a Monte Carlo algorithm that allows the estimation of the solar plant annual productivity which is a non-linear problem. This non-linearity is due to the coupling between photon transport and the metal oxyde reduction reaction. The widespread opinion is currently that "Monte Carlo methods are not generally effective for nonlinear problems mainly because expectations are linear in character"[3]. This difficulty has been overcome by Dauchet who proposed a methodology in his Phd thesis [1].

To evaluate the quantity of interest, *ie* the annual solar-to-chemical conversion rate $(\eta r\ )$, this Monte Carlo algorithm firstly computes the instantaneous power re- ceived at the entrance of the reactor at an instant $\boldsymbol{i}$, $(Pth(\boldsymbol{X}\ |\boldsymbol{i}))$. A Monte Carlo algorithm previously presented in [2] is used as a first step of the proposed algo- rithm to evaluate this quantity. The random variable $Pth(\boldsymbol{X}\ |\boldsymbol{i})$ is the contribution of optical paths $\boldsymbol{X}\ |\boldsymbol{i}$ to the thermal power collected at instant $\boldsymbol{i}$. Then, $(Pth(\boldsymbol{X}\ |\boldsymbol{i}))$, the expectation of this random variable, is the instantaneous thermal power col- lected at moment $\boldsymbol{i}$. A part of the power received by the reactor, the instantaneous useful power $Pu\ = f\,(Pth(\boldsymbol{X}\ |\boldsymbol{i}))$, represents the efficient energy used to achieve redox reactions with an instanteneaous reaction rate described by a zero-order Arrhenius-type law, strongly non-linear. Integration over time of the instanta- neous conversion rate leads to the lifetime conversion rate $(\eta r\ ((Pth(\boldsymbol{X}\ |\boldsymbol{i}))))\boldsymbol{i}$ as presented eq. (2).

$$\langle\eta_r\rangle = \int_{LifeTime} \frac{P_u(t)}{P_{th}(t)} \tag{2}$$

To treat this non-linearity, the Monte Carlo algorithm is based on a Taylor expansion of (ηr) around $Pth0 > Pth(\boldsymbol{X})$. Independent and equally distributed optical-path random variables $\boldsymbol{Xj}\,|\boldsymbol{I}$ are introduced and the infinite power series is statistically formulated thanks to the discrete random variable J whose realization

j is the order of Taylor expansion. The Taylor expansion is stopped through a Bernoulli process in order to decide whether the algorithm stops or continues without introducing a statistical bias.

Simulations are performed for the 1 MW reactor presented in [4]. A simplified model of thermal losses is applied and a comparison of the nominal performance between results obtained with the Monte Carlo algorithm and experimental results presented in [4] are presented in table 1. The difference between the measured and calculated results are due to the simplified thermal losses model as well as uncertainties of the measurements.

Table 1: Size and performance of a 1 MW reactor

Parameters	Units	Values
Aperture diameter	mm	600
Window diameter	mm	790
Cavity diameter	mm	1600
Cavity length	mm	2400
Conversion rate η_r [4]	%	54
Conversion rate estimated η_r	%	40.2

We have shown in this work that we are able to calculate the efficiency by a Monte Carlo Method, of a solar chemical reactor involving non-linearities between the heat source and the products of the reaction. A futur work could be the improvement of the heat loss model.

References

[1] J. Dauchet. *Analyse radiative des photobioracteurs*. PhD thesis, 2012.

[2] O. Farges, J. B´ezian, H. Bru, M. E. Hafi, R. Fournier, and C. Spiesser. Life- time integration using monte carlo

methods when optimizing the design of concentrated solar power plants. *Solar Energy*, 113(0):57 – 62, 2015.

[3] M. H. Kalos and P. A. Whitlock. *Monte Carlo Methods*. John Wiley & Sons,

2008.

[4] L. Schunk, W. Lipin´ski, and A. Steinfeld. Heat transfer model of a solar receiver-reactor for the thermal dissociation of zno-experimental validation at

10 kw and scale-up to 1 mw. *Chemical Engineering Journal*, 150(23):502 – 508,

2009.

[5] A. Steinfeld. Solar thermochemical production of hydrogen--a review. *Solar*

Energy, 78(5):603-615, 2005.

474861 A Model Algae Strain Extracellular Sugars: Genome Sequencing and Transcriptomic Analysis

Abstract:

Microalgae are viewed as a next generation biofuel platform. This perception is based upon rapid rates of growth, an ability to grow under a wide range of condition, and the production of a variety of different bioproducts, including lipids, proteins and carbohydrates from various microalgae species. A great deal of interest has centered on the production of lipids which could be converted into biodiesel, while the use of starch has been utilized to produce products such as ethanol in certain species. Microalgae have also been reported to produce extracellular carbohydrates, including complex polysaccharides as well as simple sugars such as glucose and maltose. Our laboratory has been studying the production of these simple sugars by a specific strain of *Chlorella* which was isolated from a symbiotic relationship with *Paramecium bursaria*. The *Chlorella* strain has been cultivated independently for several decades now, and continues to produce extracellular sugars under certain growth conditions. We have recently sequenced two strains of Chlorella species, and have completed RNA-Seq analysis of the expression profiles of this strain during growth conditions that produce sugars versus those that do not to compare the transcriptome and elucidate potential pathways toward extracellular sugar production. Results from these studies will be presented.

461039 Exploiting Models to Assess and Guide Genetic Engineering of Microalgae Strains

Abstract:

Exploiting models to assess and guide genetic engineering of microalgae strains

Andrea Bernardi[1], Andrea Meneghesso[2], Giorgio Perin[2], Tomas Morosinotto[2], Fabrizio Bezzo*[1] [1]CAPE-Lab (Computer-Aided Process Engineering Laboratory) & PAR-Lab (Padova Algae Research Laboratory), Department of Industrial Engineering, University of Padova

via Marzolo 9, 35131 Padova, Italy [2]PAR-Lab (Padova Algae Research Laboratory), Department of Biology, University of Padova, via Ugo Bassi 58B, 35131, Padova, Italy *fabrizio.bezzo@unipd.it

Microalgae are long been recognized as one of the most promising alternatives for biofuel production. The main advantages of microalgae with respect to other possible feedstock are the high potential productivity and the absence of competition with traditional crops for arable land and clean water. However, this potential is still theoretical and algae production on large scale is not profitable yet. Several issues need to be addressed to reach this objective, ranging from algae cultivation and harvesting as well as products extraction [1]. In order to be economically and energetically sustainable microalgae cultivation must use the solar irradiance, which is variable on daily and seasonal basis. One of the main problems of microalgae cultivation in large scale is the low conversion efficiency of the solar energy. In fact, there is a big gap between the conversion efficiency obtained in laboratory and in large scale photobioreactors.

Microalgae have developed a series of mechanisms to modify their photosynthetic apparatus in order optimize the light utilization in a natural environment, such as variation in pigment composition and/or spatial organization of the photosynthetic apparatus [2]. In particular, two main regulatory mechanisms can be identified: photoregulation, often referred as Non Photochemical Quenching (NPQ), and photoacclimation. Photoregulation involves the activation of dissipative pathways to convert part of the light energy in thermal energy in order to prevent photo-induced damage (photoinhibition) and acts in a time scale from minutes to hours. Photoacclimation is a slower process acting in time scales of days and induced a variation on the pigment content of the microalgal cells as consequence of a variation in the light conditions. However, certain properties of microalgae, which have been positively selected in wild-type species through evolution, can cause a significantly loss of productivity and are therefore detrimental to large-scale cultivation as the growing conditions in industrial plants are very different from the natural ones [3]. In this context, genetic improvements of microalgal strains can play a major role in the development of biofuels technology. However, addressing such a complex problem via trial-and-error, as is done currently, presents many drawbacks. First of all, evaluating the performances of a mutant is a difficult and time demanding task, as it requires to grow several microalgae cultures under different light conditions and each experiment may last several days.

First principles models could be an effective tool to reduce the experimental effort needed to select the best mutant among different candidates and they can also provide guidance for genetic engineering by identifying those modifications having the largest potential impact on productivity.

In this work we will show how a model developed by the authors [4] can be exploited to evaluate the performances of different mutants in fast and reliable way. The model describes the photoproduction, photoregulation and photoinhibition processes and has been calibrated against chlorophyll fluorescence and photosynthesis rate measurements of three strains of *Nannochloropsis gaditana*: the wild type, a mutant with reduced NPQ (mutant T3.9) and a mutant with a reduced antenna size and chlorophyll content (mutant E2). After the calibration the model has been used to evaluate the mutants performances under multiple light conditions, such as the photoperiods of different months in a typical Mediterranean country. The advantage of the proposed approach is that to calibrate the model are necessary only fast and accurate fluorescence measurements and some measurements of photosynthesis rate. In fact, as demonstrated in [4] the information contained in fluorescence experiments is sufficient to describe the dynamics of photoregulation and photoinhibition.

In Figure 1a are reported the photosynthesis rate measurements for *N. gaditana* wild type along with the model predictions for the wild type, the mutant T3.9 and the mutant E2. In Figure 1b are reported the predicted photosynthesis rate profiles under light conditions typical of July in Padova as obtained from the PVGIS database.

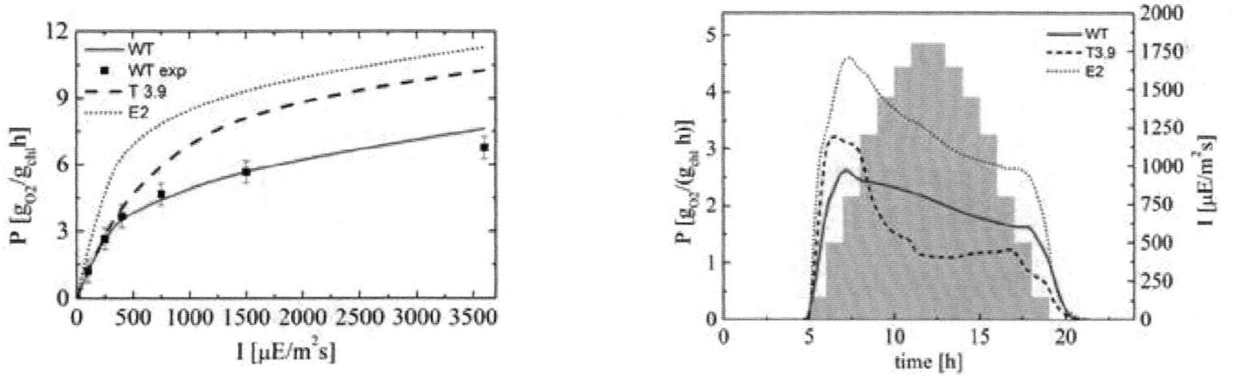

(a) (b)

Figure 1: (a) Photosynthesis rate measurements for *Nannochloropsis gaditana* wild type along with model simulation results and photosynthesis rate prediction for mutants T3.9 and E2; (b) predicted photosynthesis rate profile for July irradiantion profile for *N. gaditana* wild type, mutant T 3.9 and mutant E2.

We can observe that, even if the photosynthesis rate measurements of the NPQ-less mutants are higher than the measurements of the wild type also at high light intensities, when the culture is exposed to intense light for a prolonged amount of time, the photoinhibition lead to a dramatic reduction of the photosynthesis rate. For this reason a mutant where the NPQ activity is absent may not be a good candidate for an outdoor cultivation in the Mediterranean area and other type of mutants should be investigated. On the other hand, the mutant E2 is more promising as the model predictions show an higher photosynthesis rate both in laboratory measurements conditions and for the July photoperiod. Moreover, if a dense culture is considered the antenna-less mutant will lead to a more uniform light distribution, which will increase the gap between the wild type and the mutant.

Future efforts will be devoted to validate the model predictions for different photoperiods and to include the light attenuation and mixing effect in the modelling framework. This will allow us to investigate the mutants performances under more realistic conditions representing a real photobioreactor conditions.

References

[1] Mata, T.; Martins, A.; Caetano, N. Microalgae for biodiesel production and other applications: a review. Renewable and Sustainable Energy Rev. 2010, 14, 217-232.

[2] Williams, P. J. l. B.; Laurens, L. M. L. (2010). Microalgae as biodiesel & biomass feedstocks: Review & analysis of the biochemistry, energetics & economics. Energy & Environmental Science, 3, 554-590.

[3] Formighieri, C., Franck, F., and Bassi, R. (2012) Regulation of the pigment optical density of an algal cell: filling the gap between photosynthetic productivity in the laboratory and in mass culture. J. Biotechnol. 162, 115–23

[4] Bernardi, A.; Nikolaou, A.; Meneghesso, A.; Morosinotto, T.; Chachuat, B.; Bezzo, F. (2015). High fidelity modelling of light limited photosynthetic production in microalgae using PAM fluorescence and optimal experiment design. PlosOne. 2016, 11, e0152387

468558 Use of Unionized Ammonia to Control Zooplankton Grazers in Cultures of *Scenedesmus*

Abstract:

Algae have been targeted as a potential biomass source for scalable biofuel and bioproducts development. Unfortunately, the presence of zooplankton grazers has remained a problem in open pond systems, reducing yields and causing repeated culture crashes. At Old Dominion University's open pond, zooplankton grazers were controlled by maintaining a high pH (>9.5 pH) and total nitrogen concentration of 30 mg/L using urea. These values correspond to unionized ammonia values previously reported as toxic to the zooplankton, specifically rotifers, but not the algae in studies designed specifically for each organism. To understand this phenomenon further, particularly in the relevant combined (algae/zooplankton) system, 2.5 L cultures of zooplankton contaminated *Scenedesmus* were inoculated in bench scale photobioreactors at various pH's (8.0, 10.0 and 12.0) and unionized ammonia concentrations (0.0, 0.1, and 1.0 mM). These treatments were tested in a factorial experimental design. Trials lasting five days each were broken down by each ammonia treatment. Each trial was tested across each pH plus a control with only culturing media to ensure the algae cultures were growing normally (4 treatments per trial in duplicate). The algal and zooplankton populations were monitored by cell counts (hemocytometer) and total suspended solids (TSS) measurements (1.5 Âµm filter). Unionized ammonia was supplied in the form of ammonium chloride and the pH was daily adjusted using hydrochloric acid and sodium hydroxide additions. After the repeated crashes of the 12.0 pH cultures across all ammonia treatments, this pH treatment was removed from the experiment. To improve control over pH, phosphate and carbonate buffer systems were employed in the second set of trials, but the unbuffered systems were still tested for comparison. This resulted in 4 pH treatments plus a control in duplicate for each trial. Total ammonia was measured in filtered (1.5 Î¼m) liquid samples collected each day after TSS and frozen at -20 °C to quantify losses due to uptake by the algae and/or stripping to the atmosphere. Preliminary results indicate that 1.0 mM NH_3 concentration reduced and maintained a reduced zooplankton population for at least several days while slightly reducing the growth of the algae

population before ammonia concentrations were reduced to non-inhibitory levels.

467182 Extracellular Glycerol Production ByÂ *Dunaliella Tertiolecta* in a Membrane Photobioreactor

Abstract:

Extracellular glycerol production by *Dunaliella tertiolecta* in a membrane photobioreactor

Photosynthetic microalgae can be used as bio-factories to produce value-added chemicals from carbon dioxide. One such chemical is glycerol, which is used as a precursor in many industrial manufacturing, including in the production of food and pharmaceuticals. Many marine microalgae, especially those belonging to *Dunaliella* sp., are able to produce intracellular glycerol under saline environment to regulate cellular osmotic pressure. Some microalgae, such as *Dunaliella Tertiolecta*, are also able to produce extracellular glycerol. While the use of microalgae in glycerol production has the advantages of carbon capture and utilization, it may also generate microalgae biomass, which can act as precursors in the production of various other useful compounds, including biofuels.

The extracellular glycerol production from microalgae is a great opportunity as it avoids the usual issues related to harvesting or downstream separation of microalgae. However, the production of extracellular glycerol by microalgae is typically low and it may not be very economical to produce. On approach to improve the yield of intracellular glycerol by microalgae can be the use of membrane photobioreactor (MPBR), such that a very high biomass density can be maintained in the MBR, which may translate in high yield of glycerol.

In this research, an MPBR based on microfiltration was designed and operated for high-throughput extracellular glycerol production using *Dunaliella Tertiolecta*. The effects of various operating parameters, such as biomass concentration, CO_2 supply, hydraulic retention time, light intensity and medium salinity on biomass accumulation and glycerol production were investigated. The results indicated that the MPBR has high potential in economical and sustainable production for glycerol.

466442 Microalgal Fuels and Chemicals Production Using Kinetic Modeling and Scaled-up Experimental Studies

Abstract:

Microalgal Fuels and Chemicals Production using Kinetic Modeling and Scaled-up Experimental Studies Mesut Bekirogullari[a,b,c], Jon K. Pittman[c] and Constantinos Theodoropoulos[a,b,*]

[a]School of Chemical Engineering and Analytical Science, University of Manchester, M13 9PL, UK

[b]Centre for Process Integration, University of Manchester, M13 9PL, UK

[c]Faculty of Life Sciences, University of Manchester, M13 9PL, UK

Cultivation of microalgae for the production of biomass and lipid oils is a promising candidate in the field of renewable fuels [1,2]. The price of the microalgal biodiesel production is currently too high to compete with the fossil diesel production. Nevertheless, optimization and scaled-up control of the microalgae growth process can improve the competitiveness and sustainability of microalgal-derived biodiesel industry [3].

The aim of this work is the development of a detailed kinetic model which will be used in conjunction with in-house obtained experimental data to enable the optimization and the control of the microalgae growth in bench-scale batch systems. The proposed integrated framework will be employed to analyse the performance of microalgae growth, manifested by nutrient and environmental parameters such as substrate, nitrogen, light, temperature and pH. Successful control and optimization of such a system can lead to the utilisation of reduced excess/uncontrolled carbon, nutrient and water sources, and also to the production of high oil bodies.

A detailed ODE-based kinetic model is constructed here and it is used for the simultaneous simulation of the carbon substrate, the nitrogen, the light intensity and the pH in the system [4,5]. High fidelity experiments were conducted at a range of operating conditions to analyse the effect of substrate and nitrogen on the lipid oil production, and to complement the computational study in order to enable the estimation of system kinetic parameters of crucial importance. The constructed model is validated and it is utilised to define the optimal growth conditions, which lead to highest oil accumulation and biomass growth.

Furthermore, experiments were carried out in open raceway ponds under the obtained optimal growth conditions and they were used for kinetic model validation at a scale-up level. We believe that scaled-up process simulations using such a detailed modelling framework will enable the system industrial applicability and commercialization.

Keywords: microalgae, bench scale, raceway pond, multi-parameter quantification.

References

1. Driver, T., Bajhaiya, A. & Pitmann, J. K. 2014. Potential of Bioenergy Production from Microalgae. Current Sustainable/Renewable Energy Reports, 1, 94-103.
2. Y. Chisti, 2007; Biotechnology Advances, 25, 294-306.
3. Bechet, Q., Shilton, A. & Guieysse, B. 2013. Modeling the effects of light and temperature on algae growth: State of the art and critical assessment for productivity prediction during outdoor cultivation. Biotechnology Advances, 31, 1648-1663.
4. M. Bekirogullari, J. Pittman and C. Theodoropoulos, 2015; Computer Aided Chemical Engineering, 37, 2393-2398.
5. M. Bekirogullari, I.S. Fragkopoulos, J.K.Pittman and C. Theodoropoulos, 2016; Algal Research, *Submitted.*

* Corresponding author, *e-mail:* k.theodoropoulos@manchester.ac.uk

Tel: +44 (0)161 3064386, Fax: +44 (0)1613069321

465404 Tertiary Wastewater Treatment By *Chlorella vulgaris* in a Membrane Photobioreactor: Effects of Wastewater Composition and Light/Dark Cycle

Abstract:

Microalgae is one of the most promising and sustainable feedstock for the production of biofuels because of its high growth rate, efficient CO_2 fixation and high lipid content. These advantages had led to extensive research in the microalgae cultivation and processing. However, despite making significant progress at these fronts, production of biodiesel from microalgae remains an expensive process.

One approach to reduce the cost of microalgae cultivation is the use of urban wastewater for microalgae growth. Municipal wastewater can not only provide cheap source of organic carbon to microalgae but it also contains ample amount of nitrogen (N) and phosphorus (P) as well. The cost of microalgal biofuels can also be reduced by achieving high biomass concentration in the photobioreactor which will result in lower harvesting costs. A high biomass concentration can be achieved in continuous cultivation of microalgae in a membrane photobioreactor (MPBR), by operating it at high loading rates.

In this research, an MPBR was designed and operated for N and P removal from wastewater using *Chlorella vulgaris*. The effects of N/P ratio and light/dark cycle on MPBR performance and biomass accumulation were investigated. The MPBR achieved nearly complete N removal under all the operating conditions, although the removal efficiency/rates were lower when the MPBR was operated under alternating light/dark cycles. It was also observed that P removal improved at higher N loadings. The biomass composition was different at different wastewater composition, and higher amount of proteins were accumulation by microalgae, when N supply was increased. The results suggested that autotrophic microalgae can be used for effectively treating and recycling inorganic pollutants in tertiary wastewater.

452691 Modeling Algal Cultivation in the United Arab Emirates

Abstract:

Due to, first oil price volatility, and then climate change, Abu Dhabi has decided to become a worldwide leader in renewable sources of energy, sustainability, and carbon capture and sequestration (CCS). In 2006, Abu Dhabi launched the Masdar Initiative aimed at promoting the development and use of renewable sources of energy in Abu Dhabi [1]. This initiative includes the construction of Masdar City which will be a carbon-neutral city, the establishment of the Masdar Institute of Science and Technology, and the objective of increasing the share of renewables to at least 7% in Abu Dhabi's power generation by 2020 [1]. This has resulted in Abu Dhabi establishing a nuclear program for civilian use and the largest solar energy farm [2]. Abu Dhabi has also launched an ambitious CCS project that will have captured five million tons of CO_2 by 2012 [3], but there are concerns about the high cost of CCS. The main use of the sequestered carbon will be for Enhanced Oil Recovery (EOR), but Abu Dhabi is searching for other uses of this sequestered carbon. In addition, Abu Dhabi's city council data indicates that 82 million gallons of wastewater are dumped directly to the sea causing harmful algae blooms in the ocean.

These situations present an outstanding opportunity for algal cultivation processes because of their ability to fix atmospheric CO_2 and to remove nutrients from wastewater. Even more, biofuels and high value products can be obtained from algae reducing the cost of CCS. The United Arab Emirates (UAE) has a sizable potential for algal biofuels due to its vast desert expanses, long shorelines, year-round sunny days, availability of flue gas, seawater and wastewater, and native adapted algae species that tolerate salinity levels twice as high as those in the Gulf waters [2]. However, algal cultivation is challenging as it must be achieved in open ponds. Closed photobioreactors are not an option because they incur high capital and operating costs. Open pond systems have low biomass and lipids productivities because they are CO_2 limited and they are susceptible to invasion by undesirable microorganisms. In order to increase productivity of open pond systems, reliable and predictive process models are needed.

Here, we use the high-rate algal bacterial pond (HRAP) [4], [5] model and dynamic flux balance analysis (DFBA) [6], [7] to model algal cultivation in the UAE. The model and ideas presented in [8] are taken as a start point. In this paper microbial consortia are used for two purposes. First, they increase culture resilience by utilizing resources more efficiently compared to an algae monoculture, making pond invasion by undesirable species more difficult [9]. Second, symbiotic relationships can be established between microalgae and other microbes in which the microbes convert carbon sources into CO_2 and acetate, and algae converts the CO_2 and acetate into biomass. Therefore, algae ponds can become feasible in locations far away from flue gas sources.

Here, we present the results of different case studies concerning the cultivation of algae in Abu Dhabi. In particular, we use experimental data regarding environmental conditions, wastewater composition, and flue gas composition to test different cultivation configurations and make predictions of novel system behavior [10], [7], [11].

References

[1] D. Reiche, "Renewable Energy Policies in the Gulf countries: A case study of the carbon-neutral "Masdar City" in Abu Dhabi," *Energy Policy,* vol. 38, pp. 378-382, 2010.

[2] A. A. Jaradat, "The current potential of algae biofuels in the United Arab Emirates," *Biofuels,* vol. 4, no. 4, pp. 347-349, 2013.

[3] S. Nader, "Paths to a low-carbon economy - the Masdar example," *Energy Procedia,* vol. 1, pp. 3951-58, 2009.

[4] H. Buhr and S. Miller, "A Dynamic Model of the High-Rate Algal-Bacterial Wastewater Treatment Pond," *Water Res.,* vol. 17, pp. 29-37, 1983.

[5] A. Yang, "Modeling and Evaluation of CO2 Supply and Utilization in Algal Ponds," *Industrial & Engineering Chemistry Research,* vol. 50, pp. 11181-11192, 2011.

[6] A. Varma and B. Ø. Palsson, "Stoichiometric flux balance models quantitatively predict growth and metabolic by-product secretion in wild-type Escherichia coli W3110," *Applied and Environmental Microbiology,* vol. 60, no. 10, pp. 3724-3731, 1994.

[7] R. Mahadevan, J. Edwards and F. I. Doyle, "Dynamic flux balance analysis of diauxic growth in Escherichia coli.," *Biophysical Journal,* vol. 83, no. 3, pp. 1331-40, 2002.

[8] J. A. Gomez, K. Höffner and P. I. Barton, "From sugars to biodiesel using microalgae and yeast," *Green Chemistry,* vol. 18, no. 2, pp. 461-475, 2016.

[9] E. Kazamia, D. C. Aldridge and A. G. Smith, "Synthetic ecology - A way forward for sustainable algal biofuel production?," *Journal of Biotechnology,* vol. 162, pp. 163-169, 2012.

[10] J. D. Orth, I. Thiele and B. Ø. Palsson, "What is flux balance analysis?," *Nature Biotechnology,* vol. 28, pp. 245-248, 2010.

[11] K. Höffner and P. I. Barton, "Design of Microbial Consortia for Industrial Biotechnology," *Computer-Aided Chemical Engineering,* vol. 34, pp. 65-74, 2014.

462072 Kinetics of Coffee Oil Ethanolysis Catalyzed By an Immobilized Lipase Enzyme to Produce Biodiesel

Abstract:

As supplies of fossil fuels are depleted, the need for alternative sources of fuel becomes evident. Biodiesel is an attractive alternative to petroleum-based fuel; however, cost of the feedstock, mainly vegetable oils, makes biodiesel a costly product. In the search for alternative feedstocks, spent coffee grounds were revealed to be a largely available and economical source of vegetable oil. To design a process for biodiesel production, a kinetic rate model was necessary. Immobilized lipase was found to be an optimal catalyst for the ethanol transesterification of coffee. Tests were performed in batches, a differential packed bed reactor, and a recycle reactor. Reactor effluents were analyzed by GC-MS. Results indicated significant external mass transfer limitations and stability of the immobilized enzyme catalyst at temperatures less than or equal to 45°C. A modified Michaelis-Menten equation for two substrates provided a good fit for the reaction kinetics in the range of 30° to 45°C.

473126 Different Parameters Controlling the Biocrude Yield in Hydrothermal Liquefaction of Microalgae

Abstract:

Research into alternate sources of fuel or renewable energy reveal many promising ideas for alternative energy sources. One of these possibilities is converting biomass to biofuel. Under high pressure and temperature, microalgae can be polymerized to biofuel. There are some thermo-chemical processes available to convert the biomass like pyrolysis and gasification, but these processes require dry feedstock. Thus these processes encounter large energy loss by vaporizing the moisture. Hydrothermal Liquefaction (HTL), a high temperature, high-pressure polymerization process, can convert wet biomass, including water to biocrude oil, biochar, and nutrient rich water. The biocrude obtained from HTL are environmentally friendly and can be directly used as fuel or can be upgraded to transport fuel. Many researchers are working on increasing the efficiency of HTL. In the present study an attempt is made to optimize the HTL reactor for maximizing the biocrude yield. Batch reactions are carried out to correlate the biocrude yield with different parameters like the reaction temperature (300Â°C - 350Â°C), reaction time (30 mins â€“ 60 mins), initial solid concentration (5% - 20%) and also different species with different lipid contents. Experiments are also carried out to see the effect of different cultivating medium (wastewater and freshwater) on the crude yield after HTL. Experimental results show that different operating conditions can result in the biocrude yield to vary between 5% and 60%. These parameters are also related to the amount of heavier biocrude oil produced. This paper carefully examines and explains the relationships between the operating conditions and the biocrude yield to be able to optimize a HTL batch scale reactor for maximum crude yield.

451152 Synthesis and Characterization of Proto-Kerogen Via Simulated Diagenesis of *Litopenaeus Setiferus* and *Procambarus Clarkii* chitin

Abstract:

Petroleum formation begins with organic matter (OM). Chitin is a type of organic matter found in shellfish. Chitin is a long-chain polymer of an N-Acetyl Glucasamine, a derivative of glucose. Over geologic time and progressive burial, chitin and other OM undergo sequential catalytic modifications known as diagenesis, catagenesis, and metagenesis, resulting in the petroleum-forming intermediate known as kerogen. Catalytic activity is provided by aluminosilicate clays such as kaolin, feldspar, bentonite, and wollastonite. With progressive burial, temperatures eventually rise into the "oil window" of 100-200°C. Within the oil window, organic matter is gradually converted into complex intermediates including fulvic and humic acids, eventually forming kerogen.

For this experiment chitin is isolated from *Litopenaeus setiferus* and *Procambarus clarkii* by sequential treatment with 1M hydrochloric acid followed by 1M sodium hydroxide. Reactors are then mixed with various aluminosilicates and incubated in a temperature-controlled oven. Experimental conditions evaluated in this research include residence times ranging from 1 day to 1 month and reaction temperatures of 125°C, 155°C, and 175°C. The impact of pH on reaction products is also evaluated in this research. The products of the reaction are characterized using thermogravimetric analysis, which measures sample volatility with respect to temperature. Specific molecules produced in each reaction are then identified by NMR.

468959 The Impact of Treatment Conditions on Biomass Maturation and Conversion

Abstract:

Kerogen is the natural product of the maturation of organic matter. Over time and progressive burial, biomass undergoes sequential catalytic modifications known as diagenesis, catagenesis, and metagenesis. The aluminosilicate clays that co-deposit with organic matter provide catalytic function.

The types of biomass evaluated in this study include lignin, bagasse, yeast, chitin, and algae. Lignin was obtained by treating sugarcane bagasse with sulfuric acid to hydrolyze cellulose and hemicellulose. Chitin was isolated from crawfish and shrimp casings by sequential treatment with HCl and NaOH. *Chlorella Protothecoides* was grown phototrophically and *Saccharomyces cerevisiae* was grown heterotrophically. Biomass samples were then mixed with an aluminosilicate.

Biomass-aluminosilicate samples were loaded into isochoric reactors and incubated at elevated temperatures for the desired resident time. Different aspects of the reactions were changed to determine the various effects on pre-kerogen. The temperatures evaluated were 175°C, 150°C and 125°C. Aluminosilicate used were wollastonite, bentonite, and feldspar. Residence times ranged from 24 hours to 28 days.

The results of these variables will be presented with thermogravimetric analysis (TGA). TGA measures the volatility of a sample as temperature is increased. In this study, TGA was used to evaluate the impact of the treatment conditions outlined above on the organic products of simulated diagenesis. Characteristic volatility curves were used to determine the temperature ranges by which fractions of the biomass volatilize before and after maturation experiments.

471691 Ethanol Conversion to n-Butanol Via the Guerbet Chemistry: Role of Mixed Oxide Catalysts

Abstract:

In future ethanol is expected to be available in large quantities beyond the transportation blending needs. The excess ethanol can be converted over mixed oxide catalyst to butanol and higher alcohols via complex tandem reaction mechanism (Guerbet) which can be used as a chemical or fuel intermediate. The cascade nature of the Guerbet chemistry requires catalytic materials containing active sites with different functionalities at unique molecular distances. The hydrotalcite derived $MgO-Al_2O_3$ mixed oxide catalyst is identified as one of the efficient material to convert ethanol to n-butanol and higher alcohols. The catalyst preparation method, mixed oxide composition and the experimental conditions plays a role in directing the ethanol condensation chemistry. In this talk the conversion of ethanol to butanol over a multi-functional mixed oxide catalyst with respect to the catalyst synthesis and preparation, reaction conditions and the reaction mechanism will be discussed.

461389 Microwave-Assisted Pyrolysis of Coal-Biomass Mixturs UNDER Reactive Gases

Abstract:

Conversion of coal/biomass blends into higher value chemicals can either be achieved directly, by non-oxidative routes or indirectly, by oxidative routes. Usually, multiple chemical processes are involved in oxidative routes; first a gasification step is needed to produce syngas, followed by Fisher Tropsch reactions to produce chemicals. On the other hand, pyrolysis could be used as a single step process to convert these blends directly into chemicals. However, the main obstacle in conventional thermal pyrolysis is the formation of high molecular weight tars which is difficult to handle and require further cracking to be utilized. In this work, microwave-assisted pyrolysis was studied on samples of Mississippi lignite and PRB subB coals mixed with 25 wt% yellow pine wood sawdust. The microwave-assisted pyrolysis product distributions of the above blends were compared to those using conventional thermal pyrolysis. . Microwave energy was not only used for rapid heating but also to examine any possible kinetic enhancements of this process either under inert (N_2) or reactive hydrogen rich (H_2 and CH_4) gases. Our initial GC-MS results suggested that the oil product distributions were shifted to relatively lower molecular weight chemicals using microwave heating compared to conventional heating. It also suggested that the addition of hydrogenated gases increases the oil yield produced. The chars produced were characterized by different techniques, such as SEM and XRD to determine the morphology and crystallinity of these samples before and after pyrolysis, respectively. The porosity of fresh and char samples were determined using N_2-adsorption surface area measurements.

461151 Carbon Dioxide Reduction and Carbon Dioxide ReuseÂ in Advanced Biorefineries

Abstract:

An advanced biorefinery is effectively a polygeneration facility with integrated, efficient and flexible conversion of wastes, through a combination of physical, chemical, biochemical and thermochemical processes, into multiple products[1]. The biorefinery concept was analogous to the complex crude oil refineries adopting the process engineering principles applied in their designs, such as feedstock fractionation, multiple value-added productions, process flexibility and integration[1]. Enhancing the economic performance of biorefinery is the key challenge during the commercialisation stage. Energy products such as heat, electricity and biofuel are low value high volume products, while biochemicals and biomaterials (e.g. polymers) are high value low volume products. Biofuel or bioenergy production needs subsidies or economic incentives via supportive regulatory frameworks for feasibility. Hence, the aim of this study is to devise a strategy to realise an inherently economically competitive biorefinery system by added-value co-productions using stillage streams including carbon dioxide in offgases or exhaust gases, which can advocate a circular economy. Therefore, waste streams including those containing carbon dioxide from biofuel or bioenergy system must be reused to produce added-value products for achieving sustainability of biorefineries. Carbon dioxide reuse into biofuels, chemicals and biomaterials can be realised via three main routes: carbon dioxide reduction at cathode in microbial electrosysthesis system (MES)[2]; reusing carbon dioxide from exhaust gas of a thermochemical platform[3] and carbon dioxide capture by growing algae in algae biorefinery systems[1]. The case studies in this paper focus on the production of biopolymer, biochemical, biofuel by CO_2 reduction in MES; and biofuel production by trigeneration and Sabatier reactions. The economic performances using techno-economic modelling and environmental performances using life cycle assessment are conducted to result in integrated and advanced biorefineries that can enhance the three measures of sustainability. Producing biomaterial together with biofuels and bioenergy is the route for competitive biorefinery systems.

---

[1] Sadhukhan, J., Ng, K.S., Hernandez, E.M. 2014. *Biorefineries and Chemical Processes: Design, Integration and Sustainability Analysis.* Wiley.

[2] Sadhukhan, J., Lloyd, J., Scott, K., Premier, G.C., Yu, E., Curtis, T., Head, I. 2015. A Critical Review of Integration Analysis of Microbial Electrosynthesis (MES) Systems with Waste Biorefineries for the Production of Biofuel and Chemical from Reuse of CO_2, *Renewable & Sustainable Energy Reviews,* 56, 116-132.

[3] Ng, K.S., Zhang, N., Sadhukhan, J. 2013. Techno-economic analysis of polygeneration systems with carbon capture and storage and CO2 reuse, *Chemical Engineering Journal*, 219, 96-108.

463161 Hydrogen Storage and Carbon Dioxide Valorization through Reductive Calcination of Mineral Carbonates

Abstract:

Hydrogen storage and carbon dioxide valorization through reductive calcination of mineral carbonates

Georg Baldauf-Sommerbauer, Susanne Lux, Matthaeus Siebenhofer

Institute of Chemical Engineering and Environmental Technology, Graz University of Technology

The calcination of mineral carbonates is a widely used preparation reaction for metal oxides in industry (see Eq. 1). Carbon dioxide that has been fixed in mineral form is released, with the well-known detrimental effect on the global carbon balance and climate. This drawback of calcination can be overcome in reductive calcination. Carbon dioxide can be used as a means of storage for renewably produced hydrogen as methane and/or synthesis gas CO/H_2.

$MCO_3 = MO + CO_2$ (1)

The conversion of carbon monoxide and dioxide into methane is feasible for temperatures below 550 °C from a thermodynamic point of view. Due to kinetic hindrance, however, heterogeneous catalysis is needed to reach high yields. The Gibbs enthalpy for the reverse water gas shift reaction (RWGS: $CO_2 + H_2$ à $CO + H_2O$) decreases roughly linear from approximately +30 kJ mol^{-1} at 0 °C to approximately -6 kJ mol^{-1} at 1000 °C. The RWGS is also kinetically hindered and needs to be catalyzed for conversions near the thermodynamic equilibrium. If the carbon dioxide emitting step, namely calcination, is already performed in reductive hydrogen atmosphere, Eq. 2 can be formally derived for the direct production of carbon monoxide and methane from mineral carbonates. Calculations of

the Gibbs enthalpy revealed that methane and carbon monoxide production by reductive calcination of the industrially relevant carbonates of iron, magnesium, and other metal carbonates is possible.

$(a+b+c)MCO_3 + (b+4c)H_2 = (a+b+c)MO + aCO_2 + bCO + cCH_4 + (b+2c)H_2O$ (2)

Numerous experiments were conducted to elucidate the relationship between product gas composition, mineral carbonate, and process temperature and pressure. Experimental investigations were conducted using a tubular reactor setup with mineral iron (= siderite, 5-10 mm size fraction), magnesium (= magnesite, 5-8 mm size fraction), and magnesium/calcium (= magnesite/dolomite, 5-8 mm size fraction) carbonates as a fixed bed. The inlet gas composition was set to 90-100 vol.% hydrogen and 0-10 vol.% nitrogen at a total flow of 30 L h^{-1} at standard temperature and pressure.

Thermogravimetric experiments showed that reductive calcination proceeds at lower temperatures than calcination in inert nitrogen atmosphere. The minimum temperature needed for reductive calcination at isothermal conditions using the tubular reactor setup increases from siderite (~330 °C), through magnesite (~430 °C) to magnesite/dolomite (~475 °C).

Reductive calcination experiments with magnesite in hydrogen atmosphere (430-530 °C, 1-13 bar total pressure) revealed a maximum methane yield of ~37 % (475 °C, 13 bar). Methane formation is facilitated by increase of pressure at moderate temperatures below 475 °C. The yield of carbon monoxide ranges from 27 to 47 %. It increases from 430 to 505 °C and decreases at higher temperatures. Pressures above 1 bar slightly decrease carbon monoxide yield. When the magnesium carbonate fraction of a mixed magnesite (M) - dolomite (D) (M:D = 1.0:2.3) is calcined reductively (475-535 °C, 1-13 bar total pressure), only traces of methane could be detected in the product gas. Compared to pure magnesite, the carbon monoxide yield is enhanced. Up to 68 % of the total carbon can be released as carbon monoxide. Siderite ($FeCO_3$) differs from magnesite when the methane yield is compared. Experiments in the temperature region 330-380 °C at 1-11 bar pressure did show that the formation of methane is facilitated by low temperatures and elevated pressures. Optimization of the process parameters temperature, pressure, feed flow, size fraction, and bed length revealed that methane yields above 60 % without any additional catalyst are obtainable.

467141 Carbon Dioxide Sequestration Via Zn-Catalyzed Urea Conversion to Dimethyl Carbonate

Abstract:

Dimethyl carbonate (DMC) is emerging as an attractive electrolyte in advanced battery technologies and as a non-toxic solvent in several applications. The production of DMC via a number of pathways is currently under investigation; we report here the reaction of urea with methanol to form DMC over a Zn-based catalyst. Urea is a low-cost commodity, commonly used as nitrogen fertilizer, produced by the reaction of carbon dioxide with ammonia. The formation of DMC from urea thus represents a carbon sequestration route to a commodity chemical.

The primary pathway for urea to DMC involves two reactions through the intermediate methyl carbamate (MC). The first reaction of urea with methanol to MC is rapid and proceeds without catalyst; DMC formation is therefore kinetically limited by the Zn-catalyzed second reaction involving MC and methanol. Prior literature suggests that Zn interacts with the amino groups in urea to form a complex that is the active species in the reaction. In this work, we show that the nature of this complex changes over the course of reaction, and that a substantial fraction of the Zn complex exists as a dissolved homogeneous catalyst. The yield of DMC is further limited in reaction by the formation of byproducts including N-methyl-methyl carbamate (NMMC), formed by the methylation of MC with DMC. A kinetic model of the overall reaction network allows prediction of optimum conditions for DMC formation, and clarifies the role of the Zn catalyst in producing DMC from urea.

462623 The Stability Performance of Synthetic Ca-Based Sorbents Doped By Zirconium-Based Ceramics in Cyclic CO_2 capture Operations

Abstract:

Calcuim (Ca) based sorbents as an option for carbon dioxide (CO_2) capture process at high temperatures have displayed considerable activity loss during extended operation, due to sintering phenomena. Refractory zirconium-based ceramics have shown excellent performance as thermal barrier coatings in many high-tech applications; and, the recent doping of zirconia in Ca-based sorbents has exhibited improvement in the stability in cyclic CO_2 capture operations. In this work, sorbents were doped, at loadings of 10 wt%, with a zirconium-based ceramics, such as alkaline earth metal zirconate ($MZrO_3$, M= Mg, Ca and Ba) and aluminum zirconate ($Al_2Zr_3O_9$), which have the chemical structure of perovskite, and rare earth metal zirconates ($Ln_2Zr_2O_7$ and $LnZrO_4$, Ln = La, Ce and Sm), with the chemical structure of pyrochlore. These materials were prepared using the wet co-precipitation method. The results indicated significant sorbent stability and capacity improvement for mixed metal oxide zirconia-stabilized sorbents over those of pure CaO and natural Cadomin. Among the tested sorbents, the $CaZrO_3$-doped sorbents showed the best performance and remarkable stability due to their high resistance to sintering. The sorbent demonstrated a CO_2 capture capacity of 12 mole CO_2 / kg sorbent, with an activity loss of 19.6% through 30 cycles of carbonation at 675°C and calcination at 850°C. Among the rare earth metal zirconate ceramics, samarium zirconate ($Sm_2Zr_2O_7$) presented the best stability with an activity loss of 23.4 % loss, but at a lower uptake of 7.2 mole CO_2 / kg sorbent at the same conditions. In comparison, Cadomin (a natural calcium carbonate) at the same conditions resulted in a CO_2 capture capacity of 13 mole CO_2 / kg sorbent and an activity loss of 69%. Surface area measurements indicated that addition of the ceramics dopants to the Ca-based sorbents greatly inï¬‚uenced both the surface area and pore volume of the sorbents. X-ray diffraction results showed that the improved resistance of the sorbent is due to the formation of

structures such perovskite-type $CaZrO_3$, which provided a stable framework to the sorbent during the carbonation-calcination process.

464577 Improvement of Sn Electrocatalytic Reaction for CO_2 Conversion into Formate

Abstract:

Carbon capture and utilization is an interesting strategy to contribute to reduce CO_2 emissions and mitigate climate change, since it may involve the conversion of CO_2 into added-value products, such as liquid fuels. Among the different possibilities for CO_2 conversion, its electrochemical reduction appears as an interesting option because it could be used to store the intermittent and unpredictable energy from renewable sources, such solar or wind power, in form of chemical products. One of the most interesting products obtained from CO_2 electroreduction is formate. This product is used as raw material in several industrial processes with growing demand and it has also been proposed as a suitable fuel for fuel-cells and as a renewable energy hydrogen carrier.

One of the electrocatalysts that allow CO_2 conversion into formate on aqueous solutions is Sn [1]. This metal could be used in electrodes with different configurations, such as plate electrodes or in the form of particles of different sizes deposited on porous supports. The aim of this work is to analyze the performance of Sn electrodes with different configuration and to remark the improvements achieved using Sn nanoparticles in gas diffusion electrodes (GDEs) with respect to our previous studies, taking account the faradaic efficiency, formate concentration in the product, rate of formate production and the current density of the process.

Starting with conventional Sn plate electrodes for CO_2 electrochemical reduction, a maximum formate rate of $4.4 \cdot 10^{-4}$ mol m^{-2} s^{-1} was achieved with a 70% faradaic efficiency and formate concentration of 140 ppm at current densities around 12 mA cm^{-2} [2]. This low performance is due to the small electrocatalytic surface of the plate electrode and the mass transport limitations. To increase the specific surface and mass transport on the electrode, Sn particles were deposited on carbon paper to manufacture particulated electrodes. In these electrodes, gaseous CO_2 is directly fed to the cell and flows through the electrode. Sn commercial particles of 150 μm [3] achieved a formate concentration of 1.3 g L^{-1} with 70% faradaic efficiency working at 40 mA cm^{-2}. When the size of particle was reduced to 150 nm [4], it was possible to keep a 70% faradaic efficiency with a current density of 90 mA cm^{-2}, increasing the concentration to 1.5 g L^{-1} with a formate rate of $3.21 \cdot 10^{-3}$ mol m^{-2} s^{-1}.

Recent research efforts are focused on the development of Sn Gas Diffusion Electrodes (Sn-GDEs) which allow working at high current density with high faradaic efficiencies . Sn-GDE electrodes consist of a carbon paper as carbonaceous support, a microporous layer and the catalytic layer formed by the Sn carbon supported nanoparticles. Using GDEs we were able to achieve a 70% faradaic efficiency to formate at a current density of 150 mA cm^{-2}, obtaining a formate concentration of 2.5 g L^{-1}, which outperforms our results obtained with particulated electrodes. When the current density is increased and with a low catholyte flow, concentrations over 16 g L^{-1} were achieved, but at the expense of a 42% faradaic efficiency. However, despite the promising results obtained with Sn-GDEs, more research efforts are still required in terms of stability of the electrodes and optimization of the process.

[1] D. Kopljar, A. Inan, P. Vindayer, N. Wagner, E. Klemm., J. Appl. Electrochem (2014) 44, 1107-1116

[2] Alvarez-Guerra M, Del Castillo A, Irabien A., Chem. Eng. Res. Des. (2014) 92, 692â€“701

[3] A. Del Castillo, M. Alvarez-Guerra, A. Irabien, AIChE J., (2014) 60, 3557-3564.

[4] A. Del Castillo, M. Alvarez-Guerra, J. Solla-GullÃ³n, A. SÃ¡ez, V. Montiel, A. Irabien., Appl. Energy (2015) 157, 165-173

Innovative CO_2 Utilization by Carbonation Curing of Lightweight Concrete Made with Portland Limestone Cement
Paper Number: 451499

Hilal El-Hassan[a,*] and Yixin Shao[b]

[a] *Department of Civil and Environmental Engineering, United Arab Emirates University, Al Ain, United Arab Emirates, P.O. Box 15551*
[b] *Department of Civil Engineering and Applied Mechanics, McGill University, 817 Sherbrooke Street West, Montreal, Quebec, Canada, H3A 2K6*

Abstract
The atmospheric concentration of carbon dioxide gas has immensely increased with escalation in the world's greenhouse gas (GHG) emissions. Through CO_2 recovery, storage and utilization, emission induced global climate changes can be alleviated. The use of Portland limestone cement (PLC) as binder in lightweight concrete can effectively improve net gains in carbon emission reductions by partial replacement of cement with limestone. To further mitigate CO_2 discharge, carbonation curing was examined as a potential curing mechanism of lightweight concrete made with PLC. Slab samples were cast and initially cured at 25°C and 50% relative humidity for durations up to 18 hours. Carbonation was then carried out for 4 hours in a sealed chamber at an absolute pressure of 1 bar. The degree of reactivity was assessed by the CO_2 uptake based on PLC content and excluding preexisting limestone. The resulting uptake of fresh lightweight concretes without initial curing and after 18 hours of initial curing reached 6% and 18% respectively, corresponding to 13% and 39% reaction efficiency. The effect of carbonation curing on the mechanical properties, microstructure and reaction products of concrete was then investigated. A comparison between the optimum carbonated sample and hydrated reference was performed. While both concretes showed similar early-age compressive strength, late strength of carbonated sample was only comparable if water compensation through surface spray was performed. Hydration products were characterized as well crystalline calcium carbonates with the transformation of carbonate polymorphs into more stable calcite. In the process, PLC could replace OPC in carbonation curing of lightweight concrete while maintaining similar carbon sequestration potential.

Keywords: Carbon Capture & Storage, Materials and Sustainability & Environment.
[*] Corresponding author. Tel: +971(3)713-5156; Fax: +971(3)713-4343
E-mail address: helhassan@uaeu.ac.ae (Hilal El-Hassan)

1. Introduction

Portland limestone cement (PLC) is produced by adding limestone and calcium sulfate to Portland cement clinker in the grinding phase. It is different than ordinary Portland cement (OPC) in that cement replacement by limestone may reach 20%, thus reducing the carbon footprint associated with cement production. Such a replacement may also affect the compressive strength and hydration of concrete. Early research believed that fine limestone particles acted as nucleation sites, thus increasing the rate of hydration of the calcium silicates at early age and possibly improving the distribution of hydrates [1-2]. Ramachandran had reported that the hydration rate of C_3S also increased with an increase in the fineness and amount of $CaCO_3$ [3]. The formation of hydration products as ettringite had also been accelerated in the presence of $CaCO_3$, while C-S-H incorporated significant amount of calcium carbonates into its structure and formed calcium silicocarbonated hydrates [3]. Additionally, $CaCO_3$ will react chemically with aluminate phases to form carboaluminate phases [4]. Finally, because limestone is softer than clinker, when interground, it will achieve a finer particle size, thus refining particle packing and improving particle size distribution [5].

To further reduce the carbon footprint from cement production, CO_2 could be sequestered in PLC concrete through carbonation curing. Other work presented the physical performance and reaction products of carbonated OPC concrete [6, 7]. To investigate the effect of carbonation process parameters on the microstructure and characteristics of the reaction products of PLC concrete, fresh and air cured concretes are carbonated in a static carbonation setup. Three batches were selected based on the previous study: hydration reference without initial air curing (0a), carbonated concrete without water compensation (18a+4c), and carbonated concrete with water compensation (18+4c+sp). The reaction products of the concretes are characterized utilizing X-Ray Diffraction (XRD) and Thermogravimetry/Differential Thermogravimetry (TG/DTG). The potential replacement of OPC with PLC in manufacturing lightweight concrete by carbonation curing could allow the sequestration of CO_2 while also using lesser cement in the production.

2. Experimental Investigations

2.1. PLC concrete sample preparation and curing

Rectangular concrete samples 127 mm long, 76 mm wide, and 38 mm thick were cast for the purpose of this study. The lightweight concrete mix consisted of PLC as binder and saturated-surface dry expanded slag aggregates with 5% water content. Mixture proportions are shown in Table 1. The chemical composition of PLC was determined by infrared carbon analyzer and presented in Table 2. The limestone content was measured as 15.9%, equivalent to a CO_2 content of 7.0%. The raw materials were mixed in a pan mixer and samples were compact formed using a vibrating hammer. PLC concrete was then demolded right after casting. Three curing schemes were investigated: (1) Sealed hydration in a plastic bag served as reference (0a) without water loss; (2) Initial air curing of 18 hours in an environmental chamber of 50% relative humidity (RH) and 25°C followed by 4-hour carbonation and subsequent hydration (18a+4c); (3) Initial air curing of 18 hours in an environmental chamber of 50% relative humidity (RH) and 25°C, followed by 4-hour carbonation, water compensation and subsequent hydration (18a+4c+sp). Water compensation through spray was introduced immediately after carbonation to restore all the curing-induced water loss. Subsequent hydration was carried out in a sealed bag at a relative humidity of 80±5% and room temperature (24±1°C) until 24 hours and 28 days for compressive strength tests following ASTM C140 [8].

The static carbonation setup employed in this work is schematized in Fig. 1. Within a few seconds, the pressure is reduced to 0.7 bars below atmospheric prior to carbonation by means of a low-capacity vacuum. A heater installed at the inlet brings up the 99% pure carbon dioxide gas to room temperature as it flows into the sealed chamber. A pressure regulator is connected to the system to retain the pressure at 1 bar. The whole system is placed on a digital scale to monitor the mass increase due to the carbonation reaction.

Table 1: Mixture proportion of PLC concretes

	Slab (g)	Mass (kg/m^3)	Percent (m.-%)
Portland Limestone Cement	88	241	13
Mixing Water	35	96	5
SSD Expanded Slag	554	1502	82
Concrete Sample	677	1839	100

Table 2: Chemical composition of PLC and OPC

Cement	Constituent (%)								
	CaO	SiO_2	Al_2O_3	Fe_2O_3	MgO	Na_2O	K_2O	SO_3	CO_2
PLC	59.8	21.6	4.5	2.7	1.7	0.8	-	3.2	7.0

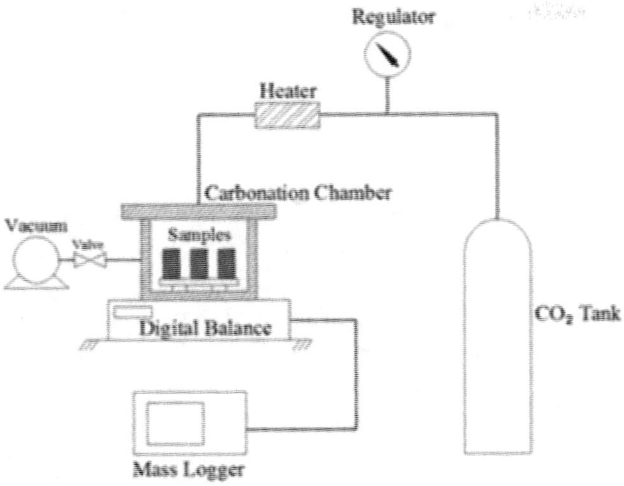

Fig. 1. Schematic of carbonation setup

2.2. Measurement of CO_2 uptake in PLC concrete

Three methods were used to quantify the carbon uptake of carbonated PLC concrete: mass gain, mass curve, and furnace decomposition. The mass gain method compares the sample's mass before and after carbonation in accordance with Equation 1. As the chemical reaction is exothermic, the increase in temperature induced water evaporation. Such water was collected and added to the final mass to ensure a closed system.

CO_2 uptake (%) = (Final mass + Mass of water loss – Initial mass) / (Mass of cement) (1)

The mass curve method was performed by placing the carbonation setup on a digital balance, which was zeroed after vacuuming the chamber. A mass curve was recorded as mass versus time until the end of the process at which CO_2 was released and the residual mass, M, was measured. The system was calibrated by repeating the tests using CO_2-insensitive styrofoam samples of the same volume to obtain a second residual mass, m. The difference between M and m represented the CO_2 uptake by concrete (Equation 2). Data collected by mass gain and mass curve methods are two simultaneous measurements from the same process and therefore should be comparable. They are independent from any CO_2 content existing before carbonation.

CO_2 uptake (%) = $(M - m)$ / (Mass of cement) (2)

An independent furnace decomposition test was employed to serve as a third check. A furnace of maximum temperature of 1100ºC was employed to test large concrete samples with mass range of 35-70 g. PLC concrete samples were ignited up to 105°C, 470°C, and 950°C to quantitatively measure the evaporable water, bound water in hydration products, and carbon dioxide in carbonates respectively [9].

2.3. Microstructure evaluation

XRD and TG/DTG were employed to characterize the microstructure of the concrete specimens. A 5-mm thick concrete surface layer were extracted using a hammer and sieved to pass a 125-μm sieve to remove coarse aggregates. No attempts were made to separate the fine aggregates from the cement. Therefore, powder samples were a mix of paste and fine aggregates. XRD and TG analysis was performed on the same concrete powder to determine the reaction products.

At first, the concrete powder was analyzed using XRD. It was performed using a Philips PW1710 Powder Diffractometer (Cu, Kα radiation, X'celerator proportional detector, scan interval 10-100°, 0.02°, and 0.5 seconds per step) to relatively quantify the amount of calcium hydroxide, calcium silicates, and calcium carbonates in carbonated and hydrated samples.

The same concrete powder to that used in XRD analysis was used in a thermogravimetry (TG/DTG) test. TG was utilized to estimate the respective contents of C-S-H and ettringite (105-420°C), calcium hydroxide (420-470°C), and calcium carbonate (470-950°C) [9]. A thermal analyzer (NETZSCH, TG 449 F3 Jupiter) with a resolution of 0.01 mg was employed with a sample mass of 120 mg for each test. The TG/DTG curves were thus obtained by plotting the mass loss between 25°C and 950°C at a heating rate of 10°C/min.

3. Experimental Results and Discussion

3.1. CO_2 uptake of PLC concrete

The CO_2 uptake and content calculated based on the mass gain, mass curve, furnace analysis, and TG are shown in Table 3. The first two methods, which are insensitive to the initial carbon content, were used to determine the carbon uptake (% by mass) of the 4-hour carbonated concrete. The results show a consistency among non-sprayed and sprayed concrete at 18.3%. In order to determine the initial carbon content at the concrete surface, the furnace analysis method was employed. The hydrated reference (0a) resulted in CO_2 content of 6.68% initially present prior carbonation. The equivalent limestone content is 15.20%. However, after 4-hour carbonation curing, the carbon content was measured as 26.14%.

The cement content of concrete powder was determined using past experimental work and equations [6]. All reaction products are expressed in terms of dry cement content by mass. The cement contents (CC) of different concrete batches served as correction factors and were on the range of 0.25. The TG analysis of the hydrated reference resulted in approximately 7% carbon content. The obtained carbonates are to be deducted from those of the carbonated samples to obtain the CO_2 uptake. The average carbon uptake is then calculated based on that of the four methods and is summarized in the last column of Table 3.

Table 3: CO_2 content and uptake (%) in PLC concretes

	CO_2 Uptake		CO_2 Content		CO_2 Uptake
Sample	Mass gain	Mass curve	Furnace	TG 28-day	Average
0a (Ref)	n/a	n/a	6.68	6.62	0.00
18a + 4c	18.34 ± 1.6	18.23	26.14	26.06	18.87
18a + 4c + sp	18.29 ± 1.7	18.09	26.92	26.54	19.14

3.2. Compressive strength of PLC concrete

Compressive strength results for 1-day and 28-day concretes are shown in Fig. 2. The 24-hour duration comprises initial air curing followed by 4-hour carbonation and subsequent hydration to 24 hours. For instance, sample 18a+4c underwent 18 hours of air curing followed by 4 hours of carbonation and 2 hours of subsequent hydration. In comparison to hydrated counterparts, carbonated concrete exhibited 13.4±0.9% higher strength after 1 day. It was obvious that carbonation was an accelerated hydration process. Even though only part of the lost water was restored, within 2 hours of subsequent hydration, water compensation improved the compressive strength by 29.5±2.1% compared to the non-sprayed carbonated concrete.

After 28 days, the carbonated concrete (18a+4c) exhibited lower compressive strength than the hydrated reference owing to the water loss in early air curing. Water spray after carbonation was therefore essential to restore the curing-induced water loss. Hence, the strength increased by 11.7±0.8% after water compensation. The results indicate that carbonation followed by water compensation is beneficial. While early carbonation does not hinder subsequent hydration, the water loss during early air curing and carbonation curing reduces the degree of hydration. With water compensation through surface spray, it is possible to make carbonation strength comparable to hydration counterpart.

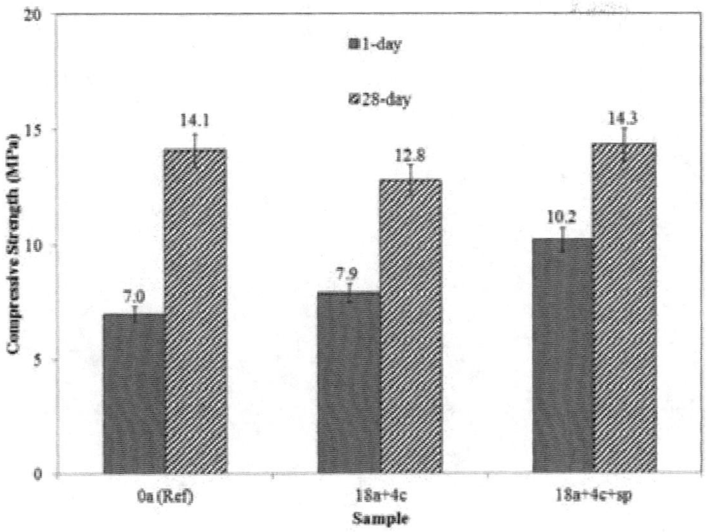

Fig. 2. Compressive strength of PLC concretes

3.2. Microstructure analysis of PLC concrete

Fig. 3 shows the 1-day XRD patterns for carbonated and hydrated PLC concretes. Dominant and strong peaks of calcium carbonate polymorphs including calcite, vaterite, and aragonite are present throughout carbonated samples. Higher carbonate peaks at 22°, 29.4° and 43° (2θ) are detected in the carbonated batches in comparison to the hydrated reference (0a). The coexistence of hydration and carbonation products is noticed in the carbonated samples. Calcium hydroxide is distinctively identified at 18° in hydrated samples. However, it is conclusive that calcium hydroxide is consumed given the absence of this peak in the carbonated PLC concrete. In addition, anhydrous calcium silicates are consumed, evidenced by the decrease in peak intensity at 34.5° and 41.5°, indicating the existence of C_3S. Ettringite has also been consumed verified by the lack of the 52.5° peak after carbonation.

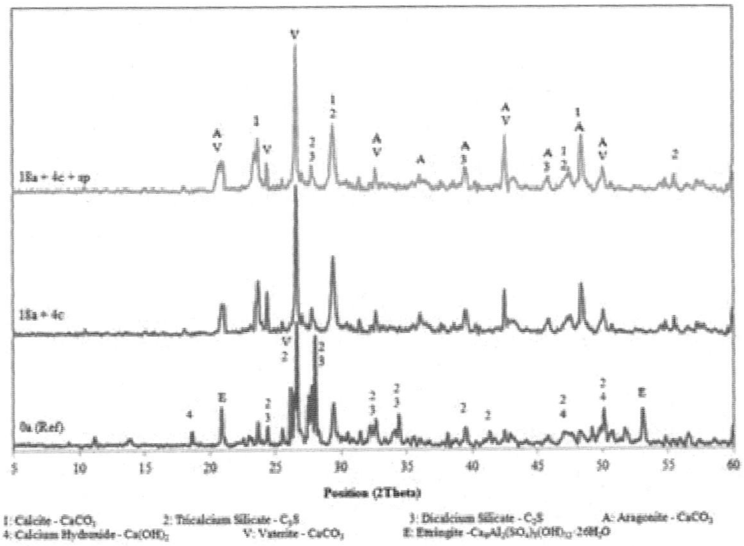

Fig. 3. XRD patterns of PLC concretes after 1 day

XRD patterns of the PLC concretes after 28-day subsequent hydration are shown in Fig. 4. The amount of vaterite and aragonite, shown at 21.5° and 43°, have decreased with reference to the patterns obtained after 1 day. While aragonite is suggested to form when the sample is allowed to dry out as in the case of 18a+4c [10], vaterite forms due to the rapid carbonation reaction and high heat release [10-11]. The hydrated reference does not differ between 1 and 28 days other than higher calcium hydroxide peaks with 27 more days of hydration. The 15% added limestone in the cement was detected as calcite peaks at 26.8° and 29.4° in the hydrated reference.

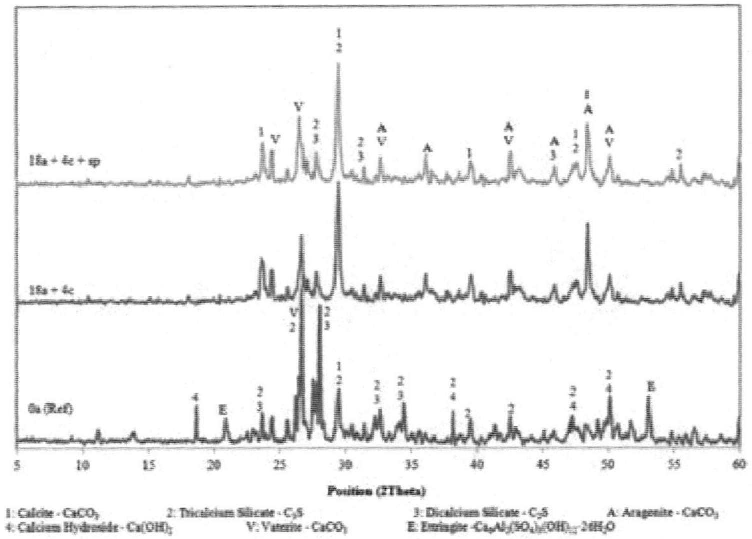

Fig. 4. XRD patterns of PLC concretes after 28 days

According to Ramachandran and Beaudoin (2001), the mass loss along thermogravimetric curves can be classified into 6 categories: Below 105°C, the mass loss is associated with the evaporable water and possibly the poorly formed C-S-H. The water loss between 105°C and 200°C represents the bound water in low temperature C-S-H and ettringite, while the loss in the range of 200°C-420°C is associated to the bound water in well-formed hydration products including C-S-H and C-A-H. Bound water in calcium hydroxide (CH) is decomposed between 420°C and 470°C [14]. While CO_2 in poorly crystalline calcium carbonates designated by aragonite and vaterite are decomposed between 470°C and 680°C, that in well crystalline calcite lies in the range of 680°C-950°C [9,11-12].

Based on TG/DTG analysis, detailed mass loss is presented in Fig. 5. Following the previously mentioned mass loss distribution, it can be noted that the hydrated reference had the lowest amounts of C-S-H and C-A-H and exceeded by the carbonated counterparts. As carbonation is an accelerated hydration process, the anhydrous calcium silicates in PLC were consumed to form C-S-H following 4-hour carbonation [14]. This explains the increase in C-S-H and C-A-H contents after carbonation. Further formation of C-S-H was identified in the sprayed carbonated sample (18a+4c+sp) after two hours of subsequent hydration. CH formation was a direct indication of the degree of hydration of any sample. The hydrated reference formed the highest amount of CH associated with no water loss during 1-day curing. The sprayed carbonated sample (18a+4c+sp) showed a comparable amount and was followed by its drier counterpart (18a+4c). Even with 0.8% loss in the range of 420-470°C, it is possible that CH was not detected by XRD possibly due to the presence of amorphous forms of CH [15]. Evidently, a sufficient water supply promoted the hydration of unreacted calcium silicates in the carbonated PLC concrete. Beyond 470°C, the hydrated reference exhibited some loss due to decarbonation of pre-existing carbonates. The respective CO_2 loss of carbonated concretes represented the CO_2 content, and included the added limestone and produced carbonates. The mass loss between 470-680°C, representing the carbonates produced from carbonation of C-S-H, showed the least thermal stability, suggesting poorly crystalline calcium carbonate designated by aragonite and vaterite [9,12,15]. The carbonates with better crystallinity, such as calcite, required higher temperatures to decompose and were lost above 680°C [10].

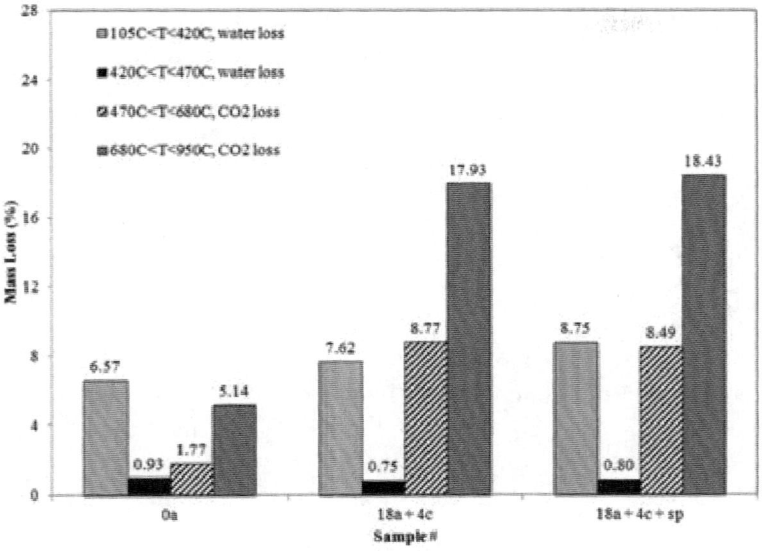

Fig. 5. Mass loss due to dehydration and decarbonation in 1-day PLC concretes

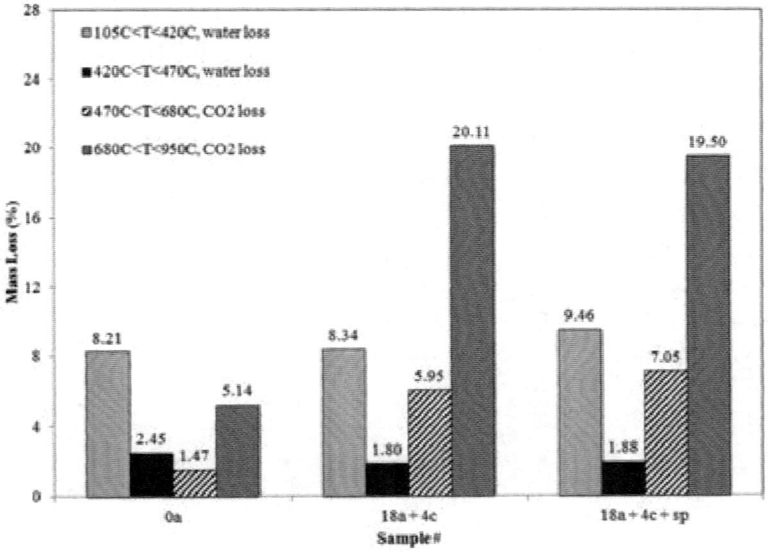

Fig. 6. Mass loss due to dehydration and decarbonation in 28-day PLC concretes

Hydration and carbonation products of 28-day PLC concretes are summarized in Fig. 6. The major change between 1 and 28 days was the hydrated reference concrete, in which increase in mass loss between 105-470ºC was indicative of increase in hydration products. Sealed curing conditions in hydration reference prevented water evaporation, which is an essential factor for subsequent hydration. The increase in compressive strength (Fig. 2) is associated with the increase in C-S-H and C-A-H content. Similar behavior was reported in the two carbonated concretes with higher losses attributed to 28-day subsequent hydration. It was suggestive that carbonation of PLC may slightly hinder the hydration of silicates and aluminates, thus affecting the development of compressive strength.

The percentage of poorly crystalline and well crystalline carbonates were calculated in order to study the stability of carbonates throughout subsequent hydration. Table 6 summarizes the poorly crystalline (470-680ºC) and well crystalline (680-950ºC) proportions of carbonation-induced carbonates in each batch. After 1 day, 33-35% of the total 4-hour reaction carbonates were poorly crystalline, while 65-67% were well crystalline. After 28 days, the percent of well crystalline increased to 72-77%. It was conclusive that unstable carbonates (aragonite and vaterite) transformed into more stable polymorphs (calcite) in PLC concretes.

Table 6: Percentage of carbonation-induced poorly and well crystalline carbonates in PLC concretes

	Poorly Crystalline		Well Crystalline	
Sample	1-day	28-day	1-day	28-day
18a+4c	35.4	23.0	64.6	77.0
18a + 4c + sp	33.6	28.0	66.4	72.0

4. Conclusions

The carbonation curing of PLC concrete was investigated. Its effect on the compressive strength and microstructure of PLC concrete was examined. Based on the findings of this study, the following conclusions may be drawn:

1. In the presence of initial air curing, a CO_2 uptake of about 19% in terms of cement mass was achieved in a lightweight PLC concrete with expanded slag aggregates. It was corresponding to a carbonation degree of 38%.
2. Compressive strength of PLC concretes varied depending on curing method. Carbonated samples exhibited higher strength at early age in comparison to hydrated counterparts. After 28-day subsequent hydration, carbonated concretes could only compare to those cured using hydration when water compensation through surface spray was employed.
3. Carbonation consumed calcium hydroxide, calcium silicates, and ettringite to produce calcium carbonates in its three polymorphs. The XRD peaks showed evidence of this phenomenon. Even though a mass loss between 420-470ºC was detected in the carbonated PLC concrete, calcium hydroxide was not identified by XRD possibly due to its amorphous nature.
4. The carbonation-induced calcium carbonates are distinguished between poorly and well crystalline. For 18-hour air cured and 4-hour carbonated PLC concrete, 33-35% of the total carbonation-induced carbonates were poorly crystalline throughout after 1 day, while 65-67% remained well crystalline. After 28 days, 72-77% of carbonates became well crystalline. It was conclusive that phase transformation of carbonates occurred in PLC concrete during subsequent hydration.

5. Acknowledgements

The authors gratefully acknowledge the support by the Natural Science and Engineering Research Council (NSERC) of Canada, Canadian Concrete Masonry Producers Association (CCMPA), and Lafarge Canada.

6. References

[1] Soroka I, Setter N. The Effect of Fillers on Strength of Cement Mortars. Cem. and Concr. Res. 1977; 7(4): 449-456.

[2] Bonavetti V, Donza H, Menéndez G, Cabrera O, Irassar, E.F. Limestone Filler Cement in Low w/c Concrete: A Rational Use of Energy. Cem. and Concr. Res. 2003; 33(6): 865-871.

[3] Ramachandran V, Zhang C. Influence of $CaCO_3$ on hydration and microstructural characteristics of tricalcium silicate. Cemento 1986; 83(3): 129-152.

[4] Matschei T, Lothenbach B, Glasser F.P. The Role of Calcium Carbonate in Cement Hydration. Cem. and Concr. Res. 2007; 37(4): 551-558.

[5] Tsivilis S, Chaniotakis E, Kakali G, and Batis G. An analysis of the properties of Portland limestone cements and concretes. Cem. and Concr. Composites 2002; 24(3): 371-378.

[6] El-Hassan H, Shao Y, Ghouleh Z. Reaction Products in Carbonation Cured Lightweight Concrete. Am. Soc. of Civ. Eng. Mater. J. 2013; 25(6): 799-809.

[7] El-Hassan H, Shao Y, Ghouleh Z. Effect of initial curing on carbonation of lightweight concrete masonry units. Am. Concr. Inst. Mater. J. 2013; 110(4): 441-450.
[8] ASTM C140. Standard Test Methods for Sampling and Testing Concrete Masonry Units and Related Units. ASTM International 2004.
[9] Ramachandran V. S, Beaudoin J. J. Handbook of Analytical Techniques in Concr. Sci. and Technol., William Andre Publishing/Noyes 2001.
[10] Goodbrake C. J. Reaction of beta-dicalcium silicate and tricalcium silicate with carbon dioxide and water. J. of the Am. Ceram. Soc. 1979; pp. 168-171.
[11] Bukowski J. M. Reactivity and strength development of CO_2 activated non-hydraulic calcium silicates. Cem. and Concr. Res. 1979; 9: 57-68.
[12] Goto S. Calcium silicate carbonation products. J. of the Am. Ceram. Soc. 1995; pp. 2867-2872.
[13] Li D, Shi, H, Deng J, Xu Y. Study on the thermal decomposition kinetics of nano-sized calcium carbonates. J. of Zhejiang University 2003; 4(3): 363-368.
[14] Chang C, Chen J. The experimental investigation of concrete carbonation depth. Cem. and Concr. Res. 2006; 36: 1760-1767.
[15] Young J. F, Berger R. L, Breese J. Accelerated Curing of Compacted Calcium Silicate Mortars on Exposure to CO_2. J. of the Am. Ceram. Soc. 1974; 57(9): 394-397.

473956 Surface Behaviors of Electrodeposited Tin Oxide on Selected Substrates and Their Effectiveness in CO_2 Electro-Reduction

Abstract:

Carbon dioxide (CO_2) effects on greener energy remain the topic of discussion in the need to direct new ways to sustainability. Electro-sequestration of CO_2 is being pursued as a promising route leading to many studies in this area. Findings have shown the effectiveness of electrocatalytic conversion of CO_2 to useful chemicals. However, high over-potentials to produce efficient current densities, coupled with a vast range of faradaic efficiencies, and issues on stability and longevity have been the major drawbacks. Post transition metal oxides have shown promise to CO_2 reduction, among which tin oxide (SnO_2) is the most cost effective and least toxic. But the reported range of current density, and faradaic efficiency using a SnO_2 catalyst is sporadic. In this work, the reduction of CO_2 is investigated by focusing on electrode materials, specifically, on surface behaviors of the anode structures. Since SnO_2 has shown great potential for CO_2 reduction, our aim is to investigate this material further. Thus, the objective is to study the cathodic electrodeposition characteristics of some notable substrates- glassy carbon, indium tin oxide, tin, and their alloys- as they interact with SnO_2 catalyst. The result of our study and the effectiveness of these materials to reduce CO_2 will be presented.

473500 Thermochemical Characterization of Sugarcane Bagasse for the Production of Combustibles Via Pyrolysis Using N2 As Carrier Gas

Abstract:

Global warming is a large-scale issue that has led humanity to seek for renewable energy sources. Biomass thermal gasification is one the new technologies to produce clean fuels using biomass as feedstock. The resultant product from this process is a renewable neutral carbon fuel. Temperature parametric results obtained from a sugarcane-bagasse (CH1.581O0.671N0.00306S0.00117) pyrolysis experimental study carried out at pyrolysis temperatures ranging between 400 C and 900 C and at a heating rate of 4.30 C/min ± 7.8%, are discussed. The experiments were developed in a static batch reactor with nitrogen as carrier gas. Sugarcane bagasse samples were thermo-chemically characterized in order to determine thermodynamic properties such as enthalpy of formation, high heating value, empirical formulas of biomass, volatile matter, etc. The chemical formula of the pyrolysis oil obtained was estimated using gas composition, given by a gas analyzer, and atom balance on components. The effect of temperature on the production of liquid and gaseous fuels are presented along with the HHV of the syngas produced. The following results were obtained: chemical empirical formulas for biomass and volatile matter were CH1.581O0.671N0.00306S0.00117 and CH1.807O0.767N0.00349S0.00133 respectively; high heating values (DAF basis) of biomass and volatile matter 18,925.08 kJ/kg and 17,472.77 kJ/kg, respectively; average chemical empirical formula of liquid fuel CH1.942O0.484N0.005S0.002.

464366 The Effect of Torrefaction Pretreatment, Temperature, and Lignin Content of Hybrid Poplar on the Quality of Fast Pyrolysis Bio-Oil

Abstract:

Introduction: Fast pyrolysis of woody biomass has been identified as a potential means for the production of advanced transportation fuels. During fast pyrolysis the three main components of biomass (cellulose, hemicellulose, and lignin) thermochemically breakdown to produce bio-oil. Several studies have isolated these individual components and looked at their bio-oil distributions. However the interactions between these three main components and how variations in feedstock composition affect product distribution are not well understood. The purpose of this work is to assess the quality of bio-oil as lignin content is varied in the hybrid poplar, how temperature of pyrolysis affects these results, and how the addition of a torrefaction pretreatment will affect the composition of bio-oil. The quality of bio-oil is assessed by calculating bio-oil yield relative to dry biomass and relative to char and gas, changes in relative abundance of oxygen in the bio-oil, and energy density of the bio-oil. Eight genetically different poplar samples with varying lignin content were pyrolyzed at 500ºC using a CDS 5200 HP micro-pyrolysis unit, which was directly connected to a ThermoFisher Trace GC coupled with a DSQII MS. Four of these hybrid poplar samples over a range of lignin content were then pyrolyzed at temperatures of 550ºC and 600ºC to determine effects of pyrolysis temperature. Finally, two of those samples, a high and low lignin concentration, were then torrefied at 300ºC for 30 minutes and then pyrolyzed at 550ºC for 20 seconds.

Results: An increase of lignin concentration, by approximately 10% relative to the low lignin content sample, decreased the abundance of phenolic species in bio-oil by 3% and decreased the relative abundance of oxygen by 8% at 500ºC. The bio-oil yield similarly decreased by 5%, due to an increase char yield at high lignin levels. With an increase in pyrolysis temperatures, there was an increase in phenolics produced overall, but with respect to increasing lignin content, it continued the trend of decreasing phenolic species in bio-oil. Again with respect to lignin content at higher temperatures, the bio-oil yield decreased as the char yield increased. The addition of torrefaction as a pretreatment step, decreased the overall yield of bio-oil produced by 10% and in agreement with previous literature, removed low molecular weight highly oxygenated species, thereby decreasing bio-oil yield but improving the quality of bio-oil produced. From these results over a limited range of lignin content, higher temperatures are needed to increase the recovery of phenolic species from pyrolysis of poplar.

465258 Effect of Operating Parameters on Ablative Pyrolysis of Wood

Abstract:

One of the main requirements for the production of bio-oil from fast pyrolysis of wood is the use of small feedstock particle sizes (1-3 mm), in order to minimize temperature gradients within the particles and therefore avoid undesired reactions that take place in the range 200-400°C. The only exception to the rule is ablative pyrolysis. In ablation, heat transfer to the wood is carried out via direct contact with a hot, moving metallic surface. This situation creates a steep temperature gradient within the particle, such that pyrolysis takes place within a thin liquid layer contacting the solid as opposed to the entire wood particle. For this reason, in principle there is no limit to the particle size in ablative pyrolysis, and this feature is of great value, since grinding costs represent 7-9 % of the total process costs in pyrolysis. Though this characteristic has been speculated in the past, there were no experimental data to back up this claim. Recently, our group developed an ablative pyrolysis unit that is able to convert entire wood chips (about 1â€™â€™ size) and even entire wood rods directly into bio-oil in one single step, with yields similar to those in a fluidized bed. In this work, we evaluated the effects of the main operating parameters involved in ablative pyrolysis of whole wood chips: the temperature of the hot plate, the relative speed between the wood chips and the hot plate, the pressure applied by the plate on the chips, and the initial thickness of the assembly of wood chips. Comparisons involve the bio-oil yield, LHV, water content, composition as detected by GC-MS, and functional groups detected by FTIR. In addition, differences in the char produced is evaluated by elemental composition and FTIR.

467137 A Combined Pyrolysis – Gasification Study to Increase Syngas Production Yield from Beer Bagasse Waste

Abstract:

Extensive resources and efforts have been dedicated to assess opportunities for using locally available biomass to produce energy, reduce CO_2emissions, and lower reliance on petroleum. The use of lignocellulosic biomass is promising since it is readily available in vast amounts and there are processes for converting different biomass sources to useful products of particularly interest in liquid transportation fuels. Energy based in biomass is a source of renewable energy from organic and inorganic matter formed in a biological or mechanical process. Although fossil fuels, such as petroleum, oil and coal, have their foundation in ancient biomass, they have been "out" of the carbon cycle for a long time. Their combustion consequently disturbs the carbon dioxide content in the atmosphere.

One promising technology that has gained interest in the past few years is pyrolysis and gasification. Although fast pyrolysis has been known and studied for many years; due the complexity of fast pyrolysis reaction mechanisms there are still a lot more to investigate to increase the understanding on fast pyrolysis of biomass.

While it is scientifically viable to exploit lignocellulosic materials and organic wastes into energy, chemicals and fuels, it is far from an economic process and the cost needs to be lowered. It also must be demonstrated that a commercial scale biomass utilization process is environmentally sound, and there is a need to find an inexpensive and widely available lignocellulosic source of biomass

The core objective of this research is to study the influence of pyrolysis as a pretreatment on spent grain from brewers' waste to improve syngas production from gasification. Biochar produced from the biomass pyrolysis was used in order to obtain compounds with higher carbon content to subsequent undergo a gasification process. In addition, commercial activated carbon was gasified to compare the yield of syngas produced in the gasification of biochar.

Brewers waste was pyrolyzed under a nitrogen environment in different operating conditions; heating rates, reaction times, and pyrolysis temperatures. Activated carbon and biochar gasification was done at 850 °C and 1 atm of pressure for two hours using water as gasifying agent in a fluidized bed reactor.

Pyrolysis at 300 °C produced a greater yield of biochar leading to a higher concentration of syngas produced during gasification of the waste biomass studied compared with the commercial activated carbon products obtained.

Finally, the processes of chemical thermo-decomposition pyrolysis and gasification applied to brewers' waste, allow to obtain products with higher added value, making possible the production of synthesis gas from the gasification of solid compounds obtained from pyrolysis of beer bagasse waste.

These studies lay the groundwork for longer-term efforts aimed at using waste biomass as a sustainable way to add value to solid waste biomass.

468387 Product Distribution and Kinetics of Hydropyrolysis of Agro Residues Via Py-GC/MS and Py-FT-IR

Abstract:

Approximately 140 billion tons/yr of agricultural waste is generated globally, which is equivalent to 50 billions tons of crude oil. (UNEP Report 2011). In India, a major fraction of the agro residues is unutilized and simply burnt in the farm, while the rest are either used as fodder or co-fired with coal in furnaces. These are large volume, yet low value solutions. Generating liquid bio-crude/biofuel from the agro residues via thermochemical techniques is a promising option to reduce the dependency on fossil-based fuels. Although fast pyrolysis of biomass residues is a promising technique that is being deployed to produce bio-oil/bio-crude, there are inherent difficulties associated with the utilization of bio-oil. These include high oxygen content, low heating value, poor storage and thermal stability, which are attributed to the presence of moisture and organics like carbonyl compounds, carboxylic acids, alcohols and phenols. While this stability issue can be partially addressed by catalytic fast pyrolysis that produces a fairly deoxygenated bio-oil, the bio-oil still requires further upgradation via catalytic hydrodeoxygenation process. Hydropyrolysis is a single step process wherein biomass is fast pyrolyzed at c.a. 500 °C in presence of hydrogen gas at moderate pressures (<35 bar). This results in the production of a high quality liquid fuel with very low O/C ratio. It is, therefore, necessary to thoroughly characterize the evolution of pyrolysates from fast hydropyrolysis and compare it with conventional fast pyrolysis process to understand the mechanism of formation of products. In this study, analytical pyrolysis (Py) of rice straw, empty palm fruit bunch (EFB) and pine wood biomass is conducted and the products are characterized using gas chromatograph/mass spectrometer (GC/MS) and Fourier transform infrared spectrometer (FT-IR).

The biomass characteristics such as volatile matter, fixed carbon, ash content, elemental C, H, N, S and O, and heating value were evaluated using standard techniques. It is important o note that the biomasses investigated in this study vary mainly in ash composition, and follow the trend: rice straw (19%) > EFB (7.5%) > pine wood (0.7%). Fast pyrolysis and hydropyrolysis experiments were conducted in a Pyroprobe® 5200 pyrolyzer in trap mode. Approximately 0.5 mg of the agro residues were pyrolyzed at a heating rate of >10,000 °C/s to the preset temperature of 400-700 °C and held for 60 seconds in hydrogen atmosphere. The vapors were characterized using Agilent two-dimensional GC/MS. The effect of hydrogen pressure from 1 to 25 bar on pyrolysate composition was investigated, and compared with pyrolysate composition in presence of He environment at atmospheric pressure. Experiments were also conducted wherein the time evolution of key functional groups of the pyrolysates was analyzed using an online FT-IR spectrometer (Agilent Cary 660). A significant reduction in the formation of oxygenated organics, especially carboxylic acids, furan derivatives and phenolics, and a concomitant increase in the production of linear and aromatic hydrocarbons were observed during fast hydropyrolysis compared to normal fast pyrolysis of Indian rice straw, even at medium pressures of hydrogen. High temperatures slightly improved the yield of linear hydrocarbons, but greatly increased the yield of aromatics during hydropyrolysis. Py-FT-IR study revealed the time evolution of the pyrolysates, and the reaction completion time. The time taken for maximum production of oxygenates varied from 10-15 s at 500 °C, which was similar for all three biomass varieties. More interesting results on (a) the evolution of individual functional group vibrations, (b) product characterization, and (c) char formation, for different biomass varieties will be discussed during the presentation. The implications of the results on the quality of bio-oil and its properties will also be addressed.

458610 Techno-Economic Assessment of CO_2 Quality Effect on Its Storage and Transport (CO2QUEST)

Abstract:

The ultimate composition of the CO_2 stream captured from fossil fuel power plants or other CO_2intensive industries and transported to a storage site using high pressure pipelines will be governed by safety, environmental and economic considerations.

So far, most of the studies performed on this topic have been limited in scope, primarily focusing on investigating the impact of the CO_2stream impurities on each part of the Carbon Capture and Sequestration (CCS) chain in isolation. This is a significant drawback given the markedly different sensitivities of the pipeline, well bore materials and storage sites to the various impurities.

For example, given the risk of water table contamination, trace elements such as Lead, Mercury and Arsenic in the CO_2 stream are of far greater concern in an aquifer storage site than compared to the pipeline. On the other hand, even small concentrations of water in the CO_2 stream are detrimental to the pipeline due to corrosion, but of benefit even at high concentrations during storage given the immobilisation effect of water on CO_2.

'*What is good for the pipeline is not necessarily good for storage'.*

It is clear that the optimum composition and concentration of the impurities in the captured CO_2 stream involves a delicate balance between the different requirements within the CCS chain, spanning capture, transportation and storage, with cost and safety implications being the over-arching factor. Pivotal to these considerations is an understanding of the impact of the impurities on the physico-chemical properties of CO and its hazard profile.

This paper presents and overview and the main findings of the FP7 European Commission CO2QUEST project. Involving the collaboration of 12 industry and academic partners in Europe, China and Canada the project addresses the fundamentally important and urgent issues regarding the impact of the typical CO_2streams impurities captured from fossil fuel power plants on its safe and economic transportation and storage.

The work programme, spanning 36 months, focuses on the development of state-of-the-art mathematical models, backed, by laboratory and industrial-scale experimentation using unique EC-funded test facilities to perform a comprehensive techno-economic, risk-based assessment of the impact of the CO_2stream impurities on the phase behaviour and the physico-chemical reactions governing the pipeline and storage site integrities.

The above involves the determination of the important CO_2 mixtures that have the most profound impact upon the pipeline pressure drop, compressor power requirements, pipeline propensity to ductile and brittle facture propagation, corrosion of the pipeline and well bore materials, geochemical interactions within the well bore and storage site, and the ensuing health and environmental hazards. Based on a cost/benefit analysis and whole system approach, the results are employed to provide recommendations for tolerance levels, mixing protocols and control measures for pipeline networks and storage infrastructure thus contributing to the development of relevant standards for the safe design and operation of CCS.

461978 Understanding the Factors That Affect CO_2 Transport, Storage, and Enhanced Oil Recovery Potential in Unconventional Reservoirs

Abstract:

Investigations on the use of subsurface geologic formations for carbon dioxide (CO_2) storage have been ongoing for many years and have traditionally focused on saline aquifers, depleted oil and gas formations, and unminable coal seams (Goodman and others, 2014; U.S. Department of Energy National Energy Technology Laboratory, 2012). In the last decade, improvements in directional drilling technology (Watney, 2016) and hydraulic fracturing (King, 2010) of organic-rich shale and other tight rock formations have opened new opportunities for CO_2 storage, with the potential cobenefit of enhancing gas and oil recovery in reservoirs that typically have less than 10% primary recovery.

Past and ongoing work at the Energy & Environmental Research Center (EERC), funded by the U.S. Department of Energy National Energy Technology Laboratory in conjunction with partners from the oil and gas industry, has been focused on better understanding the factors that affect CO_2 transport as it relates to both enhanced oil recovery (EOR) and CO_2 storage in unconventional oil reservoirs and organic-rich shales. Much of this work has been performed in several of the EERC's laboratories using rock samples from the Bakken petroleum system, the results of which can be applied to unconventional formations across the United States. The analytical work conducted to date has focused on supercritical CO_2 extraction experiments, CO_2 sorption experiments, and extensive physical and geochemical characterization using rock samples from several Bakken wells to better understand the factors that can affect CO_2 migration, oil extraction, sorption, and storage. Sample characterization has been accomplished using a variety of analytical techniques including whole-core and microcomputerized tomography imaging, conventional scanning electron microscopy (SEM) coupled with energy-dispersive x-ray spectroscopy, field emission SEM, focused ion beam SEM, x-ray fluorescence, x-ray diffraction, optical microscopy, Rock-Eval pyrolysis, vitrinite reflectance, mercury injection capillary pressure testing, and geomechanical testing. Initial characterization work focused on better understanding the occurrence and characteristics of fracture and pore networks and has expanded to the use of advanced analytical and image analysis techniques to quantify fracture and pore occurrence and mineralogical content at various resolutions. Work is ongoing to better understand and quantify the occurrence and accessibility of pores, fractures, clay minerals, and kerogen and the potential implications that those features have on CO_2 transport for EOR and on CO_2 sorption for storage. The sample characterization efforts are being used to better understand the results of supercritical CO_2 extraction experiments and ongoing CO_2 sorption experiments using Bakken samples that have undergone extensive characterization. These results are being integrated into modeling research activities to determine the fundamental mechanisms controlling fluid transport and CO_2 storage in the Bakken, resulting in the development of improved model simulations. The results of the EERC's laboratory-based work and its implications for CO_2 EOR and storage, as well as future research needs, will be discussed.

REFERENCES

Goodman, A., Fukai, I., Dilmore, R., Frailey, S., Bromhal, G., Soeder, D., Gorecki, C., Peck, W., Rodosta, T., and Guthrie, G., 2014, Methodology for assessing CO_2 storage potential of organic-rich shale formations: Energy Procedia, v. 63, p. 5178–5184.

King, G.E., 2010, Thirty years of gas shale fracturing: what have we learned? Society of Petroleum Engineers: Annual Technical Conference and Exhibition, September 19–22, Florence, Italy, SPE 133456.

U.S. Department of Energy National Energy Technology Laboratory, 2012, The United States 2012 Carbon Utilization and Storage Atlas, Fourth Edition: www.netl.doe.gov/File%20Library/Research/Coal/carbon-storage/atlasiv/Atlas-IV-2012.pdf (accessed January 2016).

Watney, W.L., 2016, Horizontal drilling – technology review, current applications, and its future in developing Kansas' petroleum resources: Kansas Geological Survey, Lawrence, Kansas, www.kgs.ku.edu/PRS/Unger/Watney-KBA-KIOGA--Horizontal_Drilling.pdf (accessed January 2016).

470324 Evaluating the Feasibility and Scale of Geologic CO_2 Storage Potential in Ohio's Appalachian Region

Abstract:

Evaluating the feasibility and scale of geologic CO_2 storage potential in Ohio's Appalachian Region

Priya Ravi Ganesh*, Autumn Haagsma, Glenn Larsen, Isis Fukai, Joel Main, Mackenzie Scharenberg, Neeraj Gupta

Battelle, Columbus, OH

Around the world, various industrial-scale carbon dioxide (CO_2) geologic storage (GS) demonstration projects have been or are being conducted in deep saline formations. To fully tap into the overall potential of CO_2 GS resources, comprehensive research focusing on the detailed geologic characterization of potential regions, such as the Appalachian Basin in the Midwestern United States, needs to be completed. The project described here is part of a long-term program to investigate CO_2 GS potential in Ohio and adjacent areas. Here, we apply an integrated multi-scale approach to evaluate the possible implications of basin-scale pressure buildup on storage capacity estimation. The reservoir analysis task is divided into two main efforts: (1) *Static capacity assessment* to help provide regional estimates as well as specific site estimates for selected saline storage site(s) and (2) *Dynamic capacity and injectivity assessments* of selected saline storage site(s) as illustrative examples for the evaluation of injectivity and capacity from the study area. The static and dynamic estimates for CO_2 storage shall be compared to highlight the effects of dynamic operational constraints such as rates, well locations, etc. typical of realistic CO_2GS site conditions.

Our study region consists of ~61,000 sq.km of complex geology in eastern Ohio, with a primary emphasis on the strata that underlie the Lower Ordovician Knox unconformity erosional surface. These strata are the Beekmantown dolomite through the Mt. Simon/basal sandstone where many of these deeper porous layers contain highly saline brines that have remained isolated from shallower zones. Key inputs to this feasibility study draw from the geologic characterization and fracture gradient assessment activities in the study area done as part of other tasks in this project. The static resource capacity estimation shall be determined with considerations for the total storage efficiency of CO_2 in these modeled saline formations. The potential storage sites determined from the geologic characterization exercise shall be evaluated to understand dynamic CO_2 GS performance metrics of interest, namely, injectivity, pressure buildup and spatial distribution of injected CO_2volume through time.

This study aims to help evaluate the feasibility and scale of CO_2 GS in regions in Ohio with respect to performance metrics of storage capacity and CO_2 injectivity. We work to evaluate different scales such as the regional or formation scale, plume or site-scale, as well as local-scale or near-well effects of CO_2 injection. This multi-scale study considers different resolution of input detail such as porosity and permeability distributions and uses key performance metrics, such distribution of injected CO_2 and pressure, to provide insights into acceptable site spacing, pressure buildup and CO_2plume distribution.

At the formation or regional level, we begin by mapping static storage resource estimates for each formation to create a roadmap to CO_2 GS in eastern Ohio. CO_2 GS resource is defined as the mass estimate of CO_2 that can be stored in a given geologic formation. We use different levels of data resolution for input formation properties (e.g., porosity and thickness) for this exercise. These range from the knowledge of average formation properties to the availability of the spatial distribution of these properties. Results show that, in descending order, the Mt. Simon/ basal sandstone, Kerbel sandstone, and Rose Run sandstone have the highest specific CO_2GS resource capacities. We investigate the subsurface dynamics by considering multiple simultaneous operational storage sites scenarios in selected study areas with high storage potential for these formations. The dynamic numerical modeling scenarios that consider multiple operational sites in a given formation or region shall provide in insights into acceptable site spacing from the extent of the pressure front to minimize pressure interference between sites.

The study areas for CO_2 GS in Cambrian formations in the eastern Ohio study area were determined from the formation maps created for the study area. Locations where the petrophysical indicators for storativity (porosity-feet) and transmissivity (permeability-feet) highs overlap are considered for further detailed study for CO_2 storage with dynamic assessment. The top three sandstone units identified from the static storage resource estimation exercise have been analyzed to identify 'sweet spots'. At the plume or local-scale examination for a given formation, these individual sites shall be modeled with possible CO_2 injection well patterns simulated. The results from this exercise test the effect of pressure buildup and potential interference issues (well spacing), as well as tracking the injected CO_2plume extent.

On a finer scale, a single well at the center of the possible well patterns for each site in a given formation are simulated for 30 years of CO_2injection, followed by 50 years of post-injection monitoring, to get insights into local-scale near-wellbore dynamics. These simplified well models with equivalent average site properties in radial configurations are being tested to help with preliminary sensitivity analysis to relate to monitoring aspects during injection operations. This exercise establishes sensitivity to well configurations, relative permeability etc. representative of local-site conditions in the chosen detailed study site(s).

The integration of results from these different scales are expected to bring together insights into the practical feasibility of commercial CO_2 GS storage in our study region. Our study aims to provide insight into the CO_2 GS potential for deep saline formations in eastern Ohio while reflecting the effects of dynamic operational constraints typical of realistic CO_2GS site conditions in the study region.

This work was supported by the Ohio Development Services Agency OCDO Grant OOE-CDO-D-13-22 and the U.S. Department of Energy through the Midwest Regional Carbon Sequestration Partnership award number DE-FC26-05NT42589. In-kind contributions and well access was provided by a number of well operators in the region.

460550 Shallow Ground Water Monitoring at the Secarb CO_2 Injection Site Near Citronelle, Alabama

Abstract:

The SECARB Anthropogenic Test is a fully integrated U.S. Department of Energy sponsored-research project located near Citronelle Alabama designed to demonstrate the safe, reliable capture, transportation and storage of CO_2. Starting in August 2012, the project team injected over 114,000 metric tonnes of CO_2 into the Paluxy Formation at a depth exceeding 2,865 m over a two-year period. Injection concluded in September 2014 and the project is currently in the post-injection site care period. Several monitoring techniques are being used to demonstrate post-injection containment of the CO_2 in the injection zone and non-endangerment of groundwater resources during the three-year post-injection monitoring effort. These techniques include time-lapse cross-well and vertical seismic profile surveys, geophysical logging, reservoir pressure and temperature measurements, soil CO_2 flux measurements, tracer surveys, and shallow groundwater sampling and analyses.

The ground water monitoring program consists of collecting water samples from three dedicated monitoring wells and one water supply well installed in the shallow surficial aquifers in close proximity to the projectâ€™s deep CO_2 injection and observation wells. A risk assessment performed during the early stages of the project identified the deep wellbores, which penetrate the confining unit overlying the injection zone, as the only potential migration pathway. No known faults are present based on available geologic and seismic data. Although the risk of CO_2 leakage is expected to be extremely small, SECARB installed the shallow ground water monitoring wells on the injection well pads to check for a CO_2 release as part of the regulatory permitting program.

Water samples are collected from each monitoring well on a quarterly schedule. Samples are analyzed on-site for time-sensitive parameters including pH, electrical conductivity, dissolved oxygen, temperature and oxidation-reduction potential and preserved for off-site analysis for 17 metals, total inorganic carbon (TIC) and other inorganic parameters by a contract laboratory. The ground water analytical results are compared to U.S. Environmental Protection Agency (EPA) drinking water standards and to background samples collected prior to CO_2 injection using EPA-approved statistical methods.

The study results indicate that shallow ground water systems are dynamic and complex, resulting in water quality results for individual naturally-occurring constituents that can fluctuate considerably with time due to natural processes or sampling-induced mobilization of inorganics. This may lead to false detection and reporting of a CO_2 release over a 40 to 100-year life span of a commercial CO_2 storage project. Therefore, multiple constituents (e.g., pH, Fe, TIC and EC), or lines of evidence, should be collectively used in the analysis to detect a release and not just one or two (e.g., pH). The duration and location of background sampling should also be carefully considered in the initial design of the storage facility due to the ubiquitous occurrence of inorganics found naturally in ground water.

471378 Probabilistic Analysis of CO_2 Storage Mechanisms in a CO_2-EOR Field Using Polynomial Chaos Expansion

Abstract:

Depleted oil fields are already used for storing CO_2 via injection of CO_2 for enhanced oil recovery (CO_2-EOR). Such storage is an outcome of CO_2-EOR, albeit not necessarily by design. A next step would be intentional storage via post-EOR CO_2 injection. Following injection for EOR or post-EOR, CO_2 may be trapped by dissolution in the oil or aqueous phases, or remain in the supercritical CO_2 phase. Forecasting the nature of trapping and the ultimate CO_2 distribution in a reservoir is hindered by uncertainty in reservoir properties. The purpose of this study is to develop and apply reduced order models (ROMs) integrated with Monte Carlo simulations to quantify the three primary types of storage mechanisms in an active CO_2-EOR field, including oil solubility trapping (oil phase), aqueous solubility trapping (aqueous phase), and hydrodynamic trapping (CO_2 in supercritical phase). The case study site for this analysis is the SACROC unit in western Texas. A Polynomial Chaos Expansion (PCE) technique was used to develop the ROMs. The sources of uncertainty considered are porosity and permeability. Model outputs of interest include dissolved mass of CO_2 in oil phase, mass of CO_2 in supercritical phase, dissolved mass of CO_2 in aqueous phase, and oil saturation in the rock formation. Reduced order models were developed for all cells in five selected layers of the model, which are adjacent to injection wells, at three specific time points of interest, including the end of a simulated CO_2-EOR period, the end of a post-EOR continuous CO_2 injection period, and the end of a post-injection monitoring period. Results of regression fit and validation analysis yield high R^2 values and low NRMSE values, indicating that the ROMs derived from PCE are able to make meaningful predictions (compared to conventional reservoir models) and effectively represent relationships between model inputs and model outputs of interest. Results of 1000 Monte Carlo simulations indicate a dominant upward transport of CO_2 during injection, driven by buoyancy, and a dominant downward transport of dissolved CO_2 after injection stops, caused by the increase in both oil and brine density. At the end of the 100-year simulation of the SACROC case study, the selected five layers store between 54% and 61% of total trapped CO_2 in the entire domain in the oil phase, between 21% and 24% in the supercritical phase, and between 11% and 15% in the aqueous phase. The selected five layers store between 89% and 98% of total trapped CO_2 in the entire domain by the end of simulation.

465751 Monitoring Ions Based on Reactive Transport Modeling of CO2 Sequestration in Ulleung Basin, Korea

Abstract:

In Carbon Capture and Storage(CCS) process geochemical monitoring technology plays important role to avoid potential CO2 leakage. In this paper, reactive transport modeling of supercritical CO2 in aquifer which is modeled after Ulleung basin; one of offshore geological storage site of Korea was performed. Supercritical CO2 have been injected into target reservoir at the rate of 30kg/s for 30 years. After injection for 30 years, long-term behavior of the CO2 by considering its geochemical reaction for 5000 years was investigated. In early stage, aqueous phase and supercritical CO2 was dominant. After 800~1000years, mineral phase became dominant phase accounts for more than 80% of injected CO2. During injection period, at monitoring points, manganese, bicarbonate ions concentration changed before reaching supercritical CO2. Ion concentration changes can give early warning of potential CO2 leakage. To determine monitoring ions for specific mineral composition, reactive transport modeling of typical mineral composition cases for geological storage reservoir all over the world was conducted.

468153 The Top-Ref Approach to Improve the Resource Efficiency of Energy Intensive Industrial Processe

Abstract:

TOP-REF project aims at developing and validating indicators, methodologies and non-invasive tools devoted to the improvement of resource efficiency in energy intensive continuous industrial processes in the agro-chemical, chemical and petrochemical sectors. These methodologies and tools will constitute to the following targets:

- To path the way for achieving the SPIRE2030 objective of reducing non-renewable, primary raw material intensity up to 20 % and fossil energy intensity up to 30%, both compared with current levels.
- Reduce production costs up to 15 % compared with the current levels.
- A significant reduction of the environmental impacts (CO2 emissions, water footprint, pollutants, hazardous emissions, etc.).

To achieve these impacts, TOP-REF develops and demonstrates a robust, resource-efficiency-focused and cross-sectorial methodology based on novel exergy-based resource indicators. This methodology is implemented in three industrial demonstration plants: solid fertilizer production, crude distillation and fractionation unit of a refinery and a naphtha steam cracker. The tools developed are validated by demonstration under real conditions in three pilots.

The methodology developed by TOP-REF consists of the following steps:

1. Development of Key Resource Indicators (KRIs) to measure the performance in resource efficiency
2. Modeling and simulation of the three industrial pilots
3. Identification of the critical process parameters (CPPs) influencing the KRIs
4. Optimization of the CPPs to improve the resource efficiency

5. Demonstration and validation under real conditions

The approach to identify the critical process parameters for the production processes investigated is based on detailed process models validated by the industrial partners. These process models are analyzed using global sensitivity analysis to identify the critical process parameters. As the full simulation models are very complex, simplified mathematical surrogate models are derived prior to global sensitivity analysis. A variance-based global sensitivity analysis is performed for all process parameters to identify the most influential operating parameters on the key resource indicators (KRIs) developed within the TOP-REF project. The use of a variance-based sensitivity analysis enables the consideration of interaction effects among the process parameters. Finally, the identified critical process parameters are optimized using the KRIs as objective. The optimized operating points will be validated under real conditions in the three production plants.

Within the contribution we will explain the TOP-REF approach developed, show its application to the three industrial processes and highlight experiences made during the application.

Acknowledgment

The research leading to the results presented in this contribution has received funding from the European Communityâ€˜s 7th Framework Program for research, technological development and demonstration under grant agreement No 604140, TOP-REF.

468313 Novel Data Envelopment Analysis Approach for Sustainability Assessment: Application to Electricity Generation Technologies

Abstract:

Moving towards sustainable development implies implementing actions, projects, programs, plans and policies, which involve the balanced integration of economic, environmental, and social objectives. However, setting sustainability goals requires some knowledge and understanding of the current level of sustainability. To this end, sustainability assessment approaches can be applied for measuring the extent of progress towards sustainability and for defining quantitative targets that ensure a more sustainable development.

In this work we explore the use of data envelopment analysis (DEA) to quantify the level of sustainability attained by a system and identify targets for improvements. DEA is a non-parametric linear programming (LP) technique that measures the efficiency of a set of entities, termed decision-making units (DMUs), each transforming multiple inputs into multiple outputs (Charnes et al., 1978). By solving the DEA LP problem, it is possible to distinguish between 'efficient units' (i.e., efficiency score of 1) and 'inefficient units' (i.e., non-frontier units with an efficiency score strictly lower than 1), for which DEA provides in turn specific guidelines to become efficient by projecting them on the efficient frontier.

In the context of sustainability assessment, DEA presents two major limitations particularly critical when a wide range of economic, environmental and social indicators are considered in the analysis. First, the standard DEA makes no distinction between the units deemed efficient, which makes it difficult to select a final alternative in the absence of a ranking scheme. Secondly, DEA is very sensitive to the set of inputs and outputs, leading to very different results depending on the sustainability criteria considered in the analysis.

To overcome the limitations of the standard DEA as applied to sustainability studies, this work proposes an enhanced DEA methodology tailored to carrying out sustainability assessments. The methodology proposed

integrates DEA with the concept of "order of efficiency", originally introduced to rank Pareto optimal solutions from a set of many Pareto points (Das, 1999). In essence, the main idea behind the new methodology consists of repeating the DEA calculations iteratively for each and every possible combination of inputs/outputs, to subsequently aggregate the results into efficiency metrics that assess each sustainability dimension separately and finally determine an overall sustainability level integrating the three scores (obtained in each dimension).

The capabilities of the methodology proposed are illustrated through its application to the assessment of the electricity-generation technologies expected to play a major role in the future electricity mix of United Kingdom (i.e., nuclear, gas, coal with and without carbon capture and storage, wind, solar photovoltaics and biomass with wood and *Miscanthus* pellets). Each technology is defined as a DMU that is characterised by a set of economic, environmental and social sustainability indicators determined using a life cycle assessment approach by Stamford and Azapagic (2014). We consider 18 sustainability indicators: three economic (capital cost, operation and maintenance cost, and fuel cost), nine environmental (freshwater ecotoxicity, marine ecotoxicity, global warming, ozone depletion, acidification, eutrophication, photochemical smog, land occupation, and land ecotoxicity) and six social (direct employment, worker injuries, human toxicity potential, radiation, depletion of elements, and depletion of fossil fuels).

The results generated with our approach show that gas, nuclear and wind technologies are efficient in each dimension of sustainability when all the indicators within each such dimension are considered. Gas is also the most economically and socially efficient (highest economic and social efficiency scores), while nuclear is the best in the environmental dimension. From an overall sustainability point of view, the gas technology presents the highest overall sustainability efficiency (mainly because of its very good performance in the social dimension), followed by nuclear and wind. On the other hand, coal is found economically inefficient; coal, solar PV and both biomass options are environmentally inefficient; and coal and biomass with *Miscanthus* are found socially inefficient. For these technologies deemed inefficient in a particular sustainability dimension, an inefficiency assessment is performed which provides targets that should be attained to reach the efficiency in each specific dimension.

The enhanced DEA proposed can facilitate the transition to a more sustainable future, since it enables ranking alternatives according to the extent to which they adhere to sustainability principles; and also pinpoints in a systematic manner the main sources of inefficiency and sets targets for improvements.

REFERENCES

Charnes, A., Cooper, W.W., Rhodes, E., 1978. Measuring the efficiency of decision making units. Eur. J. Oper. Res. 2, 429–444. doi:10.1016/0377-2217(78)90138-8

Das, I., 1999. A preference ordering among various Pareto optimal alternatives. Struct. Optim. 18, 30–35.

Stamford, L., Azapagic, A., 2014. Life cycle sustainability assessment of UK electricity scenarios to 2070. Energy Sustain. Dev. 23, 194–211. doi:10.1016/j.esd.2014.09.008

471458 Optimal Design of Sustainable Agricultural Water Networks

Abstract:

This work proposes a multi-period mixed integer non-linear programming model for the optimal design of agricultural water networks, which involves the optimal reuse of resources such as water, fertilizers and pesticides. This model is based on a new superstructure that includes attractive options for the use, reuse, recycle, regeneration and storage for the considered resources. The model incorporates simultaneously the maximization of the net annual profit and the minimization of the environmental impact. The profit is formed by the economic incomes due to the sale of crops minus the corresponding capital and operating costs associated to the fresh water, fertilizer, storage, piping, pumping and water treatment. Whereas the environmental impact is determined through the Eco-indicator 99 and the water footprint to evaluate the effect over the global warming, acidification, eutrophication, green water footprint, gray water footprint and blue water footprint. Two case studies from the Mexican state of Sinaloa were solved with the proposed model. The first one considers four crops of corn and it requires five irrigation periods; while in the second one twelve parcels are included where there are four crops of corn, four crops of bean, two crops of wheat and two crops of alfalfa, and there are needed seventeen irrigation periods. The operating, economic and design parameters are representative for the State of Sinaloa in Mexico; and the results are presented in three dimensional Pareto diagrams. From these diagrams is possible to find the optimal relationship between profit, eco-indicator and water footprint.

461551 Li-Rich Anti-Perovskite Superionic Conductor Films for All-Solid-State Li-Ion Batteries

Abstract:

Development of high-performance inorganic ionic conductors for all-solid-state lithium batteries has been considered as an ultimate solution to address the safety issues for transportation applications. The solid-state superionic conductors also enable the fabrication of thin and flexible batteries, which are urgently needed in burgeoning applications such as wearable electronics and implantable medical devices. However, the ionic conductivities of solid electrolytes are still much lower than those of the liquid electrolytes, thus limiting the applications of lithium batteries in many important areas. Here, we report the deposition of anti-perovskite Li_3OCl solid-state electrolyte films with significantly enhanced superionic conductivity via pulsed laser deposition using a composite target. A room-temperature ionic conductivity of Li3OCl films as high as 2.0 Ã— 10â^'4 S cmâ^'1 is achieved using such a processing scheme, and this value is more than two orders of magnitude higher than that of its bulk counterpart. The superior compatibility between the electrolyte film and metallic Li is revealed by cycling a Li/Li_3OCl/Li symmetric cell. The applicability of Li_3OCl film as an electrolyte for solid-state thin-film Li-ion batteries has been demonstrated for the first time by examining the full Li-ion batteries. Our findings illustrate that the anti-perovskite films hold great promise as alternative solid electrolytes for electric-energy storage devices.

466943 Carbon Foams from Polyhipe/Reduced Graphene Oxide Composites and Their Performance As Electrodes in Supercapacitor Devices

Abstract:

Carbon foams were produced from composites of macroporous poly(divinylbenzene) (poly(DVB) and reduced graphene oxide (rGO). Poly(DVB) was synthesized by polymerizing the continuous phase of water-in-oil high internal phase emulsions (HIPEs) stabilized at the interface by amphiphilic rGO, employed as a particulate emulsifier. The use of rGO as an emulsifier allows for the synthesis of stable macroporous composites of poly(DVB)HIPEs with a percolating network of rGO at the surface of the material. Both permeable and non-permeable carbon foams, or 'carboHIPEs', could be prepared by carbonization of these macroporous composites at 800 °C. The resulting carboHIPEs gave yields as high as 26 wt.% of the original material. Furthermore, carboHIPEs retain the pore structure of their macroporous precursor, while also producing a newly-formed microporous structure, leading to hierarchical porosity and huge increases in surface area upon carbonization. Surface areas of up to 1800 m^2/g and excellent electrical conductivities of up to 270 S/m are achievable, among the highest reported within the fields of both polyHIPEs and carboHIPEs. Using an rGO emulsifier when creating these structures allows for the production of true 'all-carbon' foams upon carbonization. These polyHIPE composites do not require modification, such as additional crosslinking, prior to carbonization, due to the inherently crosslinked structure of poly(DVB). It is demonstrated that the rGO derived carboHIPEs are good candidates as electrodes in supercapacitor applications such as electrical double-layer capacitor (EDLC) devices, where carboHIPEs derived from more conventional silica-stabilized HIPEs perform poorly. The use of a pourable, aqueous emulsion-template enables simple moulding, minimizes waste and avoids the strong acid treatments used to remove many conventional solid-templates. The capability of producing monolithic, porous carbon foams allows for the production of binderless devices, potentially simplifying the production process of supercapacitors. Devices demonstrated maximum specific electrode capacitance to the tune of 26 F g^{-1} at 10 mV s^{-1}, 5.2 Wh kg^{-1} of energy density, 280 W kg^{-1} of power density and coulombic efficiency of up to 99.6 %. In short, the carbonization of rGO-poly(DVB)HIPE composites enables the production of hierarchically porous carboHIPEs, suitable for a wide range of applications as sorbents and electrodes.

473013 Designing Electrolytes for Beyond Li-Ion Batteries Using Coupled High Throughput *Ab Intio* Calculations and MD Simulations

Abstract:

Transformative outcomes for transportation and electrical grid require high performance, low cost energy storage systems beyond Li-ion technology and innovations in electrodes and electrolytes, alike. Multivalent batteries, chemical transformation and redox flow batteries are promising candidates for next generation electricity storage technologies. An automatic High-throughput computational infrastructure has been constructed for the electrolyte genome project supported by the US Joint Center for Energy Storage Research (JCESR) coupled with experimental analysis with the aim to design electrolytes for beyond Li-ion battery technology.[1] A non-aqueous multivalent (e.g., Mg^{2+}, Ca^{2+} and Zn^{2+}) metal cell is one of the potential candidates for a post-lithium-ion battery. The theoretical volumetric capacity of a metal anode coupled with the lack of dendrite formation at a multivalent metal anode provide an attractive opportunity in energy storage.[1] However, development and commercialization of multivalent batteries are fraught with complications. Thus a fundamental understanding of molecular level properties of these electrolytes is required to improve the electrochemical stability and the charge transfer properties. In this work, we present a multi-scale modelling approach for multivalent salts in various solvents. We uncover a novel effect between concentration dependent ion pair formation and anion stability at reducing potential, e.g., at the metal anode.[2] We find that both Mg and Zn electrolytes are highly prone to ion pair formation, even at modest concentrations, for a wide range of solvents with different dielectric constants, which have implications for dynamics as well as charge transfer.[3] Specifically, we observe that, at Mg metal potentials, the ion pair undergoes partial reduction at the Mg cation center ($Mg^{2+} \rightarrow Mg^{+}$), which competes with the charge transfer mechanism and can activate the anion to render it susceptible to decomposition. It was observed that Zn electrolytes are also prone to similar $TFSI^-$ anion decomposition but with much higher transition state barrier, which results in better reversible Zn deposition on Zn metal anode in non-aqueous Zn electrolytes as compared to Mg electrolytes. Another category of battery that has gained much attention recently includes chemical transformation systems such as Li-S batteries, which stores energy through chemical bonds instead of intercalation. Here, we study the effect of salt anion and the solvent on the solvation structure and dynamics of Li-polysulfide in the solution. We observe that counter anion (such as TFSI, TfO, FSI & TDI) and solvent interaction strength with Li^+ is critical in controlling polysulfide solubility.[4] Also, high mobility of counter anions can cause faster capacity loss due to enhanced polysulfide solubility and SEI layer formation. The third energy storage concept is of redox flow batteries, which have shown outstanding promise for grid-scale energy storage with improved grid stability. Here we study ionic liquid tethered ferrocene catholyte used as redox center utilizing the Fc/Fc^+ reaction for improving the performance of non-aqueous redox flow battery materials. We observed that solubility of Fc/Fc^+ redox center can be increased up to eight-fold in a mixture of carbonate electrolytes through the modification Fc with a polar tetraalkylammonium function group containing $TFSI^-$anion. It was also observed that at solubility limit, the precipitation of solute is initiated through agglomeration of contact-ion pairs due to overlapping solvation shells. This works shows that the combination of multi-scale modeling with experimental techniques provides unprecedented insight into the origin of the electrochemical, structural, and transport properties of electrolytes, which is crucial in designing electrolytes for beyond Li-ion batteries.

References

[1] X. Qu, A. Jain, N. N. Rajput,Y.Zhang, S.P.Png, M. Brafman, L.Cheng, E. Maginn, L.A. Curtiss, K.A. Persson, "Electrolyte Genome Project: A Big Data Approach in Battery Materials Discovery", Computational Materials Science, 2015.

[2] N. N. Rajput, X. Qu, N. Sa, A. K. Burrell, K.A. Persson, "The Coupling between Stability and Ion Pair Formation in Magnesium Electrolytes from First-Principles Quantum Mechanics and Classical Molecular Dynamics", Journal of the American Chemical Society, 2015.

[3] S. Han, N.N. Rajput, X. Qu, B. Pan, M. He, M. Ferrandon, C.Liao, K.Persson, A. Burrell, ACS Applied Materials & Interfaces, 2016, 8(5), pp 3021-3031

[4] M. Vijaykumar, N. Govind, E. Walter, S. D. Burton, A. Shukla, A. Devraj, J. Xiao, J. Liu, C. Wang, A. Karim, S. Thevuthasan,"Molecular structure and stability of dissolved lithium polysulfide species" Physical Chemistry Chemical Physics, 2014

[5] K. S. Han, N. N. Rajput, X. Wei, W. Wang, J. Z. Hu, K. A. Persson, K. T. Mueller, "Diffusional motion of redox centers in carbonate electrolytes", The Journal of Chemical Physics, 2014

464841 The Influence of Tetrabutylammonium (TBA) Salts on the Discharge and Charge Behavior of Li-O_2 Batteries

Abstract:

The rise of carbon dioxide (CO_2) emissions from fossil fuel use due to rapid industrial growth has exacerbated changes in the climate. Alternative energy sources such as solar and wind have been developed, but extensive commercial utilization has been hindered by their intermittency. To complement these energy conversion methods, energy storage media such as batteries and fuel cells are being developed. Of the battery chemistries, Li-air (O_2) appears promising because it can allow for an order of magnitude greater gravimetric energy density than current commercial Li-ion batteries. However, numerous challenges face Li–O_2 batteries that stem from electrolyte, electrode, and salt degradation, and the limited oxygen reduction and evolution kinetics. These challenges lead to limited current rates, poor capacity retention, and lower cycle life.

Additives such as water, fluorinated solvents, and salts like $LiNO_3$ have been explored to modify the intermediate (superoxide) and O_2 solubility to increase discharge capacities. During charging, catalysts or redox mediators have also been incorporated to aid in the decomposition of the discharge product – Li_2O_2; a bulk insulator that requires charging potentials above 4 V_{Li}.

Ammonium-based salts such as tetrabutylammonium (TBA) have been shown to support oxygen reduction and evolution reactions using cyclic voltammetry, and TBA-superoxide complexes are known to form during reduction and the complexes are easily removed during oxidation. However, these salts have not been studied in actual Li–O_2 batteries.

In this work, we studied the influence of TBA on the discharge and charge behavior of Li–O_2 batteries. We fabricate Li–O_2 cells using TBA as the only salt present in the electrolyte. Despite the lack of lithium salt, a typical discharge plateau (~2.6 V_{Li}) is observed – similar to the voltage obtained with lithium salt – and lithium peroxide (Li_2O_2) is formed as the primary discharge product. The mechanism for discharge was studied, and the influence of solvent on the discharge process was also explored. The presence of TBA also affects the charging behavior, leading to lower charging overpotentials (compared to lithium salts) and Li_2O_2 oxidation. We investigated the charging mechanism and show that oxidation of TBA during charge is responsible for the subsequent Li_2O_2 oxidation. Knowledge gained from this work can spur greater understanding of the discharge and charge mechanisms of Li–O_2 batteries.

467844 Freestanding V_2O_5-PEDOT Thin Film Electrode for Rechargeable Aqueous K-Ion Energy Storage

Abstract:

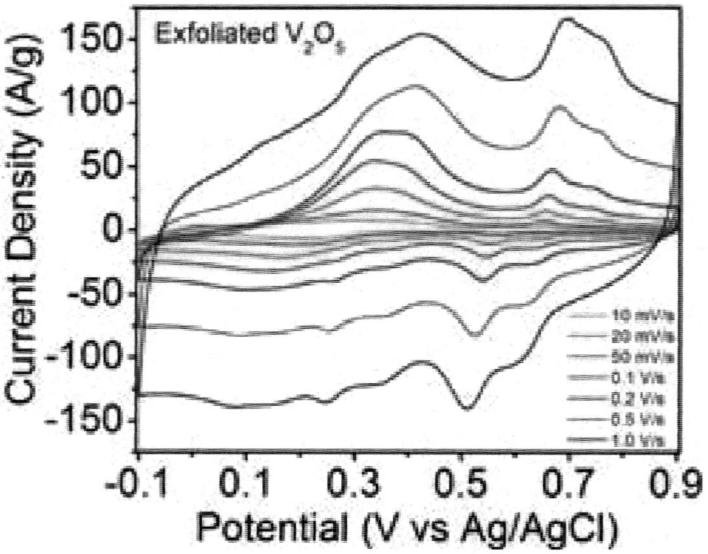

Rechargeable aqueous electrochemical energy storage (EES) devices are a promising technology for a wide range of energy storage applications due to their low cost and superior safety compared to Li-ion batteries. However, their low energy density has inhibited implementation. Materials with layered structures possess favorable characteristics for improving the energy density of aqueous EES devices. In particular, for aqueous EES using cations with larger ionic radii (e.g. Na-ions and K-ions), which are more abundant and therefore, more sustainable as well as less expensive than Li-ions. Herein we report the exfoliation of bulk vanadium pentoxide using the conductive polymer poly(3,4-ethylenedioxythiophene)-poly(styrenesulfonate) (PEDOT:PSS) via a wet chemical method. The resulting exfoliated V_2O_5-PEDOT:PSS metal oxide-polymer composite can be fabricated into freestanding thin film electrode with simple processing techniques. Structural characterization of the exfoliated V_2O_5-PEDOT:PSS composite was investigated with various techniques including synchrotron X-ray diffraction and neutron total scattering analyzed in real space with the Atomic Pair Distribution Function. A structural rearrangement from bulk vanadium pentoxide to a structure with large interlayer spacing and a coherence length on the nanoscale is observed upon exfoliation. Cyclic voltammetry measurements conducted in a 1M KCl electrolyte and 1V potential window show multiple redox features, which remain visible at all scan rates tested between 10 and 1000 mV/s. The exfoliated V_2O_5-PEDOT:PSS nanocomposite not only has the ability for high capacity K-ion storage (605 F/g at 10 mV/s, or 168 mAh/g 100s discharge time) but also the ability to operate at fast charge-discharge rates (203 F/g at 1 V/s, or 56 mAh/g at a 1s discharge time). This work demonstrates a scalable top down approach for the development of a freestanding, high capacity and high power nanocomposite electrode for aqueous K-ion energy storage.

458651 Theoretical Consideration of Nanostructured Magnetite As an Electrode Material for Li-Ion Energy Storage

Abstract:

Theoretical Consideration of Nanostructured Magnetite as an Electrode Material for Li-ion Energy Storage

C. N. Lininger[a], M. S. Hybertsen[b], A. C. West[a] [a]Department of Chemical Engineering, Columbia University, New York, NY 10027, USA

[b]Center for Functional Nanomaterials, Brookhaven National Laboratory, Upton, NY 11973, USA

The electrode material in a Li-ion battery must allow for reversible intercalation of lithium, and there are a range of materials capable of this behavior. Magnetite, Fe_3O_4, is a low cost, environmentally inert, iron oxide that allows for intercalation into its crystal lattice. Upon full reduction, magnetite has a high theoretical capacity against Li/Li^+ at 926 mAh/g as opposed to the current commercial cathode materials operating in the 100-300 mAh/g range.[1–3] Due to its low cost, low environmental impact, high theoretical capacity, and well defined nanoscale synthesis method, magnetite is a model material to push the state of the art in lithium-ion batteries.[5]

Magnetite is an inverse spinel transition metal oxide. The iron in magnetite takes on two valence states, Fe^{2+} and Fe^{3+}, spread over two coordination environments, tetrahedral 8a and octahedral 16d. The tetrahedral iron is all Fe^{3+}, while the octahedral iron centers are comprised of half Fe^{2+} and half Fe^{3+}.[7–9] The oxygen in magnetite forms a robust close packed cubic array at 32e sites, which was recently found to act as an invariant scaffold around which structural rearrangements occur during lithiation.[10] Due to the multivalent irons in magnetite, it is capable of accepting multiple electrons per iron during discharge. Multiple-electron transfer (MET) materials are of great interest to the battery community as the ability of a single cation center to accept multiple electrons greatly increases energy density.[10] However, as a result of facilitating multiple electrons, the metal center goes through oxidation state changes and structural rearrangements occur within the material. Phase transformations can be irreversible and are known to affect cyclability and limit the current rates that conversion-materials can be operated under.[10]Due to the complexity of MET materials, there exists a need for fundamental research into the electronic and ionic transport processes that are occurring in these MET materials during discharge.

Magnetite can be synthesized as a nanomaterial, with spherical nanoparticles ranging from 6-12 nm in a low-temperature synthetic method that does not involve extraneous synthetic processes.[4,5] Upon manufacture into an electrode, the magnetite nanoparticles are observed to agglomerate into micron-sized agglomerates.[6] We have developed continuum-level models for the electrochemical performance of this material including both nanoscale and agglomerate-scale physics. Neither exclusively nanoscale or agglomerate-scale simulations were able to accurately capture the experimentally observed electrochemical performance of magnetite.[6,11] In the agglomerate, ionic transport occurs through nanoporous pathways in-between the aggregated nanoparticles. On this length scale, the size of the diffusing species and the pores through which the diffusion occurs are both in the nano regime. On the nanoscale, solid-state diffusion occurs as lithium inserts into the crystalline magnetite. Depending on concentration, this causes structural rearrangements. The crystalline diffusivity can vary by orders of magnitude as a function of the state of discharge, due to these structural rearrangements that occur during lithiation.

We have used density functional theory to explore structural rearrangements that occur in magnetite during lithiation between zero and one equivalents. We have established that structural rearrangements include the movement of tetrahedral iron to an octahedral coordination environment at concentrations as low as half an equivalent of lithium. We have also established that at this concentration, the lithium has lower energy in octahedral coordinated sites as compared to tetrahedral coordination. While evidence pointed to such changes in coordination, in all previous literature, the lithium concentration at which it occurred had not been established.

Additionally, we are using DFT to explore i) the diffusivity of lithium in crystalline solid magnetite and ii) the lithium open circuit potential in the dilute concentration limit. The use of DFT to understand the kinetic resistances to the structural rearrangements that we have shown occur at low levels of lithiation will give insights into relaxation times observed for this material. Both the DFT determined solid-state diffusivity and the theoretically predicted open-circuit potential will aid in informing the limiting ionic and electronic processes occurring in magnetite for its uses as an electrode material in lithium-ion batteries.

This work was supported as part of the Center for Mesoscale Transport Properties, an Energy Frontier Research Center supported by the U. S. Department of Energy, Office of Science, Basic Energy Sciences, under award #DE-SC0012673. C. N. L. acknowledges the support of the National Science Foundation Graduate Research Fellowship under Grant No. DGE-11-44155. Any opinions, findings, and conclusions or recommendations expressed in this material are those of the authors and do not necessarily reflect the views of the National Science Foundation. This research used resources of the Center for Functional Nanomaterials, which is a U.S. DOE Office of Science Facility, at Brookhaven National Laboratory under Contract No. DE-SC0012704. The Yeti Shared HPC Cluster at Columbia University was also used, which includes support from Empire State Development's Division of Science, Technology, and Innovation under contract number C090171.

(1) Doeff, M. M. In *Encyclopedia of Sustainability Science and Technology*; Meyers, R. A., Ed.; 2013; pp. 5–49.

(2) Menard, M. C.; Takeuchi, K. J.; Marschilok, A. C.; Takeuchi, E. S. *Phys. Chem. Chem. Phys.* **2013**, *15*, 18539–18548.

(3) Ito, S.; Nakaoka, K.; Kawamura, M.; Ui, K.; Fujimoto, K.; Koura, N. *J. Power Sources* **2005**, *146*, 319–322.

(4) Zhu, S.; Marschilok, A. C.; Takeuchi, E. S.; Yee, G. T.; Wang, G.; Takeuchi, K. J. *J. Electrochem. Soc.* **2010**, *157*, 1158–1163.

(5) Zhu, S.; Marschilok, A. C.; Takeuchi, E. S.; Takeuchi, K. J. *Electrochem. Solid-State Lett.* **2009**, *12*, 91–94.

(6) Knehr, K. W.; Brady, N. W.; Cama, C. A.; Bock, D. C.; Lin, Z.; Lininger, C. N.; Marschilok, A. C.; Takeuchi, K. J.; Takeuchi, E. S.; West, A. C. *J. Electrochem. Soc.* **2015**.

(7) Thackeray, M. M.; David, W. I. F.; Goodenough, J. B. *Mater. Res. Bull.* **1982**, *17*, 785–793.

(8) Sickafus, K. E.; Wills, J. M. *J. Am. Ceram. Soc.* **1999**, *82*, 3279–3292.

(9) Menard, M. C.; Takeuchi, K. J.; Marschilok, C.; Takeuchi, E. S. **2013**, 18539–18548.

(10) Zhang, W.; Bock, D. C.; Pelliccione, C. J.; Li, Y.; Wu, L.; Zhu, Y.; Marschilok, A. C.; Takeuchi, E. S.; Takeuchi, K. J.; Wang, F. *Adv. Energy Mater.* **2016**.

(11) Knehr, K. W.; Brady, N. W.; Lininger, C. N.; Cama, C. A.; Bock, D. C.; Lin, Z.; Marschilok, A. C.; Takeuchi, K. J.; Takeuchi, E. S.; West, A. C. *ECS Trans.* **2015**, *69*, 7–19.

465161 Highly Flexible Self-Assembled V_2O_5 Cathodes Enabled By Conducting Diblock Copolymers

Abstract:

Structural energy storage materials combining load-bearing mechanical properties and high energy storage performance are desired for applications in wearable devices or flexible displays. Vanadium pentoxide (V_2O_5) is a promising cathode material for possible use in flexible battery electrodes, but it remains limited by low Li^+ diffusion coefficient and electronic conductivity, severe volumetric changes upon cycling, and limited mechanical flexibility. Here, we demonstrate a route to address these challenges by blending a diblock copolymer bearing electron- and ion-conducting blocks, poly(3-hexylthiophene)-*block*-poly(ethyleneoxide) (P3HT-*b*-PEO), with V_2O_5 to form a mechanically flexible, electro-mechanically stable hybrid electrode. V_2O_5 layers were arranged parallel in brick-and-mortar-like fashion held together by the P3HT-*b*-PEO binder. . This unique structure significantly enhances mechanical flexibility, toughness and cyclability without sacrificing capacity. Properties are probed as a function of composition using dynamic mechanical analysis, impedance spectroscopy, cyclic voltammetry, and galvanostatic cycling. It is found that optimum composition exists, where too little polymer results in brittle inflexible electrodes and too much polymer reduces the electrochemical activity. Only 5 wt % polymer is required to triple the flexibility (normal strain) of V_2O_5, and electrodes comprised of 10 wt % polymer have unusually high toughness (293 kJ/m^3) and specific energy (530 Wh/kg), both higher than reduced graphene oxide paper electrodes. Electrodes with commercial binder poly(vinylidene difluoride) shows no remarkable improvement. These results demonstrate a new approach for the development of structural energy storage materials.

472403 Combined Experimental and Theoretical Study of Oxygen Mobility in Hydrated Alpha-MnO2

Abstract:

Alpha-MnO_2 is of interest as a cathode material for conventional lithium cells and as an electrode/electrocatalyst for hybrid Li-ion/Li-O_2 cells. It has a framework structure with large 2x2 tunnels that can contain water; the H_2O component can be ion-exchanged with lithium to create a Li_2O-stablized electrode. When used in Li-O_2 cells, it is possible that the electrocatalytic process is associated with the removal from and reintroduction of lithium and oxygen into the host structure during charge and discharge, respectively. In this contribution, we present a study of the oxygen mobility via water intercalation in alpha-MnO_2. We present evidence of the dehydration of alpha-MnO_2 from temperature-resolved in-situ synchrotron powder x-ray diffraction, thermogravimetric analysis, and in-situ UV-resonance Raman spectroscopy; and structural, thermodynamic, and kinetic information from density functional theory calculations. We also present our analysis of the charge state of the hydration species from density functional theory, redox titration, and x-ray absorption near-edge spectroscopy. Finally, we discuss implications for improved sustainability in the practical usage of alpha-MnO_2 for reversible Li or Li-O_2 storage.

461208 Template Particle & Solvent Evaporation Assisted Phase Inversion: Easy and Robust Control of Filtration Membrane Morphologies for Various Applications

Abstract:

We present a robust platform allowing the controlled nanoparticle assisted formation of porous polymer membranes.[1] Applications for porous membranes range from drinking water purification,[2] gas separation,[3] biotechnological downstream processing,[4] battery separators,[5] reverse osmosis,[6] food separation processes[7] to porous scaffolds for tissue engineering.[8]

Depending on the membraneâ€™s specific application, different morphologies are required. For example, the convenient separation of particles such as tiny parvoviruses from pharmaceutical goods, requires membranes thick and flexible enough to be mechanically stable while exhibiting appropriate pore sizes. Furthermore for maximizing the flux rate, asymmetric membranes are mostly preferred, as they comprise thin selective layers containing small pores.

Our filtration membrane fabrication procedure allows the simple and exact shaping of porous polymer membranes with previously unmaintained asymmetric properties.[1] The membrane formation is enabled by solvent evaporation induced spinodal decomposition of particles containing liquids and the usage of template particles[9] as such. Membranes produced with varying concentrations of ingredients were analyzed on their morphology at different steps during membrane synthesis. This allowed deciphering mechanistic aspects of the membrane formation, as well as the key parameters for controlling the pore size and pore size gradient. By only varying the concentration ratio of non-solvent and ZnO nanoparticles within a specific polymer dispersion, selective pore sizes form sub-20 nm up to 100 nm and various pore size gradients could be reproducibly created. A selection of membranes with different pore sizes, was

finally tested on pure water flux rates as well as silica nanoparticle retention rates, and compared to commercially available high-end virus filters.

Very recently, we discovered the benefit of pore directed ZnO nanoparticles for the synthesis of MOF-polymer composite membranes. The synthesis of MOF-polymer composite membranes depends on MOF-precursors which are ideally in close contact with the membrane. For synthesis of MOF-polymer composite membranes the ZnO-nanoparticles, instead of being dissolved, were maintained within the pores and used as zinc source for the solvothermal synthesis of mechanically flexible ZIF-8 composite membranes. ZIF-8-polymer composite membranes were tested on H2, N2 and CO2 single gas permeation values.

In addition to the methodâ€™s simplicity and the large scale availability of all the ingredients, it has great upscale-potential via roll-to-roll coating.[10]

1. Hess, S. C.; Kohll, A. X.; Raso, R. A.; Schumacher, C. M.; Grass, R. N.; Stark, W. J. Template-Particle Stabilized Bicontinuous Emulsion Yielding Controlled Assembly of Hierarchical High-Flux Filtration Membranes. ACS applied materials & interfaces 2015, 7, 611-617.

2. Shannon, M. A.; Bohn, P. W.; Elimelech, M.; Georgiadis, J. G.; Marinas, B. J.; Mayes, A. M. Science and technology for water purification in the coming decades. Nature 2008, 452, 301-310.

3. Qiu, S. L.; Xue, M.; Zhu, G. S. Metal-organic framework membranes: from synthesis to separation application. Chem Soc Rev 2014, 43, 6116-6140.

4. Rathore, A. S.; Shirke, A. Recent Developments in Membrane-Based Separations in Biotechnology Processes: Review. Preparative Biochemistry & Biotechnology 2011, 41, 398-421.

5. Liu, H. Y.; Liu, L. L.; Yang, C. L.; Li, Z. H.; Xiao, Q. Z.; Lei, G. T.; Ding, Y. H. A hard-template process to prepare three-dimensionally macroporous polymer electrolyte for lithium-ion batteries. Electrochimica Acta 2014, 121, 328-336.

6. Kosutic, K.; Kastelan-Kunst, L.; Kunst, B. Porosity of some commercial reverse osmosis and nanofiltration polyamide thin-film composite membranes. J Membrane Sci 2000, 168, 101-108.

7. Makardij, A.; Chen, X. D.; Farid, M. M. Microfiltration and ultrafiltration of milk: Some aspects of fouling and cleaning. Food and Bioproducts Processing 1999, 77, 107-113.

8. Madihally, S. V.; Matthew, H. W. T. Porous chitosan scaffolds for tissue engineering. Biomaterials 1999, 20, 1133-1142.

9. Kellenberger, C. R.; Luechinger, N. A.; Lamprou, A.; Rossier, M.; Grass, R. N.; Stark, W. J. Soluble nanoparticles as removable pore templates for the preparation of polymer ultrafiltration membranes. J Membrane Sci 2012, 387, 76-82.

10. Kellenberger, C. R.; Hess, S. C.; Schumacher, C. M.; Loepfe, M.; Nussbaumer, J. E.; Grass, R. N.; Stark, W. J. Roll-to-Roll Preparation of Mesoporous Membranes by Nanoparticle Template Removal. Ind Eng Chem Res 2014, 53, 9214-9220.

466886 Intrinsically Microporous Polyimides Incorporating TröGer's Base (TB) for Membrane Gas Separation

Abstract:

Tröger's Base (TB), which is a rigid, V-shaped, and bridged bicyclic amine, has been studied during more 130 years especially in organocatalysis and stereochemistry and recently, McKeown introduced the TB units into polymers of intrinsic microporosity (PIMs) for gas separation membranes. The resultant PIM-TBs exhibited an improvement in gas transport behaviors, indicating that incorporation of the TB units into polymer backbones endows polymer rigidity and microporosity.

In this study, rigid TB moieties were incorporated in polyimides (PI-TBs) *in-situ* polycondensation reactions using dimethoxymethane (DMM). Polyimides are commonly used for membrane gas separation due to their advantages such as good thermal and chemical stabilities, and processability. Solution-processable PI-TB membranes in this work exhibited an enhancement of mechanical properties and gas transport behaviors compared to conventional polyimide membranes, indicating that TB units imparted their microporosity and improved free volume elements of polyimide chains.

Six copolyimides incorporating Tröger's Base (CoPI-TBs) were also investigated for the first time. The CoPI-TBs were prepared *in-situ* TB formation with dimethoxymethane (DMM) and two kinds of diamines, imide-containing diamines and three kinds of flourene-based diamines. Molecular architecturing of CoPI-TBs was tailored of the ratio of two diamines and species of flourene moieties and the structure-property relationship including the physical properties and gas separation properties of the corresponding CoPI-TBs was evaluated.

469370 Comparison of Performance Between Laboratory Scale and Industrial Scale Membranes

Abstract:

Abstract

Cellulose acetate (CA) membranes are among the most researched polymers for separation and barrier applications because of their many useful characteristics including relative low cost, good toughness, high biocompatibility, good desalting, high potential flux. There are different methods used to fabricate porous polymeric membranes, and these can lead to irreproducible results, especially during scale up. The performance of the membrane is greatly influenced by not only the material but also the processing methods and conditions, because they affect the morphology of the membrane. Generally, a laboratory scale approach, such as knife over blade or solution casting, is used to cast a thin sheet of the polymeric membrane solution. Then, the final form of the membrane is prepared by sintering, stretching, or phase inversion. In many studies, the primary film casting process has been limited to doctor blade extrusion, however, this method is not always the most suitable for the industrial scale, because the solution exposed to the environment during the processing phase. Alternatively, slot die extrusion is a well-developed pre-metered approach that allows for continues casting of liquid film in a controlled environment, which also minimizes changes in liquid properties, such as rheology. The slot die coating process is a commonly used process for the manufacturing of plastics and polymer films, as well as coatings and optical films for liquid crystal displays in industry. For the scale up fabrication of hollow fiber CA membranes, a hollow fiber spin-line was used. The feasibility of processing cellulose acetate using a laboratory-scale doctorâ€™s blade versus a scale up slot die extrusion vs hollow fiber membrane was examined. The effects of processing methods, conditions, and substrate on the morphology and on the flux of CA membranes were studied.Â CA membranes were tested for their performance under the same operational conditions of pressure, feed volume, and feed concentration.

472794 Interfacial Polymerization for Sugar / Ionic Liquid Separations

Abstract:

Ionic Liquids (IL) are tunable green solvents for a variety of chemical and biochemical processes. However, the economic viability of utilizing these green solvents depends on developing processes for their recycle and reuse. ILs are outstanding solvents for biomass fractionation and sugar conversion processes. However, separating saccharides from ILs is generally very difficult due to the similarity of their molecular weight. Here a selective skin layer containing 3-aminophenylboronic acid has been successfully synthesized via interfacial polymerization on the surface of asymmetric polyethersulfone (PES) membranes. The fabricated composite membranes show superior ionic liquid/sugar selectivity at competitive permeability. Moreover, the selectivity can be tuned by varying the monomer composition, the polymerization time and post treatment conditions. While [EMIM]Cl and [BMIM]Cl are able to selectively permeate through the membranes, over 90% of mono- and oligosaccharides are rejected. The compositional, chemical and morphological properties of these nanofiltration membranes are characterized and correlated with their separation characteristics.

455188 Development of High Performance Carboxylated PIM-1/P84 Blend Membranes for Pervaporation Dehydration of Isopropanol and CO_2/CH_4 Separation

Abstract:

By taking the advantages of high permeability characteristics of carboxylated PIM-1 (cPIM-1) and high selectivity properties of P84, a new type of cPIM-1/P84 blend membranes has been designed in this study for pervaporation dehydration of isopropanol and CO_2/CH_4 separation. P84 is known to be a highly selective polyimide with low permeability for alcohol dehydration and gas separation, while cPIM-1 is a highly permeable PIM-1 modified via hydrolysis reaction. Polarized light microscopy was employed to study their homogeneity. It was found that the P84/cPIM-1 blends were fully miscible when either cPIM-1 or P84 content was at or lower than 10wt%. The molecular-level interactions were further confirmed by FTIR and DSC. The cPIM-1 loading has significant impact on sorption and diffusion of water and IPA in blends, as indicated by vapor sorption tests. The pervaporation dehydration of isopropanol (IPA) shows that the incorporation of only 10wt% cPIM-1 in the P84 matrix enhances the water permeability by 87% without compromising its selectivity. It has an IPA/water selectivity of 6263 and a water permeability of 0.0864 mg m^{-1} h^{-1} Kpa^{-1}. Similarly, the addition of 30wt% cPIM-1 into P84 increases the CO_2 permeability by 430% from the original 1.3 to 6.98 Barrers. The P84/cPIM-1 membrane containing 90% cPIM-1 has the CO_2/CH_4 separation performance surpassing the Robson upper bound. The newly developed compatible cPIM-1/P84 blend membranes exhibit great potential for the applications of alcohol dehydration and gas separation.

458504 Effects of Fabrication Conditions on the Microstructures and Performances of Smart Gating Membranes with *in Situ* assembled Nanogels As Gates

Abstract:

ABSTRACT

Smart gating membranes with *in situ* assembled poly(*N*-isopropylacrylamide) (PNIPAM) nanogels as gates are successfully prepared via vapor induced phase separation (VIPS) with different exposure time periods, temperatures and relative humidities of the water vapor. Effects of the fabrication conditions on the microstructures as well as the thermo-responsive and mechanical performances of the membranes are investigated. Both the membrane microstructure and the movement of blended PNIPAM nanogels in the membrane forming solution can be controlled by adjusting the fabrication conditions. With increasing the exposure time, the membrane microstructure undergoes a transition from typical liquid-induced phase separation (LIPS) structure (unsymmetrical finger-like porous structure) to typical VIPS structure (symmetric cellular-like porous structure). The critical time periods for the microstructure transition of membranes prepared with vapor temperature and relative humidity of 25 °C/90%, 25 °C/70% and15 °C/70% are about 1.5 min, 2 min and 10 min respectively. The performances of membranes are heavily dependent on the microstructures. The membranes with unsymmetrical finger-like porous structures show large thermo-responsive factor ($R_{39/20}$, the ratio of water flux at 39 °C to that at 20 °C), with the maximum $R_{39/20}$ value being 43.2. All the membranes with symmetric cellular-like porous structures exhibit strong mechanical property and large water flux at 39 °C, which is higher than the volume phase transition temperature (VPTT) of PNIPAM nanogels. The results provide valuable guidance for rational design and fabrication of smart gating membranes with desirable performances.

453480 Novel Thermally Cross-Linked Polyimide Membranes for Ethanol Dehydration Via Pervaporation

Abstract:

In this work, two novel carboxyl-containing polyimides, 2,2′-bis(3,4-dicarboxyphenyl) hexafluoropropane dianhydride-4,4′-diaminodiphenylmethane/3,5-diaminobenzoic acid (6FDAMDA/DABA, FMD) and 3,3′,4,4′-benzophenone tetracarboxylic dianhydride-4,4′-diaminodiphenylmethane/3,5-diaminobenzoic acid (BTDA-MDA/DABA, BMD), are synthesized via chemical and thermal imidization methods, respectively, and employed as pervaporation membranes for ethanol dehydration. Chemical structures of the two polyimides are examined by FTIR and TGA to confirm the successful synthesis. A post thermal treatment of the polyimide membranes with the temperature range of 250 to 400 °C is applied, and its effects on the membrane morphology and separation performance are studied and characterized by FTIR, TGA, WXRD, solubility and sorption test. It is believed that the thermal treatment of the carboxyl-containing polyimide membrane at a relative low temperature only leads to the physical annealing, while it may cause the decarboxylation-induced crosslinking at a higher temperature. In addition, the operation temperature in pervaporation is also varied and shown to be an important factor to affect the final membrane performance. Performance benchmarking shows that the developed polyimide membranes both have superior pervaporation performance to most other flat-sheet dense membranes. This work is believed to shed useful insights on polyimide
membranes for pervaporation applications.

452769 Recent Advances in Evapoporometry for Determining the Pore-Size Distribution of Membranes

Abstract:

Evapoporometry (EP) is a novel means for characterizing the pore-size distribution (PSD) of membranes. It is based on the phenomenon described by the Kelvin equation whereby the vapor pressure is reduced owing to the interfacial curvature of a volatile wetting liquid contained in small pores. EP employs a specially designed test cell that permits establishing a partial pressure in the gas phase at the surface of the membrane that corresponds to the vapor pressure of the largest pores that are filled with the volatile wetting liquid. A small hole in the lid of the test cell controls the rate at which evaporation of the volatile wetting liquid occurs from the pores. Owing to the implications of the Kelvin equation, the gas phase adjacent to the membrane is supersaturated with respect to all pores that are smaller than those from which liquid is evaporating. This supersaturation condition causes pores to empty via evaporation progressively from the largest to the smallest. Measuring the instantaneous mass of the test cell containing the membrane sample via a high resolution microbalance permits determining the instantaneous evaporation rate. This can be related to the partial pressure of the volatile wetting liquid at the surface of the membrane via a straightforward analysis of the mass transfer in the test cell. The Kelvin equation then permits determining the instantaneous pore diameter from the partial pressure at the surface of the membrane. The limitations of the Kelvin equation imply that EP can determine the PSD for flat sheet and hollow fiber ultrafiltration membranes.

EP offers several advantages relative to other methods for determining the PSD. Since the mass can be measured with microgram accuracy, the measurement error in EP is much less than for techniques that require measuring a flow rate, pressure, volume, or heat input. Implementing EP does not require any specialized equipment or expertise. EP also has a relatively small laboratory footprint since the required equipment consists of the test cell, high resolution microbalance, environmental chamber, and an anti-vibration table. Since EP characterization is carried out at the ambient temperature and pressure, it does not alter the morphology of the membrane sample. EP also can be used to determine the PSD of fully hydrated viable biofilms that cannot be done via other PSD characterization methods. A particular advantage of EP is that it can be used to measure the effect of internal pore fouling on the PSD that cannot be done using any method that involves liquid displacement via an inert gas or immiscible liquid.

This presentation will provide an overview of the EP characterization method and discuss recent advances in this technology. The latter include an improved EP test cell design, improving the resolution of EP, extending EP to larger pore sizes, accounting for the effect of the residual adsorbed layer on the pore walls, and adapting EP for characterizing the PSD on both the outer surface and lumen side of hollow fiber membranes.

450165 Inhibitory Effect of Condensed and Non-Condensed Phenolic Moieties in Lignin on Enzymatic Hydrolysis

Abstract:

The residual lignin plays an inhibitory role in the enzymatic hydrolysis of cellulosic biomass. In this study, we examined the structure change of the isolated lignins from hot water hydrothermal pretreatment of aspen and their inhibitory effects on the enzymatic hydrolysis of Avicel. The functional groups of the isolated lignins were determined by quantitative ^{13}C, 2D HSQC and ^{31}P NMR. The increase of pretreatment temperature significantly increased the condensed and non-condensed syringyl and guaiacyl OH group contents in the isolated lignins, and decreased the aliphatic OH, p-hydroxybenzoate OH and carboxylic OH group contents. A compelling adverse association (r^2=0.998) was observed between the condensed syringyl and guaiacyl phenolic OH group contents in lignins and their inhibitory effects on enzymatic hydrolysis. Langmuir adsorption isotherms showed that higher pretreatment temperature resulted in higher binding ability between isolated lignins and enzymes, which led to more non-productive binding. It is hypothesized that condensed syringyl and guaiacyl phenolic units are mainly responsible for the inhibitory effect of lignin on enzymatic hydrolysis, in which the condensed aromatic rings enhance the hydrophobic interactions and the phenolic OH group boost the hydrogen bonding. The combination of hydrophobic interactions and hydrogen bonding can further strengthen the non-productive binding.

467057 Measurement and Evaluation of Recalcitrance Changes for Organic Solvent Pretreatment Methods

Abstract:

Biomass recalcitrance is largely responsible for the high cost of chemical and fuel production from lignocellulose. To enhance biodegradation for these applications, it is crucial to determine how a pretreatment method affects the physical characteristics of lignocellulose[1,2]. In this work, we show the efficacy of Direct Yellow 11 as a probe to measure the cellulosic accessibility to cellulase enzymes. Together with crystallinity and delignification measurements, this accessibility measurement technique provides insight into recalcitrance changes that enhance biomass susceptibility towards enzymatic conversion to glucose[2,3]. Here we elucidate how a variety of organic solvents affect the key physical characteristics of biomass. These physical characteristic changes provide an increased understanding of biomass recalcitrance and its effect on cellulase kinetics.

1. Chandra, R., et al. (**2008**). "The characterization of pretreated lignocellulosic substrates prior to enzymatic hydrolysis, part 1: a modified Simons' staining technique." Biotechnology progress 24(5): 1178-1185.
2. Kang, Y. et al. (**2013**). "SO2 catalyzed steam explosion: The effects of different severity on digestibility, accessibility, and crystallinity of lignocellulosic biomass." Biotechnology progress 29.4: 909-916.
3. Hall, M. et al. (**2010**). "Cellulose crystallinity–a key predictor of the enzymatic hydrolysis rate." FEBS journal 277.6: 1571-1582.

473939 One Step Breaking Softwood Recalcitrance By Cellulose Solvent-Based Lignocellulose Fractionation Followed By Enzymatic Saccharification

Abstract:

In our efforts to shift away from conventional petroleum-based feedstock to supply transportation fuels, plant biomass has emerged as an attractive renewable alternative. Softwoods have great potentials as renewable feedstocks for biorefineries. However, softwoods are more recalcitrant compared to hardwood counterparts. Hence, the use of softwoods in biorefineries has been viewed as limited. In the present study, we applied a cellulose solvent-based pretreatment/fractionation process on southern yellow pine and douglas fir at a mild condition: 50 ºC and 1 atm for 90 min. Pretreated softwood samples were hydrolyzed fast by cellulase (15 FPUs) and > 85% glucose yield was achieved by 24 h. Cellulose accessibility to cellulase of pretreated samples was probed by the fusion protein and was found to increase by 39 folds. After fractionation, degree of crystallinity of pretreated samples decreased and fibrous structure appeared to be disrupted as observed by solid-state ^{13}C Cross-Polarization Magic Angle Spinning (CP/MAS) nuclear magnetic resonance (NMR) and scanning electron microscopy. No changes in lignin functionalities after fractionation were observed by 2D Heteronuclear Single-Quantum Correlation (HSQC) NMR, giving rise to the potential use of lignin as a co-product in biorefineries.

472908 A Novel Acid Catalyst for Conversion of Ethanol to Butanol

Abstract:

The extensive consumption of fossil fuels contributes to a long-term threat to the ecosystem on which people highly depend. . Accordingly, people are increasingly interested in developing biorenewable fuels, among them is N-butanol. Due to its desired noncorrosivity, immiscibility with water, and relatively high energy density, preparing N-butanol has gained increasing attention. The major common challenge of the recently reported N-butanol production technologies lies in their low N-butanol selectivity. The novel acid catalyst (instead of base catalysts used in literature) developed in this work has achieved the highest selectivity to date. The effects of different factors on N-butanol selectivity and yield are evaluated. In addition, the associated reaction mechanism is understood with various material and process characterization methods.

475639 High Temperature Solar Thermochemical Energy Storage Materials

Abstract:

High temperature solar thermochemical energy storage (STCES) has promise to provide a low cost and high temperature storage solution for solar thermal power.Â However, the current status of STCES is at an early stage of development and substantial research and development is required to realize practical implementation. Two thermochemical energy storage systems with high theoretical energy densities at high temperatures have been investigated. In both the cases, we have been able to develop stable energy storage materials with energy densities in excess of 1200 MJ m^{-3} that can operate at temperatures above 1100 â„ƒ in both cases.

The first system pertains to thermochemical energy storage via decomposition and carbonation of strontium carbonate/strontium oxide in the temperature range of 1150-1235Â°C. Porous structures of pure $SrO/SrCO_3$ were studied via a fractional factorial experimental study to verify the possibility of synthesizing stable undoped $SrO/SrCO_3$ structures using the sacrificial pore formation with graphite (SPFG) method. Results of the experimental study indicate that the SPFG method is not suitable for fabricating porous structures of pure SrO or $SrCO_3$ for high temperature thermochemical energy storage. Cyclic volumetric energy density (including only the chemical reaction contribution) calculated from thermogravimetric measurements decreases substantially to values below 600 MJ m^{-3} upon repeated cycling, demonstrating that the SPFG method not viable. Further investigations revealed that doping $SrO/SrCO_3$ with MgO and BaO improved the cyclic stability significantly. Mixtures of SrO-MgO and SrO-BaO in the molar ratio of 2:3 appear to achieve cyclic stability at 860 ± 5 MJ m^{-3} (500 ± 3 kJ kg^{-1}) and 1408 ± 15 MJ m^{-3} (1001 ± 11 kJ kg^{-1}) respectively over 15 cycles. Pure BaO was also tested for thermochemical energy storage and is found to be more stable with a higher energy density of 1690 ± 24 MJ m^{-3} (1070 ± 15 kJ kg^{-1}).

A second study was performed on transition metal oxide based thermochemical energy storage systems. The purpose of the investigation was to validate the use of magnesium oxide (MgO) as a sintering inhibitor/porous structure stabilizer for high temperature solar thermochemical energy storage (STCES) using reversible redox reactions. Solid solutions of reactive metal oxides in MgO prepared using the solid state reaction method (SSR) were investigated for STCES applications. A new concept of using steam and carbon dioxide to lower the partial pressure of oxygen during thermal reduction was tested and validated. 40 and 33 mole % MnO in MgO, 20 mole % CuO in MgO and 40 mole % Fe_2O_3 in MgO were investigated for high temperature thermochemical energy storage via redox reactions. A composition of 40 mole % MnO in MgO appeared to be the most promising material among those tested here for high temperature thermochemical energy storage. The material can release heat at a temperature greater than 1200°C and has an energy storage capacity of 64212 $kJkg^{-1}$ (122236 MJm^{-3}). The material appeared stable over 20 cycles in a thermogravimetric analyzer and 12 cycles in a tube furnace bench scale reactor and is expected to maintain stability over repeated cycling.

466067 Techno-Economic Assessment of the Polygeneration of Liquid Fuels and Electricity By Solar Hybridised Gasification of Coal

Abstract:

Solar gasification offers the potential to convert a greater portion of the fuel input to syngas than is possible with a conventional gasifier, resulting in a higher production of Fischer-Tropsch liquids (FTL), by displacing the carbonaceous fuel that is otherwise burnt in O_2 to provide process heat. In this hybrid gasification process, concentrated solar thermal energy is used when it is available, while autothermal gasification is used when the incoming solar insolation is insufficient. Thus, the intermittency of the solar resource is accommodated through the use of intermediate storage of syngas and turning down the liquid fuel production rate. This paper presents a techno-economic evaluation of a coal-to-liquids process integrated with a solar hybridised, oxygen blown, atmospheric pressure vortex-flow gasifier (SCTL) located at Daggett, USA. The influence of the solar multiple (SM) and syngas storage capacity (hours) on the plant feasibility was assessed. Here, the SM is the ratio of the heliostat field area to the required heliostat field area that can generate a net annual peak solar thermal energy matching the energy required by the gasification process.

The annual specific FTL output and the specific total energetic output (FTL and electricity) of the SCTL plant relative to the conventional coal-to-liquid (CTL) were found to increase by 12% and 15%, respectively, as the SM was increased to 3. On the other hand, the annual reduction in mine-to-tank (MTT) CO_2 emission from the SCTL relative to the CTL was reduced by 17%. The utilization factor of the FT and power generation plants in the SCTL was significantly influenced by the syngas storage capacity. The utilization factor for the SCTL with no syngas storage was calculated to be 74% compared with 93% for the SCTL with 4 hours syngas storage capacity.

To produce 1500 barrels per day of FTL, the levelised cost of fuel (LCOF) for the CTL was estimated to be USD1.2/L_{2012}. The LCOF for the SCTL with no syngas storage was increased from USD1.3/L_{2012} to USD1.4/L_{2012}, and from USD1.4/L_{2012} to USD1.5/L_{2012} with 4 hours syngas storage capacity, as the SM was increased from 0.5 to 3. Despite the fact that the outputs from the SCTL were higher than that of the CTL, this could not overcome the high cost of the solar components and the syngas storage system. Furthermore, the LCOF for the SCTL is sensitive to several parameters identified in the sensitivity analysis, such as the total plant cost, the assumed cost for carbon capture and sequestration, and the operation and maintenance costs.

452450 Solar Metallothermic Production of Rare Earth Elements

Abstract:

Rare earth elements (REEs) have many applications in metal alloys, catalysts, petroleum refining, glass additives, polishing compounds, and ceramics; they also underpin many critical clean energy technologies such as PV devices, batteries, high-performance magnets, and phosphors for lighting and electronics. The demand for rare earth elements is expected to grow in the coming years, especially in uses related to the energy sector. At present, the extractive processes for REEs and rare earth oxides (REOs) are driven by fossil fuel based energy sources with very high greenhouse gas impacts and resource depletion potential. The separation and reduction stages of the REO/REE production process have the highest energy and material consumption requirements.

Cerium (Ce), lanthanum (La), neodymium (Nd), and praseodymium (Pd) REEs and REOs accounts for 54, 25, 4, and 3% (86% combined) of the global REE supply and usage. Ce, La, Nd, and Pr metals are commercially produced by electrolytic processes where the anhydrous chlorides of the metals are dissolved in molten NaCl, KCl, $CaCl_2$, or other salts. The electrolysis of these rare earth chlorides must occur at approximately 50°C above the melting point of these metals, which are 795, 920, 1021, and 931°C for cerium, lanthanum, neodymium, and praseodymium, respectively.

Energy consumption and environmental concerns, coupled with the fact that the current electrolytic production method must occur at high temperature, provides significant motivation for investigating solar thermal processing as an alternative to conventional REE production. Here, the potential for solar metallothermic production of rare earth elements, especially Ce, La, Nd, and Pr, is critically evaluated using thermodynamic analysis.

473021 Manganese As a Redox Reactant for Sustainable Thermochemical Ammonia Synthesis

Abstract:

Fixed nitrogen, as ammonia (NH_3), is the main ingredient of most fertilizers and its availability is the most common yield-limiting factor in industrial agricultures. Responsible for up to 50% of crop yields, ammonia fertilizers provide the nitrogen necessary to sustain over half of the global population. Annual global ammonia production, currently greater than 140 million metric tons, will need to expand along with the expected increase in demand for fertilizer for food production. Catalytic ammonia synthesis via the Haber-Bosch process, operates at some of the most intense process conditions for chemical production, with temperatures as high as 500°C and pressures reaching 300 atmospheres. The Haber-Bosch process consumes 0.6 tons of natural gas and emits 2.3 tons of carbon dioxide per ton of ammonia produced.

A chemical loop is experimentally demonstrated here using manganese as a cycled redox reactant for ammonia production from nitrogen, water, and methane at atmospheric pressure. Manganese is used to reduce the nitrogen which is then reacted with steam to form ammonia in the presence of a sodium promotor. The corresponding manganese oxide is reduced by a dilute methane stream also producing syngas. The role of sodium hydroxide in the promotion of ammonia liberation is investigated. The thermodynamic predictions of the system are discussed, and the reaction kinetics are evaluated.

The thermochemical cycle appears a promising method to sustainably produce ammonia at near-ambient pressures. Separating the ammonia synthesis into nitrogen reduction and subsequent protonation steps introduces an additional degree of freedom for optimizing the redox reactant and process conditions. Reduced carbon dioxide emissions, and valuable syngas coproduct are advantageous for this novel approach going forward.

467091 Highly Selective Fischer-Tropsch Synthesis for Production of JP-8 Jet Fuel from Coal and Coal/Biomass Mixtures

Abstract:

An indirect liquefaction process employing advanced process intensification approaches is being developed to reduce costs and accelerate commercialization of a coal to liquid (CTL) process for converting coal and coal-biomass mixtures to JP-8 jet fuel. In this Air Force and DOE/NETL sponsored cooperative agreement, Southern Research is leading a team consisting of IntraMicron, Inc., Southwest Research Institute, and Nexant, Inc., along with Chevron as selective Fischer-Tropsch (FT) catalyst supplier and the National Carbon Capture Center (NCCC) as the host site. The process reduces the number of parasitic syngas cleaning steps and eliminates costly upgrading of wax produced in conventional indirect liquefaction processes, resulting in improved economics and lower lifecycle greenhouse gas (GHG) emissions for CTL-based jet fuel production. These improvements directly reduce capital costs of indirect CTL processes, resulting in significantly improved economics and cost competitiveness of CTL when compared to petroleum refining. Specific objectives for FT experiments are to demonstrate a selective and active second generation cobalt-zeolite hybrid Fischer-Tropsch (FT) catalyst from Chevron that predominantly produces liquid range C5-C20 hydrocarbons with high productivity (5X conventional FT catalysts) and jet fuel selectivity while eliminating the production and associated handling of solid C21+ wax. In fall of 2015, Southern operated a bench-scale FT reactor with a 2â€ diameter heat exchange reactor tube using IntraMicron technology in a nominal 5 gallons per day skid-mounted fixed-bed installed at NCCC. In those experiments, both productivity and selectivity goals were achieved during a 300 hour experiment. In fall 2016, another experiment is planned with the reactor diameter increased to 4â€. This work will compare the results from the two long-term tests of the hybrid FT jet-selective catalyst using the skid-mounted fixed-bed FT reactor system at NCCC.

468121 Composition of Bio-Crude Oils from Hydrothermal Liquefaction of Algae in Batch Reactor

Abstract:

Due to the limitations of food, farmland, and weather, novel biofuels produced from hydrothermal liquefaction of biomass has attracted much attention. Algae-derived oils from hydrothermal liquefaction are promising because of high algae growth rates, strong CO_2-mitigation potential, and avoidance of drying requirements. In hydrothermal liquefaction, water at subcritical states (270-350°C and 80-173 bar) catalyzes depolymerization of lipids, proteins and carbohydrates in algae, leading to relatively high yields of bio-crude oil that can be valorized to transportation fuels. In this study, we investigated the influences of operating conditions (temperatures of 310-350 °C, residence times of 5-60 min, and solid algae contents of 5-10 wt.%) on the composition of bio-crude oil from hydrothermal liquefaction of *Nannochloropsis salina* and *Galdieria sulphuraria* using a variety of analytical methods. Organic molecules with molecular masses of ~150-700 Da were observed by Fourier transform ion cyclotron resonance mass spectroscopy (FT-ICR/MS) for comprehensive qualitative compositional description that employed three complimentary ionization methods. Free fatty acids in the HTL oil were quantitated by gas chromatography mass spectroscopy (GC/MS). Higher heating values and CHNS elemental content were also measured by bomb calorimeter and elemental analysis to obtain information about energy recovery relative to previous studies. Bio-crude oil characterization results will be used to optimize hydrothermal liquefaction operating conditions in a 1.8 L batch reactor and to select initial conditions for a pilot scale, continuous flow reactor.

469356 De-Alloyed Platinum-Bismuth Nanoparticles As Highly Active Electrocatalysts for Dimethyl Ether Oxidation and Oxygen Reduction

Abstract:

Dimethyl ether (DME) is a clean-burning alternative to diesel fuel that meets strict emissions standards in combustion engines. DME is currently derived from natural gas but can also be synthesized from biomass on-site and is thus a sustainable and highly distributed energy source. DME has properties similar to propane and can utilize established fueling infrastructure and handling procedures. In addition to its use in internal combustion engines, the convenience of DME generation and transport can be exploited to overcome the high-pressure infrastructure requirements of hydrogen fuel in more efficient electrochemical fuel cells. Fuel cell-driven vehicle efficiencies are among the highest of all vehicle technologies.

Relative to other direct liquid fuels envisioned for use in electrochemical cells, DME is: (1) more energy-dense than hydrogen, (2) non-toxic compared to methanol, (3) a lower dipole moment than methan which limits fuel crossover, (4) the simplest ether which can be more efficiently oxidized than ethanol through electrocatalytic C-O bond cleavage, and (5) unlike ethanol, does not interfere with world food production. A key stumbling block to the commercialization of DME fuel cells has always been identification of a suitable electrocatalyst. The most active electrocatalyst presently available for DME oxidation is highly dispersed commercial precious metal Pt/Ru nanoparticles supported on carbon. A mass activity no greater than 1.3 A/g_{Pt} is observed at peak potential and is more than an order of magnitude less than that obtained from hydrogen.

We here discuss the synthesis of a novel DME oxidation electrocatalyst that has *more than an order of magnitude* greater activity per unit mass of Pt than state-of-the art. The electrocatalyst is composed of de-alloyed $PtBi_2$ nanoparticles (NPs). The NP pre-cursor material formed without the use of toxic organic solvents, difficult-to-synthesize organic precursors, or dangerous hydrazine reducing agent typically needed to produce bismuth metal. Synthesis involves the use of stannous chloride as both reducing and stabilizing agent in aqueous suspension and is unprecedented in that Sn-Bi ligand formation was unknown prior to our work. Electrochemical and chemical de-alloying of these nanoparticles is shown to alter the surface atomic coordination relative to pure Pt and produce unusually high electrochemical areas and activities. DME electrooxidation activities greater than 30 A/g_{Pt} are observed in sulfuric acid at peak potential while the activity of the corresponding cathodic oxygen reduction reaction (ORR) increases to over 15 A/g_{Pt} at 0.9V vs. RHE in perchloric acid. ORR activity is comparable to that obtained with recent dealloyed Pt-Cu catalysts but in a one-pot synthesis approach that avoids the need for a high pressure hydrogen reduction step during synthesis. Despite the significant increases in mass-specific activity observed relative to pure Pt, the area-specific activity of the dealloyed electrocatalyst is lower. This observation is discussed in terms of the corresponding changes in surface atom coordination and electrolyte effects.

Modification of Cu-Based Catalyst System for Co-production of Methanol and Dimethylether

Lujie Ye and Sunggyu Lee

Sustainable Energy and Advanced Materials Laboratory
Department of Chemical and Biomolecular Engineering
Ohio University
Athens, OH 45701

EXTENDED ABSTRACT

Dimethylether (DME) is the simplest ether, which has outstanding physical and chemical properties in many areas including oxygenated fuels and industrial chemicals. Since DME only contains C, H, and O elements, no NO_X and SO_X emission is generated during the combustion of DME. The high oxygen content of DME also alleviates the emission of CO as well, which is a common product of incomplete combustion. It is also reported that DME has a higher cetane number (CN) than the conventional petrodiesel. Therefore, DME is regarded as a promising substitute for petrodiesel as well as a cetane enhancer. Besides the advantages of using DME as fuel, the recently developed processes of producing lower olefins like ethylene and propylene [1], gasoline range boiling hydrocarbons [2], industrial chemicals such as oxygenates [3], methyl acetate [4], and ethylene glycol intermediate like dimethoxyethane (DMET) [5] make a strong case for DME to become a promising building block chemical for the synthesis chemistry.

Co-production of methanol and DME under a dual catalyst system of $Cu/ZnO/Al_2O_3$ and γ-Al_2O_3 has been developed for years. However, deactivation of the methanol synthesis catalyst is nontrivial, thus rendering a practical applicability problem of the system. This problem is at least partially caused by the growth of the copper crystallite size of the catalyst during the synthesis reaction. It is reported that the crystallite size of Cu could increase from an average of 37 angstroms to 109 angstroms after thermal aging under CO-free or CO_2-free syngas only for short periods of time-on-stream [6]. The aged catalyst only provides 78% of the original reactivity [7]. Besides the accelerated growth of the catalyst's copper crystallite size, deactivation of the catalyst can also be triggered by accumulation (i.e., build-up) of water in the catalyst pores, which could promote the mechanical breakdown of the catalyst while reducing the overall reaction conversion. Therefore, understanding of the detailed catalytic mechanism and developing a more robust catalyst are the key challenges for improving the efficiency of the DME production. In this study, supercritical fluid treatment is performed as a pre-treatment step for the unreduced catalyst of $CuO/ZnO/Al_2O_3$ to eventually improve the catalytic reactivity of the reduced (activated) catalyst.

The supercritical treatment is carried out in a stainless steel autoclave-based reactor system. Chemically pure CO_2 is compressed by a booster pump and stored in a CO_2 storage vessel. A certain metered amount of CO_2 will be fed into the reactor from the bottom (Figure 1). Prior to the supercritical treatment, the catalyst loaded reactor will be purged and rinsed by gaseous CO_2 three times to remove the residual moisture and air. Then, the system will be pressurized to a desired supercritical condition gradually, according to the estimated P-V-T values using the Peng-Robinson equation of state. Different temperatures, pressures, and treatment times are compared to find the most effective and beneficial combinations for this treatment step.

After the supercritical fluid treatment, a portion of ZnO is transformed to $ZnCO_3$, which is reported to be beneficial to the methanol synthesis [8] [9]. The cumulative volume of pores for the treated catalyst increases about 20 percent on an average, while the internal surface area of the treated catalyst increases about 40 percent. At the same time, the crystallite size of copper is found to be decreased by about 30 percent after the supercritical CO_2 treatment.

Acknowledgment

This work was partially supported by a research grant from the Korea Institute of Science and Technology (KIST). The authors are grateful for the support.

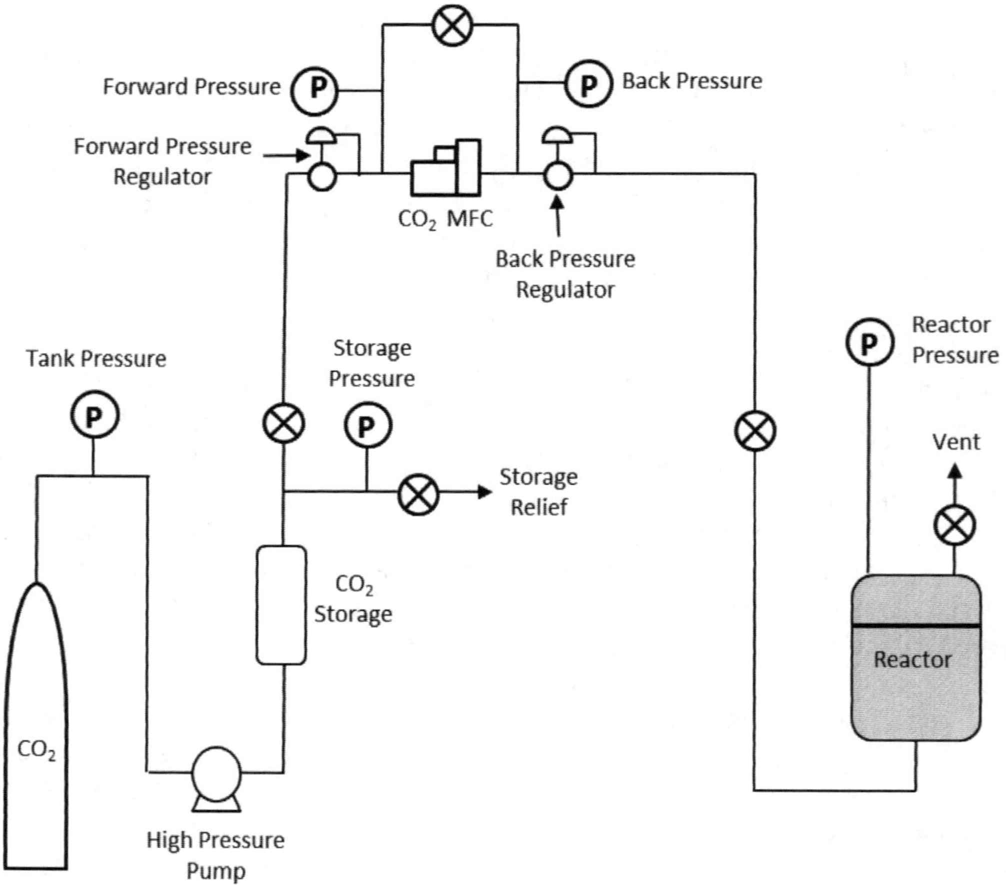

Figure 1. A process flow diagram of CO_2 supercritical treatment system (Courtesy of Chad Able, SEAM Lab, Ohio University)

References

[1] A. S. Al-Dughaither and D. L. Hugo, "Neat dimethyl ether conversion to olefins (DTO) over HZSM-5: Effect of SiO2/Al2O3 on porosity, surface chemistry, and reactivity," *Fuel,* vol. 138, pp. 52-64, 2014.

[2] P. Haro, F. Trippe, R. Stahl and E. Henrich, "Bio-syngas to gasoline and olefins via DME–a comprehensive techno-economic assessment," *Applied Energy,* vol. 108, pp. 54-65, 2013.

[3] G. Liu, B. Yan and B. Chen, "Technical review on jet fuel production," *Renewable and Sustainable Energy Reviews,* vol. 25, pp. 59-70, 2013.

[4] P. Cheung, A. Bhan, G. J. Sunley and E. Iglesia, "Selective carbonylation of dimethyl ether to methyl acetate catalyzed by acidic zeolites," *Angewandte Chemie International,* vol. 45, no. 10, pp. 1617-1620, 2006.

[5] A. Sardesai, T. Tartamella and S. Lee, "Synthesis of hydrocarbons from dimethyl ether: selectivtties towards light hydrocarbons," *Fuel Science & Technology International,* vol. 14, no. 5, pp. 703-712, 1996.

[6] S. Lee, Methanol Synthesis Technology, Boca Raton, FL: CRC Press, 1990.

[7] B. G. Lee and S. Lee, "Regeneration of liquid phase methanol synthesis catalyst," *Fuel Science and Technology Int'l,* vol. 5, no. 9, pp. 587-612, 1991.

[8] V. Parameswaran, S. Lee and S. Wender, "Roles of carbon dioxide in methanol synthesis," *Fuel Science and Technology International,* vol. 7, no. 8, pp. 1021-1057, 1989.

[9] V. Parameswaran, S. Lee and S. Wender, "The role of water in methanol synthesis," *Fuel Science and Technology International,* vol. 7, no. 7, pp. 899-918, 1989.

465206 Design of a Sabatier Reactor for CO_2 Conversion into Synthetic Methane

Abstract:

Design of a Sabatier reactor for CO_2 conversion into synthetic methane

Duo Sun and David Simakov

Department of Chemical Engineering, University of Waterloo, Waterloo, ON N2L 3G1, Canada

Converting CO_2-reach streams (e.g., biogas, fermentation off-gas, flue gas) into synthetic natural gas (SNG) via Sabatier reaction (CO_2 hydrogenation into CH_4) is an attractive route for CO_2 utilization [1-3]. There is a potential to decrease CO_2 emissions significantly and, at the same time, to reduce the consumption of fossil natural gas. To make the process sustainable, H_2 which is required for this reaction should have zero or very low carbon footprint. This requirement is achievable if H_2 is produced via water electrolysis using renewable electricity (hydro, wind, solar) or surplus (e.g., nuclear) electricity. Electrolysis can be done in a highly efficient way using polymer electrolyte membrane (PEM) electrolyzes [4]. However, a number of technological issues remain to be resolved, including catalyst deactivation and thermal management (Sabatier reaction is highly exothermic) [5]. The study presented herein focuses on the design of a highly efficient and compact Sabatier reactor for CO_2 hydrogenation into synthetic natural gas.

The general concept is a heat-exchanger type packed bed with active cooling for efficient heat removal. Compressed air, steam, or molten salt can be used as a coolant; molten salts are excellent heat transfer fluids due to their high heat capacity and thermal conductivity [6-9]. To identify the optimal configuration, we performed numerical simulations, using a transient pseudo-homogeneous model [10]. We searched for optimal operating conditions to maximize methane yield and to prevent at the same time reactor overheating and catalyst deactivation. We have identified several configurations that allow highly efficient CO_2 hydrogenation, providing CO_2 conversions and CH4 yields higher than 90%. A preliminary techno-economic evaluation demonstrates the potential of this approach for commercialization.

References

[1] Rönsch, S. et al. Review on methanation – From fundamentals to current projects. *Fuel* 2016, 166, 276-296.

[2] Schaaf, T. et al. Methanation of CO_2 – storage of renewable energy in a gas distribution system. *Energy Sustain. Soc.* 2014, 4, 29.

[3] Götz, M et al. Renewable Power-to-Gas: A technological and economic review. *Renew. Energ.* 2016, 85, 1371-1390.

[4] Gahleitner, G. Hydrogen from renewable electricity: An international review of power-to-gas pilot plants for stationary applications. *Int. J. Hydrogen Energ.* 2013, 38, 2039-2061.

[5] Kao, Y.-L.; Lee, P.-H.; Tseng, Y.-T.; Chien, I.-L.; Ward, J.-D. Design, control and comparison of fixed-bed methanation reactor systems for the production of substitute natural gas. *J. Taiwan Inst. Chem. Eng.* 2014, 45, 2346-2357.

[6] Simakov, D. S. A.; Wright, M. M.; Ahmed, A.; Mokheimer, E. M. A.; Román-Leshkov, Y. Solar thermal reforming of natural gas: a review on chemistry, catalysis and system design. *Catal. Sci. Technol.* 2015, 5, 1991-2016.

[7] Said, S. A. M.; Waseeuddin, M.; Simakov, D. S. A. A review on solar reforming systems. *Renew. Sust. Energ. Rev.* 2016, 59, 149-159.

[8] Said, S. A. M.; Simakov, D. S. A.; Mokheimer, E. M. A.; Habib, M. A.; Ahmed, S.; Waseeuddin, M.; Román-Leshkov, Y. Computational fluid dynamics study of hydrogen generation by low temperature methane reforming in a membrane reactor. *Int. J. Hydrogen Energ.* 2015, 40, 3158-3169.

[9] Said, S. A. M.; Simakov, D. S. A.; Waseeuddin, M.; Román-Leshkov, Y. Solar molten salt heated membrane reformer for natural gas upgrading and hydrogen generation: A CFD model. *Sol. Energy* 2016, 124, 163-176.

[10] Simakov, D. S. A.; Sheintuch, M. Design of a thermally balanced membrane reformer for hydrogen production. *AICHE J.* 2008, 54, 2735-2750.

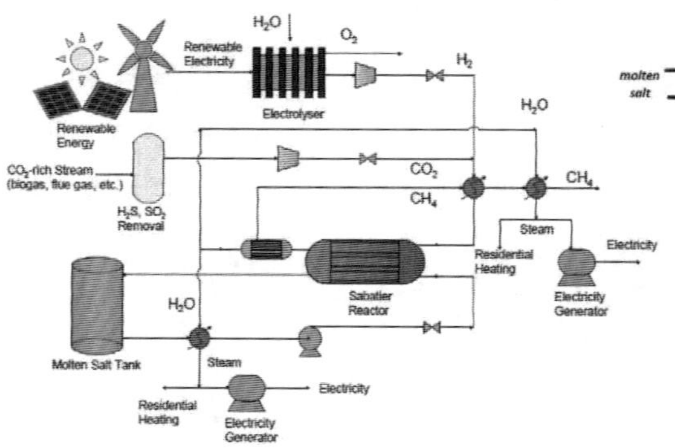

Figure 1: General concept of a system for CO_2 hydrogenation into synthetic natural gas.

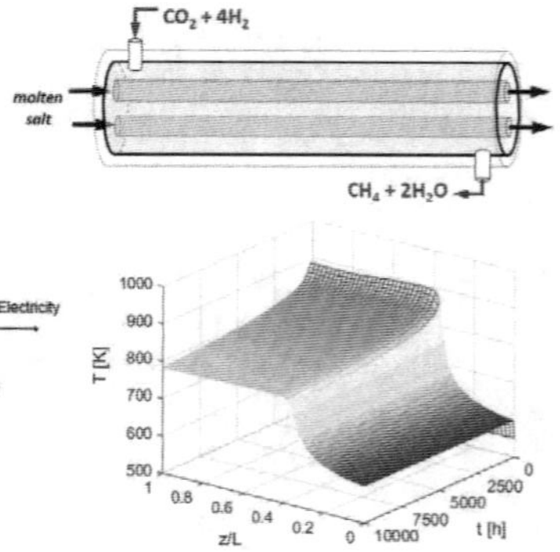

Figure 2: Reactor concept and simulated spatiotemporal temperature profile.

462795 Steam Reforming of Dimethyl Ether over a Novel Plate-Type γ-Al_2O_3/Al Monolith Supported Cu-Based Bi-Functional Catalyst

Abstract:

Nowadays, the increasing global demand of energy accelerates the depletion of fossil fuels, thus boosting the extensive exploration of new energy alternatives. Hydrogen is considered as a clean energy and steam reforming of dimethyl ether (DME) is regarded as a promising method to produce hydrogen for fuel cell applications. Unlike many other fuels (e.g. methanol, methane and bio-ethanol), DME is non-toxic and provides high H/C ratio and high energy density. Moreover, the well-developed infrastructure of LPG can readily be used for the distribution of DME due to their similar physical properties. Generally, steam reforming of dimethyl ether (DME SR) (Eq. (1)) can be expressed as a combination of DME hydrolysis (Eq. (2)) to methanol over a solid acid catalyst and steam reforming of methanol (MSR, Eq. (3)) over a metallic catalyst. Therefore, a bi-functional catalyst consisting of acid sites and MSR sites is required for overall DME SR process.

DME steam reforming: $CH_3OCH_3 + 3H_2O \leftrightarrow 6H_2 + 2CO_2 \quad H_r = +135kJ/mol$ (1)

DME hydrolysis: $CH_3OCH_3 + H_2O \leftrightarrow 2CH_3OH \quad H_r = +37kJ/mol$ (2)

MeOH steam reforming: $CH_3OH + H_2O \leftrightarrow 3H_2 + CO_2 \quad H_r = +49kJ/mol$ (3)

In our previous works,[1-3] a plate-type γ-Al_2O_3/Al monolith was prepared through anodization technology. The structured monolith was proven to have an excellent activity in DME hydrolysis. Moreover, the anodic alumina supported Cu composite catalyst exhibited a satisfied performance in DME SR system.

In this study, the self-reduction behavior of fresh Cu/γ-Al_2O_3/Al catalyst during DME SR was further studied and the self-activation mechanism was proposed based on XRD and XPS analysis. However, the Cu/γ-Al_2O_3/Al catalyst showed a poor thermal stability at 350 °C due to the sintering of copper. As an approach, nickel was added to the Cu-based catalyst, and then the effect of nickel loading and chemical state on the activity of Cu/Ni/γ-Al_2O_3/Al catalyst was investigated. It was found that the proper amount of nickel could improve the dispersion of copper species, and thus enhancing the catalytic activity. For catalysts with lower nickel loadings, the thermal stability of Cu-based catalyst could not be improved. While, for catalysts with higher nickel loadings, part of nickel species would aggregate and block the active sites of copper. Moreover, a small amount of metallic Ni appeared, which resulted in the occurrence of side-reaction of CO methanation. Finally, the durability evaluation of optimized Cu/Ni/γ-Al_2O_3/Al was carried out under critical conditions (400 °C) and the results showed that it has an excellent stability during a 120 h test, which demonstrated that the novel plate-type γ-Al_2O_3/Al monolith supported Cu and Ni composite catalyst was an excellent catalyst for DME SR. Furthermore, it would be very promising for the application of micro-channel reformer for the domestic fuel cell system as for its outstanding shape flexibility and catalytic performance.

References

[1] F.Y. Fan, Q. Zhang, J.J. Xu *et al.*, Catal. Today, 216 (2013) 194-9.

[2] Q. Zhang, F.Y. Fan, G.M. Xu *et al.*, Int. J. Hydrogen Energy, 38 (2013) 10305-14.

[3] Q. Zhang, J.J. Xu, F.Y. Fan *et al.*, Fuel Process. Technol., 119 (2014) 52-9.

461602 Computational Modeling of Electrochemical Bio-Oil Upgrading

Abstract:

Electrochemical hydrogenation (ECH) reactions are needed in the post-pyrolysis treatment of bio-oil to convert biomass into biofuels. A computational approach is taken to find the optimal catalyst material, and the conditions at which the concentration of species to react surrounding the catalyst is best. This is done to provide testable hypotheses regarding catalyst activity and selectivity for experimental work. Despite extensive research done in electrochemical reactions, the complex solvent/catalyst environment has hindered a rigorous understanding of electrocatalyst surface chemistry. Density functional theory calculations of species binding and ECH reaction energies were done as a function of voltage on the surface of numerous catalysts. Initial results show that Au, Cu, and C catalysts are best suited because they preferably bind organic molecules over hydrogen, and show a large applied potential for H_2 formation and low one for organic hydrogenation. Since only the species at the cathode vicinity are susceptible to undergo electron transfer, classical molecular dynamics simulations of solvent mixtures in an electrolytic cell were performed to assess how species concentrations differ between the solid/liquid interface and bulk regions. Simulations were performed at varying temperatures, cathode surfaces and charge. Energetic and entropic contributions were analyzed to understand competing surface binding and solvation effects. Initial results show a decrease in organic and increase in water mol fractions at the solid/liquid interface as the cathode charge in the electrolytic cell increases. This can only be compensated by a strong adhesion of the organic molecule to the cathode surface. Results are compared with concurrent experiments and implications on catalyst choice are discussed.

472694 Optimizing Catalytic Fast Pyrolysis and Hydrotreating for the Production of Biofuels

Abstract:

Liquid transportation fuels can be produced via catalytic fast pyrolysis by upgrading biomass pyrolysis vapors prior to condensation. While it is possible to obtain a completely deoxygenated hydrocarbon product by using suitable operating conditions, including high biomass-to-catalyst mass ratios, the yield of the product oil remains low. Higher yields can be obtained at lower biomass-to-catalyst ratios but the resulting oil includes oxygenates and hydrotreating is required for the final deoxygenation. With decreasing degree of deoxygenation in the catalytic pyrolysis step, higher severity is required for hydrotreating. This research discusses the optimization of the overall carbon efficiency and cost of the process for producing oils with low oxygen contents via catalytic fast pyrolysis. Pine was pyrolyzed and upgraded over ZSM-5 catalyst in a dual 2-inch fluidized bed reactor system. The catalyst-to-biomass ratio was varied to produce oils with different oxygen contents, and the oils were subsequently hydrotreated over a sulfided CoMo catalyst.

470848 In Situ Raman Analysis for the Production of Biofuel Using a Heterogeneous Layered Double Hydroxide Catalyst

Abstract:

Oral Presentation

In Situ Raman Analysis for the Production of Biofuel Using a Heterogeneous Layered Double Hydroxide Catalyst

Obakore Agbroko , Keyvan Mollaeian, William E. Holmes, Tracy J Benson

As the world's energy consumption is rapidly increasing many efforts are being made to produce renewable and sustainable fuel sources that provide independence, or at least displacement, from fossil fuels. Many of these renewable sources include sunlight, wind energy, geothermal energy, and the conversion of plant based biomass into biofuels. The conversion of vegetable oils into biodiesel is a highly viable option for supplementing fossil fuel energy. Unfortunately, most biodiesel is produced from oils stemming from row-crops, which take up viable land resources needed for food crops. One promising alternative is microbially-produced oils (MPOs) where microorganisms feed on industrial and municipal wastewaters. MPOs, however, contain high concentrations of free fatty acids (FFAs), unlike most row-crops. Under conventional biodiesel processes, high concentrations of FFAs in oil feedstocks significantly complicate conversion and separation of the fuel produced. However, if dimethyl carbonate (DMC) is used as a substitute for methanol, the formation of an unwanted glycerol byproduct is avoided. In addition, the use of a heterogeneous catalyst, in our case triazabicyclodecene bound to Mg/Al layered double hydroxide (TBD-LDH), process wash water is not required for separation of fuel components from catalyst salts.

In this study, an in situ Raman analysis method was developed to monitor the reaction conversion as well as kinetic parameters for the transesterification of lipid oils to biofuel using DMC and TBD-LDH catalyst. This in situ method, using a fiber optic probe attached to the 785 nm Raman system, does not require sample handling and workup as needed for GC analysis. The height of the C=C stretching mode of unsaturated fatty acids of the oil at a Raman shift of 1,655 cm-1 was measured to determine the conversion of the oil. A Gaussian distribution deconvolution algorithm was used to optimize the spectra for quantitative analysis. Refined canola oil and crude corn oil (4.5 wt% FFA) were used to test the robustness of this technique. A high-temperature GC-FID analysis was used to validate the Raman method.

472064 Hydroprocessing of Biocrude into Hydrocarbon Liquid Fuel

Abstract:

Catalytic hydrotreating (HDT) of bio-oil offers a strategic opportunity to produce refinery feedstocks or infrastructure-compatible finished fuels. Nevertheless, HDT of raw bio-oil has proven to be challenging due to the complexity of the reactive oxygenated compounds; the thermal instability of the raw bio-oil cause issues such as reactor plugging and catalyst coking/deactivation. RTI is developing an advanced biofuels technology that integrates catalytic biomass pyrolysis and hydrotreatingÂ to produce hydrocarbon-based biofuels. The goal is to produce a more thermally stable bio-crude with lower oxygen content compared to typical bio-oils. Results from several HDT studies on bio-crude from RTIâ€™s 1 ton per dayÂ catalytic biomass pyrolysis unit show promising results and suggest that the bio-crude oil can be successfully upgraded without a stabilization step.

The main goal of this work is to understand how bio-crude quality affects HDT by hydrotreating several well characterized bio-crude oils. The presentation will discuss in detail, the effect of biocrude composition and process condition (temperature, H_2 pressure, H_2/Oil ratio, and LHSV) on product yield/quality and hydrogen consumption. A detailed knowledge of the impact of these factors can result in innovative technology solutions to control reaction selectivity and minimize hydrogen demand.

462656 Hydrodeoxygenation and Hydroisomerization of Algae Oils to Hydrocarbon Fuels

Abstract:

Microalgae are a promising feedstock for renewable fuels and chemicals because of their potentially high growth rate and lipid content. Recent interest in converting the lipid fraction into completely deoxygenated jet- and diesel-range fuels has motivated research and development efforts in catalyst and process development for deoxygenation and isomerization steps for these lipids, but literature data for algal lipid composition with respect to its effects on DO and HI is sparse. In this work, we report detailed characterization of the extracted lipids as well as deoxygenation (DO) and hydroisomerization (HI) of the lipids extracted from mid-stage (high-carbohydrate content) *Scenedesmus* algae. Over a Pd/C catalyst, nearly complete deoxygenation is achievable, and the DO product is readily isomerized over a Pt/SAPO-11 catalyst. The isomerized product has boiling points primarily in the diesel range and cloud points < -10 °C. Minimal cleanup of the oils is required prior to catalytic upgrading, but bleaching with SiO2 or H3PO4 results in a product with lower nitrogen content.

460978 Continuous Hydrodeoxygenation of Liquid Phase Pyrolysis Oil

Abstract:

Continuous Hydrodeoxygenation of Liquid Phase Pyrolysis Oil

Liquid phase pyrolysis

Non-renewable energy resources such as oil and coal will not suffice the increasing demand of energy in the future. To fill the upcoming gap a biomass liquefaction concept has been developed. This concept consists of two main process steps. In the first step lignocellulosic biomass is converted into pyrolysis oil and pyrolysis char through the bioCRACK process[1]. In the second step the intermediate products are upgraded. Two upgrading technologies were investigated: hydrodeoxygenation (HDO) of pyrolysis oil[2] and hydrogenation of pyrolysis char. The bioCRACK process was successfully operated in pilot scale (100 kg/h lignocellulosic feed) over two years.

Continuous HDO of liquid phase pyrolysis oil

Liquid phase pyrolysis oil was upgraded by continuous hydrodeoxygenation in lab scale. Hydrotreatment was performed catalytically in a plugflow reactor. Operation parameters were 400°C and 121 bar hydrogen pressure. Lab scale results showed a carbon yield of 38% in the fuel phase with a carbon content of 86,15% and a hydrogen content of 13.35%. Fuel boiling range is shown in figure 1. The boiling range is between diesel and gasoline.

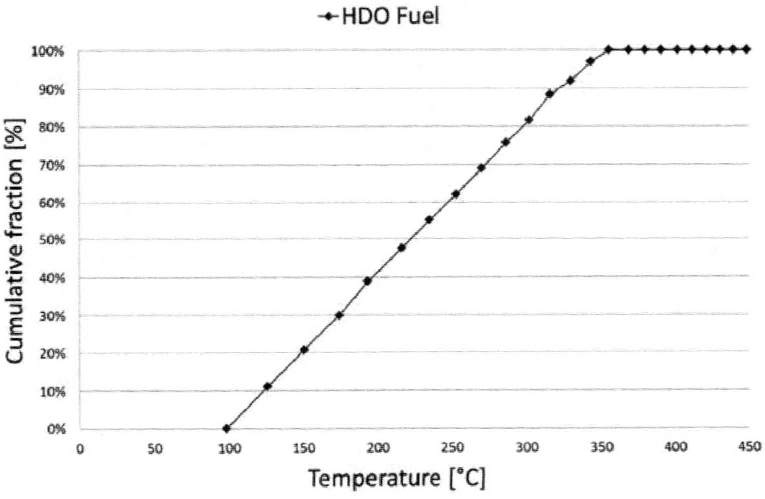

Figure SEQ Figure * ARABIC 1: boiling range of HDO fuel from liquid phase pyrolysis oil

Direct liquefaction of pyrolysis char was carried out in lab scale catalytically and non-catalytically. Non-catalytic biochar liquefaction showed a conversion of 84% and an oil yield of 72%[3], catalytic biochar liquefaction resulted in 100% conversion and 95% oil yield.

The design parameters for pilot scale of the hydrodeoxygenation technology as well as the hydrogenation technology (upgrade of pyrolysis oil as well as biochar) are under evaluation.

ADDIN Mendeley Bibliography CSL_BIBLIOGRAPHY (1) Schwaiger, N.; Elliott, D. C.; Ritzberger, J.; Wang, H.; Pucher, P.; Siebenhofer, M. *Green Chem.* **2015**, *17*, 2487–2494.

(2) Pucher, H.; Schwaiger, N.; Feiner, R.; Ellmaier, L.; Pucher, P.; Chernev, B.; Siebenhofer, M. *Green Chem.* **2015**, *17*, 1291–1298.

(3) Feiner, R.; Schwaiger, N.; Pucher, H. *RSC Adv.* **2013**, *3*, 17898–17903.

470370 Developing a Simulation Framework for Optimizing CO2-Capture Processes

Abstract:

DEVELOPING A SIMULATION FRAMEWORK

FOR OPTIMIZING CO_2-CAPTURE PROCESSES

Stuart Higgins and Y. A. Liu

PetroChina-AspenTech Center of Excellence in Process System Engineering,

Department of Chemical Engineering, Virginia Polytechnic Institute and State University,

Blacksburg, VA 24061

We have reported[1] a highly efficient, amine-based CO_2-capture unit design, achieving a thermal regeneration energy of merely 1.67 GJ/(tonne CO_2 captured). While our results look promising, the optimization of our process superstructure required a significant amount of effort, largely due to the number of optimization parameters and our use of computationally expensive rigorous models.

This experience motivated us to develop a more practical simulation framework. We focus on a modular design that allows us to use commercial modeling tools, and develop an overall simulation structure to take advantage of multi-threading, distribution, custom convergence methods, error recovery, and other features for increasing productivity.

Our new software framework enables faster, more reliable, more productive process optimization, while still using the same rigorous modeling tools validated by respected vendors. Multi-threading allows us to take advantage of the increasing core counts on modern CPUs, while distribution is similarly important as computational devices become more numerous.

In this talk, we will discuss the simulation framework and its application to optimize generalized process designs. We expect these improvements to enable optimizations that would otherwise require far too much time and effort to complete. We are particularly interested in superstructure process designs that generalize base-case process designs with potential improvements, because we can optimize these superstructures to find new process designs.

---

[1] Stuart J. Higgins and Y. A. Liu, "CO_2 Capture Modeling, Energy Savings, and Heat Pump Integration", *Industrial and Engineering Chemistry Research*, 54, 2526-2553 (2015).

461580 The Development and Demonstration of Best-Practice Guidelines for the Start-up Injection of CO_2 into Highly-Depleted Gas Fields

Abstract:

Highly-depleted gas fields represent prime potential targets for large-scale storage of captured CO_2 emitted from industrial sources and fossil-fuel power plants. Given the potentially low reservoir pressure as well as the unique thermodynamic properties of CO_2, especially in the presence of stream impurities, the injection process presents significant safety and operational challenges.

The most effective way of transporting the captured CO_2 for subsequent sequestration using high-pressure pipelines is in dense phase. The CO_2 arriving at the injection well will typically be at pressure greater than 70 bar and at temperature between 4 and 8 °C. Given the substantially lower pressure at the wellhead, the injection of CO_2 will result in its rapid, quasi-adiabatic Joule-Thomson expansion leading to significant temperature drops. This could pose several risks, including hydrate and ice formation around the wellbore and thermal shocking of the wellbore casing steel, leading to its fracture and ultimately escape of CO_2.

Despite its importance, the modelling of the flow dynamics taking place during the start-up injection of CO_2 has been devoted mainly to either steady-state operations[1] or flows where only pure CO_2 is injected[2]. The steady-state flow model is inappropriate as it ignores important phenomena, such as expansion wave propagation and bubble nucleation, occurring during the rapid depressurisation process. In addition, there is little work in the literature regarding the impact of the injection rate and the upstream inlet conditions (e.g. temperature and pressure) on the degree of cooling along the well[3]. Moreover, all previous work has been confined to pure CO_2. In practice the injected CO_2 will have a range of different impurities which will have a profound impact on the CO_2 fluid phase behaviour and hence its injectivity[4].

The aim of our work is twofold. First, we present in detail a homogeneous relaxation flow model for the numerical simulation of the highly transient phenomena taking place in the wellbore. Mass, momentum, and energy conservation equations are considered in the tubing. Wall friction, gravitational force, and heat transfer between the fluid and the surrounding formation are also taken into account. The phase equilibrium and thermodynamic properties for CO_2 are determined using the Peng-Robinson equation of state, which guarantees both accurate and computationally efficient predictions of VLE data for CO_2[5]. Operational envelopes for the injection well are obtained from data already available in the literature[3]. Second, by performing a sensitivity analysis on various injection rates and temperatures, we propose optimal transient-operation design and develop best-practice guidelines for the minimisation of the risk associated with start-up injection of CO_2 into highly-depleted gas fields.

1. Hughes D., *Carbon storage in depleted gas fields: Key challenges*, Energy Procedia, 1, 2009, 3007-3014.

2. Galic H., Cawley S. and Bishop S., *CO_2 injection into depleted gas reservoirs*, SPE 123788, 2009, 1-8.

3. Li X., Xu R., Wei L. and Jiang P., *Modeling of wellbore dynamics of CO_2 injector during transient well shut-in and start-up operations*, International Journal of Greenhouse Gas Control, 42, 2015, 602-614.

4. Brown S., Beck J., Mahgerefteh H. and Fraga E. S., *Global sensitivity analysis of the impact of impurities on CO_2 pipeline failure*, Reliability Engineering & System Safety, 115, 2013, 43-54.

5. Oosterkamp A. and Ramsen J., *State-of-the-art overview of CO_2 pipeline transport with relevance to offshore pipelines*, Technical Report, PolyTec.

464433 Application of a Systematic Methodology for Sustainable Carbon Dioxide Utilization Process Design

Abstract:

Application of a systematic methodology for sustainable carbon dioxide utilization process design

Cristina Calvera, Rebecca Frauzem*, Rafiqul Gani

Department of Chemical and Biochemical Engineering, Technical University of Denmark, Søltofts Plads Building 229, 2800 Kgs. Lyngby, Denmark

* rebfra@kt.dtu.dk

As concerns about the environment are growing, new efforts are needed to achieve more sustainable processes. One such environmental concern is global warming, which is primarily caused by the greenhouse effect or the increase in concentration of greenhouse gases [1]. The most significant greenhouse gases are carbon dioxide, methane and nitrous oxide, of which carbon dioxide is the highest constituent at 82%. Furthermore, the amount of carbon dioxide emissions is growing with time. These trends make it evident that there is a need for methods to reduce these greenhouse gases emissions. While there are two methods of reducing carbon dioxide emissions, carbon capture and storage (CCS) and carbon capture and utilization (CCU), CCU is considered promising as it makes further use of the carbon dioxide as a solvent, raw material, and reagent to produce valuable products [1]. Using such utilization processes, the emissions can be reduced as they are being utilized and profit can be obtained, or the cost of operation for the carbon dioxide treatment can be returned, through this utilization process.

In order to systematically reduce such emissions, carbon capture and utilization is considered rather than carbon capture and storage. To achieve this a methodology is developed to design sustainable carbon dioxide utilization processes. First, the information on the possible utilization alternatives is collected, including the economic potential of the process and the carbon dioxide emissions. The carbon dioxide emissions can be classified as direct and indirect emissions in a chemical process. The net carbon dioxide is determined for the utilization processes as the indirect carbon dioxide emissions minus the carbon dioxide utilized. Processes that presents zero or negative net carbon dioxide emission are desired in order to reduce the carbon dioxide emissions. Using this estimated preliminary evaluation, the top processes, with the most negative carbon dioxide emission are investigated by rigorous detailed simulation to evaluate the net carbon dioxide emissions. Once the base case design is established and evaluated, targeted improvements are made by exploiting opportunities, for example, optimization, heat integration and improved design decisions so that more sustainable and lower net carbon dioxide emission alternatives are obtained. .

This method is applied to various processes where carbon dioxide is used as raw material. First, the process data are collected and compared. The economic feasibility is evaluated. From this, five processes are selected and analyzed in detail: the production of dimethyl carbonate, succinic acid, propylene carbonate, dimethyl ethylene and methanol. Not all the studied processes could be designed for zero or negative net carbon dioxide emission. Propylene carbonate production is found to have a negative net carbon dioxide, where, implementing targeted process improvements minmized the net carbon dioxide emission to -0.389 kg of carbon dioxide per kg of propylene carbonate. On the other hand, for succinic acid production, even after targeted improvements, the net carbon dioxide remained positive, even though compared to the existing industrial processes there is a relative reduction of upto 85%.

What this study shows is that for meaningful net carbon dioxide reduction, the carbon dioxide utilization processes need to be selected very carefully to obtain the best results. However, the opportunity to potentially reduce the net carbon dioxide emissions for the production of some bulk chemicals with carbon dioxide as feedstock exists.

[1] Yuan, Z., Eden, M.R., Gani, R., 2016. Toward the Development and Deployment of Large-Scale Carbon Dioxide Capture and Conversion Processes. Ind. Eng. Chem. Res. 55, 3383–3419. doi:10.1021/acs.iecr.5b03277

466690 Dynamic Modeling and Simulation of the Carbon Capture System Using Blended Amine-Based Solvent

Abstract:

Control of greenhouse gas emission has become an important issue due to the rising concerns for global warming. Among the various potential modes of carbon capture, post-combustion capture (PCC) is an appropriate option for capturing carbon dioxide from existing coal-fired power plants because the plants could be retrofitted easily to accommodate PCC systems at commercial scale. However, extra capital and operating cost required to install and operate the PCC system is the main hurdle for its widespread application. The amine-based absorption process causes significant loss in power production due to the large amount of regeneration energies needed in the desorption stage. Therefore, the blended amine solvents, which are having bigger absorption capacity and fast reactions, have been actively searched and developed. The various efficient operation methods such as flexible operation have been also suggested in line with decreasing the cost of energy loss in the process. Furthermore, there have been some attempts to develop a dynamic model of MEA (Monoethanolamine) absorption process in order to investigate the operational issues related to the dynamics of the process. However, most dynamic models have not been validated due to lack of data.

In this research, the absorption process based on a newly developed blended-amine solvent is simulated in *gPROMS* and the process operability and dynamic issues are examined. Thermodynamics and rigorous process modeling for the absorption process using the new solvent are developed in simulator. Dynamic models are compared with pilot plant data for validation. Furthermore, dynamic behaviors with control loops are observed under some potential operational scenarios to optimize the operational strategy to reduce the regeneration energies. Important dynamics, disturbances are analyzed and control structure is selected and tested.

464107 Continuous-Time Model for Retrofitting of Power Plants and Optimal Carbon-Capture and Storage

Abstract:

Carbon capture and storage (CCS) is one of the options for mitigating the adverse effects of CO_2 emissions and the resulting greenhouse gas and climate change effects. CCS involves a two-step process of directly capturing CO_2 from different sources (such as flue gas emissions of power plants), using one of the carbon capture technologies, and then transporting and storing CO_2 in natural sinks (such as geological reservoirs or exhausted oil/gas wells). The different carbon capture technologies include pre-combustion capture in integrated gasification combined cycle (IGCC), post-combustion capture based on flue gas scrubbing, oxy-fuel combustion, and chemical looping combustion (CLC). Other alternatives for reducing CO_2 emissions include improving energy efficiency of power plants or using low-carbon fuel technologies [1-4].

CCS is essential for continued use of fossil fuel based technologies to mitigate the environmental concerns related to carbon footprint. Several approaches based on pinch analysis and mathematical programming have been developed in the literature for source-sink matching of CO_2 emissions. All the three operations: carbon capture, transportation, and storage consume some energy to perform them, again resulting in additional CO_2 emissions. On one hand we are capturing CO_2 and storing into allocated geological sink, while on the other hand we are allowing additional CO_2 emission into atmosphere in order to ensure feasibility of whole process, leading to compensatory power generation in case of power plants. Enabling of CCS requires retrofitting of existing power plants to allow for capturing of CO_2 in pure form, thus reducing the power output of these plants, which are modeled in the literature as energy penalties.

Tan et al [2] proposed a CO_2 source-sink continuous time mixed integer nonlinear programming (MINLP) model considering physical and temporal effects of planning CCS. The temporal effects include time

matching of start and end times of operating times of sources with sinks. Later they used linearization of bilinear terms and converted the model into MILP. Fixed operating lives and fixed flow rates were considered for CO_2 sources. Lee and Chen [3] presented an improved version of Tan et al [2] model with reduction in number of variables and constraints. Lee et al [4] extended this work using discrete time intervals to include decisions related to different carbon capture technologies for retrofitting of power plants, variable flow rates for CO_2 sources, and finite injection rates for CO_2 sinks.

In this work, we present an improved version of the literature models [2-3] for optimal source-sink matching in carbon capture and storage systems, without requiring bilinear terms and the resulting linearization. Hence, it directly results in an MILP model with reduced problem size in terms of number of variables and constraints. Two case studies from literature will be presented to illustrate the proposed advantages based on these modifications.

Then, we present an MILP model for retrofitting of power plants based on different carbon capture technologies using continuous time representation. The objective is to maximize CO_2 emissions reduction by retrofitting power plants and matching CO_2 sources and sinks, while keeping electricity cost within a specified limit and satisfying all temporal and physical constraints. A case study from literature [4] was solved using the proposed continuous time MILP model, yielding orders of magnitude reduction in problem size, as expected. Additionally, we will present sensitivity analysis by varying the minimum duration of viable connection between sources and sinks to study its effect on the total CO_2 captured.

References

[1] Shenoy, A. U.; Shenoy, U. V. (2012). Targeting and design of energy allocation networks with carbon capture and storage, *Chem. Eng. Sci.*, 68, 312–327.

[2] Tan, R. R.; Aviso, K.B.; Bandyopadhyay, S.; Ng, D. K. S. (2012). Continuous-time optimization model for source-sink matching in carbon capture and storage systems, *Ind. Eng. Chem. Res.*, 51, 10015–10020.

[3] Lee, J.-Y.; Chen, C.-L. (2012). Comments on 'Continuous-time optimization model for source-sink matching in carbon capture and storage systems', *Ind. Eng. Chem. Res.*, 51, 11590–11591.

[4] Lee, J.-Y.; Tan, R. R.; Chen, C.-L. (2014). A unified model for the deployment of carbon capture and storage. *App. Energy*, 121, 140–148.

467083 Heat Integration of Absorption-Based Carbon Capture Processes in the Industrial Sector

Abstract:

The industrial sector is struggling with the difficult switch to renewable feedstocks, since the generated carbon is sometimes inherent in their processes or a part of their product. Therefore, carbon capture and storage (CCS) performed in the industrial sector has gained increasing attention recently, and most research is aimed at absorption-based capture processes. The largest cost for post-combustion CCS when applied at a power plant is the cost of regeneration of the absorbent, but industrial plants often have vast amounts of excess heat that can be utilized in the capture process. The focus of this paper is, therefore, heat integration of absorption-based carbon capture in the industrial sector. Models of two capture processes, using monoethanolamine (MEA) or ammonia as absorbents, were evaluated on the basis of how suitable they are for generic, industrial applications, depending on the availability and quality of the excess heat at the industrial site, and the temperature in the stripper reboiler. Models of the two processes have been constructed in Aspen Plus using a temperature interval of 90 – 120 °C for MEA and 105 – 155 °C for ammonia, and have been presented in previous work.

The study then continues with a heat integration case study of an oil refinery situated on the west coast of Sweden, emitting 1.8 Mt CO_2/yr from the four main chimneys. Over 100 coolers, using air or water as a cooling medium, are present at the refinery, providing cooling in the proximity of 500 MW. Because a system where over 100 new units are installed would be very complex and expensive, the possibility of installing fewer heat exchangers while having an acceptable loss of the available excess heat was investigated. Both the MEA and the ammonia process were evaluated based upon the levels of heat that resulted from this investigation.

The results point at a maximum heat recovery temperature of 108 °C for ammonia, and for MEA there is a boundary solution at the lower boundary of the investigated temperature interval, 90 °C. MEA could capture more CO_2 than ammonia using only excess heat, and techno-economic calculations were then made for the implementation of the MEA process in the oil refinery, including the cost for excess heat collection. The calculations comprise of the cost of the MEA process at both boundary temperatures in the stripper-reboiler, 90 °C and (standard) 120 °C, at two different sizes, and CO_2 concentrations. Compared to estimates where excess heat is not considered, the cost for carbon capture was decreased by up to 40 %.

461901 Study of Mass Transfer and Sorption Phenomena in CO_2/Brine/Rock Systems

Abstract:

Geological sequestration of CO_2 is a promising technique to reduce the emissions of greenhouse gases and to mitigate global warming. Understanding of mass transfer and sorption characteristics of CO_2 in relevant brines, and in brine-saturated reservoir rock samples at realistic reservoir conditions is considered essential for predicting the gas distribution after injection, and for potentially estimating the cumulative amount of trapped gas. In this study, Mount Simon cores from Illinois Basin were characterized via measurements of their porosity and permeability. Their internal surface area and pore size distribution were also measured via probe-gas sorption experiments using two different gases (N_2 at 77 K and CO_2 at 195 K) in order to provide additional information about the microstructure of the cores. A series of high-pressure core flooding experiments were conducted to understand interactions between core, brine and CO_2 in terms of sorption, mass transfer and displacement characteristics. In addition, mathematical models are developed to describe the observed mass transfer and sorption behavior at the core-scale.

472020 Numerical Modeling of CO_2 Sequestration in Deep, Saline, Dolomitic-Limestone Aquifers

Abstract:

Geological sequestration of CO_2 is one of the most promising technologies for large-scale CO_2 mitigation. As per U. S. Department of Energy [1] the United States has an estimated potential of 1.8–20 trillion metric tons of underground CO_2 storage. To assess the suitability of a potential CO_2 storage site, we must understand the thermodynamics [2] and kinetics of the CO_2-brine system under the relevant conditions. Using TOUGHREACT [3], we have estimated the changes in physical and chemical properties of limestone-dolomitic aquifers over a simulated period of 100 years after injecting high pressure (160-200 bars) CO_2 for 10 years under isothermal conditions. The objectives of the numerical model are to 1) estimate the amount of CO_2 sequestered by solubility trapping, mineral trapping, and residual trapping; 2) estimate the dissolution and precipitation of different minerals in the system; 3) estimate the concentration of the major chemical species and pH change in the aquifer brine; and 4) estimate the changes in porosity and permeability of the aquifer. Also, effects of other parameters like temperature, pressure, and salinity on CO_2 sequestration have been studied for long-term feasibility of the process.

Results from the study suggest that solubility trapping accounts for the highest amount of CO_2 sequestered as compared to mineral and residual trapping. Due to formation of carbonic acid, the pH of the brine dropped to below 5. This drop in pH led to dissolution of minerals like calcite and dolomite. Also, there was precipitation of minerals like gypsum. Changes in porosity and permeability due to dissolution and precipitation of minerals have been estimated. Although the study was conducted primarily for calcite, dolomite, and gypsum, it may be applicable to other formations as well. The study gives an overall estimate of the amount of CO_2 sequestered under different conditions (salinity, pressure, temperature, and mineralogy). It also predicts the changes in chemistry and permeability of the system under different

conditions. The study thereby helps to choose the feasibility of sites for CO_2 sequestration more efficiently.

References

1. NACAP (2012) The North American Carbon Storage Atlas, The U.S. Department of Energy (DOE), Natural Resources Canada (NRCan), and the Mexican Ministry of Energy (SENER).
2. Thomas, M. W., Stewart, M., Trotz, M., and Cunningham, J. A. (2012). Geochemical modeling of CO_2 sequestration in deep, saline, dolomitic-limestone aquifers: Critical evaluation of thermodynamic sub-models. Chemical Geology, 306, 29-39.
3. Xu, T., E.L. Sonnenthal, N. Spycher, and K. Pruess, (2006) TOUGHREACT - A simulation program for non-isothermal multiphase reactive geochemical transport in variably saturated geologic media: Applications to geothermal injectivity and CO2 geological sequestration, Computers & Geosciences, v.32, p.145-165.

469496 Effects of Recycling Regenerated Heat Carrier on the Performance of an Auger Pyrolysis Reactor

Abstract:

Auger pyrolysis reactors offer unique advantages over traditional fluidized bed pyrolyzers. Some advantages include the minimal requirement of an inert gas and the flexibility of heat carrier materials with varying thermophysical properties. Previous work used a lab-scale auger pyrolyzer to test the effect of heat transfer properties of various heat carriers such as stainless steel shot, silicon carbide and sand. While high liquid yields were achieved (~65 wt. %), the long-term performance of these heat carriers is not well understood. If auger pyrolyzers are to be commercially viable, the heat carrier must be recycled to minimize the cost of system inputs and solids handling. The objective of this research was to evaluate the effect of heat carrier regeneration and recyclability on the long-term performance of an auger pyrolysis reactor. The effect on product yields and changes to the thermophysical characteristics of the heat carrier materials were of specific focus.

A twin-screw, lab-scale auger reactor was used to pyrolyze up to 1 kg/hr of red oak at 515°C. The heat carriers of interest in this study included fine (250-600 μm) and coarse sand (600-1000 μm). Pyrolysis trials using recycled heat carrier were conducted at up to 5 recycles. The heat carrier was regenerated following each trial to remove any carbon residue. A cold gas quench system was used to collect the bio-oil into two fractions: a heavy phase and an aqueous phase. Product characterization included ultimate and proximate analysis of the bio-oil and char. Gas chromatography of the bio-oil fractions was also conducted. Product yields of bio-oil (~60 wt. %), char (~17 wt. %) and gas (~18wt %) were consistent across the limited number of recycle trials. Heat carrier attrition varied from 4 to 8 wt. %, and the mean particle size decreased following the recycled trials. These results indicate that heat carrier properties may change significantly over a short number of rotations, but their impacts on biomass pyrolysis yields is small. Long-term tests would be necessary to determine heat carrier replacement rates in order to avoid significant impacts on bio-oil yields and quality.

469939 Pyrolysis of Biomass Using Blast Furnace Slag As Catalyst

Abstract:

Catalytic pyrolysis of biomass using various acidic zeolite catalysts have shown promise of producing stable biomass pyrolysis oils that can be easily upgraded to fuels and biobased products. However, zeolite catalysts are relatively expensive and tend to produce low liquid yields, large fractions of gas and aqueous phase and therefore there is a need to develop other catalysts that will improve liquid yields. Pine sawdust was pyrolyzed in a bench scale bubbling fluidized reactor using fine blast furnace slag as the pyrolysis medium and nitrogen as fluidizing gas. The pyrolysis was conducted at 500 C and feed rate of 600-700 g/h. The pyrolysis reaction showed an initial exotherm during the first 20 min and there was significant production of hydrogen. The reaction became endothermic like most conventional pyrolysis reactions and hydrogen production decreased. The organic liquid yields ranged from 28 to 48 wt% depending on the run time and were higher than those reported for acidic zeolite catalysts. The char/coke yields ranged from 14 -18 wt%; the non-condensable gas yields ranged from 9 to 13 wt% and pyrolytic water yields ranged from 8 to 18 wt%. The oxygen content of the oils was 27-30 wt%, and the total acid number (TAN) was surprisingly low (23-26 mg KOH/g oil). The oils were less viscous than the sand pyrolysis oils. Thus the blast furnace slag appears to be a promising catalyst for the catalytic pyrolysis of biomass.

471364 Recycling of Used Railroad Ties Via a Thermochemical Process Using a Semi-Pilot Scale Auger Reactor System

Abstract:

A thermochemical process, an initial thermal treatment followed by an intermediate pyrolysis step, was investigated to recover wood preservatives and produce bio-oil and biochar containing low amount of polycyclic aromatic hydrocarbons (PAHs) from used creosote-treated railroad ties. The material was first treated using a semi-pilot scale auger pyrolysis reactor at 250, 280, and 300 °C for 15min. The yield of the liquid fraction was 28, 35, and 38 wt% while the solid fraction yield was 65, 57 and 54 wt%, respectively. The liquid fraction had two phases: aqueous phase in the upper and organic phase (enriched in creosote) in the bottom. As thermal desorption temperature increased, the aqueous phase exhibited a decrease in water content from 95 to 88% and pH from 3.2 to 2.6 and an increase in total acid number (TAN) from 20 to 63 mgKOH/g. The yield of the creosote phase increased from 2.0 to 4.5 wt% of total creosote content of 7 wt% found in the wood tie. As thermal desorption temperature increased, in addition to creosote, the organic phase also contained larger amount of carbohydrates- and lignin-derived compounds, decreasing the purity of the recovered creosote. The thermally treated samples were subsequently pyrolyzed using the same semi-pilot scale auger pyrolysis system at 500 oC for 72s residence time. The yield of bio-oil produced from the untreated, 250, 280, and 300 °C-treated materials was 61, 42, 35, and 30 % while the yield of biochar was 21, 25, 32, and 33 %, respectively. The characterization of the bio-oil produced from the thermally treated ties showed that as the desorption temperature increased, water content decreased from 36 to 32 %,TAN value decreased from 127 to 113 mg KOH/g, and density increased from1.1 to 1.2 g/mL. PAHs traces found in biochar from the thermally treated wood tie were dramatically lower (less than 0.2 wt%) than that (0.8 wt%) in biochar from the untreated ties. This result demonstrates that a thermal preservative-removal step can recover valuable creosote for re-use as preservatives and subsequently supply a feedstock without significant levels of contaminant hazardous pollutants for pyrolysis to produce bio-oil and biochar.

471572 Effect of Feedstock Formatting on Gasification Performance for Hyprid Poplar, Corn Stover and Low-Quality Materials

Abstract:

This work reports the effects of moisture content and particle size distribution on the gasification performance of hybrid poplar, corn stover and municipal solid wastes (MSW) in a modular plasma-gasification system. The MSW tested includes construction wastes, plastics and refuse derived fuels. To date, there has been very little data in the literature to support the feasibility of MSW gasification. In addition, while gasification has been demonstrated at large scale (over 200 tons per day), it is unknown whether smaller scale (2-40 tons per day) deployment of gasification technologies will be economical. It is at these scales where many of the applications of MSW gasification will reside since there are relatively few areas in the US with large amounts of MSW available. The presented work may serve as a basis for techno-economic analyses and scale up research.

Hybrid poplar was tested at 9 different conditions using a 3^2 matrix with respect to particle size and moisture content. Corn stover was tested at varying particle sizes, and MSW were tested at 4-6 mm particle size and â€˜as isâ€™ moisture. In addition, the different samples were air classified and assigned a measure of buoyancy to delineate between particle size affects and buoyancy. Inorganic and organic mass balances were done for the initial feedstock materials and the resulting slag, when possible, using ICP and CHNO proximate/ultimate analyses. CO_2, CO and H_2 production were measured inline and from collected product gas samples. The presence of other hydrocarbons was tested using GC-MS, but none were detected. Results suggest that hybrid poplar produced syn gas between 30 and 40% H_2, and between 35 and 45% CO, with some dependency on formatting conditions. Less than 2% of the material was recovered as slag, and it generally consisted of silica and alkali and alkaline earth metals. Initial testing of the corn stover and MSW suggest successful gasification, but that formatting plays a more critical role in MSW gasification than in hybrid poplar gasification.

472096 Effects of Catalyst Properties on Biomass Conversion By Catalytic Fast Pyrolysis and Hydropyrolysis

Abstract:

Effects of Catalyst Properties on Biomass Conversion by Catalytic Fast Pyrolysis and Hydropyrolysis

David P. Gamliel and Julia A. Valla

Department of Chemical & Biomolecular Engineering, University of Connecticut

191 Auditorium Road, Unit 3222, Storrs, CT 06269-3222, USA,

Phone: +1-860-486 4602, e-mail: ioulia.valla@.uconn.edu

Biomass is a clean and renewable carbon source, capable of providing sustainable carbon for conversion to fuels and platform chemicals. Catalytic fast pyrolysis (CFP) of biomass is the thermochemical conversion of biomass to liquid bio-oil, gas and char under inert atmosphere at intermediate temperatures and high heating rate. Recently, catalytic pyrolysis of biomass under a hydrogen atmosphere at elevated pressure, termed hydropyrolysis, has been studied as an effective method for production of deoxygenated high-value bio-oil with plentiful aromatic and aliphatic hydrocarbons [1]. The objective of this work is to determine how catalyst properties correlate to CFP and hydropyrolysis product yields, and propose the ideal catalyst for the production of value-added products from biomass resources.

A wide variety of catalysts were prepared and characterized for this study, such that the effects of transition metal choice, metal loading, support type and support acidity could be determined. Each catalyst was completely characterized using a variety of techniques to find the surface area, morphology, pore structure, metal oxidation state and crystal structure. Pd, Ru and Ni were chosen as candidate metals due to their high activity for deoxygenation and hydrogenation [2]. These metals were deposited via dry impregnation on a variety of alumino-silicate supports including alumina, silica and ZSM-5 zeolite. Furthermore, the Si/Al ratio of the ZSM-5 support was varied in order to determine the effects of acidity on the product distribution.

CFP was performed in a pyrolysis gas chromatograph (PyGC) unit, at 600 °C, under inert conditions and at atmospheric pressure. Increasing ZSM-5 catalyst acidity was found to increase yields of mono-aromatic hydrocarbons (MAHs), naphthalenes and CO, and decrease char formation. However, once the zeolite Si/Al ratio was lower than 25, coke formation reactions became dominant. It was found that significant increase of catalyst acid site density resulted in enhanced formation of coke and char precursors, such as poly-aromatic hydrocarbons (PAHs). The results are in agreement with the theory that the primary mechanism of acid catalyzed deoxygenation of pyrolysis vapors is decarbonylation and aromatization [3]. Impregnation of 3% of each transition metal slightly favored permanent gas, primarily in the form of CO, and decreased char yield. CFP with non-zeolitic support alone was found to have high yields to oxygenates and char.

Catalytic hydropyrolysis of biomass was performed in the same unit in pure H¬2 atmosphere, at 600 °C and 450 psig pressure. When hydropyrolysis was performed in the presence of ZSM-5 zeolite, a similar product distribution to CFP was observed. When Ni was incorporated on the catalyst, high amounts of CH4 were produced at the expense of solids, and alkanes were present in the liquid product, as shown in Figure 1. As the zeolite acidity was increased naphthalenes, PAHs and carbon oxide yields increased significantly at the expense of CH4 and liquid alkanes. This demonstrates the competition between acid-catalyzed decarbonylation, decarboxylation and aromatization and metal-catalyzed methanation and hydrogenation. As with CFP, hydropyrolysis with alumina or silica produced few liquids and high char yields. When Ni was incorporated onto silica, alkanes yield totaled about 6 wt.% carbon, and CH4 yield totaled over 50 wt.% carbon. Alternatively, when Ni was incorporated on the alumina support, lower CH4 yields and almost no alkanes were observed, along with higher solid yields, proving that significant Lewis acidity and large pore structure are detrimental for value-added product formation.

Catalyst design is essential should the thermochemical conversion of biomass reach large-scale implementation. Proper catalyst acidity aids in the formation of MAH compounds, while highly acidic catalysts result in high coke production. Incorporation of metals in the hydropyrolysis environment aids in the creation of alkanes and CH4, at the expense of char. In particular, metal-catalyzed methanation and hydrogenation reactions were found to compete with acid-catalyzed deoxygenation when both were present.

Acknowledgement

The study was funded by the National Science Foundation Award CBET-1236738 and the University of Connecticut GK12 program.

References

[1] V.K. Venkatakrishnan, W.N. Delgass, F.H. Ribeiro, R. Agrawal, Green Chem. 17 (2015) 178–183.

[2] T. Prasomsri, M. Shetty, K. Murugappan, Y. Román-Leshkov, Energy Environ. Sci. 7 (2014) 2660.

[3] D.P. Gamliel, S. Du, G.M. Bollas, J.A. Valla, Bioresour. Technol. 191 (2015) 187–196.

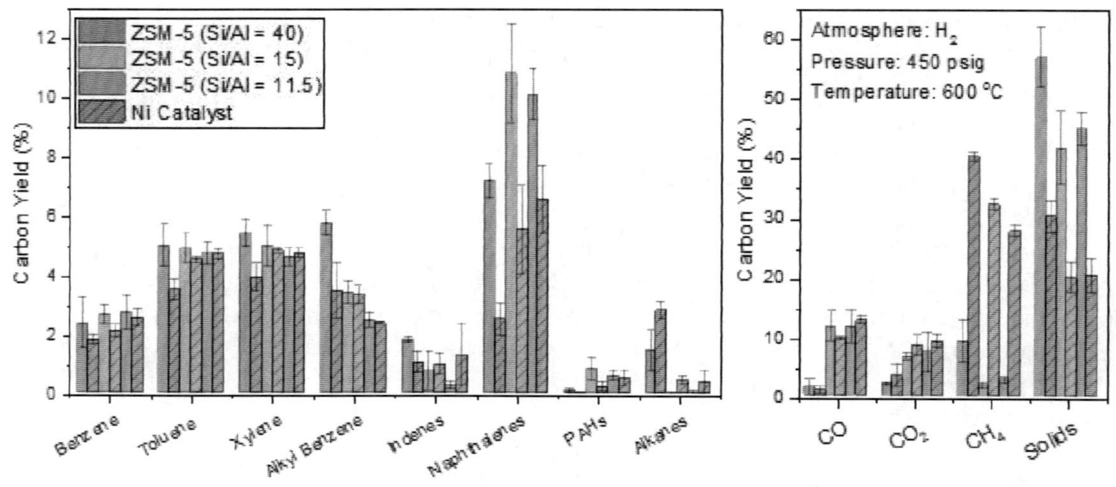

Figure 1. Bio-oil product distribution (left) and permanent gas product yields (right) for hydropyrolysis with ZSM-5 zeolites of various acidity with and without Ni incorporated.

471815 Effects of Feedstock on the Product Yields and Selectivity of Lignocellulosic Biomass Pyrolysis Using HZSM-5 Catalyst

Abstract:

Effects of Feedstock on the Product Yields and Selectivity of

Lignocellulosic Biomass Pyrolysis Using HZSM-5 Catalyst

Brendon Shea, Charles Coe, Justinus Satrio*

Department of Chemical Engineering

Villanova University, Villanova, PA 19085

One promising process of utilizing lignocellulosic biomass for the production of fuels and chemicals is fast pyrolysis. In this approach, the biomass is thermally treated at temperatures around 400-600°C in the absence of oxygen. The products formed include a solid char, a liquid bio-oil, and a mixture of non-condensable gases. The liquid bio-oil is composed of a multitude of organic compounds; many of them are oxygenated, such as ketones, phenols, and carboxylic acids. The high presence of these oxygenated compounds is not desirable since they cause bio-oil to be highly acidic, highly polar, chemically unstable, and low in heating value products. The use of solid acid catalysts, such HZSM-5, an aluminosilicate zeolite, can alter the liquid product distribution in favor of non-oxygenated compounds such as aromatics. These hydrocarbon compounds are desirable in bio-oil, especially when the bio-oil is used as intermediate feedstock for the production of liquid transportation fuels.

The product distribution is heavily affected by the feedstock of biomass that is pyrolyzed. A study on fast pyrolysis of

5 different biomass feedstocks was conducted with and without H-ZSM-5(40) as a solid acid catalyst. Feedstocks tested included pinewood, mixed hardwood, switchgrass, horse-barn bedding, and phragmites (common reeds). Fast pyrolysis experiments were conducted using a micropyrolyzer system which was connected to a GC/MS and a TCD gas analyzer for analyzing chemical composition of the pyrolytic vapors and non-condensable gases respectively. Catalytic pyrolysis experiments were performed in *ex-situ* mode, in which catalytic vapors were passed through a heated catalyst bed before being sent to the vapor and gas analyzers. The *ex-situ* catalytic pyrolysis configuration allowed multiple catalytic pyrolysis runs were done in series on the same catalyst sample to observe the effect of biomass feedstock on the catalyst activity. For completing mass balance, the amount of char formed was also measured in each run.

Experimental results indicated that biomass feedstock has a significant effect on pyrolysis product yields and selectivity as well as on the catalyst activity. Switchgrass and phragmites produced significantly more char than the other biomass feedstock. In non-catalyzed pyrolysis, these two grass-type feedstock produced less CO and more CO_2. The composition of chemical moieties of bio-oil products also was affected by the type of feedstock used for the fast pyrolysis reaction. In catalytic pyrolysis, pinewood and mixed hardwood were found to produce the most CO. while mixed hardwood and phragmites produced the most CO2. There was little correlation in the light gas formation as the catalyst deactivated. The bio-oil composition did vary as the catalyst deactivated; however the light gas formation showed very little correlation with the catalyst deactivation. For all biomass types, as the catalyst deactivated, the amounts of aromatics produced decreased while the amounts of ketones, phenols, and carboxylic acids produced increased. It appeared that pinewood had the greatest rate of change in bio-oil composition as the catalyst deactivated. While catalytic pyrolysis of pinewood showed the largest hydrocarbon yields initially, after several runs the yield difference compared to those from catalytic pyrolysis of different biomass feedstock became negligible.

*Contact: Justinus Satrio, Ph.D. Email: justinus.satrio@villanova.edu

473194 Physiochemical Characterization of Lignocellulosic Biomass Dissolution By Flowthrough Pretreatment

Abstract:

Comprehensive understanding of biomass solubilization chemistry in aqueous pretreatment such as water-only and dilute acid flowthrough pretreatment is of fundamental importance to achieve the goal of valorizing biomass to fermentable sugars and lignin for biofuels production. In this study, poplar wood was flowthrough pretreated by water-only or 0.05% (w/w) sulfuric acid at different temperatures (220-270 °C), flowrate (25 mL/min), and reaction time (8-90 min), resulting in significant disruption of the lignocellulosic biomass. Ion chromatography (IC), Fourier transform infrared (FTIR) spectroscopy, X-ray diffraction (XRD) analysis, and solid state cross polarization/magic angle spinning (CP/MAS) ^{13}C nuclear magnetic resonance (NMR) spectroscopy were applied to characterize the pretreated biomass whole slurries in order to reveal depolymerization as well as solubilization mechanism and identify unique dissolution structural features during these pretreatments.

465220 Comparison of Ethanol Yields from Fermentation of Hydrolysates Produced By CO-Solvent Enhanced Lignocellulosic Fractionation (CELF) and Dilute Acid Pretreatments of Switchgrass

Abstract:

For decades, numerous environmental, economic, and strategic concerns have motivated research into sustainable, domestic sources of fuels to replace those derived from petroleum. Lignocellulosic biomass represents the most abundant renewable resource and provides the only known route to sustainably produce liquid fuels on a large scale and low cost. Due to its complex and recalcitrant structure, raw biomass hinders biological conversion of polysaccharides, thus necessitating pretreatment to deconstruct the plant with high yields through making cellulose and hemicellulose more accessible to enzymes for release of fermentable sugars. Several pretreatment methods, including hydrothermal and dilute acid approaches, have been developed to improve enzyme accessibility, but each is limited in their ability to remove lignin, a key contributor to biomass recalcitrance, and require expensive enzyme loadings to achieve high yields. Recently, we developed a novel pretreatment called Co-solvent Enhanced Lignocellulosic Fractionation (CELF) that applies aqueous tetrahydrofuran (THF) with dilute sulfuric acid to remove a large portion of the lignin, simultaneously achieve high yields of five carbon sugars from hemicellulose during pretreatment, and subsequently hydrolyze most of the glucan at very low enzyme loadings. In this study, we applied CELF to switchgrass to determine optimum pretreatment conditions for maximum sugar release in the combined operations of pretreatment and enzymatic hydrolysis at low solids loadings and compared the results to those for DA pretreatment performed at conditions that realized maximum sugar yields from pretreatment coupled with enzymatic hydrolysis. For each pretreatment, we then established ethanol yields from fermentation of the hydrolysates produced at the optimum conditions, the degree to which fermentation was inhibited by dissolved compounds, and the effectiveness of various conditioning strategies in reducing inhibition.

466651 Understanding the Effects of Ammonia and Organosolv Pretreatments on Physicochemical Characteristics of Lignin

Abstract:

Pretreatment is an important step in biological conversion of biomass. It renders structural changes of biomass to reduce the biomass recalcitrance, thus enhancing its sugar release performance. Lignin is one of major barriers for lignocellulosic biomass utilization; therefore, various pretreatment methods have been studied to reduce the recalcitrance of biomass by removal and/or modification of lignin in the biomass. Among many pretreatment methods, two pretreatment methods using ammonium hydroxide and organic solvents were selected to understand how these methods reduces the recalcitrance and improves sugar release. In this study, the crude lignins isolated from poplar and switchgrass were subject to aqueous ammonia and organosolv pretreatments, and the effects of each pretreatment on lignin structural changes were investigated. Physicochemical characteristics of the lignins during the pretreatments were analyzed by diverse analysis methods. Chemical compositions of soluble fraction and residual solid fraction were analyzed by HPLC and GCMS. Molecular weights of both fractions were investigated using GPC. 2D ^{13}C-^{1}H HSQC NMR and ^{31}P NMR analyses were conducted to reveal the structural and chemical transformation of lignin. Adsorption of enzymes on the pretreated lignin was also investigated.

466953 How Solvent Properties Determine Pretreatment Effects on Key Lignocellulosic Characteristics

Abstract:

Chemical pretreatment efficacy is largely dependent on the solvent employed. Therefore, picking a solvent is one of the most important decisions in the early stages of chemical process design[1]. Traditionally, biomass pretreatment decisions have been more guided selection than rational design. Here we present a study of different organic solvent pretreatments on steam exploded bagasse. We demonstrate how certain chemical moieties affect the delignification capacity of a solvent. In addition to delignification, we describe the effect of organic solvent properties towards enhancing cellulose conversion to glucose. These methods provide insights into the chemistry of organic solvent pretreatments and a foundation for the molecular design of improved organic solvent pretreatment methods.

1. Shuai, L. and J. Luterbacher **(2016).** "Organic Solvent Effects in Biomass Conversion Reactions." *ChemSusChem* 9.2: 133-155.

469494 Comparison of Celf, Cu-AHP, and HS Ionic Liquid Pretreatment of Hardwoods: Hydrolysis Yields, Comprehensive Mass Balances, and Lignin Properties

Abstract:

In this work, three pretreatments under investigation at the DOE Bioenergy Research Centers (BRCs) are subjected to a side-by-side comparison to assess their performance on model bioenergy hardwoods. These will include co-solvent enhanced lignocellulosic fractionation (CELF), high-solid loading pretreatment with ionic liquid (HS-IL), and Cu-catalyzed alkaline hydrogen peroxide pretreatment (Cu-AHP). The pretreatments will be individually optimized for the feedstock and common performance metrics as assessed by a single lab will be used as a basis of comparison. The first component of this will be to assess the susceptibility of the pretreated biomass to hydrolysis by cellulolytic enzymes. This reactivity will be related to both structural and compositional changes to plant cell wall as a consequence of pretreatment. The second component will be to determine comprehensive mass balances on the pretreatments. This will include assessing the solubilization, depolymerizaiton, and conversion of cell wall biopolymers as well as the solvent requirements, their degradation, and their potential for reuse. An important feature of the pretreatments in this study is that all three have shown preliminary evidence of yielding lignins with properties amenable to further valorization. As the final component of this work, we will characterize the yields, structural properties, and reactivity of the soluble and insoluble lignin fractions generated by these pretreatments.

471752 Understanding the Influence of Hemicellulose and Lignin Removal on Deconstruction of Switchgrass By *Clostridium Thermocellum*Â Consolidated Bioprocessing *Vs.* Conventional Fungal Enzymatic Hydrolysis

Abstract:

Low cost sustainable fuels are urgently needed to support human prosperity. Currently, almost all liquid fuels are derived from petroleum that is finite in quantity and contributes significantly to climate change. According to the US Energy Information Administration, transportation accounts for about 70% of the total US oil consumption and about 95% of which is from fossil fuels consumption. Alternatively, biofuels would provide a sustainable energy source and help reduce our dependence on imported petroleum. Ethanol made from corn starch and cane sugar is presently the largest biotechnology-based product and commands a large market share of the alternative fuels market. However, currently the trend is to produce ethanol from lignocellulosic biomass that, unlike corn starch and cane sugar, does not directly compete with our food infrastructure. Ethanol is a high octane, cleaner burning fuel that, when made from lignocellulosic biomass, offers a low carbon footprint and the potential to be further upgraded to other drop-in fuels, such as butanol, jet fuel, gasoline, and diesel. Ethanol production from lignocellulosic biomass involves numerous important operations including size reduction, pretreatment, enzyme production, enzymatic hydrolysis, fermentation, and product recovery. The production of cellulolytic enzymes, typically from a fungus such as *Trichoderma reesei*, is one of the most expensive steps because of the high concentrations of enzymes needed to achieve economically attractive ethanol yields from recalcitrant lignocellulosic biomass. Consolidated bioprocessing (CBP) is a simplified, integrated bioprocess for directly converting lignocellulosic biomass into ethanol by utilizing thermophilic anaerobes that can produce their own enzyme consortium to hydrolyze biomass polysaccharide chains into fermentable sugars after which the same organism metabolizes the sugars to produce ethanol and other products. *Clostridium thermocellum,* in particular, is a very promising CBP organism. However, pretreatment may still be required to assist *C. thermocellum* in biomass deconstruction for high yields. Dilute acid and hydrothermal are leading pretreatment technologies that reduce biomass recalcitrance by removing some of the lignin and most of the hemicellulose. Alkali pretreatments, on the other hand, remove some hemicellulose and a lot of the lignin. By comparison, co-solvent enhanced lignocellulosic fractionation (CELF), a novel pretreatment technology recently invented at the University of California, Riverside, which makes use of tetrahydrofuran (THF) as a co-solvent in water, enhances the solubilization and recovery of both hemicellulose sugars and lignin from lignocellulosic biomass beyond that of conventional pretreatment technologies. Because different pretreatment technologies produce solids with different physical and compositional characteristics, it is important to evaluate how these features impact CBP performance. This study, therefore, compared the effects of CELF pretreatment combined with *C. thermocellum* CBP on total sugar release from switchgrass to those of hydrothermal, dilute acid, and alkali pretreatments. The objective was to identify pretreatment conditions that give the highest possible sugar release in pretreatment (Stage 1) and biological conversion (Stage 2) steps combined for each pretreatment technology. The pretreatments were performed over a range of temperatures and times, and the resulting washed solids were then subjected to *C. thermocellum* CBP at 5 g/L glucan loading in 50 mL bottle reactors incubated at 60ËšC with a shaking speed of 180 rpm and fungal enzymes mediated enzymatic hydrolysis at 50ËšC and 150 rpm with the same solids loading to establish overall switchgrass deconstruction and sugar recovery.

457412 Separation of Lignin from Organosolv Spent Liquor-from Phase Behavior to Continuous Processing

Abstract:

Worldwide pulping mills combust several hundred million tons of lignin every year in their recovery boilers to recycle the pulping chemicals and provide process heat [Cazacu et al. 2013]. The most applied Kraft and Sulfite pulping techniques are focused on pulp production whereas lignin is utilized as cheap energy provider.

Lignin is build up from phenylpropane monomers with hydroxyl and methoxy functional groups, making it a valuable renewable resource for the chemical and material industry. Lignocellulose biorefineries are supposed to sustainably provide renewable platform chemicals for the future by fractionating woody biomass with appropriate processes. Organosolv pulping techniques can fractionate lignocellulosic biomass into mainly pulp, lignin and (oligo-) sugars by using an organic solvent and water as pulping liquor at temperatures near 200 °C. The resulting pulp is of reasonable quality [Zhang et al. 2016] and the dissolved lignin can be separated from the spent liquor as precipitate by removing the organic solvent, which is recycled to pulping liquor. The (oligo-) sugar fraction remains dissolved in the aqueous phase. Organosolv lignin is known to be of high quality with low ash, sulfur and carbohydrate contents. The separation processes in a lignocellulose biorefinery utilizing organosolv pulping are essential for the economic viability [Viell et al. 2013].

The Fraunhofer Center for Chemical-Biotechnological Processes CBP in Leuna, Germany, operates a lignocellulose pilot plant consisting mainly of a batch forced circulation reactor for ethanol organosolv pulping, an enzymatic digestion unit for cellulose to sugar hydrolysis, as well as ethanol evaporation and lignin precipitation facilities. Fraunhofer CBP and MPI DCTS initiated a cooperation for investigation and optimization of the lignin separation process especially in terms of energy consumption, fouling prevention and downstream processability in filtration and drying.

In this contribution the path from fundamental phase behavior determination to an advanced continuous lignin precipitation process is presented [Schulze et al. 2016]. Systematic solubility measurements in the ternary lignin-ethanol-water system at different temperatures have been performed in order to elucidate attractive process possibilities. Thereby lignin was found to form a liquid phase under certain conditions as result of a decreased glass transition temperature. The liquid lignin phase potentially causes fouling on the reactor internals during separation processes if the process conditions are not carefully adjusted. On the other hand, softened lignin particles can agglomerate and coalesce to form bigger particles. A compromise between particle enlargement and incrustation formation has to be made in selection of process conditions. Enlarged lignin particles contribute to a better filterability. A (semi-) continuous process for simultaneous lignin precipitation, ethanol recovery and particle enlargement was developed on the knowledge from phase behavior. Experimental results from lab scale continuous processing of spent liquors produced at different pulping conditions indicate the influence of lignin properties (like molecular weight distribution) on the separation process. Semi-continuous precipitation experiments were performed in pilot scale confirming the scalability of the process to the next level. Inline process analytic techniques (Mettler Toledo FBRM®, PVM and ATR-FTIR) were utilized for monitoring and controlling lignin particle size distribution, lignin particle shape and aqueous phase composition. Agglomeration kinetics were derived from the monitored data. Filtration experiments with the produced lignin dispersions provided filter cake resistance data to assess filterability.

References:

Cazacu, G., Capraru, M., Popa, V. (2013). Advances concerning lignin utilization in new materials. In: Thomas, S., Visakh, P.M., Mathew, A.P. (Eds.), Advances in Natural Polymers. Springer, Berlin Heidelberg, pp. 255–312.

Schulze, P., Lorenz, H., Seidel-Morgenstern, A., Leschinsky, M., Unkelbach, G. (2014). Verfahren zur Fällung von Lignin aus Organosolv-Kochlaugen. DE priority patent application filed in 20th October 2014.

Schulze, P., Lorenz, H., Seidel-Morgenstern, A., Leschinsky, M., Unkelbach, G. (2016). Advanced process for precipitation of lignin from ethanol organosolv spent liquors. Bioresource Technology, 199, 128–134.

Viell, J., Harwardt, A., Seiler, J., Marquardt, W., (2013). Is biomass fractionation by Organosolv-like processes economically viable? A conceptual design study. Bioresource Technology, 150, 89–97.

Zhang, K., Pei, Z., Wang, D. (2016). Organic solvent pretreatment of lignocellulosic biomass for biofuels and biochemicals: a review. Bioresource Technology, 199, 21–33.

462650 Cell Lysis and Lipid Recovery from Oleaginous Yeast

Abstract:

Single-cell oils, or lipids derived from oleaginous microbes, have seen recent interest as a potential renewable source of transportation fuels. However, one of the main technical challenges involved in single-cell-oil-derived products is due to the intracellular nature of the oils: the microbial cells must be lysed and the lipids extracted from the resulting complex matrix. We have screened several approaches for cell lysis approaches using the oleaginous yeast *Lipomyces starkeyi*, and have found that a dilute acid pretreatment compares favorably to a number of alternative strategies including high-pressure homogenization, alkaline pretreatment, bead beating, enzymatic hydrolysis or thermolytic approaches. Extraction of the lysed cells with hexane allows for recovery of the lipids in yields above 90% and in high purity. The acid pretreatment is also effective on strains of *Cryptococcus curvatus* and *Rhodosporidium toruloides*, and is highly scalable.

471016 Transport and Reaction Modeling of Biomass Pretreatment and Topochemical Evolution Using Actual 3D Structure of Plant Cell Walls

Abstract:

Plant biomass is a renewable raw material for a wide variety of bio-based products which include pulp and paper, bio-plastics, biofuels, wood plastic composites etc. The process of conversion of plant biomass to other useful products is heavily dependent on our understanding of the physio-chemical structure of plant cell walls. Plant cell walls are composed of a complex network of pore spaces and fibers of cell wall components that has a significant influence on the efficiency of the biomass conversion process, the properties of the bioproducts and their end-use applications. Hence it is important to characterize the 3D plant cell wall structure. 3D characterization techniques such as X-ray computed tomography (XRCT) and Transmission electron microscopy computed tomography (TEM-CT) have been popular. Although, the X-ray CT provides micron scale visualization, it is not sufficient to get a complete understanding of the 3D structure of plant cell walls. Hence, a more powerful technique of TEM-CT is used to probe the nanoscale 3D ultrastructure. The computed tomography technique is based on capturing a series of images by successive tilts around a single axis and reconstructing the 3D image by back projection. The structural information obtained from TEM-CT can be correlated with topochemical information from Raman spectroscopy to fully understand the physiochemical structure of plant cell walls.

The first step in the conversion process is the pretreatment step, which disrupts the recalcitrant structure of lignocellulose and increases the efficiency of the subsequent hydrolysis. One or more of the cell wall components are dissolved during this process. The changes in structural characteristics such as porosity, specific surface area and 3D pore size distribution during the pretreatment process can be determined from the processed TEM-CT images. The structural information from TEM-CT can be correlated with the topochemical distribution from Raman Spectroscopy to develop a transport-reaction model based on a hybrid random walk and reaction. This model involves a stochastic dynamic approach that keeps track of the change in structure and spatial concentration of the cell wall components simultaneously. A fixed number of particles of the reagent used for pretreatment, which is representative of its concentration, diffuse through the pore spaces of the plant cell walls till it encounters a pore-fiber interface. This path can be monitored by a stochastic hybrid random walk approach. At the interface, the reagent used for pretreatment can either react with one or more of the cell wall components or undergo reflection back into the pore space, based on the reaction probability. The, spatial concentration profile of the pretreating agent at different times can be determined by keeping track of the number of particles of the pretreating agent at each position as time proceeds. The spatial concentration profiles of the cell wall components and their evolution during biomass conversion can also be determined from the corresponding rate equations.

Thus, with such a model, a complete understanding of the 3D structure and the physicochemical distribution and its evolution can be understood. This provides a fundamental insight into biomass recalcitrance and how they can be addressed in developing efficient conversion strategies.

472309 Lignin-Derived Products from Biocrude As Building Block Chemicals

Abstract:

The production of chemicals as bioproducts in an integrated biorefinery could enhance the economic viability of biomass-to-liquid fuel technologies and afford a higher return on investment. RTI International is developing an advanced biofuels technology that integrates a catalytic biomass pyrolysis step and a hydroprocessing step to produce infrastructure compatible biofuels. In recent efforts, RTI is exploring and developing processes that could be used to produce bioproducts in addition to liquid fuel from the biocrude intermediate.

The present study explores distillation as a separation method for recovery of lignin-derived chemicals from biocrude. The biocrudes used were relatively thermally stable and distillable; they were produced from catalytic pyrolysis with a non-zeolite, alumina-based catalyst. Gas chromatography-mass spectrometry (GC-MS) analysis indicated that the biocrude contained reasonable amount of lignin-derived compounds. The distillation studies were performed with a laboratory distillation unit (PETRODIST 300 CC) and the biocrude were fractionated into four distillate cuts: Cut 1 (IBP-110 °C), Cut 2 (110-200 °C), Cut 3 (200-300 ° C), and Cut 4 (300-400 °C). The yields, physicochemical properties, and chemical composition of the fractions as well as the overall distillation separation efficiency will be discussed.

461248 Continuous Butanol Extraction Using Supercritical Carbon Dioxide

Abstract:

Butanol has promise as a drop-in biofuel or as a fuel additive that can be blended with gasoline at much higher proportions than ethanol. To date, economical production of butanol has been hampered largely by its cytotoxicity, which becomes limiting at levels as low as several weight percent. In this work, we demonstrate n-butanol extraction from dilute aqueous solutions used to simulate fermentation broths using supercritical carbon dioxide ($scCO_2$). In this talk, we will focus on process engineering of the n-butanol extraction. Our process engineering work is complemented by microbiological and metabolic engineering work that has identified a $scCO_2$-tolerant bacterial strain and genetically modified it for biofuel production. Compared to other butanol recovery approaches, $scCO_2$-extraction has sterilization and potential energy balance advantages; because $scCO_2$ selectively extracts butanol instead of water, a highly concentrated butanol stream can be recovered which requires minimal post-processing purification.

Specifically in this work, we studied n-butanol extraction performance and material/energy balance analysis. For butanol extraction performance, we used batch-wise extraction to study the effects of initial n-butanol concentration, extraction vessel pressure, and $scCO_2$ volumetric flow rate on n-butanol extraction rate. Additionally, we modeled the data using a standard liquid-liquid mass transfer model to determine values for the mass transfer coefficient, $k_l a$. Figure 1 shows representative extraction data compared to the model predictions using fit mass transfer coefficients. In all cases, the mass transfer model adequately described the experimental data. Best-fit values of $k_l a$ did not vary within our estimated limits of uncertainty for variation in extraction pressure (from 10.3 to 13.7 MPa – the range over which the $scCO_2$-tolerant bacterial strain has exhibited growth); therefore, we conclude that operation at lower pressures should be favored to achieve better process economy. Similar analysis was performed to interpret the mass transfer coefficient from correlations developed for gas-liquid and liquid-liquid extraction and compared to interfacial area results obtained from $scCO_2$-droplet size analysis. In addition to n-butanol, we investigated extraction of other alcohol products (e.g., iso-butanol, pentanol, and hexanol) as these are potential metabolic targets for a modified $scCO_2$-tolerant organism. Finally, we studied the extraction of the aldehyde intermediate, butanal, to set limits on its accumulation in the fermentation products.

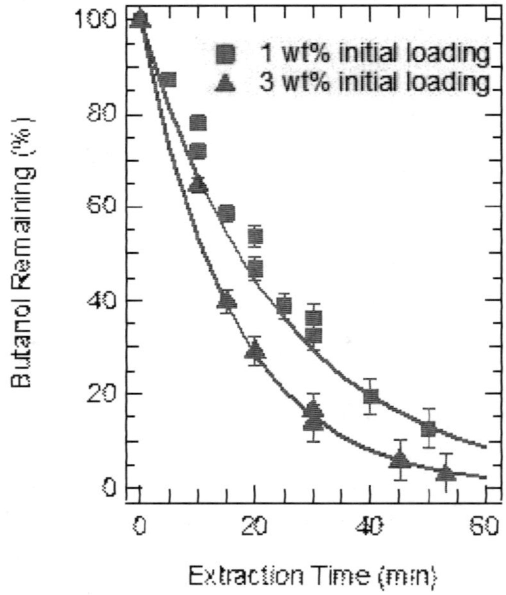

Figure 1. Representative n-butanol extraction data obtained at 10.3 MPa and 40 °C.

In addition to extraction experiments and modeling, we performed preliminary material and energy balance calculations of a conceptual continuous process. Thermodynamic and extraction data were used to model n-butanol removal from the fermenter and fermentation kinetics were modeled based on activity of similar organisms. The Peng-Robinson equation of state was used for modeling thermodynamic properties. To improve energy efficiency, the $scCO_2$ extraction stream was only partially de-pressurized from 10 to 5 MPa, resulting recovery of n-butanol at 79 wt% purity. Energy was recovered during the de-pressurization, which partly offset the energy required for CO_2 compression. Analysis was performed assuming the availability of either 10 or 0.1 MPa CO_2. Overall, we found that n-butanol with 95% purity could be produced at an energy cost of 3.9 MJ kg^{-1} (assuming that a 1 bar CO_2 source is available) or 3 MJ kg^{-1} (assuming a 10 MPa CO_2 source). These figures compare well with existing methods of butanol recovery, which range from 9 to more than 20 MJ kg^{-1}, and the lower heating value of the butanol fuel itself (33.1 MJ kg^{-1}). Further analysis is needed to account for heat loss and heat integration and to optimize process efficiency.

466873 *Development of an Integrated Aquaculture System for Clean Water and Animal Feed Production*

Abstract:

The concept of a "simulated lichen system" was recently developed by our UMN research group in which we can select different desired microalgae and fungal combinations which can be grown on the surface of specific polymers to form a biofilm. This technology can be extended to aquaculture wastewater bioremediation in order to recycle the nutrients as a proteinaceous microbial biomass feed for the reared aquatic animals. This process modification could make aquafarming more efficient as commercial feeds account for up to 50% of total production costs. The "mycoalgae biofilm" utilizes photosynthetic aeration to replace the conventional and energy intensive mechanical aeration for biological processes. By using the mycoalgae biofilm in aquaculture production, farmers will realize lowered costs for feed and reduced costs for maintenance and cleaning of the aquaculture water. Microalgae cultivation has been heavily researched to treat nutrient-rich water; however, the process is limited by problems associated with the settling and separation of small microalgae cells downstream of the treatment site. The proposed methodology of using a bioaugment of filamentous fungi in the lichen biofilm overcomes this limitation. The fungi efficiently retain algae and can also aid in the recovery and recycling of nutrients in the wastewater. The fungi retaining the algae allows for a less energy intensive harvesting process as a singular solid biofilm is produced. The strains complement each other by the exchange of required respiratory gases. The biofilm composition can be tailored based on the influent stream components, feasibility of the strains to grow together and its nutritional value as animal feed.

The effect of different initial levels of ammonia were tested in synthetic aquaculture wastewater to study the removal efficiency of nutrient pollutants by our biofilm. Starting ammonia levels were chosen based on a literature review of effluent discharge from various cultured aquatic species. Ammonia levels were reduced to undetectable limits in the medium to low initial levels of ammonia trials. At high levels of ammonia a set ammonia removal rate was observed, independent of total ammonia present in the system. We suspect the removal rate is dependent on nutrient availability and can be optimized in the future. The mycoalgae biofilm outperformed mono-cultures of fungi and algae at every level of ammonia tested in regards to total nutrients removed from the media. Combined biomass production was also higher in the co-cultures than in mono-cultures at all ranges of ammonia tested. The biomass has added value as feed for the cultured fish, thus making the process more economically advantageous. The biomass generated will be high in polyunsaturated fatty acids, proteins, and carbohydrates thus enabling robust fish growth. The biofilm's nutritional properties, low energy harvesting method, and effective nutrient removal abilities make it a promising system for wastewater remediation.

470075 Carbon Dioxide (CO_2) Capture and Utilization Pathways â€“ Assessing the Synergies Between Biological Use and Geological Sequestration of CO_2

Abstract:

As an important greenhouse gas (GHG), the emission of CO_2 to the atmosphere needs to be reduced in order to mitigate climate change. Designing of proper of CO_2 utilization pathways are imperative to reduce GHG emission. There are several ways to restrict CO_2 emission into the atmosphere. Two such promising technologies include geological sequestration of CO_2 in depleted oil and gas fields, saline aquifers and others, and biological sequestration of CO_2 via microalgae. The primary focus of the present study is to assess the integration potential of CO_2 geological sequestration and CO_2 biological sequestration via growth, harvest and conversion of microalgae to value-added bioproducts. Different ways of microalgae conversion, including lipid extraction, utilization of the spent algal biomass as well as hydrothermal liquefaction of the whole microalgal biomass, are considered for the production of value-added products. The present study aims to quantify the techno-economic tradeoffs and optimize the entire supply-chain of CO_2 utilization to ensure the most economically feasible alternative of CO_2 usage. A multi-scale supply chain optimization framework is considered which takes into account CO_2 capture, compression, pipeline transport and geologic storage as well as different processing pathways of microalgae. The state of Texas in USA is considered as the case study to evaluate the optimal design of CO_2 capture, storage and/or utilization pathways. The present study will help inform decision and policy makers the optimal pathway of CO_2 storage and utilization procedures for a particular location.

Chevron

human energy

Sustainable Development and Footprint Optimization in Shale Development

Janet Peargin, Footprint Program Manager
Hong Jin, Lead HES Engineer

Chevron Energy Technology Company

November 17, 2016

The challenge
footprint of shale/tight rock development

Many wells & moving parts

Footprint changes over time

Jobs
Community GDP
Desired energy

Land/habitat
Emissions
Trucks
Noise
Waste
Resource competition

Chevron's environmental principles

Include the environment in decision making

Reduce our environmental footprint

Operate responsibly

Steward our sites

Footprint optimization: everybody plays a role

We can't solve problems by using the same kind of thinking we used when we created them.

– Albert Einstein

Shifting your perspective is so much more powerful than being smart.

– Astro Teller, Google "X" 2016 TED Talk

The opportunity: sustainable development

SEE across the asset life to create value and protect people and the environment

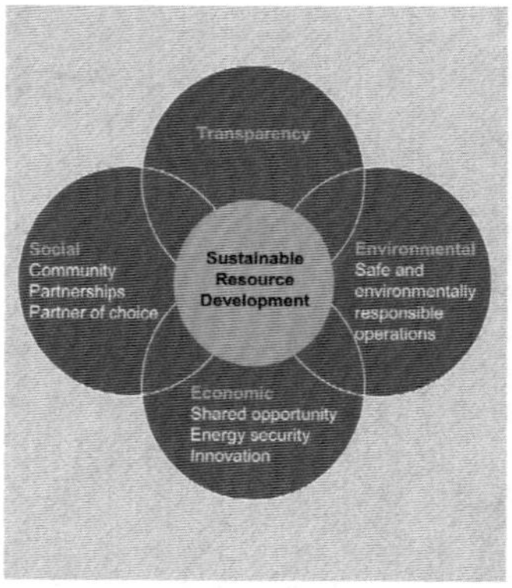

SEE triple bottom line value

SEEk win-win (all-win) ideas
Social, Environmental, Economic

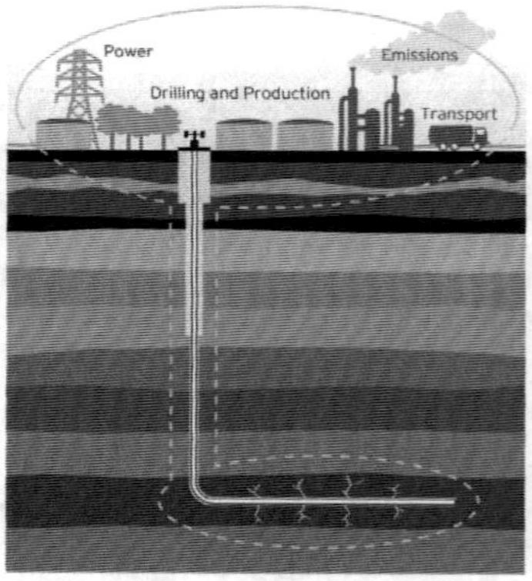

Optimize footprint

Integrated system of materials and energy that changes over time

Design with the end in mind

Decommission, restore, reuse

The win-win (all-win): optimize footprint to enhance value and reduce cost/risk

*Life Cycle Assessment Standard 14040-14044

Improve planning: SEE through the lens of sustainable development and SEE win-win (all-win) opportunities

Look at the big picture (asset, region)
Look at the long-term and change over time
Include diverse stakeholder perspectives

Take a systems view. Integrate across functions, activities, hydrocarbon/water/power
Plan and design for social, environmental, and economic (SEE) value drivers

Consider:

- Pace and scale of development
- Sequencing of events
- Degree of centralization
- Logistics and supply chain
- Ability to share infrastructure
- Power/fuel options
- Integration points

Improve decision quality: SEE more broadly and longer-term

Compliance is expected.

A traditional financial analysis is narrow and short-term.

SEE more broadly and longer term to identify risks and opportunities ... and enhance decision quality.

Compliance
- Business ethics
- Skills enhancement
- Social investments
- Safety/health

Compliance
- Biodiversity
- Emissions reduction
- Spill prevention
- Water conservation
- Waste reduction

Compliance
- Earnings
- Revenues
- Net cash flow
- Shareholder return
- Jobs created
- Local economic impacts

7

Improve decision quality: quantify environmental tradeoffs with ISO life cycle assessment (LCA)

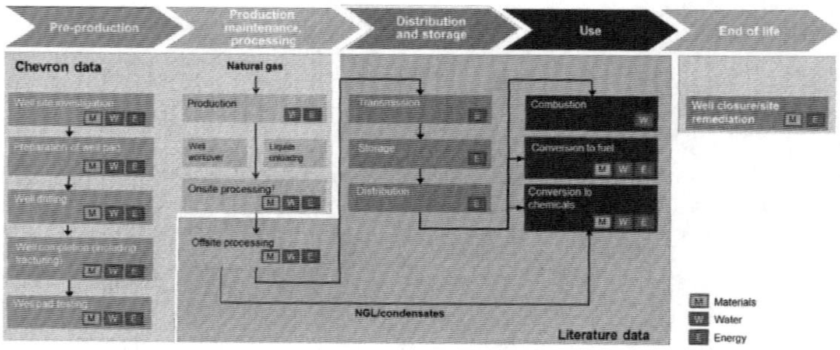

- What is our environmental footprint?
- Where are our big sources?
- What is the “best” fuel for power?
- What if I use this technology vs. that?
- What are the on and offsite impacts?

8

Improve design: apply win-win (all-win) principles

Plan/design for sustainable development	• Develop long-term integrated plans • Frame for SEE (social, environmental, economic)
Integrate and innovate	• Look across the asset life at hydrocarbons/water/power & different activities (drilling, operations, etc.)
Question business as usual	• What do stakeholders want? Is there a better way?
Do *more* with *less*	• Reduce – reuse – recycle – be (energy/materials) efficient • Eliminate no-value steps & equipment
Disturb less land	• Fewer pads; share infrastructure/right-of-ways; preserve habitat corridors
Conserve water	• Use less & lower quality; treat to quality needed for use; consider reuse
Reduce waste	• Flares/emissions, discharges, energy use, no-value steps/equipment
Green the process	• Consider non-toxic chemistry, power/fuel efficiencies, fewer steps
Leverage technology	• Look to downstream, offshore, suppliers, competitors, others • Think mobile, modular, compact, flexible, prefab, energy efficient, automated, standard
Leverage supply chain	• How can suppliers help? Can we hire and source locally?
Make informed decisions	• SEE the big picture. Consider short- *and* long-term costs and value. • Use life cycle assessment to quantify environmental aspects of alternatives

Chevron

Win-win example: electricity from field gas

Better emissions/safety/reliability/production/cash flow

Problem

Need cost-effective, low-emissions power source at remote site

Solution

Micro-turbines make electricity from field gas*

Win-Win Do *more* with *less*

MORE

Safety, flexibility, schedule certainty, reliability (8–10%), production/cash flow (sooner)

LESS

Equipment, permitting, O&M, trucks, emissions, material transfer/spill potential

** Great for ESPs. Rod pumps doable with more systems engineering*

Chevron

Everyone plays a role: ask questions

- What are our stakeholder concerns?
- What perspectives are we missing?
- What are our footprint boundaries?
- What are the priorities and tradeoffs?
- What does good (enough) look like?
- How can we minimize environmental and community impacts?
- How do we do more (product) with less (material, equipment, waste, etc.)?
- How do we design for change over time to operations, region, community?
- What can we learn from other processes?
  - LeanSigma®
  - Environmental, social, health impact assessment
  - Energy Efficiency and Waste Minimization assessments

LeanSigma is a federally registered service mark of TBM Consulting, Inc.

Key takeaways

- We all play a role in creating our company footprint – a system of materials and energy that changes over time
- Think beyond compliance. Include SEE (social, environmental, economic) drivers in planning, designs, and decisions.
- SEE through the lens of sustainable development to SEE opportunities. SEE the difference.
  - Cost savings from less equipment, optimized layout/network
  - Win-win ideas to save $ AND protect people & the environment
  - Decisions that are more robust
- A financial analysis is only part of the risk/uncertainty. We <u>can</u> quantify and include SEE drivers in decision-analysis.
- Sustainable development is good business. SEE it reduce cost and risk, add value, and improve performance.
- Everyone plays a role. You can contribute. Just change your perspective to SEE the win-win (all-win) opportunities to do more (product, energy) with less (energy, materials, waste, emissions, trucks, etc.)

References

- Hardisty, P.E. *Environmental and Economic Sustainability.* CRC Press, New York. ISBN 978-14200594891-56679-643-2, 378 pp. 2010.

- Sivapalan, M.; WorleyParsons. *Sustainable Development of Unconventional Oil & Gas Resources.* Prepared for presentation at *GasTech* 2012 Conference, London, 2012.

- Tanzil, D.; Dragan, M.; Bourque, B.; Seel, K.; Taylor, T.; Golder Associates. *Evaluating Social and Environmental Risk and Opportunities in Site Selection and Routing.* Copyright 2015, Society of Petroleum Engineers. SPE-173488-MS. Prepared for presentation at the SPE E&P Health, Safety, Security, & Environmental Conference – Americas held in Denver, Colorado, USA, 16–18 March 2015.

- Thomas, Charlotte. I*ndustrial Ecology: Moving Beyond Traditional Engineering Borders.* Society of Women Engineers magazine, spring edition, pp. 24-30. 2010

461852 Green Net Value Added As a Sustainability Metric Based on Life Cycle Assessment: An Application to Bounty ® Paper Towel

Abstract:

Sustainability measurement in economics involves evaluation of environmental and economic impact in an integrated manner. In this study, we combine system level economic data with environmental impact from a life cycle assessment (LCA) of a common product. We are exploring a costing approach that captures traditional costs and incorporates externality costs to provide a convenient, easily interpretable sustainability metric. Green Net Value Added (GNVA) is a type of full cost accounting that incorporates total revenue, the cost of materials and services, depreciation, and environmental externalities. It adjusts the traditional economic value added for environmental externality costs. Two, but not all, of the potential environmental impacts calculated by the standard LCIA method (TRACI) could be converted to externality cost values. We compute externality costs disaggregated by upstream sectors in order to accurately reflect the externality costs associated with different industries along the supply chain. We preset full cost, green profit and GNVA to evaluate the relative sustainability of Bounty® paper towels manufactured at two production facilities. We found that the longer running, more established line had a higher GNVA than the newer line. The dominant factors contributing to externality costs are to come from the stationary sources in the supply chain: electricity generation (27-35%), refineries (20-21%), pulp, and paper making (15-23%). Health related externalities from Particulate Matter (PM2.5) and Carbon Dioxide equivalent (CO_2e) emissions appear largely driven by electricity usage and emissions by the facilities, followed by pulp processing and transport. Supply chains for other facility purchases, product distribution, and end-of-life management (landfilling/incineration) contribute the least to the environmental health-related costs.

474653 Sustainability Metrics for Solar Powered Charging Infrastructure for Electric Vehicles

Abstract:

This presentation will include an analysis of sustainability metrics associated with the installation of solar panels in parking lots designed to provide electricity for commercial use and shade for those who park under the solar panels. The installation will also include electric vehicle supply equipment to charge batteries in electric vehicles. The system is connected to the electrical grid so that the electricity generated can be used efficiently and electric vehicle batteries can be charged at all times. Important sustainability metrics are cost of electricity generated, cost of vehicle charging, value of shaded parking, convenience of vehicle charging, land area, water, air quality, greenhouse gas emissions, and materials for the installation and equipment. Social metrics include impact on health status of the community, psychological well being, harmony, values, agency, and environmental justice.

458893 Evaluation of Low-Energy Recovery of N-Methyl-2-Pyrrolidone Solvent from Waste Streams in Specialty Resin Manufacture

Abstract:

N-methyl-2-pyrrolidone (NMP) is a common solvent used as a reaction medium in the manufacture of specialty polymeric resins and their precursors. NMP is used primarily because of its thermal stability, dipolar aprotic properties, and ability to produce a high molecular weight resin precursor. Since NMP is not consumed by the reaction, an aqueous hazardous waste stream is generated. A previous case study at a DuPont specialty chemical plant by the authors, shows that a distillation recovery system that can effectively purify NMP from the waste stream. The distillation recovery process reduces operating cost by 83%, but only reduces the life cycle carbon footprint by 40%. This is due to the high energy demand of the distillation column. Additionally, high NMP purity levels in the resin manufacture process and the complex multicomponent nature of the waste stream, required a more extensive design for the recovery operation.

Alternative distillation designs were explored, but none reduced energy consumption and life cycle carbon emissions as significantly as the economic life cycle impact. More sustainable low energy separations were investigated. Various membrane, adsorption, and extraction separation strategies have the potential to reduce energy demand, but each has its own limitations. An extraction operation was selected as the best candidate for separation, and the proposed process considers alternative uses for the recovered NMP waste in other manufacturing scenarios.

Results with a chloroform extraction operation to produce NMP of suitable quality for alternative use, provides the best option for solvent recovery. The new process can not only yield environmental life cycle improvements, based on various impact factors, but provide economic benefits as well. Yearly projections of waste reductions, life cycle improvements, energy utilization, and cost savings are presented. Life cycle carbon emissions are reduced by over 60% compared to the base case. Since NMP is a prominent solvent for several commercial sectors, methods aimed at reducing waste help promote broader sustainability goals. This study shows how sustainability metrics can be used to guide decision making when evaluating other processing alternatives and uses of the final product. (This work is partially supported by a grant from the EPA Pollution Prevention program).

468272 Thermoelectric Power Technology Choices Based on Water Availability

Abstract:

Thermoelectric Power Technology Choices Based on Water Availability National Energy Technology Laboratory (NETL) has developed a prototype water-energy model to investigate issues surrounding the energy-water nexus as it relates to the effect of power generation plant design choices on water availability and the overall costs of supplying water for thermoelectric power generation.Â The model utilizes the database developed by a team of researchers with the Water Security Program at Sandia National Laboratory (SNL). The database includes water availability, cost, and water use for electric power, agriculture, municipal and industrial for 38 conterminous states representing various Interconnects within the U.S. Data on current and future water use for thermoelectric power are derived from NETL databases on water use by technology and projections from the Energy Information Administrationâ€™s National Energy Modeling System (NEMS). Current and forecasted water availability data from the World Resources Institute are used to develop three water availability scenarios. The prototype water-energy model is used to project the amount of water available for power generation, to project water demand for future power generation based on a NEMS forecast, to identify regions that experience potential water shortfalls, and to estimate the cost of meeting water demand where water shortfalls exist for each water availability scenario. A case study of the Brazos River Basin in Texas is used to illustrate a specific application and to highlight the capabilities of the prototype water-energy model.

464240 Deciphering Lignin Deconstruction Mechanism of Lower Termite

Abstract:

Although fascinating, the exact process of plant cell wall degradation developed by nature and implemented in lower termites remains unclear. We have been conducting systematic studies on the lignin deconstruction mechanism over the last seven years as an effort to develop nature-inspired technologies for lignocellulose processing. A series of experimental and analytical techniques from various disciplines were used in our studies. Scientific questions explored included mircrochemical environment of the termite guts, structural and composition changes of lignocellulose along the termite digestion track, critical enzyme systems involved in lignin deconstruction, chemical reactions that modifies lignin structure, and disassociation energy levels of different chemical bonds that form lignin structure. Integrating these results makes it possible to propose a chemical reaction mechanism and associated lignin degradation pathways. The mechanism and the pathway will be presented and discussed in the proposed paper. The results of the study provide new insights for developing more sustainable processes for plant cell wall deconstruction required in biorefineries.

455682 Production of Food, Energy, and Fuel from Microalgae in Holograhic Diffractive Optic-Solar Glass Reactors

Abstract:

This project is focused on developing technology platforms for the solar-assisted production of food, fuel, and clean water from municipal wastewater sources. This novel food, energy water (FEW) platform employs new light management strategies to simultaneously: (1) generate electrical energy; and (2) optimize algae growth for food and fuel. This is accomplished by use of a holographic diffractive optic-solar glass (HDO-SG) reactor. The glass transmits the visible light spectrum that is used for photosynthesis and internal reflects the IR and UV light that is directed toward photovoltaic (PV) cells that are integrated into the reactor. The PV- generated energy can be used to operate the photobioreactor. Preliminary results show that *Chlorella* sp and *Scendesmus* sp. algae grow well in reactors containing HDO-SG. Water loss is minimized compared to open systems which is key when growing in arid climates. Growth of *Chlorella* sp on secondary and tertiary wastewater has been demonstrated, as well as growth using of flue gas as a carbon source. No observed changes in lipid productivity were found compared to control cultures grown in laboratory media.

474778 BUTANOL PRODUCTION BY *CLOSTRIDIUM ACETOBUTYLICUM* IN A SERIES OF PACKED BED BIOFILM REACTORS

Francesca Raganatia, Dipartimento di Ingegneria Chimica, dei Materiali e della Produzione Industriali - Università degli Studi di Napoli Federico II, Napoli, Italy
Alessandra Procentese, Dipartimento di Ingegneria Chimica, dei Materiali e della Produzione Industriali - Università degli Studi di Napoli Federico II, Napoli, Italy
Giuseppe Olivieri, Dipartimento di Ingegneria Chimica, dei Materiali e della Produzione Industriali - Università degli Studi di Napoli Federico II, Napoli, Italy
Maria Elena Russo, Istituto di Ricerche sulla Combustione, Consiglio Nazionale delle Ricerche, Napoli, Italy
Peter Gotz, Department of Life Sciences and Technology / Bioprocess Engineering, Beuth University of Applied Sciences Berlin, Berlin, Germany
Piero Salatino, Dipartimento di Ingegneria Chimica, dei Materiali e della Produzione Industriali - Università degli Studi di Napoli Federico II, Napoli, Italy
Antonio Marzocchella, Dipartimento di Ingegneria Chimica, dei Materiali e della Produzione Industriali - Università degli Studi di Napoli Federico II, Napoli, Italy

The reactor design plays a key role in the fermentative production of biobutanol. The high cell concentration that may be reached in confined – biofilm, membrane, recycling - cell reactors offers high conversion rates. The concentration of solvents in the broth from biofilm reactors reported in literature is not particularly high and it negatively affects the successive stages for butanol recovery (Qureshi et al., 2005). The low concentration of solvents in the produced stream is typically due to the inhibitory effect of solvents on the fermentation. Therefore, the butanol bioreactor productivity is as low as the bioreactor behaviour approaches the CSTR limit.

Another major element of the research on ABE fermentation is the mathematical modeling of this process. The experimental investigations in any typical fermentation process are not only time consuming due to slow kinetics of the process, but also involve significant cost. The performance of ABE fermentation process, in terms of selectivity and yield of solvents, is dependent on many physical and design parameters, in addition to the *Clostridium* strain involved and basic mode of fermentation operation. Thus, optimization of the process involves intense experimentation with different combinations of physical and design parameters, which is rather impractical. Simulation of the fermentation process with mathematical models gives an insight into the characteristics of the process, and also helps to identify the influence of each variable on the overall process.

This contribution reports recent advance on ABE fermentation at Napoli. The study regarded an innovative immobilized cell reactor system: an experimental campaign and the development of a model.

Clostridium acetobutylicum was adopted for the fermentation process. The conversion was carried out in 4 packed bed reactors connected in series: the first reactor of the series was kept under acidogenesis and the successive reactors were kept under solventogenesis. Tests were carried out feeding the reactor system with solutions bearing both glucose as carbon source.

A mathematical model of the PBBRs system was also formulated using kinetic data assessed experimentally. The proposed model was an unstructured-unsegregated model and summarizes biochemical as well as physiological aspects of growth and metabolite synthesis by the production strain.

Introduction

The development of biotechnological processes to produce butanol from renewable resources as eco-sustainable alternatives to the petrochemical production routes (Kumar et al., 2012) is still an open challenge. Acetone-Butanol-Ethanol (ABE) fermentation by clostridia is drawing new interest as a way to turn renewable resources into valuable base chemicals and liquid fuels. Nevertheless the high industrial potential interest for the butanol production by the biotechnological route, some features of the ABE fermentation process hinder its success on an industrial scale. Indeed, the ABE fermentation is characterized by low yield, the acid-solvent two phase feature, and low concentration of butanol due to its inhibiting effect on fermentation. Moreover, the cost of traditional feedstocks - starch and molasses – is a larger fraction of the butanol sale price. Therefore, the use of a low-cost renewable resource is a pre-requisite for the industrial success of biotechnological process to produce butanol. Key issues for the success of the ABE fermentation process include the detailed metabolic analysis of the clostridia coupled with the development of improved engineered strains too (Papoutsakis 2008).

Reactor design and operating conditions play a key role in fermentative productions. The main factors that hinder the commercial development of the traditional ABE batch fermentation processes include low cell density, cell – and relative feedstock - lost at the end of the fermentation, low reactor productivity, high down-times, nutritional limitations and severe product inhibition (Chen and Blaschek, 1999). Process intensification may be obtained by increasing cell concentration in the reactor: cell immobilized reactors and retention membrane reactors are two potential solutions (Qureshi et al., 2005). These reactors configuration takes also the advantage to be operated continuously (Qureshi et al., 2000; Procentese et al., 2015). Continuous bioconversions are characterized by several advantages with respect to batch cultures in biofilm reactors (Qureshi et al., 2000). The main advantages are: the high cell concentration, the reactor operation at high dilution rates – high throughout - without cell washout, high yield, high butanol concentration that enhance butanol recovery performances. Moreover, the biofilm support can be reused.

The reactor design and its optimization for ABE fermentation may take advantages form reactor modelling The performance of ABE fermentation process, in terms of selectivity and yield of solvents, depends on several physical and design parameters, on the Clostridium strain, and on the mode of fermentation operation. Simulation of the fermentation process with mathematical models gives an insight into the characteristics of the process, and also helps to identify the influence of each variable on the overall process.

This contribution reports recent advance on ABE fermentation at Napoli. The study regarded an innovative immobilized cell reactor system: an experimental campaign and the development of a model.

Apparatus and model

The anaerobic solventogenic commercial *Clostridium acetobutylicum* DSM 792 was used for the fermentation process. The conversion was carried out in 4 packed bed biofilm reactors (PBBRs) connected in series: the first reactor (fed with stream bearing the the carbon source) was operated under acidogenesis conditions, and the three successive reactors were operated under solventogenesis conditions. The two phases of the ABE fermentation were operated in separate vessels: acids were produced in first section of the system, the produced acids and the residual sugar were converted in solvents in the second section. The PBBR system performance was characterized in terms of final butanol concentration and productivity as a function of the dilution rate.

The apparatus was made of reactor system, liquid pumps, heating apparatus, device for pH control, and on-line diagnostics (sketch in Fig. 1). The reactor system was made of four fixed beds. Each bed was at the bottom of a 100 mL glass lined pipe (4 cm ID, 8 cm high) jacketed for the heat exchange. Water from an external circulating water bath (Julabo heating circulator MA4) was fed into the jacket of each reactor to keep the operating temperature at 37 °C. The liquid head was controlled by the overflow duct in each reactor: the working volume of each reactor was set at 40 mL. Nitrogen was sparged at the bottom of each reactor to ensure anaerobic conditions. The pH control device – one for each reactor - included a pH-meter, a peristaltic pump, a vessel with NaOH 0.3 M solution and a controller.

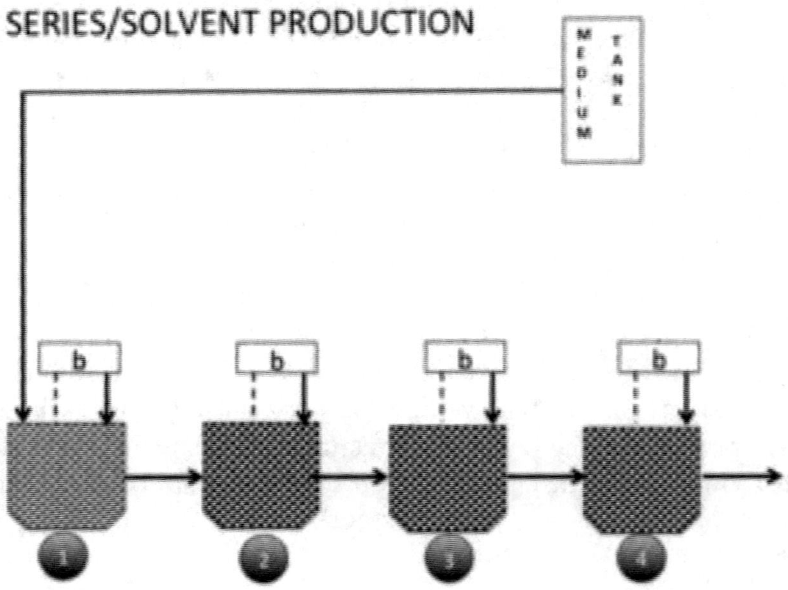

Fig.1: Outline of the apparatus used for the continuous process: A) PBBRs connected in parallel during the start-up phase; B) PBBRs connected in series during the butanol production phase. b: pH measure/control device.

Tygon rings (3/1 mm OD/ID) were chosen as biofilm carriers.
Clostridium acetobutylicum DSM 792 was used. Details regarding the reactivation and pre-inoculum procedures are reported in Raganati et al. 2013.

The composition of the medium fed to the PBBR system is reported in (Procentese et al. 2015a). Glucose (100 g/L concentration) was used as carbon source.

A mathematical model of the PBBRs system was formulated using glucose kinetic data assessed experimentally according to the procedure proposed by Procentese et al. (2015a; 2015b). The proposed model was an unstructured-unsegregated model and summarizes biochemical as well as physiological issues of growth and metabolite synthesis by the production strain. The key fermentation rates were expressed and evaluated with regard to substrate consumption and butanol end-product inhibitory effects.

Kinetic data regarding the growth rate (acidogenesis) and the butanol production rate (solventogenesis) were from specific experimental campaigns (details in Procentese et al. 2015a and 2015b). The main assumptions of the proposed model are reported hereinafter (Raganati et al., 2016).

- the PBBR system was assumed as a series of CSTR;
- the biomass present in each PBBR as free cells and immobilized cells was a heterogeneous cell population consisting of: acidogenic cells, solventogenic cells and spores ;
- the kinetics of cell growth and butanol production of biofilm-cells were assumed equal to those of the free cells;
- cells attachment and detachment processes were considered.

Experimental results

Two PBBR configurations were used (Raganati et al., 2016).

Parallel configuration) The four fixed beds were operated in parallel. This configuration was used during start-up to promote biofilm formation.

Series configuration) The four fixed beds were connected in series (Fig. 1). This configuration was used during butanol production after a biofilm layer had formed in each unit in parallel mode.

Butanol production tests were carried out with the 4 PBBRs connected in series (Fig.1) and operated at pre-set conditions. The pH of the reactor 1 (Fig.1) was set at 5.5 to promote acidogenesis conditions. The pH of the reactors 2, 3, and 4 (Fig.1) was set at 4.7 to promote solventogenesis conditions.

The overall dilution rate (D) – ratio between the feeding flow rate and the total volume of the 4 fixed beds – ranged between 0.05 and 1.4 h^{-1}. After setting the dilution rate, the reactor system was operated until steady state conditions were reached: metabolite and glucose concentration in each reactor staying constant for at least 5 times the space-time of the reactor.

Fig. 2 reports the concentration of butanol and glucose in the four PBBRs measured under steady state conditions for D=0.15 h^{-1}. It is interesting to analyse the substrate and metabolite concentration in the four reactors from the PFR behaviour point of view. The gradual decrease in glucose concentration (Fig. 2B), and the gradual increase in solvent concentration (Fig. 2A) along the PBBR series reproduce the typical behaviour of batch fermentation. As expected, the formation of a stable biofilm in the four units and the choice of the proper operating conditions lead to a successful reactor system that behaves like a PFR with continuous inoculum and constant biofilm concentration.

The PBBRs system was successfully operated for more than three months to produce butanol. The reactor performances – butanol productivity and butanol concentration – were assessed as a function of the reactor system dilution rate. By tuning the D it was possible: i) to totally convert the carbon source (D <0.15 h^{-1}); ii) to maximize the concentration of butanol in the produced stream (about 15 g/L at D = 0.65 h^{-1}); iii) to maximize the butanol productivity (about 12.6 g/Lh at D = 0.9 h^{-1}).

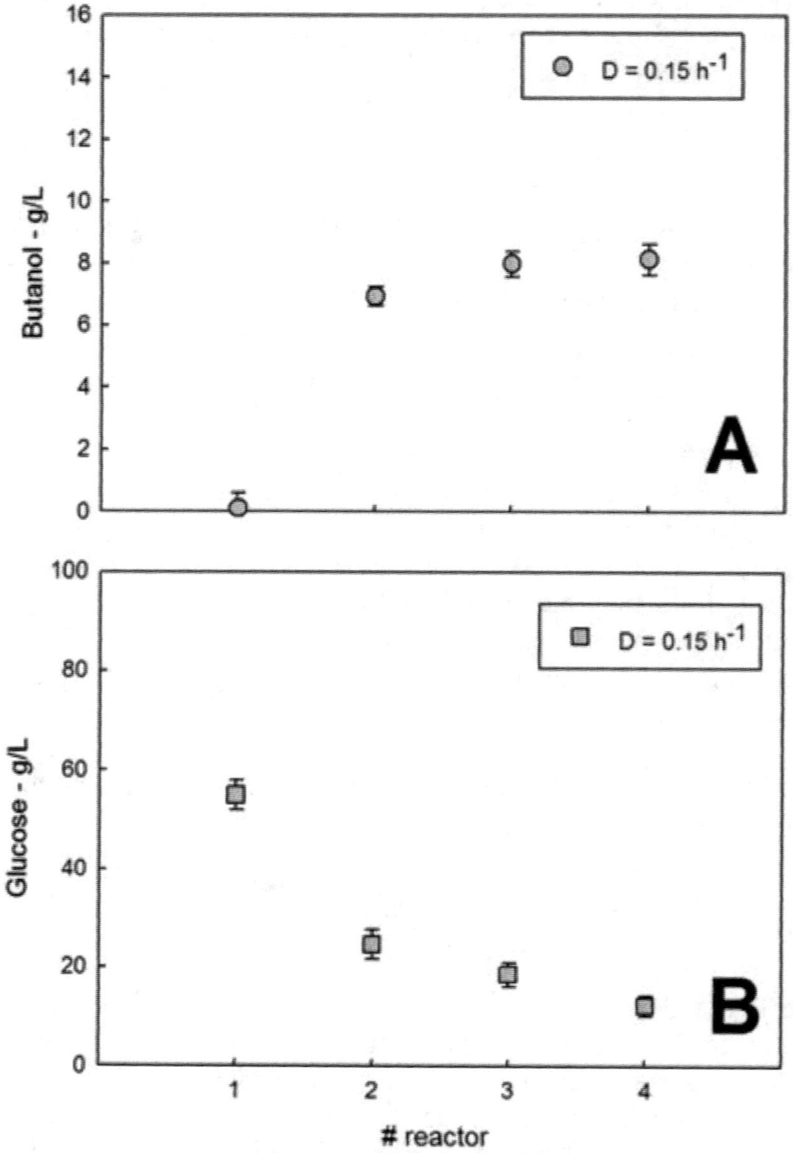

Fig.2: Butanol and glucose concentration measured along the PBBR series at D=0.15 h^{-1}.

Model results

Fig. 3 reports the estimated concentration of butanol and glucose in the four PBBRs for D=0.15 h^{-1}. It is evident that the developed model succeeded in reproducing the behavior of the PBBRs system.

Acknowledgements

The authors thank the European Union's Horizon 2020 research and innovation program for their financial support to the project Waste2Fuels (grant agreement No 654623).

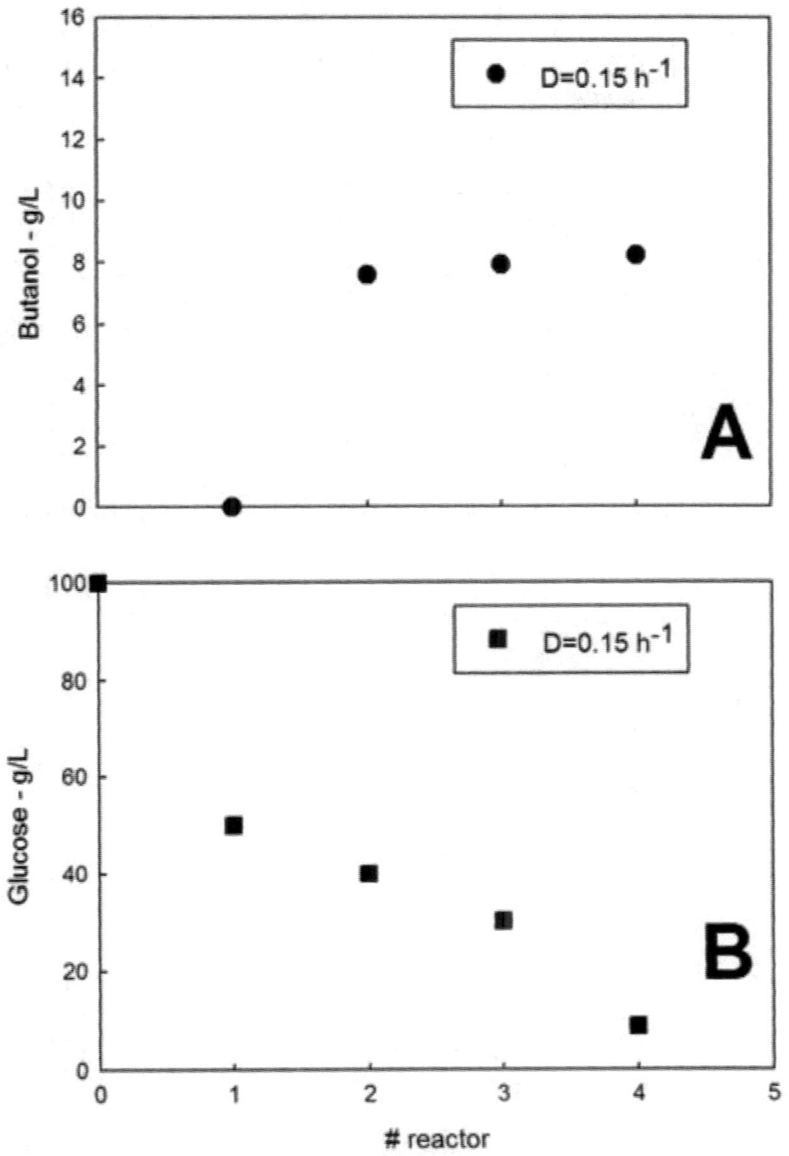

Fig.3: Estimated butanol and glucose concentration along the PBBR series at D=0.15 h^{-1}.

References

1. Chen CK, Blaschek HP. Acetate enhances solvent production and prevents degeneration in *Clostridium beijerinckii* BA101. Appl Microbiol Biotechnol 1999; 52:170–173
2. Kumar M, Goyal Y, Sarkar A, Gayen K. Comparative economic assessment of ABE fermentation based on cellulosic and non-cellulosic feedstocks. Appl Energ 2012; 93:193–204.
3. Papoutsakis, E. T. Engineering solventogenic clostridia. Current Opinion in Biotechnology 2008 19(5), 420–9.

4. Procentese, A., Raganati, F., Olivieri, G., Russo, M. E., Salatino, P., Marzocchella, A. Continuous xylose fermentation by *Clostridium acetobutylicum* – Kinetics and energetics issues under acidogenesis conditions. Bioresource Technol 2015a 164, 155–161.
5. Procentese A, Raganati F, Olivieri G, Russo ME, Salatino P, Marzocchella A. Continuous lactose fermentation by *Clostridium acetobutylicum* – Assessment of solventogenic kinetics. Bioresour Technol 2015b; 180:330–337.
6. Qureshi N, Schripsema J, Lienhardt J, Blaschek HP. Continuous solvent production by *Clostridium beijerinckii* BA101 immobilized by adsorption onto brick. World J Microb Biot 2000; 16:337-382.
7. Qureshi, N., Annous, B. A., Ezeji, T. C., Karcher, P., & Maddox, I. S. (2005). Biofilm reactors for industrial bioconversion processes : employing potential of enhanced reaction rates. Microbial Cell Factories, 21, 1–21.
8. Raganati, F., Olivieri, G., Procentese, A., Russo, M. E., Salatino, P., Marzocchella, A. Butanol production by bioconversion of cheese whey in a continuous packed bed reactor. Bioresour Technol 2013; 138:259–265.
9. Raganati, F., Procentese, A., Olivieri, G., Russo, M. E., Götz, P., Salatino, P., Marzocchella, A. Butanol production by *Clostridium acetobutylicum* in a series of packed bed biofilm reactors. Chemical Engineering Science 2016; 152; 678–688.

466294 Effect of Environmental Factors Towards Patchoulol Production By *Chlamydomonas Reinhardtii*

Abstract:

Patchoulol ($C_{15}H_{26}O$) is a sesquiterpene alcohol and a valuable compound in the fragrance and pharmaceutical industry. Traditionally, patchouli oil is extracted from the leaves of a plant, *Pogostemon cablin*. This approach faces the threat of limited and intermittent supply as these plants only grow in select regions of the world. But now with the advance of genetic engineering, it is possible to genetically modify microorganisms such as yeasts and bacteria to produce patchoulol. Photosynthetic microorganisms, such as microalgae in particular, would be of high interest as they have the advantage of directly converting carbon dioxide to high-value products through carbon fixation and photosynthesis. In this work, we have cultivated a patchoulol-producing strain of *Chlamydomonas reinhardtii* and elucidated the effect of light and temperature towards cell growth and patchoulol production. We also considered the effect of light-and-dark cycles towards patchoulol productivity. The findings from these environmental factors not only increase our understanding of patchoulol expression in microalgae, but also represent a step towards advancing this production platform to other target molecules. It is envisioned that this platform would be applicable to other terpenoids such as bisabolene and limonene, both hydrocarbon precursors to a range of industrially relevant chemicals.

470048 Nutrient Capture By a Sustainable Symbiotic Biofilm: Simultaneous Phosphorous and Nitrogen Recovery By Attached Fungi and Microalgae Biofilm

Abstract:

Nutrient pollution, a condition of excess phosphate (P) and nitrate (N) from point and non-point sources, is a major water quality concern worldwide due to the tremendous volume of urban, agricultural and industrial wastewater production [1]. The elimination of phosphorus from effluents before discharging is made mandatory as even traces of P (above 0.02 mg/L) can cause eutrophication, also P level is regulated to be lower than 0.05 mg/L, if streams discharge into lakes or reservoirs. On the other hand, the P reserves are being depleted swiftly which apparently is linked to global food security, with more than 90% of phosphorus used for agricultural purpose [2]. Currently a curative solution to combat water pollution by recuperating N and P, recycling the P for food production by closing the nutrient loop to effectively manage dwindling rock P reserve are the major challenges. The most common phosphorus removal techniques for municipal and industrial wastewater are biological treatment and chemical precipitation. Bio-valorization offers an excellent alternative to chemical treatment technologies as the biomass produced can be a source of protein and other valuable bio-chemicals also it is an eco-friendly and low cost technology.

In this present work, a novel composite biofilm called mycoalgae biofilm [3] was developed using the isolated strains of polyphosphate accumulating fungi and microalgae and we want to apply this platform for the efficient recovery and recycling of nutrients in waste stream. The effect of various process conditions (fungal strain, carbon levels, temperature, P/N ratio) on the nutrient recovery efficiency and the changes in the biofilm biomass composition was studied. Since the biofilm will be used for nutrient recovery from an array of samples a synthetic medium for the cell growth with the initial P (40 mg/L), N (200 mg/L) and other nutrients were used as the medium for the biofilm development. Depending on the nature of the waste water and its composition, and the fungi-microalgae combination intended to use for bio-remediation, the developed mycoalgae biofilm can be a sustainable microbial process for food, feed or energy source.

Methods: Water and soil samples were collected from Sarita wetland near UMN to isolate the native fungi for the biofilm formation. Fungal isolates were identified by genetic identification of fungal internal transcribed spacer (ITS) regions. Maximum Phosphorous and Nitrogen removal was observed in strain Sa7 (*Mucor sp.)* and *Chlorella sp,* combination which was considered for the biofilm formation and Nutrient recuperation. The morphology of the biofilm was also studied using CLSM and SEM at the optimized conditions to study the interaction between the cells. The nutrient removal efficiencies were calculated using the equation: $RE_i = (S_{i0}$ â€“ $S_{it})/S_{i0}$; where RE_i is the removal efficiency of the nutrients *i* (TN, TP, PO_4-P) or the pollutants (*i*) using the mycoalgae lichen type biofilm; S_{i0} and S_{it} are the initial and final concentration of component *i* after *t* days. The rate of nutrient and pollutant removal by mycoalgae biofilm is an important factor which will be calculated by; $dS/dt = r_i = (S_{i0}$ â€“ $S_{it})/$ $(t_i$ â€“ $t_0)$; Where: r_i (g/L/d) is the removal rate of substrate *i* (TN, TP, PO_4-P).

Conclusion: The composite biofilm shows promising results in waste water treatment especially in removing the N, P, suspended solids and other organics from nutrient polluted water in a single-step reactor with easy biomass recovery and high efficiency, when compared to the individual pure cell cultures. The biofilm attaches to the polymer matrix completely leaving behind the clean water for recirculation, and the composite biomass for nutrient recycle. The biofilm composition can also be tailored based on the influent stream components, feasibility of the strains to grow together and its nutrition value for further use. The commercial expansion of this biofilm system could be a step ahead in having an integrated and sustainable water treatment systems. **References**

1. E.K. Wilson, Danger from microcystins in Toledo water unclear, Chem. Eng. News, 92 (32) (2014), p. 9
2. Ye, Y., Gan, J., & Hu, B. (2015). Screening of Phosphorus-Accumulating Fungi and Their Potential for Phosphorus Removal from Waste Streams. Applied biochemistry and biotechnology, 177(5), 1127-1136.
3. Bo Hu, Aravindan Rajendran, A Novel Microalgae Cultivation Technology Using Fungi and Lichen Biofilm, UMN invention disclosure case # 20140274

461292 Incorporating Hydrodynamics into Spatiotemporal Metabolic Models of Bubble Column Gas Fermentation

Abstract:

A promising route to renewable liquid fuels and chemicals is the fermentation of carbon monoxide containing gas streams to synthesize desired products such as ethanol and 2,3-butanediol. We recently developed a spatiotemporal metabolic model for bubble column reactors with the CO fermenting bacterium *Clostridium ljungdahlii* as the microbial catalyst [1]. This first generation model was based on the simplifying assumptions of ideal plug flow for the vapor phase and plug flow plus axial dispersion for the liquid phase. However, industrial bubble column reactors are complex multiphase processes in which spatial variations in the gas superficial velocity, volumetric holdup and interfacial area can have profound e ects on fermentation performance.

We discuss the development of a second-generation bubble column model that accounts for the e ects of gas phase hydrodynamics on cellular growth, nutrient update and byproduct synthesis rates in a temporarily and spatially resolved manner. The modeling approach involves the coupling of a genome-scale metabolic reconstruction with mass and momentum balances that allow the calculation of local gas metabolite concentrations, superficial velocity, volumetric holdup and interfacial area as well as local concentrations of biomass and liquid phase metabolites. The resulting set of partial di erential equations with embedded linear programs is spatially discretized and numerically integrated using the MATLAB based code DFBAlab. Simulations results are shown for two CO fermentating organisms: the research model *C. ljungdahlii* and the industrially relevant *Clostridium autoethanogenum*. Model predictions are compared to those generated without gas phase hydrodynamics and to experimental data collected from a laboratory scale bubble column reactor with respect to CO conversion, biomass production, ethanol titer and ethanol-to-acetate ratio.

1. Chen, J., J. A. Gomez, K. Ho ner, P. I. Barton and M. A. Henson, "Metabolic Modeling of Synthesis Gas Fermentation in Bubble Column Reactors," Biotechnology for Biofuels, 8, 89, doi:10.1186/s13068-015-0272-5 (2015).

464191 A Commercial Demonstration of Biorefinery of Lipids--Coproduction of Biodiesel and 1,3-Propanediol

Abstract:

Lipase-catalyzed transesterification (also esterification) from oils or fats for biodiesel production has some advantages over chemical-catalyzed transesterification, such as environmental friendship, lower energy consumptionï¼Œwidely suitable for low quality vegetable oils. However, the low stability of the lipase due to alcohol has been being regarded as the main hurdle to the industrialization of lipase-catalyzed biodiesel production. Tsinghua University has developed a novel process to eliminate the inhibition of methanol on lipase, thus the operational life of the lipase could be improved into more than 300 cycles from several (1-5) cycles while traditional process applied. The patented novel process was successfully demonstrated in a pilot plant (200kg/day) and commercial plant (50,000ton/year) in China. It has also been being demonstrated in a pilot-plan in Brazil.

As a by-product, glycerol will be produced at about 10% of biodiesel during the process of biodiesel production. More and more concerns were pay on how to utilize crude glycerol. It could be a promising way to produce 1,3-propanediol (PDO) by fermentation with crude glycerol as feedstock. PDO is a valuable monomer to synthesize polytrimethylene terephthalate (PTT). PTT has excellent properties for textile fibre or plastic, and it was found to have some better properties than other polyesters such as PET, PBT. Tsinghua University has developed a novel flexible process for PDO production by fermentation from glycerol or glucose, and it was successfully commercialized with capacity of 20,000 tons/year PDO .

462306 Experimentally Driven Choices in the Fabrication of a 45 Kw Solar Simulator

Abstract:

Solar-thermal technology has been successfully used to drive biomass gasification, metals reduction and electrical generation in a renewable manner. The development of these applications relies on high-flux solar simulators: assemblies of high power lamps whose irradiance mimics concentrated sunlight. These platforms provide controlled environments for the evaluation solar-thermal technologies at the laboratory scale. To date more than seven high-flux solar simulators are available worldwide for academic studies.In conjunction with Alpha-Omega Power Technologies (LLC), we evaluated choices in high-flux solar simulator construction. This included the effect of different reflector coatings â€“ uncoated, silvered, and aluminized â€“ on lamp efficiency. Additionally, the importance of automated filament positioning was quantified. These results educated the installation of a new, 45 kW, solar simulator at the University of Colorado Boulder.

Carbon Nanotube (CNT) Based Multifunctional Optoelectronic Rods for Efficient Energy Generation and Transport

Sayeeda T. J. Aishee; Evelyn Gonzalez; Erika Hernandez; Glenn Grissom; H. Justin Moore, M. Jasim Uddin*

Department of Chemistry, College of Science, University of Texas at Rio Grande Valley

1 West University Blvd. Brownsville, Texas 78520

*Corresponding Author: jasim.uddin@utrgv.edu, Phone: 956-882-7772

Abstract

To keep up with the world's increasing demand for energy, a clean and perpetual energy source is vital to sustain this planet. This research activity focused on producing a flexible, inter-aligned, thermally stable, highly conductive, carbon based, and ultra-strong photovoltaic (PV) power wire from carbon nanotubes yarn (CNY) functionalized with quantum dots and N719 dye sensitizer. These innovative three dimensional PV wires, with enhanced photoconversion efficiency, have been developed using CNY as the counter electrode (CE) and TiO_2 grafted CNY as a working electrode (WE). The CEs were twisted around the WEs to collect and transmit the photogenerated electrons from the outer circuit. The optimized photovoltaic cell shows a 3.9% photo-energy conversion efficiency.

Introduction

With increasing global awareness for facilitating the energy demand of the world in which climate is constantly changing and pollution negatively affecting many areas, much research has been focused upon finding alternative sources for energy that are both inexhaustible and clean. Among these alternative sources for energy, photovoltaic cells have drawn much attention in the scientific community due to their potential for highly efficient photovoltaic energy conversion [1]. Photovoltaic power wires could revolutionize the traditional solar cells and supply a significant portion of the world's energy consumption. Much work has been done in order to increase the efficiency of the photovoltaic cells; in fact, the energy conversion efficiency of dye-sensitized solar cells has the potential to reach up to 12.1% utilizing platinized transparent conductive oxide anodes [2]. There is a particular interest in creating PV cells that have low internal resistance and low heat transport. It is important to highlight that PV cells are open to textile application, thus greater cell stability is required. Photovoltaic cells constructed from CNY are seen around the world as promising and extraordinary sources for propelling technological advancement in fields involving electronic materials and others, as their ability to convert light to energy is highly favorable and unique [5]. The research focuses on optimizing the energy conversion and improving the mechano-optoelectronic conditions of photovoltaic cells. Enhancing the efficiency of dye-sensitized carbon nanotube yarn shaped photovoltaic cells through the incorporation of MoS_2 is most definitely an innovative project with the potential to affect many fields such as those of textiles, electronics, and engineering in a greatly beneficial manner.

This paper has focused on developing a PV cell from CNY functionalized with molybdenum disulfide and cadmium sulfide as an addition to cell configurations previously reported [1-3, 4]. In order to optimize the efficiency of PV cells, we must first understand its molecular structure of the photoactive phase. The PV microwire was then coated with TiO_2 which acts as a working electrode (WE). Although most PV cells are assembled with a liquid electrolyte, the CNY was used to test the suitability and electronic potential without using any conventional liquid electrolyte. Incorporating CNY around the WE provides a greater sustainability than most traditional dye-sensitized solar cells (DSSC), and it works as a counter electrode. Indeed, the carbon nanotube yarns demonstrate excellent catalytic properties that provide high electronic performance; when the CNY was twisted around the WE, it became the primary conductor of electrons. This power microwire could lead to the widespread use of three dimensional PV wirses in daily life such as clothing, engineering structures, or electronic products.

Experimental Method

In order to develop photovoltaic cells from dye-sensitized CNY functionalized with molybdenum disulfide and attain larger fill factors with an enhanced efficiency thus leading to increased PV energy conversion, the following procedures were undertaken:

Carbon Nanotube Yarn Preparation

The CNY was prepared by removing the polymer from the CNY by soaking it in acetone, and then it was baked at 350°C in a tube furnace with light air flow to facilitate the removal of ash.

Further, the CNY was treated with nitric acid and sonicated to remove any remaining polymer ash.

Working Electrode Treatment

The WE was submerged in Triton-X and rinsed with ethanol. Five yarns were twisted together utilizing their mechanical properties; this process resulted in an assembly of 5 carbon nanotube yarns for the WE. A TiO_2 nano and macro-sol was applied and calcined at 350^0C.

The CNY was then soaked in a solution of $TiCl_4$ and ethanol. Finally, the WE underwent chemical bath deposition. The WE was functionalized with solutions containing quantum dots (molybdenum disulfide and or cadmium disulfide) and dimethyl sulfoxide, taking care to see that these processes did not ruin the integrity of the TiO_2 nano/macro coats. This was confirmed through observation in the atomic force microscope (AFM).

The WE was also placed in mixed solutions of PCBM (Phenyl-C61-butyric acid methyl ester) and P3HT (Poly(3-hexylthiophene-2,5-diyl)) (Scheme 1) , and it was then dried at RT. The PCBM and P_3HT sensitizer based photovoltaic cells facilitate a high efficiency rate in photon to electron conversion. Carrier conversion efficiency (η_c) and energy conversion efficiency (η) of CNY based photovoltaic devices are improved when semiconducting polymers incorporating Bis(1-[3-(methoxycarbonyl)propyl]-1-phenyl)-[6.6]C62 (mixture of isomers), Bis[60] PCBM (PCBM 62) or its functionalized derivatives are used in conjunction with CdS and CdSe. These processes readied the WE to facilitate photoelectron transfer. The WE was then placed in N719 with *tert*-butanol and acetonitrile solution, which is vital to the photosensitive function of the CNY. Any electrodes were kept in the dark prior to assembly, including the WE, which was ready to be assembled into the PV microwires.

Counter Electrode Preparation

The counter electrode was assembled as both single strand CNY and triple strand CNY; this readied the CE for the Pt nanoparticles deposition by sputtering. The sputtering process was done with care to minimize possible damage to the thin film of the CNY.

Electrolyte Preparation

The electrolyte was then prepared with Poly(vinylidene fluoride-hexafluoropropylene) (PVdF-HFP), I_2, and NaI using a chemical bath deposition process while making sure to obtain a smooth homogenous solution. This prepared the electrolyte to be applied to the cells by pipette.

PV Cell Construction

Finally, the PV cell was constructed with a predefined twisting (1 mm pitch distance) process and using conductive silver paste to hold electrodes connection on the slides and connect the ends of the individual electrodes [4]. Approximately thirty PV cells were developed, with half of those having quantum dot incorporated and half of them without it. These CNY cells were compared through analysis of data to confirm the significance of the use of quantum dot in enhancing the efficiency of PV cells with dye-sensitized CNY.

The materials of construction were characterized using Atomic Force Microscope (AFM), the RAMAN Spectroscopy, Infrared (IR) Spectroscopy, and Ultraviolet-Visible Spectroscopy (UV-Vis). The AFM allowed for high-resolution images of the PV cells to be developed and also allowed for sensitive force measurements to be collected. The RAMAN and IR spectroscope yielded much vital data such as exact composition of material, stress/strain state, symmetry and orientation, quality, and amount of material. The IR spectroscope analyzed the infrared light interacting with the PV cell materials by measuring the absorption, emission, and reflection. The UV-Vis yielded data relating to the wavelength and amount of light absorbed by the PV cells. A potentiostat (VERSASTAT 300) was utilized in order to control the current and voltage of the CNY powerwires.

a)

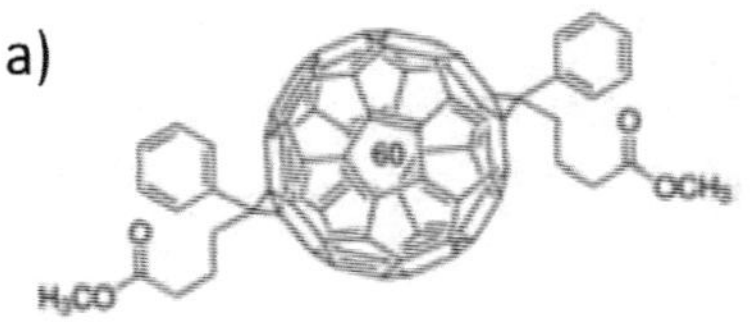

b)

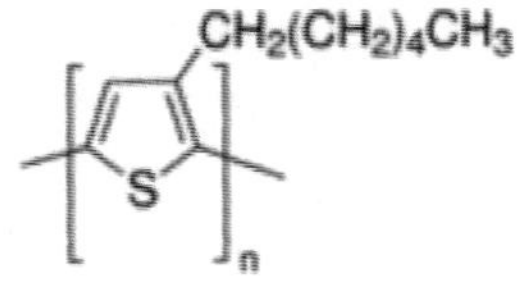

Scheme 1: a) Bis(1-[3-(methoxycarbonyl)propyl]-1-phenyl)-[6.6]C62 (mixture of isomers), Bis[60]PCBM (PCBM 62) and b) Poly(3-hexylthiophene-2,5-diyl) (P3HT).

Results and Discussion

Inventor Computer Aided Design (CAD) Modeling Software was used to better exhibit the construction of the various chemical interlayers of the PV microwires.

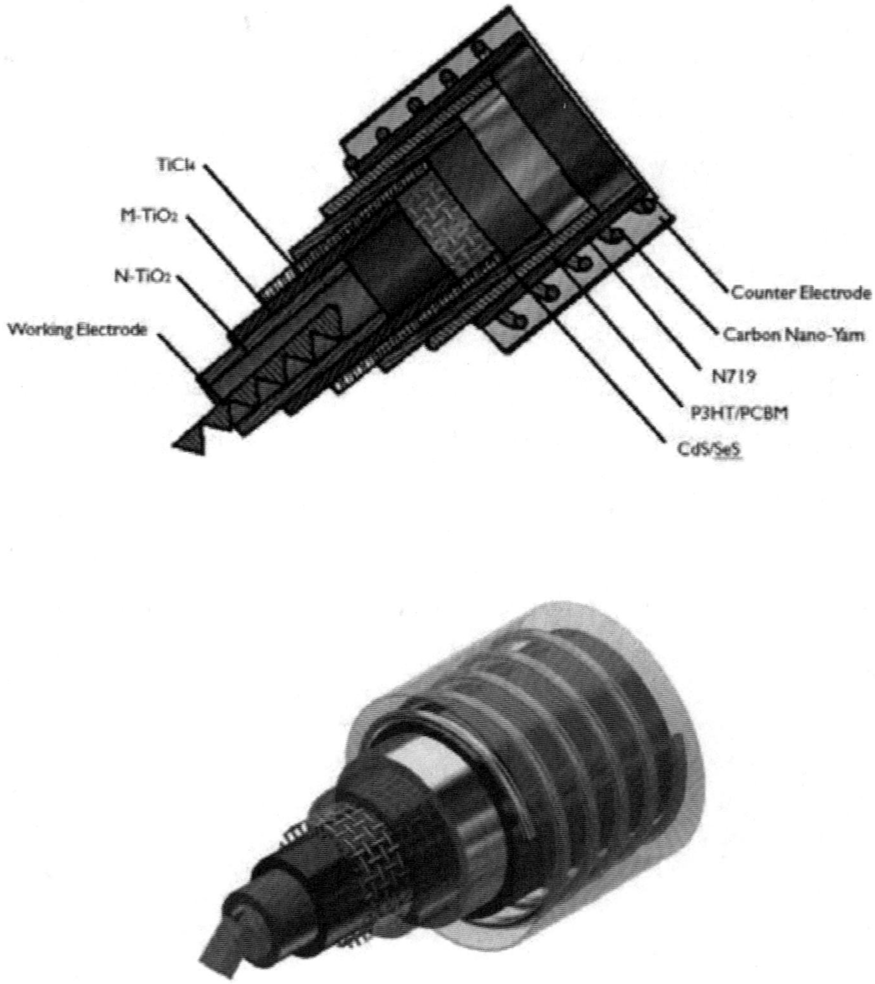

Figure 1a-b: The schematic view of the CNY PV cell.

The PV cells were constructed as shown in Figure 1 and the morphology of the CNY demonstrated with AFM images in Figure 2. The orientation of each individual layer enhanced the energy sensitivity and overall functionality of the cell. As can be seen in figures 2, the CNY clearly exhibits a homogeneous structure without agglomeration. Both images show the underlying carbon nanotube-yarn clearly despite their thick coating and difference in scale. The

CNT yarns are flexible, which enable our WEs and CEs to be twined easily with any of twisting parameters toward the WE and can maintain its structural and chemical properties unchanged after being folded/twisted or tied for multiple times.

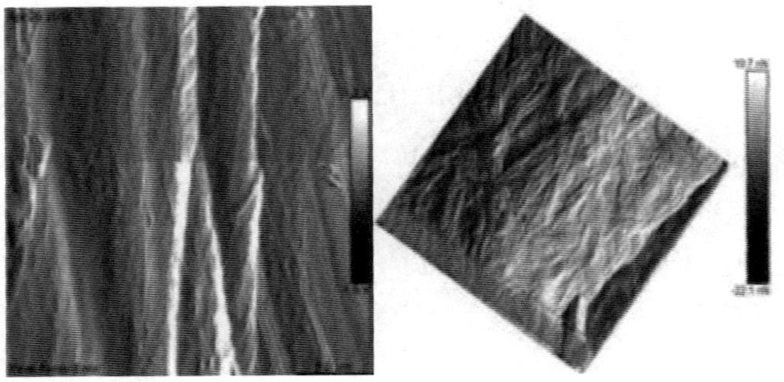

Figures 2 a-b) AFM images of Pristine CNY.

The *I-V* curve exhibits the efficiency of the photovoltaic energy conversion of the cells. This efficiency was calculated by taking the ratio of the maximum electrical power output compared to the solar power input into the PV cell. The fill factor, the measure of quality of the PV cell, was also calculated through the comparison of the maximum power to the theoretical power output at both the short circuit and open circuit voltages. The data obtained from AFM, Raman spectroscopy, and UV-Vis-DRS (Ultra Violet Visible-Diffuse Reflectance spectroscopy) are in-agreement with determining the nanostructured feature of this optoelectronic cells (*the figures are not shown here for the sake of brevity*). The RAMAN spectroscope yielded data that was analyzed in order to determine the structural phase of the WE and CE. The results infer that the titania nanophase are certainly indicative of anatase phase. Finally, data relating to the wavelength and amount of ultraviolet light absorbed by the PV cells was analyzed through the use of the UV-Vis-DRS (figure not shown for the sake of brevity.

The UV–Vis reflectance spectra were carried out at room temperature on a Perkin-Elmer UV–Vis–DRS to detect UV absorption intensities, the location of absorption edges and the quantum size effects of the nanostructured TiO_2 grafted on individual CNTs. The UV–Vis spectra have been obtained in the reflectance mode on the dried samples. This procedure is aimed to avoid systematic errors in evaluating the intensities of the specific UV–Vis absorption bands of the

TiO_2 and CNTs, because, as it is known, different fillings of the intra-crystalline voids by the adsorbed solvent molecules have an important effect on the scattering properties of the supporting phase.

The efficiency of 3D cells is largely dependent upon their fill factor, the measure of quality of the solar cell. We found the fill factor of the cells is ~0.5. Fully functioning photovoltaic cells constructed from dye-sensitized carbon nanotube yarns functionalized with CdS, and MoS_2 (QD) were developed. These PV cells have an enhanced capability for efficient conversion of solar energy, with an increase in efficiency of ~Δ1-2% from that of cells not functionalized with Cds and MoS_2.

Ultimately, the cells constructed from CNY using CdS and MoS_2 had higher fill factors than those without CdS and MoS_2. An increase in fill factor due to the addition of MoS_2 is highly beneficial because it signifies an increase in quality of the PV cells. The energy conversion efficiency of the cells constructed from CNY functionalized with MoS_2 had 3.9% photovoltaic energy conversion efficiency at 0.7 V, as shown in figure 3. Therefore, the PV powerwires were synthesized from CNY using QDs had better efficiency than those without incorporating QDs.

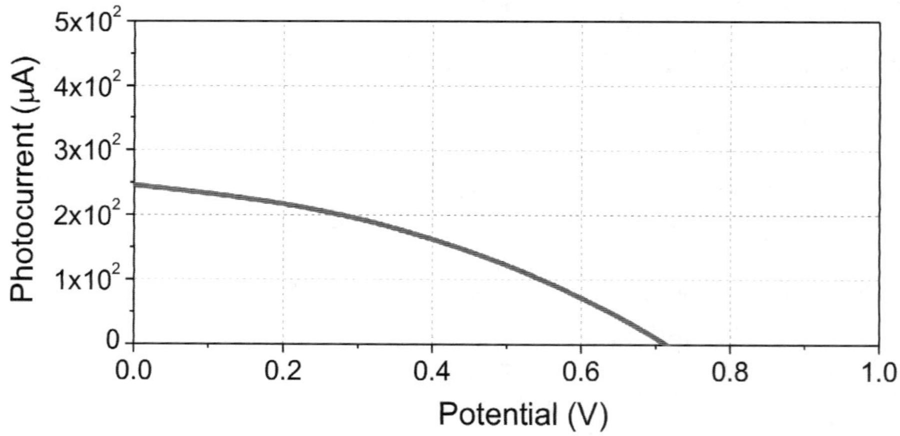

Figure 3: The *I-V* curve of the synthesized cells. The PV cell shows productive usage of power, thus indicating a high efficiency of 3.9% at 0.7 V.

Conclusion

The development of a photovoltaic cell from CNY is advantageous due to the low cost of the materials and the eco-friendly nature of the energy source. The significance of this project exists in the fact that the developed three dimensional PV cells result in much more efficient photovoltaic energy conversion, which makes the future use of photovoltaic cells in everyday life. The application of CdS, MoS_2, and CdSe enhanced the photovoltaic energy conversion productivity of the PV cell. These research outcomes helped maximize the photoenrgy conversion efficiency up to 3.90%. There are many advantages to continuing research on microware structured PV cells, since PV cells are a strong candidate for consideration for future application in a variety of fields such as those involving textiles, electricity, and other forms of engineering.

Acknowledgement:

The authors gratefully acknowledge the research grant (D01_W911SR-14-2-0001-0003) from Department of Defense through MSRDC, Washington, DC 20005.

References

[1] Yan, Jin, Uddin, M. Jasim, Dickens, Tarik J., Daramola, Deborah E., Olawale, David, and Okoli, Okenwa I. 2013. Tailoring the efficiency of 3D wire-shaped photovoltaic cells (WPVCs) by functionalization of solid-liquid interfacial properties. *Phys. Status Solidi* A, 1-7.

[2] Uddin, M.J., Davies B., Dickens, T.J., and Okoli, O.I. 2013. Self-aligned carbon nanotubes yarns (CNY) with efficient optoelectronic interface for microyarn shaped 3D photovoltaic cells. *Solar Energy Materials & Solar Cells* 115, 166-171.

[3] Uddin, M.J., Dickens, T., Yan, J., Chirayath, R., Olawale, D.O., and Okoli, O.I. 2013. Solid state dye-sensitized photovoltaic microwires (DSPMs) with carbon nanotubes yarns as counter electrode: Synthesis and characterization. *Solar Energy Materials & Solar Cells* 108, 65-69.

[4] Uddin, M. Jasim, Grissom, Glenn, Leal, Miguel, Galvez, Veronica, Trad, Tarek, Touhami, Ahmed, Islam, Nazmul, Parsons, Jason, and Moore, H. Justin. 2016. Self-aligned carbon nanotube yarns for multifunctional optoelectronic applications. *Proceedings of the ASME's International Mechanical Engineering Congress and Exposition (IMECE)* 2016-67441, 1-5.

[5] Nonoguchi, Yoshiyuki, Hata, Kenji, and Kawai, Tsuyoshi. 2015. Dispersion of Synthetic MoS_2 Flakes and Their Spontaneous Adsorption on Single-Walled Carbon Nanotubes. *ChemPlusChem* 80, 1158-1163.

464625 Thermal Ageing Characterization of Materials Under Concentrated Solar Power

Abstract:

Concentrating solar power (CSP) technologies have the potential of replacing global natural gas and liquid fuels consumptions by producing solar hydrogen. There are many factors that hinders the competitive low production price of this technology such as degradation of given properties of a solar absorber material over time due to temperature and solar radiation (Brunold *et al.*, 2004). In case of solar absorber materials, the most important factor for aging is degradation of reflectance/absorptance over time and its mechanical integrity (Boubault *et al.*, 2014a). Furthermore, sudden variation of temperature over a short period of time due to variation of solar radiation is another cause of aging (Boubault *et al.*, 2012). It is proposed (Boubault *et al.*, 2014a) that measurement of thermal performance of a material can serve as an ageing criterion (Boubault *et al.*, 2014b), which can be achieved by quantification of its absorption and heat transfer ability. In this work, we will investigate the thermal performance and the optical properties of stainless steel, as a reference material, using a high flux solar simulator and a low power light source. The testing facility consists of a high flux solar simulator which is capable of irradiating Gaussian distributed variable flux (Sarwar *et al.*, 2014). A light collimator was used to homogenize the incident flux. The flux characterization was carried out before and after homogenization using a water cooled Gardon-type circular foil radiometer. The material was exposed to the homogenized flux for given time periods, while we were monitoring continuously the temperature of the material and the heat flux at the back side of the material. The optical properties of the material such as absorption and reflectance were measured and compared before and after exposure to the high concentrated irradiance. Here we present the results of the characterization and the method validation against a total integrated reflectance/transmittance method employing a spectrometer with an integrating sphere.

Acknowledgements

This publication was made possible by a PDRA award [PDRA2-1105-14044] from the Qatar National Research Fund (a member of The Qatar Foundation). The statements made herein are solely the responsibility of the authors.

References

Boubault, A., B. Claudet, O. Faugeroux and G. Olalde, 2014a: Aging of solar absorber materials under highly concentrated solar fluxes. *Solar Energy Materials and Solar Cells*, **123**:211-219.

Boubault, A., B. Claudet, O. Faugeroux and G. Olalde, 2014b: Aging of solar absorber materials under highly concentrated solar fluxes. *Solar Energy Materials and Solar Cells*, **123**:211-219.

Boubault, A., B. Claudet, O. Faugeroux, G. Olalde and J.-J. Serra, 2012: A numerical thermal approach to study the accelerated aging of a solar absorber material. *Solar Energy*, **86**:3153-3167.

Brunold, S., M. Köhl, K. Möller and B. Carlsson, 2004. Chapter 4.4 - Accelerated Indoor Testing. In: M. K. C. J. W. Czanderna (Ed.), *Performance and Durability Assessment* (pp. 175-195). Elsevier: Amsterdam.

Sarwar, J., G. Georgakis, R. LaChance and N. Ozalp, 2014: Description and characterization of an adjustable flux solar simulator for solar thermal, thermochemical and photovoltaic applications. *Solar Energy*, **100**:179-194.

467866 Development and Thermal Test Results of a 10 Kw Porous Media Solar Reactor

Abstract:

Solar volumetric receivers are based on high porosity materials as radiation absorbers, like open cell foams or monolithic honeycombs. These structures allow the volumetric absorption of concentrated solar radiation within them. Solar heat absorbed in the porous media is then transferred to a working fluid passing through it. In this work, a 10 kW volumetric solar reactor has been developed for the reduction of metal oxides. Within the reactor there is a ceramic porous media that supports the metal oxide to be reduced. This structure is directly irradiated with concentrated solar energy in order to reach reaction temperatures above 1300 K.

The developed solar reactor was designed for experimentation with different operational conditions at the IER-UNAM Solar Furnace. The design was accomplished based on detailed ray tracing of the furnace, to obtain the concentrated flux distributions at different planes in the focal zone. The design is based on a cavity with a ceramic conical receiver, and allows for temperature measurements by thermocouples and pyrometers at different spots. Thermal simulations were carried out to determine the reactor body specifications and insulation requirements. The device has been built and preliminary tests have been carried out at the solar. These tests are aimed at evaluating the thermal performance of the system in the absence of chemical reaction. Different porous media that could be used in thermochemical processes have been used in these tests.

468047 Thermal Model of a Solar Porous Media Receiver for a Thermochemical Reactor

Abstract:

Solar receivers containing high porosity materials are a promising option due to higher efficiency compared with tubular receivers. In this system, concentrated solar radiation is absorbed on the porous media in a volumetric manner, then the absorbed heat is transferred to a working fluid when passes through the porous chanels. One critical issue when designing or optimizing volumetric solar absorbers is the knowledge of the temperatures of the solid matrix and fluid. These temperatures are linked with the efficiency and operability of the solar receiver. Hence, this work presents a 3D model of heat transfer and computational fluid dynamics in a porous media receiver subject to concentrated solar flux. This model allow us to analyze the temperature distribution both in the fluid phase and the solid phase, by applying the local thermal non- equilibrium model. The macroscopic model was then used to analyze the impact of geometric properties and operational conditions on the temperature distribution.

474675 Effect of Dust on Straight Tube Receiver of a Parabolic Trough Concentrator

Abstract:

The objective of present investigation is to enhance the performance of parabolic trough collector system. To improve its optical efficiency a dust collector /dirt accumulator has been modelled. This system is being used as parabolic trough based system for power plant application. The experiments were performed to analyze the optical efficiency and optical degradation rate at the roof top of Centre for Energy and Resources Development, Indian Institute of Technology (BHU), Varanasi (25.3176°N,82.9739°E),India. A parametric studies has also been done to estimate the effect of wind flow/air flow, pitch angle, shadow of wall, wind direction ,ambient temperature, solar radiation and collector area for high temperature, high concentration ratio application.

464032 Renewable Transportation Fuel Production from Algal Biocrude Oil Converted Via Hydrothermal Liquefaction

Abstract:

The goal of this project is to understand the properties of biocrude oil converted from Spirulina (SP), which can thrive in wastewater. Hydrothermal liquefaction (HTL) is a promising technology to convert wet biowaste into biocrude oil. Under high temperature and high pressure, HTL technology uses water in biomass as an excellent solvent and catalyst to break down and reform organic materials into biocrude oil products. Our previous study showed that the biocrude oil converted from wet biowaste via HTL contains comparable heating values (HHV) to that of petroleum crude oil. However, due to the lack of up-scaled production and proper separation of HTL biocrode oil, the application of HTL biocrude oil is limited.

In this study, the biocrude oil converted from Spirulina is separated into different fractions by the distillation process. The preliminary distillation test shows that 35-40 wt.% of HTL algal biocrude oil can be distilled at about 175 °C, indicating that the HTL algal biofuel is promising to be used as transportation fuels including kerosene and diesel. In addition, compared to other types of HTL biocrude oil, it is noticed that 25 wt.% and 10 wt.% of the biocrude oil converted from swine manure and food processing waste is respectively distilled at around 175 °C. This also suggests that the HTL algal biocrude contains compounds with lower boiling points.

Detailed physicochemical characterizations, such as acidity, viscosity, boiling point distribution, and chemical composition are examined by the viscometer, thermogravimetric analysis (TGA), CHN Analyzer, Fourier transform infrared spectroscopy (FTIR) and gas chromatography mass spectrometry (GC-MS). The physicochemical properties of the distillates are also compared to those of petroleum transportation fuel, such as jet fuel and diesel, and those of HTL biocrude oil converted from other biomass, such as swine manure and food processing waste.

465013 Biochemicals Production By Genetically Engineered Cyanobacteria Using Photobioreactors

Abstract:

The 3rd generation biorefinery research using microalgae is ongoing in many countries because fuel/chemicals production from carbon dioxide by photogenic microorganisms is essentially sustainable. Cyanobacteria is recognized as a proper platform for chemicals production due to their high growth rate, but it requires genetic modification because the original strain never produce usefulchemicals. In this presentation, I would like to introduce the current status of our research concerning the production of iso-propanol (IPA) and 1,3 propandiol (1.3PDO) by genetically engineered cyanobacteria using photobioreactor system with LED irradiation equipment, which can selectively supply various wavelength of light with low power consumption.

To achieve IPA production in cyanobacteria, we constructed a metabolically engineered Synechococcus elongatus strain TA1741 which produces IPA from acetyl-CoA by four steps. Successful cell growth was observed for blue, and blue and red LED, however, IPA was not detected for all cases. On the other hand, higher intracellular glycerol accumulation was observed, which suggests that the carbon dioxide fixed in a Calvin cycle is converted primarily into Glycogen under light condition. So we thought that, by changing from light to dark condition, accumulated glycogen may be converted to IPA. To verify this idea, Cells grown in a jar fermentor was harvested, re-suspended in new medium and purged by nitrogen gas. Then the tubes were maintained under anaerobic and dark conditions for 5-10days and analyzed. After five or 10 days cultivation, the production of IPA was confirmed for all cases showing that the genes concerning IPA biosynthesis were working in TA1741. Maximum IPA concentration was approx. 23 mg/L under anaerobic and dark conditions, however the concentration is rather low and complicated operation is required for practical IPA production.

Then, we examined another product, 1.3-Propandiol (1.3-PDO) production by genetically engineered cyanobacteria. In cyanobacteria, metabolic flux passing through acetyle-CoA is very narrow under light condition. On the other hand, PDO can be produced from DHAP which is main intermediate product in Calvin cycle. To achieve PDO production in cyanobacteria, five genes were introduced into the host strain, which produces 1.3 PDO from DHAP through glycerol. We call this strain TA2984. Using this strain, batch experiments were performed in the same way. Cell growth was successful and maximum PDO concentration was approximately 0.7 mM/L. It should be noted that the higher production of glycerol was observed. Their production was clearly associated with cell growth. This characteristics allows us to try continuous process. So we are now examining the possibility of continuous process using immobilized cyanobacteria. As a bioreactor for continuous process, an airlift fermenter with an internal draft tube was used. Successful growth was observed using an airlift bioreactor, maximum PDO concentration increased to approximately 1.8mM.

466401 Kinetic Modelling of Microalgae Growth for the Optimization of Starch and Lipid Production

Abstract:

Kinetic Modelling of microalgae growth for the optimization of starch and lipid production

Gonzalo M. Figueroa-Torres [a,b], Jon Pittman[c] and Constantinos Theodoropoulos[a,b]

[a]School of Chemical Engineering and Analytical Science, University of Manchester, M13 9PL, UK

[b]Centre for Process Integration, University of Manchester, M13 9PL, UK

[c]Faculty of Life Sciences, University of Manchester, M13 9PL, UK

Microalgal biomass has been regarded as a sustainable and renewable feedstock for biofuel production due to its ability to naturally accumulate carbohydrates - mainly in the form of starch – and lipids. These carbon-based compounds can potentially be used as raw substrates for biofuel production (Brennan & Owende, 2010). Studies have shown that carbohydrate and lipid content in microalgae cells is positively influenced under conditions of nutrient limitation. However, this increase in starch and lipid content is coupled with a decrease in biomass production (Markou, Angelidaki, & Georgakakis, 2012), an adverse effect on the overall cultivation process. An efficient approach towards improving starch and lipid production by means of a cultivation strategy involves the use of kinetic models capable of predicting the main compositional elements of microalgae growth (Bernard, 2011). However, there have been very limited modelling efforts in the literature regarding the simultaneous production of starch and lipids by microalgal biomass. Thus, the aim of this work is to develop a multi-parameter kinetic model for the optimization of starch and lipid production. The proposed model considers a set of intracellular carbon flows between two main cellular compartments: a pool of active biomass and a storage pool comprising of starch and lipids. Model parameters were fitted against experimental datasets generated from lab-scale cultivation of *Chlamydomonas reinhardtii* CCAP 11/32C in standard Tris-Acetate-Phosphate (TAP) media under different concentration regimes. Fitting was carried out through an in-house developed optimization algorithm. Validation of the model was then carried out against different experimental datasets. The model can be used to predict three carbon-based cellular pools – starch, lipids, and active biomass – as well as nutrient consumption and pH evolution, useful factors for the establishment of optimal cultivation conditions.

Bernard, O. (2011). Hurdles and challenges for modelling and control of microalgae for CO2 mitigation and biofuel production. *Journal of Process Control, 21*(10), 1378–1389. doi:10.1016/j.jprocont.2011.07.012

Brennan, L., & Owende, P. (2010). Biofuels from microalgae—A review of technologies for production, processing, and extractions of biofuels and co-products. *Renewable and Sustainable Energy Reviews, 14*(2), 557–577. doi:10.1016/j.rser.2009.10.009

Markou, G., Angelidaki, I., & Georgakakis, D. (2012). Microalgal carbohydrates: an overview of the factors influencing carbohydrates production, and of main bioconversion technologies for production of biofuels. *Applied Microbiology and Biotechnology, 96*(3), 631–645. doi:10.1007/s00253-012-4398-0

469321 Enhanced Biomass and Lipid Productivities of Outdoor Alkaliphillic Microalgae Cultures through Increased Media Alkalinity

Abstract:

Alkaliphillic microalgae have the ability to thrive under high pH and alkalinity conditions. The phototrophic biomass productivity of alkaliphillic microalgae can be improved by increasing availability of bicarbonate (HCO_3^-) at extreme pH conditions (pH >10.0) through increase in media alkalinity. The high alkalinity would allow the maintenance of sufficient â€œHCO_3^-Â bufferâ€ in solution while high media pH would allow rapid transfer of atmospheric CO_2 to compensate for HCO_3^-Â uptake by photosynthetic mechanism. Thus, growth would not get restricted by inorganic carbon limitations. Further, extreme pH conditions would permit mixotrophic cultivation with sugars present in the medium.

In the present study, phototrophic and mixotrophic cultivation strategies were implemented under high alkalinity and pH conditions for alkaliphillic *Chlorella Sp.* (strain SLA-04) in 1000L outdoor raceway ponds. Under the phototrophic conditions, when the culture concentrations were below 0.6 g/L, biomass productivities of ~20 g/m^2/day were obtained. While some of the HCO_3^- was utilized, ~ 100% of the carbon accumulated in biomass came from atmospheric CO_2. Under mixotrophic conditions, SLA-04 cultures grew without any measurable signs of contamination. In addition, due to availability of additional carbon source as well as the reducing equivalents, biomass productivity was significantly higher than phototrophic conditions (>57 g/m^2/day). Lipid productivities were also higher for cultures grown under mixotrophic conditions (7 g/m^2/day) when compared to phototrophic SLA-04 cultures (2 g/m^2/day). Thus, the use of alkaliphillic microalgae for cultivation, would result in increased carbon capture from atmosphere and thereby reduces the costs accompanied with CO_2 sparging (usually associated with mesophillic

cultures).

469333 Macroemulsion Mixture Assisted Organic Solvent Extraction of Algal Lipids from Unconcentrated Algal Culture

Abstract:

Petroleum fuels are non-renewable resources that is rapidly depleting. To meet the demand for energy while slowing the growth of petroleum fuel consumption, there is a growing interest in algal biofuels as a sustainable alternative form of energy. Algal lipid extraction is considered a major bottleneck to biofuels derived from microalgae due to their extremely small size, tough cell wall, suspended in dilute cultures, and cell density similar to that of the surrounding growth medium (*1*). Conventional methods of algal lipid extraction generally requires long extraction times, dried algal biomass with little water content, and high energy inputs (*2*). Therefore, the development of an effective and energy efficient algal lipid extraction process is critical for a successful implementation of microalgae based biofuel.

This study demonstrates a procedure for lipid extraction from microalgae culture using a mixture of macroemulsion and organic solvents at room temperature. This method does not require harvesting and dehydration of the algal biomass. The effects of extraction variables were investigated. The results showed that the macroemulsion mixture assisted solvent extraction method extracted 83-87% of total recoverable lipids at a mixing time of less than 1 min. No significant differences in terms of distribution of fatty acid profile have been obtained when compared to the Bligh and Dyer method. Evaluation of energy consumption indicates a substantial reduction in extraction costs compared to other methods.

Reference

(*1*) **N. Uduman**, **Y. Qi**, **M. K. Danquah**, and **A. F. A. Hoadley**, "Marine microalgae flocculation and focused beam reflectance measurement," *Chem. Eng. J.*, **162** (3), pp. 935–940, (2010).

(*2*) **R. Halim**, **M. K. Danquah**, and **P. A. Webley**, "Extraction of oil from microalgae for biodiesel production: A review," *Biotechnol. Adv.*, **30** (3), pp. 709–732, (2012).

469398 An Ecological and Engineering Approach to Optimizing Algal Biofuels

Abstract:

Algae are ubiquitous in natural ecosystems and have been studied extensively due to their versatility for biofuel production. Most studies to date have been conducted on the grounds of synthetic biology and process engineering with few industrial scale projects considering algal community interactions. Such interactions have often indicated the propensity of increasing overall productivity and reducing community invasability, both important characteristics for scalable projects. Over 30,000 species of algae have currently been identified with another 20,000 estimated to exist. Within this context, elucidation of these relationships remains extremely resource and time intensive. This research describes a strategy for rapid, high-throughput screening of algal community combinations using a microfluidic platform to synthesize millions of parallel, nanoliter-scale algal communities for analysis of biomass accumulation. Model communities were first studied in a bench scale flask experiment and then examined using microfluidic droplets. These experiments showed consistent results on both positively and negatively interacting algal bicultures. Specifically, these include better performance of two bicultures, *Ankistrodesmus falcatus* and *Chlorella sorokiniana*, *Chlorella sorokiniana* and *Selenastrum minutum*, and reduced performance of a biculture consisting of *Selenastrum capricornutum* and *Scenedesmus ecornis*. While the holistic success of algal biofuels hinges on an amalgamation of scientific fields, rapid screening of algal communities will prove imperative for discovering community interactions and ideally facilitating a mechanistic analysis of how such interactions arise in natural communities.

466273 Current Status and Prospects of Industrial Xylitol Production in China

Abstract:

Xylitol is an alternative sweetener that has many beneficial health properties including regulating insulin-independent metabolism in humans, preventing dental caries and a number of pediatric diseases. Based on these promising advantages, xylitol has been commercially developed as a valuable ingredient for use in food and pharmaceutical industries. In China, xylitol is mainly produced industrially by the reduction of xylose derived from hemicellulose hydrolysates of biomass materials in the presence of Raney nickel catalysts. Currently, xylitol market lies between 20,000 and 40,000 tons per year with an economic value ranging between 90 and 340 million dollars and the xylitol market is expect to increase by 15% per year. The mainly Chinese companies including FUTASTE, LONGLIVE and HUAKANG PHARMA can produce about 40,000 tons per year. While chemical synthesis of xylitol has some disadvantages including a high energy requirement, extensive purification steps, and a high cost of product. Accordingly, the biological production of xylitol has recently attracted much attention. Many efforts have been made for the biological conversion of xylitol using microorganisms from hemicellulosic hydrolyzate by Chinese scientists. It is noteworthy that the xylitol productivity of recombinant *Escherichia coli* (1.84 g l^{-1} h^{-1} and 143.8 g l^{-1} yield) by our previous work and the xylitol productivity of recombinant *Candida tropicalis* (2.01 g l^{-1} h^{-1} and 75.1 g l^{-1}yield) by Dr. Yuan's group are the highest xylitol productivity rates among recombinant bacteria and fungi from hemicellulosic hydrolyzate, respectively. Further, the industrial bio-production of xylitol and arabinose using *C.tropicalis* by Thomson Biotech from xylose mother liquor has been reported. All of above processes have been studied in an effort to provide one that is high yielding and competitive with chemical processes.

457047 Rational Substitution of Firefly Luciferase from *Photinus Pyralis* for Improvement of Thermostability

Abstract:

Firefly luciferase is widely used in many areas of biotechnology and molecular biology, but the application is restricted due to its poor thermal stability. Random mutagenesis and directed evolution paly an important role in the thermal stability of protein; nevertheless, large number of experimental attempts appear to be time- and labor-consuming. In this paper, to enhance the thermal stability of firefly luciferse, we got two mutants (E325K, E354K) by reversing the charges of individual residues at active sites with opposite interaction energies on the basis of molecular dynamic (MD) simulation. Root-mean-square-deviation(RMSD) and radius of gyration were used to evaluate the thermostability of the mutants. The residual activities of wide-type luciferase, E325K, E354K were 4.52%,7.22% and 30%, respectively, after being heated at 37 â„ƒ for 5 h, which showed the same trends with the computational predictions (E354K > E325K > WT). It was worth nothing that the T_{50} and T_m of E354K were 4.45 and 3.2 â„ƒ higher, respectively, than that of the native luciferase. Rational substitution of residue in regulation of thermostability using MD simulation would be a faster, less resource-intensive and more effective than random mutagenesis and directed evolution for instructional design to enhance thethermostability.

466981 Improving Cellular Robustness and Butanol Titers of *Clostridium Acetobutylicum* ATCC824 By Introducing the Heat Shock Proteins from Extremophilic Bacteria

Abstract:

In recent years, increasing concerns over environment, energy and climate have renewed interest in biotechnological production of butanol. However, growth inhibition by fermentation products and inhibitory components of raw biomass hindered the development of ABE fermentation. Improving cellular robustness of Clostridium acetobutylicum is of great importance for efï¬cient ABE production. In this study, we attempt to improve the robustness and butanol titers of C. acetobutylicum ATCC824 by overexpressing the groESL and dnaK from the extremely radioresistant bacteria Deinococcus wulumuqiensis R12 and C. acetobutylicum ATCC824. Three recombinants were obtained and designated as ATCC824(groESLR12), ATCC824(dnaK R12) and ATCC824(groESL824). These three recombinants were found to signiï¬cantly improve diverse types of stress-tolerances including furfural, butanol, oxidative, acid and osmotic-tolerance. Meanwhile, the butanol titers increased to 11.22g/L, 13.04g/L and 10.68g/L, which were 29.7%, 51% and 23.5% higher compared with the wild type, respectively. Especially for ATCC824(dnaK R12), the production of acetic and butyric acid decreased about 50% and 100% compared with the wild strain. These results indicated that overexpression of groESL and dnaK enhanced the metabolic rate of EMP pathway and improved acids re-uptake efficiency. In addition, ATCC824(groESLR12) was superior to ATCC824(groESL824) in diverse types of stress-tolerances and butanol titers because GroESL from an extreme bacterium could more efficiently perform their function in the heterologous host. Our study provided a new clue and a theoretical basis for improve the robustness and butanol titers of C. acetobutylicum ATCC824.

455883 Polymalic Acid Fermentation: A Platform for Production of Biopolymers and Chemicals

Abstract:

Polymalic acid (PMA) is a novel polyester polymer that has been broadly used in the medical and food industries. Its monomer, L-malic acid, is also a potential C4 platform chemical and extensively used in the food industry while it is currently supplied with the racemic mixture (DL-Malic acid) through chemical synthesis. Recently, great researches have been attempted to develop biotechnological process for L-malic acid production. In our study, a novel route for production of L-malic acid and its derivatives based on PMA platform was developed. Here, a specific fluorescence probe for detecting acidity strength was designed and used for high-throughput screening. Based on the genome-scale metabolic model *i*ZX637, Engineered the strain with overexpressed the gene of pyruvic carboxylase resulted in a PMA titer increase by 30% in shake flask. During the fermentation process, nitrogen limitation was found to be beneficial for PMA biosynthesis, which could upregulate the expression of key genes involved in the PMA biosynthesis pathway. Disruption or overexpression of the nitrogen transcriptional regulator Gat1 controlled by the TOR signaling pathway could obviously regulate PMA biosynthesis. Finally, the downstream process and progress in industry in China are presented.

476737 Engineering Yeast with Minicellulosome and Cellodextrin Pathway for Co-Utilization of Cellulose and Mixed Sugars

Abstract:

Consolidated bioprocessing (CBP), which integrates cellulase production, cellulose saccharification, and fermentation into one step, has been widely considered as the ultimate low-cost configuration for producing second-generation bioethanol from lignocellulose. However, the requirement of a microbial strain able to hydrolyze cellulosic biomass and convert the resulting sugars into high-titer ethanol limits CBP application. In this work, cellulolytic yeasts were developed by engineering ***Saccharomyces cerevisiae*** with a heterologous cellodextrin utilization pathway and bifunctional minicellulosomes. The cell-displayed minicellulosome was two-scaffoldin derived, and contained an endoglucanase and an exoglucanase, while the intracellular cellodextrin pathway consisted of a cellodextrin transporter and a β-glucosidase, which mimicked the unique cellulose-utilization system in ***Clostridium thermocellum*** and allowed ***S. cerevisiae*** to degrade and use cellulose without glucose inhibition/repression on cellulases and mixed-sugar uptake. Consequently, only a small inoculation of the non-induced yeast cells was required to efficiently co-convert both cellulose and galactose to ethanol in a single-step co-fermentation process, achieving a high specific productivity of ~62.61 mg cellulosic ethanol/g cell·h from carboxymethyl cellulose (CMC) and ~56.37 mg cellulosic ethanol/g cell·h from phosphoric acid-swollen cellulose (PASC). Our work provides a versatile engineering strategy for co-conversion of cellulose-mixed sugars to ethanol by ***S. cerevisiae***, and the achievements in this work may further promote the application of CBP in cellulosic biofuel production.

476734 Integrated Chemical and Bio-Catalysis for Production of High-Value Liquid Transportation Fuels from Agricultural Residues

Abstract:

Catalytic conversion of lignocellulosic biomass to high-valued transportation petrol, jet and diesel fuels is of great importance to develop versatile renewable energy and boost rural economy, thus receiving worldwide attention. However, the state-of-the-art processes still suffer from low efficiency in yield and productivity, which hinders the progress of its commercialization. Thus, it is essential to explore the consolidated catalytic reaction and refinery process. Here we report an advanced approach/process that could continuously produce long-chain (C5-C15) ketones from lignocellulosic biomass in a cascade mode by integrating several bio- and chemical catalysis reactions, which exhibits high efficiency and stable conversion rate (>70 %) in alkylation reaction. This technology consolidated the acetone-butanol-ethanol (ABE) fermentation, palladium catalyzed alkylation and in situproduct recovery, allowing the catalytic reaction to proceed consistently and smoothly. The remarkable conversion property is mainly attributed to excellent compatible linkage of chemical catalysis with biosynthesis and process design to stabilize the activity of Pd catalyst. This validated strategy for rational process design and integration of chemical- and bio-catalysis offers great demonstration in producing high-valued transportation biofuels derived from agricultural residues.

457314 Cholesteryl Hyaluronic Acid-Coated, Reduced Graphene Oxidenanosheets for Anti-Cancer Drug Delivery

Abstract:

Here, we report hyaluronyl reduced graphene oxide (rGO) nanosheets as a tumor-targeting delivery system for anticancer agents. Hyaluronyl-modified rGO nanosheets were prepared by synthesizing cholesteryl hyaluronic acid (CHA) and using it to coat rGO nanosheets, yielding CHA-rGO. Compared with rGO, CHA-rGO nanosheets showed increased colloidal stability under physiological conditions and improved in vivo safety, with a survival rate of 100% after intravenous administration of 40 mg/kg in mice. The doxorubicin (Dox) loading capacity of CHA-rGO was 4-fold greater than that of rGO. Uptake of Dox by CD44-overexpressing KB cells was higher for CHA-rGO than for rGO, and was decreased in the presence of hyaluronic acid through competition for CD44 receptor binding. After intravenous administration in tumor-bearing mice, CHA-rGO/Dox showed higher tumor accumulation than rGO/Dox. The in vivo antitumor efficacy of Dox delivered by CHA-rGO was significantly increased compared with free Dox or rGO/Dox. In CHA-rGO/Dox-treated mice, tumor weights were reduced to 14.1% 0.1% of those in untreated mice. Our findings indicate that CHA-rGO nanosheets possess greater stability, safety, drug loading capacity, and CD44-mediated delivery of Dox than rGO nanosheets. These beneficial properties of CHA-rGO improved the distribution of Dox to tumors and facilitated the cellular uptake of Dox by CD44-overexpressing tumor cells, resulting in enhanced anticancer effects.

454781 *Featured:* Energy Conversion Considerations in Solar-Driven Photoelectrochemical Water Splitting and Carbon Dioxide Reduction

Abstract:

A practical method to use sunlight to generate chemical fuels could dramatically change the landscape of global energy generation and storage. One specific implementation of this concept is an integrated photoelectrochemical (PEC) device that operates without added electrical bias with sunlight as the only energy input, producing either hydrogen by water splitting or carbon-based fuels via carbon dioxide reduction. Experimental demonstrations of such systems date back to the 1970s, and, since that time, there have been improvements in both the functional understanding and performance of the devices.

There are a number of design considerations which arise when considering the full solar to fuel conversion system, both for water splitting and carbon dioxide reduction. These include the type and number of semiconductor junctions used to generate the requisite photovoltage, the use of co-catalysts, the electrolyte conditions, a product separation method such as a membrane, and the geometric layout and size of components.

To date, all approaches which have a solar to hydrogen (STH) conversion efficiency of >5% employ either 2 or more semiconductor junctions arranged in a voltage-additive tandem configuration. The degree of integration varies widely, from fully integrated "artificial leaves" to systems with decoupled photovoltaics and catalysts. A small subset of devices employ a mechanism, such as a proton or anion conducting membrane, to separate products and yield a pure H_2 stream. Reported STH conversion efficiencies range from <1% to over 20% [1]. However, there are very few reports of long term operational stability, which is a clear prerequisite for a positive energy return on investment [2].

Electrochemical reduction of CO_2 to fuel molecules such as methanol and ethanol could form the basis for the production of renewable and sustainable transportation fuels, replacing the fossil fuels which are used today. There has been substantial recent progress in solar driven PEC CO2 reduction, with a number of reports of energy conversion efficiencies of over 1%. While these values are larger than those estimated for biomass generation by plants and algae (up to 1%) [3], the natural system utilizes dilute CO_2 in the atmosphere as a feed stock and generates a separated fuel product. Thus, product separation and the selective production fuels remain as outstanding challenges for the PEC technology.

This material is based upon work performed by the Joint Center for Artificial Photosynthesis, a DOE Energy Innovation Hub, supported through the Office of Science of the U.S. Department of Energy under Award Number DE-SC0004993.

[1] Ager, J. W.; Shaner, M. R.; Walczak, K. A.; Sharp, I. D.; Ardo, S. *Energy Environ. Sci.* **2015**, *8* (10), 2811–2824.

[2] Sathre, R.; Scown, C. D.; Morrow, W. R.; Stevens, J. C.; Sharp, I. D.; Ager, J. W.; Walczak, K.; Houle, F. A.; Greenblatt, J. B. *Energy Environ. Sci.* **2014**, *7*, 3264–3278.

[3] Blankenship, R. E.; Tiede, D. M.; Barber, J.; Brudvig, G. W.; Fleming, G.; Ghirardi, M.; Gunner, M. R.; Junge, W.; Kramer, D. M.; Melis, A.; Moore, T. A.; Moser, C. C.; Nocera, D. G.; Nozik, A. J.; Ort, D. R.; Parson, W. W.; Prince, R. C.; Sayre, R. T. *Science.* **2011**, *332*, 805–809.

466023 An Enabling Start-up Support System for CO_2 Utilisation: A Case Study on Barriers to Commercial Success

Abstract:

The topic of this paper is sustainable entrepreneurship focussing on 'emerging Davids' (Hockerts, Wüstenhagen 2010) and their endeavour to bring CO2 utilisation technologies to the market.

The problem is that a proof-of-concept on a technically relevant scale is essential for new technologies in order to attract external stakeholders such as investors, customers and other partners. These proofs of a scale-up production for CO2 utilisation technologies are ranging from demonstrators over pilot production facilities to large commercial plants and are most of the times very capital intensive. This capital intensity often leads to the point that new technology ventures fail to successfully commercialise their products. This results in a lack of CO2 utilisation start-ups and little radical innovation to tackle current societal challenges like resource depletion and climate change. Recent events such as the COP21 in Paris and the G7 summit in 2015 may be promising means in order to boost sustainable-oriented innovation by investing more and more money in clean technologies over the next decades, but an effective allocation and a sophisticated design for a support system are still relatively unexplored.

In order to give managerial recommendation on the design of a favourable support system for capital intensive CO2 utilisation technologies a better understanding on the barriers to those innovation is necessary. A broad portfolio of CO2 utilisation technologies are in scope of this contribution; ranging from chemicals production (e.g. intermediates and specialty chemicals from CO2) over CO2 mineralisation (e.g. building materials from CO2) to CO2 to fuels (Hendriks et al. 2013). This work builds on a multilevel perspective from current barrier research (Hueske, Guenther 2015) to investigate all relevant barriers to commercial success. The authors choose a dynamic capabilities approach (Eisenhardt, Martin 2000) in combination with stakeholder theory (Freeman 2004) to analyse both internal as well as external barriers.

A exploratory multiple case study approach (Eisenhardt, Graebner 2007; Yin 2013; Eisenhardt 1989) was used to investigate the CO2 utilisation start-up landscape worldwide. CO2 utilisation may not only be a door-opener for large emission abatement technologies but is also a potent resource efficiency technology by feeding CO2 back into the carbon-based economic system (Styring et al. 2014; Naims et al. 2015). A total of 20 semi-structured interviews have been carried out, transcribed, coded (Strauss, Corbin 2015), and triangulated with other available data sources during a 5-month-period in 2015/2016.

This work is funded by the EIT Climate-KIC enCO2re programme – 'enabling CO2 re-utilisation'.

References:

Eisenhardt, Kathleen M. (1989): Building theories from case study research. In *Academy of management review* 14 (4), pp. 532–550.

Eisenhardt, Kathleen M.; Graebner, Melissa E. (2007): Theory Building from Cases: Opportunities and Challenges. In *The Academy of Management Journal* 50 (1), pp. 25–32.

Eisenhardt, Kathleen M.; Martin, Jeffrey A. (2000): Dynamic capabilities: What are they? In *Strategic Management Journal;* 21, pp. 1105–1121.

Freeman, R. Edward (2004): The stakeholder approach revisited. In *Zfwu* 5 (3), pp. 228–241.

Hendriks, Chris; Noothout, Paul; Zakkour, Paul; Cook, Greg (2013): Implications of the Reuse of Captured CO2 for European Climate Action Policies. With assistance of Carbon Counts Ltd. ECOFYS Netherlands B.V. Utrecht. Available online at http://www.scotproject.org/sites/default/files/Carbon%20Count,%20Ecofys%20%282013%29%20Implications%20of%20the%20reuse%20of%20captured%20CO2%20-%20report.pdf, checked on 06/03/15.

Hockerts, Kai; Wüstenhagen, Rolf (2010): Greening Goliaths versus emerging Davids — Theorizing about the role of incumbents and new entrants in sustainable entrepreneurship. In *Journal of Business Venturing* 25 (5), pp. 481–492. DOI: 10.1016/j.jbusvent.2009.07.005.

Hueske, Anne-Karen; Guenther, Edeltraud (2015): What hampers innovation? External stakeholders, the organization, groups and individuals: a systematic review of empirical barrier research. In *Manag Rev Q* 65 (2), pp. 113–148. DOI: 10.1007/s11301-014-0109-5.

Naims, H.; Olfe-Kräutlein, B.; Lorente Lafuente, A. M.; Bruhn, T. (2015): CO2-Recycling – Option für Politik und Gesellschaft? Zwölf Thesen zur gesellschaftlichen und politischen Bedeutung von Carbon Capture and Utilisation(CCU)-Technologien. Institute for Advanced Sustainability Studies (IASS). Potsdam (IASS Working Paper).

Strauss, Anselm L.; Corbin, Juliet M. (2015): Basics of Qualitative Research. Techniques and Procedures for Developing Grounded Theory. 4th: Sage publications.

Styring, Peter; Quadrelli, Elsje Alessandra; Armstrong, Katy (2014): Carbon Dioxide Utilisation: Closing the Carbon Cycle. 1st edition: Elsevier Science.

Yin, Robert K. (2013): Case study research: Design and methods. 5th: Sage publications.

471447 Catalytic Dry Methane Reforming for Industrial CO_2 Capture Applications & Syngas Generation

Abstract:
Steam Methane Reforming is the most popular process for the production of H_2 rich syngas from hydrocarbon feed such as natural gas or flare gas in refineries. Generally, the process is operated to obtain a high H_2:CO ratio around 4~7, which is further enhanced via a water gas shift reactor. However, the synthesis of hydrogen gas is accompanied by the generation of a high amount of CO_2 as well, averaging around 9 tonnes of CO_2 per tonne of H_2produced.

Syngas however can be utilized for purposes other than just hydrogen generation, in applications such as the production of synthetic crude oil (via the Fischer-Tropsch process), methanol, ethanol as well as of higher products of enhanced value such as acrylates, organic acids, ether etc. In such processes, the CO molecule is as important as the H_2 molecule. Typically, for most chemical synthesis processes the ratio of consumption of H_2:CO is between 1 â€" 2.5.

Catalytic Dry Methane Reforming[1] is a process for generation of syngas (CO + H_2), utilizing CH_4 and CO_2as feed (1). The reforming reaction is accompanied by the reverse water gas shift reaction (2).

$$CH_4 + CO_2 \rightarrow 2CO + 2H_2 \quad \text{Â â€¦.(1)}$$

$$CO_2 + H_2 \rightleftharpoons CO + H_2O \quad \text{Â â€¦.(2)}$$

The product syngas, with a typical composition of H_2:CO â‰¤ 1, can be used to supplement the raw material for the above mentioned processes. The reaction however is highly endothermic, and requires operation at high temperatures (750-900 C). This can potentially lead to significant CO_2 emissions through fuel combustion to provide energy for the reaction. The reaction, however, has been found to consume nearly twice the quantity of CO_2that is produced by combustion of fuel, thus resulting in a carbon capture efficiency of over 50 %. This would make it a great tool to serve the dual purposes of emission reduction as well as valorization of waste greenhouse gases in industry.

Concerns over the increasing global temperatures have triggered an extensive search for viable technologies to mitigate industrial GHG emissions. Implementation of the Dry Methane Reforming process could be a viable solution to this problem for a variety of industries. However, the high temperature operation causes $CH_4$and CO decomposition to form carbon deposits, thus resulting in deactivation of dry reforming catalysts. The lack of a satisfactory catalyst for the DMR system has impeded the application of this process in industry.

Enerkem Inc. is a Canadian company focused on reducing the carbon footprint for cleaner chemical production. The company has developed its technology for converting municipal solid wastes to methanol, which in turn can serve as the base material for secondary chemicals, such as higher alcohols, aldehydes, acrylates etc. It is starting its first commercial operation at the Edmonton Waste Management Centre for the production of methanol. Enerkem Inc. has also developed the E20 dry reforming catalyst, which has been observed to resist deactivation when operated at over 90% CH_4 as well as CO_2conversion for more than 100 hours (Figure 1). A comparison of the emission levels for syngas production using steam methane reforming operating at equilibrium condition and the dry methane reforming using the E20 catalyst have been presented in Table 1.

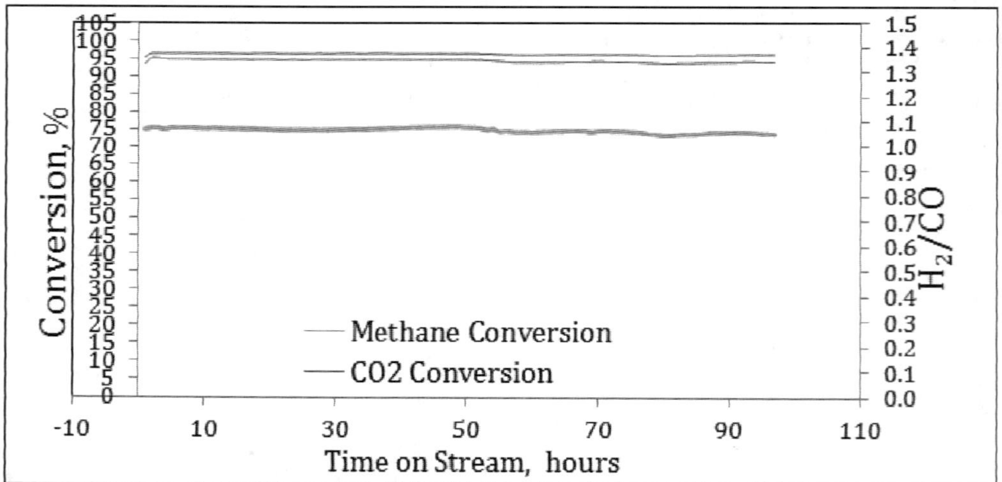

Figure 1. Performance of E20 Dry reforming catalyst for 100 hour operation with CH_4:CO_2:N_2= 40:40:20, reactor temperature of 850 ÌŠC and WHSV of 500 ml/min-g.cat

Kinetic experiments have been conducted on the E20 catalyst and a working kinetic model has been developed. Simulation studies performed on HYSYSÂ® have exhibited the CO_2capture potential of the E20 catalyst when integrated with the synthesis of high value chemicals. To showcase this potential, we have considered the case of acrylic acid synthesis, using syngas as the starting material. The most commonly utilized process for the synthesis of acrylic acid is the partial oxidation of propylene. However, the pathway for synthesis from syngas can be summarized as follows:

Syngas is used to synthesize methanol via reaction 3.

$$CO + 2H_2 \rightarrow CH_3OH \quad \text{â€¦.(3)}$$

Methanol can be utilized to synthesize formaldehyde via the Formox process[3] (4), as well as acetic acid via the methanol carbonylation process[4](5).

$$CH_3OH + \frac{1}{2}O_2 \rightarrow HCHO + H_2O$$ Â â€¦.(4)

$$CH_3OH + CO \rightarrow CH_3COOH$$ Â â€¦.(5)

The acetic acid and formaldehyde can undergo reaction via an aldol condensation route[5]to form acrylic acid (6).

$$HCHO + CH_3COOH \rightarrow CH_2(OH)CH_2COOH \rightarrow CH_2 = CHCOOH + H_2O$$ Â â€¦.(6)

The overall process would allow the valorization of waste CO_2streams into high value acrylate compounds.

	SMR	**SMR**	**DMR**	**DMR**
Reactor Temp (ÌŠC)	1000	1000	850	850
H_2O:CO_2:CH_4	5:0:1	3:0:1	0:1:1	0:1.67:1
H_2:CO	5.5	4.3	0.84	0.71
CH_4 Conversion (%)	99.2	97.2	73.8	67.8
CO_2 Conversion (%)			86.5	54.7
Energy per kg H_2 (kJ)	71933.4	61193.1	81392.6	95834.2
CO_2 produced per kg H_2 (kg)	6.5	5	-9.6	-12.6
CO_2 capture efficiency (%)			68.2	70.5
Energy per kg CO (J)	28242.8	18843.7	4896.3	4841
CO_2 produced per kg CO (kg)	2.5	1.6	-0.58	-0.64
CO_2 capture efficiency (%)			69.1	71.1

Table 1. Comparison of Energy consumption and CO_2emission of Steam Methane Reforming (SMR) and Dry Methane Reforming (DMR) processes

References

1. Bradford, M.; Vannice, M., CO2 reforming of CH4. *Catalysis Reviews* **1999,**41, (1), 1-42.

2. Pakhare, D.; Spivey, J., A review of dry (CO 2) reforming of methane over noble metal catalysts. *Chemical Society Reviews* **2014,**43, (22), 7813-7837.

3. Reuss, G.; Disteldorf, W.; Gamer, A. O.; Hilt, A., Formaldehyde. *Ullmann's Encyclopedia of Industrial Chemistry* **2005**.

4. Yoneda, N.; Kusano, S.; Yasui, M.; Pujado, P.; Wilcher, S., Recent advances in processes and catalysts for the production of acetic acid. *Applied Catalysis A: General* **2001,**221, (1), 253-265.

5. Vitcha, J. F.; Sims, V. A., Vapor Phase Aldol Reaction. Acrylic Acid by Reaction of Acetic Acid and Formaldehyde. *Industrial & Engineering Chemistry Product Research and Development* **1966,** 5, (1), 50-53.

CO_2 as a "Soft" Oxidant for Syngas Production via Chemical Looping

Amey More, Saurabh Bhavsar, Charlie Hansen and Götz Veser
Department of Chemical Engineering, University of Pittsburgh, Pittsburgh, PA

Abstract:

'Chemical Looping Combustion' is a clean combustion technology, which enables fossil fuel combustion with inherent CO_2 capture based on the cyclic oxidation and reduction of an oxygen carrier. While most efforts in chemical looping (CL) are focused on combustion, we previously demonstrated the application of the "CL principle"—the periodic oxidation and reduction of a metal oxide to couple two independent redox reactions—to the activation of CO_2 via reduction to CO [1]. In the present contribution, we investigate and compare CO_2 activation via CL in two different operating modes: In the first scheme, CO_2 reduction is coupled with CH_4 oxidation by using mixtures of Fe and Ni (as alloys or simple physical mixtures) to produce CO and syngas product streams [2]. In the second operating scheme, monometallic Ni carriers are utilized to catalytically crack CH_4, producing pure H_2 streams. The solid carbon deposits are then burnt off with CO_2, overall producing separate CO and H_2 product streams.

Supported metal oxide carriers were synthesized, and a combination of thermo-gravimetric analysis (TGA), X-ray diffraction (XRD), and electron microscopy (TEM) was used to identify structure-reactivity correlations for systematic carrier design. Gas phase conversions and selectivities were determined in periodic fixed bed reactor operation. We find that Fe-Ni alloys can indeed show good activity towards both methane activation and CO_2 reduction, and that the weak oxidant CO_2 allows controlled oxidation of Fe-Ni alloys, which enables selective oxidation of CH_4 to syngas. Remarkably, a simple physical mixture of Fe and Ni far exceeds the reactivity of an equivalent alloy carrier, which can be traced back to a synergistic gas-phase coupling between the two carrier fractions. For the second proposed scheme, we find a strong dependence on the support material, which affects particle size and distribution of the active metal species. Overall, our investigations demonstrate the potential of CO_2 as a "soft" oxidant which enables selective oxidation reactions via novel, intensified chemical looping processes.

[1] M. Najera, et al, *Chem. Eng. Res. Des.* 89 (2011) 1533; Bhavsar, et al, *Chem. Eng. Technol.* 35 (2012) 1281; S. Bhavsar and G. Veser, *Energy Fuels* 27 (2013) 2073; A. More, S. Bhavsar, and G. Veser. *Energy Technology* (2016) AOP.

464674 Process and Material Design for Micro-Encapsulated Ionic Liquids in Post-Combustion CO_2 Capture

Abstract:

Aprotic Heterocyclic Anion (AHA) Ionic Liquids (ILs) have been identified as promising new solvents for post-combustion carbon capture due to their high CO_2 uptake and the high tunability[1,2] of their binding energy with CO_2. Some of these compounds change phase (solid to liquid) on absorption of CO_2; these Phase Change ILs (PCILs)[3] offer the additional advantage that part of the heat needed to desorb the CO_2 from the absorbent is provided by the heat of fusion as the PCIL solidifies upon release of CO_2. However, the relatively high viscosity of AHA ILs and the occurrence of a phase change in PCILs present challenges for conventional absorption equipment. To overcome these challenges we are pursuing the use of new technology to micro-encapsulate the AHA ILs and PCILs. Our partners at Lawrence Livermore National Laboratory have successfully demonstrated this technology in the application of post-combustion carbon capture with sodium and potassium carbonate solutions,[4] and have recently shown the feasibility of micro-encapsulation of an AHA IL for carbon capture.[5] The large effective surface area and high CO_2 permeability of the micro-capsules is expected to offset the drawback of the high IL viscosity and to provide for a more efficient and cost-effective mass transfer operation involving AHA ILs and PCILs.

These opportunities, however, present us with both process and materials design questions. For example, what is the target CO_2 absorption strength (enthalpy of chemical absorption) for the tunable AHA IL? What is the target for micro-capsule diameter in order to obtain a high mass transfer rate and good fluidization performance? What are the appropriate temperatures and pressures for the absorber and stripper? In order to address these and other questions, we have developed a rate-based model of a post-combustion CO_2 capture process using micro-encapsulated ILs. As a performance baseline, we have also developed a rate-based model of a standard packed bed absorber using an un-encapsulated AHA IL absorbent. Using such models we can determine optimal CO_2 capture performance and investigate the sensitivity of the optimum with respect to the key thermo-physical and transport properties of the IL (e.g., CO_2 binding energy, viscosity, etc.) and the micro-capsules (e.g. diameter, CO_2 permeability, etc.). Results of these process and material design studies will be presented, and the performance of this novel micro-encapsulation technology will be assessed.

References

(1) Gurkan, B.; Goodrich, B. F.; Mindrup, E. M.; Ficke, L. E.; Massel, M.; Seo, S.; Senftle, T. P.; Wu, H.; Glaser, M. F.; Shah, J. K.; Maginn, E. J.; Brennecke, J. F.; Schneider, W. F. Molecular Design of High Capacity, Low Viscosity, Chemically Tunable Ionic Liquids for CO_2 Capture. *J. Phys. Chem. Lett.* **2010,** *1*, 3494-3499.

(2) Seo, S.; Quiroz-Guzman, M.; DeSilva, M. A.; Lee, T. B.; Huang, Y.; Goodrich, B. F.; Schneider, W. F.; Brennecke, J. F. Chemically Tunable Ionic Liquids with Aprotic Heterocyclic Anion (AHA) for CO2 Capture. *J. Phys. Chem. B* **2014,** *118*, 5740-5751.

(3) Seo, S.; Simoni, L. D.; Ma, M.; DeSilva, M. A.; Huang Y.; Stadtherr, M. A.; Brennecke, J. F. Phase-Change Ionic Liquids for Postcombustion CO_2 Capture. Energy Fuels, **2014**, *28*, 5968-5977.

(4) Vericella, J. J.; Baker, S. E.; Stolaroff, J. K.; Duoss, E. B.; Hardin, J. O.; Lewicki, J.; Glogowski, E.; Floyd, W. C.; Valdez, C. A.; Smith, W. L.; Satcher, J. H.; Bourcier, W. L.; Spadaccini, C. M.; Lewis, J. A.; Aines, R. D. Encapsulated Liquid Sorbents for Carbon Dioxide Capture, *Nature Commun.*, **2015**, *6*, 6124.

(5) Stolaroff, J. K.; Ye, C.; Oakdale, J. O.; Baker, S. E.; Smith, W. L.; Nguyen, D. T.; Spadaccini, C. M.; Aines, R. D. Microencapsulation of Advanced Solvents for CO_2 Capture. *Faraday Discuss.*, 2016, in press, DOI: 10.1039/C6FD00049E.

457289 Material Development and Assessment of a Thermochemical Energy Storage Based on $CaO/CaCO_3$

Abstract:

The usage of solar energy for heat and power supply is restricted by the fluctuation of solar radiation. To overcome this and to enable a continuous operation of concentrated solar power plants, thermal energy storages units are essential. In this context, reversible thermal decomposition reactions of solids have been proposed and studied for several years [1]: A solid salt is decomposed into another solid and a gaseous component during daytime. The composition reaction, reverse reaction, occurs during night time or at an overcast sky in order to supply continuously energy for a steam cycle. Investigated reaction schemes contain oxides, hydroxides, carbonates and hydrides [2-5]. Due to high storage densities and discharging temperatures between 400 and 1000 °C, the gas-solid reactions are discussed as a promising concept for increasing the efficiency and operation hours of solar thermal power plants.

In previous works [6], the cyclic carbonation/calcination reaction of CaO is proposed as a thermochemical storage for solar power plants. The energy is stored chemically in CaO after the endothermal calcination reaction of $CaCO_3$ (eq. R1). The formed energy-rich CaO and CO_2 are stored separately. The energy is released by initiating the exothermal carbonation reaction (eq. RR1) of CaO and CO_2 at temperatures between 600 and 750 °C.

$$CaCO_3(s) \rightarrow CaO(s) + CO_2(g) \qquad \Delta^R h^{+} = 178 \text{ kJ mol}^{-1} \qquad \text{(R1)}$$

$$CaO(s) + CO_2(g) \rightarrow CaCO_3(s) \qquad \Delta^R h^{+} = -178 \text{ kJ mol}^{-1} \qquad \text{(RR1)}$$

The involved components are cheap, non-toxic and in case of $CaCO_3$ abundantly available. Nevertheless, a profound challenge is the strong cycle-to-cycle degradation of the material due to sintering. A low cyclic stability implies a decrease of active material after several cycles and thus, the efficiency and storage density is lowered.

Within this study, Al-doped materials are synthesized with different Ca/Al molar ratios in order to stabilize the sorbent structure and prevent the mentioned sintering phenomena. The Ca/Al ratio is varied between 75/25 and 95/5. All samples are experimentally characterized in 20 calcination/carbonation cycles by a thermogravimetric analyzer measuring the CO_2 uptake at predefined conditions [6]. Furthermore, based on the results and a performed enthalpy balance of carbonation and calcination reaction the storage density and the energetic efficiency are calculated in order to evaluate the sorbents out of an energetic point of view. All samples are compared to the performance of the benchmark material (natural CaO).

The experiments show that the formed Ca/Al mixed phase stabilizes the sorbent and the CO_2 uptake behavior can thus be maintained. It could be proven that all synthesized sorbents are characterized by an improved cyclic stability and the relative loss of activity from the first to the 20th cycle is notably reduced. Whereas the benchmark material shows a loss of activity of about 70 % after the 20th cycle, the degradation of the best prepared sorbent is less than 4 %. In context of storage density and efficiency it can be seen that in general, an increase of Al-content entails a lowered maximum achievable storage density and efficiency due to a decreased initial amount of active material (CaO). Nevertheless, by taking the results of the CO_2 uptake measurements into account, all synthesized samples show an improved performance compared to natural CaO in the 20th cycle due to a constant CO_2 uptake. In terms of storage density 2 to 3.5 times higher values can be achieved. For the energetic efficiency a factor of up to 1.2 relative to the benchmark material is obtained.

Thus, based on these experimental and theoretical investigations it can be confirmed that the addition of a certain amount of initial inert material is justified in order to optimize the performance of the material. The results of the study lay the foundation for the usage of the $CaO/CaCO_3$ reaction couple as a high efficient thermochemical heat storage.

[1] W. Wentworth, E. Chen, Solar Energy, 18 (1976), 205-214

[2] A. J. Carrillo, J. Moya, A. Bayón, P. Jana, A. Víctor, M. Romero, J. Gonzalez-Aguilar, D. Serrano, P. Pizarro, J.M. Coronado, Solar Energy Materials and Solar Cells, 123 (2014), 47-57

[3] F. Schaube, A. Wörner, R. Tamme, Journal of Solar Energy Engineering, 133 (2011), 1006-1012

[4] M. Aihara, K. Tanaka, M. Watanabe, T. Takeuchi, H. Habuka, Journal of Chemical Engineering of Japan, 40 (2007), 1270-1274

[5] M. Paskevicius, D. Sheppard, K. Williamson, C. Buckley, Energy, 88 (2015), 469-477

[6] J. Obermeier, B. Müller, K. Müller, W. Arlt, Energy Technology, 4 (2016), 123 – 135

472158 Development of a Thermoreversible Medium to Culture and Efficiently Harvest Microalgae

Abstract:

With the ability to generate different types of biofuels and a variety of valuable chemicals, microalgae represent a promising alternative to fossil fuels. However, the development of high throughput cultivation and efficient harvesting techonologies is necessary to make microalgal biomass production a more viable option. Herein, a new Tris-Acetate-Phosphate-Pluronic (TAPP) medium with thermoreversible sol-gel transition properties developed for cultivation and efficient harvesting of microalgae will be presented.

This sol-gel transition phenomenon refers to the transformation from a low-viscosity fluid to an elastic solid upon heating. For the TAPP medium developed, this gelation process is completely reversible upon cooling. The thermo-rheological properties of the new pluronic-based medium as well as the resulting pluronic-microalgae matrix after cultivation are systematically characterized. moreover, a framework is proposed to efficiently harvest the microalgal biomass produced through small variations of temperature.

Microalga *Chlamydomonas reinhardtii* is successfully cultivated in the new TAPP medium and leads to equivalent production of microalgal biomass with similar lipid and carbohydrate composition than that obtained from cultivation in the traditional TAP medium. The harvesting process of the microalgal biomass produced is also highly simplified with the TAPP system. In fact, confinement of microalgal cells in the pluronic matrix leads to a cluster distribution that increases the settling velocity by a factor of ten. Through small variations of temperature, microalgae are allowed to settle, the broth subsequently jellified and ninety percent of microalgal biomass are harvested simply through scrapping off the surface.

The implications of this new cultivation and harvesting technology as well as the potential for varieties of large scale applications will be discussed.

472266 Bioprocessing Strategies for Engineered Coproduction of Biofuels and Medical Biopolymers from Diatom Microalgae

Abstract:

Photosynthetic diatom microalgae have significant capacity for biosynthesis of energy-dense biofuel molecules, as well as unique valued co-products not found in other types of algae, including metal oxide nanomaterials for advanced material applications, and glucosamine biopolymers or monomers for nutraceutical and biomedical applications. Diatoms biomineralize soluble silicon to nanostructured biosilica, and require dissolved silicon (Si) as a required substrate for cell wall biosynthesis and division. Bioprocess engineering strategies have the potential to guide the cellular biosynthesis of three aforementioned product streams, all within the same diatom cell. To exploit diatom silicon metabolism for eliciting the biosynthetic pathways of selected products, a two-stage cultivation process was developed to induce high levels of lipid and chitin production by the centric marine diatom *Cyclotella* within under tightly controlled conditions where light and CO_2 delivery are not limiting. Carbon dioxide delivery and consumption were measured in real time to determine carbon flux into biomass, lipid, and chitin biopolymer products. The two-stage cultivation process synchronized *Cyclotella* diatom cells to silicon-starved state in Stage I. In Stage II, the cells were maintained in a nominal silicon-starved state under controlled nutrient perfusion, which allowed for continued cell division, carbon dioxide assimilation, and product formation. In this cultivation mode, cell biomass, lipid, and chitin production were elicited and linearly maintained at silicon depletion without the need for nitrogen depletion. Intracellular accumulation of lipid and extracellular extrusion of chitin microfibrils were evident. Optimal stage II product yields associated with the biomass were 34 wt% total lipid and 16 wt% chitin, with 60 mol% of assimilated carbon dioxide allocated to these two products. From this information, a process flowsheet and material balance on the diatom-based photosynthetic bio-refinery for production of the nutraceutical glucosamine with co-production of biodiesel and biosilica was developed to illustrate the productivity at scale. This analysis shows that the diatom-based photosynthetic biorefinery has significant potential as a future platform for biofuels and unique, valuable co-products.

454396 Effects of Microalgal Polycultures on Quality of Biomass for Biocrude Oil Production Via Hydrothermal Liquefaction

Abstract:

Rising atmospheric carbon dioxide levels have created a continuously increasing demand for sustainable energy. Algal biomass as a feedstock for sustainable energy production is a compelling option, as algae cultivation does not necessitate arable land and algae have higher energy density than terrestrial plants [1]. Moreover, one biomass-conversion technology called hydrothermal liquefaction (HTL) can convert wet microalgae into an energy-dense biocrude oil without incurring energy penalties associated with drying the biomass [2, 3].

Raceway ponds are a primary method of growing microalgae, although the open nature of these ponds subjects algal cultures to external perturbations like light and temperature variability, predators, diseases, and invasive species, leading to crop instabilities [4, 5]. Multi-species communities, or polycultures, could potentially be more resistant to these environmental factors and thus more stable over time [6, 7]. In a recent study, we used HTL to assess the biocrude productivity and stability of 2-, 4-, and 6-species polycultures of *Ankistrodesmus falcatus, Chorella sorokiniana, Pediastrum duplex, Scenedesmus acuminatus, Scenedesmus ecornis,* and *Selenasatrum capricornutum*, compared to the monocultures of these species [8]. This study found that polycultures generally produced biocrude more stably than monocultures but not in higher amounts. However, the quality of the biomass feedstock and the resulting biocrude have yet to be considered, but are also important bases of comparison. Most importantly, these factors affect the extent of catalytic upgrading required before the biocrude can be inserted into existing refining infrastructure [9].

In this presentation, we expand upon earlier work [8] and compare the performance of polycultures and monocultures on several quality-based metrics, including biomass fatty acid content and biocrude yield, atomic ratios, and higher heating value (HHV). Microalgae richer in fatty acid content and biocrude with lower nitrogen and oxygen content and higher carbon and hydrogen content are more desirable. We found

that some polycultures met or exceeded the best monocultures for each metric we examined; moreover, the best monocultures and polycultures were generally different for each metric, suggesting that optimization of polyculture composition based on the desired biocrude specifications is possible. These metrics of quality provide additional bases of comparison for monocultures and polycultures beyond those of biocrude productivity and stability reported in earlier work by [8], allowing for a more comprehensive evaluation of the benefits of biodiversity on biocrude oil production via microalgal HTL.

References

1. Patil, V.; Tran, K. Q.; Giselrod, H. R. *International Journal of Molecular Sciences* **2008**, 9, 1188–1195.

2. Brown, T. M.; Duan, P.; Savage, P. E. *Energy & Fuels* **2010**, 24, 3639–3646.

3. Toor, S. S.; Rosendahl, L.; Rudolf, A. *Energy* **2011**, 36, 2328–2342.

4. Pienkos, P. T.; Darzins, A. Biofuels, *Bioproducts and Biorefining* **2009**, 3, 431–440.

5. Georgianna, D. R.; Mayfield, S. P. *Nature* **2012**, 488, 329–335.

6. Hooper, D. U.; Chapin, F. S.; Ewel, J. J.; Hector, A; Inchausti, P; Lavorel, S; Lawton, J. H.; Lodge, D. M.; Loreau,

   M; Naeem, S; Schmid, B; Setälä, H; Symstad, A. J.; Vandermeer, J; Wardle, D. A. *Ecological Monographs* **2005**,

   75, 3–35.

7. Cardinale, B. J.; Matulich, K. L.; Hooper, D. U.; Byrnes, J. E.; Duffy, E.; Gamfeldt, L.; Balvanera, P.; O'Connor,

   M. I.; Gonzalez, A. *American Journal of Botany* **2011**, 98, 572–592.

8. Narwani, A.; Lashaway, A.; Hietala, D.; Savage, P.; Cardinale, B. J. *Algal Research, Submitted.*

9. Duan, P.; Savage, P. E. *Bioresource technology* **2011**, 102, 1899–906.

455130 Enhanced Autotrophic Growth of *Nannochloris Sp.* with Buffer Chemicals for Sustainable Carbon Recycle

Abstract:

In this study, a potential ability of buffer chemicals for CO_2 gas capture and the resultant solutions were investigated for the autotrophic growth of *Nannochloris sp.* in a modified Shuisheng-4 culture medium. Sodium-based buffer chemicals used in this study well dissolve in water due to their high solubility and the buffer solutions can be converted to sodium bicarbonate solution after absorbing $CO_2(g)$. After absorbing $CO_2(g)$, both total inorganic carbon ($TIC=HCO_3^-$, CO_3^{2-}, H_2CO_3, $CO_2(aq)$) and dissolved inorganic carbon ($=DIC=HCO_3^-+CO_2(aq)$) in the solution significantly increased and the pH of the solution is equilibrated at pH=8.3 suitable for the growth of *Nannochloris sp.* As a result, high DIC concentration dramatically increased the autotrophic growth rate of *Nannochloris sp.* when the Na^+ concentration was less than 0.12 M. In addition, *Nannochloris sp.* was cultivated in the buffer chemicals system using simulated flue gas mixed with CO_2, N_2, SO_2, NO, and HCl gases. As a result, the buffer solutions could suppress the possible inhibition effect of acid gases such as SO_2, NO, and HCl on the growth of *Nannchloris sp.* This study suggests that the sodium-based buffer chemicals can be effectively used for biological CO_2 conversion and sustainable carbon recycle with pure CO_2 or flue gas from a coal-fired power plant.

455395 Effect of Process Parameters and Extractive Solvents on Hydrothermal Liquefaction of Microalgae

Abstract:

Hydrothermal Liquefaction processes proves to be an attractive technology from the view point of energy consumption, renewability and process integration for converting wet biomass into energy dense bio-oil. The yield and quality of bio-crude oil is influenced by process conditions like liquefaction temperature, solid-loading ratio, residence time, type of solvent for extraction. Therefore, we investigated the influence of these parameters on the hydrothermal liquefaction of microalgae *Micractinium sp*. The temperatures and solid-loading ratios were varied from 200 to 350°C and 0-20 % respectively in a stainless steel batch reactor. Experiments were also performed with residence times varied from 0 to 60 minutes (~15 minutes interval). The maximum bio-oil yield was obtained at 300°C and at a solid loading ratio of 10 % for 30 minutes residence time. Experiments were performed with different solvents including hexane, acetone and dichloromethane for the separation of bio-oil from product phase. Energy recovery calculations were performed with the analysis of the products (bio-oil, biochar, water soluble compounds, and gaseous phase) through ultimate and proximate analysis. Products were also analyzed by Gas Chromatography Mass Spectrometry (GC-MS) and calorimetric techniques.

Keywords: Microalgae, Hydrothermal Liquefaction, Renewability, Bio-oils/fuels, GC-MS

455641 Re-Use of Hydrothermal Liquefaction Bio-Char to Grow *Galdieria Sulphuraria*

Abstract:

Hydrothermal liquefaction is one of the promising methods to convert wet algal biomass into biofuels. Maximum biocrude oil was observed at 300 C when *Galdieria Sulpuraria* (GS) was liquefied using water as solvent. But a substantial amount of algal biomass was also converted into bio-char. The biochar produced at 300 C and 30 min of reaction time was analyzed for phosphorous content using inductively coupled plasma optical emission spectrometry (ICP-OES) and for nitrogen content using CHNOS elemental analyzer. It was observed that 1 gram of HTL biochar has 6.59mg of phosphorous and 6.36mg of nitrogen. The phosphorous and nitrogen present in this biochar can be used to cultivate microalgae. In this study, the macronutrients were leached using a 40 ml reactor at varying temperatures (23 C, 39 C and 55 C) and pH (2.5, 4.7 and 7.0) into cyanidium media without phosphates in it. This method was able to extract 54 mg/L of reactive phosphates and 16 mg/L of ammoniacal nitrogen at the optimum conditions. These leached phosphate and ammoniacal nitrogen were tested for the toxicity over 15 days and then used to grow GS in a Tissue Culture Roller drum apparatus with 16 mm borosilicate glass tube photo reactor and confirmed using a Microplate Assay at 40 C and 2-3 % CO in an incubator. A better growth was observed when leached nutrients were used compared to that of regular nutrients in the growth media.

467293 Engineering an Environmentally-Isolated Bacterium for Continuous Biofuel Production and Recovery Under Supercritical CO_2

Abstract:

Model organisms such as *E. coli* and *S. cerevisiae* are the workhorses of biotechnological product generation due to their well understood and easily manipulated genetics, fast doubling times, and low cost fermentation media. However, bioprocesses utilizing these hosts are prone to contamination and are negatively affected by product toxicity. Environmental strain isolation provides an opportunity to discover alternative organisms with unique growth characteristics and physiological traits to overcome these long-standing bioprocess challenges. Using a targeted bioprospecting approach by sampling fluid from a deep subsurface supercritical carbon dioxide (scCO_2) well, a bacterium was isolated that is rarely able to demonstrate consistent, robust growth in the presence of scCO_2. Due to the broad microbial lethality and solvent chemistry of scCO_2, we hypothesize that a dual-phase reactor of growth media and scCO_2 will simultaneously provide a sterile growth environment and the capacity to continuously strip off strain-produced biofuels (short chain alcohols) to alleviate product toxicity. Additionally, scCO_2 is a sustainable, labile solvent that can be separated from desired products through depressurization, leaving these products at high concentrations and providing advantages over other co-solvent extraction systems. We have begun characterizing this unique host through genome sequencing and by determining optimal growth conditions in various environments. Transformation is possible using a protoplast-based method, which has permitted the identification of promoters capable of inducible heterologous protein expression in both aerobic and anaerobic conditions. We engineered the scCO_2 tolerant strain to produce isobutanol by introducing a two-enzyme (2-ketoisovalerate decarboxylase (KivD) and alcohol dehydrogenase (Adh)) pathway. The two-step conversion occurs at approximately 70-80% from 2-ketoisovalerate when grown aerobically; however, the intermediate aldehyde was found to accumulate at short culture times. Due to the high partition coefficient for the aldehyde to the scCO_2 phase, five alternative alcohol dehydrogenases were tested. A variant was identified that lowered the build-up of isobutyraldehyde and resulted in conversion of 2-ketoisovalerate to isobutanol above 85% in aerobic cultures. Efforts to scale up this process and to develop tools to integrate the biofuel pathway into the genome are currently underway.

464608 A Whole-Plant Genome-Scale Model of Maize

Abstract:

A whole-plant metabolic model of maize was used to analyze the transport of metabolites between organs, as well as evaluate the unique metabolic properties of the major organs. The model links the root, stalk, leaf, kernel, and tassel organs using the phloem as the main tissue for metabolite transport among the organs. The network of metabolic reactions in each organ was determined using whole-plant transcriptomic data to determine organ or tissue specific gene protein reaction (O-GPR or T-GPR) relationships. 3017, 2714, 3372, 3242, and 3206 genes corresponding to metabolic reactions were expressed in the root, shoot, leaf, tassel, and kernel organs, respectively. Once reactions with known activity were allocated into each organ, the flow through the biomass reaction of each organ was restored by preferentially adding reactions known to occur in maize with no organ or tissue localization information over reactions corresponding to low expression in the focus organ. Organ-specific biomass equations were defined and ATP requirements were determined for photosynthetic vs. non-photosynthetic organs based on literature. Three growth stages of maize development representing the vegetative leaf growth, tassel development, and kernel filling stages were modeled based on the change in dry weight of each organ over time. Additionally, the transport between organs was normalized based on the proportion of dry weight at each growth stage modeled. Using parsimonious analysis, the total number of reactions required during leaf growth, tassel development, and kernel filling stages were identified with approximately 80% of the reactions in the vegetative leaf growth stage common among all growth stages. As expected, sucrose is transported to the kernel during the kernel filling stage from all other organs. Using a whole-plant model, we aim to further enhance the understanding of the metabolites transported between organs, as well as identify the reactions underpinning nitrogen utilization efficiency.

471323 Plants 2.0: Towards Metabolic Engineering of Natural Product Biosynthesis into Crops

Abstract:

By 2050, 9.7 billion people will inhabit the earth. To feed this growing population, we need to combat plant pathogens, which cause 65% of crop damage in the United States alone. One solution is to engineer production of natural defense metabolites from wild, or even domesticated, plants into crops. When attacked by pathogens, a few cruciferous plants, such as Chinese cabbage (***Brassica rapa)*** and watercress (***Nasturtium officinale***), produce a group of secondary metabolites called phytoalexins. Indole phytoalexins (brassinins) and benzyl phytoalexins (nasturlexins) are especially affective against bacterial and fungal pathogens[1]. They can improve human health as well; in vitro studies show brassinin is highly affective against cancer cell lines[2].

By engineering brassinin synthesis into a crop, we introduce not just a single anti-fungal, anti-cancer compound, but potentially over a dozen brassinin derivatives, each with unique defense and therapeutic properties. Towards this end, we recently discovered a 10 gene pathway in ***B. rapa*** for the synthesis of brassinin and two more enzymes responsible for its derivatization. Furthermore, preliminary results show this pathway is active in planta when transiently and constitutively expressed in tobacco (***Nicotiana benthamiana***). However, transient engineering is unacceptable for a crop trait that must be passed through generations and constitutive production of brassinin enzymes and intermediates appears to have detrimental effects on plant health. Therefore, our ultimate goal is creation of a stable engineered crop line which produces brassinin only under pathogen stress.

Here we present our progress towards introduction of phytoalexin biosynthesis in planta. We first focus on initial transient expression of these pathways in tobacco and the model plant ***Arabidopsis thaliana.*** Then we discuss our progress toward engineering of ***A. thaliana*** stable lines which both synthesize brassinin and, with careful promotor design, regulate brassinin biosynthesis. We chose to create an ***A. thaliana*** prototype both as a proof of concept and to study the effect of this pathway on plants and their commensal microbial communities. Finally, we present our outlook for the characterization of brassinins on the plant-microbe system and the introduction of this pathway into a crop, such as soybean (***Glycine max)***. Understanding these phenotypes and initial engineering of these pathways will be extremely important for successful integration of aromatic defense molecules into crop plants and improved food yield.

[1] Pedras, M. Soledade C., Estifanos E. Yaya, and Erich Glawischnig. "The phytoalexins from cultivated and wild crucifers: chemistry and biology."***Natural product reports*** 28.8 (2011): 1381-1405.

[2] Kim, Sung Moo, et al. "Brassinin Induces Apoptosis in PC 3 Human Prostate Cancer Cells through the Suppression of PI3K/Akt/mTOR/S6K1 Signaling Cascades." ***Phytotherapy Research*** 28.3 (2014): 423-431.

467483 Metabolic Model of *Trichodesmium Erythraeum,* a Marine Diazotroph

Abstract:

It is estimated that 42% of the annual global nitrogen fixation is performed by the marine cyanobacterium, *Trichodesmium.* It is also known to leak 30 â€“ 50 % of the nitrogen it fixes, thus acting as an important source of biologically available nitrogen in the open ocean. *Trichodesmium* is a unique diazotrophic cyanobacterium because it can simultaneously fix carbon and nitrogen by differentiating the metabolic pathways in two subsets of cells while maintaining similar cellular structure. Despite the prominence of this organism in the nitrogen cycle, little work has been done to investigate how carbon and nitrogen is directed through metabolism within and between cells. We have reconstructed the metabolic network of *T. erythraeum*, which is the one species of *Trichodesmium* with a sequenced genome. We have used this model to predict fluxes between cells based on constraints which we have measured in the lab or *in situ* data found in literature. Our model successfully predicts equilibrium cell compositions (carbon fixing versus nitrogen fixing) which have been measured in the open ocean. This model will be further refined and improved to enable more advanced modeling of cultures with distinct metabolisms.

471639 Developing a Tunable and Tightly Controlled Gene Expression System in *Synechocystis* sp. PCC6803

Abstract:

Cyanobacteria are photosynthetic microbes that can derive energy from sunlight and carbon from atmospheric carbon dioxide, as well as other sources. In the past two decades, there were numerous efforts geared towards developing cyanobacteria as bioproduction platforms. Several of these organisms are amenable to genetic manipulation, thus facilitating synthetic biology/metabolic engineering applications. The unique demands of living on light as a primary nutrient make the use of many established gene expression control systems difficult or impossible to apply. Among other approaches, researchers have used endogenous promoters such as P*psbA2* and P*cpcB* (promoters that drive expression of major photosynthetic components), and synthetic riboswitches with limited/mixed success. Hence, developing an effective genetic toolkit to regulate gene expression in cyanobacteria is of utmost importance. In this work, as a first case study, we combined a previously developed theophylline-responsive riboswitch with the T7 control system to drive expression of a yellow fluorescence protein (YFP) reporter in *Synechocystis* sp. PCC 6803. We then demonstrated, as a second case study, the application of this system by using P*nblA* to drive expression of this same reporter. Note that, under environmental stresses such as nitrogen or sulfur starvation, cyanobacteria utilize phycobilisomes (that account for 30-40% of total cellular protein) as a nutrient source through a degradation process that leads to yellowing or bleaching of the culture. The *nbl* (nonbleaching) genes are involved in this process, with *nblA* interacting directly with the phycobilisome, possibly causing rod instability and the beginning of the disassembly process. Overall, the utility of the designed gene expression control system was clearly shown via reduced baseline expression and higher ratio of induction. Thus, our work shows the utility of a tightly controlled and tunable gene expression control system that can exploit the differences in light/dark life cycle and expedite biotechnological applications in cyanobacteria.

467630 A Strain Engineering Strategy for the Containment of Genetically Modified Cyanobacteria

Abstract:

Cyanobacteria are photosynthetic microorganisms whose metabolism can be manipulated through genetic engineering for the production of many molecules of interest to industry directly from CO_2 and sunlight[1,2]. However, the costs of constructing and maintaining photobioreactors for scaled up cultivation of these genetically modified cyanobacteria are cost prohibitive[3]. A major concern for the cultivation of cyanobacteria in more affordable open pond reactors is the escape and subsequent environmental contamination by these genetically modified cyanobacteria. Modelling and preliminary experimental results suggest that a potentially useful strategy for containment of genetically modified cyanobacteria could involve abolishment of function the native CO_2-concentrating mechanism (CCM) through gene knockouts, thereby eliminating the ability to fix CO_2 in ambient air but causing no reduction in CO_2-fixation rate under an artificially elevated CO_2 concentration[4]. In the first part of this work, the efficacy of this method was studied for a model cyanobacterium, *Synechococcus sp.* PCC7002. The genes eliminated in this method have homologs across a wide diversity of cyanobacteria. Thus, a probable mechanism for subversion of containment would involve horizontal gene transfer to restore CCM function. In the second part of this work, elimination of genes associated with the natural competency of *Synechococcus sp.* PCC7002 was examined as a method to decease the likelihood of this subversion. The final result of this work is a strain of cyanobacteria engineered to have high fitness in an industrial environment containing high-CO_2, an inability to propagate in the surrounding environment, and a low likelihood of restoring viability in ambient air.

1. Angermayr, S. A., Gorchs Rovira, A. & Hellingwerf, K. J. Metabolic engineering of cyanobacteria for the synthesis of commodity products. *Trends Biotechnol.* **33,**352–361 (2015).

2. Sarkar, D. & Shimizu, K. An overview on biofuel and biochemical production by photosynthetic microorganisms with understanding of the metabolism and by metabolic engineering together with efficient cultivation and downstream processing. *Bioresour. Bioprocess.* **2,**17–36 (2015).

3. Jones, M. E. & Banholzer, W. F. Solar flux, water, and land impose limits on biology. *Biotechnol. Bioeng.* **111,**1059–61 (2014).

4. Clark, R., Cameron, J., Root, T. & Pfleger, B. Insights into the industrial growth of cyanobacteria from a model of the carbon-concentrating mechanism. *AIChE J.* **60,** 1269–1277 (2014).

466300 Numeric Optimization of Power Consumption for Simultaneous Saccharification and Fermentation of Cellulose to Ethanol

Abstract:

Using second-generation bioethanol, cellulose, the most abundant raw material in nature, can be converted to ethanol. Despite intensive research efforts at laboratory scale, the secondary effects of scale-up to large systems are still largely uncertain. A threedimensional numerical model of a laboratory scale fermenter is used to predict the effects of particulate mixing and reaction kinetics. This allows for case studies or optimizations of the reactor design to be carried out. The potential design variables are manifold ranging from agitation speed to impeller geometry while objectives are conversion rates and power consumption.

A two step reaction kinetics for simultaneous saccharification and fermentation (SSF) of Avicel (microcrystalline cellulose) particles to ethanol was presented by [van Zyl, 2012]. This separation of the two primary cellulase enzyme-kinetics gives the capability to predict the heterogeneous behavior of the enzyme-substrate interactions. This model improves the understanding of these systems while maintaining sufficient simplicity for implementation alongside a commercial computational fluid dynamics environment.

Viscosity is a mayor factor for the designing of a mixing vessel, as viscosity influences both: the shear rates and fluid dynamics of the system. To get the viscosity of the particle-fluid mixture an apparent dynamic viscosity model is used.

Cross-coupling the reaction models with computational fluid dynamic simulations provide a novel approach to capture the secondary effects of substrate conversion and particle distribution on the performance of the fermentation vessel. The first effect is the reduction in the average particle diameter due to the conversion of the cellulose. The lower drag forces of these smaller particles allow â€“ in a second effect - for a better distribution in the vessel. It is important to maintain a fully suspended particle distribution to maximize the surface contact with the rest of the mixing fluid.

These effects enable a reduced impeller rotation rate at later stages of the conversion decreasing the power consumption.

Difficulties in the modeling arise due to the different time scales in the simulation. The mixing with rotation rates of 150 rpm compared to the fermentation lasting for several hours. An approach to couple these phenomena in one application is presented.

In this contribution we show how to model this analysis efficiently using the commercial CFD software STAR-CCM+by CD-adapco.

References:
van Zyl, Josebus Maree. *Three-dimensional modelling of simultaneous saccharification and fermentation of cellulose to ethanol.* Diss. Stellenbosch University, 2012.

467208 Furfural Production By Continuous Reactive Extraction

Abstract:

1. Introduction

The bio-based economy grows rapidly during the 21st century. So far, oil and natural gas dominate the industrial world and are widely considered as raw materials. During the last few years, the use of biomass as raw material has increased significantly (1).

One important renewable bio-based chemical is furfural. It is widely used as solvent for saturated hydrocarbons because of its chemical structure. Moreover, it has the ability to dissolve aromatics and other unsaturated olefins (2) and is extensively used in oil and gas industry. Finally, furfural has a high potential as a platform chemical.

The objective of this study is to investigate the reaction mechanism and kinetics of xylose conversion towards furfural production by a continuous reactive extraction process. Moreover, the reaction pathway will be further elucidated and the kinetic rates of the intermediate steps are determined. Finally, the kinetic model for xylose dehydration to furfural is used to maximize furfural yield.

2. Experimental section

2.1. Reaction mechanism

The conversion of xylose to furfural is quite complex. Several studies (2), (3) have been performed in order to analyze and interpret the mechanism of the xylose dehydration. The scheme below shows the suggested mechanisms.

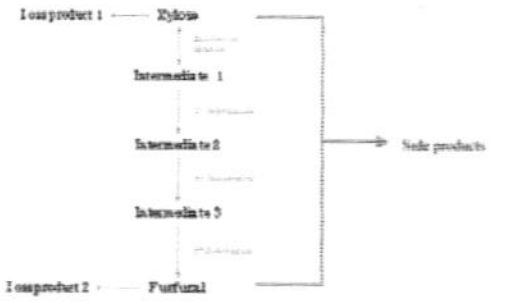

Scheme 1 Reaction mechanisms of xylose dehydration (2), (3)

According to Scheme 2 (blue), xylose is in equilibrium with intermediate I1. I1 undergoes three dehydration reactions to furfural. Byproducts are formed directly from either xylose or furfural (orange). The furfural degradation to byproduct 2 is significant and influences the selectivity. Increasing the temperature and acidity will further catalyze the degradation. An additional reaction is mentioned in (3), where xylose and furfural react and form side products (green).

The formation of byproducts from furfural can be reduced by in situ removal of furfural from the aqueous solution, by evaporation (4) or by extraction. Various studies (3), (4) have been performed to investigate which solvent performs best. Toluene and MIBK seem the most promising candidates with high affinity to furfural. Therefore, in this study, toluene is selected.

2.2. Experimental Setup

The experiments are performed in a biphasic continuous reactor as shown in Scheme 2. The configuration includes 3 pumps for sulfuric acid, xylose solution, and toluene. The maximum total flow is 15 mL/min and the minimum flow is 4 mL/min, with the flow ratio organic/aqueous of 2:1.

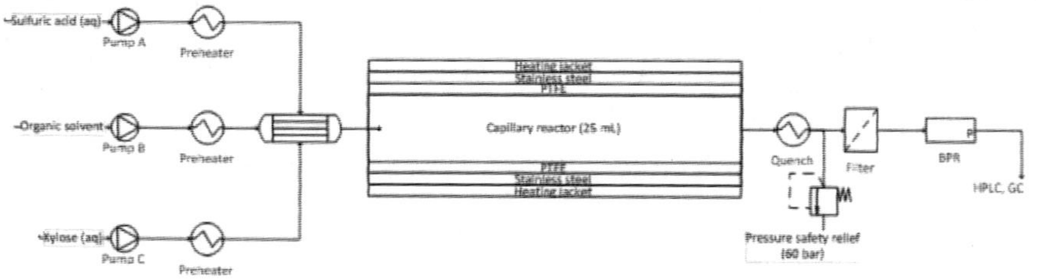

Scheme 2 Experiment lay out

The three streams flow separately in stainless steel tubes of 1.5 m length and 0.5 mm inner diameter through a heat exchanger to acquire the desired temperature. The preheat system is followed by two consecutive T-joints that allow the capillaries to be inserted in each other. Therefore the three streams are injected and mixed at the reactor's entrance (Scheme 3).

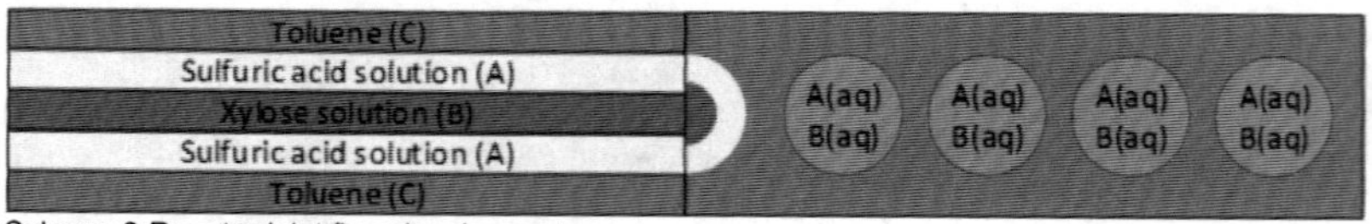

Scheme 3 Reactor inlet flow development

The reactor of 1.5 m in length and an inner diameter of 3.1 mm consists of a PTFE wall and a stainless steel mesh jacket to provide structural integrity. The PTFE wall is hydrophobic and wetted by toluene only, which prevents solid deposition at the reactor wall. The mixture is then quenched followed by a pressure safety relief and a filter with 2 μm pores that protects the back-pressure regulator (BPR) from blocking by solids formed during the reaction. Samples are taken after the BPR at atmospheric pressure and ambient temperature and analyzed with HPLC (aqueous phase) and GC (toluene phase).

The temperature range of the experiment is 100-170°C and the pressure is 30 bar so the presence of any vapor or gas at the working temperature is avoided. The selected catalyst is sulfuric acid (98%, Emsure) at 0.1 M concentration. The xylose inlet concentration is 12 wt %.

3. Results

3.1. Product characterization

In Figures 1 and 2 examples of the water phase HPLC are presented at τ=100s and 500s, respectively. Two signals are presented; (a) from the RI detector and (b) from the UV-vis detector. The compounds known from the calibration are indicated in Figures 1 and 2. In the RI signal, two unknown peaks are present: I1 and I2. In the UV signal more peaks are present. The most significant are termed A, B, C and D. Peaks B_a and B_b have a similar retention time as peak I1, who appears to have a shoulder and probably consists of two components. Component C is close to the retention time of I2. However, its concentration increases while I2 decreases in concentration.

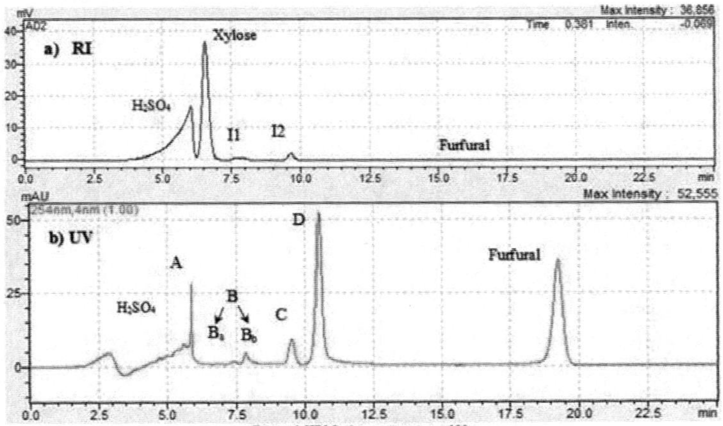

Figure 1 HPLC chromatogram; τ=100 s

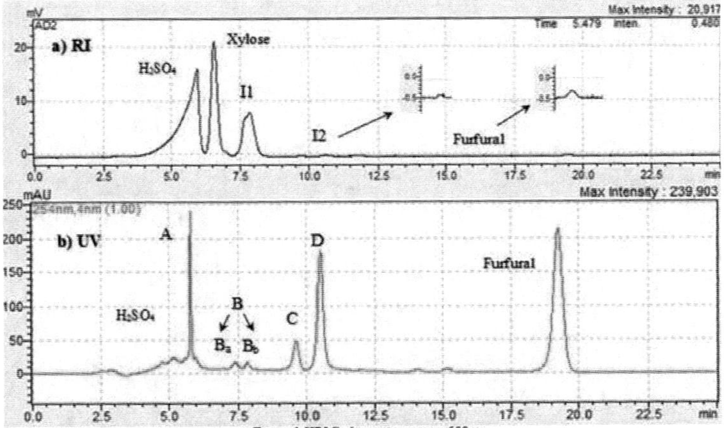

Figure 2 HPLC chromatogram; τ=500 s

Regarding UV chromatograms (Figures 1(b) and 2(b)), peaks corresponding to compounds B_a, B_b increase from 2 and 4 mAu to 10 and 8 mAu respectively with residence time. Moreover, peak of compound C rises from 10 mAu to 50 at the end of the residence time. Finally, peak of D has a raise from 52 to 180 mAu at 500s.

3.2. Reactor performance

In Figure 3 the reactor outlet concentration of furfural and xylose are presented. The highest xylose conversion is 44.52 % and the overall furfural yield is 14.45 %. The selectivity towards furfural appears to be relatively constant at 35% with increasing residence time.

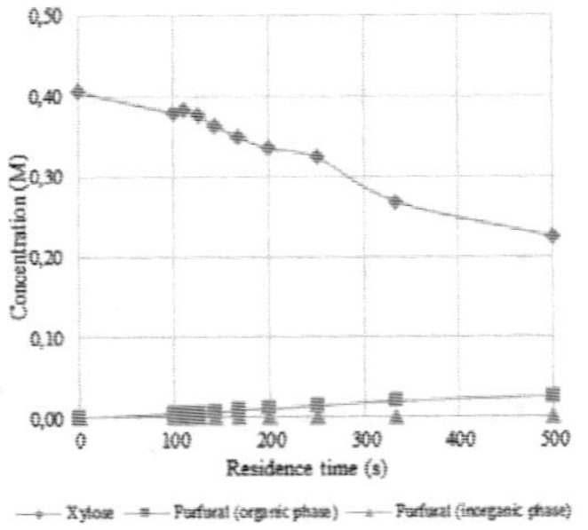

Figure 3 Concentration Profiles

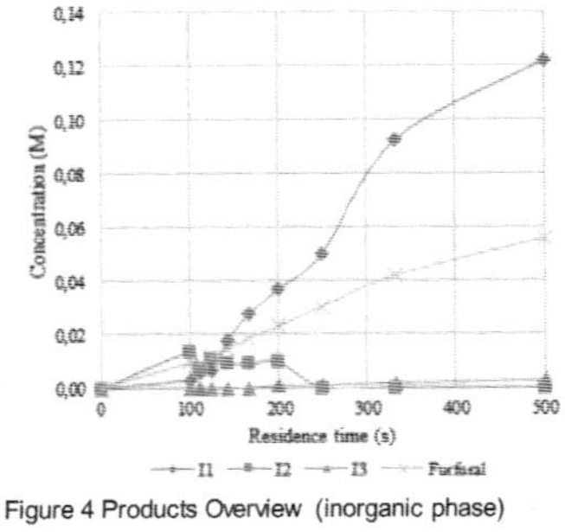

Figure 4 Products Overview (inorganic phase)

In Figure 4, the concentrations of the unknown compounds are shown as a function of residence time, with the concentration of furfural. The intermediates' concentrations are based on the RI chromatograms (Figures 1(a) and 2(a)). However, the intermediates have not been isolated, characterized, and calibrated in the HPLC. Their concentrations are based on an estimate from their RI peak areas and overall mole balance. The RI sensitivity of I1, I2, and I3 are assumed equal and is estimated from the mole balance (Eq. (1)) for the highest conversion. The balance accuracy is more than 99.9%.

$$C_i^{aq} = \left(C^{aq}_{xylose,in} - C^{aq}_{xylose,out} - C^{aq}_{furfural,out} - 2C^{org}_{furfural,out}\right)\frac{A_i}{\sum_i A_i}$$

Byproducts that are not detected by the HPLC are neglected, although a very small amount of solid residue is noticed in the samples with higher residence times.

4. Discussion

It can be observed from Figures 1(a), 2(a) and 4 that component I1 shows a similar trend as furfural at higher concentration. This cannot be explained by the irreversible consecutive reaction scheme 1. It is very likely that I1 corresponds to intermediate 3 in Scheme 1 and is in equilibrium with furfural. Alternatively, all dehydration reactions (Scheme 1) are equilibrium reactions and I1 corresponds to any of the three intermediates. In Figures 1 and 2 the RI does not respond to components C and D and they are presented only in UV. Thus, it can be assumed that their concentration is low. Possibly, C and D correspond to the intermediates

2 and 3 (Scheme 1).

The concentration of component I2 increases before the lowest residence time and decreases to zero after 250s. I2 is probably a precursor to the solid residue formed and reacts with itself and the solid residue. Thus, the declining concentration with residence time can be explained. The concentration of I3 remains low and follows the same trend as furfural, which would indicate that it is a byproduct in equilibrium with I1.

5. Conclusions

By the continuous reactive extraction at 150°C, 44.52 % xylose conversion is achieved and 14.45 % total furfural yield. Compounds in equilibrium with furfural have been detected validating the theory about reaction mechanism presented earlier (Scheme 1). Moreover, additional compounds are presented in the system potentially in equilibrium with furfural intermediates. Finally, solids are formed from furfural degradation in aqueous phase even in short residence times.

6. Outlook

Future work includes additional experiments in the temperature range of 100°C to 170°C and shorter and higher residence times. Moreover, the influence of the catalyst will be investigated. Regarding the extraction part, variations will also be made on the nature and the ratio of the solvent. Finally, a kinetic model will be developed to describe the reaction rates as a function of temperature and concentrations.

7. References

1. B. Kamm, P. R. Gruber, M. Kamm. *Biorefineries-Industrial processes and Products.* s.l. : Wiley, 2006.

2. *The Chemistry and technology of furfural production in modern lignocellulose-feedstock biorefineries.* Marcotullio, Gianluca. s.l. : Delft University of technology, 2011.

3. *Kinetics of furfural production by dehydration of xylose in a biphasic reactor with microwave heating.* R. Weingarten, J. Cho, Wm. C. Conner, Jr. and G.W. Huber. 1423-1429, s.l. : Green Chemistry, 2010, Vol. 12.

4. *Extraction of furfural from aqueous solution uding Alcohols.* J. L. Cabezas, L. A. Barcena, J. Coca, M, Cockrem. 435-437, s.l. : J. Chem. Eng. Data, 1983, Vol. 33.

5. *The chemistry and technology of furfural and its many by-products.* Zeitch, K. J. s.l. : Elsevier, 2000.

469286 An Investigation of the Degradation of Lignin in Sub-Super Critical Water in a Continuous Flow Reactor

Abstract:

Lignin, after cellulose, is the most abundant organic compound on the planet and its conversion has significant potential as a source for the sustainable production of fuels and bulk chemicals. However, Lignin conversion into more meaningful products is a challenge due to the inherent complexity of lignin molecule. Despite many efforts in this field, degradation of lignin has not been entirely successful especially at large scale due to the plugging of catalysts with tars due to unwanted oligomerization reactions. To avoid this phenomenon, continuous flow systems appear to be a better option since they allow for reduced residence times, which may inhibit tar formation and because they are more likely to lead to viable industrial scale processes.

In this research, the degradation of lignin was investigated in a continuous flow reactor in super/sub-critical water at 25, 35 and 45 ml/min and 250, 300 and 320 . Solid, liquid, and gas phase products were collected and analyzed to determine the near optimum reaction conditions and performance results that can be expected from this type of system.

456337 Interaction Between Model Compounds in Lignocellulosic Biomass Under Supercritical Water Conditions

Abstract:

Supercritical water gasification (SCWG) is a conversion technique to transform biomass into energy without necessity of drying process because water in supercritical state (temperature and pressure above 374 °C and 22.1 MPa, respectively) behaves as an organic solvent so that a high conversion can be achieved. Plant biomass or lignocellulosic biomass is a good candidate for a feedstock utilized in this process due to its large availability on Earth. It consists of cellulose, hemicellulose, which are easy to be gasified and lignin, which inhibits gasification. Previously, reaction pathways of cellulose, hemicellulose and lignin were individually proposed by many researchers. However, the study of the reaction mechanisms of the overall biomass is still rare. Therefore, this research is to study the interaction between these compounds in supercritical water aiming for better understanding in a behavior of lignocellulosic biomass in the gasification. To simplify the reaction mechanisms, model compounds for each component are employed instead of the complex structure of those. The interaction between glucose that is a model compound of cellulose, xylose that is a model compound of hemicellulose and guaiacol that is a model compound of lignin is investigated using tubular reactor system composed of high-pressure pumps, a preheater, a reactor made of stainless steel SS316 with inner and outer diameters of 1 mm and 1.59 mm (1/16 inch), respectively, heat exchangers, a back-pressure regulator, and a gas- and liquid-sampling system. Molten salt is used as heating medium for heating pre-heated water and the reactor. Mixture solutions of glucose, xylose and guaiacol were prepared in designated concentrations: 0.3 % wt.: 0.0 % wt.: 0.1% wt., 0.2 % wt.: 0.2 % wt.: 0.0 % wt., 0.0 % wt.: 0.3 % wt.: 0.1 % wt. and 0.15 % wt.: 0.15 % wt.: 0.1 % wt. Reaction temperatures are above 300 °C. A pressure of the system is 25 MPa for all conditions. The effect of each compound in the feedstock, temperature and residence time on product yields as well as interaction networks of glucose, xylose and guaiacol will be presented.

464614 A Novel Single Microsphere-Microreactor Approach Modeling and Experimental for Study of Biomass Fast Pyrolysis

Abstract:

Many technologies are being investigated and developed for chemicals and fuel production from renewable resources. Biomass fast pyrolysis is emerging as a promising approach. Practical question still remain, however, regarding the kinetics of fast pyrolysis; mainly due to the very fast, three-phase reactions that occur. It appears that the pyrolysis heating rate, solid/gas phase residence time at temperature and biomass particle size have an important impact on char, condensable and gas yields and selectivity. In an effort to elucidate reaction kinetics, a novel micro-particle micro-reactor system has been developed allowing reaction and quantification of rates for single manufactured spherical particles of nominally 10 microgram. The very fast rate of reaction for pyrolysis makes all pyrolysis experiments dynamic at some level. This new approach, wherein a single 10 microgram particle is injected into a hot furnace, achieves heating rates as high as 5000 ^{o}C/s. Experimental data is captures in the time-domain wherein time and gas-phase response (using fast flame ionization) are trapped at rates as high as 10,000 Hz. Translation of this time-domain data to temperature-domain can only be done by a model-based approach. This presentation focuses on modeling the reactor dynamics and correlation of time-temperature-reaction rate datasets. Prior studies show that for fast pyrolysis of single microspheres, the amount of hydrocarbon and char production increases linearly as the biomass particle size increases, however the reaction rate decreases. Furthermore, the amount of volatilized hydrocarbon appears to increase linearly as the temperature increases. Finally, the rate of reaction decreases as the size of biomass particle increases and increases as the temperature increases.

467108 Computational Fluid Dynamics Study of Full-Scale Aerobic Bioreactors:Â Evaluation of Dynamic Oxygen Distribution, Gas-Liquid Mass Transfer, and Reaction

Abstract:

If advanced biofuels are to measurably displace fossil fuels in the near term, they will have to operate at levels of scale, efficiency, and margin unprecedented in the current biotech industry. For aerobically-grown products in particular, scale-up is complex and the practical size, cost, and operability of extremely large reactors is not well understood. Put simply, the problem of how to attain fuel-class production scales comes down to cost-effective delivery of oxygen at high mass transfer rates and low capital and operating costs. To that end, very large reactor vessels (>500 m^3) are proposed in order to achieve favorable economies of scale. Additionally, techno-economic evaluation indicates that bubble-column reactors are more cost-effective than stirred-tank reactors in many low-viscosity cultures.

In order to advance the design of extremely large aerobic bioreactors, we have performed computational fluid dynamics (CFD) simulations of bubble-column reactors. A multiphase Euler-Euler model is used to explicitly account for the spatial distribution of air (i.e., gas bubbles) in the reactor. Expanding on the existing bioreactor CFD literature (typically focused on the hydrodynamics of bubbly flows), our simulations include interphase mass transfer of oxygen and a simple phenomenological reaction representing the uptake and consumption of dissolved oxygen by submerged cells.

The simulations reproduce the expected flow profiles, with net upward flow in the center of column and downward flow near the wall. At high simulated oxygen uptake rates (OUR), oxygen-depleted regions can be observed in the reactor. By increasing the gas flow to enhance mixing and eliminate depleted areas, a maximum oxygen transfer

(OTR) rate is obtained as a function of superficial velocity. These insights regarding minimum superficial velocity and maximum reactor size are incorporated into NRELâ€™s larger techno-economic models to supplement standard reactor design equations.

AUTHOR INDEX

AUTHOR INDEX

AUTHOR INDEX

AUTHOR INDEX

AUTHOR INDEX

AUTHOR INDEX

AUTHOR INDEX

AUTHOR INDEX

AUTHOR INDEX

AUTHOR INDEX

AUTHOR INDEX

AUTHOR INDEX

AUTHOR INDEX

AUTHOR INDEX

AUTHOR INDEX

AUTHOR INDEX

AUTHOR INDEX

AUTHOR INDEX

AUTHOR INDEX

AUTHOR INDEX

AUTHOR INDEX

AUTHOR INDEX

AUTHOR INDEX

AUTHOR INDEX

AUTHOR INDEX

AUTHOR INDEX

AUTHOR INDEX

AIChE
120 Wall Street, FL 23
New York, NY 10005-4020

ISBN 978-1-5108-3420-0

9 781510 834200

CURRAN ASSOCIATES INC.
proceedings
.com